VOCABULARY

VOCABULARY PREVIEW *Check the box that applies.* | Got It | Must Study

Multiplication: Multiplication is a quick way to perform repeated addition.

Product: A **product** is the result of multiplying.

Factors: Factors are the numbers that are multiplied

Commutative Property of Multiplication: The orde the product. Example: $2 \cdot 3 = 3 \cdot 2$

Study these terms when they appear in the text. At th ing the Vocabulary Review in the section exercises.

INTERACTIVE DEFINITION Product and Factor

A **product** is the result of multiplying.
Factors are the numbers that are multiplied to give a product.

EXAMPLE	GUIDED PRACTICE
1. Identify the product and factors in $8 \cdot 6 = 48$.	**1.** Identify the product and factors in $63 = 7 \cdot 9$.
Product → $8 \cdot 6 = 48$ ← Factors	☐ → $63 = 7 \cdot 9$ ← ☐
48 is the product of 6 and 8. 8 and 6 are factors of 48.	____ is the product of 7 and 9. ____ and ____ are factors of 63.

DO YOU UNDERSTAND how to identify the product in a multiplication exercise? | Got It | Get Help

DO YOU UNDERSTAND how to identify the factors in a multiplication exercise? | Got It | Get Help

VOCABULARY REVIEW *Review the Vocabulary you can check* Got It *for every term.*

Use these words to complete each sentence.

multiplication • product • factors • Commutati

	Got It	Get Help
18. A __________ is the result of multiplying.		
19. Numbers that are multiplied are called __________.		
20. The __________ allows you to reorder factors without changing the product.		
21. In the spaces below, write *factor* or *product* as appropriate. __________ · __________ = __________		
22. __________ can be performed using repeated addition.		

How will you get help for any vocabulary about which you are unsure?

Your instructor ____ MyMathLab ____ A classmate ____ A tutor ____ Other ____

CONCEPT

Concept Check

B1. Using money, explain why a 3 needs to be carried when multiplying $9 \cdot \$24$.

B2. Why do we multiply the smallest place values first? *Hint:* Think about a problem in which we would need to carry.

EXERCISE PRACTICE

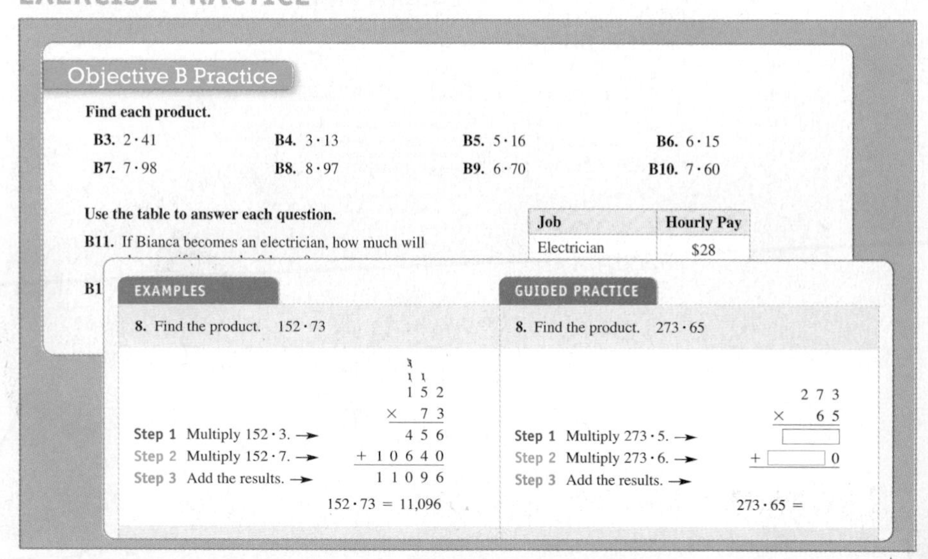

Objective B Practice

Find each product.

B3. $2 \cdot 41$	**B4.** $3 \cdot 13$	**B5.** $5 \cdot 16$	**B6.** $6 \cdot 15$
B7. $7 \cdot 98$	**B8.** $8 \cdot 97$	**B9.** $6 \cdot 70$	**B10.** $7 \cdot 60$

Use the table to answer each question.

B11. If Bianca becomes an electrician, how much will

Job	Hourly Pay
Electrician	$28

B1

EXAMPLES	GUIDED PRACTICE
8. Find the product. $152 \cdot 73$	**8.** Find the product. $273 \cdot 65$
Step 1 Multiply $152 \cdot 3$. → 456	Step 1 Multiply $273 \cdot 5$. → ☐
Step 2 Multiply $152 \cdot 7$. → + 10640	Step 2 Multiply $273 \cdot 6$. → + ☐0
Step 3 Add the results. → 11096	Step 3 Add the results. →
$152 \cdot 73 = 11{,}096$	$273 \cdot 65 =$

Resources to Help You

MyMathLab®

MyMathLab provides students with a personalized interactive learning environment, where they can learn at their own pace and gain immediate feedback and help. MyMathLab engages students in active learning—it's modular, self-paced, accessible anywhere with Web access, and adaptable to each student's learning style. In addition, MyMathLab provides instructors with a rich and flexible set of text-specific resources, including course management tools, to support online, hybrid, or traditional courses. MyMathLab is available to qualified adopters. For more information, visit our website at www.mymathlab.com or contact your Pearson representative.

MathXL®

MathXL is a powerful online homework, tutorial, and assessment system. With MathXL, instructors can create, edit, and assign online homework and tests using algorithmically generated exercises for infinite practice opportunities.

Take a tour of MathXL today! Visit www.mathxl.com.

Twitter®

Follow Goetz/Smith/Tobey Basic Math on Twitter (twitter.com/gstbasicmath)! The authors have provided a tweet for each exercise set in every section of the text. These tweets provide hints and suggestions to help you with specific exercises. Sign up to follow Goetz/Smith/Tobey Basic Math on Twitter (**@gstbasicmath**) and receive helpful hints and suggestions from the authors all semester.

Basic Mathematics

Brian Goetz
Kellogg Community College

Graham Smith
Kellogg Community College

John Tobey
North Shore Community College

Boston Columbus Indianapolis New York San Francisco Upper Saddle River
Amsterdam Cape Town Dubai London Madrid Milan Munich Paris Montréal Toronto
Delhi Mexico São Paulo Sydney Hong Kong Seoul Singapore Taipei Tokyo

Editorial Director, Mathematics: Christine Hoag
Editor in Chief: Paul Murphy
Executive Project Manager: Kari Heen
Project Editor: Courtney Slade
Associate Editor: Christine Whitlock
Editorial Assistant: Kristin Rude
Editor in Chief, Development: Carol Trueheart
Senior Development Editor: Elaine Page
Development Editor: Anne Scanlan-Rohrer
Senior Managing Editor: Karen Wernholm
Production Manager: Ron Hampton
Production Management/Composition: Pre-Press PMG
Senior Designer/Cover Art Direction: Beth Paquin
Cover Art Creation: Studio Montage
Interior Design: ARK Design
Digital Assets Manager: Marianne Groth
Manager, Visual Research: Elaine Soares
Production Coordinator: Katherine Roz
Associate Media Producer: Shana Rosenthal
Executive Manager, Course Production: Peter Silvia
Senior Content Developer: Mary Durnwald
Executive Project Manager, MathXL: Eileen Moore
Executive Marketing Manager: Michelle Renda
Marketing Manager: Adam Goldstein
Associate Marketing Manager: Tracy Rabinowitz
Marketing Assistant: Margaret Wheeler
Senior Author Support/Technology Specialist: Joe Vetere
Permissions Project Supervisor: Michael Joyce
Senior Manufacturing Buyer: Evelyn Beaton

Cover images: ***Mini:*** Lee Waters/Transtock/Jupiter Images; ***Illustration of buildings:*** Albert Campbell/Shutterstock Image; ***Illustration of trees:*** Robert Adrian Hillman/Shutterstock Images; ***Highway sign:*** Carsten Reisinger/Shutterstock Images

Library of Congress Cataloging-in-Publication Data

Goetz, Brian,
Basic mathematics / Brian Goetz, Graham Smith, John Tobey.—1st ed.
p. cm.
Includes bibliographical references and index.
ISBN-10: 0-13-229611-X (student edition)
1. Arithmetic—Textbooks. I. Smith, Graham, II. Tobey, John, III. Title.
QA107.2.G64 2011
510—dc22 2009031251

2 3 4 5 6 7 8 9 10—WC—14 13 12 11 10

www.pearsonhighered.com

ISBN-13: 978-0-13-229611-3
ISBN-10: 0-13-229611-X

CONTENTS

about the authors

Brian Goetz has helped students of all levels achieve success in mathematics for more than sixteen years. He believes that an active and supportive environment is needed for students to succeed. Brian has been teaching mathematics at Kellogg Community College in Michigan since 2001. Prior to this, as a curriculum specialist for the Grand Rapids Area Precollege Engineering Program, he created materials to motivate and inspire underserved populations. Brian also ran a math learning center at Bay de Noc Community College, where he helped students exceed their expectations. When he isn't teaching or writing, Brian spends time with his family and friends, as well as mountain biking and kayaking. He dreams of spending a summer kayaking around Lake Superior.

Graham Smith has spent his entire life immersed in education. He was raised in a family of six teachers, where dinner conversations often centered on public education. The majority of Graham's professional life has been focused on education and the success of underprepared students. Graham has 16 years of classroom experience, including the time he spent teaching with Americorps from 1993–1995. Currently, Graham is the Developmental Mathematics Coordinator at Kellogg Community College, where has been teaching since 2000. Graham's training and experience in mathematics and developmental education (Kellogg Institute at Appalachian State University and the National Center for Developmental Education) inspire both his teaching and writing. In his spare time, Graham enjoys spending time with his wife, Amy, catching big fish, playing the guitar, and tinkering with his car that runs on recycled vegetable oil.

Dr. John Tobey currently teaches mathematics at North Shore Community College in Danvers, Massachusetts, where he has taught for 39 years and served as mathematics department chair for 5 years. Previously, Dr. Tobey taught calculus at the United States Military Academy at West Point. He holds a doctorate from Boston University and a master's degree from Harvard University. He has authored and coauthored 7 college mathematics textbooks with Pearson/Prentice Hall. He is a past president of New England Mathematics Association of Two-Year Colleges (NEMATYC) and is an active member of the American Mathematics Association of Two-Year Colleges (AMATYC). In 1993, Dr. Tobey received the NISOD award for excellence in teaching.

PREFACE

This text is designed to provide an engaging and accessible book for a new generation of basic mathematics students who, due in part to the rise of instant communication, thrive in an active environment. While students in this generation have a strong desire and potential for academic success, we have seen many students that are not ready for college-level mathematics. This book bridges the gap between the skills of today's basic math students and the rigor of college-level mathematics through the process of active learning. Three primary goals guided the development of this text:

- Provide accessible instruction for an academically diverse population.
- Maintain the standards of current basic mathematics courses.
- Embed study skills and strategies into the text, modeling an active mathematics learning system for this course and beyond.

The result is a learning system structured as an on-ramp to mastery. By providing increased levels of support and interactivity within the text, students experience many small successes, building confidence before encountering more involved concepts. Students take an active role while using the text. It is rare for more than half a page to pass without students being asked to make a decision, solve an exercise, or provide an answer to a question. This interactive structure and support (on-ramp) allows students to get up to speed within their own comfort level, building success as they go.

Below you will find a summary of the text's pedagogical framework and its distinct features. We adhered to 5 simple principles while developing these features:

- Present one topic at a time in small, manageable objectives.
- Reinforce content with immediate guided practice.
- Promote active learning through constant and varied interaction.
- Develop student's conceptual understanding.
- Emphasize visual instruction.

Providing Accessible Instruction for an Academically Diverse Population

The following features are designed to focus and develop student understanding in small, manageable, objectives. Students are provided ample opportunity to develop understanding and build elementary skills before extending and combining concepts. Stronger students will accelerate through these features quickly to meet the more rigorous material at the end of each section.

Introduction

The Concept: Explains the content for each objective utilizing visuals, easily understood patterns, and references to previously mastered material. Through the *Concept*, students learn the ideas and principles vital to the mathematics they are studying.

Procedure: Explicitly states the steps used to perform the mathematics associated with each objective. This is important at the developmental level for two reasons. First and foremost, a procedure demonstrates that there is a logical and sequential reasoning behind mathematics. Secondly, through an explicit listing of steps, students are provided a convenient feature for study and review.

Vocabulary Development

Vocabulary Preview: Provided on the first page of every section, students are encouraged to become familiar with the words they are going to need to communicate mathematics.

Interactive Definitions: Many vocabulary terms that are critical to success or difficult to understand are developed in *Interactive Definitions.* Here, students develop understanding through *Examples* and *Guided Practice.*

Vocabulary Review: Provided at the start of exercise sets, students are encouraged to briefly assess their working knowledge of important vocabulary. The *review* is linked to the *preview* and *Interactive Definitions,* encouraging students to reflect on prior learning.

Instruction

Examples: Demonstrate how each *Procedure* is implemented. Within each *Example*, the procedural steps are restated to help the student commit the procedure to memory. To imitate classroom instruction, explanation is placed either above or to the left of the mathematics being performed. Students read the instruction and anticipate what is going to happen before seeing the mathematical step.

Guided Practice: Mimic a time-honored instruction technique used in the classroom. Following the model from the *Example*, students are guided through the process for finding a solution. ***The Guided Practice exercises encourage students to consult, utilize, and become actively involved with the text.*** Through consistent use of this feature, students will learn how to read example problems in mathematics texts.

Why It Works: Found after suitable procedures, this feature utilizes easily recognized patterns to present students with an informal proof so they can understand why a procedure works.

Twitter@gstbasicmath: Students and instructors can follow the authors via Twitter (@gstbasicmath). The authors provide helpful hints and teaching tips for each exercise set in the text via tweets (140 characters or less). The tweets are also available via MyMathLab (Instructor Resources) so they can be modified and distributed by instructors through their own Twitter accounts.

Practice

Concept Check: These exercises are assignable and precede any practice problems. By focusing student attention to the main ideas and important facts presented in the *Concept*, students consider the meaning and ideas behind the procedures that they are learning.

Objective Practice: Used to practice basic skills immediately after the guided practice. This is analogous to desk work in the classroom. These exercises are numbered so they can be assigned and are formatted to be as prominent as the traditional end-of-section exercises. These exercises are especially important because they provide an opportunity for students to achieve mastery, one skill at a time.

Section Exercises: Occur at the end of each section. They are similar to exercise sets in traditional texts in that they review all basic skills. *Section Exercises* also incorporate higher-level thinking skills including the application, analysis, and synthesis of basic concepts.

Design

The design and layout of the text are structured to promote student interaction, ease of use and readability.

- To maximize the amount of space for student interaction with the features of the text, the margin column, seen in most mathematics texts, was eliminated.
- By using the full width of the page, written procedural steps are aligned with Examples and the corresponding steps of Guided Practice exercises.
- Color-coded procedural steps are coordinated with the corresponding mathematics in examples, allowing students to locate exactly where a given step is performed.
- The writing is presented in small manageable paragraphs, separated with white space. This makes it easy for students to locate information and can keep students from feeling overwhelmed by large blocks of text.
- Features are distinctly identified with color-coded tabs to aid student navigation within the text. Practice and section exercises have a different background color making them easy to locate.
- To ease eye strain, a light yellow background color is used throughout the text.

Maintain the Current Standards of Basic Mathematics Courses

Recognizing and solving multiple problem types, application exercises, and understanding how concepts tie together are skills critical to students' continued success. The following features are designed to deepen the students' understanding of the material and to help them develop the skills necessary to be successful in college mathematics.

Combining Concepts and Applications: Found at the end of appropriate sections to present students with multiple-skill and application exercises. Here, multistep exercises and more difficult concepts/extensions are introduced. Students will have practiced an abundance of skill-based problems prior to encountering these higher level examples.

Focus On: Present topics that are often deeper or more involved than the surrounding objectives. Topics include estimation, proportional reasoning, foundations for algebra, etc. These are presented without objective titles, so they can be assigned as instructors see fit.

Exercises: End-of-section exercises are written with a level of rigor comparable to traditional basic mathematics texts. However, by working *Objective Practice* exercises first, students are better prepared to tackle rigorous concepts at the end of a section.

Chapter Review **and** ***Chapter Test:*** These features provide students with an effective tool for self-assessment. They offer an opportunity for students to recognize and solve different problem types out of the context of a section heading.

Embed Study Skills and Strategies in the Text, Modeling an Active System for Learning Mathematics

Good study habits and strategies are crucial elements for success in college courses and are most effectively learned when integrated into a course. Accordingly, we have embedded study skills and strategies into the text so students can gain proficiency with various learning techniques. Several elements of the text are designed to build mathematics study skills.

How to Study Mathematics: The text is designed to teach students an effective system for studying a subject that builds on itself. Students learn, practice, and receive feedback one topic at a time. Through this process, students learn the importance of understanding a mathematical skill before moving onto the next and realize their own potential to understand the subject.

Self-Assessment: Students are provided constant feedback as they work through a section and are encouraged to frequently assess their knowledge. Self-assessment check boxes appear throughout the text so students can record that they "Got It" or need to "Get Help."

Question Logs: Each section ends with an organized table where students can record their questions and keep track of exercises that they need to have answered.

Chapter Organizer: Each chapter contains an organizer that summarizes the vocabulary and procedures with examples from every section of the chapter. Students can use this feature to study for a test or preview a chapter.

In Closing

This text has been developed to provide accessible, interactive instruction that resonates with the diverse learning needs of today's students. While the individual features are designed to meet these goals, it is the union of these features that truly represents the power of the learning system.

- ***Unprecedented vocabulary development:*** The vocabulary preview, interactive definitions, and vocabulary review combine to form a powerful learning tool that is without parallel in other basic mathematics texts.
- ***Balanced conceptual and procedural development:*** Within each objective, students simultaneously develop conceptual and procedural knowledge, helping them to master and retain information.
- ***One topic at a time:*** The objectives are designed to present material in small, manageable parts that contain instruction, immediate practice, and feedback. Students are guided step by step through each example as they concurrently work a *Guided Practice* exercise. Assignable objective practice immediately follows so students can reinforce what they have learned before tackling a new objective. Experiencing one small success after another enables many students to overcome their anxiety and realize that success in math is attainable.
- ***On-Ramp:*** The collection of objectives in a section acts as an on-ramp to success. By incorporating learning support in a one topic at a time approach, the system meets students where they are, allowing them to quickly get up to speed. The on-ramp empowers students and builds the confidence that comes from independent learning. Most importantly, more students are successful with higher level exercises that require application, analysis, and synthesis of concepts.

The primary goal of this text is to provide a framework from which more of the current generation of students can be successful. Based on the performance and response we have received from our students at Kellogg Community College and the students at the 15 other colleges and universities who have class tested the manuscript to date, we modestly submit that this text helps today's basic mathematics learners as well, if not better than, any other basic math text.

Resources for the Student

Student Solutions Manual

ISBN 0-13-229613-6, 978-0-132-29613-7

- Solutions to all odd-numbered section exercises
- Solutions to every (even and odd) Guided Practice exercise, Concept Check exercise, and Objective Practice exercise
- Solutions to every exercise (even and odd) in the Extra Practice sections, Mid-Chapter Reviews, Chapter Reviews, and Chapter Tests

Lecture Videos

ISBN 0-13-229621-7, 978-0-13-229621-2

- Organized by section, contain problem-solving techniques and examples from the textbook
- Step-by-step solutions to selected exercises from each textbook section
- Available in MyMathLab®

Resources for the Instructor

Annotated Instructor's Edition

ISBN 0-13-229612-8, 978-0-13-229612-0

- Complete student text with answers to all Guided Practice exercises, Concept Check exercises, Objective Practice exercises, Extra Practice sections, Mid-Chapter Reviews, Chapter Reviews, and Chapter Tests

Instructor's Solutions Manual

ISBN 0-13-229614-4, 978-0-13-229614-4

- Detailed step-by-step solutions to all exercises (even and odd), including Guided Practice, Concept Check, Objective Practice, Extra Practice sections, Mid-Chapter Reviews, Chapter Reviews, and Chapter Tests

Instructor's Resource Manual with Tests

ISBN 0-13-229615-2, 978-0-13-229615-1

For each section there is

- One minilecture with key learning objectives, classroom examples, and teaching notes.
- Two short group activities per chapter are provided in a convenient ready-to-use handout format.
- Answers are included for all items.
- Alternate test forms with answers:
  - Six Chapter Tests per chapter (3 free response, 3 multiple choice)
  - Two Final Exams (1 free response, 1 multiple choice)

TestGen®

TestGen (www.pearsoned.com/testgen) enables instructors to build, edit, and print tests using a computerized bank of questions developed to cover all the objectives of the text. TestGen is algorithmically based, allowing instructors to create multiple but equivalent versions of the same question or test with the click of a button. Instructors can also modify test bank questions or add new questions. The software and testbank are available for download from Pearson Education's online catalog.

Pearson Math Adjunct Support Center

The **Pearson Math Adjunct Support Center** (http://www.pearsontutorservices.com/math-adjunct.html) is staffed by qualified instructors with more than 100 years of combined experience at both the community college and university levels. Assistance is provided for faculty in the following areas:

- Suggested syllabus consultation
- Tips on using materials packed with your book
- Book-specific content assistance
- Teaching suggestions, including advice on classroom strategies

Media Resources

MathXL® Online Course (access code required)

MathXL is a powerful online homework, tutorial, and assessment system that accompanies Pearson Education's textbooks in mathematics or statistics.

With MathXL, instructors can

- create, edit, and assign online homework and tests using algorithmically generated exercises correlated at the objective level to the textbook.
- create and assign their own online exercises and import TestGen tests for added flexibility.
- maintain records of all student work tracked in MathXL's online gradebook.

With MathXL, students can

- take chapter tests in MathXL and receive personalized study plans based on their test results.
- use the study plan to link directly to tutorial exercises for the objectives they need to study and retest.
- access supplemental animations and video clips directly from selected exercises.

MathXL is available to qualified adopters. For more information, visit our website at www.mathxl.com, or contact your Pearson sales representative.

MyMathLab® Online Course (access code required)

MyMathLab provides students a personalized interactive learning environment, where they can learn at their own pace and gain immediate feedback and help. MyMathLab engages students in active learning—it's modular, self-paced, accessible anywhere with Web access, and adaptable to each student's learning style. In addition, MyMathLab provides instructors with a rich and flexible set of text-specific resources, including course management tools, to support online, hybrid, or traditional courses. MyMathLab is available to qualified adopters. For more information, visit our website at www.mymathlab.com or contact your sales representative.

Acknowledgments

We offer our heartfelt thanks to the following individuals for their vital contributions to the project. To Jack Mayleben, for bringing the authors together and getting the project off the ground; Paul Murphy, who dreams big and has the talent to bring ideas into reality; Courtney Slade, for understanding our work and making it much better; Kari Heen, for her guidance over the team throughout the process; Dona Kenly for her work with class testers and focus groups; Doris Lewis, for her tireless efforts on answer keys through innumerable versions; Christine Whitlock for her coordination and creative ideas for connecting with students; Anne Scanlan-Rohrer for her encouragement and ability to find a common direction between seemingly disparate comments; Marlana Voerster, Adam Goldstein, Tracy Rabinowitz, and Michelle Renda for the innovative ways they have spread the message about the text; Marianne Miller for her remarkable attention to detail; Beth Paquin for her design and creative ideas to help students use the text; Ron Hampton for his ability to realize and improve our vision of the layout and graphics; Laura Hakala; for her attentive eyes throughout production; Mark Naber for his assistance with the exercise sets; Deana Richmond and Elaine Page for insightful editorial comments.

We are grateful to the many instructors from across the country that devoted their time and talent to the development of this text. Their insights, ideas and improvements have been invaluable in making this a better book. We will attempt to thank all the instructors who class tested, reviewed, or participated in a focus group below. *Note:* At the time this text went to press, new class testers were beginning to contribute to the project. We apologize for any omissions.

Class Testers

Gabriel Attar, Monroe County Community College
Sam Bazzi, Henry Ford Community College
Elena Bogardus, Camden County College
Christine Brady, Suffolk County Community College
Gwen Braun, Montana State University–Billings
Carolyn Chapel, Western Technical College
Jerry Chen, Suffolk County Community College
Anna Cox, Kellogg Community College
Cynthia Erbacher, Jefferson Community College
Roshele Friudenberg, Lee College
Eric Gilbertson, Montana State University–Billings
Mathew Hudock, St. Philip's College
Kelly Jackson, Camden Community College
Dan Kleinfelter, College of the Desert
Mary Levis, Montana State University–Billings
Doris Lewis, Kellogg Community College
Ilva Mariani, Cerritos College
Janice McCue, College of Southern Maryland
Mark Moses, Southwest Tennessee Community College
Judy Reed, Kellogg Community College
Suzanne Rosenberger, Harrisburg Area Community College
Kathleen Shepherd, Monroe County Community College
Lisa Stop, Monroe County Community College
Jane Timper, Rockingham Community College
Ghazi Zouaui, Monroe County Community College

Reviewers and Focus Group Attendees

Ali Ahmad, Doña Ana Community College of New Mexico State University
Khadija Ahmed, Monroe County Community College
Karen Anglin, Blinn College–Brenham
Sam Bazzi, Henry Ford Community College
Joseph Brenkert, Front Range Community College
Beverly Broomell, Suffolk County Community College
Andrew Burch, Estrella Mountain Community College
Torrey Burden, North Carolina A&T University
Kashonda S. Bynum, North Carolina A&T University
Mark Chapman, Baker College
Kristin Chatas, Washtenaw Community College
Phong Chau, Glendale Community College
Mark Crawford, Waubonsee Community College
Julie DePree, University of New Mexico
Heather Foes, Rock Valley College
Suellen Gifford, Ferris State University
Gay Grubbs, Griffin Technical College
Bobbie Hill, Coastal Bend College
Sandee House, Georgia Perimeter College
Glenn Jablonski, Triton College
Cameron Kishel, Columbus State Community College
Dan Kleinfelter, College of the Desert
Pat Kopf, Kellogg Community College
Chris Lorencen, Kellogg Community College
Annette Magyar, Southwestern Michigan College
Ilva Mariani, Cerritos College
Tammi Marshall, Cuyamaca College
Jerry Mason, Monroe County Community College
Janice McCue, College of Southern Maryland
Michelle Moravec, McLennan Community College
Jeffrey Morford, Henry Ford Community College
Mark Moses, Southwest Tennessee Community College
Mark Naber, Monroe County Community College
William Parker, Greenville Technical College
Jose Rico, Laredo Community College South
Lisa Rombes, Washtenaw Community College
Suzanne Rosenberger, Harrisburg Area Community College
Victoria Seals, Gwinnett Technical College
Bob Secrist, Kellogg Community College
Mark Sigfrids, Kalamazoo Valley College
Lynn Smolarkiewicz, Kellogg Community College
Sue Stetler, Kellogg Community College
Sean Stewart, Owen State Community College
Shirley Taylor, Trident Technical College
Cassonda Thompson, York Technical College
Todd Troutman, Lansing Community College
Susie Tummers, El Camino College
Alma Wlazinski, McLennan Community College
Jen Wood, Mesa Community College
Carol Zavarella, Hillsborough Community College

We offer special thanks to our colleagues at Kellogg and North Shore Community Colleges who contributed greatly to the development of this book. Thanks to Doris Lewis who, in addition to providing answers and writing the *Instructor's Resource Manual*, helped us develop many of the fundamental features within the text. Thanks to Mary Weller for her expertise and ideas regarding readability and study skills. Thanks to Anna Cox, Pat Kopf, Saeed Sabouni, Bob Secrist, and Sue Stetler for their input and feedback over many hallway conversations. Thanks to Ed Williams, for his photographic expertise. Most importantly, we wish to thank our students who, over the last five years, have used the manuscript in the classroom. It was their feedback that contributed most to the direction of the text.

Writing this book would have been impossible for us without the support of our family and friends. Our deepest thanks and love go to Laura Easter, Amy Young, and Nancy Tobey. Your understanding, love, and patience have been a source of great encouragement.

Brian Goetz
Graham Smith
John Tobey

We dedicate this book
to Jack Mayleben.
—B.G.
—G.S.

CHAPTER

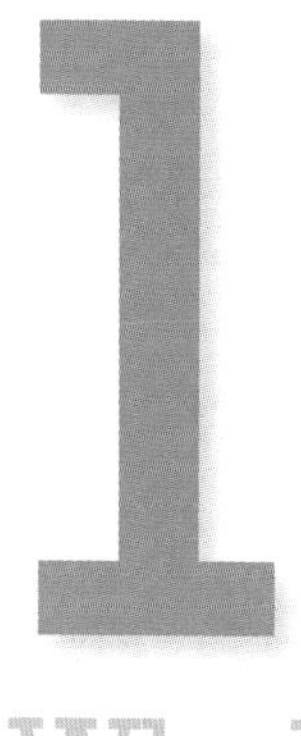

Whole Numbers

In this chapter, you will study whole numbers and the four basic operations: addition, subtraction, multiplication, and division. As you work through this chapter, focus on increasing your mental math skills. Mental math skills will help you perform calculations and understand the properties of numbers.

Planning a wedding is a difficult job, one that requires both mental estimation and accurate calculations. In the early stages of planning, estimation is used to balance a couple's wedding wish list with an affordable budget. Once a rough plan is agreed upon, actual calculations must be used to ensure that the wedding doesn't exceed the budget.

Several exercises in this chapter relate to planning a wedding, including the costs of a catered meal in Exercise 82, Section 1.6.

Section 1.1 Understanding Whole Numbers

Whole numbers are the building blocks of mathematics. They are the numbers in the list 0, 1, 2, 3, 4, 5, 6, 7, 8, 9, 10, 11, 12... The three dots indicate that the list goes on forever. Your goal in this section is to understand how to write whole numbers and to determine the value of each digit in a whole number.

The **Objectives** in Section 1.1 will help you

- **A** Understand digits, whole numbers, and place value.
- **B** Write whole numbers in expanded form.
- **C** Graph whole numbers on a number line.
- **D** Round whole numbers.
- **E** Read whole numbers on a bar graph.

VOCABULARY PREVIEW *Check the box that applies.*

	Got It	Must Study
Digits: The **digits** in the whole number system are 0, 1, 2, 3, 4, 5, 6, 7, 8, and 9.		
Whole Numbers: Whole numbers are the numbers in the list 0, 1, 2, 3, 4, 5, 6 ...		
Place Value: Place value uses the place of a digit in a whole number to indicate that digit's value.		
Periods: Periods divide the whole numbers into groups of three digits. The first five periods are ones, thousands, millions, billions, and trillions.		
Rounding: Rounding is the process of choosing the best approximation for a number at a given place value.		

Study these terms when they appear in the text. At the end of the section, test your understanding by completing the Vocabulary Review in the section exercises.

Objective A Understand Digits, Whole Numbers, and Place Value

The Concept In this section, you will study digits, whole numbers, and place value.

- **The whole numbers are 0, 1, 2, 3, 4, 5, 6, 7, 8, 9, 10, 11, 12, 13...**
  The three dots indicate that the list goes on forever.

Whole numbers are written using digits. There are exactly ten digits in the whole number system. Because we write any whole number using these ten digits, we can think of digits as the alphabet of the whole numbers.

- **The digits are 0, 1, 2, 3, 4, 5, 6, 7, 8, and 9.**

The whole number two hundred thirty-four is written with three digits.

Whole Number

234

3 digits

To write "two hundred thirty-four," we used the digits 2, 3, and 4 and the concept of place value. Place value uses the position or placement of the digits in a number to indicate the value of the digit.

$234 = $200 + $30 + $4

- The 2 is in the hundreds place, meaning that there are two hundreds in 234.
- The 3 is in the tens place, meaning that there are three tens in 234.
- The 4 is in the ones place, meaning that there are four ones in 234.

Commas are used to make large numbers easier to read.

1234567890	1,234,567,890
Without commas, this is not easy to read.	Commas make this easy to read.

Each group of three digits is called a period. The name of each period is written as shown in the following chart:

These are the periods.	Trillions			Billions			Millions			Thousands			Ones		
The digits go here.			,			**1,**	**2**	**3**	**4,**	**5**	**6**	**7,**	**8**	**9**	**0**
Each digit has the place value shown.	Hundred trillions	Ten trillions	Trillions	Hundred billions	Ten billions	Billions	Hundred millions	Ten millions	Millions	Hundred thousands	Ten thousands	Thousands	Hundreds	Tens	Ones

To read 1,234,567,890, we can read the number in each period followed by the name of the period.

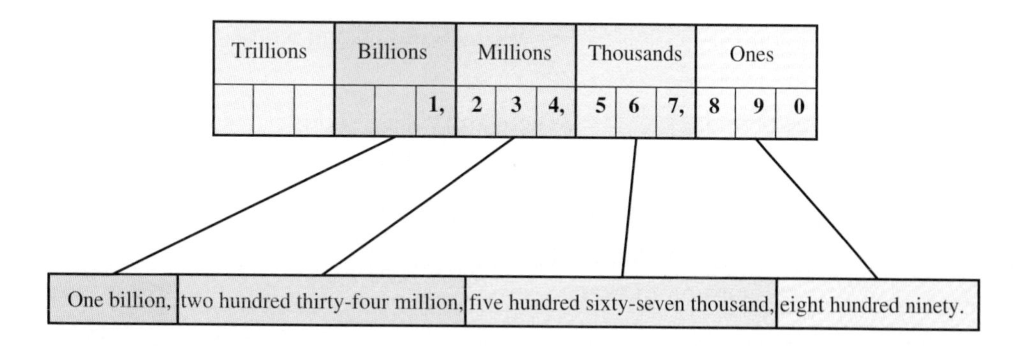

You do not include the name of the ones period when writing whole numbers in words.

Once you have learned the periods in the place value chart, you can read or write any number using words.

Procedure Write a Whole Number Using Words

Step 1 Determine the largest period of the number.

Count the periods: "ones, thousands, millions, billions, trillions . . ."

Step 2 From the left, write the number in each period followed by the name of the period.

EXAMPLES

1. Write 56,302 using words.

Step 1 Determine the largest period. — There are two periods.

Think ones, thousands ... — 56 is in the thousands period.

Step 2 From the left, write the number in each period followed by the name of the period. — The number is written as fifty-six thousand, three hundred two.

It is incorrect to write three hundred AND two. *And* is used to indicate a decimal. We will study decimals in Chapter 3.

2. Write 5,000,000,023 using words.

Step 1 Determine the largest period. — There are four periods.

Think ones, thousands, millions, billions ... — 5 is in the billions period.

Step 2 From the left, write the number in each period followed by the name of the period. — The number is written as five billion, twenty-three.

Do not write *five billion, zero million, zero thousand, twenty-three.* When a period has all zeros, you do not write the name of the period.

3. Write 34,120,000 using words.

Step 1 Determine the largest period. — There are three periods.

Think ones, thousands, millions ... — 34 is in the millions period.

Step 2 From the left, write the number in each period followed by the name of the period. — The number is written as thirty-four million, one hundred twenty thousand.

GUIDED PRACTICE

1. Write 412,706 using words.

There are ____ periods.

412 is in the ________ period.

The number is written as ______________________________.

2. Write 12,000,003 using words.

There are ____ periods.

12 is in the ________ period.

The number is written as ________________.

3. Write 17,400 using words.

There are ____ periods.

17 is in the ________ period.

The number is written as ______________________.

We can also use the place value chart to determine the place value of a digit.

Procedure Determine the Place Value of a Digit

Step 1 Determine the period of the digit.

Step 2 Determine the value of the digit in that period.

Is it in the ones, tens, or hundreds place of the period?

EXAMPLES		GUIDED PRACTICE
4. Determine the place value of the digit 2 in the whole number 456,423,578.		**4.** Determine the place value of the digit 2 in the whole number 274,565,890.
Step 1 Determine the period of the digit. Think ones, thousands, millions ...	2 is in the thousands period.	2 is in the ______ period.
Step 2 Determine the value of the digit in that period.	The number in that period is 423.	The number in that period is 274.
	2 is in the tens place of 423.	2 is in the ______ place of 274.
	2 is in the ten thousands place.	2 is in the ________ place.
5. Determine the place value of the digit 6 in the whole number 456,423,578.		**5.** Determine the place value of the digit 8 in the whole number 274,565,890.
Step 1 Determine the period of the digit. Think ones, thousands, millions ...	6 is in the millions period.	8 is in the ____ period.
Step 2 Determine the value of the digit in that period.	The number in that period is 456.	The number in that period is ____.
	6 is in the ones place of 456.	8 is in the ______ place of ____.
	6 is in the millions place.	8 is in the ______ place.
	When a digit is in the ones place of a period, we do not include the label "ones" with the place value name. 6 is said to be in the millions place, not the one millions place.	

Concept Check

A1. List the digits in the whole number system.

A2. What is the largest place value of a whole number with five digits?

A3. List the first five periods in the whole number system.

Objective A Practice

Write each whole number using words.

A4. 153 **A5.** 492 **A6.** 7,005 **A7.** 3,080

A8. 100,309 **A9.** 50,003 **A10.** 4,000,000,005 **A11.** 8,000,005,000

Determine the place value of 4 in each whole number.

A12. 234,115 **A13.** 564,789 **A14.** 123,435 **A15.** 2,435,789

A16. 768,988,456 **A17.** 435,261,678 **A18.** 678,945,555 **A19.** 24,756,900,132

Objective B Write Whole Numbers in Expanded Form

The Concept Understanding expanded form (also known as expanded notation) is important because it is used to explain arithmetic and estimation in this chapter. Expanded notation uses addition to write the meaning of each digit in a whole number.

Standard Notation		Expanded Notation
234	=	200 + 30 + 4

The digits in 234 stand for 200, 30, and 4.

Procedure Write a Whole Number in Expanded Form

Write the sum of what each digit represents.

Use place value to determine the meaning of each digit.
You do not need to write the meaning of zeros. (See Example 2.)

EXAMPLES

6. Write 75,321 in expanded form.

Think of the meaning of each digit.

75,321
1
20
300
5,000
70,000

Write the sum of what each digit represents.
75,321 = 70,000 + 5,000 + 300 + 20 + 1

7. Write 400,020,006 in expanded form.

Write the sum of what each digit represents.
400,020,006 = 400,000,000 + 20,000 + 6

GUIDED PRACTICE

6. Write 451,386 in expanded form.

Think of the meaning of each digit.

451,386
__
__0
__00
__,000
__0,000
__00,000

Write the sum of what each digit represents.
451,386 = _____ + _____ + ____ + ___ + __ + __

7. Write 5,300,010 in expanded form.

Write the sum of what each digit represents.
5,300,010 = _____ + _____ + _____

Concept Check

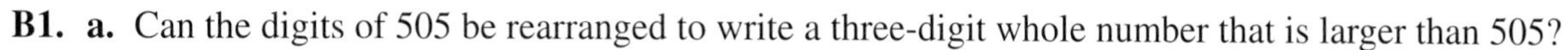

B1. **a.** Can the digits of 505 be rearranged to write a three-digit whole number that is larger than 505?
b. If so, write 505 and the larger number using expanded form.
c. Use the expanded form of the larger number to explain why the number is greater than 505.

B2. **a.** Can the digits of 560 be rearranged to write a three-digit whole number that is smaller than 560?
b. If so, write 560 and the smaller number using expanded form.
c. Use the expanded form of the smaller number to explain why the number is less than 560.

Objective B Practice

Write each whole number in expanded form.

B3. 1,534 **B4.** 5,492 **B5.** 6,002 **B6.** 3,040

B7. 205,100,309 **B8.** 420,050,003 **B9.** 4,000,000,005 **B10.** 8,000,005,000

Objective C Graph Whole Numbers on a Number Line

The Concept The number line is used to graph numbers and will be used throughout the text to demonstrate concepts. In this section, you will use a number line to understand the relative sizes of whole numbers.

Procedure Graph Whole Numbers on a Number Line

Step 1 Draw a number line with arrows at each end.
Step 2 Indicate the scale, using labels.
Space the labels in a convenient way based on the numbers to be graphed.
Step 3 Graph each whole number on the number line.

DETAILED EXAMPLES Graphing Whole Numbers on a Number Line

Graph 3, 6, and 10 on a number line.

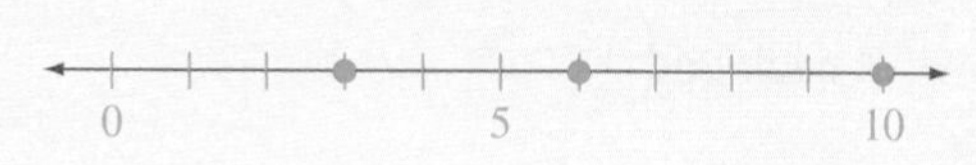

Step 1 Draw a number line with arrows at each end.
Step 2 Indicate the scale, using labels.

- Because the numbers range from 3 to 10, we draw the tick marks for each whole number.
- We did not write a number below each tick mark. When the tick marks are evenly spaced, we can determine the value they represent by counting.

Step 3 Graph each whole number on the number line.

(Continued)

Graph 10, 30, and 83 on a number line.

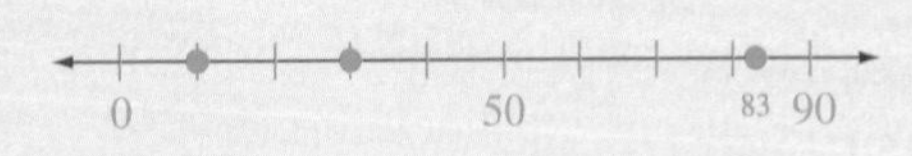

Step 1 Draw a number line with arrows at each end.

Step 2 Indicate the scale, using labels.

- Because the numbers range from 10 to 83, draw the tick marks every 10 units and label them at 0, 50, and 90.

Step 3 Graph each whole number on the number line.

- Because there isn't a tick mark at 83, we included a label for 83.

EXAMPLES

8. Graph 5, 7, 9, and 11 on a number line.

Step 1 Draw a number line with arrows at each end.

Step 2 Indicate the scale, using labels.

Step 3 Graph each whole number on the number line.

9. Graph 20, 40, 55, and 99 on a number line.

Step 1 Draw a number line with arrows at each end.

Step 2 Indicate the scale, using labels.

Step 3 Graph each whole number on the number line.

10. Graph 150, 300, and 890 on a number line.

Step 1 Draw a number line with arrows at each end.

Step 2 Indicate the scale, using labels.

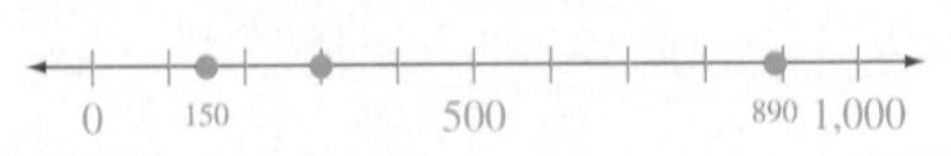

Step 3 Graph each whole number on the number line.

GUIDED PRACTICE

8. Graph 2, 6, 9, and 13 on a number line.

9. Graph 10, 50, 43, and 67 on a number line.

10. Graph 200, 410, and 700 on a number line.

Concept Check

C1. To graph 33, 150, and 728 on the same number line, why is it a good idea to put tick marks for every 100 units instead of for every 1 unit?

C2. If a number is graphed on a tick mark, why don't we have to include a label for it?

Objective C Practice

Graph each set of whole numbers on a number line.

C3. 3, 7, 9, 13

C4. 2, 5, 10, 14

C5. 20, 70, 88, 105

C6. 5, 30, 60, 107

C7. 120, 300, 1,100

C8. 400, 570, 1,200

Objective D Round Whole Numbers

The Concept To demonstrate the concept of rounding, we will use an example that involves money. Imagine that you owe a friend $128. You have $200 in your pocket. Your friend needs cash now, but you don't have the exact change. What should you do? The answer depends on what bills you have.

INTERACTIVE DEFINITION Rounding

Rounding is the process of choosing the best approximation for a number at a given place value.

EXAMPLE

11. You owe a friend $128, and you have only $100 bills. What is the fairest amount to pay your friend at this time?

This question requires that you round $128 to the hundreds place.

1 is in the hundreds place.

128

Your payment options with $100 bills are as follows:

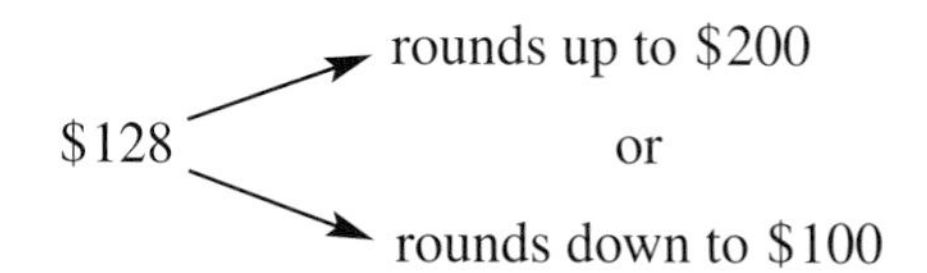

128 is closer to 100, so 100 is the better approximation of 128.

$128 rounded to the hundreds place is $100.

GUIDED PRACTICE

11. You owe a friend $128, and you have only $10 bills. What is the fairest amount to pay your friend at this time?

This question requires that you round $128 to the tens place.

_____ is in the tens place.

128

Your payment options with $10 bills are as follows:

$128 rounds up to _____

or

rounds down to _____

128 is closer to _____, so _____ is the better approximation of 128.

$128 rounded to the tens place is _____.

DO YOU UNDERSTAND how to approximate a whole number by rounding? Got It | Get Help

A whole number rounded to a given place value is the best approximation of the number at that place value. For instance, 128 rounds to whichever of 120 and 130 is the better approximation. After you identify the two choices, decide which number is closer to the number being rounded.

EXAMPLES

12. Round 487 to the nearest hundred.

487 will round up to 500 or down to 400.

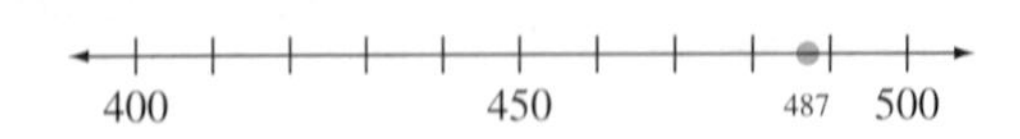

487 is closer to 500, so it is the better approximation. 487 rounds up to 500.

13. Round 3,173 to the nearest thousand.

3,173 will round up to 4,000 or down to 3,000.

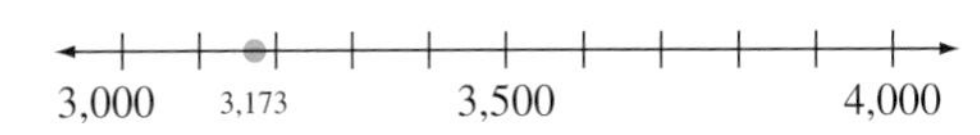

3,173 is closer to 3,000, so it is the better approximation. 3,173 rounds down to 3,000.

14. Round 285 to the nearest ten.

285 will round up to 290 or down to 280.

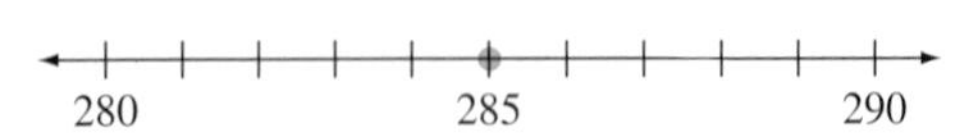

Because 285 is exactly halfway between 280 and 290, 285 rounds up to 290.

15. Round 98,129 to the nearest ten thousand.

98,129 will round up to 100,000 or down to 90,000.

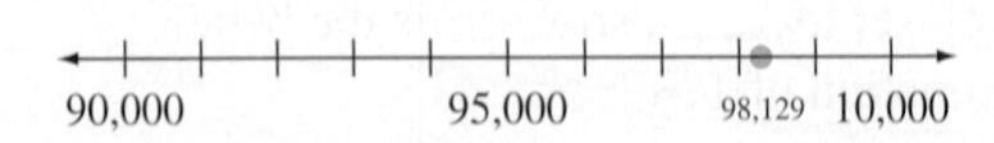

98,129 is closer to 100,000, so it is the better approximation. 98,129 rounds up to 100,000.

GUIDED PRACTICE

12. Round 812 to the nearest hundred.

812 will round up to _____ or down to _____.

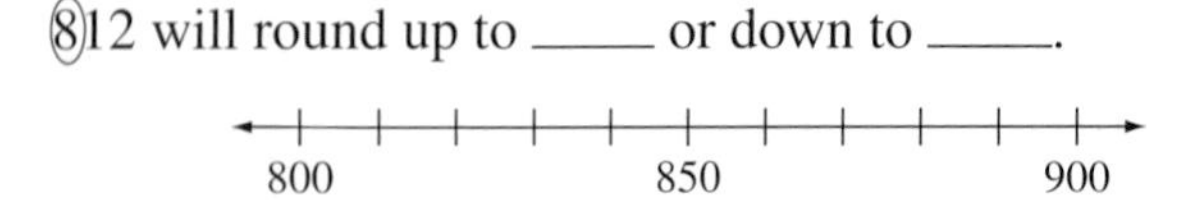

812 is closer to _____, so it is the better approximation. 812 rounds up/down to _____.

13. Round 5,632 to the nearest thousand.

5,632 will round up to _____ or down to _____.

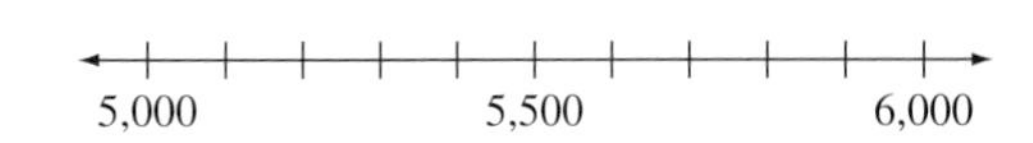

5,623 is closer to _____, so it is the better approximation. 5,632 rounds up/down to _____.

14. Round 135 to the nearest ten.

135 will round up to _____ or down to _____.

Because 135 is _______ between _____ and _____, 135 rounds up/down to _____.

When a whole number is halfway between the two possible approximations, round up to the next whole number. "Ties go to the large number."

15. Round 9,423 to the nearest thousand.

9,423 will round up to ____,000 or down to ____,000.

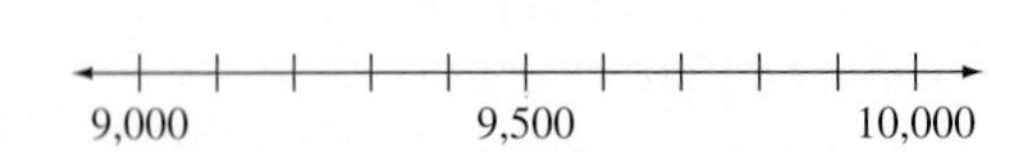

9,423 is closer to _____, so it is the better approximation. 9,423 rounds up/down to _____ .

The formal procedure for rounding uses the concepts from Guided Practices 11 through 15. Keep those concepts in mind as you look at the next example.

Procedure Round a Whole Number

Step 1 Identify the digit to be rounded.

Step 2 Decide if the digit to be rounded increases by one or stays the same.

Increase the digit when the digit to the right is 5 through 9.
The digit stays the same when the digit to the right is 0 through 4.

Step 3 All of the digits to the right of the rounded digit will be zero.

DETAILED EXAMPLE Rounding a Whole Number

Round 735,082 to the nearest ten thousand.

Step 1 Identify the digit to be rounded.

7(3)5,082

Three is in the ten thousands place.

Step 2 Decide if the digit to be rounded increases by one or stays the same.

- Increase the digit when the digit to the right is 5 through 9.
- The digit stays the same when the digit to the right is 0 through 4.

7(3)5,082

Because the next digit to the right is 5, the digit 3 will increase by one, 7(3)5,082 rounds up to 7(4)0, 000.

Step 3 All of the digits to the right of the rounded digit will be zero.

7(4)0,000

All of the place values to the right of 4 are replaced with zeros, 740,000 is the best approximation for 735,082.

735,082 rounds up to 740,000.

Notice the following:

- We say that a number "rounds up" when the rounded digit increases by one.
- We say that a number "rounds down" when the rounded digit stays the same.
- Whether we are rounding up or down, all digits to the right of the rounded digit will become zero.

EXAMPLES

16. Round 5,590 to the hundreds place.

Step 1 Identify the digit in the hundreds place.

5,(5)90

5,590 will round to 5,500 or 5,600.

Step 2 Because the tens digit is 9, we round up.

Step 3 5,590 rounds up to 5,600.

GUIDED PRACTICE

16. Round 34,821 to the thousands place.

Step 1 Identify the digit in the thousands place.

34,821

34,821 will round to ______ or ______.

Step 2 Because the hundreds digit is _____, we round up/down.

Step 3 34,821 rounds up/down to _____.

(Continued)

17. Round 834,724 to the ten thousands place.	**17.** Round 17,683 to the hundreds place.
Step 1 Identify the digit in the ten thousands place. 834,724 834,724 will round to 830,000 or 840,000. **Step 2** Because the thousands digit is 4, we round down. **Step 3** 834,724 rounds down to 830,000.	**Step 1** Identify the digit in the hundreds place. 17,683 17,683 will round to _____ or _____. **Step 2** Because the tens digit is ____, we round up/down. **Step 3** 17,683 rounds up/down to _____.
18. Round 487,380 to the hundred thousands place.	**18.** Round 52,908 to the ten thousands place.
Step 1 Identify the digit in the hundred thousands place. 487,380 487,380 will round to 400,000 or 500,000. **Step 2** Because the ten thousands digit is 8, we round up. **Step 3** 487,380 rounds to 500,000.	**Step 1** Identify the digit in the ten thousands place. 52,908 52,908 will round to _____ or _____. **Step 2** Because the thousands digit is ____, we round up/down. **Step 3** 52,908 rounds to _____

Concept Check

D1. **a.** What whole number is halfway between 11,000 and 12,000?
b. How can you decide if a whole number is closer to 12,000 than 11,000?

D2. How can you decide if a whole number is closer to 0 or 1,000,000?

D3. Describe the concept of rounding.

Objective D Practice

Choose the best approximation for each whole number.

D4. Is 1,000 or 2,000 a better approximation for 1,587?

D5. Is 1,000 or 2,000 a better approximation for 1,298?

D6. Is 320 or 330 a better approximation for 321?

D7. Is 170 or 180 a better approximation for 172?

D8. Is 432,000 or 433,000 a better approximation for 432,501?

D9. Is 237,000 or 238,000 a better approximation for 237,499?

Round each whole number to the given place value.

D10. 63 to the tens place

D11. 87 to the tens place

D12. 673 to the hundreds place

D13. 843 to the hundreds place

D14. 15,732 to the thousands place

D15. 64,382 to the thousands place

D16. 946,093 to the hundred thousands place

D17. 97,000 to the ten thousands place

D18. A new house costs $186,542. Round its cost to the nearest ten thousand dollars.

D19. An airplane costs $3,839,512. Round its cost to the nearest hundred thousand dollars.

D20. If you have $1,483 in the bank, how much money do you have, to the nearest hundred dollars?

D21. A used car costs $12,499. How much does the car cost, to the nearest thousand dollars?

Objective E Read Whole Numbers on a Bar Graph

The Concept Bar graphs (or charts) can be used to present information in a way that is easy to read and understand. The following bar graph relates temperature to the time of day.

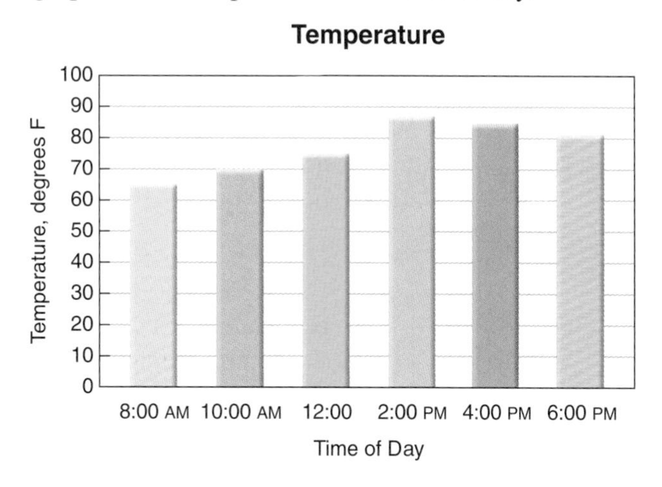

Using the graph, we can quickly approximate the temperature at different times of day. To estimate the temperature at 10 o'clock, find the bar that represents 10 A.M. and read the temperature at the top of the bar, 70°F.

Often you will have to estimate values on a bar graph. The temperature at 2 P.M. is between 80° and 90°F. 85°F might seem like a good guess. However, the bar is above the halfway point, so 87°F is a better approximation.

EXAMPLES

Plato's parents measured his height as he was growing up and made a bar graph using the data they collected. Read the following bar graph to answer the questions below. Answers should be accurate to the nearest inch.

19. How tall was Plato when he was 16?

We must read the height from the top of the bar labeled "16."

Plato was 67 inches tall when he was 16.

GUIDED PRACTICE

On a rainy day, Mr. Descartes became curious about the value of a car as it ages. To investigate, he made a bar graph. Read the following bar graph to answer the questions below. Answers should be accurate to the nearest thousand dollars.

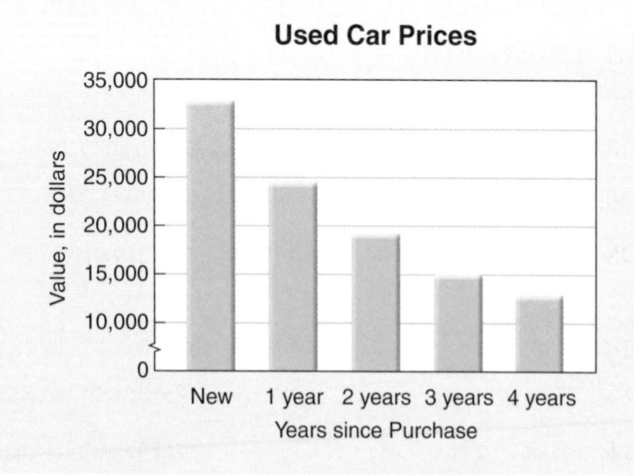

19. How much was the car worth after 3 years?

We must read the height from the top of the bar labeled "_______".

Mr. Descartes' car was worth $ _______ after 3 years.

(Continued)

20. How tall was Plato when his parents started measuring his height?

We need to look at the first bar.

Plato was 55 inches tall when his parents started measuring his height.

21. Between what ages does it look as though Plato grew the most?

We need to look at the top of each bar for the biggest step up between ages.

It looks as though Plato grew the most between ages 15 and 16.

22. How tall was Plato when his parents stopped recording the data?

We need to look at the last bar.

Plato was about 73 inches tall when his parents stopped recording the data.

20. How much did Mr. Descartes pay for his car?

The purchase price of the car is represented by the _____ bar.

Mr. Descartes paid $_______ for his car.

21. In what one-year period does it look as though the car lost the most value?

Between what one-year periods do the bars make the biggest step down?

Mr. Descartes' car lost the most value between _____ and _____.

22. How much was the car worth when Mr. Descartes stopped recording the data?

The last piece of data shown is above the bar labeled "_____ years."

Mr. Descartes' car was worth _______ after _____ years.

Concept Check

E1. What type of graph can be used to show data that relates two pieces of information?

E2. Describe the meaning of *data* in your own words.

Objective E Practice

Use the following bar graph to answer each question. Estimate each value as accurately as you can.

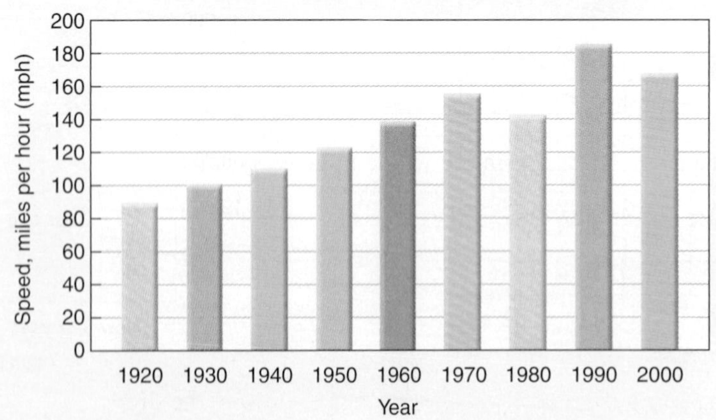

E3. What was the average winning speed in 1950?

E4. What was the average winning speed in 2000?

E5. What was the slowest winning speed from 1970 to 2000?

E6. What was the fastest winning speed from 1950 to 1980?

E7. Over how many decades did the average winning speed increase?

E8. Over how many decades did the average winning speed decrease?

Section 1.1 Exercises

FOR EXTRA HELP     

To understand digits, whole numbers, and place value:

1. Answer the Objective A Concept Checks.
2. Answer the odd Objective A Practice Exercises.
3. Answer the even Objective A Practice Exercises.

To write whole numbers in expanded form:

4. Answer the Objective B Concept Checks.
5. Answer the odd Objective B Practice Exercises.
6. Answer the even Objective B Practice Exercises.

To graph whole numbers on a number line:

7. Answer the Objective C Concept Checks.
8. Answer the odd Objective C Practice Exercises.
9. Answer the even Objective C Practice Exercises.

To round whole numbers:

10. Answer the Objective D Concept Checks.
11. Answer the odd Objective D Practice Exercises.
12. Answer the even Objective D Practice Exercises.

To understand how to read whole numbers on a bar graph:

13. Answer the Objective E Concept Checks.
14. Answer the odd Objective E Practice Exercises.
15. Answer the even Objective E Practice Exercises.

VOCABULARY REVIEW

Review the Vocabulary Preview for Section 1.1. Study the definitions until you can check Got It *for every term.*

Use these words to complete each sentence.

digits • whole numbers • place value • periods • rounding

	Got It	Get Help
16. 17, 23, 9, and 2 are examples of ____________.		
17. There are exactly ten ____________ in the whole number system.		
18. ____________ divide whole numbers into groups of three digits. The first four are ____________, ____________, ____________, and ____________.		
19. The process of choosing the best approximation for a number at a given place value is called ____________.		
20. The value represented by a digit in a whole number depends on the digit's ____________.		
21. List the first ten whole numbers. ____________		

How will you get help for any vocabulary about which you are unsure?

Your instructor____ MyMathLab____ A classmate____ A tutor____ Other____

Write each whole number using words.

22. 12,348 **23.** 67,456 **24.** 126,785 **25.** 3,127

26. 215,854 **27.** 315 **28.** 456,734 **29.** 1,125,568

30. 309 **31.** 560,200,107 **32.** 8,070,000,012 **33.** 12,005,000

Fill in the empty blanks in each check.

34.

Johann Doh
1620 Plymouth Rock
Heavy, Massachusetts 98067
1168
DATE *Jan. 13* 20*10*
PAY to the ORDER of *Dave Osborne* $ ____
Three thousand four hundred fifty DOLLARS
United Bank of Trusts
1 Round Circle
Diameter, PI 31415
MEMO *Being super!* *Johann Doh*

35.

Johann Doh
1620 Plymouth Rock
Heavy, Massachusetts 98067
1168
DATE *March 14* 20*10*
PAY to the ORDER of *Dorothy Hammel* $ ____
Eight hundred eight thousand eighty-eight DOLLARS
United Bank of Trusts
1 Round Circle
Diameter, PI 31415
MEMO *A lifetime of figure eights!* *Johann Doh*

36.

Johann Doh
1620 Plymouth Rock
Heavy, Massachusetts 98067
1168
DATE *Feb. 29* 20*10*
PAY to the ORDER of *Yogi Berra* $ *10,065*
____ DOLLARS
United Bank of Trusts
1 Round Circle
Diameter, PI 31415
MEMO ____ *Johann Doh*

37.

Johann Doh
1620 Plymouth Rock
Heavy, Massachusetts 98067
1168
DATE *Dec. 25* 20*10*
PAY to the ORDER of *Rod Price* $ *201,041*
____ DOLLARS
United Bank of Trusts
1 Round Circle
Diameter, PI 31415
MEMO ____ *Johann Doh*

Write each whole number in expanded form.

38. 28 **39.** 37 **40.** 2,600 **41.** 3,800

42. 91,730 **43.** 85,290 **44.** 7,000,000,300 **45.** 8,000,100,000

46. A bank cashier has only \$1's, \$10's, and \$100's in her drawer. Write \$233 in expanded form.

47. A supermarket cashier has \$1's, \$5's, \$10's, \$20's, \$50's, and \$100's in his drawer. Write \$222 in expanded form.

Write each whole number in standard form.

48. $200 + 60$ **49.** $300 + 40$ **50.** $500 + 40 + 3$ **51.** $600 + 50 + 4$

52. $4{,}000 + 300 + 50$ **53.** $2{,}000 + 700 + 20$ **54.** $8{,}000 + 10$ **55.** $3{,}000 + 9$

Graph each set of whole numbers on a number line.

56. 2, 14, 6, 10 **57.** 27, 21, 22, 30 **58.** 5, 30, 20, 40

59. 10, 70, 50, 20 **60.** 200, 1,000, 500 **61.** 1,200, 1,500, 1,600

Round each whole number to the given place value.

62. 67 to the tens place **63.** 86 to the tens place **64.** 623 to the hundreds place

goetz When rounding numbers, remember that you are finding a 'best approximation.' Rounded numbers are easier to work with.
For more tips and tweets, go to twitter.com/gstbasicmath

65. 868 to the hundreds place

66. 16,532 to the thousands place

67. 84,523 to the thousands place

68. 234,542 to the ten thousands place

69. 689,512 to the hundred thousands place

70. 740,001 to the hundred thousands place

71. A scrap dealer loaded 5,329 pounds of aluminum onto a trailer. How many pounds did he load, to the nearest hundred pounds?

72. A scrap dealer received $11,872 for a truckload of aluminum. How much money did he receive, to the nearest thousand dollars?

73. Zeus took 78 hours to complete a quilt for a charity auction. How many hours is that, to the nearest ten hours?

74. Round $54,635 to the nearest:

a. Hundred dollars.
b. Ten dollars.
c. Ten thousand dollars.

75. Round $123,564 to the nearest:

a. Hundred dollars.
b. Hundred thousand dollars.
c. Ten thousand dollars.

76. Round 6,567,347 to the nearest:

a. Thousand.
b. Hundred thousand.
c. Million.

77. Round 3,564,326 to the nearest:

a. Hundred.
b. Hundred thousand.
c. Million.

Use the bar graphs to answer each question.

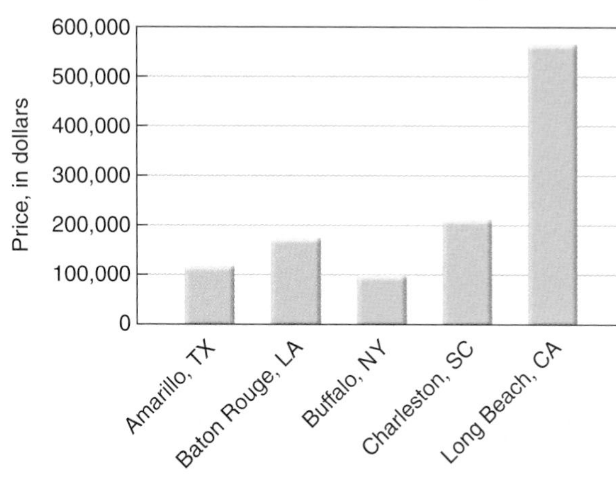

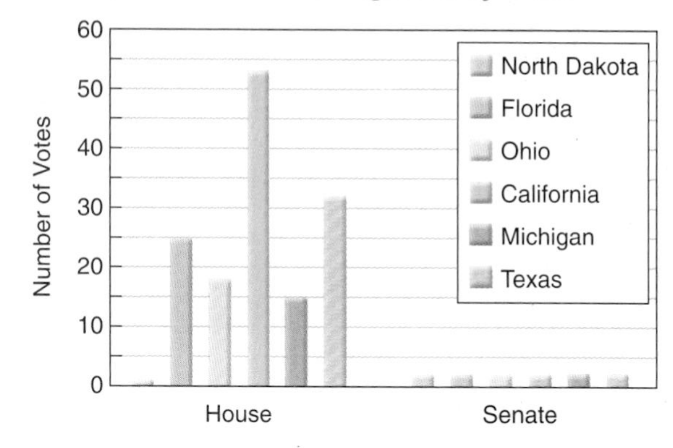

78. The bar graph shows how the median (middle) price of a home varies in the United States.

a. What is the median price of a home in Buffalo, NY?
b. What is the median price of a home in Charleston, SC?
c. Approximately how many homes in Amarillo, TX, could you buy with the money that it takes to purchase one home in Long Beach, CA?

79. The bar graph shows how many votes each state had in both the U.S. House of Representatives and the U.S. Senate during 2005.

a. Which state had the most votes in the House?
b. Order the states from most to least number of votes in the House. Indicate both the state and the number of votes the state had.
c. Which states, if any, have more votes in the Senate than in the House?

Section 1.1 Question Log

Use this space to write questions. Make sure you get these answered and revisit them when you prepare for your exam.

Page ____ Answered ☐	Page ____ Answered ☐
Page ____ Answered ☐	Page ____ Answered ☐

Section 1.2 Adding Whole Numbers

The addition procedure is based on counting to get a total. To use it, you must learn the basic facts and understand how place value is used in each step.

The **Objectives** in Section 1.2 will help you

- A Understand the basic addition facts.
- B Carry to regroup a number.
- C Add large numbers.

VOCABULARY PREVIEW *Check the box that applies.*

	Got It	Must Study
Operations: The four elementary **operations** are addition, subtraction, multiplication, and division.		
Sum: A **sum** is the result of adding.		
Carrying: Carrying lets you regroup a 10 from one place value into a 1 in the next larger place value.		
Commutative Property of Addition: The order in which you add numbers will not change the sum. Example: $2 + 3 = 3 + 2$		
Perimeter: The distance around a figure is its **perimeter.**		
Variable: A **variable** is a symbol, usually a letter of the alphabet, that represents a number.		

Study these terms when they appear in the text. At the end of the section, test your understanding by completing the Vocabulary Review in the section exercises.

Objective A Understand the Basic Addition Facts

The Concept Addition is developed from counting. For example, think about the sum $5 + 4$. If someone did not know this basic fact, he or she could count to the answer using a number line.

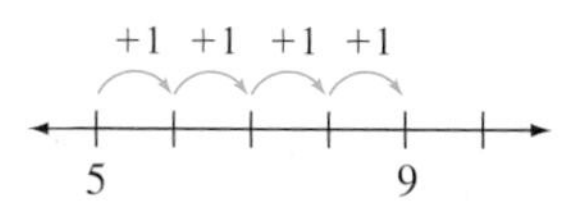

▶ **NOTE** Being able to visualize what is happening in mathematics is a useful skill. Even if you know all of the basic facts, you will benefit from drawing number lines to demonstrate addition.

The information in this objective will help you understand addition using a number line and improve your knowledge of basic addition facts.

Procedure Use a Number Line for Addition

Step 1 Draw a number line with a dot at the first number.

Step 2 Move the second number of units to the right.

EXAMPLE

1. Use a number line to find the sum. $7 + 4$

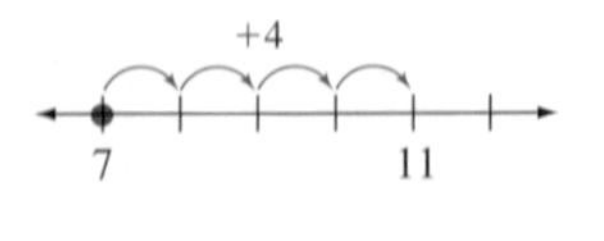

$7 + 4 = 11$

GUIDED PRACTICE

1. Use a number line to find the sum. $9 + 3$

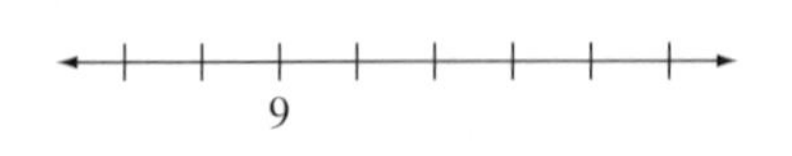

$9 + 3 =$ ____

The basic operations, including addition, have properties that let us work with numbers more easily. For instance, the Commutative Property of Addition tells us that $2 + 3 = 3 + 2$.

INTERACTIVE DEFINITION Commutative Property of Addition

The **Commutative Property of Addition** states that the order in which numbers are added does not change the sum.

EXAMPLE

2. Demonstrate the Commutative Property of Addition using $4 + 3$.

$4 + 3 = 7$
$3 + 4 = 7$

The order in which numbers are added does not change the sum.

GUIDED PRACTICE

2. Demonstrate the Commutative Property of Addition using $6 + 5$.

$6 + 5 =$ ____
$5 + 6 =$ ____

The ____ in which numbers are added does not change the sum.

DO YOU UNDERSTAND the Commutative Property of Addition? Got It Get Help

Concept Check

A1. In your own words, how can you use a number line to demonstrate the concept of addition?

Objective A Practice

Draw a number line to visualize each sum.

A2. $4 + 2$ **A3.** $6 + 2$ **A4.** $7 + 8$ **A5.** $9 + 6$

Complete each addition table. The outlined box is 12 because 5 + 7 = 12.

A6.

+	7	3	9	8
5	12			
7				
8				
6				

A7.

+	4	9	5	6
4				
9				
3				
6				

Writing only the answers, try to complete each column in fewer than 30 seconds. Accuracy is important. Take enough time to answer each exercise correctly.

A8. Start Time: _____
a. 0 + 6
b. 1 + 2
c. 3 + 8
d. 7 + 2
e. 6 + 7
f. 8 + 5
g. 5 + 2
h. 4 + 9
i. 9 + 7
j. 8 + 8
Finish Time: _____
Total Time: _____

A9. Start Time: _____
a. 9 + 1
b. 6 + 9
c. 5 + 2
d. 2 + 8
e. 9 + 0
f. 7 + 6
g. 3 + 9
h. 8 + 4
i. 7 + 5
j. 8 + 7
Finish Time: _____
Total Time: _____

A10. Start Time: _____
a. 5 + 6
b. 8 + 5
c. 3 + 7
d. 4 + 9
e. 0 + 8
f. 9 + 5
g. 9 + 8
h. 1 + 4
i. 9 + 9
j. 6 + 8
Finish Time: _____
Total Time: _____

A11. Start Time: _____
a. 8 + 9
b. 5 + 1
c. 6 + 8
d. 2 + 6
e. 7 + 8
f. 8 + 1
g. 7 + 7
h. 6 + 5
i. 9 + 0
j. 9 + 4
Finish Time: _____
Total Time: _____

▶ **NOTE** Don't worry if it takes you longer than 30 seconds. However, if it does, make flash cards to study your basic facts.

Objective B Carry to Regroup a Number

The Concept When we add digits with the same place value, we may get an answer that is greater than 9. When this happens, we must regroup the answer by carrying a digit to its correct place value. To help you understand the concept of carrying, we will look at examples that use sums of money. To add money in these examples, we will do the following:

- Use only \$1 bills and \$10 bills
- Change or regroup the bills to keep the fewest number of bills possible

As you follow the examples, keep in mind that the bills are used to indicate place value.

DETAILED EXAMPLE Carrying to Regroup a Number

A Sum That Does Not Require Carrying

Find the sum of $14 and $23.

$$\begin{array}{r} 1\ 4 \\ +\ 2\ 3 \\ \hline 7 \end{array}$$ Basic fact: $4 + $3 = $7

Bacause we have fewer than 10 ones, we do not need to change any $1 bills for $10 bills. In other words, we do not need to carry.

$$\begin{array}{r} 1\ 4 \\ +\ 2\ 3 \\ \hline 3\ 7 \end{array}$$ Basic fact: $1 + $2 = $3

$14 + $23 = $37

A Sum That Requires Carrying

Find the sum of $24 and $38.

$$\begin{array}{r} ^{1} \\ 2\ 4 \\ +\ 3\ 8 \\ \hline 2 \end{array}$$ Basic fact: $4 + $8 = $12

Because we have more than 10 ones, we change ten $1 bills for one $10 bill. This leaves one $10 bill and two $1 bills. The $10 bill is regrouped by carrying the digit 1 to the tens place.

$$\begin{array}{r} ^{1} \\ 2\ 4 \\ +\ 3\ 8 \\ \hline 6\ 2 \end{array}$$ $1 + $2 + $3 = $6

$24 + $38 = $62

Procedure Add Whole Numbers

Step 1 Add the digits in the ones place. Carry if necessary.

Step 2 Add the digits in the tens place. Carry if necessary.

EXAMPLES

3. Find the sum of 36 and 23.

Step 1 Add the digits in the ones place.

Step 2 Add the digits in the tens place.

$$\begin{array}{r} 3\ 6 \\ +\ 2\ 3 \\ \hline 5\ 9 \end{array}$$

4. Find the sum of 39 and 7.

Step 1 Add the digits in the ones place and carry the one.

Step 2 Add the digits in the tens place, including the carried one.

$$\begin{array}{r} ^{1} \\ 3\ 9 \\ +\ \ \ 7 \\ \hline 4\ 6 \end{array}$$

GUIDED PRACTICE

3. Find the sum of 42 and 17.

$$\begin{array}{r} 4\ 2 \\ +\ 1\ 7 \\ \hline \end{array}$$

4. Find the sum of 58 and 5.

$$\begin{array}{r} 5\ 8 \\ +\ \ \ 5 \\ \hline \end{array}$$

5. Find the sum of 85 and 49.	**5.** Find the sum of 73 and 38.
Step 1 Add the digits in the ones place and carry the one. **Step 2** Add the digits in the tens place and carry the one to the hundreds place. $\begin{array}{r} ^{1\,1} \\ 85 \\ +\ 49 \\ \hline 134 \end{array}$	$\begin{array}{r} 73 \\ +\ 38 \\ \hline \end{array}$
6. Find the sum of 71, 52, and 13.	**6.** Find the sum of 92, 23, and 11.
Step 1 Add the digits in the ones place. **Step 2** Add the digits in the tens place and carry the one to the hundreds place. $\begin{array}{r} ^{1} \\ 71 \\ 52 \\ +\ 13 \\ \hline 136 \end{array}$	$\begin{array}{r} 92 \\ 23 \\ +\ 11 \\ \hline \end{array}$

Concept Check

B1. Use \$5 + \$8 to explain the concept of carrying.

Objective B Practice

Find each sum.

B2. 32 + 54 **B3.** 61 + 38 **B4.** 42 + 39 **B5.** 65 + 28

B6. 93 + 28 **B7.** 48 + 81 **B8.** 65 + 78 **B9.** 83 + 58

B10. 17 + 42 + 83 **B11.** 65 + 21 + 83

Objective C Add Large Numbers

The Concept The procedure to add large numbers is the same as the procedure to add small numbers. However, we will have more columns of digits to add.

Procedure Add Large Numbers

Step 1 Add the digits in the ones place. Carry if necessary.
Step 2 Add the digits in the tens place. Carry if necessary.
Step 3 Repeat the procedure for any higher place values.

EXAMPLES

7. Find the sum of 1,852 and 427.

Step 1 Add the digits in the ones place.
Step 2 Add the digits in the tens place.
Step 3 Repeat the procedure for any higher place values.

```
   1
   1 8 5 2
+    4 2 7
   2 2 7 9
```

GUIDED PRACTICE

7. Find the sum of 3,687 and 1,032.

```
   3 6 8 7
+  1 0 3 2
```

EXAMPLES

8. Find the sum of 390, 785, and 629.

Step 1 Add the digits in the ones place.
Step 2 Add the digits in the tens place.
Step 3 Repeat the procedure for any higher place values.

```
   1 2 1
     3 9 0
     7 8 5
+    6 2 9
   1 8 0 4
```

To add three or more digits, use small steps. To add the tens digits in the example, we thought "1 + 9 is 10, 10 + 8 is 18, 18 + 2 is 20."

GUIDED PRACTICE

8. Find the sum of 482, 813, and 992.

```
   4 8 2
   8 1 3
+  9 9 2
```

Concept Check

C1. Use small steps to add 2 + 3 + 5 + 7 mentally. See the note from Example 8 and Guided Practice 8.

Objective C Practice

Find each sum.

C2. 345 + 621

C3. 893 + 104

C4. 393 + 478

C5. 263 + 479

C6. 4,582 + 3,987

C7. 7,306 + 1,837

C8. 73,498 + 18,395

C9. 32,875 + 45,950

Use the table to answer each question.

C10. What were the combined receipts on Wednesday and Thursday?

C11. What were the combined receipts on Friday and Saturday?

Cash Register Receipts

Day	Amount
Wednesday	$492
Thursday	$634
Friday	$2,342
Saturday	$1,837

FOCUS ON Adding Several Numbers Efficiently

To add several numbers efficiently, we can use the Commutative Property of Addition.

Commutative Property of Addition

$$3 + 5 = 5 + 3$$

The order in which we add numbers does not change the sum.

When we are adding many digits, it is often useful to mentally group numbers that add to 10. Consider the sum, $4 + 6 + 3 + 9 + 7$. Using the commutative property, we mentally group the numbers to get sums of 10.

$$4 + 6 + 3 + 9 + 7$$

The colored pairs add to 10. Two "tens" and a 9 are left over. The sum is 29. Sometimes, three numbers can be grouped to form a sum of 10.

$$8 + 3 + 4 + 7 + 5 + 1$$

(The 3 and 7 are grouped to make 10; the 4 and 5 + 1 are grouped to make 10.)

The 3 can be paired with the 7. We look for a 6 to be paired with the 4. There isn't a 6, but the 5 and 1 can be added with 4. Two "tens" and an 8 are left over, so the sum is 28.

PRACTICE

Determine each sum by grouping digits to form "tens."

1. $6 + 4 + 9 + 1 + 8$
2. $7 + 3 + 5 + 5 + 4$
3. $1 + 7 + 4 + 2 + 6 + 5$
4. $2 + 6 + 3 + 1 + 4 + 5$
5. $4 + 7 + 6 + 2 + 5 + 1 + 4$
6. $5 + 6 + 3 + 2 + 4 + 1 + 3$

Combining Concepts and Applications

CONCEPT I Finding the Perimeter of a Figure The **perimeter** of a figure is the distance around the figure. To find a perimeter, add the lengths of all of the sides.

EXAMPLE

9. To plant flowers, Pierre must find the perimeter of his backyard.

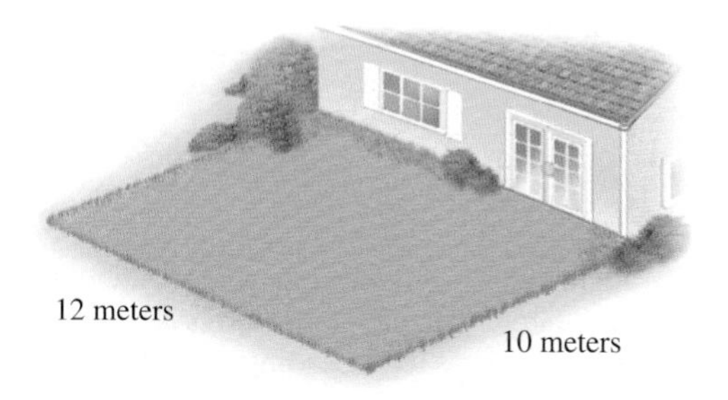

GUIDED PRACTICE

9. In a triathlon, Arianna must swim the course shown below. Find the perimeter of the triangle to see how far Arianna must swim.

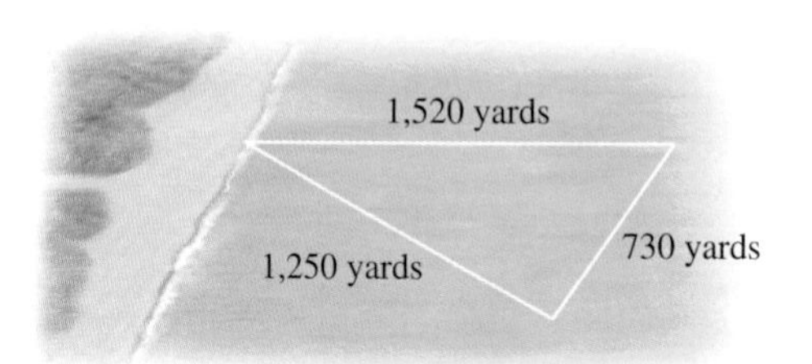

(Continued)

To find the perimeter of a figure, add the lengths of all of the sides.

$$\begin{aligned} \text{Perimeter} &= \underbrace{12 + 10} + 12 + 10 \\ &= \underbrace{22 + 12} + 10 \\ &= \underbrace{34 + 10} \\ &= 44 \text{ meters} \end{aligned}$$

$$\begin{aligned} \text{Perimeter} &= \underbrace{\quad + \quad} + \quad \\ &= \underbrace{\quad + \quad} \\ &= \quad \text{yards} \end{aligned}$$

CONCEPT II Solving for an Unknown Number Often you will need to solve a problem that has an unknown quantity. In mathematics, we use variables to represent unknown numbers. A **variable** is a symbol, usually a letter of the alphabet, that represents a number. For instance, if you are 8 miles into a 15-mile race, you may wonder how far you have to go. To answer that question, you could ask, "What number added to 8 gives 15?"

$$8 + x = 15$$
$$x = 7$$

Use the basic fact 8 + 7 = 15.

Two methods for solving this type of exercise are shown in the following examples:

EXAMPLES

10. What number added to 7 gives 12?

Use basic facts to find the unknown number.

$$7 + x = 12$$
$$x = 5$$

Use the basic fact 7 + 5 = 12.

11. What number added to 21 gives 53?

Count to find the unknown number.

$$21 + y = 53$$

Starting at 21, count by tens. Don't go over 53.

21, 31, 41, 51

3 tens

We are still 2 short of 53, so we need to add 2 ones.

21, 31, 41, 51, 52, 53

3 tens *2 ones*

$$3 \text{ tens} + 2 \text{ ones} = 32$$
$$y = 32$$

GUIDED PRACTICE

10. What number added to 4 gives 10?

Use basic facts to find the unknown number.

$$4 + y = 10$$
$$y =$$

Use the basic fact 4 + = .

11. What number added to 35 gives 68?

Count to find the unknown number.

$$35 + x = 68$$

Starting at 35, count by tens. Don't go over 68.

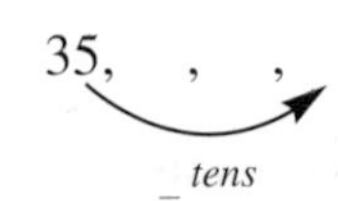

We are still _____ short of 68, so we need to add _____ ones.

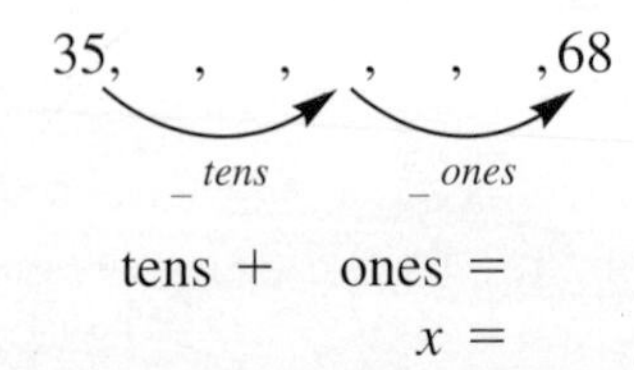

tens + ones =

$$x =$$

When counting to an answer, it is often convenient to use your fingers. You may be surprised to learn that professional mathematicians sometimes use their fingers to do arithmetic.

Section 1.2 Exercises

FOR EXTRA HELP      

To understand the basic addition facts:

1. Answer the Objective A Concept Check.
2. Answer the odd Objective A Practice Exercises.
3. Answer the even Objective A Practice Exercises.

To carry to regroup a number:

4. Answer the Objective B Concept Check.
5. Answer the odd Objective B Practice Exercises.
6. Answer the even Objective B Practice Exercises.

To add large numbers:

7. Answer the Objective C Concept Check.
8. Answer the odd Objective C Practice Exercises.
9. Answer the even Objective C Practice Exercises.

To add several numbers efficiently:

10. Answer the odd Focus On Adding Several Numbers Efficiently Practice Exercises.
11. Answer the even Focus On Adding Several Numbers Efficiently Practice Exercises.

VOCABULARY REVIEW *Review the Vocabulary Preview for Section 1.2. Study the definitions until you can check* Got It *for every term.*

Use these words to complete each sentence.

operations • carrying • perimeter • sum • Commutative Property of Addition • variable

	Got It	Get Help
12. ____________ lets you regroup a 10 from one place value into a 1 in the next larger place value.		
13. The ____________ allows you to reorder numbers that are being added.		
14. The distance around a figure is the figure's ____________.		
15. A letter of the alphabet that is used to represent a number is called a ____________.		
16. Addition, subtraction, multiplication, and division are the four elementary ____________.		
17. The result of adding is called a ____________.		

How will you get help for any vocabulary about which you are unsure?

Your instructor ___ MyMathLab ___ A classmate ___ A tutor ___ Other ___

Draw a number line to visualize each sum.

18. $3 + 5$ **19.** $5 + 2$ **20.** $2 + 8$

21. $9 + 3$ **22.** $10 + 6$ **23.** $3 + 19$

24. $0 + 6$

25. $8 + 0$

Complete each addition table.

26.

+	7	5	6	9
4				
8				
7				
6				

27.

+	9	7	5	4
3				
6				
5				
9				

28.

+	4	7	8	2
7				
9				
5				
4				

29.

+	8	4	9	0
2				
7				
6				
9				

Writing only the answers, try to complete each column in fewer than 30 seconds. Accuracy is important. Take enough time to work through each exercise correctly.

30. Start Time: ____
- **a.** $0 + 6$
- **b.** $1 + 2$
- **c.** $3 + 1$
- **d.** $2 + 6$
- **e.** $7 + 3$
- **f.** $3 + 8$
- **g.** $5 + 4$

Finish Time: ____

Total Time: ____

31. Start Time: ____
- **a.** $9 + 1$
- **b.** $6 + 9$
- **c.** $5 + 2$
- **d.** $6 + 6$
- **e.** $9 + 8$
- **f.** $4 + 9$
- **g.** $5 + 8$

Finish Time: ____

Total Time: ____

32. Start Time: ____
- **a.** $5 + 6$
- **b.** $8 + 5$
- **c.** $3 + 4$
- **d.** $5 + 6$
- **e.** $7 + 8$
- **f.** $4 + 8$
- **g.** $4 + 7$

Finish Time: ____

Total Time: ____

33. Start Time: ____
- **a.** $8 + 9$
- **b.** $5 + 1$
- **c.** $6 + 8$
- **d.** $3 + 6$
- **e.** $9 + 6$
- **f.** $8 + 7$
- **g.** $7 + 9$

Finish Time: ____

Total Time: ____

Find each sum.

34. $\begin{array}{r} 23 \\ +\,14 \\ \hline \end{array}$

35. $\begin{array}{r} 31 \\ +\,38 \\ \hline \end{array}$

36. $\begin{array}{r} 42 \\ +\,31 \\ \hline \end{array}$

37. $\begin{array}{r} 55 \\ +\,24 \\ \hline \end{array}$

38. $\begin{array}{r} 84 \\ +\,28 \\ \hline \end{array}$

39. $\begin{array}{r} 48 \\ +\,45 \\ \hline \end{array}$

40. $57 + 78$

41. $83 + 76$

42. $\begin{array}{r} 8 \\ 7 \\ +\,4 \\ \hline \end{array}$

43. $\begin{array}{r} 5 \\ 6 \\ +\,3 \\ \hline \end{array}$

44. $12 + 14 + 13$

45. $15 + 21 + 17$

46. $\begin{array}{r} 332 \\ +\;\;54 \\ \hline \end{array}$

47. $\begin{array}{r} 561 \\ +\;\;38 \\ \hline \end{array}$

48. $742 + 39$

49. $265 + 28$

50. $493 + 328$

51. $648 + 281$

52. $765 + 478$

53. $583 + 658$

Answer each question.

54. You sold three items at a yard sale: baseball cards for \$15, a dresser for \$22, and a weight bench for \$81. How much did you earn?

55. You are paying bills. Your electric bill is \$45, the phone bill is \$71, and the cable bill is \$37. How much money do you need in the bank to cover your checks?

56. In three straight games, a running back ran for 65 yards, 52 yards, and 63 yards. How many total yards did he run for all three games?

57. Sabrina is on a road trip. She has driven 75 miles to pick up a friend, 41 miles to get to the beach, and 66 miles to get to a better beach. What is the total distance she has driven?

Find each sum.

58. 245 + 321

59. 873 + 158

60. 293 + 468

61. 251 + 479

62. $\begin{array}{r} 1522 \\ +\ 4581 \\ \hline \end{array}$

63. $\begin{array}{r} 9406 \\ +\ 1733 \\ \hline \end{array}$

64. 23,698 + 48,395

65. 22,875 + 44,950

66. 654,293 + 28,764

67. 975,482 + 81,635

68. 651,789 + 278,675

69. 838,738 + 584,539

70. 678,456 + 34,567

Determine each sum by grouping digits to form "tens."

71. 7 + 4 + 3 + 6

72. 8 + 5 + 3 + 2

73. 4 + 3 + 2 + 6 + 5 + 3

74. 2 + 7 + 3 + 1 + 4 + 5

75. 4 + 7 + 4 + 2 + 1 + 2

76. 5 + 4 + 3 + 2 + 4 + 1 + 2

Find the perimeter of each figure.

77.

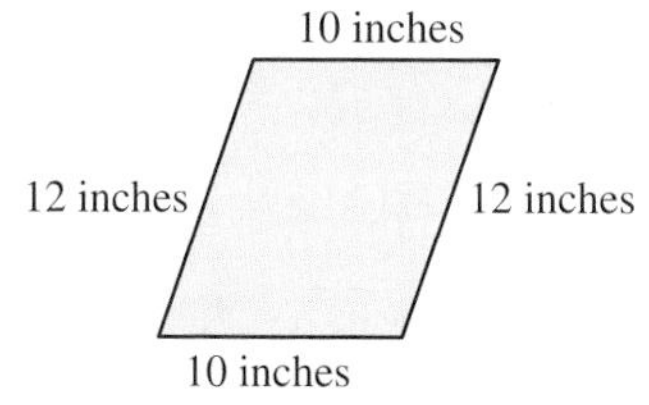

78.

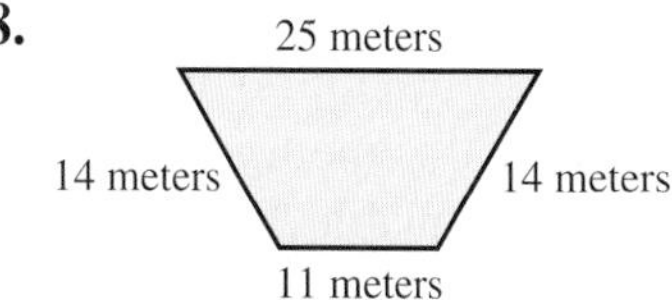

79.

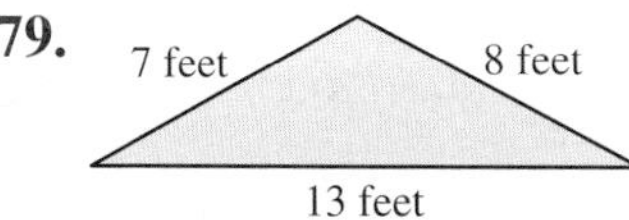

80.

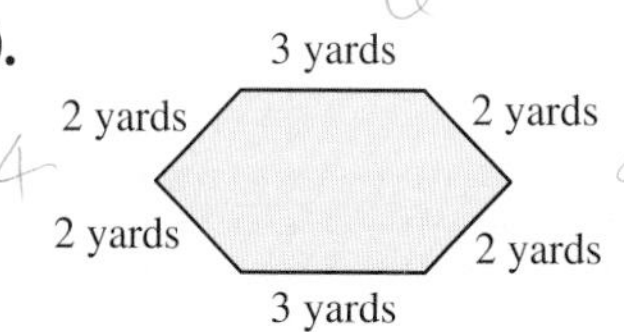

81. Workers from the building in the following picture are on a coffee break. They decide to have a shopping cart race inside the building, along the building's perimeter. About how long is the race course?

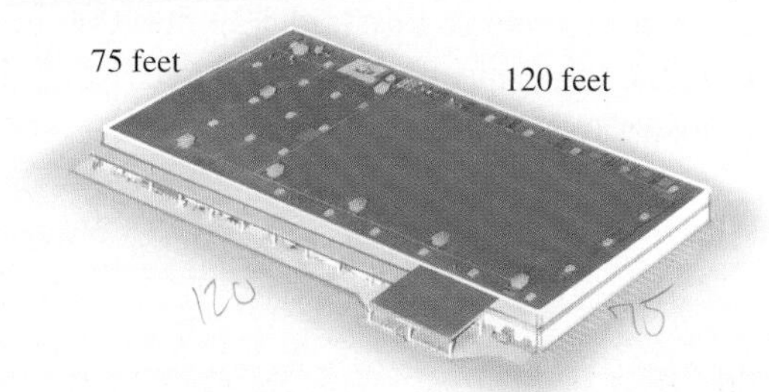

82. Gutters will be installed at the lower edge of each roof in the following picture. How many feet of gutter must be installed?

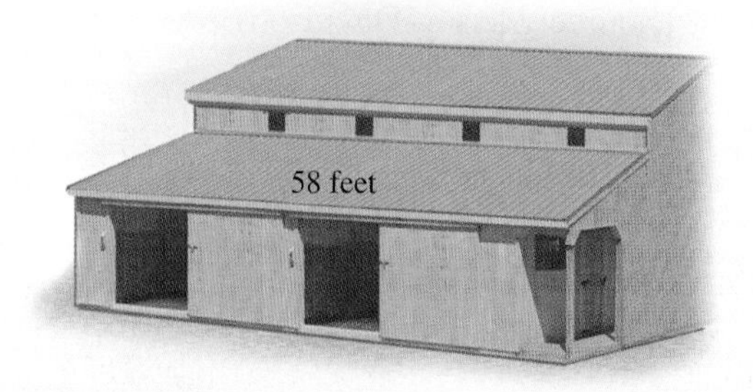

83. This month Pat spent \$120, \$85, \$202, and \$85 at the grocery store.

a. How much has she spent?

b. If the grocer gives a \$20 discount card to its customers when they spend more than \$500 in a month, will Pat get the card?

84. A retail store is placing an order for 50 pairs of pants, 125 shirts, 70 pair of socks, and 15 purses.

a. How many items did the store order?

b. A 10% discount is provided for all orders of at least 250 items. Will the store receive the discount?

Use the map provided to answer each question. Each of the line segments drawn around the upper and lower peninsulas of Michigan is approximately 150 miles long.

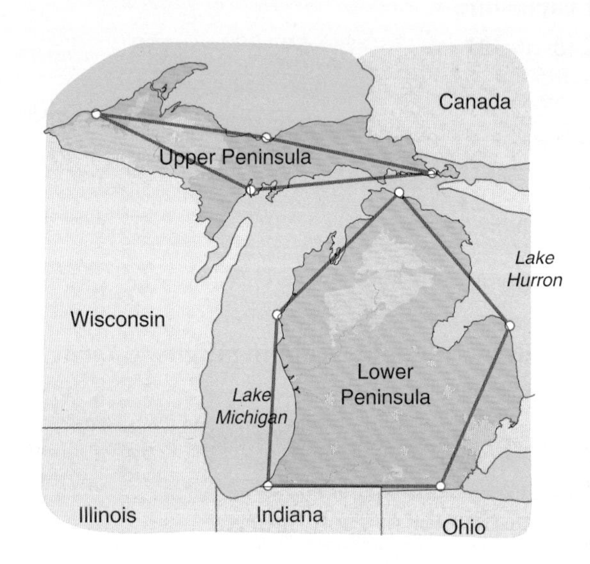

85. Approximately how long is the coastline of the upper peninsula?

86. Approximately how long is the coastline of the lower peninsula?

87. Do you believe the estimate in Exercise 85 is too high or too low? Why?

88. Do you believe the estimate in Exercise 86 is too high or too low? Why?

89. Even though the estimates above are not accurate, why might someone say that this method of finding a coastline's length is reasonable?

Solve each exercise.

90. Find the sum of 235 and 469.

91. Find the sum of 356 and 635.

92. Find the total of 54, 67, and 43.

93. Find the total of 58, 78, and 53.

94. What is 576 increased by 123?

95. What is 764 increased by 236?

96. In 2005, Harley-Davidson earned \$4,183,515,000 from motorcycle sales and \$815,678,000 from sales of parts and accessories. What were the total sales? (Source: Harley-Davidson)

97. Recording artist Jay-Z's album sales are fast approaching the album sales of Earth, Wind & Fire. By August 2006, Jay-Z had \$23,000,000 in sales and Earth, Wind & Fire had \$23,500,000 in sales. Find the total sales for both. (Source: Recording Industry Association of America)

98. As part of their training, a college basketball team runs wind sprints on the court. To complete a wind sprint, the players run from the base line, to the center line, back to the base line, down to the opposite base line, and back. What is the distance of one wind sprint?

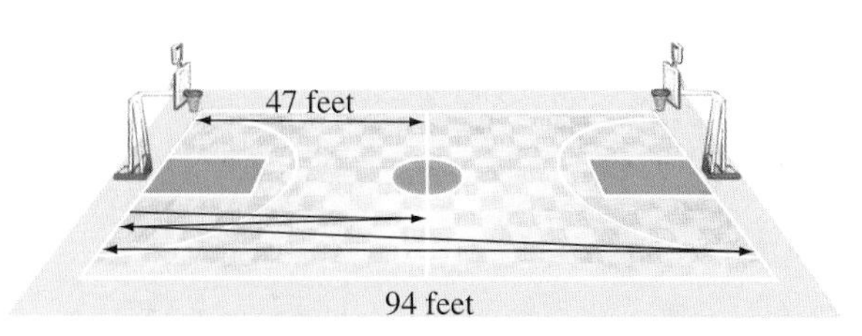

99. As part of their training, a college basketball team runs laps around the court. Find the total distance of three laps.

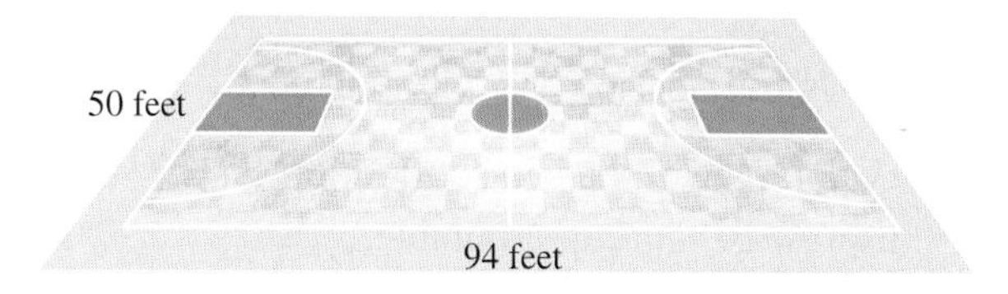

Solve for the unknown number.

100. $5 + x = 15$ **101.** $10 + x = 17$ **102.** $8 + y = 13$ **103.** $7 + y = 16$

104. $8 = 2 + z$ **105.** $11 = 4 + z$ **106.** $28 = 11 + x$ **107.** $38 = 29 + x$

108. $56 = 22 + y$ **109.** $24 + y = 45$ **110.** $52 = 14 + z$ **111.** $29 + z = 65$

Smith Use basic facts to identify the value of x that will make a true statement.

For more tips and tweets, go to twitter.com/gstbasicmath

Section 1.2 Question Log

Use this space to write questions. Make sure you get these answered and revisit them when you prepare for your exam.

Page ____ Answered	Page ____ Answered

Section 1.2 Extra Practice

See the Appendix A for extra practice on Section 1.2.

Section 1.3 Subtracting Whole Numbers

The subtraction procedure is based on counting backward to find a difference between two numbers. As with addition, you must learn the basic facts and understand how place value is used in each step.

The **Objectives** in Section 1.3 will help you

- A Understand the basic subtraction facts.
- B Borrow to regroup a number.
- C Subtract large numbers.
- D Subtract several numbers.

VOCABULARY PREVIEW *Check the box that applies.*	Got It	Must Study
Inverse Operations: Two operations are called **inverse operations** if they "undo" each other. Addition and subtraction are inverse operations.		
Difference: A **difference** is the result of subtracting.		
Borrowing: Borrowing lets you regroup a 1 from one place value into a 10 in the next smaller place value.		

Study these terms when they appear in the text. At the end of the section, test your understanding by completing the Vocabulary Review in the section exercises.

Objective A Understand the Basic Subtraction Facts

The Concept To develop an understanding of subtraction, we can use addition. Addition and subtraction are called **inverse operations** because they "undo" each other. To add, we count forward. To subtract, we count backward. This relationship can be visualized using a number line.

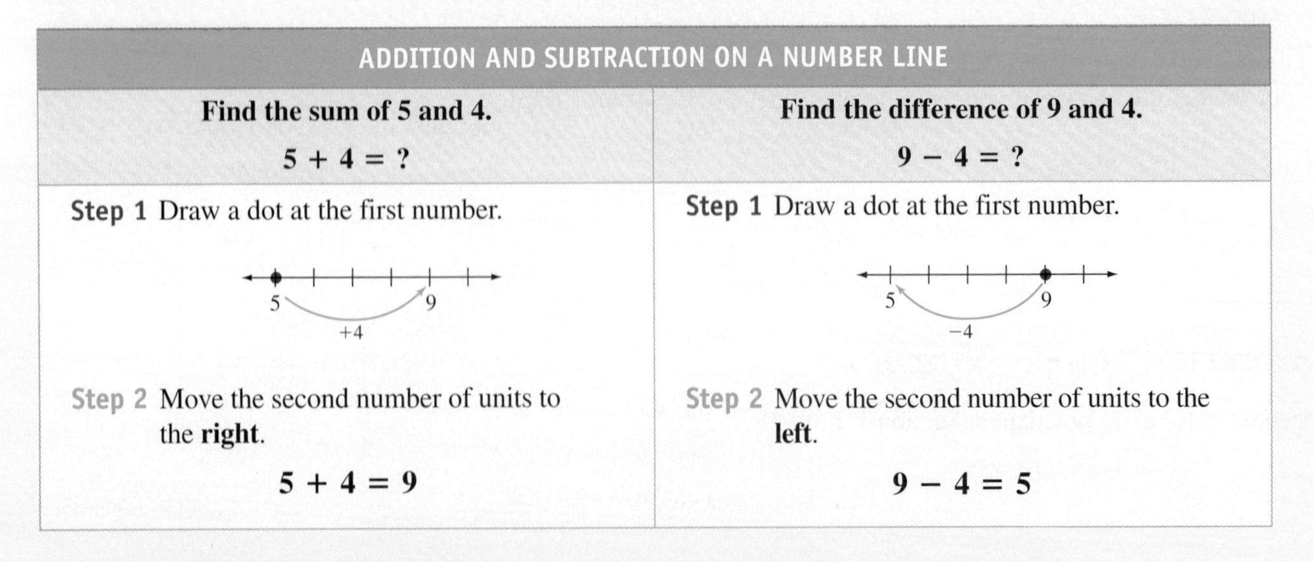

ADDITION AND SUBTRACTION ON A NUMBER LINE	
Find the sum of 5 and 4. $5 + 4 = ?$	**Find the difference of 9 and 4.** $9 - 4 = ?$
Step 1 Draw a dot at the first number.	**Step 1** Draw a dot at the first number.
Step 2 Move the second number of units to the **right**. $5 + 4 = 9$	**Step 2** Move the second number of units to the **left**. $9 - 4 = 5$

These two graphs can be combined to see why addition and subtraction "undo" each other.

Addition and Subtraction "Undo" Each Other

Because addition and subtraction are inverse operations, they "undo" each other. If we add 4 and then subtract 4, we'll end up where we started.

Starting with 5, add 4 and then subtract 4 to return to 5.

$5 + 4 = 9 \qquad 9 - 4 = 5$

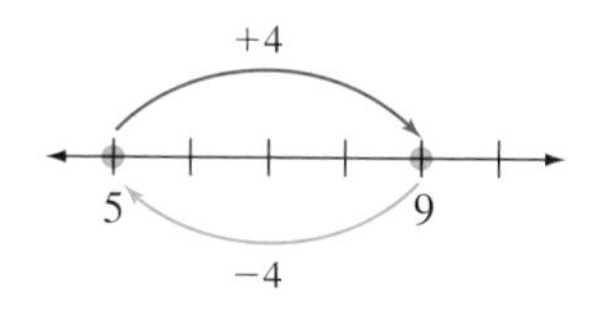

Procedure **Use a Number Line for Subtraction**

Step 1 Draw a number line with a dot at the first number.

Step 2 Move the second number of units to the left.

EXAMPLE

1. Use a number line to find the difference. $8 - 3$

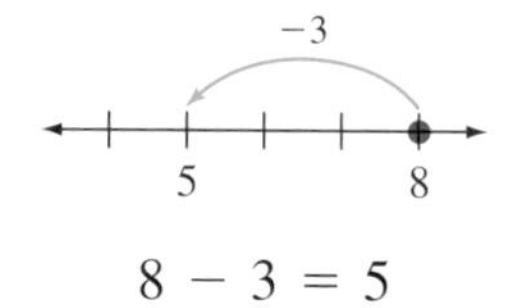

$8 - 3 = 5$

GUIDED PRACTICE

1. Use a number line to find the difference. $10 - 6$

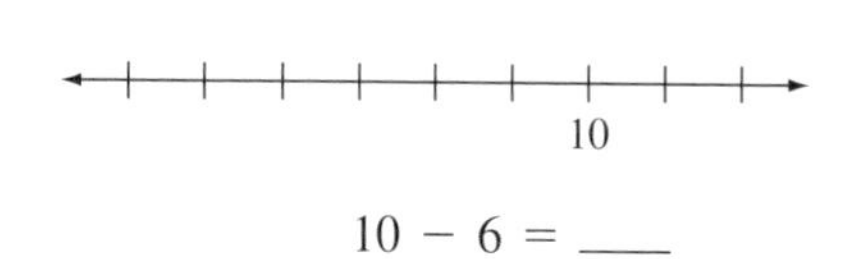

$10 - 6 =$ ____

Concept Check

A1. In your own words, how can you use a number line to demonstrate the concept of subtraction?

Objective A Practice

Draw a number line to visualize each difference.

A2. $4 - 2$

A3. $6 - 3$

A4. $15 - 8$

A5. $14 - 6$

Complete each addition table. You will need to use subtraction to find the numbers in some of the white boxes. The outlined box is 5 because 12 − 7 = 5.

A6.

+	7		8	
5	12			
				12
			12	
9		13		19

A7.

+				
	7	11		
6			13	
		13		10
	10		15	

Writing only the answers, try to complete each column in fewer than 30 seconds. Accuracy is important. Take enough time to work through each exercise correctly.

A8. Start Time: ____
a. 11 − 6
b. 13 − 5
c. 4 − 1
d. 9 − 2
e. 13 − 7
f. 10 − 8
g. 7 − 2
h. 8 − 4
i. 16 − 9
j. 14 − 8
Finish Time: ____
Total Time: ____

A9. Start Time: ____
a. 10 − 1
b. 15 − 9
c. 12 − 5
d. 13 − 9
e. 9 − 0
f. 13 − 6
g. 16 − 8
h. 12 − 4
i. 12 − 9
j. 9 − 7
Finish Time: ____
Total Time: ____

A10. Start Time: ____
a. 11 − 5
b. 13 − 8
c. 7 − 4
d. 14 − 6
e. 8 − 8
f. 15 − 6
g. 12 − 8
h. 15 − 8
i. 18 − 9
j. 11 − 9
Finish Time: ____
Total Time: ____

A11. Start Time: ____
a. 17 − 8
b. 6 − 1
c. 14 − 5
d. 11 − 8
e. 15 − 7
f. 10 − 1
g. 14 − 7
h. 11 − 4
i. 9 − 7
j. 13 − 4
Finish Time: ____
Total Time: ____

Objective B Borrow to Regroup a Number

The Concept When we subtract numbers, often we need to borrow from larger place values. To help you understand the concept of borrowing, we will look at examples that use differences of money. To subtract money in these examples, we will do the following:

- Use only \$1 bills and \$10 bills
- Change or regroup the bills to keep the fewest number of bills possible

If you understand how a \$10 bill can be "broken" into ten \$1 bills, then you understand how borrowing works.

Objective C Subtract Large Numbers

The Concept The procedure to subtract large numbers is the same as the procedure to subtract small numbers. However, we will have more columns of digits to subtract.

Procedure Subtract Large Numbers

Step 1 Subtract the digits in the ones place. Borrow if necessary.
Step 2 Subtract the digits in the tens place. Borrow if necessary.
Step 3 Repeat the procedure for any higher place values.

EXAMPLE

6. Find the difference. 1,852 − 427

Step 1 Subtract the digits in the ones place. Borrow if necessary.
Step 2 Subtract the digits in the tens place. Borrow if necessary.
Step 3 Repeat the procedure for any higher place values. Borrow if necessary.

$$\begin{array}{r} 1\ 8\ \overset{4}{\cancel{5}}\ {}^{1}2 \\ -\ \ \ 4\ 2\ 7 \\ \hline 1\ 4\ 2\ 5 \end{array}$$

GUIDED PRACTICE

6. Find the difference. 3,635 − 462

$$\begin{array}{r} 3\ 6\ 3\ 5 \\ -\ \ \ 4\ 6\ 2 \\ \hline \end{array}$$

DETAILED EXAMPLE Borrowing from Two or More Places Away

Find the difference. 307 − 128

At first, we cannot borrow from the tens place.

$$\begin{array}{r} 3\ 0\ 7 \\ -\ 1\ 2\ 8 \\ \hline \end{array}$$

7 − 8 is not a basic fact, and we can't borrow from 0 tens.

But we can borrow from the hundreds place.

$$\begin{array}{r} \overset{2}{\cancel{3}}\ {}^{1}0\ 7 \\ -\ 1\ \ 2\ 8 \\ \hline \end{array}$$

We can borrow 1 hundred, which gives us 10 tens.

This will let us borrow from the tens place.

$$\begin{array}{r} \overset{2}{\cancel{3}}\ \overset{9}{\cancel{10}}\ {}^{1}7 \\ -\ 1\ \ 2\ \ 8 \\ \hline \end{array}$$

Now we can borrow 1 ten, leaving 9 tens and 10 ones.

Together the solution will look like this:

$$\begin{array}{r} \overset{2}{\cancel{3}}\ \overset{9}{\cancel{10}}\ {}^{1}7 \\ -\ 1\ \ 2\ \ 8 \\ \hline 1\ \ 7\ \ 9 \end{array}$$

EXAMPLE	GUIDED PRACTICE
7. Find the difference. 2,304 − 1,208	**7.** Find the difference. 8,507 − 3,319

Step 1 Subtract the digits in the ones place. Borrow from two or more places if necessary.

$$\begin{array}{r} \overset{2}{\cancel{3}}\,\overset{9}{\cancel{0}}\,{}^{1}4 \\ 2\ \ \ \ \ \ \ \ \\ -1\,2\,0\,8 \\ \hline 1\,0\,9\,6 \end{array}$$

Step 2 Subtract the digits in the tens place.

Step 3 Repeat the procedure for any higher place values.

$$\begin{array}{r} 8\,5\,0\,7 \\ -3\,3\,1\,9 \\ \hline \end{array}$$

Concept Check

C1. When borrowing from the hundreds place to get more ones, why is it important to know that 100 is equal to 9 tens and 10 ones?

C2. Compare the two subtraction problems 123 − 89 and 543,983 − 145,320. Why might someone say the second problem is not harder than the first, just longer?

Objective C Practice

Find each difference.

C3. 607 − 337 **C4.** 402 − 174 **C5.** 1,848 − 753 **C6.** 2,787 − 697

C7. 500 − 356 **C8.** 600 − 278 **C9.** 2,000 − 521 **C10.** 30,000 − 2,331

Use the table to answer each question.

C11. JP wants to purchase a used Mini. If he receives $8,750 to trade in his Hummer, how much cash will the dealer give him?

C12. Alexis is purchasing a PT Cruiser. The dealer has offered $6,500 for her current car. Unfortunately, she still owes $7,395. How much does she owe in excess of what her car is worth?

Car	Cost
Lexus	$21,873
PT Cruiser	$9,342
Mini	$4,800

Objective D Subtract Several Numbers

Concept When subtracting several numbers, perform the operations one at a time from left to right. Organize your work to match the examples. This will help in later chapters and in future classes.

Procedure Subtract Several Numbers

Perform one subtraction at a time from left to right.

EXAMPLES

8. Subtract. $19 - 4 - 3$

Perform one subtraction at a time from left to right.

$$\underbrace{19 - 4}_{first} - 3 = 15 - 3$$
$$= 12$$

The second = is in line with the first. 12 is written to the right of the second =.

9. Subtract. $24 - 1 - 8 - 9$

Perform one subtraction at a time from left to right.

$$\underbrace{24 - 1}_{first} - 8 - 9 = \underbrace{23 - 8}_{second} - 9$$
$$= \underbrace{15 - 9}_{third}$$
$$= 6$$

10. Subtract. $356 - 109 - 52$

Perform one subtraction at a time from left to right. Use scratch paper if necessary.

$$\underbrace{356 - 109}_{first} - 52 = \underbrace{247 - 52}_{second}$$
$$= 195$$

$$\begin{array}{r} 3\overset{4}{\cancel{5}}{}^{1}6 \\ -\,109 \\ \hline 247 \end{array} \qquad \begin{array}{r} \overset{1}{\cancel{2}}{}^{1}47 \\ -\,52 \\ \hline 195 \end{array}$$

GUIDED PRACTICE

8. Subtract. $38 - 2 - 5$

$$\underbrace{38 - 2}_{first} - 5 = \qquad - 5$$
$$=$$

Line up a second equal sign. Write your answer to the right of the second =.

9. Subtract. $47 - 5 - 6 - 9$

$$\underbrace{47 - 5}_{first} - 6 - 9 = \underbrace{\qquad - 6}_{second} - 9$$
$$= \underbrace{\qquad - 9}_{third}$$
$$=$$

10. Subtract. $976 - 428 - 35$

$$\underbrace{976 - 428}_{first} - 35 = \underbrace{\qquad - 35}_{second}$$
$$=$$

$$\begin{array}{r} 976 \\ -\,428 \\ \hline \end{array}$$

Concept Check

D1. In your own words, explain the procedure for subtracting several numbers.

Objective D Practice

Subtract.

D2. $20 - 5 - 2$

D3. $30 - 7 - 3$

D4. $69 - 4 - 3$

D5. $48 - 5 - 1$

D6. $732 - 258 - 309$

D7. $851 - 302 - 217$

(Continued)

An account ledger is a document used to record the deposits and withdrawals from a checking account. Use the account ledger to answer each question.

Trans Type	Date	Description of Transaction	Payment/ Debit (−)	Fee (if any)	Deposit/ Credit (+)	Balance
	1-Jan	Balance Forward				$490
ATM	1-Jan	Needed Cash	$20	$2		
Ck 1254	2-Jan	Rent Check	$350			
Deposit	4-Jan	Paycheck			$250	

D8. What was the balance at the end of January 1?

D9. What was the balance at the end of January 2?

D10. What was the balance at the end of January 4?

D11. Was there enough money on January 5 to write a check for $369?

Combining Concepts and Applications

CONCEPT I Finding Differences Between Graphical Data

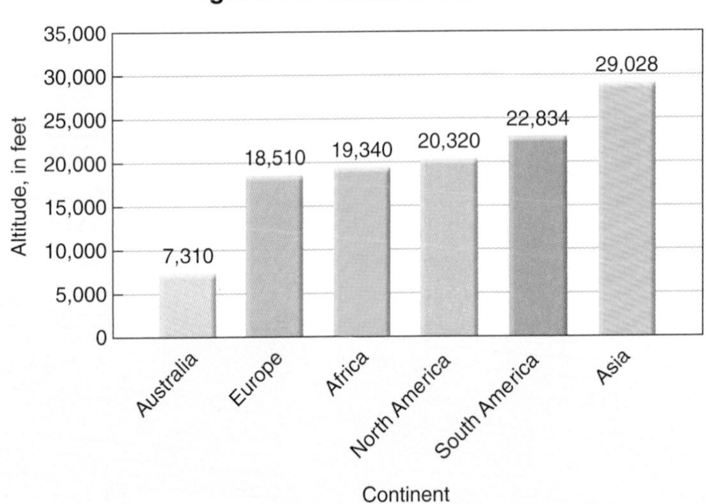

EXAMPLE

11. One person has climbed the highest mountain in North America. A second person has climbed the highest mountain in South America. How much higher has the second person climbed?

Step 1 Identify the altitude for each mountain.

North America = 20,320
South America = 22,834

Step 2 A difference in altitude indicates subtraction.

Difference = Larger − Smaller
= 22,834 − 20,320
= 2,514

Step 3 State the answer in words.

The second climber has climbed 2,514 feet higher than the first climber.

GUIDED PRACTICE

11. What is the largest difference between the highest mountains on any two continents highest mountains?

Step 1 Identify the altitude for each mountain.

Step 2 A difference in altitude indicates subtraction.

Step 3 State the answer in words.

CONCEPT II Solving for an Unknown Number

In the last section, you used basic facts and counting to solve for an unknown number. You can also use an algebra technique. The fact that addition and subtraction "undo" each other can be used to solve for an unknown number.

EXAMPLES

12. Solve for the unknown number. $x + 3 = 9$

$$x + 3 = 9$$
$$x + 3 - (3) = 9 - (3)$$
$$x = 6$$

Subtracting 3 from both sides of the equal sign will "undo" the addition of 3.

Replace x with 6 and use addition to check the answer.

$$x + 3 = 9$$
$$6 + 3 = 9$$
$$9 = 9$$

13. Solve for the unknown number. $y - 2 = 6$

$$y - 2 = 6$$
$$y - 2 + 2 = 6 + 2$$
$$y = 8$$

Adding 2 to both sides of the equal sign will "undo" the subtraction of 2.

Replace y with 8 and use subtraction to check the answer.

$$y - 2 = 6$$
$$8 - 2 = 6$$
$$6 = 6$$

GUIDED PRACTICE

12. Solve for the unknown number. $x + 12 = 15$

$$x + 12 = 15$$
$$x + 12 - (\quad) = 15 - (\quad)$$
$$x =$$

Subtracting _____ from both sides of the equal sign will "undo" the addition of 12.

Replace x with _____ and use addition to check the answer.

$$x + 12 = 15$$
$$\quad + 12 = 15$$
$$= 15$$

13. Solve for the unknown number. $y - 15 = 19$

$$y - 15 = 19$$
$$y - 15 + (\quad) = 19 + (\quad)$$
$$y =$$

Adding _____ to both sides of the equal sign will "undo" the subtraction of 15.

Replace _____ with _____ and use subtraction to check the answer.

$$y - 15 = 19$$
$$\quad - 15 = 19$$
$$= 19$$

(Continued)

14. Arianna has finished the first two legs of a swim, as shown below. The total swim is 3,400 yards. How far, x, must she still swim?

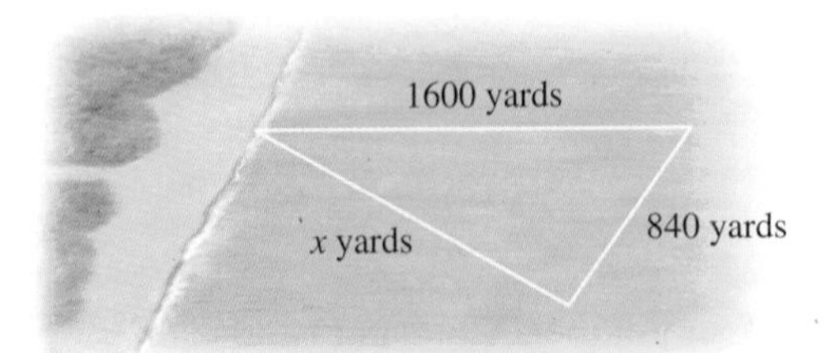

Write a word equation.

$$\text{Total} = \text{Portion finished} + \text{Portion left}$$

Substitute the numbers into the equation and solve.

$$\begin{aligned} \text{Total} &= \text{Portion finished} + \text{Portion left} \\ 3{,}400 &= 2{,}440 + x \\ 3{,}400 - 2{,}440 &= 2{,}440 - 2{,}440 + x \\ 960 &= x \end{aligned}$$

Arianna still must swim 960 yards.

14. To earn a 10% discount, a retail store must purchase 300 items at once. If the store has an order consisting of 190 shirts and 75 pants, how many pairs of socks, x, must be ordered to earn the discount?

Write a word equation.

$$\underline{\qquad\qquad} = \text{Items ordered} + \underline{\qquad\qquad}$$

Substitute the numbers into the equation and solve.

$$\begin{aligned} \underline{\qquad\qquad} &= \text{Items ordered} \quad + x \\ &= \qquad\qquad + x \\ -(\qquad) &= \quad -(\qquad) + x \\ &= x \end{aligned}$$

The store must order ____ more items to earn the discount.

Section 1.3 Exercises

FOR EXTRA HELP MyMathLab MathXL PRACTICE WATCH DOWNLOAD READ REVIEW

To understand the basic subtraction facts:

1. Answer the Objective A Concept Check.
2. Answer the odd Objective A Practice Exercises.
3. Answer the even Objective A Practice Exercises.

To borrow to regroup a number:

4. Answer the Objective B Concept Checks.
5. Answer the odd Objective B Practice Exercises.
6. Answer the even Objective B Practice Exercises.

To subtract large numbers:

7. Answer the Objective C Concept Checks.
8. Answer the odd Objective C Practice Exercises.
9. Answer the even Objective C Practice Exercises.

To subtract several numbers:

10. Answer the Objective D Concept Check.
11. Answer the odd Objective D Practice Exercises.
12. Answer the even Objective D Practice Exercises.

VOCABULARY REVIEW *Review the Vocabulary Preview for Section 1.3. Study the definitions until you can check* Got It *for every term.*

Use these words to complete each sentence.

inverse operations • difference • borrow

	Got It	Get Help
13. To regroup a 1 from one place value into a 10 in the next smaller place value, one must ____________.		
14. ____________ undo each other.		
15. A ____________ is the result of subtracting.		

How will you get help for any vocabulary about which you are unsure?

Your instructor ____ MyMathLab ____ A classmate ____ A tutor ____ Other ____

Draw a number line to visualize each difference.

16. 8 − 5 **17.** 9 − 7 **18.** 12 − 8 **19.** 14 − 6

Complete each addition table. You will need to use subtraction to find the numbers in some of the white boxes.

20.

+	5		8	
				11
	13			
		10	17	16
	12			

21.

+				
	11		15	
	12	18		
7			14	
		13		9

22.

+			7	
	16			14
		12		15
	13		12	
		6		

23.

+			9	
			13	
6	11	14		12
			16	
			10	

Writing only the answers, try to complete each column in fewer than 30 seconds. Accuracy is important. Take enough time to work through each exercise correctly.

24. Start Time: ____
- **a.** 13 − 6
- **b.** 14 − 8
- **c.** 12 − 9
- **d.** 10 − 7
- **e.** 7 − 3
- **f.** 14 − 5
- **g.** 16 − 9

Finish Time: ____

Total Time: ____

25. Start Time: ____
- **a.** 12 − 6
- **b.** 15 − 8
- **c.** 14 − 6
- **d.** 10 − 7
- **e.** 11 − 5
- **f.** 8 − 6
- **g.** 18 − 9

Finish Time: ____

Total Time: ____

26. Start Time: ____
- **a.** 17 − 8
- **b.** 15 − 6
- **c.** 11 − 8
- **d.** 12 − 7
- **e.** 10 − 5
- **f.** 13 − 7
- **g.** 14 − 7

Finish Time: ____

Total Time: ____

27. Start Time: ____
- **a.** 12 − 8
- **b.** 15 − 7
- **c.** 16 − 9
- **d.** 13 − 8
- **e.** 11 − 4
- **f.** 13 − 5
- **g.** 17 − 9

Finish Time: ____

Total Time: ____

Find each difference.

28. 26 − 14

29. 36 − 34

30. 42 − 34

31. 55 − 28

32. 84 − 28

33. 48 − 19

34. 78 − 59

35. 83 − 76

36. 18 − 7 − 4

37. 25 − 6 − 3

38. 32 − 14 − 13

39. 55 − 21 − 17

40. 332 − 54

41. 561 − 78

42. 742 − 79

43. 265 − 88

44. The balance in Quigly's bank account was $765 before he wrote a $478 check. What is the new balance?

45. Johann was paid $648 to install a new door on a client's house. Then he had to pay $281 in expenses. How much money is left over?

46. Lindsey invited 493 guests to her wedding. Only 328 showed up. How many invited guests did not attend?

47. Goran grew a giant pumpkin last summer. On August 1, the pumpkin weighed 523 pounds. By August 21, it weighed 658 pounds. How much did the pumpkin increase in weight?

48. 85 − 22 − 31

49. 75 − 11 − 37

50. 165 − 32 − 13

51. 175 − 41 − 22

52. 234 − 54 − 56

53. 455 − 65 − 45

54. 300 − 24 − 63

55. 500 − 32 − 37

56. 745 − 368

57. 873 − 188

58. 200 − 168

59. 400 − 279

60. 6022 − 4581

61. 9406 − 1738

62. 20,198 − 18,395

63. 22,075 − 14,957

64. 654,293 − 28,764

65. 975,482 − 81,635

66. 651,389 − 278,375

67. 838,738 − 584,539

tobey Watch out! You must work from left to right, one subtraction at a time. Don't try to change the order.

For more tips and tweets, go to twitter.com/gstbasicmath

Solve.

68. Subtract 8 from 25.

69. Subtract 6 from 31.

70. What is the difference of 52 and 30?

71. What is the difference of 63 and 20?

72. What is 76 decreased by 23?

73. What is 64 decreased by 36?

74. Subtract 25 from 89.

75. Subtract 54 from 78.

76. Find 45 less than 65.

77. Find 36 less than 76.

78. Harvey and Yolanda took a road trip to Yosemite National Park. When they started their trip, the odometer read 45,678 miles. When they reached the park, the odometer read 46,534 miles. How far had they driven?

79. Kao and Tina took a road trip to the Rock and Roll Hall of Fame. When Tina started driving, the odometer read 124,347 miles. When she reached the Rock and Roll Hall of Fame, the odometer read 124,700 miles. How far had Tina driven?

80. Rick has $1,254 in his checking account. If he writes a check for $765, what is his balance?

81. Amy is reading *The Fountainhead*. If the book has 1,345 pages and she has read 689 pages, how many pages does she need to read to finish the book?

82. Nick is trading in his pickup to buy a new car. The car he wants to buy costs $23,465. If Nick gets $4,670 for his old truck, how much more must he pay for the new car?

83. Tara is shopping for a used car. She finds the Blue Book value of the car she wants to buy. The Blue Book value is $11,670, and the private seller value is $9,866. How much will she save if she buys the car from the private seller?

84. On *The Biggest Loser*, six team members start with a combined weight of 1,482 pounds. They lose 31, 65, 18, 42, 57, and 48 pounds. What is their new combined weight?

85. On *The Greatest Race*, the fourth-place team finishes a leg in 8 hours, 57 minutes. The third-place team takes 18 minutes less than the fourth-place team. The second-place team finishes 21 minutes before the third-place team. The winning team beats the second-place team by 3 minutes. How long does it take the winning team to finish?

Use the following checking account ledger to answer each question.

Trans Type	Date	Description of Transaction	Payment/ Debit (−)	Fee (if any)	Deposit/ Credit (+)	Balance
	1-April	Balance Forward				$490
Ck 1783	1-April	Rent	$325			
ATM	3-April	Needed cash	$40	$3		
Ck 1784	5-April	Cable Bill	$40			
Deposit	12-April	Paycheck			$420	
Ck 1785	15-April	Electricity Bill	$50			

86. What was the balance before the paycheck was deposited?

87. What was the balance after the paycheck was deposited?

88. What was the total of all payments/debits and fees?

89. What was the balance after April 15?

Use the following bar graph to respond to each exercise.

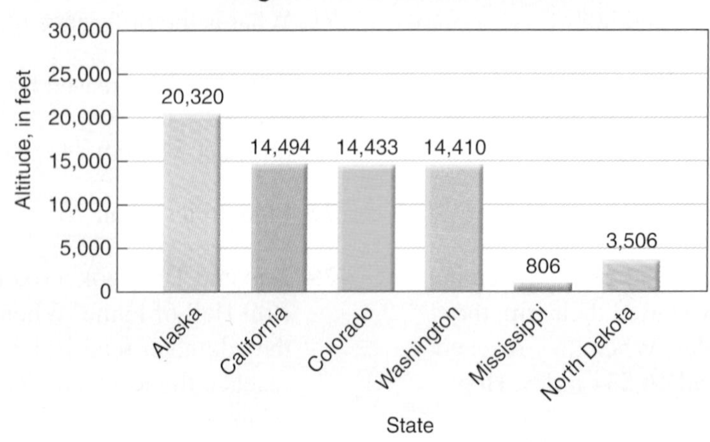

90. Clint has climbed the highest mountain in Alaska and Washington. How much higher is the highest point in Alaska compared to the highest point in Washington?

91. Find the difference in altitude of the highest points in Colorado and North Dakota.

92. Rafael has climbed the highest mountain in California, Alaska, and Mississippi. He thinks that the difference in height between the mountains in California and Mississippi is greater than the difference in height between the mountains in Alaska and California. Compare the differences in height to see if Rafael is correct.

Solve for the unknown number.

93. $x - 16 = 12$

94. $x - 10 = 7$

95. $y - 28 = 13$

96. $y - 11 = 16$

97. $8 = z - 22$

98. $11 = z - 24$

99. $23 = x + 4$

100. $28 = x + 9$

101. $34 = y - 8$

102. $y - 24 = 5$

103. $64 = z + 12$

104. $89 = 65 + z$

Section 1.3 Question Log

Use this space to write questions. Make sure to get these answered and revisit them when you prepare for your exam.

Page ____ Answered	Page ____ Answered

Section 1.3 Extra Practice

See Appendix A for extra practice on Section 1.3.

Section 1.4 Multiplying Whole Numbers

The multiplication procedure comes from the idea of repeatedly adding the same number. As with other operations, you must learn the basic facts and understand how place value is used in each step.

The **Objectives** in Section 1.4 will help you

- A Understand the basic multiplication facts.
- B Multiply a one-digit number and a two-digit number.
- C Multiply by a power of ten.
- D Multiply large numbers.
- E Multiply a list of factors.

VOCABULARY PREVIEW *Check the box that applies.*

	Got It	Must Study
Multiplication: Multiplication is a quick way to perform repeated addition.		
Product: A **product** is the result of multiplying.		
Factors: Factors are the numbers that are multiplied to give a product.		
Commutative Property of Multiplication: The order in which you multiply factors will not change the product. Example: $2 \cdot 3 = 3 \cdot 2$		

Study these terms when they appear in the text. At the end of the section, test your understanding by completing the Vocabulary Review in the section exercises.

Objective A Understand the Basic Multiplication Facts

The Concept Repeated addition of the same number can be performed quickly using **multiplication**. Imagine you are having a car wash to raise money for charity. You charge $6 to wash one car. After washing eight cars, you wonder how much money you have raised.

Repeated Addition	**Multiplication**
$\underbrace{6+6+6+6+6+6+6+6}_{\text{8 sixes}} = 48$	$8 \cdot 6 = 48$

You should be comfortable with the four ways of writing multiplication. The following notations all mean "three times two."

$$3 \cdot 2 \qquad 3 \times 2 \qquad 3 \cdot (2) \qquad 3(2) \qquad (3)(2)$$

The notation $3 \cdot 2$ is often used instead of 3×2 because $\times$ can be confused with the variable x in algebra. The dot, while harder to see, does not cause that confusion.

The words *product* and *factors* are used to describe the numbers in a multiplication exercise.

INTERACTIVE DEFINITION Product and Factor

A **product** is the result of multiplying.

Factors are the numbers that are multiplied to give a product.

EXAMPLE

1. Identify the product and factors in $8 \cdot 6 = 48$.

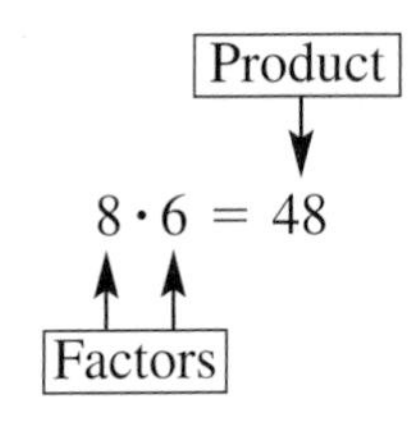

48 is the product of 6 and 8.

8 and 6 are factors of 48.

GUIDED PRACTICE

1. Identify the product and factors in $63 = 7 \cdot 9$.

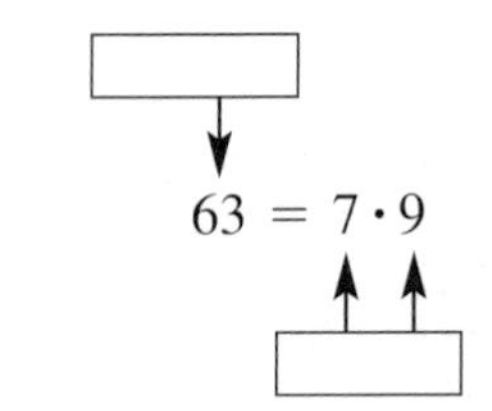

_____ is the product of 7 and 9.

_____ and _____ are factors of 63.

DO YOU UNDERSTAND how to identify the product in a multiplication exercise? Got It Get Help

DO YOU UNDERSTAND how to identify the factors in a multiplication exercise? Got It Get Help

Sometimes it will be useful to reorder factors to find a product. The Commutative Property of Multiplication allows us to reorder factors.

INTERACTIVE DEFINITION Commutative Property of Multiplication

The **Commutative Property of Multiplication** states that changing the order in which factors are multiplied does not change the product.

EXAMPLE

2. Demonstrate the Commutative Property of Multiplication with $3 \cdot 2$.

$$3 \cdot 2 = 6$$
$$2 \cdot 3 = 6$$

The order in which factors are multiplied does not change the product.

GUIDED PRACTICE

2. Demonstrate the Commutative Property of Multiplication with $4 \cdot 5$.

$$4 \cdot 5 =$$
$$5 \cdot 4 =$$

The _____ in which factors are multiplied does not change the product.

DO YOU UNDERSTAND the Commutative Property of Multiplication? Got It Get Help

Concept Check

A1. In your own words, explain how multiplication is related to repeated addition.

A2. Use the factors 2 and 5 to show that multiplication is commutative.

A3. Identify the factors and the product in $3 \cdot 4 = 12$.

Objective A Practice

Find each product using repeated addition.

A4. $3 \cdot 5$ **A5.** $4 \cdot 7$ **A6.** $3 \cdot 6$ **A7.** $4 \cdot 8$

A8. Complete the multiplication table by multiplying each factor in the top row by each factor in the left column. The outlined box is 15 because $3 \cdot 5 = 15$.

×	1	2	3	4	5	6	7	8	9	10
1										
2										
3										
4										
5			15							
6										
7										
8										
9										
10										

Study the patterns in the multiplication table to increase your speed and accuracy. Notice that each row or column can be filled in using repeated addition. This can help you memorize the basic facts.

Writing only the answers, try to complete each column in fewer than 30 seconds. Accuracy is important. Take enough time to work through each exercise correctly.

A9. Start Time: _____
- **a.** $0 \cdot 5$
- **b.** $6 \cdot 8$
- **c.** $9 \cdot (6)$
- **d.** 1×3
- **e.** $5 \cdot 6$
- **f.** $7 \cdot 6$
- **g.** $4 \cdot 7$
- **h.** 10×4
- **i.** $8(8)$
- **j.** $4 \cdot 4$

Finish Time: _____

Total Time: _____

A10. Start Time: _____
- **a.** $4 \cdot 1$
- **b.** 6×9
- **c.** $3 \cdot 0$
- **d.** $7(7)$
- **e.** 5×10
- **f.** $8 \cdot 7$
- **g.** $5 \cdot 9$
- **h.** $4 \cdot (5)$
- **i.** $9 \cdot 8$
- **j.** $3 \cdot 2$

Finish Time: _____

Total Time: _____

A11. Start Time: _____
- **a.** $4 \cdot 3$
- **b.** $10 \cdot 3$
- **c.** 8×9
- **d.** $4 \cdot 8$
- **e.** $5 \cdot (7)$
- **f.** $9 \cdot 7$
- **g.** $0 \cdot 10$
- **h.** $7(8)$
- **i.** 3×6
- **j.** $6 \cdot 7$

Finish Time: _____

Total Time: _____

A12. Start Time: _____
- **a.** 3×3
- **b.** $8 \cdot 6$
- **c.** $5 \cdot 8$
- **d.** $0(0)$
- **e.** $7 \cdot 9$
- **f.** 6×6
- **g.** $4 \cdot 10$
- **h.** $4 \cdot (9)$
- **i.** $9 \cdot 9$
- **j.** $4 \cdot 2$

Finish Time: _____

Total Time: _____

Objective B Multiply a One-Digit Number and a Two-Digit Number

The Concept To multiply numbers with more than one digit, it helps to look at an example using money. To multiply money in this example, we will do the following:

- Use only \$1 bills and \$10 bills
- Separate the money into \$1 bills and \$10 bills before we multiply

As you read the examples, keep in mind that the bills are used to indicate place value.

DETAILED EXAMPLE Demonstrating the Multiplication Procedure

Multiply. $3 \cdot \$14$

Example with Money

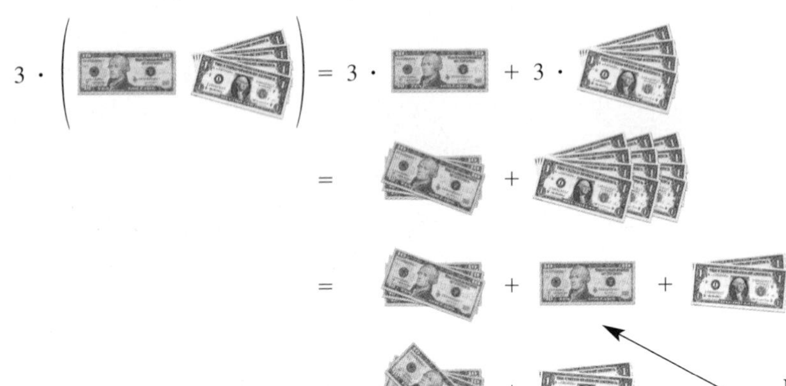

Formal Procedure

First, multiply the ones digits. $3 \cdot 4 = 12$
Write the 2 in the ones place and carry 1 ten.

$$\begin{array}{r} {}^{1}\\ 14 \\ \times\ \ 3 \\ \hline 2 \end{array}$$

Second, multiply the tens digits. $3 \cdot 1 = 3$
Add the carried digit. $3 + 1 = 4$

Making 1 ten from 10 ones

$$\begin{array}{r} {}^{1}\\ 14 \\ \times\ \ 3 \\ \hline 42 \end{array}$$

Procedure Multiply a One-Digit Number and a Two-Digit Number

Step 1 Multiply the ones place of the two-digit number by the one-digit number. Carry if necessary.

Step 2 Multiply the tens place of the two-digit number by the one-digit number. Add any carried number.

EXAMPLES

3. Find the product. $4 \cdot 21$

Step 1 Multiply the ones place of the two-digit number by the one-digit number. Carry if necessary.

Step 2 Multiply the tens place of the two-digit number by the one-digit number. Carry if necessary.

$$\begin{array}{r} 21 \\ \times\ \ 4 \\ \hline 84 \end{array}$$

GUIDED PRACTICE

3. Find the product. $3 \cdot 32$

$$\begin{array}{r} 32 \\ \times\ \ 3 \\ \hline \end{array}$$

(Continued)

4. Find the product. $6 \cdot 87$ **Step 1** Multiply the ones place of the two-digit number by the one-digit number. Carry if necessary. **Step 2** Multiply the tens place of the two-digit number by the one-digit number. Carry if necessary. $\begin{array}{r} \scriptstyle 4 \\ 87 \\ \times \quad 6 \\ \hline 522 \end{array}$	**4.** Find the product. $7 \cdot 53$ $\begin{array}{r} 53 \\ \times \quad 7 \\ \hline \end{array}$
5. Find the product. $5 \cdot 40$ **Step 1** Multiply the ones place of the two-digit number by the one-digit number. Carry if necessary. **Step 2** Multiply the tens place of the two-digit number by the one-digit number. Carry if necessary. $\begin{array}{r} 40 \\ \times \quad 5 \\ \hline 200 \end{array}$	**5.** Find the product. $6 \cdot 90$ $\begin{array}{r} 90 \\ \times \quad 6 \\ \hline \end{array}$

Concept Check

B1. Using money, explain why a 3 needs to be carried when multiplying $9 \cdot \$24$.

B2. Why do we multiply the smallest place values first? *Hint:* Think about a problem in which we would need to carry.

Objective B Practice

Find each product.

B3. $2 \cdot 41$

B4. $3 \cdot 13$

B5. $5 \cdot 16$

B6. $6 \cdot 15$

B7. $7 \cdot 98$

B8. $8 \cdot 97$

B9. $6 \cdot 70$

B10. $7 \cdot 60$

Use the table to answer each question.

Job	Hourly Pay
Electrician	\$28
Nurse's aide	\$14
Mechanic	\$18

B11. If Bianca becomes an electrician, how much will she earn if she works 8 hours?

B12. If James becomes a nurse's aide, how much will he earn if he works 8 hours?

Objective C Multiply by a Power of 10

The Concept Whole numbers that are "powers of 10" begin with a one and have no digits other than zero written to the right. The first seven powers of ten are written as follows:

Numeral	Name	Number of Zeros
1	One	0
10	Ten	1
100	One Hundred	2
1,000	One Thousand	3
10,000	Ten Thousand	4
100,000	One Hundred Thousand	5
1,000,000	One Million	6

DETAILED EXAMPLES Multiplying by a Power of Ten

Multiply. $54 \cdot 100$

To understand the concept, think of multiplication as repeated addition.

Think "repeated addition."

$$54 \cdot 100 = \text{Fifty-four 100s}$$
$$= \text{Fifty-four hundred}$$
$$= 5{,}400$$

To work efficiently, count the zeros.

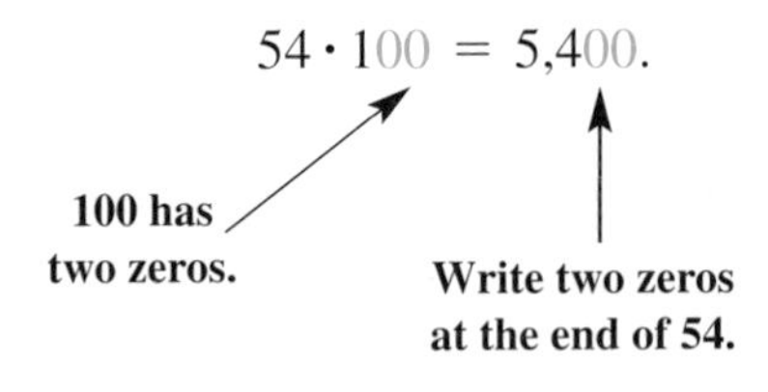

Multiply. $200 \cdot 1{,}000$

Again, think of multiplication as repeated addition.

Think "repeated addition."

$$200 \cdot 1{,}000 = \text{Two hundred 1,000's}$$
$$= \text{Two hundred thousand}$$
$$= 200{,}000$$

To work efficiently, count the zeros.

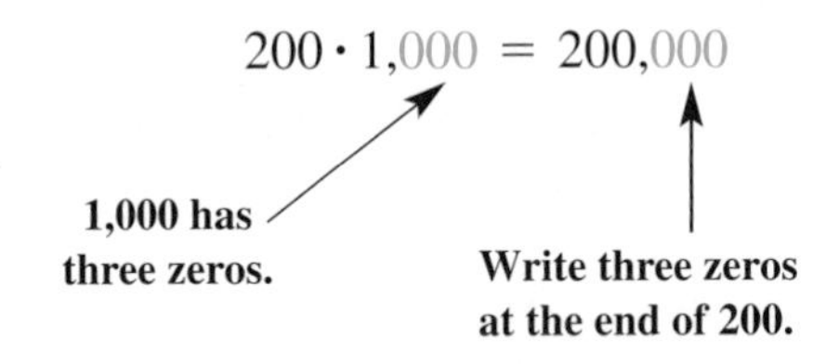

Procedure Multiply by a Power of Ten

Step 1 Count the number of zeros in the power of ten.

Step 2 Write that many zeros at the end of the other factor.

EXAMPLES	GUIDED PRACTICE
6. Find the product. $1{,}000 \cdot 73$	**6.** Find the product. $10{,}000 \cdot 63$
1,000 has 3 zeros. $1{,}000 \cdot 73 = 73{,}000$	10,000 has ____ zeros. $10{,}000 \cdot 63 = 63$
7. Find the product. $298 \cdot 100{,}000$	**7.** Find the product. $6513 \cdot 10{,}000{,}000$
100,000 has 5 zeros. $298 \cdot 100{,}000 = 29{,}800{,}000$	10,000,000 has ____ zeros. $6{,}513 \cdot 10{,}000{,}000 =$

Concept Check

C1. Performing the multiplication, we can see that $45 \cdot 100 = 4{,}500$. Why are we allowed to write two zeros after 45 to get the answer? *Hint:* Think about how many 100's there are.

C2. What property states that $38 \cdot 1{,}000 = 1{,}000 \cdot 38$?

Objective C Practice

Find each product.

C3. $14 \cdot 10$ **C4.** $23 \cdot 10$ **C5.** $523 \cdot 1{,}000$ **C6.** $12 \cdot 100$

C7. $100 \cdot 73$ **C8.** $1{,}000 \cdot 879$ **C9.** $10{,}000 \cdot 987$ **C10.** $100{,}000 \cdot 87$

Objective D Multiply Large Numbers

The Concept To multiply by numbers with more than one digit, multiply by one digit at a time. First, multiply by the digit in the ones place and then multiply by the digit in the tens place.

$$\begin{array}{r} 5\ 8 \\ \times\ 3\ 2 \\ \hline \end{array}$$

First, multiply 58 by 2.
Second, multiply 58 by 3.

Procedure Multiply Large Numbers

Step 1 Multiply by the digit in the ones place.

Step 2 Multiply by the digit in the tens place. Write the result in the tens place.

Step 3 Continue multiplying by any digits in higher place values. Write each result in the correct place value.

Step 4 Add the results.

DETAILED EXAMPLE Multiplying Large Numbers

Multiply. $78 \cdot 32$

Multiply by the digit in the ones place.

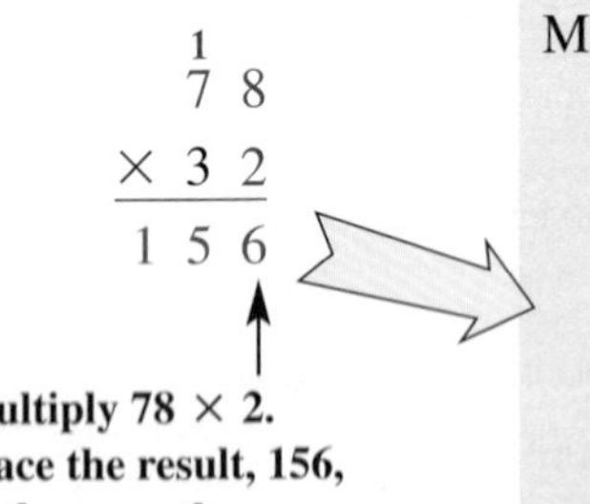

Multiply 78 × 2.
Place the result, 156,
in the ones place.
78 × 2 ones = 156 ones

Multiply by the digit in the tens place.

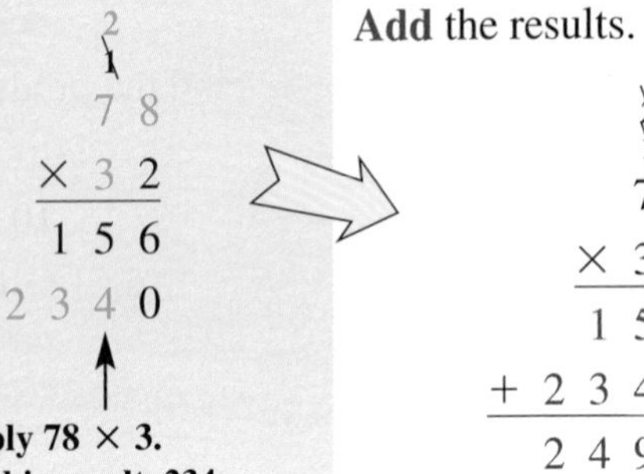

Multiply 78 × 3.
Place this result, 234,
in the tens place.
78 × 3 tens = 234 tens
= 2,340

Add the results.

$$\begin{array}{r} 78 \\ \times\ 32 \\ \hline 156 \\ +\ 2340 \\ \hline 2496 \end{array}$$

Together, the steps look like this:

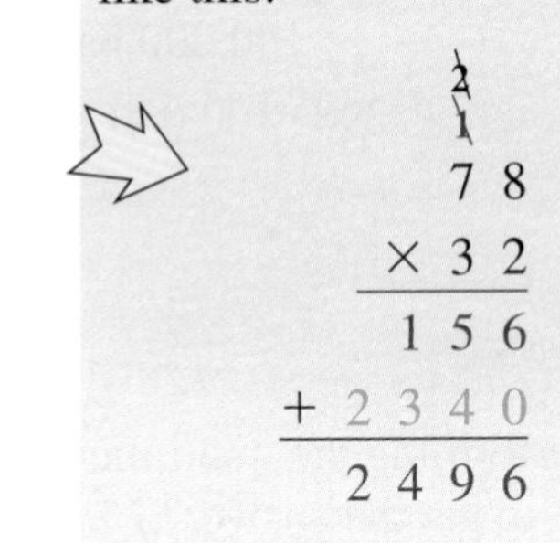

In step 2, we multiplied $78 \cdot 3 = 234$ and included a placeholder zero by writing 2340. In the following Guided Practice exercises, placeholder zeros are written for you. Make sure you write placeholder zeros when multiplying large numbers.

EXAMPLES

8. Find the product. $152 \cdot 73$

Step 1 Multiply $152 \cdot 3$. →
Step 2 Multiply $152 \cdot 7$. →
Step 3 Add the results. →

$$\begin{array}{r} 152 \\ \times\ \ 73 \\ \hline 456 \\ +\ 10640 \\ \hline 11096 \end{array}$$

$152 \cdot 73 = 11{,}096$

9. Find the product. $839 \cdot 723$

Step 1 Multiply $839 \cdot 3$. →
Step 2 Multiply $839 \cdot 2$. →
Step 3 Multiply $839 \cdot 7$. →
Step 4 Add the results. →

$$\begin{array}{r} 839 \\ \times\ 723 \\ \hline 2517 \\ 16780 \\ +\ 587300 \\ \hline 606597 \end{array}$$

$839 \cdot 723 = 606{,}597$

GUIDED PRACTICE

8. Find the product. $273 \cdot 65$

Step 1 Multiply $273 \cdot 5$. →
Step 2 Multiply $273 \cdot 6$. →
Step 3 Add the results. →

$$\begin{array}{r} 273 \\ \times\ \ 65 \\ \hline \square \\ +\ \square 0 \\ \hline \end{array}$$

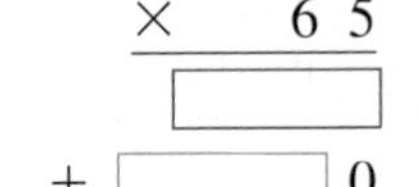

$273 \cdot 65 =$

9. Find the product. $936 \cdot 287$

Step 1 Multiply $936 \cdot 7$. →
Step 2 Multiply $936 \cdot 8$. →
Step 3 Multiply $936 \cdot 2$. →
Step 4 Add the results. →

$$\begin{array}{r} 936 \\ \times\ 287 \\ \hline \square \\ \square 0 \\ +\ \square 00 \\ \hline \end{array}$$

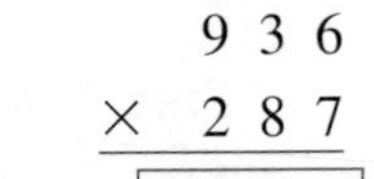

$936 \cdot 287 =$

Concept Check

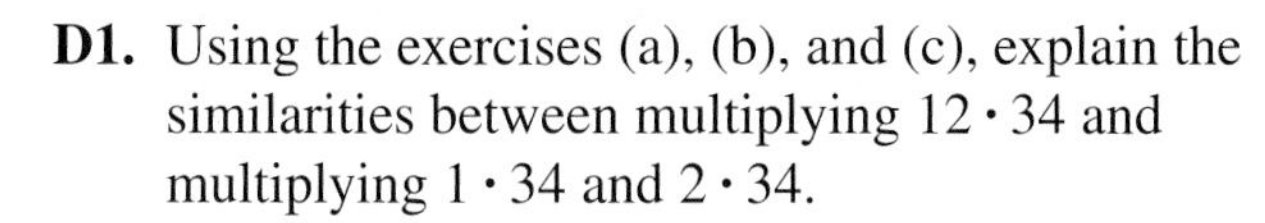

D1. Using the exercises (a), (b), and (c), explain the similarities between multiplying $12 \cdot 34$ and multiplying $1 \cdot 34$ and $2 \cdot 34$.

D2. Using the exercises (a), (b), and (c), explain the differences between multiplying $12 \cdot 34$ and multiplying $1 \cdot 34$ and $2 \cdot 34$.

a. $\begin{array}{r} 34 \\ \times 12 \\ \hline 68 \\ +340 \\ \hline 408 \end{array}$ **b.** $\begin{array}{r} 34 \\ \times 2 \\ \hline 68 \end{array}$ **c.** $\begin{array}{r} 34 \\ \times 1 \\ \hline 34 \end{array}$

Objective D Practice

Find each product.

D3. $423 \cdot 12$ **D4.** $142 \cdot 21$ **D5.** $365 \cdot 18$ **D6.** $734 \cdot 24$

D7. $152 \cdot 312$ **D8.** $164 \cdot 731$ **D9.** $943 \cdot 739$ **D10.** $186 \cdot 274$

D11. A hot dog and bun together have approximately 195 calories. How many calories does a contestant consume if she eats 37 hot dogs?

D12. A hot dog and bun together have approximately 195 calories. Kobayashi won the contest by eating 53 hot dogs. How many calories did he consume?

FOCUS ON ESTIMATING Products of Large Numbers

Estimation can help you check the accuracy of a multiplication exercise.

If someone tells you that $673 \cdot 59 = 27{,}707$, you can use estimation to quickly check the answer. Example 10 shows us that this answer is too low to be correct because $673 \cdot 59 \approx 42{,}000$. The symbol $\approx$ means that the result is equal to approximately 42,000.

Procedure Estimate the Product of Large Numbers

Step 1 Round each number to its largest place value.

Step 2 Multiply the rounded numbers.

(Continued)

EXAMPLES

10. Estimate. $673 \cdot 59$

Step 1 Round each number to its largest place value.

$673 \approx 700$
$59 \approx 60$

Step 2 Multiply the rounded numbers.

$673 \cdot 59 \approx \underline{700} \cdot \underline{60}$
$= 42,\underline{000}$

$6 \cdot 7 = 42$

700 has 2 zeros. 60 has 1 zero. Write 3 zeros after the 42.

GUIDED PRACTICE

10. Estimate. $212 \cdot 285$

$212 \approx$
$285 \approx$

$212 \cdot 285 \approx$
$=$

Count the zeros to multiply quickly.

EXAMPLES

11. Estimate. $1,829 \cdot 431$

Step 1 Round each number to its largest place value.

$1,829 \approx 2,000$
$431 \approx 400$

Step 2 Multiply the rounded numbers.

$1,829 \cdot 431 \approx 2,\underline{000} \cdot 4\underline{00}$
$= 8\underline{00,000}$

$2 \cdot 4 = 8$

2,000 has 3 zeros. 400 has 2 zeros. Write 5 zeros after the 8.

GUIDED PRACTICE

11. Estimate. $1,183 \cdot 299$

$1,183 \approx$
$299 \approx$

$1,183 \cdot 299 \approx$
$=$

PRACTICE

Many of the following products are incorrect. Use estimation to find any answers that you think are incorrect. Then perform the multiplication to find the actual answer.

1. $65 \cdot 73 = 4,745$ **2.** $58 \cdot 93 = 5,394$ **3.** $1,258 \cdot 68 = 58,544$ **4.** $1,893 \cdot 42 = 97,506$

5. $359 \cdot 34 = 21,306$ **6.** $483 \cdot 67 = 29,306$ **7.** $4,863 \cdot 192 = 830,466$ **8.** $7,412 \cdot 967 = 716,404$

Objective E Multiply a List of Factors

The Concept When multiplying a list of factors, we can find the product by multiplying the factors from left to right.

$$\underbrace{2 \cdot 7}_{\text{First}} \cdot 5 \cdot 10 = \underbrace{14 \cdot 5}_{\text{Second}} \cdot 10$$
$$= \underbrace{70 \cdot 10}_{\text{Third}}$$
$$= 700$$

Procedure Multiply a List of Factors

Perform the multiplications from left to right.

EXAMPLES

12. Multiply. $4 \cdot 5 \cdot 9 \cdot 2$

Perform the multiplications from left to right.

$$\underbrace{4 \cdot 5}_{\text{1st}} \cdot 9 \cdot 2 = \underbrace{20 \cdot 9}_{\text{2nd}} \cdot 2 = \underbrace{180 \cdot 2}_{\text{3rd}} = 360$$

13. Multiply. $5 \cdot 4 \cdot 2 \cdot 6 \cdot 2$

Perform the multiplications from left to right.

$$\underbrace{5 \cdot 4}_{\text{1st}} \cdot 2 \cdot 6 \cdot 2 = \underbrace{20 \cdot 2}_{\text{2nd}} \cdot 6 \cdot 2 = \underbrace{40 \cdot 6}_{\text{3rd}} \cdot 2 = \underbrace{240 \cdot 2}_{\text{4th}} = 480$$

GUIDED PRACTICE

12. Multiply. $5 \cdot 3 \cdot 7 \cdot 2$

$$\underbrace{5 \cdot 3}_{\text{1st}} \cdot 7 \cdot 2 = \underbrace{\_\_\_ \cdot 7}_{\text{2nd}} \cdot 2 = \underbrace{\_\_\_ \cdot 2}_{\text{3rd}} = \_\_\_$$

13. Multiply. $2 \cdot 12 \cdot 3 \cdot 5 \cdot 2$

$$\underbrace{2 \cdot 12}_{\text{1st}} \cdot 3 \cdot 5 \cdot 2 = \underbrace{\_\_\_ \cdot 3}_{\text{2nd}} \cdot 5 \cdot 2 = \underbrace{\_\_\_ \cdot 5}_{\text{3rd}} \cdot 2 = \underbrace{\_\_\_ \cdot 2}_{\text{4th}} = \_\_\_$$

Concept Check

E1. In your own words, explain how to multiply a list of factors.

Objective E Practice

Find each product.

E2. $2 \cdot 9 \cdot 5$

E3. $5 \cdot 7 \cdot 2$

E4. $2 \cdot 9 \cdot 2 \cdot 3$

E5. $8 \cdot 3 \cdot 7 \cdot 2$

E6. $14 \cdot 6 \cdot 5$

E7. $16 \cdot 7 \cdot 5$

E8. $2 \cdot 8 \cdot 3 \cdot 3 \cdot 2$

E9. $3 \cdot 4 \cdot 2 \cdot 9 \cdot 3$

Combining Concepts and Applications

CONCEPT I Finding the Area of a Rectangle To find the area of a rectangle, find the number of square units that will cover it.

This is one square inch, a square that is 1 in. by 1 in.

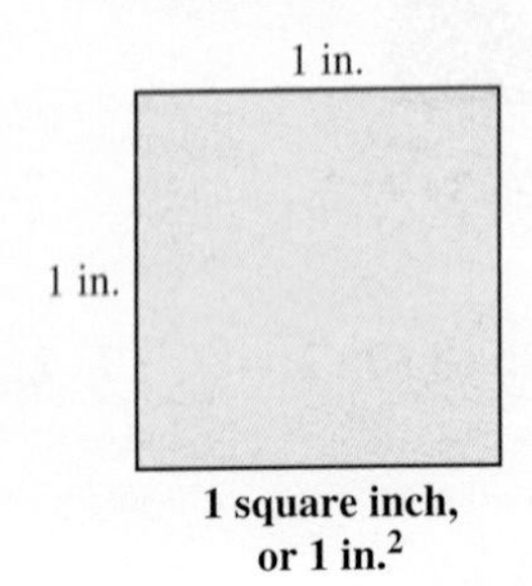

1 square inch, or 1 in.^{2}

This is one square centimeter, a square that is 1 cm by 1 cm.

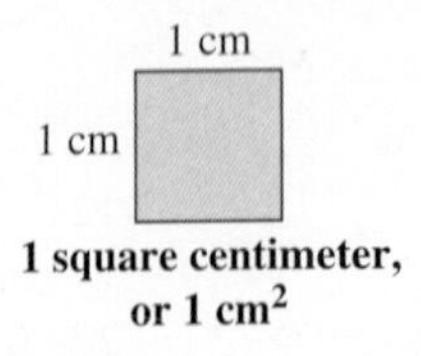

1 square centimeter, or 1 cm^2

To find the area of a rectangle, determine how many square units are needed to cover the rectangle. The rectangle below has an area of 12 square centimeters because it takes 12 square centimeters to cover the figure.

12 square centimeters

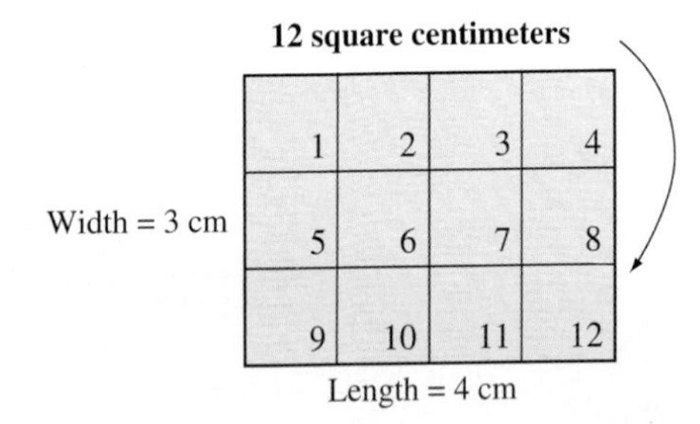

We can count the squares to find the area of the rectangle. However, it is easier to use the formula $A = l \cdot w$ to calculate the area of a rectangle.

$$\begin{aligned} A &= l \cdot w \\ &= 3 \text{ cm} \cdot 4 \text{ cm} \\ &= 12 \text{ square centimeters} \end{aligned}$$

EXAMPLE	GUIDED PRACTICE
14. Find the area of the rectangle.	**14.** Find the area of the rectangle.
34 ft; 54 ft	15 in.; 16 in.
$A = l \cdot w$ $= 34 \cdot 54$ $= 1{,}836$ square feet	$A = l \cdot w$ $=$ $=$

CONCEPT II Solving for an Unknown Number

SOLVING FOR AN UNKNOWN NUMBER	
In this section, we will use basic facts or repeated addition when solving for an unknown number.	
Solving Using Basic Facts	**Solving Using Repeated Addition**
Solve $6 \cdot x = 42$ for the unknown number.	Solve $y \cdot 13 = 65$ for the unknown number.
$6 \cdot x = 42$ $x = 7$ **Basic fact:** $6 \cdot 7 = 42$	$y \cdot 13 = 65$ Counting by 13, we have 13, 26, 39, 52, $\boxed{65}$. 65 is the fifth number in the list, so $5 \cdot 13 = 65$ $y = 5$.

EXAMPLES

15. Solve for the unknown number. $9 \cdot x = 72$

Is there a basic fact "9 times something = 72"? Yes, there is a basic fact.

$$9 \cdot x = 72$$
$$x = 8$$

Basic fact: $9 \cdot 8 = 72$

16. Solve for the unknown number. $14 \cdot y = 56$

Multiplication of 14 is not a basic fact. Repeatedly add 14 until we reach 56.

14, 28, 42, 56

Four 14's add to 56.

$$14 \cdot y = 56$$
$$y = 4$$

GUIDED PRACTICE

15. Solve for the unknown number. $8 \cdot x = 48$

Is there a basic fact "8 times something = 48"? Yes, there is a basic fact.

$$8 \cdot x = 48$$
$$x = $$

16. Solve for the unknown number. $41 \cdot y = 205$

Multiplication of 41 is not a basic fact. Repeatedly add ____ until we reach ____.

____, ____, ____, ____, ____

____ 41's add to 205.

$$41 \cdot y = 205$$
$$y = $$

Section 1.4 Exercises

FOR EXTRA HELP

To understand the basic multiplication facts:

1. Answer the Objective A Concept Checks.
2. Answer the odd Objective A Practice Exercises.
3. Answer the even Objective A Practice Exercises.

To multiply a one-digit number and a two-digit number:

4. Answer the Objective B Concept Checks.
5. Answer the odd Objective B Practice Exercises.
6. Answer the even Objective B Practice Exercises.

To multiply by a power of 10:

7. Answer the Objective C Concept Checks.
8. Answer the odd Objective C Practice Exercises.
9. Answer the even Objective C Practice Exercises.

To multiply large numbers:

10. Answer the Objective D Concept Checks.
11. Answer the odd Objective D Practice Exercises.
12. Answer the even Objective D Practice Exercises.

To estimate products of large numbers:

13. Answer the odd Focus On Estimation Practice Exercises.
14. Answer the even Focus On Estimation Practice Exercises.

To multiply a list of factors:

15. Answer the Objective E Concept Check.
16. Answer the odd Objective E Practice Exercises.
17. Answer the even Objective E Practice Exercises.

VOCABULARY REVIEW *Review the Vocabulary Preview for Section 1.4. Study the definitions until you can check* Got It *for every term.*

Use these words to complete each sentence.

multiplication • product • factors • Commutative Property of Multiplication

	Got It	Get Help
18. A ____________ is the result of multiplying.		
19. Numbers that are multiplied are called ____________.		
20. The ____________ allows you to reorder factors without changing the product.		
21. In the spaces below, write *factor* or *product* as appropriate. ____________ · ____________ = ____________		
22. ____________ can be performed using repeated addition.		

How will you get help for any vocabulary about which you are unsure?

Your instructor ____ MyMathLab ____ A classmate ____ A tutor ____ Other ____

Find each product using repeated addition.

23. $2 \cdot 4$ **24.** $3 \cdot 4$ **25.** $4 \cdot 2$ **26.** $3 \cdot 6$

Complete each multiplication table by multiplying the factors in the top row by the factors in the left column. The outlined box is 15 because 3 × 5 = 15.

27.

×	7	8	4	3
5				15
8				
6				
9				

28.

×	9	7	3	8
4				
7				
6				
9				

Writing only the answers, try to complete each column in fewer than 30 seconds. Accuracy is important. Take enough time to work through each exercise correctly.

29. Start Time: _____
- **a.** $6 \cdot 6$
- **b.** $3 \cdot (6)$
- **c.** 9×4
- **d.** $9 \cdot 7$
- **e.** $4(7)$
- **f.** 8×9
- **g.** $6 \cdot 5$

Finish Time: _____
Total Time: _____

30. Start Time: _____
- **a.** 3×4
- **b.** $8 \cdot (7)$
- **c.** $9 \cdot 5$
- **d.** 7×6
- **e.** $3(8)$
- **f.** $4 \cdot 9$
- **g.** $6 \cdot 9$

Finish Time: _____
Total Time: _____

31. Start Time: _____
- **a.** $5 \cdot 4$
- **b.** $(4) \cdot 6$
- **c.** 7×8
- **d.** $6 \cdot 3$
- **e.** $8(4)$
- **f.** 9×9
- **g.** $8 \cdot 5$

Finish Time: _____
Total Time: _____

32. Start Time: _____
- **a.** $5 \cdot 6$
- **b.** $(7) \cdot 3$
- **c.** 9×6
- **d.** $8 \cdot 6$
- **e.** $7(9)$
- **f.** 4×8
- **g.** $7 \cdot 7$

Finish Time: _____
Total Time: _____

Find each product.

33. $\begin{array}{r} 23 \\ \times\ 4 \\ \hline \end{array}$ **34.** $\begin{array}{r} 31 \\ \times\ 8 \\ \hline \end{array}$ **35.** $\begin{array}{r} 42 \\ \times\ 7 \\ \hline \end{array}$ **36.** $\begin{array}{r} 55 \\ \times\ 9 \\ \hline \end{array}$

37. $\begin{array}{r} 543 \\ \times\ 5 \\ \hline \end{array}$ **38.** $\begin{array}{r} 621 \\ \times\ 6 \\ \hline \end{array}$ **39.** $\begin{array}{r} 745 \\ \times\ 3 \\ \hline \end{array}$ **40.** $\begin{array}{r} 389 \\ \times\ 9 \\ \hline \end{array}$

41. Louise sold 3 ATVs last week. Each ATV sold for $2,463. What was her total in ATV sales last week?

42. Tickets to a rugby game cost $8 each. If 3,351 people attended the game, what was the total in ticket sales?

43. To help victims of a typhoon, a merchant asked for $5 donations from her customers. 51,472 people made a donation. What was the total amount collected?

44. To replace the car inventory recently sold, a car dealership purchased 7 Porsches for $56,745 each. What is the value of the new car inventory?

Multiply by the given power of 10.

45. $\begin{array}{r} 84 \\ \times\ 10 \\ \hline \end{array}$ **46.** $\begin{array}{r} 56 \\ \times\ 10 \\ \hline \end{array}$ **47.** $\begin{array}{r} 854 \\ \times\ 100 \\ \hline \end{array}$ **48.** $\begin{array}{r} 732 \\ \times\ 100 \\ \hline \end{array}$

49. $561 \cdot 1{,}000$ **50.** $489 \cdot 1{,}000$ **51.** $100 \cdot 5{,}789$ **52.** $100 \cdot 1{,}453$

Find each product.

53. $\begin{array}{r} 45 \\ \times\ 20 \\ \hline \end{array}$ **54.** $\begin{array}{r} 93 \\ \times\ 20 \\ \hline \end{array}$ **55.** $\begin{array}{r} 926 \\ \times\ 300 \\ \hline \end{array}$ **56.** $\begin{array}{r} 527 \\ \times\ 300 \\ \hline \end{array}$

goetz Have difficulty with basic fact accuracy/speed? Try flash cards—every minute you study basic facts will save you lots of future time.
For more tips and tweets, go to twitter.com/gstbasicmath

57. As part of an economic stimulus package, 400 people each received $300 from the IRS. What was the total amount given to the 400 people?

58. As part of an economic stimulus package, 850 people each received $600 from the IRS. What was the total amount given to the 850 people?

59. To help its employees improve their skills, a company contributed $5,000 to a scholarship fund every week for a year (52 weeks). How much money was given to the scholarship fund?

60. An appliance store has the following offer. Any customer donating 20 cans of food will receive $100 off the price of any appliance. If 87 people take advantage of the offer, how much food will be collected?

Multiply.

61. $\begin{array}{r} 84 \\ \times\ 28 \\ \hline \end{array}$

62. $\begin{array}{r} 48 \\ \times\ 45 \\ \hline \end{array}$

63. $57 \cdot 78$

64. $83 \cdot 76$

65. $8 \cdot 7 \cdot 4$

66. $5 \cdot 6 \cdot 3$

67. $12 \cdot 14 \cdot 13$

68. $15 \cdot 21 \cdot 17$

69. $\begin{array}{r} 332 \\ \times\ 54 \\ \hline \end{array}$

70. $\begin{array}{r} 561 \\ \times\ 38 \\ \hline \end{array}$

71. $742 \cdot 39$

72. $265 \cdot 28$

73. 581 people donated, on average, $1,522 each to a politician's campaign fund. How much was donated in total?

74. An egg farm has 9,406 chickens. Last year every chicken laid, on average, 733 eggs. How many eggs did the farm produce last year?

75. What is the product of 3,698 and 8,005?

76. What is the product of 2,875 and 4,050?

77. $\begin{array}{r} 465 \\ \times\ 657 \\ \hline \end{array}$

78. $\begin{array}{r} 583 \\ \times\ 734 \\ \hline \end{array}$

79. $765 \cdot 478$

80. $843(658)$

81. $245 \cdot 321$

82. $873 \cdot 158$

83. $293 \cdot 405$

84. $201 \cdot 409$

Find the area of each rectangle.

85.

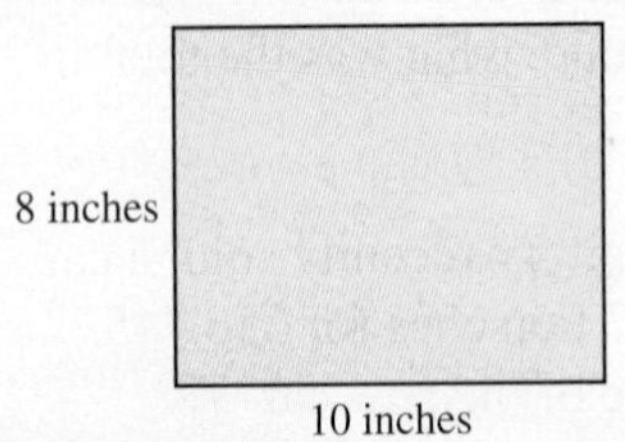

86.

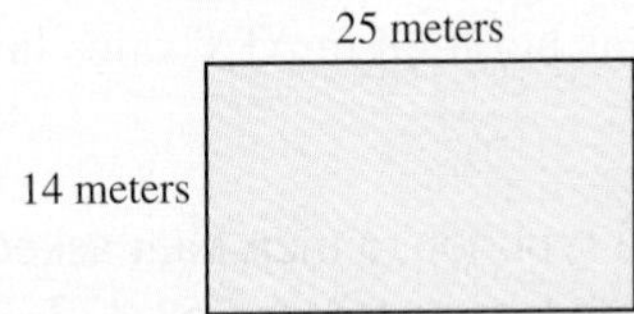

87. The rectangular reflecting pool in front of the Washington Monument measures 2,029 feet by 167 feet. What is its area?

88. A garden's rectangular reflecting pool is about 12 feet by 27 feet. Estimate its area.

Solve.

89. Multiply 35 and 46.

90. Multiply 62 and 24.

91. Find the product of 54 and 53.

92. Find the product of 78 and 73.

93. What is 6 times 123?

94. What is 4 times 236?

95. Juan worked 15 hours during the first 2 days of the week. If Juan makes $31 per hour, how much money did he earn?

96. Dusty worked 22 hours during the first 3 days of the week. If Dusty makes $18 per hour, how much money did he earn?

97. Mia rents a house to Jeff for 6 months. If the house rents for $850 a month, how much income does Mia receive from Jeff?

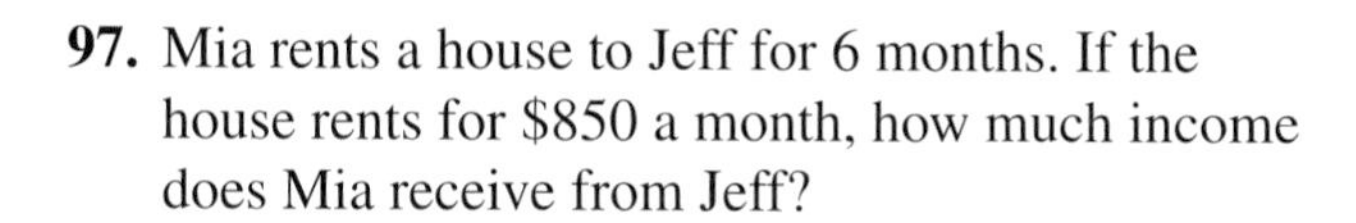

98. Doug buys an apartment complex. The complex has 6 apartments that will rent for $945 a month each. What is Doug's monthly income if every apartment is rented?

99. The distance between bases in a baseball field is 90 feet. The infield is the area inside the bases. What is the area of the infield?

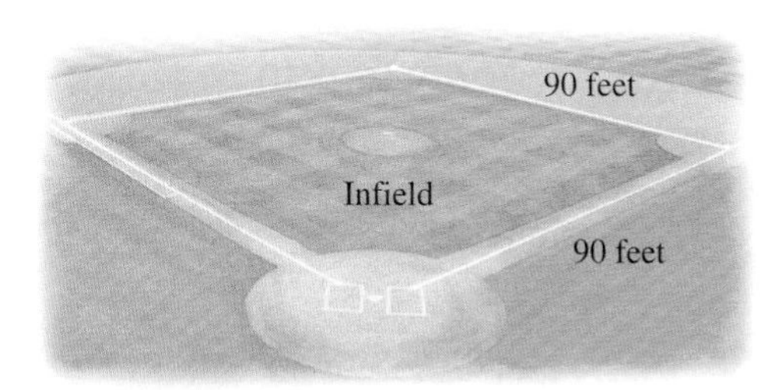

100. The game of tennis requires that each player defend the entire side of his or her court. What is the area that a player must defend?

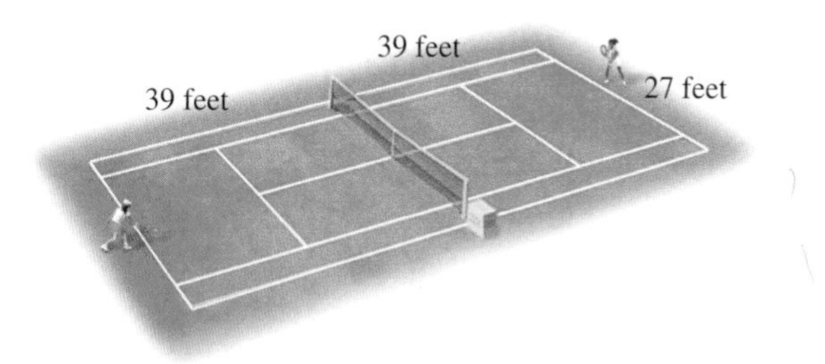

101. Terrell's Volkswagen car gets 42 miles per gallon during highway driving. If he has 13 gallons of gas, how far can he travel on the highway?

102. Darnel's Volkswagen SUV gets 11 miles per gallon during highway driving. If he has 23 gallons of gas, how far can he travel on the highway?

103. Jeaubert bought a used car for $9,000. He has made 14 payments of $250.

a. How much does he still owe on the car?

b. The car has depreciated and is now worth only $5,200. Is the car worth more or less than what is still owed?

104. Celeste purchased 288 eggs to cook omelets at a convention.

a. Celeste has cooked 53 omelets, using three eggs for each omelet. How many eggs does she have left?

b. Does she have enough eggs left to make 45 more omelets?

105. Epoxy is used to refinish a basketball court.

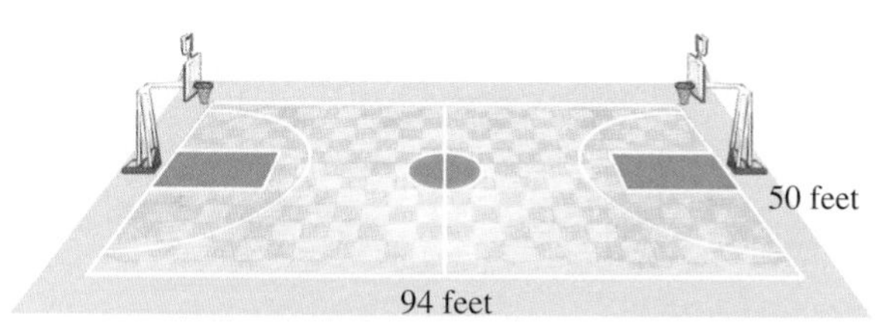

a. What is the area of the basketball court?

b. One gallon of epoxy can coat 400 square feet. How many square feet can be coated with 10 gallons of epoxy?

c. Will 10 gallons be enough to refinish the court?

106. A Zamboni machine is used to resurface the rink after each period in a hockey game.

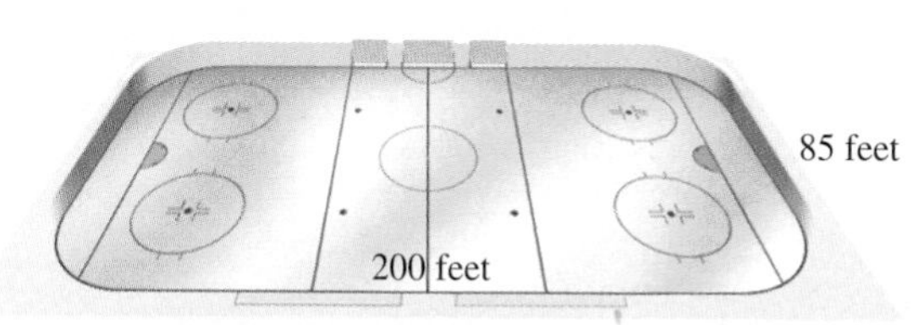

a. What is the approximate area of the rink? *Hint:* A rectangle gives a good approximation for the area of the hockey rink.

b. One gallon of water is needed to resurface 400 square feet. How many square feet can be resurfaced with 20 gallons of water?

c. Will 20 gallons be enough to resurface the rink?

To determine which electrician is cheaper, a contractor uses the following table to determine how much each electrician charges.

107. How much will Herman charge to wire a room with 8 outlets, 2 switches, and 6 lights?

108. How much will Pedro charge to wire a room with 8 outlets, 2 switches, and 6 lights?

	Herman's Costs	Pedro's Costs
Outlets	$45	$50
Switches	$35	$40
Lights	$70	$60

Solve for the unknown number.

109. $5 \cdot x = 15$

110. $4 \cdot x = 24$

111. $8 \cdot y = 32$

112. $7 \cdot y = 42$

113. $80 = 10 \cdot z$

114. $110 = 11 \cdot z$

115. $660 = 220 \cdot x$

116. $840 = 420 \cdot y$

Section 1.4 Question Log

Use this space to write questions. Make sure you get these answered and revisit them when you prepare for your exam.

Page ____ **Answered** ☐	**Page** ____ **Answered** ☐
Page ____ **Answered** ☐	**Page** ____ **Answered** ☐

Section 1.4 Extra Practice

See the Appendix A for extra practice on Section 1.4.

Section 1.5 Dividing Whole Numbers

Division can be viewed in two different ways. We can interpret "6 ÷ 3 = 2" to mean that

- There are two 3's inside one 6.
- If 6 is divided into 3 pieces, then each piece will have a size of 2.

To use the division procedure, you must learn the basic facts and understand how place value is used in each step.

The **Objectives** in Section 1.5 will help you

- **A** Understand the basic division facts.
- **B** Perform long division with a remainder.
- **C** Estimate quotients.
- **D** Divide by a large number.

VOCABULARY PREVIEW *Check the box that applies.*

The following four definitions use the example **7 ÷ 3 = 2 with remainder 1.**	Got It	Must Study
Dividend: The number being divided is the **dividend.** In the example above, 7 is the dividend.		
Divisor: The number that does the dividing is the **divisor.** In the example above, 3 is the divisor.		
Quotient: A **quotient** is the result of dividing. It is the whole number of times that the divisor goes into the dividend. In the example above, 2 is the quotient.		
Remainder: The **remainder** is the number that is left over when the divisor does not divide the dividend evenly. In the example above, 1 is the remainder.		
Division Notation: The following **notation** will be used throughout this section. $6 \div 3 = 2$ ↕ dividend ÷ divisor = quotient; $3\overline{)6}$ with quotient 2 ↕ $\text{divisor}\overline{)\text{dividend}}$ with quotient; $\frac{6}{3} = 2$ ↕ $\frac{\text{dividend}}{\text{divisor}} = \text{quotient}$		

Study these terms when they appear in the text. At the end of the section, test your understanding by completing the Vocabulary Review in the section exercises.

Objective A Understand the Basic Division Facts

The Concept We can use multiplication to help us learn the basic division facts. For example, to find $8 \div 2 = \boxed{?}$, we can ask ourselves, $\boxed{?} \cdot 2 = 8$. Because $\boxed{4} \cdot 2$ is 8, we know that $8 \div 2$ is $\boxed{4}$. The following examples use multiplication to solve division exercises.

Division	⇒	Corresponding Multiplication
$6 \div 3 = x$	⇒	$x \cdot 3 = 6$
$15 \div 5 = y$	⇒	$y \cdot 5 = 15$
Dividend ÷ Divisor = z	⇒	$z \cdot$ Divisor = Dividend

Reverse the order of the numbers to write a corresponding multiplication exercise.

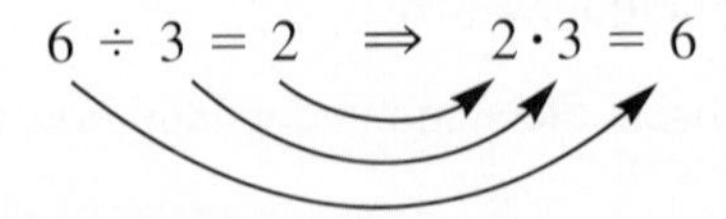

Procedure **Solve a Division Exercise Using Multiplication**

Solve the corresponding multiplication exercise.

EXAMPLES

1. Write the corresponding multiplication exercise for $8 \div 2$.

Reverse the order of the numbers to write the corresponding multiplication exercise.

$8 \div 2 = \boxed{?} \Rightarrow \boxed{?} \cdot 2 = 8$

2. Solve the corresponding multiplication exercise for $15 \div 3$.

Solve the corresponding multiplication exercise.

$15 \div 3 = \boxed{?} \Rightarrow \boxed{?} \cdot 3 = 15$

$5 \cdot 3 = 15$

Therefore, $15 \div 3 = 5$.

GUIDED PRACTICE

1. Write the corresponding multiplication exercise for $10 \div 5$.

Reverse the order of the numbers to write the corresponding multiplication exercise.

$10 \div 5 = \boxed{?} \Rightarrow \underline{\boxed{?}} \cdot \_\_\_\_ = \_\_\_\_$

2. Solve the corresponding multiplication exercise for $24 \div 6$.

Solve the corresponding multiplication exercise.

$24 \div 6 = \boxed{?} \Rightarrow \boxed{?} \cdot \_\_\_\_ = \_\_\_\_$

$\_\_\_\_ \cdot \_\_\_\_ = \_\_\_\_$

Therefore, $24 \div 6 = \_\_\_\_$.

FOCUS ON Division Involving Zero

Be very careful when a 0 is in a division exercise.

- When zero is divided by another number, the result is zero.
- When a number is divided by zero, the result is undefined. It is undefined because no answer makes sense.

Because it is the best way to understand why division by zero is undefined, you should solve the corresponding multiplication exercise.

(Continued)

Division of Zero Is Zero

$0 \div 4 = 0,\ 4\overline{)0} = 0,$ and $\frac{0}{4} = 0.$

Explanation Using Multiplication

$0 \div 4 = \boxed{?} \Rightarrow \boxed{?} \cdot 4 = 0$

$\boxed{0} \cdot 4 = 0$

Because $0 \cdot 4 = 0,\ 0 \div 4 = 0.$

Division by Zero Is Undefined

$4 \div 0,\ 0\overline{)4},$ and $\frac{4}{0}$ are undefined.

Explanation Using Multiplication

$4 \div 0 = \boxed{?} \Rightarrow \boxed{?} \cdot 0 = 4$

$\boxed{} \cdot 0 = 4$

Every number times 0 is zero, not 4.

Because we can't get a product of 4, $4 \div 0$ is undefined.

PRACTICE

For each exercise:

a. Write the corresponding multiplication exercise.
b. Try to solve the corresponding multiplication exercise to see if the answer is *0* or *undefined.*

1. $0\overline{)10}$

2. $\frac{10}{0}$

3. $10\overline{)0}$

4. $10 \div 0$

5. $\frac{0}{10}$

6. $0 \div 10$

Concept Check

A1. Write the corresponding multiplication exercise for $45 \div 9 = \boxed{?}$

A2. Use multiplication to explain why the answer to $6 \div 0$ is undefined.

Objective A Practice

For each division exercise, write and solve the corresponding multiplication exercise.

A3. $30 \div 3$

A4. $40 \div 4$

A5. $60 \div 10$

A6. $48 \div 8$

A7. $63 \div 7$

A8. $54 \div 6$

A9. $80 \div 20$

A10. $100 \div 50$

Complete each multiplication table. You will need to use division to find the numbers in some of the empty boxes. The outlined box is 5 because 40 ÷ 8 = 5.

A11.

×	7		8	
5			40	10
9				18
		36	48	
	28			

A12.

×		5	9	
			81	
		35		42
4				
	24		72	

Writing only the answers, try to complete each column in fewer than 30 seconds. Accuracy is important. Take enough time to work through each exercise correctly.

A13. Start Time: _____

a. $10 \div 5$
b. $48 \div 8$
c. $\frac{54}{6}$
d. $3 \div 3$
e. $6\overline{)30}$
f. $42 \div 6$
g. $\frac{28}{7}$
h. $40 \div 0$
i. $64 \div 8$
j. $4\overline{)16}$

Finish Time: _____

Total Time: _____

A14. Start Time: _____

a. $4 \div 1$
b. $54 \div 9$
c. $\frac{0}{3}$
d. $49 \div 7$
e. $0\overline{)7}$
f. $56 \div 7$
g. $\frac{54}{9}$
h. $20 \div 5$
i. $72 \div 8$
j. $2\overline{)6}$

Finish Time: _____

Total Time: _____

A15. Start Time: _____

a. $12 \div 3$
b. $30 \div 3$
c. $\frac{72}{9}$
d. $32 \div 8$
e. $7\overline{)35}$
f. $63 \div 7$
g. $\frac{0}{7}$
h. $56 \div 8$
i. $6 \div 0$
j. $7\overline{)42}$

Finish Time: _____

Total Time: _____

A16. Start Time: _____

a. $9 \div 3$
b. $48 \div 6$
c. $\frac{40}{8}$
d. $25 \div 5$
e. $9\overline{)63}$
f. $36 \div 6$
g. $\frac{40}{10}$
h. $36 \div 9$
i. $81 \div 9$
j. $0\overline{)8}$

Finish Time: _____

Total Time : _____

▶ **NOTE** Don't worry if it takes you longer than 30 seconds. However, if it does, make flash cards to study your basic facts.

Objective B Perform Long Division with a Remainder

The Concept Often a divisor will not go into a dividend evenly. This results in a remainder. Imagine that two people have only dollar bills and they must divide a $13 restaurant check. Each person can pay $6, for a total of $12, resulting in $1 left (remaining) to be paid.

$$13 \div 2 = 6 \text{ Remainder } 1$$
$$13 \div 2 = 6\text{R}1$$

To perform a division exercise with a remainder, you need to understand the notation and vocabulary used in long division.

INTERACTIVE DEFINITION Long Division Vocabulary

$$\begin{array}{r}2\\3\overline{)6}\end{array} \Leftrightarrow \begin{array}{r}\text{quotient}\\\text{divisor}\overline{)\text{dividend}}\end{array}$$

EXAMPLE

3. Identify the dividend, divisor, and quotient in $18 \div 6 = 3$.

$$18 \div 6 = 3 \Rightarrow \begin{array}{r}3\\6\overline{)18}\end{array}$$

The dividend is 18.
The divisor is 6.
The quotient is 3.

GUIDED PRACTICE

3. Identify the dividend, divisor, and quotient in $12 \div 3 = 4$.

$$12 \div 3 = 4 \Rightarrow \overline{)\quad}$$

The dividend is _____.
The divisor is _____.
The quotient is _____.

DO YOU UNDERSTAND how to identify the dividend, divisor, and quotient in a division exercise? Got It Get Help

Procedure **Division with a Remainder**

Step 1 Determine how many times the divisor goes into the dividend.
Step 2 Multiply the divisor by that value.
Step 3 Subtract the result from the dividend.

The remainder is what is left over.

EXAMPLE

4. Divide. $3\overline{)17}$

Step 1 Determine how many times 3 goes into 17.

$3 \cdot 5 = 15$
$3 \cdot 6 = 18$
17

3 goes into 17 five times. →

Step 2 Multiply the divisor by that value.
$3 \cdot 5 = 15$

Step 3 Subtract the result from the dividend.
$17 - 15 = 2$

$$\begin{array}{r} 5 \\ 3\overline{)17} \\ -15 \\ \hline 2 \end{array}$$

Because 3 does not go into 2, the remainder is 2.
To show that 3 goes into 17 five times with a remainder of two, we write $17 \div 3 = 5R2$.

GUIDED PRACTICE

4. Divide. $6\overline{)29}$

Step 1 Determine how many times 6 goes into 29.

$6 \cdot$ ____ $= 24$
$6 \cdot$ ____ $= 30$
29

6 goes into 29 ____ times. →

Step 2 Multiply the divisor by that value.
$6 \cdot$ ____ $=$ ____

Step 3 Subtract the result from the dividend.
$29 -$ ____ $=$ ____

$$\begin{array}{r} 6\overline{)29} \\ - \quad \\ \hline \end{array}$$

Because 6 does not go into ____, the remainder is ____. To show that 6 goes into ____ ____ times with a remainder of ____, we write $29 \div 6 =$ ____.

The steps outlined in Example 4 and Guided Practice 4 are the keys for performing long division. These steps are repeated for each place value in the quotient.

DETAILED EXAMPLE **The Structure of a Long Division Exercise**

Divide. $63 \div 5$

Set up the problem.

$$\begin{array}{r} \text{R} \\ 5\overline{)6\ 3} \end{array}$$

Draw the lines.

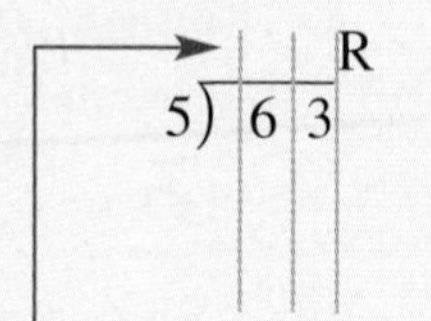

Solve the exercise.

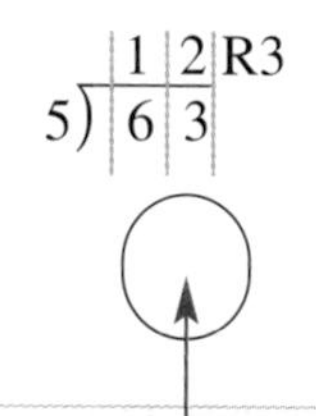

Every "box" on the top of the division symbol will have a digit in it when you are done.

Your work will be shown here.

Procedure Perform Long Division with a Remainder

Start with the largest place value of the dividend and work toward the smallest place value.

Step 1 Determine how many times the divisor *goes into* the dividend.
This value will be a one-digit number, possibly zero.

Step 2 *Multiply* the divisor by that value.
Write the answer below the dividend.

Step 3 *Subtract* that quantity.

Step 4 *Bring down* the next digit of the divisor.

Repeat until every digit of the dividend has been divided.

In the following examples, the steps are abbreviated as *goes into, multiply, subtract,* and *bring down.* These steps are repeated until every digit in the dividend has a corresponding digit in the quotient, even if the digit is 0. The four steps are designed to subtract as much of the divisor as possible in each step and then see what is left over.

Use the following advice to answer more division exercises correctly.

1. Use graph paper to stay organized.
2. If you don't have graph paper, draw the lines to set up a grid.
3. If you are using lined paper, turn your paper sideways to use the lines.

EXAMPLES

5. Divide. 63 ÷ 4

Begin by dividing 4 into 6.

4 goes into 6 one time.
Multiply 1 and 4.
Subtract 4.
Bring down the 3.

```
     1
4)  6 3
  - 4 ↓
    2 3
```

Divide 4 into what's left.

4 goes into 23 five times.
Multiply 4 and 5.
Subtract 20.
We are at the end of the dividend. The remainder is 3.

```
     1 5 R3
4)  6 3
  - 4 ↓
    2 3
  - 2 0
      3
```

63 ÷ 4 = 15R3

GUIDED PRACTICE

5. Divide. 73 ÷ 3

Draw the lines

Key steps
- Goes into
- Multiply
- Subtract
- Bring down

Helpful hints
- Write big to avoid little mistakes.
- Draw the lines.
- Every "box" gets a digit.

6. Divide. $312 \div 7$

7 will not go into 3, so we write a 0 above the 3.

Now divide 7 into 31.

7 goes into 31 four times.
Multiply 7 and 4.
Subtract 28.
Bring down the 2.

$$\begin{array}{r} 04 \\ 7\overline{)312} \\ -28\downarrow \\ \hline 32 \end{array}$$

Divide 7 into what's left.

7 goes into 32 four times.
Multiply 7 and 4.
Subtract 28.
We are at the end of the dividend. The remainder is 4.

$$\begin{array}{r} 044 \text{ R4} \\ 7\overline{)312} \\ -28 \\ \hline 32 \\ -28 \\ \hline 4 \end{array}$$

$312 \div 7 = 44R4$

6. Divide. $459 \div 6$

Draw the lines.

$$\overline{)}$$

Key steps
- Goes into
- Multiply
- Subtract
- Bring down

Helpful hints
- Write big to avoid little mistakes.
- Draw the lines.
- Every "box" gets a digit.

Concept Check

B1. Why might drawing the lines help you solve a division problem?

B2. In your own words, describe a remainder.

Objective B Practice

Divide.

B3. $75 \div 4$ **B4.** $83 \div 2$ **B5.** $67 \div 8$ **B6.** $48 \div 7$

B7. $\frac{654}{4}$ **B8.** $\frac{845}{3}$ **B9.** $\frac{523}{6}$ **B10.** $\frac{625}{7}$

B11. A group of 5 basketball players won a tournament and its prize, \$1,500. If they split the prize equally, how much will each player get?

B12. If the basketball players from Exercise B11 share the prize with their coach, how much will each person get?

FOCUS ON The Meaning of Remainders

Recall that a quotient is the result of dividing. In an earlier example, two people split a $13 restaurant bill. To divide the bill, we said that $13 \div 2 = 6R1$. The remainder of $1 still needs to be split, or divided, to pay the bill. When the $1 remainder is divided, it becomes part of the quotient. The divided bill could be expressed using either a decimal ($6.50) or a mixed number $\left(\$6\frac{1}{2}\right)$. Notice how we can write the answer using a mixed number.

$$13 \div 2 = 6R1 \quad = 6\frac{1}{2}$$

Mixed numbers are made of a whole number and a fraction. They are covered in detail in chapter 2.

Procedure Use a Mixed Number to Write a Quotient and Remainder

Step 1 Divide the dividend by the divisor.

$$\text{divisor}\overline{)\text{dividend}} \quad \text{quotient and remainder}$$

Step 2 Write the mixed number.

$$\text{quotient}\,\frac{\text{remainder}}{\text{divisor}}$$

EXAMPLES

7. $13 \div 5 = 2R3$.
Write the answer as a mixed number.

For $13 \div 5 = 2R3$, identify the quotient, remainder, and divisor.

Quotient = 2
Remainder = 3
Divisor = 5

Write the mixed number.

$$\text{quotient}\,\frac{\text{remainder}}{\text{divisor}} = 2\frac{3}{5}$$

8. Write the answer to $22 \div 9$ as a mixed number.

Step 1 Divide the dividend by the divisor.

Use scratch paper to perform the division.

$$22 \div 9 = 2R4.$$

Step 2 Write the mixed number.

$$22 \div 9 = 2\frac{4}{9}$$

GUIDED PRACTICE

7. $17 \div 3 = 5R2$.
Write the answer as a mixed number.

For $17 \div 3 = 5R2$, identify the quotient, remainder, and divisor.

Quotient =
Remainder =
Divisor =

Write the mixed number.

$$\text{quotient}\,\frac{\text{remainder}}{\text{divisor}} = \underline{\quad}$$

8. Write the answer to $25 \div 6$ as a mixed number.

Step 1 Divide the dividend by the divisor.

Use scratch paper to perform the division.

$$25 \div 6 = \underline{\quad R \quad}$$

Step 2 Write the mixed number.

$$25 \div 6 =$$

(Continued)

PRACTICE

Write the answer to each division exercise as a mixed number.

1. $14 \div 3 = 4R2$ **2.** $21 \div 8 = 2R5$ **3.** $26 \div 3$ **4.** $17 \div 8$

5. $32 \div 5$ **6.** $29 \div 7$ **7.** $41 \div 5$ **8.** $53 \div 7$

Objective C Estimate Quotients

The Concept Being able to estimate a quotient is a valuable skill. It will allow you to check whether an answer is reasonable. Estimation is a practical tool for using division in everyday life. Often a quick estimate is all that is needed.

DETAILED EXAMPLE Estimating a Quotient in an Application Exercise

A class of 31 students is raising money to go on a research trip. The total cost of the trip will be $8,768. Estimate the amount that each student must raise.

To estimate, we round both numbers to their largest place value.

$$31\overline{)8768} \approx 30\overline{)9000}$$

Rounded Values
31 ≈ 30
8,768 ≈ 9,000

The rounded values make the division easier.

$$\begin{array}{r} 0300 \\ 30\overline{)9000} \\ -90 \\ \hline 000 \end{array}$$

Answer: Each student must raise about $300.

The actual amount is $282.84. The estimate, $300, is useful because the students will understand that they need to raise about $300.

Procedure Estimate Quotients

Step 1 Round the dividend and divisor to their largest place values.

Step 2 Divide with the rounded values.

Ignore the remainder when stating the estimate.

EXAMPLE

9. Estimate the quotient. $465 \div 21$

Step 1 Round each number to its largest place value.

$21 \approx 20$ (Rounded to the tens place)

$465 \approx 500$ (Rounded to the hundreds place)

Step 2 Divide with the rounded values.

$$\begin{array}{r} 025 \\ 20\overline{)500} \\ -40\downarrow \\ \hline 100 \\ -100 \\ \hline 0 \end{array}$$

$465 \div 21 \approx 25$

(Actual answer: 22R3)

GUIDED PRACTICE

9. Estimate the quotient. $193 \div 11$

Step 1 Round each number to its largest place value.

$11 \approx$ _____ (Rounded to the _____ place)

$193 \approx$ _____ (Rounded to the _______ place)

Step 2 Divide with the rounded values.

$\overline{)\quad}$

$193 \div 11 \approx$

(Actual answer: 17R6)

Is an estimate close enough? If you believe that an actual answer is not close enough to your estimate, then it's a good idea to double-check your calculations.

How inaccurate can an estimate be? If we estimate $44\overline{)1520}$, we get 50. The actual calculation is 34R24. Both the answer and the estimate are correct even though they are very different. This is a case where we would definitely double-check our calculations!

EXAMPLE

10. Estimate the quotient. $341 \div 42$

Step 1 Round each number to its largest place value.

$42 \approx 40$ (Rounded to the tens place)

$341 \approx 300$ (Rounded to the hundreds place)

Step 2 Divide with the rounded values.

$$\begin{array}{r} 007 \\ 40\overline{)300} \\ -280 \\ \hline 20 \end{array}$$

$341 \div 42 \approx 7$

(Actual answer: 8R5)

GUIDED PRACTICE

10. Estimate the quotient. $571 \div 31$

Step 1 Round each number to its largest place value.

$31 \approx$ _____ (Rounded to the _____ place)

$571 \approx$ _____ (Rounded to the _______ place)

Step 2 Divide with the rounded values.

$\overline{)\quad}$

$571 \div 31 \approx$ _____

(Actual answer: 18R13)

Concept Check

C1. In your own words, what is the advantage of estimating an answer to a division exercise?

C2. Round each number to its largest place value.
a. 4,832 **b.** 92,387 **c.** 23 **d.** 850

Objective C Practice

Round each number to its largest place value to estimate each quotient.

C3. $387 \div 16$ **C4.** $212 \div 12$ **C5.** $973 \div 48$ **C6.** $912 \div 33$

C7. $\dfrac{5{,}342}{17}$ **C8.** $\dfrac{7{,}921}{23}$ **C9.** $\dfrac{6{,}185}{58}$ **C10.** $\dfrac{3{,}182}{95}$

C11. Twenty-three campers are working together to build a 9,400-pound monument on top of a mountain. If each person moves the same amount of rock, about how many pounds of rock will each person move?

C12. If four additional campers help build the monument from Exercise C11, about how many pounds of rock will each person move?

Objective D Divide by a Large Number

The Concept Often we will have to find the quotient of large numbers such as $1{,}724 \div 32$. When we divide by a large number, we use the same steps that we have already learned. However, because the multiples of 32 are not basic facts, we use estimation to find the digits of the quotient.

EXAMPLE Using Estimation to Find One Digit of a Quotient

Divide. $1{,}724 \div 32$. Do only the first step.

Set up the exercise.

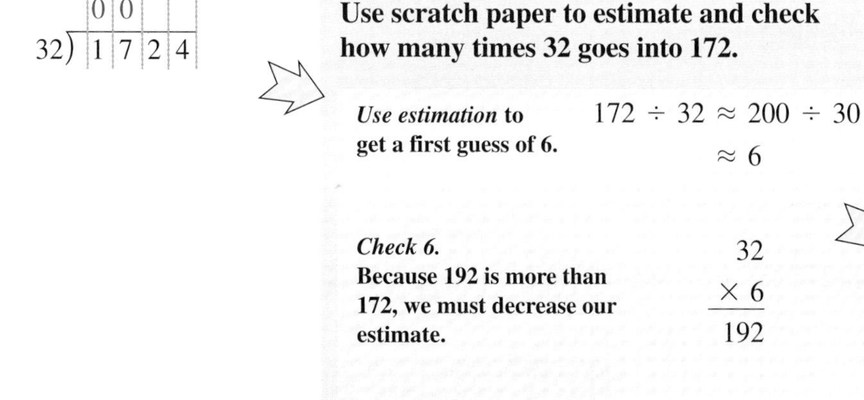

This pattern is continued to finish the division exercise.

$$\begin{array}{r} 005 \\ 32\overline{)1724} \\ -160 \\ \hline 124 \end{array}$$

Our next step would be to divide 124 by 32.

In the previous example, our initial estimate of 6 was too large. It is useful to recognize when initial estimates are too large or too small.

RECOGNIZING WHEN AN ESTIMATE IS TOO SMALL OR TOO LARGE

When Is an Estimate Too Small?	When Is an Estimate Too Large?
Find the quotient. 53 ÷ 17	**Find the quotient.** 58 ÷ 12
Initial Estimate: We estimate 53 ÷ 17 using 50 ÷ 20 to guess that 17 goes into 53 two times.	**Initial Estimate:** We estimate 58 ÷ 12 using 60 ÷ 10 to guess that 12 goes into 58 six times.
$\begin{array}{r} 02 \\ 17\overline{)53} \\ -34 \\ \hline 19 \end{array}$	$\begin{array}{r} 06 \\ 12\overline{)58} \\ -72 \\ \hline \end{array}$
Because 19 is greater than 17, we know that our estimate was too small.	Because 72 is greater than 58, we know that our estimate was too large.
Correction: Try increasing the estimate to 3.	**Correction:** Try decreasing the estimate to 4.
$\begin{array}{r} 03 \\ 17\overline{)53} \\ -51 \\ \hline 2 \end{array}$	$\begin{array}{r} 04 \\ 12\overline{)58} \\ -48 \\ \hline 10 \end{array}$
This is the correct value.	This is the correct value.

Procedure Divide by a Large Number

Start with the largest place value of the dividend and work toward the smallest place value.

Step 1 Determine how many times the divisor *goes into* the dividend.

Step 2 *Multiply* the divisor by that value.

Step 3 *Subtract* that quantity.

Step 4 *Bring down* the next digit of the divisor to see what is left over.

Repeat until every digit of the dividend has been divided.

▶ **NOTE** In the examples that follow, each color in the scratch paper corresponds to one cycle of "goes into, multiply, subtract, bring down."

EXAMPLES

11. Divide. 423 ÷ 31

Calculate 31 into 42.
31 goes into 42 once.

Calculate 31 into 113.
Estimate: 100 ÷ 30 = 3

Check:
31 · 3 = 93.
31 goes into 113 three times.

$$\begin{array}{r} 013\text{R20} \\ 31\overline{)423} \\ -31 \\ \hline 113 \\ -93 \\ \hline 20 \end{array}$$

423 ÷ 31 = 13R20

GUIDED PRACTICE

11. Divide. 372 ÷ 21

Calculate 21 into 37.

372 ÷ 21 =

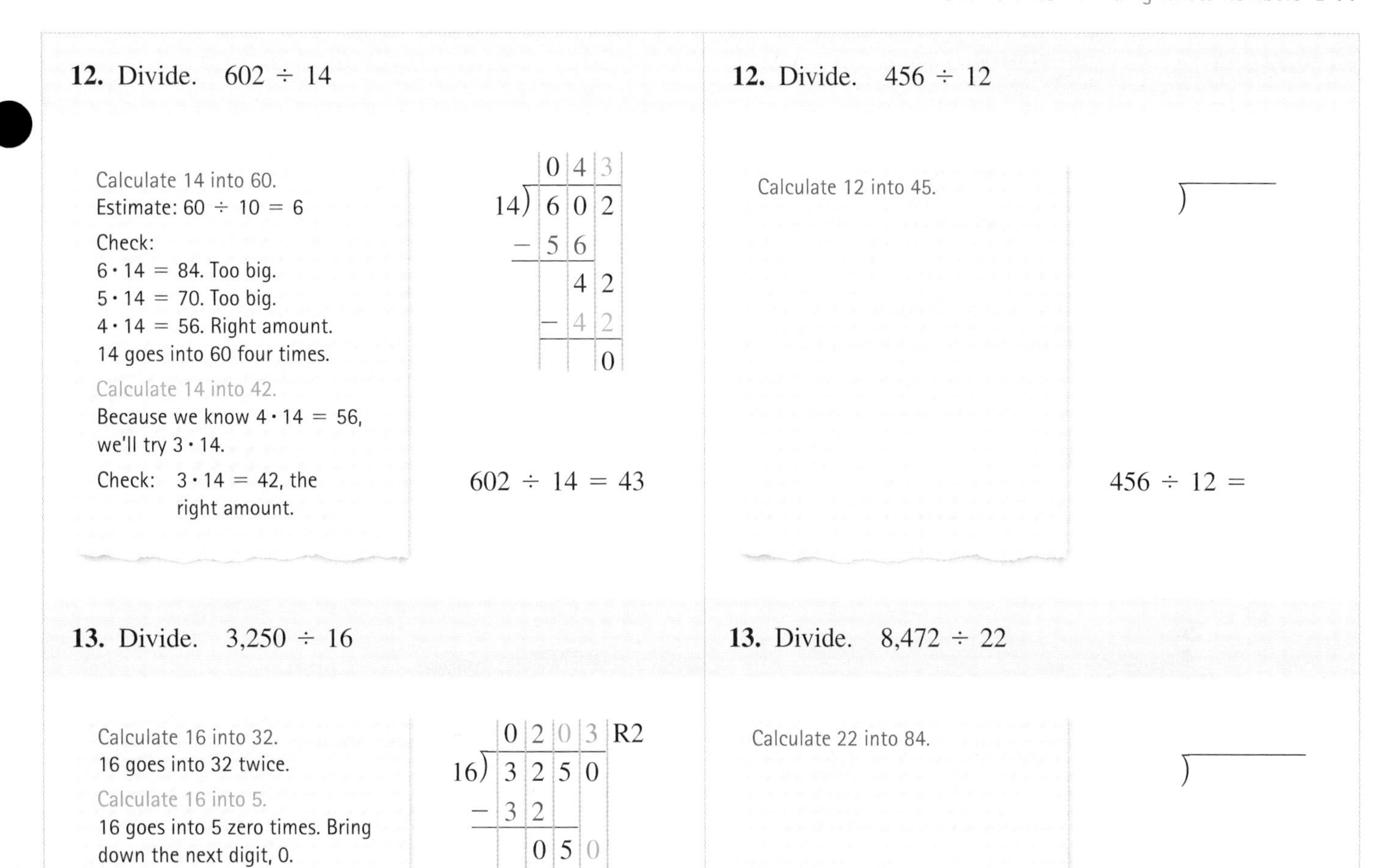

Concept Check

D1. Use estimation to make a first guess at how many times 31 goes into 271.

D2. Use multiplication to show that 31 goes into 271 less than 9 times.

D3. Use the concept of place value to explain what is wrong in the following problem:

$$\begin{array}{r} 2R3 \\ 12\overline{)243} \\ \underline{-24} \\ 03 \end{array}$$

D4. In your own words, how would drawing the lines have helped the student find the quotient in Exercise D3?

Objective D Practice

Divide.

D5. $321 \div 53$ **D6.** $483 \div 62$ **D7.** $956 \div 22$

D8. $837 \div 31$ **D9.** $2{,}550 \div 16$ **D10.** $3{,}973 \div 21$

D11. Answer each question.

a. Deana has asked to place the gas bill for her business on a budget plan. To calculate her monthly payment, the gas company divides last year's bill, \$7,584, by 12 months. Find her monthly gas bill.

b. After having her gas bill placed on a budget plan, Deana's business used \$9,516 in gas over the next year. What will the new monthly bill be if she remains on a budget plan?

Combining Concepts and Applications

CONCEPT I Solving for an Unknown Number To solve for an unknown number in a division exercise, solve the corresponding multiplication exercise.

EXAMPLES	GUIDED PRACTICE
14. Solve for the unknown number. $54 \div x = 6$	**14.** Solve for the unknown number. $42 \div y = 7$
Here the corresponding multiplication exercise is a basic fact. $54 \div x = 6$ Corresponding multiplication $6 \cdot x = 54$ Basic fact: $6 \cdot 9 = 54$ $x = 9$	Here the corresponding multiplication exercise is a basic fact. $42 \div y = 7$ Corresponding multiplication ____ • ____ = 42 Basic fact: ____ • ____ = 42 $y =$
15. Solve for the unknown number. $96 \div y = 12$	**15.** Solve for the unknown number. $65 \div x = 13$
Here the corresponding multiplication exercise is not a basic fact. Keep adding 12 until you reach 96. $96 \div y = 12 \Rightarrow 12 \cdot y = 96$ Multiples of 12 are 12, 24, 36, 48, 60, 72, 84, 96. Eight 12's add to 96. $y = 8$	Here the corresponding multiplication exercise is not a basic fact. Keep adding 13 until you reach 65. $65 \div x = 13 \Rightarrow$ ____ $\cdot x =$ ____ Multiples of 13 are ____, ____, ____, ____, ____. ____ 13's add to 65. $x =$

Section 1.5 Exercises

FOR EXTRA HELP

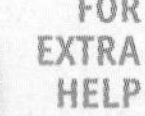

MathXL PRACTICE WATCH DOWNLOAD READ REVIEW

To understand the basic division facts:

1. Answer the Objective A Concept Checks.
2. Answer the odd Objective A Practice Exercises.
3. Answer the even Objective A Practice Exercises.

To understand division involving zero:

4. Answer the odd Focus On Division Involving Zero Practice Exercises.
5. Answer the even Focus On Division Involving Zero Practice Exercises.

To perform long division with a remainder:

6. Answer the Objective B Concept Checks.
7. Answer the odd Objective B Practice Exercises.
8. Answer the even Objective B Practice Exercises.

To understand the meaning of remainders:

9. Answer the odd Focus On the Meaning of Remainders Practice Exercises.
10. Answer the even Focus On the Meaning of Remainders Practice Exercises.

To estimate quotients:

11. Answer the Objective C Concept Checks.
12. Answer the odd Objective C Practice Exercises.
13. Answer the even Objective C Practice Exercises.

To understand how to divide by a large number:

14. Answer the Objective D Concept Checks.
15. Answer the odd Objective D Practice Exercises.
16. Answer the even Objective D Practice Exercises.

VOCABULARY REVIEW

Review the Vocabulary Preview for Section 1.5. Study the definitions until you can check Got It *for every word.*

Use these words to complete each sentence.

dividend • divisor • quotient • remainder

	Got It	Get Help
17. A ____________ is the result of dividing.		
18. The number being divided is the ____________.		
19. The number that does the dividing is the ____________.		
20. When a divisor does not divide the dividend exactly, the portion that is left over is called the ____________.		
Fill in the parts of each division exercise with the word *dividend*, *divisor*, or *quotient*. Assume that there is no remainder.		
21. ____________ ÷ ____________ = ____________		
22. $\left(\frac{}{}\right)$ = ____________		
23. ____________ $\overline{)}$ ____________		

How will you get help for any vocabulary about which you are unsure?

Your instructor ____ MyMathLab ____ A classmate ____ A tutor ____ Other ____

For each exercise:
a. Write the corresponding multiplication exercise.
b. Find the unknown number.

24. $8 \div 2 = x$ **25.** $16 \div 4 = y$ **26.** $25 \div 5 = z$ **27.** $28 \div 7 = x$

28. $42 \div 6 = y$ **29.** $54 \div 9 = z$ **30.** $32 \div 4 = x$ **31.** $60 \div 12 = y$

Complete each multiplication table. You will need to use division to find the numbers in some of the empty boxes. The outlined box is 2 because $14 \div 7 = 2$.

32.

×	7	6	4	8
2	14	12	8	16
5	35	30	20	40
8	56	48	32	64
7	49	42	28	56

33.

×	6	8	9	3
5	30	40	45	15
4	24	32	36	12
7	42	56	63	21
8	48	64	72	24

Writing only the answers, try to complete each column in fewer than 30 seconds. Accuracy is important. Take enough time to work through each exercise correctly.

34. Start Time: ____
- **a.** $32 \div 8$
- **b.** $\dfrac{48}{6}$
- **c.** $9\overline{)54}$
- **d.** $\dfrac{56}{7}$
- **e.** $4 \div 0$
- **f.** $3\overline{)24}$
- **g.** $36 \div 4$

Finish Time: ____
Total Time: ____

35. Start Time: ____
- **a.** $9\overline{)36}$
- **b.** $\dfrac{12}{4}$
- **c.** $8\overline{)64}$
- **d.** $\dfrac{24}{3}$
- **e.** $49 \div 7$
- **f.** $5\overline{)45}$
- **g.** $0 \div 1$

Finish Time: ____
Total Time: ____

36. Start Time: ____
- **a.** $3\overline{)21}$
- **b.** $\dfrac{32}{4}$
- **c.** $9\overline{)45}$
- **d.** $\dfrac{35}{7}$
- **e.** $0 \div 7$
- **f.** $6\overline{)30}$
- **g.** $81 \div 9$

Finish Time: ____
Total Time: ____

37. Start Time: ____
- **a.** $56 \div 8$
- **b.** $\dfrac{18}{3}$
- **c.** $8\overline{)72}$
- **d.** $\dfrac{8}{0}$
- **e.** $36 \div 6$
- **f.** $3\overline{)21}$
- **g.** $63 \div 9$

Finish Time: ____
Total Time: ____

Divide.

38. $84 \div 7$ **39.** $104 \div 8$ **40.** $\dfrac{45}{3}$ **41.** $\dfrac{48}{4}$

42. $5\overline{)125}$ **43.** $8\overline{)128}$ **44.** $7\overline{)196}$ **45.** $6\overline{)324}$

46. $3\overline{)19}$ **47.** $31 \div 6$ **48.** $28 \div 5$ **49.** $7\overline{)24}$

50. $6\overline{)104}$ **51.** $195 \div 9$ **52.** $8\overline{)424}$ **53.** $9\overline{)288}$

54. Seven people are going to share $294 equally. How much will each person get?

55. Five people are working to inventory items on 185 shelves in a store. How many shelves will each person inventory?

56. A rotisserie costing $364 can be yours for just 4 easy payments. How much is each payment?

57. Andrew bought a weight set for $648. He will pay for it in 3 payments. How much is each payment?

Divide.

58. $6\overline{)518}$

59. $7\overline{)431}$

60. $6{,}024 \div 8$

61. $2{,}349 \div 9$

62. $3\overline{)3{,}462}$

63. $2\overline{)56{,}872}$

64. $8{,}084 \div 6$

65. $15{,}467 \div 8$

66. $\dfrac{3{,}429}{9}$

67. $\dfrac{1{,}644}{4}$

68. $\dfrac{0}{67{,}548}$

69. $\dfrac{76{,}483}{0}$

70. $23\overline{)1{,}150}$

71. $42\overline{)1{,}974}$

72. $715 \div 55$

73. $588 \div 28$

Solve each word exercise.

74. Allison leg-pressed a total of 11,250 pounds by doing 75 reps. How many pounds are in each rep?

75. How many boxes that are 54 centimeters tall can be stacked in a room that is 300 centimeters high?

76. Sebastian rode his bike 7,644 miles last year. Divide the number of miles by 12 to determine the average number of miles he rode each month.

77. Leo took a two-week cruise across the Atlantic. On the trip, he read 1,302 pages of novels. On average, how many pages did he read each day?

For each exercise:

a. Estimate each quotient.

b. Perform the division.

78. $30\overline{)57{,}900}$

79. $20\overline{)41{,}720}$

80. $896 \div 128$

81. $2{,}112 \div 132$

82. $16\overline{)34{,}582}$

83. $18\overline{)67{,}843}$

84. $124\overline{)635{,}252}$

85. $142\overline{)105{,}790}$

Solve each word exercise.

86. Divide 400 by 16.

87. Divide 576 by 18.

88. Find the quotient of 954 and 9.

89. Find the quotient of 657 and 8.

90. What is 386 divided by 12?

91. What is 528 divided by 11?

92. John worked a total of 10 hours on the first 2 days of the week. If John made $310 for this work, how much money did he earn per hour?

93. Dewey worked a total of 20 hours on the first 3 days of the week. If Dewey made $1,200 for this work, how much money did he earn per hour?

94. A landscaping truck carried 104 cubic yards of topsoil in 13 trips. How many cubic yards of topsoil did the truck carry in each trip?

95. A group of 11 friends invested their money in a Costa Rican vacation home. If the vacation home cost $61,600, how much money did each person invest?

96. If Anne and Amy drove 455 miles in 7 hours, what speed (miles per hour) did they drive?

97. If Tyson and Cheryl drove 186 miles at a speed of 62 miles per hour, how many hours did they drive?

smith Keep your work neat and organized. Draw the lines and use scratch paper to perform the calculations.

For more tips and tweets, go to twitter.com/gstbasicma

98. **a.** Terry's home is small and well insulated. This year she paid $984 to heat her house. What was her average monthly bill?
b. Terry saved $1,640 for heating expenses. For how many months will this amount cover her heating bills?

99. **a.** Darrell's home is large and poorly insulated. This year he paid $3,936 to heat his house. What was his average monthly bill?
b. Darrell saved $1,640 for heating expenses. For how many months will this amount cover his heating bills?

100. A group of 8 students volunteered to "get out the vote." They need to visit 176 houses in a neighborhood.
a. How many houses must each student visit?
b. If the coordinator asks them to visit the houses in pairs, how many houses will each pair of students visit?

101. As a bonus, a company is going to distribute $21,000 in profits to its 15 employees.
a. How much will each employee receive?
b. Before the checks are written, one of the employees is fired for inappropriate conduct. How much will the remaining employees receive?

102. Jamal and his 3 children went to the movies. Tickets and snacks cost $16 per person.
a. How much did he spend at the movies?
b. If he had $80 in cash, did he have $10 to buy the children ice cream after the movie?

103. Jeanette and her 4 children went to a baseball game. Tickets and snacks cost $37 per person.
a. How much did she spend at the game?
b. If she had $190 in cash, did she have enough money to pay $15 for a foam finger?

Solve for the unknown number.

104. $15 \div x = 5$

105. $24 \div x = 4$

106. $8 = 32 \div y$

107. $7 = 42 \div y$

108. $80 \div z = 10$

109. $110 \div z = 11$

110. $600 = 2 \cdot x$

111. $800 = 4 \cdot x$

112. To find $6 \div 3$, we can find the unknown number in $n \cdot 3 = 6$. Use this concept to explain why $0 \div 6 = 0$ but $6 \div 0$ is undefined.

Section 1.5 Question Log

Use this space to write questions. Make sure you get these answered and revisit them when you prepare for your exam.

Page ____ **Answered**	**Page** ____ **Answered**

Mid-Chapter 1 Review Exercises

Before starting Section 1.6, solve the exercises in Appendix A.

Section 1.6 Exponents, Groupings, and the Order of Operations

When an exercise has several operations, it is important for everyone to perform the calculations in the same order. If not, people will get different answers to the same problem. To help us perform calculations in the correct order, we follow the order of operations.

The **Objectives** in Section 1.6 will help you

- A Evaluate expressions with exponents.
- B Simplify expressions with groupings.
- C Evaluate expressions using the order of operations.

VOCABULARY PREVIEW *Check the box that applies.*

	Got It	Must Study
Exponents: Exponents are used to show repeated multiplication.		
Base: A number with an exponent is the **base** of the exponential expression.		
Exponent: An **exponent** indicates how many of the base must be multiplied.		
Grouping Symbols: There are many **grouping symbols,** including parentheses ( ), brackets [ ], and fraction bars —.		
Grouping: A **grouping** is a collection of numbers within a grouping symbol.		
Order of Operations **First:** Perform operations in groupings. **Second:** Perform operations with exponents. **Third:** Perform multiplication and division as they occur from left to right. **Fourth:** Perform addition and subtraction as they occur from left to right.		

Study these terms when they appear in the text. At the end of the section, test your understanding by completing the Vocabulary Review in the section exercises.

Objective A Evaluate Expressions with Exponents

The Concept An expression with an exponent has two parts: a base and an exponent. The expression 4^3 is an exponential expression. The exponent, 3, is written above and to the right of the base, 4. The base is the number or symbol that appears immediately before the exponent.

base → 4^3 ← exponent

Example	Base	Exponent
2^3	2	3
$8 + 15^2$	15	2

EXAMPLE

1. Identify the base and the exponent in 7^5.

base → 7^5 ← exponent

The base is 7.
The exponent is 5.

GUIDED PRACTICE

1. Identify the base and the exponent in 9^8.

9^8

The base is ____.
The exponent is ____.

INTERACTIVE DEFINITION Exponent

An **exponent** indicates how many of the base must be multiplied.

EXAMPLES

2. Evaluate. 4^3

4^3 has a base of 4 and an exponent of 3.

To evaluate 4^3, we must multiply 3 factors of the base, 4.

$$\begin{aligned} 4^3 &= 4 \cdot 4 \cdot 4 \\ &= 16 \cdot 4 \\ &= 64 \end{aligned}$$

3. Write the following using exponential notation.
$2 \cdot 2 \cdot 2 \cdot 2 \cdot 2 \cdot 2$

There are 6 factors of 2.
The base is 2.
The exponent is 6.

$$2 \cdot 2 \cdot 2 \cdot 2 \cdot 2 \cdot 2 = 2^6$$

GUIDED PRACTICE

2. Evaluate. 2^4

2^4 has a base of ____ and an exponent of ____.

To evaluate 2^4, we must multiply ____ factors of the base, ____.

$$\begin{aligned} 2^4 &= \\ &= \\ &= \\ &= \end{aligned}$$

3. Write the following using exponential notation.
$3 \cdot 3 \cdot 3 \cdot 3 \cdot 3 \cdot 3 \cdot 3$

There are ____ factors of ____.
The base is ____.
The exponent is ____.

$$3 \cdot 3 \cdot 3 \cdot 3 \cdot 3 \cdot 3 \cdot 3 =$$

DO YOU UNDERSTAND what an exponent indicates? Got It | Get Help

Procedure Evaluate Expressions with Exponents

Step 1 Write the repeated multiplication.
Write the base as a factor as many times as the exponent.

Step 2 Multiply.

EXAMPLES

4. Evaluate. 2^3

Step 1 Write the repeated multiplication.

An exponent indicates how many of the base must be multiplied.

Step 2 Multiply.

Write the base, 2, three times.

$$2^3 = 2 \cdot 2 \cdot 2$$
$$= 4 \cdot 2$$
$$= 8$$

5. Evaluate. 15^2

Step 1 Write the repeated multiplication.

An exponent indicates how many of the base must be multiplied.

Step 2 Multiply.

Write the base, 15, two times.

$$15^2 = 15 \cdot 15$$
$$= 225$$

GUIDED PRACTICE

4. Evaluate. 3^4

Write the base, _____, _____ times.

$3^4 =$

$=$

$=$

$=$

5. Evaluate. 12^2

Write the base, _____, _____ times.

$12^2 =$

$=$

INTERACTIVE DEFINITION Exponent of Zero

Any number* with an **exponent of zero** is 1.

EXAMPLE

6. Evaluate. 5^0

Any number with an exponent of zero is 1.

$$5^0 = 1$$

GUIDED PRACTICE

6. Evaluate. $3{,}589^0$

Any number with an exponent of _____ is _____.

$3{,}589^0 =$

DO YOU UNDERSTAND how to evaluate a number with an exponent of zero? Got It | Get Help

WHY IT WORKS Any Number with an Exponent of Zero Is 1

To understand why a number to the zero power is 1, study the following pattern for $2^0 = 1$.

The exponents decrease by one going down the column.

$$2^4 = 16$$
$$2^3 = 8$$
$$2^2 = 4$$

The results are divided by the base, 2, as we go down the column.

Continuing with this pattern, we see why $2^0 = 1$.

$$2^2 = 4$$
$$2^1 = 2$$

When the exponent is decreased by one, **we divide by the base.** $2 \div 2 = 1$

$$2^0 = 1$$

*The only exception to this rule is that 0^0 is undefined.

Concept Check

A1. In your own words, how do we identify the base and the exponent in an exponential expression?

A2. How do the values of the base and the exponent relate to repeated multiplication?

Objective A Practice

For each expression, identify the base and the exponent. Do not multiply.

A3. 3^7 **A4.** 7^2 **A5.** $2 + 13^2$ **A6.** $5 + 4^{10}$

Write each exponential expression using repeated multiplication. Then evaluate it.

A7. 6^3 **A8.** 5^3 **A9.** 3^4 **A10.** 3^5

A11. 16^2 **A12.** 21^2 **A13.** 7^0 **A14.** 9^0

Objective B Simplify Expressions with Groupings

The Concept Groupings are used to collect operations that need to be performed first. We use parentheses, brackets, and fraction bars to group expressions.

TYPES OF GROUPINGS		
Parentheses	**Brackets**	**Fraction Bars**
$2 + (4 \cdot 3)$ Grouping: $(4 \cdot 3)$	$[8 - 3] \cdot 4$ Grouping: $[8 - 3]$	$\frac{4 + 6}{3 - 1}$ Two groupings: $4 + 6$ and $3 - 1$
To simplify each expression, we simplify the groupings first and then simplify whatever is left.		
$2 + (4 \cdot 3) = 2 + 12$ $= 14$	$[8 - 3] \cdot 4 = 5 \cdot 4$ $= 20$	$\frac{4 + 6}{3 - 1} = \frac{10}{3 - 1}$ $= \frac{10}{2}$ $= 10 \div 2$ $= 5$

Procedure Simplify Expressions with Groupings

Step 1 Perform operations in the groupings first.

Step 2 Perform other operations second.

EXAMPLES

7. Simplify. $10 \div (12 - 7)$

Step 1 Perform operations in the grouping first. $10 \div (12 - 7) = 10 \div 5$

Step 2 Perform the other operation second. $= 2$

8. Simplify. $[14 + 2] \div 8$

Step 1 Perform operations in the grouping first. $[14 + 2] \div 8 = 16 \div 8$

Step 2 Perform the other operation second. $= 2$

9. Simplify. $\dfrac{9 + 12}{9 - 2}$

Step 1 Numerators and denominators are groupings. $\dfrac{9 + 12}{9 - 2} = \dfrac{21}{9 - 2}$

Step 2 Fraction bars indicate division. $= \dfrac{21}{7}$

$= 21 \div 7$

$= 3$

10. LeBron James had a great game, scoring twice as many points as the two leading scorers on the other team combined. If those two players scored 15 and 11 points, how many points did LeBron score?
a. Write an expression that models the situation.
b. Evaluate the expression to answer the question.

a. Write an expression.
The first sentence tells us what LeBron scored:

$$2 \cdot (\text{1st highest} + \text{2nd highest})$$

Substituting the scores for the two other players:

$$\text{LeBron's points} = 2(15 + 11)$$

b. Evaluate the expression.
Following the order of operations, LeBron scored:

$$\text{LeBron's points} = 2 \cdot (15 + 11)$$
$$= 2(26)$$
$$= 52$$

LeBron scored 52 points.

GUIDED PRACTICE

7. Simplify. $15 - (12 \div 4)$

$15 - (12 \div 4) =$

$=$

8. Simplify. $3 \cdot [15 - 9]$

$3 \cdot [15 - 9] =$

$=$

9. Simplify. $\dfrac{42 - 32}{15 \div 3}$

$\dfrac{42 - 32}{15 \div 3} = \dfrac{\_\_\_\_}{15 \div 3}$

$= \dfrac{\_\_}{\_\_}$

$=$

$=$

10. Candace Parker had a great game, scoring three times as many points as the two leading scorers on the other team combined. If those two players scored 12 and 9 points, how many points did Candace score?
a. Write an expression that models the situation.
b. Evaluate the expression to answer the question.

a. Write an expression.
The first sentence tells us what Candace scored:

____ · (________ + ________)

Substituting the scores for the two other players:

Candace's points = ____ · (___ + __)

b. Evaluate the expression.
Following the order of operations, Candace scored:

Candace's points =

=

=

Candace scored ____ points.

Concept Check

B1. In your own words, what does a grouping indicate?

B2. What should be calculated first, operations within parentheses or exponential expressions?

Objective B Practice

Simplify each expression.

B3. $5 + (6 - 2)$

B4. $13 - (2 \cdot 6)$

B5. $28 \div [14 - 10]$

B6. $[5 + 4] \cdot 3$

B7. $\dfrac{7 + 8}{8 - 7}$

B8. $\dfrac{4 \cdot 4}{3 + 5}$

B9. $\dfrac{3 + (4 - 2)}{6 - 1}$

B10. $\dfrac{4 - 2}{12 \div (5 + 1)}$

B11. A college theater group sold 893 tickets for $12 each. If the total expenses for the show came to $9,312, how much money did the theater group make/lose?

B12. Danika earns $7 per hour after taxes. Today she worked 8 hours but spent $12 in gas for her commute to and from work. How much money did she have at the end of the day?

Objective C Evaluate Expressions Using the Order of Operations

The Concept The order in which two or more tasks are done can make a huge difference in the result.

The Correct Order	We put our socks on before we put on our shoes.	Multiplication is performed before addition. $2 + 3 \cdot 4 = 2 + 12$ $= 14$
The Wrong Order	Putting our shoes on before our socks is ridiculous.	We will not get the right answer if we perform addition before multiplication. $2 + 3 \cdot 4 = 5 \cdot 4$ $= 20$

The **order of operations** tells us the order in which we must perform each arithmetic step in an expression. It guarantees that people will get the same answer for the same exercise. Whenever you perform arithmetic, you must follow the order of operations.

Procedure

First: Perform operations in parentheses/groupings.

Second: Perform operations with exponents.

Third: Perform multiplication and division as they occur from left to right.
- Multiplication and division are tied in the order of operations.

Fourth: Perform addition and subtraction as they occur from left to right.
- Addition and subtraction are tied in the order of operations.

The order of operations is very important. You must memorize this procedure. Students often remember the order of operations using the first letters of the operations.

Memorizing the Order of Operations

Parentheses/Groupings	Please
Exponents	Excuse
Multiply and Divide → In a tie, work from left to right.	My Dear → Tie
Add and Subtract → In a tie, work from left to right	Aunt Sally → Tie

To organize an order of operations exercise, use the three C's.

Choose **Copy** **Calculate**

DETAILED EXAMPLE How to Organize an Order of Operations Exercise

Evaluate. $20 \cdot 3 - 7 \cdot (4 + 2)$. Do only the first step.

Choose the operation to be performed.

$20 \cdot 3 - 7 \cdot (4 + 2)$

The grouping must be calculated first.

Copy everything in the expression that will not change.

$20 \cdot 3 - 7 \cdot (4 + 2) = 20 \cdot 3 - 7 \cdot (\ \ )$

Copy everything except the calculation in the grouping.

Calculate the operation.

$20 \cdot 3 - 7 \cdot (4 + 2) = 20 \cdot 3 - 7 \cdot (6)$

Fill in the calculation from the grouping.

EXAMPLES

11. Evaluate. $5 + 6 \cdot 5$

Groupings (none)
Exponents (none)
Multiply and divide, left to right.
Add and subtract, left to right.

$$5 + \underbrace{6 \cdot 5}_{\text{1st}} = \underbrace{5 + 30}_{\text{2nd}}$$
$$= 35$$

12. Evaluate. $(6 \cdot 4) - 2 \cdot 5$

Groupings
Exponents (none)
Multiply and divide, left to right.
Add and subtract, left to right.

$$\underbrace{(6 \cdot 4)}_{\text{1st}} - 2 \cdot 5 = (24) - \underbrace{2 \cdot 5}_{\text{2nd}}$$
$$= \underbrace{24 - 10}_{\text{3rd}}$$
$$= 14$$

13. Evaluate. $50 - 20 + 5 \cdot 2$

Groupings (none)
Exponents (none)
Multiply and divide, left to right.
Add and subtract, left to right.

$$50 - 20 + \underbrace{5 \cdot 2}_{\text{1st}} = \underbrace{50 - 20}_{\text{2nd}} + 10$$
$$= \underbrace{30 + 10}_{\text{3rd}}$$
$$= 40$$

14. Evaluate. $24 - 12 \div 2 \cdot 3$

Groupings (none)
Exponents (none)
Multiply and divide, left to right.
Add and subtract, left to right.

$$24 - \underbrace{12 \div 2}_{\text{1st}} \cdot 3 = 24 - \underbrace{6 \cdot 3}_{\text{2nd}}$$
$$= \underbrace{24 - 18}_{\text{3rd}}$$
$$= 6$$

15. Evaluate. $(5 - 3)^2 - 3 + 5$

Groupings
Exponents
Multiply and divide (none)
Add and subtract, left to right.

$$(5 - 3)^2 - 3 + 5 = (2)^2 - 3 + 5$$
$$= 4 - 3 + 5$$
$$= 1 + 5$$
$$= 6$$

GUIDED PRACTICE

11. Evaluate. $12 + 6 \cdot 3$

$$12 + \underbrace{6 \cdot 3}_{\text{1st}} = \underbrace{12 + \quad}_{\text{2nd}}$$
$$=$$

12. Evaluate. $30 - (2 + 3) \cdot 4$

$$30 - \underbrace{(2 + 3)}_{\text{1st}} \cdot 4 = 30 - \underbrace{(\quad) \cdot 4}_{\text{2nd}}$$
$$= \underbrace{30 - \quad}_{\text{3rd}}$$
$$=$$

13. Evaluate. $30 + 2 \cdot 9 - 8$

$$30 + \underbrace{2 \cdot 9}_{\text{1st}} - 8 = \underbrace{30 + \quad}_{\text{2nd}} - 8$$
$$= \underbrace{\quad - 8}_{\text{3rd}}$$
$$=$$

14. Evaluate. $16 + 30 \div 3 \cdot 2$

$$16 + 30 \div 3 \cdot 2 =$$
$$=$$
$$=$$

15. Evaluate. $5 + (2 + 3)^2 - 20$

$$5 + (2 + 3)^2 - 20 =$$
$$=$$
$$=$$
$$=$$

The next example may look difficult because it is long. However, if you work one step at a time, you will see that the problem is just longer, not more difficult. Remember this helpful hint: Choose what operation to do first, copy everything else, and then calculate the operation.

EXAMPLES

16. Evaluate. $3 \cdot (8 - 6)^3 - 4 \cdot 3 + 1$

Groupings	$3 \cdot (8 - 6)^3 - 4 \cdot 3 + 1$
Exponents	$= 3 \cdot (2)^3 - 4 \cdot 3 + 1$
Multiply and divide, left to right.	$= 3 \cdot (8) - 4 \cdot 3 + 1$
	$= 24 - 4 \cdot 3 + 1$
Add and subtract, left to right	$= 24 - 12 + 1$
	$= 12 + 1$
	$= 13$

GUIDED PRACTICE

16. Evaluate. $27 \div (9 - 6)^3 + 8 \cdot 2 - 7$

$27 \div (9 - 6)^3 + 8 \cdot 2 - 7$

=

=

=

=

=

Concept Check

C1. Why is it important for everyone to follow the same order when simplifying expressions?

C2. If one expression has both multiplication and division, how do you determine which to do first?

C3. If one expression has both addition and subtraction, how do you determine which to do first?

Objective C Practice

Simplify each expression by following the order of operations.

C4. $6 - 2 + 3$

C5. $12 - 6 + 3$

C6. $20 \div 5 \cdot 2$

C7. $30 \div 5 \cdot 3$

C8. $12 - 4 \cdot 2$

C9. $12 + 8 \div 4$

C10. $3 + 4^2 \div 8$

C11. $6^2 - 3 \cdot 8$

C12. $25 - (6 - 2)^2 + 9$

C13. $8^2 - (12 + 4) \cdot 2$

C14. $3 \cdot (8 - 4) + 2 \cdot 3^2$

C15. $3 \cdot 2^3 - 8(15 - 13)$

Combining Concepts and Applications

CONCEPT I Finding an Average An *average* is used to describe a typical value from a list of many values. Most college instructors use averages to assign grades. By calculating your average in a class, you can determine your grade at any point during the semester.

Formula Used to Find an Average:

$$\text{Average} = \frac{\text{Sum of the data values}}{\text{Number of data values}}$$

EXAMPLE

17. The following bar graph shows a student's exam scores.

a. Estimate the average exam score using the graph.

The scores range from 75 to 95.
The average must be above 75 and below 95.

Draw a horizontal line that looks as though it goes through the middle of the values.

Use the scale on the left to guess the average value. It looks as though the average exam score is about 82.

Since this is an estimate, answers may vary.

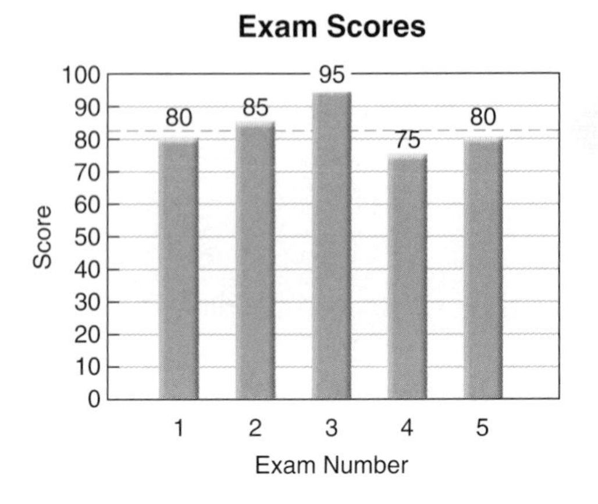

b. Calculate the average exam score using the formula.

$$\text{Average} = \frac{\text{Sum of the data values}}{\text{Number of data values}}$$

Add the exam scores.
Divide by the number of exams, 5.

$$= \frac{80 + 85 + 95 + 75 + 80}{5}$$

Add the test scores first.

$$= \frac{415}{5}$$

Divide.

$$= 83$$

GUIDED PRACTICE

17. The following bar graph shows daily high temperatures.

a. Estimate the average daily high temperature using the graph.

The temperatures range from _____ to _____.
The average value must be above _____ and below _____.

Draw a horizontal line that looks as though it goes through the middle of the values.

Use the scale on the left to guess the average value. It looks as though the average daily high temperature is about _____ degrees F.

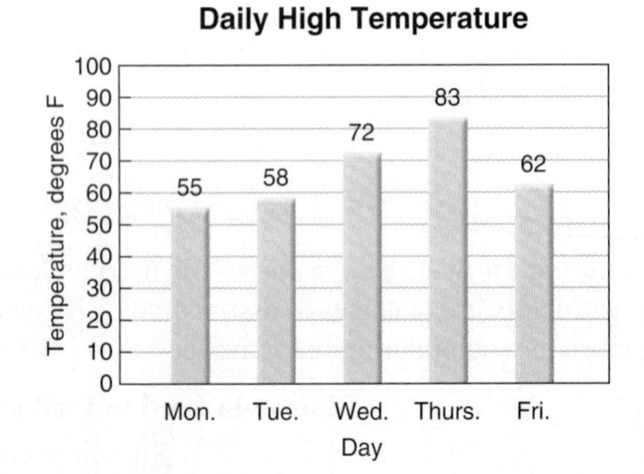

b. Calculate the average daily high temperature using the formula.

$$\text{Average} = \frac{\text{Sum of the data values}}{\text{Number of data values}}$$

Add the high temperatures.
Divide by the number of days, ____.

$$= \frac{\quad + \quad + \quad + \quad + \quad}{\quad}$$

Add the temperatures first.

$$= \frac{\quad}{\quad}$$

Divide.

$$=$$

Section 1.6 Exercises

FOR EXTRA HELP

Math XL PRACTICE

WATCH

DOWNLOAD

READ

REVIEW

To evaluate expressions with exponents:

1. Answer the Objective A Concept Checks.
2. Answer the odd Objective A Practice Exercises.
3. Answer the even Objective A Practice Exercises.

To simplify expressions with groupings:

4. Answer the Objective B Concept Checks.
5. Answer the odd Objective B Practice Exercises.
6. Answer the even Objective B Practice Exercises.

To evaluate expressions using the order of operations:

7. Answer the Objective C Concept Checks.
8. Answer the odd Objective C Practice Exercises.
9. Answer the even Objective C Practice Exercises.

VOCABULARY REVIEW *Review the Vocabulary Preview for Section 1.6. Study the definitions until you can check* Got It *for every word.*

Use these words to complete each sentence.

exponents • base • exponent • grouping symbols • grouping • order of operations

	Got It	Get Help
10. A(n) ____________ indicates how many of the ____________ must be multiplied.		
11. ____________ are used to show repeated multiplication.		
12. Three types of ____________ are used in this section. They are ____________, ____________, and ____________.		
13. A collection of numbers inside parentheses is called a(n) ____________.		

(Continued)

	Got It	Get Help
14. List the order of operations. 1st: 2nd: 3rd: 4th:		

How will you get help for any vocabulary about which you are unsure?

Your instructor ____ MyMathLab ____ A classmate ____ A tutor ____ Other ____

15. If an expression has both multiplication and division, which should you perform first?

16. If an expression has both addition and subtraction, which should you perform first?

Write each expression using exponents. Do not evaluate the expressions.

17. $4 \cdot 4 \cdot 4$

18. $3 \cdot 3 \cdot 3 \cdot 3 \cdot 3$

19. $6 \cdot 6 \cdot 6 \cdot 6 \cdot 6$

20. $7 \cdot 7 \cdot 7 \cdot 7 \cdot 7 \cdot 7$

Use repeated multiplication to rewrite each expression. Do not evaluate the expressions.

21. 6^5

22. 7^4

23. 9^3

24. 3^9

Writing only the answers, try to complete each column in fewer than 30 seconds. Accuracy is important. Take enough time to work through each exercise correctly.

25. Start Time: ____
- **a.** 1^2
- **b.** 3^2
- **c.** 2^3
- **d.** 4^2
- **e.** 7^2
- **f.** 6^2
- **g.** 9^2

Finish Time: ____

Total Time: ____

26. Start Time: ____
- **a.** 8^2
- **b.** 1^3
- **c.** 2^2
- **d.** 10^2
- **e.** 9^0
- **f.** 2^4
- **g.** 4^2

Finish Time: ____

Total Time: ____

27. Start Time: ____
- **a.** 5^2
- **b.** 2^2
- **c.** 5^3
- **d.** 3^2
- **e.** 6^2
- **f.** 3^3
- **g.** 9^2

Finish Time: ____

Total Time: ____

28. Start Time: ____
- **a.** 8^2
- **b.** 6^2
- **c.** 7^2
- **d.** 4^2
- **e.** 5^2
- **f.** 1^2
- **g.** 6^0

Finish Time: ____

Total Time: ____

Evaluate.

29. 3^4

30. 6^3

31. 5^1

32. 9^1

33. 1^8

34. 1^{13}

35. 10^3

36. 10^4

37. 2^4

38. 9^3

39. 5^2

40. 12^2

41. 4^3

42. 5^4

43. 2^6

44. 2^7

Evaluate.

45. $(16 + 4) \cdot 5$ **46.** $(12 - 3) \cdot 2$ **47.** $8(24 - 15)$ **48.** $7(5 + 7)$

49. $(24 \div 6) \cdot 7$ **50.** $15 \div 3 \cdot 6$ **51.** $19 - (33 \div 11)$ **52.** $24 + (63 \div 7)$

53. $\dfrac{21 - 3}{7 + 2}$ **54.** $\dfrac{35 - 7}{8 - 4}$

55. Two groups of people worked together collecting morel mushrooms. A group of 3 people collected 29 mushrooms. A second group of 4 people collected 48 mushrooms. If all of the mushrooms were shared equally, how many mushrooms did each person get?

56. A grocer sold 320 chips and salsa combos for a total of $2,080. To make the combos, it costs the grocer $2 for each bag of chips and $3 for each jar of salsa. How much money did the grocer earn?

Simplify, following the order of operations.

57. $12 \div 3 + 4 \cdot 9$ **58.** $16 \div 2 - 4 \cdot 2$ **59.** $81 \div 3^2 - 6$ **60.** $36 \div 2^2 - 7$

61. $\dfrac{4 + 3 \cdot 6}{7 + 4}$ **62.** $\dfrac{7 \cdot 8 - 20}{17 - 8}$ **63.** $5(4 + 3^2)$ **64.** $6(5^2 - 9)$

65. $8^2 - (4^2 - 6)$ **66.** $2^5 - (15 - 3^2)$

For each exercise, perform only the first step according to the order of operations. Match your answer to one of the lettered answers (a) through (h). The lettered answers can be used more than once.

67. $12 \div 3^2$	**a.** Answer not shown.	**68.** $12 \div 3 \cdot 3$
	b. $12 \div 6$	
69. $12 \div 2 \cdot 3$	**c.** $6 \cdot 3$	**70.** $(2 + 4) \cdot 3$
	d. $12 \div 9$	
71. $8 - 2 + 4$	**e.** $4 \cdot 3$	**72.** $4^2 - 6$
	f. $9 + 4$	
73. $3^3 + 4$	**g.** $6 + 4$	**74.** $3^2 + 4$
	h. $8 - 6$	

Simplify.

75. $40 - 10 \times 2 + 3^2$ **76.** $15 + 2 \times 3 - 4^2$ **77.** $5 - (9 - 7)^2 + 1$

78. $75 \div (8 - 3)^2 \times 3$ **79.** $24 \div 8 \times 3 - (5 - 3)^2$ **80.** $12 \div 3 \times 2 + (5 + 1)^2$

81. Tickets at a basketball game cost $17 for the upper deck and $35 for the lower deck. There are 8,500 seats in the upper deck and 6,100 in the lower deck. How much money was earned if all of the tickets were sold?

82. A caterer charges $17 for a chicken meal and $15 for vegetarian meal. At a wedding reception, 186 people ordered chicken and 31 people ordered vegetarian. How much did the caterer charge?

tobey First do the grouping. Then the exponent. Then multiply and divide from left to right. Finally, add and subtract from left to right.

For more tips and tweets, go to twitter.com/gstbasicmath

Calculate the average for each list of numbers.

83. 9, 16, 23

84. 17, 13, 24

85. 100, 200, 240

86. 200, 300, 232

87. 0, 90, 100, 110

88. 70, 80, 90, 100

89. 120, 240, 180, 100

90. a. Estimate the average exam score by drawing a horizontal line on the graph that goes through the "middle" of the values.

b. Calculate the average exam score for the five tests.

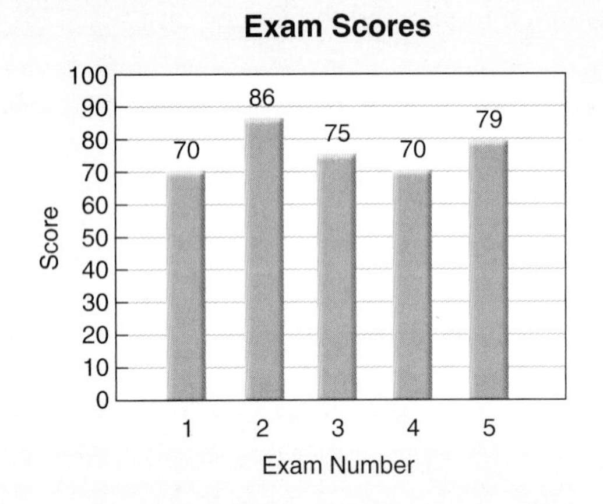

91. a. Estimate the average daily high temperature by drawing a horizontal line on the graph that goes through the "middle" of the values.

b. Calculate the average daily high temperature for these 5 days.

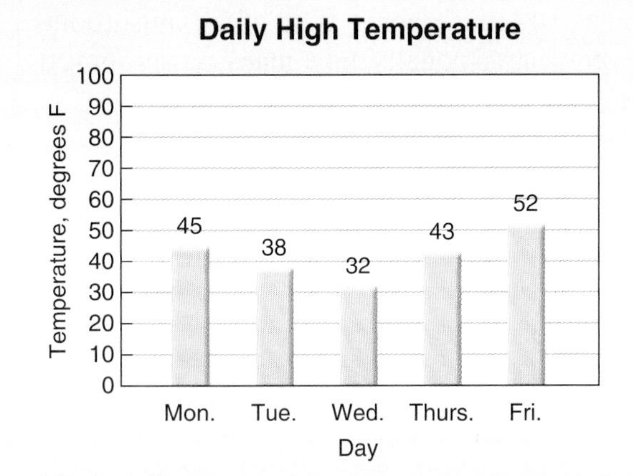

92. Find the average ticket price for all four football games.

AVERAGE TICKET PRICES FOR SELECT GAMES FROM STUBHUB'S TOP 25 RIVALRY RANKINGS, 2006	
Ticket Price	**Football Game**
$890	Michigan Wolverines vs. Ohio State Buckeyes, 11/18/2006
$419	Oklahoma vs. Texas (Red River Shootout), 11/18/2006
$388	Notre Dame Fighting Irish vs. USC Trojans, 11/25/2006
$191	USC Trojans vs. UCLA Bruins, 12/02/2006

Source: StubHub

93. What was the average number of top movies with a PG-13 rating between 2002 and 2006?

TOP MOVIES BY RATING				
Year	**G**	**PG**	**PG-13**	**R**
2006	1	4	13	2
2005	1	4	13	2
2004	1	6	10	3
2003	1	3	11	5
2002	1	6	13	0

Source: National Association of Theatre Owners

94. What is the average number of 1st place medals that the top four countries won?

TEAM MEDALS WON IN THE WORLD CUP SOCCER COMPETITION			
Pos.	**Team**	**1st Place**	**2nd Place**
1	Brazil	4	2
2	West Germany	3	3
3	Italy	3	2
4	Argentina	2	2
5	Uruguay	2	0
6	France	1	0
7	England	1	0
8	Czechoslovakia Holland Hungary	0	2

Source: www.world-cup-info.com

Section 1.6 Question Log

Use this space to write questions. Make sure you get these answered and revisit them when you prepare for your exam.

Page ______ **Answered**

Page ______ **Answered**

Page ______ **Answered**

Page ______ **Answered**

Section 1.7 Properties of Whole Numbers

We often use the concepts in this section when working with fractions and algebra. Learning these topics will be helpful for later chapters and future classes.

The **Objectives** in Section 1.7 will help you

- **A** Understand prime numbers.
- **B** Find the factorizations of a number.
- **C** Find multiples of a number.

VOCABULARY PREVIEW *Check the box that applies.*	Got It	Must Study
Divisible: A number is **divisible** by any number that divides into it evenly with no remainder.		
Prime Number: A **prime number** is divisible by only two whole numbers: itself and 1.		
Composite Number: A **composite number** is divisible by a whole number other than itself and 1.		
Factor (noun): A **factor** of a number divides that number evenly.		
Factor Pair: A **factor pair** of a number is made up of any two numbers that multiply to give the number.		
Factorization: A **factorization** is used to rewrite a number as a product of two or more factors.		
Prime Factorization: When a factorization of a number is written using only prime numbers, it is called the number's **prime factorization**.		
Factor Tree: A **factor tree** can be used to help find a number's prime factorization.		
Multiples: Multiplying a number by whole numbers $(1, 2, 3, 4 \ldots)$ gives **multiples** of the number.		

Study these terms when they appear in the text. At the end of the section, test your understanding by completing the Vocabulary Review in the section exercises.

Objective A Understand Prime Numbers

The Concept Every whole number greater than 1 is either a prime number or a composite number. You can tell whether a whole number is prime or composite by identifying the numbers that will divide it evenly.

INTERACTIVE DEFINITION Prime and Composite Numbers

A **prime number** is divisible by only two whole numbers: itself and 1.
A **composite number** is divisible by a whole number other than itself and 1.

EXAMPLES

1. Is 5 prime or composite?

Neither 2, 3, nor 4 divide it evenly.
Because 5 is not divisible by any number other than 1 and itself, 5 is a prime number.

2. Is 8 prime or composite?

2 and 4 divide 8 evenly.

$$8 \div 2 = 4$$
$$8 \div 4 = 2$$

Because 8 is divisible by a whole number other than itself and one, 8 is a composite number.

GUIDED PRACTICE

1. Is 7 prime or composite?

Neither ___, ___, ___, ___, nor ___ divide it evenly.
Because 7 is not divisible by any number other than ____ and itself, 7 is a _____ number.

2. Is 15 prime or composite?

____ and ____ divide 15 evenly.

$$15 \div \_\_ = \_\_$$
$$15 \div \_\_ = \_\_$$

Because 15 is divisible by a whole number other than ____ and ____, 15 is a _______ number.

DO YOU UNDERSTAND the definitions for prime and composite numbers? Got It | Get Help

▶ **NOTE** One and zero are the only whole numbers that are neither prime nor composite.

To decide if a number is prime or composite, check for divisibility by the prime numbers.

The first seven prime numbers are 2, 3, 5, 7, 11, 13, 17.

To help you decide if a number is divisible by 2, 3, or 5, use the following tests for divisibility.

TESTS FOR DIVISIBILITY
2: A number is divisible by 2 if it is even.
• Even numbers end in 0, 2, 4, 6, or 8. • 176 is divisible by 2 because 176 is even. • 133 is not divisible by 2 because 133 is odd.
3: A number is divisible by 3 if the sum of its digits is divisible by 3.
• 132 is divisible by 3 because $1 + 3 + 2 = 6$ and 6 is divisible by 3. • 452 is not divisible by 3 because $4 + 5 + 2 = 11$ and 11 is not divisible by 3.
5: A number is divisible by 5 if it ends in 0 or 5.
• 455 is divisible by 5 because it ends in 5. • 34 is not divisible by 5 because it does not end in 0 or 5.

▶ **NOTE** There are divisibility tests for many other numbers. If you are curious, you can do an online search using Google for *divisibility tests* for 7, 11, 13, and so on.

EXAMPLE

3. For the numbers 629, 860, and 375, answer the following questions:
a. Which numbers are divisible by 2?
b. Which numbers are divisible by 3?
c. Which numbers are divisible by 5?

a. A number is divisible by 2 if it is even.
629 is odd, so it is not divisible by 2.
860 is even, so it is divisible by 2.
375 is odd, so it is not divisible by 2.

b. A number is divisible by 3 if the sum of its digits is divisible by 3.
629: $6 + 2 + 9 = 17$; 17 is not divisible by 3, so 629 is not divisible by 3.
860: $8 + 6 + 0 = 14$; 14 is not divisible by 3, so 860 is not divisible by 3.
375: $3 + 7 + 5 = 15$; 15 is divisible by 3, so 375 is divisible by 3.

c. A number is divisible by 5 if it ends in 0 or 5.
629's last digit is 9, so it is not divisible by 5.
860's last digit is 0, so it is divisible by 5.
375's last digit is 5, so it is divisible by 5.

GUIDED PRACTICE

3. For the numbers 915, 720, and 837, answer the following questions:
a. Which numbers are divisible by 2?
b. Which numbers are divisible by 3?
c. Which numbers are divisible by 5?

a. A number is divisible by 2 if it is _____.
915 is even/odd, so it is/is not divisible by 2.
720 is even/odd, so it is/is not divisible by 2.
837 is even/odd, so it is/is not divisible by 2.

b. A number is divisible by 3 if the sum of its digits is divisible by _____.
915: 9 + 1 + 5 = _____; _____ is/is not divisible by 3, so 915 is/is not divisible by 3.
720: 7 + _____ + _____ = _____; _____ is/is not divisible by 3, so 720 is/is not divisible by 3.
837: _____ + _____ + _____ = _____; _____ is/is not divisible by 3, so 837 is/is not divisible by 3.

c. A number is divisible by 5 if it ends in _____ or _____.
915's last digit is _____, so it is/is not divisible by 5.
720's last digit is _____, so it is/is not divisible by 5.
837's last digit is _____, so it is/is not divisible by 5

Procedure Determine Whether a Number Is Prime or Composite: Part 1

Check for divisibility by the prime numbers (2, 3, 5 . . .).
The number is prime if it is divisible only by 1 and itself.
The number is composite if it is divisible by a whole number other than 1 and itself.

EXAMPLES

4. Is 7 prime or composite?

Check for divisibility by the prime numbers.
Check 2
7 is odd, so 2 does not divide it.
Check 3
7 is not divisible by 3.
Check 5
7 ends in 7, so it is not divisible by 5.
Check 7
7 is divisible by itself and 1.

Prime or Composite?
7 is a prime number because it is divisible only by itself and 1.

GUIDED PRACTICE

4. Is 11 prime or composite?

Check for divisibility by the prime numbers.
Check 2
11 is even/odd, so 2 does/does not divide it.
Check 3
11 is/is not divisible by 3.
Check 5
11 ends in _____, so it is/is not divisible by 5.
Check 7
11 is/is not divisible by 7.
Check 11
11 does/does not divide itself evenly.

Prime or Composite?
11 is prime/composite because _____.

5. Is 57 prime or composite?

Check for divisibility by the prime numbers.

Check 2

57 is odd, so it is not divisible by 2.

Check 3

5 + 7 = 12, which is divisible by 3.
Therefore, 3 divides 57 evenly.

Prime or Composite?

57 is a composite number because it is divisible by a number other than itself and 1.

5. Is 35 prime or composite?

Check for divisibility by the prime numbers.

Check 2

35 is even/odd, so it is/is not divisible by 2.

Check 3

3 + 5 = ____, which is/is not divisible by 3.
Therefore, 3 does/does not divide 35 evenly

Check 5

The last digit of 35 is ____, so 5 does/does not divide 35 evenly.

Prime or Composite?

35 is prime/composite because ________.

Fortunately, we don't need to check for divisibility by all of the primes to see if a number is prime or composite. To check if a number is prime or composite, check for divisibility by the primes whose square is less than the number being checked.

A PARTIAL LIST OF PRIME NUMBERS AND THEIR SQUARES						
Prime Numbers	2	3	5	7	11	13
Their Squares	$2^2 = 4$	$3^2 = 9$	$5^2 = 25$	$7^2 = 49$	$11^2 = 121$	$13^2 = 169$

EXAMPLES

6. Which prime numbers must be checked to see if 47 is prime or composite?

Check the primes whose square is less than 47.

$2^2 = 4, 3^2 = 9, 5^2 = 25,$ ~~$7^2 = 49$~~

We must check 2, 3, and 5.

7. Which prime numbers must be checked to see if 133 is prime or composite?

Check the primes whose square is less than 133.

$2^2 = 4, 3^2 = 9, 5^2 = 25, 7^2 = 49, 11^2 = 121,$

~~$13^2 = 169$~~

We must check 2, 3, 5, 7, and 11.

GUIDED PRACTICE

6. Which prime numbers must be checked to see if 87 is prime or composite?

Check the primes whose square is less than ____.

$2^2 =$ ____, $3^2 =$ ____, $5^2 =$ ____,
$7^2 =$ ____, $11^2 =$ ____, $13^2 =$ ____

We must check ________.

7. Which prime numbers must be checked to see if 109 is prime or composite?

Check the primes whose square is less than ____.

$2^2 =$ ____, $3^2 =$ ____, $5^2 =$ ____,

$7^2 =$ ____, $11^2 =$ ____, $13^2 =$ ____

We must check ________.

Procedure Determine Whether a Number Is Prime or Composite: Part 2

Step 1 List the prime numbers whose square is less than the number being checked.

Step 2 Check for divisibility by these prime numbers.

- If the number is divisible, it is composite.
- If it is divisible only by itself and 1, it is prime.

DETAILED EXAMPLE Determining Whether a Number Is Prime or Composite

Is 37 prime or composite?

Step 1 List the prime numbers whose square is less than the number being checked.

$2^2 = 4$
$3^2 = 9$
$5^2 = 25$
~~$7^2 = 49$~~

These primes have squares less than 37. We will check these prime numbers.

Stop! 49 is larger than 37. We will not check prime numbers greater than 5.

Step 2 Check for divisibility by these prime numbers.

Check 2
37 is odd, so 2 does not divide 37.

Check 3
37 is not divisible by 3.

Check 5
37 is not divisible by 5.

37 is a prime number.

EXAMPLE

8. Is 41 prime or composite?

Step 1 List the prime numbers that must be checked.

$2^2 = 4$, $3^2 = 9$, $5^2 = 25$, ~~$7^2 = 49$~~

Because 49 is more than 41, we must check through 5.

Step 2 Check for divisibility by these prime numbers.
Check 2
41 is odd, so 41 is not divisible by 2.

Check 3
$4 + 1 = 5$ is not divisible by 3, so 41 is not divisible by 3.

Check 5
41's last digit is 1, so 41 is not divisible by 5.

41 is prime.

GUIDED PRACTICE

8. Is 53 prime or composite?

Step 1 List the prime numbers that must be checked.

$2^2 =$ ____, $3^2 =$ ____, $5^2 =$ ____, $7^2 =$ ____, $11^2 =$ ____

Because ____ is more than 53, we must check through ____.

Step 2 Check for divisibility by these prime numbers.
Check 2
53 is odd/even, so 53 is/is not divisible by 2.

Check 3
____ + ____ = ____ is/is not divisible by 3, so 53 is/is not divisible by 3.

Check 5
The last digit of 53 is ____, so 53 is/is not divisible by 5.

Check 7
53 is/is not divisible by 7.

53 is prime/composite.

Concept Check

A1. What is a prime number?

A2. What is a composite number?

Use the following list of numbers to answer Exercises A3, A4, and A5.

12, 34, 41, 75, 87, 111, 213, 220

A3. Use divisibility tests to determine which numbers are divisible by 2.

A4. Use divisibility tests to determine which numbers are divisible by 3.

A5. Use divisibility tests to determine which numbers are divisible by 5.

A6. What is the largest prime number that must be checked for divisibility to determine if 23 is prime or composite?

Objective A Practice

For each exercise:

a. List the prime numbers that you must check for divisibility.

b. Determine whether the number is prime or composite.

A7. 63 **A8.** 75 **A9.** 93 **A10.** 97

A11. 133 **A12.** 141 **A13.** 89 **A14.** 57

Objective B Find the Factorizations of a Number

The Concept A factorization of a number will make many calculations easier. The goal of this objective is to help you learn both the vocabulary and the procedures used to find the prime factorization of a number.

INTERACTIVE DEFINITION Factor

A **factor** of a number divides that number evenly.

EXAMPLE	GUIDED PRACTICE
9. Identify the factors of 18.	**9.** Identify the factors of 12.
There are exactly six numbers that divide 18 evenly.	There are exactly six numbers that divide 12 evenly.
$18 \div 1 = 18$ $18 \div 2 = 9$ $18 \div 3 = 6$	$12 \div$ ___ = ___ $12 \div$ ___ = ___ $12 \div$ ___ = ___
$18 \div 18 = 1$ $18 \div 9 = 2$ $18 \div 6 = 3$	$12 \div$ ___ = ___ $12 \div$ ___ = ___ $12 \div$ ___ = ___
The six factors of 18 are	The six factors of 12 are
1, 2, 3, 6, 9, 18.	___, ___, ___, ___, ___, ___

DO YOU UNDERSTAND how to identify the factors of a number? Got It Get Help

INTERACTIVE DEFINITION Factor Pair

A **factor pair** of a number is made up of any two numbers that multiply to give the number.

EXAMPLE

10. Identify the factor pairs of 18.

Use Example 9 to determine the factor pairs of 18.

The three factor pairs of 18 are
1 and 18, 2 and 9, 3 and 6.

GUIDED PRACTICE

10. Identify the factor pairs of 12.

Use Guided Practice 9 to identify the factor pairs of 12.

The three factor pairs of 12 are
____ and ____, ____ and ____, ____ and ____.

DO YOU UNDERSTAND how to identify the factor pairs of a number? Got It Get Help

INTERACTIVE DEFINITION Two-Number Factorizations

The **two-number factorizations** of a number are all of the factor pairs that multiply to give the number.

EXAMPLE

11. Write the two-number factorizations of 18.

To find the two-number factorizations, use the factor pairs of 18.

The two-number factorizations of 18 are
$1 \cdot 18, \quad 2 \cdot 9, \quad 3 \cdot 6.$

GUIDED PRACTICE

11. Write the two-number factorizations of 12.

To find the two-number factorizations, use the factor pairs of 12.

The two-number factorizations of 12 are
____, ____, ____

DO YOU UNDERSTAND how to find two-number factorizations? Got It Get Help

INTERACTIVE DEFINITION Prime Factorizations

When a factorization of a number is written using only prime numbers, it is called the number's **prime factorization**.

EXAMPLE

12. Find the prime factorization of 18.

Start with any two-number factorization of 18.

$$18 = 2 \cdot 9$$

GUIDED PRACTICE

12. Find the prime factorization of 12.

Start with any two-number factorization of 12.

$$12 = 2 \cdot 6$$

This is not a prime factorization of 18 because 9 is composite. However, we can write 9 as a product of primes, $9 = 3 \cdot 3$.

$$18 = 2 \cdot 9$$
$$18 = 2 \cdot 3 \cdot 3$$

The prime factorization of 18 is

$$18 = 2 \cdot 3 \cdot 3.$$

Each factor is prime, and the product is 18.

This is not a prime factorization of 12 because ____ is composite. However, we can write ____ as a product of primes, ____ = ____ · ____

$$12 = 2 \cdot 6$$
$$12 = 2 \cdot \_\_\_\_ \cdot \_\_\_\_$$

The prime factorization of 12 is

$$12 = \_\_\_\_ \cdot \_\_\_\_ \cdot \_\_\_\_.$$

Each factor is _____, and the product is ____.

DO YOU UNDERSTAND how to find the prime factorization of a number? Got It | Get Help

To help you find a number's prime factorization, you may want to build a factor tree. A **factor tree** can be used to find a number's prime factorization. The process for making a factor tree is shown in the following example.

DETAILED EXAMPLE Using a Factor Tree

Use a factor tree to find the prime factorization of 30.

Step 1 The first branches can be any two-number factorization of 30.

Step 2 Because 6 is composite, write its two-number factorization in the second branch. This gives the prime factorization.

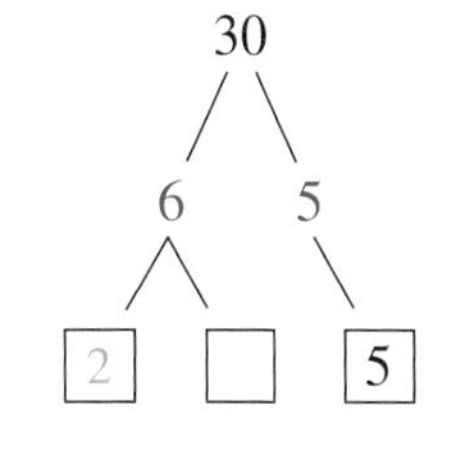

$$30 = 2 \cdot 3 \cdot 5$$

Keep factoring all composites until every number is prime.

▶ **NOTE** When we write prime factorizations, we will order the factors from least to greatest, $30 = 2 \cdot 3 \cdot 5$. It is okay to list factors in other orders, $30 = 3 \cdot 5 \cdot 2$.

Procedure Find a Prime Factorization

Step 1 Write the number using any of its two-number factorizations.

Step 2 If any of those factors are composite, perform step 1 on them. Repeat this process until every factor is prime.

Fact: There is only one prime factorization for every number.
Result: When finding the prime factorization, we can start with any two-number factorization. As long as we keep factoring all of the composite factors, the prime factorization will be correct.

Below are three factor trees that we could use to find the prime factorization of 150. There are at least two more possible factor trees for 150. Can you find them?

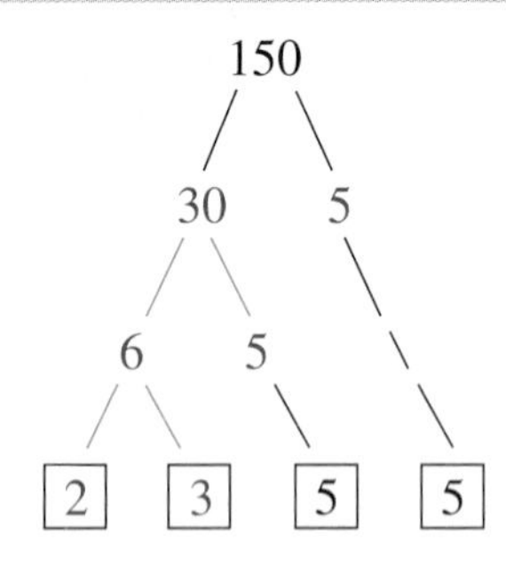

$150 = 2 \cdot 3 \cdot 5 \cdot 5$

150
75 2
25 3
5 5 3 2

$150 = 2 \cdot 3 \cdot 5 \cdot 5$

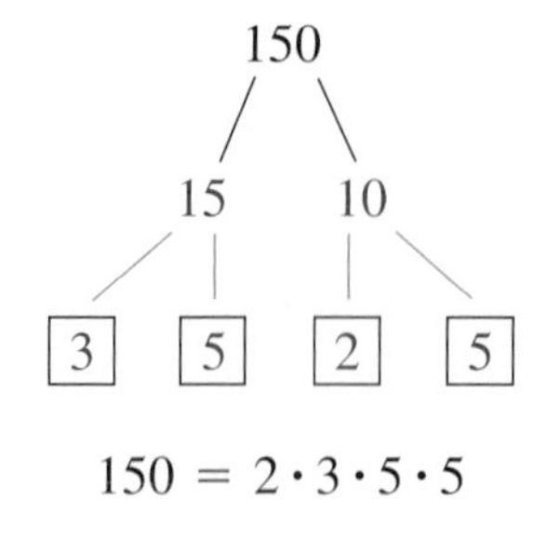

$150 = 2 \cdot 3 \cdot 5 \cdot 5$

EXAMPLES

13. Find the prime factorization of 50. Write any repeated factors using exponents.

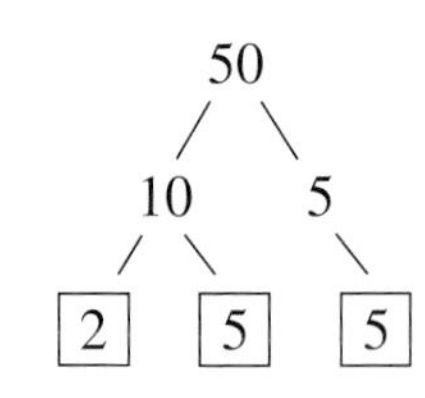

$50 = 2 \cdot 5 \cdot 5$

$50 = 2 \cdot 5^2$

14. Find the prime factorization of 81. Write any repeated factors using exponents.

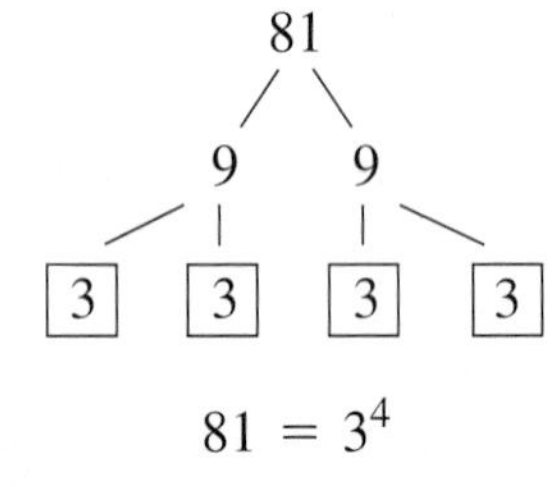

$81 = 3^4$

GUIDED PRACTICE

13. Find the prime factorization of 20. Write any repeated factors using exponents.

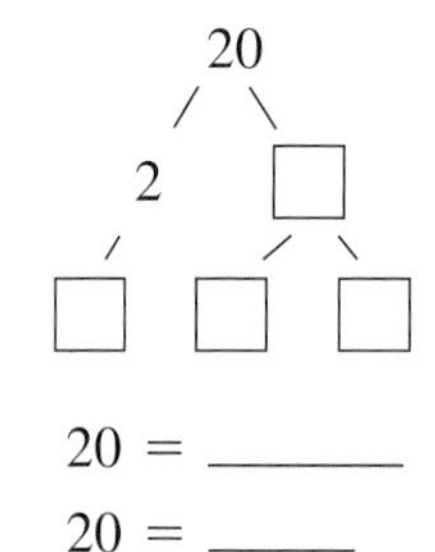

20 = ______

20 = ______

14. Find the prime factorization of 16. Write any repeated factors using exponents.

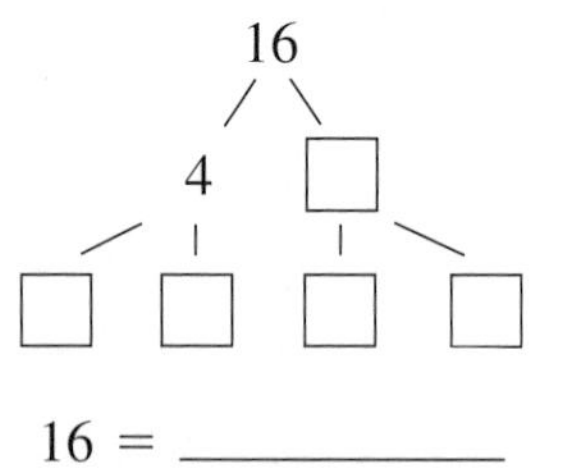

16 = ______________

Concept Check

B1. In your own words, explain what it means to write a prime factorization.

B2. We found the prime factorization of 50 by writing $50 = 2 \cdot 5 \cdot 5$. We then wrote $50 = 2 \cdot 5^2$. Use the prime factorization of 81 to explain why using exponents makes writing a prime factorization easy.

Objective B Practice

B3. **a.** Find all factors of 36.
b. Write every two-number factorization of 36.
c. Write the prime factorization of 36.

B4. **a.** Find all factors of 40.
b. Write every two-number factorization of 40.
c. Write the prime factorization of 40.

Find the prime factorization of each number.

B5. 35 **B6.** 33 **B7.** 50 **B8.** 45

B9. 48 **B10.** 100 **B11.** 72 **B12.** 56

Objective C Find Multiples of a Number

The Concept In Objective B, you learned about the factors of a number. For example, the factors of 6 are 2 and 3 because 2 and 3 divide into 6 evenly. In this objective, you will find **multiples**. To find multiples of 6, you can either multiply 6 by whole numbers or "count by sixes." The first four multiples of 6 are 6, 12, 18, and 24.

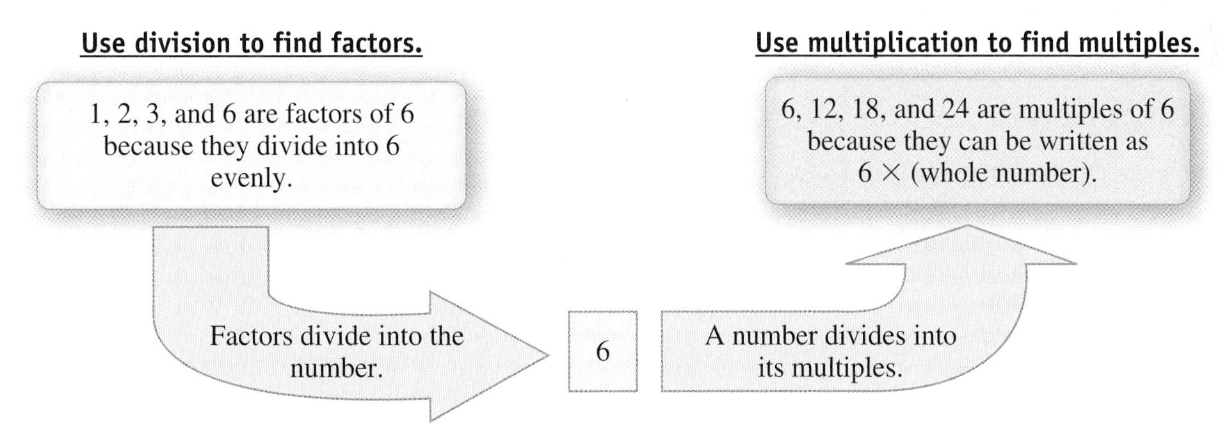

It is useful to remember two facts:

- Factors of a number cannot be greater than the number.
- Multiples of a number cannot be less than the number.

INTERACTIVE DEFINITION Multiples

Multiplying a number by whole numbers $(1, 2, 3, 4 \ldots)$ will give **multiples** of the number.

EXAMPLES

15. Identify the first five multiples of 3.

First five multiples of 3:

$$1 \cdot 3 = 3$$
$$2 \cdot 3 = 6$$
$$3 \cdot 3 = 9$$
$$4 \cdot 3 = 12$$
$$5 \cdot 3 = 15$$

GUIDED PRACTICE

15. Identify the first five multiples of 4.

First five multiples of 4:

____ · 4 = ____
____ · 4 = ____
____ · 4 = ____
____ · 4 = ____
____ · 4 = ____

(Continued)

16. Identify the first five multiples of 15.	16. Identify the first five multiples of 12.
First five multiples of 15: $1 \cdot 15 = 15$ $2 \cdot 15 = 30$ $3 \cdot 15 = 45$ $4 \cdot 15 = 60$ $5 \cdot 15 = 75$	First five multiples of 12: ____ $\cdot$ 12 = ____ ____ $\cdot$ 12 = ____ ____ $\cdot$ 12 = ____ ____ $\cdot$ 12 = ____ ____ $\cdot$ 12 = ____

DO YOU UNDERSTAND how to find the multiples of a number? Got It | Get Help

Concept Check

C1. List four multiples of 10.

C2. List four factors of 10.

C3. Which can be less than a number, its factors or its multiples?

C4. Which can be greater than a number, its factors or its multiples?

Objective C Practice

List the first five multiples of each number.

C5. 8 **C6.** 7 **C7.** 30 **C8.** 20

C9. 16 **C10.** 18 **C11.** 24 **C12.** 32

Section 1.7 Exercises

FOR EXTRA HELP  

To understand prime numbers:

1. Answer the Objective A Concept Checks.
2. Answer the odd Objective A Practice Exercises.
3. Answer the even Objective A Practice Exercises.

To find the factorizations of a number:

4. Answer the Objective B Concept Checks.
5. Answer the odd Objective B Practice Exercises.
6. Answer the even Objective B Practice Exercises.

To find multiples of a number:

7. Answer the Objective C Concept Checks.
8. Answer the odd Objective C Practice Exercises.
9. Answer the even Objective C Practice Exercises.

VOCABULARY REVIEW *Review the Vocabulary Preview for Section 1.7. Study the definitions until you can check* Got It *for every term.*

Use these words to complete each sentence.

divisible • factor (noun) • factor tree • prime number • factor pair • multiple • composite number • prime factorization

	Got It	Get Help
10. A ____________ of a number is made up of any two numbers that multiply to give the number.		
11. A ____________ is divisible by only two whole numbers: itself and 1.		
12. A ____________ can be used to help find the prime factorization of a number.		
13. A ____________ of a number is found by multiplying the number by any whole number.		
14. A ____________ of a number divides that number evenly.		
15. A number is ____________ by any number that divides into it evenly with no remainder.		
16. A ____________ is divisible by a whole number other than itself and 1.		
17. When a factorization of a number is written using only prime numbers, it is called the number's ____________.		

How will you get help for any vocabulary, about which you are unsure?

Your instructor ____ MyMathLab ____ A classmate ____ A tutor ____ Other____

Using your own words, complete each divisibility rule.

18. A number is divisible by 2 if . . .

19. A number is divisible by 3 if . . .

20. A number is divisible by 5 if . . .

Determine whether each number is prime or composite.

21. 17 **22.** 21 **23.** 32 **24.** 29

25. 39 **26.** 46 **27.** 121 **28.** 91

List every factor of each number.

29. 6 **30.** 15 **31.** 11 **32.** 13

33. 28 **34.** 18 **35.** 32 **36.** 24

37. 75 **38.** 100 **39.** 34 **40.** 39

For each exercise:

a. List the prime numbers that you must check for divisibility.

b. Determine whether the number is prime or composite.

41. 10 **42.** 16 **43.** 23 **44.** 29

45. 39 **46.** 37 **47.** 43 **48.** 41

49. 73 **50.** 91 **51.** 93 **52.** 87

Draw a factor tree for each number. Use the factor tree to write the number's prime factorization.

53. 18 **54.** 20 **55.** 42 **56.** 75

Find the prime factorization of each number. If the number is prime, write that it is prime.

57. 12 **58.** 28 **59.** 31 **60.** 47

61. 18 **62.** 56 **63.** 42 **64.** 54

65. 72 **66.** 90 **67.** 17 **68.** 27

69. 100 **70.** 38 **71.** 60 **72.** 64

List the first five multiples of each number.

73. 6 **74.** 8 **75.** 25 **76.** 30

77. 14 **78.** 11 **79.** 17 **80.** 13

81. 150 **82.** 120 **83.** 35 **84.** 45

For each exercise, use the numbers shown in the middle column.

- **The answer to each exercise can use several of the numbers in the middle column.**
- **The numbers in the middle column are used for several exercises, so do not eliminate them.**

85. Which numbers are factors of 6?	1 2 3	**86.** Which numbers are multiples of 6?
87. Which numbers are multiples of 4?	5 6 8	**88.** Which numbers are factors of 4?
89. Which numbers are factors of 12?	9 10 12	**90.** Which numbers are multiples of 12?
91. Which numbers are multiples of 5?	11 14 16	**92.** Which numbers are factors of 7?
93. Which numbers are factors of 8?	20 21 24	**94.** Which numbers are multiples of 9?
95. Which numbers are multiples of 20?	47 75 80	**96.** Which numbers are factors of 20?

Fill in each blank with the word *multiple* or *factor*, as appropriate.

97. A box of 36 pencils can be divided evenly between 3 people because 3 is a _____ of 36.

98. Paper comes bundled in packages of 250 sheets. Roshawna can purchase 2,500 sheets of paper because 2,500 is a _______ of 250.

99. Stamps come in sheets of 20. Liam can sell 400 stamps to Sam because 400 is a _______ of 20.

100. A box of 24 envelopes can be divided evenly between 4 people because 4 is a _______ of 24.

goetz The mnemonic "multiples more, factors fewer" may help you answer these questions.

For more tips and tweets, go to twitter.com/gstbasicmath

Section 1.7 Question Log

Use this space to write questions. Make sure you get these answered and revisit them when you prepare for your exam.

Page ______ **Answered**	**Page** ______ **Answered**
Page ______ **Answered**	**Page** ______ **Answered**

Section 1.8 Greatest Common Factor and Least Common Multiple

To work efficiently with fractions in Chapter 2, it is important to understand the concepts related to the greatest common factor (GCF) and least common multiple (LCM). This section presents two methods for determining both the GCF and LCM.

The **Objectives** in Section 1.8 will help you

- A Find the greatest common factor by listing factors.
- B Find the least common multiple by listing multiples.
- C Find the GCF and LCM by factoring.

VOCABULARY PREVIEW *Check the box that applies.*

	Got It	Must Study
Factor: A **factor** of a number divides that number evenly.		
Common Factor: A **common factor** of two numbers divides into both numbers evenly.		
Greatest Common Factor (of several numbers): The largest number that divides the two or more numbers evenly. The GCF is less than or equal to the smallest of the numbers.		
GCF: The **G**reatest **C**ommon **F**actor is abbreviated **GCF**.		
Multiples: Multiplying a number by whole numbers (1, 2, 3, 4 . . .) gives **multiples** of the number.		
Common Multiple: A **common multiple** of two numbers is a multiple of both numbers.		
Least Common Multiple (of several numbers): The smallest multiple that is common to several numbers is the **least common multiple** of the numbers. The LCM is greater than or equal to the largest of the numbers.		
LCM: The **L**east **C**ommon **M**ultiple is abbreviated **LCM**.		
Venn Diagram: A **Venn diagram** uses overlapping circles to represent the relationship between two or more items.		

Study these terms when they appear in the text. At the end of the section, test your understanding by completing the Vocabulary Review in the section exercises.

Objective A Find the Greatest Common Factor by Listing Factors

The Concept The greatest common factor (GCF) of two numbers is used frequently when multiplying, dividing, and simplifying fractions.

When finding a greatest common factor, it helps to think about the parts of speech in the term *greatest common factor*.

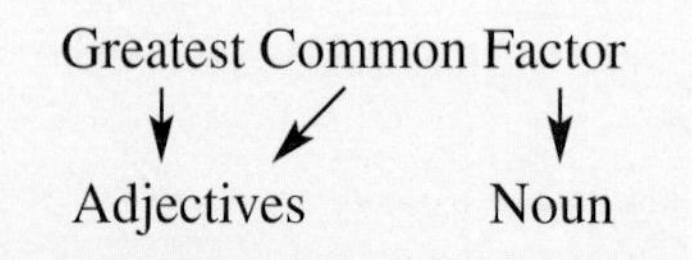

The adjectives *greatest* and *common* describe specific types of factors. Starting with lists of factors, we'll narrow the lists using the adjectives *common* and *greatest*.

- Find the common factors of the numbers.
- From the list of common factors, determine the greatest common factor.

INTERACTIVE DEFINITION Common Factors of Numbers

Common Factors: Any number that divides two or more numbers evenly is a **common factor** of those numbers.

EXAMPLE

1. Find the common factors of 16 and 24.

List the factors of each number and circle the common factors.

Factors of 16: 1, 2, 4, 8, 16

Factors of 24: 1, 2, 3, 4, 6, 8, 12, 24

1, 2, 4, and 8 are common factors.

The common factors are 1, 2, 4, and 8 because they divide both 16 and 24 evenly.

GUIDED PRACTICE

1. Find the common factors of 12 and 18.

List the factors of each number and circle the common factors.

Factors of 12:

Factors of 18:

____, ____, ____, and ____ are common factors.

The common factors are ____, ____, ____, and ____ because they divide both 12 and 18 evenly.

DO YOU UNDERSTAND how to find common factors? Got It Get Help

INTERACTIVE DEFINITION Greatest Common Factor (GCF)

The **greatest common factor (GCF)** is the largest number that divides two or more numbers evenly.

EXAMPLE

2. Find the greatest common factor (GCF) of 16 and 24.

From Example 1, the common factors of 16 and 24 are

1, 2, 4, 8.

The greatest of the common factors is 8.

The GCF of 16 and 24 is 8.

GUIDED PRACTICE

2. Find the greatest common factor (GCF) of 12 and 18.

From Guided Practice 1, the common factors of 12 and 18 are

____, ____, ____, ____.

The greatest of the common factors is ____.

The GCF of 12 and 18 is ____.

DO YOU UNDERSTAND how to find the greatest common factor? Got It Get Help

Procedure Find the Greatest Common Factor by Listing Factors

Step 1 List the factors.
Step 2 List the common factors.
Step 3 Find the greatest of the common factors.

EXAMPLES

3. Find the greatest common factor of 12 and 16.

Step 1 Factors of 12: [1], [2], 3, [4], 6, 12
Factors of 16: [1], [2], [4], 8, 16
Step 2 1, 2, and 4 are common factors.
Step 3 4 is the greatest of the common factors.
GCF = 4

4. Find the greatest common factor of 30 and 40.

Step 1 Factors of 30: [1], [2], 3, [5], 6, [10], 15, 30
Factors of 40: [1], [2], 4, [5], 8, [10], 20, 40
Step 2 1, 2, 5, and 10 are common factors.
Step 3 10 is the greatest of the common factors.
GCF = 10

5. Find the greatest common factor of 10 and 21.

Step 1 Factors of 10: [1], 2, 5, 10
Factors of 21: [1], 3, 7, 21
Step 2 1 is the only common factor.
Step 3 1 is the greatest of the common factors.
GCF = 1

GUIDED PRACTICE

3. Find the greatest common factor of 20 and 8.

Step 1 Factors of 20:
Factors of 8:
Step 2 ___, ___, and ___ are common factors.
Step 3 ___ is the greatest of the common factors.
GCF = ___

4. Find the greatest common factor of 14 and 42.

Step 1 Factors of 14:
Factors of 42:
Step 2 ___, ___, ___, and ___ are common factors.
Step 3 ___ is the greatest of the common factors.
GCF = ___

5. Find the greatest common factor of 8 and 27.

Step 1 Factors of 8:
Factors of 27:
Step 2 ___ is the only common factor.
Step 3 ___ is the greatest of the common factors.
GCF = ___

Concept Check

A1. Explain why 8 is not a common factor of 16 and 18.

A2. Explain why 5 is a common factor but not the greatest common factor (GCF) of 10 and 20.

A3. Find two numbers such that 3 is a common factor but not the greatest common factor.

Objective A Practice

A4. 6 and 12

a. List the factors of 6 and 12.
b. List the common factors of 6 and 12.
c. Find the greatest common factor of 6 and 12.

A5. 12 and 36

a. List the factors of 12 and 36.
b. List the common factors of 12 and 36.
c. Find the greatest common factor of 12 and 36.

List factors to find the greatest common factor (GCF) of each set of numbers.

A6. 6 and 4 **A7.** 10 and 15 **A8.** 9 and 12 **A9.** 8 and 10

A10. 6 and 5 **A11.** 7 and 10 **A12.** 28 and 56 **A13.** 32 and 48

Objective B Find the Least Common Multiple by Listing Multiples

The Concept The least common multiple (LCM) of two numbers is used frequently when adding or subtracting fractions. When finding the least common multiple, it helps to think about the parts of speech in the term *least common multiple*.

Least Common Multiple

Adjectives Noun

The adjectives *least* and *common* describe specific types of multiples. Starting with lists of multiples, we'll narrow the lists using the adjectives *common* and *least*.

- Find the common multiples of the numbers.
- From the list of common multiples, determine the least common multiple.

INTERACTIVE DEFINITION Common Multiple

A **common multiple** of two numbers is a multiple of both numbers.

EXAMPLE

6. Find the first two common multiples of 2 and 3.

First six multiples of 2:

$1 \cdot 2 = 2$
$2 \cdot 2 = 4$
$3 \cdot 2 = \boxed{6}$
$4 \cdot 2 = 8$
$5 \cdot 2 = 10$
$6 \cdot 2 = \boxed{12}$

First four multiples of 3:

$1 \cdot 3 = 3$
$2 \cdot 3 = \boxed{6}$
$3 \cdot 3 = 9$
$4 \cdot 3 = \boxed{12}$

The first two common multiples of 2 and 3 are 6 and 12.

GUIDED PRACTICE

6. Find the first two common multiples of 4 and 6.

First six multiples of 4:

$1 \cdot 4 =$
$2 \cdot 4 =$
$3 \cdot 4 =$
$4 \cdot 4 =$
$5 \cdot 4 =$
$6 \cdot 4 =$

First four multiples of 6:

$1 \cdot 6 =$
$2 \cdot 6 =$
$3 \cdot 6 =$
$4 \cdot 6 =$

The first two common multiples of 4 and 6 are ____ and ____.

DO YOU UNDERSTAND how to find common multiples? Got It Get Help

INTERACTIVE DEFINITION Least Common Multiple

The **least common multiple (LCM)** is the smallest multiple that is common to two or more numbers.

EXAMPLE	GUIDED PRACTICE
7. Find the least common multiple (LCM) of 2 and 3.	**7.** Find the least common multiple (LCM) of 4 and 6.
From Example 6, the first two common multiples of 2 and 3 are 6 and 12. The least of the common multiples is 6. The LCM of 2 and 3 is 6.	From Guided Practice 6, the first two common multiples of 4 and 6 are _____ and _____. The least of the common multiples is _____. The LCM of 4 and 6 is _____.

DO YOU UNDERSTAND how to find the least common multiple? Got It Get Help

To find the least common multiple of two numbers, list multiples for each number until you find the first common multiple. That number is the LCM.

Procedure Find the Least Common Multiple by Listing Multiples

List multiples of each number and stop at the first common multiple.

EXAMPLES	GUIDED PRACTICE
8. Find the least common multiple of 5 and 3.	**8.** Find the least common multiple of 5 and 6.
List multiples of each number and stop at the first common multiple. Multiples of 5: 5, 10, [15], 20 Multiples of 3: 3, 6, 9, 12, [15] The LCM is 15.	List multiples of each number and stop at the first common multiple. Multiples of 5: Multiples of 6: The LCM is _____.
9. Find the least common multiple of 12 and 15.	**9.** Find the least common multiple of 12 and 16.
List multiples of each number and stop at the first common multiple. Multiples of 12: 12, 24, 36, 48, [60], 72 Multiples of 15: 15, 30, 45, [60] The LCM is 60.	List multiples of each number and stop at the first common multiple. Multiples of 12: Multiples of 16: The LCM is _____.

Concept Check

B1. What is the least common multiple of two numbers?

B2. What is the greatest common factor of two numbers?

B3. When finding the GCF and LCM of two numbers, which can be greater than the two numbers, the GCF or LCM?

B4. When finding the GCF and LCM of two numbers, which can be less than the two numbers, the GCF or LCM?

Objective B Practice

List multiples to find the least common multiple of each set of numbers.

B5. 4 and 16 **B6.** 5 and 25 **B7.** 20 and 15 **B8.** 9 and 12

B9. 50 and 75 **B10.** 30 and 20 **B11.** 600 and 400 **B12.** 800 and 600

Fill in each blank with *GCF* or *LCM* to make a true statement.

B13. 10 is the ____________ of 2 and 10.

B14. 15 is the ____________ of 15 and 30.

B15. 9 is the ____________ of 9 and 27.

B16. 12 is the ____________ of 12 and 6.

Objective C Find the GCF and LCM by Factoring

The Concept It is often inefficient to find the GCF or LCM using the listing methods. Fortunately, there is a second method you can use to find the GCF and LCM. This method uses the prime factorization of two numbers and a **Venn diagram.**

INTERACTIVE DEFINITION Venn Diagram

A **Venn diagram** uses overlapping circles to represent the relationship between two or more items.

- The characteristics of each item are written in the corresponding circle.
- Characteristics common to both items are written where the circles overlap.

(Continued)

EXAMPLE

10. The items on two sub sandwiches are shown using the following Venn diagram.
a. What items are on the cold cut sub?
b. What items are on the veggie sub?
c. What items are common to both subs?

a) Items on the cold cut sub:
Ham, turkey, lettuce

b) Items on the veggie sub:
Lettuce, cucumber, tomato, peppers

Cold Cut: ham, turkey
Veggie: cucumber, tomato, peppers
Overlap: lettuce

c) Common item:
Lettuce

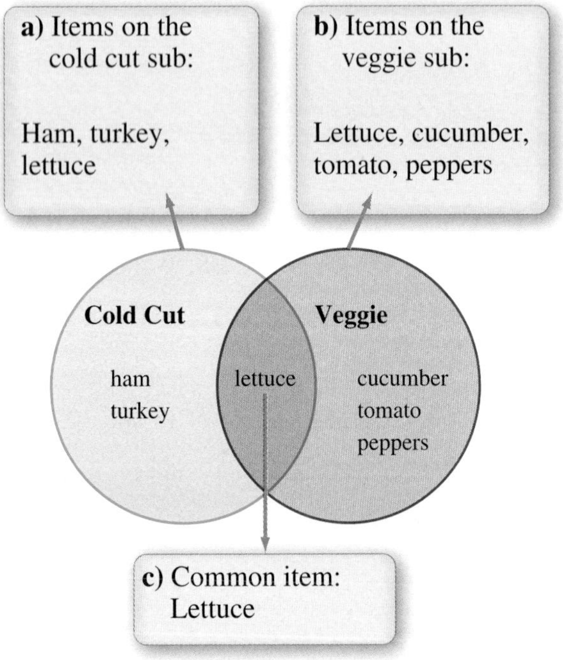

GUIDED PRACTICE

10. The items on two sub sandwiches are shown using the following Venn diagram.
a. What items are on the turkey sub?
b. What items are on the ham sub?
c. What items are common to both subs?

a) Items on the turkey sub:

b) Items on the ham sub:

Turkey: turkey, tomato
Ham: ham, lettuce, peppers
Overlap: swiss

c) Common item:

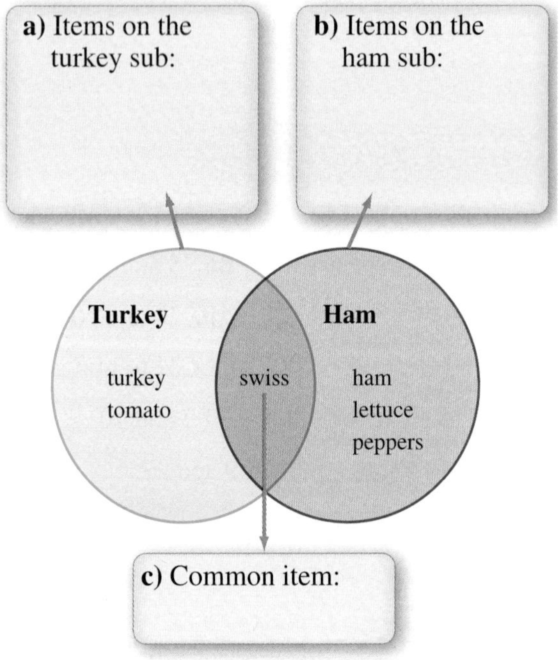

DO YOU UNDERSTAND how to read a Venn diagram? Got It Get Help

When finding the GCF or LCM, we will use a Venn diagram to sort each number's factors. Each circle will contain a number's prime factorization. The overlapping section will contain the prime factors common to both numbers.

Procedure Draw a Venn Diagram for the Factors of Two Numbers

Step 1 Write the prime factorization of each number and circle each pair of common factors.

Step 2 Write the common factors where the circles overlap.

Step 3 Write the remaining factors in the corresponding circles.

DETAILED EXAMPLE Drawing a Venn Diagram

Step 1 Write the prime factorization of each number and circle each pair of common factors.

$$18 = 2 \cdot 3 \cdot 3$$
$$30 = 2 \cdot 3 \cdot 5$$

Step 2 Write the common factors where the circles overlap.

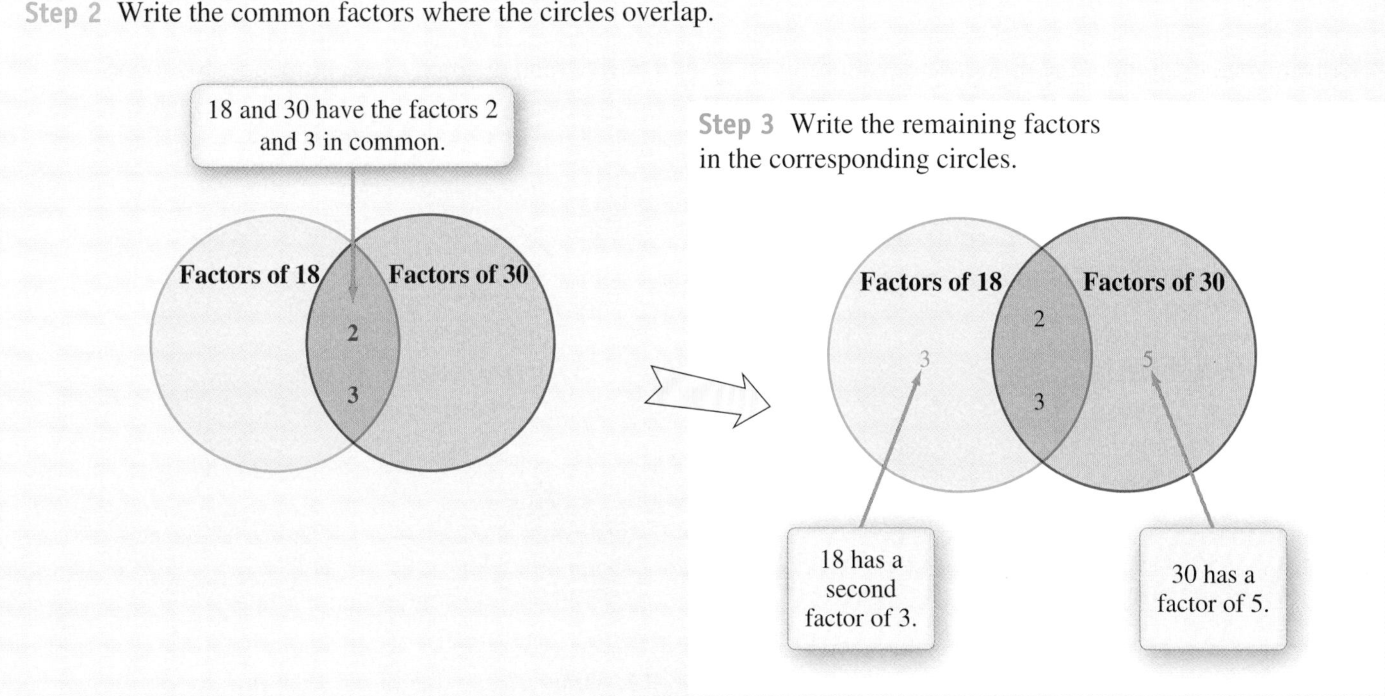

After making a Venn diagram, cover each circle with your hand to make sure each circle has the correct prime factors.

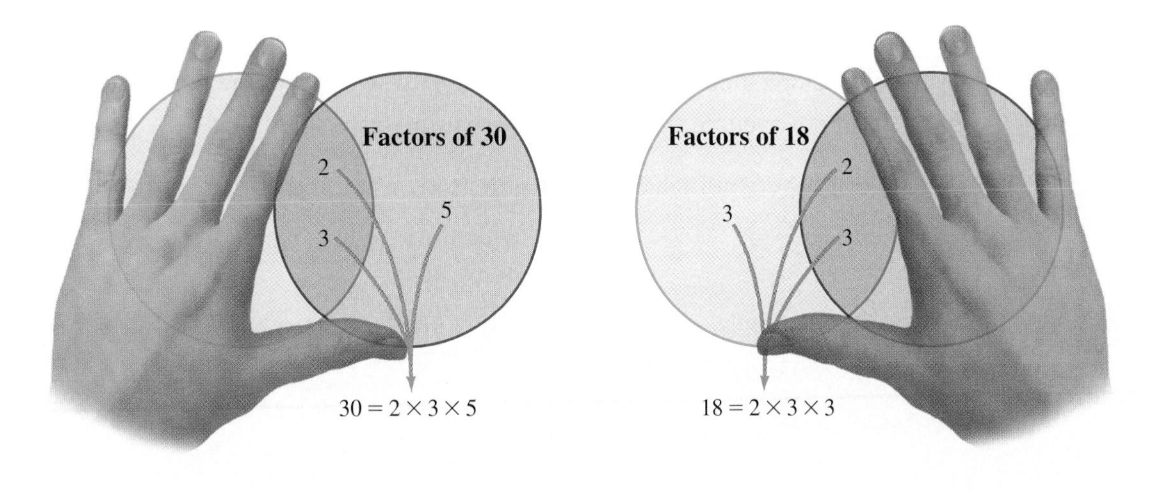

EXAMPLES

11. Draw a Venn diagram for the prime factorizations of 10 and 15.

Write the prime factorization of each number.

$$10 = 2 \cdot 5$$
$$15 = 3 \cdot 5$$

Draw a Venn diagram of the prime factorizations.

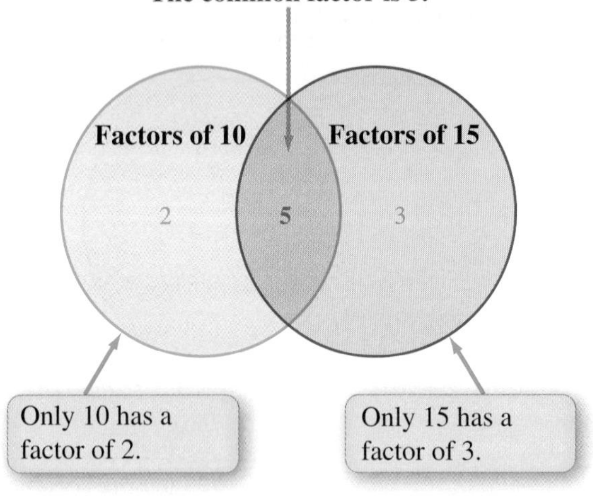

GUIDED PRACTICE

11. Drawa Venn diagram for the prime factorizations of 4 and 6.

Write the prime factorization of each number.

4 =

6 =

Draw a Venn diagram of the prime factorizations.

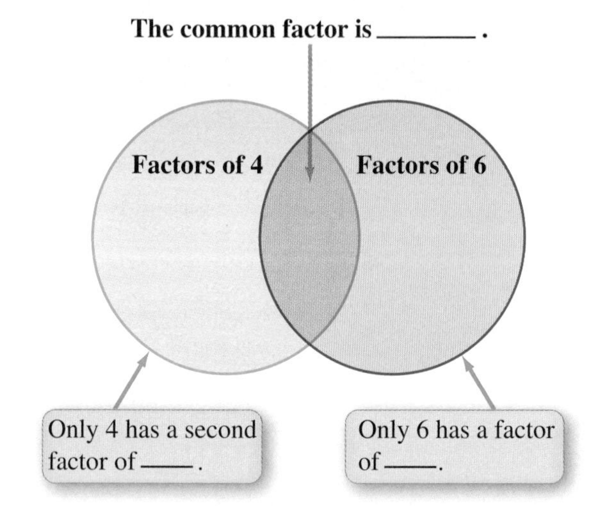

EXAMPLES

12. Draw a Venn diagram for the prime factorizations of 18 and 24.

Write the prime factorization of each number.

$$18 = 2 \cdot 3 \cdot 3$$
$$24 = 2 \cdot 2 \cdot 2 \cdot 3$$

Draw a Venn diagram of the prime factorizations.

The common factors, 2 and 3, go in the overlap.

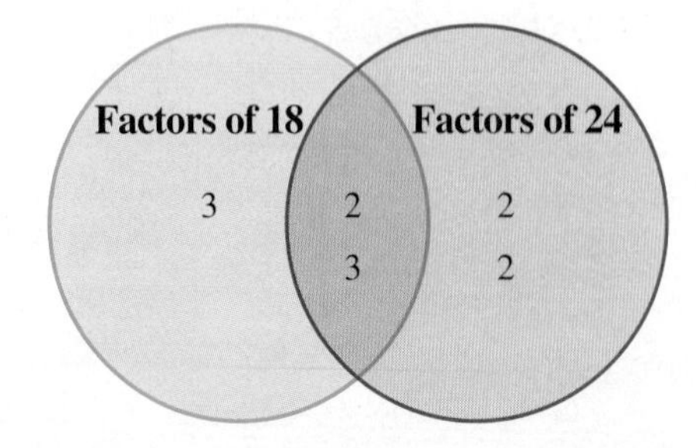

GUIDED PRACTICE

12. Draw a Venn diagram for the prime factorizations of 20 and 12.

Write the prime factorization of each number.

20 =

12 =

Draw a Venn diagram of the prime factorizations.

The common factors, ____ and ____, go in the overlap.

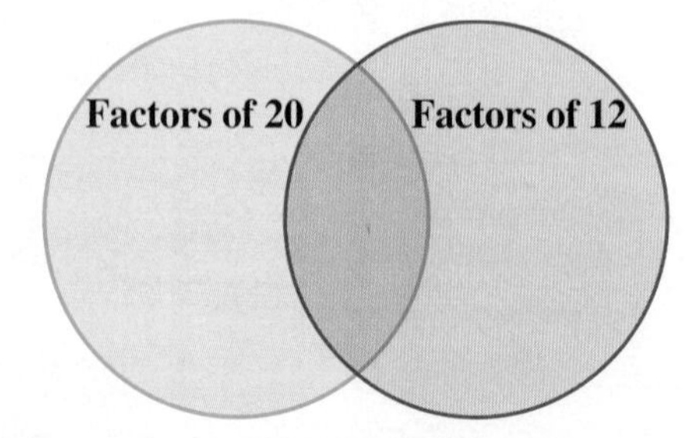

Using a Venn diagram and the prime factorizations of numbers, you can quickly find the GCF and LCM.

Procedure Find the GCF and LCM by Factoring

Step 1 Write the prime factorization of each number.
Step 2 Draw a Venn diagram of the prime factorizations.
Step 3 The GCF is the product of the factors inside the overlap.
Step 4 The LCM is the product of all of the factors in the circles.

EXAMPLES

13. Find the GCF and LCM of 12 and 30.

Step 1 Write the prime factorization of each number.

$$12 = 2 \cdot 2 \cdot 3$$
$$30 = 2 \cdot 3 \cdot 5$$

Step 2 Draw a Venn diagram of the prime factorizations.
The common factors, 2 and 3, go in the overlap.

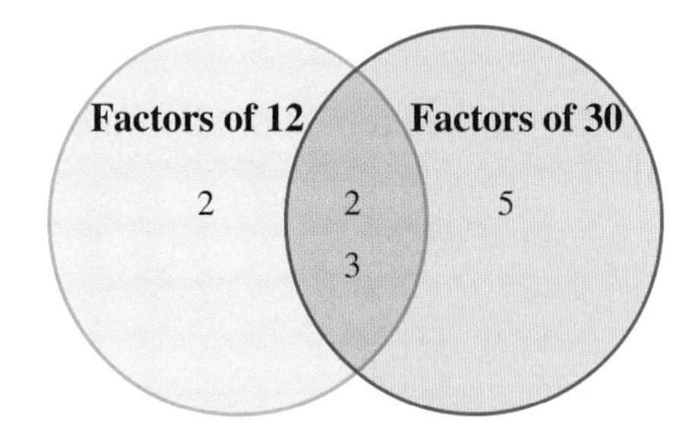

Step 3 The GCF is the product of the factors inside the overlap.

$$\text{GCF} = 2 \cdot 3$$
$$= 6$$

Step 4 The LCM is the product of all of the factors in the circles.

$$\text{LCM} = 2 \cdot 2 \cdot 3 \cdot 5$$
$$= 60$$

GUIDED PRACTICE

13. Find the GCF and LCM of 50 and 20.

Step 1 Write the prime factorization of each number.

$$50 =$$
$$20 =$$

Step 2 Draw a Venn diagram of the prime factorizations.
The common factors, _____ and _____, go in the overlap.

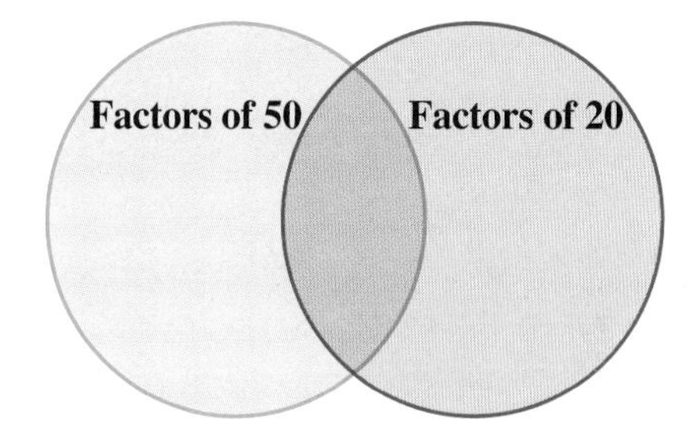

Step 3 The GCF is the product of the factors inside the overlap.

$$\text{GCF} =$$
$$=$$

Step 4 The LCM is the product of all of the factors in the circles.

$$\text{LCM} =$$
$$=$$

(Continued)

14. Find the GCF and LCM of 14 and 28.

Step 1 Write the prime factorization of each number.

$$14 = 2 \cdot 7$$
$$28 = 2 \cdot 2 \cdot 7$$

Step 2 Draw a Venn diagram of the prime factorizations.

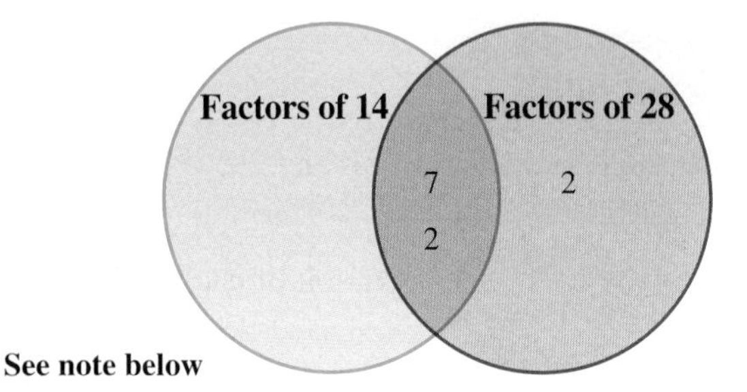

See note below

Step 3 The GCF is the product of the factors inside the overlap.

$$\text{GCP} = 2 \cdot 7$$
$$= 14$$

Step 4 The LCM is the product of all of the factors in the circles.

$$\text{LCM} = 2 \cdot 2 \cdot 7$$
$$= 28$$

14. Find the GCF and LCM of 18 and 54.

Step 1 Write the prime factorization of each number.

$$18 =$$
$$54 =$$

Step 2 Draw a Venn diagram of the prime factorizations.

Step 3 The GCF is the product of the factors inside the overlap.

$$\text{GCF} =$$
$$=$$

Step 4 The LCM is the product of all of the factors in the circles.

$$\text{LCM} =$$
$$=$$

▶ **NOTE** When we draw a Venn diagram of prime factorizations, each part of the diagram is understood to contain a factor of 1. Because of this, we could have written the Venn diagram from Example 14 as shown.

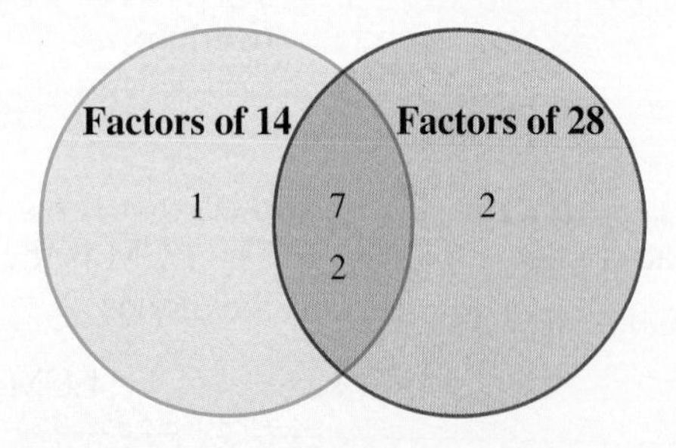

Concept Check

C1. The number 14 is either the GCF or LCM of 28 and 42. Without performing any calculations, which is it? Why?

C2. The number 84 is either the GCF or LCM of 12 and 14. Without performing any calculations, which is it? Why?

C3. Choose between LCM and GCF to make each statement true.

a. If all of the numbers inside a Venn diagram are multiplied, the result will be the LCM/GCF.

b. If the numbers inside the overlapping portion of a Venn diagram are multiplied, the result will be the LCM/GCF.

C4. What factor is understood to be in the overlapping section of the Venn diagram below?

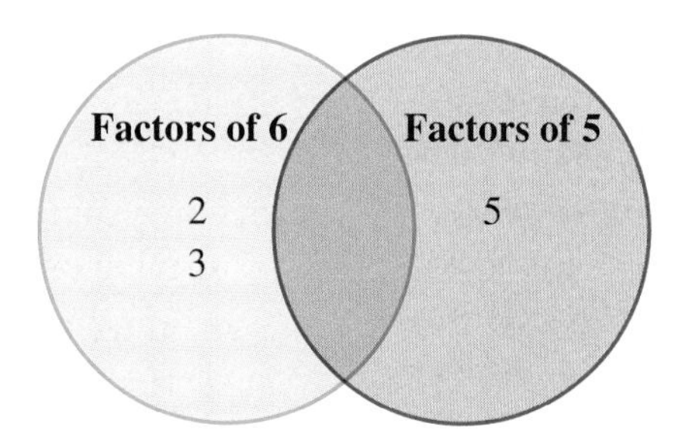

Answer the questions for each Venn diagram.

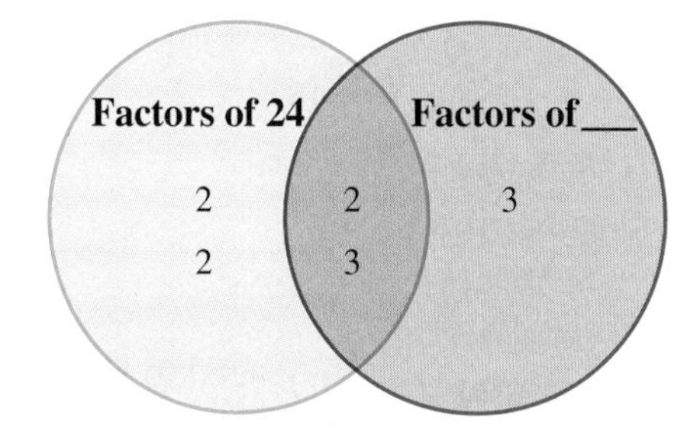

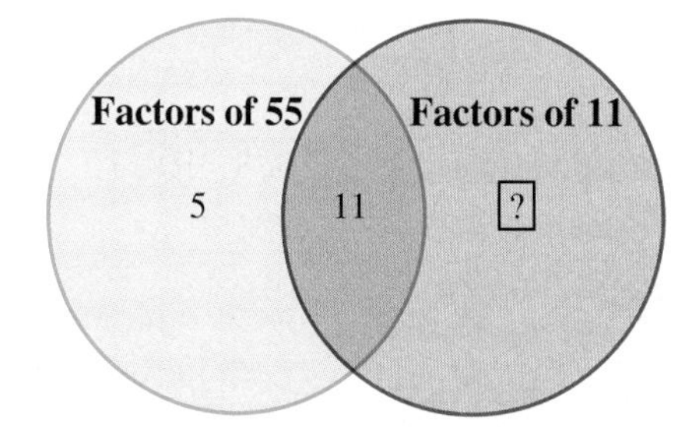

C5. a. What number is represented by the circle on the right?

b. What is the greatest common factor of the numbers?

c. What is the least common multiple of the numbers?

C6. a. What factor of 11 is the [?] equal to?

b. What is the greatest common factor of 11 and 55?

c. What is the least common multiple of 11 and 55?

Objective C Practice

Use a Venn diagram to find the greatest common factor (GCF) and least common multiple (LCM) for each set of numbers.

C7. 6 and 24

C8. 9 and 36

C9. 15 and 25

C10. 21 and 35

C11. 22 and 55

C12. 36 and 48

C13. 64 and 14

C14. 34 and 12

C15. 9 and 16

C16. 25 and 12

C17. 20 and 70

C18. 50 and 90

Combining Concepts and Applications

CONCEPT I LCM Applications

EXAMPLE

15. Kendrick has worked out the following schedule for buying groceries:

- Every two weeks he buys milk.
- Every five weeks he buys peanut butter.

If he just bought both of those items today, how many weeks will it be before he buys both items on the same shopping trip again?

Find the LCM of the numbers.

Multiples of 2: 2, 4, 6, 8, $\boxed{10}$, 12

Multiples of 5: 5, $\boxed{10}$, 15

LCM = 10

It will be 10 weeks until he buys both items on the same trip again.

GUIDED PRACTICE

15. Terry and her husband are in different car pools, driving on the following schedule:

- Terry takes her turn driving every five days.
- Her husband takes his turn driving every six days.

On the days that both of them drive, they borrow a car. If they both drove today, how many days will it be until they need to borrow a car again?

Find the LCM of the numbers.

Multiples of 5:

Multiples of 6:

LCM =

It will be _____ days until they need to borrow a car again.

Section 1.8 Exercises

FOR EXTRA HELP      

To find the greatest common factor by listing factors:

1. Answer the Objective A Concept Checks.
2. Answer the odd Objective A Practice Exercises.
3. Answer the even Objective A Practice Exercises.

To find the least common multiple by listing multiples:

4. Answer the Objective B Concept Checks.
5. Answer the odd Objective B Practice Exercises.
6. Answer the even Objective B Practice Exercises.

To find the GCF and LCM by factoring:

7. Answer the Objective C Concept Checks.
8. Answer the odd Objective C Practice Exercises.
9. Answer the even Objective C Practice Exercises.

VOCABULARY REVIEW *Review the Vocabulary Preview for Section 1.8. Study the definitions until you can check* Got It *for every word.*

Use these words to complete each sentence.

factor • common factor • greatest common factor • multiple • common multiple • least common multiple

	Got It	Get Help
10. A __________ of a number is found by multiplying the number by any whole number.		
11. A __________ of a number divides that number evenly.		
12. A number that divides two numbers evenly is a __________ of the numbers.		
13. The smallest multiple that is common to several numbers is the __________ of the numbers.		
14. A __________ of two numbers is a multiple of each of the numbers.		
15. The largest number that divides two or more numbers evenly is the __________ of the numbers.		

How will you get help for any vocabulary about which you are unsure?

Your instructor ____ MyMathLab ____ A classmate ____ A tutor ____ Other ____

List factors to find the greatest common factor (GCF) of each set of numbers.

16. 6 and 8

17. 6 and 9

18. 20 and 15

19. 30 and 18

20. 9 and 45

21. 7 and 42

22. 60 and 80

23. 40 and 50

List multiples to find the least common multiple (LCM) of each set of numbers.

24. 6 and 3

25. 5 and 10

26. 16 and 8

27. 33 and 11

28. 16 and 3

29. 3 and 22

30. 12 and 15

31. 30 and 18

Use the following Venn diagrams to answer each question.

32.

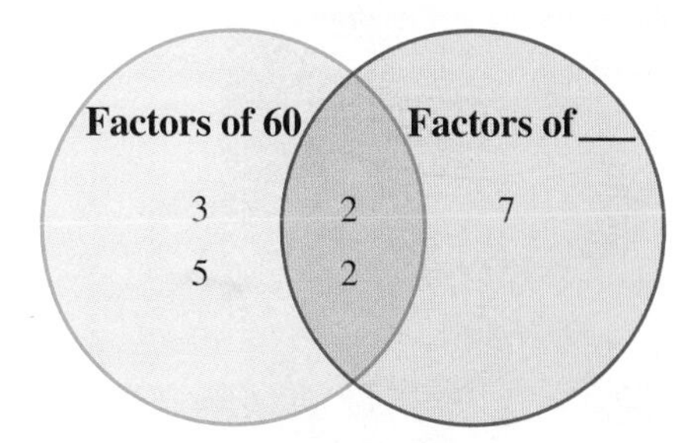

a. What number is represented by the circle on the right?

b. What is the greatest common factor of the two numbers?

c. What is the least common multiple of the two numbers?

33.

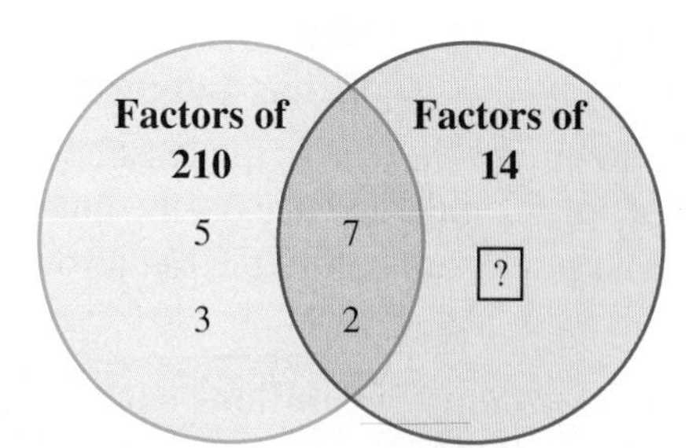

a. What factor of 14 is [?] equal to?

b. What is the greatest common factor of the two numbers?

c. What is the least common multiple of the two numbers?

For each set of numbers:
a. Draw a Venn diagram for each number's prime factorization.
b. Determine the GCF.
c. Determine the LCM.

34. 16 and 30
35. 15 and 25
36. 14 and 21
37. 12 and 18
38. 49 and 21
39. 35 and 25
40. 22 and 20
41. 26 and 24
42. 10 and 45
43. 21 and 35
44. 20 and 70
45. 45 and 18
46. 24 and 40
47. 54 and 72

48. Cleo and Andy have a crush on each other. Cleo volunteers with Habitat for Humanity every four days. Andy volunteers every three days. The last day they worked together was the fifth of the month. When is the next time they will work together?

49. From Exercise 48, Cleo decides to volunteer every two days to increase the number of times she will work with Andy. The fifth of the month was the last day they worked together. When is the next time they will work together?

Complete each multiplication table. You will need to use common factors and division to fill in some of the empty boxes. The factors in the white boxes are not arranged in any order.

50. **To start this exercise, find the GCF of the numbers in this column.** ↓ (second column)

×	7	3	8	6
4	28	12	32	24
5	35	15	40	30
9	56	27	72	54
7	49	35	56	42

51. **To start this exercise, find the GCF of the numbers in this column.** ↓ (second column)

×				
	63			56
		35		40
	54		24	
		21		

For Exercises 52 and 53, the gears at the right have the following number of teeth:

A has 50 teeth. **B has 125 teeth.**
C has 50 teeth. **D has 140 teeth.**

52. a. For gears A and B, what is the least common multiple of 50 and 125? This is the number of teeth that must pass before the red arrows realign.

To determine how many revolutions each gear pair must make before the red or green arrows line up, divide the LCM by the number of teeth on the gear.

b. How many revolutions must A make before the red arrows on A and B line up again?

c. How many revolutions must B make before the red arrows on A and B line up again?

Challenge Question: If gears B and C are attached, how many revolutions must A make before both the red and green arrows line up at the same time? Answer: 35 revolutions.

smith It's easy to confuse the GCF and LCM. Remember, the GCF is less than the LCM. You can always check an answer using this fact. *For more tips and tweets, go to twitter.com/gstbasicmath*

53. a. For gears C and D, what is the least common multiple of 50 and 140? This is the number of teeth that must pass before the green arrows realign.

To determine how many revolutions each gear pair must make before the red or green arrows line up, divide the LCM by the number of teeth on the gear.

b. How many revolutions must C make before the green arrows on C and D line up again?

c. How many revolutions must D make before the green arrows on C and D line up again?

Section 1.8 Question Log

Use this space to write questions. Make sure you get these answered and revisit them when you prepare for your exam.

Page ____ **Answered**	**Page** ____ **Answered**
Page ____ **Answered**	**Page** ____ **Answered**

Section 1.9 Applications with Whole Numbers

When solving applied exercises, it is important that you have a good strategy. The following strategy will help you solve the applied exercises in this section.

Strategy for Solving Applied Exercises

1. Read and understand the word exercise.
2. Change the word problem to a math problem. Write a corresponding math exercise.
3. Solve the math.
4. Interpret the results.

The **Objectives** in Section 1.9 will help you

A Translate words into mathematics.
B Solve word exercises with one operation.
C Solve word exercises with more than one operation.

VOCABULARY PREVIEW *Check the box that applies.*	Got It	Must Study
I can list eight terms that are used for **addition**.		
I can list nine terms that are used for **subtraction**.		
I can list six terms that are used for **multiplication**.		
I can list six terms that are used for **division**.		
I can list six terms that are used for **equality**.		

Study these terms when they appear in the text. At the end of the section, test your understanding by completing the Vocabulary Review in the section exercises.

Objective A Translate Words into Mathematics

The Concept The first step in solving a word exercise is to translate the words into mathematics. To do this efficiently, you must recognize terms that indicate addition, subtraction, multiplication, division, and equality.

KEY WORDS IN MATHEMATICS

Addition	Subtraction	Multiplication	Division	Equality
Sum	Difference	Product	Quotient	Equals
Add	Subtract	Times	Divided by	Is/Was
Total	Minus	Of	Divided into	Results in
Plus	Less	Doubled	Goes into	Is the same as
More than	Fewer than	Tripled	Per	Gives
Increased by	Less than	Multiplied by	Ratio	Amounts to
Deposit ($)	Decreased by			
Greater than	Loss of			
	Withdraw ($)			

Procedure Translate Words into Mathematics

Step 1 Identify the term that indicates an operation.

Step 2 Write a corresponding math exercise.

In the following Examples and Guided Practice Exercises, focus only on translating words into math. Do not solve the exercises.

EXAMPLES

Translate each question into mathematics. Do not answer the questions.

1. What is the product of 15 and 20?

"Product" indicates multiplication.
Multiply the two numbers that follow.

The product $= 15 \cdot 20$

2. What is the quotient of 20 and 5?

"Quotient" indicates division.
Divide the first number by the second number.

The quotient $= \frac{20}{5}$

3. At the zoo, the number of wallabies, 85, was decreased by 8. How many wallabies are left?

"Decrease" indicates subtraction.
Subtract the second number from the first number.

Wallabies $= 85 - 8$

GUIDED PRACTICE

1. What is the result of tripling 40?

"Tripling" indicates ____________.

The result = ____

2. How many 4's go into 32?

"Go into" indicates ________.

The result = _____

3. $45 was deposited into an account that already had $50 in it. What is the new balance?

"Deposit" indicates ________.

Balance = ________

Concept Check

A1. Try to do the following without looking at the Key Words in Mathematics.

a. List at least six terms that indicate addition.

b. List at least six terms that indicate subtraction.

c. List at least four terms that indicate multiplication.

d. List at least four terms that indicate division.

e. List at least four terms that indicate equality.

A2. Write a word exercise that requires adding 10 and 20.

A3. Write a word exercise that requires dividing 20 by 10.

Objective A Practice

For each exercise, translate words into mathematics. Do not answer the questions.

A4. What is the difference between 15 and 7?

A5. What is the sum of 8 and 9?

A6. What is the result of doubling 15?

A7. How many times does three divide into 42?

A8. What is the balance if $40 is withdrawn from an account with $80 in it?

A9. Pierre purchased 5 packs of soda with each pack containing 6 cans. How many cans of soda did he purchase?

A10. What is the product of 3, 4, and 5?

A11. What is the total of 3, 4, and 5?

Objective B Solve Word Exercises with One Operation

The Concept Once you set up a word exercise, perform the mathematics necessary to solve it. After you solve a word exercise, write what your answer represents using words.

Procedure Solve Word Exercises with One Operation

Step 1 Read and understand the word exercise.

Step 2 Write a corresponding math exercise.
Write an intermediate word equation if necessary.

Step 3 Solve the math.

Step 4 Interpret the result.

EXAMPLES

4. Joyce had $42 in her account on Monday. She deposited $34 on Tuesday. What was her new balance?

Step 1 Read and understand the word exercise.

"Deposited" indicates addition. We must add the two values.

Step 2 Write a corresponding math exercise.

Step 3 Solve the math.

New balance = Old balance + Deposit
= 42 + 34
= 76

Write a word equation to guide you.

Step 4 Interpret the result.

Joyce's new balance is $76.

5. Umberto bought soda for a wedding reception. The soda comes in packs of 18 cans. How many cans did he purchase if he bought 50 packs?

Step 1 Read and understand the word exercise.

There are 50 *of* the packs. *Of* often means multiplication. Multiply the number of packs by the number of cans in one pack.

Step 2 Write a corresponding math exercise.

Step 3 Solve the math.

cans = (# packs) · (# cans per pack)
= 50 · 18
= 900

Word exercises should have word answers.

Step 4 Interpret the result.

Umberto purchased 900 cans of soda.

6. Eco had 792 empty soda cans in his pantry. To return the cans, he repackaged them in cases that hold 18 cans in each case. How many cases did he fill?

Step 1 Read and understand the word exercise.

To find out how many cases are filled, we must determine how many 18's go into 792. This is a division exercise.

Step 2 Write a corresponding math exercise.

Step 3 Solve the math.

cases = (# cans) ÷ (# cans per case)
= 792 ÷ 18
= 44

Step 4 Interpret the result.

Eco filled 44 cases.

GUIDED PRACTICE

4. Jim had $76 in his account on Tuesday. He withdrew $25 on Wednesday. What was his new balance?

"________" indicates ________.

We must ______________.

New balance =
=
=

Jim's new balance is _____.

5. Buggs can make 8 carrot cakes in an hour. If he works for 4 hours, how many cakes can he make?

6. Daniel is packaging carrots into 3-pound bags. If he has a total of 45 pounds of carrots, how many bags will he fill?

Concept Check

B1. Write a word exercise in which $50 is multiplied by 12 months.

B2. Write a word exercise in which $600 is divided by 12.

Objective B Practice

Solve each word exercise.

B3. Sergio can paint 15 birdhouses in an hour. If 75 birdhouses are ordered, how long will it take Sergio to paint them?

B4. The quantity *1 gross* is equal to 144. If Quinten ordered five gross of pens for a bookstore, how many pens did he order?

B5. Claudia purchased a house for $75,000. With repairs, she spent $84,312. How much did the repairs cost?

B6. Anita is in a chemistry class with 183 other students. There are 212 other students in her history class. What is the difference in the sizes of the two classes?

B7. Gumby's gumball business collected $45, $63, $21, and $12 from four machines. What was the total amount collected?

B8. After collecting $835 in payment for landscaping work, Terrence had to pay a repair bill of $245 and a fuel bill of $120. How much money was left?

B9. An acre of land has an area of 43,560 square feet. What is the area of 23 acres?

B10. Together four employees of a company earn $176,000. What is their average income?

Objective C Solve Word Exercises with More Than One Operation

The Concept Often word exercises require several steps to get an answer. Once you recognize the different steps, you can follow the order of operations to solve the word exercise. Make sure you take extra time to read and understand these exercises.

Procedure Solve Word Exercises with More Than One Operation

Step 1 Read and understand the word exercise.

Step 2 Write a corresponding math exercise.
Write an intermediate word equation if necessary.

Step 3 Solve the math.

Step 4 Interpret the result.

EXAMPLES

7. Juanita received the following scores on her math tests: 85, 78, 90, 83. If the minimum average for a B is 83%, did she earn a B?

Step 1 Read and understand. To find an average, add the data values and divide by the number of values.

Step 2 Write an intermediate word equation.

$$\text{Avg} = \frac{\text{Sum of scores}}{\text{\# of scores}}$$

Substitute the known values.

$$= \frac{85 + 78 + 90 + 83}{4}$$

Step 3 Solve the math.

$$= \frac{336}{4}$$

$$= 84$$

Step 4 Interpret the result. Because Juanita averaged 84%, she earned a B.

8. Kwon took a trip to visit his grandmother. At the start of the trip, his car's odometer read 90,444 miles. At the end of the trip, it read 91,308 miles. If the car used 36 gallons of gas, how many miles per gallon did the car get on the trip?

Step 1 Read and understand. To find the mileage (miles per gallon), divide the miles driven by the gas used.

Step 2 Write an intermediate word equation.

$$\text{Mileage} = \frac{\text{Miles driven}}{\text{Gas used}}$$

$$= \frac{\text{Change in odometer}}{\text{Gas used}}$$

Substitute the known values.

$$= \frac{91{,}308 - 90{,}444}{36}$$

Step 3 Solve the math.

$$= \frac{864}{36}$$

$$= 24$$

Step 4 Interpret the result. Include the units. Kwon's car averaged 24 miles per gallon on the trip.

GUIDED PRACTICE

7. Over 4 days, the daily high temperatures were 64, 61, 72, and 59 degrees. Was the average daily high temperature on those days higher or lower than 65 degrees?

8. Sergei noticed that his sink was leaking. To determine how much water he was wasting, he checked his water meter. When he checked the meter, it read 12,983 gallons. Two hours later the meter read 13,037 gallons. If no other water was used, what was the average amount of water wasted per hour?

Concept Check

C1. Why might writing an intermediate word equation help you solve a word exercise?

C2. Why is it important to write the units in answers to word exercises?

Objective C Practice

Use the following bar graph to answer each question.

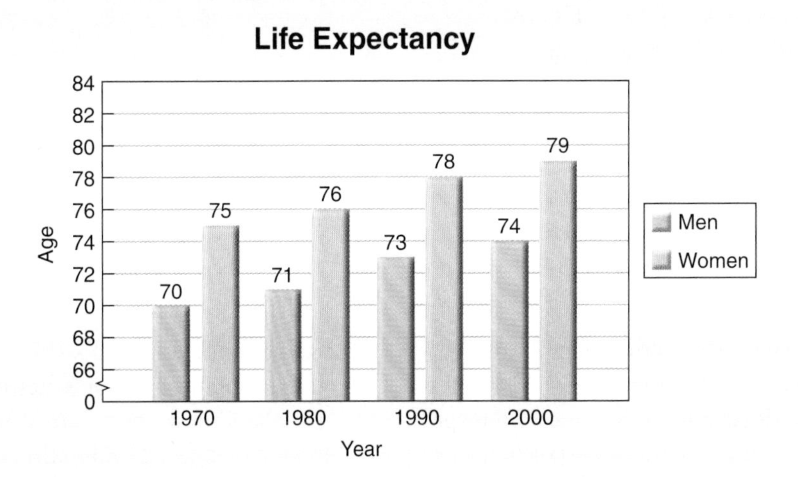

C3. What is the average life expectancy of men for the years shown?

C4. What is the average life expectancy of women for the years shown?

Answer each question.

C5. A local school district sponsored a high school math competition. 450 students entered the competition. 212 of them were seniors, and 175 were juniors. How many contestants were either freshmen or sophomores?

C6. Yanis purchased 3 gallons of milk at \$3 each, 4 bags of bread at \$2 each, and 16 cans of tuna at \$2 each. How much was Yanis' bill?

C7. To remodel a bathroom, JeJuan paid \$150 for a vanity, \$85 for a sink, \$39 for a mirror, and \$43 for towel bars. After he used a \$75 gift card, what was his total bill?

C8. After purchasing \$350 in materials to remodel a bedroom, Janise returned three items that cost \$45, \$52, and \$35. After the returns, how much money did Janise spend on the project?

Section 1.9 Exercises

FOR EXTRA HELP   PRACTICE  WATCH  DOWNLOAD  READ REVIEW

To translate words into mathematics:

1. Answer the Objective A Concept Checks.
2. Answer the odd Objective A Practice Exercises.
3. Answer the even Objective A Practice Exercises.

To solve word exercises with one operation:

4. Answer the Objective B Concept Checks.
5. Answer the odd Objective B Practice Exercises.
6. Answer the even Objective B Practice Exercises.

To solve word exercises with more than one operation:

7. Answer the Objective C Concept Checks.
8. Answer the odd Objective C Practice Exercises.
9. Answer the even Objective C Practice Exercises.

VOCABULARY REVIEW *Review the Vocabulary Preview for Section 1.9. Study the definitions until you can check* Got It *for every term.*

Use these words to indicate what the words below represent.

addition • subtraction • multiplication • division • equality

	Got It	Get Help		Got It	Get Help		Got It	Get Help
10. Deposit ____			**11.** Goes into ____			**12.** Sum ____		
13. Quotient ____			**14.** Product ____			**15.** Gives ____		
16. Tripled ____			**17.** Increased by ____			**18.** Of ____		
19. Difference ____			**20.** Withdraw ____			**21.** Fewer than ____		
22. Is ____			**23.** Amounts to ____			**24.** Per ____		

How will you get help for any vocabulary about which you are unsure?

Your instructor ____ MyMathLab ____ A classmate ____ A tutor ____ Other ____

Answer each question.

25. Find 45 multiplied by 19.

26. Find the quotient of 145 and 5.

27. Find 27 less than 38.

28. Find the product of 11 and 56.

29. Calculate the total of 7, 17, and 22.

30. Calculate 26 fewer than 42.

31. Calculate 10 more than the product of 6 and 8.

32. Calculate 9 less than the product of 7 and 5.

33. Dexter withdrew $245 from an account with a balance of $1,236. What is the new balance?

34. Tandy deposited $387 into an account with a balance of $3,476. What is the new balance?

35. The owner's manual recommends changing the oil in a 2008 Pontiac Grand Am every 5,500 miles. If the car's odometer read 76,473 miles when the oil was changed, at what odometer reading should the next oil change occur?

36. Germaine is creating an MP3 player of music for an upcoming party. The player holds 700 megabytes. The songs he has chosen take up 678 megabytes. Which of the following songs also will fit on the MP3 player?

a. Pollination (32 megabytes)
b. Heaviest Pigeon (26 megabytes)
c. Herb Roberts (21 megabytes)

37. Karman Auto Sales is offering \$350 off the purchase of a new carport. If the original price of a carport is \$834, what is the sale price?

38. Eric purchased a car for \$12,950. The 6% sales tax was \$777. What was the total price of the car including sales tax?

39. A case of eggs contains 12 cartons. If 12 eggs are in each carton, how many eggs are in a case?

40. Seven friends are splitting \$1,211 evenly. How much will each person receive?

41. The temperature was 43 degrees at 7 A.M. It increased 16 degrees by 11 A.M. and then decreased 7 degrees by 5 P.M. What was the temperature at 5 P.M.?

42. The treasurer for a college volleyball team deposited \$1,560 into an account that already had \$12,785 in it. Later on he withdrew \$967. What was the new account balance?

43. It takes a gallon of water to resurface 8 square feet of ice in a hockey rink. How many gallons of water does it take to resurface the entire hockey rink? An ice rink is shaped approximately like a rectangle that measures 200 feet by 85 feet.

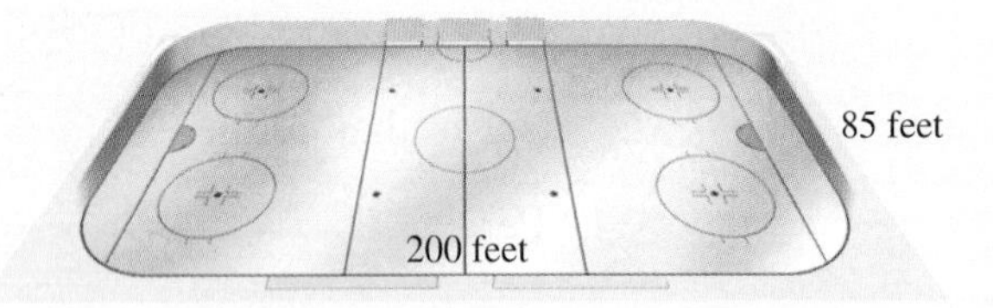

44. One quart of epoxy is needed to finish 25 square feet of flooring on a basketball court. How many quarts of epoxy will it take to finish the entire basketball court?

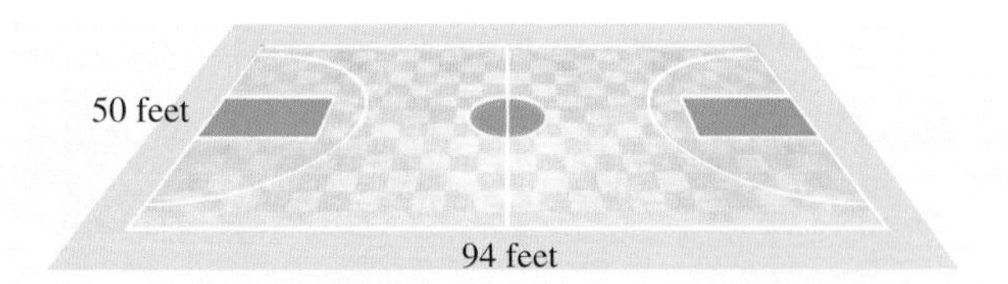

45. Ward is calculating his yearly expenses for playing in a recreational soccer league. He spent \$940 during the last four years traveling to and from the games. What was his average cost to travel per year?

46. Chantel is calculating the year's travel expenses for her company. If her company spent \$58,212 on travel this year, what was the average cost per month?

47. The last 3 days have had low temperatures of 55, 61, and 52 degrees. What is the average low temperature for the last 3 days?

48. The last 5 days have had high temperatures of 75, 68, 64, 64, and 54 degrees. What is the average high temperature for the last 5 days?

49. Thomas bought a 52-acre organic fruit farm. If there are 43,560 square feet in an acre, how many square feet does his farm cover?

50. Thomas's 52-acre fruit farm has 23 trees per acre. If Thomas makes a profit of \$62 per tree, how much profit should he expect from the farm?

51. Kendra drove her car on a 364-mile trip. On the trip, her car used 14 gallons of gasoline. How many miles per gallon did her car get?

52. Corrine drove her car on a 372-mile trip. On the trip, her car used 12 gallons of gasoline. How many miles per gallon did her car get?

53. Find the total cost of purchasing two pairs of shoes at $26 each and three shirts at $14 each.

54. Find the total cost of purchasing two pairs of gloves at $9 each and three hats at $11 each.

55. Kao's car usually gets 34 miles per gallon when driven on the highway. At the start of a recent highway trip, his odometer read 114,456 miles. When he stopped, his odometer read 114,744 miles. The trip used 9 gallons of gas. What was his gas mileage on the trip?

56. The EPA estimates that the 2006 Honda Insight gets 66 miles per gallon when driven on the highway. At the start of a recent highway trip, the odometer read 9,345 miles. At the end of the trip, the odometer read 10,089. The car used 12 gallons of gas. What was the gas mileage on the trip?

57. Courtney rents a house to a family for $950 per month. Her expenses are $639 every month for the mortgage and $1,968 in taxes, once per year. Calculate Courtney's yearly profit.

58. Branford rents a house to college students for $775 per month. His expenses are $534 every month for the mortgage and $1,234 in taxes, once per year. Calculate Branford's yearly profit.

tobey How's the car doing? Subtract the odometer readings. Divide miles driven by gallons of gas. Is the result close to the expected MPG?

For more tips and tweets, go to twitter.com/gstbasicmath

Section 1.9 Question Log

Use this space to write questions. Make sure you get these answered and revisit them when you prepare for your exam.

Page ____ Answered	Page ____ Answered
Page ____ Answered	Page ____ Answered

Chapter 1 Organizer

VOCABULARY

Use the following steps to review the vocabulary for Chapter 1.

1. *Write the definition for each word from memory.*
2. *Compare the definitions you have written with the definitions in the Vocabulary Preview.*
3. *Study any definitions that you could not remember or that you defined incorrectly.*

1.1
Digits • Whole Numbers • Place Value • Periods • Rounding

1.2
Operations • Sum • Carrying • Commutative Property of Addition • Perimeter • Variable

1.3
Inverse Operations • Difference • Borrowing

1.4
Multiplication • Product • Factors • Commutative Property of Multiplication

1.5
Dividend • Divisor • Quotient • Remainder • Division Notation

1.6
Exponents • Base • Exponent • Grouping Symbols • Grouping • Order of Operations

1.7
Divisible • Prime Number • Composite Number • Factor (noun) • Factor Pair • Factorization • Prime Factorization • Factor Tree • Multiples

1.8
Factor • Common Factor • Greatest Common Factor (of several numbers) • GCF • Multiples • Common Multiple • Least Common Multiple (of several numbers) • LCM • Venn Diagram

1.9
Addition Terms • Subtraction Terms • Multiplication Terms • Division Terms • Equality Terms

PROCEDURES

Procedure/Topic	Steps	Example
Write a Whole Number Using Words (Section 1.1)	**Step 1** Determine the largest period of the number. Count the periods: "ones, thousands, millions, billions, trillions . . ." **Step 2** From the left, write the number in each period followed by the name of the period.	Write 123,456 using words. One hundred twenty-three thousand, four hundred fifty-six
Determine the Place Value of a Digit (Section 1.1)	**Step 1** Determine the period of the digit. **Step 2** Determine the value of the digit in that period. Is it in the ones, tens, or hundreds place of the period?	Determine the place value of the digit 2 in the whole number 123,456. 2 is in the tens place of the thousands period, which means that 2 is in the ten thousands place.
Write a Whole Number in Expanded Form (Section 1.1)	Write the sum of what each digit represents. Use place value to determine the meaning of each digit. You do not need to write the meaning of zeros.	Write 20,306 in expanded form. $20{,}306 = 20{,}000 + 300 + 6$

Procedure/Topic	Steps	Example
Graph Whole Numbers on a Number Line (Section 1.1)	**Step 1** Draw a number line with arrows at each end. **Step 2** Indicate the scale, using labels. Space the labels in a convenient way based on the numbers to be graphed. **Step 3** Graph each whole number on the number line.	Graph 3, 8, and 15 on a number line. 0 3 5 8 10 15
Round a Whole Number (Section 1.1)	**Step 1** Identify the digit to be rounded. **Step 2** Decide if the digit to be rounded increases by one or stays the same. Increase the digit when the digit to the right is 5–9. The digit stays the same when the digit to the right is 0–4. **Step 3** All of the digits to the right of the rounded digit will be zero.	Round 76,497 to the thousands place. 7⑥,497 rounds down to 76,000.
Use a Number Line for Addition (Section 1.2)	**Step 1** Draw a number line with a dot at the first number. **Step 2** Move the second number of units to the right.	Use a number line to find the sum. $3 + 12$ 0 3 5 10 15 $3 + 12 = 15$
Add Whole Numbers (Section 1.2)	**Step 1** Add the digits in the ones place. Carry if necessary. **Step 2** Add the digits in the tens place. Carry if necessary.	Find the sum. $84 + 57$ $\begin{array}{r} {}^{1}\,{}^{1} \\ 8\,4 \\ +\,5\,7 \\ \hline 1\,4\,1 \end{array}$
Add Large Numbers	**Step 1** Add the digits in the ones place. Carry if **Step 2** Add the digits in the tens place. Carry if necessary. **Step 3** Repeat the procedure for any higher place values.	Find the sum. $2{,}432 + 385$ $\begin{array}{r} {}^{1} \\ 2\,4\,3\,2 \\ +\,3\,8\,5 \\ \hline 2\,8\,1\,7 \end{array}$
Use a Number Line for Subtraction (Section 1.3)	**Step 1** Draw a number line with a dot at the first number. **Step 2** Move the second number of units to the left.	Use a number line to find the difference. $15 - 12$ 0 3 5 10 15 $15 - 12 = 3$
Subtract Whole Numbers (Section 1.3)	**Step 1** Subtract the digits in the ones place. Borrow if necessary. **Step 2** Subtract the digits in the tens place. Borrow if necessary. **Step 3** Repeat the procedure for any higher place values.	Find the difference. $574 - 439$ $\begin{array}{r} 5\,\overset{6}{\not7}\,{}^{1}4 \\ -\,4\,3\,9 \\ \hline 1\,3\,5 \end{array}$

Procedure/Topic	Steps	Example
Subtract Large Numbers (Section 1.3)	**Step 1** Subtract the digits in the ones place. Borrow if necessary. **Step 2** Subtract the digits in the tens place. Borrow if necessary. **Step 3** Repeat the procedure for any higher place values.	Find the difference. 4,259 − 724 $$\begin{array}{r} \overset{3}{\cancel{4}}\,{}^{1}2\,5\,9 \\ -\ \ 7\,2\,4 \\ \hline 3\,5\,3\,5 \end{array}$$
Subtract Several Numbers (Section 1.3)	Perform one subtraction at a time from left to right.	Subtract. 50 − 10 − 7 $50 - 10 - 7 = 40 - 7$ $= 33$
Multiply a One-Digit Number and a Two-Digit Number (Section 1.4)	**Step 1** Multiply the ones place of the two-digit number by the one-digit number. Carry if necessary. **Step 2** Multiply the tens place of the two-digit number by the one-digit number. Add and carry if necessary.	Find the product. 25 × 9 $$\begin{array}{r} \overset{4}{2}\,5 \\ \times\ \ 9 \\ \hline 2\,2\,5 \end{array}$$
Multiply by a Power of Ten (Section 1.4)	**Step 1** Count the number of zeros in the power of ten. **Step 2** Write that many zeros at the end of the other factor.	Find the product. 318 · 1,000 $318 \cdot 1{,}000 = 318{,}000$ 1,000 has 3 zeros. Write 3 zeros at the end of 318.
Multiply Large Numbers (Section 1.4)	**Step 1** Multiply by the digit in the ones place. **Step 2** Multiply by the digit in the tens place. Write the result in the tens place. **Step 3** Continue multiplying by any digits in higher place values. Write each result in the correct place value. **Step 4** Add the results.	Find the product. 214 × 21 $$\begin{array}{r} 2\,1\,4 \\ \times\ 2\,1 \\ \hline 2\,1\,4 \\ +\ 4\,2\,8\,0 \\ \hline 4\,4\,9\,4 \end{array}$$ **Step 1** Multiply 214 · 1. → 214 **Step 2** Multiply 214 · 2. → + 4280 **Step 4** Add the results. → 4494
Estimate the Product of Large Numbers (Section 1.4)	**Step 1** Round each number to its largest place value. **Step 2** Multiply the rounded numbers.	Estimate. 592 · 33 $592 \cdot 33 \approx 600 \cdot 30$ $\approx 18{,}000$
Multiply a List of Factors (Section 1.4)	Perform the multiplications from left to right.	Multiply. 2 · 3 · 5 · 4 $\underbrace{2 \cdot 3}_{1st} \cdot 5 \cdot 4 = \underbrace{6 \cdot 5}_{2nd} \cdot 4$ $= \underbrace{30 \cdot 4}_{3rd}$ $= 120$

Procedure/Topic	Steps	Example
Solve a Division Exercise Using Multiplication (Section 1.5)	Solve the corresponding multiplication exercise. Figure out what number must multiply the divisor to result in the dividend.	Write the corresponding multiplication exercise for $42 \div 6$. Corresponding Multiplication $42 \div 6 = x \Rightarrow \boxed{x} \cdot 6 = 42$ $\boxed{7} \cdot 6 = 42$ Therefore, $42 \div 6 = 7$.
Division with a Remainder (Section 1.5)	**Step 1** Determine how many times the divisor goes into the dividend. **Step 2** Multiply the divisor by that value. **Step 3** Subtract the result from the dividend. The remainder is what is left over.	Divide. $23 \div 5$ $\begin{array}{r} 04\text{ R }3 \\ 5\overline{)23} \\ -20 \\ \hline 3 \end{array}$ $23 \div 5 = 4R3$
Perform Long Division with a Remainder (Section 1.5)	Start with the largest place value of the dividend and work toward the smallest place value. **Step 1** Determine how many times the divisor *goes into* the dividend. This value will be a one-digit number, possibly zero. **Step 2** *Multiply* the divisor by that value. Write the answer below the dividend. **Step 3** *Subtract* that quantity. **Step 4** *Bring down* the next digit of the divisor. Repeat until every digit of the dividend has been divided.	Divide. $3{,}250 \div 16$ $\begin{array}{r} 0203\text{ R }2 \\ 16\overline{)3250} \\ -32 \\ \hline 050 \\ -48 \\ \hline 2 \end{array}$ $3{,}250 \div 16 = 203R2$
Use a Mixed Number to Write a Quotient and Remainder (Section 1.5)	**Step 1** Divide the dividend by the divisor. **Step 2** Write the mixed number.	$42 \div 5 = 8R2$. Write the answer as a mixed number. $42 \div 5 = 8\frac{2}{5}$
Estimate Quotients (Section 1.5)	**Step 1** Round the dividend and divisor to their largest place values. **Step 2** Divide with the rounded values.	Estimate the quotient. $485 \div 53$ $485 \div 53 \approx 500 \div 50$ $= 10$
Divide by a Large Number (Section 1.5)	Start with the largest place value of the dividend and work toward the smallest place value. **Step 1** Determine how many times the divisor *goes into* the dividend. **Step 2** *Multiply* the divisor by that value. **Step 3** *Subtract* that quantity. **Step 4** *Bring down* the next digit of the divisor to see what is left over. Repeat until every digit of the dividend has been divided.	Divide. $461 \div 21$ $\begin{array}{r} 021\text{ R }20 \\ 21\overline{)461} \\ -42 \\ \hline 41 \\ -21 \\ \hline 20 \end{array}$

Procedure/Topic	Steps	Example
Evaluate Expressions with Exponents (Section 1.6)	**Step 1** Write the repeated multiplication. Write the base as a factor as many times as the exponent. **Step 2** Multiply.	Evaluate. 2^3 $2^3 = 2 \cdot 2 \cdot 2$ $= 4 \cdot 2$ $= 8$
Simplify Expressions with Groupings (Section 1.6)	**Step 1** Perform the operations in the groupings first. **Step 2** Perform other operations second.	Simplify. $4 \div (6 - 4)$ $4 \div (6 - 4) = 4 \div 2$ $= 2$
Use the Order of Operations (Section 1.6)	**First:** Perform operations in groupings. **Second:** Perform operations with exponents. **Third:** Perform multiplication and division as they occur from left to right. **Fourth:** Perform addition and subtraction as they occur from left to right.	Evaluate. $25 - (6 - 3)^2 \div 3 \cdot 2 + 4$ $25 - (6 - 3)^2 \div 3 \cdot 2 + 4$ $= 25 - (3)^2 \div 3 \cdot 2 + 4$ $= 25 - 9 \div 3 \cdot 2 + 4$ $= 25 - 3 \cdot 2 + 4$ $= 25 - 6 + 4$ $= 19 + 4$ $= 23$
Determine Whether a Number Is Prime or Composite (Section 1.7)	**Step 1** List the prime numbers whose square is less than the number being checked. **Step 2** Check for divisibility by these prime numbers. If the number is divisible, it is composite. If it is divisible only by itself and 1, it is prime.	Is 59 prime or composite? $2^2 = 4$ $3^2 = 9$ $5^2 = 25$ $7^2 = 49$ $7^2 = 49$ ~~$11^2 = 121$~~ Check 2, 3, 5, 7 59 is prime because it is not divisible by 2, 3, 5, or 7.
Find a Prime Factorization (Section 1.7)	**Step 1** Write the number using any of its two-number factorizations. **Step 2** If any of those factors are composite, perform Step 1 on them. Repeat this process until every factor is prime.	Find the prime factorization of 90. Write any repeated factors using exponents. 90 → 18, 5; 18 → 9, 2; 5 → 5; 9 → [3], [3]; 2 → [2]; 5 → [5] $90 = 2 \cdot 3 \cdot 3 \cdot 5$ $90 = 2 \cdot 3^2 \cdot 5$
Find the Greatest Common Factor by Listing Factors (Section 1.8)	**Step 1** List the factors. **Step 2** List the common factors. **Step 3** Find the greatest of the common factors.	Find the greatest common factor of 12 and 30. 12: 1, [2], [3], 4, [6], 12 30: 1, [2], [3], 5, [6], 10, 15, 30 Because 6 is the largest factor in common, it is the greatest common factor of 12 and 30.

Procedure/Topic	Steps	Example
Find the Least Common Multiple by Listing Multiples (Section 1.8)	List multiples of each number and stop at the first common multiple.	Find the least common multiple of 6 and 4. 6: 6, [12], 18, [24], 30, [36], 42 . . . 4: 4, 8, [12], 16, 20, [24], 32, [36] . . . Because 12 is the smallest multiple in common, it is the least common multiple of 6 and 4.
Draw a Venn Diagram for the Factors of Two Numbers (Section 1.8)	**Step 1** Write the prime factorization of each number and circle each pair of common factors. **Step 2** Write the common factors where the circles overlap. **Step 3** Write the remaining factors in the corresponding circles.	Draw a Venn diagram for the prime factorizations of 18 and 30. $18 = 2 \cdot 3 \cdot 3$ $30 = 2 \cdot 3 \cdot 5$ Factors of 18: 3; Both: 2, 3; Factors of 30: 5
Find the GCF and LCM by Factoring (Section 1.8)	**Step 1** Write the prime factorization of each number. **Step 2** Draw a Venn diagram of the prime factorizations. **Step 3** The GCF is the product of the factors inside the overlap. **Step 4** The LCM is the product of all of the factors in the circles.	Find the GCF and LCM of 18 and 30. Using the Venn diagram above: • The GCF is 6, the product of 2 and 3. • The LCM is the product of all of the factors. $\text{LCM} = 2 \cdot 3 \cdot 3 \cdot 5$ $= 90$
Translate Words into Mathematics (Section 1.9)	**Step 1** Identify the terms that indicate an operation. **Step 2** Write a corresponding math exercise.	What is the product of 10 and 7? "Product" means multiplication. The product $= 10 \times 7$
Solve Word Exercises with One Operation (Section 1.9)	**Step 1** Read and understand the word exercise. **Step 2** Write a corresponding math exercise. Write an intermediate word equation if necessary. **Step 3** Solve the math. **Step 4** Interpret the result.	Balloon bunches cost \$5 each. If Jarred purchases 8 bunches of balloons, how much does he spend? $\$\text{Spent} = \text{Cost} \cdot \text{Bunches}$ $= 5 \times 8$ $= 40$ Jarred spends \$40 on balloons.
Solve Word Exercises with More Than One Operation	**Step 1** Read and understand the word exercise. **Step 2** Write a corresponding math exercise. Write an intermediate word equation if necessary. **Step 3** Solve the math. **Step 4** Interpret the result.	Jarred paid \$5 for each of 8 bunches of balloons. If he paid with a \$50 bill, what was his change? $\text{Change} = \$\text{ Paid} - \text{Cost}$ $= 50 - 5 \cdot 8$ $= 50 - 40$ $= 10$ Jarred received \$10 in change.

Chapter 1 Review Exercises

1.1

Answer each question

1. What is the place value of the digit 7 in 272,492?

2. What is the place value of the digit 2 in 2,953,973?

3. Write $1,098,342 in words, as you would on a check.

4. Write $34,908 in words, as you would on a check.

5. Write 30,402 in expanded form.

6. Write 105,030 in expanded form.

7. Round 938,412 to the nearest thousand.

8. Round 87 to the nearest hundred.

9. Graph the given set of numbers on a number line. Use the number line to order the numbers from greatest to least. $\{16, 32, 25, 43\}$

10. Graph the given set of numbers on a number line. Use the number line to order the numbers from least to greatest. $\{125, 80, 175, 50\}$

The bar graph shows the number of McDonald's restaurants within driving distance of Southfield, Michigan.

11. How many McDonald's are within 10 miles of Southfield? Round your answer to the nearest 10 restaurants.

12. If someone in Southfield wanted to visit 300 different McDonald's in one year, what is the longest distance he or she would need to travel to get to any one of the restaurants?

▶ **NOTE** By contrast, only 11 McDonald's are within 100 miles of Ontonogan, Michigan, and none are closer than 35 miles.

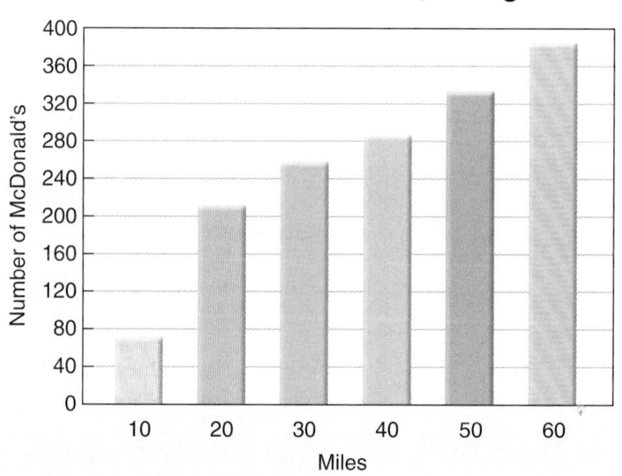

Source: www.foodio.com

1.2

Find each sum.

13. 45 + 13

14. 21 + 76

15. 127 + 45

16. 385 + 87

17. 1,548 + 531

18. 2,632 + 433

19. 8,382 + 7,152

20. 7,392 + 9,285

Dante is going to buy an engagement ring for his girlfriend, whose birthstone is the ruby. Use the bar graph to answer each question.

21. How much will a ring with a 0.4-carat ruby and a 0.2-carat diamond cost? Can Dante afford this ring if he has $750?

22. How much will a ring with a 0.4-carat diamond and a 0.2-carat ruby cost? Can Dante afford this ring if he has $1,000?

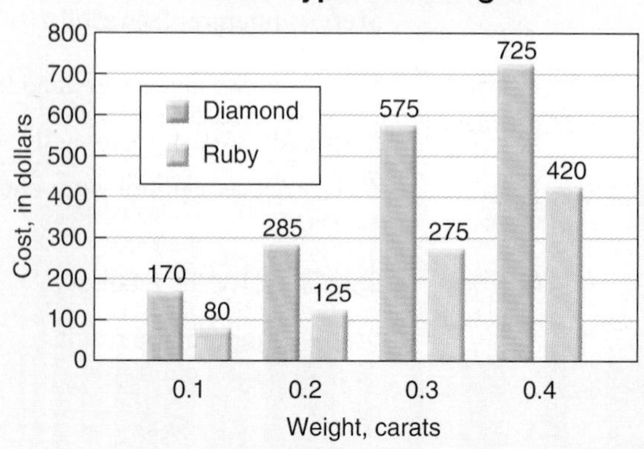

1.3

Find each difference

23. 75 − 43 **24.** 56 − 23 **25.** 172 − 38 **26.** 492 − 55

27. 1,275 − 183 **28.** 5,326 − 417 **29.** 8,057 − 3,265 **30.** 5,307 − 289

The bar graph shows how much money Elaine and Blair earned in the first through fourth years after graduating from college. Use the graph to answer each question.

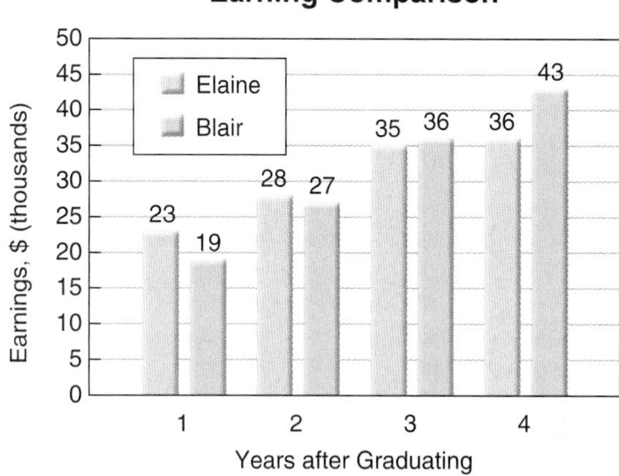

31. In what year was the difference in salaries the greatest? How much more/less did Elaine earn than Blair in that year?

32. What is the total amount that each person earned in the four years?

33. Who earned the most money in the four years? How much more did that person earn over those years?

34. Fill in the blanks. Doing so will help you recognize the similarities and differences between addition and subtraction.

Addition		Subtraction
When adding on a number line, start at the first number and move to the ____.	← similar →	When subtracting on a number line, start at the first number and move to the ____.
Sometimes we need to ____ a 1 to the next highest place value.	← similar →	Sometimes we need to ____ a 1 from the next highest place value.
To check an addition answer, we can ______.	← similar →	To check a subtraction answer, we can ____.

1.4

Estimate each product by rounding each factor to its largest place value. Then find the product.

35. $21 \cdot 41$ **36.** $32 \cdot 27$ **37.** $67 \cdot 24$

38. $38 \cdot 79$ **39.** $184 \cdot 53$ **40.** $285 \cdot 76$

Find each product.

41. $2 \cdot 6 \cdot 7 \cdot 5$ **42.** $8 \cdot 3 \cdot 7 \cdot 2$

Answer each question.

43. What is the area of each of the following rectangles? By how many square feet is the short rectangle larger/smaller than the tall rectangle?

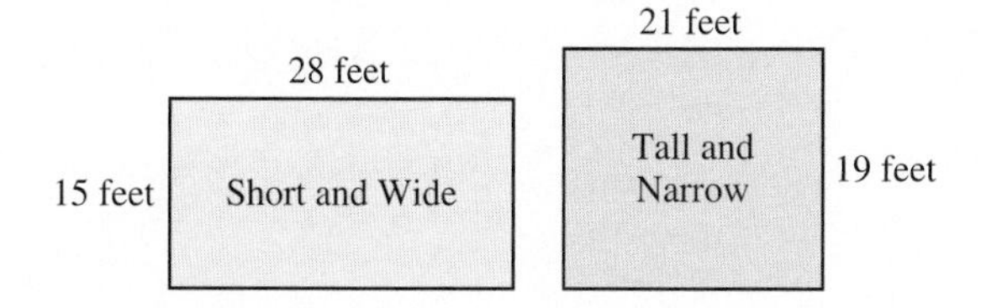

44. Alli has made 15 payments of $372 on her car. What is the total of her payments? If she paid $12,856 for the car, how much does she still owe?

45. In football, a touchdown is worth 6 points and a field goal is worth 3 points. How many points did the Bears score if they made 4 field goals and 3 touchdowns? Exclude extra points.

46. Carlos and Jan can rent a house at \$850 per month for 12 months. If they sign a 24-month lease, they can rent the house for \$817 per month. How much will Carlos and Jan save in the first 12 months if they sign a 24-month lease?

1.5

Find each quotient by solving the corresponding multiplication exercise.

47. $72 \div 9$

48. $56 \div 8$

Estimate each quotient. Then perform the division.

49. $73 \div 7$

50. $58 \div 3$

51. $171 \div 14$

52. $237 \div 18$

53. $6{,}572 \div 21$

54. $7{,}540 \div 36$

Answer each question.

55. A local lumberyard sells bundles of wood, each containing 200 two-by-fours. A carpenter builds sheds for homeowners. Each shed uses 85 two-by-fours. How many sheds can the carpenter build using the two-by-fours from four bundles? Hint: ignore any remainder.

56. At a grocery store, a typical checkout lane uses 85 bags in one hour. A bagger just stocked one lane with 3 bundles, each containing 100 bags. How many hours will pass before more bags are needed? Hint: ignore any remainder.

1.6

57. State the order of operations.

Simplify each expression, following the order of operations.

58. 2^3

59. 4^2

60. $2 + 3^3$

61. $100 - 7^2$

62. $15 - 3 + 2$

63. $21 \div 7 \cdot 3$

64. $(5 - 3)^2 + 5$

65. $25 - (8 - 7)^3$

66. $15 \div (8 - 3)$

67. $42 - (52 - 33) + 5$

68. $\dfrac{7 - (5 - 2)}{12 \div (8 - 2)}$

69. $\dfrac{40 - 5 \cdot 6}{4^2 - 11}$

Use the bar graph to answer each question.

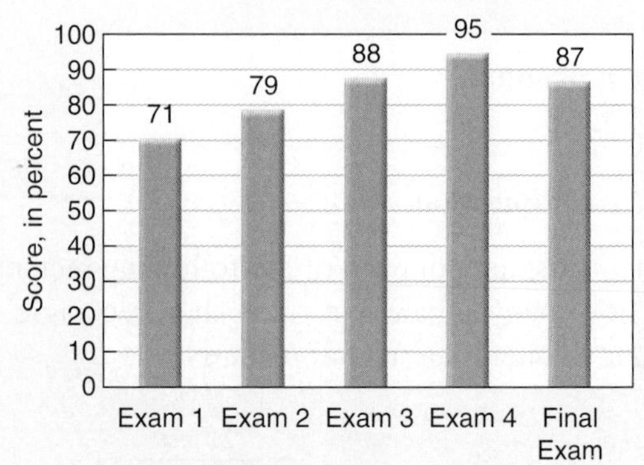

70. Estimate Alex's average exam score by drawing a horizontal line through the "middle" of the column tops.

71. Calculate Alex's average exam score. Use it to determine his grade from the following scale:

A = 93 to 100, A− = 90 to 92
B+ = 88 to 89, B = 83 to 87, B− = 80 to 82
C+ = 78 to 79, C = 73 to 77, C− = 70 to 72

72. By looking at the bar graph, how can you tell that it isn't necessary to list grades lower than C− in the scale?

1.7

Answer each question.

73. Is 39 prime or composite?

74. Is 41 prime or composite?

75. List all of the factor pairs of 20.

76. List all of the factor pairs of 36.

77. What is the prime factorization of 40?

78. What is the prime factorization of 66?

79. List the first 8 multiples of 6.

80. List the first 6 multiples of 8.

1.8

The numbers 33 and 198 are related to two other numbers. One is the greatest common factor, and the other is the least common multiple.

81. Which of the two numbers is the greatest common factor? How do you know?

82. Which of the two numbers is the least common multiple? How do you know?

83. Find the greatest common factor of 24 and 16 by listing factors.

84. Find the greatest common factor of 28 and 21 by listing factors.

85. Find the least common multiple of 12 and 15 by listing multiples.

86. Find the least common multiple of 12 and 9 by listing multiples.

Find the least common multiple and greatest common factor of each pair of numbers. Use a Venn diagram and the factoring technique.

87. 20 and 24

88. 32 and 28

89. 20 and 40

90. 14 and 15

1.9

Answer each question.

91. What is the product of 16 and 24?

92. What is the quotient of 256 and 64?

93. The quotient of 72 and 8 is increased by 17. What is the result?

94. The product of 8 and 7 is decreased by 30. What is the result?

95. Angela was checking out of the grocery store. Before the clerk rang up 3 boxes of granola bars, Angela's total was $76. After the clerk rang up the granola bars, the total was $97. Angela thinks that the final total is incorrect. How much was she charged for each box?

96. In the last three months, Par had three equal car payments deducted from her savings account. Her account had a starting balance of $2,598 and an ending balance of $1,944. What was the amount of each payment?

97. A farmer built 5 windmills on his property. In one day, the windmills produced 115, 118, 117, 126, and 99 kilowatt hours of electricity. What is the average output for each windmill? For the farmer to break even on his investment, the windmills must average 110 kilowatt hours per day. Is the farmer making or losing money?

98. A new subdivision will purchase electricity from a local wind farm. A typical house in the subdivision will use 12 kilowatt hours per day. If 73 houses are in the subdivision, how much electricity is needed for the subdivision? If the wind farm can generate only 600 kilowatt hours per day, how much more electricity will the subdivision need?

Chapter 1 Practice Test

Answer each question.

1. What is the place value of the digit 8 in 263,876,276?

2. Write \$60,809 in words, as you would on a check.

3. Round 827,398 to the nearest ten-thousand.

4. Write 107,060 in expanded form.

5. Graph $\{23, 30, 10, 18\}$ on a number line.

6. Using the bar graph, identify the largest gem that Dante can purchase for \$300. Your answer should state the type of gem and its weight.

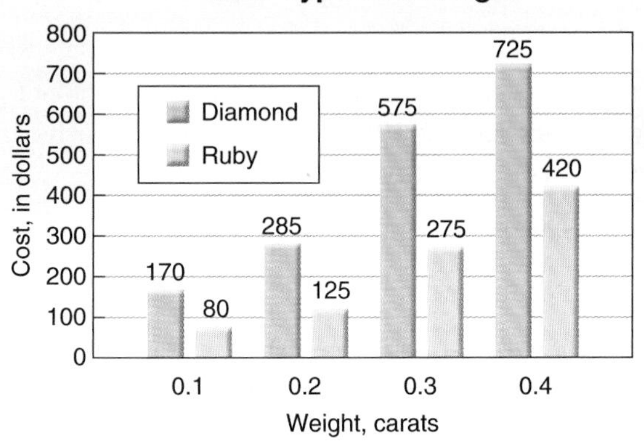

Find each sum or difference.

7. 137 + 42

8. 293 − 52

9. 475 − 186

10. 398 + 275

11. 4,235 + 783

12. 7,032 − 671

13. 5,045 − 352

14. 6,956 + 372

The following bar graph shows how much money a company's CEO was paid each year. Use the information in the bar graph to answer each question.

15. The CEO was the same person in 2004, 2005, and 2006. How much money, in total, did that CEO earn during those three years?

16. By how much did the CEO's earnings increase from 2004 to 2006?

17. In 2007, the company hired a new CEO. How much more did the new CEO earn than the previous CEO earned in 2006?

18. On average, how much money did the company pay its CEO over these four years?

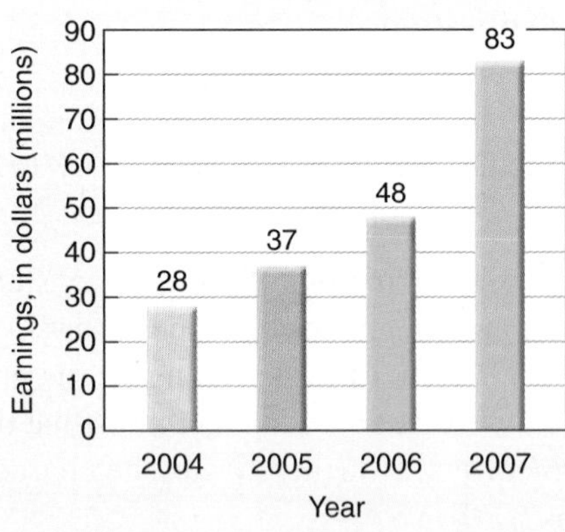

Estimate each product or quotient. Then find the exact answer.

19. $48 \cdot 72$

20. $423 \div 11$

21. $392 \div 18$

22. $162 \cdot 67$

23. $5{,}723 \cdot 492$

24. $6{,}834 \div 52$

Answer each question.

25. Every square inch of an inflatable jack exerts 15 pounds of force on the bottom of a truck. If the jack makes 190 square inches of contact with the truck, how much force (in pounds) is acting on the truck?

26. An inflatable jack is lifting a 3,096-pound truck. The jack creates 18 pounds of force for every square inch of contact it makes with the truck. How many square inches of contact is the jack making?

27. An author gave a public reading of her work. After the reading, she sold copies of her book for $28 each. If she sold a total of $896 in books, how many books did she sell?

28. Recently, 48 people attended an author reading. 28 of those who attended decided to purchase a signed copy of the book. If each book sold for $19, how much income did the author earn?

Simplify each expression, following the order of operations.

29. $3^4 - 31$

30. $53 - 24 + 6$

31. $3 + 2 \cdot 5$

32. $40 \div 2 \cdot 5$

33. $2 \cdot (5 - 3)^2$

34. $\dfrac{24 - 4 \cdot 2}{2^3}$

Answer each question.

35. What are all of the factor pairs of 40?

36. What is the prime factorization of 42?

37. What is the least common multiple of 25 and 15?

38. What is the greatest common factor of 30 and 20?

Use a Venn diagram to find the least common multiple and greatest common factor for each pair of numbers.

39. 42 and 24

40. 16 and 28

Answer each question.

41. Jarett purchased 20 microphones for $18 each. If he had $583 in his expense account before his purchase, how much is left in his account?

42. Al's truck gets 15 miles per gallon when not towing a load and 10 miles per gallon when towing a load. He drove 75 miles to get a load. He then drove 100 miles towing the load. How many gallons of gas did his truck use?

CHAPTER

2 Fractions

Fractions are numbers that represent portions of a whole. This chapter is designed to help you understand what a fraction represents and how to add, subtract, multiply and divide fractions.

Dave Grohl of the Foo Fighters performed for the 2007 Live Earth concert at Wembley Stadium in London. Live Earth was a series of concerts that included politicians, musicians, and celebrities drilling home the dangers of global warming and urging people to "go green." To hold this concert, a portion of the stadium seats had to be blocked off.

In Exercise 63, Section 2.1, we will discover the fraction of seats available in Wembley Stadium for this concert.

Section 2.1 Visualizing Fractions

In this section, you will visualize fractions using the **part–whole concept** of a fraction. This will allow you to understand fractions and learn about their relative sizes.

The **Objectives** in Section 2.1 will help you

- A Represent a picture as a fraction.
- B Represent a fraction as a picture.
- C Relate improper fractions to their pictures.
- D Compare fractions by visualizing them.
- E Compare fractions with the same numerator or denominator.

VOCABULARY PREVIEW *Check the box that applies.*	Got It	Must Study
Fraction: A **fraction** is a number that represents a portion of a whole quantity.		
Numerator: The **numerator** is the top number in a fraction. It indicates how many equal-sized pieces are shaded. The numerator of $\frac{3}{5}$ is 3.		
Denominator: The **denominator** is the bottom number in a fraction. It indicates the total number of equal-sized pieces in one whole object. The denominator of $\frac{3}{5}$ is 5.		
Fraction Bar: A **fraction bar** is the line that is drawn between the numerator and the denominator. A fraction bar indicates division.		
Part–Whole Concept: The **part–whole concept** of a fraction states that a fraction such as $\frac{3}{8}$ represents three parts of a whole that has been divided into eight equal-sized pieces.		
Proper Fraction: A **proper fraction** has a numerator that is less than the denominator. The fraction $\frac{3}{5}$ is proper because 3 is less than 5.		
Improper Fraction: An **improper fraction** has a numerator that is equal to or greater than the denominator. The fraction $\frac{5}{3}$ is improper because 5 is greater than 3.		

Study these words when they appear in the text. After the section, test your understanding by completing the Vocabulary Review in the section exercises.

Objective A Represent a Picture as a Fraction

The Concept A **fraction** is a number that represents a portion of a whole quantity. Visualizing a fraction will help you understand the portion of a whole that is being represented. Figure 2.1 shows a visual comparison of some commonly used fractions.

1 ONE WHOLE									
$\frac{1}{2}$	$\frac{1}{2}$								
$\frac{1}{3}$	$\frac{1}{3}$	$\frac{1}{3}$							
$\frac{1}{4}$	$\frac{1}{4}$	$\frac{1}{4}$	$\frac{1}{4}$						
$\frac{1}{5}$	$\frac{1}{5}$	$\frac{1}{5}$	$\frac{1}{5}$	$\frac{1}{5}$					
$\frac{1}{6}$	$\frac{1}{6}$	$\frac{1}{6}$	$\frac{1}{6}$	$\frac{1}{6}$	$\frac{1}{6}$				
$\frac{1}{8}$	$\frac{1}{8}$	$\frac{1}{8}$	$\frac{1}{8}$	$\frac{1}{8}$	$\frac{1}{8}$	$\frac{1}{8}$	$\frac{1}{8}$		
$\frac{1}{10}$	$\frac{1}{10}$	$\frac{1}{10}$	$\frac{1}{10}$	$\frac{1}{10}$	$\frac{1}{10}$	$\frac{1}{10}$	$\frac{1}{10}$	$\frac{1}{10}$	$\frac{1}{10}$

Figure 2.1

From the row showing eighths in Figure 2.1, we can build fractions such as $\frac{3}{8}$, shown in Figure 2.2. There are three shaded portions. Each portion is $\frac{1}{8}$ of the whole. Therefore, the figure shows $\frac{3}{8}$ of a whole.

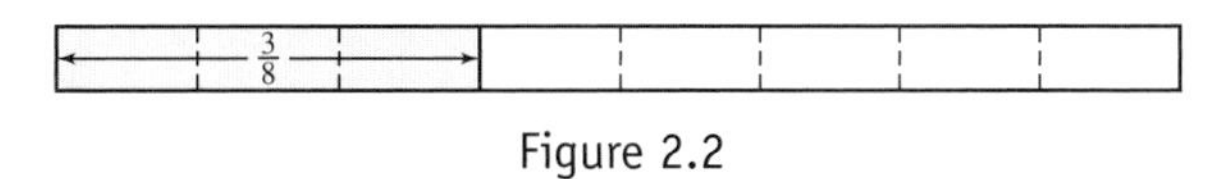

Figure 2.2

When working with fractions, it is important to understand the words *numerator* and *denominator*.

INTERACTIVE DEFINITION Numerator and Denominator

The **numerator** is the top number in a fraction. It indicates how many equal-sized pieces are shaded.

The **denominator** is the bottom number in a fraction. It indicates the total number of equal-sized pieces in one whole object.

EXAMPLE

1. What fraction is represented below?

There are five shaded pieces.
The numerator is 5.

There are six pieces in all.
The denominator is 6.

The fraction is $\frac{\text{numerator}}{\text{denominator}}$ or $\frac{5}{6}$.

GUIDED PRACTICE

1. What fraction is represented below?

There are ______ shaded pieces.
The numerator is ______.

There are ______ pieces in all.
The denominator is ______.

The fraction is $\frac{\text{numerator}}{\text{denominator}}$ or ——

DO YOU UNDERSTAND how to determine the numerator from a picture?	Got It	Get Help
DO YOU UNDERSTAND how to determine the denominator from a picture?	Got It	Get Help

▶ **NOTE** A denominator cannot be zero. If you see a fraction such as $\frac{3}{0}$, the fraction is undefined. Try drawing a whole object with zero pieces!

Procedure Represent a Picture as a Fraction

Step 1 The denominator is the total number of equal-sized parts in one whole object.

Step 2 The numerator is the number of shaded parts in one whole object.

EXAMPLES

2. Represent the picture as a fraction.

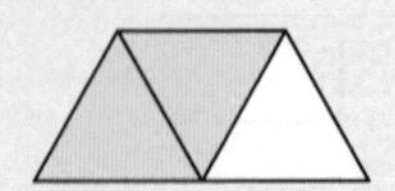

Step 1 Since there are three equal-sized parts in all, the denominator is 3.

$$\text{Fraction} = \frac{}{3}$$

Step 2 Since there are two shaded parts, the numerator is 2.

$$\text{Fraction} = \frac{2}{3}$$

3. Using the chart, what fraction of a work day does Jain spend on the phone?

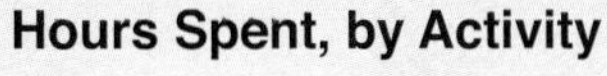

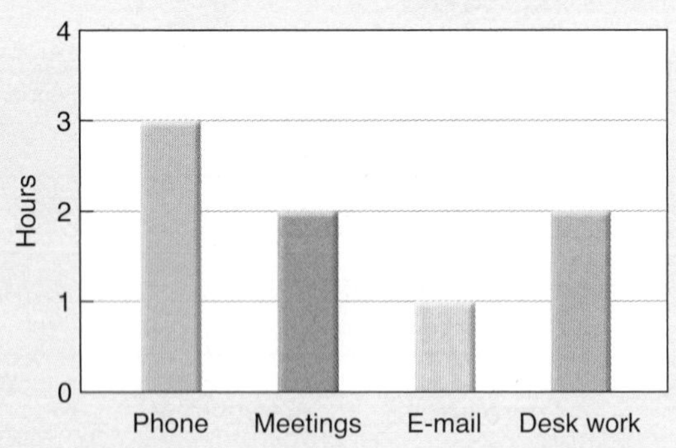

Step 1 The denominator is the total number of hours in a work day.

$$\text{Denominator} = 3 + 2 + 1 + 2$$
$$= 8$$

Step 2 The numerator is the number of phone hours spent in a work day.

$$\text{Numerator} = 3$$

$$\text{Fraction} = \frac{\text{Numerator}}{\text{Denominator}}$$
$$= \frac{3}{8}$$

GUIDED PRACTICE

2. Represent the picture as a fraction.

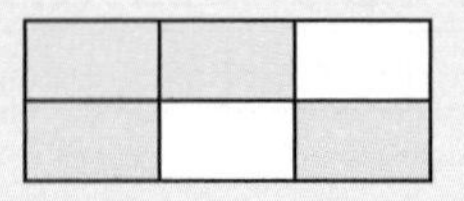

Step 1 Since there are ______ equal-sized parts in all, the denominator is ______.

$$\text{Fraction} = \frac{}{}$$

Step 2 Since there are ______ shaded parts, the numerator is ______.

$$\text{Fraction} = \frac{}{}$$

3. Using the chart, what fraction of a school day does Xailu spend reading?

Hours Spent, by Activity

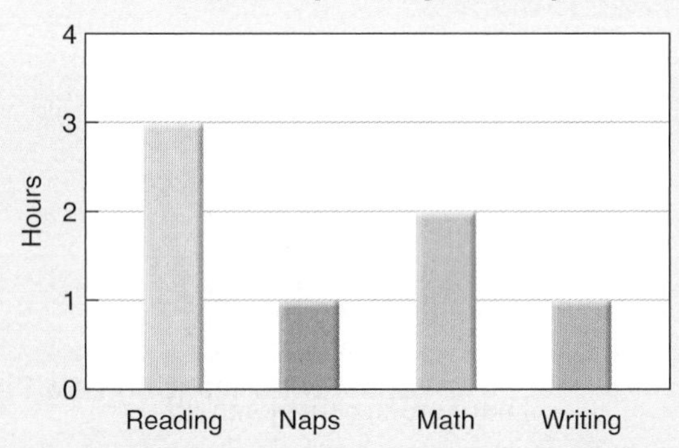

Step 1 The denominator is the total number of hours in a school day.

$$\text{Denominator} = \quad + \quad + \quad +$$
$$=$$

Step 2 The numerator is the number of ______ hours spent in a school day.

$$\text{Numerator} =$$

$$\text{Fraction} = \frac{\text{Numerator}}{\text{Denominator}}$$
$$= \frac{}{}$$

Concept Check

A1. For the fraction $\frac{2}{7}$, the numerator is ____ and the denominator is ____ .

A2. When you represent a picture as a fraction, the number of equal-sized pieces in a whole becomes the __________ and the number of shaded pieces becomes the __________.

Objective A Practice

Represent each picture as a fraction.

A3.

A4.

A5.

A6.

A7.

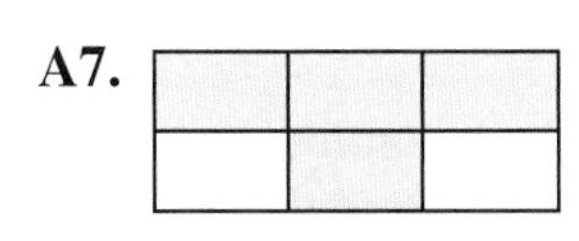

A8.

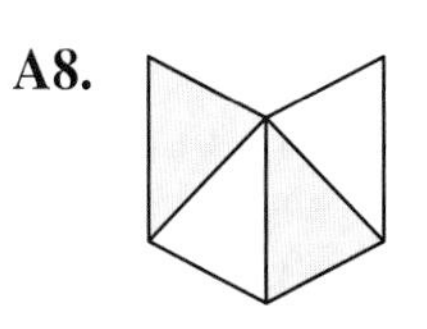

A9. Using the chart, what fraction of the clothes in Jan's closet are shoes?

Clothes in Jan's Closet

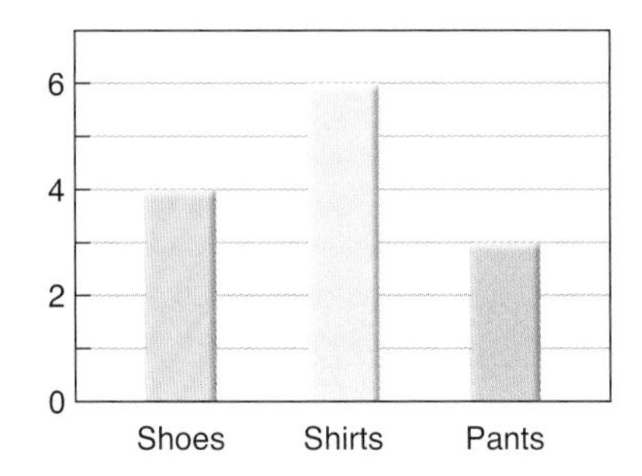

A10. Using the chart, what fraction of the bags of fruit in Alana's cart are grapes?

Bags of Fruit in Cart

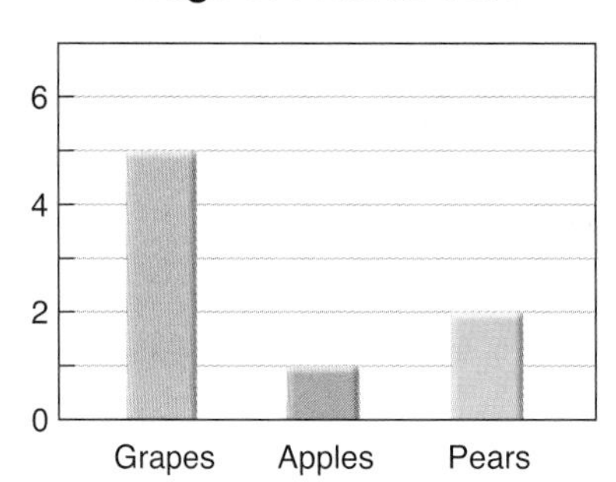

Objective B **Represent a Fraction as a Picture**

The Concept It is useful to visualize the portion of a whole that a fraction represents. The denominator indicates the total number of parts in one whole. The numerator indicates how many of those parts are represented.

> **Procedure Represent a Fraction as a Picture**
>
> **Step 1** Draw a shape with equal-sized parts matching the number in the denominator.
>
> **Step 2** Shade as many parts as the number in the numerator.

EXAMPLE

4. Represent $\frac{5}{7}$ as a picture.

Step 1 Since the denominator = 7, draw an object with 7 equal-sized parts.

Step 2 Since the numerator = 5, shade 5 of those parts.

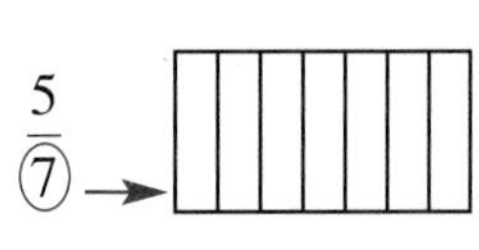

GUIDED PRACTICE

4. Represent $\frac{3}{8}$ as a picture.

Step 1 Since the denominator = ____, draw an object with ____ equal-sized parts.

Step 2 Since the numerator = ____, shade ____ of those parts.

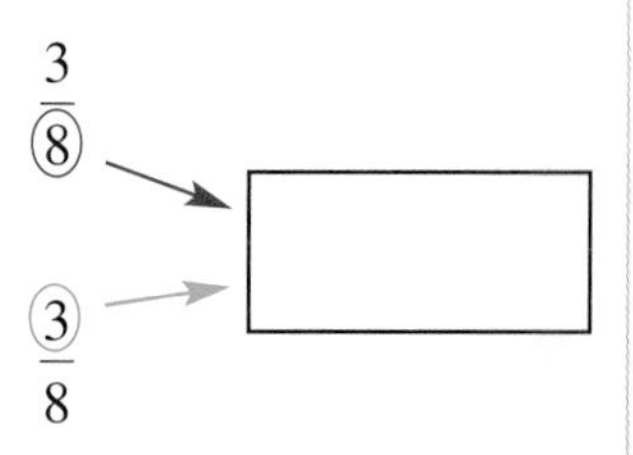

Concept Check

B1. When representing a fraction as a picture, what does the numerator indicate?

B2. When representing a fraction as a picture, what does the denominator indicate?

Objective B Practice

Match each fraction with its corresponding picture.

B3. $\frac{3}{6}$

B4. $\frac{5}{6}$

B5. $\frac{3}{5}$

B6. $\frac{6}{7}$

a. 

b.

c.

d.

Represent each fraction as a picture.

B7. $\frac{2}{3}$ **B8.** $\frac{5}{8}$ **B9.** $\frac{0}{4}$ **B10.** $\frac{0}{2}$

Draw a picture for each statement below. Have fun with the drawings and don't worry about artistic quality.

B11. There are five cowboys. Four-fifths of them are wearing cowboy hats.

B12. Four vehicles are at a gas station. Three-fourths of them are pickup trucks.

Objective C Relate Improper Fractions to Their Pictures

The Concept In Objectives A and B, we worked with proper fractions. **Proper fractions** have numerators that are less than the denominators. In this objective, we will work with **improper fractions**.

INTERACTIVE DEFINITION Improper Fractions

An **improper fraction** has a numerator that is equal to or greater than the denominator.

EXAMPLE

5. Identify each picture or fraction as a proper fraction or an improper fraction.

a.

b.

c. $\frac{3}{7}$

d.

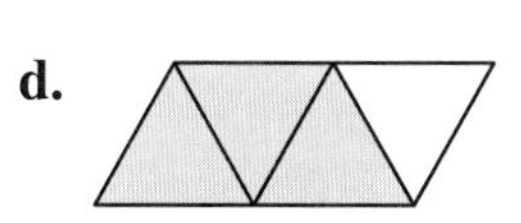

e. $\frac{9}{5}$

f. $\frac{4}{4}$

Proper Fractions
Fraction c is proper because the numerator is less than the denominator.

Picture d is a proper fraction because less than one whole object is shaded.

Improper Fractions
Fractions e and f are improper because the numerator is equal to or greater than the denominator.

Pictures a and b show improper fractions because at least one whole object is shaded.

GUIDED PRACTICE

5. Identify each picture or fraction as a proper fraction or an improper fraction.

a. $\frac{5}{8}$

b.

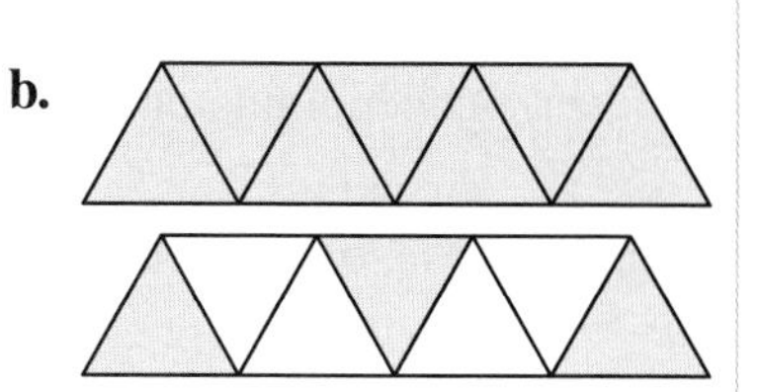

c.

d. $\frac{13}{13}$

e.

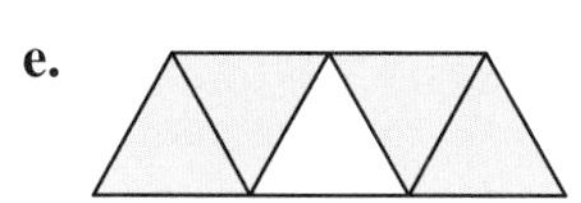

f. $\frac{14}{3}$

Proper Fractions
Fraction _____ is proper because the numerator is _______ than the denominator.

Picture _____ is a proper fraction because _____ than one whole object is shaded.

Improper Fractions
Fractions _____ and _____ are improper because the numerator is _______ or _________ the denominator.

Pictures _____ and _____ show improper fractions because ______ one ______ object is shaded.

DO YOU UNDERSTAND how to identify improper fractions? Got It Get Help

To represent a fraction that has a numerator larger than the denominator, we shade more parts than can be drawn using a single object. Therefore, we draw more than one object.

Procedure Relate Improper Fractions to Their Pictures

Step 1 Match the denominator with the number of equal-sized parts in one whole object.

Step 2 Match the numerator with the total number of shaded parts.

EXAMPLES

6. Represent the picture as an improper fraction.

Step 1 Since there are 5 equal-sized parts in each object, the denominator is 5. Fraction $= \frac{\quad}{5}$

Step 2 Since there are 8 shaded parts, the numerator is 8. Fraction $= \frac{8}{5}$

7. Represent $\frac{9}{4}$ as a picture.

Step 1 Since the denominator is 4, we draw objects with 4 equal-sized parts.

Step 2 The numerator is 9, so we must shade 9 parts.

$\frac{9}{4} =$ 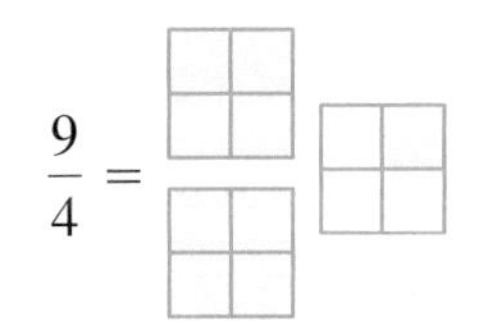

A total of 3 whole objects must be drawn to shade 9 parts.

GUIDED PRACTICE

6. Represent the picture as an improper fraction.

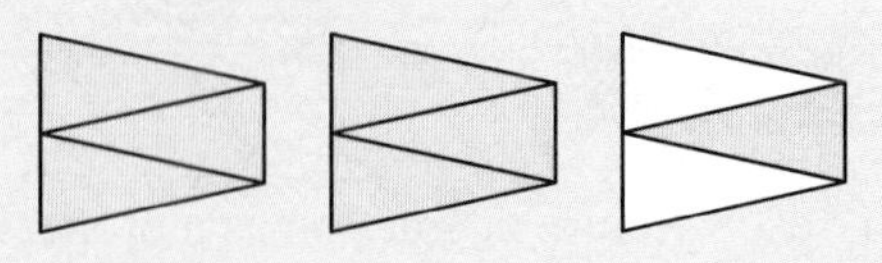

Step 1 Since there are _____ equal-sized parts in each object, the denominator is _____. Fraction $= \frac{\quad}{\quad}$

Step 2 Since there are _____ shaded parts, the numerator is _____. Fraction $= \frac{\quad}{\quad}$

7. Represent $\frac{11}{5}$ as a picture.

Step 1 Since the denominator is _____, we draw objects with _____ equal-sized parts.

Step 2 The numerator is _____, so we must shade _____ parts.

$\frac{11}{5} =$

A total of _____ whole objects must be drawn to shade _____ parts.

Concept Check

C1. Circle the phrase that makes the following sentence correct.
Since an improper fraction can represent more than one whole object, we may need to draw (more than)/(less than) one object to represent the fraction as a picture.

C2. What is true of an improper fraction that can be drawn with exactly one object?

Objective C Practice

Match each improper fraction with its corresponding picture.

C3. $\frac{6}{3}$ **C5.** $\frac{5}{5}$

C4. $\frac{6}{5}$ **C6.** $\frac{6}{1}$

a.

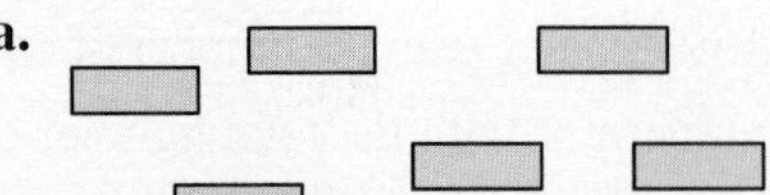

c.

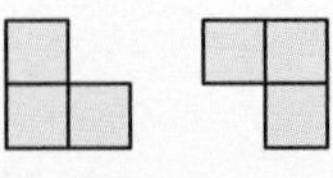

b.

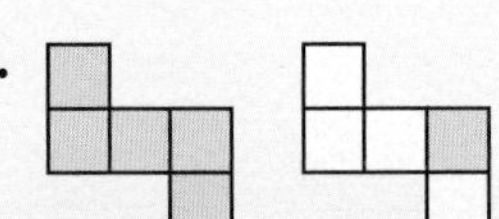

d.

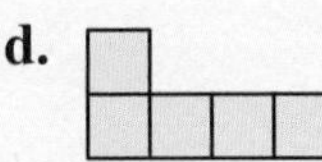

Represent each improper fraction as a picture.

C7. $\frac{4}{2}$ **C8.** $\frac{10}{5}$ **C9.** $\frac{7}{5}$ **C10.** $\frac{5}{3}$

Represent each picture as a fraction. State whether the fraction is proper or improper.

C11.

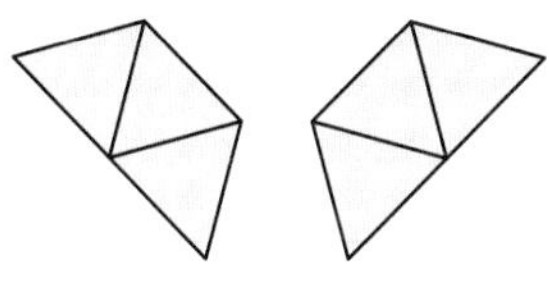

C12.

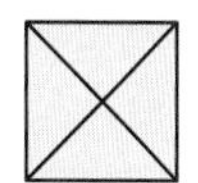 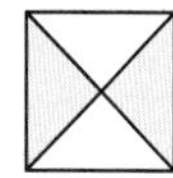

C13. What fraction of the ice cream is vanilla?

C14. What fraction of the milk is chocolate?

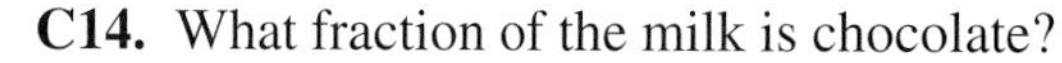

C15. What is wrong with trying to draw a picture that represents $\frac{2}{0}$?

Objective D Compare Fractions by Visualizing Them

The Concept Being able to visualize a fraction is a skill that can help you understand how large one fraction is relative to another. When you go to the store, you may see signs saying "Sale! $\frac{1}{3}$ off!" or "Sale! $\frac{1}{4}$ off!" Being able to compare fractions will help you understand that a "$\frac{1}{3}$ off" sale is a better value than a "$\frac{1}{4}$ off" sale.

The chart in Figure 2.3 can be used to quickly compare the sizes of some common fractions.

1 ONE WHOLE									
$\frac{1}{2}$	$\frac{1}{2}$								
$\frac{1}{3}$	$\frac{1}{3}$	$\frac{1}{3}$							
$\frac{1}{4}$	$\frac{1}{4}$	$\frac{1}{4}$	$\frac{1}{4}$						
$\frac{1}{5}$	$\frac{1}{5}$	$\frac{1}{5}$	$\frac{1}{5}$	$\frac{1}{5}$					
$\frac{1}{6}$	$\frac{1}{6}$	$\frac{1}{6}$	$\frac{1}{6}$	$\frac{1}{6}$	$\frac{1}{6}$				
$\frac{1}{7}$	$\frac{1}{7}$	$\frac{1}{7}$	$\frac{1}{7}$	$\frac{1}{7}$	$\frac{1}{7}$	$\frac{1}{7}$			
$\frac{1}{8}$	$\frac{1}{8}$	$\frac{1}{8}$	$\frac{1}{8}$	$\frac{1}{8}$	$\frac{1}{8}$	$\frac{1}{8}$	$\frac{1}{8}$		
$\frac{1}{10}$	$\frac{1}{10}$	$\frac{1}{10}$	$\frac{1}{10}$	$\frac{1}{10}$	$\frac{1}{10}$	$\frac{1}{10}$	$\frac{1}{10}$	$\frac{1}{10}$	$\frac{1}{10}$

Figure 2.3

Procedure Compare Two or More Fractions

Visualize the picture of each fraction and then compare their sizes.

EXAMPLES

8. Which is larger, $\frac{2}{3}$ or $\frac{3}{5}$?

Use the chart to create a visual representation of each fraction.

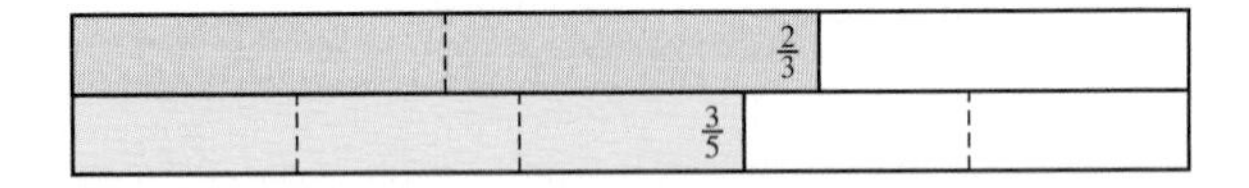

Since $\frac{2}{3}$ represents a larger portion of a whole, $\frac{2}{3}$ is larger than $\frac{3}{5}$.

9. Graph $\frac{3}{5}, \frac{2}{7}$, and $\frac{1}{8}$ on a number line.

Use the chart to create a visual representation of each fraction.

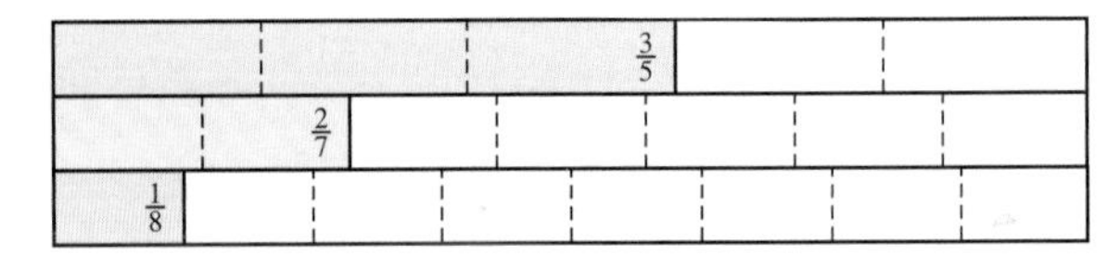

Order the fractions from smallest to largest.

$$\frac{1}{8}, \frac{2}{7}, \text{ and } \frac{3}{5}$$

Graph the fractions on the same number line.

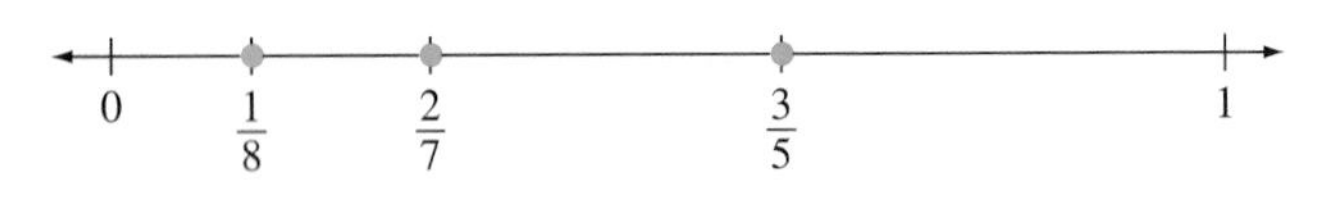

GUIDED PRACTICE

8. Which is larger, $\frac{3}{4}$ or $\frac{4}{7}$?

Use the chart to create a visual representation of each fraction.

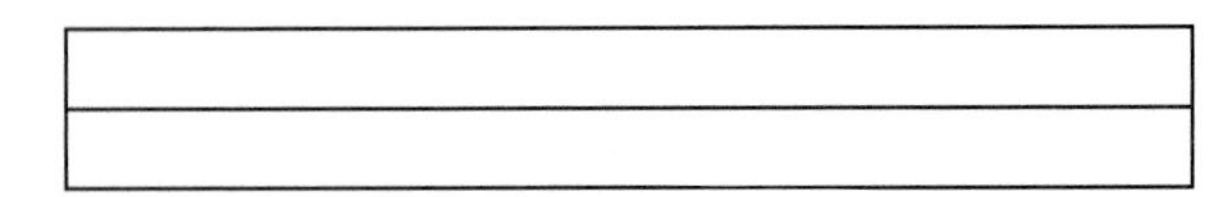

Since ―― represents a larger portion of a whole, ―― is larger than ――.

9. Graph $\frac{5}{8}, \frac{4}{6}$, and $\frac{3}{4}$ on a number line.

Use the chart to create a visual representation of each fraction.

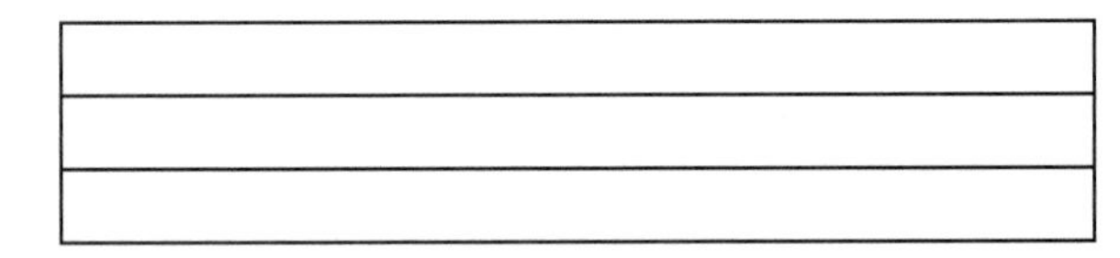

Order the fractions from smallest to largest.

Graph the fractions on the same number line.

Concept Check

D1. What will the picture of a larger fraction look like when compared to the picture of a smaller fraction of the same object?

Objective D Practice

Use Figure 2.3 on the previous page to determine which of the fractions is larger.

D2. $\frac{1}{7}, \frac{1}{5}$

D3. $\frac{1}{2}, \frac{4}{6}$

D4. $\frac{5}{8}, \frac{3}{4}$

D5. $\frac{5}{7}, \frac{6}{8}$

Order the fractions from smallest to largest and graph them on a number line.

D6. $\frac{1}{3}, \frac{1}{4}, \frac{1}{5}$

D7. $\frac{3}{4}, \frac{2}{3}, \frac{5}{7}$

D8. $\frac{3}{7}, \frac{3}{6}, \frac{3}{5}, \frac{3}{8}$

D9. $\frac{2}{4}, \frac{2}{6}, \frac{2}{5}, \frac{2}{7}$

Objective E Compare Fractions with the Same Numerator or Denominator

The Concept Using a chart or pictures to compare fractions is a great way to determine when one fraction is greater than another. However, most people don't carry fraction charts with them wherever they go. It helps to know other techniques to compare fractions. Below we compare fractions that have the same denominator or the same numerator.

▶ **NOTE** Even though people don't carry fraction charts with them, they can represent two fractions as pictures to compare them.

Comparing Fractions with the Same Denominator

When two fractions have the same denominator, their pictures will have pieces that are the same size. The greater fraction will be the one with more pieces. Which is greater, $\frac{5}{8}$ or $\frac{3}{8}$?

$\frac{5}{8}$ →	1	2	3	4	5			
$\frac{3}{8}$ →	1	2	3					

← **A larger numerator indicates more pieces.**

Each fraction is made of eighths. The fraction $\frac{5}{8}$ is greater than $\frac{3}{8}$ since it has more of the same-sized pieces.

Comparing Fractions with the Same Numerator

When two fractions have the same numerator, their pictures will have the same number of shaded pieces. The greater fraction will be the one whose pieces are larger in size. Which is greater, $\frac{5}{8}$ or $\frac{5}{6}$?

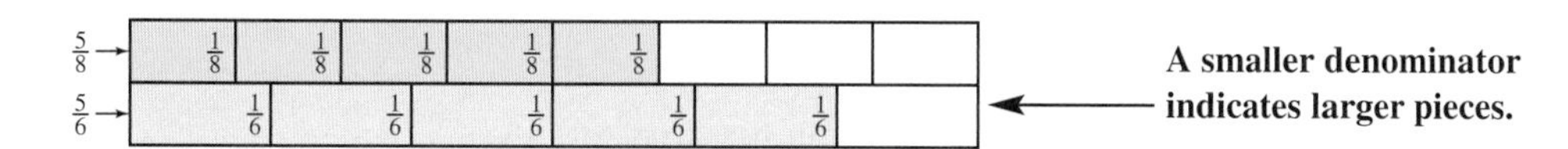

← **A smaller denominator indicates larger pieces.**

Both fractions have five shaded pieces. The fraction $\frac{5}{6}$ is greater than $\frac{5}{8}$ because sixths are larger than eighths.

To help us compare and order fractions, we will use the inequality symbols $<$ and $>$.

INTERACTIVE DEFINITION Inequality Symbols

The symbol $<$ means "is less than."

The symbol $>$ means "is greater than."

EXAMPLE

10. Use $<$ or $>$ to translate the written sentence into a mathematical sentence.

a. One-fourth *is less than* one-half.

$$\frac{1}{4} \boxed{<} \frac{1}{2}$$

An inequality symbol always opens toward the larger number.

b. Three-fifths *is greater than* two-fifths.

$$\frac{3}{5} \boxed{>} \frac{2}{5}$$

GUIDED PRACTICE

10. Use $<$ or $>$ to translate the written sentence into a mathematical sentence.

a. One-third *is less than* one-half.

$$\frac{1}{3} \square \frac{1}{2}$$

b. Two-sixths *is greater than* one-sixth.

$$\frac{2}{6} \square \frac{1}{6}$$

DO YOU UNDERSTAND how to use inequality symbols? Got It Get Help

Procedure Compare Fractions with the Same Numerator or Denominator

Visualize the fractions.

If the denominators are the same, the fraction with the larger numerator (more pieces) is greater.

If the numerators are the same, the fraction with the smaller denominator (larger pieces) is greater.

EXAMPLES

11. Which is greater, $\frac{7}{8}$ or $\frac{7}{10}$?

Visualize the fractions: Since the numerators are the same, the fraction with larger pieces is greater.

Since 8ths are larger than 10ths,

$$\frac{7}{8} > \frac{7}{10}.$$

12. Which is less, $\frac{10}{3}$ or $\frac{8}{3}$?

Visualize the fractions: Since the denominators are the same, the fraction with fewer pieces is less than the other.

$$\frac{8}{3} < \frac{10}{3}$$

GUIDED PRACTICE

11. Which is greater, $\frac{4}{6}$ or $\frac{4}{5}$?

Visualize the fractions: Since the numerators are the same, the fraction with ________________ is greater.

Since ______ths are larger than ______ths,

$$\frac{4}{\_\_} > \frac{4}{\_\_}.$$

12. Which is less, $\frac{4}{5}$ or $\frac{3}{5}$?

Visualize the fractions: Since the denominators are the same, the fraction with ________________ is less than the other.

$$\frac{\_\_}{5} < \frac{\_\_}{5}$$

Concept Check

E1. When comparing fractions that have the same denominator, why does a larger numerator indicate a larger fraction? Draw two pictures to demonstrate.

E2. When comparing fractions that have the same numerator, why does a smaller denominator indicate a larger fraction? Draw two pictures to demonstrate.

Objective E Practice

Use the symbols > or < to indicate the relative sizes of the fractions.

E3. $\frac{9}{7} \square \frac{10}{7}$ **E4.** $\frac{3}{3} \square \frac{2}{3}$ **E5.** $\frac{15}{31} \square \frac{15}{19}$ **E6.** $\frac{21}{52} \square \frac{21}{25}$

E7. $\frac{6}{2} \square \frac{6}{4}$ **E8.** $\frac{7}{10} \square \frac{7}{4}$ **E9.** $\frac{8}{90} \square \frac{10}{90}$ **E10.** $\frac{70}{120} \square \frac{20}{120}$

Combining Concepts and Applications

CONCEPT I Recognizing Fractions in Application Exercises Many exercises in this text will require you to translate a word statement into a fraction. To correctly translate the words in an exercise into the numerator and denominator of a fraction, use the part–whole concept of fractions.

Procedure Translate Word Statements That Contain Fractions

Identify the part and whole in the word statement. $\frac{\text{Part}}{\text{Whole}} = \frac{\text{Numerator}}{\text{Denominator}}$

EXAMPLES

13. Sarah answered 17 out of 20 questions correctly on her American history exam. What fraction of the questions did she answer correctly?

Identify the part and whole: 17 is the part of the test answered correctly, and 20 is the number of questions on the whole test.

$$\frac{\text{Part}}{\text{Whole}} = \frac{\text{Numerator}}{\text{Denominator}} = \frac{17}{20}$$

14. There are 62 female students and 55 male students in a class. What fraction of the students in the class are female?

Identify the part and whole: 62 is the part of students that are female. $62 + 55 = 117$ is the number of students in the whole class.

$$\frac{\text{Part}}{\text{Whole}} = \frac{\text{Numerator}}{\text{Denominator}} = \frac{62}{117}$$

GUIDED PRACTICE

13. Luigi washed 8 of the 13 cars in a lot. What fraction of the cars did he wash?

Identify the part and whole: ____ is the part of the cars that have been washed, and ____ is the number of cars in the whole lot.

$$\frac{\text{Part}}{\text{Whole}} = \frac{\text{Numerator}}{\text{Denominator}} = \frac{\quad}{\quad}$$

14. There are 25 sedans and 38 SUVs in a car lot. What fraction of the cars are sedans?

Identify the part and whole: ____ is the part of the cars that are sedans. ____ + ____ = ____ is the number of cars in the whole lot.

$$\frac{\text{Part}}{\text{Whole}} = \frac{\text{Numerator}}{\text{Denominator}} = \frac{\quad}{\quad}$$

CONCEPT II Using a Ruler Rulers for measuring inches are usually made by repeatedly cutting 1 inch in half. This process is detailed below.

Not to scale.

A whole cut into halves.

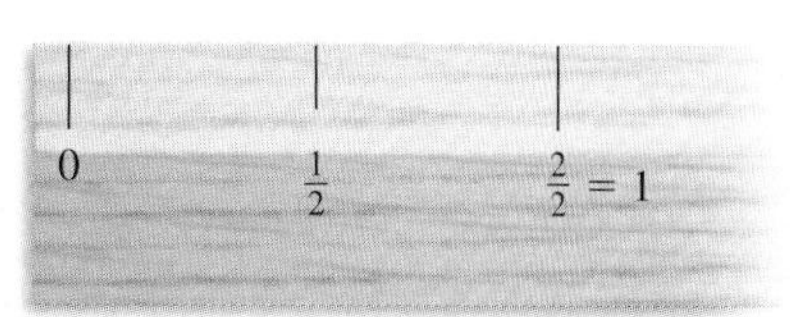

Not to scale.

Then each half is cut in half to show fourths.

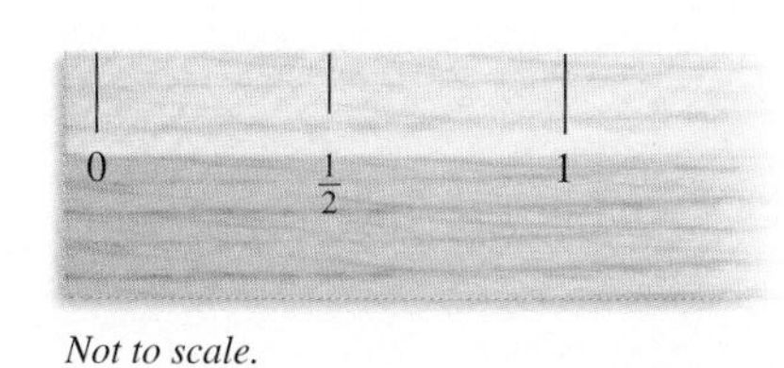

Not to scale.

Halves cut into fourths.

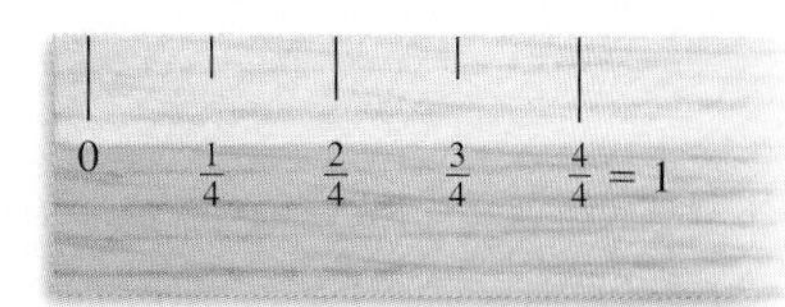

Not to scale.

This is repeated to show eighths and sixteenths.

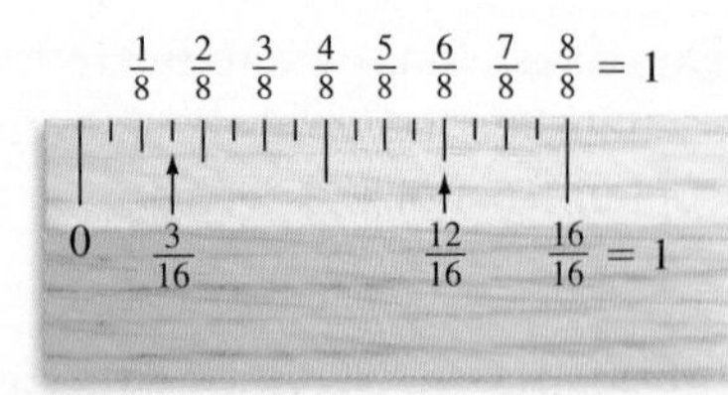

Not to scale.

Notice that equivalent fractions can be measured on the same mark of the ruler.

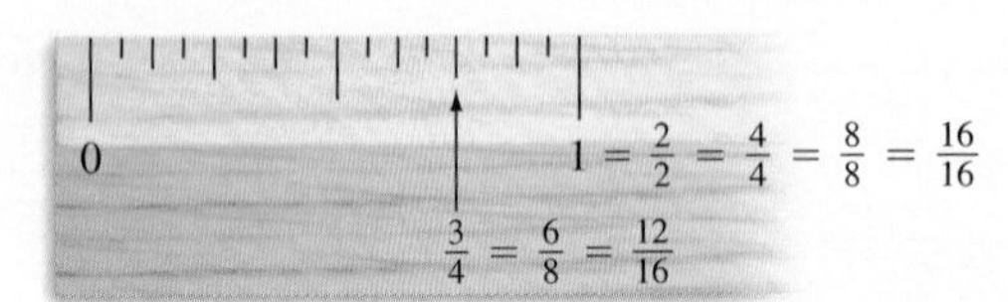

Not to scale.

EXAMPLES

15. Mark every half on the ruler.

To mark halves, the ruler must be cut into two equal-sized pieces.

Cut this line into two equal pieces.

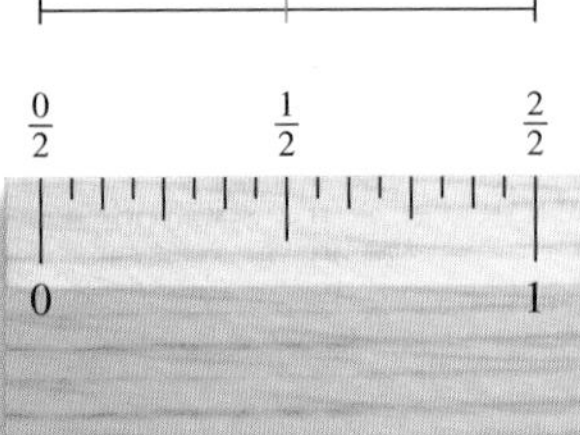

Not to scale.

16. Graph $\frac{2}{2}$, $\frac{0}{4}$, and $\frac{1}{4}$ on the same ruler.

Decide which tick marks represent halves and which represent fourths.

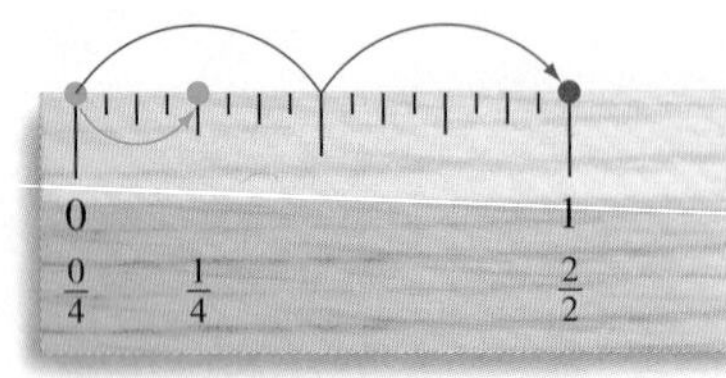

Not to scale.

Count halves from the left and graph $\frac{2}{2}$.

Count fourths from the left and graph $\frac{0}{4}$ and $\frac{1}{4}$.

GUIDED PRACTICE

15. Mark every fourth on the ruler.

To mark fourths, the ruler must be cut into _____ equal-sized pieces.

Cut this line into ___ equal pieces.

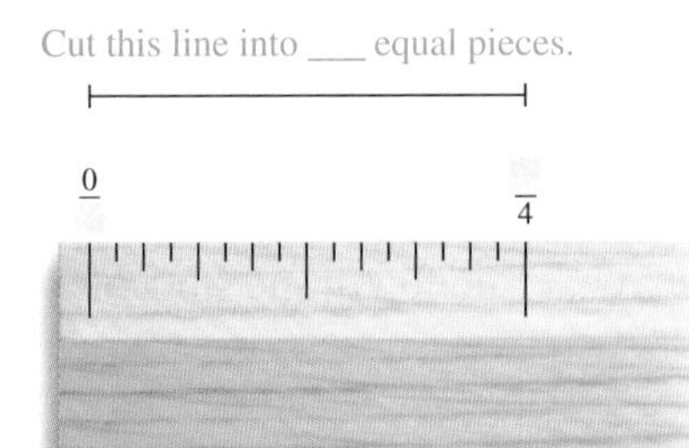

Not to scale.

16. Graph $\frac{1}{2}$, $\frac{3}{4}$, and $\frac{4}{4}$ on the same ruler.

Decide which tick marks represent halves and which represent fourths.

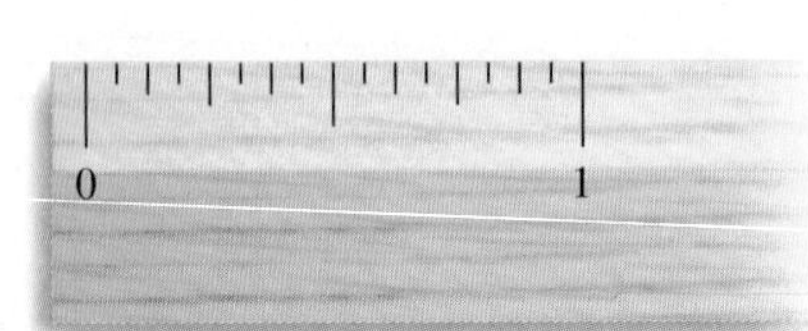

Not to scale.

Count halves from the left and graph $\frac{1}{2}$.

Count fourths from the left and graph $\frac{3}{4}$ and $\frac{4}{4}$.

17. Graph $\frac{7}{8}$ and $\frac{3}{16}$ on the same ruler.

Decide which tick marks represent eighths and which represent sixteenths.

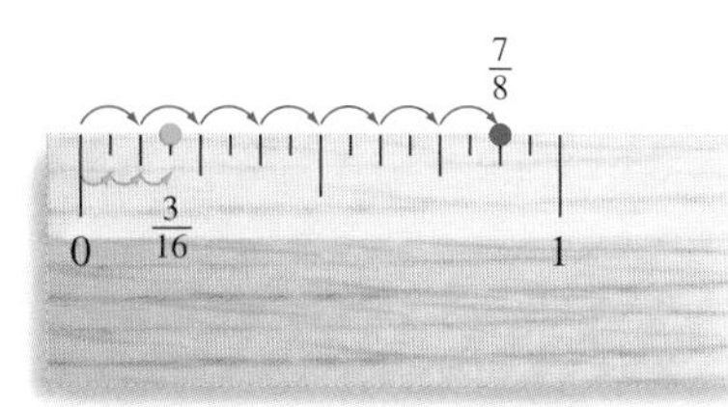

Not to scale.

Count eighths from the left and graph $\frac{7}{8}$.

Count sixteenths from the left and graph $\frac{3}{16}$.

17. Graph $\frac{3}{8}$ and $\frac{9}{16}$ on the same ruler.

Decide which tick marks represent eighths and which represent sixteenths.

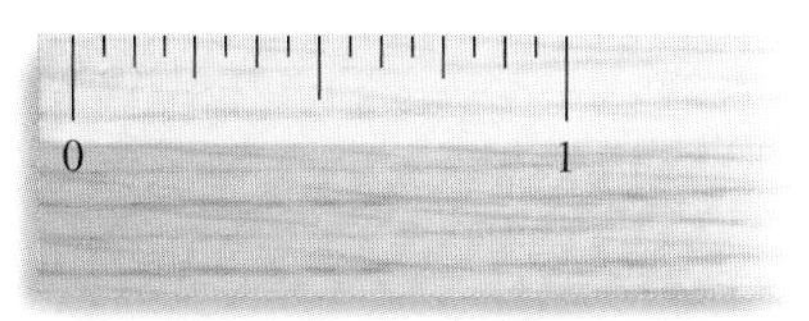

Not to scale.

Count eighths from the left and graph $\frac{3}{8}$.

Count sixteenths from the left and graph $\frac{9}{16}$.

Section 2.1 Exercises

FOR EXTRA HELP     

MyMathLab MathXL PRACTICE WATCH DOWNLOAD READ REVIEW

To represent a picture as a fraction:

1. Answer the Objective A Concept Checks.
2. Answer the odd Objective A Practice Exercises.
3. Answer the even Objective A Practice Exercises.

To represent a fraction as a picture:

4. Answer the Objective B Concept Checks.
5. Answer the odd Objective B Practice Exercises.
6. Answer the even Objective B Practice Exercises.

To relate improper fractions to their pictures:

7. Answer the Objective C Concept Checks.
8. Answer the odd Objective C Practice Exercises.
9. Answer the even Objective C Practice Exercises.

To compare fractions by visualizing them:

10. Answer the Objective D Concept Check.
11. Answer the odd Objective D Practice Exercises.
12. Answer the even Objective D Practice Exercises.

To compare fractions with the same numerator or denominator:

13. Answer the Objective E Concept Checks.
14. Answer the odd Objective E Practice Exercises.
15. Answer the even Objective E Practice Exercises.

VOCABULARY REVIEW *Review the Vocabulary Preview for Section 2.1. Study the definitions until you can check* Got It *for every word.*

Use these words to complete each sentence.

fraction • fraction bar • numerator • proper fraction • denominator • improper fraction

	Got It	Get Help
16. The ____________ of a fraction indicates how many equal-sized parts of the whole are represented.		
17. A(n) ____________ is a number that represents a portion of a whole quantity.		
18. A fraction that represents more than one whole is called a(n) ____________.		
19. A fraction that represents less than one whole is called a(n) ____________.		
20. The line that is drawn between the numerator and the denominator is called a(n) ____________.		
21. The ____________ of a fraction indicates how many parts are in the whole.		

How will you get help for any vocabulary that you are unsure about?

Your instructor ____ MyMathLab ____ A classmate ____ A tutor ____ Other ____

22. What does the denominator of a fraction tell about its picture?

23. What does the numerator of a fraction tell about its picture?

Represent each fraction as a picture or represent the picture as a fraction as appropriate.

24. 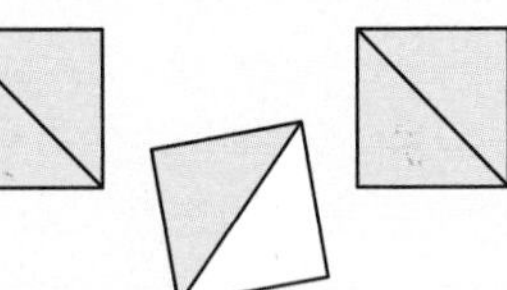

25. 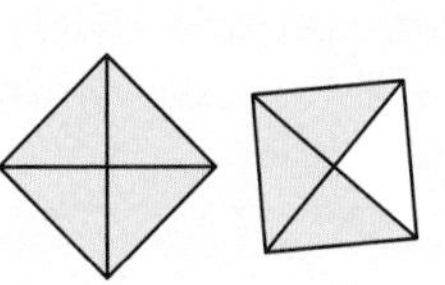

26. $\frac{2}{5}$

27. $\frac{1}{6}$

28. $\frac{5}{4}$

29. $\frac{4}{2}$

30. What fraction of the babies are wearing pink shirts?

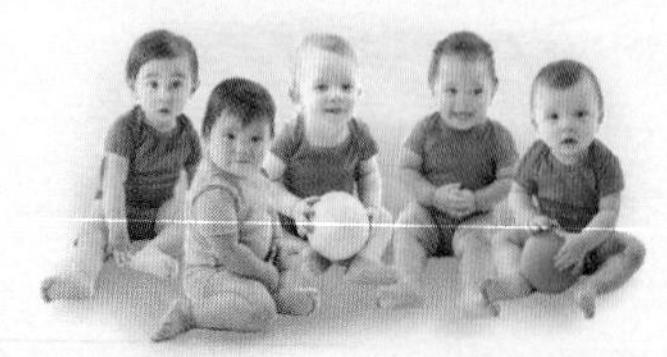

31. What fraction of the children are wearing pink shirts?

32. $\frac{0}{2}$

33. 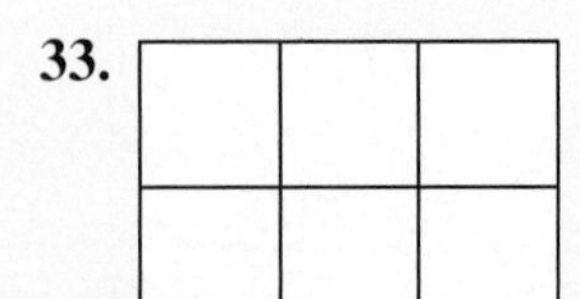

34. 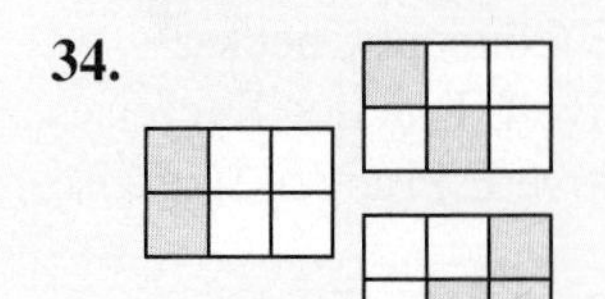

35. 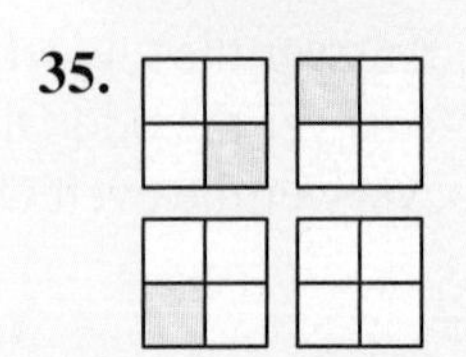

goetz Careful! When two or more objects are shown, the denominator is the number of pieces in one object!

For more tips and tweets, go to twitter.com/gstbasicmath

For each exercise:
a. Draw lines on the objects to make each of its pieces equal in size.
b. Determine the fraction represented by the picture.

36. The figure should have 12 equal-sized parts.

37. The figure should have 8 equal-sized parts.

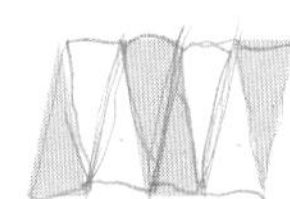

38. The figure should have 9 equal-sized parts.

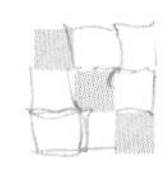

39. The figure should have 4 equal-sized parts.

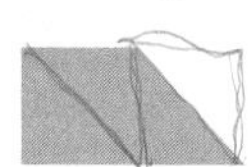

40. The figure should have 6 equal-sized parts.

41. The figure should have 3 equal-sized parts.

42. Why do pictures of improper fractions often require more than one whole object?

Each picture represents one of the four elementary arithmetic operations. Identify the operation represented and explain how you know that it is this operation.

43.

44.

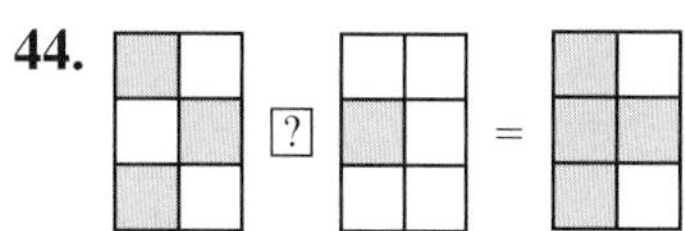

For each exercise:
a. Represent the three fractions as pictures of the same object.
b. Use the pictures from (a) to determine which of the fractions are equal in size.

45. $\frac{1}{2}, \frac{1}{5}, \frac{2}{4}$ **46.** $\frac{1}{3}, \frac{3}{5}, \frac{2}{6}$ **47.** $\frac{3}{5}, \frac{3}{4}, \frac{6}{8}$ **48.** $\frac{5}{6}, \frac{2}{3}, \frac{10}{12}$

Use < or > to compare the fractions. Give reasons to support your answer.

49. $\frac{4}{7} \square \frac{3}{7}$ **50.** $\frac{3}{5} \square \frac{2}{5}$ **51.** $\frac{1}{20} \square \frac{1}{100}$ **52.** $\frac{7}{15} \square \frac{7}{13}$

53. $\frac{5}{12} \square \frac{5}{20}$ **54.** $\frac{2}{7} \square \frac{4}{7}$ **55.** $\frac{12}{11} \square \frac{13}{11}$ **56.** $\frac{4}{9} \square \frac{4}{10}$

Graph each fraction on the ruler provided.

57. $\frac{3}{4}, \frac{5}{16}, \frac{11}{16}, \frac{5}{8}, \frac{3}{16}$

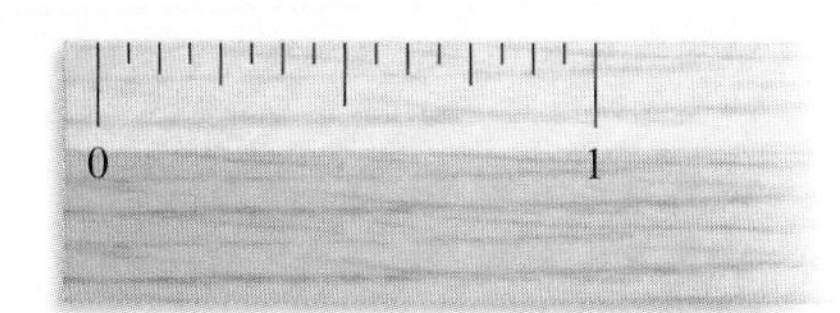

Not to scale.

58. $\frac{1}{2}, \frac{3}{8}, \frac{5}{16}, \frac{11}{16}, \frac{5}{8}$

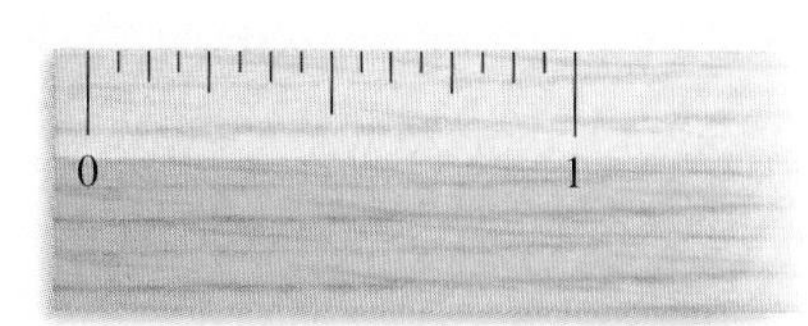

Not to scale.

59. $\frac{1}{4}, \frac{1}{16}, \frac{5}{16}, \frac{3}{8}, \frac{3}{16}$

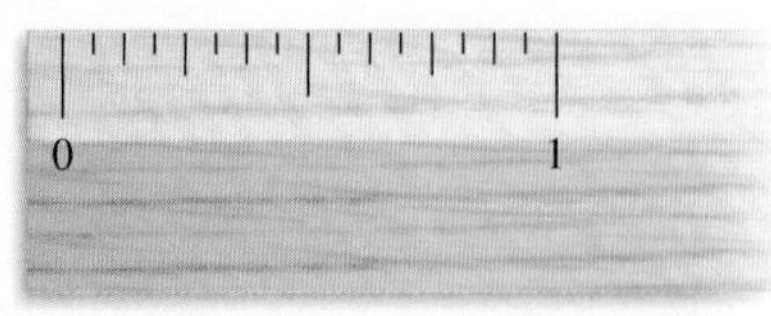

Not to scale.

60. $\frac{1}{16}, \frac{1}{8}, \frac{1}{4}, \frac{1}{2}$

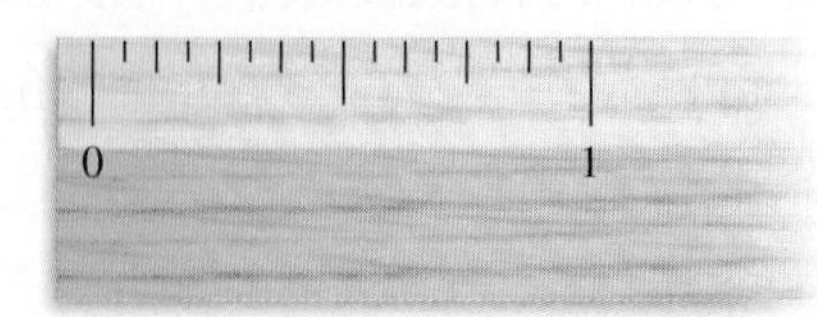

Not to scale.

61. Two students are debating whether the techniques learned in this section can be used to interpret the picture provided using a fraction.

a. Connie says that you cannot use the techniques learned because of the way the picture is drawn. What is her argument?

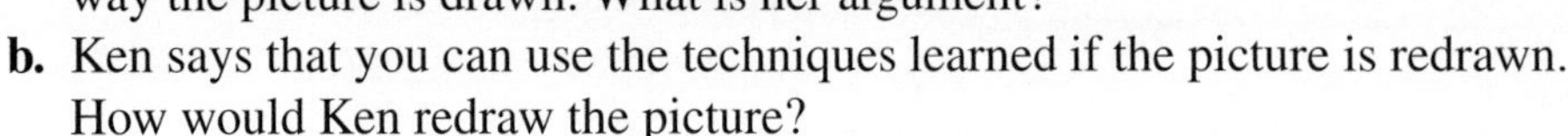

b. Ken says that you can use the techniques learned if the picture is redrawn. How would Ken redraw the picture?

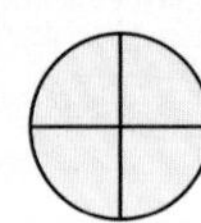

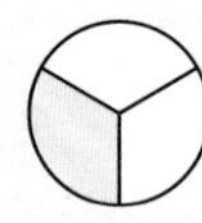

62. Write a set of steps that someone else could use to answer this question: What fraction of the people in the room are wearing jeans?

63. Wembley Stadium, which seats 90,000 people, has been sectioned off for the Live Earth concert. If each section of the stadium seats approximately the same number of people, what fraction of the seats (indicated in green) in the stadium are being used?

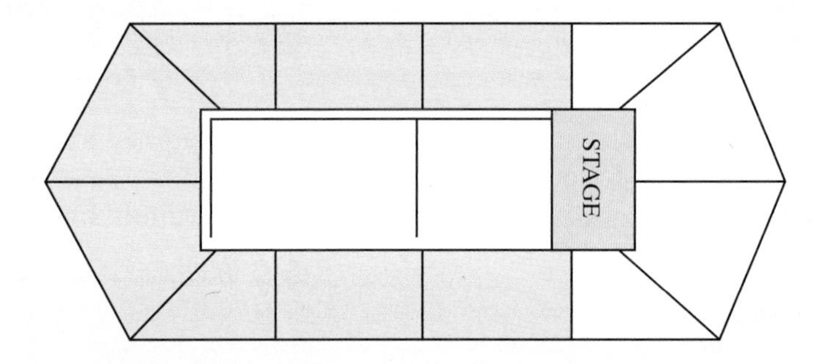

64. What fractional part of a day is 11 hours?

65. At a party, 14 guests were men and 11 guests were women.

a. How many people were at the party?

b. What fraction of the partygoers were men?

c. What fraction of the partygoers were women?

66. Marvin spent \$231 of his \$435 paycheck on an external hard drive for his computer.

a. What fraction of his paycheck did he spend?

b. What fraction of his paycheck did he have left?

67. A pizza was cut into ten equal-sized pieces. Seven of the pieces were eaten.

a. What fraction of the pizza was eaten?

b. What fraction of the pizza was not eaten?

68. Archie and Veronica are debating how to compare fractions. Archie thinks that fractions with bigger numbers are larger. Veronica thinks that fractions with smaller numbers are larger. Who is right? Could they both be partially right? Explain your answer.

69. A local pet store specializes in selling tropical fish. The store's showcase tank has 12 angelfish, 4 oscar fish, and 15 discus fish.

a. What fraction of the fish in this tank are angelfish?

b. What fraction of the fish in this tank are discus fish?

c. What fraction of the fish in this tank are either angelfish or discus fish?

70. Use the markings on the fuel gauge to determine what fraction of fuel is left and what fraction of fuel has been used.

a. The fuel gauge is on A.

b. The fuel gauge is on B.

c. The fuel gauge is halfway between C and D.

71. A jet plane has two fuel tanks: a main tank and an auxiliary tank.

a. What fraction of the main tank's fuel remains?

b. What fraction of the auxiliary tank's fuel remains?

c. If both tanks are combined, what fraction of a single tank remains?

Main Tank Auxiliary Tank

72. Answer the following questions about the quilt square shown.

a. What fraction of the total area is red?

b. What fraction of the total area is made of red triangles?

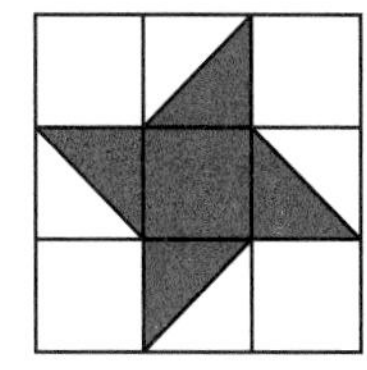

73. Use the diagram below to answer each question.

a. What fraction of the shapes are red?

b. What fraction of the shapes are square?

c. What fraction of the shapes are both red and square?

d. What fraction of the shapes are red or square (one or the other or both)?

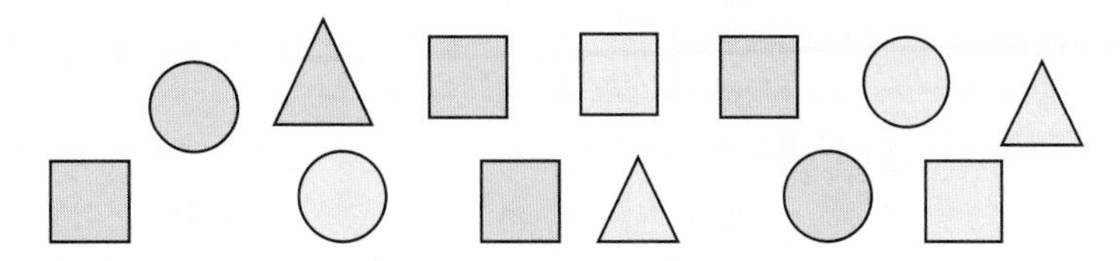

SELF-ASSESSMENT

	YES	NO
1. On your last test, were you satisfied with your score? *If you answered yes, good job. Do not continue here.* *If you answered no, respond to the following:*		
2. Did you preview each section of the textbook before your instructor taught it in class?		
3. Did you complete all of the homework before it was due?		
4. Did you get help with the homework exercises that you didn't understand?		

5. Set two goals that will help you become better prepared for your next test.

Goal 1: ______________________________

Goal 2: ______________________________

Section 2.1 Question Log

Use this space to write questions. Be sure to get these answered and revisit them when you prepare for your exam.

Page ____ **Answered** ☐	**Page** ____ **Answered** ☐
Page ____ **Answered** ☐	**Page** ____ **Answered** ☐

Section 2.2 Multiplying Fractions

In everyday life, it is important to be able to simplify and multiply fractions.

Simplifying fractions makes them easier to understand.

Instead of saying that $\frac{57,932}{66,208}$ of the fans wore red, we say that $\frac{7}{8}$ of the fans wore red.

Multiplying by a fraction calculates part of a whole.

To calculate $\frac{3}{4}$ of the 4 cars shown, we multiply $\frac{3}{4} \cdot 4$.

Multiplying by units fractions converts units of measure.

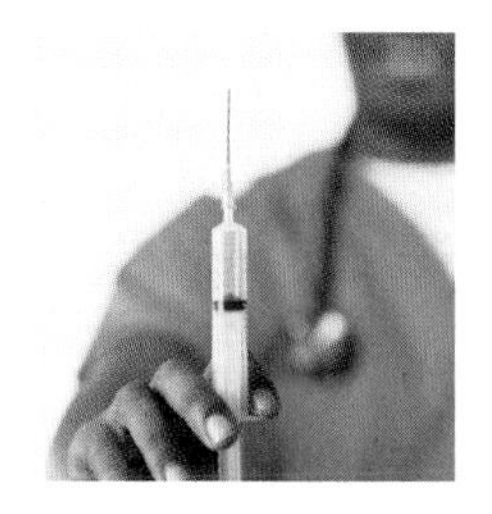

Nurses often multiply by units fractions to calculate a dose of medication.

The **Objectives** in Section 2.2 will help you

- A Simplify a fraction.
- B Multiply two fractions.
- C Build equivalent fractions.
- D Use a fraction to convert units of measure.

VOCABULARY PREVIEW *Check the box that applies.*	Got It	Must Study
Quotient Concept: The **quotient concept** of a fraction states that a fraction such as $\frac{6}{2}$ can be interpreted as $6 \div 2$.		
Factor: To **factor** a number, write it as a product of numbers. Factoring 6 gives $2 \cdot 3$.		
Simplest Form: A fraction is in **simplest form** when its numerator and denominator have no common factors other than 1.		
Simplify/Reduce a Fraction: To **simplify or reduce** a fraction, write it in simplest form.		
Greatest Common Factor: The largest number that will divide several numbers evenly is the **greatest common factor (GCF)** of those numbers.		
Equivalent Fractions: Two fractions that are written differently but represent the same quantity are **equivalent** fractions. $\frac{1}{2}$ and $\frac{3}{6}$ are equivalent fractions.		
Units Fraction: A **units fraction** contains units of equal measure in the numerator and denominator. Units fractions are equal to 1.		
Missing Multiplier: A **missing multiplier** is a factor that multiplies the numerator and denominator of a fraction to build the desired equivalent fraction.		

Study these words when they appear in the text. After the section, test your understanding by completing the Vocabulary Review in the section exercises.

Objective A Simplify a Fraction

The Concept Simplifying or reducing a fraction involves writing a fraction in an equivalent but more easily understood form. Compare the pictures for $\frac{9}{18}$ and $\frac{1}{2}$ below.

$\frac{9}{18}$ →

$\frac{1}{2}$ →

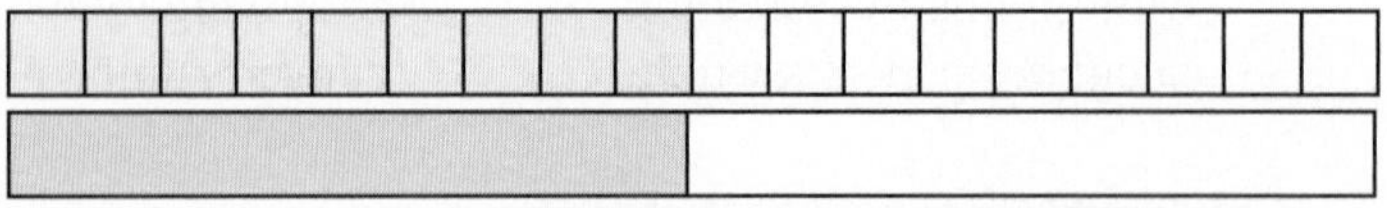

Each fraction represents the same portion of a whole. $\frac{9}{18}$ and $\frac{1}{2}$ are **equivalent fractions** because they represent the same quantity but are written differently. Since each fraction represents $\frac{1}{2}$, an easy fraction to understand, we should call both of them $\frac{1}{2}$. That is, we should write $\frac{9}{18}$ in its **simplest form** as $\frac{1}{2}$.

INTERACTIVE DEFINITION Equivalent Fractions and Simplest Form

Equivalent fractions are written differently but represent the same quantity.

A fraction is in **simplest form** when its numerator and denominator have no common factors other than 1.

EXAMPLE

1. Do the following for the fractions pictured below.
a. Identify the two fractions that are equivalent.
b. Identify which equivalent fraction is in simplest form.

$\frac{1}{2} =$

$\frac{2}{3} =$

$\frac{2}{4} =$

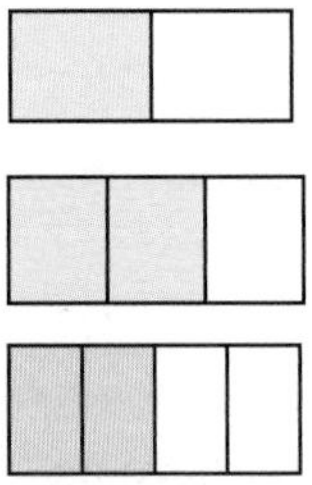

a. Since $\frac{1}{2}$ and $\frac{2}{4}$ are written differently but represent the same quantity, they are equivalent fractions.

b. Of the equivalent fractions $\frac{1}{2}$ and $\frac{2}{4}$, $\frac{1}{2}$ is in simplest form.

GUIDED PRACTICE

1. Do the following for the fractions pictured below.
a. Identify the two fractions that are equivalent.
b. Identify which equivalent fraction is in simplest form.

$\frac{4}{6} =$

$\frac{2}{3} =$

$\frac{2}{4} =$

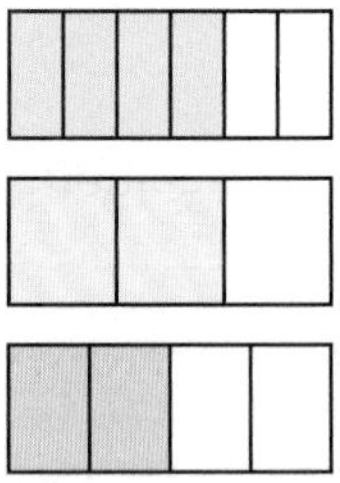

a. Since ___ and ___ are written differently but represent the same quantity, they are ___________ fractions.

b. Of the ___________ fractions ___ and ___, ___ is in simplest form.

DO YOU UNDERSTAND the concept of equivalent fractions? Got It Get Help

DO YOU UNDERSTAND the picture of a fraction that is in simplest form? Got It Get Help

EXAMPLE

2. Identify the simplest form for $\frac{6}{8}$ by using a picture and by factoring.

$\frac{6}{8} \Rightarrow$ ▭▭▭▭▭▭▭▭

Picture
The greatest common factor of 6 and 8 is 2. Combine every two pieces into a single piece.

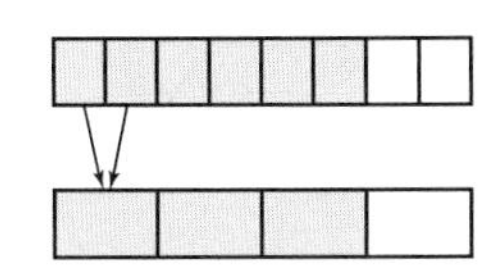

Factoring
Grouping every two pieces together is equivalent to factoring a 2 out of the numerator and denominator.

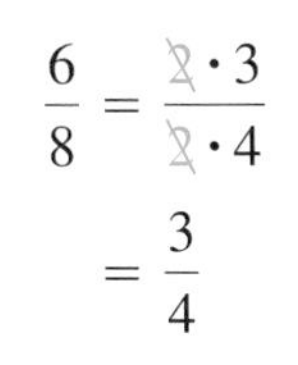

$$\frac{6}{8} = \frac{2 \cdot 3}{2 \cdot 4} = \frac{3}{4}$$

the same portions

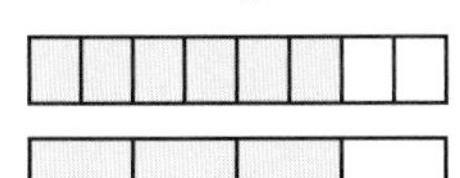

equivalent fractions

$$\frac{6}{8} = \frac{3}{4}$$

GUIDED PRACTICE

2. Identify the simplest form for $\frac{6}{9}$ by using a picture and by factoring.

$\frac{6}{9} \Rightarrow$ ▭▭▭▭▭▭▭▭▭

Picture
The greatest common factor of 6 and 9 is _____. Combine every _____ pieces into a single piece.

Factoring
Grouping every _____ pieces together is equivalent to factoring a _____ out of the numerator and denominator.

$$\frac{6}{9} = \frac{\quad}{\quad} = \frac{\quad}{\quad}$$

the same portions

equivalent fractions

$$\frac{6}{9} = \frac{\quad}{\quad}$$

Procedure Simplify a Fraction

Step 1 Factor the numerator completely.
Step 2 Factor the denominator completely.
Step 3 Divide out common factors.

▶ **NOTE** You may already know a method for simplifying/reducing fractions that seems different from this procedure. Make sure you learn the three-step procedure for simplifying a fraction. When you take algebra, this will be the *only way* to simplify some fractions.

EXAMPLES

3. Write $\frac{10}{15}$ in simplest form.

Factor the numerator and denominator.

10 → 2, 5 15 → 3, 5

GUIDED PRACTICE

3. Write $\frac{4}{10}$ in simplest form.

Factor the numerator and denominator.

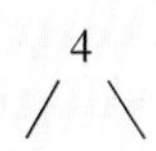

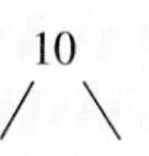

(Continued)

Step 1 Factor the numerator.
Step 2 Factor the denominator.
Step 3 Divide out the common factors.

$$\frac{10}{15} = \frac{2 \cdot 5}{3 \cdot 5} = \frac{2 \cdot \cancel{5}}{3 \cdot \cancel{5}} = \frac{2}{3}$$

the same portions

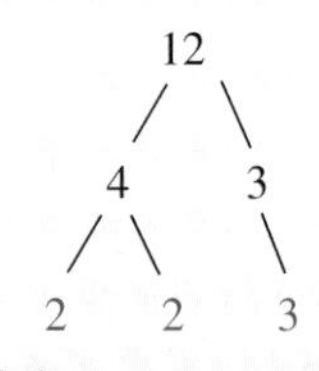

=

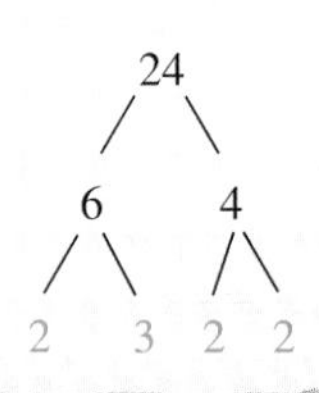

equivalent fractions

$$\frac{10}{15} = \frac{2}{3}$$

Step 1 Factor the numerator.
Step 2 Factor the denominator.
Step 3 Divide out the common factors.

$$\frac{4}{10} = \frac{\quad}{\quad} = \frac{\quad}{\quad} = \frac{\quad}{\quad}$$

the same portions

=

equivalent fractions

$$\frac{4}{10} = \frac{\quad}{\quad}$$

4. Write $\frac{12}{24}$ in simplest form.

Factor the numerator and denominator.

12 → 4, 3; 4 → 2, 2; 3 → 3

24 → 6, 4; 6 → 2, 3; 4 → 2, 2

Step 1 Factor the numerator.
Step 2 Factor the denominator.
Step 3 Divide out the common factors.

$$\frac{12}{24} = \frac{2 \cdot 2 \cdot 3}{2 \cdot 2 \cdot 2 \cdot 3} = \frac{{}^1\cancel{2} \cdot \cancel{2}^1 \cdot \cancel{3}^1}{{}^1\cancel{2} \cdot \cancel{2}^1 \cdot 2 \cdot \cancel{3}^1} = \frac{1}{2}$$

$$\frac{12}{24} = \frac{1}{2}$$

If all the factors divide out of the numerator, the result is a numerator of 1.

4. Write $\frac{18}{36}$ in simplest form.

Factor the numerator and denominator.

18

36

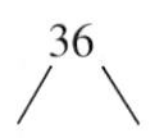

$$\frac{18}{36} = \frac{\quad}{\quad} = \frac{\quad}{\quad} = \frac{\quad}{\quad}$$

$$\frac{18}{36} = \frac{\quad}{\quad}$$

5. Simplify. $\frac{120}{12}$

Step 1 Factor the numerator.
Step 2 Factor the denominator.
Step 3 Divide out the common factors.

$$\frac{120}{12} = \frac{2 \cdot 2 \cdot 2 \cdot 3 \cdot 5}{2 \cdot 2 \cdot 3} = \frac{{}^1\cancel{2} \cdot \cancel{2}^1 \cdot 2 \cdot \cancel{3}^1 \cdot 5}{{}^1\cancel{2} \cdot \cancel{2}^1 \cdot \cancel{3}^1} = \frac{10}{1} = 10$$

A fraction with 1 in the denominator is a whole number

This fraction could have been simplified using the quotient concept of a fraction.

$$\frac{120}{12} = 120 \div 12 = 10$$

5. Simplify. $\frac{240}{16}$

$$\frac{240}{16} = \frac{\quad}{\quad} = \frac{\quad}{\quad} = \frac{\quad}{\quad} =$$

$$\frac{240}{16} = 240 \div 16 =$$

When a denominator is a factor of a numerator, it is often convenient to use the **quotient concept of a fraction:** $\frac{\text{numerator}}{\text{denominator}}$ = numerator ÷ denominator. This can help you simplify some fractions more quickly.

$$\frac{10}{5} \text{ is the same as } 10 \div 5 = 2, \text{ so } \frac{10}{5} = 2$$

Concept Check

A1. Why is it important to simplify fractions? Use an example of a fraction and its simplified form (such as $\frac{9}{18}$ and $\frac{1}{2}$) to explain your answer.

Objective A Practice

Do the following for the fractions pictured below.
a. Identify the two fractions that are equivalent.
b. Identify which equivalent fraction is in simplest form.

A2. $\frac{2}{6}$ →

$\frac{2}{4}$ →

$\frac{1}{3}$ →

A3. $\frac{3}{9}$ →

$\frac{1}{3}$ →

$\frac{4}{9}$ →

Identify the simplest form for each fraction by using a picture and by factoring.

A4. $\frac{3}{6}$ =

A5. $\frac{2}{6}$ =

Write each fraction in simplest form.

A6. $\frac{4}{6}$

A7. $\frac{6}{8}$

A8. $\frac{12}{30}$

A9. $\frac{16}{40}$

A10. $\frac{60}{45}$

A11. $\frac{50}{35}$

A12. $\frac{96}{16}$

A13. $\frac{84}{12}$

A14. Johann sold a camera for \$120. He had to pay \$15 in fees to eBay. What fraction of the sales price were fees?

A15. Sebastian purchased Johann's camera for \$120 plus \$10 for shipping. What fraction of the total did the shipping cost represent?

FOCUS ON Estimating Fractions

Estimating fractions will help you make quick calculations when you encounter fractions in everyday life. Estimation can also help you realize when you have made a mistake in your calculations.

In this section and in the previous section, you practiced visualizing and simplifying fractions. Both skills are used here. When estimating the size of a fraction, do not worry about making a perfect picture. Rather, draw an approximation of the actual fraction using rounded values.

(Continued)

Procedure Estimate a Fraction

Step 1 Round either the numerator or the denominator to its largest place value.

Step 2 Simplify.

Step 3 Compare the estimate with the original fraction and create a picture of each.

EXAMPLES

Estimate each fraction. Compare the estimate with the original fraction, then create a picture of each.

6. $\frac{33}{60}$

$$\frac{33}{60} \approx \frac{30}{60} = \frac{1}{2}$$

Because 33 is a little more than 30, we shade a little more than half the object.

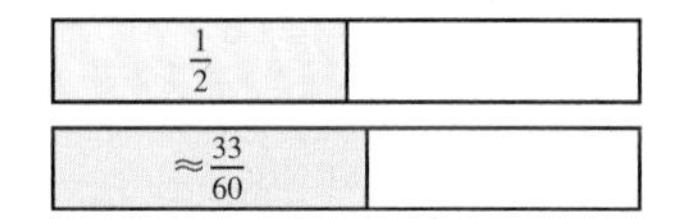

$\frac{33}{60}$ is a little more than $\frac{1}{2}$.

7. $\frac{17}{80}$

$$\frac{17}{80} \approx \frac{20}{80} = \frac{1}{4}$$

Because 17 is less than 20, we shade a little less than a fourth of the object.

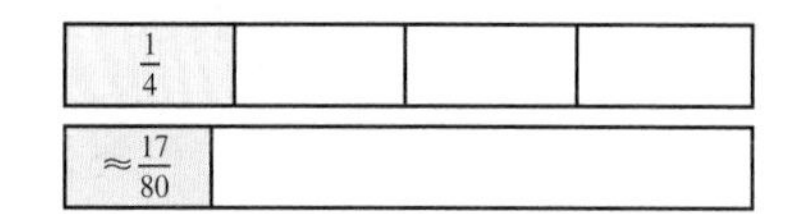

$\frac{17}{80}$ is a little less than $\frac{1}{4}$.

8. $\frac{90}{101}$

$$\frac{90}{101} \approx \frac{90}{100} = \frac{9}{10}$$

Since 101^{ths} are smaller than 100^{ths}, we shade a little less than nine-tenths of the object.

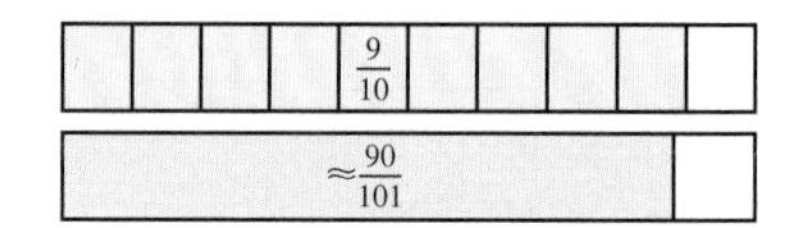

$\frac{90}{101}$ is a little less than $\frac{9}{10}$.

Could we have done better? Probably. But the fractions visualized above are good enough for us to get an idea of the portion that the actual fraction represents.

PRACTICE

Estimate each fraction. Compare the estimate with the original fraction and create a picture of each.

1. $\frac{11}{30}$

2. $\frac{9}{40}$

3. $\frac{20}{35}$

4. $\frac{60}{94}$

5. $\frac{60}{83}$

6. $\frac{18}{20}$

Objective B Multiply Two Fractions

The Concept Reading the expression $\frac{2}{3} \cdot \frac{5}{7}$ aloud, we say, "two-thirds of five-sevenths." To find two-thirds of five-sevenths, we can start with $\frac{5}{7}$ and determine $\frac{2}{3}$ of it. We begin by interpreting $\frac{5}{7}$ as a picture.

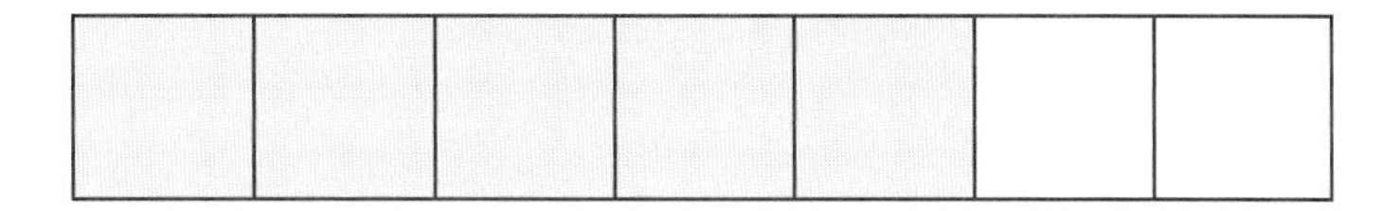

Our answer will be $\frac{2}{3}$ of the shaded region. To determine that portion, we divide each piece of $\frac{5}{7}$ (whether shaded or not) into three pieces.

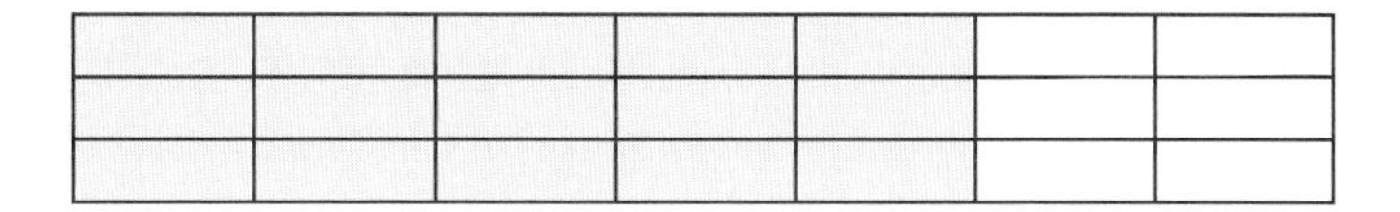

To keep $\frac{2}{3}$ of the original, we will keep two out of every three shaded parts or "erase" one out of every three shaded parts.

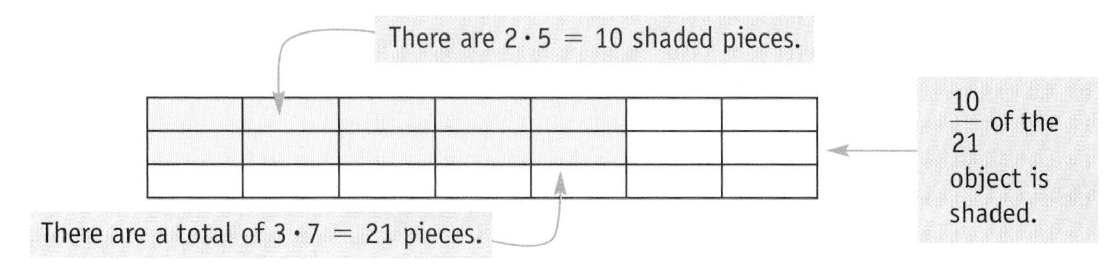

That gives us $\frac{2}{3} \cdot \frac{5}{7} = \frac{10}{21}$. Two very important things just happened. When we multiplied the two fractions, the total number of shaded parts doubled from 5 to 10; we multiplied the numerators, $2 \cdot 5 = 10$. Also, the total number of parts, whether shaded or not, tripled from 7 to 21; we multiplied the denominators, $3 \cdot 7 = 21$. Thus, we have demonstrated the procedure for multiplying two fractions.

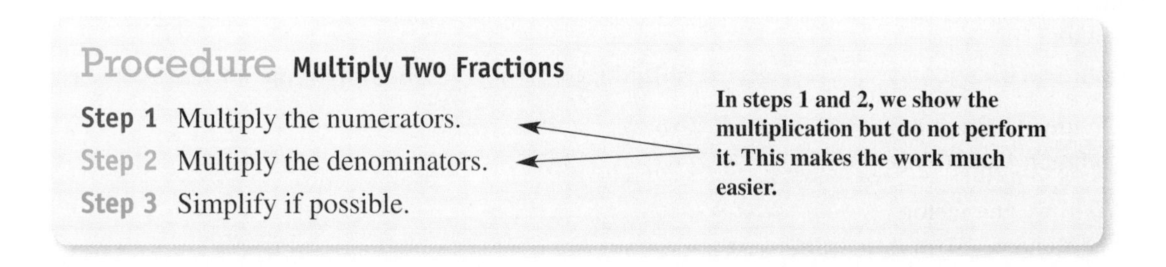

Procedure Multiply Two Fractions

Step 1 Multiply the numerators.

Step 2 Multiply the denominators.

Step 3 Simplify if possible.

In steps 1 and 2, we show the multiplication but do not perform it. This makes the work much easier.

DETAILED EXAMPLE Multiplying Fractions

Multiply. $\frac{6}{5} \cdot \frac{15}{8}$

Step 1 Multiply the numerators.
Write the numerators multiplied together, but don't perform the multiplication.

$$\frac{6}{5} \cdot \frac{15}{8} = \frac{6 \cdot 15}{}$$

Step 2 Multiply the denominators.
Write the denominators multiplied together, but don't perform the multiplication.

$$= \frac{6 \cdot 15}{5 \cdot 8}$$

Because we didn't perform the multiplication in steps 1 and 2, it is easier to factor in step 3.

Step 3 Simplify.

- Factor the numerator and denominator.

$$= \frac{\overbrace{2 \cdot 3}^{6} \cdot \overbrace{3 \cdot 5}^{15}}{5 \cdot \underbrace{2 \cdot 2 \cdot 2}_{8}}$$

- Divide out the common factors.

$$= \frac{\cancel{2} \cdot 3 \cdot 3 \cdot \cancel{5}}{\cancel{5} \cdot \cancel{2} \cdot 2 \cdot 2}$$

- Multiply any factors that remain.

$$= \frac{9}{4}$$

EXAMPLE

9. Multiply. $\frac{4}{5} \cdot \frac{9}{8}$

Step 1 Multiply the numerators.
Step 2 Multiply the denominators.

$$\frac{4}{5} \cdot \frac{9}{8} = \frac{4 \cdot 9}{5 \cdot 8}$$

Step 3 Simplify.
- Factor.

$$= \frac{\overbrace{2 \cdot 2}^{4} \cdot \overbrace{3 \cdot 3}^{9}}{5 \cdot \underbrace{2 \cdot 2 \cdot 2}_{8}}$$

- Divide out the common factors.

$$= \frac{\cancel{2} \cdot \cancel{2} \cdot 3 \cdot 3}{5 \cdot \cancel{2} \cdot \cancel{2} \cdot 2}$$

- Multiply any factors that remain.

$$= \frac{9}{10}$$

GUIDED PRACTICE

9. Multiply. $\frac{12}{5} \cdot \frac{10}{9}$

$$\frac{12}{5} \cdot \frac{10}{9} = \frac{\quad}{\quad}$$

$$= \frac{\overbrace{\quad}^{12} \cdot \overbrace{\quad}^{10}}{5 \cdot \underbrace{\quad}_{9}}$$

$$= \frac{\quad}{\quad}$$

$$= \frac{\quad}{\quad}$$

Keep two things in mind when multiplying fractions:

- Combine the numerators and denominators without performing the multiplication.
- Divide any common factors out before you multiply. Work smarter, not harder!

EXAMPLES

10. Multiply. $\frac{5}{14} \cdot \frac{21}{10}$

Step 1 Multiply the numerators.
Step 2 Multiply the denominators.
Step 3 Simplify.
- Factor.
- Divide out common factors.
- Multiply the remaining factors.

$$\frac{5}{14} \cdot \frac{21}{10} = \frac{5 \cdot 21}{14 \cdot 10}$$
$$= \frac{\cancel{5} \cdot 3 \cdot \cancel{7}}{2 \cdot \cancel{7} \cdot 2 \cdot \cancel{5}}$$
$$= \frac{3}{4}$$

Simplify before multiplying.

11. Multiply. $\frac{7}{27} \cdot 9$

Write the whole number over 1.

Step 1 Multiply the numerators.
Step 2 Multiply the denominators.
Step 3 Simplify. Common factors that are not prime can also be divided out.

$$\frac{7}{27} \cdot 9 = \frac{7}{27} \cdot \frac{9}{1}$$
$$= \frac{7 \cdot 9}{27 \cdot 1}$$
$$= \frac{7 \cdot \cancel{9}}{3 \cdot \cancel{9} \cdot 1}$$
$$= \frac{7}{3}$$

Simplify before multiplying.

12. Carlos and Clara are members of a club that is raising money to study abroad. They will receive $\frac{3}{10}$ of the \$1,200 that the club has raised. How much money will they receive?

To find a fractional part of something, multiply by the fraction.

Step 1 Multiply the numerators.
Step 2 Multiply the denominators.
Step 3 Simplify. Common factors that are not prime can also be divided out.

$$\frac{3}{10} \cdot 1{,}200 = \frac{3}{10} \cdot \frac{1{,}200}{1}$$
$$= \frac{3 \cdot 1{,}200}{10 \cdot 1}$$
$$= \frac{3 \cdot \cancel{10} \cdot 120}{\cancel{10} \cdot 1}$$
$$= \frac{360}{1}$$
$$= 360$$

Simplify before multiplying.

They will receive \$360.

GUIDED PRACTICE

10. Multiply. $\frac{14}{9} \cdot \frac{3}{35}$

$$\frac{14}{9} \cdot \frac{3}{35} = \underline{\qquad}$$
$$= \underline{\qquad\qquad}$$
$$= \underline{\qquad}$$

11. Multiply. $6 \cdot \frac{5}{12}$

$$6 \cdot \frac{5}{12} = \frac{\quad}{1} \cdot \underline{\qquad}$$
$$= \underline{\qquad}$$
$$= \underline{\qquad}$$
$$= \underline{\quad}$$

12. Victoria and Toma have written a grant to start a recycling program at their college. If they are awarded $\frac{3}{5}$ of the \$5,000 they requested, how much money will they receive?

$$\underline{\quad} \cdot 5{,}000 = \underline{\quad} \cdot \underline{\qquad}$$
$$= \underline{\qquad}$$
$$= \underline{\qquad\qquad}$$

$$=$$

They will receive ______.

Concept Check

Each picture shows a multiplication problem. Complete the picture of the fraction that is the product.

B1. $\frac{1}{3} \cdot$ [ ] = [ ]

B2. $\frac{3}{4} \cdot$ [ ] = [ ]

Objective B Practice

Multiply the fractions. Write your answer in simplest form. Remember to simplify before you perform the multiplication.

B3. $\frac{3}{4} \cdot \frac{5}{7}$

B4. $\frac{2}{5} \cdot \frac{9}{11}$

B5. $7 \cdot \frac{1}{84}$

B6. $\frac{5}{6} \cdot 24$

B7. $\frac{5}{3} \cdot 12$

B8. $3 \cdot \frac{5}{9}$

B9. $\frac{6}{35} \cdot \frac{21}{10}$

B10. $\frac{3}{5} \cdot \frac{10}{17}$

B11. Without looking back at the previous pages, write the procedure for multiplying fractions.

B12. An inheritance will be split so that each member of the family receives $\frac{1}{5}$ of the estate. If the estate is worth $40,000, how much will each family member receive?

B13. Moises invested $\frac{1}{15}$ of his salary last year in a retirement account. If his salary last year was $60,000, how much did he invest in the retirement account?

Objective C Build Equivalent Fractions

The Concept Two basic facts will help you understand the concepts used in this objective.

Fact 1: Any fraction with the same numerator and denominator is equal to one.

$$= \frac{5}{5} = 1 \quad \text{because} \quad \frac{5}{5} = \frac{1 \cdot \cancel{5}}{1 \cdot \cancel{5}} = 1$$

$$\frac{675}{675} = 1 \quad \text{because} \quad \frac{675}{675} = (675) \div (675) = 1$$

This is the quotient concept of a fraction.

▶ **NOTE** The only exception is when the fraction is $\frac{0}{0}$. Using the quotient concept of a fraction, $\frac{0}{0}$ indicates division by zero. $\frac{0}{0}$ is undefined.

Fact 2: Any number, multiplied by one, is the same number.

$$98 \cdot 1 = 98 \quad \text{and} \quad \frac{3}{5} \cdot 1 = \frac{3}{5}$$

Using Facts 1 and 2, we can multiply $\frac{3}{5}$ by a fraction equal to one, $\frac{2}{2}$, to build the equivalent fraction $\frac{6}{10}$.

$$\frac{3}{5} = \frac{3}{5} \cdot \frac{2}{2} = \frac{6}{10}$$

Multiplying by $\frac{2}{2}$ will not change the value of $\frac{3}{5}$.

Procedure Build an Equivalent Fraction

Step 1 Set up the framework.

Step 2 Determine the missing multiplier.

Step 3 Multiply the numerator and denominator by the missing multiplier.

The **missing multiplier** is the factor that must multiply the numerator and denominator of the fraction to build the desired equivalent fraction.

DETAILED EXAMPLE Building Equivalenct Fractions

Fill in the framework to build equivalent fractions. $\frac{5}{8} \cdot \left(\frac{}{}\right) = \frac{}{40}$

Step 2 Determine the missing multiplier. Use basic facts to complete the multiplication in the denominator.

The missing multiplier is 5 because $8 \cdot 5 = 40$.

$$\frac{5}{8} \cdot \left(\frac{?}{5}\right) = \frac{}{40}$$

Step 3 Multiply the numerator and denominator by the missing multiplier.

$$\frac{5}{8} \cdot \left(\frac{5}{5}\right) = \frac{25}{40}$$

$\frac{5}{8}$ and $\frac{25}{40}$ are equivalent fractions.

EXAMPLE

13. Fill in the framework to build equivalent fractions. $\frac{3}{4} \cdot \left(\frac{}{}\right) = \frac{}{8}$

Step 2 Determine the missing multiplier.

The missing multiplier is 2 because $4 \cdot 2 = 8$.

Step 3 Multiply the numerator and denominator by the missing multiplier.

$$\frac{3}{4} \cdot \left(\frac{2}{2}\right) = \frac{6}{8}$$

$\frac{3}{4}$ and $\frac{6}{8}$ are equivalent fractions.

GUIDED PRACTICE

13. Fill in the framework to build equivalent fractions. $\frac{3}{5} \cdot \left(\frac{}{}\right) = \frac{}{35}$

The missing multiplier is _____ because _____

$$\frac{3}{5} \cdot \left(\frac{}{}\right) = \frac{}{35}$$

$\frac{3}{5}$ and $\frac{}{}$ are __________ fractions.

Using this framework to build equivalent fractions will be helpful. Use this framework even if you know another way to build equivalent fractions.

EXAMPLE

14. Build $\frac{5}{6}$ into an equivalent fraction with a denominator of 42.

The missing multiplier is 7 because $6 \cdot 7 = 42$.

Step 1 Set up the framework.
Step 2 Determine the missing multiplier.
Step 3 Multiply the numerator and denominator by the missing multiplier.

$$\frac{5}{6} \cdot \left(\frac{7}{7}\right) = \frac{35}{42}$$

$\frac{5}{6}$ and $\frac{35}{42}$ are equivalent fractions.

GUIDED PRACTICE

14. Build $\frac{3}{7}$ into an equivalent fraction with a denominator of 21.

The missing multiplier is ____ because ________.

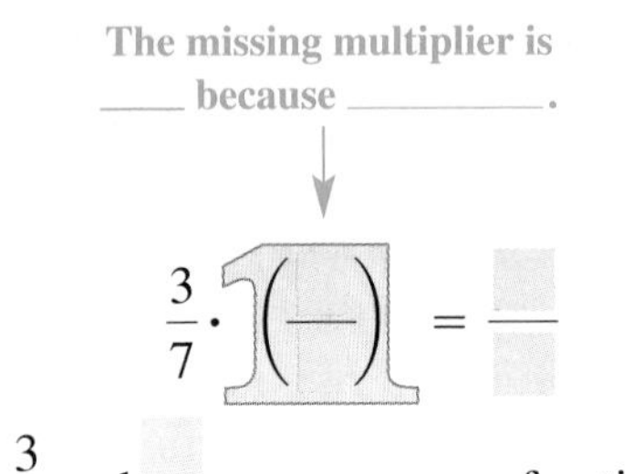

$\frac{3}{7}$ and $\frac{\square}{\square}$ are ________ fractions.

Concept Check

C1. Write three different fractions equal to one.

C2. Use the part–whole concept of fractions to describe why the fractions from Exercise C1 are equal to one.

C3. Use the quotient concept of fractions (division) to describe why the fractions from Exercise C1 are equal to one.

Answer each question without multiplying the fractions.

C4. $\frac{3}{7} \cdot \frac{5}{5} = \frac{3}{7}$ True or False? Why?

C5. $7 \cdot \frac{9}{5} = 7$ True or False? Why?

Objective C Practice

Fill in the framework to build an equivalent fraction with the specified denominator.

C6. $\frac{1}{4} \cdot \left(\frac{\square}{\square}\right) = \frac{\square}{12}$

C7. $\frac{1}{3} \cdot \left(\frac{\square}{\square}\right) = \frac{\square}{15}$

Create an equivalent fraction with the specified denominator.

C8. $\dfrac{3}{4} = \dfrac{\quad}{16}$ **C9.** $\dfrac{5}{3} = \dfrac{\quad}{21}$ **C10.** $\dfrac{7}{9} = \dfrac{\quad}{27}$ **C11.** $\dfrac{11}{12} = \dfrac{\quad}{24}$

C12. $\dfrac{16}{21} = \dfrac{\quad}{84}$ **C13.** $\dfrac{5}{23} = \dfrac{\quad}{46}$ **C14.** $\dfrac{10}{13} = \dfrac{\quad}{39}$ **C15.** $\dfrac{15}{16} = \dfrac{\quad}{64}$

C16. Six people will share the remaining brownies in the pan.

a. If they split the remaining brownies equally, what portion of a whole pan is each person's piece?

b. Based on part a, we know that $\frac{1}{2}$ is equivalent to $\dfrac{6}{\text{what number?}}$.

C17. Two-thirds of Martha's strawberry pie remains when four friends come to visit.

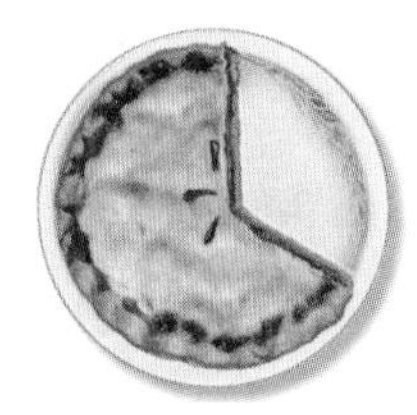

a. If they split the remaining pie equally, what portion of a whole pie is each person's piece?

b. Based on part a, we know that $\frac{2}{3}$ is equivalent to $\dfrac{4}{\text{what number?}}$.

Objective D Use a Fraction to Convert Units of Measure

The Concept Many applications of mathematics require conversions from one unit of measure to another. To convert units of measure, we multiply by a **units fraction**. A units fraction has equal measures in the numerator and denominator.

Commonly Used Equal Measures

Time	Volume
60 seconds (sec) = 1 minute (min)	8 ounces (oz) = 1 cup (c)
60 minutes = 1 hour (h)	2 cups = 1 pint (pt)
24 hours = 1 day (d)	2 pints = 1 quart (qt)
7 days = 1 week (wk)	4 quarts = 1 gallon (gal)
365 days = 1 year (yr)	
Length	**Weight**
12 inches (in.) = 1 foot (ft)	16 ounces (oz) = 1 pound (lb)
3 feet = 1 yard (yd)	2,000 pounds = 1 ton
5,280 feet = 1 mile (mi)	

Because the units of measure are equal in a units fraction, units fractions are equal to one. For example, seven days and one week are equal measures of time. If we make a fraction with these measures, the fraction is equal to one and is a units fraction.

$$\frac{7 \text{ days}}{1 \text{ week}} = 1$$

INTERACTIVE DEFINITION Units Fraction

A **units fraction** contains units of equal measure in the numerator and denominator. Units fractions are equal to 1.

EXAMPLE

15. Identify the two units fractions that convert between feet and yards.

Since 3 feet = 1 yard, the two units fractions are

$$\frac{3 \text{ ft}}{1 \text{ yd}} = 1 \quad \text{and} \quad \frac{1 \text{ yd}}{3 \text{ ft}} = 1.$$

GUIDED PRACTICE

15. Identify the two units fractions that convert between inches and feet.

Since ____ inches = ____ foot, the two units fractions are

$$\frac{}{} = 1 \text{ and } \frac{}{} = 1$$

DO YOU UNDERSTAND how to build units fractions? Got It Get Help

In the previous example, we identified the two units fractions that convert between feet and yards.

$$\frac{3 \text{ ft}}{1 \text{ yd}} \quad \text{or} \quad \frac{1 \text{ yd}}{3 \text{ ft}}$$

The units fraction that we use depends on whether we are converting from yards to feet or feet to yards. This is demonstrated in the following example.

EXAMPLE

16. Convert 12 yards to feet.

Step 1 Identify equal measures for the units.

$$1 \text{ yard} = 3 \text{ feet}$$

Step 2 Identify the two units fractions.

$$\frac{3 \text{ feet}}{1 \text{ yard}} = 1 \quad \text{and} \quad \frac{1 \text{ yard}}{3 \text{ feet}} = 1$$

Step 3 Choose the units fraction that is needed for this conversion. This units fraction will have $\frac{\text{new units}}{\text{original units}}$.

$$\frac{\text{New units}}{\text{Original units}} = \frac{3 \text{ feet}}{1 \text{ yard}}$$

Step 4 Multiply by the units fraction to perform the conversion.

$$\frac{12 \cancel{\text{yards}}}{1} \cdot \left(\frac{3 \text{ feet}}{1 \cancel{\text{yard}}}\right) = \frac{36 \text{ feet}}{1}$$
$$= 36 \text{ feet}$$

GUIDED PRACTICE

16. Convert 5 pints to cups.

Step 1 Identify equal measures for the units.

___ pint = ___ cups

Step 2 Identify the two units fractions.

$$\frac{}{} = 1 \quad \text{and} \quad \frac{}{} = 1$$

Step 3 Choose the units fraction that is needed for this conversion. This units fraction will have $\frac{\text{new units}}{\text{original units}}$.

$$\frac{\text{New units}}{\text{Original units}} = \frac{}{}$$

Step 4 Multiply by the units fraction to perform the conversion.

$$\frac{5 \cancel{\text{pints}}}{1} \cdot \left(\frac{}{}\right) = \frac{}{1}$$
$$=$$

Procedure Use a Fraction to Convert Units of Measure

Step 1 Write the original quantity as a fraction.

Step 2 Multiply by the units fraction that will divide out the original units and introduce the new units.

Step 3 Simplify to convert the units.

EXAMPLES

17. Convert 3 tons to pounds (lb).

Step 1 Identify equal measures for the units. 1 ton = 2,000 lb

Step 2 Multiply by a units fraction that will divide out the original units.

$$\frac{3 \text{ tons}}{1} \cdot \left(\frac{2{,}000 \text{ lb}}{1 \text{ ton}}\right) = \frac{6{,}000 \text{ lb}}{1}$$

$$= 6{,}000 \text{ lb}$$

Step 3 Simplify the fraction to convert the units.

18. Convert 60 inches (in.) to feet (ft).

Step 1 Write the original quantity as a fraction.

$$60 \text{ in.} = \frac{60 \text{ in.}}{1}$$

Step 2 Multiply by a units fraction that will divide out the original units.

$$= \frac{60 \text{ in.}}{1} \cdot \left(\frac{1 \text{ ft}}{12 \text{ in.}}\right)$$

Step 3 Simplify.

$$= \frac{60 \text{ ft}}{12}$$

$$= 5 \text{ feet}$$

19. Can 2 gallons (gal) completely fill a 6-quart (qt) container?

Convert 2 gallons into quarts.

Step 1 Write the original quantity as a fraction.

$$2 \text{ gal} = \frac{2 \text{ gal}}{1}$$

Step 2 Multiply by a units fraction that will divide out the original units.

$$= \frac{2 \text{ gal}}{1} \cdot \left(\frac{4 \text{ qt}}{1 \text{ gal}}\right)$$

$$= 8 \text{ quarts}$$

Step 3 Simplify.

Yes, 2 gallons will fill a 6-quart container.

GUIDED PRACTICE

17. Convert 4 miles to feet.

___ mile = ______ feet

$$\frac{4 \text{ miles}}{1} \cdot \left(\frac{\qquad}{\qquad}\right) = \frac{\qquad \text{feet}}{1}$$

$$=$$

18. Convert 28 cups to pints.

$$28 \text{ cups} = \frac{\qquad}{1}$$

$$= \frac{\qquad}{1} \cdot \frac{\qquad}{\qquad}$$

$$=$$

19. Is 6 pounds (lb) of chocolate enough to complete a recipe that calls for 100 ounces (oz) of chocolate?

Convert ________ into ________.

$$6 \text{ lb} = \frac{\qquad}{1}$$

$$= \frac{\qquad}{1} \cdot$$

$$=$$

Yes/No, 6 pounds of chocolate will/will not be enough to complete the recipe.

Concept Check

D1. Write the two units fractions that can be used to convert between hours and minutes.

D2. Write the two units fractions that can be used to convert between quarts and gallons.

D3. Simplify.

$$\frac{8\text{ qt}}{1} \cdot \frac{2\text{ pints}}{1\text{ quart}}$$

D4. Simplify.

$$\frac{4{,}000\text{ lb}}{1} \cdot \frac{1\text{ ton}}{2{,}000\text{ pounds}}$$

Objective D Practice

Write the appropriate units fraction and perform the conversion.

D5. $\frac{5\text{ ft}}{1} \cdot \left(\frac{\quad\text{ in.}}{\quad}\right) = \quad$ inches

D6. $\frac{4\text{ yd}}{1} \cdot \left(\frac{\quad\text{ ft}}{\quad}\right) = \quad$ feet

D7. $\frac{8\text{ cups}}{1} \cdot \left(\frac{\quad\text{ pint}}{\quad}\right) = \quad$ pints

D8. $\frac{12\text{ pints}}{1} \cdot \left(\frac{\quad\text{ quarts}}{\quad}\right) = \quad$ quarts

Use a units fraction to perform each conversion.

D9. How many feet are in 7 yards?

D10. How many days are in 9 weeks?

D11. 192 ounces is equal to how many pounds?

D12. 336 hours is equal to how many days?

D13. Can 64 ounces of soda completely fill a 6-cup container?

D14. A recipe calls for 7 cups of milk. Will 3 pints of milk be enough to complete the recipe?

Combining Concepts and Applications

CONCEPT I Evaluating an Exponent on a Fraction To evaluate an exponent, use repeated multiplication.

EXAMPLE	GUIDED PRACTICE
20. Simplify. $\left(\frac{2}{5}\right)^3$	**20.** Simplify. $\left(\frac{2}{3}\right)^4$
Step 1 Write the repeated multiplication. $\left(\frac{2}{5}\right)^3 = \frac{2}{5} \cdot \frac{2}{5} \cdot \frac{2}{5}$ **Step 2** Multiply the numerators and denominators. $= \frac{2 \cdot 2 \cdot 2}{5 \cdot 5 \cdot 5}$ $= \frac{4 \cdot 2}{25 \cdot 5}$ $= \frac{8}{125}$	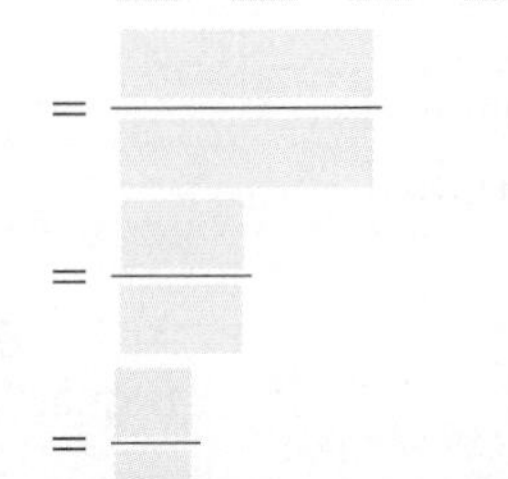

CONCEPT II Evaluating Expressions with More Than One Operation When there is more than one operation in an expression, you must follow the order of operations.

First: Perform operations in groupings.

Second: Perform operations with exponents.

Third: Perform multiplication and division as they occur from left to right.

Fourth: Perform addition and subtraction as they occur from left to right.

EXAMPLE

21. Simplify. $\frac{1}{2}\cdot\left(\frac{5}{8}\cdot\frac{16}{25}\right)^2$

Groupings $\frac{1}{2}\cdot\left(\frac{5}{8}\cdot\frac{16}{25}\right)^2 = \frac{1}{2}\cdot\left(\frac{\cancel{5}\cdot\cancel{8}\cdot 2}{\cancel{8}\cdot\cancel{5}\cdot 5}\right)^2$

Exponents $= \frac{1}{2}\cdot\left(\frac{2}{5}\right)^2$

$= \frac{1}{2}\cdot\left(\frac{2}{5}\cdot\frac{2}{5}\right)$

Multiply $= \frac{1\cdot\cancel{2}\cdot 2}{\cancel{2}\cdot 5\cdot 5}$

$= \frac{2}{25}$

GUIDED PRACTICE

21. Simplify. $\frac{2}{3}\cdot\left(\frac{9}{8}\cdot\frac{4}{3}\right)^3$

$\frac{2}{3}\cdot\left(\frac{9}{8}\cdot\frac{4}{3}\right)^3 =$

$=$

$=$

$=$

$=$

CONCEPT III Application Exercises To find a fractional part of a quantity, we multiply by that fraction. For example, to find one-twelfth of a quantity, we multiply the quantity by $\frac{1}{12}$. Recall that the word *of* often indicates multiplication when written in a word exercise.

EXAMPLE

22. Terrance has a rectangular backyard that is 60 ft wide by 70 ft long. He installed a swimming pool that occupies $\frac{1}{12}$ of that area. What is the area of the swimming pool?

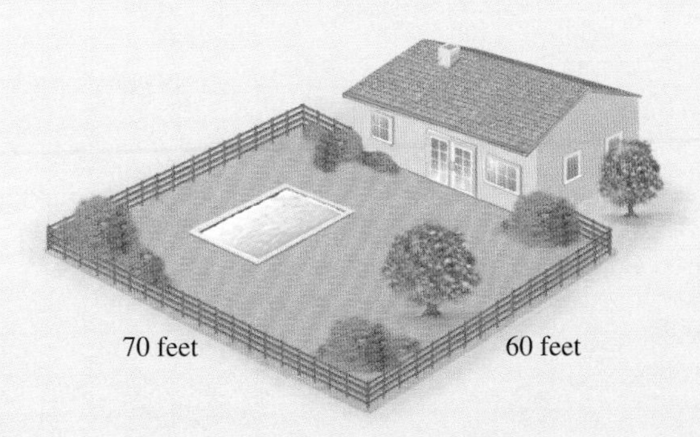

(Continued)

Step 1 Determine the area of the backyard. The area of a rectangle is $A = l \cdot w$.

The area of the yard $= 60 \text{ ft} \cdot 70 \text{ ft}$
$= 4{,}200 \text{ ft}^2$

Step 2 Multiply the area of the yard by the fractional part that the pool occupies. To find $\frac{1}{12}$ of the yard's area, multiply the area by $\frac{1}{12}$.

The area of the pool = "One-twelfth" of "the backyard"

$$= \frac{1}{12} \cdot \frac{4{,}200 \text{ ft}^2}{1}$$

$$= \frac{4{,}200 \text{ ft}^2}{12}$$

Step 3 Simplify.

$$= \frac{\cancel{3} \cdot \cancel{4} \cdot 350 \text{ ft}^2}{\cancel{3} \cdot \cancel{4}}$$

$$= 350 \text{ ft}^2$$

The area of the pool is 350 ft^2.

GUIDED PRACTICE

22. Jill has a rectangular backyard that is 30 ft wide by 60 ft long. She installed a garden that occupies $\frac{1}{9}$ of that area. What is the area of the garden?

Step 1 Determine the area of the backyard. The area of a rectangle is $A = l \cdot w$.

Area of the yard = __________
= __________ ft^2

Step 2 Multiply by the area of the yard by the fractional part that the garden occupies. To find $\frac{1}{9}$ of the yard's area, multiply the area by $\frac{1}{9}$.

Area of the garden = "__________" of "__________"

Area of the garden =

=

=

Step 3 Simplify.

=

The area of the garden is ______ ft^2.

This flowchart is presented to help you organize your thoughts about how to multiply fractions.

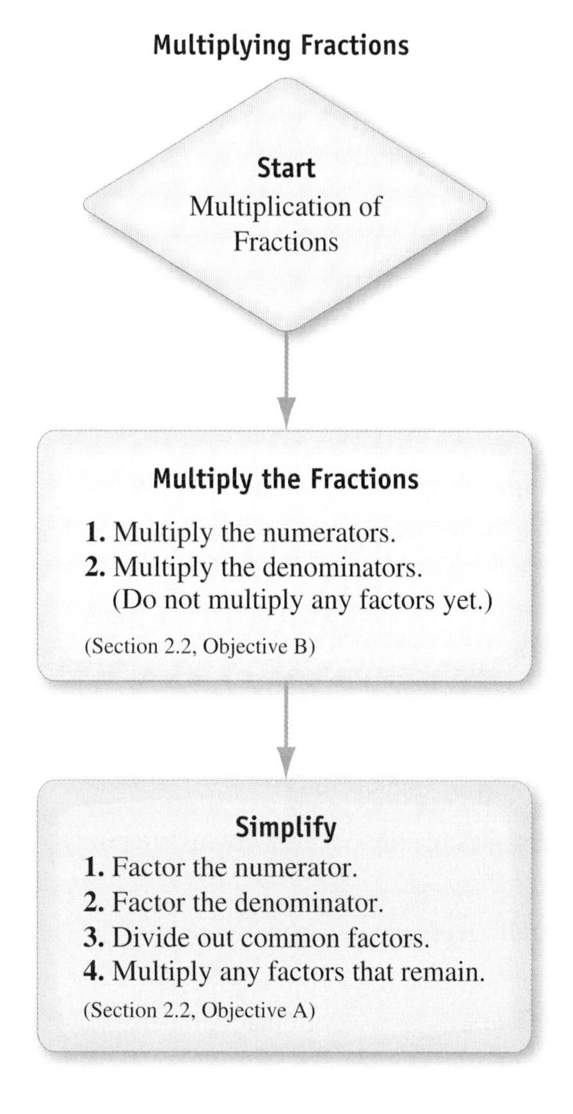

Section 2.2 Exercises

REVIEW

To simplify a fraction:

1. Answer the Objective A Concept Check.
2. Answer the odd Objective A Practice Exercises.
3. Answer the even Objective A Practice Exercises.

To estimate a fraction:

4. Answer the odd Focus on Estimating Exercise.
5. Answer the even Focus on Estimating Exercises.

To multiply two fractions:

6. Answer the Objective B Concept Checks.
7. Answer the odd Objective B Practice Exercises.
8. Answer the even Objective B Practice Exercises.

To build equivalent fractions:

9. Answer the Objective C Concept Checks.
10. Answer the odd Objective C Practice Exercises.
11. Answer the even Objective C Practice Exercises.

To use a fraction to convert units of measure:

12. Answer the Objective D Concept Checks.
13. Answer the odd Objective D Practice Exercises.
14. Answer the even Objective D Practice Exercises.

VOCABULARY REVIEW *Review the Vocabulary Preview for Section 2.2. Study the definitions until you can check* Got It *for every word.*

Use these words to complete each sentence.

factor • simplify/reduce • greatest common factor • missing multiplier • simplest form • equivalent • units fraction • quotient concept

	Got It	Get Help
15. A fraction is in ____________ when it has no common factors other than one in the numerator and denominator.		
16. To ____________ a fraction, write it in simplest form.		
17. Two fractions that represent the same quantity but are written differently are ____________.		
18. A fraction that is written with equivalent measures and is equal to 1 is a ____________.		
19. To ____________ a number, write it as a product of other numbers.		
20. When building equivalent fractions, you must multiply the numerator and denominator of one fraction by the ____________ to build the equivalent fraction.		

How will you get help for any vocabulary that you are unsure about?

Your instructor ____ MyMathLab ____ A classmate ____ A tutor ____ Other ____

Simplify each fraction.

21. $\frac{6}{9}$
22. $\frac{4}{8}$
23. $\frac{25}{35}$
24. $\frac{10}{40}$
25. $\frac{36}{54}$
26. $\frac{49}{77}$
27. $\frac{10}{210}$
28. $\frac{5}{65}$
29. $\frac{99}{117}$
30. $\frac{85}{135}$
31. $\frac{42}{85}$
32. $\frac{143}{187}$

33. In your own words, what steps are used to simplify a fraction?

34. Why is it important to simplify fractions? Use an example of a fraction and its simplified form to explain your answer.

Multiply and simplify the result. Remember to divide out common factors before performing any multiplication.

35. $\frac{1}{5}\cdot\frac{3}{4}$ **36.** $\frac{3}{8}\cdot\frac{1}{7}$ **37.** $\frac{5}{18}\cdot\frac{9}{2}$ **38.** $\frac{4}{3}\cdot\frac{7}{12}$

39. $\frac{6}{5}\cdot\frac{9}{12}$ **40.** $\frac{3}{18}\cdot\frac{12}{5}$ **41.** $\frac{73}{18}\cdot\frac{9}{146}$ **42.** $\frac{4}{21}\cdot\frac{7}{20}$

43. $\frac{9}{4}\cdot\frac{10}{15}$ **44.** $\frac{3}{12}\cdot\frac{15}{9}$ **45.** $8\cdot\frac{1}{4}$ **46.** $9\cdot\frac{1}{3}$

47. $\frac{3}{16}\cdot\frac{4}{2}\cdot\frac{2}{9}$ **48.** $\frac{6}{10}\cdot\frac{5}{36}\cdot\frac{5}{3}$ **49.** $\frac{14}{25}\cdot\frac{15}{21}\cdot\frac{3}{4}$ **50.** $\frac{12}{15}\cdot\frac{10}{15}\cdot\frac{5}{8}$

51. In your own words, what steps are used to multiply two fractions?

Write the repeated multiplication for each exponent. Do not perform the multiplication.

52. $\left(\frac{3}{5}\right)^4$ **53.** $\left(\frac{2}{7}\right)^3$

Follow the order of operations to simplify each expression.

54. $\left(\frac{4}{5}\right)^2$ **55.** $\left(\frac{6}{7}\right)^2$ **56.** $\left(\frac{1}{2}\right)^3\cdot\frac{16}{21}$ **57.** $\left(\frac{1}{3}\right)^3\cdot\frac{9}{20}$

58. $\left(\frac{7}{3}\cdot\frac{5}{21}\right)^2$ **59.** $\left(\frac{5}{6}\cdot\frac{3}{20}\right)^2$ **60.** $\frac{9}{15}\cdot\left(\frac{15}{64}\cdot4^3\right)$ **61.** $\frac{9}{12}\cdot\left(\frac{15}{27}\cdot3^3\right)$

Create an equivalent fraction with the given denominator.

62. $\frac{7}{5}=\frac{?}{25}$ **63.** $\frac{2}{9}=\frac{?}{81}$ **64.** $\frac{12}{5}=\frac{?}{60}$ **65.** $\frac{14}{16}=\frac{?}{64}$

66. $\frac{21}{11}=\frac{?}{99}$ **67.** $\frac{11}{12}=\frac{?}{48}$ **68.** $\frac{13}{14}=\frac{?}{28}$ **69.** $\frac{7}{16}=\frac{?}{48}$

70. How can you identify a fraction equal to one?

Write the appropriate units fraction to complete the conversion.

71. $\frac{10\text{ ft}}{1}\cdot\left(\frac{\text{in.}}{}\right)=$ inches **72.** $\frac{14\text{ yd}}{1}\cdot\left(\frac{\text{ft}}{}\right)=$ feet

73. $\frac{98\text{ days}}{1}\cdot\left(\frac{\text{week}}{}\right)=$ weeks **74.** $\frac{146}{}\cdot\left(\frac{\text{quart}}{}\right)=$ quarts

Convert each quantity to the desired unit.

75. Convert 3 pounds to ounces.

76. Convert 6 days to hours.

77. How many ounces are in 15 cups?

78. How many seconds are in 32 minutes?

79. How many pounds is 320 ounces equal to?

80. How many days is 432 hours equal to?

Use fractions to answer each question.

81. Dewey recently purchased a computer and a wireless router for his home. The computer cost $675. The wireless router cost $75. What fractional part of the total cost was the wireless router? Your answer should be in simplest form.

82. Alexandria purchased 200 new books for the library. If the books had been purchased individually, their total cost would have been $4,500. However, the library was given a $1,000 discount. As a fraction in simplest form, what discount did the library receive?

83. In a recent survey of 216 college students, the results showed that $\frac{1}{3}$ of the students prefer rap music; $\frac{1}{6}$ of the students prefer country music, and $\frac{1}{2}$ of the students prefer rock music.

a. How many students prefer rap music?
b. How many students prefer country music?
c. How many students prefer rock music?

84. When a pair of six-sided dice are rolled, a sum of 7 is likely to occur $\frac{1}{6}$ of the time; a sum of 9 is likely to occur $\frac{1}{9}$ of the time, and a sum of 11 is likely to occur $\frac{1}{18}$ of the time. Answer the following questions assuming that James rolled a pair of six-sided dice 36 times.

a. How many 7s did James likely roll?
b. How many 9s did James likely roll?
c. How many 11s did James likely roll?

85. Tiny Tom's Garden Center is having a $\frac{1}{3}$-off sale. Before the discount, your purchase totaled $231.

a. How much did you save on your purchase?
b. How much did you spend on your purchase?

86. Crazy Kurt's Fireworks is going out of business. To liquidate inventory, the store is having a $\frac{1}{4}$-off sale. Before the discount, you have put $148 worth of merchandise in your cart.

a. How much will you save on your purchase?
b. How much will you spend on your purchase?

87. Amy is making a rectangular planter box for growing fresh herbs. The planter box is 2 yards by $\frac{1}{2}$ yard. What is the area of the planter box in square yards?

88. You answered 16 of the 17 questions on an English test. If $\frac{3}{4}$ of the questions you answered were correct, what fraction of all of the questions did you answer correctly?

89. Last month John earned $2,000. He spent $\frac{2}{5}$ of this amount on housing, $\frac{1}{5}$ on transportation, and $\frac{1}{10}$ on a credit card payment.

a. In total, how much did John spend on these expenses?
b. How much did John have left for other expenses?

90. To approve a real estate development project, the city council needs to have a $\frac{2}{3}$-majority vote from the council members. There are 15 members on the city council.

a. How many members must vote for the development project to approve it?
b. How many members must vote against the development project to stop it?

Think carefully about part b.

Section 2.2 Question Log

Use this space to write questions. Be sure to get these answered and revisit them when you prepare for your exam.

Page ______ **Answered**	**Page** ______ **Answered**
Page ______ **Answered**	**Page** ______ **Answered**

Section 2.3 Dividing Fractions

If you can multiply fractions, you can divide fractions. It's that simple. What is nice about dividing fractions is that there is a procedure that changes a division exercise into a multiplication exercise.

The **Objectives** in Section 2.3 will help you

A Find the reciprocal of a number.
B Divide one fraction by another.

VOCABULARY PREVIEW *Check the box that applies.*

	Got It	Must Study
Reciprocal: Two numbers are **reciprocals** if their product is 1. Example: $\frac{3}{5}$ and $\frac{5}{3}$ are reciprocals since $\frac{3}{5} \cdot \frac{5}{3} = \frac{15}{15} = 1$.		
Dividend: The number being divided is the **dividend.**		
Divisor: The number that does the dividing is the **divisor.**		

Study these words when they appear in the text. After the section, test your understanding by completing the Vocabulary Review in the section exercises.

Objective A Find the Reciprocal of a Number

The Concept To understand why we need reciprocals, let's look ahead to division with fractions. To divide by a fraction, we multiply by the reciprocal of the fraction. Since we need the reciprocal to perform division, we must understand how to find it.

INTERACTIVE DEFINITION Reciprocal

Two numbers are **reciprocals** if their product is 1.

EXAMPLE	GUIDED PRACTICE
1. Find the reciprocal of $\frac{3}{4}$.	**1.** Find the reciprocal of $\frac{5}{7}$.
What fraction, when multiplied by $\frac{3}{4}$, gives a product of one? $\frac{3}{4} \cdot \frac{?}{?} = 1$	$\frac{5}{7} \cdot \frac{?}{?} = 1$

The reciprocal's numerator is the original fraction's denominator. The reciprocal's denominator is the original fraction's numerator.

$\frac{3}{4} \cdot \frac{4}{3} = 1$

$\frac{5}{7} \cdot \frac{\quad}{\quad} = 1$

The reciprocal of a fraction is found by interchanging the numerator and the denominator, or "flipping" the fraction.

The reciprocal of $\frac{3}{4}$ is $\frac{4}{3}$.

The reciprocal of $\frac{5}{7}$ is $\frac{\quad}{\quad}$.

DO YOU UNDERSTAND how to find a reciprocal? Got It Get Help

Procedure Find a Reciprocal

Step 1 If the number is a whole number, write it as a fraction.

Example: $3 = \frac{3}{1}$ (3 can be written as 3 over 1.)

Step 2 Interchange the numerator and denominator. "Flip" the fraction.

▶ **NOTE** Zero does not have a reciprocal because division by zero is not defined. Think about what the reciprocal of zero would look like: $\frac{1}{0}$, or $1 \div 0$, which is undefined.

EXAMPLES

2. Find the reciprocal of $\frac{8}{3}$.

Step 1 If the number is a whole number, write it as a fraction.

$\frac{8}{3}$ is already a fraction.

Step 2 Interchange the numerator and denominator. "Flip" the fraction.

$\frac{8}{3} \Rightarrow \frac{3}{8}$

The reciprocal of $\frac{8}{3}$ is $\frac{3}{8}$.

GUIDED PRACTICE

2. Find the reciprocal of $\frac{65}{23}$.

$\frac{65}{23}$ is already a fraction.

$\frac{65}{23} \Rightarrow \frac{\quad}{\quad}$

The reciprocal of $\frac{65}{23}$ is $\frac{\quad}{\quad}$.

(Continued)

3. Find the reciprocal of 17.	**3.** Find the reciprocal of 96.
Step 1 If the number is a whole number, write it as a fraction. $17 = \frac{17}{1}$ **Step 2** Interchange the numerator and denominator. "Flip" the fraction. $\frac{17}{1} \Rightarrow \frac{1}{17}$ The reciprocal of 17 is $\frac{1}{17}$.	$96 = \frac{\quad}{\quad}$ The reciprocal of 96 is $\frac{\quad}{\quad}$.

Concept Check

A1. Any number times ____________ is equal to one.

A2. Multiply the following fractions. Once you see the pattern, you will be able to do all of these problems in your head. In part f, neither a nor b are zero.

a. $\frac{6}{7} \cdot \frac{7}{6}$ **b.** $\frac{54}{12} \cdot \frac{12}{54}$ **c.** $\frac{123}{321} \cdot \frac{321}{123}$

d. $\frac{9}{10} \cdot \frac{10}{9}$ **e.** $\frac{164}{73} \cdot \frac{73}{164}$ **f.** $\frac{a}{b} \cdot \frac{b}{a}$ ← This problem may look difficult but it's really quite easy.

Objective A Practice

Find the reciprocal of each number. Simplify if possible.

A3. $\frac{9}{7}$ **A4.** $\frac{5}{2}$ **A5.** $\frac{3}{13}$ **A6.** $\frac{21}{52}$

A7. $\frac{1}{3}$ **A8.** $\frac{1}{9}$ **A9.** 22 **A10.** 16

Objective B Divide One Fraction by Another

The Concept The following pattern demonstrates the procedure used to divide one fraction by another. To understand this pattern, remember that we can perform the division exercise $4 \div 2$ by identifying how many 2's go into 4.

Evaluate $4 \div 2 = 2$ by finding how many 2's go into 4.

4

2 2

Two 2's go into 4.

$$4 \div 2 = 2 \quad \text{is the same as} \quad \frac{4}{1} \cdot \frac{1}{2} = 2.$$

To divide by 2, we can multiply by the reciprocal of 2.

Evaluate $4 \div \frac{1}{2}$ by finding how many $\frac{1}{2}$'s go into 4.

4

$\frac{1}{2}$ $\frac{1}{2}$ $\frac{1}{2}$ $\frac{1}{2}$ $\frac{1}{2}$ $\frac{1}{2}$ $\frac{1}{2}$ $\frac{1}{2}$

Eight $\frac{1}{2}$'s go into 4.

$$4 \div \frac{1}{2} = 8 \quad \text{is the same as} \quad 4 \cdot 2 = 8.$$

To divide by $\frac{1}{2}$, we can multiply by the reciprocal of $\frac{1}{2}$.

Evaluate $4 \div \frac{1}{4}$ by finding how many $\frac{1}{4}$'s go into 4.

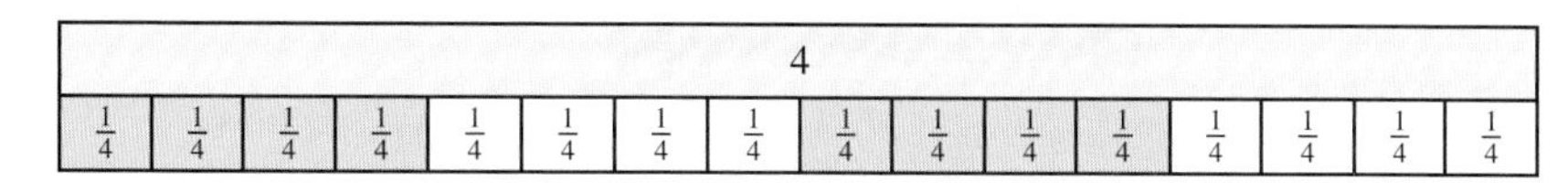

Sixteen $\frac{1}{4}$'s go into 4.

$$4 \div \frac{1}{4} = 16 \quad \text{is the same as} \quad 4 \cdot 4 = 16.$$

To divide by $\frac{1}{4}$, we can multiply by the reciprocal of $\frac{1}{4}$.

Based on the pattern, to divide by a fraction, you can multiply by the reciprocal of the fraction.

Procedure Divide One Fraction by Another

Step 1 (Skip) Rewrite the first fraction.

Step 2 (Flip) Write the reciprocal of the second fraction.

Step 3 (Multiply) Change the division symbol to multiplication.

Step 4 Simplify and perform the multiplication.

DETAILED EXAMPLE Dividing One Fraction by Another

Divide. $\frac{5}{3} \div \frac{7}{4}$

Step 1 (Skip) Rewrite the first fraction.

$$\frac{5}{3} \div \frac{7}{4} = \frac{5}{3}$$

Step 2 (Flip) Write the reciprocal of the second fraction.

$$\frac{5}{3} \div \frac{7}{4} = \frac{5}{3} \quad \frac{4}{7}$$

Step 3 (Multiply) Change the division symbol to multiplication.

$$\frac{5}{3} \div \frac{7}{4} = \frac{5}{3} \cdot \frac{4}{7}$$

Simplify and perform the multiplication.

$$\frac{5}{3} \div \frac{7}{4} = \frac{5}{3} \cdot \frac{4}{7} = \frac{20}{21}$$

EXAMPLES

4. Divide. $\frac{3}{5} \div \frac{9}{10}$

Skip, flip, and multiply. $\frac{3}{5} \div \frac{9}{10} = \frac{3}{5} \cdot \frac{10}{9}$

Simplify. $= \frac{\cancel{3} \cdot 2 \cdot \cancel{5}}{\cancel{5} \cdot \cancel{3} \cdot 3}$

$= \frac{2}{3}$

Simplify before mutiplying.

5. Divide. $\frac{1}{6} \div \frac{4}{9}$

Skip, flip, and multiply. $\frac{1}{6} \div \frac{4}{9} = \frac{1}{6} \cdot \frac{9}{4}$

Simplify. $= \frac{1 \cdot \cancel{3} \cdot 3}{2 \cdot \cancel{3} \cdot 4}$

$= \frac{3}{8}$

Simplify before mutiplying.

GUIDED PRACTICE

4. Divide. $\frac{4}{7} \div \frac{16}{35}$

$\frac{4}{7} \div \frac{16}{35} = \frac{4}{7} \cdot \frac{\quad}{\quad}$

$= \frac{\quad}{\quad}$

$= \frac{\quad}{\quad}$

5. Divide. $\frac{3}{8} \div \frac{5}{2}$

$\frac{3}{8} \div \frac{5}{2} = \frac{\quad}{\quad} \cdot \frac{\quad}{\quad}$

$= \frac{\quad}{\quad}$

$= \frac{\quad}{\quad}$

6. Divide. $\frac{6}{5} \div 2$

Write any whole numbers as fractions. $\frac{6}{5} \div 2 = \frac{6}{5} \div \frac{2}{1}$

Skip, flip, and multiply. $= \frac{6}{5} \cdot \frac{1}{2}$

Simplify. $= \frac{\cancel{2} \cdot 3 \cdot 1}{5 \cdot \cancel{2}}$

$= \frac{3}{5}$

6. Divide. $4 \div \frac{3}{5}$

$4 \div \frac{3}{5} = \frac{}{} \div \frac{3}{5}$

$=$

$=$

$=$

7. If a carpenter needs $\frac{1}{2}$ bottle of glue to make a table, how many tables can she make from 6 bottles of glue?

This question asks us to find how many $\frac{1}{2}$'s fit into 6.

Use division to answer the question.

Write the whole number as a fraction. $6 \div \frac{1}{2} = \frac{6}{1} \div \frac{1}{2}$

Skip, flip, and multiply. $= \frac{6}{1} \cdot \frac{2}{1}$

$= \frac{12}{1}$

$= 12$

She can make 12 tables from 6 bottles of glue.

7. If a child grows $\frac{3}{4}$ of an inch each year, how long will it take the child to grow 3 inches?

This question asks us to find how many —'s fit into ____.

Use ____________ to answer the question.

$3 \quad \frac{3}{4} =$

$=$

$=$

$=$

It will take the child ____ years to grow 3 inches.

Concept Check

B1. Show that each pair of exercises gives the same answer.

a. $8 \div 2$ and $8 \cdot \frac{1}{2}$

b. $12 \div 4$ and $12 \cdot \frac{1}{4}$

c. $36 \div 9$ and $36 \cdot \frac{1}{9}$

B2. Based on Exercise B1, division by a number and multiplication by the number's ________ are equivalent.

B3. If you "flip" a fraction, you are finding its ________.

B4. A friend called you on the phone, asking for help with the exercise $\frac{4}{3} \div \frac{5}{6}$. Write every step your friend should use to solve this exercise. To see if your steps are correct, read them to a classmate and have him or her solve the exercise following your steps. Did your steps lead your classmate to a correct solution?

Objective B Practice

Divide.

B5. $\frac{1}{2} \div \frac{3}{4}$

B6. $\frac{3}{7} \div \frac{3}{5}$

B7. $\frac{12}{16} \div \frac{9}{4}$

B8. $\frac{15}{8} \div \frac{35}{6}$

B9. $\frac{36}{11} \div 18$

B10. $\frac{5}{18} \div 15$

B11. $5 \div \frac{1}{2}$

B12. $8 \div \frac{2}{5}$

B13. If a tree grows $\frac{1}{8}$ inch in diameter each year, how many years will it take the tree to grow 5 inches in diameter?

B14. Mackenzie needs $\frac{1}{3}$ can of spray paint to paint a chair. If she has two cans of spray paint, can she paint her five kitchen chairs?

Combining Concepts and Applications

CONCEPT I Using the Order of Operations

The Order of Operations

First: Perform operations in groupings.
Second: Perform operations with exponents.
Third: Perform multiplication and division as they occur from left to right.
Fourth: Perform addition and subtraction as they occur from left to right.

To increase your accuracy, follow this process for *each step* of your solution.

- Choose the operation to be performed.
- Copy everything in the expression that will not change.
- Calculate the operation.
- Repeat this process for each operation in the expression.

EXAMPLE

8. Simplify. $\left(\frac{1}{2} \div \frac{2}{3}\right)^2 \cdot 5$

Groupings $\left(\frac{1}{2} \div \frac{2}{3}\right)^2 \cdot 5 = \left(\frac{1}{2} \cdot \frac{3}{2}\right)^2 \cdot 5$

Exponents $= \left(\frac{3}{4}\right)^2 \cdot 5$

$= \left(\frac{3}{4}\right) \cdot \left(\frac{3}{4}\right) \cdot 5$

Multiplication and division from left to right $= \left(\frac{9}{16}\right) \cdot \frac{5}{1}$

$= \frac{45}{16}$

GUIDED PRACTICE

8. Simplify. $\left(\frac{1}{3} \div \frac{3}{2}\right)^2 \cdot 7$

$\left(\frac{1}{3} \div \frac{3}{2}\right)^2 \cdot 7 = \left(\frac{\square}{\square} \cdot \frac{\square}{\square}\right)^2 \cdot 7$

$= \left(\frac{\square}{\square}\right)^2 \cdot 7$

=

=

=

CONCEPT II Application Exercises One way of thinking about division is to find how many times the divisor fits into the dividend. Many professionals, including carpenters and caterers, apply this concept of division to perform their jobs.

EXAMPLE

9. A carpenter wants to cut an 8-foot board into pieces that are $\frac{2}{3}$ foot long. How many $\frac{2}{3}$-foot boards can be made? Do not consider waste made by the saw.

Identify the arithmetic operation needed to solve the problem.

To find how many $\frac{2}{3}$-foot boards fit in an 8-foot board, divide 8 by $\frac{2}{3}$.

Perform the arithmetic operation.

$$8 \div \frac{2}{3} = \frac{8}{1} \cdot \frac{3}{2}$$
$$= \frac{4 \cdot \cancel{2} \cdot 3}{\cancel{2}}$$
$$= 12$$

Interpret the results.

The carpenter can make twelve $\frac{2}{3}$-foot boards from the 8-foot board.

GUIDED PRACTICE

9. A caterer provides seven cakes for a dinner party. If each piece of cake is $\frac{1}{16}$ of an entire cake, how many pieces are there?

Identify the arithmetic operation needed to solve the problem.

To find how many ______'s are in ______ cakes, divide ______ by ______.

Perform the arithmetic operation.

$$\quad \div \quad =$$
$$=$$
$$=$$

Interpret the results.

The 7 cakes have a total of ______ pieces.

This flowchart is presented to help you organize your thoughts about how to divide one fraction by another.

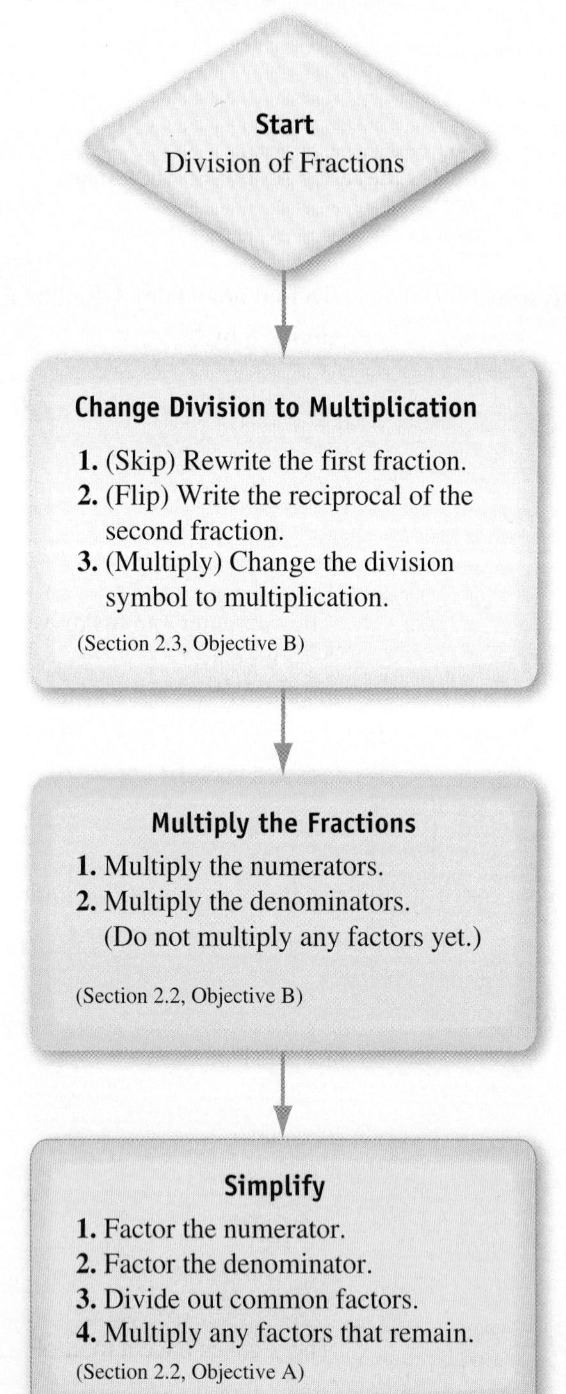

Section 2.3 Exercises

FOR EXTRA HELP **MyMathLab**  Math XL PRACTICE  WATCH  DOWNLOAD  READ REVIEW

To find the reciprocal of a number:

1. Answer the Objective A Concept Checks.
2. Answer the odd Objective A Practice Exercises.
3. Answer the even Objective A Practice Exercises.

To divide one fraction by another:

4. Answer the Objective B Concept Checks.
5. Answer the odd Objective B Practice Exercises.
6. Answer the even Objective B Practice Exercises.

VOCABULARY REVIEW *Review the Vocabulary Preview for Section 2.3. Study the definitions until you can check* Got It *for every word.*

Use these words to complete each sentence.

reciprocal • dividend • divisor

	Got It	Get Help
7. The number being divided is called the ____________.		
8. The number that does the dividing is called the ____________.		
9. A ____________ is formed by interchanging the numerator and denominator.		

How will you get help for any vocabulary that you are unsure about?

Your instructor ____ MyMathLab ____ A classmate ____ A tutor ____ Other ____

Divide. Write your answer in simplest form.

10. $\frac{7}{3} \div \frac{5}{6}$ **11.** $\frac{1}{5} \div \frac{3}{10}$ **12.** $\frac{1}{15} \div \frac{2}{45}$ **13.** $\frac{3}{2} \div \frac{9}{5}$

14. $\frac{30}{16} \div \frac{60}{32}$ **15.** $\frac{20}{9} \div \frac{50}{45}$ **16.** $\frac{7}{3} \div \frac{0}{9}$ **17.** $\frac{7}{9} \div \frac{0}{3}$

18. $\frac{0}{5} \div 6$ **19.** $\frac{0}{5} \div 3$ **20.** $\frac{4}{9} \div \frac{9}{4}$ **21.** $\frac{7}{12} \div \frac{12}{7}$

22. $10 \div \frac{5}{6}$ **23.** $20 \div \frac{5}{8}$ **24.** $\frac{27}{100} \div \frac{3}{20}$ **25.** $\frac{25}{64} \div \frac{5}{32}$

26. In your own words, how do you divide one fraction by another?

Simplify each expression, following the order of operations. Write your answer in simplest form.

27. $\frac{12}{7} \div \frac{24}{35}$ **28.** $\frac{9}{7} \div \frac{5}{14}$ **29.** $\frac{1000}{49} \cdot \frac{7}{100}$ **30.** $\frac{18}{14} \cdot \frac{7}{36}$

tobey When dividing by fractions and the # 0 is involved, we must be careful. What do you notice–what math property do we need to use? *For more tips and tweets, go to twitter.com/gstbasicmath*

31. $\frac{5}{3} \div \frac{3}{5} \cdot \frac{27}{10}$

32. $\frac{8}{7} \div \frac{7}{8} \cdot \frac{7}{8}$

33. $\left(\frac{5}{9} \cdot \frac{3}{10}\right)^2$

34. $\left(\frac{6}{15} \cdot \frac{5}{18}\right)^2$

35. $\left(\frac{1}{4} \div \frac{2}{3}\right)^2$

36. $\left(\frac{2}{9} \div \frac{4}{3}\right)^2$

37. $\left(\frac{7}{9}\right)^2 \div 49$

38. $\left(\frac{2}{3}\right)^2 \div 12$

39. $\frac{4}{9} \div 2^2$

40. $\frac{12}{15} \div 4^2$

41. $\frac{3}{5} \div \left(\frac{3}{5}\right)^2 \cdot 1$

42. $\frac{3}{5} \div \left(\frac{3}{2}\right)^2 \cdot 1$

43. $\frac{0}{3} \cdot \frac{7}{5}$

44. $\frac{6}{9} \cdot \frac{0}{6}$

45. $\left(4 \div \frac{1}{2}\right)^2 \div 8$

46. $\left(\frac{12}{5} \div 3\right)^2 \div 4$

47. Why is it wrong to try to divide $\frac{8}{3}$ by $\frac{0}{1}$?

48. Why is it okay to divide $\frac{0}{15}$ by $\frac{1}{3}$?

In each worked exercise, a student made a common mistake.
a. Identify the location of the mistake and explain why it is a mistake.
b. Identify the correct answer.

49. $\frac{3}{4} \div \frac{2}{9} = \frac{3}{4} \cdot \frac{2}{9}$
$= \frac{6}{36}$

50. $\frac{4}{7} \div \frac{21}{8} = \frac{\cancel{4}^{1}}{\cancel{7}_{1}} \div \frac{\cancel{21}^{3}}{_{2}\cancel{8}}$
$= \frac{1}{1} \div \frac{2}{3}$
$= \frac{2}{3}$

51. $3 \cdot \frac{3}{4} = \frac{3 \cdot 3}{3 \cdot 4}$
$= \frac{9}{12}$
$= \frac{3}{4}$

52. $\frac{5}{6} \div 6 = \frac{5}{6} \cdot \frac{6}{1}$
$= \frac{5}{\cancel{6}} \cdot \frac{\cancel{6}}{1}$
$= \frac{5}{1}$

Without performing any calculations, use the order of operations to answer each exercise.

53. a. Identify one pair of expressions that have identical answers.
b. Identify a second pair of expressions that have identical answers.

A. $\frac{1}{2} \div \left(\frac{3}{4} \cdot \frac{5}{6}\right)$ **B.** $\left(\frac{1}{2} \div \frac{3}{4}\right) \cdot \frac{5}{6}$ **C.** $\frac{1}{2} \cdot \left(\frac{3}{4} + \frac{5}{6}\right)$

D. $\left(\frac{1}{2} \cdot \frac{3}{4}\right) + \frac{5}{6}$ **E.** $\frac{1}{2} \div \frac{3}{4} \cdot \frac{5}{6}$ **F.** $\frac{1}{2} \cdot \frac{3}{4} + \frac{5}{6}$

54. a. Identify one pair of expressions that have identical answers.
b. Identify a second pair of expressions that have identical answers.

A. $\frac{1}{3} \div \left(\frac{1}{4} \cdot \frac{5}{6}\right)$ **B.** $\left(\frac{1}{3} \div \frac{1}{4} \cdot \frac{5}{6}\right)$ **C.** $\frac{1}{3} \cdot \frac{1}{4} + \frac{5}{6}$

D. $\left(\frac{1}{3} \cdot \frac{1}{4}\right) + \frac{5}{6}$ **E.** $\left(\frac{1}{3} \div \frac{1}{4}\right) \cdot \frac{5}{6}$ **F.** $\frac{1}{3} \cdot \left(\frac{1}{4} + \frac{5}{6}\right)$

Use multiplication or division to answer each question. Write any fractional answers in simplest form.

55. A nurse must determine how many patients can be treated with a prescription that requires a $\frac{3}{8}$-grain dosage. (Grains are a measure of weight.) The hospital has 21 grains in stock. How many patients can be treated?

56. A carpenter needs to build a 3-inch-thick board by stacking individual boards. How many $\frac{3}{4}$-inch-thick boards must be stacked to make a 3-inch board?

57. Ms. Crocker brought 13 pies to a picnic. She cut each pie into pieces that are $\frac{1}{8}$ of a pie. Will there be enough pie for 98 people?

58. Each player on a football team will be given a quarter cup of pickle juice at halftime to boost his electrolyte levels. If the trainer has 14 cups of pickle juice, is there enough juice for the 54 players on the team?

59. In a furniture factory, a machine can put a $\frac{1}{32}$-inch-thick layer of polyurethane on a table. How many layers must be applied to make the polyurethane $\frac{1}{8}$ inch thick?

60. The local supermarket packages green beans in $\frac{3}{4}$-pound packages. If green beans come in 21-pound boxes, how many packages can be made from one box?

61. A machinist is making parts from a steel rod. Each part requires $\frac{3}{8}$ inch of rod. What length of rod will make 24 parts?

62. A quilter has 50 yards of fabric. Each square requires $\frac{3}{8}$ yard of fabric. How much fabric is needed to make 20 squares for a quilt?

63. To determine the mileage of his car, Ingmar divides the number of miles traveled by the gallons of gas used. What is his car's mileage if he traveled 48 miles using $\frac{24}{10}$ gallons of gas? Make sure you state the units.

64. The manager of a bowling alley is oiling the lanes. It takes $\frac{5}{8}$ cup of oil for one lane. How many cups will it take to oil 40 lanes?

65. An orthodontist uses $\frac{7}{8}$ inch of wire to connect each tooth in a set of braces. How much wire is needed to connect 24 teeth?

66. A dentist uses $\frac{3}{16}$ ounce of ceramic for every filling. If the dentist has 6 ounces of ceramic, how many fillings can she make?

67. Geoff can run around a $\frac{3}{8}$-mile track in 2 minutes. How many laps must Geoff run to reach a total of 3 miles?

68. What happens when you divide one fraction by an equivalent fraction? Assume that neither fraction is equal to zero. Use the following as an example: $\frac{15}{24}$ and $\frac{5}{8}$ are equivalent fractions.

Section 2.3 Question Log

Use this space to write questions. Be sure to get these answered and revisit them when you prepare for your exam.

Page ____ Answered ☐	Page ____ Answered ☐
Page ____ Answered ☐	Page ____ Answered ☐

Section 2.4 Adding and Subtracting Fractions

In this section, you will learn why fractions need to have like denominators to be added and subtracted. Since fractions often have different denominators, you will learn how to build like fractions with a least common denominator (LCD). After learning this process, you will be able to add and subtract unlike fractions.

The **Objectives** in Section 2.4 will help you

- A Add and subtract like fractions.
- B Build like fractions by listing multiples.
- C Add and subtract unlike fractions: Apply Objective B.
- D Build like fractions by factoring.
- E Add and subtract unlike fractions: Apply Objective D.

VOCABULARY PREVIEW *Check the box that applies.*	Got It	Must Study
Like Fractions: Like fractions have the same denominator.		
Unlike Fractions: Unlike fractions have different denominators.		
Least Common Denominator (LCD): The **least common denominator (LCD)** of several fractions is the least common multiple of their denominators. (Review LCM in Section 1.8.)		
Build Like Fractions: Building like fractions is the act of rewriting two fractions as equivalent fractions with a common denominator.		
Missing Multiplier: A **missing multiplier** is a factor that multiplies the numerator and denominator of a fraction to build a desired equivalent fraction.		

Study these words when they appear in the text. After the section, test your understanding by completing the Vocabulary review in the section exercises.

Objective A Add and Subtract Like Fractions

The Concept When adding and subtracting fractions, we need to make sure that the fractions are like fractions. **Like fractions** have the same denominator.

$\frac{1}{4}$ and $\frac{2}{4}$ are like fractions because they have identical denominators.

Fractions with the same denominator represent objects that are divided into pieces of the same size, such as fourths. We can visualize the addition of like fractions by combining pieces.

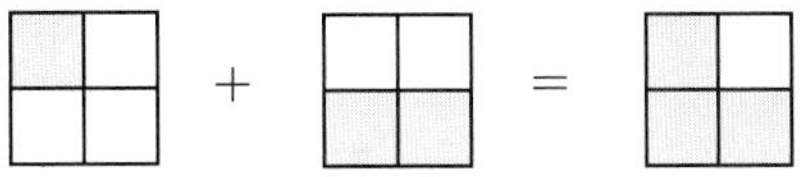

(1 fourth) + (2 fourths) = (3 fourths)

Notice that when the pieces are added, their size, fourths of a whole, doesn't change. This demonstrates how to add like fractions. To add like fractions, we add the numerators and keep the denominator the same.

$$\frac{1}{4} + \frac{2}{4} = \frac{1 + 2}{4}$$ Adding the numerators tells us how many fourths we have in total.

$$= \frac{3}{4}$$

To better understand why the denominator does not change, compare adding like fractions to adding like things, such as dollar bills.

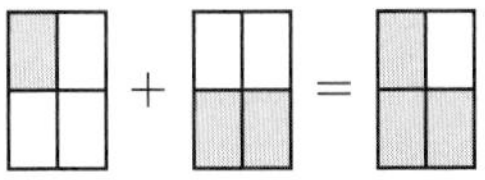

$\frac{1}{4} + \frac{2}{4} = \frac{3}{4}$ the total number of fourths

\$1 + \$2 = \$3 the total number of dollars

The concept is the same when we subtract like fractions.

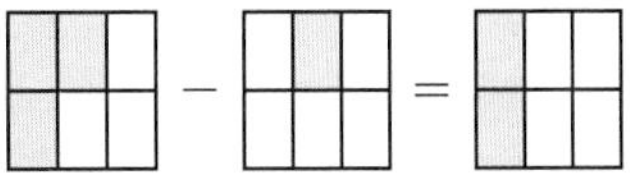

$\frac{3}{6} - \frac{1}{6} = \frac{2}{6}$ the total number of sixths

\$3 − \$1 = \$2 the total number of dollars

When combining like things, the type of thing does not change; only the quantity changes.

When combining dollars, the result will be the total number of dollars.

When combining like fractions, the denominator remains the same.

Procedure Add or Subtract Like Fractions

Step 1 Add or subtract the numerators.

Step 2 Keep the denominator the same.

Step 3 Simplify if possible.

DETAILED EXAMPLE Adding and Subtracting Like Fractions

Subtract. $\frac{5}{8} - \frac{3}{8}$

Step 1 Subtract the numerators.

$$\frac{5}{8} - \frac{3}{8} = \frac{2}{}$$

Step 2 Keep the denominator the same.

$$\frac{5}{8} - \frac{3}{8} = \frac{2}{8}$$

Step 3 Simplify.

$$= \frac{2}{8}$$

$$= \frac{1 \cdot \cancel{2}}{4 \cdot \cancel{2}}$$

$$= \frac{1}{4}$$

The steps written together.

$$\frac{5}{8} - \frac{3}{8} = \frac{2}{8}$$

$$= \frac{1 \cdot \cancel{2}}{4 \cdot \cancel{2}}$$

$$= \frac{1}{4}$$

EXAMPLES

1. Add. $\frac{4}{9} + \frac{2}{9}$

Step 1 Add the numerators.

Step 2 Keep the denominator the same.

$$\frac{4}{9} + \frac{2}{9} = \frac{4+2}{9}$$

$$= \frac{6}{9}$$

Step 3 Simplify.

$$= \frac{2 \cdot \cancel{3}}{3 \cdot \cancel{3}}$$

$$= \frac{2}{3}$$

2. Subtract. $\frac{13}{16} - \frac{9}{16}$

Step 1 Subtract the numerators.

Step 2 Keep the denominator the same.

$$\frac{13}{16} - \frac{9}{16} = \frac{13-9}{16}$$

$$= \frac{4}{16}$$

Step 3 Simplify.

$$= \frac{1 \cdot \cancel{4}}{4 \cdot \cancel{4}}$$

$$= \frac{1}{4}$$

GUIDED PRACTICE

1. Add. $\frac{1}{6} + \frac{7}{6}$

$$\frac{1}{6} + \frac{7}{6} = \frac{\quad + \quad}{\quad}$$

$$= \frac{\quad}{\quad}$$

$$= \frac{\quad \cdot \quad}{\quad \cdot \quad}$$

$$= \frac{\quad}{\quad}$$

2. Subtract. $\frac{9}{12} - \frac{5}{12}$

$$\frac{9}{12} - \frac{5}{12} = \frac{\quad - \quad}{\quad}$$

$$= \frac{\quad}{\quad}$$

$$= \frac{\quad}{\quad}$$

$$= \frac{\quad}{\quad}$$

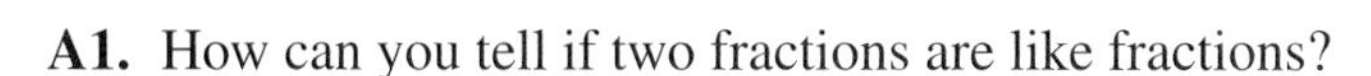

Concept Check

A1. How can you tell if two fractions are like fractions?

A2. List three pairs of like fractions.

A3. With respect to the fraction 3 tenths, what is the numerator? What is the denominator?

A4. When adding or subtracting like objects, the number of objects changes, not the type of object. With that in mind, add or subtract the like objects listed below.

a. 10 oranges + 4 oranges
b. 1 fifth + 3 fifths
c. 12 gumballs − 8 gumballs
d. 8 tenths − 3 tenths

Objective A Practice

Add or subtract the like fractions. Simplify if possible.

A5. $\frac{1}{3} + \frac{1}{3}$ **A6.** $\frac{3}{11} + \frac{4}{11}$ **A7.** $\frac{2}{5} - \frac{1}{5}$ **A8.** $\frac{6}{7} - \frac{4}{7}$

A9. $\frac{5}{8} + \frac{1}{8}$ **A10.** $\frac{15}{24} + \frac{3}{24}$ **A11.** $\frac{15}{26} - \frac{2}{26}$ **A12.** $\frac{17}{22} - \frac{6}{22}$

Fill in the missing parts of each exercise.
- **Shade the appropriate portions of each picture.**
- **Complete the written operations.**
- **Follow the completed example shown to the right.**

Completed Example:

$\frac{1}{4} + \frac{2}{4} = \frac{3}{4}$

A13. $\frac{4}{5} - \frac{\quad}{\quad} = \frac{\quad}{\quad}$

A14. $\frac{\quad}{\quad} - \frac{1}{4} = \frac{\quad}{\quad}$

A15. $\frac{5}{6} + \frac{1}{6} = \frac{\quad}{\quad}$

A16. $\frac{3}{4} + \frac{\quad}{\quad} = \frac{\quad}{\quad}$

A17. $\frac{1}{2} + \frac{\quad}{\quad} =$

A18. $\frac{1}{3} + \frac{\quad}{\quad} = \frac{\quad}{\quad}$

Objective B Build Like Fractions by Listing Multiples

The Concept **Unlike fractions** are fractions that have different denominators. To add or subtract unlike fractions, we must build them into like fractions. The following pictures demonstrate two things.

1. Figure 2.4 shows unlike fractions being added.
2. Figure 2.5 gives a visualization of what it means to build like fractions.

$\frac{1}{2}$ of one pizza and $\frac{1}{3}$ of another were put into a single box.

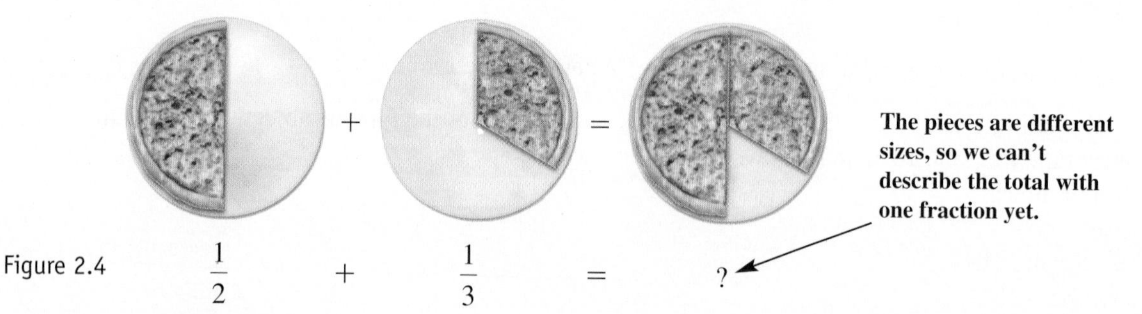

Figure 2.4 $\frac{1}{2} + \frac{1}{3} = ?$

Because the portions "halves" and "thirds" are not like portions, we cannot describe their sum right away. However, if we cut each piece into sixths of a whole pizza, the sum can be given as a fraction.

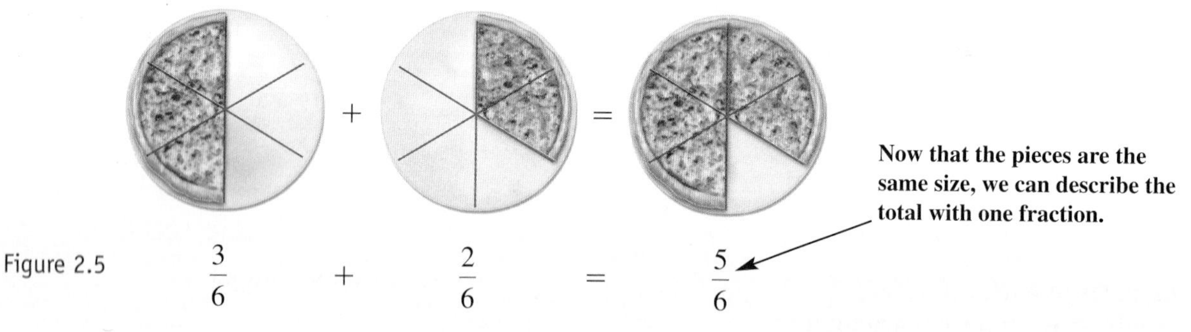

Figure 2.5 $\frac{3}{6} + \frac{2}{6} = \frac{5}{6}$

The number of pieces of leftover pizza, 5, gives the numerator. The size of the pieces, sixths, gives the denominator.

To add unlike fractions, such as the ones above, we must build them into like fractions. This will require that we find the least common denominator (LCD). Often it will be convenient to find the LCD by listing multiples.

INTERACTIVE DEFINITION Least Common Denominator (LCD)

The **least common denominator (LCD)** of several fractions is the least common multiple of their denominators.

EXAMPLE

3. Identify the LCD of $\frac{1}{6}$ and $\frac{5}{9}$ by listing multiples.

List multiples of the larger denominator, 9, until the smaller denominator, 6, divides a multiple evenly.

$9 \cdot 1 = 9$ . . . Keep going.

$9 \cdot 2 = 18$ . . . Stop. 6 divides 18 evenly.

LCD $= 18$

GUIDED PRACTICE

3. Identify the LCD of $\frac{1}{8}$ and $\frac{3}{10}$ by listing multiples.

List multiples of the larger denominator, ____, until the smaller denominator, ____, divides a multiple evenly.

$10 \cdot 1 =$ Keep going/Stop

$10 \cdot 2 =$ Keep going/Stop

$10 \cdot 3 =$ Keep going/Stop

$10 \cdot 4 =$ Keep going/Stop

LCD = ____

DO YOU UNDERSTAND how to identify the LCD of two fractions? Got It Get Help

Procedure Build Like Fractions by Listing Multiples

Step 1 Determine the LCD.

List multiples of the larger denominator until you find a multiple of the smaller denominator.

Step 2 Create the framework to build like fractions.

The framework here is the same as that used for building equivalent fractions in Section 2.2.

Step 3 Multiply the numerators and denominators by the missing multipliers.

DETAILED EXAMPLE Building Like Fractions by Listing Multiples

Build like fractions for $\frac{3}{8}$ and $\frac{2}{6}$.

Step 1 Determine the LCD.

$8 \cdot 1 = 8 \ldots$ Keep going.
$8 \cdot 2 = 16 \ldots$ Keep going.
$8 \cdot 3 = 24 \ldots$ Stop. 6 divides 24 evenly.
LCD $= 24$

Step 2 Create the framework.
Write the LCD, 24, in the framework.

$$\frac{3}{8} \cdot \left(\frac{\quad}{\quad}\right) = \frac{\quad}{24}$$
$$\frac{2}{6} \cdot \left(\frac{\quad}{\quad}\right) = \frac{\quad}{24}$$

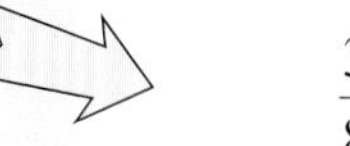

Step 3 Multiply by the missing multipliers.

$$\frac{3}{8} \cdot \left(\frac{3}{3}\right) = \frac{9}{24}$$
$$\frac{2}{6} \cdot \left(\frac{4}{4}\right) = \frac{8}{24}$$

When you build like fractions, you should use the framework shown in the following examples. To find the missing multiplier, ask yourself what number must multiply each denominator to get the LCD.

EXAMPLES

4. Build like fractions for $\frac{2}{5}$ and $\frac{1}{4}$.

Step 1 Find the LCD.
List multiples of 5 until 4 divides a multiple evenly.

~~5~~, ~~10~~, ~~15~~, 20
4 divides 20 evenly.
LCD $= 20$

$$\frac{2}{5} \cdot \left(\frac{4}{4}\right) = \frac{8}{20}$$

Step 2 Create the framework.
Step 3 Multiply by the missing multipliers.

$$\frac{1}{4} \cdot \left(\frac{5}{5}\right) = \frac{5}{20}$$

GUIDED PRACTICE

4. Build like fractions for $\frac{3}{4}$ and $\frac{1}{6}$.

Step 1 Find the LCD.
List multiples of 6 until 4 divides a multiple evenly.

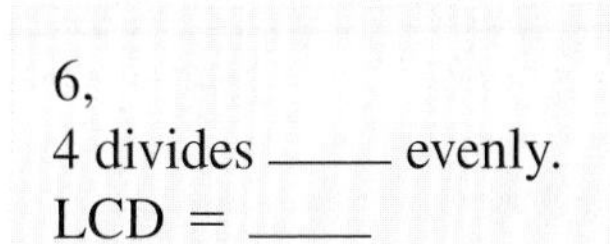
6,
4 divides ____ evenly.
LCD = ____

$$\frac{3}{4} \cdot \left(\frac{\quad}{\quad}\right) = \frac{\quad}{\quad}$$

Step 2 Create the framework.
Step 3 Multiply by the missing multipliers.

$$\frac{1}{6} \cdot \left(\frac{\quad}{\quad}\right) = \frac{\quad}{\quad}$$

(Continued)

5. Build like fractions for $\frac{3}{10}$ and $\frac{1}{6}$.

Step 1 Find the LCD.

~~10~~, ~~20~~, 30
6 divides 30 evenly.
LCD = 30

$$\frac{3}{10} \cdot \left(\frac{3}{3}\right) = \frac{9}{30}$$

Step 2 Create the framework.
Step 3 Multiply by the missing multipliers.

$$\frac{1}{6} \cdot \left(\frac{5}{5}\right) = \frac{5}{30}$$

5. Build like fractions for $\frac{4}{7}$ and $\frac{2}{5}$.

Step 1 Find the LCD.

7,
5 divides ____ evenly.
LCD = ____

$$\frac{4}{7} \cdot \left(\frac{}{}\right) = \frac{}{}$$

Step 2 Create the framework.
Step 3 Multiply by the missing multipliers.

$$\frac{2}{5} \cdot \left(\frac{}{}\right) = \frac{}{}$$

6. Build like fractions for $\frac{2}{3}$ and $\frac{5}{4}$.

Step 1 Find the LCD.

~~4~~, ~~8~~, 12
3 divides 12 evenly.
LCD = 12

$$\frac{2}{3} \cdot \left(\frac{4}{4}\right) = \frac{8}{12}$$

Step 2 Create the framework.
Step 3 Multiply by the missing multipliers.

$$\frac{5}{4} \cdot \left(\frac{3}{3}\right) = \frac{15}{12}$$

6. Build like fractions for $\frac{1}{12}$ and $\frac{11}{9}$.

Step 1 Find the LCD.

12,
9 divides ____ evenly.
LCD = ____

$$\frac{1}{12} \cdot \left(\frac{}{}\right) = \frac{}{}$$

Step 2 Create the framework.
Step 3 Multiply by the missing multipliers.

$$\frac{11}{9} \cdot \left(\frac{}{}\right) = \frac{}{}$$

Concept Check

B1. In Figure 2.4 on page 2-60, why couldn't we describe the fraction after adding the pieces of pizza together?

B2. In Figure 2.5 on page 2-60, why could we describe the fraction after adding the pieces of pizza together?

B3. Answer each of the following.

a. Identify the value of the fraction. $\frac{4}{4} =$

b. Using the answer from part a, explain why $\frac{2}{5} \cdot 1$ is equal to $\frac{2}{5} \cdot \left(\frac{4}{4}\right)$.

c. Identify each answer as true or false.

$\frac{2}{5} \cdot 1 = \frac{2}{5}$ True/False $\qquad$ $\frac{2}{5} \cdot \left(\frac{4}{4}\right) = \frac{8}{20}$ True/False

d. Using the answers from parts a, b, and c, explain why $\frac{2}{5} = \frac{8}{20}$.

e. When building like fractions, why must we multiply both the numerator and denominator by the same number?

Objective B Practice

List multiples to identify the LCD and build like fractions.

B4. $\frac{1}{2}\cdot\left(\frac{}{}\right)=\frac{}{}$ $\frac{1}{6}\cdot\left(\frac{}{}\right)=\frac{}{}$

B5. $\frac{2}{5}\cdot\left(\frac{}{}\right)=\frac{}{}$ $\frac{4}{15}\cdot\left(\frac{}{}\right)=\frac{}{}$

B6. $\frac{3}{8}\cdot\left(\frac{}{}\right)=\frac{}{}$ $\frac{7}{10}\cdot\left(\frac{}{}\right)=\frac{}{}$

B7. $\frac{3}{8}\cdot\left(\frac{}{}\right)=\frac{}{}$ $\frac{2}{3}\cdot\left(\frac{}{}\right)=\frac{}{}$

Build like fractions.

B8. $\frac{3}{20}, \frac{3}{5}$ **B9.** $\frac{1}{6}, \frac{2}{3}$ **B10.** $\frac{4}{9}, \frac{5}{12}$ **B11.** $\frac{5}{9}, \frac{3}{4}$

B12. $\frac{17}{25}, \frac{1}{3}$ **B13.** $\frac{5}{8}, \frac{3}{12}$ **B14.** $\frac{11}{26}, \frac{1}{4}$ **B15.** $\frac{1}{6}, \frac{3}{14}$

Objective C Add and Subtract Unlike Fractions: Apply Objective B

The Concept In the previous objective, we learned how to build like fractions. Building like fractions allows us to add and subtract fractions that do not represent like objects.

We can't add these fractions because they are unlike; the pieces aren't the same size.

$$\frac{1}{3} + \frac{1}{4} = ?$$

We build like fractions by dividing the objects into twelfths. Then the like fractions can be added by combining the pieces.

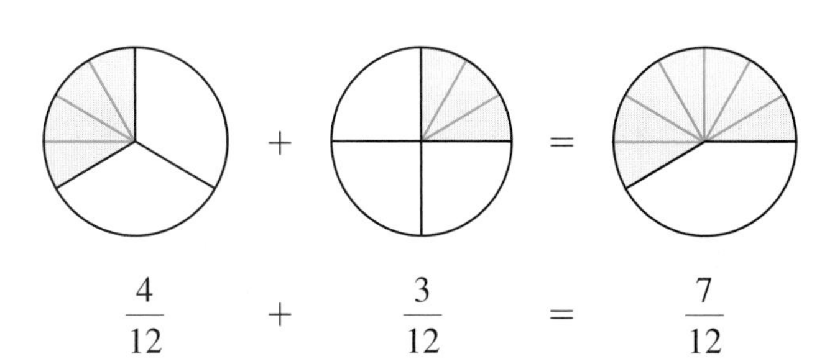

$$\frac{4}{12} + \frac{3}{12} = \frac{7}{12}$$

Procedure Add or Subtract Unlike Fractions

Step 1 Build like fractions.

Step 2 Add or subtract the like fractions.

Step 3 Simplify if possible.

EXAMPLE

7. Add. $\frac{1}{4} + \frac{2}{3}$

Step 1 Build like fractions. $\frac{1}{4} \cdot \left(\frac{3}{3}\right) = \frac{3}{12}$

~~4~~, ~~8~~, 12
LCD = 12

$\frac{2}{3} \cdot \left(\frac{4}{4}\right) = \frac{8}{12}$

Step 2 Add the like fractions. $\frac{3}{12} + \frac{8}{12} = \frac{11}{12}$

Step 3 Simplify. $\frac{11}{12}$ is in simplest form.

8. Subtract. $\frac{9}{12} - \frac{4}{8}$

Determine the LCD. LCD = ~~12~~, 24

Step 1 Build like fractions. $\frac{9}{12} - \frac{4}{8} = \frac{9}{12} \cdot \left(\frac{2}{2}\right) - \frac{4}{8} \cdot \left(\frac{3}{3}\right)$

Step 2 Subtract the like fractions. $= \frac{18}{24} - \frac{12}{24}$

$= \frac{6}{24}$

Step 3 Simplify. $= \frac{1 \cdot \cancel{6}}{\cancel{6} \cdot 4}$

$= \frac{1}{4}$

9. Subtract. $\frac{5}{6} - \frac{4}{5}$

Determine the LCD. LCD = ~~6~~, ~~12~~, ~~18~~, ~~24~~, 30

Step 1 Build like fractions. $\frac{5}{6} - \frac{4}{5} = \frac{5}{6} \cdot \left(\frac{5}{5}\right) - \frac{4}{5} \cdot \left(\frac{6}{6}\right)$

Step 2 Subtract the like fractions. $= \frac{25}{30} - \frac{24}{30}$

Step 3 Simplify. $= \frac{1}{30}$

$\frac{1}{30}$ is in simplest form.

GUIDED PRACTICE

7. Add. $\frac{1}{4} + \frac{3}{5}$

Step 1 Build like fractions. $\frac{1}{4} \cdot \left(\frac{\ }{\ }\right) = \frac{\ }{\ }$

5,
LCD =

$\frac{3}{5} \cdot \left(\frac{\ }{\ }\right) = \frac{\ }{\ }$

Step 2 Add the like fractions. $\frac{\ }{\ } + \frac{\ }{\ } = \frac{\ }{\ }$

Step 3 Simplify.

8. Subtract. $\frac{7}{15} - \frac{1}{6}$

LCD =

$\frac{7}{15} - \frac{1}{6} = \frac{7}{15} \cdot \left(\frac{\ }{\ }\right) - \frac{1}{6} \cdot \left(\frac{\ }{\ }\right)$

$= \frac{\ }{\ } - \frac{\ }{\ }$

$= \frac{\ }{\ }$

$= \frac{\ }{\ }$

$= \frac{\ }{\ }$

9. Subtract. $\frac{4}{9} - \frac{1}{12}$

LCD =

$\frac{4}{9} - \frac{1}{12} =$

10. Add. $5 + \frac{2}{3}$

Write the whole number as a fraction. $5 + \frac{2}{3} = \frac{5}{1} + \frac{2}{3}$

Determine the LCD. LCD = 3

Step 1 Build like fractions. $\frac{5}{1} + \frac{2}{3} = \frac{5}{1} \cdot \frac{3}{3} + \frac{2}{3}$

$= \frac{15}{3} + \frac{2}{3}$

Step 2 Add the like fractions. $= \frac{17}{3}$

10. Add. $6 + \frac{1}{4}$

$6 + \frac{1}{4} = \frac{6}{\square} + \frac{1}{4}$

LCD =

$\frac{6}{\square} + \frac{1}{4} = \frac{6}{\square} \cdot \frac{\square}{\square} + \frac{\square}{\square}$

$= \frac{\square}{\square} + \frac{\square}{\square}$

$= \frac{\square}{\square}$

Concept Check

C1. A classmate believes incorrectly that $\frac{1}{3} + \frac{1}{2} = \frac{2}{5}$.

a. Represent each of the three fractions, using a picture.

$\frac{1}{2} =$

$\frac{1}{3} =$

$\frac{2}{5} =$

b. Does the picture for $\frac{2}{5}$ look like it has the same area as the shaded areas for $\frac{1}{3}$ and $\frac{1}{2}$ combined?

c. Based on your answer for part b, is it possible that $\frac{1}{3} + \frac{1}{2} = \frac{2}{5}$?

d. What mistake(s) did your classmate make when adding these fractions?

C2. In your own words, why must we build like fractions to add fractions that have different denominators?

Objective C Practice

Add or subtract the unlike fractions.

C3. $\frac{1}{2} + \frac{1}{4}$ **C4.** $\frac{2}{3} - \frac{2}{9}$ **C5.** $\frac{3}{5} - \frac{1}{6}$ **C6.** $\frac{3}{4} + \frac{1}{5}$

C7. $\frac{3}{10} + \frac{2}{15}$ **C8.** $\frac{2}{9} + \frac{1}{6}$ **C9.** $\frac{3}{8} - \frac{1}{6}$ **C10.** $\frac{3}{4} - \frac{1}{6}$

C11. In your own words, what steps are used to build like fractions from unlike fractions?

C12. In your own words, what steps are used to add or subtract like fractions?

Objective D Build Like Fractions by Factoring

The Concept Listing multiples to build like fractions can be very time-consuming when the fractions have large denominators. For many fractions, it is easier to find the LCD by factoring. This method uses the Venn diagram technique for finding the LCM from Section 1.8.

The factoring technique is important because it is the only way to find the LCD for fractions called *rational expressions*.

▶ **NOTE** A rational expression is a fraction that contains algebraic expressions, such as $\frac{1}{x + 3}$. You'll learn more about rational expressions if you take algebra.

Procedure Identify a LCD by Factoring

Step 1 Write the prime factorization for each denominator.

Step 2 Create a Venn diagram of the prime factorizations.

Step 3 The LCD is the product of all of the factors in the circles.

Below we compare the two methods by identifying the LCD for $\frac{1}{26}$ and $\frac{1}{14}$.

Listing Multiples

$1 \cdot 26 = 26$ not divisible by 14
$2 \cdot 26 = 52$ not divisible by 14
$3 \cdot 26 = 78$ not divisible by 14
$4 \cdot 26 = 104$ not divisible by 14
$5 \cdot 26 = 130$ not divisible by 14
$6 \cdot 26 = 156$ not divisible by 14
$7 \cdot 26 = 182$ is divisible by 14.

This looks like a lot of work, and it doesn't show the division that was performed.

Factoring

Step 1 Write the prime factorization for each denominator.

$$14 = 2 \cdot 7$$
$$26 = 2 \cdot 13$$

Step 2 Create a Venn diagram of the prime factorizations.

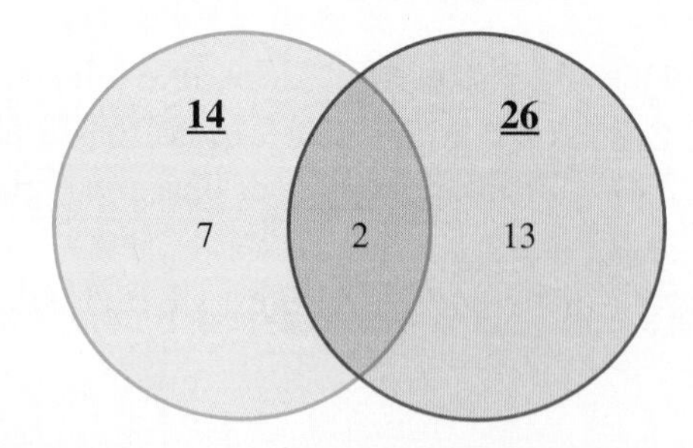

Step 3 The LCD is the product of all the factors in the circles.

$$\begin{aligned} \text{LCD} &= 7 \cdot 2 \cdot 13 \\ &= 14 \cdot 13 \\ &= 182 \end{aligned}$$

This is much easier, and it shows the missing multipliers.

EXAMPLE

11. Identify the LCD of $\frac{9}{12}$ and $\frac{4}{16}$ by factoring.

Step 1 Write the prime factorization for each denominator.

$$12 = 2 \cdot 2 \cdot 3$$
$$16 = 2 \cdot 2 \cdot 2 \cdot 2$$

12 and 16 have two factors of 2 in common.

Step 2 Create a Venn diagram of the prime factorizations.

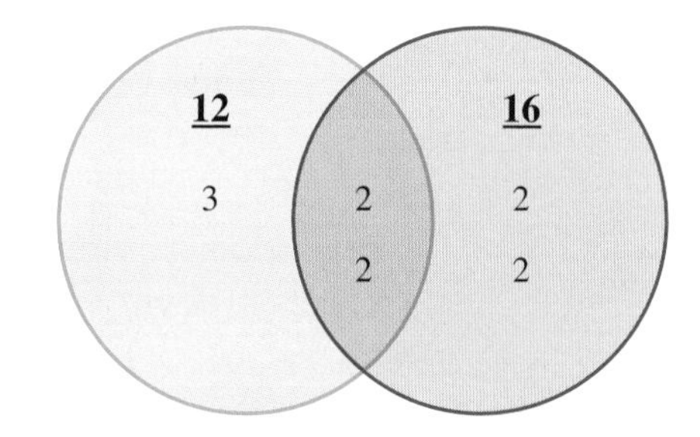

Step 3 The LCD is the product of all of the factors in the circles.

$$\text{LCD} = 3 \cdot 2 \cdot 2 \cdot 2 \cdot 2$$
$$= 48$$

GUIDED PRACTICE

11. Identify the LCD of $\frac{7}{18}$ and $\frac{4}{27}$ by factoring.

Step 1 Write the prime factorization for each denominator.

18 =

27 =

18 and 27 have two factors of ______ in common.

Step 2 Create a Venn diagram of the prime factorizations.

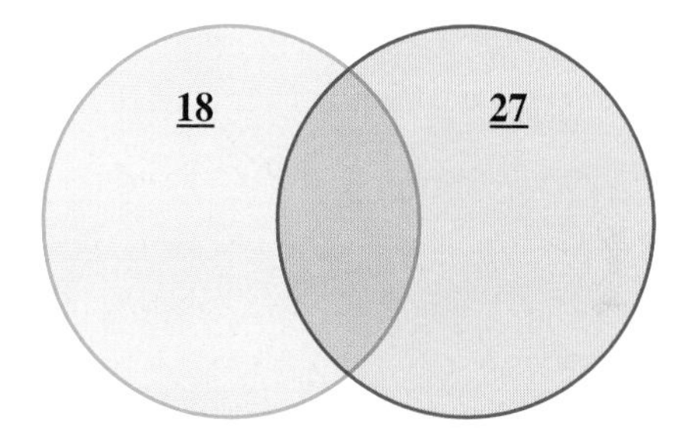

Step 3 The LCD is the product of all of the factors in the circles.

LCD =

=

Once you create the Venn diagram, you can use it to identify the missing multipliers and build like fractions.

Procedure Build Like Fractions by Factoring

Step 1 Identify the least common denominator (LCD) by factoring.

Step 2 Identify the missing multipliers from the factorizations.

A number's missing multipliers are the factors outside the number's Venn diagram circle.

Step 3 Create the framework and build like fractions.

DETAILED EXAMPLE Building Like Fractions by Factoring

Build like fractions for $\frac{1}{20}$ and $\frac{5}{24}$.

Step 1 Identify the LCD by factoring.

$20 = 2 \cdot 2 \cdot 5$

$24 = 2 \cdot 2 \cdot 2 \cdot 3$

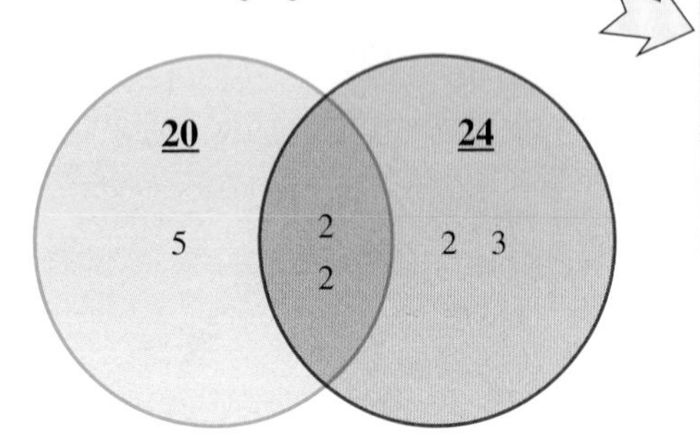

$\text{LCD} = 5 \cdot 2 \cdot 2 \cdot 2 \cdot 3$
$= 120$

Step 2 Identify the missing multipliers. The missing multipliers are the factors outside a number's circle.

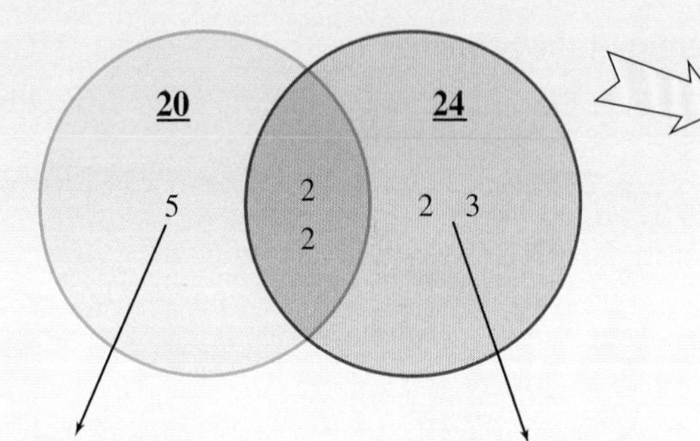

5 is outside 24's circle, so 5 is 24's missing multiplier.

2 · 3 = 6 is outside 20's circle, so 6 is 20's missing multiplier.

Step 3 Create the framework and build like fractions.

$$\frac{5}{24} \cdot \left(\frac{5}{5}\right) = \frac{25}{120}$$

$$\frac{1}{20} \cdot \left(\frac{6}{6}\right) = \frac{6}{120}$$

EXAMPLES

12. Build like fractions for $\frac{1}{21}$ and $\frac{2}{9}$.

Step 1 Identify the LCD.

$21 = 7 \cdot 3$

$9 = 3 \cdot 3$

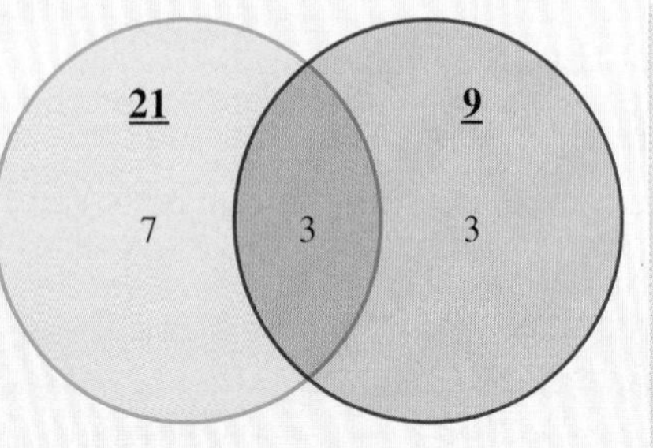

$\text{LCD} = 7 \cdot 3 \cdot 3$
$= 63$

Step 2 Identify the missing multipliers.

21 is missing a 3.

9 is missing a 7.

GUIDED PRACTICE

12. Build like fractions for $\frac{3}{14}$ and $\frac{1}{10}$.

Step 1 Identify the LCD.

14 =

10 =

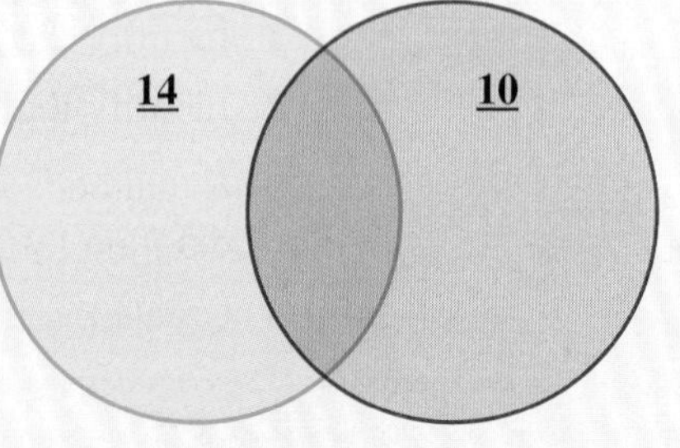

LCD =   ·   ·
=

Step 2 Identify the missing multipliers.

14 is missing a ______.

10 is missing a ______.

Step 3 Build like fractions.
Use the missing multipliers identified on the previous page.

$$\frac{1}{21}\cdot\left(\frac{3}{3}\right)=\frac{3}{63}$$
$$\frac{2}{9}\cdot\left(\frac{7}{7}\right)=\frac{14}{63}$$

13. Build like fractions for $\frac{5}{27}$ and $\frac{7}{36}$.

Step 1 Identify the LCD.

$27 = \phantom{2\cdot 2\cdot{}} 3\cdot 3\cdot 3$
$36 = 2\cdot 2\cdot 3\cdot 3$

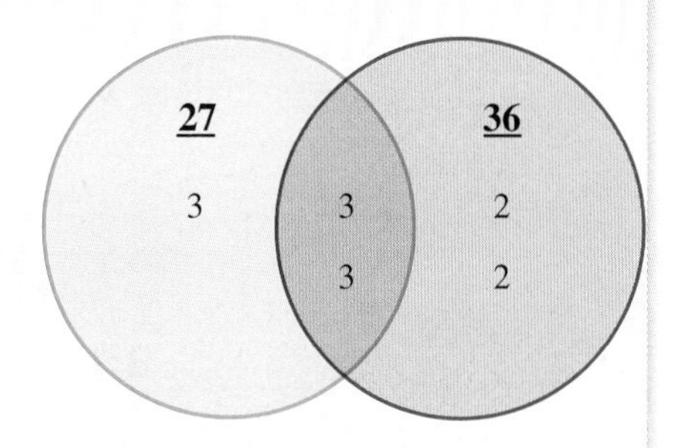

$$\text{LCD} = 3\cdot 3\cdot 3\cdot 2\cdot 2 = 108$$

Step 2 Identify the missing multipliers.

27 is missing $2\cdot 2 = 4$.
36 is missing a 3.

Step 3 Build like fractions.
Use the missing multipliers identified above.

$$\frac{5}{27}\cdot\left(\frac{4}{4}\right)=\frac{20}{108}$$
$$\frac{7}{36}\cdot\left(\frac{3}{3}\right)=\frac{21}{108}$$

Step 3 Build like fractions.
Use the missing multipliers identified on the previous page.

$$\frac{3}{14}\cdot\left(\frac{}{}\right)=\frac{}{}$$
$$\frac{1}{10}\cdot\left(\frac{}{}\right)=\frac{}{}$$

13. Build like fractions for $\frac{13}{12}$ and $\frac{17}{54}$.

Step 1 Identify the LCD.

12 =
54 =

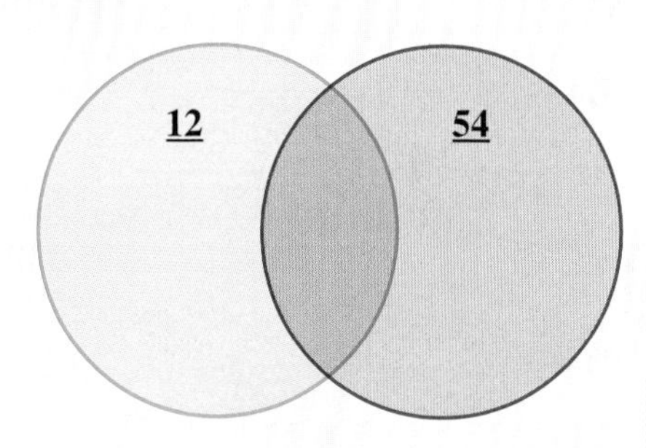

LCD =
=

Step 2 Identify the missing multipliers.

12 is missing _____ · _____ = _____.
54 is missing _____.

Step 3 Build like fractions.
Use the missing multipliers identified above.

$$\frac{13}{12}\cdot\left(\frac{}{}\right)=\frac{}{}$$
$$\frac{17}{54}\cdot\left(\frac{}{}\right)=\frac{}{}$$

Concept Check

Before you begin these exercises, make sure you understand the factoring method for finding the LCD.

D1. Using the listing multiples method, time how long it takes you to find the least common multiple for 21 and 24.

D2. Using the factoring method, time how long it takes you to find the least common multiple for 21 and 24.

D3. Which technique was quicker?

Objective D Practice

Build like fractions. Use the factoring method to determine the LCD.

D4. $\frac{7}{13}\left(\frac{\quad}{\quad}\right) = \frac{\quad}{\quad}$

$\frac{5}{11}\left(\frac{\quad}{\quad}\right) = \frac{\quad}{\quad}$

D5. $\frac{3}{10}\left(\frac{\quad}{\quad}\right) = \frac{\quad}{\quad}$

$\frac{5}{16}\left(\frac{\quad}{\quad}\right) = \frac{\quad}{\quad}$

D6. $\frac{3}{14}\left(\frac{\quad}{\quad}\right) = \frac{\quad}{\quad}$

$\frac{1}{6}\left(\frac{\quad}{\quad}\right) = \frac{\quad}{\quad}$

D7. $\frac{7}{8}\left(\frac{\quad}{\quad}\right) = \frac{\quad}{\quad}$

$\frac{5}{22}\left(\frac{\quad}{\quad}\right) = \frac{\quad}{\quad}$

D8. $\frac{3}{20}, \frac{6}{35}$

D9. $\frac{1}{21}, \frac{2}{49}$

D10. $\frac{5}{18}, \frac{5}{42}$

D11. $\frac{1}{31}, \frac{1}{19}$

Objective E Add and Subtract Unlike Fractions: Apply Objective D

The Concept To add and subtract like fractions in this objective, we use the factoring method to identify the LCD and missing multipliers. Because this method will be necessary if you take algebra, we repeat something we said earlier. It is important for you to practice the factoring technique for building like fractions. If you happen to see the LCD, take the time to verify it by factoring.

Procedure Add or Subtract Unlike Fractions

Step 1 Build like fractions.

Use the factoring method to find the LCD.

Step 2 Add or subtract the like fractions.

Step 3 Simplify if possible.

EXAMPLE

14. Add. $\frac{5}{14} + \frac{10}{21}$

Identify the LCD and the missing multipliers.

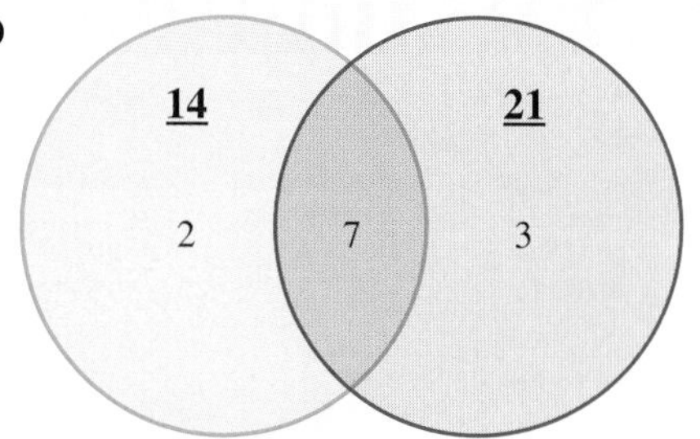

LCD = $2 \cdot 7 \cdot 3$
= 42

14 is missing a 3.
21 is missing a 2.

Step 1 Build like fractions. $\frac{5}{14} + \frac{10}{21} = \frac{5}{14}\left(\frac{3}{3}\right) + \frac{10}{21}\left(\frac{2}{2}\right)$

$= \frac{15}{42} + \frac{20}{42}$

Step 2 Add the like fractions. $= \frac{35}{42}$

Step 3 Simplify. $= \frac{5 \cdot \cancel{7}}{6 \cdot \cancel{7}}$

$= \frac{5}{6}$

15. Subtract. $\frac{11}{21} - \frac{5}{12}$

Identify the LCD and the missing multipliers.

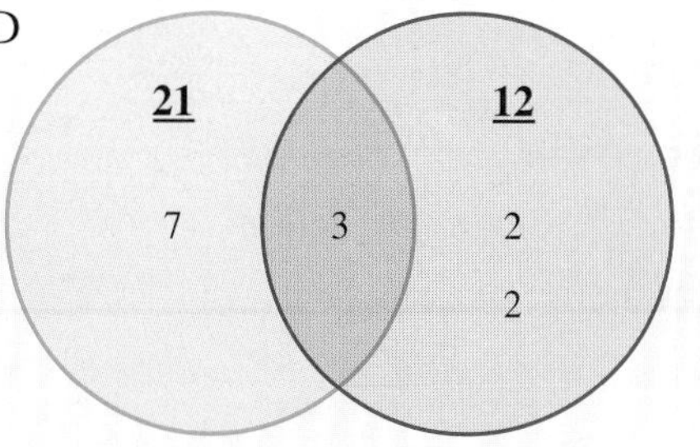

LCD = $3 \cdot 7 \cdot 2 \cdot 2$
= 84

21 is missing $2 \cdot 2 = 4$.
12 is missing a 7.

GUIDED PRACTICE

14. Add. $\frac{1}{6} + \frac{3}{50}$

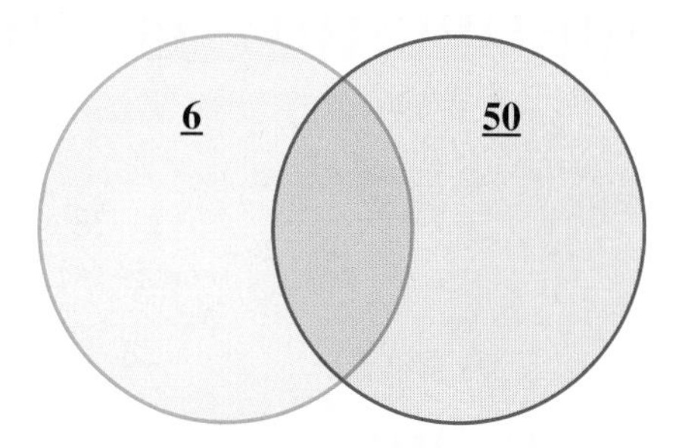

LCD =
=

6 is missing ______.
50 is missing ______.

$\frac{1}{6} + \frac{3}{50} = \frac{1}{6}\left(\frac{\quad}{\quad}\right) + \frac{3}{50}\left(\frac{\quad}{\quad}\right)$

$= \frac{\quad}{\quad} + \frac{\quad}{\quad}$

$= \frac{\quad}{\quad}$

$= \frac{\quad}{\quad}$

$= \frac{\quad}{\quad}$

15. Subtract. $\frac{17}{24} - \frac{3}{18}$

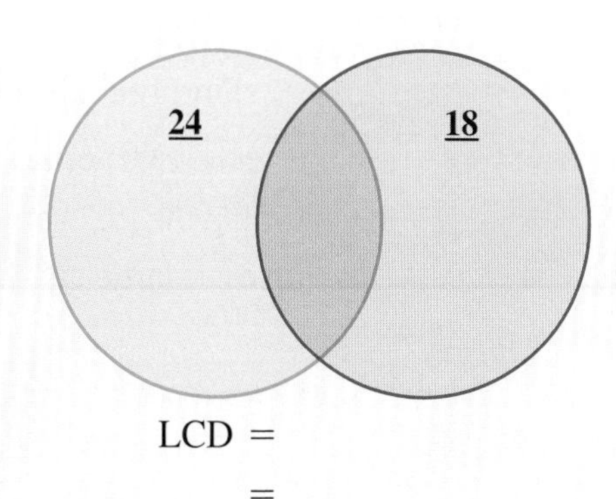

LCD =
=

24 is missing ______.
18 is missing ______.

(Continued)

Step 1 Build like fractions.	$\frac{11}{21} - \frac{5}{12} = \frac{11}{21}\left(\frac{4}{4}\right) - \frac{5}{12}\left(\frac{7}{7}\right)$ $= \frac{44}{84} - \frac{35}{84}$	$\frac{17}{24} - \frac{3}{18} =$ $=$
Step 2 Subtract the like fractions.	$= \frac{9}{84}$	$=$
Step 3 Simplify.	$= \frac{\not{3} \cdot 3}{\not{3} \cdot 28}$ $= \frac{3}{28}$	$=$ $=$

Concept Check

E1. Why is it important to practice building like fractions by factoring?

Objective E Practice

Add or subtract the unlike fractions. Use the factoring method to build like fractions.

E2. $\frac{7}{20} - \frac{1}{24}$

E3. $\frac{11}{20} - \frac{5}{12}$

E4. $\frac{5}{21} + \frac{9}{28}$

E5. $\frac{4}{55} + \frac{7}{22}$

E6. $\frac{1}{8} + \frac{7}{18}$

E7. $\frac{7}{12} + \frac{5}{26}$

E8. $\frac{8}{15} - \frac{5}{12}$

E9. $\frac{11}{24} - \frac{5}{16}$

Combining Concepts and Applications

CONCEPT I Adding and Subtracting More Than Two Unlike Fractions When adding and subtracting more than two unlike fractions, apply the order of operations. Perform addition or subtraction as they occur from left to right. To work efficiently, find the LCD of all three fractions, using two steps.

Procedure Find the LCD of Three Fractions

Step 1 First, find the LCD of the first two fractions.

Step 2 Second, find the LCD of that denominator and the third fraction.

This process can be continued for four or more fractions.

EXAMPLE

16. Perform the operations. $\frac{3}{5} - \frac{1}{6} + \frac{1}{20}$

Step 1 Find the LCD of the first two fractions.
The LCD for 5ths and 6ths is 30.

Step 2 Find the LCD of that denominator and the third fraction.
The LCD for 30ths and 20ths is 60.

Build like fractions with a denominator of 60.

$$\frac{3}{5} - \frac{1}{6} + \frac{1}{20} = \frac{3}{5}\left(\frac{12}{12}\right) - \frac{1}{6}\left(\frac{10}{10}\right) + \frac{1}{20}\left(\frac{3}{3}\right)$$
$$= \frac{36}{60} - \frac{10}{60} + \frac{3}{60}$$
$$= \frac{26}{60} + \frac{3}{60}$$
$$= \frac{29}{60}$$

GUIDED PRACTICE

16. Perform the operations. $\frac{3}{4} - \frac{1}{3} + \frac{1}{8}$

Step 1 Find the LCD of the first two fractions.
The LCD for 4ths and 3rds is ______.

Step 2 Find the LCD of that denominator and the third fraction.
The LCD for ______ and ______ is ______.

Build like fractions with a denominator of ______.

$$\frac{3}{4} - \frac{1}{3} + \frac{1}{8} = \frac{3}{4}\left(\frac{\quad}{\quad}\right) - \frac{1}{3}\left(\frac{\quad}{\quad}\right) + \frac{1}{8}\left(\frac{\quad}{\quad}\right)$$
$$= \frac{\quad}{\quad} - \frac{\quad}{\quad} + \frac{\quad}{\quad}$$
$$= \frac{\quad}{\quad} + \frac{\quad}{\quad}$$
$$= \frac{\quad}{\quad}$$

CONCEPT II Application Exercises

EXAMPLE

17. Shilo is making blueberry pie for dessert. The recipe requires $\frac{1}{3}$ cup of sugar for the pie and $\frac{2}{5}$ cup of sugar for the whipped topping. How much sugar will she need to complete the recipe?

Step 1 Determine whether the problem requires addition or subtraction to solve.

Shilo needs sugar for both the pie and the whipped topping, so we add the amounts together.

Step 2 Add or subtract the fractions. Simplify if possible.

$$\frac{1}{3} + \frac{2}{5} = \frac{1}{3}\left(\frac{5}{5}\right) + \frac{2}{5}\left(\frac{3}{3}\right)$$
$$= \frac{5}{15} + \frac{6}{15}$$
$$= \frac{11}{15}$$

Step 3 Check your solution with the original problem.

$\frac{11}{15}$ is greater than both $\frac{1}{3}$ and $\frac{2}{5}$, so our answer seems reasonable.

Step 4 Interpret the results.

Shilo needs a total of $\frac{11}{15}$ cups of sugar.

GUIDED PRACTICE

17. Rachel is driving a $\frac{3}{4}$-inch nail through a board that is $\frac{5}{8}$ inch thick. How far will the nail stick out on the other side of the board if she drives the nail all the way in?

This flowchart is presented to help you organize your thoughts about adding and subtracting fractions.

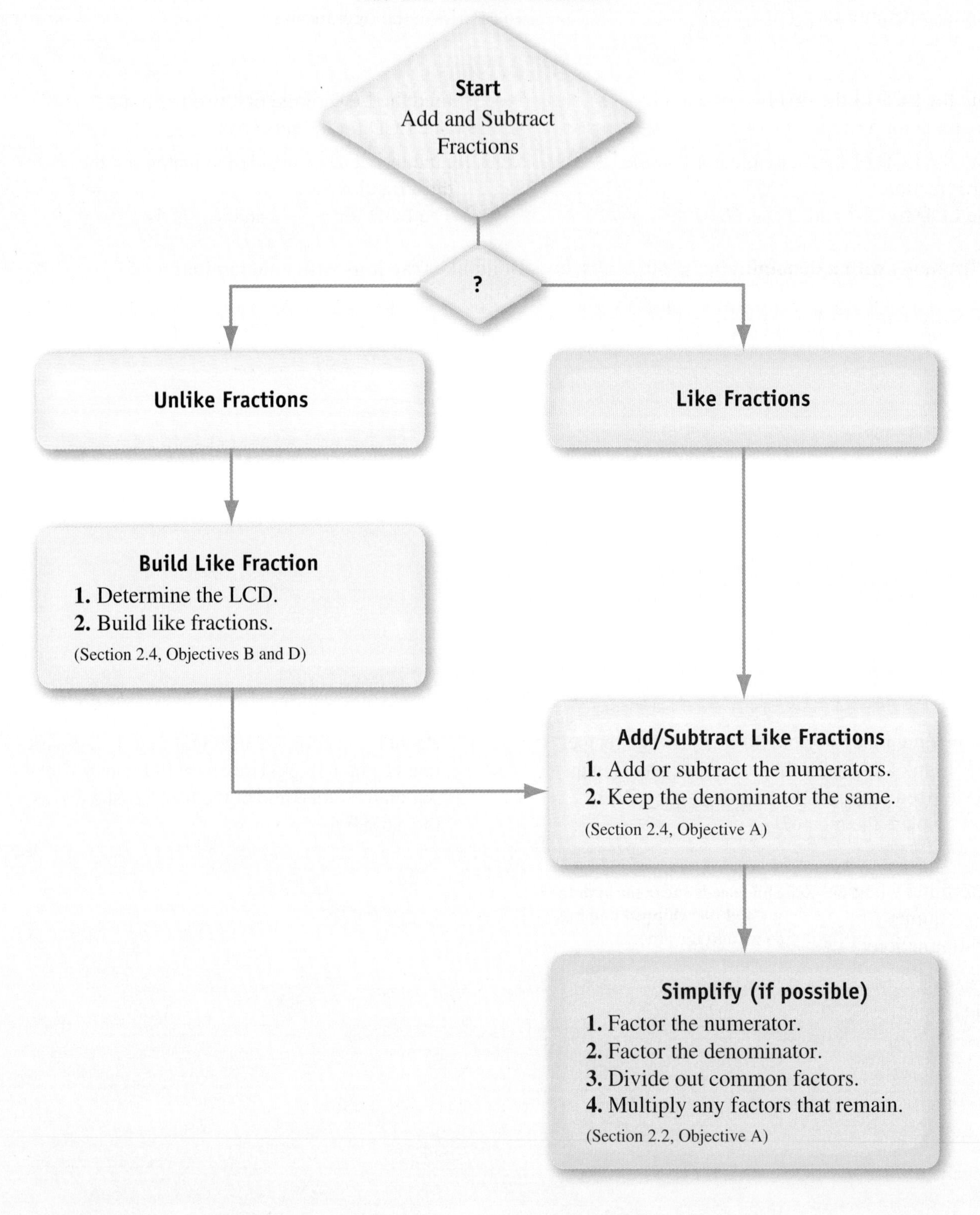

Section 2.4 Exercises

FOR EXTRA HELP     

To add and subtract like fractions:

1. Answer the Objective A Concept Checks.
2. Answer the odd Objective A Practice Exercises.
3. Answer the even Objective A Practice Exercises.

To build like fractions by listing multiples:

4. Answer the Objective B Concept Checks.
5. Answer the odd Objective B Practice Exercises.
6. Answer the even Objective B Practice Exercises.

To add and subtract unlike fractions by applying Objective B:

7. Answer the Objective C Concept Checks.
8. Answer the odd Objective C Practice Exercises.
9. Answer the even Objective C Practice Exercises.

To build like fractions by factoring:

10. Answer the Objective D Concept Checks.
11. Answer the odd Objective D Practice Exercises.
12. Answer the even Objective D Practice Exercises.

To add and subtract unlike fractions by applying Objective D:

13. Answer the Objective E Concept Checks.
14. Answer the odd Objective E Practice Exercises.
15. Answer the even Objective E Practice Exercises.

VOCABULARY REVIEW *Review the Vocabulary Preview for Section 2.4. Study the definitions until you can check* Got It *for every word.*

Use these words to complete each sentence.

like fractions • building like fractions • unlike fractions • missing multiplier • least common denominator (LCD)

	Got It	Get Help
16. Two fractions with different denominators are called ____________.		
17. The act of rewriting two fractions as equivalent fractions with a common denominator is called ____________.		
18. Fractions with identical denominators are called ____________.		
19. To build an equivalent fraction, multiply both the numerator and denominator by the fraction's ____________.		
20. To add unlike fractions, first you must find the ____________.		

How will you get help for any vocabulary that you are unsure about?

Your instructor ____ MyMathLab ____ A classmate ____ A tutor ____ Other ____

Add or subtract the like fractions and simplify your answer.

21. $\frac{3}{7} + \frac{5}{7}$ **22.** $\frac{2}{9} + \frac{5}{9}$ **23.** $\frac{9}{10} - \frac{3}{10}$ **24.** $\frac{5}{12} - \frac{1}{12}$

25. $\frac{8}{15} + \frac{4}{15}$ **26.** $\frac{9}{10} + \frac{3}{10}$ **27.** $\frac{13}{14} - \frac{5}{14}$ **28.** $\frac{5}{6} - \frac{2}{6}$

Fill in the missing parts of each exercise.

- **Shade the appropriate portion of each picture.** (Draw lines around the pieces if necessary.)
- **Complete the written operations.**

▶ **NOTE** All the fractions are like fractions in these exercises.

29.

$\frac{1}{3} + \frac{1}{\ } = \frac{\ }{\ }$

30.

$\frac{\ }{\ } + \frac{\ }{\ } = \frac{5}{6}$

31.

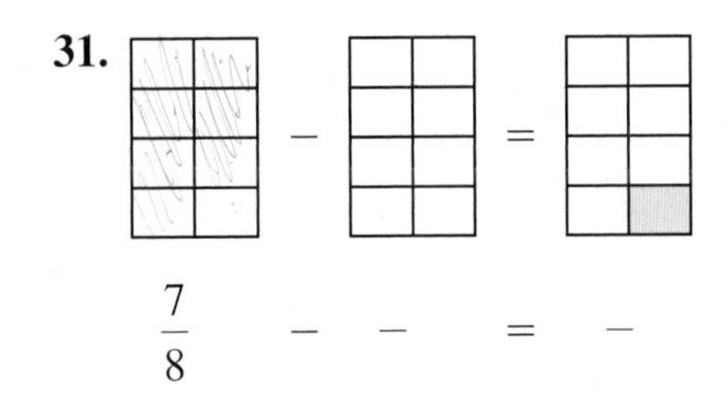

$\frac{7}{8} - \frac{\ }{\ } = \frac{\ }{\ }$

32.

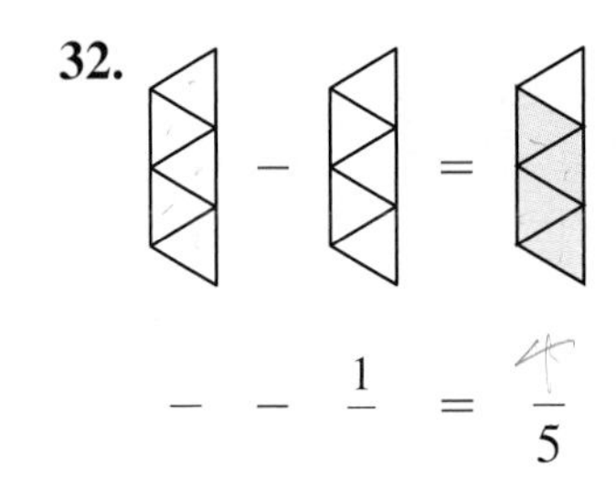

$\frac{\ }{\ } - \frac{1}{\ } = \frac{\ }{5}$

33.

$\frac{\ }{2} - \frac{1}{\ } = \frac{\ }{2}$

34.

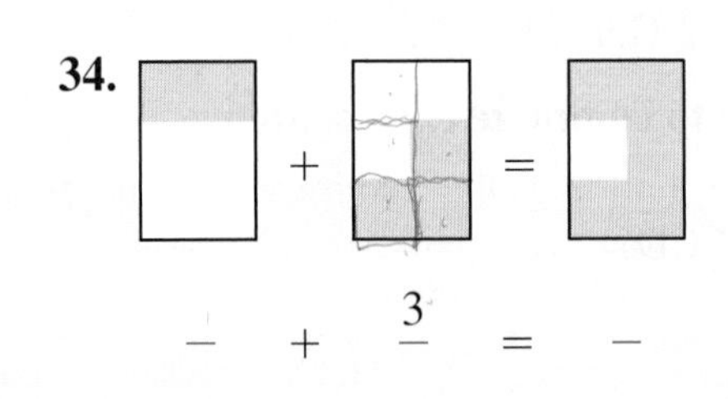

$\frac{\ }{\ } + \frac{3}{\ } = \frac{\ }{\ }$

List multiples to identify the LCD for each pair of fractions.

35. $\frac{3}{5}, \frac{1}{2}$ **36.** $\frac{1}{3}, \frac{2}{7}$ **37.** $\frac{4}{9}, \frac{1}{6}$ **38.** $\frac{5}{8}, \frac{5}{6}$

Use the factoring method to identify the LCD for each pair of fractions.

39. $\frac{1}{15}, \frac{2}{21}$ **40.** $\frac{4}{25}, \frac{6}{15}$ **41.** $\frac{1}{7}, \frac{5}{12}$ **42.** $\frac{1}{30}, \frac{3}{40}$

Add or subtract and simplify your answer. As you look for the least common denominator, try listing multiples first. If you can't determine the LCD quickly by listing multiples, use the factoring method.

43. $\frac{3}{4}+\frac{5}{12}$ **44.** $\frac{5}{12}+\frac{5}{6}$ **45.** $\frac{7}{10}-\frac{1}{15}$ **46.** $\frac{5}{6}-\frac{4}{9}$

47. $\frac{8}{15}-\frac{2}{9}$ **48.** $\frac{3}{8}+\frac{5}{18}$ **49.** $\frac{13}{14}-\frac{4}{21}$ **50.** $\frac{5}{12}+\frac{7}{30}$

51. $\frac{17}{20}-\frac{1}{6}$ **52.** $\frac{7}{20}+\frac{4}{25}$ **53.** $\frac{5}{21}+\frac{3}{14}$ **54.** $\frac{14}{55}-\frac{5}{22}$

55. $\frac{1}{18}+\frac{16}{27}$ **56.** $\frac{10}{17}+\frac{5}{34}$ **57.** $\frac{8}{15}-\frac{5}{12}$ **58.** $\frac{23}{24}-\frac{7}{16}$

59. $\frac{3}{16}+\frac{7}{18}$ **60.** $\frac{8}{21}-\frac{8}{35}$ **61.** $\frac{7}{20}-\frac{7}{24}$ **62.** $\frac{9}{18}+\frac{1}{24}$

Perform the indicated operations, following the order of operations.

63. $\frac{3}{4}+\frac{1}{3}-\frac{1}{2}$ **64.** $\frac{9}{2}+\frac{3}{4}-\frac{5}{3}$ **65.** $\frac{3}{4}-\frac{1}{8}-\frac{1}{2}$ **66.** $\frac{8}{5}-\frac{1}{4}-\frac{2}{5}$

67. $9-\frac{1}{3}+\frac{1}{5}$ **68.** $4-\frac{5}{6}+\frac{3}{5}$ **69.** $\frac{1}{6}+\frac{1}{8}+\frac{1}{12}$ **70.** $\frac{3}{10}+\frac{2}{5}+\frac{7}{10}$

For each exercise:
a. Build like fractions.
b. Order the fractions from least to greatest.

Hint: When fractions have the same denominator, the fraction with the larger numerator indicates more pieces and a larger fraction.

71. $\frac{3}{4}, \frac{5}{8}$ **72.** $\frac{2}{3}, \frac{3}{6}$ **73.** $\frac{13}{16}, \frac{3}{4}$ **74.** $\frac{55}{100}, \frac{1}{2}$

75. $\frac{5}{6}, \frac{4}{5}$ **76.** $\frac{5}{12}, \frac{7}{16}$ **77.** $\frac{2}{3}, \frac{7}{10}$ **78.** $\frac{3}{7}, \frac{2}{5}$

Answer each question.

79. Hardware Plus sells plywood in thicknesses of $\frac{11}{32}, \frac{5}{8}$, and $\frac{1}{2}$ inch. Order the plywood from smallest to largest.

80. Cloverdale Hardware sells dowels in diameters of $\frac{9}{16}, \frac{1}{2}, \frac{3}{8}$, and $\frac{5}{8}$ inch. Order the dowels from smallest to largest.

81. Three buttons have diameters of $\frac{11}{32}, \frac{3}{8}$, and $\frac{5}{16}$ inch. Order the buttons from largest to smallest.

82. A machinist has three bolts with diameters of $\frac{1}{2}, \frac{7}{8}$, and $\frac{3}{4}$ inch. Order the bolts from smallest to largest.

goetz Don't forget. Common denominators are always needed when adding or subtracting fractions.
For more tips and tweets, go to twitter.com/gstbasicmath

83. When like fractions are added, why does the denominator stay the same?

84. When adding like fractions, why are you allowed to add the numerators?

85. Ginger and Mark did not coordinate their shopping lists. Ginger bought $\frac{2}{3}$ lb of grapes. Mark bought $\frac{1}{4}$ lb of grapes. How many pounds of grapes did they purchase together?

86. How much olive oil should be added to $\frac{1}{6}$ oz of vinegar to make $\frac{3}{4}$ oz of salad dressing?

87. Vanessa is a carpenter. Her current project, installing kitchen cupboards, is almost complete. All she needs to do is put the knobs on the cupboard doors. If the cupboard doors are $\frac{3}{4}$ inch thick and the knobs require $\frac{5}{12}$ inch of exposed bolt thread, how long do the bolts need to be?

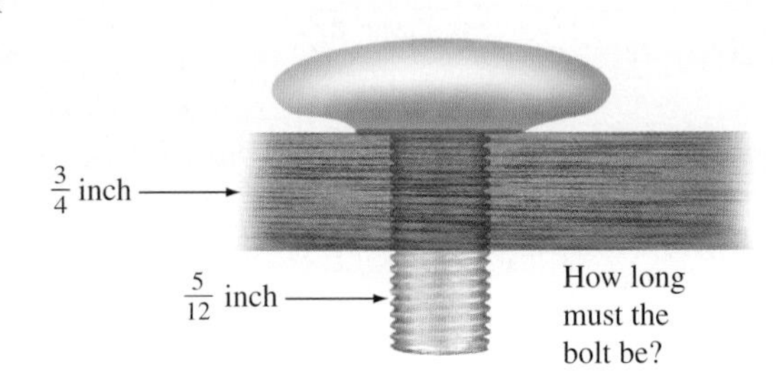

88. During an experiment, a chemist mixed $\frac{1}{9}$ oz of one solution with $\frac{5}{12}$ oz of another solution. What is the total volume of the combined solution?

89. On a 50-question quiz, you answered 84%, or $\frac{84}{100}$, of the questions correctly. How many questions did you answer correctly?

90. Cecilia is screwing $\frac{11}{32}$-inch plywood boards over her windows to prepare her house for a hurricane. She is using screws that are $\frac{7}{4}$ inches long. After passing through the plywood, how far will the screws go into the window frames?

91. Thomas and Rosie are getting married next summer. They plan on buying a house soon after their wedding. Rosie has saved $\frac{3}{7}$ of the down payment on the house. Thomas has saved $\frac{1}{5}$ of the down payment.

a. Who has saved more for the house?

b. Together, what fractional part of the down payment do they have?

c. What fractional part do they still need to save?

92. A drill press has a reservoir that holds 1 gallon of oil. Over one week, a machinist carefully monitored how much oil was used and added. The reservoir was full at the start of his shift on Monday, when he used $\frac{16}{64}$ gal of oil. On Tuesday, he used $\frac{20}{64}$ gal of oil. On Wednesday, he used $\frac{25}{64}$ gal of oil. How much oil was in the reservoir at the end of Wednesday?

93. An orthodontist tightened a child's braces periodically for several months by turning a screw that connects the braces. Two months ago, the orthodontist turned the screw $\frac{2}{16}$ of a turn. Last month, she turned the screw $\frac{3}{16}$ of a turn, and this month, she turned the screw another $\frac{3}{16}$ of a turn. What fraction of a complete turn did the orthodontist make over the last three months?

94. What fraction of a dollar is 2 quarters, 1 dime, and 1 nickel?

95. What fraction of a dollar is 1 quarter, 3 dimes, and 4 pennies?

96. Andy and Aziza are going to race around the rectangular plaza shown below. Find its perimeter to determine the distance of the race.

97. A town has agreed to allow the construction of the park shown below. The only condition is that the park must have a fence around it. Find the park's perimeter, in kilometers, to determine how much fencing is needed.

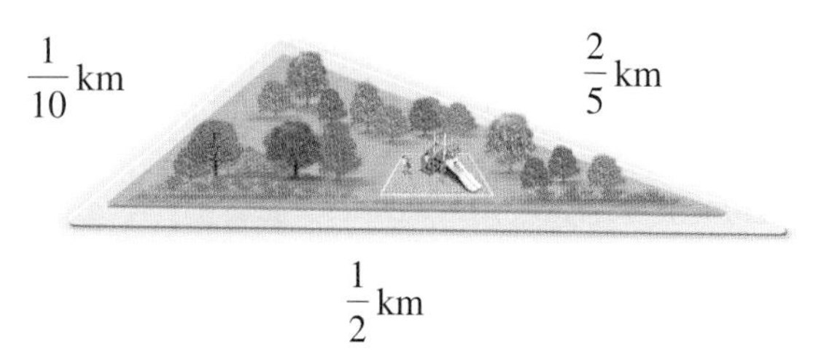

Use the graph to answer each question.

98. What fraction of the day did Steve spend doing school-related activities?

99. By what fraction of the day did Steve's TV time go beyond his homework time?

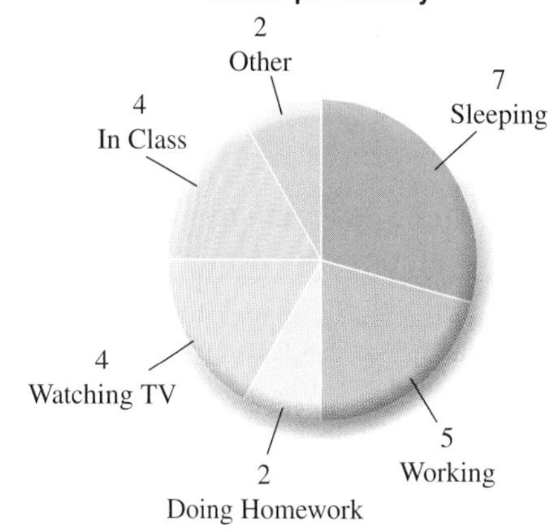

Section 2.4 Question Log

Use this space to write questions. Be sure to get these answered and revisit them when you prepare for your exam.

Page _____ **Answered**	**Page** _____ **Answered**
Page _____ **Answered**	**Page** _____ **Answered**

Mid-Chapter 2 Review Exercises

Before starting Section 2.5, solve the exercises in Appendix A.

Section 2.5 Fractions and the Order of Operations

In this section, you will use the order of operations with fractions. As with whole numbers, it is important for everyone to perform operations in the same order. If people do not follow the order of operations, they may get different answers to identical problems.

Because the steps for adding, subtracting, multiplying, and dividing fractions may still feel new to you, be sure to reference the flowchart for working with fractions as you work in this section. It is located on page 2-86.

The **Objective** in Section 2.5 will help you

A Use the order of operations with fractions.

VOCABULARY PREVIEW *Check the box that applies.*	Got It	Must Study
Order of Operations: **First:** Perform operations in groupings. **Second:** Perform operations with exponents. **Third:** Perform multiplication and division as they occur from left to right. **Fourth:** Perform addition and subtraction as they occur from left to right.		

Study the order of operations until you can recite it, word for word, from memory.

Objective A Use the Order of Operations with Fractions

The Concept If the order of operations is not followed, you can get different values for identical exercises. For instance, look at the examples below.

Following the Order of Operations

$$\frac{5}{8}+\frac{1}{4}\cdot\frac{1}{2}=\frac{5}{8}+\frac{1}{8}$$
$$=\frac{6}{8}$$
$$=\frac{3}{4}$$

***Not* Following the Order of Operations**

$$\frac{5}{8}+\frac{1}{4}\cdot\frac{1}{2}=\frac{5}{8}+\frac{2}{8}\cdot\frac{1}{2}$$
$$\neq\frac{7}{8}\cdot\frac{1}{2}$$
$$\neq\frac{7}{16}$$

The multiplication should have been performed *before* the addition.

The Order of Operations

Groupings: Numerators and denominators are groupings.
Exponents: Write the repeated multiplication; then multiply.
Multiplication and Division: Perform them as they occur from left to right.
Addition and Subtraction: Perform them as they occur from left to right.

Since the numerator and denominator of a fraction are groupings, they should be simplified first.

DETAILED EXAMPLES Numerators and Denominators Are Groupings

Simplify. $\frac{1}{10} + \frac{1}{2+3}$

The denominator is a grouping. $\frac{1}{10} + \frac{1}{2+3} = \frac{1}{10} + \frac{1}{5}$

Build like fractions to add the fractions. $= \frac{1}{10} + \frac{2}{10}$

$= \frac{3}{10}$

Simplify. $\frac{1}{4} \cdot \frac{3+5}{3}$

The numerator is a grouping. $\frac{1}{4} \cdot \frac{3+5}{3} = \frac{1}{4} \cdot \frac{8}{3}$

Multiply and simplify by dividing out common factors. $= \frac{1}{\cancel{4}} \cdot \frac{2 \cdot \cancel{4}}{3}$

$= \frac{2}{3}$

Procedure **Use the Order of Operations**

First: Perform operations in groupings.
Second: Perform operations with exponents.
Third: Perform multiplication and division as they occur from left to right.
Fourth: Perform addition and subtraction as they occur from left to right.

The next two examples look more difficult because they are longer. To make long exercises easier, do the following:

- Choose the operation to be performed.
- Copy everything that will not change.
- Calculate the operation.
- Repeat this process for every operation in the expression.

EXAMPLES

1. Perform the operations. $\frac{5-4}{7} \div \frac{3}{14}$

Groupings $\frac{5-4}{7} \div \frac{3}{14} = \frac{1}{7} \div \frac{3}{14}$

Divide. $= \frac{1}{7} \cdot \frac{14}{3}$

Simplify. $= \frac{1 \cdot 2 \cdot \cancel{7}}{\cancel{7} \cdot 3}$

$= \frac{2}{3}$

GUIDED PRACTICE

1. Perform the operations. $\frac{5}{8} \div \frac{3}{9-5}$

$\frac{5}{8} \div \frac{3}{9-5} = \frac{5}{8} \div \frac{3}{\_\_}$

$= \frac{\_\_}{\_\_} \cdot \frac{\_\_}{\_\_}$

$= \frac{\_\_}{\_\_}$

$= \frac{\_\_}{\_\_}$

(Continued)

2. Perform the operations. $\frac{7}{8} - \left(\frac{1}{4} + \frac{1}{2}\right)$

Groupings
Build like fractions to add the fractions in the grouping.

$$\frac{7}{8} - \left(\frac{1}{4} + \frac{1}{2}\right) = \frac{7}{8} - \left(\frac{1}{4} + \frac{2}{4}\right)$$

$$= \frac{7}{8} - \left(\frac{3}{4}\right)$$

Subtract.
Build like fractions to subtract.

$$= \frac{7}{8} - \frac{6}{8}$$

$$= \frac{1}{8}$$

2. Perform the operations. $\left(\frac{3}{5} + \frac{1}{10}\right) - \frac{7}{15}$

$$\left(\frac{3}{5} + \frac{1}{10}\right) - \frac{7}{15} = \left(\frac{\quad}{\quad} + \frac{1}{10}\right) - \frac{7}{15}$$

$$= \left(\frac{\quad}{\quad}\right) - \frac{\quad}{\quad}$$

$$= \frac{\quad}{\quad} - \frac{\quad}{\quad}$$

$$= \frac{\quad}{\quad}$$

3. Perform the operations. $\frac{5}{6} - \frac{7-3}{15} \cdot \frac{5}{2}$

Groupings

$$\frac{5}{6} - \frac{7-3}{15} \cdot \frac{5}{2} = \frac{5}{6} - \frac{4}{15} \cdot \frac{5}{2}$$

Multiply and simplify.

$$= \frac{5}{6} - \frac{2 \cdot \cancel{2} \cdot \cancel{5}}{3 \cdot \cancel{5} \cdot \cancel{2}}$$

Subtract.
Build like fractions to subtract.

$$= \frac{5}{6} - \frac{2}{3}$$

$$= \frac{5}{6} - \frac{4}{6}$$

$$= \frac{1}{6}$$

3. Perform the operations. $\frac{13}{20} - \frac{3}{4} \cdot \frac{9-5}{5}$

$$\frac{13}{20} - \frac{3}{4} \cdot \frac{9-5}{5} = \frac{13}{20} - \frac{3}{4} \cdot \frac{\quad}{5}$$

$$= \frac{\quad}{\quad} - \frac{\quad}{\quad}$$

$$= \frac{\quad}{\quad} - \frac{\quad}{\quad}$$

$$= \frac{\quad}{\quad} - \frac{\quad}{\quad}$$

$$= \frac{\quad}{\quad}$$

4. Perform the operations. $\left(\frac{5}{3} \cdot \frac{5}{4} + \frac{9}{12}\right) \cdot \frac{3}{17}$

Groupings
Multiply; then add.

$$\left(\frac{5}{3} \cdot \frac{5}{4} + \frac{9}{12}\right) \cdot \frac{3}{17} = \left(\frac{25}{12} + \frac{9}{12}\right) \cdot \frac{3}{17}$$

Multiply.

$$= \left(\frac{34}{12}\right) \cdot \frac{3}{17}$$

Simplify.

$$= \frac{\cancel{2} \cdot \cancel{17} \cdot \cancel{3}}{\cancel{2} \cdot 2 \cdot \cancel{3} \cdot \cancel{17}}$$

$$= \frac{1}{2}$$

4. Perform the operations. $\left(\frac{11}{6} \cdot \frac{1}{5} + \frac{17}{30}\right) \cdot \frac{6}{7}$

$$\left(\frac{11}{6} \cdot \frac{1}{5} + \frac{17}{30}\right) \cdot \frac{6}{7} = \left(\frac{\quad}{\quad} + \frac{\quad}{\quad}\right) \cdot \frac{6}{7}$$

$$= \left(\frac{\quad}{\quad}\right) \cdot \frac{\quad}{\quad}$$

$$= \frac{\quad}{\quad}$$

$$= \frac{\quad}{\quad}$$

Concept Check

Each of the following exercises was solved incorrectly. Find the first error in each exercise and identify it as either a computation mistake or a violation of the order of operations. The first two examples are done for you.

Computation mistake. The first two fractions are not like, so they can't be added yet.

$$\frac{7}{8}+\frac{3}{4}-\frac{1}{6}=\frac{10}{12}-\frac{1}{6}=\frac{5}{6}-\frac{1}{6}=\frac{4}{6}=\frac{2}{3}$$

Order of operations violation: The division needs to be performed first since it is left of the multiplication.

$$\frac{1}{2}\div\frac{3}{4}\cdot\frac{5}{6}=\frac{1}{2}\div\frac{15}{24}=\frac{1}{2}\div\frac{5}{8}=\frac{1}{2}\cdot\frac{8}{5}=\frac{4}{5}$$

A1. $$\frac{3}{4}-\frac{1}{2}+\frac{1}{6}=\frac{9}{12}-\frac{6}{12}+\frac{2}{12}=\frac{9}{12}-\frac{8}{12}=\frac{1}{12}$$

A2. $$\left(\frac{3}{5}\right)^2\div\left(\frac{9}{14}\right)=\frac{6}{10}\div\frac{9}{14}=\frac{6}{10}\cdot\frac{14}{9}=\frac{\cancel{2}\cdot\cancel{3}}{\cancel{2}\cdot 5}\cdot\frac{\cancel{2}\cdot 7}{\cancel{3}\cdot\cancel{3}}=\frac{7}{5}$$

A3. In an order of operations exercise, why should the numerator and denominator of a fraction be simplified first?

Objective A Practice

Perform the operations.

A4. $\frac{19}{20}-\left(\frac{1}{2}+\frac{2}{5}\right)$

A5. $\frac{7}{21}+\left(\frac{5}{7}-\frac{1}{3}\right)$

A6. $\frac{2}{3}\cdot\left(\frac{3}{4}\right)^2$

A7. $\left(\frac{1}{5}\right)^2\div\frac{3}{5}$

A8. $\frac{2}{3}-\frac{5}{12}+\frac{3}{4}$

A9. $\frac{7}{5}-\frac{3}{10}+\frac{1}{5}$

A10. $\frac{4}{5}\div\frac{8}{15}\cdot\frac{2}{9}$

A11. $\frac{4}{5}\div\frac{1}{3}\cdot\frac{25}{6}$

A12. $\frac{13}{21}-\frac{5}{7}\cdot\left(\frac{9-7}{3}\right)$

A13. $\frac{7}{24}-\frac{1}{5}\div\left(\frac{6}{2+3}\right)$

A14. $\frac{25}{16}-\left(\frac{3}{4}\right)^2+\frac{5}{8}$

A15. $\frac{9}{25}+\left(\frac{4}{5}\right)^2+\frac{1}{10}$

Combining Concepts and Applications

CONCEPT I Application Exercises Each of the word exercises in this section requires two or more steps. When solving a word exercise, remember to check your answer and make sure it is reasonable.

EXAMPLE

5. Alice has finished writing $\frac{3}{4}$ of the 200 thank-you cards from her wedding reception yesterday. If she finished another $\frac{1}{8}$ of the cards today, how many more cards does she have to write?

First Add to determine the fractional portion that is completed.

$$\frac{3}{4}+\frac{1}{8}=\frac{6}{8}+\frac{1}{8}$$
$$=\frac{7}{8}$$

Second Subtract the completed portion from one to determine the portion that is not completed.

$$1-\frac{7}{8}=\frac{8}{8}-\frac{7}{8}$$
$$=\frac{1}{8}$$

Third Multiply the total amount by the portion that is not completed.

$$200\cdot\left(\frac{1}{8}\right)=\frac{25\cdot\cancel{8}}{1}\cdot\frac{1}{\cancel{8}}$$
$$=25$$

Does the answer make sense? Since $\frac{3}{4}$ plus $\frac{1}{8}$ represents almost all of the cards, only a small portion should remain. 25 is a small portion of 200.

State your answer in words. Alice still has 25 letters to write.

GUIDED PRACTICE

5. Between 8 A.M. and 11 A.M., Alex set up $\frac{3}{5}$ of the 300 chairs needed for a wedding. If he sets up another $\frac{3}{10}$ of the chairs before lunch, how many more chairs will he need to set up after lunch?

$$\frac{3}{5}+\frac{3}{10}=\frac{\quad}{\quad}+\frac{\quad}{\quad}$$
$$=\frac{\quad}{\quad}$$

$$1-\frac{\quad}{\quad}=\frac{\quad}{\quad}-\frac{\quad}{\quad}$$
$$=\frac{\quad}{\quad}$$

$$300\cdot\left(\frac{\quad}{\quad}\right)=\frac{\quad}{\quad}\cdot\frac{\quad}{\quad}$$
$$=$$

FOCUS ON ALGEBRA Solving an Equation for an Unknown Value

In algebra, many equations can be solved by undoing an operation.

- Addition and subtraction undo each other.
- Multiplication and division undo each other.

To solve each equation, undo the operation that is being done to the variable *n*.

EXAMPLES

6. What value of *n* makes the equation true?

$n - \frac{1}{3} = \frac{3}{4}$

$$n - \frac{1}{3} = \frac{3}{4}$$

Step 1 To undo subtraction, add the same number to each side.

$$n - \frac{1}{3} + \frac{1}{3} = \frac{3}{4} + \frac{1}{3}$$

Step 2 Build like fractions to add.

$$n = \frac{9}{12} + \frac{4}{12}$$

$$n = \frac{13}{12}$$

7. What value of *n* makes the equation true? $\frac{3}{5} \cdot n = \frac{1}{4}$

$$\frac{3}{5} \cdot n = \frac{1}{4}$$

Step 1 To undo multiplication by a fraction, multiply both sides by the reciprocal of that fraction.

$$\frac{\cancel{5}}{\cancel{3}} \cdot \frac{\cancel{3}}{\cancel{5}} \cdot n = \frac{5}{3} \cdot \frac{1}{4}$$

Step 2 Multiply the product on the right.

$$n = \frac{5}{12}$$

GUIDED PRACTICE

6. What value of *n* makes the equation true?

$n - \frac{2}{3} = \frac{1}{2}$

$$n - \frac{2}{3} = \frac{1}{2}$$

$$n - \frac{2}{3} + \frac{\square}{\square} = \frac{1}{2} + \frac{\square}{\square}$$

$$n = \frac{\square}{\square} + \frac{\square}{\square}$$

$$n = \frac{\square}{\square}$$

7. What value of *n* makes the equation true? $\frac{3}{2} \cdot n = \frac{4}{5}$

$$\frac{3}{2} \cdot n = \frac{4}{5}$$

$$\frac{\square}{\square} \cdot \frac{3}{2} \cdot n = \frac{\square}{\square} \cdot \frac{4}{5}$$

$$n = \frac{\square}{\square}$$

PRACTICE

What value of *n* makes the equation true?

1. $n + \frac{1}{4} = \frac{3}{4}$

2. $n + \frac{5}{8} = \frac{9}{8}$

3. $n - \frac{1}{2} = \frac{3}{2}$

4. $n - \frac{1}{10} = \frac{4}{10}$

5. $\frac{1}{5} \cdot n = \frac{1}{8}$

6. $\frac{1}{4} \cdot n = \frac{1}{9}$

7. $\frac{5}{6} \cdot n = 2$

8. $\frac{7}{3} \cdot n = 6$

9. $n + \frac{1}{3} = \frac{11}{18}$

10. $n + \frac{1}{6} = \frac{3}{12}$

11. $n - \frac{1}{3} = \frac{5}{6}$

12. $n - \frac{1}{5} = \frac{3}{10}$

13. $n + \frac{5}{6} = \frac{6}{5}$

14. $n + \frac{8}{9} = \frac{3}{2}$

15. $\frac{4}{5} \cdot n = \frac{3}{20}$

16. $\frac{9}{5} \cdot n = \frac{18}{7}$

This flowchart is designed to help you remember the correct steps for the correct problem. Study this flowchart and practice writing it on a sheet of paper.

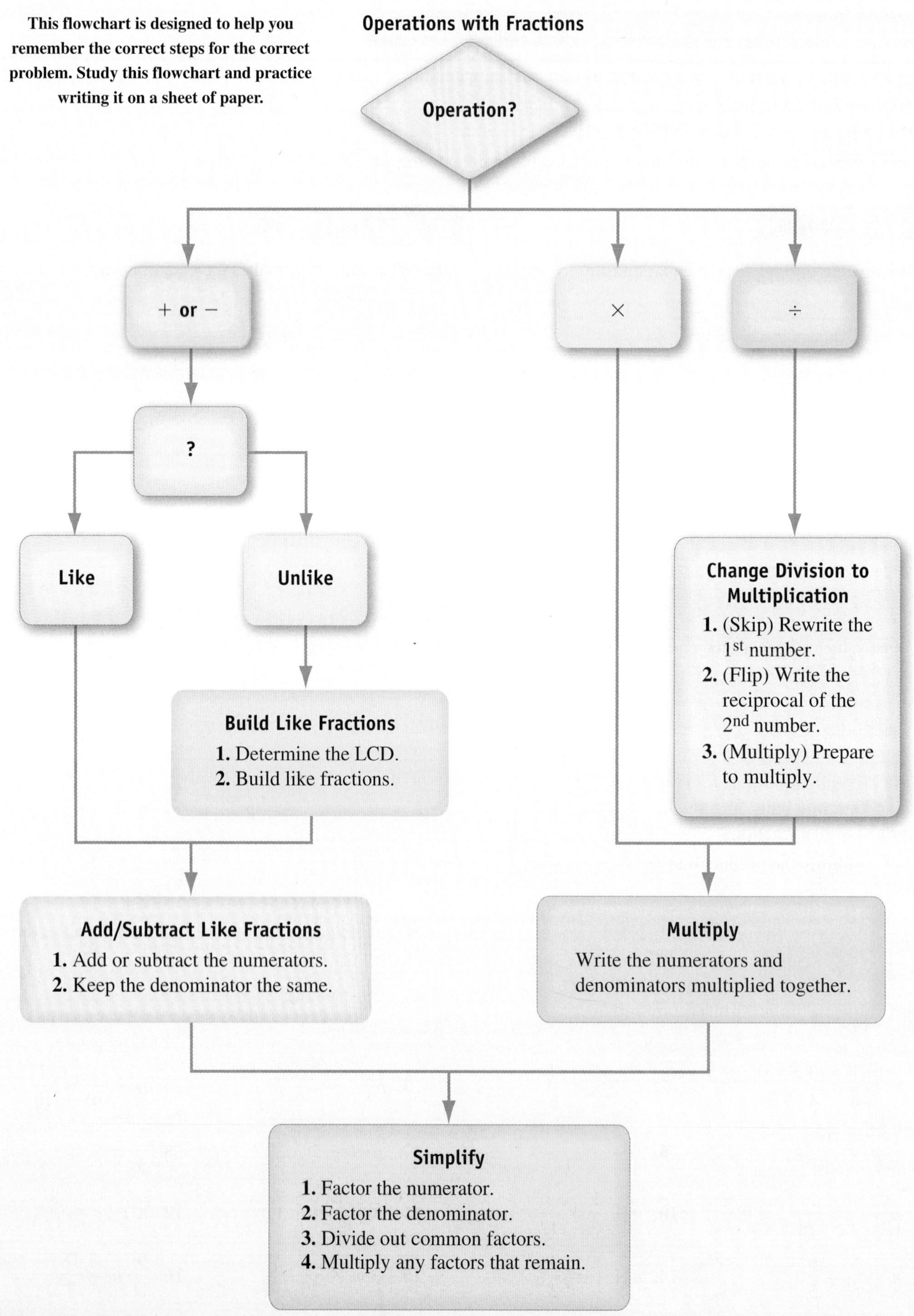

▶ **NOTE** An LCD is needed only when you are adding or subtracting. Do not find the LCD when you are multiplying or dividing.

Section 2.5 Exercises

FOR EXTRA HELP     

To use the order of operations with fractions:

1. Answer the Objective A Concept Checks.
2. Answer the odd Objective A Practice Exercises.
3. Answer the even Objective A Practice Exercises.

To solve an equation for an unknown value:

4. Answer the odd Focus on Algebra Practice Exercises.
5. Answer the even Focus on Algebra Practice Exercises.

Follow the order of operations to simplify each expression.

6. $\frac{1}{2} - \frac{1}{3} + \frac{1}{4}$

7. $\frac{1}{5} - \frac{1}{6} + \frac{1}{8}$

8. $\frac{3}{10} \cdot \frac{4}{9} \div \frac{1}{3}$

9. $\frac{5}{12} \cdot \frac{9}{20} \div \frac{3}{10}$

10. $\frac{16}{35} - \frac{3}{7} \cdot \frac{2}{5}$

11. $\frac{27}{32} - \frac{3}{8} \cdot \frac{1}{4}$

12. $\frac{1}{5} \div \frac{1}{6} \cdot \frac{1}{8}$

13. $\frac{1}{2} \div \frac{1}{3} \cdot \frac{1}{4}$

14. $\left(\frac{5}{7} - \frac{3}{7}\right) \cdot \frac{2}{5}$

15. $\left(\frac{7}{8} - \frac{3}{8}\right) \cdot \frac{1}{4}$

16. $\left(\frac{4}{5}\right)^2 \cdot \frac{15}{8}$

17. $\left(\frac{3}{7}\right)^2 \cdot \frac{14}{9}$

18. $\frac{3}{4} - \left(\frac{1}{4}\right)^2 + \frac{3}{8}$

19. $\frac{2}{3} - \left(\frac{2}{3}\right)^2 + \frac{1}{9}$

20. $\frac{3}{10} \cdot \left(\frac{4}{9} \div \frac{1}{3}\right)$

21. $\frac{5}{12} \cdot \left(\frac{9}{20} \div \frac{3}{10}\right)$

22. $\left(\frac{3}{4}\right)^2 - \frac{3}{8}$

23. $\left(\frac{4}{5}\right)^2 - \frac{3}{5}$

24. $\left(2 - \frac{8}{5}\right)^2$

25. $\left(\frac{9}{2} - 4\right)^2$

26. $\frac{4}{9} \div \left(\frac{5-3}{7-4}\right)$

27. $\left(\frac{5}{8} - \frac{1}{4}\right) \cdot \frac{16}{21}$

28. $\left(\frac{1}{5} + \frac{1}{15}\right) \cdot \left(\frac{7}{8} - \frac{1}{4}\right)$

29. $\left(\frac{1}{3} + \frac{1}{6}\right) \cdot \left(\frac{3}{4} - \frac{1}{2}\right)$

30. $\frac{12-3}{10} - \frac{4}{5} \div \frac{8}{3}$

31. $\frac{9}{2} - \frac{6-1}{6} \div \frac{5}{9}$

32. $\left(\frac{1}{6-4} - \frac{1}{8}\right) + \frac{3}{4}$

33. $\left(\frac{3}{16-9} - \frac{3}{14}\right) + \frac{2}{5}$

34. Andrew walked $\frac{3}{4}$ of a mile. Jolina walked twice as far and $\frac{1}{8}$ of a mile farther than Andrew. How far did Jolina walk?

35. Andréa had 4 spools of white thread. If each wedding dress she makes requires $\frac{1}{3}$ spool of thread and she has sewn 4 dresses, how much thread does she have left?

smith The numerators and denominators are groupings. Perform any calculations in the numerator or denominator first.
For more tips and tweets, go to twitter.com/gstbasicmath

What value of n makes the equation true?

36. $n + \frac{5}{8} = \frac{7}{8}$

37. $n - \frac{1}{8} = \frac{5}{8}$

38. $n - \frac{1}{4} = \frac{3}{8}$

39. $n + \frac{1}{3} = \frac{1}{2}$

40. $\frac{3}{4} \cdot n = \frac{6}{1}$

41. $\frac{4}{3} \cdot n = \frac{2}{3}$

42. $n - \frac{3}{5} = \frac{2}{7}$

43. $n + \frac{7}{9} = \frac{8}{5}$

Answer each question.

44. Julia made guacamole using five avocado halves. When she was done, she had three avocado halves left over.

a. Explain why the following equation tells us how many avocados Julia had before she made the guacamole.

$$n - \frac{5}{2} = \frac{3}{2}$$

b. How many avocados did Julia start with?

45. Victor made orange juice using seven orange halves. When he was done, he had five orange halves left over.

a. Explain why the following equation tells us how many oranges Victor had before he made the orange juice.

$$n - \frac{7}{2} = \frac{5}{2}$$

b. How many oranges did Victor start with?

46. A building code for railings requires that connecting bolts have a diameter that is *more* than $\frac{1}{4}$ the thickness of the attaching plate. An attaching plate for a new railing is $\frac{1}{2}$ inch thick.

a. What is the minimum allowable diameter of the bolt?

b. If the bolts are available only in increments of 16ths of an inch, what is the smallest diameter bolt that may be used?

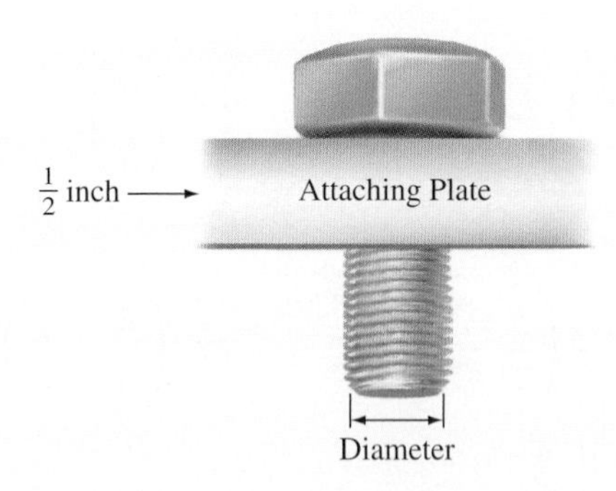

47. Antwan wants to build a fence around his backyard that measures 80 feet by 60 feet. His 15-foot-by-20-foot garden is already fenced. The unfenced portion of his yard is shown with a dashed line on the diagram.

a. What is the perimeter of Antwan's entire backyard?

b. How much fencing must Antwan buy?

c. Before he puts up the new fence, what fraction of his project is already complete?

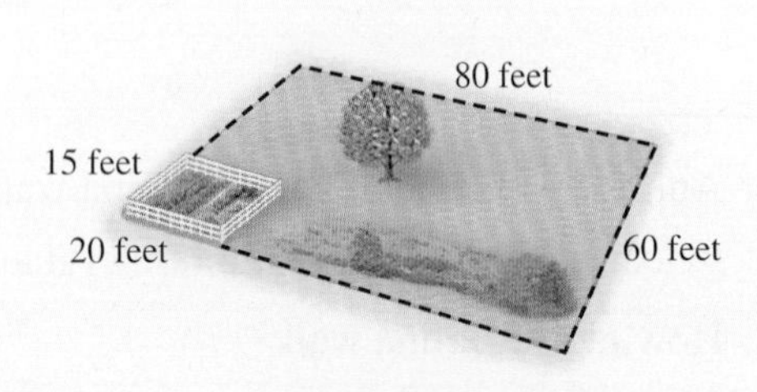

Section 2.5 Question Log

Use this space to write questions. Be sure to get these answered and revisit them when you prepare for your exam.

Page ______ **Answered**	**Page** ______ **Answered**
Page ______ **Answered**	**Page** ______ **Answered**

Section 2.6 Mixed Numbers

The mixed number $1\frac{1}{4}$ is equivalent to the improper fraction $\frac{5}{4}$.

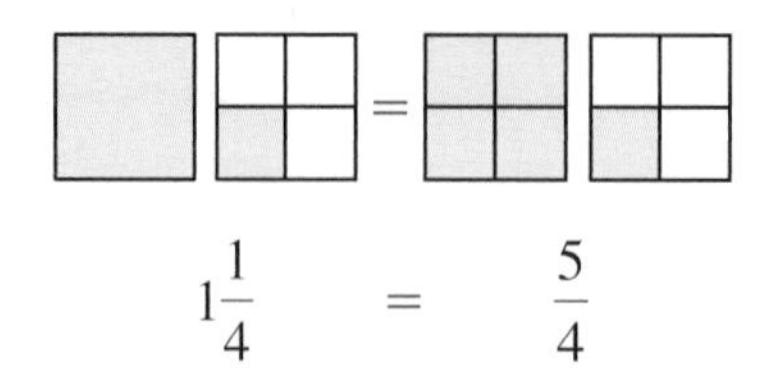

A logical question to ask is, Why do we have both mixed numbers and improper fractions?

There are several advantages of mixed numbers.

- Many people feel that is easier to identify the value of a number when it is written as a mixed number. The value of the mixed number $1\frac{1}{4}$ might be easier to understand than the value of the improper fraction $\frac{5}{4}$.
- By writing the whole number and fractional parts separately, it is often easier to estimate mixed number calculations.

There are several advantages of improper fractions.

- It is often easier to perform calculations using improper fractions.
- The concepts used with improper fractions are used in algebra.

The **Objectives** in Section 2.6 will help you

- **A** Convert an improper fraction to a mixed number.
- **B** Convert a mixed number to an improper fraction.
- **C** Multiply and divide mixed numbers.
- **D** Add mixed numbers.
- **E** Subtract mixed numbers.

VOCABULARY PREVIEW *Check the box that applies.*	Got It	Must Study
Improper Fraction: An **improper fraction** has a numerator that is greater than or equal to the denominator.		
Mixed Number: The sum of a whole number greater than zero and a proper fraction is a **mixed number.**		
Whole Number: A number from the list 0, 1, 2, 3, 4 ... is a **whole number.**		
Quotient: A **quotient** is the result of dividing. In $3\overline{)17}$ 5R2, 5 is the quotient.		
Remainder: The **remainder** is the number that is "left over" when the divisor does not divide the dividend evenly. In $3\overline{)17}$ 5R2, 2 is the remainder.		

Study these words when they appear in the text. After the section, test your understanding by completing the Vocabulary Review in the section exercises.

Objective A Convert an Improper Fraction to a Mixed Number

The Concept Mixed numbers are used when both a whole number and a fraction are needed to represent something.

INTERACTIVE DEFINITION Mixed Number

The sum of a whole number greater than zero and a proper fraction is a **mixed number.**

EXAMPLE

1. Represent the picture as a mixed number.

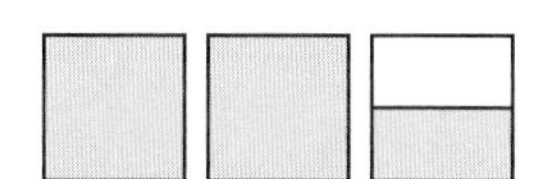

Since there are 2 whole objects and $\frac{1}{2}$ of another, the mixed number $2\frac{1}{2}$ represents this picture.

GUIDED PRACTICE

1. Represent the picture as a mixed number.

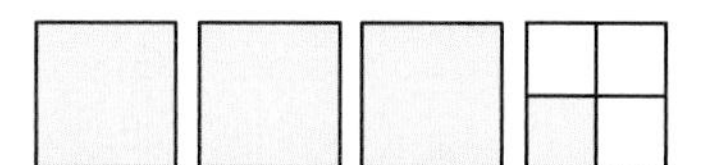

Since there are _____ whole objects and $\frac{\quad}{\quad}$ of another, the mixed number $\frac{\quad}{\quad}$ represents this picture.

DO YOU UNDERSTAND mixed numbers? Got It Get Help

For every mixed number, there is an equivalent improper fraction. The mixed number and improper fractions pictured below represent an equivalent number of oranges.

Equivalent Number of Oranges

$2\frac{1}{2}$ oranges

$\frac{5}{2}$ oranges

A picture can help you understand how an improper fraction can be converted to a mixed number.

EXAMPLES

2. Use a picture to convert $\frac{7}{4}$ to a mixed number.

Step 1 Represent the improper fraction as a picture. $\frac{7}{4} =$

Step 2 Draw 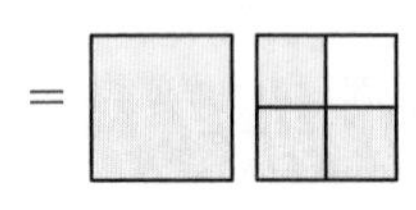 as one whole object. $=$

Step 3 Describe the picture as a mixed number. $= 1\frac{3}{4}$

3. Use division to convert the same improper fraction, $\frac{7}{4}$, to a mixed number.

Step 1 Divide the numerator by the denominator. $\frac{7}{4} \rightarrow 4\overline{)7}$ 1R3

Quotient = 1

Remainder = 3

Denominator = 4

Step 2 Write the mixed number as quotient $\frac{\text{remainder}}{\text{denominator}}$. $\frac{7}{4} = 1\frac{3}{4}$

GUIDED PRACTICE

2. Use a picture to convert $\frac{3}{2}$ to a mixed number.

Step 1 Represent the improper fraction as a picture. $\frac{3}{2} =$

Step 2 Draw  as one whole object. $=$

Step 3 Describe the picture as a mixed number. $=$

3. Use division to convert the same improper fraction, $\frac{3}{2}$, to a mixed number.

Step 1 Divide the numerator by the denominator. $\frac{3}{2} \rightarrow 2\overline{)3}$

Quotient = ______

Remainder = ______

Denominator = ______

Step 2 Write the mixed number as quotient $\frac{\text{remainder}}{\text{denominator}}$. $\frac{3}{2} =$

Procedure Convert an Improper Fraction to a Mixed Number

Step 1 Divide the numerator by the denominator.

$$\text{denominator}\overline{)\text{numerator}} \quad \text{quotient and remainder}$$

Step 2 Write the mixed number.

$$\text{quotient}\ \frac{\text{remainder}}{\text{denominator}}$$

Step 3 Simplify the fraction if necessary.

EXAMPLES

4. Convert $\frac{9}{4}$ to a mixed number.

Step 1 Divide the numerator by the denominator. $9 \div 4 = 2$ Remainder 1

Step 2 Write the mixed number. $\frac{9}{4} = 2\frac{1}{4}$

quotient $\frac{\text{remainder}}{\text{denominator}}$

5. Convert $\frac{13}{6}$ to a mixed number.

Step 1 Divide the numerator by the denominator. $13 \div 6 = 2$ Remainder 1

Step 2 Write the mixed number. $\frac{13}{6} = 2\frac{1}{6}$

6. Convert $\frac{27}{3}$ to a mixed number.

Step 1 Divide the numerator by the denominator. $27 \div 3 = 9$ Remainder 0

Step 2 Write the mixed number. $\frac{27}{3} = 9$

GUIDED PRACTICE

4. Convert $\frac{17}{6}$ to a mixed number.

$17 \div 6 =$ ______ Remainder ______

$\frac{17}{6} =$

5. Convert $\frac{15}{8}$ to a mixed number.

$15 \div 8 =$ ______ Remainder ______

$\frac{15}{8} =$

6. Convert $\frac{84}{7}$ to a mixed number.

$84 \div 7 =$ ______ Remainder ______

$\frac{84}{7} =$ ______

When the remainder is zero, the answer is a whole number.

Concept Check

A1. In your own words, explain how to use the quotient concept of a fraction to convert an improper fraction to a mixed number.

Objective A Practice

For each exercise:

a. What improper fraction is represented?

b. What mixed number is represented?

A2. 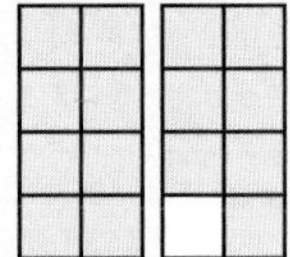

A3. 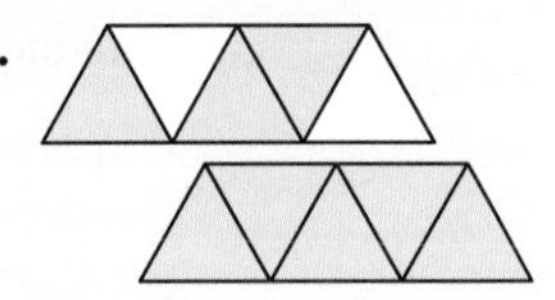

A4. 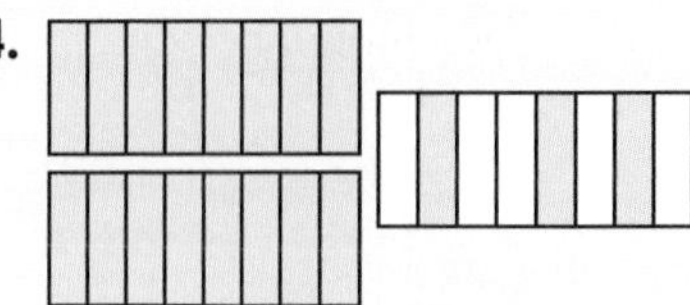

A5. 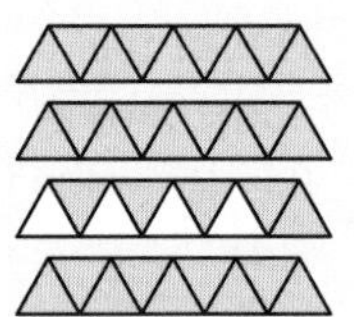

Answer each question.

A6. Write a mixed number and an improper fraction to represent the number of limes.

A7. Write a mixed number and an improper fraction to represent the number of dollars.

Convert each improper fraction to a mixed number or whole number. Write your answer in simplest form.

A8. $\frac{12}{5}$

A9. $\frac{17}{3}$

A10. $\frac{42}{10}$

A11. $\frac{39}{9}$

A12. $\frac{48}{8}$

A13. $\frac{26}{2}$

Objective B Convert a Mixed Number to an Improper Fraction

The Concept Visualizing a picture can help you understand the procedure used to convert mixed numbers to improper fractions.

EXAMPLE

7. Use a picture to convert $2\frac{1}{4}$ to an improper fraction.

$$2\frac{1}{4} = \frac{4}{4} + \frac{4}{4} + \frac{1}{4} = \frac{8}{4} + \frac{1}{4} = \frac{9}{4}$$

Each whole is divided to match the denominator.

GUIDED PRACTICE

7. Use a picture to convert $2\frac{1}{3}$ to an improper fraction.

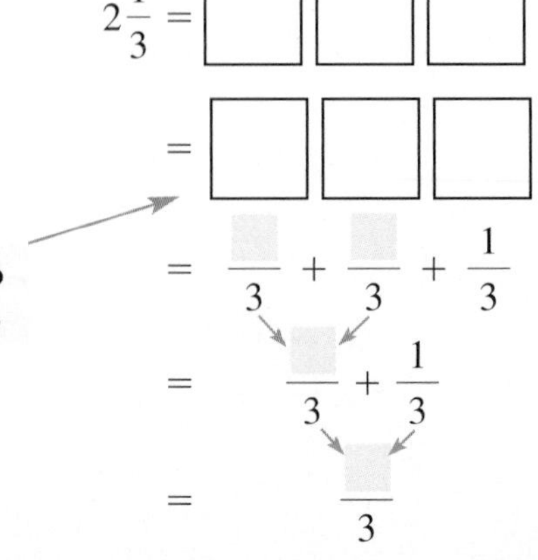

The mixed numbers from Example 7 and Guided Practice 7 are used again for Example 8 and Guided Practice 8. Use those pictures to understand the procedure that follows.

EXAMPLES

8. Answer each question to convert $2\frac{1}{4}$ to an improper fraction.

a. How many fourths does the "whole" portion of $2\frac{1}{4}$ represent?

Multiply the denominator and the whole number.
There are $4 \cdot 2 = 8$ fourths in 2.

b. How many fourths are represented by the fractional part of $2\frac{1}{4}$?

There is 1 fourth in the fractional part of $2\frac{1}{4}$.

c. In total, how many fourths are there?
Add the results. There are $8 + 1 = 9$ fourths.

$$2\frac{1}{4} = \frac{9}{4}$$

9. Convert $1\frac{3}{4}$ to an improper fraction.

a. How many fourths does the "whole" portion of $1\frac{3}{4}$ represent?

Multiply the denominator and the whole number.
There are $4 \cdot 1 = 4$ fourths in 1.

b. How many fourths are represented by the fractional part of $1\frac{3}{4}$?

There are 3 fourths in the fractional part of $1\frac{3}{4}$.

c. In total, how many fourths are there?

Add the results. There are $4 + 3 = 7$ fourths.

$$1\frac{3}{4} = \frac{7}{4}$$

GUIDED PRACTICE

8. Answer each question to convert $2\frac{1}{3}$ to an improper fraction.

a. How many thirds does the "whole" portion of $2\frac{1}{3}$ represent?

Multiply the denominator and the whole number.
There are ______ • ______ = ______ thirds in 2.

b. How many thirds are represented by the fractional part of $2\frac{1}{3}$?

There is ______ third in the fractional part of $2\frac{1}{3}$.

c. In total, how many thirds are there?
Add the results. There are ___ + ___ = ___ thirds.

$$2\frac{1}{3} = \frac{}{3}$$

9. Convert $5\frac{7}{8}$ to an improper fraction.

a. How many eighths does the "whole" portion of $5\frac{7}{8}$ represent?

Multiply the denominator and the whole number.
There are ______ • ______ = ______ eighths in 5.

b. How many eighths are represented by the fractional part of $5\frac{7}{8}$?

There are ___ eighths in the fractional part of $5\frac{7}{8}$.

c. In total, how many eighths are there?
Add the results. There are ___ + ___ = ___ eighths.

$$5\frac{7}{8} = \frac{}{8}$$

Procedure Convert a Mixed Number to an Improper Fraction

Step 1 New numerator = Denominator · Whole number + Numerator
Visualize a loop to help determine the new numerator.

Step 2 The denominator stays the same.

DETAILED EXAMPLE Converting a Mixed Number to an Improper Fraction

Convert $4\frac{2}{3}$ to an improper fraction.

Step 1 Visualize a loop.
- Multiply the denominator and the whole number.
- Add that product to the numerator.

Step 2 The denominator stays the same.

$4\frac{2}{3} = 4\frac{+2}{\cdot 3}$

$= \frac{14}{3}$ ← **The new numerator is $3 \cdot 4 + 2 = 14$.**

EXAMPLES

10. Convert $2\frac{5}{8}$ to an improper fraction.

Step 1 Determine the new numerator. $8 \cdot 2 + 5 = 21$

Denominator · Whole number + Numerator of fraction

Step 2 The denominator stays the same. $2\frac{5}{8} = \frac{21}{8}$

11. Convert $3\frac{1}{6}$ to an improper fraction.

Step 1 Determine the new numerator. $6 \cdot 3 + 1 = 19$

Denominator · Whole number + Numerator of fraction

Step 2 The denominator stays the same. $3\frac{1}{6} = \frac{19}{6}$

GUIDED PRACTICE

10. Convert $3\frac{1}{4}$ to an improper fraction.

_____ · _____ + _____ = _____

$3\frac{1}{4} = \frac{\square}{\square}$

11. Convert $5\frac{7}{9}$ to an improper fraction.

_____ · _____ + _____ = _____

$5\frac{7}{9} = \frac{\square}{\square}$

Concept Check

B1. This question relates to the procedure for converting a mixed number to an improper fraction.
a. Why do you multiply the denominator and the whole number?
b. Why is the numerator of the improper fraction equal to denominator · (whole number) + (numerator of proper fraction)?

Objective B Practice

B2. Convert $6\frac{7}{8}$ to an improper fraction.

a. How many eighths are in 6?

b. How many eighths are in $\frac{7}{8}$?

c. How many eighths are in $6\frac{7}{8}$?

B3. Convert $9\frac{2}{3}$ to an improper fraction.

a. How many thirds are in 9?

b. How many thirds are in $\frac{2}{3}$?

c. How many thirds are in $9\frac{2}{3}$?

Convert each mixed number to an improper fraction.

B4. $1\frac{6}{7}$ **B5.** $3\frac{2}{5}$ **B6.** $2\frac{1}{2}$ **B7.** $6\frac{3}{5}$

B8. $10\frac{1}{3}$ **B9.** $3\frac{1}{10}$ **B10.** $9\frac{1}{9}$ **B11.** $7\frac{5}{8}$

FOCUS ON Rounding Mixed Numbers

When performing calculations with mixed numbers, it is a good idea to estimate each answer. To estimate an answer, round any mixed numbers to the nearest whole number. Consider rounding $3\frac{7}{13}$. Notice that $3\frac{7}{13}$ is between 3 and 4. Which is a better approximation for $3\frac{7}{13}$? If the fractional part is $\frac{1}{2}$ or more, 4 is the better approximation. If the fractional part is less than $\frac{1}{2}$, 3 is the better approximation.

Comparing a fraction to $\frac{1}{2}$.

To compare $\frac{7}{13}$ to $\frac{1}{2}$, we will build like fractions.

$$\frac{7}{13} \square \frac{1}{2}$$

$$\frac{7}{13}\cdot\frac{2}{2} \square \frac{1}{2}\cdot\frac{13}{13}$$

$$\frac{14}{26} \boxed{>} \frac{13}{26}$$

Since $\frac{7}{13} \boxed{>} \frac{1}{2}$, $3\frac{7}{13}$ rounds up to 4.

Note: If you are comfortable with decimals, you can use decimals to compare $\frac{7}{13}$ to $\frac{1}{2}$.

Half of 13 is 6.5; so we can write $\frac{1}{2}$ as $\frac{6.5}{13}$.

$$\frac{7}{13} > \frac{6.5}{13}$$

Since $\frac{7}{13} > \frac{1}{2}$, $3\frac{7}{13}$ rounds up to 4.

EXAMPLES

12. Round $4\frac{1}{3}$ to the nearest whole number.

Since $\frac{1}{3} < \frac{1}{2}$, $4\frac{1}{3}$ rounds down to 4.

GUIDED PRACTICE

12. Round $5\frac{9}{15}$ to the nearest whole number.

Since $\frac{9}{15} \square \frac{1}{2}$, $5\frac{9}{15}$ rounds up/down to ______.

(Continued)

13. Round $2\frac{5}{10}$ to the nearest whole number.	**13.** Round $12\frac{3}{7}$ to the nearest whole number.
Since $\frac{5}{10} = \frac{1}{2}$, $2\frac{5}{10}$ rounds up to 3.	Since $\frac{3}{7} \square \frac{1}{2}$, $12\frac{3}{7}$ rounds up/down to ____.

PRACTICE

Round each mixed number to the nearest whole number.

1. $5\frac{3}{5}$ **2.** $3\frac{2}{4}$ **3.** $7\frac{4}{9}$ **4.** $17\frac{9}{20}$

5. $12\frac{3}{4}$ **6.** $1\frac{5}{8}$ **7.** $6\frac{9}{20}$ **8.** $43\frac{2}{3}$

9. $7\frac{58}{100}$ **10.** $1\frac{1}{1000}$ **11.** $12\frac{16}{33}$ **12.** $13\frac{32}{63}$

Objective C Multiply and Divide Mixed Numbers

The Concept To multiply and divide mixed numbers, estimate the product or quotient and then calculate the answer. Estimation is useful because it will help you catch mistakes.

DETAILED EXAMPLE Estimating a Product or Quotient

Estimate. $1\frac{7}{8} \cdot 3\frac{2}{5}$

Round each number to the nearest whole number. $1\frac{7}{8} \approx 2$ and $3\frac{2}{5} \approx 3$

Estimate using the rounded numbers.

$$1\frac{7}{8} \cdot 3\frac{2}{5} \approx 2 \cdot 3$$
$$\approx 6$$

When we perform the calculation $1\frac{7}{8} \cdot 3\frac{2}{5}$, we check to see that the answer is close to 6.

Procedure Multiply or Divide a Mixed Number

Step 1 Estimate the product or quotient.
Step 2 Convert each mixed number to an improper fraction.
Step 3 Multiply or divide the fractions.
Step 4 Write the result as a mixed number.

EXAMPLES

14. Multiply. $1\frac{2}{3} \cdot 3\frac{1}{5}$

Step 1 Estimate the product.

$$1\frac{2}{3} \cdot 3\frac{1}{5} \approx 2 \cdot 3$$
$$\approx 6$$

Step 2 Convert to improper fractions.

$$1\frac{2}{3} \cdot 3\frac{1}{5} = \frac{5}{3} \cdot \frac{16}{5}$$

Step 3 Multiply the fractions.

$$= \frac{\cancel{5} \cdot 16}{3 \cdot \cancel{5}}$$
$$= \frac{16}{3}$$

Step 4 Write as a mixed number.

$$= 5\frac{1}{3}$$

This is close to the estimate, so we trust our answer.

15. Divide. $5\frac{2}{3} \div 2\frac{1}{3}$

Step 1 Estimate the quotient.

$$5\frac{2}{3} \div 2\frac{1}{3} \approx 6 \div 2$$
$$\approx 3$$

Step 2 Convert to improper fractions.

$$5\frac{2}{3} \div 2\frac{1}{3} = \frac{17}{3} \div \frac{7}{3}$$

Step 3 Change to multiplication and simplify.

$$= \frac{17}{3} \cdot \frac{3}{7}$$
$$= \frac{17}{\cancel{3}} \cdot \frac{\cancel{3}}{7}$$
$$= \frac{17}{7}$$

Step 4 Write as a mixed number.

$$= 2\frac{3}{7}$$

This is close to the estimate, so we trust our answer.

GUIDED PRACTICE

14. Multiply. $7\frac{1}{7} \cdot 6\frac{3}{5}$

$$7\frac{1}{7} \cdot 6\frac{3}{5} \approx$$
$$\approx$$
$$7\frac{1}{7} \cdot 6\frac{3}{5} = \frac{\quad}{\quad} \cdot \frac{\quad}{\quad}$$
$$= \frac{\quad}{\quad}$$
$$= \frac{\quad}{\quad}$$
$$=$$

Is the answer close to the estimate?

15. Divide. $16\frac{1}{4} \div 8\frac{1}{3}$

$$16\frac{1}{4} \div 8\frac{1}{3} \approx$$
$$\approx$$
$$16\frac{1}{4} \div 8\frac{1}{3} = \frac{\quad}{\quad} \div \frac{\quad}{\quad}$$
$$= \frac{\quad}{\quad} \cdot \frac{\quad}{\quad}$$
$$= \frac{\quad}{\quad}$$
$$= \frac{\quad}{\quad}$$
$$= \quad \frac{\quad}{\quad}$$

Is the answer close to the estimate?

Often an estimate will not be a whole number. For example, consider $12\frac{3}{8} \div 6\frac{7}{9}$.

$$12\frac{3}{8} \div 6\frac{7}{9} \approx 12 \div 7$$
$$\approx 1\frac{5}{7}$$

Here we would expect our answer to be a little less than 2.

Concept Check

C1. In your own words, how do you estimate the value of a mixed number?

C2. In your own words, why is estimating mixed numbers useful?

Objective C Practice

Estimate and then multiply or divide the mixed numbers as indicated.

C3. $3\frac{3}{7} \cdot 4\frac{1}{8}$

C4. $1\frac{1}{2} \cdot 2\frac{2}{3}$

C5. $9\frac{1}{3} \cdot 7\frac{3}{4}$

C6. $5\frac{1}{7} \cdot 6\frac{1}{3}$

C7. A brownie recipe calls for $2\frac{1}{4}$ cups of flour. If you prepare $\frac{1}{2}$ of this recipe, how many cups of flour will you need?

C8. A chicken soup recipe calls for $\frac{3}{4}$ pound of chicken. If you want to prepare $1\frac{1}{2}$ times this recipe, how many pounds of chicken will you need?

C9. $5\frac{5}{8} \div 1\frac{2}{7}$

C10. $3\frac{7}{9} \div 3\frac{1}{5}$

C11. $\frac{7}{8} \div 2\frac{2}{3}$

C12. $\frac{3}{4} \div 2\frac{2}{3}$

Objective D Add Mixed Numbers

The Concept To multiply and divide mixed numbers, change each mixed number to an improper fraction. We could follow a similar procedure to add mixed numbers, but it often makes more work than is necessary.

Rather, we can rewrite the sum $3\frac{9}{10} + 2\frac{8}{10}$ as $3 + \frac{9}{10} + 2 + \frac{8}{10}$.

Now to find the sum, we can use the commutative property to add the fractional portions and then add the whole numbers separately.

DETAILED EXAMPLE Adding Mixed Numbers

Add. $3\frac{9}{10} + 2\frac{8}{10}$

Step 1 Reorder and add the fractions.

$$3 + \frac{9}{10} + 2 + \frac{8}{10} = (3 + 2) + \left(\frac{9}{10} + \frac{8}{10}\right)$$

$$= 3 + 2 + \frac{17}{10}$$

The commutative property lets us reorder numbers that are being added. $a + b = b + a$

Step 2 Convert $\frac{17}{10}$ to a mixed number.

$$= 3 + 2 + 1\frac{7}{10}$$

Step 3 Add the whole numbers.

$$= 6\frac{7}{10}$$

Notice that we waited to add the whole numbers until after we added the fractions. When adding mixed numbers, add the fractions first in case you need to carry. This technique is similar to the procedure you used when adding whole numbers in Chapter 1.

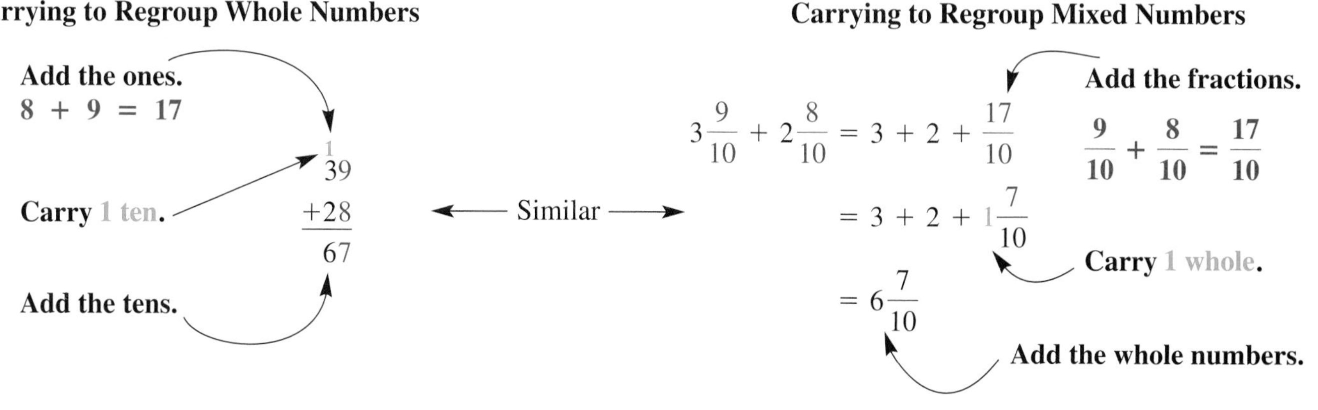

Procedure Add Mixed Numbers

Step 1 Estimate the sum.

Step 2 Add the fractions.
Convert any improper fractions to mixed numbers.

Step 3 Add the whole numbers.

EXAMPLES

16. Add. $1\frac{2}{3} + 3\frac{1}{5}$

Step 1 Estimate the sum. $1\frac{2}{3} + 3\frac{1}{5} \approx 2 + 3$

≈ 5

Step 2 Build like fractions and add. $1\frac{2}{3} + 3\frac{1}{5} = 1\frac{10}{15} + 3\frac{3}{15}$

$= 1 + 3 + \frac{13}{15}$

Step 3 Add the whole numbers $= 4\frac{13}{15}$

This is close to the estimate, so we trust our answer.

17. Add. $4\frac{4}{5} + 5\frac{1}{4}$

Step 1 Estimate the sum. $4\frac{4}{5} + 5\frac{1}{4} \approx 5 + 5$

≈ 10

GUIDED PRACTICE

16. Add. $3\frac{1}{8} + 4\frac{3}{4}$

$3\frac{1}{8} + 4\frac{3}{4} \approx$

$\approx$

$3\frac{1}{8} + 4\frac{3}{4} = 3\frac{\square}{\square} + 4\frac{\square}{\square}$

$= 3 + 4 + \frac{\square}{\square}$

$= \square\frac{\square}{\square}$

Is the answer close to the estimate?

17. Add. $7\frac{5}{6} + 5\frac{2}{3}$

$7\frac{5}{6} + 5\frac{2}{3} \approx$

$\approx$

(Continued)

Step 2 Build like fractions and add.

$$4\frac{4}{5} + 5\frac{1}{4} = 4\frac{16}{20} + 5\frac{5}{20}$$

Convert the improper fraction to a mixed number.

$$= 4 + 5 + \frac{21}{20}$$

$$= 4 + 5 + 1\frac{1}{20}$$

Step 3 Add the whole numbers.

$$= 10\frac{1}{20}$$

This is close to the estimate, so we trust our answer.

$$7\frac{5}{6} + 5\frac{2}{3} = 7\frac{\square}{\square} + 5\frac{\square}{\square}$$

$$= 7 + 5 + \frac{\square}{\square}$$

$$= \square + \square + \frac{\square}{\square}$$

$$= \square\frac{\square}{\square}$$

Is the answer close to the estimate?

Concept Check

D1. When performing the addition problem 37 + 19, first you add the digits in the ones column and then you add the digits in the tens column. With that in mind, why is it often a good idea to add the fractional portions of mixed numbers before adding the whole number portions?

Objective D Practice

Add.

D2. $3\frac{3}{7} + 4\frac{1}{8}$

D3. $1\frac{1}{2} + 2\frac{1}{3}$

D4. $9\frac{1}{3} + 7\frac{3}{5}$

D5. $10\frac{3}{5} + 6\frac{3}{10}$

D6. Knowing that the scale is balanced, how much does the object on the right weigh?

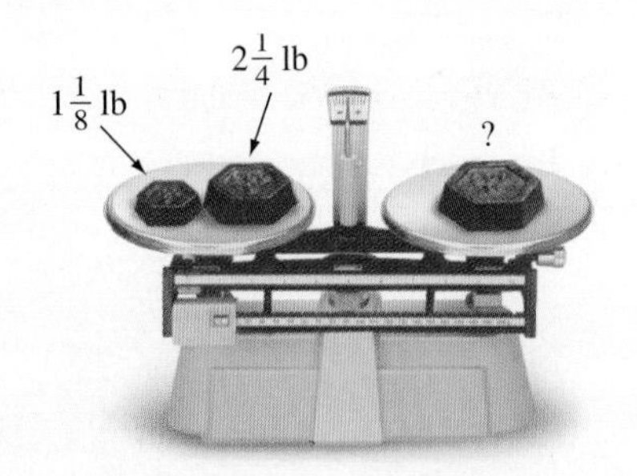

D7. Knowing that the scale is balanced, how much does the object on the right weigh?

$5\frac{1}{3}$ lb $3\frac{1}{12}$ lb ?

D8. $4\frac{8}{9} + 3\frac{1}{6}$

D9. $9\frac{7}{9} + 10\frac{5}{9}$

D10. $13\frac{13}{14} + 6\frac{6}{7}$

D11. $98\frac{3}{5} + 1\frac{8}{10}$

Objective E Subtract Mixed Numbers

The Concept When you subtract mixed numbers, subtract the fractions first and then subtract the whole numbers.

EXAMPLE

18. Subtract. $3\frac{2}{3} - 1\frac{1}{5}$

Step 1 Estimate the difference.

$$3\frac{2}{3} - 1\frac{1}{5} \approx 4 - 1$$
$$\approx 3$$

Step 2 Build like fractions to subtract.

$$3\frac{2}{3} - 1\frac{1}{5} = 3\frac{10}{15} - 1\frac{3}{15}$$

Step 3 Subtract the fractions and whole numbers separately.

$$= 2\frac{7}{15}$$

This is close to the estimate, so we trust our answer.

GUIDED PRACTICE

18. Subtract. $5\frac{5}{8} - 2\frac{1}{16}$

$$5\frac{5}{8} - 2\frac{1}{16} \approx$$
$$\approx$$

$$5\frac{5}{8} - 2\frac{1}{16} = 5\frac{\quad}{\quad} - 2\frac{\quad}{\quad}$$
$$= \frac{\quad}{\quad}$$

Is the answer close to the estimate?

When subtracting mixed numbers, you may need to borrow to regroup a number.

INTERACTIVE DEFINITION **Borrow to Regroup a Mixed Number**

To **borrow** within a mixed number, regroup a 1 from the whole number and add it to the fraction.

EXAMPLES

19. Answer parts a, b, and c to subtract a mixed number where borrowing is necessary.

a. Why must we borrow to find $5\frac{1}{8} - 2\frac{3}{8}$?

We cannot subtract the fractions because $\frac{3}{8}$ is greater than $\frac{1}{8}$. To subtract, we need to borrow a 1 from the whole number 5.

b. Borrow a 1 to regroup the mixed number. $5\frac{1}{8}$

Step 1 Borrow a 1 from 5.

$$5\frac{1}{8} = 4 + 1 + \frac{1}{8}$$

Step 2 Write the 1 as the fraction $\frac{8}{8}$.

$$= 4 + \frac{8}{8} + \frac{1}{8}$$

Step 3 Add the fractions to write the equivalent mixed number.

$$= 4\frac{9}{8}$$

GUIDED PRACTICE

19. Answer parts a, b, and c to subtract a mixed number where borrowing is necessary.

a. Why must we borrow to find $7\frac{1}{6} - 3\frac{5}{6}$?

We cannot subtract the fractions because ___ is greater than ___. To subtract, we need to borrow a 1 from the whole number ___.

b. Borrow a 1 to regroup the mixed number. $7\frac{1}{6}$

Step 1 Borrow a 1 from 7.

$$7\frac{1}{6} = 6 + \quad + \frac{1}{6}$$

Step 2 Write the 1 as the fraction $\frac{6}{6}$.

$$= 6 + \frac{\quad}{\quad} + \frac{1}{6}$$

Step 3 Add the fractions to write the equivalent mixed number.

$$= 6\frac{\quad}{6}$$

(Continued)

c. Replace $5\frac{1}{8}$ with the new equivalent mixed number and subtract.	**c.** Replace $7\frac{1}{6}$ with the new equivalent mixed number and subtract.
Step 1 Replace $5\frac{1}{8}$ with its equivalent. $5\frac{1}{8} - 2\frac{3}{8} = 4\frac{9}{8} - 2\frac{3}{8}$ **Step 2** Subtract. $= 2\frac{6}{8}$ **Step 3** Simplify. $= 2\frac{3}{4}$	**Step 1** Replace $7\frac{1}{6}$ with its equivalent. $7\frac{1}{6} - 3\frac{5}{6} = \_\frac{\_}{\_} - 3\frac{5}{6}$ **Step 2** Subtract. $= \_\frac{\_}{\_}$ **Step 3** Simplify. $=$

DO YOU UNDERSTAND when you must borrow to regroup a mixed number?	Got It	Get Help
DO YOU UNDERSTAND how to borrow to regroup a mixed number?	Got It	Get Help

EXAMPLE

20. Subtract. $6\frac{1}{3} - 2\frac{2}{3}$

We cannot subtract $\frac{1}{3} - \frac{2}{3}$. We need to borrow a 1 from the 6. Write the borrowed 1 as $\frac{3}{3}$ so we have common denominators.

Write 6 as 5 + $\frac{3}{3}$

$$6\frac{1}{3} - 2\frac{2}{3} = 5 + \frac{3}{3} + \frac{1}{3} - 2\frac{2}{3}$$
$$= 5\frac{4}{3} - 2\frac{2}{3}$$
$$= 3\frac{2}{3}$$

GUIDED PRACTICE

20. Subtract. $3\frac{1}{5} - 1\frac{2}{5}$

We cannot subtract $\frac{\_}{\_} - \frac{\_}{\_}$. We need to borrow a 1 from the _____. Write the borrowed 1 as $\frac{\_}{\_}$ so we have common denominators.

Write 3 as _____ + $\frac{5}{5}$

$$3\frac{1}{5} - 1\frac{2}{5} = \_ + \frac{\_}{\_} + \frac{1}{5} - 1\frac{2}{5}$$
$$= \_\frac{\_}{\_} - 1\frac{2}{5}$$
$$=$$

Procedure **Subtract Mixed Numbers**

Step 1 Estimate the difference.
Step 2 Build like fractions.
Step 3 If necessary, borrow to regroup.
Step 4 Subtract the fractions and whole numbers separately.

EXAMPLES

21. Subtract. $5\frac{1}{4} - 2\frac{3}{4}$

Estimate the difference. $5\frac{1}{4} - 2\frac{3}{4} \approx 5 - 3$
≈ 2

Borrow 1 from the 5 to subtract the fractions. $5\frac{1}{4} - 2\frac{3}{4} = 4 + \frac{4}{4} + \frac{1}{4} - 2\frac{3}{4}$
$= 4\frac{5}{4} - 2\frac{3}{4}$

Subtract the fractions and whole numbers separately. $= 2\frac{2}{4}$

Simplify. $= 2\frac{1}{2}$

This is close to the estimate, so we trust our answer.

22. Subtract. $6\frac{1}{3} - 2\frac{1}{2}$

Estimate the difference. $6\frac{1}{3} - 2\frac{1}{2} \approx 6 - 3$
≈ 3

Build like fractions. $6\frac{1}{3} - 2\frac{1}{2} = 6\frac{2}{6} - 2\frac{3}{6}$

Borrow 1 to subtract the fractions. $= 5 + \frac{6}{6} + \frac{2}{6} - 2\frac{3}{6}$
$= 5\frac{8}{6} - 2\frac{3}{6}$

Subtract the fractions and whole numbers separately. $= 3\frac{5}{6}$

This is close to the estimate, so we trust our answer.

23. Subtract. $3 - 1\frac{1}{7}$

Estimate the difference. $3 - 1\frac{1}{7} \approx 3 - 1$
≈ 2

GUIDED PRACTICE

21. Subtract. $5\frac{5}{8} - 2\frac{7}{8}$

$5\frac{5}{8} - 2\frac{7}{8} \approx$
$\approx$

$5\frac{5}{8} - 2\frac{7}{8} = 4 + \frac{\quad}{\quad} + \frac{5}{8} - 2\frac{7}{8}$
$= 4\frac{\quad}{\quad} - 2\frac{7}{8}$
$=$
$=$

Is the answer close to the estimate?

22. Subtract. $9\frac{1}{4} - 4\frac{1}{3}$

$9\frac{1}{4} - 4\frac{1}{3} \approx$
$\approx$

$9\frac{1}{4} - 4\frac{1}{3} = 9\frac{\quad}{\quad} - 4\frac{\quad}{\quad}$
$= 8 + \frac{\quad}{\quad} + \frac{\quad}{\quad} - 4\frac{\quad}{\quad}$
$= 8\frac{\quad}{\quad} - 4\frac{\quad}{\quad}$
$= \frac{\quad}{\quad}$

Is the answer close to the estimate?

23. Subtract. $4 - 2\frac{1}{5}$

$4 - 2\frac{1}{5} \approx$
$\approx$

(Continued)

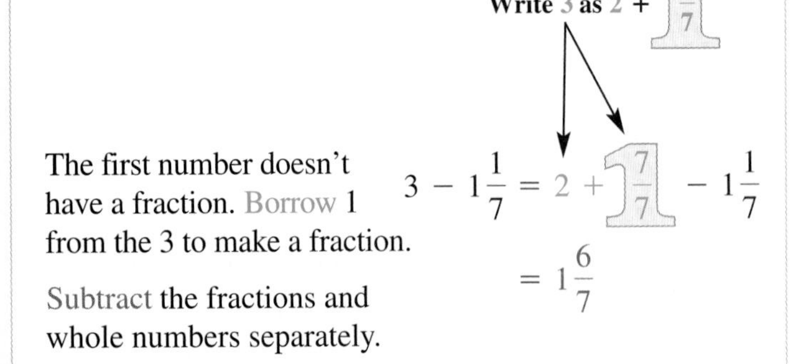

The first number doesn't have a fraction. Borrow 1 from the 3 to make a fraction.

$$3 - 1\frac{1}{7} = 2 + \frac{7}{7} - 1\frac{1}{7}$$

Subtract the fractions and whole numbers separately.

$$= 1\frac{6}{7}$$

This is close to the estimate, so we trust our answer.

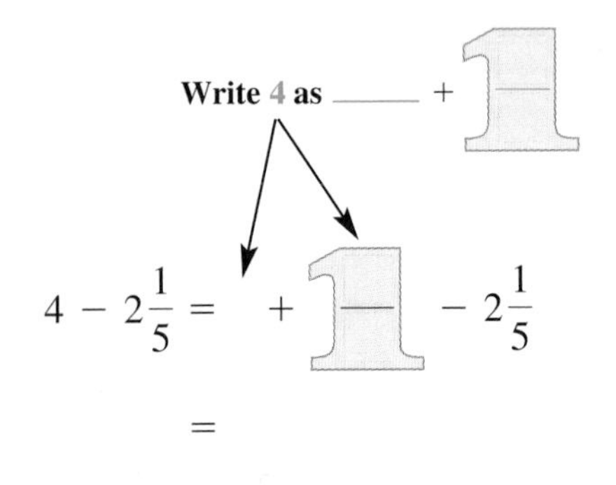

$$4 - 2\frac{1}{5} = \quad + \frac{\quad}{\quad} - 2\frac{1}{5}$$

$$=$$

Is the answer close to the estimate?

▶ **NOTE** Many students write these problems in a vertical format so that the whole numbers and fractional parts line up. That method is also correct.

Concept Check

E1. If we borrow from the 5, what is the new numerator?

$$5\frac{1}{6} = 4\frac{?}{6}$$

E2. If we borrow from the 8, what is the new numerator?

$$8\frac{2}{7} = 7\frac{?}{7}$$

E3. In the following two expressions, one is borrowed from the whole number to perform a subtraction exercise.

$$3\frac{2}{5} = 2\frac{7}{5} \quad \text{and} \quad 4\frac{1}{8} = 3\frac{9}{8}$$

In both cases, the numerator of the mixed number increased by the value of the denominator. Explain why this must happen.

Objective E Practice

Subtract. Borrow or regroup, as necessary.

E4. $7\frac{1}{5} - 3\frac{1}{10}$ **E5.** $12\frac{5}{8} - 3\frac{3}{16}$ **E6.** $4\frac{2}{3} - 3\frac{3}{5}$ **E7.** $10\frac{7}{12} - 6\frac{1}{4}$

E8. $9\frac{1}{9} - 3\frac{2}{3}$ **E9.** $13\frac{2}{7} - \frac{10}{21}$ **E10.** $23\frac{5}{6} - 10\frac{9}{10}$ **E11.** $20\frac{1}{5} - \frac{2}{3}$

E12. $8 - 3\frac{1}{3}$ **E13.** $7 - \frac{1}{6}$ **E14.** $6\frac{1}{6} - 4$ **E15.** $8\frac{4}{5} - 5$

Combining Concepts and Applications

CONCEPT I Using the Order of Operations If more than one operation is present, we must follow the order of operations.

Procedure **Use the Order of Operations**

First: Perform operations in groupings.
Second: Perform operations with exponents.
Third: Perform multiplication and division as they occur from left to right.
Fourth: Perform addition and subtraction as they occur from left to right.

EXAMPLES

24. Simplify. $\left(1\frac{1}{4}-\frac{1}{2}\right)^2\cdot\left(\frac{1}{3}\right)$

Groupings $\left(1\frac{1}{4}-\frac{1}{2}\right)^2\cdot\left(\frac{1}{3}\right)=\left(\frac{5}{4}-\frac{2}{4}\right)^2\cdot\left(\frac{1}{3}\right)$

Exponents $=\left(\frac{3}{4}\right)^2\cdot\left(\frac{1}{3}\right)$

Multiplication and division $=\frac{9}{16}\cdot\left(\frac{1}{3}\right)$

Simplify. $=\frac{3\cdot\cancel{3}\cdot1}{2\cdot2\cdot2\cdot2\cdot\cancel{3}}$

$=\frac{3}{16}$

GUIDED PRACTICE

24. Simplify. $\left(3\frac{1}{6}-1\frac{1}{3}\right)^2\div\left(\frac{1}{2}\right)$

$\left(3\frac{1}{6}-1\frac{1}{3}\right)^2\div\left(\frac{1}{2}\right)=$

CONCEPT II Application Exercises

EXAMPLE

25. Hector is making a birdhouse. It requires two boards that are $1\frac{1}{4}$ feet long, two boards that are 1 foot long, and three boards that are $\frac{3}{4}$ foot long. The boards will be cut from a piece of lumber that is 8 feet long. After the boards are cut, how long is the piece of lumber that is left over?

Step 1 Determine the total used.
- Multiply each length or weight by its quantity.
- Add.
- Simplify.

Wood used $=2\cdot\left(1\frac{1}{4}\right)+2\cdot(1)+3\cdot\left(\frac{3}{4}\right)$

$=2\cdot\frac{5}{4}+2+\frac{9}{4}$

$=\frac{10}{4}+\frac{8}{4}+\frac{9}{4}$

$=\frac{27}{4}$

$=6\frac{3}{4}$

GUIDED PRACTICE

25. Stephanie is repackaging coffee from a tub that contains 10 pounds of coffee. If she fills five $\frac{3}{4}$-pound bags, one $1\frac{1}{2}$-pound bag, and two 1-pound bags, how much coffee will be left over?

Total packaged =

=

=

=

=

(Continued)

Step 2 Subtract the total used from the available material to see what is left over.	Leftover lumber $= 8 - 6\frac{3}{4}$ $= 7\frac{4}{4} - 6\frac{3}{4}$ $= 1\frac{3}{4}$	Leftover coffee = = =
Step 3 Interpret the results.	The piece of lumber that is left over is $1\frac{3}{4}$ feet long.	lb of coffee will be left over.

Section 2.6 Exercises

FOR EXTRA HELP

To convert an improper fraction to a mixed number:

1. Answer the Objective A Concept Check.
2. Answer the odd Objective A Practice Exercises.
3. Answer the even Objective A Practice Exercises.

To convert a mixed number to an improper fraction:

4. Answer the Objective B Concept Check.
5. Answer the odd Objective B Practice Exercises.
6. Answer the even Objective B Practice Exercises.

To round mixed numbers:

7. Answer the odd Focus on Rounding Mixed Numbers Practice Exercises.
8. Answer the even Focus on Rounding Mixed Numbers Practice Exercises.

To multiply and divide mixed numbers:

9. Answer the Objective C Concept Checks.
10. Answer the odd Objective C Practice Exercises.
11. Answer the even Objective C Practice Exercises.

To add mixed numbers:

12. Answer the Objective D Concept Check.
13. Answer the odd Objective D Practice Exercises.
14. Answer the even Objective D Practice Exercises.

To subtract mixed numbers:

15. Answer the Objective E Concept Checks.
16. Answer the odd Objective E Practice Exercises.
17. Answer the even Objective E Practice Exercises.

VOCABULARY REVIEW *Review the Vocabulary Preview for Section 2.6. Study the definitions until you can check* Got It *for every word.*

Use these words to complete each sentence.

mixed number • whole number • quotient • remainder • proper fraction • improper fraction

	Got It	Get Help
18. A(n) ____________ has a numerator that is greater than or equal to the denominator.		
19. A mixed number is the sum of a whole number greater than zero and a(n) ____________.		
20. A(n) ____________ is the whole number of times the denominator goes into the numerator.		
21. The sum of a whole number greater than zero and a proper fraction is a(n) ____________.		

How will you get help for any vocabulary that you are unsure about?

Your instructor _____ MyMathLab _____ A classmate _____ A tutor _____ Other _____

Describe each picture using (a) an improper fraction and (b) a mixed number.

22.

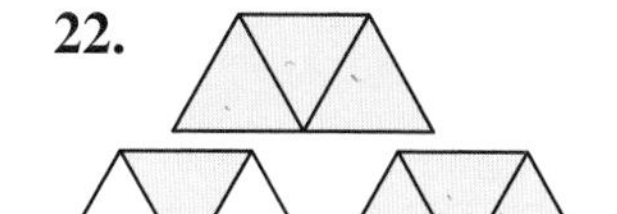

23.

24.

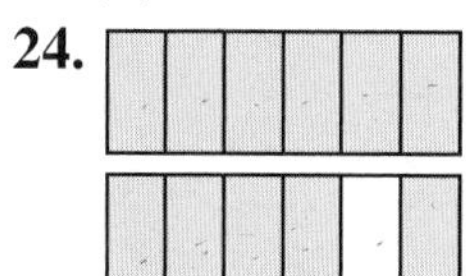

25.

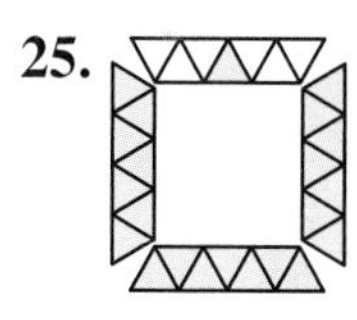

26.

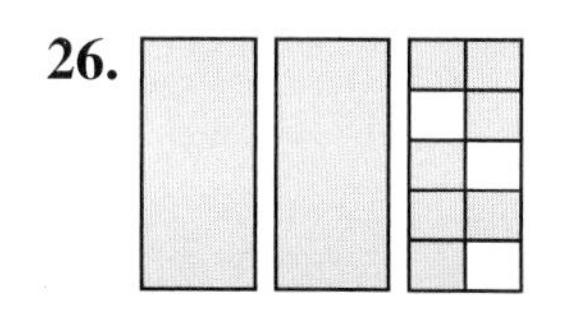

27.

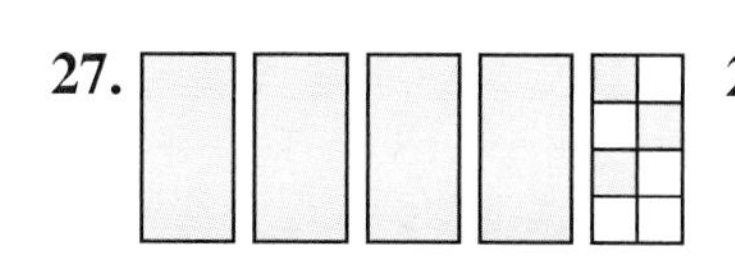

28.

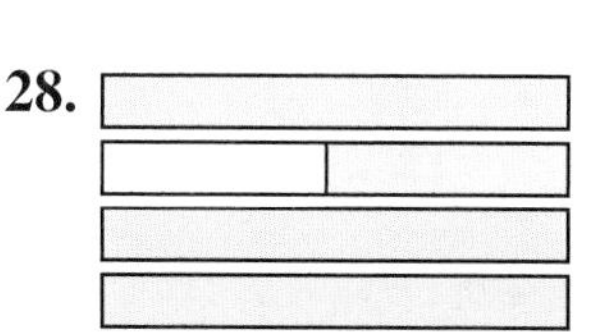

29.

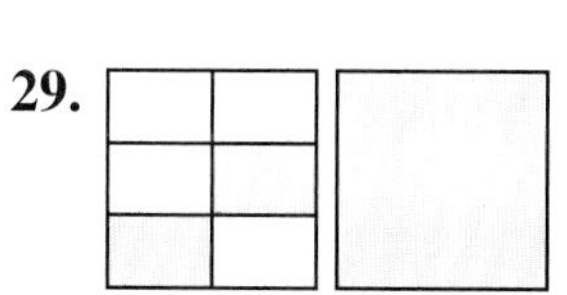

Convert each number to a mixed number or an improper fraction as appropriate.

30. $4\frac{5}{6}$ **31.** $3\frac{1}{8}$ **32.** $\frac{12}{5}$ **33.** $\frac{16}{3}$ **34.** $21\frac{1}{2}$

35. $13\frac{1}{3}$ **36.** $\frac{99}{10}$ **37.** $\frac{102}{7}$ **38.** $41\frac{1}{4}$ **39.** $\frac{41}{14}$

For each exercise:

a. Estimate the product or quotient.

b. Multiply or divide.

40. $1\frac{3}{5} \cdot 2\frac{1}{7}$ **41.** $2\frac{3}{4} \cdot 3\frac{2}{3}$ **42.** $8\frac{1}{3} \div 2\frac{1}{4}$ **43.** $2\frac{2}{3} \div 4\frac{1}{3}$

44. $3\frac{3}{4} \div 6\frac{1}{2}$ **45.** $10\frac{3}{5} \div 2\frac{4}{5}$ **46.** $4\frac{2}{3} \cdot 5\frac{1}{7}$ **47.** $5\frac{2}{3} \cdot 2\frac{1}{7}$

For each exercise:
a. Estimate the sum or difference.
b. Add or subtract.

48. $18\frac{3}{5} + 12\frac{1}{5}$

49. $21\frac{5}{9} + 4\frac{1}{9}$

50. $7\frac{3}{10} - 4\frac{2}{10}$

51. $62\frac{9}{11} - 52\frac{3}{11}$

52. $11\frac{3}{8} - 9\frac{3}{4}$

53. $50\frac{1}{10} - 25\frac{2}{5}$

54. $9 - 7\frac{1}{3}$

55. $10 - 5\frac{1}{10}$

56. Knowing that the scale is balanced, how much does the unknown object on the left weigh?

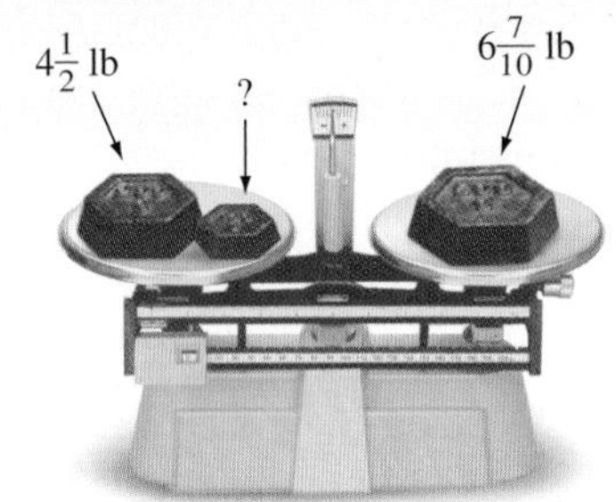

57. Knowing that the scale is balanced, how much does the unknown object on the left weigh?

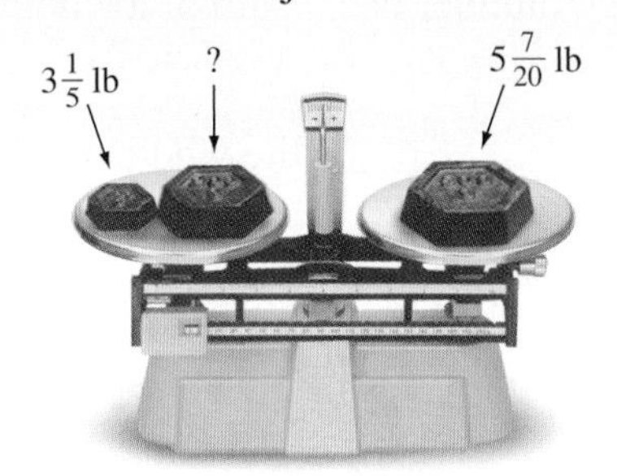

For each exercise:
a. Estimate the result of performing the indicated operation.
b. Perform the indicated operation.

58. $4\frac{1}{3} + 8$

59. $8\frac{7}{10} + 4\frac{3}{5}$

60. $12 - 4\frac{5}{8}$

61. $53 - 26\frac{3}{5}$

62. $76\frac{4}{5} - 3\frac{1}{5}$

63. $7\frac{2}{3} \div 4$

64. $10\frac{1}{8} \div 4\frac{3}{4}$

65. $100\frac{15}{16} + 25\frac{3}{4}$

66. $5\frac{1}{4} + 8\frac{7}{8}$

67. $1\frac{2}{3} \cdot 3\frac{4}{5}$

68. $4\frac{1}{16} \cdot 3\frac{7}{18}$

69. $7\frac{2}{5} - 6\frac{13}{30}$

70. $5 \cdot 6\frac{3}{8}$

71. $2\frac{3}{4} \cdot 3\frac{2}{3}$

72. $24\frac{3}{4} \div 5\frac{1}{4}$

73. $9\frac{1}{4} \div 5$

Answer each question.

74. Hector was training for a marathon and ran the following distances (in miles) over the course of four days: $5\frac{1}{2}, 10\frac{7}{10}, 5\frac{1}{2}, 15\frac{3}{5}$.

a. What was the total distance that Hector ran?
b. What was the average distance over the four days?

75. Vladimir skied cross country for $17\frac{3}{4}$ km on Monday, $21\frac{1}{2}$ km on Tuesday, and $12\frac{1}{2}$ km on Wednesday.

a. What was the total distance that Vladimir skied?
b. What was the average daily distance that Vladimir skied from Monday to Wednesday?

Evaluate each expression, following the order of operations.

76. $\left(2\frac{1}{6}\right) - \left(1\frac{2}{3}\right) + \left(\frac{5}{6}\right)$

77. $\left(4\frac{1}{8}\right) - \left(2\frac{3}{8}\right) + \left(\frac{3}{4}\right)$

78. $1\frac{1}{8} \div 1\frac{4}{5} \cdot 2\frac{1}{5}$

tobey Be clear on 2 concepts. When adding/subtracting fractions, find the LCD. When multiplying/dividing fractions, do NOT find the LCD. *For more tips and tweets, go to twitter.com/gstbasicmath*

79. $\left(4\frac{1}{5}\right) \div \left(1\frac{2}{5}\right) \cdot \left(3\frac{1}{2}\right)$

80. $\left(1\frac{1}{5}\right) + \left(1\frac{3}{5}\right) \cdot \left(2\frac{1}{2}\right)$

81. $\left(2\frac{1}{3}\right) + \left(1\frac{1}{3}\right) \cdot \left(8\frac{1}{4}\right)$

82. $\left(1\frac{5}{8}\right) \div \left(3\frac{1}{8} - 1\frac{1}{2}\right)$

83. $16\frac{3}{4} \div \left(1\frac{1}{8} - \frac{3}{8}\right)$

84. $\left(2\frac{1}{5}\right) \cdot \left(\frac{3}{11}\right) - \left(\frac{2}{5}\right)$

85. $\left(1\frac{1}{8}\right) \cdot \left(3\frac{1}{4}\right) - \left(\frac{7}{8}\right)$

86. $\frac{1}{5} \div \left(3 - \left(\frac{1}{2}\right)^2\right)$

87. $\frac{14}{23} \div \left(2 - \left(\frac{2}{3}\right)^2\right)$

Answer each question.

88. Kim purchased $2\frac{7}{8}$ pounds of hamburger. That night she made five burgers, each weighing $\frac{1}{4}$ pound. How much hamburger was left over?

89. Alfonzo had 3 pounds of coffee beans. He gave $1\frac{1}{8}$ pounds to Sally and another $\frac{3}{4}$ pound to Caleb. How much coffee did Alfonzo have left?

90. Julius and Cleopatra are building a dog run for their bloodhound. The run will measure $38\frac{1}{3}$ feet by $52\frac{3}{4}$ feet. How much fencing wire will they need if the wire is wrapped around the perimeter and an additional $2\frac{1}{3}$ feet is needed to secure one end of the fence to the other?

91. Patrick and J.R. were making cookies but had only 4 cups of flour. Patrick put $2\frac{1}{4}$ cups of flour in his bowl. Using the remaining flour, J.R. noticed that he didn't have the $2\frac{1}{4}$ cups he needed. When Patrick wasn't looking, J.R. took enough flour from Patrick's bowl to complete his recipe. How much flour did J.R. take?

92. Bill was paid $1,200 to build a porch for his neighbor. He gave $\frac{1}{10}$ of this money to his brother to pay back a debt. If $\frac{1}{3}$ of it was used to pay for supplies and $\frac{1}{6}$ was used to pay his helper, how much of the $1,200 did Bill have left?

93. Miranda was given $1,000 to invest for her daughter to go to college. If Miranda put $\frac{1}{4}$ of the money in a CD and $\frac{1}{5}$ of the money in a money market account, how much money was left over to purchase a savings bond?

Section 2.6 Question Log

Use this space to write questions. Make sure you get these answered and revisit them when you prepare for your exam.

Page ____ Answered	Page ____ Answered
Page ____ Answered	Page ____ Answered

Chapter 2 Organizer

VOCABULARY

Use the following steps to review the vocabulary for Chapter 2.

1. *Write the definition for each word from memory.*
2. *Compare the definitions you have written with the definitions in the Vocabulary Preview.*
3. *Study any definitions that you could not remember or that you defined incorrectly.*

2.1
Fraction • Numerator • Denominator • Fraction Bar • Part–Whole Concept • Quotient Concept • Proper Fraction • Improper Fraction

2.2
Factor • Simplest Form • Simplify/Reduce a Fraction • Greatest Common Factor • Equivalent Fractions • Units Fraction • Missing Multiplier

2.3
Reciprocal • Dividend • Divisor

2.4
Like Fractions • Unlike Fractions • Least Common Denominator (LCD) • Build Like Fractions • Missing multiplier

2.5
Order of Operations

2.6
Improper Fraction • Mixed Number • Whole Number • Quotient • Remainder

PROCEDURES

Procedure/Topic	Steps	Example
Part–Whole Concept; Represent a Picture as a Fraction (Section 2.1)	**Step 1** The denominator is the total number of equal-sized parts in one whole object. **Step 2** The numerator is the number of shaded parts in one whole object.	Represent the picture as a fraction. $= \frac{7}{4}$ ← Total shaded parts / Number of equal parts in each object.
Represent a Fraction as a Picture (Section 2.1)	**Step 1** Draw a shape with equal-sized parts matching the number in the denominator. **Step 2** Shade as many parts as the number in the numerator.	Represent $\frac{1}{6}$ as a picture. $\frac{1}{6} =$
Compare Fractions with the Same Numerator (Section 2.1)	The fraction with the larger pieces (smaller denominator) is greater.	Which is larger $\frac{5}{6}$ or $\frac{5}{8}$? $\frac{5}{6}$ is larger than $\frac{5}{8}$
Compare Fractions with the Same Denominator (Section 2.1)	The fraction with more pieces (larger numerator) is greater.	Which is greater $\frac{4}{5}$ or $\frac{2}{5}$? $\frac{4}{5}$ is greater than $\frac{2}{5}$.

Procedure/Topic	Steps	Example
Simplify a Fraction (Section 2.2)	**Step 1** Factor the numerator completely. **Step 2** Factor the denominator completely. **Step 3** Divide out common factors.	Simplify. $\frac{18}{30}$ $\frac{18}{30} = \frac{\cancel{2}\cdot\cancel{3}\cdot 3}{\cancel{2}\cdot\cancel{3}\cdot 5}$ $= \frac{3}{5}$
Multiply Two Fractions (Section 2.2)	**Step 1** Multiply the numerators. **Step 2** Multiply the denominators. **Step 3** Simplify if possible.	Multiply. $\frac{5}{12}\cdot\frac{4}{15}$ $\frac{5}{12}\cdot\frac{4}{15} = \frac{{}^{1}\cancel{5}\cdot\cancel{4}^{1}}{3\cdot{}^{1}\cancel{4}\cdot 3\cdot\cancel{5}^{1}}$ ← **Simplify *before* multiplying** $= \frac{1}{9}$
Build Equivalent Fractions (Section 2.2)	**Step 1** Set up the framework. **Step 2** Determine the missing multiplier. **Step 3** Multiply the numerator and denominator by the missing multiplier.	Build $\frac{5}{8}$ into an equivalent fraction with a denominator of 24. $\frac{5}{8} = \frac{?}{24} \Rightarrow \frac{5}{8}\left(\frac{3}{3}\right) = \frac{15}{24}$ $\frac{5}{8}$ and $\frac{15}{24}$ are equivalent.
Use a Fraction to Convert Units of Measure (Section 2.2)	**Step 1** Write the original quantity as a fraction. **Step 2** Multiply by the units fraction that will divide out the original units and introduce the new units. **Step 3** Simplify to convert the units.	Convert 3 feet to inches. $3\text{ ft} = \frac{3\cancel{\text{ft}}}{1}\cdot\frac{12\text{ inches}}{1\cancel{\text{ft}}}$ $= 36$ inches
Find a Reciprocal (Section 2.3)	**Step 1** If the number is a whole number, write it as a fraction. **Step 2** Interchange the numerator and denominator. "Flip" the fraction.	Find the reciprocal of $\frac{3}{4}$. $\frac{3}{4} \Rightarrow \frac{4}{3}$ Remember, 0 has no reciprocal.
Divide One Fraction by Another (Section 2.3)	**Step 1** (Skip) Rewrite the first fraction. **Step 2** (Flip) Write the reciprocal of the second fraction. **Step 3** (Multiply) Change the division symbol to multiplication. **Step 4** Simplify and perform the multiplication.	Divide. $\frac{8}{3} \div \frac{2}{5}$ $\frac{8}{3} \div \frac{2}{5} = \frac{8}{3}\cdot\frac{5}{2}$ $= \frac{2\cdot 2\cdot\cancel{2}}{3}\cdot\frac{5}{\cancel{2}}$ $= \frac{20}{3}$

Procedure/Topic	Steps	Example
Add or Subtract Like Fractions (Section 2.4)	**Step 1** Add or subtract the numerators. **Step 2** Keep the denominator the same. **Step 3** Simplify if possible.	Add. $\frac{3}{8}+\frac{1}{8}$ $\frac{3}{8}+\frac{1}{8}=\frac{4}{8}$ $=\frac{{}^{1}\cancel{2}\cdot\cancel{2}^{1}}{\cancel{2}\cdot\cancel{2}\cdot 2}$ $=\frac{1}{2}$
Build Like Fractions by Listing Multiples (Section 2.4)	**Step 1** Determine the LCD. **Step 2** Create the framework to build like fractions. **Step 3** Multiply the numerators and denominators by the missing multipliers.	Build like fractions for $\frac{3}{8}$ and $\frac{2}{6}$ by listing multiples. $\cancel{8}\cdot\cancel{16}, 24 \quad \text{LCD} = 24$ $\frac{3}{8}\cdot\left(\frac{3}{3}\right)=\frac{9}{24}$ $\frac{2}{6}\cdot\left(\frac{4}{4}\right)=\frac{8}{24}$
Add or Subtract Unlike (Section 2.4)	**Step 1** Build like fractions. **Step 2** Add or subtract the like fractions. **Step 3** Simplify if possible.	Add. $\frac{3}{4}+\frac{7}{10}$ $\frac{3}{4}+\frac{7}{10}=\frac{3}{4}\cdot\left(\frac{5}{5}\right)+\frac{7}{10}\cdot\left(\frac{2}{2}\right)$ $=\frac{15}{20}+\frac{14}{20}$ $=\frac{29}{20}$
Build Like Fractions by Factoring (Section 2.4)	**Step 1** Identify the least common denominator (LCD) by factoring. **Step 2** Identify the missing multipliers from the factorizations. A number's missing multipliers are the factors outside the number's Venn diagram circle. **Step 3** Create the framework and build like fractions.	Build like fractions for $\frac{2}{21}$ and $\frac{4}{9}$ by factoring. 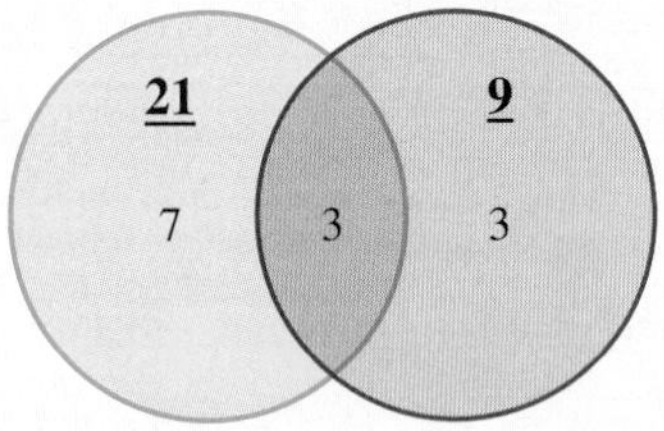 21: 7, 3 (shared) — 9: 3, 3 (shared) $\text{LCD} = 7\cdot 3\cdot 3$ $= 63$ 9 is missing a 7. 21 is missing a 3. $\frac{2}{21}\cdot\left(\frac{3}{3}\right)=\frac{6}{63}$ $\frac{4}{9}\cdot\left(\frac{7}{7}\right)=\frac{28}{63}$

Procedure/Topic	Steps	Example
Use the Order of Operations (Sections 2.5 and 2.6)	**First:** Groupings **Second:** Exponents **Third:** Multiplication and division, left to right **Fourth:** Addition and subtraction, left to right	Perform the operations. $\frac{7}{8} - \left(\frac{1}{4} + \frac{1}{2}\right)$ $\frac{7}{8} - \left(\frac{1}{4} + \frac{1}{2}\right) = \frac{7}{8} - \left(\frac{1}{4} + \frac{2}{4}\right)$ $= \frac{7}{8} - \left(\frac{3}{4}\right)$ $= \frac{7}{8} - \frac{6}{8}$ $= \frac{1}{8}$
Convert an Improper Fraction to a Mixed Number (Section 2.6)	**Step 1** Divide the numerator by the denominator. **Step 2** Write the mixed number. $\text{quotient}\,\frac{\text{remainder}}{\text{denominator}}$ **Step 3** Simplify the fraction if necessary.	Convert $\frac{7}{3}$ to a mixed number. $\frac{7}{3} \Rightarrow 3\overline{)7}\ \text{2R1} \Rightarrow 2\frac{1}{3}$ $\frac{7}{3} = 2\frac{1}{3}$
Convert a Mixed Number to an Improper Fraction (Section 2.6)	**Step 1** New numerator = Denominator · Whole number + Numerator **Step 2** The denominator stays the same.	Convert $2\frac{1}{3}$ to an improper fraction. $2\frac{1}{3} \Rightarrow 3 \cdot 2 + 1 = 7 \Rightarrow \frac{7}{3}$ $2\frac{1}{3} = \frac{7}{3}$
Multiply or Divide Mixed Numbers (Section 2.6)	**Step 1** Estimate the product or quotient. **Step 2** Convert each mixed number to an improper fraction. **Step 3** Multiply or divide the fractions. **Step 4** Write the result as a mixed number.	Multiply. $2\frac{1}{4} \cdot 1\frac{1}{5}$ $2\frac{1}{4} \cdot 1\frac{1}{5} \approx 2 \cdot 1 = 2$ $2\frac{1}{4} \cdot 1\frac{1}{5} = \frac{9}{4} \cdot \frac{6}{5}$ $= \frac{3 \cdot 3 \cdot \cancel{2} \cdot 3}{2 \cdot \cancel{2} \cdot 5}$ $= \frac{27}{10}$ $= 2\frac{7}{10}$
Add Mixed Numbers (Section 2.6)	**Step 1** Estimate the sum **Step 2** Add the fractions. Convert any improper fractions to mixed numbers. **Step 3** Add the whole numbers.	Add. $3\frac{4}{5} + 1\frac{2}{5}$ $3\frac{4}{5} + 1\frac{2}{5} \approx 4 + 1 = 5$ $3\frac{4}{5} + 1\frac{2}{5} = 3 + 1 + \frac{6}{5}$ $= 3 + 1 + 1\frac{1}{5}$ $= 5\frac{1}{5}$

Procedure/Topic	Steps	Example
Subtract Mixed Numbers (Section 2.6)	**Step 1** Estimate the difference. **Step 2** Build like fractions. **Step 3** If necessary, borrow to regroup. **Step 4** Subtract the fractions and whole numbers separately.	Subtract. $3 - 1\frac{1}{7}$ $3 - 1\frac{1}{7} \approx 4 - 1$ ≈ 3 **Write 4 as** $3\frac{7}{7}$ $4 - 1\frac{1}{7} = 3\frac{7}{7} - 1\frac{1}{7}$ $= 2\frac{6}{7}$

Chapter 2 Review Exercises

2.1

Represent each fraction as a picture or each picture as a fraction.

1.

2. 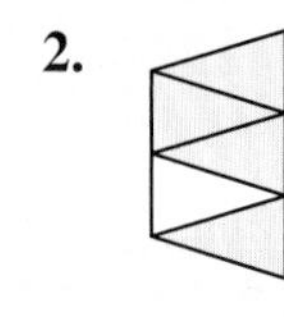 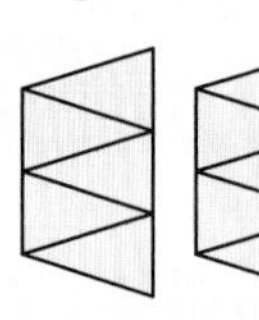

3. $\frac{2}{5}$

4. $\frac{7}{3}$

In Exercises 5 and 6:

a. Draw lines on each object to make the pieces equal.

b. Determine the fraction represented by the picture.

5. This figure should have 9 equal parts.

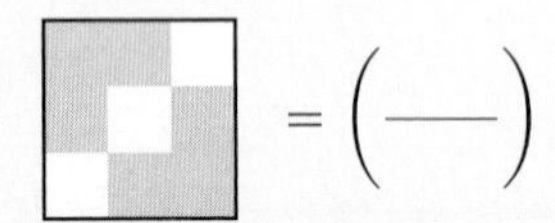

6. This figure should have 4 equal parts.

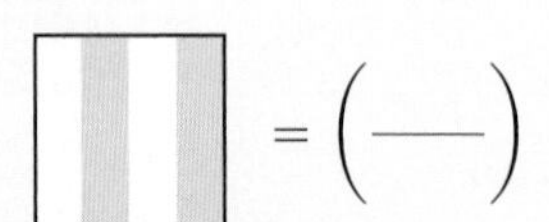

7. Use < or > to make a true statement.

$$\frac{5}{8} \square \frac{5}{9}$$

Explain your answer.

8. Use < or > to make a true statement.

$$\frac{5}{12} \square \frac{7}{12}$$

Explain your answer.

9. Adalina has two jobs. She works 6 hours per day at Snicker's and 7 hours per day at Chuckles. What fraction of her work is at Snicker's?

10. The Bumblebees have 7 players, and the Flowers have 8 players. What fraction of the players are on the Flowers?

2.2

11. a. Which two pictures represent equivalent fractions?
b. Which of the two equivalent fractions is drawn in simplest form?

A.

B.

C.

a. ____ and ____ are equivalent.

b. Of those, ____ is in simplest form.

12. Redraw the following fraction in simplest form.

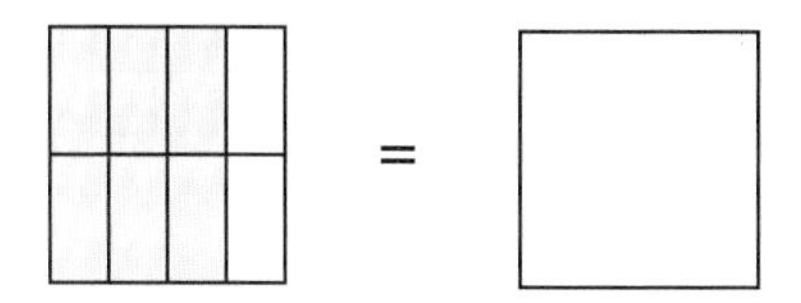

13. Simplify. $\frac{15}{20}$

14. Simplify. $\frac{21}{49}$

Perform the operations. Write your answer in simplest form.

15. $\frac{7}{2} \cdot \frac{3}{2}$

16. $\frac{1}{3} \cdot \frac{9}{5}$

17. $\frac{4}{7} \cdot \frac{21}{5}$

18. $\frac{7}{10} \cdot \frac{15}{14}$

19. $\frac{3}{6} \cdot \frac{18}{7} \cdot \frac{14}{3}$

20. $\left(\frac{5}{3}\right)^2$

21. Antonio purchased a set of tires for \$200. He had a coupon for \$40. What fraction of the original cost did Antonio pay?

22. Falicia's new clothes totaled \$240 before the clerk applied a "one-fourth" discount. How much did Falicia pay?

23. Convert 15 feet to inches.

24. Convert 11 cups to ounces.

2.3

Find the reciprocal of each number.

25. $\frac{5}{9}$

26. $\frac{1}{45}$

27. 3

Perform the operations. Write your answer in simplest form.

28. $\frac{5}{6} \div \frac{2}{3}$

29. $\frac{6}{7} \div \frac{3}{14}$

30. $\frac{5}{6} \div \frac{6}{5}$

31. $\frac{14}{5} \div \frac{7}{10}$

32. $\frac{3}{7} \div \frac{3}{7}$

33. $\frac{0}{2} \div \frac{8}{5}$

34. $\left(\frac{3}{5} \div \frac{6}{5}\right)^2$

35. $\left(\frac{3}{2}\right)^2 \div \frac{15}{8}$

Answer each question.

36. A machinist needs to cut a 4-meter-long plate of steel into pieces that are $\frac{8}{10}$ meter long. How many $\frac{8}{10}$-meter pieces can be made?

37. Jelly beans are going to be put into bags that weigh $\frac{4}{5}$ pound. How many bags of jelly beans will equal 14 pounds?

38. To calculate the mileage of a go-cart, Thaddeus drove $\frac{3}{8}$ of a mile using $\frac{1}{64}$ gallon of gas. How many miles per gallon did the go-cart get?

Fill in the missing parts of each exercise.
- **Shade the appropriate portions of each picture.**
- **Complete the written operations.**

39.

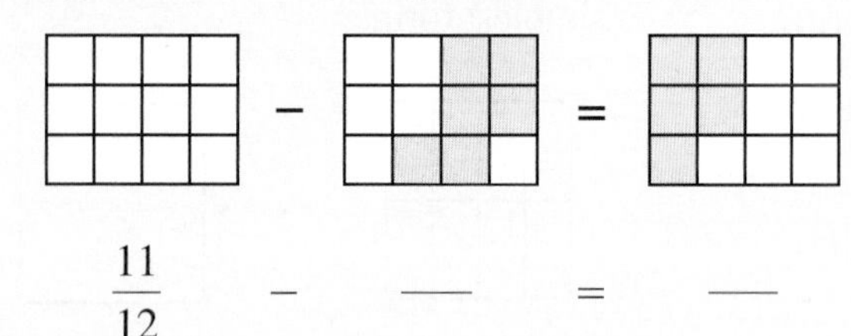

$\frac{11}{12} - \underline{\quad} = \underline{\quad}$

40.

$\underline{\quad} + \frac{1}{\underline{\quad}} = \underline{\quad}$

For each exercise:

a. Identify the least common denominator.

b. Build the two fractions into equivalent like fractions.

41. $\frac{5}{12}, \frac{1}{8}$

42. $\frac{7}{8}, \frac{5}{6}$

Perform the operations. Write your answer in simplest form.

43. $\frac{7}{9} + \frac{1}{9}$

44. $\frac{6}{7} - \frac{2}{7}$

45. $\frac{7}{18} + \frac{2}{9}$

46. $\frac{3}{5} - \frac{1}{3}$

47. $\frac{3}{14} - \frac{2}{21}$

48. $\frac{7}{10} + \frac{3}{25}$

49. $\frac{4}{5} - \frac{3}{5} + \frac{3}{10}$

50. $\frac{5}{8} + \frac{3}{4} - \frac{1}{8}$

51. Order the fractions from largest to smallest.

$\frac{7}{16}, \frac{1}{4}, \frac{3}{8}$

52. Order the fractions from smallest to largest.

$\frac{4}{5}, \frac{5}{6}, \frac{9}{15}$

53. A chemist needs to make $1\frac{1}{2}$ gallons of solution. How much water must be added to $\frac{3}{8}$ gallon of concentrate to make the solution?

2.5

Perform the operations. Write your answer in simplest form.

54. $\frac{7}{8} - \frac{3}{4} \cdot \frac{1}{2}$

55. $\frac{1}{7-3} + \frac{1}{5}$

56. $\frac{6}{7} \cdot \left(\frac{2}{3}\right)^2$

57. $\frac{7}{9} \div \frac{14}{3} \cdot \frac{2}{3}$

58. $\frac{4}{5} + \frac{1}{7} - \frac{1}{3}$

59. $\left(\frac{1}{3} + \frac{1}{2}\right)^2$

60. $\frac{5-2}{6} + \frac{4}{5-2}$

61. $\left(\frac{7}{2} \div \frac{3}{7}\right) \div 49$

What value of *n* makes the equation true?

62. $n - \frac{1}{2} = \frac{1}{3}$

63. $n + \frac{4}{5} = \frac{13}{15}$

64. $\frac{1}{3} \cdot n = \frac{5}{9}$

65. $\frac{3}{5} \cdot n = \frac{7}{10}$

Answer each question.

66. Muame has been asked to inventory an aisle at a store. There are 50 display cabinets in the aisle, each of which has 7 shelves. If he has completed $\frac{4}{7}$ of the job, how many shelves must he still inventory?

67. To make one batch of Calypso Cake, $\frac{3}{4}$ cup of milk is needed for the cake and $\frac{1}{8}$ cup of milk is needed for the frosting. How much milk is needed to make 5 batches of cake with frosting?

2.6

Describe each picture using (a) an improper fraction and (b) a mixed number.

68.

69. 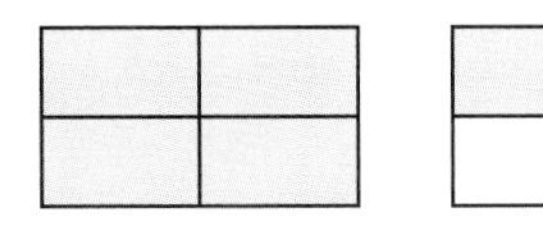

Answer each question.

70. Convert $7\frac{1}{6}$ to an improper fraction.

71. Convert $\frac{53}{8}$ to a mixed number.

72. Estimate the difference without finding the actual answer.

$13\frac{3}{5} - 5\frac{1}{2}$

73. Estimate the product without finding the actual answer.

$5\frac{1}{4} \cdot 8\frac{7}{9}$

Perform each operation. Write any improper fractions as mixed numbers.

74. $4\frac{5}{7} + 1\frac{1}{14}$

75. $3\frac{1}{2} \div 5\frac{3}{7}$

76. $9\frac{1}{8} - 2\frac{5}{8}$

77. $4\frac{2}{3} \cdot 2\frac{1}{7}$

78. $3\frac{2}{3} + 4\frac{7}{9}$

79. $8 - 2\frac{5}{6}$

80. $3\frac{3}{7} \div \frac{5}{14}$

81. $3\frac{3}{4} \div 3$

82. $100\frac{1}{5} - 10\frac{1}{3}$

83. $\left(3\frac{1}{4}\right)^2$

Chapter 2 Practice Test

1. Represent $\frac{5}{6}$ using a picture.

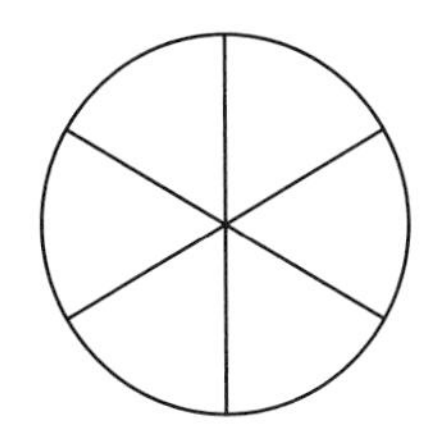

2. a. Which two pictures represent equivalent fractions?
b. Which of the two equivalent fractions is drawn in simplest form?

A. 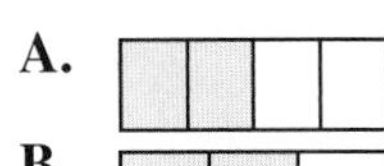

B.

C. 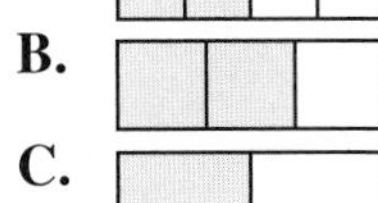

a. ____ and ____ are equivalent.

b. Of those, ____ is in simplest form.

Fill in the missing parts of each exercise.
- **Shade the appropriate portions of each picture.**
- **Complete the written operations.**

3. 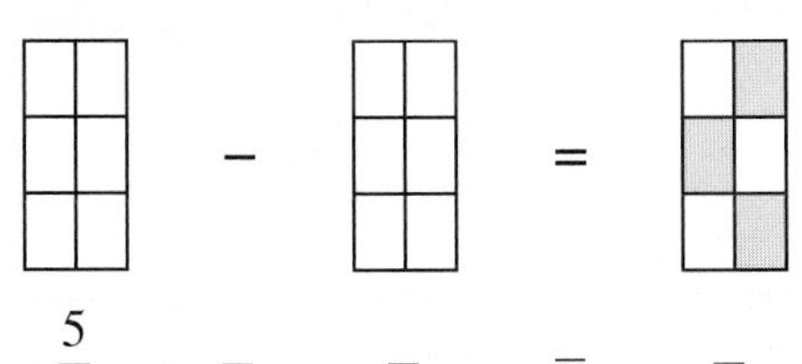

$\frac{5}{6} - \frac{\quad}{6} = \frac{\quad}{6}$

4. 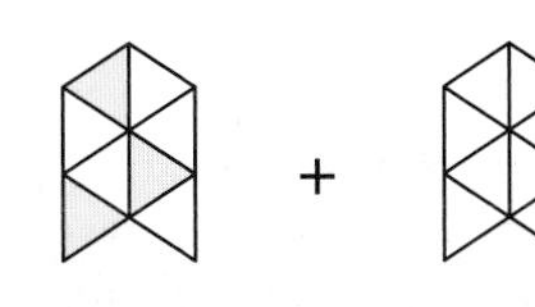 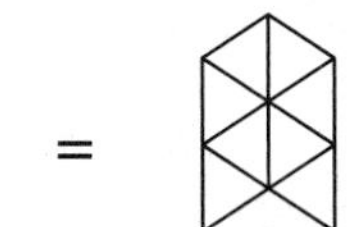

$\frac{\quad}{\quad} + \frac{2}{\quad} = \frac{\quad}{\quad}$

Perform the operations. Write your answer in simplest form.

5. $\frac{1}{5} \cdot \frac{1}{3}$

6. $\frac{5}{6} - \frac{2}{6}$

7. $\frac{5}{8} \div \frac{3}{4}$

8. $\frac{7}{12} + \frac{1}{3}$

9. $\frac{7}{12} + \frac{1}{14}$

10. $\frac{3}{14} \div \frac{2}{7}$

11. $\frac{7}{20} - \frac{1}{16}$

12. $\frac{5}{6} \cdot 3$

13. $\frac{2}{7} - \frac{1}{14} + \frac{3}{7}$

14. $\frac{8}{5} + \frac{1}{3} \cdot \frac{2}{5}$

15. $\frac{3}{8} - \frac{1}{14}$

16. $\frac{8}{5} \cdot \frac{1}{2} \cdot \frac{10}{4}$

17. $\left(\frac{5}{4}\right)^2 - \frac{7}{16}$

18. $\frac{4}{3} \cdot \frac{9}{24} + \frac{3}{2}$

19. Order the fractions from largest to smallest.

$\frac{3}{4}, \frac{15}{24}, \frac{7}{12}$

20. If 36 out of 60 guests have replied to a party invitation, what fraction of the guests have not replied?

21. Judge Pineda has read $\frac{2}{3}$ of the documents submitted for a court case. If there are a total of 99 documents, how many more must she still read?

22. Mariah is testing blood samples for malaria. She has determined that 2 out of every 1,000 samples test positive for malaria. If she tests 7,000 people in her town, how many people will not test positive for malaria?

Convert each mixed number to an improper fraction or each improper fraction to a mixed number.

23. $\frac{11}{4}$

24. $3\frac{1}{7}$

25. $3\frac{2}{6}$

26. $\frac{33}{5}$

Perform the operations. Write any improper fractions as mixed numbers.

27. $\frac{3}{4} - \frac{1}{4} \cdot \frac{1}{4}$

28. $3\frac{1}{8} + 1\frac{1}{2}$

29. $\left(\frac{1}{2} + \frac{1}{4}\right) \cdot \frac{10}{9}$

30. $4\frac{1}{2} \cdot 3\frac{1}{5}$

31. $5\frac{3}{4} - 3\frac{5}{8}$

32. $\frac{8 - 3}{3} \cdot \frac{1}{12 + 3}$

33. $6\frac{1}{4} - 2\frac{7}{8}$

34. $2\frac{4}{6} \div 5\frac{1}{3}$

35. $2 \div \left(\frac{3}{4} - \frac{1}{2}\right)$

36. $\frac{1}{2} \div \frac{2}{3} \cdot \frac{3}{4}$

37. A carnival game gives a $\frac{1}{4}$ ticket for every point scored. How many points must Mila win if she wants a stuffed gorilla that costs 150 tickets?

CHAPTER

3 Decimals

As with fractions, decimal numbers, also known as decimals, are used to represent portions of a whole. Decimal numbers are sometimes called decimal fractions because they give us another way to represent fractions and mixed numbers.

Auto races can last hours and can be won by thousandths of a second. Traveling at nearly 190 miles per hour, a car leading by a few feet will win a race by only a few thousandths of a second. Kevin Harvick won the closest race in NASCAR history by 0.002 seconds. His average speed was 149.335 miles per hour.

In Exercise 95 of Section 3.1, place value is used to order other winning speeds for the Daytona 500.

Section 3.1 Understanding Decimal Numbers

The place value of each digit to the right of the decimal point represents a fraction.

A dime is $\frac{1}{10}$ of a dollar. Usually, we describe a dime with a decimal.

$$\$\frac{1}{10} = \mathbf{\$0.10}$$

A penny is $\frac{1}{100}$ of a dollar. We also describe a penny with a decimal.

$$\$\frac{1}{100} = \mathbf{\$0.01}$$

In the following diagram, the same quantities are shown as a picture, a fraction, and a decimal.

A picture	A fraction	A decimal
	$= \frac{3}{4}$	$= 0.75$
	$= 2\frac{3}{4}$	$= 2.75$

In this section, you will learn the value of each digit in a decimal number and will learn how to approximate a decimal number by rounding it to a certain place value.

The **Objectives** in Section 3.1 will help you

- **A** Understand the place values of decimal numbers.
- **B** Write decimal numbers in words.
- **C** Convert between decimals and fractions.
- **D** Round decimal numbers.
- **E** Order decimal numbers.

VOCABULARY PREVIEW *Check the box that applies.*

	Got It	Must Study
Digits: The **digits** in our number system are 0, 1, 2, 3, 4, 5, 6, 7, 8, and 9.		
Decimal Point: The **decimal point** separates the digits that represent whole numbers from those that represent fractions. Often the decimal point is called the decimal.		
Decimal Places: The number of **decimal places** is the number of digits written to the right of the decimal point.		
Place Value: Place value uses the place of a digit in a number to indicate that digit's value.		
Like Decimals: Like decimals have the same number of decimal places.		

Study these words when they appear in the text. After the section, test your understanding by completing the Vocabulary Review in the section exercises.

Objective A Understand the Place Values of Decimal Numbers

The Concept Decimal numbers are made up of digits. Each digit represents some value based on its place value. Digits to the left of the decimal point represent whole numbers. Digits to the right of the decimal point represent parts of a whole.

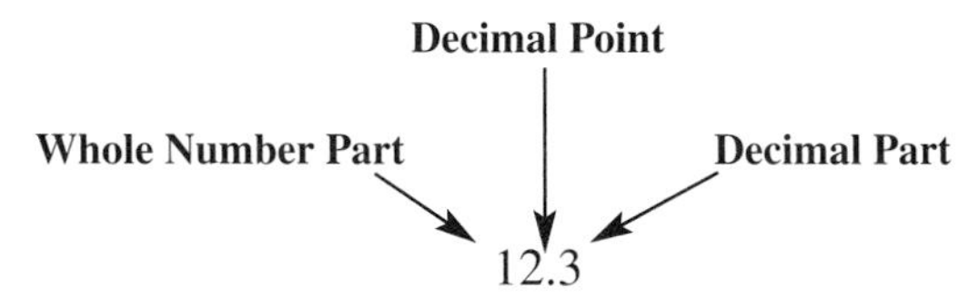

The decimal point separates "wholes" from "portions of a whole."

The decimal part of 12.3 is read as "three tenths" because .3 represents the fraction $\frac{3}{10}$. Place values to the right of the decimal point can be written as fractions. For example, each of the last three digits of the decimal number 25.<u>379</u> represents a fraction. Expanded notation can help you understand the fractions that the digits 3, 7, and 9 represent in the decimal number 25.379.

25.379 Written in Expanded Form:

$$25.379 = 20 + 5 + \frac{3}{10} + \frac{7}{100} + \frac{9}{1{,}000}$$

25.379 Written in the Place Value Chart

Thousands			Ones			Fractions Written in Decimal Form					
				2	5	.	3	7	9		
Hundred thousands 100,000's	Ten thousands 10,000's	Thousands 1,000's	Hundreds 100's	Tens 10's	Ones 1's		Tenths $\frac{1}{10}$ths	Hundredths $\frac{1}{100}$ths	Thousandths $\frac{1}{1{,}000}$ths	Ten thousandths $\frac{1}{10{,}000}$ths	Hundred thousandths $\frac{1}{100{,}000}$ths

Notice that the whole number place values end in '*s*.' The decimal place values end in '*ths*.'

Recognizing patterns in the place value chart will help you identify the place values. In our number system, each place value is one-tenth of the place value to its left. To identify the next place value to the right, multiply by $\frac{1}{10}$.

$$\cdots \quad \underset{\text{Tens}}{10} \quad 10 \cdot \frac{1}{10} = 1 \longrightarrow \underset{\text{Ones}}{1} \quad 1 \cdot \frac{1}{10} = \frac{1}{10} \longrightarrow \underset{\text{Tenths}}{\frac{1}{10}} \quad \frac{1}{10} \cdot \frac{1}{10} = \frac{1}{100} \longrightarrow \underset{\text{Hundredths}}{\frac{1}{100}} \quad \cdots$$

To help you to identify a digit's place value, it is important to notice that place values are symmetric on either side of the ones place. For example, hundreds and hundredths are the same number of places on either side of the ones place.

Whole number place values end in '*s*.' | Decimal place values end in '*ths*.'

					.			
10,000's	1,000's	100's	10's	1's	10ths	100ths	1,000ths	10,000ths

Procedure Find the Place Values of Digits in Decimal Numbers

Step 1 Count the number of places that the digit is from the decimal.

Step 2 List that number of place values. "tenths, hundredths . . . "

EXAMPLES

1. What is the place value of the digit 3 in the number 568.495392?

Step 1 The digit 3 is in the fourth place to the right of the decimal.

Step 2
- tenths
- hundredths
- thousandths
- ten thousandths

3 is in the ten thousandths place.

2. What is the place value of the digit 2 in the number 568.95392?

Step 1 The digit 2 is in the fifth place to the right of the decimal.

Step 2
- tenths
- hundredths
- thousandths
- ten thousandths
- hundred thousandths

2 is in the hundred thousandths place.

GUIDED PRACTICE

1. What is the place value of the digit 5 in the number 84,988.435821?

Step 1 The digit 5 is in the ______ place to the right of the decimal.

Step 2
- ____________
- ____________
- ____________

Filling in these blanks will help you to remember the place values for decimal numbers.

5 is in the ____________ place.

2. What is the place value of the digit 1 in the number 88.93821?

Step 1 The digit 1 is in the ______ place to the right of the decimal.

Step 2
- ____________
- ____________
- ____________
- ____________
- ____________

1 is in the ____________ place.

Procedure Write a Decimal Number in Expanded Form

Write the sum of what each digit represents.

Be sure to multiply each digit by its place value.

EXAMPLES

3. Write 4.35 in expanded form.

$$4.35 = 4 + \frac{3}{10} + \frac{5}{100}$$

4. Write 53.8013 in expanded form.

$$53.8013 = 50 + 3 + \frac{8}{10} + \frac{1}{1{,}000} + \frac{3}{10{,}000}$$

GUIDED PRACTICE

3. Write 8.19 in expanded form.

$$8.19 = \quad + \frac{\quad}{10} + \frac{\quad}{100}$$

4. Write 300.40012 in expanded form.

$$300.40012 =$$

Concept Check

A1. Using multiplication, explain why the first digit to the right of the ones place is the tenths place.

A2. Fill in the blanks: Because of the symmetry of place values, "thousands" and "thousandths" are ______ places to the left and right, respectively, of the digit in the ______ place.

A3. Write 4.56 in expanded form. Be sure to represent the decimal part using fractions.

A4. Make a place value chart that shows the place values from ten thousands to ten thousandths.

Objective A Practice

For each number, find the place value of the digit 4.

A5. 3.8742 **A6.** 8.942 **A7.** 9.4827 **A8.** 98.000498

A9. 987.73145 **A10.** 23.87894 **A11.** 6789.4236 **A12.** 0.004

Write each decimal number in expanded form.

A13. 0.8 **A14.** 0.9 **A15.** 0.768 **A16.** 0.475

A17. 0.0087 **A18.** 0.0567 **A19.** 4.909 **A20.** 7.076

Objective B Write Decimal Numbers in Words

The Concept To write a decimal such as 0.32 in words, it is useful to think of the number in expanded form.

Write 0.32 in expanded form. $0.32 = \frac{3}{10} + \frac{2}{100}$

Build like fractions and add. $= \frac{30}{100} + \frac{2}{100}$

Read the resulting fraction. $= \frac{32}{100}$

0.32 is read as "thirty-two hundredths."

To read a decimal number, say the number represented by the digits followed by the place value of the last digit on the right.

Procedure **Write a Decimal Number in Words**

Step 1 Write the words for the digits to the right of the decimal point.

Step 2 Write the place value of the last digit.

EXAMPLES	GUIDED PRACTICE
5. Write 0.321 in words.	**5.** Write 0.43 in words.
Write the words for 321 and the place value of the 1. 0.321 is three hundred twenty-one thousandths.	Write the words for ___ and the place value of the ___. 0.43 is ________________.
6. Write 0.0542 in words.	**6.** Write 0.00987 in words.
Write the words for 542 and the place value of the 2. 0.0542 is five hundred forty-two ten thousandths.	Write the words for ___ and the place value of the ___. 0.00987 is ____________________________.

We can also describe "mixed" decimal numbers in words. We use the word *and* to indicate the decimal point.

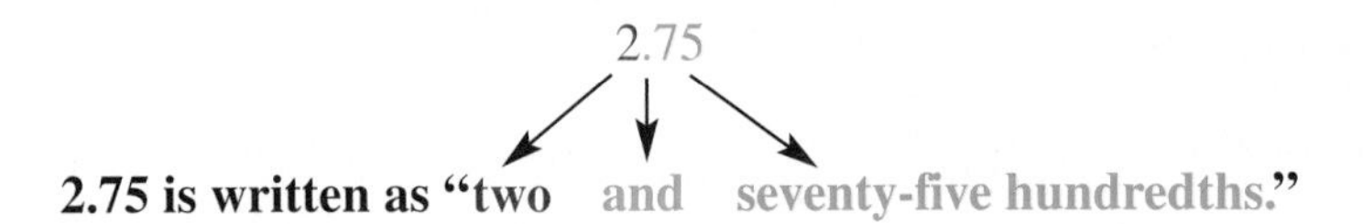

Procedure **Write a "Mixed" Decimal Number in Words**

Step 1 Write the whole number portion in words.

Step 2 Write *and* for the decimal point.

Step 3 Write the decimal portion in words.

EXAMPLES	GUIDED PRACTICE
7. Write 6.892 in words.	**7.** Write 13.094 in words.
Write: Whole portion and decimal portion Six and eight hundred ninety-two thousandths	Write: Whole portion and decimal portion __________ and ________________
8. Write 300.0204 in words.	**8.** Write 503.1006 in words.
Write: Whole portion and decimal portion Three hundred and two hundred four ten thousandths	Write: Whole portion and decimal portion ____________ and ______________________

Concept Check

B1. Match each number with its form in words.

1. 0.08	**a.** eight thousand
2. 8,000	**b.** eight thousandths
3. 0.008	**c.** eight hundred
4. 800	**d.** eight hundredths

B2. When writing a "mixed" decimal number, what word is used to separate the whole number portion from the decimal portion?

B3. When you write 1.5 and $1\frac{5}{10}$ in words, what do you notice?

Objective B Practice

Write each decimal number in words.

B4. 0.8 **B5.** 0.9 **B6.** 0.0087 **B7.** 0.0567

B8. 1.53 **B9.** 5.49 **B10.** 60.02 **B11.** 55.05

B12. 5.834 **B13.** 4.892 **B14.** 5.9006 **B15.** 6.90003

Objective C Convert Between Decimals and Fractions

The Concept To write a decimal as a fraction or mixed number, think of the decimal written in words.

A decimal number writen in words describes the equivalent mixed number.

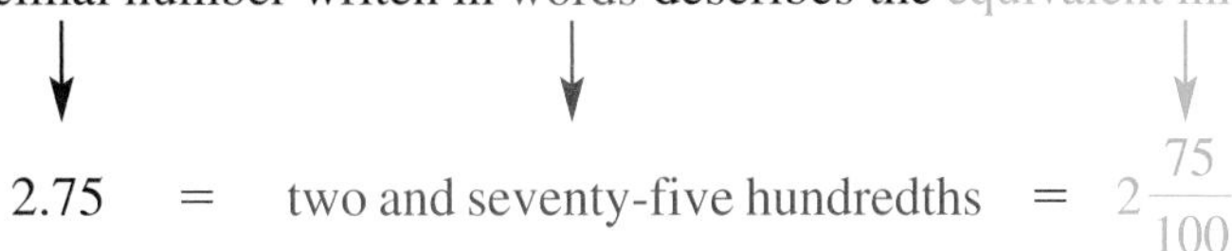

$$2.75 = \text{two and seventy-five hundredths} = 2\frac{75}{100}$$

Procedure Write a Decimal Number as a Fraction or Mixed Number

Step 1 Read the decimal number in words.

Step 2 Convert the words to a fraction or mixed number.

EXAMPLES

9. Write 4.56 as a fraction or mixed number.

4.56 = four and fifty-six hundredths

$$4.56 = 4\frac{56}{100}$$

2 decimal places — 2 zeros

GUIDED PRACTICE

9. Write 6.839 as a fraction or mixed number.

6.839 = ___ and ______________________

$$6.839 = (\quad)\frac{(\quad)}{1\quad}$$

The number of zeros in the denominator will be the same as the number of decimal places.

10. Write 67.00008 as a fraction or mixed number.

67.00008
= sixty-seven and eight hundred-thousandths

$$67.00008 = 67\frac{8}{100,000}$$

5 decimal places — 5 zeros

10. Write 320.0004 as a fraction or mixed number.

320.0004
= ______________ and ______________

$$320.0004 = \quad \frac{\quad}{1\quad}$$

If the denominator of a fraction is a power of 10, we can reverse the procedure to find the equivalent decimal number. We will discuss how to convert any fraction to a decimal number in Section 3.4.

Procedure Write a Fraction or Mixed Number as a Decimal

The denominator must be a power of 10 for this procedure to be used.

Step 1 Read the fraction or mixed number in words.

Step 2 Convert the words to a decimal number.

EXAMPLES

11. Write $\frac{365}{1,000}$ as a decimal.

$$\frac{365}{1,000} = 365 \text{ thousandths}$$

$$\frac{365}{1,000} = 0.365$$

3 zeros — 3 decimal places

12. Write $14\frac{2}{10,000}$ as a decimal.

$$14\frac{2}{10,000} = 14 \text{ and two ten thousandths}$$

$$14\frac{4}{10,000} = 14.\,0002$$

4 zeros — 4 decimal places

GUIDED PRACTICE

11. Write $\frac{23}{100}$ as a decimal.

$$\frac{23}{100} =$$

$$\frac{23}{100} =$$

12. Write $74\frac{103}{100,000}$ as a decimal.

$$74\frac{103}{100,000} =$$

$$74\frac{103}{100,000} =$$

Concept Check

C1. Write 310.67 in words.

C2. What is wrong with writing a check for $310.67 using the words "three hundred and ten $\frac{67}{100}$?"

C3. A decimal number is converted to the mixed number $3\frac{4}{10,000}$. What is the place value of the digit 4 in the decimal number?

Objective C Practice

Write each number as a fraction or mixed number. Do not simplify (reduce) the fraction.

C4. 4.09 **C5.** 3.12 **C6.** 54.4 **C7.** 23.2

C8. 3,000.003 **C9.** 4,000.004 **C10.** 230.00043 **C11.** 837.0439

Write each fraction or mixed number as a decimal.

C12. $\frac{3}{10}$ **C13.** $\frac{7}{10}$ **C14.** $\frac{37}{100}$ **C15.** $\frac{73}{100}$

C16. $5\frac{1}{100}$ **C17.** $3\frac{13}{1{,}000}$ **C18.** $36\frac{27}{10{,}000}$ **C19.** $25\frac{29}{100{,}000}$

Objective D Round Decimal Numbers

The Concept Many real-life decimal numbers must be rounded to make them useful. For example, a 6% sales tax on an item that costs $25.60 is calculated to be $1.536. Since pennies are the smallest coin, $1.536 needs to be rounded to the nearest cent or hundredth of a dollar.

▶ **NOTE** It might be useful for you to briefly review rounding whole numbers in Section 1.1, Objective D.

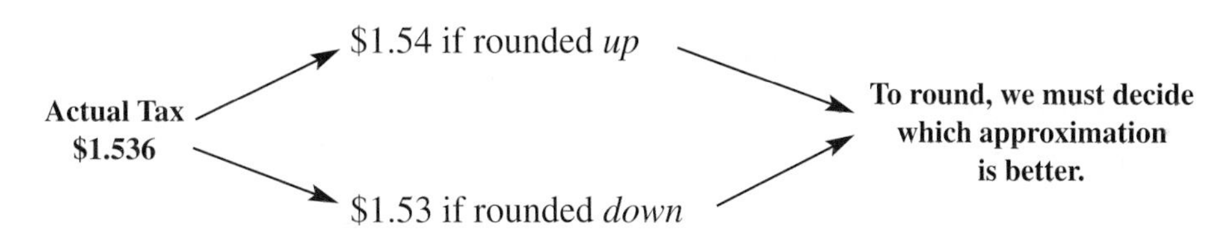

Since $1.54 is closer to the actual tax, $1.536 rounds up to $1.54.

Procedure Choose the Best Approximation

Choose the rounded number that is closer to the original.

Remember, you are finding the best approximation written to the desired place value.

EXAMPLES

13. To round 12.61 to the tenths place, we
round up to 12.7
or
round down to 12.6.
Which is a better approximation?

Since 12.6 is closer to 12.61, it is the best approximation. Round the number down.

GUIDED PRACTICE

13. To round 1.346 to the hundredths place, we
round up to 1.35
or
round down to 1.34.
Which is a better approximation?

Since ______ is closer to 1.346, it is the best approximation. Round the number ______.

14. To round 2.567 to the hundredths place, we round up to 2.57 or round down to 2.56. Which is a better approximation?	**14.** To round 3.44 to the tenths place, we round up to 3.5 or round down to 3.4. Which is a better approximation?
Since 2.57 is closer to 2.567, it is the best approximation. Round the number up.	Since ______ is closer to 3.44, it is the best approximation. Round the number ______ .
15. To round 2.65 to the tenths place, we round up to 2.7 or round down to 2.6. Which is a better approximation?	**15.** To round 9.935 to the hundredths place, we round up to 9.94 or round down to 9.93. Which is a better approximation?
It's a tie. Neither 2.7 nor 2.6 is closer to 2.65. We always round up when a tie occurs. Round the number up.	It's a ______. Neither 9.94 nor 9.93 is closer to 9.935. We always ________ when a tie occurs. Round the number ______.

Procedure **Round Decimal Numbers**

Step 1 Identify the digit to be rounded.

Step 2 Decide if the digit to be rounded increases by one or stays the same.

Increase the digit when the digit to the right is 5 through 9.
Keep the digit the same when the digit to the right is 0 through 4.

Step 3 All the digits to the right of the rounded digit will be zero.

DETAILED EXAMPLE Rounding a Decimal Number

Round 5.37082 to the thousandths place.

Step 1 Identify the digit to be rounded.

5.37(0)82

Zero is in the thousandths place.

Step 2 Decide if the digit to be rounded increases by one or stays the same.

Increase the digit when the digit to the right is 5 through 9.
Keep the digit the same when the digit to the right is 0 through 4.

5.37(0)82

Since the next digit is 8, we increase (0) to (1).

Step 3 All the digits to the right of the rounded digit will be zero.

5.37(1)

Decimal numbers ending in 0's do not need those 0's to be written. See the following note for an explanation.

5.37082 rounds up to 5.371.

▶ **NOTE** Decimal numbers ending in 0's do not need those 0's to be written. 0.50 = 0.5 because $0.50 = \frac{50}{100}$, which simplifies to $\frac{5}{10}$, or 0.5.

EXAMPLES

16. Round 5.873 to the hundredths place.

5.873

Step 1 The digit 7 is in the hundredths place.
Round up to 5.88 or down to 5.87.

Step 2 The digit to the right is 3, so we round down.
5.873 ≈ 5.87 is the best approximation.

17. Round 40.0989 to the tenths place.

40.0989

Step 1 The digit 0 is in the tenths place.
Round up to 40.1 or down to 40.0.

Step 2 The digit to the right is 9, so we round up.
40.0989 ≈ 40.1 is the best approximation.

18. A NASCAR driver had an average speed of 171.5087 miles per hour during a race. Round this speed to the nearest ten miles per hour.

171.5087

Step 1 The digit 7 is in the tens place.
Round up to 180 or down to 170.

Step 2 The digit to the right is a 1, so we round down.
171.5087 ≈ 170 miles per hour is the best approximation.

GUIDED PRACTICE

16. Round 6.8726 to the thousandths place.

6.8726

Step 1 The digit ____ is in the thousandths place.
Round up to ______ or down to ______.

Step 2 The digit to the right is ____, so we round up/down.
6.8726 ≈ ______ is the best approximation.

17. Round 89.07239 to the hundredths place.

89.07239

Step 1 The digit ____ is in the hundredths place.
Round up to ______ or down to ______.

Step 2 The digit to the right is ____, so we round up/down.
89.07293 ≈ ________ is the best approximation.

18. In a bicycle race, the winner averaged 18.523 miles per hour. Round this speed to the nearest tenth of a mile per hour.

18.523

Step 1 The digit ____ is in the tenths place.
Round up to ______ or down to ______.

Step 2 The digit to the right is a ____, so we round up/down.
18.523 ≈ ________ miles per hour is the best approximation.

Concept Check

D1. What is wrong with saying that 310.143 rounded to the hundredths place is 300?

D2. Rounding to the nearest hundredth, 3.9273 ≈ 3.93. Use this example to explain why the procedure for rounding a decimal number gives the best approximation for the number to a given place value.

D3. When rounding 241 to the tens place, we must write the 0 in 240. When rounding 2.41 to the tenths place, we do not write the 0 in 2.40. Why must the zero be included in one case but not the other? *Hint:* The answer relates to holding place value.

Objective D Practice

Choose the best approximation.

D4. To round 5.678 to the hundredths place, we
round up to 5.68
or
round down to 5.67.
Which is a better approximation?

D5. To round 2.567 to the tenths place, we
round up to 2.6
or
round down to 2.5.
Which is a better approximation?

D6. To round 2.456 to the ones place, we
round up to 3
or
round down to 2.
Which is a better approximation?

D7. To round 0.3567 to the tenths place, we
round up to 0.4
or
round down to 0.3.
Which is a better approximation?

Round each decimal number to the given place value.

D8. 8.987; hundredths

D9. 37.369; tenths

D10. 100.00907; thousandths

D11. 782.12861; ten thousandths

D12. 12,679.6728; thousands

D13. 54,789.782; hundreds

D14. 10.978; ones

D15. 20.432; ones

D16. At Cedar Point Amusement Park, the Top Thrill Dragster roller coaster reaches a speed of 119.57 miles per hour. Round this speed to the nearest mile per hour.

D17. At Six Flags Great Adventure, the Kingda Ka roller coaster reaches a speed of 127.53 miles per hour. Round this speed to the nearest ten miles per hour.

Objective E Order Decimal Numbers

The Concept To order numbers, we must make sure they are like quantities. To order fractions, we build like fractions. Consider the fractions $\frac{1}{10}$ and $\frac{9}{100}$.

$$\frac{1}{10} \square \frac{9}{100}$$

$$\frac{10}{10} \cdot \frac{1}{10} \square \frac{9}{100}$$

$$\frac{10}{100} > \frac{9}{100}$$

$$\text{Therefore, } \frac{1}{10} > \frac{9}{100}.$$

To order decimal numbers, we build **like decimals**. That is, we write each decimal number to the same place value. Like decimals are easier to compare and order than unlike decimals. To order 0.1 and 0.09, we add a "missing" zero to 0.1 so that both numbers are written as hundredths.

Add a missing zero to build like decimals.

0.1 $\square$ 0.09

0.10 $\boxed{>}$ 0.09

The number 0.1 is greater than 0.09 since ten hundredths is more than nine hundredths. Because of the place value of the digits, 0.1 is greater than 0.09 even though 1 is less than 9.

▶ **NOTE** Fractions are "like" when they have the same denominator. Decimals are "like" when they are written to the same place value.

Procedure Order Decimal Numbers

Step 1 Build like decimals.
Line up the decimal points.
Add missing zeros.

Step 2 Compare the numbers.

EXAMPLES

19. Order 0.3 and 0.04 from largest to smallest.

Step 1 Line up the decimal points.

$$0.3 = 0.30 \leftarrow \text{Add one zero.}$$
$$0.04 = 0.04$$

Step 2 Since both numbers are hundredths, we can easily compare them.

$30 > 4$, so $0.3 > 0.04$.

20. Order 0.90, 0.0952, and 0.897 from smallest to largest.

Step 1 Line up the decimal points.

$$0.9 = 0.9000$$
$$0.0952 = 0.0952$$
$$0.897 = 0.8970$$

Add missing zeros.

Step 2 Since the numbers are all ten thousandths, we can easily compare them. $952 < 8{,}970 < 9{,}000$, so

$0.0952 < 0.897 < 0.9$.

GUIDED PRACTICE

19. Order 0.073 and 0.2 from largest to smallest.

Step 1 Line up the decimal points.

Add missing zeros.

Step 2 Since both numbers are ____________, we can easily compare them.

____ > ____, so ____________.

20. Order 0.4, 0.039, and 0.38 from smallest to largest.

Step 1 Line up the decimal points.

Add missing zeros.

Step 2 Since the numbers are all ____________, we can easily compare them. ____ < ____ < ____, so

____________________.

Concept Check

E1. **a.** Write the following decimal numbers as fractions: 0.003, 0.04, and 0.2.

b. Build each of the fractions from part a into like fractions.

c. Use the fractions in part b to write 0.003, 0.04, and 0.2 in order from largest to smallest.

E2. **a.** Build $\frac{2}{10}$ and $\frac{5}{100}$ into like fractions and compare their values. Use <, >, or = to make a true statement.

$\frac{2}{10} \square \frac{5}{100}$

b. Build 0.2 and 0.05 into like decimals and compare their values. Use <, >, or = to make a true statement.

$0.2 \square 0.05$

c. When comparing quantities, building like fractions is equivalent to writing each decimal number to the same ____________.

Objective E Practice

Use <, >, or = to make each statement true.

E3. 0.4 ☐ 0.06 **E4.** 0.01 ☐ 0.004 **E5.** 0.07 ☐ 0.6 **E6.** 0.024 ☐ 0.15

Order each list of decimal numbers from largest to smallest.

E7. 0.04, 0.042, 0.12

E8. 0.087, 0.53, 0.056

Answer each question.

E9. Laptop computers are being made smaller than ever. Some are thinner than the diameter of a penny. Which laptop is the thinnest?

Laptop	Thickness
MacBook Air	0.76 inches
Mobile Metro	0.7 inches
Lenovo ThinkPad	0.73 inches

E10. Mobile phones are being made smaller than ever. Which mobile phone is the thickest?

Phone	Thickness
Motorola L6	0.428 inches
Nokia 6500	0.3733 inches
Samsung SGH	0.275 inches

Section 3.1 Exercises

FOR EXTRA HELP

To understand the place values of decimal numbers:

1. Answer the Objective A Concept Checks.
2. Answer the odd Objective A Practice Exercises.
3. Answer the even Objective A Practice Exercises.

To write decimal numbers in words:

4. Answer the Objective B Concept Checks.
5. Answer the odd Objective B Practice Exercises.
6. Answer the even Objective B Practice Exercises.

To convert between decimals and fractions:

7. Answer the Objective C Concept Checks.
8. Answer the odd Objective C Practice Exercises.
9. Answer the even Objective C Practice Exercises.

To round decimal numbers:

10. Answer the Objective D Concept Checks.
11. Answer the odd Objective D Practice Exercises.
12. Answer the even Objective D Practice Exercises.

To order decimal numbers:

13. Answer the Objective E Concept Checks.
14. Answer the odd Objective E Practice Exercises.
15. Answer the even Objective E Practice Exercises.

VOCABULARY REVIEW *Review the Vocabulary Preview for Section 3.1. Study the definitions until you can check* Got It *for every word.*

Use these words to complete each sentence.

digits • decimal point • decimal places • place value • like decimals

	Got It	Get Help
16. The __________ of a digit indicates that digit's value in a number.		
17. Decimal numbers are __________ when they are written to the same place value.		
18. The number of __________ is the number of digits written to the right of the decimal point.		
19. The __________ in our number system are 0, 1, 2, 3, 4, 5, 6, 7, 8, and 9.		
20. The __________ is a period that separates the digits representing whole numbers from the digits representing fractions.		

How will you get help for any vocabulary that you are unsure about?

Your instructor ___ MyMathLab ___ A classmate ___ A tutor ___ Other ___

Find the place value of the digit 4 in each number.

21. 70.0042 **22.** 53.40982 **23.** 21.9452 **24.** 80.89423

25. 867.423 **26.** 87.8423 **27.** 134.87 **28.** 249.872

Write each decimal number in expanded form.

29. 1,004.0123 **30.** 2,900.0015 **31.** 0.03204 **32.** 0.10013

Write each decimal number in words.

33. 0.65 **34.** 0.89 **35.** 203.129 **36.** 408.853

37. 9.0002 **38.** 4.0005 **39.** 12,000.0012 **40.** 87,000.0087

Write each decimal number in words, as you would on a check.

41.

Joseph M. Johnson
Linda J. Johnson
123 Maple Street
Your City, California 94000

1168

DATE __________ 20____

PAY to the ORDER of *Pat Kopf* $ 15.67

__________ DOLLARS

New West Bank
San Francisco California 94106

MEMO __________

42.

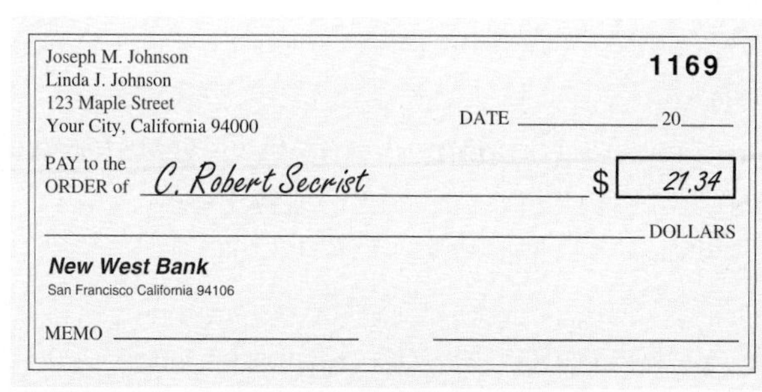

Joseph M. Johnson
Linda J. Johnson
123 Maple Street
Your City, California 94000

1169

DATE __________ 20____

PAY to the ORDER of *C. Robert Secrist* $ 21.34

__________ DOLLARS

New West Bank
San Francisco California 94106

MEMO __________

goetz Don't forget: Decimal place values end in "ths," whole number place values do not.

For more tips and tweets, go to twitter.com/gstbasicmath

43.

Joseph M. Johnson
Linda J. Johnson
123 Maple Street
Your City, California 94000

1170

DATE ______ 20___

PAY to the ORDER of *Doris Lewis* $ 102.45

______ DOLLARS

New West Bank
San Francisco California 94106

MEMO ______

44.

Joseph M. Johnson
Linda J. Johnson
123 Maple Street
Your City, California 94000

1171

DATE ______ 20___

PAY to the ORDER of *Saeed Saboni* $ 1,004.04

______ DOLLARS

New West Bank
San Francisco California 94106

MEMO ______

45.

Joseph M. Johnson
Linda J. Johnson
123 Maple Street
Your City, California 94000

1172

DATE ______ 20___

PAY to the ORDER of *Sue Stetler* $ 504.98

______ DOLLARS

New West Bank
San Francisco California 94106

MEMO ______

46.

Joseph M. Johnson
Linda J. Johnson
123 Maple Street
Your City, California 94000

1173

DATE ______ 20___

PAY to the ORDER of *Anna Cox* $ 300.09

______ DOLLARS

New West Bank
San Francisco California 94106

MEMO ______

Write each decimal as a fraction or mixed number. Do not simplify the fractions.

47. 0.12 **48.** 0.43 **49.** 1.2 **50.** 4.8

51. 12.009 **52.** 13.04 **53.** 102.908 **54.** 342.987

55. 92.12902 **56.** 34.13534 **57.** 0.1004 **58.** 0.04002

Write each fraction or mixed number as a decimal.

59. $\frac{3}{10}$ **60.** $\frac{7}{10}$ **61.** $2\frac{5}{10}$ **62.** $8\frac{1}{10}$

63. $14\frac{15}{100}$ **64.** $17\frac{98}{100}$ **65.** $3\frac{7}{100}$ **66.** $2\frac{3}{100}$

67. $54\frac{5}{1{,}000}$ **68.** $98\frac{3}{1{,}000}$ **69.** $10{,}000\frac{10}{1{,}000}$ **70.** $100\frac{1}{1{,}000}$

71. $\frac{1}{100{,}000}$ **72.** $\frac{1}{10{,}000}$ **73.** $165\frac{1{,}674}{10{,}000}$ **74.** $153\frac{3{,}002}{10{,}000}$

Round each decimal number to the given place value.

75. 163.45; tenths **76.** 157.21; tenths **77.** 163.45; tens **78.** 157.21; tens

79. 236.979; hundredths **80.** 967.863; hundredths **81.** 236.979; hundreds **82.** 467.863; hundreds

83. 872.8260; thousandths **84.** 8,279.27354; hundreds **85.** 7,327.8277; thousandths **86.** 2,794.2947; thousandths

87. 

The average temperature on Venus is 864.54 degrees Fahrenheit. Round this temperature to the nearest tenth of a degree.

88. 

The tallest skyscraper in the world is the Burj Dubai tower, with a height of 2,087 feet. Round this height to the nearest hundred feet.

89. A bank charges interest to the nearest cent. If the interest charges on a mortgage are $345.6872, how much will the bank charge?

90. Ocean currents flowing under the Golden Gate Bridge can exceed 5.6 miles per hour. What is this speed to the nearest mile per hour?

Order each list of numbers from largest to smallest.

91. 0.45, 0.103, 0.08, 0.7

92. 0.098, 0.10, 0.3, 0.298

93. 1.38, 1.83, 1.083, 1.308, 1.8

94. 4.089, 4.98, 4.009, 4.098, 4.9

95. The average speeds for the Daytona 500 winning drivers have been recorded for many years.

Order these speeds from fastest to slowest.

Year	Driver	Speed
2007	Harvick	149.335 mph
1997	Gordon	148.295 mph
1970	Hamilton	149.601 mph

96. Earthquakes are measured on the Richter scale. The larger the magnitude, the stronger the earthquake, as registered on the Richter scale.

Rank these earthquakes from largest to smallest magnitude.

Year	Location	Magnitude
1965	Alaska	8.7
1960	Chile	9.2
1952	Kamchatka	9.0

97. Mohs scale of mineral hardness starts with the softest $(1 = \text{talc})$ and ends with the hardest $(10 = \text{diamond})$.

Organize these items from softest to hardest.

Hardness	Item
2.2	Fingernail
5.5	Glass
2	Gypsum

98. In the women's 100-meter dash, the top four times were 12.034 seconds, 13.204 seconds, 12.4 seconds, and 12.204 seconds. Order these times to identify first place through fourth place.

Answer each question.

99. Brittany's final exam score is 87.493%. To receive a B+ on her final, she needs to get 87.50%. If her instructor rounds grades to the hundredths place, will she get a B+ on her final exam?

100. Chris calculates her grade point average (GPA) to be 3.1752987. If the registrar at her college reports GPAs to the nearest hundredth, what is Chris's GPA?

101. A digital scale shows that a wrestler weighs 154.798 pounds. What is the wrestler's weight to the nearest pound?

102. Gasoline at one station sells for $3.259 per gallon. Round this price to the nearest cent.

SELF-ASSESSMENT

	YES	NO
1. On your last test, were you satisfied with your score? *If you answered yes, good job. Do not continue here.* *If you answered no, answer the following.*		
2. Did you preview each section of the textbook before your instructor taught it in class?		
3. Did you complete all of the homework before it was due?		
4. Did you get help with the homework exercises that you didn't understand?		

5. Set two goals that will help you become better prepared for your next test.

Goal 1: ______

Goal 2: ______

Section 3.1 Question Log

Use this space to write questions. Be sure to get these answered and revisit them when you prepare for your exam.

Page ____ **Answered** ☐	**Page** ____ **Answered** ☐
Page ____ **Answered** ☐	**Page** ____ **Answered** ☐

Section 3.2 Adding and Subtracting Decimal Numbers

You have likely encountered addition and subtraction with decimal numbers before.

The way that we add and subtract decimal numbers is similar to how we add and subtract whole numbers. When adding decimal numbers, we must make sure that we combine the digits that are in the same place value.

Decimal numbers are added to get a total and subtracted to make change.

The **Objective** in Section 3.2 will help you

A Add and subtract decimal numbers.

VOCABULARY PREVIEW *Check the box that applies.*	Got It	Must Study
Operations: The four arithmetic **operations** are addition, subtraction, multiplication, and division.		
Sum: A **sum** is the result of adding.		
Difference: A **difference** is the result of subtracting.		
Carrying: Carrying lets you regroup a 10 from one place value into a 1 in the next larger place value.		
Borrowing: Borrowing lets you regroup a 1 from one place value into a 10 in the next smaller place value.		

Study these words when they appear in the text. After the section, test your understanding by completing the Vocabulary Review in the section exercises.

Objective A Add and Subtract Decimal Numbers

The Concept To add and subtract fractions, we have to build like fractions. To add and subtract decimals, we build like decimals by writing each decimal number to the same place value. To do this, we align the decimal points and add zeros to fill any missing place values. Once each decimal number has the same place value, we can add or subtract the digits from right to left.

Procedure Add or Subtract Decimal Numbers

Step 1 Build like decimals.

Line up the decimal points.
Add zeros to fill in any missing place values.

Step 2 Add or subtract the digits, from right to left.

EXAMPLES

1. Find the sum. 4.87 + 3.8

Line up the decimal points.

$$\begin{array}{r} \overset{1}{4}.87 \\ +\ 3.80 \\ \hline 8.67 \end{array}$$

Add one missing zero to build like decimals.

To review carrying, go to Section 1.2.

2. Find the sum. 5.00302 + 0.979

Line up the decimal points.

$$\begin{array}{r} \overset{1}{5}.00302 \\ +\ 0.97900 \\ \hline 5.98202 \end{array}$$

Add two zeros to build like decimals.

3. Find the difference. 3.108 − 2.3

Line up the decimal points.

$$\begin{array}{r} \overset{2}{\cancel{3}}.\,{}^{1}1\ 0\ 8 \\ -\ 2.\ 3\ 0\ 0 \\ \hline 0.\ 8\ 0\ 8 \end{array}$$

Add two zeros to build like decimals.

To review borrowing, go to Section 1.3.

4. Find the difference. 90 − 8.45

Line up the decimal points.

$$\begin{array}{r} \overset{8}{\cancel{9}}\ \overset{9}{\cancel{{}^{1}0}}.\ \overset{9}{\cancel{{}^{1}0}}\ {}^{1}0 \\ -\quad 8.\ 4\ 5 \\ \hline 8\ 1.\ 5\ 5 \end{array}$$

Add two zeros to build like decimals.

GUIDED PRACTICE

1. Find the sum. 3.879 + 5.71

Line up the decimal points. Add one zero and then add.

+ ______

2. Find the sum. 0.002378 + 3.392

Line up the decimal points. Add three zeros and then add.

+ ______

3. Find the difference. 6.002378 − 3.392

Line up the decimal points. Add ______ zeros and then subtract.

− ______

4. Find the difference. 40 − 7.32

Line up the decimal points. Add ______ zeros and then subtract.

Be careful when borrowing from more than one place away.

− ______

Concept Check

A1. When adding decimal numbers, what should you line up to make sure the correct digits are added together?

A2. In your own words, why is adding 0.3 and 0.4 similar to adding like fractions?

A3. Use fractions to show that 0.3 and 0.30 are the same number.

A4. Why are you allowed to add zeros on the far right of a decimal?

Objective A Practice

Find each sum.

A5. 1.3 + 2.5 **A6.** 2.5 + 3.3 **A7.** 14.91 + 2.07 **A8.** 3.08 + 2.91

A9. 4 + 3.4 **A10.** 8 + 9.21

Find each difference.

A11. 12.56 − 3.24 **A12.** 11.93 − 4.71 **A13.** 9.12 − 5.04 **A14.** 14.36 − 3.17

A15. 8.912 − 4.1912 **A16.** 75.789 − 6.0789 **A17.** 4 − 0.14 **A18.** 7 − 0.87

Answer each question.

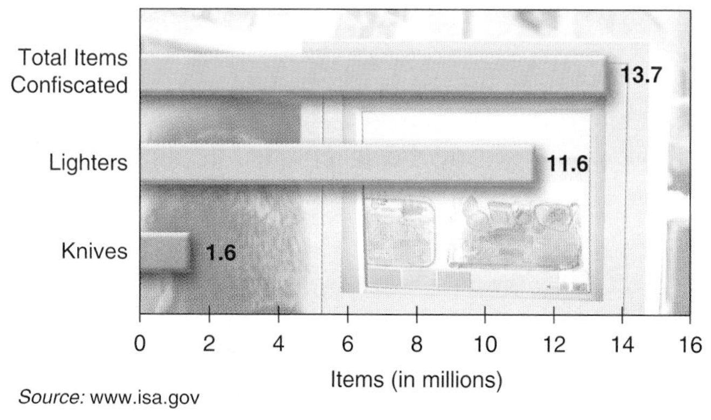

A19. What is the total number of knives and lighters that the Transportation Security Administration confiscated in 2006?

A20. How many of the intercepted items were neither knives nor lighters?

Combining Concepts and Applications

CONCEPT I Applications with the Order of Operations When adding and subtracting decimal numbers, remember to follow the order of operations. Addition and subtraction is performed from left to right.

EXAMPLE

5. On Monday, Angela's bank account had a starting balance of $355.74. Over the next week, she wrote a check for $54.21, withdrew $40.00 from an ATM machine, and deposited her paycheck of $89.32. What is her new balance?

Decide if each transaction adds to or subtracts from her starting balance.

- Writing a check uses subtraction.
- Making an ATM withdrawal uses subtraction.
- Depositing money uses addition.

Add or subtract the quantities from the starting balance.

$$\begin{aligned}\text{New balance} &= \text{Starting balance} - \text{Check} \\ &\quad - \text{Withdrawal} + \text{Deposit} \\ &= 355.74 - 54.21 - 40 + 89.32 \\ &= 301.53 - 40 + 89.32 \\ &= 261.53 + 89.32 \\ &= 350.85\end{aligned}$$

A word equation can help you understand which arithmetic operation to use.

Angela's new balance is $350.85.

GUIDED PRACTICE

5. On Monday morning, the stock market was at 13,562.33. Over the next 3 days, it rose 129.43, decreased by 39.87, and then fell by 67.13. What is the new value?

Decide if each change adds to or subtracts from the starting value.

- A rise represents ____________.
- A decrease represents ____________.
- A fall represents ____________.

Add or subtract the quantities from the starting value.

New value = Starting value

=

=

=

=

The new value is ____________.

CONCEPT II Interpreting Graphical Data

EXAMPLE

6. Stephan plotted his retirement account's value over 6 years on the following bar graph.

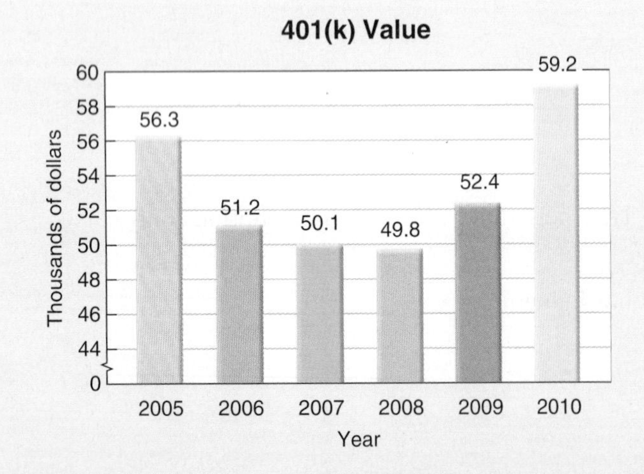

In which years did the value of Stephan's retirement account increase and decrease the most? By what amounts?

GUIDED PRACTICE

6. Stephan plotted the amount of money saved for his child's college tuition on the following bar graph.

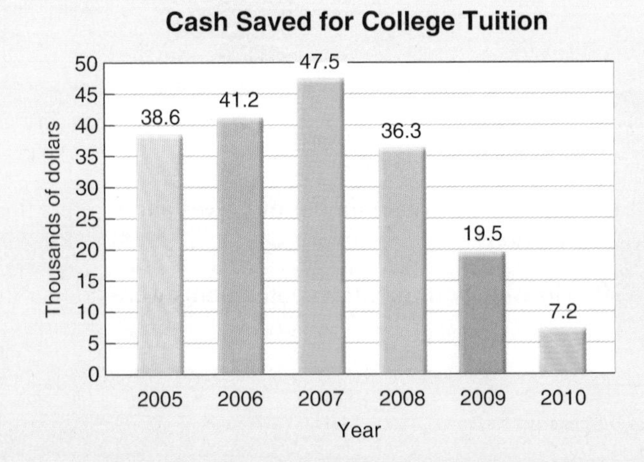

In which years did the amount saved increase and decrease the most? By what amounts?

The greatest increase occurred from 2009 to 2010. The increase is equal to the difference in values: ($59.2 - 52.4 = 6.8$). In that year, Stephan's account gained \$6.8 thousand or \$6,800 in value.

The greatest decrease occurred from 2005 to 2006. The decrease is equal to the difference in values: ($56.3 - 51.2 = 5.1$). In that year, Stephan's account lost \$5.1 thousand or \$5,100 in value.

The greatest increase occurred from ______ to ______. The increase is equal to the ________ in values: (______ +/− ______ = ____). In that year, Stephan's account increased by ______ thousand or ______ in value.

The greatest decrease occurred from ______ to ______. The decrease is equal to the ________ in values: (______ +/− ______ = ____). In that year, Stephan's account decreased by ______ thousand or ______ in value.

Section 3.2 Exercises

FOR EXTRA HELP MyMathLab Math XL PRACTICE WATCH DOWNLOAD  READ REVIEW

To add and subtract decimal numbers:

1. Answer the Objective A Concept Checks.
2. Answer the odd Objective A Practice Exercises.
3. Answer the even Objective A Practice Exercises.

VOCABULARY REVIEW *Review the Vocabulary Preview for Section 3.2. Study the definitions until you can check* Got It *for every word.*

Use these words to complete each sentence.

operations • sum • difference • carrying • borrowing

	Got It	Get Help
4. __________ lets you regroup a 1 from one place value into a 10 in the next smaller place value.		
5. A __________ is the result of adding.		
6. __________ lets you regroup a 10 from one place value into a 1 in the next larger place value.		
7. A __________ is the result of subtracting.		
8. The four arithmetic __________ are addition, subtraction, multiplication, and division.		

How will you get help for any vocabulary that you are unsure about?

Your instructor ___ MyMathLab ___ A classmate ___ A tutor ___ Other ___

Find each sum.

9. $7.4 + 4.3$

10. $7.3 + 5.2$

11. $67.12 + 2.93$

12. $12.54 + 6.76$

13. $\begin{array}{r} 2.9 \\ +\ 5.42 \\ \hline \end{array}$

14. $\begin{array}{r} 4.97 \\ +\ 1.3 \\ \hline \end{array}$

15. $7.982 + 2.73$

16. $19.679 + 3.87$

17. Sergei charged a customer $61.57 for labor and $17.34 for parts. What was the total bill?

18. Ian collected 48.45 pounds of wool from his sheep on Monday and another 73.52 pounds of wool on Tuesday. How much wool did Ian collect in all?

19. The technology club is collecting returnable bottles to raise money. In the first month, club members raised $45.80. In the second month, they raised $39.55. How much money did they raise in total?

20. Mesilla purchased spaghetti for $3.45 and bread for $4.28. How much did she spend in total?

Find each difference.

21. $\begin{array}{r} 6.4 \\ -\ 2.3 \\ \hline \end{array}$

22. $\begin{array}{r} 8.3 \\ -\ 4.1 \\ \hline \end{array}$

23. $\begin{array}{r} 15.65 \\ -\ 6.82 \\ \hline \end{array}$

24. $\begin{array}{r} 21.14 \\ -\ 8.32 \\ \hline \end{array}$

t **25.** $\begin{array}{r} 19.2 \\ -\ 5.75 \\ \hline \end{array}$

t **26.** $\begin{array}{r} 21.4 \\ -\ 3.92 \\ \hline \end{array}$

t **27.** $83.8 - 1.983$

t **28.** $4.2 - 1.032$

29. Your grocery bill was $95.34 before the clerk rang up $23.85 in coupons. What is your new grocery bill?

30. Samantha and Alex are roommates. Their phone bill is $28.18. If Alex's portion is $17.63, what is Samantha's portion of the bill?

31. Aaron purchased $86.24 in lumber. If he gave the cashier $100, how much change should he get back?

32. Amy paid $154.78 for landscaping materials. If she gave the cashier $160, how much change should she get back?

Find each sum or difference.

33. $\begin{array}{r} 423.1784 \\ +\ \ \ 4.64 \\ \hline \end{array}$

34. $\begin{array}{r} 456.902 \\ +\ \ \ 4.2344 \\ \hline \end{array}$

35. $\begin{array}{r} 924 \\ -\ \ 0.123 \\ \hline \end{array}$

36. $\begin{array}{r} 245 \\ -\ \ 0.153 \\ \hline \end{array}$

37. $128 + 8.83$

38. $200 - 8.99$

39. $315 - 7.45$

40. $71.8918 + 5.892$

41. $54.9237 - 3.901$

42. $87.92 - 12.9073$

43. $78.99 + 43.87$

44. $21.973 + 32.242$

45. Mr. McFily weighs a package at 7.27 pounds. If he removes an item that weighs 2.3 pounds, what is the new weight of the package?

46. If Jolanas starts the week with 698.2 gallons of heating oil in his tank and uses 83.9 gallons, how much oil is left in the tank?

t **smith** Be careful! You will need to add the missing zeros and borrow to perform the subtraction.

For more tips and tweets, go to twitter.com/gstbasicmath

47. Anita's Grocery is 5.2 miles from Bob's Boats. Bob's Boats is 2.1 miles from Cal's Pancakes. Cal's Pancakes is 3.15 miles from the intersection.

a. How far is Bob's Boats from the intersection?
b. How far is Anita's Grocery from the intersection?

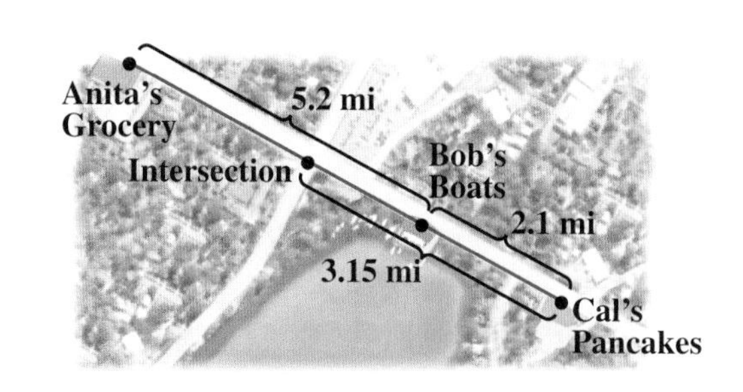

Find each sum or difference.

48. 786.92 − 34.091 **49.** 23.76 − 4.9724 **50.** 34.962 + 65.038 **51.** 45.972 + 54.028

Answer each question.

52. Your shopping cart contains the following items:

Soup, $1.89
Cereal, $3.89
Soap, $1.15
Milk, $2.94
Carrots, $1.50
Broccoli, $1.74
Yogurt, $2.00

a. Round each price to the nearest dollar and estimate the total cost of your groceries.
b. You only have $20. Based on your estimate from part a, can you afford a bag of coffee that costs $4.50?
c. Determine the actual cost of the groceries already in the cart.
d. Can you afford the bag of coffee?

53. You need to pay the monthly bills listed below:

Rent, $325.00
Car payment, $212.78
Gas, $37.87
Electric, $16.21
Phone/cable, $39.99
Credit card, $35.00

a. Round each bill to the nearest ten dollars and estimate your total bills.
b. You have $698.43 in the bank. Based on your estimate from part a, will you be able to pay any extra on your credit card this month?
c. Determine the actual total of all the bills.
d. How much money will be left in your account after you pay the bills?

Balance the following checkbooks by finding the balance after each transaction.

54.

Trans Type	Date	Description of Transaction	Payment/ Debit (−)	Fee (if any)	Deposit/ Credit (+)	Balance
	1-Dec	Balance Forward				$420.85
Check 220	1-Dec	Rent	$350.00			
ATM	6-Dec	Needed Cash	$40.00	$3.00		
	10-Dec	Paycheck			$331.24	
Check 221	10-Dec	Phone Bill	$85.32			
ATM	11-Dec	Needed Cash	$20.00	$2.50		

55.

Trans Type	Date	Description of Transaction	Payment/ Debit (−)	Fee (if any)	Deposit/ Credit (+)	Balance
	5-Feb	Balance Forward				$438.45
Check 5930	6-Feb	Food	$74.32			
Check 5931	7-Feb	Rent	$275.00			
ATM	7-Feb	Movie Night	$25.00	$3.00		
Check 5932	15-Feb	Cable Bill	$49.95			
	15-Feb	Cashed in Tip Money			$256.98	

Use the following bar graph to answer each question.

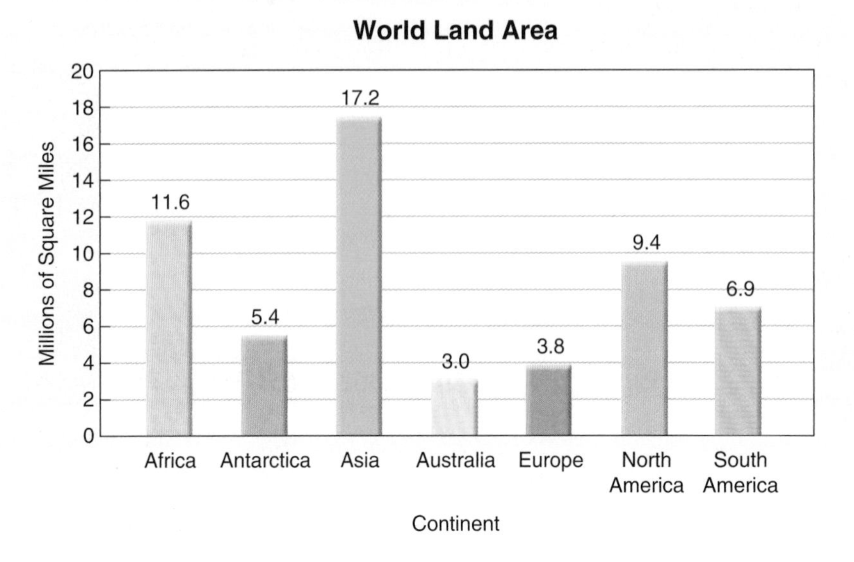

56. What is the total land area of the world?

57. How much more/less area is there in North and South America combined compared to Asia?

58. How much more/less area is there on the four smallest continents combined compared to the largest continent?

59. How much more area would Africa need for it to be the largest continent?

Section 3.2 Question Log

Use this space to write questions. Be sure to get these answered and revisit them when you prepare for your exam.

Page ____ **Answered** ☐	**Page** ____ **Answered** ☐
Page ____ **Answered** ☐	**Page** ____ **Answered** ☐

Section 3.3 Multiplying Decimal Numbers

You have likely multiplied decimal numbers in your everyday life.

Multiplication is used to calculate a total when many of the same items are purchased.

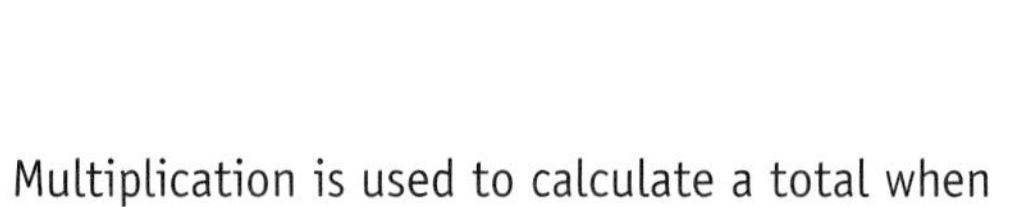

Multiplication is used to calculate a total when something is rented by the hour.

The **Objectives** in Section 3.3 will help you

- A Multiply one-digit decimal numbers.
- B Multiply any two decimal numbers.
- C Multiply by powers of 10.
- D Estimate products.
- E Use pi to find the circumference and area of a circle.

VOCABULARY PREVIEW *Check the box that applies.*	Got It	Must Study
Decimal Places: The number of **decimal places** is the number of digits written to the right of the decimal point.		
Product: A **product** is the result of multiplying.		
Factors: Factors are the numbers that are multiplied to give a product.		
Diameter: The **diameter** of a circle is the distance across a circle through its center.		
Radius: The **radius** of a circle is the distance from the circle's center to the edge of the circle.		
Pi (π): Pi (π) is a number that relates the diameter of a circle to its circumference. $\pi \approx 3.14$		
Circumference: The **circumference** of a circle is its perimeter.		

Study these words when they appear in the text. After the section, test your understanding by completing the Vocabulary Review in the section exercises.

Objective A Multiply One-digit Decimal Numbers

The Concept To multiply decimal numbers:

- Multiply the digits in the same way that you multiply whole numbers.
- The number of decimal places in the product must match the total number of decimal places in the factors.

2 decimal places	→	.04
+ 1 decimal place	→	× .2
3 decimal places	→	.008

We can multiply decimal numbers and their fraction equivalents to understand the reasoning behind this rule.

WHY IT WORKS Counting Place Values to Multiply Decimals

To find the product of 0.7 × 0.5, we can use fractions or decimals.

Multiply Using Fractions

$$\frac{7}{10} \cdot \frac{5}{10} = \frac{35}{100}$$
$$= 35 \text{ hundredths}$$
$$= 0.35$$

Multiply Using Decimals

0.7	←	**1 decimal place**
× 0.5	←	**1 decimal place**
0.3 5	←	**Must have 1 + 1 = 2 decimal places**

Procedure Multiply Decimal Numbers

Step 1 Multiply the numbers, ignoring the decimal points.

Step 2 Add the number of decimal places in each factor.

Step 3 Write the product so that it matches the number of decimal places in step 2.

EXAMPLES

1. Find the product. 0.06 × 0.04

2 decimal places	→	0.06
2 decimal places	→	× 0.04
2 + 2 = 4 decimal places	→	0.0024

4 decimal places

2. Find the product. 0.005 × 0.8

3 decimal places	→	0.005
1 decimal place	→	× 0.8
3 + 1 = 4 decimal places	→	0.0040

4 decimal places

GUIDED PRACTICE

1. Find the product. 0.003 × 0.02

___ decimal places →

___ decimal places → × ______

___ + ___ = ___ decimal places →

2. Find the product. 0.7 × 0.009

___ decimal places →

___ decimal places → × ______

___ + ___ = ___ decimal places →

3. Find the product. 0.000003 × 7	**3.** Find the product. 5 × 0.0000003
6 decimal places → 0.000003 0 decimal places → × 7 6 + 0 = 6 decimal places → 0.000021 (6 decimal places)	__ decimal places → __ decimal places → ______ __ + __ = __ decimal places →

Concept Check

A1. a. Write 0.3 and 0.5 as fractions.

b. How does the number of decimal places in 0.3 and 0.5 affect the denominators of their fractional forms?

c. Multiply the fractions from part a.

d. Based on the answer from part c, what is 0.3 × 0.5 written as a decimal number?

e. How do the number of decimal places in 0.3 and 0.5 determine the number of decimal places in the product 0.3 × 0.5?

Objective A Practice

Find each product.

A2. 0.003 × 0.04

A3. 0.0004 × 0.04

A4. 0.006 × 0.00008

A5. 0.1 × 0.003

A6. 9 × 0.0004

A7. 5 × 0.0006

A8. 8.0 × 9.0

A9. 8.0 × 2.0

A10. The thickness of one sheet of copy paper is 0.004 inch. What is the total thickness of 80 sheets of paper?

A11. The thickness of one ceramic floor tile is 0.3 inch. What is the total thickness of 40 stacked tiles?

Objective B Multiply Any Two Decimal Numbers

The Concept To multiply any two decimal numbers, follow the same procedure that you used to multiply decimal numbers with one digit.

Procedure Multiply Any Two Decimal Numbers

Step 1 Multiply the numbers, ignoring the decimal points.

Step 2 Add the number of decimal places in each factor.

Step 3 Write the product so that it matches the number of decimal places in step 2.

EXAMPLES

4. Find the product. 3.72×0.4

$$\begin{array}{lr} \text{2 decimal places} \rightarrow & 3.72 \\ \text{1 decimal place} \rightarrow & \times\ 0.4 \\ 2 + 1 = \text{3 decimal places} \rightarrow & 1.488 \end{array}$$

3 decimal places

$3.72 \times 0.4 = 1.488$

5. Find the product. 0.184×1.5

$$\begin{array}{lr} \text{3 decimal places} \rightarrow & 0.184 \\ \text{1 decimal place} \rightarrow & \times\ 1.5 \\ & 920 \\ & +184 \\ 3 + 1 = \text{4 decimal places} \rightarrow & 0.2760 \end{array}$$

4 decimal places

$0.184 \times 1.5 = 0.276$

6. Find the product. 250×0.0045

$$\begin{array}{lr} \text{0 decimal places} \rightarrow & 250 \\ \text{4 decimal places} \rightarrow & \times\ 0.0045 \\ & 1250 \\ & +\ 1000 \\ 0 + 4 = \text{4 decimal places} \rightarrow & 1.1250 \end{array}$$

4 decimal places

$250 \times 0.0045 = 1.125$

GUIDED PRACTICE

4. Find the product. 50.67×0.03

__ decimal places →

__ decimal places → × ______

__ + __ = __ decimal places →

$50.67 \times 0.03 =$ _____

5. Find the product. 0.00345×27

__ decimal places →

__ decimal places → × ______

______

__ + __ = __ decimal places →

$0.00345 \times 27 =$ _____

6. Find the product. $1{,}500 \times 0.00073$

__ decimal places →

__ decimal places → × ______

______

__ + __ = __ decimal places →

$1{,}500 \times 0.0073 =$ _____

Concept Check

B1. In your own words, how do you decide where to put the decimal point when multiplying two decimal numbers?

B2. In Example 6, why is it okay to drop the zero in 1.1250 and write the answer as 1.125?

Objective B Practice

Find each product.

B3. 0.573×0.07

B4. 0.0689×0.02

B5. 3.65×8.2

B6. 4.93×2.4

(Continued)

B7. 45.965 × 0.0032 **B8.** 85.403 × 0.023 **B9.** 7,600 × 1.06 **B10.** 3,800 × 2.12

B11. In 2004, how much money could a person deduct on her tax return if she drove her car 700 miles for business?

B12. In 2008, how much money could a person deduct on his tax return if he drove his car 900 miles for business?

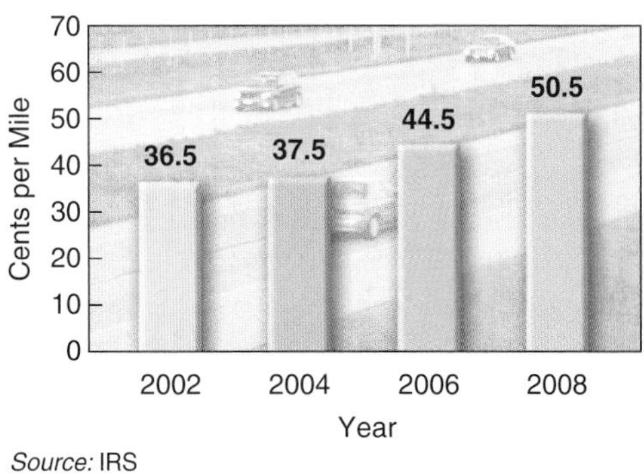

The values in this chart show the amount, per mile, that someone can deduct from his or her 2008 taxes for driving "business miles."

Objective C Multiply by Powers of 10

The Concept Knowing how to multiply by a power of 10 will allow you to estimate the products of numbers quickly. The following table shows how to move a decimal point when multiplying by a power of 10.

HOW TO MULTIPLY BY POWERS OF TEN QUICKLY			
The Number	**Key Characteristic**	**Effect on the Decimal Point**	**The Idea**
0.001 or smaller	Many decimal places	Moves that many places to the left	**Multiply by a small number, and the product gets smaller.**
0.01	Two decimal places	Moves 2 places to the left	
0.1	One decimal place	Moves 1 place to the left	
1	Has no decimal places Has no zeros	No effect	**Multiplying by 1 doesn't change the value.**
10	One zero	Moves 1 place to the right	**Multiply by a large number, and the product gets larger.**
100	Two zeros	Moves 2 places to the right	
1,000 or larger	Many zeros	Moves that many places to the right.	

If you are unsure about which direction to move the decimal point, perform the multiplication using the usual procedure.

Procedure Multiply by Decimal Powers of 10

Step 1 Count the number of decimal places in the power of ten.

Step 2 Move the decimal point the counted number of places to the left.

Multiplying a number by 0.1, 0.01, 0.001... results in a product that is smaller.

Procedure Multiply by Whole Number Powers of 10

Step 1 Count the number of zeros in the power of 10.

Step 2 Move the decimal point the counted number of zeros to the right.

Multiplying a number by 10, 100, 1,000... results in a product that is larger.

EXAMPLES

7. Find the product. 45.6×0.001

Multiplying by 0.001 will make the product smaller than 45.6.

Move the decimal point three places to the left.

$$45.6 \times 0.\underbrace{001}_{\text{3 decimal places}} = 0.0456 \quad (\leftarrow \text{3 places})$$

8. Find the product. 98.001×0.01

Even though 98.001 ends in .001, it is not a power of 10. Look at 0.01 to see how to move the decimal. Multiplying by 0.01 will make the product smaller than 98.001.

Move the decimal point two places to the left.

$$98.001 \times 0.\underbrace{01}_{\text{2 decimal places}} = 0.98001 \quad (\leftarrow \text{2 places})$$

9. Find the product. 15.73×100

Multiplying by 100 will make the product larger than 15.73.

Move the decimal point two places to the right.

$$15.73 \times 1\underbrace{00}_{\text{2 zeros}} = 1573 \quad (\rightarrow \text{2 places})$$

10. Find the product. $0.00032 \times 10{,}000$

Multiplying by 10,000 will make the product larger than 0.00032.

Move the decimal point four places to the right.

$$0.00032 \times 1\underbrace{0{,}000}_{\text{4 zeros}} = 3.2 \quad (\rightarrow \text{4 places})$$

GUIDED PRACTICE

7. Find the product. 501×0.01

Multiplying by ____ will make the product smaller/larger than 501.

Move the decimal point ____ places to the ____.

$501 \times 0.01 =$ ________

8. Find the product. 3.01×0.001

Even though 3.01 ends in .01, it is ____ a power of 10. Look at 0.001 to see how to move the decimal. Multiplying by ____ will make the product smaller/larger than 3.01.

Move the decimal point ____ places to the ____.

$3.01 \times 0.001 =$ ________

9. Find the product. $23.2 \times 100{,}000$

Multiplying by ________ will make the product smaller/larger than 23.2.

Move the decimal point ____ places to the ____.

$23.2 \times 100{,}000 =$ ________

10. Find the product. $3.76 \times 1{,}000{,}000$

Multiplying by __________ will make the product smaller/larger than 3.76.

Move the decimal point ____ places to the ____.

$3.76 \times 1{,}000{,}000 =$ ________

Concept Check

C1. **a.** Write any decimal number.
b. Rewrite the number, moving the decimal point one place to the left.
c. If a decimal point is moved to the left in a number, will the result be larger or smaller than the original number? Use the numbers from part a and b in your explanation.

C2. Moving the decimal point three places to the right in 45.9 is the same as multiplying 45.9 by ________.

Objective C Practice

Find each product.

C3. 13.6×100

C4. $3.71 \times 1{,}000$

C5. 0.001×10.001

C6. 0.1×101

C7. $0.00098 \times 1{,}000{,}000$

C8. $854{,}982 \times 0.00001$

C9. 987×0.0000001

C10. $87 \times 100{,}000$

Objective D Estimate Products

The Concept An estimate of a product can be found by rounding each number to its largest place value and multiplying.

DETAILED EXAMPLE Estimating a Product

Estimate. 352×0.0089

Step 1 Round each number to its largest place value. $352 \times 0.0089 \approx 400 \times 0.009$

Step 2 Multiply the rounded numbers. ≈ 3.6 **(Actual answer = 3.1328)**

We could have used a mental calculation

1. After rounding, multiply the nonzero digits.
   $4 \times 9 = 36$
2. The 4 came from the hundreds place.
   Move the decimal point two places to the right.
   $36 \rightarrow 3{,}600$
3. The 9 came from the thousandths place.
   Move the decimal point three places to the left.
   $3{,}600 \rightarrow 3.6$

Procedure Estimate Products

Step 1 Round each number to its largest place value.

Step 2 Multiply the rounded numbers.

EXAMPLES

11. Estimate the product. 0.0045×0.00312

Step 1 Round each number to its largest place value. $0.0045 \times 0.00312 \approx 0.005 \times 0.003$

Step 2 Multiply the rounded numbers. $\approx .000015$

Mental Calculation

1. After rounding, multiply the nonzero digits.
   $5 \cdot 3 = 15$
2. The 5 came from the thousandths place.
   Move the decimal point three places to the left.
   $15 \rightarrow 0.015$
3. The 3 came from the thousandths place.
   Move the decimal point three more places to the left.
   $0.015 \rightarrow 0.000015$

12. Estimate the product. 89.456×0.675

Step 1 Round each number to its largest place value. $89.456 \times .675 \approx 90 \times 0.7$

Step 2 Multiply the rounded numbers. ≈ 63

Mental Calculation

1. After rounding, multiply the nonzero digits.
   $9 \cdot 7 = 63$
2. The 9 came from the tens place.
   Move the decimal point one place to the right.
   $63 \rightarrow 630$
3. The 7 came from the tenths place.
   Move the decimal point one place to the left.
   $630 \rightarrow 63$

13. Estimate the product. $2{,}295 \times 0.0457$

Step 1 Round each number to its largest place value. $2{,}295 \times 0.0457 \approx 2{,}000 \times 0.05$

Step 2 Multiply the rounded numbers. ≈ 100

Mental Calculation

1. After rounding, multiply the nonzero digits.
   $2 \cdot 5 = 10$
2. The 2 came from the thousands place.
   Move the decimal point three places to the right.
   $10 \rightarrow 10{,}000$
3. The 5 came from the hundredths place.
   Move the decimal point two places to the left.
   $10{,}000 \rightarrow 100$

GUIDED PRACTICE

11. Estimate the product. 0.00373×0.065

$0.00373 \times 0.065 \approx \quad \times$

$\approx$

Mental Calculation

1. After rounding, multiply the nonzero digits.
   ____ • ____ = ____
2. The first digit came from the ____________ place.
   Move the decimal point ____ places to the ____.
   ____ → ________
3. The second digit came from the ____________ place.
   Move the decimal point ____ places to the ____.
   ______ → ________

12. Estimate the product. 8.953×0.076

$8.953 \times 0.076 \approx \quad \times$

$\approx$

Mental Calculation

1. After rounding, multiply the nonzero digits.
   ____ • ____ = __________
2. The first digit came from the ________ place.
   Move the decimal point ____ places to the ____.
   ____ → ____
3. The second digit came from the __________ place.
   Move the decimal point ____ places to the ____.
   ____ → ____

13. Estimate the product. $4{,}723 \times 0.0523$

$4{,}723 \times 0.0523 \approx \quad \times$

$\approx$

Mental Calculation

1. After rounding, multiply the nonzero digits.
   ____ • ____ = __________
2. The first digit came from the _________ place.
   Move the decimal point ____ places to the ____.
   ____ → ________
3. The second digit came from the __________ place.
   Move the decimal point ____ places to the ____.
   ______ → ____

Concept Check

D1. $4 \times 5 = 20$. If you were to multiply 0.004×0.5, would you move the decimal point in 20 to the left or right? How many places would you move it?

D2. $4 \times 5 = 20$. If you were to multiply 40×500, would you move the decimal point in 20 to the left or right? How many places would you move it?

D3. $4 \times 5 = 20$. If you were to multiply 40×0.5, would you move the decimal point in 20 to the left or right? How many places would you move it?

Objective D Practice

Estimate each product.

D4. 0.423×0.622

D5. 0.842×0.21

D6. 36.5×0.74

D7. 85.4×0.39

D8. $26{,}984 \times 0.0034$

D9. $54{,}987 \times 0.00074$

D10. 0.00034×0.56347

D11. 0.67843×0.00387

D12. As of January 2007, the most expensive four-year U.S. college was George Washington University. One college credit at George Washington University averages $1,260.67. Estimate the cost of 8 college credits at George Washington University.

D13. As of January 2007, the least expensive four-year U.S. college was Northern New Mexico College. One college credit at Northern New Mexico College averages $34.33. Estimate the cost of 8 college credits at Northern New Mexico College.

Objective E Use Pi to Find the Circumference and Area of a Circle

The Concept One of the most famous numbers in all of mathematics is pi, often written as π. The number π cannot be written exactly using decimals, so we use a symbol, π.

$$\pi = 3.141592653589\ldots \text{ and on and on and on}$$

$$\pi \approx 3.14$$

If we use the approximation $\pi \approx 3.14$, our calculations will be close but not exact. For many purposes, this is okay. If we need more accurate calculations, we can use a better approximation for pi, such as 3.14159.

The number π is used to calculate many things, including the area and circumference of a circle. To find area or circumference using π, first we need to understand the relationship between a circle's diameter and a circle's radius.

INTERACTIVE DEFINITION Diameter and Radius

- The **diameter** of a circle is the distance across a circle through its center.
- The **radius** of a circle is the distance from the circle's center to the edge of the circle.

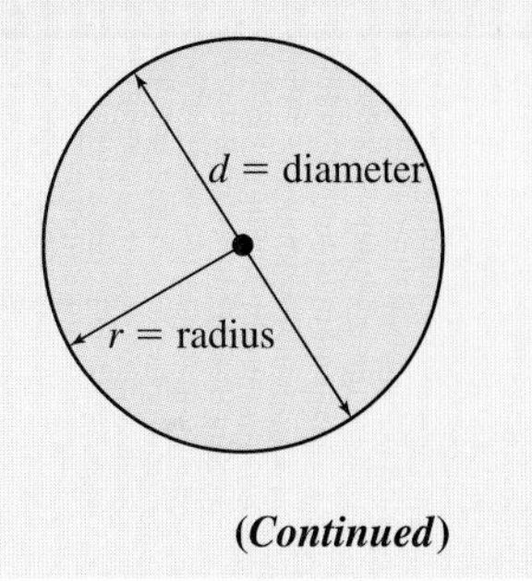

(Continued)

EXAMPLES	GUIDED PRACTICE
14. What is the radius of the circle?	**14.** What is the radius of the circle?
The radius is half the diameter. Radius = Diameter ÷ 2 = 3.2 ÷ 2 = 1.6 The radius is 1.6 m. 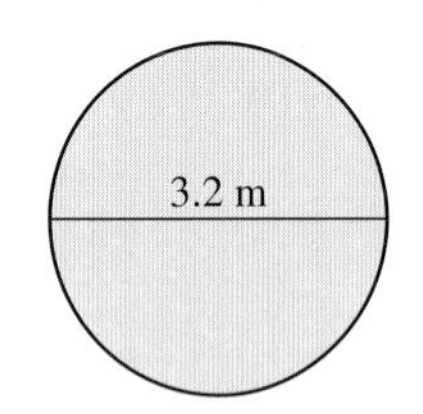	The radius is ______ the diameter. Radius = Diameter ÷ ____ = ÷ = The radius is ____ cm. 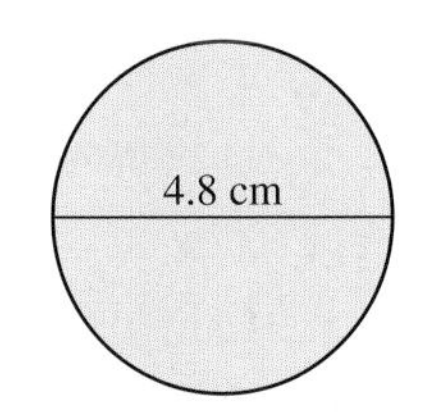
15. What is the diameter of the circle?	**15.** What is the diameter of the circle?
The diameter is twice the radius. Diameter = Radius × 2 = 1.25 × 2 = 2.5 The radius is 2.5 ft. 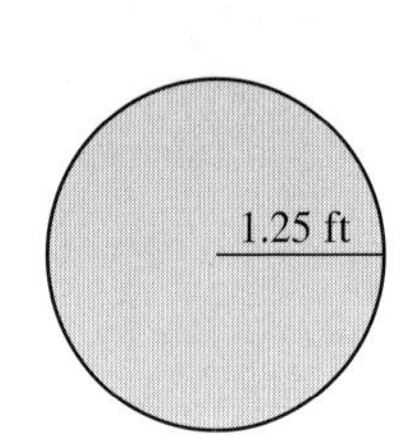	The diameter is ______ the radius. Diameter = Radius × ____ = × = The diameter is ____ in. 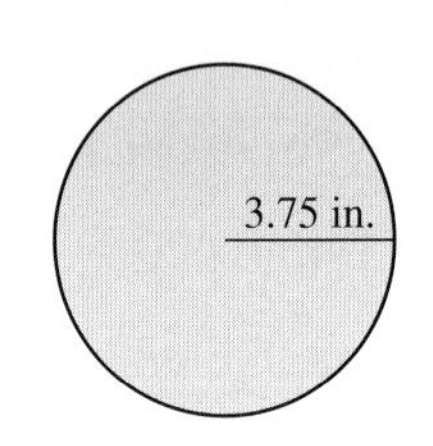

DO YOU UNDERSTAND how to find a circle's radius if you know its diameter?	Got It	Get Help
DO YOU UNDERSTAND how to find a circle's diameter if you know its radius?	Got It	Get Help

Circumference and Area of a Circle

Circumference of a Circle	Area of a Circle
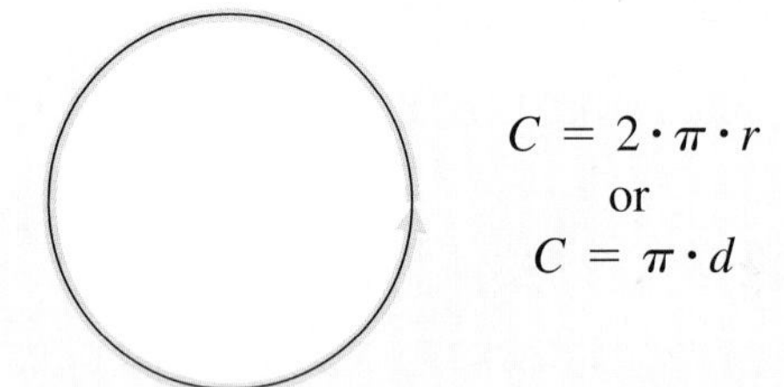 $C = 2 \cdot \pi \cdot r$ or $C = \pi \cdot d$	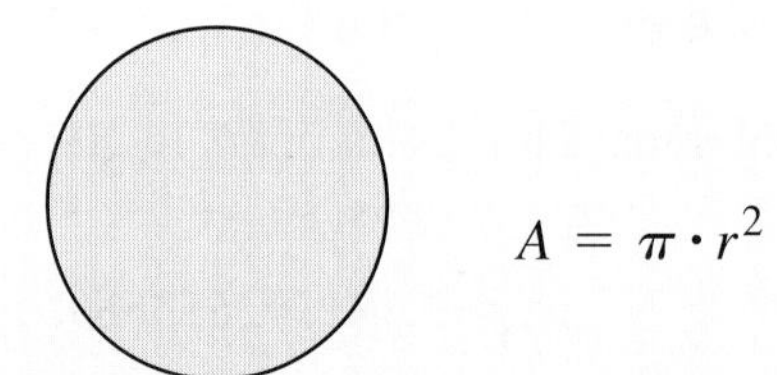 $A = \pi \cdot r^2$
The length of a string wrapped around a circle will give the circumference. The **circumference** of a circle is its perimeter.	The shaded region inside a circle is the circle's area. The area of a circle is equal to the number of "square units" that are needed to completely cover the circle.

Procedure Find the Circumference or Area of a Circle

Step 1 Write the appropriate formula.

Step 2 Substitute the values for each variable.

Step 3 Simplify.

EXAMPLES

16. What is the circumference of a circle with a radius of 3 inches?

Substitute
$\pi \approx 3.14$
$r = 3$ inches

$C = 2\pi r$
$C = 2(3.14)(3 \text{ inches})$
$C = 6.28(3 \text{ inches})$
$C = 18.84 \text{ inches}$

17. What is the area of a circle with a diameter of 10 inches?

Substitute
$\pi \approx 3.14$
$r = 5$ inches
Remember, the radius is half the diameter.

$A = \pi r^2$
$A = (3.14)(5 \text{ inches})^2$
$A = 3.14(25 \text{ square inches})$
$A = 78.5 \text{ square inches}$

GUIDED PRACTICE

16. What is the circumference of a circle with a radius of 20 meters?

$C = 2\pi r$
$C = 2(\quad)(\quad)$
$C =$
$C =$

Substitute
$\pi \approx 3.14$
$r =$ ____

17. What is the area of a circle with a diameter of 8 centimeters?

$A = \pi r^2$
$A = (\quad)(\quad)^2$
$A =$
$A =$

Substitute
$\pi \approx 3.14$
$r =$ ____
Remember, the radius is half the diameter.

Concept Check

E1. How are circumference and perimeter related?

E2. From the list below, identify which measurements could be used to describe the area of a circle.
a. 54.5 inches
b. 18.86 square centimeters
c. 78.5 ft^2
d. 34.7

E3. If you know the diameter, how can you find the radius?

Objective E Practice

Find the area or circumference as indicated. Use $\pi \approx 3.14$.

E4. What is the area of a circle with a radius of 3 kilometers?

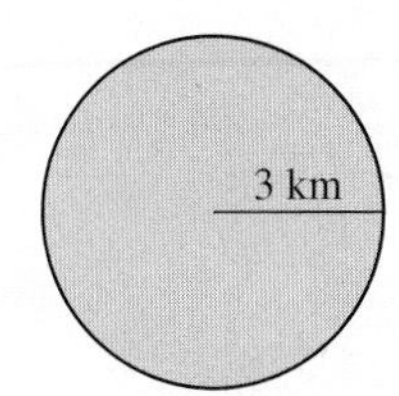

E5. What is the area of a circle with a radius of 2 miles?

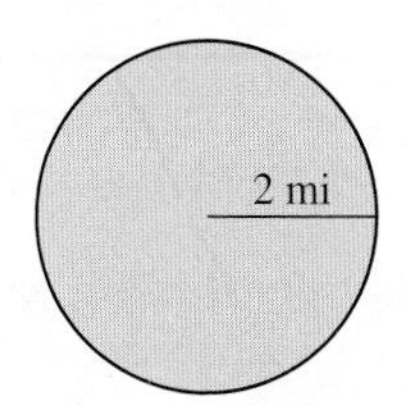

(Continued)

E6. What is the circumference of a circle with a diameter of 42 inches?

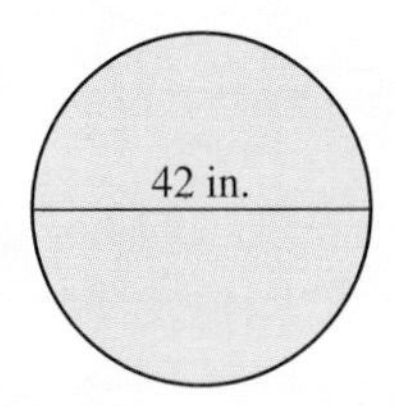

E7. What is the circumference of a circle with a diameter of 58 centimeters?

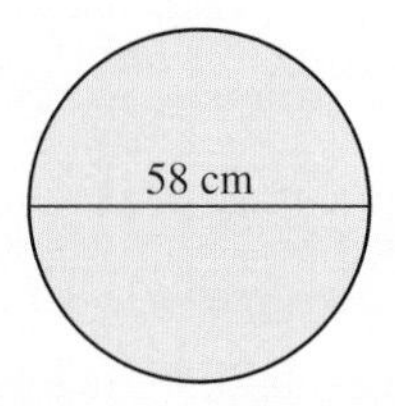

E8. What is the area of a circle with a diameter of 14 inches?

E9. What is the area of a circle with a diameter of 28 meters?

E10. What is the circumference of a circle with a radius of 3.8 millimeters?

E11. What is the circumference of a circle with a radius of 2.5 inches?

Section 3.3 Exercises

FOR EXTRA HELP

To multiply one-digit decimal numbers:

1. Answer the Objective A Concept Check.
2. Answer the odd Objective A Practice Exercises.
3. Answer the even Objective A Practice Exercises.

To multiply any two decimal numbers:

4. Answer the Objective B Concept Checks.
5. Answer the odd Objective B Practice Exercises.
6. Answer the even Objective B Practice Exercises.

To multiply by powers of 10:

7. Answer the Objective C Concept Checks.
8. Answer the odd Objective C Practice Exercises.
9. Answer the even Objective C Practice Exercises.

To estimate products:

10. Answer the Objective D Concept Checks.
11. Answer the odd Objective D Practice Exercises.
12. Answer the even Objective D Practice Exercises.

To use pi to find the circumference and area of a circle:

13. Answer the Objective E Concept Checks.
14. Answer the odd Objective E Practice Exercises.
15. Answer the even Objective E Practice Exercises.

VOCABULARY REVIEW *Review the Vocabulary Preview for Section 3.3. Study the definitions until you can check* Got It *for every word.*

Use these words to complete each sentence.

product • factors • diameter • radius • pi (π) • circumference • decimal places

(Continued)

	Got It	Get Help
16. The ____________ of a circle is the distance from the circle's center to the edge of the circle.		
17. ____________ is a number that relates the diameter of a circle to its circumference. It is approximately equal to 3.14.		
18. The distance across a circle, measured through the center, is called the ____________ of a circle.		
19. A ____________ is the result of multiplying factors.		
20. ____________ are the numbers that are multiplied to give a product.		
21. The ____________ of a circle is its perimeter.		

How will you get help for any vocabulary that you are unsure about?

Your instructor ___ MyMathLab ___ A classmate ___ A tutor ___ Other ___

Find each product.

22. 0.004×0.3

23. 0.005×0.005

24. 0.3×0.008

25. 0.9×0.0007

26. 1.12×0.8

27. 2.31×0.4

28. 2.3×3.4

29. 4.2×1.7

30. 350.1×3.1

31. 212.6×5.3

32. 0.045×0.234

33. 0.946×0.051

34. At the Pittsburgh baseball stadium, what is the price of three hot dogs?

35. At the Los Angeles baseball stadium, what is the price of two hot dogs?

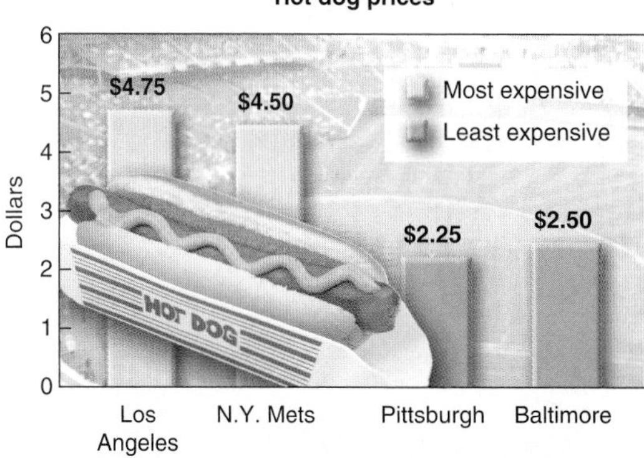

36. How much will it cost to park in downtown New York for 4.5 days?

37. How much will it cost to park in Honolulu for 3.5 days?

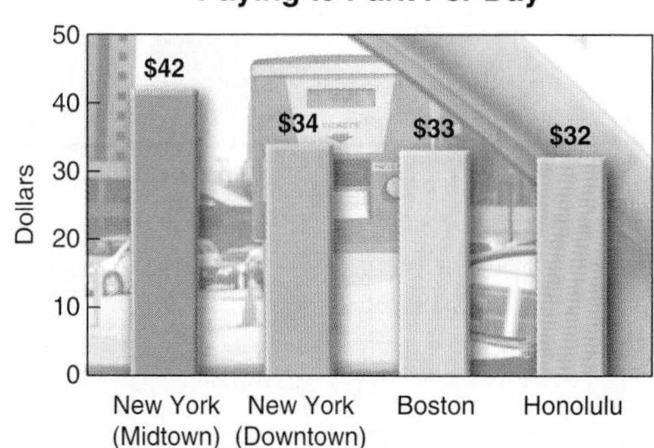

Multiply by the indicated power of 10.

38. 3.45 × 10 **39.** 43.2 × 0.1 **40.** 89.09 × 0.01 **41.** 12.65 × 100

42. 0.00078 × 1,000 **43.** 0.00078 × 0.001 **44.** 0.021 × 0.01 **45.** 0.021 × 100

For each exercise:

a. Round the factors to their largest place value.
b. Estimate each product by multiplying the rounded numbers from part a.
c. Calculate the actual product. Round the answer to the same place value as the estimate in part b.

46. 523 × 2.7 **47.** 125 × 3.2 **48.** 47,000 × 0.003 **49.** 43.2 × 0.006

50. 9.008 × 0.0052 **51.** 2.0008 × 0.0067 **52.** 213.9 × 0.00352 **53.** 879.9 × 0.0452

Given the measurement in each circle, find the indicated quantity. Round your answers to the tenths place.

54. Circumference

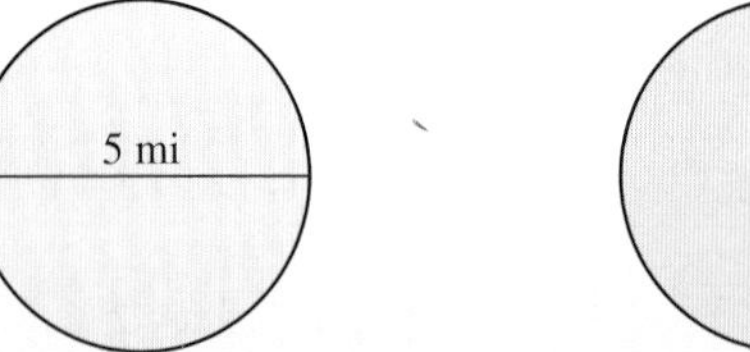

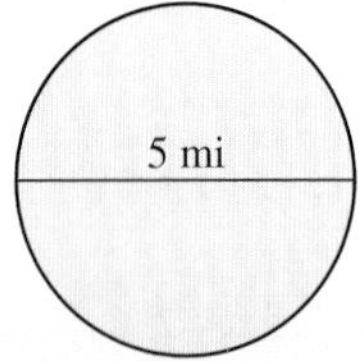

55. Circumference

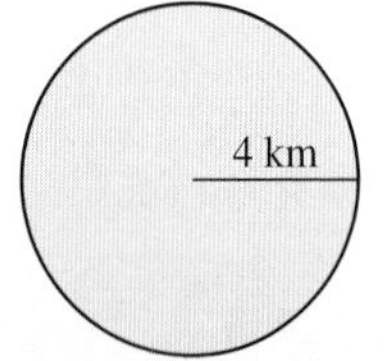

56. Area

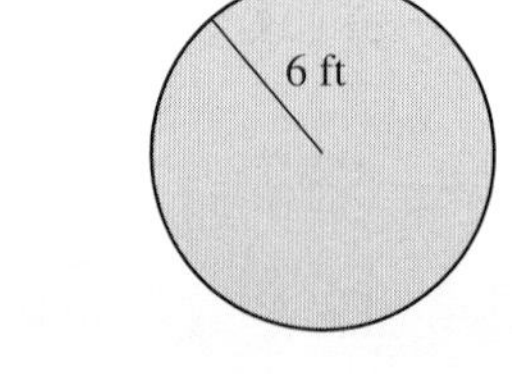

57. Area

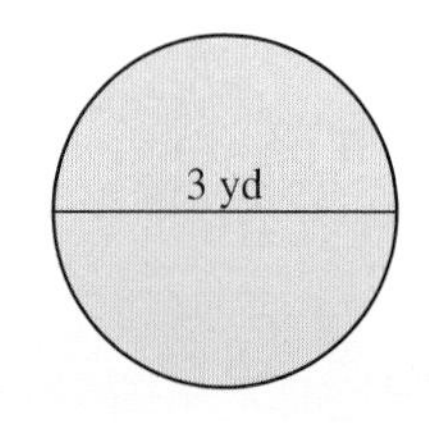

58. Calculate the circumference of the rim.

$d = 17.25$ in.

59. Calculate the circumference of the decorative plate.

$r = 4.5$ in.

60. Calculate the area of the irrigation circle.

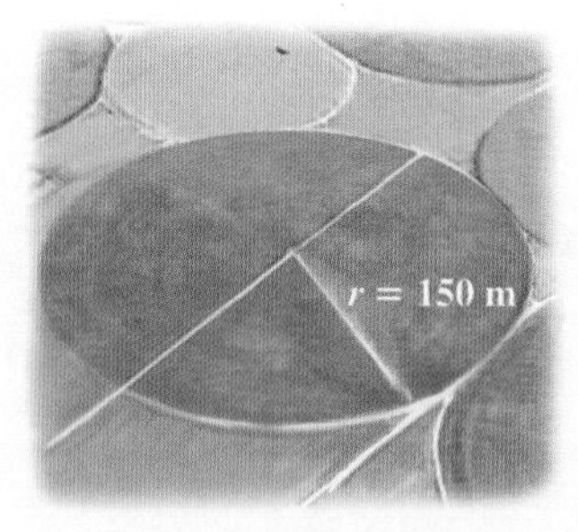

$r = 150$ meters

61. Calculate the area of the lid on a 55-gallon barrel.

$d = 60$ centimeters

Answer each question.

62. A company is going to ship 144 pulleys. Each pulley weighs 3.219 kilograms. What is the total weight?

63. You have just pumped 8.54 gallons of gas at a cost of \$3.549 per gallon. What is the bill rounded to the nearest cent?

64. If you make \$9.87 per hour and work for 33.2 hours, how much will you get paid in total?

65. On a gas bill, the charge is \$0.713 per thermal unit. If you used 71.1 thermal units last month, what was the total charge rounded to the nearest cent?

Answer each question.

66. A cell phone plan is offered at $19.99 per month, plus $0.05 per minute. If you talk for 850 minutes:

a. What is the per minute charge for talking 850 minutes?

b. How much is the entire bill for talking 850 minutes?

c. Would it be cheaper to get a plan with unlimited minutes for $59.99?

67. You rent a truck to move across town. The truck costs $29.99 plus $14.99 per hour. If you rent the truck for 3.5 hours:

a. What is the per hour charge for renting a truck for 3.5 hours?

b. How much is the entire bill for a 3.5 hour truck rental?

c. Would it be cheaper to rent a truck for a daily rate of $89.99?

tobey Which is cheaper? Cell phone plans charging by the minute or by the month? You can save a lot by solving these types of problems. *For more tips and tweets, go to twitter.com/gstbasicmath*

Section 3.3 Question Log

Use this space to write questions. Be sure to get these answered and revisit them when you prepare for your exam.

Page ____ Answered	Page ____ Answered
Page ____ Answered	Page ____ Answered

Section 3.4 Dividing Decimal Numbers

At some point in your life, you may have encountered division with decimals.

You may have used division with decimals to divide a restaurant check.

You may have used division to determine the time it takes to drive a certain distance.

As with other operations, dividing decimals is very similar to dividing whole numbers.

The **Objectives** in Section 3.4 will help you

- A Set up division exercises with decimals.
- B Divide by a one-digit divisor.
- C Write a fraction or mixed number as a decimal.
- D Estimate quotients.
- E Divide by a multidigit divisor.

VOCABULARY PREVIEW *Check the box that applies.*

	Got It	Must Study
Dividend: The number being divided is the **dividend**.		
Divisor: The number that does the dividing is the **divisor**.		
Quotient: The **quotient** is the number of times the divisor goes into the dividend.		
Division Notation: $6 \div 3 = 2$ ↕ dividend ÷ divisor = quotient; $3\overline{)6}$ with quotient 2 ↕ $\text{divisor}\overline{)\text{dividend}}$ with quotient on top; $\frac{6}{3} = 2$ ↕ $\frac{\text{dividend}}{\text{divisor}} = \text{quotient}$		
Annexing Zeros: Inserting zeros after the last digit in the dividend is called **annexing zeros.** Zeros can be annexed only to the right of the decimal point and only after the last digit on the right.		

Study these words when they appear in the text. After the section, test your understanding by completing the Vocabulary Review in the section exercises.

Objective A Set Up Division Exercises with Decimals

The Concept It is easier to divide by a whole number than by a decimal. Moving the decimal point in the divisor will allow you to divide by a whole number. To demonstrate why the decimal point can be moved, consider 12.34 ÷ 0.5 in its fraction form.

WHY IT WORKS Moving the Decimal Point to Divide by a Whole Number

Set up the division exercise. $12.34 \div 0.5$

Write the division exercise using a fraction.	$12.34 \div 0.5 = \dfrac{12.34}{0.5}$
Multiply by $\frac{10}{10}$ to move the decimal points.	$= \dfrac{12.34}{0.5} \cdot \dfrac{10}{10}$
Simplify. The result is a divisor that is a whole number.	$= \dfrac{123.4}{5}$
We can rewrite the expression.	$12.34 \div 0.5 = 123.4 \div 5$

Notice that the decimal point has moved one place to the right in both the numerator and the denominator.

When you divide with decimals, set up the problem carefully. To keep your work organized, use the following steps.

DETAILED EXAMPLE Setting up a Division Exercise with Decimals

Divide. $12.34 \div 0.5$

Step 1 Move both decimal points to divide by a whole number.

$0\,5.\overline{)1\,2\,3.\,4}$

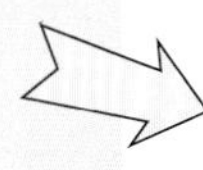

Step 2 Draw the lines to organize your work.

$05\overline{)1\,2\,3.\,4}$

Step 3 Write the decimal point in the quotient directly above the decimal point in the dividend.

$05\overline{)1\,2\,3.\,4}$

Procedure Set Up Division Exercises with Decimals

Step 1 Move the decimal points to divide by a whole number.

Step 2 "Draw the lines."

Organizing your work like this will help you to get the correct place values in your answers.

Step 3 Write the decimal point in the quotient directly above the decimal point in the dividend.

EXAMPLES

1. Set up the division exercise. $10.3 \div 0.2$

The divisor, 0.2, has one decimal place. Move both decimal points one place to the right.

Step 1 Move the decimal points.
Step 2 Draw the lines.
Step 3 Write the decimal point above.

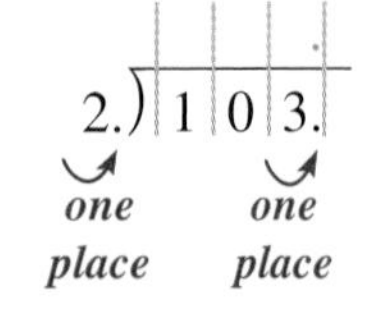

2. Set up the division exercise. $4.012 \div 1.03$

The divisor, 1.03, has two decimal places. Move both decimal points two places to the right.

Step 1 Move the decimal points.
Step 2 Draw the lines.
Step 3 Write the decimal point above.

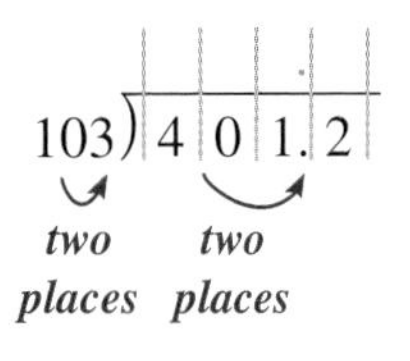

3. Set up the division exercise. $51 \div 0.4$

To move the decimal point to the right in 51, we must write a placeholder zero.

$$0.4\overline{)5\ 1} \rightarrow 0.4\overline{)5\ 1\ .\ 0} \rightarrow 4\overline{)5\ 1\ 0.}$$

Step 1 Move the decimal points.
Step 2 Draw the lines.
Step 3 Write the decimal point above.

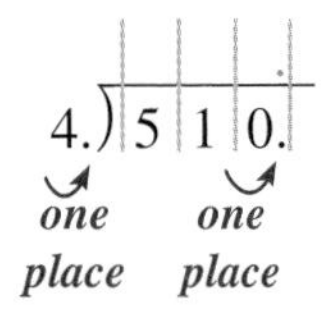

GUIDED PRACTICE

1. Set up the division exercise. $34.86 \div 0.91$

The divisor, ____, has ____ decimal place(s). Move both decimal points ____ place(s) to the right.

Step 1 Move the decimal points.
Step 2 Draw the lines.
Step 3 Write the decimal point above.

2. Set up the division exercise. $3.21 \div 1.2$

The divisor, ____, has ____ decimal place(s). Move both decimal points ____ place(s) to the right.

Step 1 Move the decimal points.
Step 2 Draw the lines.
Step 3 Write the decimal point above.

3. Set up the division exercise. $12.3 \div 0.012$

To move the decimal points ____ places to the right in 12.3, we must write two ____.

Step 1 Move the decimal points.
Step 2 Draw the lines.
Step 3 Write the decimal point above.

Concept Check

A1. For the exercise $1.2 \div 0.06$:
  a. Which number is the dividend?
  b. Which number is the divisor?

A2. If you move a decimal three places in the divisor, how many places should you move the decimal point in the dividend?

Objective A Practice

Set up but don't solve each division exercise. Be sure to draw the lines and write the decimal point in the quotient.

A3. $12.5 \div 0.4$ **A4.** $13.2 \div 0.5$ **A5.** $71.45 \div 2$ **A6.** $10.97 \div 3$

A7. $3.1 \div 0.002$ **A8.** $4.3 \div 0.002$ **A9.** $3.56 \div 1.332$ **A10.** $4.26 \div 2.672$

Objective B Divide by a One-Digit Divisor

The Concept To divide by a one-digit divisor, set up the problem and then divide. The procedure is similar to dividing whole numbers except that there is a decimal point in the quotient.

Procedure Divide by a One-Digit Divisor

Step 1 Set up the problem.

Step 2 Perform long division.

Stop when every digit of the dividend has been divided evenly.

or

Stop when you have enough digits to round to the desired place value.

Step 3 Round the quotient if necessary.

EXAMPLES

4. Find the quotient. $21 \div 0.5$

Step 1 Move both decimal points. Include a placeholder zero, if necessary.

$0.5\overline{)2\,1.} \rightarrow 5\overline{)2\,1\,0.}$

Step 2 Perform long division.

$$\begin{array}{r} 0\,4\,2. \\ 5\overline{)\,2\,1\,0.} \\ -2\,0 \\ 1\,0 \\ -1\,0 \\ 0 \end{array}$$

Step 3 Since the quotient is exact, we don't need to round.

GUIDED PRACTICE

4. Find the quotient. $22 \div 0.4$

$\overline{)}$

When dividing with decimal numbers, do not use remainders. Instead, continue dividing until you reach an exact answer or can round the quotient to a given place value. To round to a specific place value, carry out the division to the next smaller place value. For example, to round an answer to the hundredths place, carry out the division to the thousandths place.

DETAILED EXAMPLE Rounding a Quotient to a Specific Place Value

Find the quotient $1.161 \div 8$, rounding the answer to the hundredths place.

Step 1 Set up the exercise.

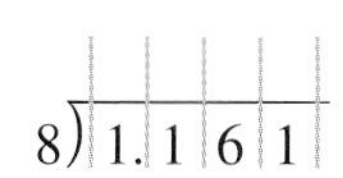

$8\overline{)1.1\,6\,1}$

Step 2 Carry out the division to the thousandths place.

$$\begin{array}{r} 0.1\,4\,5 \\ 8\overline{)1.1\,6\,1} \end{array}$$

To round the hundredths digit, we must identify the thousandths digit of the quotient.

Step 3 Round to the hundredths place.

Since 0.145 rounds to 0.15,

$1.161 \div 8 \approx 0.15.$

It is often necessary to insert zeros when carrying out division to a desired place value. We call this process "annexing zeros."

INTERACTIVE DEFINITION Annexing Zeros

Inserting zeros after the last digit in the dividend is called **annexing zeros**. Zeros can be annexed only to the right of the decimal point and only after the last digit on the right.

EXAMPLE

5. Divide $6\overline{)23}$. Annex enough zeros to round the answer to the hundredths place.

To round to the hundredths place, continue dividing until the thousandths place.

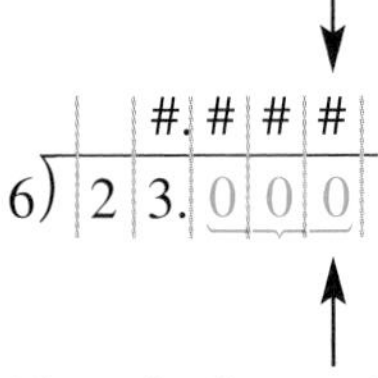

To divide to the thousandths place, annex three zeros.

GUIDED PRACTICE

5. Divide $3\overline{)14}$. Annex enough zeros to round the answer to the tenths place.

To round to the tenths place, continue dividing until the ______________ place.

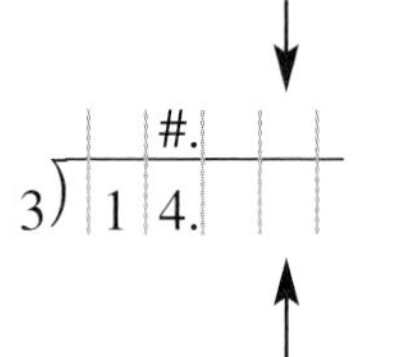

To divide to the ______________ place, annex ______________ zeros.

DO YOU UNDERSTAND how to annex zeros to round to a desired place value? Got It Get Help

DETAILED EXAMPLE Dividing When Annexing Zeros and Rounding

Find the quotient of $2 \div 0.7$. Round the answer to the tenths place.

Step 1 Set up the problem. Move both decimal points. Annex two zeros.

$7\overline{)20.}$ ⟶ $7\overline{)20.\,00}$ (quotient #. # #)

We moved both decimal points one place.

Annex two zeros to carry out the division to the hundredths place.

Step 2 Perform the division.

$$\begin{array}{r} 2.85 \\ 7\overline{)20.00} \\ -14. \\ 60 \\ -56 \\ 40 \end{array}$$

The hundredths digit is needed to round to the tenths place.

Step 3 Round to the correct place value. $2.85 \approx 2.9$

State your answer. $20 \div 7 \approx 2.9$

EXAMPLES

6. Find the quotient, accurate to the tenths place.
0.003 ÷ 0.02

Step 1 Move both decimal points. Annex enough zeros to divide to the hundredths place.

```
    #.##
2)  0.30
```

Step 2 Perform the division.

```
    0.15
2)  0.30
   -2
     10
    -10
```

Step 3 Round to the tenths place.

$0.003 \div 0.02 \approx 0.2$

7. Find the quotient, accurate to the hundredths place.
3 ÷ 0.7

Step 1 Move both decimal points. Annex enough zeros to divide to the thousandths place.

```
    ##.###
7)  30.000
```

Step 2 Perform the division.

```
    04.285
7)  30.000
   -28
     20
    -14
      60
     -56
       40
```

Step 3 Round to the hundredths place.

$3 \div 0.7 \approx 4.29$

GUIDED PRACTICE

6. Find the quotient, accurate to the tenths place.
0.006 ÷ 0.05

7. Find the quotient, accurate to the hundredths place.
7 ÷ 0.6

Concept Check

B1. What place value of the quotient must you find to round to the tenths place?

B2. a. In your own words, what does it mean to annex a zero?
b. Why is it that zeros can be annexed only to the right of the last digit to the right of the decimal point?

Objective B Practice

Find each quotient. Round each answer to the thousandths place when necessary.

B3. $2.3 \div 0.04$

B4. $5.3 \div 0.008$

B5. $0.00045 \div 0.003$

B6. $0.00073 \div 0.06$

B7. $12 \div 0.02$

B8. $14 \div 0.005$

B9. $17.683 \div 0.9$

B10. $42.876 \div 0.08$

B11. The top shelf of a display cabinet is 33.75 inches wide. If the cases that hold DVDs are 0.4 inch wide, how many DVD cases can fit on the top shelf?

B12. If six people split a $43.20 bill for a picnic, how much did each person pay?

Objective C Write a Fraction or Mixed Number as a Decimal

The following figure gives a visual representation of common decimals and their fraction equivalents.

Decimals and fractions are both used to describe portions of a whole. Because they are used to represent the same thing, decimals and fractions can be interchanged. To write a fraction or mixed number as a decimal, use the quotient concept of a fraction.

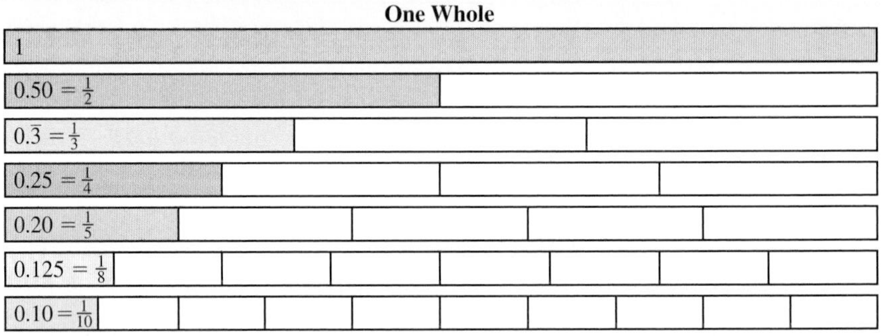

One Whole

1

$0.50 = \frac{1}{2}$

$0.\overline{3} = \frac{1}{3}$

$0.25 = \frac{1}{4}$

$0.20 = \frac{1}{5}$

$0.125 = \frac{1}{8}$

$0.10 = \frac{1}{10}$

Figure 3.1

Procedure Write a Fraction or Mixed Number as a Decimal

Divide the numerator by the denominator.

In a mixed number, the whole number remains the same.

EXAMPLES

8. Write $\frac{3}{8}$ as a decimal.

Divide the numerator by the denominator. $\frac{3}{8} \Rightarrow 8\overline{)3}$

$$
\begin{array}{r}
0.375 \\
8\overline{)3.000} \\
-2\,4 \\
\hline
60 \\
-56 \\
\hline
40 \\
-40 \\
\hline
0
\end{array}
$$

$\frac{3}{8} = 0.375$

GUIDED PRACTICE

8. Write $\frac{3}{4}$ as a decimal.

$\frac{3}{4} \Rightarrow \overline{)}$

$\frac{3}{4} =$

9. Write $\frac{5}{6}$ as a decimal.

Divide the numerator by the denominator. $\frac{5}{6} \Rightarrow 6\overline{)5}$

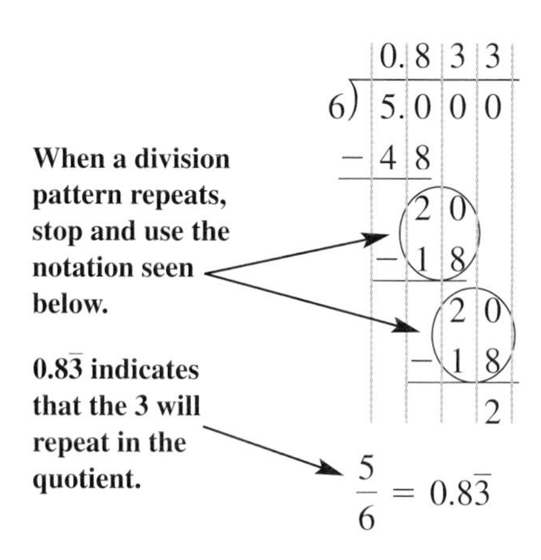

$\frac{5}{6} = 0.8\overline{3}$

9. Write $\frac{1}{6}$ as a decimal.

$\frac{1}{6} \Rightarrow \overline{)}$

$\frac{1}{6} =$

10. When $\frac{1}{7}$ as a decimal. Round to the nearest hundredth.

Divide the numerator by the denominator. $\frac{1}{7} \Rightarrow 7\overline{)1}$

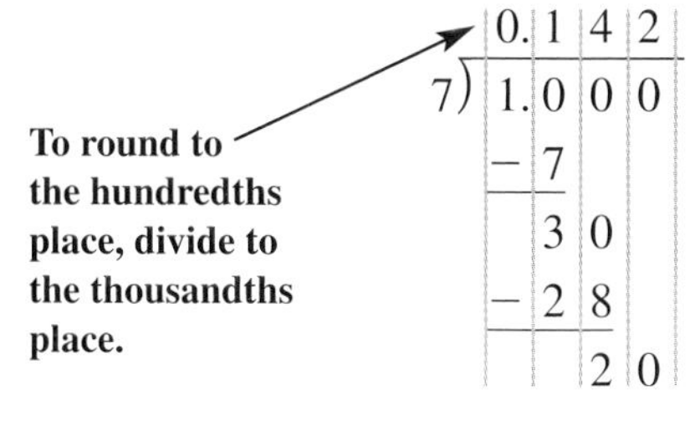

$\frac{1}{7} \approx 0.14$

10. Write $\frac{3}{7}$ as a decimal. Round to the nearest hundredth.

$\frac{3}{7} \Rightarrow \overline{)}$

$\frac{3}{7} \approx$

11. Write $3\frac{1}{4}$ as a decimal.

Divide the numerator by the denominator. $\frac{1}{4} \Rightarrow 4\overline{)1}$

$$\begin{array}{r} 0.25 \\ 4\overline{)1.00} \\ -8 \\ \hline 20 \\ -20 \\ \hline 0 \end{array}$$

$3\frac{1}{4} = 3.25$

The whole number remains the same in a mixed number or a mixed decimal.

11. Write $4\frac{1}{5}$ as a decimal.

$\frac{1}{5} \Rightarrow \overline{)}$

$4\frac{1}{5} =$

Concept Check

C1. Use Figure 3.1 at the beginning of this objective to identify the decimal equivalents of these fractions.

a. $\frac{1}{2} = $ ______ **b.** $\frac{1}{4} = $ ______ **c.** $\frac{1}{5} = $ ______ **d.** $\frac{1}{10} = $ ______

C2. Identify the pattern to convert from mixed numbers to mixed decimals.

a. $1\frac{1}{4} = 1.25$ **b.** $2\frac{1}{4} = 2.25$ **c.** $3\frac{1}{4} = $ ______ **d.** ___$\frac{1}{4} = 4.25$

C3. When converting a mixed number to a decimal, what happens to the whole number portion?

Objective C Practice

Write each fraction or mixed number as a decimal.

C4. $\frac{2}{5}$ **C5.** $\frac{7}{10}$ **C6.** $\frac{5}{8}$ **C7.** $\frac{1}{8}$

C8. $\frac{1}{3}$ **C9.** $\frac{5}{6}$ **C10.** $7\frac{3}{5}$ **C11.** $8\frac{3}{10}$

Write each fraction or mixed number as a decimal. Round each answer to the hundredths place.

C12. $\frac{3}{8}$ **C13.** $\frac{2}{3}$ **C14.** $1\frac{5}{12}$ **C15.** $3\frac{7}{8}$

Objective D Estimate Quotients

The Concept To estimate quotients with decimals, use the same method you used to estimate quotients with whole numbers.
To estimate a quotient:

- Round each number to its largest place value.
- Divide.

DETAILED EXAMPLE Estimating a Quotient

Estimate the quotient. $0.0276 \div 0.215$

Step 1 Round each number to its largest place value.

$0.0276 \approx 0.030$
$0.215 \approx 0.2$

Step 2 Set up the division.

$2\overline{)0.30}$

Perform the division.

$\begin{array}{r} 0.15 \\ 2\overline{)0.30} \end{array}$

Perform the division mentally. 2 into 30 is 15. Align the last digits in 15 and 30 to identify the place values in the quotient.

$0.0276 \div 0.215 \approx 0.15$

Estimates are useful because

- Exact answers are not always required; you can make an estimate quickly and move on.
- Estimating will help you catch arithmetic mistakes.

Procedure Estimate Quotients

Step 1 Round each number to its highest place value.

Step 2 Divide the resulting numbers.

You do not need the estimate to be exact; just find the first one or two nonzero digits.

EXAMPLES

12. Estimate the quotient. $6.2 \div 0.0512$

Step 1 Round each number. $6.2 \approx 6.0$, $0.0512 \approx 0.05$

Step 2 Divide. $5\overline{)600}$ = 120

5 goes into 60 twelve times.

$6.2 \div 0.0512 \approx 120$

(The actual answer is 121.09.)

13. Estimate the quotient. $0.0278 \div 34.97$

Step 1 Round each number. $0.0278 \approx 0.03$, $34.97 \approx 30$

Step 2 Divide. $30\overline{)0.030}$ = 0.001

30 goes into 30 one time.

$0.0278 \div 34.97 \approx 0.001$

(The actual answer is 0.00079.)

14. Estimate the quotient. $0.000794 \div 0.09127$

Step 1 Round each number. $0.000794 \approx 0.0008$, $0.09127 \approx 0.09$

Step 2 Move the decimals and divide. $9\overline{)0.080}$ = 0.009

$9 \cdot 9 = 81$, which is close to 80. For an estimate, this is close enough.

$0.000794 \div 0.09127 \approx 0.009$

(The actual answer is 0.0087.)

GUIDED PRACTICE

12. Estimate the quotient. $9.32 \div 0.0425$

$9.32 \approx$ _____

$0.0425 \approx$ _____

$9.32 \div 0.0425 \approx$

(The actual answer is 219.29.)

13. Estimate the quotient. $0.0784 \div 8.726$

$0.0784 \approx$ _____

$8.726 \approx$ _____

$0.0784 \div 8.726 \approx$

(The actual answer is 0.00898.)

14. Estimate the quotient. $0.004643 \div 0.0835$

$0.004643 \approx$ _____

$0.0835 \approx$ _____

$0.004643 \div 0.0835 \approx$

(The actual answer is 0.056.)

Concept Check

D1. In your opinion, which estimate from Guided Practices 12 through 14 is the best approximation of the actual answer? Why?

D2. In your opinion, which estimate from Guided Practices 12 through 14 is the worst approximation of the actual answer? Why?

D3. Using estimation, why must answer (c) be the correct answer to the following multiple choice question? Do not perform the division.

Find the quotient. $0.00312 \div 0.342$

a. 0.912 **b.** 0.0912 **c.** 0.00912 **d.** 0.000912

Objective D Practice

Estimate each quotient.

D4. $87.56 \div 6.32$

D5. $45.765 \div 2.35$

D6. $0.00657 \div 23.87$

D7. $0.0252 \div 11.96$

D8. $0.04845 \div 0.0185$

D9. $0.00857 \div 0.0642$

D10. $739.28 \div 43.6$

D11. $867.623 \div 23.7$

D12. Your speedometer reads 72.3 miles per hour. Your GPS system shows that you are 221.7 miles from your destination. Use estimation to approximate the time it will take to complete your trip to the nearest hour.

D13. You are earning $11.67 per hour at your job. Your car payment for the month is $226.32. Approximately how many hours per month must you work just to pay for your car?

Objective E Divide by a Multidigit Divisor

The Concept Dividing by a multidigit divisor uses the same procedure as dividing by a one-digit divisor. Whenever you are asked to find a quotient that is accurate to a given place value, you must calculate the quotient to the next smallest place value.

Procedure Divide by a Multidigit Divisor

Step 1 Set up the problem and estimate the answer.

Step 2 Perform long division.

Stop when every digit of the dividend has been divided evenly.

or

Stop when you have determined enough digits to round.

Step 3 Round the quotient if necessary.

Step 4 Check your answer.

You may find it useful to use scratch paper to perform calculations and estimations in an exercise.

DETAILED EXAMPLE Dividing by a Multidigit Divisor

Find the quotient of 369.5 ÷ 4.3. Round the answer to the tens place.

Step 1 Set up the problem and estimate the answer.

$43 \approx 40$

$3{,}695 \approx 4{,}000$

$4\,3.\overline{)3\,6\,9\,5.} \longrightarrow 40\overline{)4000}$ (estimate: 100)

Step 2 Perform long division.
To round to the tens place, continue dividing to the ones place.

		8	5
43.) 3	6	9	5.
−3	4	4	
	2	5	5
−	2	1	5

Stop here! Since we know the digit in the ones place, we do not need to continue.

Step 3 Round the quotient.
The digit to the right is 5, so the 8 rounds up.

$85 \approx 90$

Step 4 Check your answer.
The solution is close to the estimate, so we have confidence in our answer.

$369.5 \div 4.3 \approx 90$

▶ **NOTE** To review division with a multidigit divisor, see Section 1.5.

EXAMPLES

15. Find the quotient. $6.3 \div 0.12$

Set up and estimate. $12\overline{)630} \longrightarrow 10\overline{)600}$ (60)

Estimate. $6.3 \div 0.12 \approx 60$

Perform the division.
Annex one zero to complete the division.

	0	5	2.	5
12)	6	3	0.	0
−	6	0		
		3	0	
	−	2	4	
			6	0
		−	6	0
				0

Check your answer. $6.3 \div 0.12 = 52.5$

The quotient is close to our estimate, so we trust our work.

GUIDED PRACTICE

15. Find the quotient. $2.8 \div 0.16$

Set up and estimate. $16\overline{)280} \longrightarrow$) ______

Estimate. $2.8 \div 0.16 \approx$

) ______

$2.8 \div 0.16 =$

Is your answer close to your estimate?

16. Find the quotient, accurate to the tenths place.
$6.235 \div 0.24$

Set up and estimate. $24\overline{)623.5} \rightarrow 20\overline{)600}$ (quotient 30)

Estimate. $6.235 \div 0.24 \approx 30$

Perform the division.
We divide to the hundredths place to round to the tenths place.

$$\begin{array}{r} 025.97 \\ 24\overline{)623.50} \\ -48 \\ 143 \\ -120 \\ 235 \\ -216 \\ 190 \\ -168 \end{array}$$

Stop here! Since we know the digit in the hundredths place, we do not need to continue.

Round to the tenths place. $25.97 \approx 26.0$

Check your answer. $6.235 \div 0.24 \approx 26.0$

The quotient is close to our estimate, so we trust our work.

16. Find the quotient, accurate to the tenths place.
$8.634 \div 0.52$

Set up and estimate. $52\overline{)863.4} \rightarrow \overline{)}$

Estimate. $8.634 \div 0.52 \approx$

$\overline{)}$

$8.634 \div 0.52 \approx$

Is your answer close to your estimate?

17. Find the quotient, accurate to the thousandths place.
$1.524 \div 61.2$

Set up and estimate. $612\overline{)15.24} \rightarrow 600\overline{)15.000}$ (quotient .025)

Estimate. $1.524 \div 61.2 \approx 0.025$

Perform the division.

$$\begin{array}{r} 00.0249 \\ 612\overline{)15.2400} \\ -1224 \\ 3000 \\ -2448 \\ 5520 \end{array}$$

Stop here! We can now round to the thousandths place.

Round to the thousandths place. $0.0249 \approx 0.025$

Check your answer. $1.524 \div 61.2 \approx 0.025$

The quotient is close to our estimate, so we trust our work.

17. Find the quotient, accurate to the thousandths place. $3.51 \div 77$

Set up and estimate. $77\overline{)3.51} \rightarrow \overline{)}$

Estimate. $3.51 \div 77 \approx$

$\overline{)}$

$3.51 \div 77 \approx$

Is your answer close to your estimate?

Concept Check

E1. To round an answer to the hundredths place, continue dividing to the ________ place.

Set up and estimate each quotient. Do not solve.

E2. $3.8\overline{)42.0}$ **E3.** $.77\overline{)3.70}$ **E4.** $57\overline{)6.3}$ **E5.** $28\overline{)2.5}$

Objective E Practice

Find each quotient. Round your answer to the hundredths place when necessary.

E6. $0.045 \div 1.4$ **E7.** $0.032 \div 0.63$ **E8.** $5.47 \div 0.34$ **E9.** $9.85 \div 0.72$

Find each quotient. Round your answer to the tenths place when necessary.

E10. $7.15 \div 0.055$ **E11.** $6.93 \div 0.63$ **E12.** $0.4858 \div 0.042$ **E13.** $7.026 \div 0.45$

E14. Your speedometer reads 64 miles per hour. Your GPS system shows that you are 431.7 miles from the Grand Canyon. At this speed, how long will it take you to reach the Grand Canyon? Round your answer to the nearest tenth of an hour.

E15. You are saving for the down payment on your first house. If you save about $320 a month, how long will it take you to save $15,000? Round your answer to an appropriate place value.

Section 3.4 Exercises

FOR EXTRA HELP

REVIEW

To set up division exercises with decimals:

1. Answer the Objective A Concept Checks.
2. Answer the odd Objective A Practice Exercises.
3. Answer the even Objective A Practice Exercises.

To divide by a one-digit divisor:

4. Answer the Objective B Concept Checks.
5. Answer the odd Objective B Practice Exercises.
6. Answer the even Objective B Practice Exercises.

To write a fraction or mixed number as a decimal:

7. Answer the Objective C Concept Checks.
8. Answer the odd Objective C Practice Exercises.
9. Answer the even Objective C Practice Exercises.

To estimate quotients:

10. Answer the Objective D Concept Checks.
11. Answer the odd Objective D Practice Exercises.
12. Answer the even Objective D Practice Exercises.

To divide by a multidigit divisor:

13. Answer the Objective E Concept Checks.
14. Answer the odd Objective E Practice Exercises.
15. Answer the even Objective E Practice Exercises.

VOCABULARY REVIEW *Review the Vocabulary Preview for Section 3.4. Study the definitions until you can check* Got It *for every word.*

Use these words to complete each sentence.

dividend • divisor • quotient • annexing zeros

	Got It	Get Help
16. The ____________ is the number of times the divisor goes into the dividend.		
17. The number that does the dividing is the ____________.		
18. The number being divided is the ____________.		
19. Division Notation: $6 \div 3 = 2$: 6 is the ____________. 3 is the ____________. 2 is the ____________. $5\overline{)35}$ with quotient 7: 5 is the ____________. 35 is the ____________. 7 is the ____________. $\frac{16}{8} = 2$: 16 is the ____________. 8 is the ____________. 2 is the ____________.		
20. Writing zeros to the right of the last nonzero decimal place is called ____________.		

How will you get help for any vocabulary that you are unsure about?

Your instructor ___ MyMathLab ___ A classmate ___ A tutor ___ Other ___

Set up each division exercise. Do not perform the division.

21. $3.45 \div 2.1$ **22.** $8.34 \div 4.8$ **23.** $0.0034 \div 0.012$ **24.** $0.0043 \div 0.023$

Divide. Round each answer to the thousandths place when necessary.

25. $2.34 \div 6$ **26.** $3.12 \div 7$ **27.** $2.19 \div 0.2$ **28.** $5.98 \div 0.3$

29. $0.00763 \div 0.03$ **30.** $0.00653 \div 0.09$ **31.** $5 \div 0.06$ **32.** $7 \div 0.009$

33. $20.005 \div 0.004$ **34.** $32.009 \div 0.003$ **35.** $13.2 \div 0.06$ **36.** $12.8 \div 0.008$

37. Five friends agree to split a $112.23 restaurant bill evenly. How much does each person pay? Round to the nearest cent.

38. Ward was riding his motorcycle and recording his mileage. If he traveled 341 miles using 7 gallons of gas, how many miles per gallon did his motorcycle get? Round to the nearest tenth.

Write each fraction or mixed number as a decimal. Round to the hundredths place when necessary.

39. $\frac{6}{5}$ **40.** $\frac{13}{10}$ **41.** $\frac{5}{9}$ **42.** $\frac{4}{9}$

43. $4\frac{1}{4}$ **44.** $2\frac{1}{5}$ **45.** $\frac{7}{16}$ **46.** $\frac{9}{16}$

47. $1\frac{4}{25}$ **48.** $12\frac{1}{25}$ **49.** $3\frac{2}{7}$ **50.** $4\frac{5}{7}$

goetz Stay organized for accuracy. Draw the lines and make sure every box on the quotient line gets a digit, even if it's a zero.
For more tips and tweets, go to twitter.com/gstbasicmath

Divide. Round each answer to the hundredths place when necessary.

51. 4.16 ÷ 1.8
52. 3.87 ÷ 1.2
53. 84 ÷ 1.7
54. 32 ÷ 3.8

55. 32.973 ÷ 0.8
56. 43.821 ÷ 0.6
57. 0.0034 ÷ 0.027
58. 0.0071 ÷ 0.048

59. 0.2168 ÷ 0.0024
60. 0.981 ÷ 0.0046
61. 25.87 ÷ 8.3
62. 48.97 ÷ 9.7

63. To pay for a new transmission, Luigi has set up a payment plan with his mechanic. If Luigi pays his mechanic $3,855 in 12 monthly installments, how much will each payment be? Round your answer to the nearest penny.

64. Deborah took a 325-mile trip in her SUV. If she used 42 gallons of gas, how many miles did her car travel on one gallon of gas? Round your answer to the nearest tenth of a mile per gallon.

Divide. Round each answer to the tenths place when necessary.

65. 39.6 ÷ 3.1
66. 94.9 ÷ 1.7
67. 12.8 ÷ 0.63
68. 78.2 ÷ 0.37

69. 315 ÷ 0.24
70. 268 ÷ 0.12
71. 756 ÷ 0.89
72. 193 ÷ 0.52

Divide. Round each answer to the hundredths place when necessary.

73. 33.68 ÷ 0.019
74. 83.54 ÷ 0.041
75. 0.0162 ÷ 0.089
76. 0.0452 ÷ 0.052

Answer each question.

77. A package is labeled "Contents: 36 iPods." A single iPod in its box weighs 1.3 pounds. If the contents of the package weigh 45.5 pounds, how great an error was made in filling the package?

78. A chef has ordered a 50-pound bag of baby carrots for her signature buttered-honey carrots. If each serving of carrots uses 0.18 pound of carrots, how many servings can she make?

79. If a machinist can make a part using 1.4 inches of material, how many complete parts can be made from an 8-foot length of material? Be sure to convert to inches. Round your answer to the largest number of complete parts that can be made.

80. A third-grade class is raising money to take a field trip. If they earn $1.50 for each candy bar they sell, how many candy bars must they sell to raise $550? Round your answer so that the students raise at least $550.

Section 3.4 Question Log

Use this space to write questions. Be sure to get these answered and revisit them when you prepare for your exam.

Page ____ Answered ☐	Page ____ Answered ☐
Page ____ Answered ☐	Page ____ Answered ☐

Chapter 3 Organizer

VOCABULARY

Use the following steps to review the vocabulary for Chapter 3.

1. *Write the definition for each word from memory.*
2. *Compare the definitions you have written with the definitions in the Vocabulary Preview.*
3. *Study any definitions that you could not remember or that you defined incorrectly.*

3.1
Digits • Decimal Point • Decimal Places • Place Value • Like Decimals

3.2
Operations • Sum • Difference • Carrying • Borrowing

3.3
Decimal Places • Product • Factors • Diameter • Radius • Pi (π) • Circumference

3.4
Dividend • Divisor • Quotient • Division Notation • Annexing Zeros

PROCEDURES

Procedure/Topic	Steps	Example
Find the Place Values of Digits in Decimal Numbers (Section 3.1)	**Step 1** Count the number of places the digit is from the decimal point. **Step 2** List that number of place values. "tenths, hundredths..."	What is the place value of the digit 6 in 45.74264? **Step 1** The digit 6 is in the fourth place to the right of the decimal point. **Step 2** • tenths • hundredths • thousandths • ten thousandths 6 is in the ten thousandths place.
Write a Decimal Number in Expanded Form (Section 3.1)	Write the sum of what each digit represents. Be sure to multiply each digit by its place value.	Write 462.359 in expanded form. $400 + 60 + 2 + \frac{3}{10} + \frac{5}{100} + \frac{9}{1{,}000}$
Write a Decimal Number in Words ("Mixed" Decimal Numbers, too) (Section 3.1)	**Step 1** Write the whole number portion in words. **Step 2** Write *and* for the decimal point. **Step 3** Write the decimal portion in words.	Write 5.112 in words. Whole portion and decimal portion Five and one hundred twelve thousandths
Write a Decimal Number as a Fraction or Mixed Number (Section 3.1)	**Step 1** Read the decimal number in words. **Step 2** Convert the words to a fraction or mixed number.	Write 6.97 as a fraction or mixed number. 6.97 = six and ninety-seven hundredths $6.97 = 6\frac{97}{100}$ 2 decimal places → 2 zeros
Write a Fraction or Mixed Number as a Decimal (Section 3.1)	The denominator must be a power of 10 for this procedure to be used. **Step 1** Read the fraction or mixed number in words. **Step 2** Convert the words to a decimal number.	Write $2\frac{361}{1{,}000}$ as a decimal. $2\frac{361}{1{,}000}$ = 2 and three hundred sixty-one thousandths = 2.361

Procedure/Topic	Steps	Example
Choose the Best Approximation (Section 3.1)	Choose the rounded number that is closer to the original. Remember, you are finding the best approximation written to the desired place value.	To round 7.846 to the hundredths place, round up to 7.85 or round down to 7.84. Which is the best approximation? Since 7.846 is closer to 7.85, 7.85 is the best approximation.
Round Decimal Numbers (Section 3.1)	**Step 1** Identify the digit to be rounded. **Step 2** Decide if the digit to be rounded increases by one or stays the same. Increase the digit when the digit to the right is 5 through 9. Keep the digit the same when the digit to the right is 0 through 4. **Step 3** All the digits to the right of the rounded digit will be zero.	Round 7.491 to the hundredths place. 7.4⑨1 **Step 1** The digit 9 is in the hundredths place. Round up to 7.50 or down to 7.49. **Step 2** The digit to the right is 1, so we round down. $7.491 \approx 7.49$ is the best approximation.
Order Decimal Numbers (Section 3.1)	**Step 1** Build like decimals. Line up the decimal points. Add missing zeros. **Step 2** Compare the numbers.	Order 0.6 and 0.09 from largest to smallest. **Step 1** Line up the decimal points. $0.6 = 0.60$ ← Add one zero. $0.09 = 0.09$ **Step 2** Since both numbers are hundredths, we can compare them easily. $60 > 9$, so $0.6 > 0.09$.
Add or Subtract Decimal Numbers (Section 3.2)	**Step 1** Build like decimals. Line up the decimal points. Add zeros to fill in any missing place values. **Step 2** Add or subtract the digits, from right to left.	Find the sum. $5.65 + 2.7$ Line up the decimal points. $\begin{array}{r} \overset{1}{5}.65 \\ +2.70 \\ \hline 8.35 \end{array}$ ← Add one missing zero to build like decimals.
Multiply Any Two Decimal Numbers (Section 3.3)	**Step 1** Multiply the numbers, ignoring the decimal points. **Step 2** Add the number of decimal places in each factor. **Step 3** Write the product so that it matches the number of decimal places in step 2.	Find the product. 0.07×0.06 2 decimal places → 0.07 2 decimal places → × 0.06 2 + 2 = 4 decimal places → 0.0042 (4 decimal places)
Multiply by Decimal Powers of 10 (Section 3.3)	**Step 1** Count the number of decimal places in the power of 10. **Step 2** Move the decimal point the counted number of places to the left. Multiplying a number by 0.1, 0.01, 0.001... results in a product that is smaller.	Find the product. 75.2×0.001 Multiplying by 0.001 will make the product smaller than 75.2. Move the decimal point three places to the left. $75.2 \times 0.001 = 0.0752$ 3 decimal places; 3 places

Procedure/Topic	Steps	Example
Multiply by Whole Number Powers of 10 (Section 3.3)	**Step 1** Count the number of zeros in the power of 10. **Step 2** Move the decimal point the counted number of zeros to the right. Multiplying a number by 10, 100, 1,000... results in a product that is larger.	Find the product. $3.24 \times 1{,}000$ Multiplying by 1,000 will make the product larger than 3.24. Move the decimal point three places to the right. $3.24 \times \underbrace{1000}_{3\text{ zeros}} = 3\,2\,4\,\_$ (3 places)
Estimate Products (Section 3.3)	**Step 1** Round each number to its largest place value. **Step 2** Multiply the rounded numbers.	Estimate the product. 0.0072×0.0049 $0.0072 \times 0.0049 \approx 0.007 \times 0.005$ ≈ 0.000035
Find the Circumference or Area of a Circle (Section 3.3)	**Step 1** Write the appropriate formula. **Step 2** Substitute the values for each variable. **Step 3** Simplify.	What is the circumference of a circle with a radius of 4 inches? $C = 2\pi r$ $C = 2(3.14)(4\text{ inches})$ $C = 6.28(4\text{ inches})$ $C = 25.12\text{ inches}$
Set Up Division Exercises with Decimals (Section 3.4)	**Step 1** Move the decimal points to divide by a whole number. **Step 2** "Draw the lines." Organizing your work like this will help you get the correct place values in your answers. **Step 3** Write the decimal point in the quotient directly above the decimal point in the dividend.	Set up the division exercise. $4.32 \div 0.3$ $4.32 \div 0.3 \Rightarrow 3.\overline{)\,4\,3.2}$
Divide by a One-Digit Divisor (Section 3.4)	**Step 1** Set up the problem. **Step 2** Perform long division. Stop when every digit of the dividend has been divided evenly. or Stop when you have enough digits to round to the desired place value. **Step 3** Round the quotient, if necessary	Find the quotient. $4.32 \div 0.3$ $3\overline{)\,43.2}$ = 14.4 43.2 − 3 → 13; − 12 → 12; − 12 → 0
Write a Fraction or Mixed Number as a Decimal (Section 3.4)	Divide the numerator by the denominator. In a mixed number, the whole number remains the same.	Write $4\frac{3}{8}$ as a decimal. $\frac{3}{8} \Rightarrow 8\overline{)\,3}$ $8\overline{)\,3.000}$ = 0.375 − 24 → 60; − 56 → 40; − 40 → 0 $4\frac{3}{8} = 4.375$

Procedure/Topic	Steps	Example
Estimate Quotients (Section 3.4)	**Step 1** Round each number to its highest place value. **Step 2** Divide the resulting numbers.	Estimate the quotient. $8.9 \div 0.0342$ $8.9 \approx 9.0$ $0.0342 \approx 0.03$ $3\overline{)900}$ = 300 $8.9 \div 0.0342 \approx 300$
Divide by a Multidigit Divisor (Section 3.4)	**Step 1** Set up the problem and estimate the answer. **Step 2** Perform long division. Stop when every digit of the dividend has been divided evenly. or Stop when you have determined enough digits to round. **Step 3** Round the quotient if necessary. **Step 4** Check your answer.	Find the quotient. $5.58 \div 1.8$ $1.8\overline{)5.58} \longrightarrow 2\overline{)6.00}$ = 3 $\longrightarrow 5.58 \div 1.8 \approx 3$ $1.8\overline{)5.58}$ = 03.1 -54 18 -18 0 **Check** 3.1 is close to our estimate of 3, so we trust our work.

Chapter 3 Review Exercises

3.1

Answer the following exercises.

1. Write 1.2 and $1\frac{2}{10}$ in words. Explain why the numbers are equal.

2. Identify the missing place values.

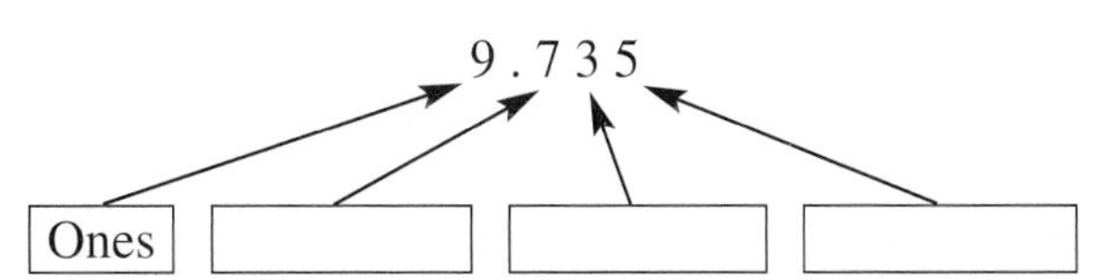

3. Finding the best approximation for a number to a certain place value is called ____________ the number.

4. When a decimal number is written in words, the word ____________ separates the whole number and decimal portions.

Identify the place value of the digit 3 in each number.

5. 3,456 **6.** 45.013 **7.** 5.007839 **8.** 0.0153

Write each decimal number in words.

9. 76.34 **10.** 2.057 **11.** 12.6003 **12.** 5.07002

Write each decimal number as a fraction and each fraction as a decimal number.

13. 0.035 **14.** 0.45 **15.** $\frac{14}{1,000}$ **16.** $\frac{8}{10,000}$

Round each decimal number to the given place value.

17. 67.841; Tenths **18.** 578.8762; Hundredths **19.** 3.56721; Thousandths **20.** 1,356.57; Hundreds

Order each list of decimal numbers from smallest to largest.

21. 0.0435, 0.14, 0.0436 **22.** 0.567, 0.0987, 7.003

3.2

Answer each question.

23. When decimal numbers are subtracted, what should be aligned to make sure like place values get subtracted?

24. Why can zeros be written on the right side of the number 2.35 without its value changing?

Find each sum or difference.

25. 12.56 − 9.45 **26.** 768.9 − 34.5 **27.** 26.89 + 12.571 **28.** 123.4 + 5.89

29. 19.72 + 9 **30.** 345.9 + 15 **31.** 24 − 11.31 **32.** 5 − 2.89

33. Use the sign to answer each question.
  a. What is the distance from McDonald's to Wendy's?
  b. What is the distance from the Holiday Inn to McDonald's?

34. Rinaldo wrote one check for $345.65 and a second check for $57.80.
  a. How much did he spend?
  b. If his account had $1,456.25 in it before he wrote the checks, what was the new balance in his account?

3.3

Answer the following concept exercises.

35. When decimal numbers are multiplied, the sum of the decimal places in the factors must match the number of decimal places in the ___________.

36. Moving the decimal point 3 places to the right in 4.57 is the same as multiplying 4.57 by ___________.

37. If you know the radius of a circle, how can you find the diameter of the circle?

Multiply.

38. 0.006 × 0.03 **39.** 0.09 × 0.0002 **40.** 8 × 0.04 **41.** 7 × 0.005

42. 623 × 0.01 **43.** 70 × 0.0001 **44.** 1,000 × 0.041 **45.** 10,000 × 0.0087

46. An artist created a sculpture out of construction paper. One piece of construction paper has a thickness of 0.008 inch. What is the total thickness of part of a sculpture that is 60 pieces of construction paper thick?

47. A fourth-grade teacher wants to buy Laffy Taffy for her class. The cost is $0.15 per piece of candy. How much will it cost the teacher to buy one piece of candy for each of 35 students?

For each exercise:

a. Estimate the product.

b. Find the product.

48. 0.568×0.0324

49. 0.684×0.00026

50. 235×0.0386

51. 478×0.06739

Find the area or circumference as indicated.

52. What is the circumference of a circle with a radius of 8 m?

53. What is the circumference of a circle with a diameter of 6 mm?

54. Find the area.

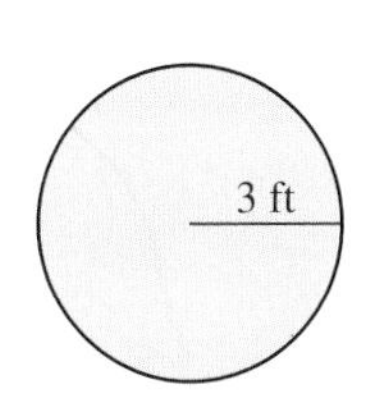

55. Find the area.

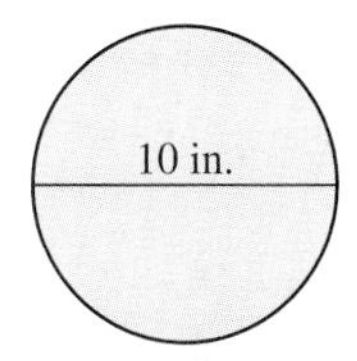

3.4

Answer the following concept exercises.

56. When setting up the division exercise $1.2 \div 0.15$, you should move the decimal point in the divisor and dividend ____________ places to the ____________.

57. To round a quotient to the thousandths place, the digit in the ____________ place must be known.

Divide. Round to the nearest hundredth when necessary.

58. $6 \div 0.03$

59. $8 \div 0.0002$

60. $0.016 \div 0.04$

61. $0.07 \div 0.005$

62. $0.623 \div 0.014$

63. $0.576 \div 0.012$

64. $18.91 \div 0.71$

65. $25.87 \div 0.58$

66. Your speedometer shows 70.8 miles per hour. Your GPS system shows that you are 235.3 miles from Venice Beach. At this speed, how long will it take you to get to Venice Beach? Round to the nearest tenth of an hour.

67. Three friends agree to split a $56.87 restaurant bill evenly. How much does each person pay? Round to the nearest cent.

Write each fraction or mixed number as a decimal number.

68. $\frac{3}{5}$

69. $\frac{5}{8}$

70. $\frac{7}{6}$

71. $\frac{11}{6}$

72. $5\frac{1}{3}$

73. $7\frac{2}{3}$

Chapter 3 Practice Test

Answer the following exercises.

1. Identify the missing place values. 45.8765

Tenths | ______ | ______ | ______

2. When multiplying decimal numbers, how is the number of decimal places in the product determined?

3. When setting up the division exercise $4.5 \div 0.678$, you should move the decimal point in the divisor and dividend ____________ places to the ____________.

4. To round a quotient to the tenths place, you must know the digit in the ____________ place of the quotient.

Write the value of the check in words.

5.

Joseph M. Johnson
Linda J. Johnson
123 Maple Street
Your City, California 94000

1173

DATE __________ 20____

PAY to the ORDER of *Shun Young* $ 216.67

____________________________ DOLLARS

New West Bank
San Francisco California 94106

MEMO ____________ ____________

Write each decimal number as a fraction and each fraction as a decimal number.

6. 0.16

7. $\frac{9}{1{,}000}$

Round each decimal number to the given place value.

8. 145.261; tenths

9. 5.8729; hundredths

10. 10.54781; thousandths

11. 1,375.9872; thousands

Order the decimal numbers from smallest to largest.

12. 0.0655, 0.101, 0.0499

Find each sum or difference.

13. $33.56 + 10.45$

14. $76 - 54.6$

15. $2{,}689 + 19.371$

16. $342 - 0.089$

17. Use the sign to answer each question.

a. What is the distance from Cracker Barrel to Taco Bell?

b. What is the distance from American Inn to Taco Bell?

18. Steve and Renatta are paying their home mortgage. The *principal* is the balance left to pay on the mortgage, without interest. The current principal is \$87,560. Each month, the payment is \$850.

a. This month, the bank charged \$452.33 in interest. How much of the \$850 payment was left to apply towards the principal?

b. Based on your answer to part (a), what is the new principal?

For each exercise:
a. Estimate the product.
b. Find the product.

19. 1.83×0.55

20. 7.35×0.089

21. 25.56×0.0578

Find the exact answer:

22. How much money could a person deduct on her 2008 tax return if she drove her car 568 miles for business?

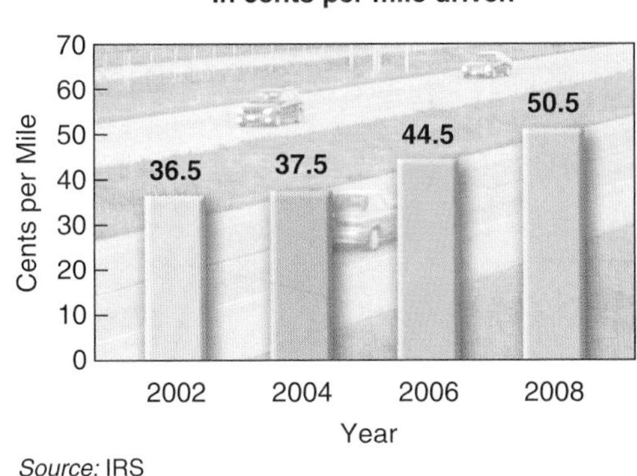

The values in this chart show the amount, per mile, that someone can deduct from her taxes for driving "business miles."

Find the area or circumference, as indicated.

23. Find the circumference.

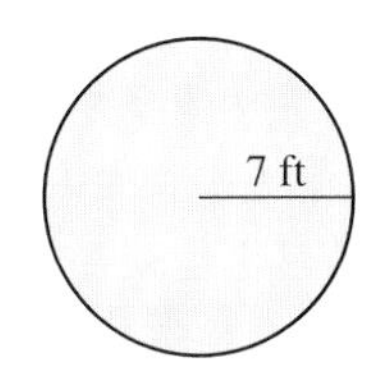

24. Find the area.

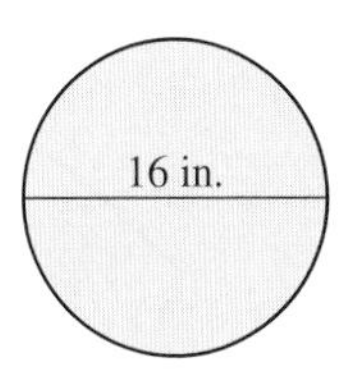

Divide. Round to the nearest thousandth when necessary.

25. $0.075 \div 0.041$

26. $0.065 \div 0.152$

27. You have purchased a \$1,356.87 laptop. If you pay \$75 a week toward the cost of the laptop, how long will it take you to pay for the laptop?

28. In the 2008 regular season, Kobe Bryant scored 2,320 points. If he played 82 games, how many points per game did he average? Round to the nearest tenth.

Write each fraction or mixed number as a decimal number.

29. $\frac{3}{8}$

30. $4\frac{2}{5}$

31. $\frac{5}{6}$

32. $\frac{9}{4}$

CHAPTER 4

Ratios, Rates, and Proportions

Ratios, rates, and proportions are used to describe the relationship between quantities. This chapter is designed to help you understand how to use ratios, rates, and proportions.

The long-running television show *ER* has done more than entertain its viewers. In a survey performed by the Kaiser Family Foundation, 55% of viewers said that in addition to being entertained by *ER,* they also learned about important health issues from watching the show.

In Exercise 41, Section 4.3, we will use a proportion to reproduce the survey technique used to get this information about *ER* viewers.

Section 4.1 Ratios and Rates

Ratios and rates are used to compare the relative amounts of two quantities.

Ratios compare quantities with the same units:	**Rates compare quantities with different units:**
Earning \$3 for every \$100 invested.	Earning \$9.50 for every hour worked.

In this section, you will learn how to compare quantities using ratios and rates and perform calculations with them.

The **Objectives** for Section 4.1 will help you

- **A** Write ratios.
- **B** Write rates and unit rates.

VOCABULARY PREVIEW *Check the box that applies.*

	Got It	Must Study
Ratio: A **ratio** is a quotient that compares two quantities with the same units. Units do not need to be written in a ratio.		
Rate: A **rate** is a quotient that compares two quantities with different units. Units must be written in a rate.		
Unit Rate: A **unit rate** is a rate with a denominator of 1 unit.		
Per: The word ***per*** indicates division. In ratios and rates, division is written as a fraction. $45 \text{ miles per hour} = \frac{45 \text{ miles}}{1 \text{ hour}}$		
Reference Fraction: A **reference fraction** is a fraction that demonstrates how the units should be organized in a ratio or rate.		

Study these words when they appear in the text. After the section, test your understanding by completing the Vocabulary Review in the section exercises.

Objective A Write Ratios

The Concept Ratios are used to compare two quantities with the same units. You may have encountered ratios when shopping for a high-definition television (HDTV).

HDTVs have a 16 : 9 aspect ratio. This ratio gives the screen a wide-screen display.

Standard TVs have a 4 : 3 aspect ratio. This ratio gives the screen a display that is almost a square.

There are three different ways to write a ratio. The aspect ratio of an HDTV can be written as 16:9, 16 to 9, or $\frac{16}{9}$. Each form of a ratio is useful in certain situations, and you may encounter all three forms in your everyday life.

INTERACTIVE DEFINITION Ratio

A **ratio** is a quotient that compares two quantities with the same units. Units do not need to be written in a ratio.

EXAMPLES

1. Compare 3 to 5 using the three forms of a ratio.

Use the word *to* to write the ratio as 3 to 5.

Use a fraction to write the ratio as $\frac{3}{5}$.

Use a colon to write the ratio as 3 : 5.

2. Write a ratio comparing a \$5 bill to a \$1 bill. Write this ratio as a fraction.

\$5 → (five-dollar bill)
to →
\$1 → (one-dollar bill)

$= \frac{\$5}{\$1} = \frac{5}{1}$

Since the bills have the same units, dollars, this fraction is a ratio. The common units, dollars, divide out, giving the ratio $\frac{5}{1}$.

GUIDED PRACTICE

1. Compare 5 to 7 using the three forms of a ratio.

Use the word *to* to write the ratio as ______.

Use a fraction to write the ratio as —.

Use a colon to write the ratio as ______.

2. Write a ratio comparing a dime to a penny. Write this ratio as a fraction.

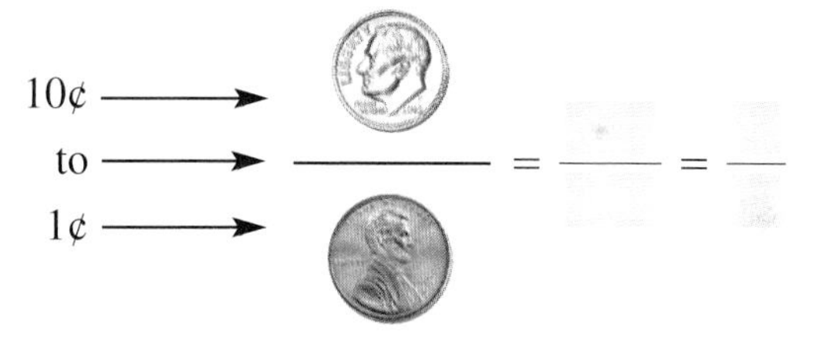

Since the coins have the same units, ______, this fraction is a ______. The common units, ______, divide out, giving the ratio —.

DO YOU UNDERSTAND how to write the three forms of a ratio?	Got It	Get Help
DO YOU UNDERSTAND how to write a ratio as a fraction?	Got It	Get Help

When using a ratio in a calculation, write the ratio as a fraction. It might be tempting to rewrite a ratio like $\frac{5}{1}$ as 5. However, it is important to write both the numerator and denominator when working with ratios. The ratio $\frac{5}{1}$ is read as "five to one," not "five."

Procedure Write a Ratio of *A* to *B* as a Fraction

Step 1 Write *A* (the first number) in the numerator.

Step 2 Write *B* (the second number) in the denominator.

EXAMPLES

3. Write the ratio "7 to 8" as a fraction.

The ratio of 7 to 8 is $\frac{7}{8}$. ← **First number** (7) ← **Second number** (8)

4. Write the ratio "9 : 10" as a fraction.

The ratio of 9 : 10 is read as "9 to 10."

$$9:10 = \frac{9}{10}$$

5. Write a ratio for the situation: Lance Armstrong rides 15 miles to every 2 miles that Ian rides.

Write a reference fraction to identify the ratio.

Lances' ride to Ian's ride

$$= \frac{\text{Lance's ride}}{\text{Ian's ride}}$$

$$= \frac{15 \not{mi}}{2 \not{mi}}$$

$$= \frac{15}{2}$$

The ratio of Lance's ride to Ian's ride is $\frac{15}{2}$.

6. Bruce's paycheck was $400 after taxes. If he paid $111 in taxes, what was the ratio of the taxes he paid to the amount of his paycheck after taxes?

$$\text{Taxes to paycheck} = \frac{\text{Taxes}}{\text{Paycheck}}$$ ← **Write a reference fraction to identify the ratio.**

$$= \frac{\not{\$}111}{\not{\$}400}$$

$$= \frac{111}{400}$$

The ratio of Bruce's taxes paid to his paycheck after taxes is $\frac{111}{400}$.

GUIDED PRACTICE

3. Write the ratio "3 to 7" as a fraction.

The ratio of 3 to 7 is $\frac{\square}{\square}$.

4. Write the ratio "6 : 13" as a fraction.

The ratio of 6 : 13 is read as "6 ____ 13."

$$6:13 = \frac{\square}{\square}$$

5. Write a ratio for the situation: Julian ate 40 hot dogs to every 13 hot dogs that Mikito ate.

Write a reference fraction to identify the ratio.

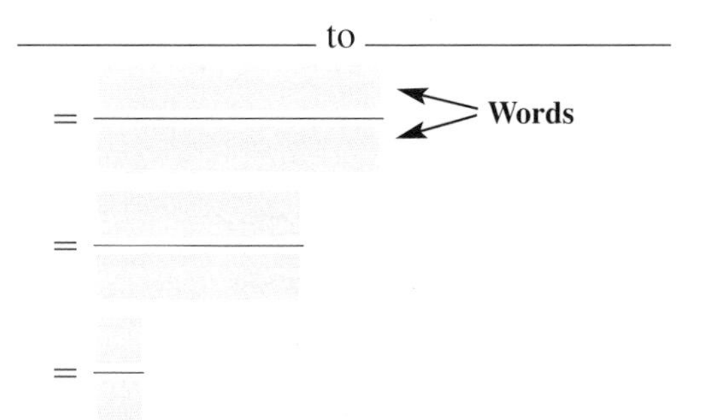

$= \frac{\square}{\square}$

$= \frac{\square}{\square}$

The ratio of Julian eating hot dogs to Mikito eating hot dogs is $\frac{\square}{\square}$.

6. The taxes that Bruce paid included $88 in federal taxes and $23 in state taxes. What was the ratio of state taxes to federal taxes in Bruce's paycheck?

→ ________ to ________ $= \frac{\square}{\square}$

$= \frac{\square}{\square}$

$= \frac{\square}{\square}$

The ratio of state taxes to federal taxes in Bruce's paycheck is $\frac{\square}{\square}$.

In Example 6 and Guided Practice 6, we had to be careful when writing the ratio because the first number was not the numerator of the ratio. Writing a reference fraction helped us set up the ratio correctly. Always read the question carefully and write a reference fraction before setting up a ratio.

When writing a ratio as a fraction, write it in simplest form.

- A ratio in simplest form does not have decimal numbers.
- A ratio in simplest form is a fraction written in lowest terms.

Procedure Write a Ratio as a Fraction in Simplest Form

Step 1 Write the ratio as a fraction.

Step 2 Clear any decimals.

Step 3 Simplify the fraction.

EXAMPLES

7. Write the ratio of 55 pounds to 40 pounds as a fraction in simplest form.

Write the ratio as a fraction. $\text{Ratio} = \dfrac{55 \text{ pounds}}{40 \text{ pounds}}$

Include the units.

Simplify the fraction. Divide out common factors. $= \dfrac{\cancel{5} \cdot 11}{\cancel{5} \cdot 8}$

$= \dfrac{11}{8}$

8. Write the ratio of 0.23 gallons to 0.7 gallons as a fraction in simplest form.

Write the ratio as a fraction. $\text{Ratio} = \dfrac{0.23 \text{ gallons}}{0.7 \text{ gallons}}$

Include the units.

Clear the decimals. To move the decimal point two places to the right, multiply by $\frac{100}{100}$. $= \dfrac{0.23}{0.7} \cdot \dfrac{100}{100}$

$= \dfrac{23}{70}$

GUIDED PRACTICE

7. Write the ratio of 42 kilograms to 30 kilograms as a fraction in simplest form.

Ratio = ———

= ———

= ——

8. Write the ratio of 0.13 liters to 0.5 liters as a fraction in simplest form

Ratio = ———

= ——— · ———

= ——

(Continued)

9. Write the ratio of $1\frac{1}{3}$ to $2\frac{2}{3}$ in simplest form.

Write the ratio as a fraction.

$$\text{Ratio} = \frac{1\frac{1}{3}}{2\frac{2}{3}}$$

Simplify the fraction.
Write the mixed numbers as improper fractions.

$$= \frac{\frac{4}{3}}{\frac{8}{3}}$$

Interpret the fraction bar as division.

$$= \frac{4}{3} \div \frac{8}{3}$$

Change division to multiplication.

$$= \frac{4}{3} \cdot \frac{3}{8}$$

Divide out common factors.

$$= \frac{\cancel{4} \cdot \cancel{3}}{\cancel{3} \cdot 2 \cdot \cancel{4}}$$

$$= \frac{1}{2}$$

9. Write the ratio of $1\frac{1}{2}$ to $2\frac{1}{4}$ in simplest form.

$$\text{Ratio} = \frac{\_}{\_}$$

$$= \frac{\_\_}{\_\_}$$

$$= \_\_ \div \_\_$$

$$= \_\_ \cdot \_\_$$

$$= \_\_ \quad \_\_\_$$

$$= \_\_$$

10. Write the fraction $\frac{7.2 \text{ inches}}{3 \text{ feet}}$ as a ratio in simplest form.

Inches and feet cannot form a ratio because the units are different. Convert 3 ft to inches so that the units in the numerator and denominator match.

$$3 \text{ feet} = \frac{3 \cancel{\text{feet}}}{1} \cdot \frac{12 \text{ inches}}{1 \cancel{\text{foot}}}$$

$$= 36 \text{ inches}$$

Write the ratio as a fraction.
Include the units.

$$\text{Ratio} = \frac{7.2 \cancel{\text{inches}}}{36 \cancel{\text{inches}}}$$

Clear the decimals.
To move the decimal point one place to the right, multiply by $\frac{10}{10}$.

$$= \frac{7.2}{36} \cdot \frac{10}{10}$$

$$= \frac{72}{36 \cdot 10}$$

Simplify the fraction.
Divide out common factors.

$$= \frac{\cancel{36} \cdot \cancel{2}}{\cancel{36} \cdot \cancel{2} \cdot 5}$$

$$= \frac{1}{5}$$

10. Write the fraction $\frac{14.4 \text{ hours}}{2 \text{ days}}$ as a ratio in simplest form.

Hours and days cannot form a ratio because the units are ________. Convert 2 days to hours so that the units in the numerator and denominator match.

$$2 \text{ days} = \frac{2 \text{ days}}{1} \cdot \_\_\_\_$$

$$= \quad \text{hours}$$

$$\text{Ratio} = \_\_\_\_$$

$$= \_\_\_\_ \cdot \_\_$$

$$= \_\_\_\_$$

$$= \_\_\_\_$$

$$= \_\_$$

Concept Check

A1. A ratio is a quotient that compares two quantities with the ____ units.

A2. Why isn't $\frac{30 \text{ miles}}{1 \text{ gallon}}$ a ratio?

Objective A Practice

Write each ratio as a fraction in simplest form.

A3. 4 to 6

A4. 5 to 10

A5. 33 : 15

A6. 40 : 54

A7. $3\frac{1}{2}$ minutes to $4\frac{1}{4}$ minutes

A8. $1\frac{1}{6}$ tons to $2\frac{1}{3}$ tons

A9. 1.2 centimeters to 71 centimeters

A10. 3.6 inches to 53 inches

A11. In NASCAR races, drivers gauge their speed by looking at the engine's revolutions per minute (rpm). When an engine is running at 2,000 rpm, the wheels turn at 600 rpm. What is the ratio of the engine's rpm to the wheel's rpm?

A12. The highest point in North Dakota is White Butte, with an elevation of 3,506 feet. The lowest point in North Dakota is Red River, with an elevation of 750 feet. What is the ratio of White Butte's elevation to Red River's elevation?

Convert the units and write each rate as a ratio in simplest form.

A13. 45 minutes to 1.5 hours

A14. 3.5 hours to 20 minutes

A15. One inch on a blueprint represents 25 feet at an actual building site. What is the ratio of measurements at the building site to measurements on the blueprint?

A16. A child's toy football field is 1 yard long. A real football field is 100 yards long. What is the ratio of the size of a real football field to the size of the toy football field?

Objective B Write Rates and Unit Rates

The Concept The difference between a ratio and a rate is subtle.

- **Ratios** compare quantities with the **same units.**
- **Rates** compare quantities with **different units.**

Rates can be used to describe how fast you are traveling (miles per hour), how expensive things are (dollars per pound), and many other everyday comparisons.

Speedometer

$$50 \text{ miles per hour} = \frac{50 \text{ miles}}{1 \text{ hour}}$$

This rate could be used to determine how long it will take to travel 120 miles.

Tomatoes: $3 for 2 pounds

$$\$3 \text{ for } 2 \text{ pounds} = \frac{\$3}{2 \text{ pounds}}$$

This rate could be used to determine how many pounds of tomatoes you could buy for $5.

Classified Ad

Typesetter: Progressive printing company seeks typesetter. $12 per hour, health ins., 401(k)

$$\$12 \text{ per hour} = \frac{\$12}{1 \text{ hour}}$$

This rate could be used to determine how much money a typesetter would make in a week.

You will often encounter the word *per* when working with rates. The word ***per*** indicates division. In ratios and rates, use a fraction to represent division.

Procedure Write a Rate

Step 1 Write a fraction with the two quantities.
You must include the units.

Step 2 Simplify the fraction if possible.

EXAMPLES

11. Write a rate for driving 351 miles in 7 hours.

Step 1 Write a fraction with the two quantities.

$$\text{Rate} = \frac{351 \text{ miles}}{7 \text{ hours}}$$

Include the units.

12. Write a rate for earning $360 in 50 hours.

Step 1 Write a fraction with the two quantities.

Step 2 Simplify the fraction. Divide out common factors.

$$\text{Rate} = \frac{\$360}{50 \text{ hours}} = \frac{\$36 \cdot \cancel{10}}{5 \cdot \cancel{10} \text{ hours}} = \frac{\$36}{5 \text{ hours}}$$

Include the units.

13. Write a rate for prescribing 250 milligrams (mg) of medicine to a 200-pound (lb) person.

Step 1 Write a fraction with the two quantities.

Step 2 Simplify the fraction. Divide out common factors.

$$\text{Rate} = \frac{250 \text{ mg}}{200 \text{ lb}} = \frac{5 \cdot \cancel{50} \text{ mg}}{4 \cdot \cancel{50} \text{ lb}} = \frac{5 \text{ mg}}{4 \text{ lb}}$$

Include the units.

GUIDED PRACTICE

11. Write a rate for running 5 miles in 42 minutes.

Rate = ______

12. Write a rate for earning $200 in 15 hours.

Rate = ______

= $\frac{\ \cdot\ }{\ \cdot\ }$

= ______

13. Write a rate for prescribing 40 milligrams of medicine to a 120-pound person.

Rate = ______

= ______

= ______

When a rate has a denominator of one, it is called a unit rate.

INTERACTIVE DEFINITION Unit Rate

A **unit rate** is a rate with a denominator of 1 unit. A unit rate can be calculated by dividing the numerator by the denominator.

EXAMPLE

14. Keshena earns $36 in 5 hours. State her rate of pay as a unit rate.

$$\$36 \text{ in } 5 \text{ hours} = \frac{\$36}{5 \text{ hours}} = \frac{\$7.20}{1 \text{ hour}} = \$7.20 \text{ per hour}$$

$$5\overline{)36.0} = 7.2$$

GUIDED PRACTICE

14. Tyler earns $42 in 5 hours. State his rate of pay as a unit rate.

$$\$42 \text{ in } 5 \text{ hours} = \frac{\$\_\_}{\_\_ \text{ hours}} = \frac{\$\_\_}{\text{hour}} = \$\_\_\_\_ \text{ per hour}$$

DO YOU UNDERSTAND how to write a rate as a unit rate? Got It Get Help

Unit rates make it easy to compare two rates. For example, we can use unit rates to compare the gas mileage of two cars.

Earl drove 162 miles using 5 gallons.

$$\frac{162 \text{ miles}}{5 \text{ gallons}} = \frac{32.4 \text{ miles}}{1 \text{ gallon}}$$

Earl's car gets 32.4 miles per gallon.

Bob drove 110 miles using 4 gallons.

$$\frac{110 \text{ miles}}{4 \text{ gallons}} = \frac{27.5 \text{ miles}}{1 \text{ gallon}}$$

Bob's car gets 27.5 miles per gallon.

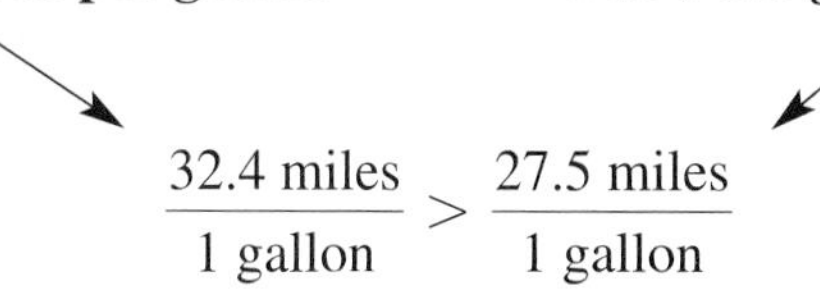

Earl's car gets better gas mileage than Bob's car.

Because unit rates often have decimals, they may not be in simplest form. However, the benefits of a unit rate justify the more complicated appearance. As you study mathematics, you will find that what is considered "simplified" in one instance is not considered "simplified" in another. Like so much in life, what is simplest depends on where you are and what you are doing.

Procedure Write a Unit Rate to Compare Two Quantities

Step 1 Write a rate that compares the two quantities.

Be sure to include the units.

Step 2 Divide the numerator by the denominator.

EXAMPLES

15. Write a unit rate for driving 300 miles using 15 gallons of gas.

Step 1 Write a rate.

$$\text{Rate} = \frac{300 \text{ miles}}{15 \text{ gallons}}$$

Step 2 Divide the numerator by the denominator.

$$\text{Unit rate} = \frac{20 \text{ miles}}{1 \text{ gallon}}$$

$$15\overline{)300} = 20$$

The unit rate is 20 miles per gallon.

16. Write a unit rate for earning $390 in 40 hours.

Step 1 Write a rate.

$$\text{Rate} = \frac{\$390}{40 \text{ hours}}$$

Step 2 Divide the numerator by the denominator.

$$\text{Unit rate} = \frac{\$9.75}{1 \text{ hour}}$$

$$40\overline{)390.} = 9.75$$

The unit rate is $9.75 per hour.

17. Write a unit rate for prescribing 75 milligrams of medicine to a 150-pound person.

Step 1 Write a rate.

$$\text{Rate} = \frac{75 \text{ milligrams}}{150 \text{ pounds}}$$

Step 2 Divide the numerator by the denominator.

$$\text{unit rate} = \frac{0.5 \text{ milligram}}{1 \text{ pound}}$$

$$150\overline{)75.} = 0.5$$

The unit rate is 0.5 milligram per pound.

GUIDED PRACTICE

15. Write a unit rate for driving 160 miles using 4 gallons of gas.

Include the units

Rate = ______

Unit rate = ______

The unit rate is ______ per ______.

16. Write a unit rate for earning $138 in 12 hours.

Include the units

Rate = ______

Unit rate = ______

The unit rate is ______ per ______.

17. Write a unit rate for prescribing 20 milligrams of medicine to a 100-pound person.

Rate = ______

Unit rate = ______

The unit rate is ______ per ______.

Concept Check

B1. In your own words, what is the difference between a ratio and a rate?

B2. What is a unit rate?

B3. Trevor's truck can travel 315 miles using 15 gallons of gas. Valerie's van can travel 138 miles using 6 gallons of gas.

a. Express each rate as a unit rate.

b. Compare the unit rates to determine which vehicle has the better gas mileage.

c. Why is it useful to express the fuel consumption of a vehicle as a unit rate?

Objective B Practice

Write each comparison as a rate in simplest form.

B4. Running 7 miles in 42 minutes

B5. Driving 48 miles in 15 minutes

B6. Earning $64 in 6 hours

B7. Earning $46 in 4 hours

Write a unit rate for each comparison.

B8. Earning $250 in 40 hours

B9. Earning $165 in 30 hours

B10. Cereal costing $3.90 for 15 ounces

B11. Cereal costing $4.80 for 20 ounces

B12. Prescribing 750 milligrams of medicine to a 150-pound person

B13. Prescribing 400 milligrams of medicine to a 80-pound person

B14. The speed of a car that has traveled 318 miles in 6 hours

B15. The speed of a car that has traveled 360 miles in 8 hours

Combining Concepts and Applications

CONCEPT I Comparison Shopping to Find the Better Buy When shopping, you can determine a best buy using a unit rate. Usually, shoppers want one of two things:

People want to get a larger quantity for each dollar spent.	**People want to pay fewer dollars for the same quantity.**
Comparing with units of "quantity per dollar," the unit rate will be written as $\frac{\text{quantity}}{\text{1 dollar}}$.	Comparing with units of "dollar per quantity," the unit rate will be written as $\frac{\text{dollars}}{\text{quantity}}$.
As a consumer, you want to *get more* per dollar. A larger rate is a better buy.	As a consumer, you want to *pay less* for the same quantity. A smaller rate is a better buy.

Procedure Determine the Best Buy

Compare the unit rates for each product.

EXAMPLES

18. Write two unit rates with units of $\frac{\text{ounces}}{1 \text{ dollar}}$ to determine the better buy.

a. 15 ounces (oz) of potato chips for $3.00
b. 21 ounces (oz) of potato chips for $3.50

Compare the unit rates. Write the unit rates in the form of "quantity per dollar." Use scratch paper to perform the division.

a. $\frac{15 \text{ oz}}{\$3} = 5$ oz per dollar

b. $\frac{21 \text{ oz}}{\$3.50} = 6$ oz per dollar

$3\overline{)15}$ = 5 $3.5\overline{)21}$ = 6

A larger rate is the better buy because you get more per dollar.

Purchasing the larger bag will give you more chips for every dollar. Therefore, the 21-ounce bag is the better buy.

19. Write two unit rates with units of $\frac{\text{dollars}}{1 \text{ item}}$ to determine the better buy.

a. Harvey's fresh fruit stand sells melons at "5 for $3."
b. Hector's fresh fruit stand sells melons at "$0.75 cents each."

Compare the unit rates. Write the unit rates in the form of "dollars per quantity." Use scratch paper to perform the division.

a. $\frac{\$3.00}{5 \text{ melons}} = \$0.60$ per melon

b. $\frac{\$0.75}{1 \text{ melon}} = \$0.75$ per melon

$5\overline{)3.00}$ = 0.60

A smaller rate is the better buy because you pay less for the same quantity.

Since the "per melon" cost is less at Harvey's stand, it is offering the better buy.

GUIDED PRACTICE

18. Write two unit rates with units of $\frac{\text{ounces}}{1 \text{ dollar}}$ to determine the better buy.

a. 20 oz of cereal for $5.00
b. 27 oz of cereal for $5.40

a. $\frac{\quad}{\$\quad} =$

b. $\frac{\quad}{\$\quad} =$

Purchasing ______ box will give you more ______ for every ______. Therefore, the ______ box is the better buy.

19. Write two unit rates with units of $\frac{\text{dollars}}{1 \text{ item}}$ to determine the better buy.

a. Hector's fresh fruit stand sells rutabagas at "6 for $1.50."
b. Harvey's fresh fruit stand sells rutabagas at "10 for $2.60."

a. $\frac{\$\quad}{\quad} =$

b. $\frac{\quad}{\quad} =$

Since the "per ______" cost is less at ______ stand, it is offering the better buy.

Section 4.1 Exercises

FOR EXTRA HELP       

To write ratios:

1. Answer the Objective A Concept Checks.
2. Answer the odd Objective A Practice Exercises.
3. Answer the even Objective A Practice Exercises.

To write rates and unit rates:

4. Answer the Objective B Concept Checks.
5. Answer the odd Objective B Practice Exercises.
6. Answer the even Objective B Practice Exercises.

VOCABULARY REVIEW *Review the Vocabulary Preview for Section 4.1. Study the definitions until you can check* Got It *for every word.*

Use these words to complete the sentences below.

ratio • per • rate • reference fraction • unit rate

	Got it	Get help
7. A rate with a denominator of one unit is called a ____________.		
8. A ____________ is a quotient that compares two quantities with different units.		
9. A quotient that compares two quantities with the same units is called a ____________.		

How will you get help for any vocabulary that you are unsure about?

Your instructor ____ MyMathLab ____ A classmate ____ A tutor ____ Other ____

For each exercise:

a. Identify each comparison as a rate or ratio.
b. Explain why it is a rate or ratio.
c. Write the corresponding rate or ratio in simplest form.

10. 4:12

11. 7:35

12. 6:9

13. 8:14

14. 13 boys to 14 girls

15. 14 girls to 27 students

16. 8 ounces per dollar

17. 5 yards per dollar

18. $225 to 3 credit hours

19. $135 to 3 credit hours

20. Earning $93 for 15 hours of work

21. Earning $84 for 15 hours of work

22. Using $2\frac{1}{2}$ lb of sugar to every $4\frac{1}{4}$ lb of flour

23. Spinning $3\frac{1}{3}$ rpm on a merry-go-round to every $6\frac{1}{6}$ rpm that your kid sister spins

Convert the units and write each rate as a ratio in simplest form.

24. 8 cents per dollar

25. 6 cents per dollar

26. Studying 4 hours every day

27. 3 cloudy days in a week

Answer each statement.

28. Write three rates that are useful in your life.

29. Write three ratios that are useful in your life.

Write each comparison as a unit rate. Round each answer to the nearest whole number unless otherwise stated.

30. Driving 480 miles using 25 gallons of gas

31. Driving 380 miles in 6.5 hours

32. 312 students to 16 instructors

33. 508 students in 21 classes

34. Paying $1.53 in sales tax for a $25.50 bill (Round to the nearest cent.)

35. Tipping a server $8.50 for a $42.35 bill (Round to the nearest cent.)

36. Painting 1,950 square feet of wall with 5 gallons of paint

37. Painting 500 square feet of wall every 2.5 hours

38. 375 mg of medication given to a 200-pound patient (Round to the nearest tenth.)

39. 120 mg of medicine given every 14 minutes (Round to the nearest tenth.)

40. 33 cookies shared between 11 children

41. 440 people riding to work on 11 buses

Answer each question.

42. What is the English translation of the math phrase *4 : 1*?

43. What steps should be used to write the ratio $\frac{32}{56}$ in simplest form?

44. What steps should be used to write the ratio $\frac{1.2}{5}$ in simplest form?

45. What steps should be used to write the rate $\frac{63 \text{ miles}}{2 \text{ gallons}}$ as a unit rate?

Calculate a unit rate for each situation to determine the better buy.

46. Tires designed to last 60,000 miles are selling for $250 a set. Tires designed to last 40,000 miles are selling for $200 a set. Which is the better buy?

47. A subscription to *Adbusters* costs $35 for 6 issues and $48 for 12 issues. Which is the better buy?

smith If you write each unit rate with money in the numerator, you can easily compare the cost of each item.

For more tips and tweets, go to twitter.com/gstbasicmath

48. Bottles of dish soap have the following prices:
- 12.6 oz for $1.39
- 38 oz for $2.59

a. For each bottle of dish soap, write a unit rate with units of $\frac{\$}{1 \text{ ounce}}$ to determine the better buy.

b. How much money will you save per ounce if you purchase the better buy?

49. Boxes of oat cereal have the following prices:
- 10 ounces for $2.12
- 15 ounces for $2.32
- 20 ounces for $4.89

a. Which is the best buy?

b. Which is the worst buy?

50. Packages of soap have the following prices:
- 12-bar package for $3.99
- 3-bar package for $1.19

a. Based on the number of bars in each package, which is the better buy?

b. Suppose that each bar in the 12-pack is larger than each bar in the 3-pack. Does this fact impact your answer to part a?

The tables below show miles driven, fuel used, and time of trip. Complete each table by filling out the indicated rates.

51.

Miles Driven	Time of Trip	Fuel Used	Rate of Travel (miles per hour)	Rate of Fuel Used (miles per gallon)
60 miles	1.2 hours	2.5 gal		
42 miles	2.1 hours	2.8 gal		

52.

Miles Driven	Time of Trip	Fuel Used	Rate of Travel (miles per hour)	Rate of Fuel Used (miles per gallon)
60 miles	1.5 hours	3 gal		
84 miles	2.4 hours	3.5 gal		

Determine the appropriate ratio or rate for each situation.

53. A retailer purchased 120 personal digital assistants (PDAs) for $32,160. The PDAs were then sold for a total of $35,880. Find the retailer's profit per PDA.

54. A bookstore purchased 75 calculators for a total of $6,750. If the total sales for the calculators were $7,875, what was the bookstore's profit per calculator?

55. A tiger shark grows approximately 24,000 teeth over a 10-year period. Approximately how many teeth per year does a tiger shark grow?

56. Some species of bamboo can grow over 36 inches in a day. How many inches per hour is this?

57. An army general with 26 years of experience earns $13,769 per month. A private with the same years of service earns $1,385 per month. Round each number to its largest place value and estimate the ratio of an army general's salary to a private's salary. (*Source*: *NYT Almanac*)

58. In 2005, the average chief executive officer (CEO) of a company in the S&P 500 earned $11,750,000, while an army general earned $165,228. Round each number to its largest place value and estimate the ratio of a CEO's salary to an army general's salary. (*Source:* www.aflcio.org)

Answer each question by finding and comparing unit rates.

59. **a.** Who was more likely to hit a home run: Mark McGwire, who hit 583 home runs in 6,187 attempts, or Babe Ruth, who hit 714 home runs in 8,399 attempts?

b. Who was more likely to strike out: Mark McGwire, who struck out 1,596 times in 6,187 attempts, or Babe Ruth, who struck out 1,330 times in 8,399 attempts?

c. Based solely on your answers to parts a and b, who was the better hitter—Babe Ruth or Mark McGwire? There is no right answer to this question, but you must use evidence from parts a and b to support your answer.

60. **a.** An investor bought a commercial building for $150,000 and sold it one year later for $165,000. What was the investor's rate of return (profit per dollar invested)?

b. An investor bought a residential home for $100,000 and sold it two years later for $116,000. What was the investor's rate of return (profit per dollar invested)?

c. Based solely on your answers to parts a and b, which was the better investment—the commercial building or the residential home? There is no right answer to this question, but you must use evidence from parts a and b to support your answer.

61. **a.** A worker produced 240 parts over an 8-hour shift. How many parts per hour did the worker produce?

b. Including benefits, the worker is paid $45 per hour. How much money does the worker earn per part produced? (Round to the nearest cent.)

62. **a.** A cook prepared meals for 225 people in 5 hours. For how many people per hour did the cook prepare food?

b. The cook is paid $15 per hour. How much money does the cook earn per meal cooked? (Round to the nearest cent.)

Use the following table of NBA scoring records to answer each question or statement.
Each * represents an NBA record.

	Total Points in Career	Career Points per Game	Best Season Points per Game	Games Played
Kareem Abdul-Jabbar	38,387*	24.61	34.8	1,560*
Michael Jordan	32,292	30.12*	37.1	1,072
Wilt Chamberlain	31,419	30.06	50.4*	1,045

Source: www.rauzulusstreet.com

63. Which player scored the most points in his career?

64. Which player scored the most points per game for a season?

65. Which player scored the most points per game over his career?

66. Use the concept of rate to explain why someone could argue that Kareem Abdul-Jabbar was the best scorer of all time.

67. Use the concept of rate to explain why someone could argue that Wilt Chamberlain was the best scorer of all time.

68. Use the concept of rate to explain why someone could argue that Michael Jordan was the best scorer of all time.

69. In your opinion, which of the three players was the best scorer of all time? There is no right answer to this question, but you must use evidence from the table to support your answer.

SELF-ASSESSMENT

	YES	NO
1. On your last test, were you satisfied with your score? *If you answered yes, good job. Do not continue here.* *If you answered no, respond to the following:*		
2. Did you preview each section of the textbook before your instructor taught it in class?		
3. Did you complete all of the homework before it was due?		
4. Did you get help with the homework exercises that you didn't understand?		

5. Set two goals that will help you become better prepared for your next test.

Goal 1: ______

Goal 2: ______

Section 4.1 Question Log

Use this space to write questions. Be sure to get these answered and revisit them when you prepare for your exam.

Page ____ Answered	Page ____ Answered
Page ____ Answered	Page ____ Answered

Section 4.2 Writing and Solving Proportions

In many instances, a ratio or rate is useful by itself.

A speed limit is a rate.

Knowing that you are driving 65 miles per hour is useful if you don't want to get a speeding ticket.

$$\text{Rate of speed} = \frac{65 \text{ miles}}{1 \text{ hour}}$$

A safe dose of a medication is a rate.

DOSAGE MAY BE ADMINISTERED WITHOUT REGARD TO MEALS.
Usual dose: Children: 30–50 mg/kg/day in divided doses. See enclosure for adult dose and full prescribing information.

A nurse must make sure that a prescription does not exceed the safe dosage rate.

$$\text{Safe dosage rate} = \frac{50 \text{ milligrams}}{1 \text{ kilogram}}$$

However, it is often valuable to write ratios and rates in a proportion. A **proportion** is an equation stating that two ratios or rates are equal. Proportions allow us to solve many real-life problems.

A proportion can be used to determine how long it will take to travel 195 miles when driving 65 miles per hour.

$$\frac{65 \text{ miles}}{1 \text{ hour}} = \frac{195 \text{ miles}}{n \text{ hours}}$$

When this proportion is solved, we would see that it takes 3 hours to travel 195 miles.

A proportion can be used to determine a safe dose of medication for an 18-kilogram child if the dosage rate is 50 milligrams per kilogram.

$$\frac{50 \text{ milligrams}}{1 \text{ kilogram}} = \frac{n \text{ milligrams}}{18 \text{ kilograms}}$$

When this proportion is solved, we would see that the safe dose should not exceed 900 milligrams.

In this section, you will learn how to write and solve proportions similar to those shown above.

The **Objectives** in Section 4.2 will help you

- **A** Write a proportion.
- **B** Determine if a statement is a proportion.
- **C** Solve a proportion.

VOCABULARY PREVIEW *Check the box that applies.*

	Got It	Must Study
Proportion: A **proportion** is an equation stating that two ratios or rates are equal.		
Reference Fraction: A **reference fraction** is a fraction that demonstrates how the units should be organized in a ratio, rate, or proportion.		

Study these words when they appear in the text. After the section, test your understanding by completing the Vocabulary Review in the section exercises.

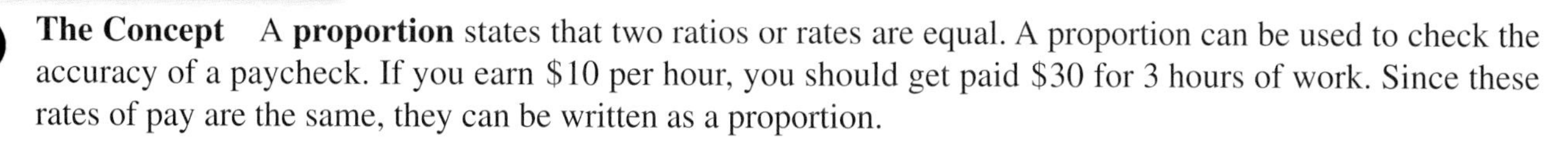

Write a Proportion

The Concept A **proportion** states that two ratios or rates are equal. A proportion can be used to check the accuracy of a paycheck. If you earn \$10 per hour, you should get paid \$30 for 3 hours of work. Since these rates of pay are the same, they can be written as a proportion.

A Proportion

$$\frac{\$10}{1\text{ hour}} = \frac{\$30}{3\text{ hours}}$$

first rate = second rate

There are three ways to write a proportion:

1. $\frac{\$10}{1\text{ hour}} = \frac{\$30}{3\text{ hours}}$
2. \$10 is to one hour as \$30 is to three hours
3. \$10 : 1 hour : : \$30 : 3 hours.

The symbol : : means "as." In a proportion, the units must match regardless of its form.

- In the fraction form of a proportion, the units of both numerators and both denominators are the same, $\frac{\text{dollars}}{\text{hours}} = \frac{\text{dollars}}{\text{hours}}$.
- In the other two forms of a proportion, the order of the units is the same, dollars : hours or dollars to hours.

Procedure Write a Proportion

Step 1 Write a reference fraction, showing the units.
Choose the units that will go in the numerators and denominators.

Step 2 Write the proportion.

EXAMPLES

1. Yesterday, 20 of 500 parts failed inspection and were labeled defective. Today, 25 of 625 failed inspection. Write a proportion for these events.

Step 1 Write a reference fraction that compares the defective parts to the total parts.

$$\text{Reference fraction} = \frac{\text{Defective parts}}{\text{Total parts}}$$

Step 2 Write the proportion.

$$\frac{20\text{ defective parts}}{500\text{ total parts}} = \frac{25\text{ defective parts}}{625\text{ total parts}}$$

Make sure the units in the proportion match the units in the reference fraction.

GUIDED PRACTICE

1. Yesterday, 480 of 500 parts passed inspection and were labeled good. Today, 384 of 400 parts passed inspection. Write a proportion for these events.

Step 1 Write a reference fraction that compares the ________ to the ________.

Reference fraction = ________

Step 2 Write the proportion.

Good parts

$$\frac{\_\_\_}{\_\_\_} = \frac{\_\_\_}{\_\_\_}$$

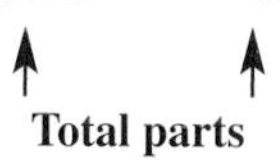

(*Continued*)

2. I drove 40 miles per hour. In 3 hours, I drove 120 miles. Write a proportion for these events.

Step 1 Write a reference fraction that compares the miles to the hours.

$$\text{Reference fraction} = \frac{\text{Miles}}{\text{Hours}}$$

Step 2 Write the proportion.

$$\frac{40 \text{ miles}}{1 \text{ hour}} = \frac{120 \text{ miles}}{3 \text{ hours}}$$

2. One box of cereal costs $1.50. The cereal is advertised as "3 boxes for $4.50." Write a proportion for these costs.

Step 1 Write a reference fraction that compares the ________ to the ____________.

Reference fraction = ————

Step 2 Write the proportion.

———— = ————

3. A 40-pound child received 1.2 ounces of medication. A 50-pound child received x ounces of medication. Write a proportion for these events.

Step 1 Write a reference fraction that compares the ounces to the pounds.

$$\text{Reference fraction} = \frac{\text{Ounces}}{\text{Pounds}}$$

Step 2 Write the proportion.

$$\frac{1.2 \text{ ounces}}{40 \text{ pounds}} = \frac{x \text{ ounces}}{50 \text{ pounds}}$$

3. A 1,500-pound bull requires 4,000 milligrams of vaccine. A 1,200-pound bull requires x milligrams of vaccine. Write a proportion for these events.

Step 1 Write a reference fraction that compares the ____________ to the ____________.

Reference fraction = ————

Step 2 Write the proportion.

———— = ————

Concept Check

A1. What type of mathematical expression states that two ratios or rates are equal?

A2. In your own words, how is a rate different from a proportion?

A3. Why isn't the following statement a proportion? 14 pounds is to 4 bricks as 8 bricks is to 28 pounds.

Objective A Practice

Write a proportion for the following events. Do not solve the proportion.

A4. 5 cups is to 3 pounds as 15 cups is to 9 pounds.

A5. 8 liters is to 7 kilograms as 32 liters is to 28 kilograms.

A6. When building a wall, a carpenter used the proportion 9 boards are to a 10-foot wall as 18 boards are to a 20-foot wall.

A7. When creating sculptures for an art fair, an artist used the proportion 3 pounds of clay is to 8 vases as 9 pounds of clay is to 24 vases.

A8. Five copies of a pamphlet were made using 45 pieces of paper. Later, 450 pieces of paper were used to make 50 copies of the pamphlet.

A9. When planting a garden, a gardener uses 1 packet of seeds to plant a row of vegetables. Since the garden has 30 rows, 30 seed packets are needed.

A10. Natalie earns \$8 per hour. In 7.5 hours, she will earn n dollars.

A11. Jaime earns \$18 per hour. In 16.4 hours, he will earn n dollars.

Objective B Determine If a Statement Is a Proportion

The Concept To determine if a statement is a proportion, check to see if the ratios or rates are equal.

This statement is a proportion.

A vendor is selling peas for \$0.50 a pound, and you pay \$1.00 for 2 pounds. These rates are equal, so they can form a proportion.

$$\frac{\$0.50}{1\text{ pound}} = \frac{\$1.00}{2\text{ pounds}}$$

This statement is not a proportion.

A vendor is selling peas for \$0.50 a pound, and you pay \$3.00 for 3 pounds. These rates are not equal, so they cannot form a proportion.

$$\frac{\$0.50}{1\text{ pound}} \neq \frac{\$3.00}{3\text{ pounds}}$$

To see if a statement is a proportion, we can build the ratios or rates into like fractions. We can also calculate the cross products and check to see if they are equal.

The cross products of a proportion are found by multiplying along each line of a cross written over a proportion.

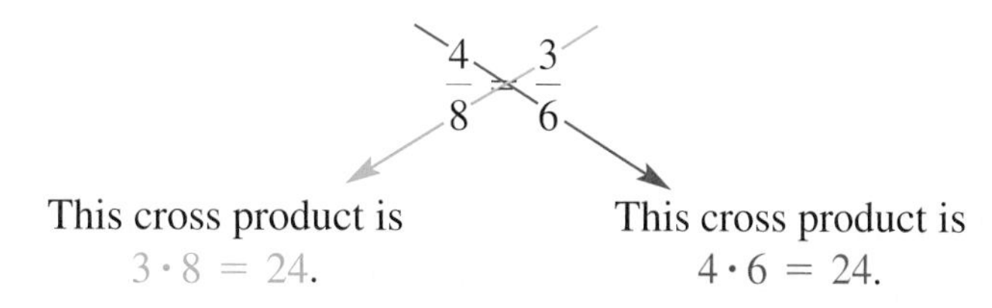

Since the cross products are equal, the statement *is* a proportion.

Procedure **Determine If a Statement Is a Proportion**

Step 1 Calculate the cross products.

Step 2 Check to see if the cross products are equal.

DETAILED EXAMPLE Using the Cross Products to Determine If a Statement Is a Proportion

Is $\frac{\$1.75}{3\text{ pounds}} \stackrel{?}{=} \frac{\$1}{2\text{ pounds}}$ a proportion?

$$\frac{1.75}{3} \stackrel{?}{=} \frac{1}{2}$$

Step 1 Multiply the numbers on each line of the cross.

$$3 \cdot 1 \stackrel{?}{=} 1.75 \cdot 2$$

Step 2 Check to see if the cross products are equal.

$$3 \neq 3.5$$

Since the cross products are not equal, the statement is not a proportion.

$$\frac{\$1}{2\text{ pounds}} \neq \frac{\$1.75}{3\text{ pounds}}$$

WHY IT WORKS Cross Products Are the Numerators of Like Fractions

Show that $\frac{\$1}{2\text{ pounds}} \neq \frac{\$1.75}{3\text{ pounds}}$ by building like fractions.

$$\frac{1}{2} \stackrel{?}{=} \frac{1.75}{3}$$

Step 1 Build like fractions with an LCD of 6.

$$\frac{3}{3} \cdot \frac{1}{2} \stackrel{?}{=} \frac{1.75}{3} \cdot \frac{2}{2}$$

$$\frac{3 \cdot 1}{6} \stackrel{?}{=} \frac{1.75 \cdot 2}{6}$$

Step 2 Notice that the numerators are the same as the cross products. The fractions are not equal, so the statement is not a proportion.

$$\frac{3}{6\text{ pounds}} \neq \frac{3.5}{6\text{ pounds}}$$

Comparing the cross products is the same as comparing the numerators of like fractions.

However we approach the problem, 3 ≠ 3.5, the statement is not a proportion.

Many students are tempted to use cross products when adding, subtracting, multiplying, or dividing fractions. A cross product can be used only when you begin with an equation. Even then, you must be very careful! Until you learn more about equations, we suggest that you use cross products only when checking to see if a statement is a proportion.

EXAMPLES

4. Is the statement a proportion?

$$\frac{72}{78} \stackrel{?}{=} \frac{12}{13}$$

Step 1 Calculate the cross products.

$$12 \cdot 78 \stackrel{?}{=} 13 \cdot 72$$
$$936 = 936$$

Step 2 The statement is a proportion because the cross products are equal.

5. Is the statement a proportion?

$$\frac{400 \text{ milligrams}}{180 \text{ pounds}} \stackrel{?}{=} \frac{120 \text{ milligrams}}{50 \text{ pounds}}$$

Step 1 Calculate the cross products.

$$120 \cdot 180 \stackrel{?}{=} 400 \cdot 50$$
$$21{,}600 \neq 20{,}000$$

Step 2 The statement isn't a proportion because the cross products aren't equal.

6. Is the statement a proportion?

$$\frac{166 \text{ miles}}{8.3 \text{ gallons}} \stackrel{?}{=} \frac{120 \text{ miles}}{6 \text{ gallons}}$$

Step 1 Calculate the cross products.

$$120 \cdot 8.3 \stackrel{?}{=} 166 \cdot 6$$
$$996 = 996$$

Step 2 The statement is a proportion because the cross products are equal.

GUIDED PRACTICE

4. Is the statement a proportion?

$$\frac{14}{15} \stackrel{?}{=} \frac{98}{105}$$

Step 1 Calculate the cross products.

____ · ____ $\stackrel{?}{=}$ ____ · ____

____ = or ≠ ____
which?

Step 2 The statement is/isn't a proportion because the cross products are/aren't equal.

5. Is the statement a proportion?

$$\frac{50 \text{ milligram}}{40 \text{ pounds}} \stackrel{?}{=} \frac{175 \text{ milligram}}{140 \text{ pounds}}$$

Step 1 Calculate the cross products.

____ · ____ $\stackrel{?}{=}$ ____ · ____

____ = or ≠ ____
which?

Step 2 The statement is/isn't a proportion because the cross products are/aren't equal

6. Is the statement a proportion?

$$\frac{280 \text{ miles}}{8 \text{ gallons}} \stackrel{?}{=} \frac{360 \text{ miles}}{10.1 \text{ gallons}}$$

Step 1 Calculate the cross products.

____ · ____ $\stackrel{?}{=}$ ____ · ____

____ = or ≠ ____
which?

Step 2 The statement is/isn't a proportion because the cross products are/aren't equal.

Concept Check

B1. In your own words, how can you determine if a statement is a proportion?

B2. Write a statement that is a proportion. Calculate the cross products to see if your statement is indeed a proportion.

Objective B Practice

Determine if each statement is a proportion.

B3. $\frac{7}{13} \overset{?}{=} \frac{2}{3}$

B4. $\frac{14}{18} \overset{?}{=} \frac{35}{45}$

B5. $\frac{10}{18} \overset{?}{=} \frac{25}{45}$

B6. $\frac{42}{100} \overset{?}{=} \frac{22}{55}$

B7. $\frac{1\frac{1}{5}}{8} \overset{?}{=} \frac{1\frac{1}{2}}{10}$

B8. $\frac{1\frac{1}{3}}{9} \overset{?}{=} \frac{1\frac{7}{9}}{12}$

B9. $\frac{3.1}{2} \overset{?}{=} \frac{9.4}{6}$

B10. $\frac{2.5}{3} \overset{?}{=} \frac{15}{18}$

B11. Wilton is the highest-scoring player on the college basketball team.

	Points Scored	Minutes Played
Game 1	24	10
Game 2	50	25

a. Determine Wilton's scoring rate in the units of $\frac{\text{points scored}}{\text{minutes played}}$ for each game.

b. Calculate the cross products to see if the two rates are proportional.

B12. Jesse earns extra money by plowing snow in the winter.

	Driveways Plowed	Hours of Work
Day 1	5	2
Day 2	7	3

a. Determine Jesse's plowing rate in the units of $\frac{\text{drives plowed}}{\text{hours worked}}$ for each day.

b. Calculate the cross products to see if the two rates are proportional.

FOCUS ON Proportional Reasoning

It is useful to develop proportional reasoning skills. The techniques that you learn here will help you

- Estimate the solution to a proportion.
- Solve many proportions in your head.
- Check to see if the solution of a proportion is reasonable.

Build the proportions so that they have the same numerator or denominator.

- If the numerators are equal, then the denominators must be equal.
- If the denominators are equal, then the numerators must be equal.

EXAMPLE

7. Solve. $\frac{7}{20} = \frac{35}{n}$

Build a common numerator.

$$\frac{7}{20} = \frac{35}{n}$$

Step 1 Determine the missing multiplier between the ratios.

$$1\left(\frac{?}{?}\right) \cdot \frac{7}{20} = \frac{35}{n}$$

Step 2 Multiply.

$$1\left(\frac{5}{5}\right) \cdot \frac{7}{20} = \frac{35}{n}$$

Step 3 Compare the fractions to solve for n.

$$\frac{35}{100} = \frac{35}{n}$$

$$100 = n$$

GUIDED PRACTICE

7. Solve. $\frac{3}{6} = \frac{n}{12}$

Build a common denominator.

$$\frac{3}{6} \cdot = \frac{n}{12}$$

$$1\left(\frac{?}{?}\right) \cdot \frac{3}{6} = \frac{n}{12}$$

$$1\left(\frac{\quad}{\quad}\right) \cdot \frac{3}{6} = \frac{n}{12}$$

$$\frac{\quad}{12} = \frac{n}{12}$$

$$\quad = n$$

PRACTICE

Solve each proportion.

1. $\frac{5}{6} = \frac{n}{18}$ 2. $\frac{5}{4} = \frac{n}{24}$ 3. $\frac{n}{8} = \frac{36}{48}$ 4. $\frac{n}{7} = \frac{9}{63}$ 5. $\frac{15}{n} = \frac{3}{4}$

6. $\frac{6}{7} = \frac{42}{n}$ 7. $\frac{6}{9} = \frac{2}{n}$ 8. $\frac{10}{45} = \frac{2}{n}$ 9. $\frac{n}{4} = \frac{8}{1}$ 10. $\frac{n}{2} = \frac{3}{1}$

In a proportion, numerators and denominators must be the same multiple of each other.

- Each numerator must be the same multiple of its denominator.
- Each denominator must be the same multiple of its numerator.

EXAMPLE

8. Solve. $\frac{10}{5} = \frac{n}{7}$

Step 1 Compare the values in the known fraction.

$$10 = 2 \cdot 5$$

$$\text{Numerator} = 2 \cdot \text{denominator}$$

Step 2 Use the same relationship in the fraction with the unknown.

$$\text{Numerator} = 2 \cdot \text{denominator}$$

$$n = 2 \cdot 7$$

$$n = 14$$

GUIDED PRACTICE

8. Solve. $\frac{3}{24} = \frac{13}{n}$

Step 1 Compare the values in the known fraction.

24 = ______ · 3

Denominator = ______ · numerator

Step 2 Use the same relationship in the fraction with the unknown.

Denominator = ______ · numerator

n = ______ · 13

n =

PRACTICE

Solve each proportion.

11. $\frac{27}{9} = \frac{n}{12}$ 12. $\frac{20}{4} = \frac{n}{11}$ 13. $\frac{n}{15} = \frac{2}{6}$ 14. $\frac{n}{20} = \frac{8}{32}$ 15. $\frac{3}{12} = \frac{7}{n}$

16. $\frac{6}{n} = \frac{2}{12}$ 17. $\frac{12}{n} = \frac{20}{5}$ 18. $\frac{72}{8} = \frac{18}{n}$ 19. $\frac{n}{\frac{1}{2}} = \frac{9}{3}$ 20. $\frac{24}{4} = \frac{n}{\frac{1}{4}}$

Use Estimation to Check the Solution of a Proportion

How can your instructor tell if an answer is incorrect without performing any computations? He or she knows how to estimate the unknown value in a proportion. In each problem on the next page, estimation is used to show that the red number has been calculated incorrectly.

(Continued)

Example

$$\frac{12}{78} \neq \frac{169}{26}$$

12 is less than 78, so
169 should be less than 26.

A fraction less than one cannot be equal to a fraction greater than one.

Example

$$\frac{15}{16} \neq \frac{14}{98}$$

15 is close to 16, so
14 should be close to 98.

A fraction a little less than one cannot be equal to a fraction much less than one.

Example

$$\frac{3}{6} \neq \frac{9}{20}$$

6 is twice 3,
20 is not twice 9.

Each fraction's denominator is not the same multiple of the fraction's numerator.

PRACTICE

Explain why each of the red numbers is incorrect.

Use estimation to show that each of these proportions is false. There are many ways to show that each proportion is false, so don't be surprised if your answer does not match the answer in the back of the book.

21. $\frac{43}{32} = \frac{15}{23}$ **22.** $\frac{5}{15} = \frac{13}{38}$ **23.** $\frac{88}{90} = \frac{41}{63}$ **24.** $\frac{20}{100} = \frac{6}{36}$

25. $\frac{1}{3} = \frac{2}{4}$ **26.** $\frac{4}{10} = \frac{2}{7}$ **27.** $\frac{53}{30} = \frac{62}{60}$ **28.** $\frac{1}{1{,}000} = \frac{2}{500}$

Objective C Solve a Proportion

The Concept Proportions are useful for solving problems where something occurs with a known ratio or rate. Consider the following examples:

- The rate 60 mph can be used in a proportion to determine that a 300-mile trip will take 5 hours to complete.
- The ratio $\frac{16}{9}$ can be used in a proportion to determine that a 27-inch-tall HDTV display will have a width of 48 inches.

The proportion exercises in this section look similar to the following exercise:

$$\underbrace{\frac{16}{9} = \frac{n}{27}}_{\text{variable in the numerator}} \quad \text{or} \quad \underbrace{\frac{9}{16} = \frac{27}{n}}_{\text{variable in the denominator}}$$

The two proportions are equivalent to each other. Each proportion has a known ratio or rate on one side of the equation and a single unknown quantity on the other side. When solving a proportion, get the variable in the numerator and then get the variable alone on one side of the equal sign.

Procedure Solve a Proportion

Step 1 Get the variable in the numerator.
Rewrite the proportion if necessary.

Step 2 Undo the division to get the variable alone.

Step 3 Simplify each side.

Step 4 Check the answer.
Substitute the solution into the original proportion. Then calculate and compare the cross products.

DETAILED EXAMPLE Solving a Proportion

Solve the proportion. $\frac{5}{3} = \frac{10}{n}$

$$\frac{5}{3} = \frac{10}{n}$$

Step 1 Rewrite the proportion to get the variable in the numerator. $\frac{3}{5} = \frac{n}{10}$

Step 2 In $\frac{n}{10}$, n is divided by 10. Multiply both sides by $\frac{10}{1}$ to undo this. $\frac{3}{5} \cdot \frac{10}{1} = \frac{n}{10} \cdot \frac{10}{1}$

Step 3 Simplify each side. $\frac{3}{\cancel{5}} \cdot \frac{2 \cdot \cancel{5}}{1} = \frac{n}{\cancel{10}} \cdot \frac{\cancel{10}}{1}$

$$\frac{6}{1} = n$$

$$6 = n$$

Step 4 Check the answer. $\frac{5}{3} = \frac{10}{n}$

Replace n with the solution $n = 6$. $\frac{5}{3} \stackrel{?}{=} \frac{10}{6}$

Calculate the cross products of the proportion. $3 \cdot 10 \stackrel{?}{=} 5 \cdot 6$

Since the cross products are equal, the solution is correct. $30 = 30$

EXAMPLES

9. Solve the proportion. $\frac{n}{12} = \frac{5}{3}$

Step 1 The variable is already in the numerator. $\frac{n}{12} = \frac{5}{3}$

Step 2 A number is dividing the variable. To undo division, multiply both sides by the number. $\frac{\cancel{12}}{1} \cdot \frac{n}{\cancel{12}} = \frac{12}{1} \cdot \frac{5}{3}$

$n = \frac{\cancel{3} \cdot 4}{1} \cdot \frac{5}{\cancel{3}}$

Step 3 Simplify. $n = 20$

Step 4 Check the answer.

Replace n with the solution. $\frac{20}{12} \stackrel{?}{=} \frac{5}{3}$

Calculate and compare the cross products of the proportion. $20 \cdot 3 \stackrel{?}{=} 5 \cdot 12$

$60 = 60$

10. Solve the proportion. $\frac{3}{n} = \frac{6}{5}$

Step 1 Rewrite the proportion to get the variable in the numerator. $\frac{3}{n} = \frac{6}{5}$

$\frac{n}{3} = \frac{5}{6}$

Step 2 A number is dividing the variable. To undo division, multiply both sides by the number. $\frac{\cancel{3}}{1} \cdot \frac{n}{\cancel{3}} = \frac{3}{1} \cdot \frac{5}{6}$

$n = \frac{\cancel{3}}{1} \cdot \frac{5}{\cancel{3} \cdot 2}$

Step 3 Simplify. $n = \frac{5}{2}$

Step 4 Check the answer. $\frac{3}{\left(\frac{5}{2}\right)} \stackrel{?}{=} \frac{6}{5}$

Replace n with the solution.

Calculate and compare the cross products of the proportion. $3 \cdot 5 \stackrel{?}{=} \frac{5}{2} \cdot 6$

$3 \cdot 5 \stackrel{?}{=} \frac{5}{2} \cdot \frac{6}{1}$

Simplify. $15 \stackrel{?}{=} \frac{5}{\cancel{2}} \cdot \frac{\cancel{2} \cdot 3}{1}$

$15 = 15$

GUIDED PRACTICE

9. Solve the proportion. $\frac{n}{28} = \frac{3}{14}$

$\frac{n}{28} = \frac{3}{14}$

$\frac{\quad}{1} \cdot \frac{n}{28} = \frac{\quad}{1} \cdot \frac{3}{14}$

$n = \frac{\quad}{\quad} \cdot \frac{\quad}{\quad}$

$n =$

Step 4 Check the answer.

$\frac{\quad}{28} \stackrel{?}{=} \frac{3}{14}$

____ · ____ $\stackrel{?}{=}$ ____ · ____

$=$

10. Solve the proportion. $\frac{4}{n} = \frac{16}{7}$

$\frac{4}{n} = \frac{16}{7}$

$\frac{\quad}{\quad} = \frac{\quad}{\quad}$

$\frac{\quad}{1} \cdot \frac{\quad}{\quad} = \frac{\quad}{1} \cdot \frac{\quad}{\quad}$

$n = \frac{\quad}{1} \cdot \frac{\quad}{\quad}$

$n = \frac{\quad}{\quad}$

Step 4 Check the answer.

$\frac{4}{\quad} \stackrel{?}{=} \frac{16}{7}$

$\stackrel{?}{=}$

$\stackrel{?}{=}$

$\stackrel{?}{=}$

$=$

11. Write and solve the proportion: 1.5 is to 5 as n is to 10.

Step 1 The variable is already in the numerator. $\frac{1.5}{5} = \frac{n}{10}$

Step 2 A number is dividing the variable. To undo division, multiply both sides by the number. $\frac{1.5}{5} \cdot \frac{10}{1} = \frac{n}{\cancel{10}} \cdot \frac{\cancel{10}}{1}$

$\frac{1.5}{\cancel{5}} \cdot \frac{2 \cdot \cancel{5}}{1} = n$

Step 3 Simplify. $3 = n$

Step 4 Check the answer.

Replace n with the solution. $\frac{1.5}{5} \stackrel{?}{=} \frac{3}{10}$

Calculate and compare the cross products of the proportion. $3 \cdot 5 \stackrel{?}{=} 1.5 \cdot 10$

$15 = 15$

11. Write and solve the proportion: 1.5 is to 4 as n is to 12.

$\frac{\quad}{\quad} = \frac{n}{\quad}$

$\frac{\quad}{\quad} \cdot \frac{\quad}{\quad} = \frac{\quad}{\quad} \cdot \frac{\quad}{\quad}$

$\frac{\quad}{\quad} \cdot \frac{\quad}{\quad} = n$

$\quad = n$

Step 4 Check the answer.

$\frac{1.5}{4} \stackrel{?}{=} \frac{\quad}{12}$

$\stackrel{?}{=}$

$=$

Concept Check

C1. Write a story problem for the proportion.

$$\frac{1 \text{ cookie}}{250 \text{ calories}} = \frac{3 \text{ cookies}}{x \text{ calories}}$$

C2. Rewrite the proportion above with the variable in the numerator.

C3. In the 30 days of November, Vic noticed that 4 out of every 5 days were cloudy. To figure out the number of cloudy days in November, he wrote the following proportion. How do you know that the proportion is not written correctly?

$$\frac{4}{5} = \frac{30}{C}$$

Objective C Practice

Solve each proportion.

C4. $\frac{3}{5} = \frac{n}{15}$

C5. $\frac{7}{12} = \frac{n}{24}$

C6. $\frac{2}{5} = \frac{7}{n}$

C7. $\frac{13}{11} = \frac{2}{n}$

C8. $\frac{5}{24} = \frac{n}{16}$

C9. $\frac{3}{35} = \frac{n}{21}$

C10. $\frac{1.2}{7} = \frac{n}{10}$

C11. $\frac{3.8}{3} = \frac{n}{5}$

Set up a proportion to solve each exercise.

C12. To make lemonade, Cassandra follows a recipe that calls for 3 lemons to make 2 quarts of lemonade. If she follows this recipe and uses 24 lemons, how many quarts of lemonade can she make?

C13. Echo lives 40 miles from her college. She drove the first 16 miles of her commute in 20 minutes. If her rate of speed stays the same, how long will her entire commute take?

Section 4.2 Exercises

FOR EXTRA HELP   PRACTICE  WATCH  DOWNLOAD  READ  REVIEW

To write a proportion:

1. Answer the Objective A Concept Check questions.
2. Answer the odd Objective A Practice Exercises.
3. Answer the even Objective A Practice Exercises.

To determine if a statement is a proportion:

4. Answer the Objective B Concept Check questions.
5. Answer the odd Objective B Practice Exercises.
6. Answer the even Objective B Practice Exercises.

To develop proportional reasoning skills:

7. Answer the odd Focus on Proportional Reasoning Skills Practice Exercises.
8. Answer the even Focus on Proportional Reasoning Skills Practice Exercises.

To solve a proportion:

9. Answer the Objective C Concept Check questions.
10. Answer the odd Objective C Practice Exercises.
11. Answer the even Objective C Practice Exercises.

VOCABULARY REVIEW *Review the Vocabulary Preview for Section 4.2. Study the definitions until you can check* Got It *for every word.*

Use these words to complete each sentence.

ratio • rate • reference fraction • proportion

	Got It	Get Help
12. A ____________ is a quotient that compares two quantities with different units.		
13. A ____________ states that two ratios or rates are equal.		
14. A quotient that compares two quantities with the same units is called a ____________.		

How will you get help for any vocabulary that you are unsure about?

Your instructor ____ MyMathLab ____ A classmate ____ A tutor ____ Other ____

Write a proportion for each statement. Do not solve the proportion.

15. Five is to three as n is to seven.

16. Four is to nine as n is to three.

17. Eight is to eleven as six hundredths is to n.

18. Six is to thirteen as seven thousandths is to n.

19. Fifty-three out of every one hundred people in town are opposed to a proposition. If 4,300 people live in the town, how many people are opposed to the proposition?

20. At a manufacturing company, four out of every one thousand parts made are defective. If thirteen parts are defective, how many parts were made?

Write a proportion for each statment. Do not solve the proportion.

21. Pavona's *Hammerhead* was purchased at a local art show for a unit cost of about $0.261 per square inch. The painting has an area of 766 square inches. Approximately what was the purchase price?

22. Picasso's *Boy with a Pipe* sold at auction for a price of about $1,093.75 per square inch. The painting has an area of 1,280 square inches. Approximately what was the purchase price?

23. Sansui earns $8.55 per hour as a librarian's assistant. If Sansui worked 30 hours last week, how much did she earn?

24. Quincy paid $25 for four hours of babysitting. What hourly rate did Quincy pay the babysitter?

25. Jessica's report is 3,000 words long. If each page has 250 words, how many pages is the report?

26. Arnold's report has 300 words per page. If the report is 8 pages long, how many words are in the report?

27. A chili recipe required $\frac{1}{4}$ cup of jalapeño peppers. Heinrich used $1\frac{1}{2}$ cups of jalepenos to make several batches of chili. How many batches did Heinrich make?

28. Emilio is making a soup called borscht from a recipe that requires $3\frac{1}{2}$ cups of chopped cabbage. If Emilio uses 14 cups of cabbage, how many batches of borscht can he make?

Determine if each statement is a proportion.

29. $\frac{15}{7} \stackrel{?}{=} \frac{60}{28}$

30. $\frac{4}{12} \stackrel{?}{=} \frac{20}{60}$

31. $\frac{15}{25} \stackrel{?}{=} \frac{21}{34}$

32. $\frac{6}{9} \stackrel{?}{=} \frac{11}{15}$

33. $\frac{2\frac{1}{3}}{3} \stackrel{?}{=} \frac{7}{15}$

34. $\frac{7\frac{1}{3}}{3} \stackrel{?}{=} \frac{23}{9}$

35. $\frac{3.5}{\frac{1}{3}} \stackrel{?}{=} \frac{61}{6}$

36. $\frac{2.5}{4\frac{3}{4}} \stackrel{?}{=} \frac{8}{20}$

37. $\frac{82 \text{ miles}}{2 \text{ hours}} \stackrel{?}{=} \frac{123 \text{ miles}}{3 \text{ hours}}$

38. $\frac{95 \text{ miles}}{6 \text{ hours}} \stackrel{?}{=} \frac{105 \text{ miles}}{7 \text{ hours}}$

39. $\frac{38 \text{ people}}{2 \text{ buses}} \stackrel{?}{=} \frac{19 \text{ buses}}{1 \text{ person}}$

40. $\frac{14 \text{ hours}}{2 \text{ days}} \stackrel{?}{=} \frac{49 \text{ days}}{7 \text{ hours}}$

Follow the instructions to answer each question.

41. Is windchill proportional to wind speed?

Windchill relates the wind speed to an apparent temperature that your skin feels. During a winter day, the wind was blowing at 30 mph and the windchill was 5°F. Later that day, the wind was blowing at 5 mph and the windchill was 33°F.

a. Write a proportion for the two rates.
b. Calculate the cross products to determine if these two events are proportional.

42. Is the measured rainfall proportional to time?

During a rainstorm, it rained 0.125 inch the first hour. After 4 hours, it had rained a total of 0.5 inch.

a. Write a proportion for the two rates.
b. Calculate the cross products to determine if these two events are proportional.

43. Is the area of a square proportional to the length of its sides?

A square with side lengths of 3 inches has an area of 9 square inches. A square with side lengths of $\frac{1}{3}$ inch has an area equal to $\frac{1}{9}$ square inch.

a. Write a ratio for the side lengths and another ratio for the areas of the squares.
b. Write a proportion for the two ratios.
c. Calculate the cross products to determine whether the two ratios are proportional.

44. Is the area of a triangle proportional to the length of its base and height?

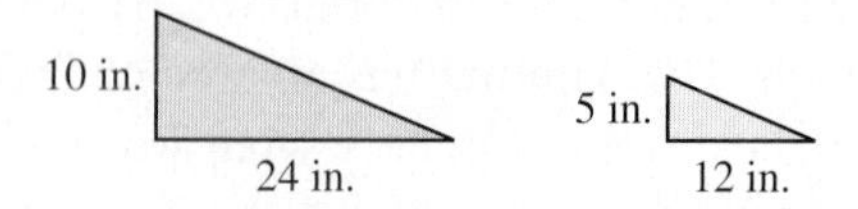

These triangles are called *similar triangles* because the corresponding sides are the same multiple of each other. In the diagram, the corresponding sides are the sides that have the same relative position. The area of the first triangle is 120 square inches. The area of the second triangle is 30 square inches.

a. Write a ratio for one set of corresponding side lengths and another ratio for the areas of the triangles.
b. Write a proportion for the two ratios.
c. Calculate the cross products to determine whether the two ratios are proportional.

45. Is postage proportional to weight? Elfreide mailed letters to her two grandchildren. The first letter weighed 3 ounces and required \$0.87 in stamps. The second letter weighed 8 ounces and cost \$2.07 to mail. Was the postage proportional to the weight of the letters?

46. Is a phone bill proportional to the minutes used? Last month, Joseph had 50 minutes of long-distance calls and a bill of \$11. This month, he had 85 minutes of calls and a bill of \$13.45. Is his bill proportional to the number of minutes of calls?

47. Is Giuseppe's model proportional to the actual ship?

Giuseppe is building a scale model of a boat. The actual ship has a length of 320 feet. The model is 9.6 inches long. The actual ship is 60 feet wide. The model is 1.8 inches wide.

a. Write a proportion for the two rates.
b. Calculate the cross products to determine whether the model is proportional.

48. Is the length of a shadow proportional to the height of an object?

Aretha wants to show that the length of an object's shadow is proportional to its height. She is 5 feet tall, and her shadow is 4.5 feet long. A flagpole is 30 feet tall and has a shadow that is 27 feet long.

a. Write a proportion for the two ratios.
b. Calculate the cross products to determine whether the shadows are proportional to the heights.

Solve each proportion. State the units of your answer when appropriate.

49. $\frac{n}{5} = \frac{6}{15}$

50. $\frac{n}{4} = \frac{3}{12}$

51. $\frac{1}{6} = \frac{2}{n}$

52. $\frac{22}{n} = \frac{11}{45}$

53. $\frac{n}{20} = \frac{3.2}{10}$

54. $\frac{6.4}{8} = \frac{n}{16}$

55. $\frac{450}{n} = \frac{5}{3}$

56. $\frac{20}{520} = \frac{40}{n}$

57. $\frac{11}{27} = \frac{n}{3}$

58. $\frac{n}{6} = \frac{8}{9}$

59. $\frac{3.5}{7} = \frac{10.5}{n}$

60. $\frac{0.75}{2} = \frac{n}{8}$

61. $\frac{\$n}{40 \text{ hours}} = \frac{\$100}{8 \text{ hours}}$

62. $\frac{36 \text{ inches}}{1 \text{ yard}} = \frac{n \text{ inches}}{5 \text{ yards}}$

63. $\frac{3 \text{ kilograms}}{60 \text{ pounds}} = \frac{5 \text{ kilograms}}{n \text{ pounds}}$

64. $\frac{21 \text{ teachers}}{546 \text{ students}} = \frac{1 \text{ teacher}}{n \text{ students}}$

65. $\frac{16 \text{ credit hours}}{\$1{,}000} = \frac{1 \text{ credit hour}}{n \text{ dollars}}$

66. $\frac{24 \text{ staplers}}{1 \text{ case}} = \frac{n \text{ staplers}}{15 \text{ cases}}$

Answer each question using a proportion.

67.

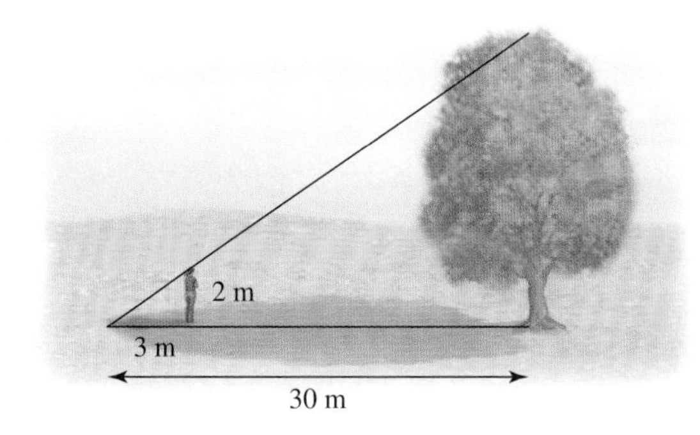

When sunlight casts a shadow on an object, the length of the shadow is proportional to the height of the object. If a 2-meter person casts a shadow that is 3 meters long, how tall is a tree that has a 30-meter shadow?

68. An arctic explorer must eat 8,000 calories a day to maintain her weight. If her diet consists entirely of Chesuncook cookies that have 950 calories each, how many cookies must she eat per day to maintain her weight? Round your answer to the nearest whole cookie.

69. A jet travels 850 miles in two hours. How long will it take the plane to travel 1,275 miles if it continues to fly at the same rate?

70. A blueprint is drawn so that 1 inch on the diagram represents 5 feet on a building. If the blueprint shows the height of a wall with a 4.6-inch line, what is the actual height of the wall?

tobey This method was used by Lewis & Clark during their 1804 explorations of the U.S. Have you needed to measure a distance outdoors? *For more tips and tweets, go to twitter.com/gstbasicmath*

Section 4.2 Question Log

Use this space to write questions. Make sure you get these answered and revisit them when you prepare for your exam.

Page ______ **Answered**	**Page** ______ **Answered**
Page ______ **Answered**	**Page** ______ **Answered**

Section 4.3 Applications of Ratios, Rates, and Proportions

Many occupations require an understanding of proportions. This section presents three applications you may find useful.

Scale drawings are important to people in technical fields.

Statistical sampling is used by news organizations to present information about our world.

Dosage calculations are critical for people in medical fields.

The **Objectives** for Section 4.3 will help you

- A Interpret a scale drawing.
- B Use a sampling method.
- C Compare a prescription dose to the usual dose.
- D Determine how much medication to give.

VOCABULARY PREVIEW *Check the box that applies.*	Got It	Must Study
Scale: A **scale** describes how the dimensions of a model relate to the dimensions of an actual object.		
Population: A **population** is an entire collection of objects to be studied.		
Sample: A **sample** is a small portion of a population used to estimate a quantity for the entire population.		
Prescribed Dose: The **prescribed dose** is the amount of medication prescribed to a patient.		
Drug Concentration: The **drug concentration** is a rate that indicates the amount of drug in a specific volume of solution.		
Usual Dose: The **usual dose** is the amount of medication typically given to a patient.		

Study these words when they appear in the text. After the section, test your understanding by completing the Vocabulary Review in the section exercises.

Objective A Interpret a Scale Drawing

The Concept Many jobs require the ability to use a scale drawing. While the example in this objective is a scale drawing of a cabin, the concepts apply to a scale drawing of any object.

A blueprint is a scale or proportional drawing. Scale drawings are used to design buildings, ships, mechanical parts, and cars. Every blueprint should have a scale printed on it. The **scale** describes how the dimensions of the drawing relate to the actual object.

Scale Drawing of a Cabin
Assume the gridlines are spaced 1 inch apart

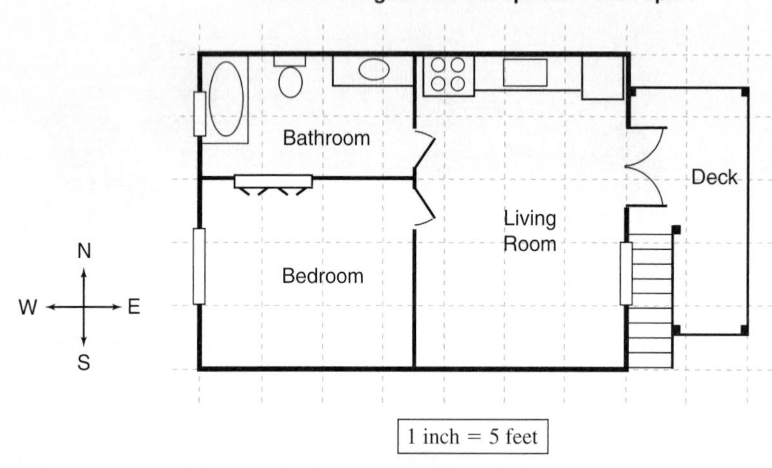

1 inch = 5 feet

The scale on the drawing of the cabin indicates that 1 inch represents 5 feet. This scale can be written as a rate.

$$\frac{5 \text{ feet}}{1 \text{ inch}} \text{ or } \frac{1 \text{ inch}}{5 \text{ feet}}$$

Writing a proportion using either of these rates, we can determine:

- A measurement in the cabin based on the drawing.
- A measurement in the drawing based on the cabin.

$$\frac{5 \text{ feet}}{1 \text{ inch}} = \frac{\text{Measure of cabin}}{\text{Measure in the drawing}}$$

Given one of these measures, we can solve the proportion for the other.

▶ **IMPORTANT** When writing a proportion, it is critical that the units in each rate be the same. Notice that the unit in each numerator is feet and the unit in each denominator is inches.

Procedure **Interpret a Scale Drawing**

Step 1 Set up a proportion using the rate defined by the scale.

Step 2 Solve the proportion.

EXAMPLES

1. Measuring from west to east, including the deck, what is the overall length of the cabin?

On the blueprint, the length of the cabin is 7 inches.

Step 1 Set up. $\frac{5 \text{ ft}}{1 \text{ in.}} = \frac{n \text{ ft}}{7 \text{ in.}}$

Step 2 Solve. $\frac{5 \text{ ft}}{1 \text{ in.}} \cdot \frac{7 \text{ in.}}{1} = \frac{n \text{ ft}}{7 \text{ in.}} \cdot \frac{7 \text{ in.}}{1}$

$35 \text{ ft} = n$

The length of the cabin is 35 feet.

2. Find the length and width of the bedroom to calculate its area.

Find the actual dimensions of the bedroom.

Length of the bedroom (west to east):

Step 1 Set up. $\frac{5 \text{ ft}}{1 \text{ in.}} = \frac{n \text{ ft}}{3.5 \text{ in.}}$

Step 2 Solve. $\frac{5 \text{ ft}}{1 \text{ in.}} \cdot \frac{3.5 \text{ in.}}{1} = \frac{n \text{ ft}}{3.5 \text{ in.}} \cdot \frac{3.5 \text{ in.}}{1}$

$17.5 \text{ ft} = n$

Width of the bedroom (north to south):

Step 1 Set up. $\frac{5 \text{ ft}}{1 \text{ in.}} = \frac{n \text{ ft}}{3 \text{ in.}}$

Step 2 Solve. $\frac{5 \text{ ft}}{1 \text{ in.}} \cdot \frac{3 \text{ in.}}{1} = \frac{n \text{ ft}}{3 \text{ in.}} \cdot \frac{3 \text{ in.}}{1}$

$15 \text{ ft} = n$

Use the actual dimensions to find the area.

Area of the bedroom:

$A = \text{Length} \cdot \text{Width}$

$= 17.5 \text{ ft} \cdot 15 \text{ ft}$

$= 262.5 \text{ ft}^2$

GUIDED PRACTICE

1. Measuring from north to south, what is the overall width of the cabin?

On the blueprint, the width of the cabin is ____ inches.

Step 1 Set up. $\frac{5 \text{ ft}}{1 \text{ in.}} = \frac{n \text{ ft}}{\quad}$

Step 2 Solve. $\frac{5 \text{ ft}}{1 \text{ in.}} \cdot \frac{\quad}{\quad} = \frac{n \text{ ft}}{\quad} \cdot \frac{\quad}{\quad}$

$= n$

The width of the cabin is ______.

2. Find the length and width of the bathroom to calculate its area.

Find the actual dimensions of the bathroom.

Length of the bathroom (west to east):

Step 1 Set up. $\frac{5 \text{ ft}}{1 \text{ in.}} = \frac{n \text{ ft}}{\quad}$

Step 2 Solve. $=$

$=$

Width of the bathroom (north to south):

Step 1 Set up. $\frac{5 \text{ ft}}{1 \text{ in.}} = \frac{\quad}{\quad}$

Step 2 Solve. $=$

$=$

Use the actual dimensions to find the area.

Area of the bathroom:

$A = \text{Length} \cdot \text{Width}$

$=$

$=$

Concept Check

A1. If a scale drawing shows a 2-inch line that represents an 8-foot wall, the scale is 1 in. = _____.

A2. If the scale on a drawing is 1 in. = 6 ft, what is the actual length of an object represented by a 5-inch line on the drawing?

Objective A Practice

Use the diagram of the cabin on page 4-36 to answer each question.

A3. Measuring from north to south, what is the actual width of the living room?

A4. Measuring from west to east, what is the actual length of the living room?

A5. If installed carpet costs \$1.50 per square foot, how much will it cost to carpet the living room of the cabin?

A6. If installed hardwood floor costs \$2.50 per square foot, how much will it cost to install a hardwood floor in the bedroom of the cabin?

Answer each question.

A7. A map's scale says that 1 inch = 25 miles. If you are 4.5 inches from your destination on the map, how far do you still have to travel?

A8. A model train was made to the scale of 1 inch to 87 inches. That is, it was designed using the proportion $\frac{\text{model length}}{\text{train length}} = \frac{1\text{ in.}}{87\text{ in.}}$.
If the model is 8 inches long, how long is the actual locomotive?

A9. An action figure was made to the scale of 1 inch to 1.25 feet for an actual person. If the action figure is 5.2 inches tall, how tall is the person?

A10. **a.** How tall are you in feet?
b. If someone made a model of you using the scale of 1 inch = 1.25 feet, how tall would the model be?

Objective B Use a Sampling Method

The Concept Sampling methods are used to estimate information about a large group or **population**. When sampling a population, information is gathered from a small group, called a **sample**. This information is then extended to the entire population using a proportion. Three steps are typically used to sample a population:

1. A question is asked about a population.
2. A small group is sampled to establish a rate.
3. The rate is applied to the entire population using a proportion.

For example, a college president wants to answer this question: How many students change their majors? It would be difficult to ask each of the 3,000 students attending the college, so a sampling technique is used. Fifty students are asked if they changed their majors. Of the 50 asked, 21 said yes. This sample forms a rate that can be used to approximate the total number of students that changed their majors.

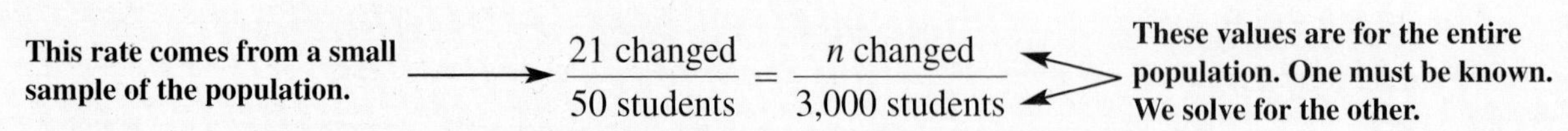

Solving this proportion, we learn that approximately 1,260 students changed majors.

Procedure Use a Sampling Method

Step 1 Use a sample to establish a rate.

Step 2 Use the rate to write a proportion for the entire population.

Step 3 Solve the proportion and answer the question.

EXAMPLE

3. Jeanie purchased a 100-pound (lb) box of mixed jelly beans (red, orange, and white). She scooped out a small sample of jelly beans that weighed 1.3 pounds and contained 103 red jelly beans. Approximately how many red jelly beans are in the box?

Step 1 Use a sample to establish a rate.

$$\text{The rate} = \frac{103 \text{ red jelly beans}}{1.3 \text{ lb}}$$

Step 2 Use the rate to write and solve a proportion.

$$\frac{103 \text{ red jelly beans}}{1.3 \text{ lb}} = \frac{n}{100 \text{ lb}}$$

$$\frac{103 \text{ red jelly beans}}{1.3 \not{\text{lb}}} \cdot \frac{100 \not{\text{lb}}}{1} = \frac{n}{\cancel{100 \text{ lb}}} \cdot \frac{\cancel{100 \text{ lb}}}{1}$$

$$\frac{103 \text{ red jelly beans} \cdot 100}{1.3} = n$$

$$7{,}923 \text{ red jelly beans} \approx n$$

Step 3 Answer the question.

There are approximately 7,923 red jelly beans in the 100 lb box.

GUIDED PRACTICE

3. Jeannie's coworker wonders how many white jelly beans are in the same 100-pound box. If the same 1.3-pound sample has 86 white jelly beans, approximately how many white jelly beans are in the box?

Step 1 Use a sample to establish a rate.

$$\text{The rate} = \frac{\underline{\quad} \text{ white jelly beans}}{1.3 \text{ lb}}$$

Step 2 Use the rate to write and solve a proportion.

$$\frac{\underline{\quad} \text{ white jelly beans}}{1.3 \text{ lb}} = \frac{\quad}{\quad}$$

$$=$$

$$\approx$$

Step 3 Answer the question.

There are approximately ______ white jelly beans in the 100 lb box.

The use of sampling to count the number of jelly beans in a box is silly. However, the same concept can be used to answer more important questions.

Which party will win the county election?

Population = All voters in the county

Sample = 500 voters selected at random

Is the city's water safe?

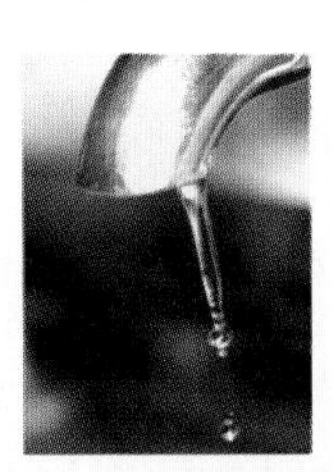

Population = City's water supply

Sample = 1 cup of water from 300 homes

Is the road wide enough?

Population = Amount of yearly traffic on the road

Sample = Amount of weekly traffic on the road

Are there too many elk in Yellowstone National Park?

Population = Number of elk in Yellowstone

Sample = Number of elk in one location of Yellowstone

Concept Check

B1. The college registrar wants to know how many students take the bus to campus. She polled 300 students in the dining hall and concluded that approximately 650 out of 4,500 students ride the bus.
a. What is the population being studied?
b. What is the sample being used?

B2. A farmer wanted to estima3te the value of the upcoming apple harvest. He looked closely at 20 of his 500 trees and concluded that he would harvest about 10,000 bushels of apples.
a. What is the population being studied?
b. What is the sample being used?

B3. When using a sampling technique to answer a question, Sheila said, "There are approximately 5,291 fish in the tank." Why didn't she say, "There are exactly 5,291 fish in the tank?"

Objective B Practice

B4. There are 4,500 students on a college campus. A pollster found that 120 out of 300 students take night classes. Approximately how many students take night classes?

B5. After checking his mail for several days, Brian found that 14 out of 20 pieces of mail were junk mail. If he receives about 1,600 pieces of mail per year, approximately how much junk mail does he receive?

B6. A biologist watched fish swimming upstream. She estimated that 40 fish swam by her in 3 minutes. Approximately how many fish swam upstream in a 24-hour period?

B7. To estimate the number of words in a dictionary, Kendrick counted 68 words on one page. If there are 1,540 pages in the dictionary, approximately how many words are there?

Objective C Compare a Prescription Dose to the Usual Dose

The Concept Before medication is given to patients, two important questions must be answered:

1. Is the prescribed dose safe?
2. What volume of medication must be given?

Failure to answer either question accurately can result in illness or even death of a patient. Because these questions are so important, many nursing programs have a gateway exam for dosage calculations. Students often need to score 100% on the exam to pass the course.

How to read a label on a prescription bottle:
The label below is for the drug Maclor, an antibiotic. The label includes critical information that must be used and interpreted before the drug is given to a patient.

100 mL Maclor MAFACLOR
FOR ORAL SUSPENSION, USP
375 mg per 5 mL. Oversize bottle provides extra room for shaking prior to consumption. May be kept for 14 days without significant lose of potency. Store tightly closed. Discard unused portion 14 days after opening.
SHAKE WELL BEFORE USING

USUAL DOSE: Children: 20 mg per kg per day (40 mg per kg in otitis media) in two divided doses. **Adults:** 375 mg two times per day. See literature for complete dosage details.
Prior to Mixing: Store at a controlled room temperature of 59 to 86°F (15° to 30°C).
Each 5 mL (approx. 1 measured teaspoonful) will contain Mafaclor monohydrate equivalent to 375 mg Maclor.
Mfd. by Roseco Pharmaceuticals in U.S.A.

100 mL when mixed
Roseco Pharmaceuticals
Maclor®
MAFACLOR
for Oral Suspension, USP
375 mg per 5 mL
Caution: Federal (U.S.A.) law prohibits dispensing without prescription.

Usual Dose The **usual dose** is a rate that indicates the appropriate dose for a patient. For safety, it should be compared to the **prescribed dose**. For a child, the rate $\frac{20 \text{ mg}}{1 \text{ kg}}$ indicates that each day, the child should receive 20 milligrams times his or her weight in kilograms. The label also tells us that the dose must be split into two doses per day.

Drug Concentration The **drug concentration** indicates the strength of the medication or amount of drug in a specific volume of solution. It will be used in a proportion to determine how many milliliters of medication must be given to a patient. This bottle's concentration indicates that there are 375 milligrams of drug in every 5 milliliters of solution.

▶ **NOTE** It is easy to confuse the meanings of a *prescribed dose*, the *usual dose*, and the *drug concentration*. Be sure that you can describe the differences between these phrases before your next test!

Procedure Compare a Prescription Dose to the Usual Dose

Step 1 Determine the usual dose from the label.
Read the label to find this rate.

Step 2 Determine the prescribed dose by reading the prescription.
The prescribed dose will be a rate.

Step 3 Compare the two rates.
If the prescribed dose is more than the usual dose, don't give the prescribed dose to the patient.

EXAMPLE

4. A prescription for a 15-kilogram (kg) child reads "Maclor, 300 milligrams (mg), twice daily." Should the drug be given at this dose?

USUAL DOSE: Children: 20 mg per kg per day (40 mg per kg in otitis media) in two divided doses. Adults: 375 mg two times per day. See literature for complete dosage details.
Prior to Mixing: Store at a controlled room temperature of 59 to 86°F (15° to 30°C).
Each 5 mL (approx. 1 measured teaspoonful) will contain Mafaclor monohydrate equivalent to 375 mg Maclor.
Mfd. by Roseco Pharmaceuticals in U.S.A.
100 mL when mixed
Roseco Pharmaceuticals
Maclor®
MAFACLOR for Oral Suspension, USP
375 mg per 5 mL
Caution: Federal (U.S.A.) law prohibits dispensing without prescription.

Step 1 Determine the usual dose from the label.

A maximum of "$\frac{20 \text{ mg}}{1 \text{ kg}}$ a day in two divided doses"

Divide by 2 to get $\frac{10 \text{ mg}}{1 \text{ kg}}$ for a single dose, twice daily.

Step 2 Write the prescribed dose as a unit rate.

$$\text{Dose} = \frac{300 \text{ mg}}{15 \text{ kg}} = \frac{20 \text{ mg}}{1 \text{ kg}}$$

Step 3 Compare the prescribed dose to the usual dose.

$$\frac{20 \text{ mg}}{1 \text{ kg}} > \frac{10 \text{ mg}}{1 \text{ kg}}$$

Since the prescribed dose is more than the usual dose, don't administer the drug to the patient.

GUIDED PRACTICE

4. A prescription for a 5-kilogram (kg) child reads "RyPed 200, 300 milligrams (mg), twice daily." Should the drug be given at this dose?

USUAL DOSE: Children: 33–50 mg/kg per day in divided doses. See literature for complete adult dosage details.
Prior to Mixing: Store at a controlled room temperature below 86°F (30°C).
May be administered without regard to meals.
Each 5 mL (approx. 1 measured teaspoonful) will contain Rythamyacin equivalent to 200 mg RyPed 200.
A product of Drummond Labs, U.S.A.
PEEL
100 mL when mixed
RyPed® 200
RYTHAMYACIN FOR ORAL SUSPENSION, USP
200 mg per 5 mL
Drummond Labs
Caution: Federal (U.S.A.) law prohibits dispensing without prescription.

Step 1 Determine the usual dose from the label.

A maximum of "______ a day in two divided doses"

Divide by 2 to get ______ for a single dose, twice daily.

Step 2 Write the prescribed dose as a unit rate.

Dose = ______

= ______

Step 3 Compare the prescribed dose to the usual dose.

Since the prescribed dose is/isn't more than the usual dose, do/don't administer the drug to the patient.

Concept Check

C1. A medication label reads "Usual Dose—60 mg per kg a day in three divided doses." What is the single dose rate for the medication?

Objective C Practice

Use the prescription labels to answer each question.

USUAL DOSE: Children: 33–50 mg/kg per day in divided doses. See literature for complete adult dosage details.

Prior to Mixing: Store at a controlled room temperature below 86°F (30°C).

May be administered without regard to meals.

Each 5 mL (approx. 1 measured teaspoonful) will contain Rythamyacin equivalent to 200 mg RyPed 200.

A product of Drummond Labs, U.S.A.

PEEL

100 mL when mixed

RyPed® 200

RYTHAMYACIN FOR ORAL SUSPENSION, USP

200 mg per 5 mL

Drummond Labs

Caution: Federal (U.S.A.) law prohibits dispensing without prescription.

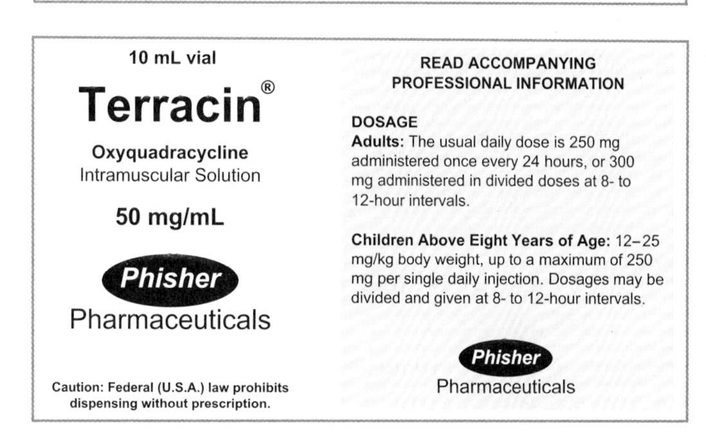

C2. What is the usual dose of RyPed 200 for children?

C3. What is the usual dose of Terracin?

C4. a. Write the prescribed dose, "300 milligrams per day of RyPed for a 12-kilogram child," as a unit rate.

b. Is this prescribed dose safe?

C5. a. Write the prescribed dose, "225 milligrams per day of Terracin for a 15-kilogram child," as a unit rate.

b. Is this prescribed dose safe?

Objective D Determine How Much Medication to Give

The Concept After determining that a prescribed dose is safe, the correct amount of medicine must be given. If the medicine is in liquid form, only part of the medicine is the actual drug. The **drug concentration** is a rate that indicates the amount of drug in a volume of medicine. The drug concentration can be found on the medication label.

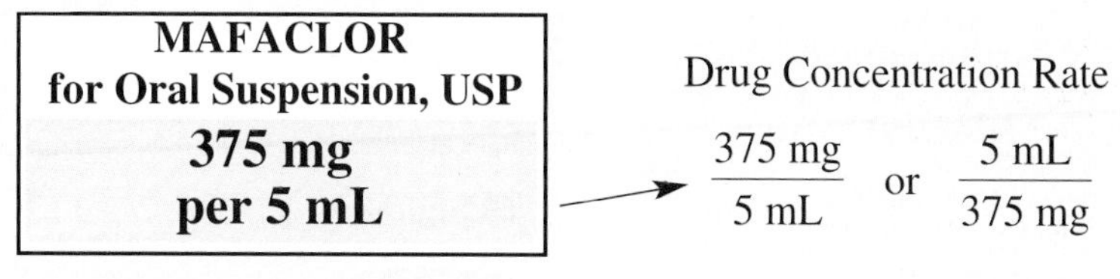

The drug concentration is found on the label.

Once we know the drug concentration, we can write a proportion to find the volume of medicine that must be given to the patient.

Procedure Determine How Much Medication to Give

Step 1 Determine if the prescribed dose is safe.

If the prescribed dose is not safe, draw an X through the syringe and stop.
If the prescribed dose is safe, continue.

Step 2 Determine the volume of medication to give to the patient.

Use the drug concentration to set up and solve a proportion.

Step 3 Shade the appropriate volume on a syringe.

DETAILED EXAMPLE Determining How Much Medication to Give

How many milliliters (mL) are needed to give 750 milligrams (mg) of Mafaclor?

Step 2 Determine the volume of medicine that must be given to the patient.

The drug concentration comes from the label. → $\frac{5 \text{ mL}}{375 \text{ mg}} = \frac{n}{750 \text{ mg}}$ ← **This is the volume to be given.** ← **This is the amount prescribed.**

$$\frac{5 \text{ mL}}{375 \text{ mg}} \cdot \frac{750 \text{ mg}}{1} = \frac{n}{750 \text{ mg}} \cdot \frac{750 \text{ mg}}{1}$$

$$\frac{5 \text{ mL}}{\cancel{375 \text{ mg}}} \cdot \frac{2 \cdot \cancel{375 \text{ mg}}}{1} = \frac{n}{\cancel{750 \text{ mg}}} \cdot \frac{\cancel{750 \text{ mg}}}{1}$$

$$10 \text{ mL} = n$$

Step 3 Using a syringe or dropper, shade the volume that will be given to the patient.

Solving this proportion, we determine that 10 mL of Mafaclor must be used. We show this volume of medication by shading it on an oral syringe.

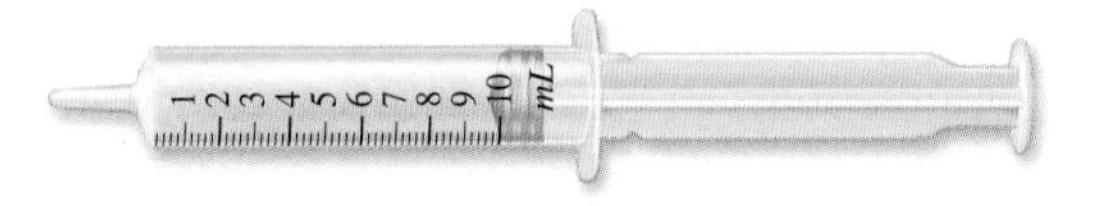

EXAMPLE

5. A prescription for an adult reads a "Maclor, 300 mg, twice daily." How much medicine must be given?

USUAL DOSE: Children: 20 mg per kg per day (40 mg per kg in otitis media) in two divided doses. **Adults:** 375 mg two times per day. See literature for complete dosage details.

Prior to Mixing: Store at a controlled room temperature of 59 to 86°F (15° to 30°C).

Each 5 mL (approx. 1 measured teaspoonful) will contain Mafaclor monohydrate equivalent to 375 mg Maclor.

Mfd. by Roseco Pharmaceuticals in U.S.A.

100 mL when mixed

Roseco Pharmaceuticals

Maclor®

MAFACLOR for Oral Suspension, USP

375 mg per 5 mL

Caution: Federal (U.S.A.) law prohibits dispensing without prescription.

GUIDED PRACTICE

5. A prescription for an adult reads a "Terracin, 125 mg, twice daily." How much medicine must be given?

10 mL vial

Terracin®

Oxyquadracycline
Intramuscular Solution

50 mg/mL

Pharmaceuticals

Caution: Federal (U.S.A.) law prohibits dispensing without prescription.

READ ACCOMPANYING PROFESSIONAL INFORMATION

DOSAGE
Adults: The usual daily dose is 250 mg administered once every 24 hours, or 300 mg administered in divided doses at 8- to 12-hour intervals.

Children Above Eight Years of Age: 12–25 mg/kg body weight, up to a maximum of 250 mg per single daily injection. Dosages may be divided and given at 8- to 12-hour intervals.

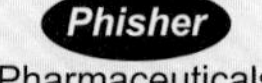

Pharmaceuticals

Step 1 Determine if the prescribed dosage is safe.

Since the usual adult dose is 375 mg taken twice daily, the dose is safe.

Step 2 Determine the volume of medication to give to the patient.

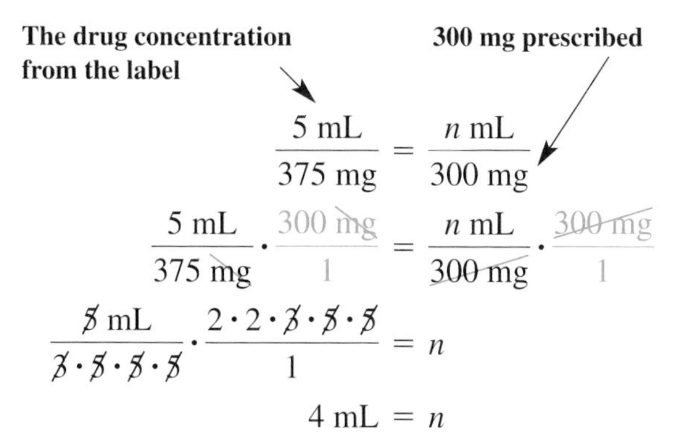

$$\frac{5\text{ mL}}{375\text{ mg}} = \frac{n\text{ mL}}{300\text{ mg}}$$

$$\frac{5\text{ mL}}{375\text{ mg}} \cdot \frac{300\text{ mg}}{1} = \frac{n\text{ mL}}{300\text{ mg}} \cdot \frac{300\text{ mg}}{1}$$

$$\frac{\cancel{5}\text{ mL}}{\cancel{3}\cdot\cancel{5}\cdot\cancel{5}\cdot\cancel{5}} \cdot \frac{2\cdot 2\cdot\cancel{3}\cdot\cancel{5}\cdot\cancel{5}}{1} = n$$

$$4\text{ mL} = n$$

4 mL must be given to the patient.

Step 3 Shade the volume on the oral syringe.

Step 1 Determine if the prescribed dosage is safe.

Since the usual adult dose is ____ mg, the dose is/isn't safe.

Step 2 Determine the volume of medication to give to the patient.

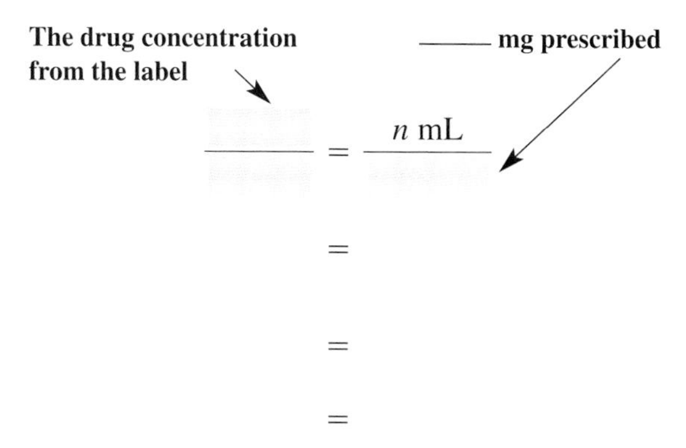

$$\frac{\_\_\_\_}{\_\_\_\_} = \frac{n\text{ mL}}{\_\_\_\_}$$

$$=$$

$$=$$

$$=$$

____ mL must be given to the patient.

Step 3 Shade the volume on the syringe.

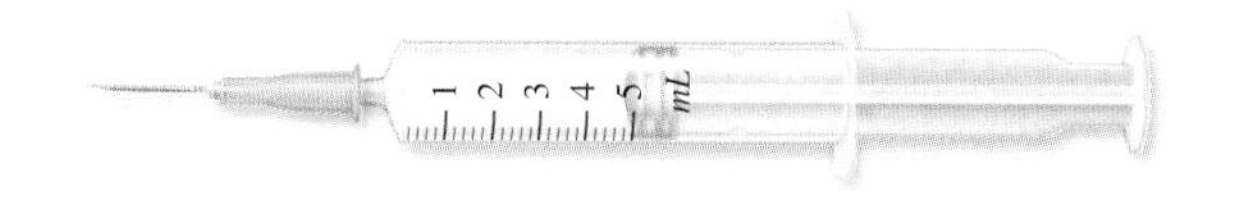

Concept Check

D1. The drug concentration for Terracin is $\frac{50\text{ mg}}{1\text{ mL}}$.

a. How much drug is in 2 mL of Terracin?

b. How much drug is in 3 mL of Terracin?

c. In your own words, explain a drug concentration of $\frac{125\text{ mg}}{2\text{ mL}}$.

Objective D Practice

Use the following labels to answer each question.

USUAL DOSE: Children: 33–50 mg/kg per day in divided doses. See literature for complete adult dosage details.

Prior to Mixing: Store at a controlled room temperature below 86°F (30°C).

May be administered without regard to meals.

Each 5 mL (approx. 1 measured teaspoonful) will contain Rythamyacin equivalent to 200 mg RyPed 200.

A product of Drummond Labs, U.S.A.

PEEL

100 mL when mixed

RyPed® 200

RYTHAMYACIN
FOR
ORAL SUSPENSION, USP

200 mg
per 5 mL

Drummond Labs

Caution: Federal (U.S.A.) law prohibits dispensing without prescription.

10 mL vial

Terracin®

Oxyquadracycline
Intramuscular Solution

50 mg/mL

Pharmaceuticals

Caution: Federal (U.S.A.) law prohibits dispensing without prescription.

READ ACCOMPANYING PROFESSIONAL INFORMATION

DOSAGE
Adults: The usual daily dose is 250 mg administered once every 24 hours, or 300 mg administered in divided doses at 8- to 12-hour intervals.

Children Above Eight Years of Age: 12–25 mg/kg body weight, up to a maximum of 250 mg per single daily injection. Dosages may be divided and given at 8- to 12-hour intervals.

Pharmaceuticals

D2. **a.** What is the usual dose of RyPed?
b. Is the prescribed dose, "350 mg of RyPed per day for a 8 kg child," safe?
c. What is the drug concentration of RyPed?
d. To administer 350 mg of RyPed, how many milliliters must be given?
e. Shade the amount on the oral syringe.

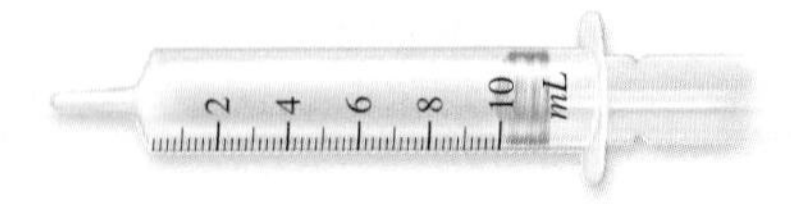

D3. **a.** What is the usual dose of Terracin?
b. Is the prescribed dose, "440 mg of Terracin per day for a 30 kg child," safe?
c. What is the drug concentration of Terracin?
d. To administer 440 mg of Terracin, how many milliliters must be given?
e. Shade the amount on the syringe.

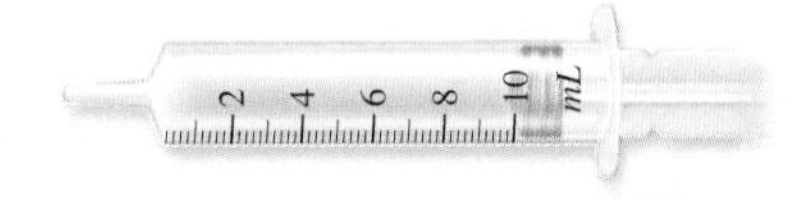

Section 4.3 Exercises

FOR EXTRA HELP

MathXL PRACTICE

WATCH

DOWNLOAD

READ

REVIEW

To interpret a scale drawing:

1. Answer the Objective A Concept Checks.
2. Answer the odd Objective A Practice Exercises.
3. Answer the even Objective A Practice Exercises.

To use a sampling method:

4. Answer the Objective B Concept Checks.
5. Answer the odd Objective B Practice Exercises.
6. Answer the even Objective B Practice Exercises.

To compare a prescription dose to the usual dose:

7. Answer the Objective C Concept Check.
8. Answer the odd Objective C Practice Exercises.
9. Answer the even Objective C Practice Exercises.

To determine how much medication to give:

10. Answer the Objective D Concept Check.
11. Answer the odd Objective D Practice Exercise.
12. Answer the even Objective D Practice Exercise.

VOCABULARY REVIEW *Review the Vocabulary Preview for Section 4.3. Study the definitions until you can check* Got It *for every word.*

Use these words to complete each sentence.

scale • population • sample • prescribed dose • drug concentration • usual dose

	Got It	Get Help
13. The __________ is the amount of medication that is prescribed to a patient.		
14. A __________ is an entire collection of objects to be studied.		
15. The __________ is the amount of medication that is typically given to a patient.		
16. A __________ describes how the dimensions of a model relate to the dimensions of an actual object.		
17. The __________ is a rate that indicates the amount of drug in a volume of medicine.		
18. A __________ is a small portion of a population used to estimate a quantity for the entire population.		

How will you get help for any vocabulary that you are unsure about?

Your instructor ____ MyMathLab ____ A classmate ____ A tutor ____ Other ____

Use the blueprint to answer the following questions. Round each measurement on the model to the nearest quarter square.

Assume the gridlines are 1 inch apart

1 inch = 1.5 feet

19. What is the actual length of the truck?

20. The wheelbase is the distance between the centers of the front and rear wheels. What is the wheelbase of this truck?

21. What is the outer diameter of the tires?

22. How tall is the truck?

23. What is the length of the bed of the truck? The bed is the rear portion and is used for hauling.

24. How far off the ground is the top of the truck bed?

25. How far off the ground is the bottom of the rear bumper?

26. How far off the ground is the bottom of the front bumper?

Use the blueprint and scale to answer the following questions.

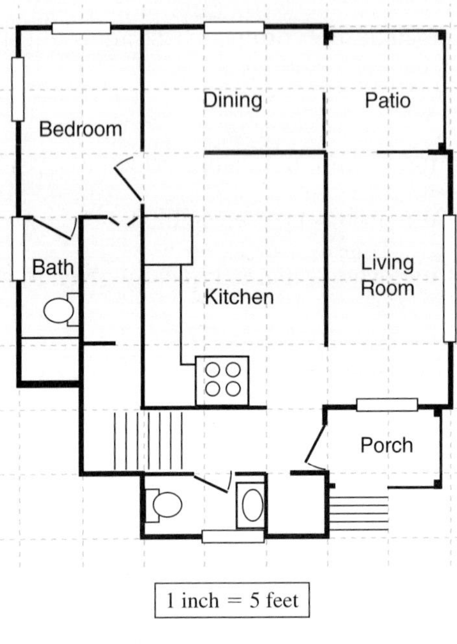

27. What is the length and width of the kitchen?

28. What is the length and width of the living room?

29. How many square feet of flooring is needed in the kitchen? (Area = Length · Width)

30. How many square feet of carpet is needed in the living room? (Area = Length · Width)

31. If every window is 4 feet tall, what is the total area of all the windows?

▶ **NOTE** Not all windows have the same width.

32. Crown molding is the decorative trim installed along the top of a wall. If crown molding is to be installed around the entire perimeter of the dining room, kitchen, and living room, how much crown molding is needed?

33. Baseboard is the decorative trim installed along the bottom of a wall. How much baseboard is needed for the dining room and living room? Do not include doorways.

34. Aluminum flashing is to be installed around the perimeter of the house. How much flashing is needed?

35. The golden ratio is a famous number that is often referenced in architecture. Structures that have a length-to-width ratio equal to the golden ratio are visually pleasing. The golden ratio is approximately 1.6 to 1. Is the ratio of length to width of the living room close to the golden ratio?

Set up and solve a proportion to answer the following questions.

36. A university has a total of 12,000 students. A sample of 180 students were surveyed to see if they approve of the basketball coach's performance. If 137 students from the sample approve, about how many students at the university approve of the coach's performance?

37. The Department of Fisheries released 10,000 lake trout with electronic tags into Lake Superior. Over the course of a season, fishermen caught a sample of 1,000,000 lake trout. Of the lake trout caught, 250 had electronic tags. What is the approximate total population of lake trout in Lake Superior?

38. A citizen of Richland, Florida, complained that "300 quarry trucks a day" were driving past his house during business hours (a total of 9 hours daily). To decide if the citizen's claim was true, a reporter watched the road for half an hour. She observed 17 trucks in that time. Does the citizen's claim appear to be true?

39. When nails are manufactured, 15 of every 10,000 nails are defective. If a company makes 500,000 nails a day, about how many of these nails are defective?

40. A health official wants to know how many people in Jonesville contracted a severe cold last winter. The health worker surveyed 200 people, 83 of whom reported having a severe cold. If 24,000 people live in Jonesville, about how many people had a severe cold?

41. The Kaiser Family Foundation found that 165 out of 300 ER viewers surveyed said that they learned important health information from the television show *ER*. If *ER* had 9,000,000 viewers last week, about how many of them would say that they learned important health information from the show?

42. Four thousand people in Jonesville got a flu vaccine last winter. A health worker polled 150 of them and found that 65 of them had contracted a severe cold. About how many people who got vaccinated ended up with a severe cold?

Apply your knowledge of sampling to the following exercise.

43. Design your own sampling experiment by following the steps below.

a. Ask a question that interests you about the students at your college.

b. Establish a sample by asking a small group of people to answer your question.

c. Look online to determine how many students attend your college.

d. Write and solve a proportion to apply the rate from the sample to the school's population.

e. Did your sample include a good mix of students? (Day vs. evening students, math class vs. history class, traditional students vs. nontraditional students)

f. Why will it make a difference if your sample includes 2 people or 200 people?

g. Do you believe your results are accurate?

h. How could you improve your sampling technique to get a better result?

Use the following drug labels for Exercises 44 through 53.

USUAL DOSE: Children: 20 mg per kg per day (40 mg per kg in otitis media) in two divided doses. **Adults:** 375 mg two times per day. See literature for complete dosage details.

Prior to Mixing: Store at a controlled room temperature of 59 to 86°F (15° to 30°C).

Each 5 mL (approx. 1 measured teaspoonful) will contain Mafaclor monohydrate equivalent to 375 mg Maclor.

Mfd. by Roseco Pharmaceuticals in U.S.A.

100 mL when mixed

Roseco Pharmaceuticals

Maclor®

MAFACLOR
for Oral Suspension, USP

375 mg
per 5 mL

Caution: Federal (U.S.A.) law prohibits dispensing without prescription.

10 mL vial

Terracin®

Oxyquadracycline
Intramuscular Solution

50 mg/mL

Pharmaceuticals

Caution: Federal (U.S.A.) law prohibits dispensing without prescription.

READ ACCOMPANYING PROFESSIONAL INFORMATION

DOSAGE

Adults: The usual daily dose is 250 mg administered once every 24 hours, or 300 mg administered in divided doses at 8- to 12-hour intervals.

Children Above Eight Years of Age: 12–25 mg/kg body weight, up to a maximum of 250 mg per single daily injection. Dosages may be divided and given at 8- to 12-hour intervals.

Pharmaceuticals

USUAL DOSE: Children: 33–50 mg/kg per day in divided doses. See literature for complete adult dosage details.

Prior to Mixing: Store at a controlled room temperature below 86°F (30°C).

May be administered without regard to meals.

Each 5 mL (approx. 1 measured teaspoonful) will contain Rythamyacin equivalent to 200 mg RyPed 200.

A product of Drummond Labs, U.S.A.

PEEL

100 mL when mixed

RyPed® 200

RYTHAMYACIN
FOR
ORAL SUSPENSION, USP

200 mg
per 5 mL

Drummond Labs

Caution: Federal (U.S.A.) law prohibits dispensing without prescription.

120 mL

Florazine®

Hydrofloride Intensate

(Oral Concentrate, USP)

Usual Dosage: 75 to 400 mg daily. See package insert for complete prescribing information. Protect from light, dilute before using.

SUZANNE Laboratories

LOT
EXP.

For each exercise:

a. **Use the medication's usual dose to determine if the prescription is safe. If the prescription is not safe, put an X through the syringe and do not answer parts b and c.**
b. **Determine the volume of medication that should be given for one dose.**
c. **Shade the appropriate volume on the drawing of the syringe.**

44. Florazine: 75 mg per dose, three times daily, to an adult patient

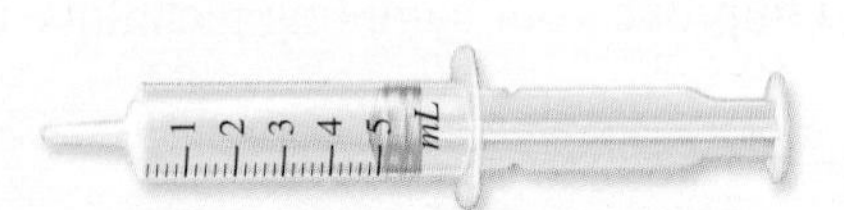

45. RyPed: 500 mg per dose, three times daily, to a 25 kg child

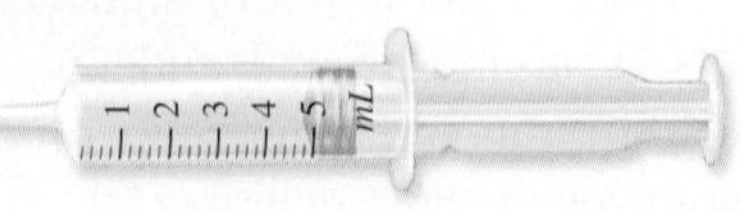

46. Terracin: 80 mg per dose, three times daily, to a 30 kg child

goetz Be sure to study the difference between a prescribed dose, usual dose, and drug concentration to avoid mistakes with dosage exercises. *For more tips and tweets, go to twitter.com/gstbasicmath*

47. Maclor: 300 mg per dose, twice daily, to an adult patient

48. Florazine: 200 mg, three times daily, to an adult patient

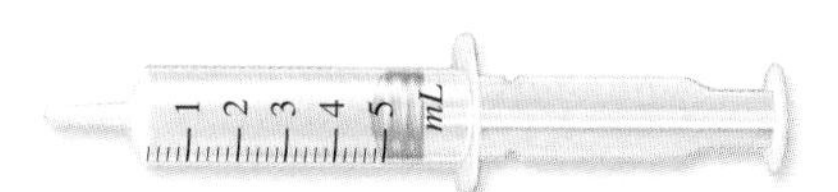

49. Terracin:100 mg per dose, every 8 hours, to a 30-year-old male patient

50. Maclor: 200 mg per dose, twice daily, to a 20 kg child

51. RyPed: 300 mg per dose, three times daily, to a 30 kg child

Section 4.3 Question Log

Use this space to write questions. Be sure to get these answered and revisit them when you prepare for your exam.

Page ____ **Answered**	**Page** ____ **Answered**
Page ____ **Answered**	**Page** ____ **Answered**

Chapter 4 Organizer

VOCABULARY

Use the following steps to review the vocabulary for Chapter 4.

1. *Write the definition for each word from memory.*
2. *Compare the definitions you have written with the definitions in the section Vocabulary Preview.*
3. *Study any definitions that you could not remember or that you defined incorrectly.*

4.1

Ratio • Rate • Unit Rate • Per • Reference Fraction

4.2

Proportion • Reference Fraction

4.3

Scale • Population • Sample • Prescribed Dose • Drug Concentration • Usual Dose

PROCEDURES

Procedure/Topic	Steps	Example
Write a Ratio of *A* to *B* as a Fraction (Section 4.1)	**Step 1** Write *A* (the first number) in the numerator. **Step 2** Write *B* (the second number) in the denominator.	The ratio of 5 to 7 is $\frac{5}{7}$.
Write a Ratio as a Fraction in Simplest Form (Section 4.1)	**Step 1** Write the ratio as a fraction. **Step 2** Clear any decimals. **Step 3** Simplify the fraction.	Write the ratio of 2.2 to 4 in simplest form. $\frac{2.2}{4} = \frac{2.2}{4} \cdot \frac{10}{10}$ $= \frac{22}{40}$ $= \frac{\cancel{2} \cdot 11}{\cancel{2} \cdot 20}$ $= \frac{11}{20}$
Write a Rate (Section 4.1)	**Step 1** Write a fraction with the two quantities. You must include the units. **Step 2** Simplify the fraction if possible.	Write a rate for earning \$50 for 8 hours of work. The rate is $\frac{\$50}{8 \text{ hours}} = \frac{\$25}{4 \text{ hours}}$.
Write a Unit Rate to Compare Two Quantities (Section 4.1)	**Step 1** Write a rate that compares the two quantities. Be sure to include the units. **Step 2** Divide the numerator by the denominator.	Write a unit rate for earning \$50 for 8 hours of work. The rate is $\frac{\$50.00}{8 \text{ hours}} = \frac{\$6.25}{1 \text{ hour}}$
Write a Proportion (Section 4.2)	**Step 1** Write a reference fraction, showing the units. **Step 2** Write the proportion.	Earlier, we drove 60 miles in 1 hour. Later, driving at the same pace, we drove 120 miles in two hours. The reference fraction is $\frac{\text{miles}}{\text{hours}}$. $\frac{60 \text{ miles}}{1 \text{ hour}} = \frac{120 \text{ miles}}{2 \text{ hours}}$

Procedure/Topic	Steps	Example
Determine If a Statement Is a Proportion (Section 4.2)	**Step 1** Calculate the cross products. **Step 2** Check to see if the cross products are equal.	Is $\frac{180 \text{ minutes}}{3 \text{ hours}} = \frac{120 \text{ minutes}}{2 \text{ hours}}$ a proportion? $120 \cdot 3 = 180 \cdot 2$ $360 = 360$ It is a proportion.
Solve a Proportion (Section 4.2)	**Step 1** Get the variable in the numerator. Rewrite the proportion if necessary. **Step 2** Undo the division to get the variable alone. **Step 3** Simplify each side. **Step 4** Check the answer. Substitute the solution into the original proportion. Then calculate and compare the cross products.	Solve $\frac{5}{n} = \frac{10}{7}$. **Solution** $\frac{n}{5} = \frac{7}{10}$ $\frac{5}{1} \cdot \frac{n}{5} = \frac{7}{10} \cdot \frac{5}{1}$ $n = \frac{7 \cdot 5}{2 \cdot 5}$ $n = \frac{7}{2}$ **Check** $\frac{n}{5} = \frac{7}{10}$ $\frac{\frac{7}{2}}{5} = \frac{7}{10}$ $\frac{7}{2} \cdot \frac{10}{1} = 5 \cdot 7$ $35 = 35$
Interpret a Scale Drawing (Section 4.3)	**Step 1** Set up a proportion using the rate defined by the scale. **Step 2** Solve the proportion.	A map's legend says that 1 inch = 25 miles. If you are 4 inches from your destination on the map, how far do you still have to travel? $\frac{25 \text{ mi}}{1 \text{ in.}} = \frac{n \text{ mi}}{4 \text{ in.}}$ $\frac{25 \text{ mi}}{1 \text{ in.}} \cdot \frac{4 \text{ in.}}{1} = \frac{n \text{ mi}}{4 \text{ in.}} \cdot \frac{4 \text{ in.}}{1}$ $100 = n$
Use a Sampling Method (Section 4.3)	**Step 1** Use a sample to establish a rate. **Step 2** Use the rate to write a proportion for the entire population. **Step 3** Solve the proportion and answer the question.	There are 4,500 students on campus. A pollster found that 120 of 300 students take night classes. Approximately how many students on campus take night classes? $\frac{120}{300}$ students take night classes. $\frac{120}{300} = \frac{n}{4{,}500}$ $\frac{120}{300} \cdot \frac{4{,}500}{1} = \frac{n}{4{,}500} \cdot \frac{4{,}500}{1}$ $1{,}800 = n$ Approximately 1,800 students on campus take night classes.

Procedure/Topic	Steps	Example
Compare a Prescription Dose to the Usual Dose (Section 4.3)	**Step 1** Determine the usual dose from the label. Read the label to find this rate. **Step 2** Determine the prescribed dose by reading the prescription. The prescribed dose will be a rate. **Step 3** Compare the two rates. If the prescribed dose is more than the usual dose, do not give the prescribed dose to the patient.	A prescription for an adult reads "Maclor, 300 mg, twice daily." Is it safe to administer the prescription? **Maclor®** **375 mg per 5 mL** **USUAL DOSE: Children:** 20 mg per kg per day (40 mg per kg in otitis media) in two divided doses. **Adults:** 375 mg two times per day. See literature for complete dosage details. Each 5 mL (approx. 1 measured teaspoonful) will contain Mafaclor monohydrate equivalent to 375 mg Maclor. The usual adult dose is 375 mg. The prescription is for 300 mg. Since the prescribed dose is less than the usual dose, it is safe.
Determine How Much Medication to Give (Section 4.3)	**Step 1** Determine if the prescribed dose is safe. **Step 2** Determine the volume of medication to give to the patient. Use the drug concentration to set up and solve a proportion. **Step 3** Shade the appropriate volume on a syringe.	This example uses the same prescription as the preceding example. How much medicine must be given? From the label / From the prescription $\frac{5\text{ mL}}{375\text{ mg}} = \frac{n\text{ mL}}{300\text{ mg}}$ $\frac{5\text{ mL}}{375\text{ mg}} \cdot \frac{300\text{ mg}}{1} = \frac{n\text{ mL}}{300\text{ mg}} \cdot \frac{300\text{ mg}}{1}$ $\frac{5\text{ mL}}{3 \cdot 5 \cdot 5 \cdot 5} \cdot \frac{2 \cdot 2 \cdot 3 \cdot 5 \cdot 5}{1} = n$ $4\text{ mL} = n$ 4 mL must be given to the patient.

Chapter 4 Review Exercises

4.1

1. A ratio is a quotient that compares two quantities with ___________ units.

Write each ratio in simplest form.

2. 21 miles to 15 miles

3. 30 mph to 64 mph

4. 1.4 cm to 50 cm

5. 3.5 inches to 40 inches

6. What is the difference between a ratio and a rate?

7. What is a unit rate?

Write each comparison as a unit rate. Round to the hundredths place when necessary.

8. Earning \$28 in 3 hours

9. Earning \$36 in 5 hours

10. Cereal at \$2.90 for 20 ounces

11. Cereal at \$3.80 for 30 ounces

Write the cost of each product as a unit rate and determine which is the better buy. Round to the hundredths place when necessary.

12. Which is the better buy?

12 oz for \$1.00 or 68 oz for \$1.56

13. Which is the better buy?

128 oz for \$2.55 or 100 oz for \$2.10

14. Calculate the unit price in dollars per ounce. How much will someone save per ounce if he or she purchases the better buy?

128 oz for \$3.25

58 oz for \$1.25

15. Calculate the unit price in dollars per ounce. How much will someone save per ounce if he or she purchases the better buy?

10 oz for \$2.25

24 oz for \$3.45

4.2

16. What type of mathematical equation states that two rates are equal?

17. Why isn't the following statement a proportion? \$12 is to 14 burgers as 28 burgers is to \$24.

Write a proportion for each exercise.

18. 15 is to 45 as 1 is to 3.

19. 11 is to 55 as 22 is to 110.

20. Mohammed is building a 100-meter fence for his backyard. If it took him 4 days to complete 60 meters of fence, it will take him x days to complete the 100-meter fence.

21. Shelia is typing an 8-page paper. If it takes her 3 hours to type 5 pages, it will it take her h hours to type the 8-page paper.

Determine if each statement is a proportion.

22. $\frac{6}{10} \stackrel{?}{=} \frac{2}{3}$

23. $\frac{4}{11} \stackrel{?}{=} \frac{77}{28}$

24. $\frac{3.5}{3} \stackrel{?}{=} \frac{14}{12}$

25. $\frac{1\frac{1}{2}}{9} \stackrel{?}{=} \frac{2\frac{1}{9}}{16}$

Solve each proportion.

26. $\frac{2}{5} = \frac{n}{30}$

27. $\frac{7}{20} = \frac{n}{80}$

28. $\frac{3}{8} = \frac{7}{n}$

29. $\frac{9}{11} = \frac{2}{n}$

30. Sarah earned \$85 last week working 10 hours. If she works 14 hours this week and is paid at the same rate, how much will she earn?

31. Chanta and her husband went out to dinner this week. They hired a babysitter for 3 hours and paid him \$25. If they go out to dinner for 4 hours next week and pay the babysitter the same rate, how much will they pay him?

4.3

Use a proportion to answer each question about scale drawings.

32. A map's legend says that 1 inch = 25 miles. If you are 2.5 inches from your destination on the map, how far do you still have to travel?

33. A map's legend says that 1 inch = 10 miles. If you are 5.6 inches from your destination on the map, how far do you still have to travel?

Use the following scale drawing to answer each question.

Western Model's Boeing B-314 Clipper

Assume each gridline is 1 inch

1 inch = 4 inches

34. What is the length of the model plane?

35. What is the wingspan, or width, of the model plane?

36. What is the length from the center of the plane to the tip of one wing?

37. What is the width of the tail?

Use a proportion to answer each question.

38. There are 3,000 students on a college campus. A pollster found that 150 out of 400 students were taking math. About how many students at the college are taking math?

39. After checking her mail several days, Brenda found that 12 out of 26 pieces of mail were junk mail. If she receives about 1,500 pieces of mail per year, about how much junk mail does she get?

40. A biologist observed salmon swimming upstream. She estimated that 35 fish swam by her in 2 minutes. About how many fish swam upstream in an hour?

41. Ron estimated that there were 580 words on one page of a novel. If there are 425 pages in this novel, about how many words are there in the novel?

Use the labels to answer parts a, b, and c for each exercise.

a. Use the medication's usual dose to determine if the prescription is safe. If the prescription is not safe, put an X through the syringe and do not answer parts b and c.

b. Determine the volume of medication to give for one dose.

c. Shade the appropriate volume on the drawing of the syringe.

120 mL

LOT
EXP.

Florazine®
Hydrofloride Intensate
(Oral Concentrate, USP)

30 mg per mL

Usual Dosage: 75 to 400 mg daily. See package insert for complete prescribing information. Protect from light, dilute before using.

10 mL vial

Terracin®
Oxyquadracycline
Intramuscular Solution

50 mg/mL

Phisher
Pharmaceuticals

Caution: Federal (U.S.A.) law prohibits dispensing without prescription.

READ ACCOMPANYING PROFESSIONAL INFORMATION

DOSAGE
Adults: The usual daily dose is 250 mg administered once every 24 hours, or 300 mg administered in divided doses at 8- to 12-hour intervals.

Children Above Eight Years of Age: 12–25 mg/kg body weight, up to a maximum of 250 mg per single daily injection. Dosages may be divided and given at 8- to 12-hour intervals.

Pharmaceuticals

USUAL DOSE: Children: 33–50 mg/kg per day in divided doses. See literature for complete adult dosage details.

Prior to Mixing: Store at a controlled room temperature below 86°F (30°C).

May be administered without regard to meals.

Each 5 mL (approx. 1 measured teaspoonful) will contain Rythamyacin equivalent to 200 mg RyPed 200.

A product of Drummond Labs, U.S.A.

PEEL

100 mL when mixed

RyPed® 200
RYTHAMYACIN
FOR
ORAL SUSPENSION, USP

200 mg per 5 mL

Drummond Labs

Caution: Federal (U.S.A.) law prohibits dispensing without prescription.

USUAL DOSE: Children: 20 mg per kg per day (40 mg per kg in otitis media) in two divided doses. **Adults:** 375 mg two times per day. See literature for complete dosage details.

Prior to Mixing: Store at a controlled room temperature of 59 to 86°F (15° to 30°C).

Each 5 mL (approx. 1 measured teaspoonful) will contain Mafaclor monohydrate equivalent to 375 mg Maclor.

Mfd. by Roseco Pharmaceuticals in U.S.A.

100 mL when mixed

Roseco Pharmaceuticals

Maclor®

MAFACLOR
for Oral Suspension, USP

375 mg per 5 mL

Caution: Federal (U.S.A.) law prohibits dispensing without prescription.

42. Florazine: 750 mg per dose, twice daily, to an adult male patient

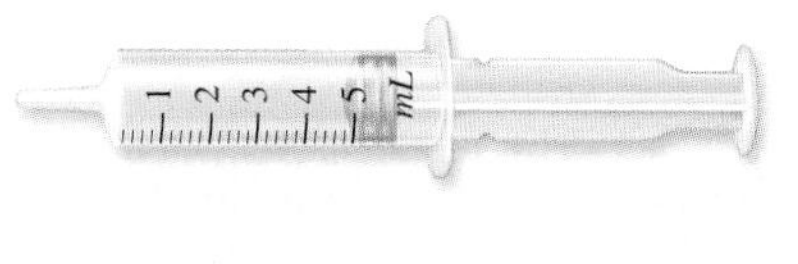

43. RyPed 200 : 250 mg per dose, twice daily, to a 10 kg child

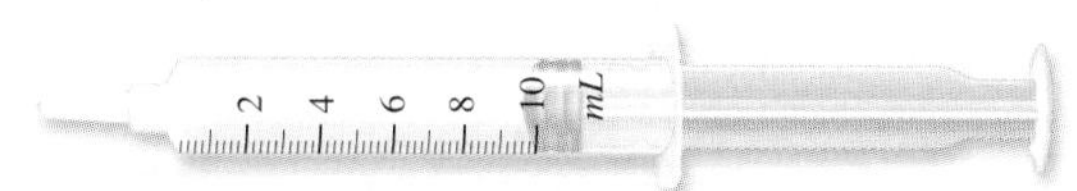

44. Terracin: 70 mg per dose, three times daily, to a 20 kg child

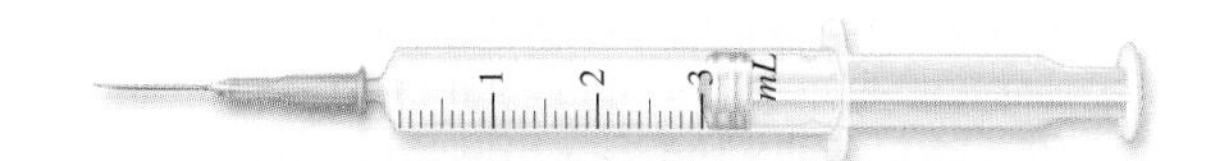

Chapter 4 Practice Test

Describe each situation with a ratio or rate. Explain why your answer is a ratio or rate.

1. James needs 2 cups of raisins for 3 loaves of bread.

2. Alexis runs 3 miles for every 2 miles that her daughter runs.

3. The tax was $1.24 on a $41.27 bill.

4. The printer uses two ink cartridges for every box of paper.

Write each rate as a unit rate.

5. $\frac{\$1.28}{8 \text{ ounces}}$

6. $\frac{144 \text{ miles}}{6 \text{ gallons}}$

7. Driving 145 miles in 2.5 hours

8. Earning $328 for working 40 hours

Write the price of each product as a unit rate to determine the better buy.

9. Which is the better buy?

$3.85 for 20 oz

$2.69 for 15 oz

10. Which is the better buy?

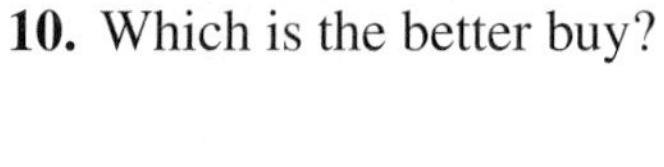

$1.64 for 4 lb

$1.99 for 5 lb

Set up a proportion to answer each question. Do not solve the proportion.

11. A mechanic earns $45 for a two-hour job. How much does she earn for a three-hour job?

12. A nurse gives 300 mg of Aspinozin to an 80 kg patient.
a. What is this dose as a rate in mg per kg?
b. Based on the rate from part a, how much Aspinozin should a 50 kg patient receive?

Determine if each statement is a proportion.

13. $\frac{10}{14} \stackrel{?}{=} \frac{5}{7}$

14. $\frac{13.6 \text{ kg}}{8 \text{ m}} \stackrel{?}{=} \frac{15.3 \text{ kg}}{9 \text{ m}}$

15. $\frac{12}{47} \stackrel{?}{=} \frac{3}{11}$

16. $\frac{42 \text{ feet}}{17 \text{ hours}} \stackrel{?}{=} \frac{33 \text{ feet}}{13 \text{ hours}}$

Solve each proportion. Round to the tenths place when necessary.

17. Twenty-four is to twenty as fifteen is to n.

18. $\frac{70 \text{ meters}}{16 \text{ seconds}} = \frac{n}{24 \text{ seconds}}$

19. A painter used 3 gallons of paint to paint a wall that is 1,050 square feet. How many gallons of paint will he need to paint a wall that is 1,400 square feet?

20. Janice finished two-thirds of her report in three hours. At this rate, how long will it take her to finish the entire report?

Use a proportion to answer each question.

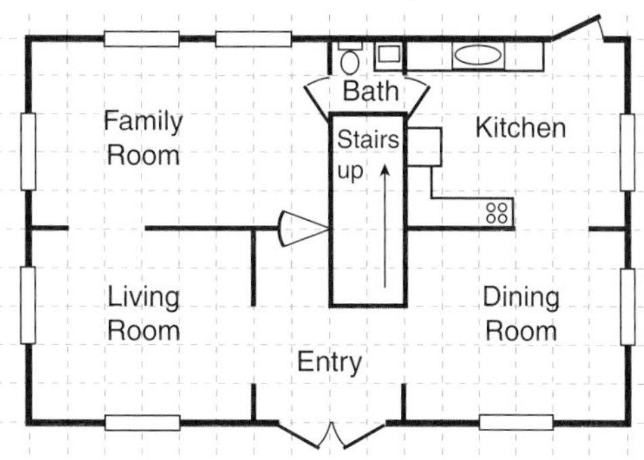

21. Using the blueprint, find the area of the family room and living room.

22. The family's sofa measures 7 feet by 3 feet. If the sofa were drawn in the family room, how many grid-lines long and wide would the sofa be?

23. An architect polled 20 of his clients and found that 13 of them preferred beige paint. If the architect has 325 clients, about how many prefer beige paint?

24. A biologist put tags on 500 fish in a pond. Later, the biologist caught 50 fish, 8 of which had been tagged. Approximately how many fish are in the pond?

200 mg Rythamyacin
FOR ORAL SUSPENSION, USP
200 mg per 5 mL. Oversize bottle provides extra room for shaking prior to consumption. May be kept for 14 days without significant lose of potency. Store tightly closed. Discard unused portion 14 days after opening.
SHAKE WELL BEFORE USING

USUAL DOSE: Children: 33–50 mg/kg per day in divided doses. See literature for complete adult dosage details.

Prior to Mixing: Store at a controlled room temperature below 86°F (30°C).

May be administered without regard to meals.

Each 5 mL (approx. 1 measured teaspoonful) will contain Rythamyacin equivalent to 200 mg RyPed 200.

A product of Drummond Labs, U.S.A.

PEEL

100 mL when mixed

RyPed® 200

RYTHAMYACIN
FOR
ORAL SUSPENSION, USP

200 mg
per 5 mL

Drummond Labs

Caution: Federal (U.S.A.) law prohibits dispensing without prescription.

25. Is a prescription of RyPed 200, 500 mg per day, safe for a 20 kg child?

26. What volume of RyPed 200 must be given to administer 500 mg of medication? Shade the appropriate volume on the syringe.

CHAPTER 5

Percents

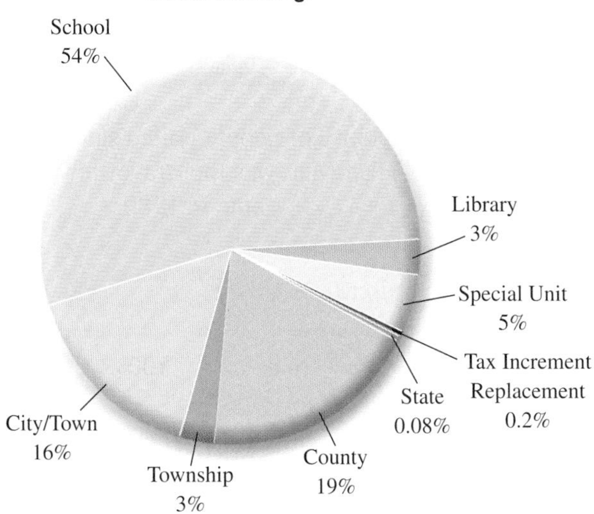

Percents describe portions of a quantity. For example, we pay a percent of every paycheck in taxes. These taxes are used to pay for many community, state, and federal services.

This circle graph uses percents to show how local taxes are shared among government agencies.

In Exercises 25 and 26 of Section 5.1, you will use this circle graph to answer questions about how taxes are used.

Section 5.1 Percents, Fractions, and Decimals

A **percent** is a portion of a whole.

40% of the U.S. Population

40% of the U.S. population

40% of the U.S. States

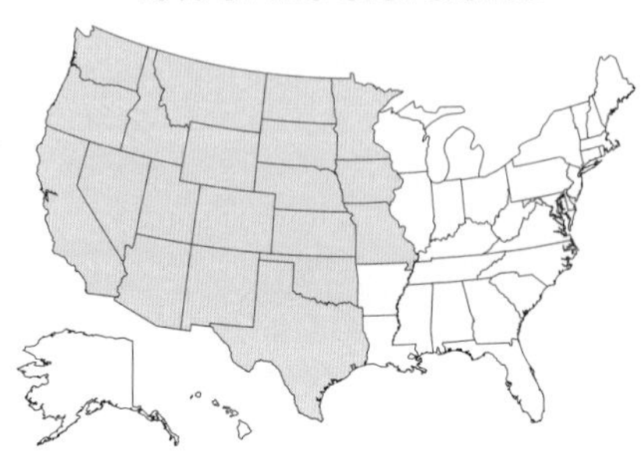

40% of the United States

Most Popular Video Game, in 1980

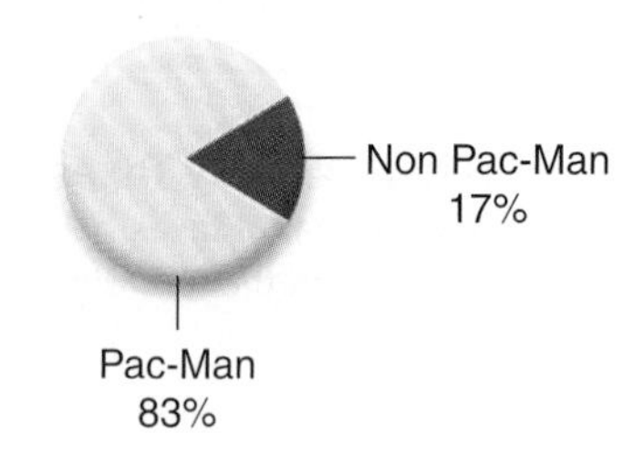

83% of people polled said that Pac-Man was their favorite video game in 1980.

The word *percent* means "per hundred," so 1% represents "one part per hundred."

Cent Sense	
Century	100 years
1 cent	$\frac{1}{100}$ of a dollar
Centimeter	$\frac{1}{100}$ meter
Centennial	100-year anniversary

1%

1% of something is $\frac{1}{100}$, or 0.01, of that thing.

50%

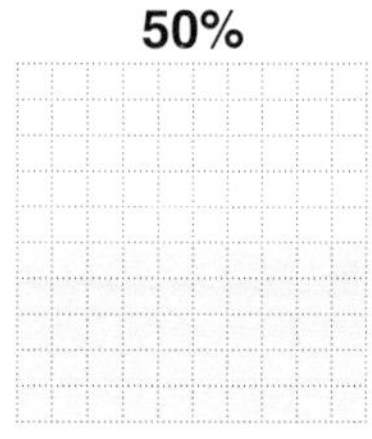

50% of something is $\frac{50}{100}$, or 0.50, of that thing.

100%

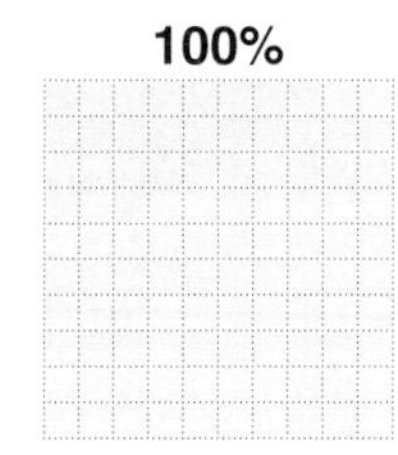

100% of something is $\frac{100}{100}$, or 1.00, of that thing.

Recall that decimals and fractions are also portions of a whole. Because decimals, fractions, and percents can represent the same thing, they can be interchanged. In this section, you will learn how to convert between decimals, fractions, and percents.

The **Objectives** in Section 5.1 will help you

- **A** Visualize percents.
- **B** Convert a percent to a fraction or mixed number.
- **C** Convert a fraction or mixed number to a percent.
- **D** Convert between a percent and a decimal.

VOCABULARY PREVIEW *Check the box that applies.*

	Got It	Must Study
Per: Per is a prefix that indicates division.		
Cent: The word ***cent*** means "one hundred."		
%: The symbol % is read as "percent."		
Percent: Percent is used in place of the phrases *per hundred* and *out of one hundred.*		
Units Fraction: A **units fraction** contains units of equal measure in the numerator and denominator. Units fractions are equal to 1.		
Decimal–Percent–Fraction Triangle: The **decimal–percent–fraction triangle** shows how to convert between decimals, percents, and fractions.		

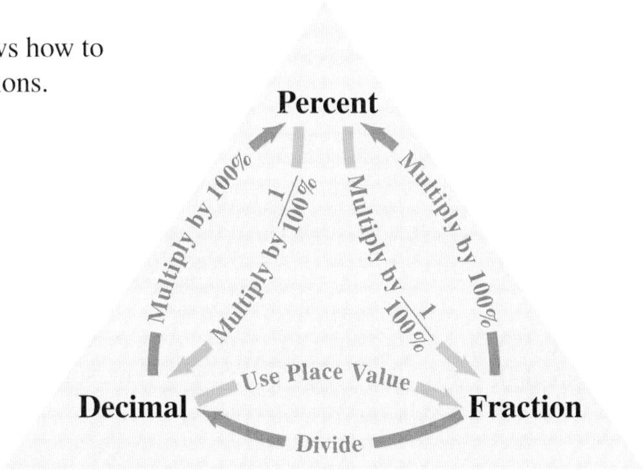

Study these words when they appear in the text. After the section, test your understanding by completing the Vocabulary Review in the section exercises.

Objective A Visualize Percents

The Concept The following chart gives a visual representation of commonly used percents, fractions, and decimals.

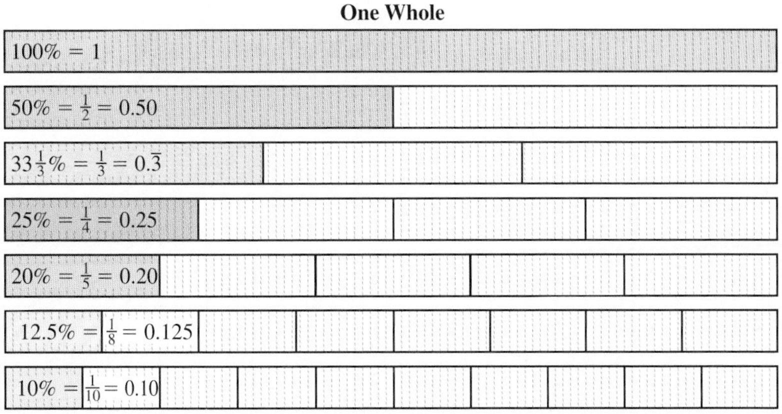

A percent is easy to visualize when it is considered as a ratio. The percent 53% indicates the ratio of 53 percent, or 53 per 100.

Visualize 53%	Visualize 121%
$53\% = 53 \text{ per } 100$ $= \frac{53}{100}$	$121\% = 121 \text{ per } 100$ $= \frac{121}{100}$ $= 1\frac{21}{100}$
53%	**121%**
A little more than half the box is shaded.	Any percent that is greater than 100% represents more than one whole.

Procedure Visualize Percents

Step 1 Interpret the percent as a ratio.

Step 2 Shade the appropriate portion of a square.

EXAMPLES

1. Shade 41% of a square.

Step 1 Interpret the percent as a ratio.

$$41\% = \frac{41}{100}$$

Step 2 Shade the appropriate portion of a square.

41%

GUIDED PRACTICE

1. Shade 37% of a square.

$$37\% = \frac{\square}{100}$$

37%

2. Shade 87% of a square.

Step 1 Interpret the percent as a ratio. $87\% = \frac{87}{100}$

Step 2 Shade the appropriate portion of a square.

87%

2. Shade 63% of a square.

$63\% = \frac{\quad}{100}$

63%

3. Shade 113% of a square.

Step 1 Interpret the percent as a ratio. $113\% = \frac{113}{100} = 1\frac{13}{100}$

Step 2 Shade the appropriate portion of a square.

113%

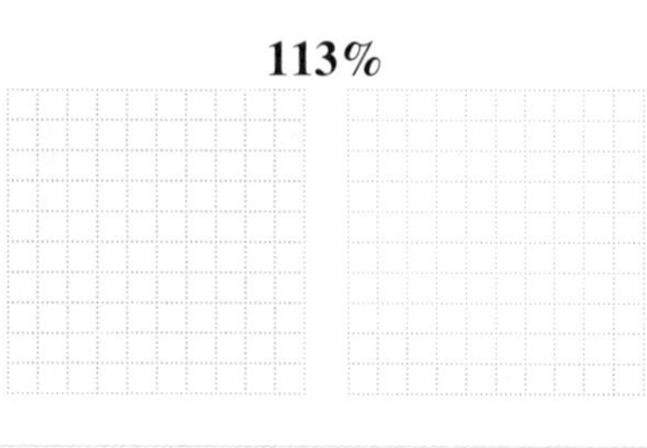

3. Shade 171% of a square.

$171\% = \frac{\quad}{100} = 1\frac{\quad}{100}$

171%

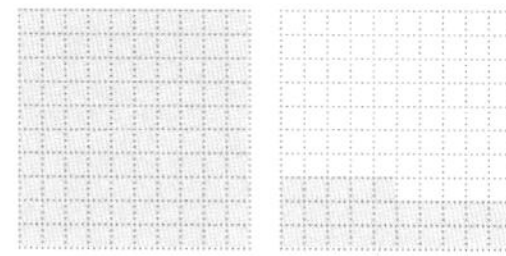

Concept Check

Match each percent with its corresponding picture. Not every picture will have a match.

A1. 90%

A2. 50%

A3. 20%

A4. 80%

A5. 225%

A6. 325%

a. **b.** **c.** **d.** **e.** **f.** **g.** **h.** **i.**

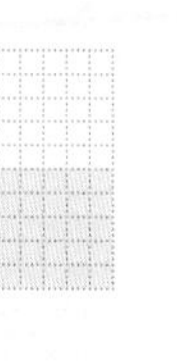

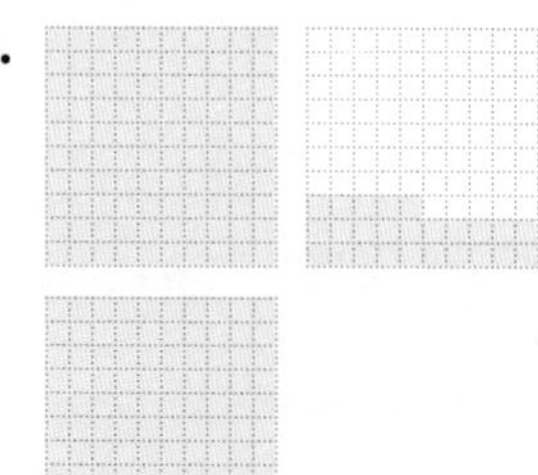

Objective A Practice

Visualize each percent.

A7. Shade 10% of a square.

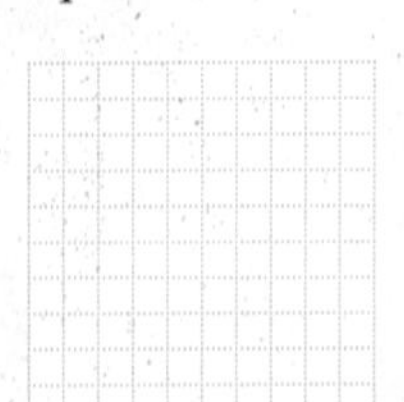

A8. Shade 33% of a square.

A9. Shade 50% of a square.

A10. Shade 75% of a square.

A11. Shade 180% of a square.

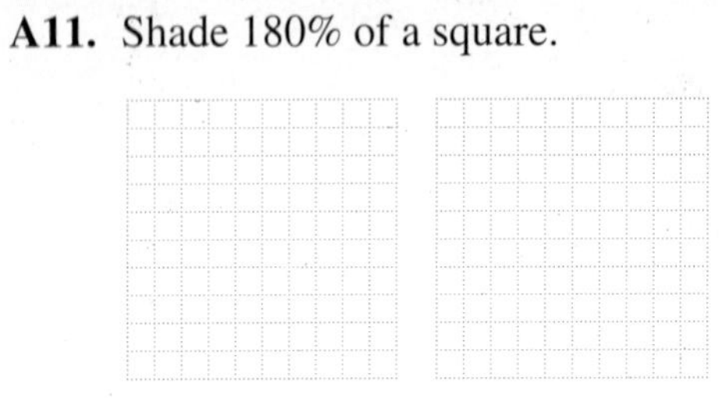

A12. Shade 125% of a square.

A13. Shade 290% of a square.

A14. Shade 300% of a square.

For each exercise, determine the percent that is shaded.

A15.

A16.

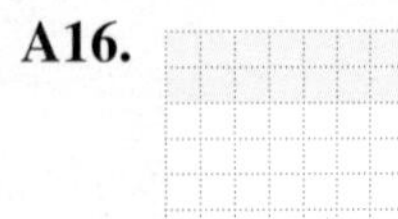

A17.

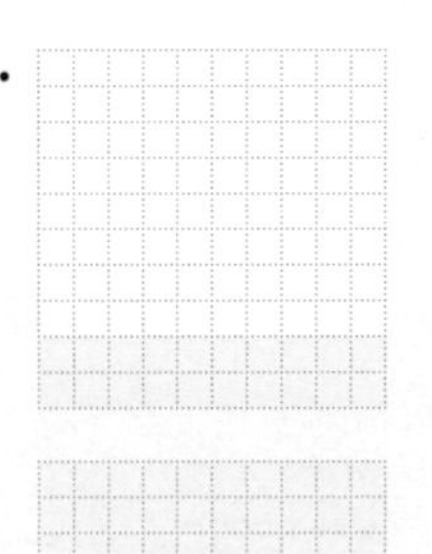

A18.

Objective B Convert a Percent to a Fraction or Mixed Number

The Concept To convert a percent to a fraction or mixed number, we multiply the percent by a fraction that is equal to one and has the percent symbol (%) in the denominator. This is the same procedure used to convert units of measure with units fractions in Section 2.2.

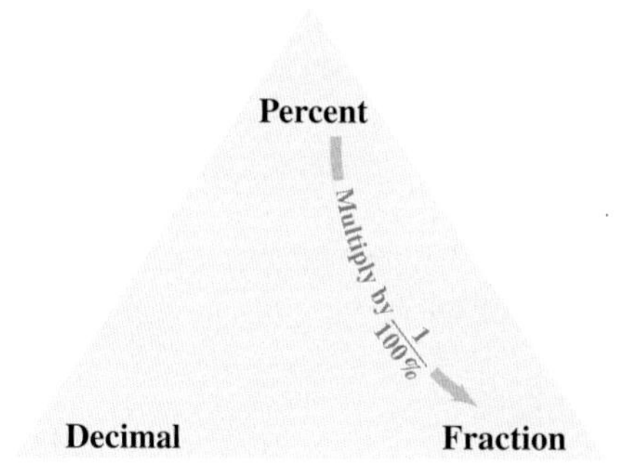

To convert between percents and fractions, use the fact that $1 = 100\%$ to make a units fraction.

Since $1 = 100\%$, $\dfrac{1}{100\%}$ is a units fraction.

Using this units fraction, we can change percents to fractions.

Notice the similarities between converting units and converting percents.

CONVERTING UNITS

Convert 30 inches to feet.

Step 1 To convert units, use a fraction equal to one.

$$1 = \frac{1 \text{ foot}}{12 \text{ inches}}$$

Step 2 Multiply the original quantity by this form of 1.

$$30 \text{ inches} = \left(\frac{30 \text{ in.}}{1}\right) \cdot \frac{1 \text{ ft}}{12 \text{ in.}}$$
$$= \frac{30}{12} \text{ ft}$$
$$= \frac{5 \cdot 6}{2 \cdot 6} \text{ ft}$$
$$= \frac{5}{2} \text{ ft}$$

CONVERTING PERCENTS

Convert 32% to a fraction.

Step 1 To convert percents, use a fraction equal to one.

$$1 = \frac{1}{100\%}$$

Step 2 Multiply the original quantity by this form of 1.

$$32\% = \left(\frac{32\%}{1}\right) \cdot \frac{1}{100\%}$$
$$= \frac{32}{100}$$
$$= \frac{4 \cdot 8}{4 \cdot 25}$$
$$= \frac{8}{25}$$

Procedure Convert a Percent to a Fraction or Mixed Number

Step 1 Multiply by 1 in the form of $\dfrac{1}{100\%}$ to divide out the percent symbol.

Step 2 Simplify.
If the result is an improper fraction, convert the fraction to a mixed number.

EXAMPLES

4. Convert 45% to a fraction or mixed number.

Step 1 Multiply by $\frac{1}{100\%}$ to divide out the % symbol.

$$45\% = \frac{45\%}{1} \cdot \frac{1}{100\%}$$

$$= \frac{45}{100}$$

Step 2 Simplify.

$$= \frac{9 \cdot 5}{20 \cdot 5}$$

$$= \frac{9}{20}$$

5. Convert $12\frac{1}{2}\%$ to a fraction or mixed number.

Convert the mixed number to an improper fraction first.

$$12\frac{1}{2}\% = \frac{25\%}{2}$$

Step 1 Multiply by $\frac{1}{100\%}$.

$$= \frac{25\%}{2} \cdot \frac{1}{100\%}$$

$$= \frac{25}{2} \cdot \frac{1}{100}$$

Step 2 Simplify.

$$= \frac{25}{2} \cdot \frac{1}{4 \cdot 25}$$

$$= \frac{1}{8}$$

6. Convert 140% to a fraction or mixed number.

Step 1 Multiply by $\frac{1}{100\%}$.

$$140\% = \frac{140\%}{1} \cdot \frac{1}{100\%}$$

$$= \frac{140}{100}$$

Step 2 Simplify.

$$= \frac{7 \cdot 20}{5 \cdot 20}$$

$$= \frac{7}{5} \text{ or } 1\frac{2}{5}$$

When a percent is more than 100%, the result will be an improper fraction, which should be converted to a mixed number.

GUIDED PRACTICE

4. Convert 60% to a fraction or mixed number.

$$60\% = \frac{60\%}{1} \cdot \frac{}{}$$

$$= \frac{}{}$$

$$= \frac{}{}$$

$$= \frac{}{}$$

5. Convert $16\frac{2}{3}\%$ to a fraction or mixed number.

$$16\frac{2}{3}\% = \frac{}{}$$

$$= \frac{}{} \cdot \frac{}{}$$

$$=$$

$$=$$

$$=$$

6. Convert 220% to a fraction or mixed number.

$$220\% = \frac{220\%}{1} \cdot \frac{}{}$$

$$= \frac{}{}$$

$$= \frac{}{}$$

$$= \frac{}{} \text{ or } \frac{}{}$$

Concept Check

B1. In your own words, what is a percent?

B2. In your own words, how are fractions and percents related?

Objective B Practice

Convert each percent to a fraction or mixed number.

B3. 75% **B4.** 20% **B5.** 89% **B6.** 8%

B7. $33\frac{1}{3}\%$ **B8.** $37\frac{1}{2}\%$ **B9.** 160% **B10.** 115%

Objective C Convert a Fraction or Mixed Number to a Percent

The Concept To convert a fraction or mixed number to a percent, we introduce the percent symbol (%.) To do this, we multiply the fraction or mixed number by a fraction that is equal to one and has the symbol % in the numerator.

$$\text{Multiply by } 1 = \frac{100\%}{1}.$$

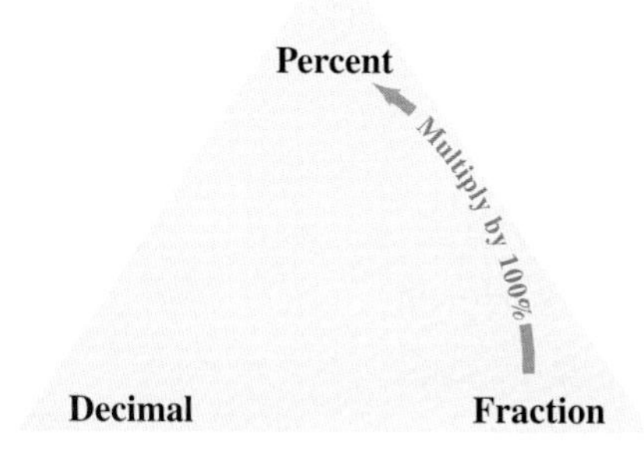

Procedure Convert a Fraction or Mixed Number to a Percent

Step 1 Multiply by 1 in the form of $\frac{100\%}{1}$.

Step 2 Simplify.
If your result is an improper fraction, convert the fraction to a mixed number.

DETAILED EXAMPLE Converting a Fraction or Mixed Number to a Percent

Convert $\frac{3}{2}$ to a percent.

Step 1 Multiply by 1 in the form of $\frac{100\%}{1}$.

Step 2 Simplify.

$$\frac{3}{2} = \frac{3}{2}\cdot\frac{100\%}{1}$$
$$= \frac{3}{\cancel{2}}\cdot\frac{\cancel{2}\cdot 50\%}{1}$$
$$= 150\%$$

Since $\frac{3}{2}$ is greater than a whole, 150% is greater than 100%.

The fraction $\frac{3}{2}$ = 150%.

=

EXAMPLES

7. Convert $\frac{7}{20}$ to a percent.

Step 1 Multiply by $\frac{100\%}{1}$.

Step 2 Simplify.

$$\frac{7}{20} = \frac{7}{20}\cdot\frac{100\%}{1} = \frac{7}{\cancel{20}}\cdot\frac{\cancel{20}\cdot 5\%}{1} = \frac{35\%}{1} = 35\%$$

8. Convert $\frac{2}{3}$ to a percent.

Step 1 Multiply by $\frac{100\%}{1}$.

Step 2 Simplify.

$$\frac{2}{3} = \frac{2}{3}\cdot\frac{100\%}{1} = \frac{200\%}{3} = 66\frac{2}{3}\%$$

66R2
3)200

Write the improper fraction as a mixed number.

9. Convert $3\frac{1}{16}$ to a percent.

Convert the mixed number to an improper fraction first.

Step 1 Multiply by $\frac{100\%}{1}$.

Step 2 Simplify.

$$3\frac{1}{16} = \frac{49}{16} = \frac{49}{16}\cdot\frac{100\%}{1} = \frac{49}{4\cdot\cancel{4}}\cdot\frac{\cancel{4}\cdot 25\%}{1} = \frac{1{,}225\%}{4} = 306\frac{1}{4}\%$$

306R1
4)1225

GUIDED PRACTICE

7. Convert $\frac{18}{25}$ to a percent.

$$\frac{18}{25} = \frac{18}{25}\cdot\frac{\quad}{\quad} = \frac{18}{25}\cdot\frac{\quad}{\quad} = \frac{\quad}{\quad} =$$

8. Convert $\frac{3}{8}$ to a percent.

$$\frac{3}{8} = \frac{3}{8}\cdot\frac{\quad}{\quad} = \frac{\quad\%}{8} =$$

8)

9. Convert $1\frac{1}{12}$ to a percent.

$$1\frac{1}{12} = \frac{\quad}{\quad} = \frac{\quad}{\quad}\cdot\frac{\quad}{\quad} = \frac{\quad}{\quad}\cdot\frac{\quad}{\quad} = \frac{\quad\%}{\quad} = \quad\%$$

Concept Check

C1. The key step used to convert a number to a percent is to multiply by 100%. This step does not change the value of the number because multiplying by 100% is the same as multiplying by ______.

C2. Which is greater, a quantity multiplied by 100% or a quantity multiplied by a mixed number?

C3. Examine the mixed numbers that have been converted to percents.

a. Identify the pattern and complete the list.

$$\frac{1}{4} = 25\%, \quad 1\frac{1}{4} = 125\%, \quad 2\frac{1}{4} = 225\%, \quad 3\frac{1}{4} = \_\_\_\%, \quad 4\frac{1}{4} = \_\_\_\%$$

b. Based on the list in part a, make a prediction about what happens to the whole-number part of a mixed number when you convert it to a percent.

Objective C Practice

Convert each fraction or mixed number to a percent.

C4. $\frac{5}{12}$ **C5.** $\frac{17}{40}$ **C6.** $\frac{1}{8}$ **C7.** $\frac{5}{6}$

C8. $1\frac{3}{10}$ **C9.** $2\frac{3}{5}$ **C10.** $1\frac{1}{2}$ **C11.** $3\frac{7}{20}$

FOCUS ON Converting Fractions or Mixed Numbers to Common Percents

Sometimes a fraction has properties that make it easy to convert to a percent. If you see something that makes your calculations easy, use it. If not, use the usual method of multiplying by 100% and simplifying the result.

If the denominator is a factor of 100, much of the arithmetic can be done mentally. Build a like fraction with a denominator of 100 and convert it to a percent.

Procedure Convert a Fraction to a Percent by Building a Like Fraction

Step 1 Determine the missing multiplier that gives a denominator of 100.

Step 2 Multiply by a fraction equal to one.

Step 3 Convert the fraction to a percent.

(Continued)

EXAMPLE

10. Convert $\frac{7}{4}$ to a percent.

Step 1 Determine the missing multiplier that gives a denominator of 100.

The missing multiplier is 25 because $4 \cdot 25 = 100$.

Step 2 Multiply by a fraction equal to one.

$$\frac{7}{4} \cdot \left(\frac{25}{25}\right) = \frac{175}{100} \rightarrow \frac{n}{100} = n\%$$

Step 3 Convert the fraction to a percent.

$$= 175\%$$

GUIDED PRACTICE

10. Convert $\frac{4}{5}$ to a percent.

The missing multiplier is ____ because $5 \cdot$ ____ $= 100$.

$$\frac{4}{5} \cdot \left(\frac{\quad}{\quad}\right) = \frac{\quad}{100} \rightarrow \frac{n}{100} = n\%$$

$$= \quad\%$$

When converting a mixed number to a percent, the whole number portion will correspond to "hundreds of percent." The fractional part will correspond to a portion less than 100%. These facts make converting a mixed number to a percent easier to do.

Procedure Convert a Mixed Number to a Percent

Step 1 Convert the whole number to a percent.

Step 2 Convert the fraction to a percent.

Step 3 Add the two percents.

EXAMPLE

11. Convert $3\frac{2}{5}$ to a percent.

$$3\frac{2}{5} = 3 + \frac{2}{5}$$

Convert the whole number to a %.

$$3 = 3 \cdot 100\%$$
$$= 300\%$$

Convert the fraction to a %.

$$\frac{2}{5} = \frac{2}{5} \cdot \frac{20}{20}$$
$$= \frac{40}{100}$$
$$= 40\%$$

Add the two percents.

$$3\frac{2}{5} = 300\% + 40\%$$
$$= 340\%$$

GUIDED PRACTICE

11. Convert $2\frac{6}{25}$ to a percent.

$$2\frac{6}{25} = \quad + \quad$$

Convert the whole number to a %.

____ = ____ $\cdot$ 100%

= ____%

Convert the fraction to a %.

$$\frac{\quad}{\quad} = \frac{\quad}{\quad} \cdot \frac{\quad}{\quad}$$
$$= \frac{\quad}{\quad}$$
$$= \quad\%$$

Add the two percents.

$2\frac{6}{25}$ = ____ + ____

= ____%

PRACTICE

Convert each fraction or mixed number to a percent.

1. $\frac{9}{10}$
2. $\frac{3}{10}$
3. $\frac{1}{5}$
4. $\frac{4}{5}$
5. $2\frac{3}{4}$
6. $3\frac{1}{4}$
7. $2\frac{1}{20}$
8. $1\frac{1}{20}$
9. 2
10. 5
11. $\frac{13}{2}$
12. $\frac{7}{2}$
13. When converting to a percent, how does the whole number portion of a mixed number relate to the value of a percent?

Objective D Convert between a Percent and a Decimal

The Concept Converting a percent to a decimal and converting a decimal to a percent use similar procedures. Therefore, we present both procedures side by side. Read each explanation individually, then go back and compare the two procedures.

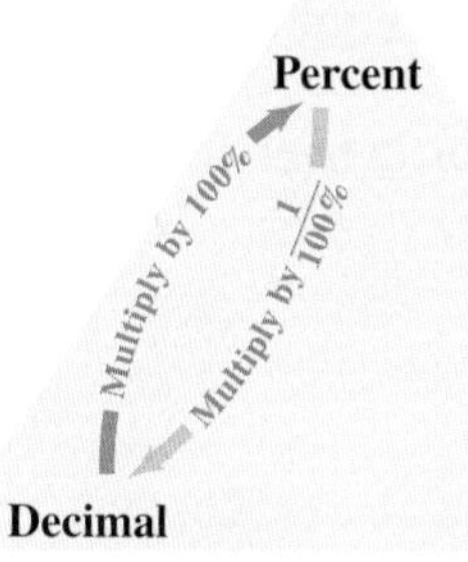

Notice the similarities between converting percents to decimals and converting decimals to percents.

CONVERTING PERCENTS TO DECIMALS

Convert 12.5% to a decimal.

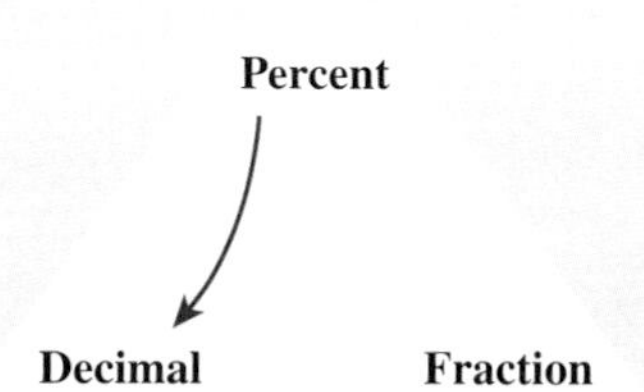

To convert a percent to a decimal, multiply by $\frac{1}{100\%}$. The percent symbol will divide out.

Step 1 Multiply by $\frac{1}{100\%}$.

$$12.5\% = \frac{12.5\%}{1} \cdot \frac{1}{100\%}$$
$$= \frac{12.5}{100}$$

Step 2 Simplify.

$$= 0.125$$

Division by 100 moves the decimal point two places to the left.

CONVERTING DECIMALS TO PERCENTS

Convert 0.375 to a percent.

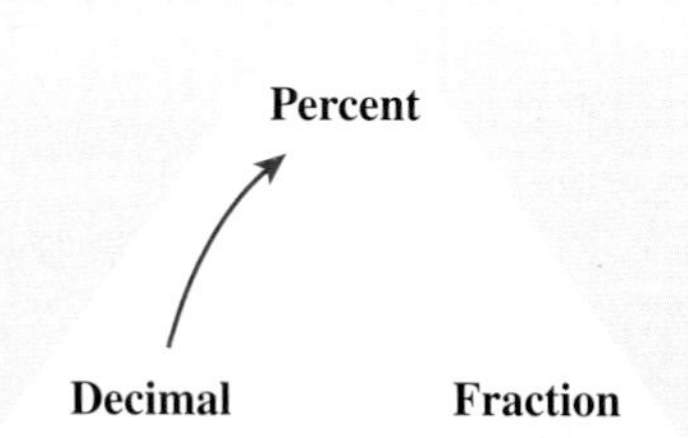

To convert a decimal to a percent, multiply by 100%.

The percent symbol will be introduced.

Step 1 Multiply by 100%.

$$0.375 = 0.375 \cdot 100\%$$

Step 2 Simplify.

$$= 37.5\%$$

Multiplication by 100 moves the decimal point two places to the right.

The **decimal–percent–fraction triangle** can help you remember how to move the decimal point when converting between decimals and percents.

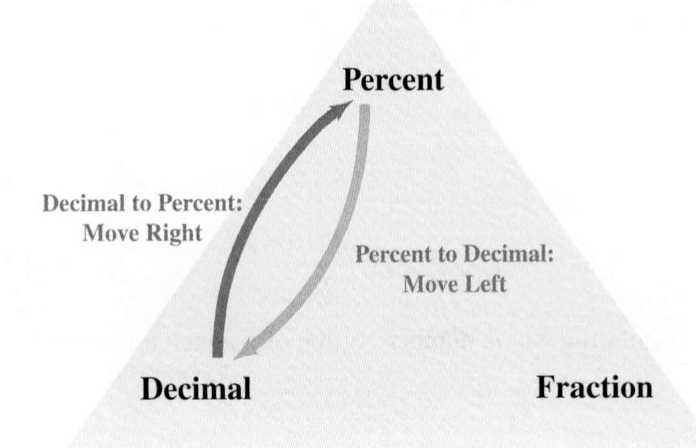

When converting between decimals and percents, remember to multiply by 1 in the form of 100% or $\frac{1}{100\%}$. Think about which form will introduce the percent symbol and which form will divide out the percent symbol.

▶ **NOTE** The key steps for converting between percents and fractions and converting between percents and decimals are the same.

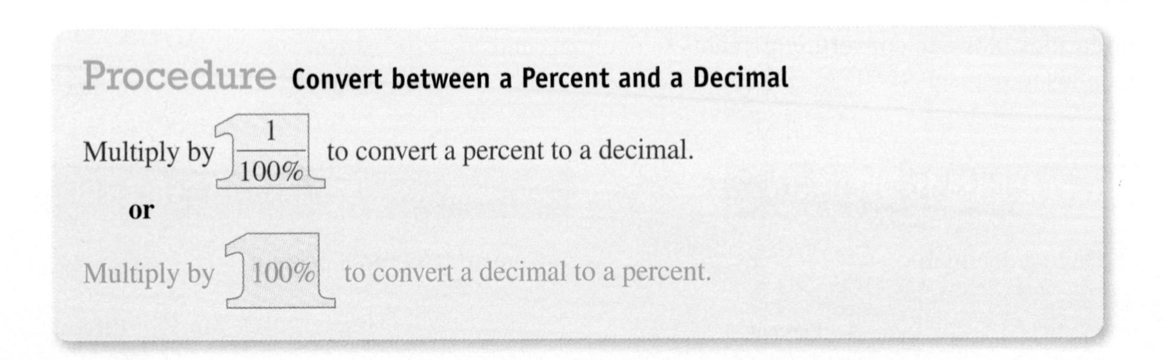

Procedure **Convert between a Percent and a Decimal**

Multiply by $\frac{1}{100\%}$ to convert a percent to a decimal.

or

Multiply by 100% to convert a decimal to a percent.

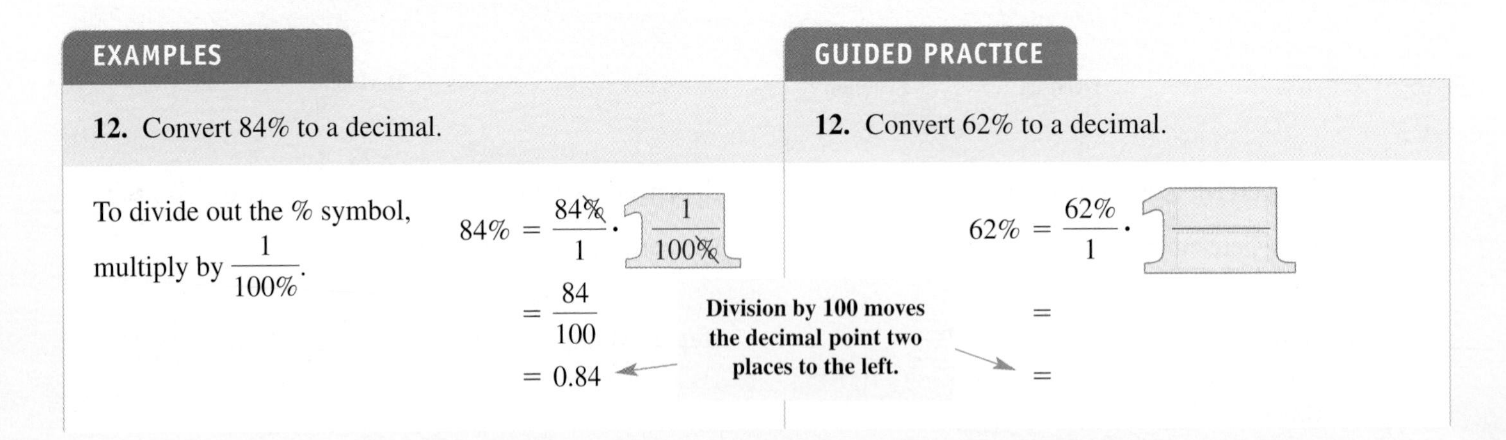

EXAMPLES

12. Convert 84% to a decimal.

To divide out the % symbol, multiply by $\frac{1}{100\%}$.

$$84\% = \frac{84\%}{1} \cdot \frac{1}{100\%}$$
$$= \frac{84}{100}$$
$$= 0.84$$

Division by 100 moves the decimal point two places to the left.

GUIDED PRACTICE

12. Convert 62% to a decimal.

$$62\% = \frac{62\%}{1} \cdot \frac{\quad}{\quad}$$
$$=$$
$$=$$

13. Convert 0.0812 to a percent.

To introduce the % symbol, multiply by 100%.

$0.0812 = 0.0812 \cdot 100\%$

$= 8.12\%$

Multiplication by 100 moves the decimal point two places to the right.

13. Convert 0.0031 to a percent.

$0.0031 = 0.0031 \cdot$ ____

$=$

14. Convert 250% to a decimal.

Multiply by $\frac{1}{100\%}$.

$250\% = \frac{250\%}{1} \cdot \frac{1}{100\%}$

$= \frac{250}{100}$

$= 2.5$

14. Convert 340% to a decimal.

$340\% = \frac{340\%}{1} \cdot$ ____

$=$

$=$

15. Convert 1.8 to a percent.

Multiply by 100%.

$1.8 = 1.8 \cdot 100\%$

$= 180\%$

15. Convert 10.3 to a percent.

$10.3 = 10.3 \cdot$ ____

$=$

Concept Check

D1. a. Choose the correct form of one used to convert 43% to a decimal.

100% or $\frac{1}{100\%}$

b. Explain why your answer to part a will perform the conversion correctly.

D2. When converting 6.5% to a decimal, the decimal point must be moved two places to the right or two places to the left.

a. Which of the following conversions is correct?

i. $6.5\% = 650$ **ii.** $6.5\% = 0.065$

b. Explain why your answer shows the correct conversion.

Objective D Practice

Convert each number to a decimal or a percent as appropriate.

D3. 85% **D4.** 36% **D5.** 0.92 **D6.** 0.12

D7. 823% **D8.** 555% **D9.** 6.5 **D10.** 4.2

D11. 0.3% **D12.** 0.0087 **D13.** 0.0012 **D14.** 0.5%

D15. 1,000% **D16.** 4,000 **D17.** 1,000 **D18.** 4,000%

Combining Concepts and Applications

CONCEPT I The Decimal–Percent–Fraction Triangle Being flexible in how you solve a problem can improve your success in mathematics. The following examples will help you develop different techniques to convert between fractions, decimals, and percents.

DETAILED EXAMPLES Converting a Percent to a Fraction

Convert 12% to a fraction.

Solution 1: Two Steps

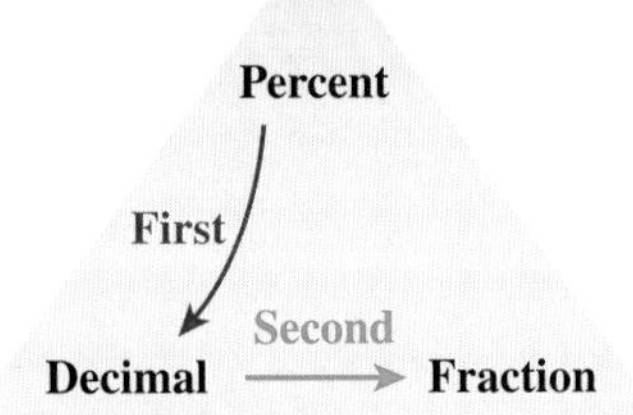

Convert 12% to a decimal. Then convert the decimal to a fraction.

Step 1 Move the decimal point two places to the left. $12\% = 0.12$

Step 2 Use place value to convert to a fraction. $= \frac{12}{100}$

Step 3 Simplify. $= \frac{3 \cdot \cancel{4}}{\cancel{4} \cdot 25} = \frac{3}{25}$

Solution 2: One Step

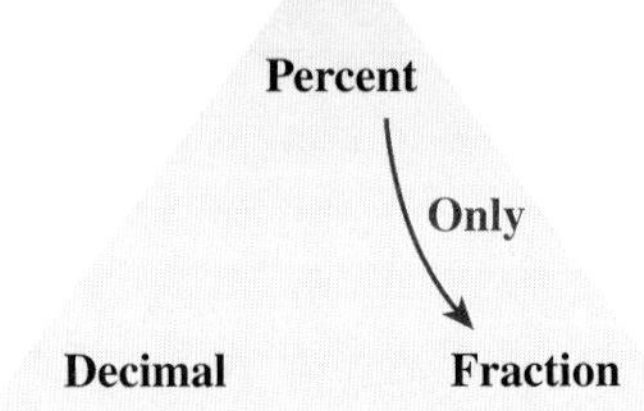

Convert 12% to a fraction in one step.

Step 1 Multiply by $\frac{1}{100\%}$ to convert to a fraction. $12\% = \frac{12\cancel{\%}}{1} \cdot \frac{1}{100\cancel{\%}} = \frac{12}{100}$

Step 2 Simplify. $= \frac{3 \cdot \cancel{4}}{\cancel{4} \cdot 25} = \frac{3}{25}$

Since decimals, percents, and fractions are interchangeable forms, we get the same answer by taking either path.

To help organize your thoughts, the decimal–percent triangle shows the key steps for decimal–percent–fraction conversions.

The Decimal–Percent–Fraction Triangle

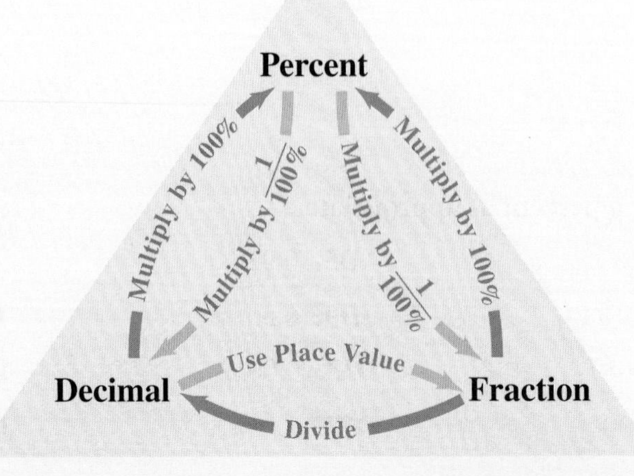

EXAMPLES

16. Convert $\frac{5}{4}$ to a percent in two steps.

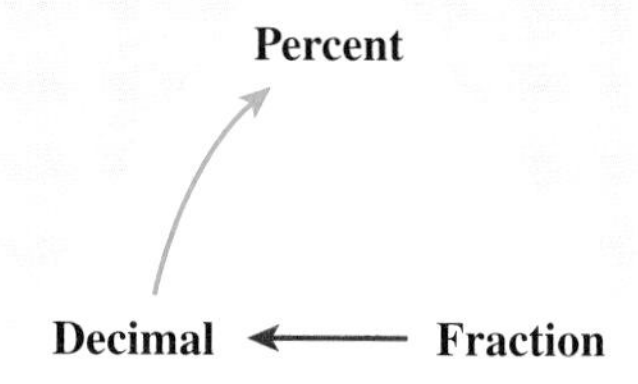

- Convert $\frac{5}{4}$ to a decimal.
- Convert the decimal to a percent.

Step 1 Divide the numerator by the denominator. $\frac{5}{4} = 1.25$

Step 2 Multiply by 100%. $= 1.25 \cdot 100\%$
$= 125\%$

17. Convert $\frac{5}{4}$ to a percent in one step.

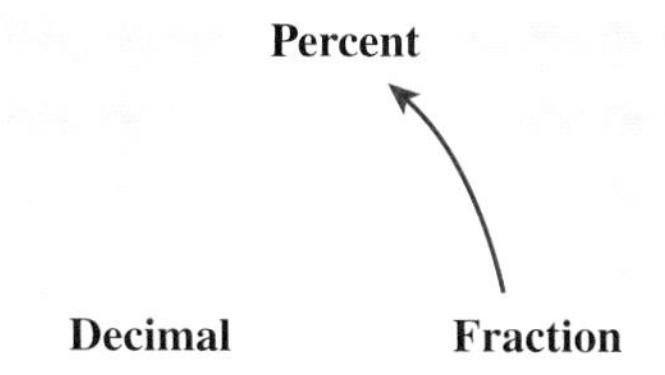

Multiply by 100%.

$$\frac{5}{4} = \frac{5}{4} \cdot \frac{100\%}{1}$$
$$= \frac{5 \cdot \cancel{4} \cdot 25\%}{\cancel{4}}$$
$$= 125\%$$

GUIDED PRACTICE

16. Convert $\frac{3}{5}$ to a percent in two steps.

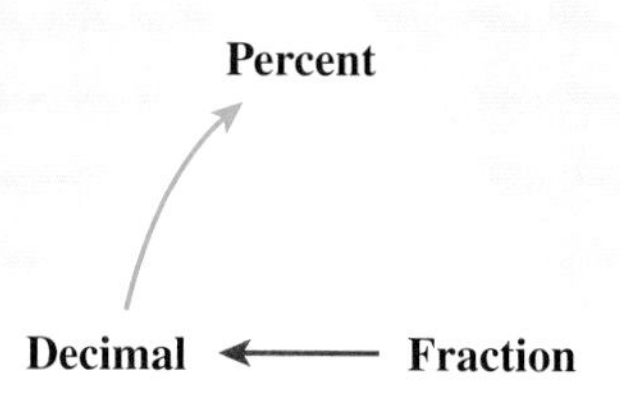

- Convert $\frac{3}{5}$ to a decimal.
- Convert the decimal to a percent.

$\frac{3}{5} =$

$=$

$=$

17. Convert $\frac{3}{5}$ to a percent in one step.

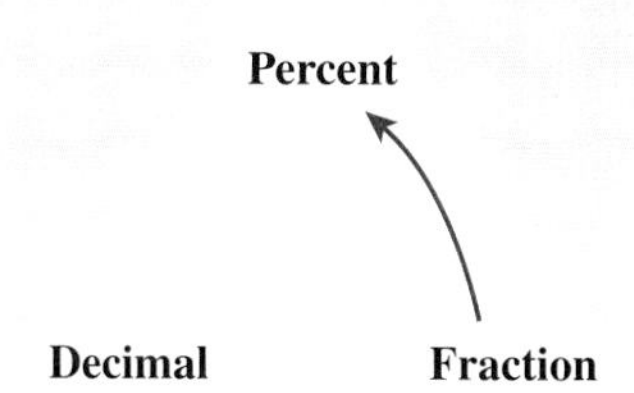

$\frac{3}{5} = \frac{3}{5} \cdot$

$= \frac{\quad}{\quad}$

$=$

18. Convert 0.064 to a percent in two steps.

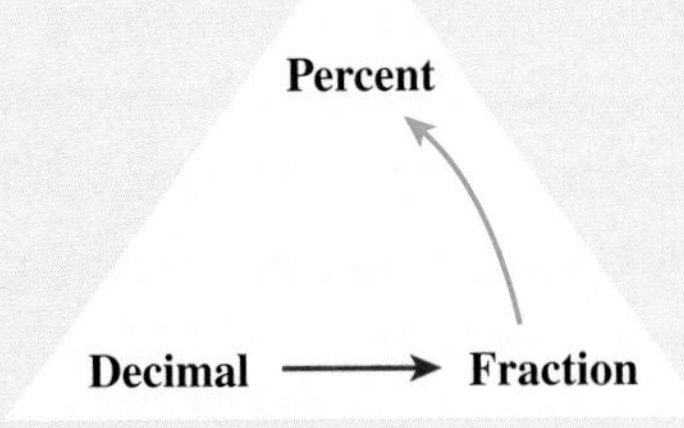

- Convert 0.064 to a fraction.
- Convert the fraction to a percent.

Step 1 Use place value to convert the decimal to a fraction. $0.064 = \dfrac{64}{1{,}000}$

Step 2 Multiply by 100%. $= \dfrac{64}{1{,}000} \cdot \dfrac{100\%}{1}$

Step 3 Simplify. $= \dfrac{64}{10 \cdot \cancel{100}} \cdot \dfrac{\cancel{100}\%}{1}$

Step 4 Divide the numerator by the denominator. $= \dfrac{64}{10}\%$

$= 6.4\%$

18. Convert 0.045 to a percent in two steps.

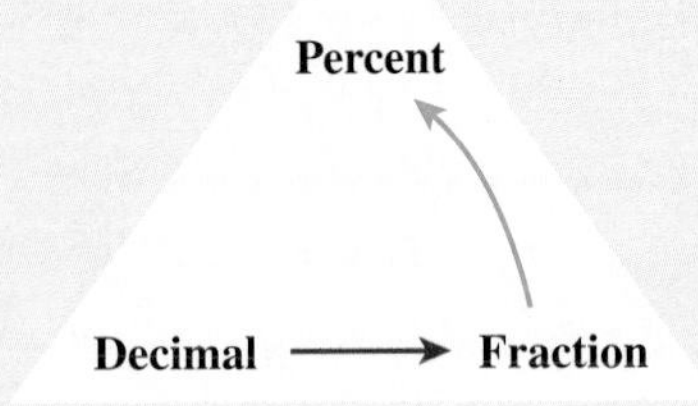

- Convert 0.045 to a fraction.
- Convert the fraction to a percent.

$0.045 = \underline{\qquad}$

$= \underline{\qquad} \cdot \dfrac{100\%}{1}$

$= \underline{\qquad} \cdot \dfrac{100\%}{1}$

$= \underline{\qquad}\%$

$= \qquad\%$

19. Convert 0.064 to a percent in one step.

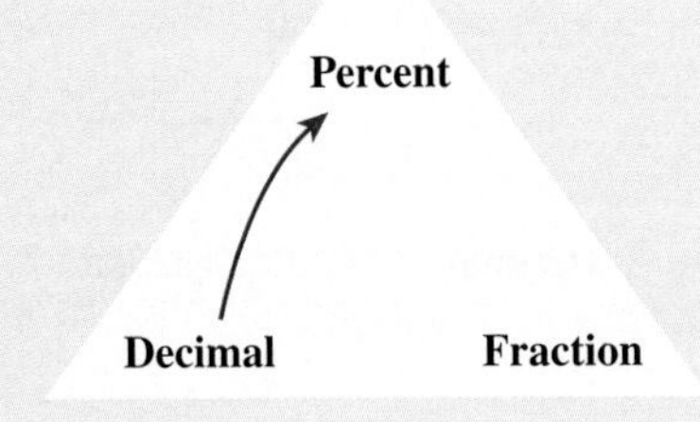

Multiply by 100%. $0.064 = 0.064 \cdot 100\%$

$= 6.4\%$

19. Convert 0.045 to a percent in one step.

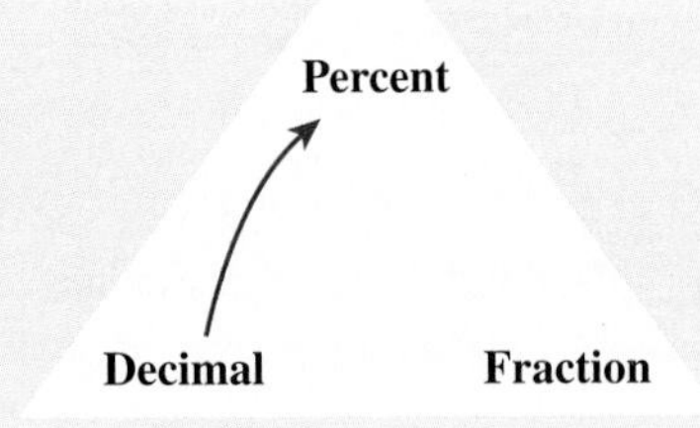

$0.045 = 0.045 \cdot \underline{\qquad}\%$

$= \qquad\%$

CONCEPT II Interpreting Graphical Data Graphs are useful in displaying large amounts of data in a form that is easy to interpret. That is why newspapers, magazines, and television programs frequently use graphs to present data. It is important to learn how to estimate portions of a whole that are represented in a graph. The following circle graph for weather patterns is used in Example 20.

Local Weather Pattern

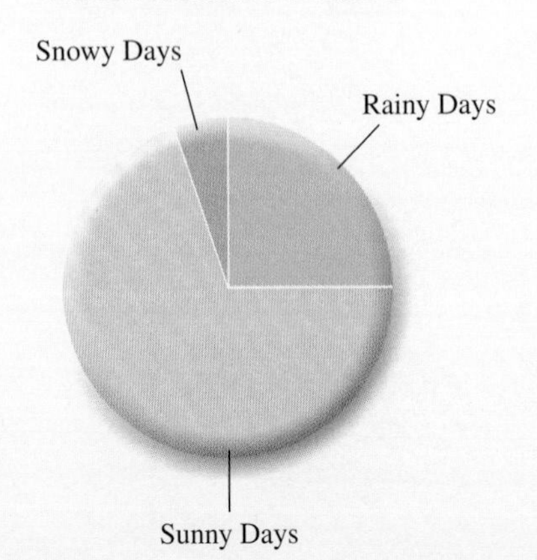

EXAMPLE Interpreting Graphical Data

20. a. Approximately what percent of the days were rainy?

Step 1 Estimate the area of interest with a fraction.

The blue sector represents rainy days. $\frac{1}{4}$ is a good estimate for the blue portion.

Step 2 Convert the fraction to a percent.

$$\frac{1}{4}\cdot\frac{100\%}{1} = \frac{1}{\cancel{4}}\cdot\frac{\cancel{4}\cdot 25\%}{1}$$
$$= 25\%$$

Approximately 25% of the days were rainy.

20. b. Approximately what percent of the days were sunny?

Step 1 Estimate the area of interest with a fraction.

The gold sector looks like it is more than $\frac{2}{3}$ but less than $\frac{3}{4}$.
The estimate should be between these two fractions.

Low Estimate as a Fraction	High Estimate as a Fraction
$\frac{2}{3}$	$\frac{3}{4}$

Step 2 Convert the fractions to percents and pick a reasonable estimate between the two values.

Low Estimate as a Percent

$$\frac{2}{3} = \frac{2}{3}\cdot\frac{100\%}{1} = \frac{200}{3}\% = 66\tfrac{2}{3}\%$$

High Estimate as a Percent

$$\frac{3}{4} = \frac{3}{4}\cdot\frac{100\%}{1} = \frac{3}{\cancel{4}}\cdot\cancel{4}\cdot 25\% = 75\%$$

70% is a good estimate because it is between these values.

Approximately 70% of the days were sunny.

20. c. Approximately what percent of the days were snowy?

Subtract the known percentages from 100%.
The percentages of the rainy, sunny, and snowy days must add to 100%.
Subtract the percents for rainy and sunny days from 100% to find the percent of snowy days.

$$100\% - 25\% - 70\% = 75\% - 70\%$$
$$= 5\%$$

Approximately 5% of the days were snowy.

The circle graph for a college budget is used in Guided Practice 20.

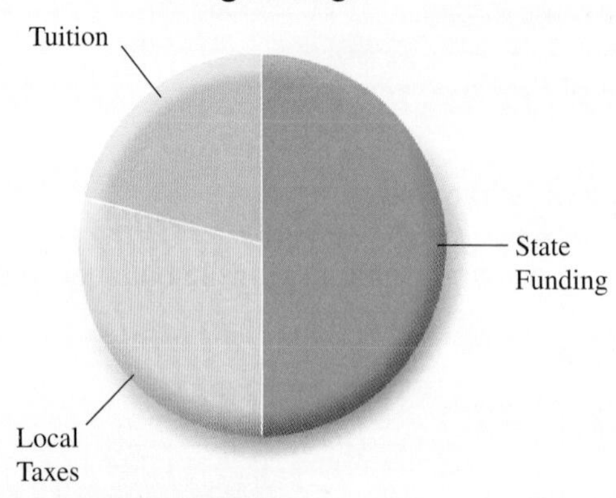

GUIDED PRACTICE Interpreting Graphical Data

20. a. Approximately what percent of the college budget comes from state funding?

Step 1 Estimate the area of interest with a fraction.

The ______ sector represents state funding. $\left(\frac{}{}\right)$ is a good estimate for this portion.

Step 2 Convert the fraction to a percent.

$$\frac{}{} \cdot \frac{100\%}{1} = $$
$$=$$

Approximately ______ % of the college budget comes from state funding.

20. b. Approximately what percent of the college budget comes from local taxes?

Step 1 Estimate the area of interest with a fraction.

The sector for local taxes looks like it is more than $\left(\frac{}{}\right)$ but less than $\left(\frac{}{}\right)$.

The final estimate should be between these two values.

Low Estimate as a Fraction

High Estimate as a Fraction

$\left(\frac{}{}\right)$

Step 2 Convert the fractions to percents and pick a reasonable estimate between the two values.

Low Estimate as a Percent

$$\frac{}{} \cdot \frac{100\%}{1} =$$
$$=$$

High Estimate as a Percent

$$\frac{}{} \cdot \frac{100\%}{1} =$$
$$=$$

______% is a good estimate because it is between these values.

Approximately ______% of the college budget comes from local taxes. Look back at the graph to make sure your answer is reasonable.

20. c. Approximately what percent of the college budget comes from tuition?

Subtract the known percentages from 100%.
The percentages of the local taxes, state funding, and tuition must add to ______%. Subtract the known percents from ______% to find the percent of the college budget that comes from tuition.

$$100\% - \_\_\_\_\% - \_\_\_\_\% = \_\_\_\_\% - \_\_\_\_\%$$
$$= \_\_\_\_\%$$

Approximately ______% of the college budget comes from tuition.

Section 5.1 Exercises

FOR EXTRA HELP  

To visualize percents:

1. Answer the Objective A Concept Checks.
2. Answer the odd Objective A Practice Exercises.
3. Answer the even Objective A Practice Exercises.

To convert a percent to a fraction or mixed number:

4. Answer the Objective B Concept Checks.
5. Answer the odd Objective B Practice Exercises.
6. Answer the even Objective B Practice Exercises.

To convert a fraction or mixed number to a percent:

7. Answer the Objective C Concept Checks.
8. Answer the odd Objective C Practice Exercises.
9. Answer the even Objective C Practice Exercises.

To convert fractions or mixed numbers to common percents:

10. Answer the even Focus On Converting Fractions or Mixed Numbers to Common Percents Exercises.
11. Answer the odd Focus On Converting Fractions or Mixed Numbers to Common Percents Exercises.

To convert between a percent and a decimal:

12. Answer the Objective D Concept Checks.
13. Answer the odd Objective D Practice Exercises.
14. Answer the even Objective D Practice Exercises.

VOCABULARY REVIEW *Review the Vocabulary Preview for Section 5.1. Study the definitions until you can check* Got It *for every word.*

Use these words to complete each sentence.

per • cent • percent

	Got It	Get Help
15. ____________ is used in place of the phrases *per hundred* and *out of one hundred.*		
16. The word ____________ means one hundred.		
17. ____________ is a prefix that indicates division.		

How will you get help for any vocabulary that you are unsure about?

Your instructor ____ MyMathLab ____ A classmate ____ A tutor ____ Other ____

18. Describe each picture using a percent, a fraction, and a decimal.

a. 

b. 

c.  

Visualize each percent.

19. Shade 45% of a square.

20. Shade 23% of a square.

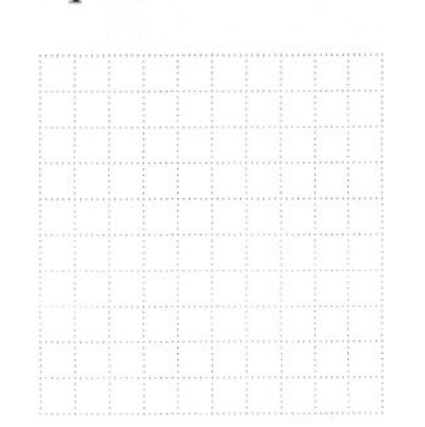

21. Shade 78% of a square.

22. Shade 1% of a square.

23. Shade 99% of a square.

24. Shade 52% of a square.

Use the circle graph to answer each question.

25. What percent of tax dollars are used to fund the library and school combined?

26. What percent of tax dollars go to the city/town, county, and township combined?

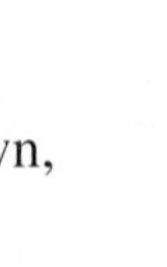

Where Are My Taxes Spent?
Each dollar of taxes paid is spent on the following:

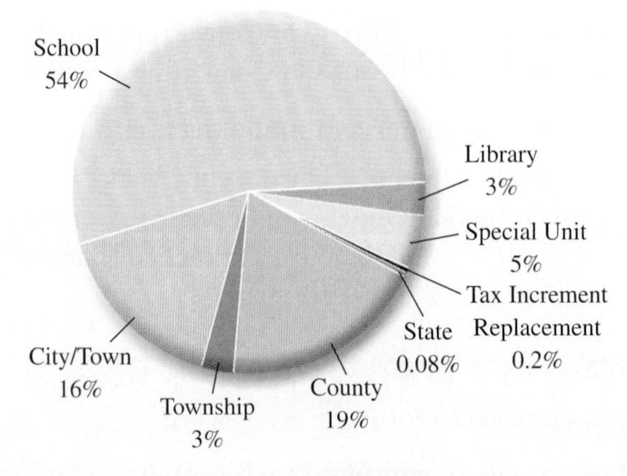

Convert each percent to a fraction or mixed number. Write your answer in simplest form.

27. 51%

28. 33%

29. 80%

30. 90%

31. 100%

32. 200%

33. $33\frac{1}{3}\%$

34. $166\frac{2}{3}\%$

35. 87.5%

36. 62.5%

37. 175%

38. 325%

39. 3,000%

40. 4,500%

41. 0.02%

42. 0.012%

For each statement, convert the percent to a fraction or mixed number.

43. An optimist said that the glass was fifty-five percent full.

44. A pessimist said that the glass was forty-five percent empty.

45. An accountant said that the stock's new value was 110% of the old value.

46. An *American Idol* judge told the contestant "You're going to Hollywood, 110%!"

47. In Exercise 45, the account's new value is 110% of the original value. In Exercise 46, a contestant is "going to Hollywood, 110%." Why is it possible to have 110% of a value, but impossible to "go to Hollywood, 110%"?

Convert each fraction or mixed number to a percent.

48. $\frac{1}{2}$ **49.** $\frac{3}{10}$ **50.** $\frac{5}{7}$ **51.** $\frac{4}{9}$

52. $\frac{5}{4}$ **53.** $\frac{5}{3}$ **54.** $2\frac{1}{2}$ **55.** $3\frac{1}{5}$

56. $\frac{33}{50}$ **57.** $\frac{17}{20}$ **58.** $\frac{1}{100}$ **59.** $\frac{100}{100}$

60. $5\frac{1}{8}$ **61.** $3\frac{3}{10}$ **62.** 7 **63.** 6

Convert each fraction to a percent.

64. Nineteen out of twenty students completed their homework on time.

65. In the United States, fourteen out of fifty-six homes have cable television.
Source: TV-Free America

66. In the United States, three out of five people can identify the Three Stooges.
Source: TV-Free America

67. The value of the stock account was four-ninths of the original value.

68. Of the nine gorillas in the room, zero of them weighed eight hundred pounds.

69. In a Hawaiian police department, there were five detectives, none of whom wore Bermuda shorts.

Convert between decimals and percents as appropriate.

70. 1 **71.** 3 **72.** 40% **73.** 60% **74.** 0.23 **75.** 0.81

76. 115% **77.** 250% **78.** 0.0016 **79.** 0.0076 **80.** 0.03% **81.** 0.08%

82. 2.2 **83.** 4.5 **84.** 80% **85.** 450% **86.** 0.04 **87.** 4.00

88. 0.06% **89.** 6% **90.** 82 **91.** 0.082 **92.** 1,000% **93.** 0.001%

smith Because a percent is out of one hundred, it looks larger than its decimal equivalent. You can always check an answer using this fact.
For more tips and tweets, go to twitter.com/gstbasicmath

Perform each conversion. Round answers to the hundredths place as appropriate.

94. a. Convert $\frac{1}{5}$ to a decimal. Then convert the decimal to a percent.
b. Convert $\frac{1}{5}$ to a percent in one step.

95. a. Convert $\frac{1}{20}$ to a decimal. Then convert the decimal to a percent.
b. Convert $\frac{1}{20}$ to a percent in one step.

96. a. Convert 0.056 to a fraction. Then convert the fraction to a percent.
b. Convert 0.056 to a percent in one step.

97. a. Convert 0.034 to a fraction. Then convert the fraction to a percent.
b. Convert 0.034 to a percent in one step.

98. a. Convert 145% to a decimal. Then convert the decimal to a fraction.
b. Convert 145% to a fraction in one step.

99. a. Convert 220% to a decimal. Then convert the decimal to a fraction.
b. Convert 220% to a fraction in one step.

Fill in the missing values on each decimal–percent–fraction triangle.

100.

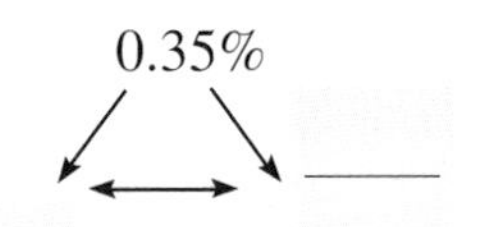

Decimal Fraction

101.

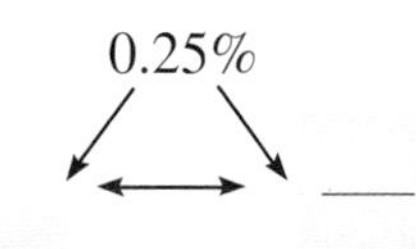

Decimal Fraction

102.

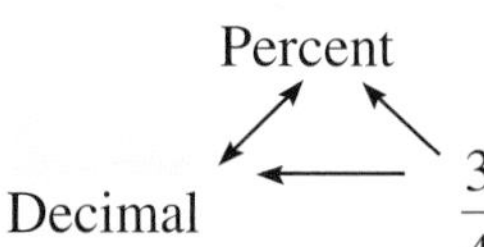

Decimal

103.

Percent

Decimal $\frac{17}{20}$

104. Identify the steps used to convert a fraction to a decimal.

105. Identify the steps used to convert a percent to a decimal.

106. Identify the steps used to convert a percent to a fraction.

107. Identify the steps used to convert a fraction to a percent.

Use the part–whole concept of a fraction to estimate the percent of each sector in the following circle graphs.

Andy's Paycheck:
Taxes vs. Take-home Pay

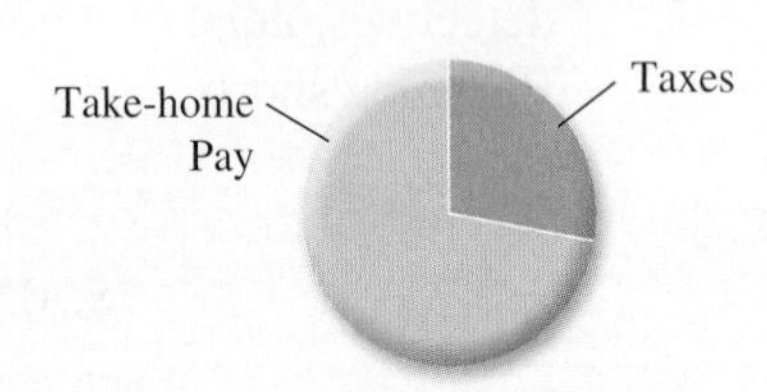

Credit Card
Minimum monthly payment

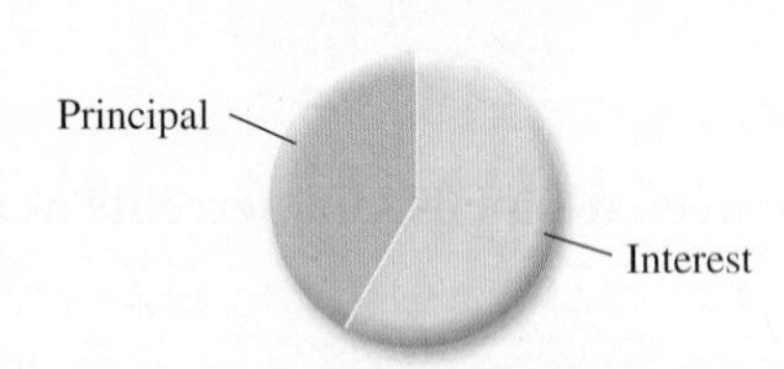

108. a. About what percent of Andy's paycheck goes to taxes?
b. About what percent of Andy's paycheck does he take home?
c. Why must the percents from parts a and b add to 100%?

109. a. About what percent of the minimum monthly payment goes to pay for the principal?
b. About what percent of the minimum monthly payment goes to pay for interest?
c. Why must the percents from parts a and b add to 100%?

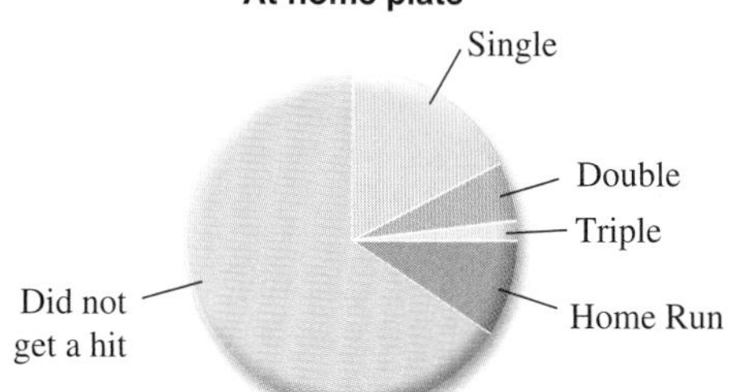

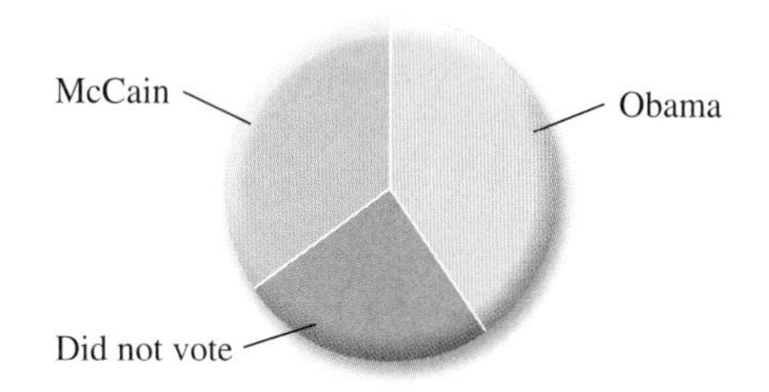

110. **a.** About what percent of the time did Babe Ruth not get a hit?
b. About what percent of the time did Babe Ruth hit a home run?
c. Why don't the percents from parts a and b add to 100%?

111. **a.** About what percent of eligible voters did not vote in the 2008 presidential election?
b. About what percent of eligible voters voted for Barack Obama?
c. Why don't the percents from parts a and b add to 100%?

112. When he enrolled in school in January, Chapal had four sources of income, as shown in the circle graph.
a. What percent of Chapal's January income came from his job?
b. What percent of Chapal's January income came from financial aid?
c. What percent of Chapal's January income came from his job and eBay sales combined?
d. What percent of Chapal's January income came from financial aid and gifts from his parents combined?

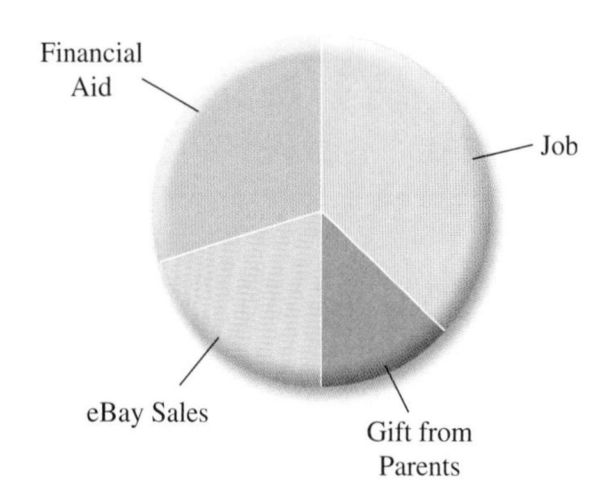

Complete the following tables by writing the equivalent fraction, decimal, or percent in each empty box. When you are finished, study the patterns in the tables to become familiar with common conversions.

113.

	Halves		
	Decimal	**Fraction**	**Percent**
a.	0.5		
b.		$\frac{2}{2}$	

114.

	Thirds		
	Decimal	**Fraction**	**Percent**
a.		$\frac{1}{3}$	
b.	$0.\overline{6}$		
c.			100%

115.

	Fourths		
	Decimal	Fraction	Percent
a.	0.25		
b.		$\frac{2}{4}$	
c.			75%
d.	1.0		

116.

	Tenths		
	Decimal	Fraction	Percent
a.	0.1		
b.	0.2		
c.		$\frac{3}{10}$	
d.			40%
e.	0.5		
f.		$\frac{6}{10}$	
g.	0.7		
h.			80%
i.	0.9		
j.		$\frac{10}{10}$	

117.

	Fifths		
	Decimal	Fraction	Percent
a.			20%
b.		$\frac{2}{5}$	
c.	0.6		
d.		$\frac{4}{5}$	
e.			100%

Complete the following tables by writing the equivalent fraction, decimal, or percent in each empty box.

118.

	Decimal	Fraction	Percent
a.		$\frac{1}{7}$	
b.			35%
c.		$\frac{1}{15}$	
d.	1.24		

119.

	Decimal	Fraction	Percent
a.		$\frac{5}{9}$	
b.			128%
c.	2.9		
d.		$\frac{1}{32}$	

SELF-ASSESSMENT

	YES	NO
1. On your last test, were you satisfied with your score? *If you answered yes, good job. Do not continue here.* *If you answered no, answer the following.*		
2. Did you preview each section of the textbook before your instructor taught it in class?		
3. Did you complete all of the homework before it was due?		
4. Did you get help with the homework exercises that you didn't understand?		

5. Set two goals that will help you become better prepared for your next test.

Goal 1: ____________________

Goal 2: ____________________

Section 5.1 Question Log

Use this space to write questions. Be sure to get these answered and revisit them when you prepare for your exam.

Page _____ **Answered**	**Page** _____ **Answered**
Page _____ **Answered**	**Page** _____ **Answered**

Section 5.2 Use Proportions to Solve Percent Exercises

By reading the newspaper or watching TV, you have likely seen or heard statements that use percents.

29% of Californians have a college degree

If you know California's population, you can use the percent, 29%, to determine how many Californians have a college degree.

This section presents techniques that will help you answer everyday questions about percents. For example, you will learn how to translate written statements into percent proportions and solve these proportions for an unknown quantity.

The **Objectives** in Section 5.2 will help you

- **A** Translate a written statement into a percent proportion.
- **B** Solve a percent exercise.

VOCABULARY PREVIEW *Check the box that applies.*

Term	Got It	Must Study
Of: The word ***of*** is often used to represent multiplication in word exercises.		
Is: The words ***is*** and ***are*** can be used to represent equality in word exercises.		
Percent Proportion (Math): The standard form of a **percent proportion** is $\frac{P}{100} = \frac{A}{B}$, where • P is the **percent.** It is written as a ratio of "P to 100." • A is the **amount.** • B is the **base.**		
Percent Proportion (Words): The standard form of a **percent proportion** written in words is *The percent is to 100 as the amount is to the base.*		

Study these words when they appear in the text. After the section, test your understanding by completing the Vocabulary Review in the section exercises.

Objective A Translate a Written Statement into a Percent Proportion

The Concept A **percent proportion** is a mathematical statement that two ratios are equal. To write a percent proportion, you must identify the percent, base, and amount.

INTERACTIVE DEFINITION Percent Proportion

The standard form of a **percent proportion** is $\frac{P}{100} = \frac{A}{B}$.

- The percent is written as a ratio, $\frac{P}{100}$.
- The amount and base form the second ratio, $\frac{A}{B}$.

EXAMPLE

1. Identify the percent, base, and amount in the percent proportion.

$$\frac{50}{100} = \frac{15}{30}$$

The percent is 50%.
The base is 30.
The amount is 15.

GUIDED PRACTICE

1. Identify the percent, base, and amount in the percent proportion.

$$\frac{10}{100} = \frac{2}{20}$$

The percent is ______%.
The base is ______.
The amount is ______.

DO YOU UNDERSTAND how to identify the percent, base, and amount in a percent proportion? Got It Get Help

To write a proportion for a percent exercise, determine the values to substitute into the proportion $\frac{P}{100} = \frac{A}{B}$.

Two of these values will be given, while the third will be represented with a variable.

Procedure Translate a Written Statement into a Percent Proportion

Step 1 Identify the percent, base, and amount.
Choose a variable for the unknown value.

Step 2 Write the percent proportion.

Students who look for key phrases when translating word exercises are usually more successful. Often, percent exercises are written with the phrase *a percent of the base is the amount.* Look for these words to help you identify the percent, base, and amount.

EXAMPLE

2. Translate the written statement into a percent proportion. "What percent of 30 is 21?"

Step 1 Identify the percent, base, and amount.

To identify the percent, look for the "percent."

The percent is unknown.
Let P = percent.

To identify the base B, look for "percent (%) of."

$B = 30$

To identify the amount A, look for the word *is*.

$A = 21$

Step 2 Write the percent proportion.
Substitute the values for P, B, and A.

$$\frac{P}{100} = \frac{21}{30}$$

GUIDED PRACTICE

2. Translate the written statement into a percent proportion. "What percent of 20 is 50?"

Step 1 Identify the percent, base, and amount.

To identify the percent, look for the "percent."

The percent is unknown.
Let P = _______.

To identify the base B, look for "percent (%) of."

B = _____

To identify the amount A, look for the word *is*.

A = _____

Step 2 Write the percent proportion.
Substitute the values for P, B, and A.

$$\frac{\square}{100} = \frac{\square}{\square}$$

In the first step of Example 2, we identified the percent as the unknown quantity. We identified the variable by writing "Let P = Percent." Identifying the variable in this way is called "choosing the variable."

EXAMPLES

3. Translate the written statement into a percent proportion. "What is 40% of 15?"

Step 1 Identify the percent, base, and amount.

To identify the percent, look for the "percent."

$P = 40$

To identify the base, look for "percent (%) of."

$B = 15$

To identify the amount, look for the word *is*.

The amount is unknown.
Let A = amount.

Step 2 Write the percent proportion.
Substitute the values for P, B, and A.

$$\frac{40}{100} = \frac{A}{15}$$

GUIDED PRACTICE

3. Translate the written statement into a percent proportion "What is 120% of 35?"

Step 1 Identify the percent, base, and amount.

To identify the percent, look for the "percent."

P = _____

To identify the base, look for "percent (%) of."

B = _____

To identify the amount, look for the word *is*.

The amount is unknown.
Let A = _______.

Step 2 Write the percent proportion.
Substitute the values for P, B, and A.

$$\frac{\square}{100} = \frac{\square}{\square}$$

4. Translate the written situation into a percent proportion. Carlos's commission, $16, is 10% of his sales. How much were his sales?

Step 1 Identify the percent, base, and amount.

To identify the percent, look for the "percent."

$$P = 10$$

To identify the base, look for "percent (%) of."

The base is unknown.
Let B = Carlos's sales.

To identify the amount, look for the word *is.*

$$A = 16$$

Step 2 Write the percent proportion.
Substitute the values for P, B, and A.

$$\frac{10}{100} = \frac{16}{B}$$

4. Translate the written situation into a percent proportion. 45 percent of Angelina's time at work was spent calling 10 customers. At this rate, how many customers can she call using all her time at work?

Step 1 Identify the percent, base, and amount.

To identify the percent, look for the "percent."

$P =$ ____

To identify the base, look for "percent (%) of."

The base is unknown.
Let ____ = ____________________.

To identify the amount, look for the word *is.* In this case, the past tense *was* indicates the amount.

$A =$ ____

Step 2 Write the percent proportion.
Substitute the values for P, B, and A.

$$\frac{\quad}{100} = \frac{\quad}{\quad}$$

5. Translate the written situation into a percent proportion. Julia works for a marketing firm in New Orleans. She paid 25% of her salary in taxes last year. If she earned $38,752 in salary, how much are her taxes?

The important information is highlighted in the paragraph above.

Step 1 Identify the percent, base, and amount.

To identify the percent, look for the "percent."

$$P = 25$$

To identify the base, look for "percent (%) of."

$$B = 38{,}752$$

To identify the amount, look for the word *is.* In this case, the key word is *are.*

The amount is unknown.
Let A = last year's taxes.

Step 2 Write the percent proportion.
Substitute the values for P, B, and A.

$$\frac{25}{100} = \frac{A}{38{,}752}$$

5. Translate the written situation into a percent proportion. Henry and Anais are each writing a 40-page term paper. After printing a draft, Anais ran out of paper. Henry gave her 232 pages, which was 58% of his paper. How much paper did Henry have before he gave paper to Anais?

Underline or highlight the important information in the paragraph above.

Step 1 Identify the percent, base, and amount.

To identify the percent, look for the "percent."

$P =$ ____

To identify the base, look for "percent (%) of."

The base is unknown.
Let ____ = ____________________.

To identify the amount, look for the word *is.* In this case, the key word is *was.*

$A =$ ____

Step 2 Write the percent proportion.
Substitute the values for P, B, and A.

$$\frac{\quad}{100} = \frac{\quad}{\quad}$$

Concept Check

A1. Use the written statement "40% of the horses are black" to answer each question.
  a. Which part of the statement indicates the base?
  b. Which part of the statement indicates the amount?

A2. When translating a written statement into a percent proportion, certain words or phrases can help you identify the percent, base, and amount.
  a. To identify the percent, look for the word ______ or the symbol ____.
  b. To identify the base, look for the phrase ______.
  c. To identify the amount, look for the words ______.

Objective A Practice

Translate each written situation or question into a percent proportion. Use *P* for percent, *B* for base, and *A* for amount. Identify the variable that you chose for the unknown. Do not solve the proportions.

A3. What percent of 30 is 18?

A4. What percent of 40 is 12?

A5. What is 80% of 20?

A6. What is 20% of 50?

A7. 250% of what is 90?

A8. 125% of what is 50?

A9. Saeed had to pay 4% of his textbook cost in sales tax. If the sales tax is $8.80, what was the cost of his textbook before the tax was added?

A10. Kao sells used cars. He earns 5% of the sale price for each car that he sells. If Kao earned $3,750 last month, what was the total value of the cars he sold?

A11. Krystal has to read a 150-page novel for her English class. After reading 110 pages, what percent of the novel has she read?

A12. Keisha just missed setting the record for the most circuit boards assembled in an hour. If she assembled 79 circuit boards and the record was 80, what percent of the record amount did she assemble?

FOCUS ON Writing Percent Proportions

For each exercise, identify the correct percent proportion located in the center column. You do not need to solve for the unknown value. Each percent proportion is used twice. *P* = percent, *B* = base, *A* = amount.

1. What percent of 50 is 20?

2. 50 is what percent of 20?

3. What number is 20% of 50?

4. 50% of 20 is what number?

5. 50% of what number is 20?

6. 50 is 20% of what number?

7. How much money is fifty percent of twenty dollars?

8. The twenty cars with magwheels represent fifty percent of how many cars?

9. To poll twenty percent of the students, fifty students must be questioned.

10. The number of students that earned A's is twenty percent of the fifty students enrolled.

11. An account balance of fifty dollars is what percent of an initial investment of twenty dollars?

12. If twenty out of fifty students earned B's, what percent of the students earned B's?

a. $\frac{20}{100} = \frac{A}{50}$

b. $\frac{P}{100} = \frac{20}{50}$

c. $\frac{50}{100} = \frac{20}{B}$

d. $\frac{50}{100} = \frac{A}{20}$

e. $\frac{P}{100} = \frac{50}{20}$

f. $\frac{20}{100} = \frac{50}{B}$

Objective B Solve a Percent Exercise

The Concept When solving percent exercises, you will solve a proportion in the form

$$\frac{P}{100} = \frac{A}{B}.$$

Two out of three variables will be known, and we must solve for the value of the third. The following examples show how to solve the percent proportion for each unknown variable.

Procedure Solve a Percent Exercise

Step 1 Translate the written statement into a percent proportion.

Step 2 Solve the proportion.

- Rewrite the proportion with the unknown variable in the numerator.
- Multiply both sides of the equation to isolate the variable.

EXAMPLES

6. What percent of 20 is 30?

Step 1 Identify P, B, and A. The percent is unknown.
Let P = the percent.
$B = 20$
$A = 30$

Write a percent proportion. $\frac{P}{100} = \frac{30}{20}$

Step 2 Multiply both sides by 100 to isolate P. $\frac{P}{\cancel{100}} \cdot \frac{\cancel{100}}{1} = \frac{30}{20} \cdot \frac{100}{1}$

$P = \frac{30 \cdot 100}{20}$

Simplify. $P = \frac{30 \cdot \cancel{20} \cdot 5}{\cancel{20}}$

$= 150$

150% of 20 is 30.

GUIDED PRACTICE

6. What percent of 40 is 16?

The percent is unknown.
Let P = ______.
$B = 40$
$A = 16$

$\frac{\quad}{100} = \frac{\quad}{\quad}$

$\frac{\quad}{100} \cdot \frac{\quad}{1} = \frac{\quad}{\quad} \cdot \frac{\quad}{1}$

$P =$ ______

$P =$ ______

$P =$

____% of 40 is ____.

7. 8 is 80% of what number?

Identify P, B, and A.

$P = 80$
The base is unknown.
Let $B =$ the number.
$A = 8$

Step 1 Translate the written statement.

$$\frac{80}{100} = \frac{8}{B}$$

Step 2 Rewrite the proportion with the variable in the numerator.

$$\frac{100}{80} = \frac{B}{8}$$

$$\frac{100}{80} \cdot \frac{8}{1} = \frac{B}{8} \cdot \frac{8}{1}$$

Use multiplication to isolate B.

$$\frac{\cancel{10} \cdot 10 \cdot \cancel{8}}{\cancel{8} \cdot \cancel{10} \cdot 1} = B$$

Simplify.

$$10 = B$$

8 is 80% of 10.

7. 64 is 16% of what number?

$P =$ ____
The base is unknown.
Let $B =$ ________.
$A =$ ____

$$\frac{\quad}{100} = \frac{\quad}{\quad}$$

$$\frac{100}{\quad} = \frac{\quad}{\quad}$$

$$\frac{100}{\quad} \cdot \frac{\quad}{1} = \frac{\quad}{\quad} \cdot \frac{\quad}{1}$$

$$\frac{\quad}{\quad} =$$

$$=$$

____ is ____% of ____.

8. Andrea paid a 25% tip for a \$40 bill. What was the tip?

Identify P, B, and A.

$P = 25$
$B = 40$
The amount is unknown
Let $A =$ the tip.

Step 1 Translate the written statement.

$$\frac{25}{100} = \frac{A}{40}$$

Step 2 Use multiplication to isolate A.

$$\frac{25}{100} \cdot \frac{40}{1} = \frac{A}{\cancel{40}} \cdot \frac{\cancel{40}}{1}$$

Simplify.

$$\frac{5 \cdot \cancel{5} \cdot 2 \cdot \cancel{20}}{\cancel{20} \cdot \cancel{5} \cdot 1} = A$$

$$10 = A$$

The tip was \$10.

8. Kato must write 40% of a 20-page paper by tomorrow. How many pages must he write by tomorrow?

$P =$ ____
$B =$ ____
The amount is unknown.
Let $A =$ ____________.

$$\frac{\quad}{100} = \frac{\quad}{\quad}$$

$$\frac{\quad}{100} \cdot \frac{\quad}{1} = \frac{\quad}{\quad} \cdot \frac{\quad}{1}$$

$$\frac{\quad}{\quad} =$$

$$=$$

Kato must write ____ pages by tommorow.

Concept Check

B1. Translate the percent proportion into a written statement: $\frac{22}{100} = \frac{24}{n}$

____ is ____ percent of ________.

B2. a. In the percent proportion $\frac{120}{100} = \frac{A}{30}$, why must the amount A be more than 30?

b. In the percent proportion $\frac{P}{100} = \frac{20}{25}$, why must the percent P be less than 100%?

Objective B Practice

Set up and solve a percent proportion.

B3. What is 30% of 60?

B4. What is 110% of 40?

B5. 100 is what percent of 80?

B6. 90 is what percent of 45?

B7. 75 is 25% of what number?

B8. 300 is 15% of what number?

B9. Dee was making 180 candles for a customer. Because Dee enjoyed her work, she accidentally made 105% of the candles required. How many candles did Dee make?

B10. Officer Samson pulled over 175 drivers in July. 140 of these drivers were wearing seat belts. What percent of the drivers were wearing seat belts?

B11. Ed took a test in his managerial finance class. If he answered 80% of 80 questions correctly, how many questions did Ed answer correctly?

B12. Denise wanted to meet with each of her 150 employees. If she met 90 of the employees, what percent of the employees did she meet?

Combining Concepts and Applications

CONCEPT I Solving Multistep Applications Now that you can translate written statements and solve percent proportions, you can solve more complex applications. The following examples demonstrate a three-step technique for organizing and solving multistep applications.

Procedure Solve Multistep Applications

Step 1 Identify the question in the application.

Step 2 Identify the information needed to answer the question.

Step 3 Use the information to set up and solve a percent proportion.

EXAMPLES

9. In a class of 35 students, 21 have a job. What percent of the students do not have a job?

Step 1 Identify the question in the application.

"What percent of the students do not have a job?"

Step 2 Identify the information needed to answer the question.

$$\begin{aligned} \text{No job} &= \text{Total students} - \text{Students with jobs} \\ &= 35 - 21 \\ &= 14 \end{aligned}$$

GUIDED PRACTICE

9. A bike manufacturer makes 125 bikes per day. If 5 of these bikes are racing bikes, what percent of the bikes are nonracing bikes?

Step 1 Identify the question in the application.

"What ______________________

Step 2 Identify the information needed to answer the question.

Nonracing bikes = ______ − ______

= ______ − ______

= ______

(Continued)

Step 3 Use the information to set up and solve a percent proportion.

Let P = the percent.
$B = 35$
A = students without jobs = 14

$$\frac{P}{100} = \frac{14}{35}$$

$$\frac{\cancel{100}}{1} \cdot \frac{P}{\cancel{100}} = \frac{100}{1} \cdot \frac{14}{35}$$

$$P = \frac{2 \cdot \cancel{5} \cdot 2 \cdot 5 \cdot 2 \cdot \cancel{7} \cdot}{\cancel{5} \cdot \cancel{7}}$$

$$P = 40$$

40% of the students do not have a job.

Step 3 Use the information to set up and solve a percent proportion.

Let ____ = __________.
B = ____
A = ______________ = ____

$$\frac{}{100} = \frac{}{}$$

____ are nonracing bikes.

10. Last year, Jeff earned \$35,000 and paid 31% of his income in taxes. How much money did Jeff earn after taxes?

Step 1 Identify the question in the application.
"How much money did Jeff earn after taxes?"

Step 2 Identify the information needed to answer the question.

Percent after taxes = 100% − percent in taxes
= 100% − 31%
= 69%

Step 3 Use the information to set up and solve a percent proportion.

$P = 69$
$B = 35{,}000$
Let A = amount earned.

$$\frac{69}{100} = \frac{A}{35{,}000}$$

$$\frac{69}{100} \cdot \frac{35{,}000}{1} = \frac{A}{\cancel{35{,}000}} \cdot \frac{\cancel{35{,}000}}{1}$$

$$\frac{69 \cdot \cancel{100} \cdot 350}{\cancel{100}} = A$$

$$69 \cdot 350 = A$$

$$24{,}150 = A$$

Jeff earned \$24,150 after taxes.

10. To buy a \$15,000 car, Samantha needs \$1,500 for the down payment. So far, she has saved 60% of the down payment. How much more does she need to save?

Step 1 Identify the question in the application.
"How ______________________________"

Step 2 Identify the information needed to answer the question.

Percent to save = 100% − percent saved so far
= ____% − ____%
= ____%

Step 3 Use the information to set up and solve a percent proportion.

P = ______
B = ______
Let A = __________.

$$\frac{}{100} = \frac{}{}$$

=

=

=

=

Samantha still needs to save __________.

Section 5.2 Exercises

FOR EXTRA HELP     

To translate a written statement into a percent proportion:

1. Answer the Objective A Concept Checks.
2. Answer the odd Objective A Practice Exercises.
3. Answer the even Objective A Practice Exercises.

To write percent proportions:

4. Answer the odd Focus On Writing Percent Proportions Exercises.
5. Answer the even Focus On Writing Percent Proportions Exercises.

To solve percent exercises:

6. Answer the Objective B Concept Checks.
7. Answer the odd Objective B Practice Exercises.
8. Answer the even Objective B Practice Exercises.

VOCABULARY REVIEW *Review the Vocabulary Preview for Section 5.2. Study the definitions until you can check* Got It *for every word.*

Use these words to complete each sentence.

of • is • percent proportion

	Got It	Get Help
9. The standard form of a __________ is $\frac{P}{100} = \frac{A}{B}$.		
10. In a percent exercise, the word __________ is often written in front of the base.		
11. The word __________ can mean "equals" in word exercises.		
12. Identify the percent, base, and amount in "twenty is forty percent of fifty." Percent = _____ Base = _____ Amount = _____		

How will you get help for any vocabulary that you are unsure about?

Your instructor _____ MyMathLab _____ A classmate _____ A tutor _____ Other _____

For each exercise, set up but do not solve the percent proportion.

13. 35% of 70 is what number?

14. 70% of 35 is what number?

15. 40% of what number is 60?

16. 25% of what number is 14?

17. 33 is what percent of 50?

18. 24 is what percent of 25?

19. 1.2% of 300 is what number?

20. 3.3% of 20 is what number?

21. What percent of 450 is 2?

22. What percent of 6 is 1?

23. 40% of 300 people each ate a hamburger at a party. How many people ate a hamburger?

24. In an order of French fries, 48% of the calories come from fat. If the fries have 250 total calories, how many calories are from fat?

Solve each percent proportion. Round to the nearest hundredth as appropriate.

25. What is 300% of 4?

26. What is 400% of 5?

27. 0.01% of what number is 7?

28. 0.02% of what number is 10?

29. 90% of what number is 70?

30. 300% of what mixed-number is 4?

31. What percent of 2 is 4?

32. What percent of 5 is 25?

33. A medication was tested on 3,000 people. Of those 3,000 people, 2,910 had no side effects. What percent of the people had no side effects?

34. A new hybrid corn seed was tested. If 3,880 out of 4,000 seeds grew to become mature plants, what percent of the seeds became mature plants?

35. Bessie the Wonder Cow produced 35 pounds of milk per day in March. In April, Bessie produced 42 pounds of milk per day. Her milk production in March is what percent of her milk production in April?

36. Andy delivered 1,200 newspapers in May. By July, that number had fallen to 1,032 newspapers. His deliveries in May are what percent of his deliveries in July?

37. The sales tax on a new car is $720. If the price of the car is $12,000 what percent of the price is the tax?

38. Ingrid works for a watch manufacturer. If she sold $900,000 in watches and earned $45,000 for these sales, what percent of the sales were her earnings?

The following circle graph shows the energy sources used to generate 4,000 gigawatt-hours of electricity. Use the graph to answer each question. Round to the nearest percent.

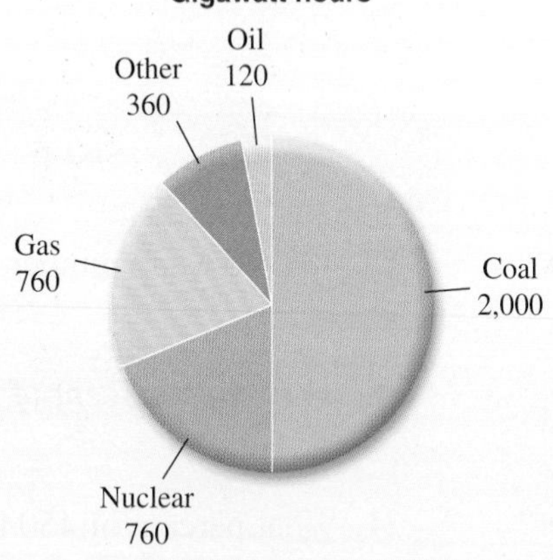

39. Oil, gas, and coal are fossil fuels. What percent of energy for electricity comes from fossil fuels?

40. The amount of electricity made from coal is what percent of electricity made from gas?

41. The amount of electricity made from nuclear sources is what percent of electricity made from oil ?

42. What percent of energy for electricity comes from other sources?

tobey Make sure you are clear how to solve each type of percent proportion. Take extra time to master each type.

For more tips and tweets, go to twitter.com/gstbasicmath

The bar graph shows the salaries of public school workers. Use the graph to answer each question. Round answers to the nearest whole percent.

43. What percent of a teacher's salary is a superintendent's salary?

44. What percent of a superintendent's salary is a teacher's salary?

45. What percent of a librarian's salary is a principal's salary?

46. What percent of a teacher's salary is a secretary's salary?

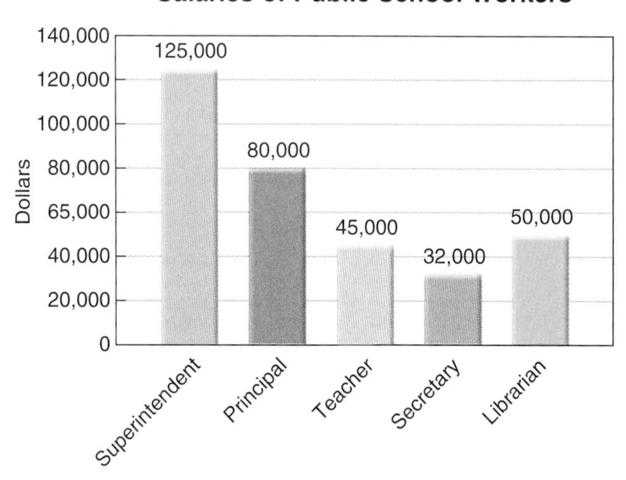

Source: *Statistical Abstract of the United States*, 2006

The bar graph shows income levels based on the highest degree earned. Use the graph to answer each question. Round to the nearest percent.

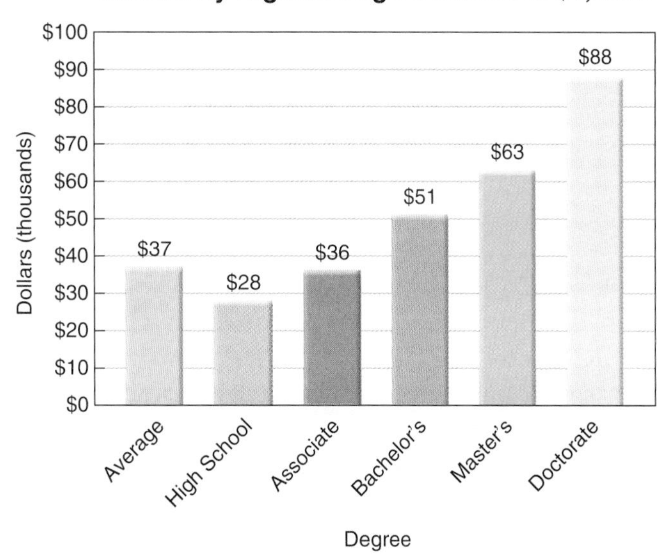

47. A person with an associate degree earns what percent of the income of a person with a high school degree?

48. A person with an associate degree earns what percent of the income of a person with a bachelor's degree?

49. What percent of the average income does someone make if she has a doctorate degree?

50. What percent of the average income does someone make if she has a high school degree?

Use the following nutrition label to answer each question about Oatloops. Round answers to the nearest whole number or percent.

Amount per Serving Cereal, No Milk		Oatloops	With 0.5 Cup Skim Milk
Calories		110	150
		% Daily Recommended Amount	
Fat	2 g	3%	3%
Sodium	210 mg	9%	12%
Potassium	200 mg	6%	12%
Fiber	3 g	11%	11%

51. Based on the nutrition label for Oatloops, what is the daily recommended amount of fat, in grams?

52. What percent of the total calories does 0.5 cup of skim milk contribute to a serving of Oatloops with skim milk?

53. What percent of the total calories does a serving of Oatloops contribute to a serving of Oatloops with skim milk?

54. How many milligrams of potassium does 0.5 cup of skim milk contain?

55. How many milligrams of sodium does 0.5 cup of skim milk contain?

56. What is the recommended daily amount of fiber, to the nearest gram?

57. How many grams of fiber are in 0.5 cup of skim milk? *Hint:* Determine the difference in percents for the Oatloops and the Oatloops with skim milk.

Solve each multistep application.

58. John Paul earned $40,000 last year. Before taxes were taken out, he bought an IRA for $4,000.Then he paid 21% of the remainder of his earnings in taxes. How much did John Paul pay in taxes?

59. Together, Greg and Marsha earn $52,000 a year. They pay 25% of their income in taxes. Then they put 5% of their remaining earnings in a savings account. How much did they put in the savings account?

60. Jean bought a rental house for $140,000. He paid $56,000 in cash and got a mortgage for the balance. If each mortgage payment is 1% of the balance, how much is each mortgage payment?

61. Ben and Mary's bill at a restaurant came to $45. They tipped the waitress 20%. If Ben and Mary each paid 50% of the bill, how much did each person pay, including tip?

62. Kwame invested $10,000 in his favorite stock. The value of the stock went up 14%. He then sold $3,000 of the stock to put a down payment on a car. What is the value of Kwame's remaining stock?

63. Pai purchased a car with a sticker price of $15,000. The dealer lowered the price by 10% and then subtracted $1,750 in incentives. How much did Pai pay for the car?

64. What is 80% of 125% of 36?

65. What is 200% of 50% of 76?

66. Write 80% and 125% as fractions in simplest from. Why do these fractions help explain the answer for Exercise 64?

67. Write 200% and 50% as fractions in simplest form. Why do these fractions help explain the answer for Exercise 65?

Section 5.2 Question Log

Use this space to write questions. Be sure to get these answered and revisit them when you prepare for your exam.

Page ____ **Answered**	**Page** ____ **Answered**
Page ____ **Answered**	**Page** ____ **Answered**

Section 5.3 Use Equations to Solve Percent Exercises

Percent exercises can be solved using either proportions or equations. In this section, you will learn how to solve percent exercises using the standard form of a percent equation.

Standard Form of a Percent Equation

Percent · Base = Amount

The **Objectives** in Section 5.3 will help you

- A Translate a written statement into a percent equation.
- B Solve a percent equation.

VOCABULARY PREVIEW *Check the box that applies.*	Got It	Must Study
Of: The word ***of*** is often used to represent multiplication in word exercises.		
Is: The words ***is*** and ***are*** can be used to represent equality in word exercises.		
Percent Equation: The standard form of a **percent equation** is as follows: In words: The percent of the base is the amount. In an Equation: Percent · Base = Amount		
Percent: In a percent equation, the **percent** describes the portion of the base that is equal to the amount.		
Base: In a percent equation, the **base** is the whole quantity.		
Amount: In a percent equation, the **amount** is the portion that results from multiplying the percent and the base.		

Study these words when they appear in the text. After the section, test your understanding by completing the Vocabulary Review in the section exercises.

Objective A Translate a Written Statement into a Percent Equation

The Concept The statement "50% of 10 is 5" is a **percent equation.**

50% of 10 is 5.

↓ ↓

$50\% \cdot 10 = 5$

Recall that an equation is a mathematical statement with an equal sign. The following equation indicates that two times *B* is 12:

$$2 \cdot B = 12$$

To use percents in practical applications, you must translate between word phrases and the percent equation. For example, a tip at a restaurant can be found using the statement *twenty percent of the bill is the tip*. This statement can be translated to the equation 20% · bill = tip.

INTERACTIVE DEFINITION Percent Equation

The standard form of a **percent equation** is

The percent of the base is the amount.

↓ ↓

Percent · Base = Amount

- The **percent** describes the portion of the base that is equal to the amount.
- The word ***of*** indicates multiplication.
- The **base** is the whole quantity that is multiplied by the percent.
- The word ***is*** indicates equality.
- The **amount** is the portion that results from multiplying the percent and the base.

EXAMPLES

1. Identify the percent, base, and amount in the percent equation.

$$15\% \cdot 50 = 7.5$$

The percent is 15%.

The base is 50 because it is multiplied by the percent.

The amount is 7.5 because it is the result of multiplying the percent and the base.

2. Translate "Twenty percent of the bill is six dollars" into a percent equation.

Identify the five parts of the percent equation.

1. The percent is 20%.
2. *Of* indicates multiplication.
3. The "bill" is the base since it is multiplied by the percent. It is unknown, so we choose a variable.

   Let B = the bill.
4. *Is* indicates an equal sign.
5. The amount is $6.

Write the percent equation.

Percent · Base = Amount

↓ ↓ ↓

$20\% \cdot B = \$6$

GUIDED PRACTICE

1. Identify the percent, base, and amount in the percent equation.

$$25\% \cdot 200 = 50$$

The percent is ____%.

The base is ____ because it is ________ by the percent.

The amount is ____ because it is the result of

________________________.

2. Translate "Sixteen percent of the points scored is eight points" into a percent equation.

Identify the five parts of the percent equation.

1. The percent is ____%.
2. *Of* indicates ____________.
3. The __________ is the base since it is ________ by the ________. It is unknown, so we choose a variable.

   Let B = ____________________.
4. *Is* indicates an ____________.
5. The amount is ____________.

Write the percent equation.

Percent · Base = Amount

↓ ↓ ↓

=

DO YOU UNDERSTAND how to identify the percent, base, and amount in a percent equation?	Got It	Get Help
DO YOU UNDERSTAND how to translate a written statement into a percent equation?	Got It	Get Help

The third step of Example 2 required us to represent an unknown quantity with a variable. When we identified the variable, we wrote "Let B = the bill." Identifying the variable in this way is called "choosing the variable."

Procedure Translate a Written Statement into a Percent Equation

Step 1 Identify the percent, base, and amount.
Choose P, B, or A as a variable for the unknown value.

Step 2 Write the percent equation.
Substitute the values for P, B, and A.

EXAMPLES

3. Translate the written statement into a percent equation.
"What number is 13% of 130?"

Step 1 Identify the percent, base, and amount.
To identify the percent, look for the % symbol.

$$P = 13$$

To identify the base, look for "percent (%) of base."

$$B = 130$$

To identify the amount, look for the word *is*.

The amount is unknown.
Let A = the number.

Step 2 Write the percent equation.
Substitute the values for P, B, and A.

$$P \cdot B = A$$
$$13\% \cdot 130 = A$$

4. Translate the written statement into a percent equation.
"What percent of 20 is 500?"

Step 1 Identify the percent, base, and amount.
To identify the percent, look for the "percent."

The percent is unknown.
Let P = percent.

To identify the base, look for "percent (%) of base."

$$B = 20$$

To identify the amount, look for the word *is*.

$$A = 500$$

Step 2 Write the percent equation.
Substitute the values for P, B, and A.

$$P \cdot B = A$$
$$P \cdot 20 = 500$$

GUIDED PRACTICE

3. Translate the written statement into a percent equation.
"What number is 40% of 200?"

Step 1 Identify the percent, base, and amount.
To identify the percent, look for the % symbol.

$P =$ ____

To identify the base, look for "percent (%) of base."

$B =$ ____

To identify the amount, look for the word *is*.

The amount is unknown.
Let $A =$ ________.

Step 2 Write the percent equation.
Substitute the values for P, B, and A.

$$P \cdot B = A$$
$$=$$

4. Translate the written statement into a percent equation.
"What percent of 700 is 35?"

Step 1 Identify the percent, base, and amount.
To identify the percent, look for the "percent."

The percent is unknown.
Let $P =$ _______.

To identify the base, look for "percent (%) of base."

$B =$ ____

To identify the amount, look for the word *is*.

$A =$ ____

Step 2 Write the percent equation.
Substitute the values for P, B, and A.

$$P \cdot B = A$$
$$=$$

5. Translate the written statement into a percent equation.

"Twenty percent of the computer's price is eighty-three dollars."

Step 1 Identify the percent, base, and amount.

To identify the percent, look for the "percent."

$P = 20$

To identify the base, look for "percent (%) of base."

The base is unknown.
Let B = the price of the computer.

To identify the amount, look for the word *is*.

$A = 83$

Step 2 Write the percent equation.

Substitute the values for P, B, and A.

$P \cdot B = A$

$20\% \cdot B = 83$

5. Translate the written statement into a percent equation.

"Sixty-one pounds is fifteen percent of the tiger's weight."

Step 1 Identify the percent, base, and amount.

To identify the percent, look for the "percent."

$P =$ ____

To identify the base, look for "percent (%) of base."

The base is unknown.
Let ____ = __________.

To identify the amount, look for the word *is*.

$A =$ ____

Step 2 Write the percent equation.

Substitute the values for P, B, and A.

$P \cdot B = A$

=

Some word exercises have extra information that is not needed to write a percent equation. While this extra information is not important to the math, it makes the problem meaningful. When solving a word exercise, focus on the necessary information first. Then use the extra information to interpret the answer in a meaningful way.

EXAMPLE

6. During the 2005–2006 NBA season, Ben Wallace won the award for defensive player of the year. He made only 41.55% of his free throws. If he attempted 296 free throws, how many free throws did he make?

The important information is highlighted in the paragraph above.

Step 1 Identify the percent, base, and amount.

To identify the percent, look for the % symbol.

$P = 41.55$

To identify the base, look for "percent (%) of base."

The base is the number of free throw attempts.

$B = 296$

GUIDED PRACTICE

6. Oprah Winfrey earns more money than almost any other celebrity. In 2005, she earned about 668% of Shaquille O'Neal's earnings. If Shaquille O'Neal earned $33.4 million in 2005, how much did Oprah earn that year?

Underline or highlight the important information in the paragraph above.

Step 1 Identify the percent, base, and amount.

To identify the percent, look for the % symbol.

$P =$ ____

To identify the base, look for "percent (%) of base."

The base is ________________.

$B =$ ________

(Continued)

Since we know the percent and the base, the amount is the number of free throws made.

The amount is unknown.

Let A = the number of free throws made.

Step 2 Write the percent equation.

Substitute the values for P, B, and A.

$$P \cdot B = A$$

$$41.55\% \cdot 296 = A$$

Ben Wallace made only 123 free throws.

Since we know the percent and the base, the amount is ______________.

The amount is unknown.

Let _____ = _______'s earnings.

Step 2 Write the percent equation.

Substitute the values for P, B, and A.

$$P \cdot B = A$$

$$=$$

Oprah earned about $223 million.

Concept Check

A1. What word is often used to represent multiplication in word exercises?

A2. What words can be used to represent equality in word exercises?

A3. What is the standard form of a percent equation?

Objective A Practice

For each exercise, set up a percent equation.
Choose *P, B,* or *A* to represent the unknown value.
Do not solve the equation.

A4. What percent of 15 is 12?

A5. What percent of 20 is 7?

A6. What is 30% of 70?

A7. What is 70% of 30?

A8. 35% of what number is 60?

A9. 25% of what number is 60?

A10. Allison purchased a new car in Michigan, where a 6% sales tax is charged. If the sales tax was $900, what was the cost of the car before the tax?

A11. Jamica ate dinner at Food-Dance, a local eatery, and had excellent service from the server, Sham. If Jamica tipped Sham 20% of the $45.50 cost of her dinner, how much was the tip?

A12. Lebron bought new basketball shoes. As part of his endorsement, he was given a discount on the original price. The regular price was $120. He had to pay $90. What percent of the regular price did Lebron pay?

A13. Dale wants to improve his lap times before a race. In the morning, his lap time was 40 seconds. In the afternoon, his lap time dropped to 37 seconds. What percent of the morning lap time was the afternoon lap time?

FOCUS ON Writing Percent Equations

For each exercise, identify the correct percent equation located in the center column. You do not need to solve for the unknown value. Each percent equation is used twice. P = percent, B = base, A = amount

a. $P \cdot 40 = 50$

b. $0.40 \cdot 50 = A$

c. $40 \cdot B = 50$

d. $P \cdot 50 = 40$

e. $0.50 \cdot B = 40$

f. $0.50 \cdot 0.40 = A$

1. Fifty percent of forty-hundredths is what?

2. What percent of 40 is 50?

3. Forty is what percent of fifty?

4. Fifty percent of what number is forty?

5. What number is forty percent of fifty?

6. Four thousand percent of what number is fifty?

7. Fifty kilograms is four thousand percent of what weight?

8. A baseball pitcher normally has forty pitches in a game. What percent of her normal number of pitches is fifty pitches?

9. A runner ran forty miles in a week. If that number is fifty percent of his normal distance, how far does he normally run in a week?

10. Originally, there were fifty pamphlets in a display. After two hours, there were forty. What percent of the original amount is the new amount?

11. A prescription calls for 0.4 gram of medicine. What is fifty percent of the daily amount?

12. After breaking a neighbor's window, a child had to clean the neighbor's fifty windows. If he has cleaned forty percent of the windows so far, how many has he cleaned?

Objective B Solve a Percent Equation

The Concept When solving percent exercises, you will solve equations in the form

$$P \cdot B = A.$$

Two out of three variables will be known, and we must solve for the value of the third. The following examples show how to solve the percent equation for each unknown variable.

Procedure Solve a Percent Equation

Step 1 Identify the values of P, B, and A.

- If the percent is given, convert it to a decimal.

Step 2 Write the percent equation.

Step 3 Substitute the values.

Step 4 Isolate the variable.

- If the variable is not alone, divide both sides of the equation to isolate it.
- If the variable is the percent, convert the decimal answer to a percent.

EXAMPLES

7. What number is 15% of 40?

Step 1 Identify the values of P, B, and A.

P is 15% = 0.15. **Convert the percent to a decimal.**
$B = 40$
The amount is unknown.
Let A = "what number."

Step 2 Write the percent equation. $P \cdot B = A$

Step 3 Substitute the values. $0.15 \cdot 40 = A$

Step 4 Simplify. $6 = A$

The number is 6.

8. 8% of what number is 5.6?

Step 1 Identify the values of P, B, and A.

P is 8% = 0.08. **Convert the percent to a decimal.**
The base is unknown.
Let B = "what number."
$A = 5.6$

Step 2 Write the percent equation. $P \cdot B = A$

Step 3 Substitute the values. $0.08 \cdot B = 5.6$

Step 4 Isolate B. Divide both sides by the number that multiplies B.

$$\frac{\cancel{0.08} \cdot B}{\cancel{0.08}} = \frac{5.6}{0.08}$$

$$B = 70$$

The number is 70.

9. What percent of 60 is 6?

Step 1 Identify the values of P, B, and A.

The percent is unknown.
Let P = "what percent."
$B = 60$
$A = 6$

Step 2 Write the percent equation. $P \cdot B = A$

Step 3 Substitute the values. $P \cdot 60 = 6$

Step 4 Isolate P. Divide both sides by the number that multiplies P.

$$\frac{P \cdot \cancel{60}}{\cancel{60}} = \frac{6}{60}$$

$$P = \frac{1}{10}$$

$$P = 0.1$$ **Convert the percent to a decimal.**

The percent is 10%.

GUIDED PRACTICE

7. What number is 20% of 50?

P is ____ % = ____.
B = ____
The amount is unknown.
Let A = "________."

$P \cdot B = A$
=
=

The number is ____.

8. 6% of what number is 4.2?

P is ____ % = ____.
The base is unknown.
Let B = "________."
A = ____.

$P \cdot B = A$
=
=
=

The number is ____.

9. What percent of 55 is 11?

The percent is unknown.
Let P = "________."
B = ____
A = ____

$P \cdot B = A$
=
=
=
=

The percent is ____.

10. 25% of 16 is what number?

Step 1 Identify the values of P, B, and A.

P is $25\% = 0.25$.
$B = 16$
The amount is unknown.
Let $A =$ "what number."

Step 2 Write the percent equation.

$$P \cdot B = A$$

Step 3 Substitute the values.

$$0.25 \cdot 16 = A$$

Step 4 Simplify.

$$4 = A$$

The number is 4.

10. What number is 40% of 10?

P is ____ % = ____.
$B =$ ____
The amount is unknown.
Let $A =$ "__________."

$$P \cdot B = A$$

$=$

$=$

The number is ____.

11. 175 is 35% of what number?

Step 1 Identify the values of P, B, and A.

$P = 35\% = 0.35$
The base is unknown.
Let $B =$ "what number."
$A = 175$

Step 2 Write the percent equation.

$$P \cdot B = A$$

Step 3 Substitute the values.

$$0.35 \cdot B = 175$$

Step 4 Isolate B.

$$\frac{\cancel{0.35} \cdot B}{\cancel{0.35}} = \frac{175}{0.35}$$

$$B = 500$$

$$0.35\overline{)175} \quad 500$$

The number is 500.

11. 104 is 26% of what number?

P is ____ % = ____.
The base is unknown.
Let $B =$ "__________."
$A =$ ____

$$P \cdot B = A$$

$=$

$=$

$=$

The number is ____.

12. Armando owed a friend \$34. If Armando paid his friend \$17, what percent of his debt did Armando repay?

Step 1 Identify the values of P, B, and A.

The percent is unknown.
Let $P =$ "what percent."
$B = 34$
$A = 17$

Step 2 Write the percent equation.

$$P \cdot B = A$$

Step 3 Substitute the values.

$$P \cdot 34 = 17$$

Step 4 Isolate P.

$$\frac{P \cdot \cancel{34}}{\cancel{34}} = \frac{17}{34}$$

$$P = \frac{\cancel{17}}{2 \cdot \cancel{17}}$$

$$P = 0.5$$

$$2\overline{)1.0} \quad 0.5$$

Armando repaid 50% of his debt.

12. Angel owed a friend \$52. If Angel paid her friend \$13, what percent of her debt did Angel repay?

The percent is unknown.
Let $P =$ "__________."
$B =$ ____
$A =$ ____

$$P \cdot B = A$$

$=$

$=$

$=$

$=$

Angel repaid ____ of her debt.

Concept Check

B1. Whenever a percent equation is solved, a conversion between a number and a percent must occur. Each calculation below requires converting either a percent to a decimal or a decimal to a percent. Fill in the blanks accordingly.

a. To perform the calculations for $40\% \cdot B = 60$, convert from a ____________ to a ____________.

b. To perform the calculations for $30\% \cdot 20 = A$, convert from a ____________ to a ____________.

c. After solving the percent equation $P \cdot 30 = 90$, convert from a ____________ to a ____________.

B2. When given the equation $0.5 \cdot B = 10$, what operation must you perform to solve for B?

B3. Answer each question.

a. An amount is 110% of the base. Is the amount more or less than the base? Explain.

b. An amount is 90% of the base. Is the amount more or less than the base? Explain.

Objective B Practice

Set up and solve each percent equation.

B4. What is 12% of 30?

B5. What is 90% of 250?

B6. 20 is 40% of what number?

B7. 40 is 20% of what number?

B8. 30 is what percent of 50?

B9. 40 is what percent of 120?

B10. 32% of Lourdes's paycheck goes to taxes. If her paycheck is $380 before taxes, how much does she pay in taxes?

B11. Amber, an ambulance driver, spends 40% of her shift waiting for calls. If her shift lasts 9 hours, how long does she spend waiting for calls?

Combining Concepts and Applications

CONCEPT I Determining If a Solution to a Percent Equation Is Reasonable When finding an answer to a percent equation, always check to see if it is a reasonable answer by asking the question "Is the percent of the base close to the amount?"

Relationships between Amount, Percent, and Base

The amount shaded at right is equal to the percent of one base.

Base	Percent	Amount
1 Base	$\cdot\ 200\% =$	The amount is twice the base.
1 Base	$\cdot\ 100\% =$	The amount is equal to the base.
1 Base	$\cdot\ 50\% =$	The amount is half the base.
1 Base	$\cdot\ 25\% =$	The amount is one-fourth the base.
1 Base	$\cdot\ 10\% =$	The amount is one-tenth the base.

Procedure Determine If a Solution to a Percent Equation Is Reasonable

Step 1 Draw an object to represent the base.

Step 2 Use the percent to shade a portion of the base.
Draw additional objects if the percent is more than 100%.

Step 3 Determine if the shaded region approximates the amount.

EXAMPLES

13. Is "13% of 120 is 100" reasonable?

Step 1 The base is 120.

Step 2 Shade approximately 13% of 120.

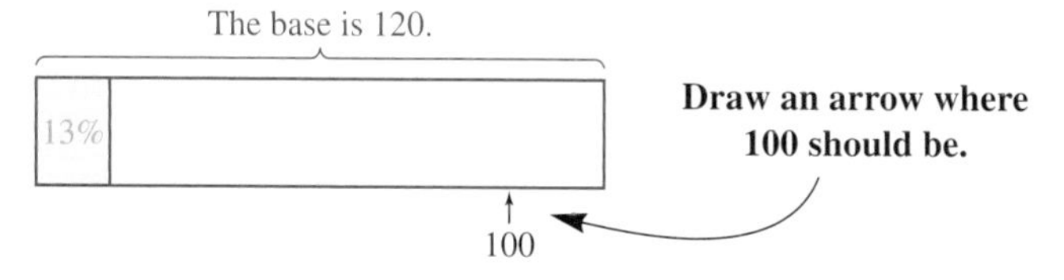

Step 3 Since 100 is not close to the shading for 13% of 120, the answer is not reasonable.

14. Is "312% of 50 is 156" reasonable?

Step 1 The base is 50.

Step 2 For each 100%, we must shade one base. To draw 312%, shade a little more than three bases.

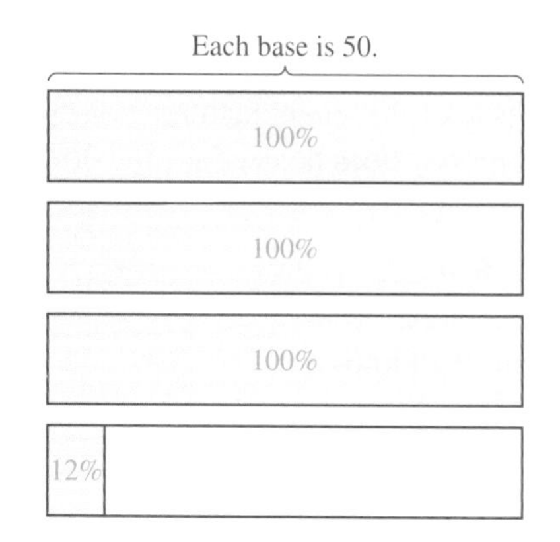

Step 3 Since each base is 50, we have $3 \cdot 50 = 150$ plus a little more. The amount 156 is reasonable.

15. Is "56% of 80 is 35" reasonable?

Step 1 The base is 80.

Step 2 Shade approximately 56% of 80.

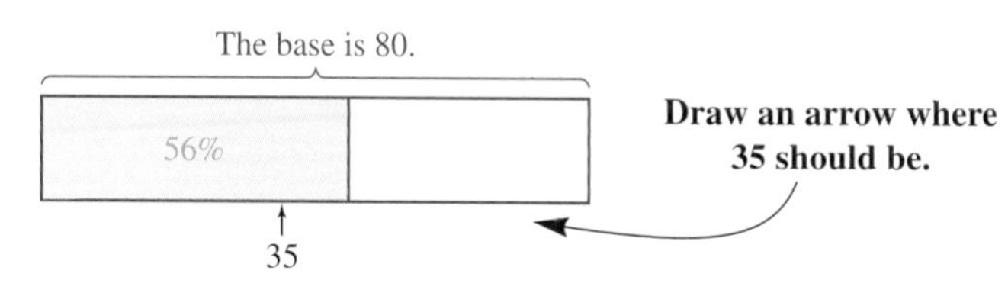

Step 3 35 is less than half of 80. 56% is more than half. The statement is not reasonable.

GUIDED PRACTICE

13. Is "10% of 30 is 25" reasonable?

Step 1 The base is _____.

Step 2 Shade approximately _____% of _____.

The base is _____ .

Draw an arrow where 25 should be.

Step 3 Since _____ is/is not close to the shading for _____% of _____, the answer is/is not reasonable.

14. Is "230% of 120 is 276" reasonable?

Step 1 The base is _____.

Step 2 For each _____%, we must shade one base. To draw _____%, shade a little more than _____ bases.

Each base is _____ .

Step 3 Using your picture, explain why the answer is reasonable or unreasonable.

15. Is "45% of 60 is 40" reasonable?

Step 1 The base is _____.

Step 2 Shade approximately _____% of _____.

The base is _____ .

Draw an arrow where 40 should be.

Step 3 Using your picture, explain why the answer is reasonable or unreasonable.

16. Is "40% of 70 is 28" reasonable?

Step 1 The base is 70.

Step 2 Shade approximately 40% of 70.

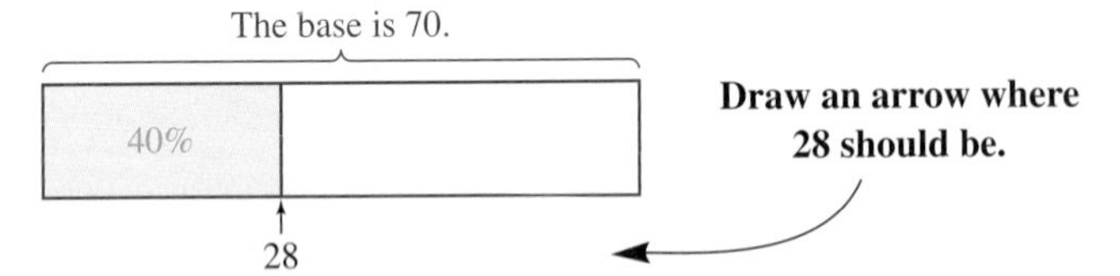

Step 3 28 is a little less than half of 70. 40% is a little less than half. The answer is reasonable.

16. Is "30% of 90 is 27" reasonable?

Step 1 The base is ____.

Step 2 Shade approximately ____% of ____.

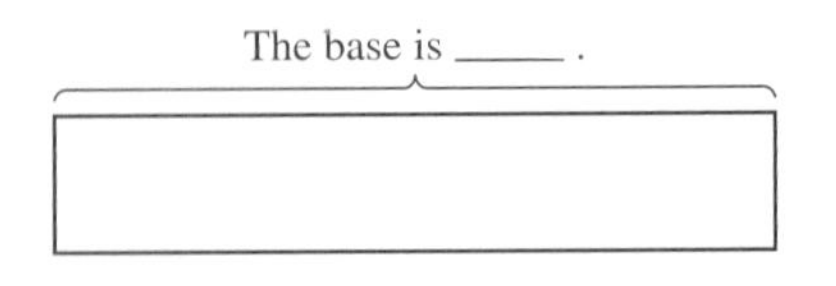

Step 3 Using your picture, explain why the answer is reasonable or unreasonable.

CONCEPT II Solving Multistep Applications To solve some applications, you need to perform addition or subtraction before setting up the percent equation.

EXAMPLE

17. Aria wants to buy a car and will make a 20% down payment. If the car's price is \$12,400 and there is a \$1,500 manufacturer's discount, how much money does she need for the down payment?

P is 20% = 0.20.
$B = 12{,}400 - 1{,}500 = 10{,}900$
The amount is unknown.
Let A = the down payment.

Step 1 Write the percent equation. $P \cdot B = A$

Step 2 Substitute the values. $0.20 \cdot 10{,}900 = A$

Step 3 Simplify. $2{,}180 = A$

She needs \$2,180 for the down payment.

GUIDED PRACTICE

17. David wants to buy a boat and will make a 30% down payment. If the boat's price is \$9,200 and there is a \$1,400 manufacturer's discount, how much money does he need for the down payment?

P is ____ % = ____.
B = ____ − ____ = ____
The amount is unknown.
Let A = ____________.

$P \cdot B = A$
=
=

He needs \$____ for the down payment.

Section 5.3 Exercises

FOR EXTRA HELP       

To translate a written statement into a percent equation:

1. Answer the Objective A Concept Checks.
2. Answer the odd Objective A Practice Exercises.
3. Answer the even Objective A Practice Exercises.

To write percent equations:

4. Answer the odd Focus On Writing Percent Equations Exercises.
5. Answer the even Focus On Writing Percent Equations Exercises.

To solve a percent equation:

6. Answer the Objective B Concept Checks.
7. Answer the odd Objective B Practice Exercises.
8. Answer the even Objective B Practice Exercises.

VOCABULARY REVIEW *Review the Vocabulary Preview for Section 5.3. Study the definitions until you can check* Got It *for every word.*

Use these words to complete each sentence.

of • is • percent equation • percent • base • amount

	Got It	Get Help
9. In a percent equation, the ____________ describes the portion of the base that is equal to the amount.		
10. An equation of the form percent · base = amount is called a(n) ____________.		
11. The word ____________ often used to represent equality in word exercises.		
12. The word ____________ is often used to represent multiplication in word exercises.		
13. In a percent equation, the ____________ is the whole quantity.		
14. In a percent equation, the ____________ is the portion that results from multiplying the percent and the base.		

How will you get help for any vocabulary that you are unsure about?

Your instructor ____ MyMathLab ____ A classmate ____ A tutor ____ Other ____

For each exercise, set up a percent equation. Choose *P, B,* or *A* to represent the unknown value. Do not solve the equation.

15. 13% of 312 is what number?

16. 41% of 73 is what number?

17. 54 is what percent of 20?

18. 21 is what percent of 2?

19. 30 is 30% of what number?

20. 21 is 21% of what number?

21. What is 0.05% of 10,000?

22. What is 0.08% of 12,000?

23. 2 is what percent of 10,000?

24. 4 is what percent of 100,000?

25. 135% of what number is 297?

26. 18% of what number is 45?

27. 150% of 70 is what number?

28. 80% of what number is 30?

29. What percent of 600 is 120?

30. What percent of 300 is 2,000?

31. 350% of what number is 21?

32. 45% of 30 is what number?

goetz If you're having a hard time writing percent equations, restudy Objective A, then do the Focus On exercises on page 5-47.
For more tips and tweets, go to twitter.com/gstbasicmath

Set up and solve each percent equation. Round to the nearest hundredth when necessary. Determine if your answer is reasonable.

33. What percent of 40 is 50?

34. What percent of 50 is 40?

35. What is 75% of 212?

36. What is 130% of 50?

37. 175% of 7 is what number?

38. 60% of 24 is what number?

39. What is 90% of 80?

40. What is 110% of 300?

41. 70 is 40% of what number?

42. 82 is 200% of what number?

43. What percent of 75 is 90?

44. What percent of 30 is 12?

45. 150% of 70 is A.

46. 80% of B is 30.

47. What percent of 600 is 120?

48. What percent of 2,000 is 1,100?

49. 350% of B is 21.

50. 45% of 30 is A.

Determine if the given solution is reasonable. Explain why or why not. Do not calculate the actual answer.

51. 13% of 312 is A.
Solution: $A = 42$

52. 45% of 20 is A.
Solution: $A = 9$

53. 54 is what percent of 20?
Solution: $P = 37\%$

54. 21 is what percent of 2?
Solution: $P = 10\%$

55. 30 is 30% of B.
Solution: $B = 100$

56. 21 is 21% of B.
Solution: $B = 4$

57. A is 0.05% of 10,000.
Solution: $A = 500$

58. A is 0.08% of 12,000.
Solution: $A = 10$

59. 2 is what percent of 10,000?
Solution: $P = 5{,}000\%$

60. 4 is what percent of 100,000?
Solution: $P = 10\%$

61. 135% of B is 297.
Solution: $B = 220$

62. 18% of B is 45.
Solution: $B = 250$

For each exercise:
Use a percent equation to answer the question. Use estimation to check if your answer is reasonable.

63. What is 40% of 80?

64. What is 60% of 90?

65. What percent of 40 is 15?

66. What percent of 80 is 120?

67. 60% of what number is 48?

68. 90% of what number is 72?

69. 0.5 is what percent of 2?

70. 0.9 is what percent of 4.5?

71. 10% of what number is 1.7?

72. 1% of what number is 2.5?

73. 10% of what number is 3?

74. 30% of what number is 4?

Solve each percent equation. Round to the nearest penny or percent if necessary. Determine if your answer is reasonable.

75. Veronica needs a new harmonica. The cost of a new harmonica is \$16.60. How much sales tax is charged if the state has a 5% sales tax?

76. Rebecca purchased a notebook for \$2.50. The bookstore charged her 8% in sales tax. How much sales tax did Rebecca pay?

77. Kaylia watched $3\frac{1}{2}$ hours of TV yesterday. In that time, she watched 1 hour of commercials. What percent of her time watching TV was spent watching commercials?

78. Alli and Jason drove to get their favorite doughnuts. They spent a total of \$10.50 on the trip, including \$8.00 for gas to drive 60 miles and \$2.50 for the doughnuts. What percent of this trip's cost was spent on gas?

Solve each multistep application. Round to the nearest penny or percent.

79. Andreas printed photographs. He used \$14.00 in photographic paper and \$3.50 in ink. What percent of the total cost goes toward paper?

80. Kelvin watched his favorite hour-long TV show last night. The actual show was on for 38 minutes, with commercials taking up the rest of the hour. What percent of the hour did Kelvin spend watching commercials?

81. Kesta wants to buy an airline ticket to Tanzania. He has \$1,100 in the bank. The ticket cost \$1,020. In addition to the ticket cost, there is an extra 6% fuel charge. Does Kesta have enough money in the bank to buy the ticket?

82. Arianna is buying a car and wants to make a 10% down payment. If the car's price is \$16,480 and there is a \$1,500 manufacturer's discount, how much money does she need for the down payment?

Section 5.3 Question Log

Use this space to write questions. Be sure to get these answered and revisit them when you prepare for your exam.

Page ____ **Answered**	**Page** ____ **Answered**
Page ____ **Answered**	**Page** ____ **Answered**

Section 5.4 Percent Applications

In everyday situations, it is common for one quantity to be calculated as a percent of another.

- Salespeople are often paid a percent of their sales in commission.
- Many investments earn a percent of the amount invested in interest.
- Urban planners use information about the percent increase or percent decrease in populations to guide their decisions.

In this section, you will learn how each of these situations relates to the standard percent equation percent · base = amount.

The **Objectives** in Section 5.4 will help you

A Solve a commission exercise.
B Solve a simple interest exercise.
C Solve percent increase and percent decrease exercises.

VOCABULARY PREVIEW *Check the box that applies.*

	Got It	Must Study
Commission: A **commission** is the amount of money that is earned as a percent of sales.		
Commission Rate: A **commission rate** is the percent of sales that determines the commission earned.		
Commission Equation: The **commission equation** comes from the percent equation but uses new terminology. Percent of base is amount. **Commission rate of sales is commission.** **Commision rate · Sales = Commision**		
Principal: When money is invested, the **principal** is the initial investment.		
Balance: The **balance** is the amount of money in an account at any time.		
Interest: The amount of money that is earned as a percent of a balance is called **interest.**		
Interest Rate: An **interest rate** is a percent of the balance that is earned over a period of time.		
Simple Interest Formula: Interest = Principal · Interest rate · Time $I = P \cdot r \cdot t$ where I is the interest, P is the principal, r is the interest rate, and t is the time period of the investment.		
Percent Increase: A **percent increase** is the percent by which the base is increased.		
Percent Decrease: A **percent decrease** is the percent by which the base is decreased.		
Compound Interest: With **compound interest,** the interest earned is added to the balance and included in the next interest calculation.		

Study these words when they appear in the text. After the section, test your understanding by completing the Vocabulary Review in the section exercises.

Objective A Solve a Commission Exercise

The Concept A **commission** is the amount of money that is earned as a percent of sales. If someone sells \$200 of merchandise and earns a 10% **commission rate**, he or she will earn 10% of \$200, or \$20, in commission. The **commission equation** comes from the standard percent equation but uses new terminology.

Amount = **Percent** · **Base**

↕ ↕ ↕

Commission = **Commission rate** · **Sales**

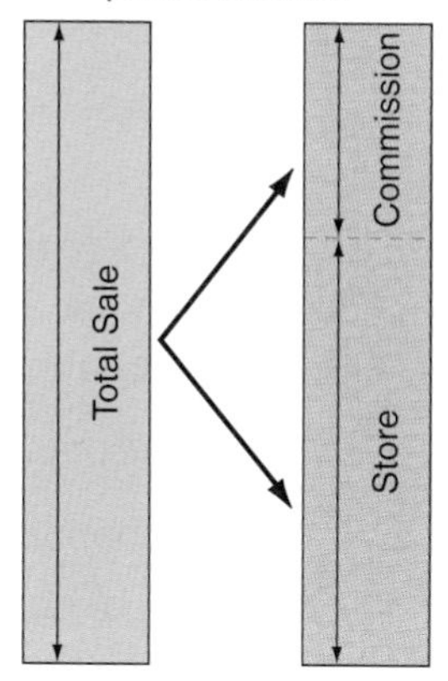

Procedure Solve a Commission Exercise

Step 1 Identify the commission earned, commission rate, and sales. Choose a variable for the unknown value.

Step 2 Write and solve a commission equation.

EXAMPLES

1. Merle is selling his house. He expects to close the deal on Tuesday. His real estate agent's commission rate is 3.5% of the sale price. If the sale price is \$120,000, what is the agent's commission?

Step 1 Identify the commission earned, commission rate, and sales.

(Amount) Let c = the commission.
(%) The commission rate is 3.5% = 0.035.
(Base) Sales = \$120,000

Step 2 Write and solve a commission equation.

Commission = Commission rate · Sales

$c = 0.035 \cdot 120{,}000$

$c = 4{,}200$

Merle's agent made \$4,200 in commission.

GUIDED PRACTICE

1. Kwame is purchasing a car. His salesperson earns a 5% commission rate. If the price of the car is \$19,000, what commission will the salesperson earn?

Step 1 Identify the commission earned, commission rate, and sales.

(Amount) Let c = the commission.
(%) The commission rate is____% = ____.
(Base) Sales = \$_____

Step 2 Write and solve a commission equation.

Commission = Commission rate · Sales

$c =$

$c =$

The salesperson made \$____ in commission.

2. Apu earns 16% commission on his propane sales. If he earned $40,000 last year, how much propane did he sell?

Step 1 Identify the commission earned, commission rate, and sales.

Commission = $40,000
The commission rate is 16% = 0.16.
Let s = propane sales.

Step 2 Write and solve a commission equation.

$$\text{Commission} = \text{Commission rate} \cdot \text{Sales}$$
$$40{,}000 = 0.16 \cdot s$$
$$\frac{40{,}000}{0.16} = \frac{0.16 \cdot s}{0.16}$$
$$250{,}000 = s$$

Apu sold $250,000 in propane.

2. Hank pays himself 5% commission on all sales at the Quick-E-Mart. If Hank earned $45,000 last year, what were his total sales at the Quick-E-Mart?

Step 1 Identify the commission earned, commission rate, and sales.

Commission = ______
The commission rate is _____% = _____.
Let s = ____________________.

Step 2 Write and solve a commission equation.

Commission = Commission rate · Sales

=

______ = ______

=

Hank's total sales were ________.

3. Khuang earned $3,200 in commission after selling $200,000 in appliances. What commission rate did Khuang earn?

Step 1 Identify the commission earned, commission rate, and sales.

Commission = $3,200
Let r = the commission rate.
Sales = $200,000

Step 2 Write and solve a commission equation.

$$\text{Commission} = \text{Commission rate} \cdot \text{Sales}$$
$$3{,}200 = r \cdot 200{,}000$$
$$\frac{3{,}200}{200{,}000} = \frac{r \cdot 200{,}000}{200{,}000}$$
$$.016 = r$$
$$1.6\% = r$$

Khuang earned a 1.6% commission rate.

3. Gloria earned a $12,500 bonus last year. Her bonus was given as a percentage of the $2,500,000 she saved the company. What was the commission rate of Gloria's bonus?

Step 1 Identify the commission earned, commission rate, and sales.

Commission = $12,500
Let _____ = the commission rate.
Savings = $________

Step 2 Write and solve a commission equation.

Commission = Commission rate · Savings

=

______ = ______

=

=

Gloria earned a _____ commission rate.

Concept Check

A1. Rewrite the standard form of the percent equation "amount = percent · base" using the terminology for the commission equation.

__________ = __________ · __________

A2. If your friend just solved a commission problem and concluded that "Pat earned a commission rate of 105%," why should you suspect that your friend's answer is incorrect?

Objective A Practice

Solve each commission exercise.

A3. Hector sells lighting fixtures. He earns a commission rate of 9%. What commission will he earn on an $800 light fixture?

A4. Shari hired a manager for her business. The manager earns a commission rate of 1% of the business's gross sales. If sales totaled $2,100,000 last year, what was the manager's commission?

A5. Lester sold $42,500 in hay last month. If he received $2,125 in commission for those sales, what commission rate did Lester earn?

A6. Tama sold $30,000 in skis last month, earning $4,500 in commission. What commission rate did she earn?

A7. An investor pays an advisor a commission rate of 0.25% of his portfolio's value. The investor's portfolio was valued at $1,260,000. What was the advisor's commission?

A8. Farr earns a 6% commission rate for running an eBay store. If she earned $1,500 last month, what were the total sales for the month?

Objective B Solve a Simple Interest Exercise

The Concept When people invest money, the initial investment is called the **principal.** In a **simple interest** account, the principal earns money, or **interest,** for the owner after a period of time. For example, an account with a principal of $50 earning 10% simple interest per year will earn $5 after each year the money has been invested. Simple interest is calculated using the simple interest formula.

Simple Interest Formula

Interest = Principal • Interest rate • Time

$$I = P \cdot r \cdot t$$

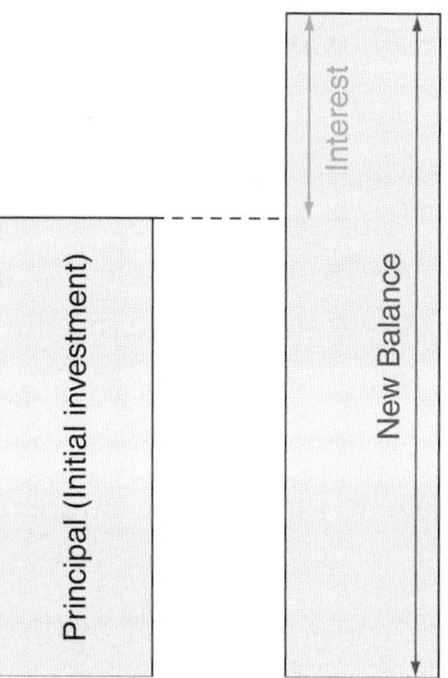

Procedure Solve a Simple Interest Exercise

Step 1 Identify the interest, principal, interest rate, and time. Choose a variable for the unknown value.

Step 2 Write and solve a simple interest equation.

EXAMPLES

4. Chuck borrowed $175 to buy books for school. He was charged 3% simple interest per month. How much interest was he charged if he paid the loan back after 3 months?

Step 1 Identify the interest, principal, interest rate, and time.

Principal = $175
Let I = interest.
Interest rate is 3% = 0.03 per month.
Time = 3 months

The units of time must be the same.

Step 2 Write and solve a simple interest equation.

$$\text{Interest} = \text{Principal} \cdot \text{Interest rate} \cdot \text{Time}$$
$$I = P \cdot r \cdot t$$
$$I = 175 \cdot 0.03 \cdot 3$$
$$I = 5.25 \cdot 3$$
$$I = 15.75$$

Chuck was charged $15.75 interest.

5. Latoya set up a mutual fund to pay for college. The principal at the beginning of the year was $12,000. At the end of 1 year, the account had earned $1,080. What interest rate did the account earn if it paid simple interest?

Step 1 Identify the interest, principal, interest rate, and time.

Principal = $12,000
Interest = $1,080
Let r = interest rate per year.
Time = 1 year

The units match.

Step 2 Write and solve a simple interest equation.

$$I = P \cdot r \cdot t$$
$$1{,}080 = 12{,}000 \cdot r \cdot 1$$
$$1{,}080 = 12{,}000 \cdot r$$
$$\frac{1{,}080}{12{,}000} = \frac{\cancel{12{,}000} \cdot r}{\cancel{12{,}000}}$$
$$0.09 = r$$
$$9\% = r$$

Latoya earned an interest rate of 9%.

GUIDED PRACTICE

4. Antonio bought $3,500 of merchandise with his credit card, which charges 1.5% simple interest per month. How much interest was Antonio charged after 2 months?

Step 1 Identify the interest, principal, interest rate, and time.

Principal = $_____
Let I = interest.
Interest rate is _____% = _____ per _____.
Time = _____

The units of time must be the same.

Step 2 Write and solve a simple interest equation.

$$\text{Interest} = \text{Principal} \cdot \text{Interest rate} \cdot \text{Time}$$
$$I = P \cdot r \cdot t$$
=
=
=

Antonio was charged _____ interest.

5. Samantha put $12,000 into a savings account to pay for college. At the end of 1 year, the account had $12,420, an increase of $420. If the account paid simple interest each year, what interest rate did this account earn?

Step 1 Identify the interest, principal, interest rate, and time.

Principal = $_____
Interest = _____
Let r = _____ per _____.
Time = _____

Do the units match?

Step 2 Write and solve a simple interest equation.

$$I = P \cdot r \cdot t$$
=
=
=
=
=

Samantha earned an interest rate of _____.

6. The simple interest rate on Alberto's credit card is 18% per year. If the principal is $2,500, how much interest is Alberto charged after 1 month?

Step 1 Identify the interest, principal, interest rate, and time.

Principal or balance = $2,500
Let I = interest.
Interest rate = 18% per year

$= \frac{18\%}{12}$ ← **18% per year gets divided over 12 months.**

= 1.5%

= 0.015 per month

Time = 1 month

The units match.

Step 2 Write and solve a simple interest equation.

$I = P \cdot r \cdot t$

$I = 2{,}500 \cdot 0.015 \cdot 1$

$I = 2{,}500 \cdot 0.015$

$I = 37.5$

Alberto is charged $37.50 interest after 1 month.

6. Andrea has a home equity loan with a simple interest rate of 9% per year. If the principal is $9,500, how much interest does Andrea owe after 1 month?

Step 1 Identify the interest, principal, interest rate, and time.

Principal or balance = _____
Let _____ = _____.
Interest rate = _____% per year

$= \frac{\_\_\%}{12}$ ← **9% per year gets divided over 12 months.**

= _____%

= _____ per month

Time = _____

Do the units match?

Step 2 Write and solve a simple interest equation.

$I = P \cdot r \cdot t$

=

=

=

Andrea owes _____ interest after 1 month.

Concept Check

B1. How do you change a monthly interest rate to a yearly interest rate?

B2. How do you change a yearly interest rate to a monthly interest rate?

Objective B Practice

Set up and solve each simple interest exercise.

B3. Barry invests $200,000 in a savings account. If he earns $15,000 each year in simple interest, what is the interest rate?

B4. Jada invests $500,000 in a bond account. If she earns $30,000 each year in simple interest, what is the interest rate?

B5. Edwina's checking account had an initial investment of $630. The account earned 1.3% simple interest each year. How much interest did the account earn in 3 years?

B6. Gladwin invested $1,580 in a money market account. If the account earns 2.5% simple interest each year, how much interest will the account earn in 4 years?

B7. Hiro's credit card company charges 15% simple interest each year. If he was charged $50 in interest last month, what was his balance?

B8. Saiesha's credit card company charges 20% simple interest each year. If she was charged $27 in interest last month, what was her balance?

Objective C Solve Percent Increase and Percent Decrease Exercises

The Concept To describe a change to a quantity, we use the formula

New quantity = Original quantity + Change

or

New quantity = Original quantity − Change

For example, Maria earned $10 per hour at work. She did great work, so her boss gave her a 10% raise. What was her new salary?

New salary = Original salary + Raise
= $10 + 10% of $10
= $10 + $1
= $11

In previous sections, we would say that Maria's new salary is a percent of her old salary. Specifically, her new salary is 110% of her old salary.

In this section, we will focus on percent *change* and how it relates to the original quantity.

Percent Change Equation

$$\text{Percent change} = \frac{\text{Change in quantity}}{\text{Original quantity}}$$

Percent Increase

Original Quantity + % Change = New Quantity

Percent Decrease

Original Quantity − % Change = New Quantity

INTERACTIVE DEFINITION Percent Increase and Percent Decrease

A **percent increase** is the percent by which the base is increased.

A **percent decrease** is the percent by which the base is decreased.

EXAMPLE

7. Identify each situation as a percent increase or a percent decrease.

a. If a $250 dress is marked 30% off, the discount is a percent decrease.

b. A bacteria population that is 200% of its original size has experienced a 100% increase.

GUIDED PRACTICE

7. Identify each situation as a percent increase or a percent decrease.

a. When you buy an iPod for $200, you are charged sales tax. The sales tax is a percent ______.

b. A bacteria population that is 75% of its original size has experienced a 25% ______.

DO YOU UNDERSTAND how to identify a percent increase and a percent decrease? Got It Get Help

Procedure Solve Percent Increase and Percent Decrease Exercises

Step 1 Identify the percent change, the change in quantity, and the original quantity. Choose a variable for the unknown value.

Step 2 Write and solve a percent change equation.

EXAMPLES

8. Last year, season tickets sold for \$800. The cost increased 3% this year. What is the change in cost?

Step 1 Identify the percent change, the change in quantity, and the original quantity.

Percent change is 3% = 0.03.
Let c = the change in cost.
Original cost = \$800

Step 2 Write and solve a percent change equation.

$$\text{Percent change} = \frac{\text{Change in cost}}{\text{Original cost}}$$

$$0.03 = \frac{c}{800}$$

$$0.03 \cdot 800 = \frac{c}{\cancel{800}} \cdot \frac{\cancel{800}}{1}$$

$$24 = c$$

The cost of season tickets increased by \$24.

9. A car's original cost was \$14,000. After some negotiation and a rebate, the cost was lowered by \$2,100. What percent discount did the customer get?

Step 1 Identify the percent change, the change in quantity, and the original quantity.

Let p = the percent discount.
Change in cost = a decrease of \$2,100
Original cost = \$14,000

Step 2 Write and solve a percent change equation.

$$\text{Percent change} = \frac{\text{Change in cost}}{\text{Original cost}}$$

$$p = \frac{2,100}{14,000}$$

$$p = 0.15$$

$$p = 15\%$$

The customer received a 15% discount.

GUIDED PRACTICE

8. Thirty-six inches of snow was on the ground before a storm. The storm added 30% more snow. What was the change in the amount of snow?

Step 1 Identify the percent change, the change in quantity, and the original quantity.

Percent change is ____% = ____.
Let ____ = the change in snow.
Original amount = ____

Step 2 Write and solve a percent change equation.

Percent change = ______________

= ____

= ____ · ____

=

The amount of snow ________ by ________.

9. A company produced 70,000 fewer tennis balls than last year. If the company produced 500,000 tennis balls last year, what percent decrease occurred?

Step 1 Identify the percent change, the change in quantity, and the original quantity.

Let ____ = ____________.
Change in production = a ________ of ____________
Original amount = ____________.

Step 2 Write and solve a percent change equation.

Percent change = ______________

=

=

=

Tennis ball production ________ by ____.

(*Continued*)

10. An engineer tracks quality control at a plant. Last month, 600 defective parts were made. This month, the number increased to 615. What percent increase in defects occurred?

Step 1 Identify the percent change, the original quantity, and the change in quantity.

Let p = the percent increase.
Change in defective parts $= 615 - 600$
$= 15$
Original defective parts $= 600$

Step 2 Write and solve a percent change equation.

$$\text{Percent change} = \frac{\text{Change in defective parts}}{\text{Original defective parts}}$$

$$p = \frac{15}{600}$$

$$p = 0.025$$

$$p = 2.5\%$$

The number of defects increased by 2.5%.

10. Daniel earned $32,000 last year. He took a pay cut to get a more interesting position. His new salary is $28,000. What percent decrease in salary did Daniel take?

Step 1 Identify the percent change, the original quantity, and the change in quantity.

Let ____ = ______________.
Change in salary = ________ − ________
= ________
Original salary = ________

Step 2 Write and solve a percent change equation.

Percent change = ______________
=
=
=

Daniel's salary ______________.

Concept Check

C1. A company owner is happy with the performance of a department, so she gives everyone in that department a 5% raise. If employees in this department earn different salaries, does everyone in the department receive the same raise, in dollars?

Objective C Practice

Set up and solve each percent increase or percent decrease exercise.

C2. A person earning $35,000 is given a 2.1% cost-of-living increase. What is the change in salary?

C3. Due to budget cuts, a person earning $38,000 is given a 3% pay cut. What is the change in salary?

C4. Last year, Arejo's car insurance cost $900. This year, his insurance increased by $207. By what percent did Arejo's insurance go up?

C5. Four years ago, the average cost of gas was $2.40 a gallon. This year, the average cost of gas is $3.84. What is the percent increase in gas price?

C6. After a very unprofitable year, a company employee earning $40,000 was given a pay cut of $5,000. What percent decrease in salary does this represent?

C7. A town had 80 police officers last year. Due to budget cuts, 3 officers are laid off this year. What is the percent decrease in the number of police officers?

Combining Concepts and Applications

CONCEPT I Solving a Compound Interest Exercise While the simple interest formula is useful, it does not reflect how interest values usually accumulate over time. With simple interest, the interest you earn never earns interest itself. With **compound interest,** the interest earned is added to the balance and the new balance (old balance plus interest) is used to determine the next interest calculation. The difference in money earned over time can be dramatic, as shown in the following graph:

With compound interest, $1,000 will grow to $17,000 in the same time it takes a simple interest account to grow from $1,000 to $4,000.

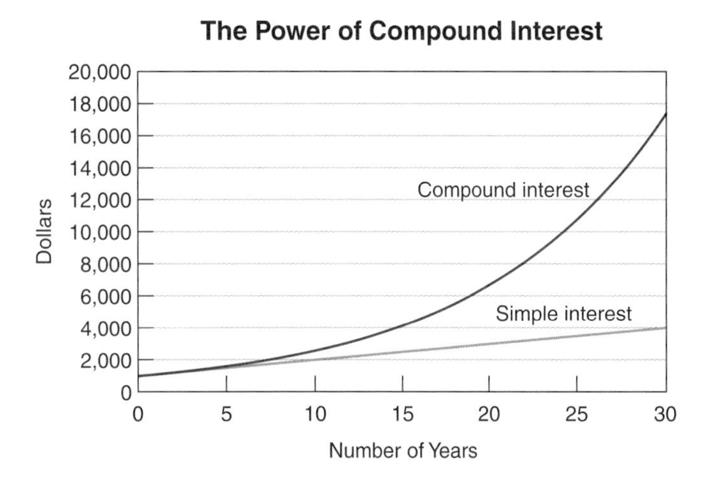

In the preceding graph, $1,000 is invested in an account that earns 10% compound interest. To understand the way compound interest is calculated, look at the value of this account for the first 2 years.

1st Year	2nd Year
$\begin{aligned} 1^{st}\text{ Year's Balance} &= (1 + r)\cdot\text{Principal} \\ &= (1 + 0.10)\cdot 1{,}000 \\ &= (1.10)\cdot 1{,}000 \\ &= \$1{,}100 \end{aligned}$	$\begin{aligned} 2^{nd}\text{ Year's Balance} &= (1 + r)\cdot 1^{st}\text{ Year's Balance} \\ &= (1.10)\cdot 1{,}100 \\ &= \$1{,}210 \end{aligned}$

Multiplying the balance by a factor of $(1 + r)$, the new amount becomes the balance increased by the interest.

- When the "1" multiplies the current balance, it maintains the amount of money in the account.
- When the "$+ r$" multiplies the current balance, it adds the earned interest to the account.

DETAILED EXAMPLE Calculating Compound Interest

If $1,000 is invested in an account that pays 5% interest compounded annually, calculate the account balance for the first three years.

$(1 + r) \cdot \text{Balance} = \text{New balance}$

1st year
Calculate the interest for the 1st year using the principal of $1,000.

$(1.05) \cdot 1{,}000.00 = \$1{,}050.00$

2nd year
Use the balance from the end of the 1st year to find the 2nd year's balance.

$(1.05) \cdot \$1{,}050.00 = \$1{,}102.50$

3rd year
Use the balance from the end of the 2nd year to find 3rd year's balance.

$(1.05) \cdot \$1{,}102.50 = \$1{,}157.63$

EXAMPLES

11. If $200 is invested in an account that pays 6% interest compounded annually, calculate the account balance for the first three years. Round your answers to the nearest cent.

Step 1 Identify the principal and the interest rate.

Principal = $200
The interest rate is 6% = 0.06.
$1 + r = 1 + 0.06 = 1.06$

Step 2 Calculate the balance of the account.

$(1 + r) \cdot \text{Balance} = \text{New balance}$

$(1.06) \cdot \$200.00 = \$212.00$

$(1.06) \cdot \$212.00 = \$224.72$

$(1.06) \cdot \$224.72 = \$238.20$

The account balance after three years is $238.20.

GUIDED PRACTICE

11. If $100 is invested in an account that pays 2% interest compounded annually, calculate the account balance for the first three years. Round your answers to the nearest cent.

Step 1 Identify the principal and the interest rate.

Principal = ____
The interest rate is ____ = ____.
$1 + r = 1 +$ ____ = ____

Step 2 Calculate the balance of the account.

$(1 + r) \cdot \text{Balance} = \text{New balance}$

(1.__) · $_____ = $_____

(1.__) · $_____ = $_____

(1.__) · $_____ = $_____

The account balance after three years is ______.

12. Use the following bar graph to compare a simple interest investment and a compound interest investment. Each account had a starting principal of $3,000 invested at 9% interest.

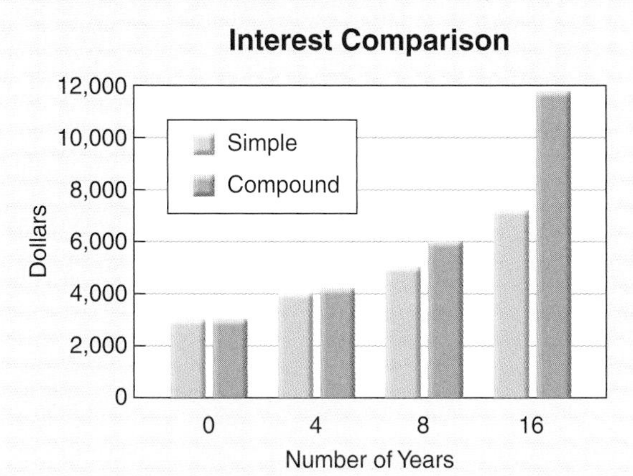

a. Estimate the balance difference between the accounts after 4 years.
b. Estimate the balance difference between the accounts after 16 years.

a. Balance difference after 4 years:

$$\text{Compound} - \text{Simple} \approx \$4{,}200 - \$4{,}000$$
$$\text{Difference} \approx \$200$$

b. Balance difference after 16 years:

$$\text{Compound} - \text{Simple} \approx \$12{,}000 - \$7{,}500$$
$$\text{Difference} \approx \$4{,}500$$

12. Use the following bar graph to compare a simple interest investment and a compound interest investment. Each account had a starting principal of $5,000 invested at 10% interest.

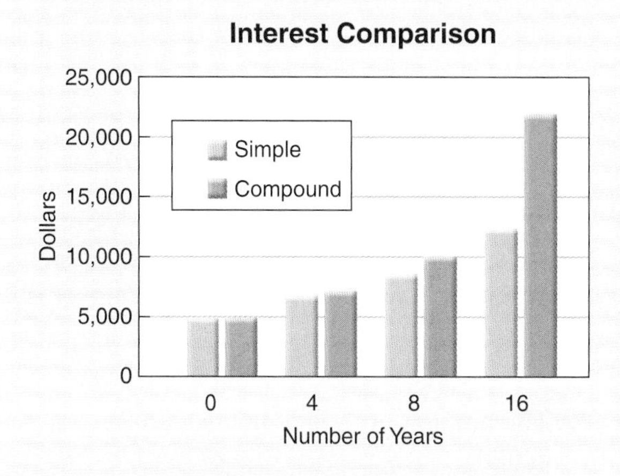

a. Estimate the balance difference between the accounts after 4 years.
b. Estimate the balance difference between the accounts after 16 years.
c. If two accounts earn the same interest rate, explain why an account earning compound interest will earn more money than an account earning simple interest.

a. Balance difference after 4 years:

$$\text{Compound} - \text{Simple} \approx$$
$$\text{Difference} \approx$$

b. Balance difference after 16 years:

$$\text{Compound} - \text{Simple} \approx$$
$$\text{Difference} \approx$$

c. A compound interest account will earn more money than a simple interest account if both have the same interest rate because

____

______________.

Section 5.4 Exercises

FOR EXTRA HELP      

To solve a commission exercise:

1. Answer the Objective A Concept Checks.
2. Answer the odd Objective A Practice Exercises.
3. Answer the even Objective A Practice Exercises.

To solve a simple interest exercise:

4. Answer the Objective B Concept Checks.
5. Answer the odd Objective B Practice Exercises.
6. Answer the even Objective B Practice Exercises.

To solve percent increase and percent decrease exercises:

7. Answer the Objective C Concept Check.
8. Answer the odd Objective C Practice Exercises.
9. Answer the even Objective C Practice Exercises.

VOCABULARY REVIEW *Review the Vocabulary Preview for Section 5.4. Study the definitions until you can check* Got It *for every word.*

Use these words to complete each sentence.

commission • commission rate • commission equation • principal • balance • interest • interest rate • simple interest • percent increase • percent decrease • compound interest

	Got It	Get Help
10. The amount of money that is earned as a percent of a balance is called ____________.		
11. A(n) ____________ is the percent of the base that is added to the base to get a new amount.		
12. A(n) ____________ is the percent of the base that is subtracted from the base to get a new amount.		
13. A ____________ is the percent of sales that determines the commission earned.		
14. When money is invested, the ____________ is the initial investment.		
15. A(n) ____________ is a percent of the balance that is earned over a period of time.		
16. With ____________, the interest earned is added to the balance to be included in the next interest calculation.		
17. ____________ is the amount of money that is earned as a percent of sales.		
18. The amount of money in an account at any time is called the ____________.		

How will you get help for any vocabulary that you are unsure about?

Your instructor ____ MyMathLab ____ A classmate ____ A tutor ____ Other ____

Set up and solve each commission exercise.

19. Sigmund earns 11% commission on all of the Mother's Day cards that his company sells. If his company sells $575,000 in cards, what is Sigmund's commission?

20. George earned 15% commission on all lumber sold to a furniture factory. If George sold $4,240 of lumber, how much commission did he earn?

21. Trudy earned $3,000 in commission for selling a $120,000 house. What was her commission rate?

22. Linda sold $480,000 in parts to an auto company. If she earned $3,600 in commission, what was her commission rate?

goetz With percent word problems, always check if your answer makes sense by comparing the sizes of the amount & base against the percent.

For more tips and tweets, go to twitter.com/gstbasicmath

23. Marcos sells cars at a dealership. His commission rate is 2%. Last month, Marcos earned $5,500 in commission. What were Marcos's sales last month?

24. Ayden sells shoes at the mall. She earns 12% in commission. If she earned $435 last week, what were her sales?

Use the table to answer each question.

A car company has a variable commission rate to reward high volumes of sales.

Car Sale Commissions	
Total Sales	**Commission Rate**
$0–$19,999	2%
$20,000–$74,999	3%
$75,000–$124,999	4%
$125,000 or more	5%

25. If Germaine sells a car for $15,000, what will he earn in commission?

26. If Latoya sells a car for $24,000, what will she earn in commission?

27. On the last day of the month, Elaine increased her sales from $124,000 to $125,000. By how much did her commission increase?

28. On the last day of the month, Bayani increased his sales from $74,000 to $75,000. By how much did his commission increase?

29. If Eddie earned $4,000 in commission, what was the amount of his car sales?

30. If Laura earned $900 in commission, what was the amount of her car sales?

Set up and solve each simple interest exercise.

31. Ziggy put $4,500 in a checking account that earns 2% per year in simple interest. How much interest will the account earn in two years?

32. Hugo purchased a $250 savings bond that earns 4.2% simple interest. How much interest will the bond earn in three years?

33. Ashley has a $2,580 balance on her credit card. If the credit card company charges 15% simple interest per year, what is her interest charge for one month?

34. Vanessa has a $7,400 balance on her credit card. If the credit card company charges 24% simple interest per year, what is her interest charge for one month?

35. Keith's credit card charges 12% annual interest. He was charged $18.53 in simple interest last month. What was his balance?

36. Christy's credit card charges 15% annual simple interest. She was charged $14 in simple interest last month. What was her balance?

Set up and solve each percent increase or percent decrease exercise. Round to the nearest hundredth of a percent when necessary.

37. A shirt that costs $25.00 was marked down to $8.50. What percent decrease in price is this?

38. A dress that costs $50.00 was marked down to $25.50. What percent decrease in price is this?

39. A store purchases hats for $15 each and sells them for $45 each. What percent increase is this?

40. A restaurant purchases wine for $8 per bottle and sells it for $32 per bottle. What percent increase is this?

41. From 2004 to 2005, the interest paid on the national debt rose from $160 billion to $178 billion. What percent increase is this?

42. From 2004 to 2005, the national debt increased from $7.3 trillion to $8 trillion. What percent increase is this?

43. Someone earning $45,000 is given a pay raise of 4%. What is the change in salary?

44. Due to budget cuts, a person earning $56,000 is given a 2% pay cut. What is the change in salary?

45. If a store sells a $5 product for $9, what percent increase is this?

46. If a store sells a $30 product for $45, what percent increase is this?

The bar graph shows the value of LaQuesia's retirement account over four years.

47. What percent of the 2007 value was the 2008 value?

48. What percent of the 2008 value was the 2010 value?

49. By what percent did the value decrease from 2007 to 2008?

50. By what percent did the value increase from 2008 to 2010?

51. What percent of the 2007 value was the 2010 value?

52. By what percent did the value increase from 2007 to 2010?

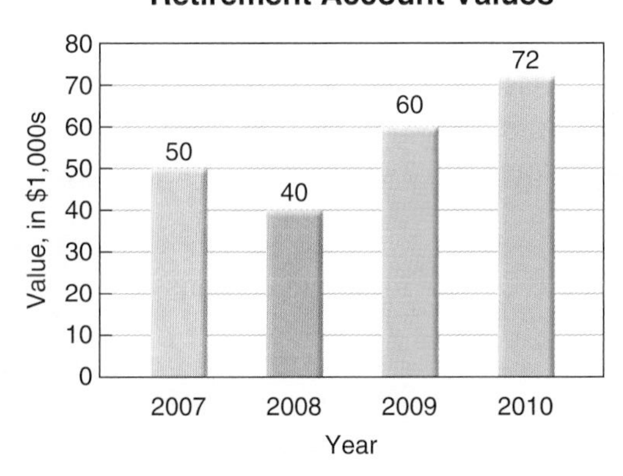

Solve each exercise. Some will require two or more steps to solve.

53. If $1,230 in property taxes is charged for a property with a 0.75% tax rate, what is the property worth?

54. If $800 in property taxes is charged on a property with a 1.6% tax rate, what is the property worth?

55. By decreasing the weight of a race car by 200 pounds, a pitcrew decreased the lap time by 3%. If the car originally had a lap time of 17.8 seconds, what was the new lap time?

56. By working extra hours, a salesperson increased her sales to $12,000. If she earns a 5% commission, how much of the $12,000 goes to the store?

57. Ben and Jennifer went out to dinner. Each paid half of the $38 bill. Jennifer gave the waiter a 20% tip on her half, and Ben gave a 15% tip on his half. What was the waiter's total tip, in dollars?

58. Of 3,000 people polled, 1,832 worked full-time, 751 worked part-time, 327 did not work, and the rest were looking for work. What percentage of people polled were looking for work?

59. A real estate agent earns 2.5% in commission. The agent had $120,000 in sales. After paying 28% in taxes, how much money did the agent have left?

60. Carly had $5,000 in an account that earned 7% interest last year. After paying 24% in taxes, how much interest did Carly keep?

61. A philanthropist was worth $10,000,000 at the start of the year. Her net worth went up 8% over the course of the year. At the end of the year, she donated $1,080,000 to charity. What was her new net worth?

62. A car had an original price of $15,000. A customer negotiated a 10% discount. After the discount, the customer also received a $500 rebate. If the customer paid $780 in taxes, what was the total cost of the car?

63. The area of a square is found using the formula $A = s^2$, where A is the area and s is the length of a side. If the 2-cm sides are increased by 100%, what is the percent increase in area?

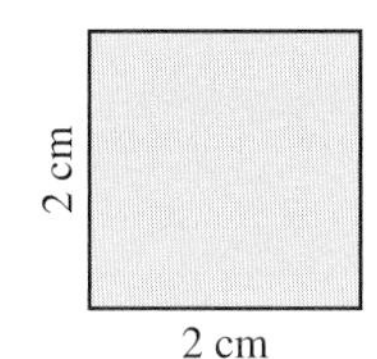

64. The area of a circle is found using the formula where A is the area and r is the length of the radius. If a circle's radius of 2 in. decreases by 50%, what is the percent decrease in area?

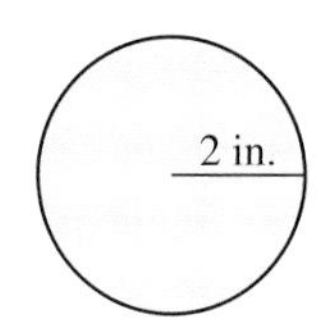

Use the figure to answer each question. There are a total of 20 shapes in this figure.

65. What percent of the shapes are circles but are not red?

66. What percent of the shapes are triangles?

67. What percent of the shapes are either red or blue squares?

68. What percent of the shapes are neither brown triangles nor blue triangles?

69. What percent of the shapes are brown circles?

70. What percent of the shapes are not red triangles?

Section 5.4 Question Log

Use this space to write questions. Be sure to get these answered and revisit them when you prepare for your exam.

Page ______ Answered

Page ______ Answered

Page ______ Answered

Page ______ Answered

Chapter 5 Organizer

VOCABULARY

Use the following steps to review the vocabulary for Chapter 5.

1. *Write the definition for each word from memory.*
2. *Compare the definitions you have written to the definitions in the Vocabulary Preview.*
3. *Study any definitions that you could not remember or that you defined incorrectly.*

5.1

Per • Cent • % • Percent • Units Fraction • Decimal–Percent–Fraction Triangle

5.2

Of • Is • Percent Proportion (Math and Words)

5.3

Of • Is • Percent Equation • Percent • Base • Amount

5.4

Commission • Commission Rate • Commission Equation • Principal • Balance • Interest • Interest Rate
Simple Interest Formula • Percent Increase • Percent Decrease • Compound Interest

PROCEDURES

Procedure/Topic	Steps	Example
Visualize Percents (Section 5.1)	**Step 1** Interpret the percent as a ratio. **Step 2** Shade the appropriate portion of a square.	Shade 47% of a square. $47\% = \frac{47}{100} \Rightarrow$
Convert a Percent to a Fraction or Mixed Number (Section 5.1)	**Step 1** Multiply by 1 in the form of $\frac{1}{100\%}$ to divide out the percent symbol. **Step 2** Simplify.	Convert 75% to a fraction. $75\% = \frac{75\%}{1} \cdot \frac{1}{100\%}$ $= \frac{75}{100} = \frac{3 \cdot 25}{1} \cdot \frac{1}{4 \cdot 25}$ $= \frac{3}{4}$
Convert a Fraction or Mixed Number to a Percent (Section 5.1)	**Step 1** Multiply by 1 in the form of $\frac{100\%}{1}$. **Step 2** Simplify. If the result is an improper fraction, convert the fraction to a mixed number.	Convert $\frac{3}{10}$ to a percent. $\frac{3}{10} = \frac{3}{10} \cdot \frac{100\%}{1}$ $= \frac{3}{10} \cdot \frac{10 \cdot 10\%}{1}$ $= 30\%$
Convert Between a Percent and a Decimal (Section 5.1)	Multiply by $\frac{1}{100\%}$ to convert a percent to a decimal. **or** Multiply by 100% to convert a decimal to a percent.	Convert each number to a percent or decimal as appropriate. $120\% = \frac{120\%}{1} \cdot \frac{1}{100\%}$ $= \frac{120}{100}$ $= 1.2$ or $0.05 = 0.05 \cdot 100\%$ $= 5\%$

Procedure/Topic	Steps	Example
Translate a Written Statement into a Percent Proportion (Section 5.2)	**Step 1** Identify the percent, base, and amount. Choose a variable for the unknown value. **Step 2** Write the percent proportion.	Translate. "What percent of 10 is 20?" Let P = percent, B = 10, and A = 20. $\frac{P}{100} = \frac{20}{10}$
Solve a Percent Exercise (Section 5.2)	**Step 1** Translate the written statement into a percent proportion. **Step 2** Solve the proportion. • Rewrite the proportion with the unknown variable in the numerator. • Multiply both sides of the equation to isolate the variable.	5 is 50% of what number? $P = 50$; let B = the number; $A = 5$ $\frac{50}{100} = \frac{5}{B}$ $\frac{100}{50} = \frac{B}{5}$ $\frac{100}{50} \cdot \frac{5}{1} = \frac{B}{5} \cdot \frac{5}{1}$ $\frac{10 \cdot 10 \cdot 5}{5 \cdot 10} = B$ $10 = B$ 5 is 50% of 10.
Translate a Written Statement into a Percent Equation (Section 5.3)	**Step 1** Identify the percent, base, and amount. Choose P, B, or A as a variable for the unknown value. **Step 2** Write the percent equation. Substitute the values for P, B, and A.	Translate. "What number is 20% of 50?" $P = 20\%$; $B = 50$; let A = amount. $P \cdot B = A$ $20\% \cdot 50 = A$
Solve a Percent Equation (Section 5.3)	**Step 1** Identify the values of P, B, and A. • If the percent is given, convert it to a decimal. **Step 2** Write the percent equation. **Step 3** Substitute the values. **Step 4** Isolate the variable. • If the variable is not alone, divide both sides of the equation to isolate it. • If the variable is the percent, convert the decimal answer to a percent.	What is 40% of 60? $P = 0.40$; $B = 60$; let A = the amount. $P \cdot B = A$ $0.40 \cdot 60 = A$ $24 = A$
Solve a Commission Exercise (Section 5.4)	**Step 1** Identify the commission earned, commission rate, and sales. Choose a variable for the unknown value. **Step 2** Write and solve a commission equation.	What were the sales if $2,000 was earned in commission with a commission rate of 5%? Commission earned = $2,000 The commission rate is 5% = 0.05. Let s = sales. $2{,}000 = 0.05 \cdot s$ $\frac{2{,}000}{0.05} = \frac{0.05 \cdot s}{0.05}$ $40{,}000 = s$ The sales were $40,000.

Procedure/Topic	Steps	Example
Solve a Simple Interest Exercise (Section 5.4)	**Step 1** Identify the interest, principal, interest rate, and time. Choose a variable for the unknown value. **Step 2** Write and solve a simple interest equation.	How much simple interest is earned in 3 months if \$2,000 is invested at 2% monthly interest? Principal or balance = \$2,000 Let I = 2% = 0.02. Time = 3 months $I = P \cdot r \cdot t$ $I = 2{,}000 \cdot 0.02 \cdot 3$ $I = 120$ \$120 is earned.
Solve Percent Increase and Percent Decrease Exercises (Section 5.4)	**Step 1** Identify the percent change, the change in quantity, and the original quantity. Choose a variable for the unknown value. **Step 2** Write and solve a percent change equation.	A tree was 120 feet tall last year. This year, it is 3% taller. What was the change in height? Percent change is 3% = 0.03. Let c = the change in height. Original height = 120 $0.03 = \frac{c}{120}$ $0.03 \cdot 120 = \frac{c}{\cancel{120}} \cdot \frac{\cancel{120}}{1}$ $3.6 = c$ The tree has grown 3.6 feet.

Chapter 5 Review Exercises

5.1

Fill in the blanks.

1. _____ is used in place of the phrases *per hundred* or *out of one hundred*.

2. The word _____ means "one hundred."

3. **a.** $\frac{1}{2} = 50\%,\ 1\frac{1}{2} = 150\%,\ 2\frac{1}{2} = 250\%,\ 3\frac{1}{2} =$ _____%, $4\frac{1}{2} =$ _____%

   b. 1 = _____%, 2 = _____%, 3 = _____%, 9 = _____%, 10 = _____%

Visualize each percent.

4. Shade 20% of a square.

5. Shade 83% of a square.

6. Shade 100% of a square.

7. Shade 1% of a square.

8. Shade 160% of a square.

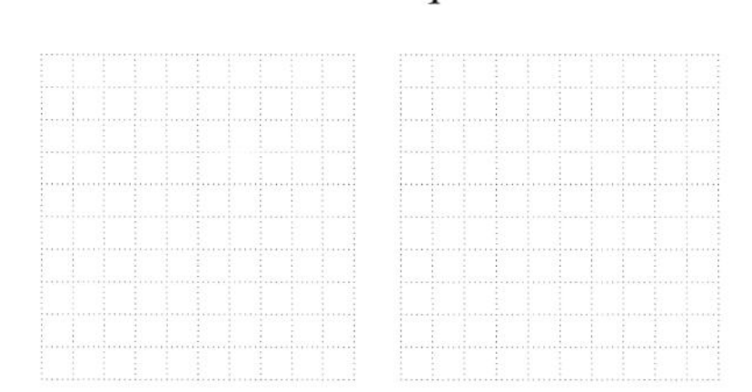

9. Shade 115% of a square.

Convert each percent to a fraction or mixed number. Write your answer in simplest form.

10. 25% **11.** 40% **12.** 78% **13.** 82%

14. $50\frac{1}{4}\%$ **15.** $40\frac{1}{2}\%$ **16.** 180% **17.** 120%

Convert each fraction or mixed number to a percent. Round to the nearest tenth of a percent.

18. $\frac{7}{10}$ **19.** $\frac{11}{20}$ **20.** $\frac{3}{8}$ **21.** $\frac{5}{8}$

22. $4\frac{1}{10}$ **23.** $5\frac{1}{5}$ **24.** $2\frac{1}{9}$ **25.** $3\frac{5}{6}$

Convert between decimals and percents as appropriate.

26. 57% **27.** 48% **28.** 0.67 **29.** 0.52

30. 235% **31.** 3.6 **32.** 8.7 **33.** 434%

34. 0.0054 **35.** 0.2% **36.** 0.9% **37.** 0.0025

38. 6,000% **39.** 7,000 **40.** 3,000 **41.** 3,000%

For each exercise, write the equivalent decimal, fraction, or percent in the empty boxes of each row.

	Decimal	Fraction	Percent
42.	0.75		
43.			$33\frac{1}{3}\%$
44.			20%
45.		$\frac{3}{3}$	

	Decimal	Fraction	Percent
46.		$\frac{1}{8}$	
47.			122%
48.	2.6		
49.		$\frac{7}{10}$	

5.2

Answer each concept question.

50. Write the standard form of a percent proportion.

51. The word ____ is often used to represent equality in word exercises.

52. The word ____ is often used to represent multiplication in word exercises.

Solve each percent exercise using a proportion. Round to the nearest percent when necessary.

53. What is 30% of 80? **54.** What is 20% of 50? **55.** 100 is what percent of 40?

56. 70 is what percent of 35?

57. 18 is 25% of what number?

58. 15 is 15% of what number?

59. In 2008, Apple increased the production of the iPhone from 10 million to 24 million. By what percent did Apple increase iPhone production?

60. If you answered 85% of 60 questions correctly, how many questions did you answer correctly?

61. Caleb made a down payment of $5,300 on a new car. This was 20% of the purchase price. What was the purchase price of the car?

62. Denise wanted to meet with each of her 150 employees. If she was able to meet with only 90 of her employees, what percent of her employees did she meet?

Answer each question about Puffy Flakes breakfast cereal. Round to the nearest percent when necessary.

63. Based on the given nutrition facts, what is the daily recommended amount of sodium, in grams?

64. In a bowl of cereal with milk, what percent of the calories come from the milk?

65. How many milligrams of carbohydrate does a half cup of milk contain?

Nutrition Facts

Serving Size 1 cup (35g)
Servings Per Container 10

	Cereal alone	Cereal with ½ cup milk
Calories	130	170
Calories from Fat	0	0
	% Daily Value**	
Total Fat 0 g*	0%	0%
Saturated Fat 0 g	0%	0%
Cholesterol 0 mg	0%	0%
Sodium 200 mg	8%	11%
Total Carbohydrate 30 g	10%	12%
Dietary Fiber 4 g	16%	16%

5.3

Answer each concept question.

66. What word is often used to indicate multiplication in a percent problem?

67. What words are often used to indicate equality in a percent problem?

68. What is the general form of a percent equation?

69. In a percent equation, the ____________ describes the portion of the base that is equal to the amount.

70. In a percent equation, the _______ is the whole quantity.

71. In a percent equation, the ____________ is the portion that results from multiplying the percent and the base.

Solve each percent exercise using an equation. Round to the nearest percent when necessary.

72. What is 15% of 35?

73. What is 80% of 22?

74. 25 is 40% of what number?

75. 46 is 20% of what number?

76. 65 is what percent of 15?

77. 84 is what percent of 12?

78. Your computer has 50 GB of memory available. If you use 42 GB, what percent of the remaining storage capacity have you used?

79. Manuel needs a new set of guitar strings for a performance he is playing in this weekend. When he purchased the strings, the 8% sales tax came to $0.96. How much did the strings cost before the sales tax was added?

80. Rebecca purchased a house for $109,000. If she made a down payment of 20%, how much was her down payment?

5.4

Answer each concept question.

81. ____________ is the amount of money that a salesperson earns as a percent of sales.

82. A(n) ____________ multiplies the balance and the time period of the investment to determine how much interest is earned.

83. When money is invested, the original investment is called the ____________.

84. ____________ is the amount of money earned as a percent of the balance.

Solve each commission exercise. Round to the nearest percent when necessary.

85. Victor sells appliances at a home improvement store. He earns a commission rate of 8%. If he sells a refrigerator for $1,200, what commission will he earn?

86. A manager's commission rate is 2% of the business's gross sales. If gross sales totaled $1,100,000, what was the manager's commission?

87. Veronica sold a car for $25,000. If she received $2,250 for this sale, what commission rate did she earn?

88. Tamica sells skin care products. This month, she sold $14,000 in products. If she earned $1,680 for these sales, what was her commission rate?

89. An investor pays an advisor a commission rate of 0.25% of his portfolio's value each year. If the advisor earns $650 this year, what is the value of the portfolio?

90. Kao earns a 4% commission rate for running a friend's e-Bay store. If Kao earned $580, what were the sales?

Solve each simple interest exercise. Round to the nearest percent when necessary.

91. Edward's checking account had a principal investment of $860. If the account earned 1.5% in simple interest each year, how much interest did the account earn in 2 years?

92. Greg's money market account had a principal investment of $1,080. If the account earned 2.5% in simple interest each year, how much interest did the account earn in 3 years?

93. Kayla invests $20,000 in a savings account. If she earns $500 a year in simple interest, what is the interest rate?

94. Yolanda invests $50,000 in a bond account. If she earns $800 a year in simple interest, what is the interest rate?

95. Rocco's credit card company charges 24% simple interest each year. If he charges $1,700 on his card, how much interest will he be charged for the first month?

96. Annatto's credit card company charges 18% simple interest each year. If she charges $1,500 on her card, how much will she be charged for the first month?

Solve each percent increase or percent decrease exercise. Round to the nearest percent when necessary.

97. If someone with a $45,000 salary receives a 3% cost-of-living raise, what is the change in salary?

98. If someone with a $36,000 salary must take a 2% pay cut, what is the change in salary?

99. Last year, Alli's car insurance cost $900. This year, her insurance decreased by $171. By what percent did Alli's insurance decrease?

100. If someone with a $36,000 salary earns a $4,000 bonus, what percent increase of salary does this bonus represent?

Chapter 5 Practice Test

Fill in the blanks.

1. ____________ is used in place of the phrases *per hundred* or *out of one hundred.*

2. In a percent equation, the ____________ tells what portion of the base gives the amount.

3. In a percent equation, the ____________ is the whole quantity.

4. In a percent equation, the ________ is the portion that results from multiplying the percent and the base.

5. The word ____ is often used to represent equality in word exercises.

6. The word ____ is often used to represent multiplication in word exercises.

7. A ____________ is the amount of money a salesperson earns as a percent of sales.

8. When money is invested, the original investment is called the ____________.

9. A percent ____________ is a percent of the base that is subtracted to make a new decreased amount.

Visualize each percent.

10. Shade 68% of a square.

11. Shade 142% of a square.

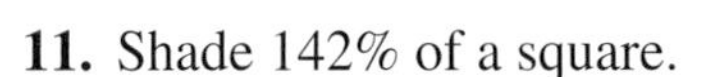

Convert each percent to a fraction or mixed number. Write your answer in simplest form.

12. 28%

13. 160%

14. $40\frac{1}{4}\%$

Convert each fraction or mixed number to a percent. Round to the nearest tenth of a percent.

15. $\frac{9}{10}$

16. $\frac{1}{8}$

17. $3\frac{2}{5}$

Convert between decimals and percents as appropriate.

18. 72%

19. 0.078

20. 0.45

21. 123%

Solve each percent exercise using a proportion. Round to the nearest percent when necessary.

22. What is 60% of 50?

23. 100 is what percent of 80?

24. 300 is 15% of what number?

25. Microsoft decreased the production of its Xbox game from 18 million to 12 million. By what percent did Microsoft decrease Xbox game production?

26. If you answered 95% of 40 questions correctly, how many questions did you answer correctly?

Solve each percent exercise using an equation. Round to the nearest percent when necessary.

27. What is 14% of 44?

28. 50 is 40% of what number?

29. 65 is what percent of 20?

30. Your computer can store 160 GB of information. If you use 100 GB, what percent of the storage capacity have you used?

31. When Maria purchased a new microphone, the 8% sales tax came to $5.44. How much did the microphone cost before the sales tax was added?

Solve each commission exercise. Round to the nearest percent when necessary.

32. Chantal sells appliances at a home improvement store. She earns a commission rate of 4%. If she sells a washing machine for $1,100, what commission will she earn?

33. Ken sells hair care products. This month, he sold $25,000 in products. If he earned $1,750 for these sales, what was his commission rate?

34. Kaleko earns a 5% commission rate for running a friend's e-Bay store. If Kaleko earned $680, what were the sales?

Solve each simple interest exercise. Round to the nearest percent when necessary.

35. Daryl's checking account had a principal investment of $15,450. If the account earned 1.5% in simple interest each year, how much interest did the account earn in 3 years?

36. Yolko invests $40,000 in a bond account. If she earns $800 a year in simple interest, what is the interest rate?

37. Ringo's credit card company charges 24% simple interest each year. If he spends $1,300 on his card, how much interest will he be charged for the first month?

Solve each percent increase or percent decrease exercise. Round to the nearest percent when necessary.

38. If someone earning a $26,000 salary receives a 2% cost-of-living raise, what is the change in salary?

39. Last year, Elli's car insurance cost $600. This year, her insurance decreased by $150. By what percent did Elli's insurance decrease?

CHAPTER 6

Measurement

The U.S. and metric systems of measure are the two major measurement systems in the world. You have likely encountered both measurement systems. Since most of the world uses the metric system, it is important that you learn both the U.S. system and the metric system.

GPS (Global Positioning System) directions can be given in miles or kilometers. Miles is a unit of the U.S. system, while kilometers is a unit of the metric system.

In Exercise 72, Section 6.3, we will convert miles to kilometers to make sure that a car is driving within the speed limit on a Canadian highway.

Section 6.1 U.S. System Units of Measure

In the United States, we use the U.S. system of measure. You are probably familiar with many measurements in the U.S. system: miles, cups, gallons, pounds, and tons. In this section, you will learn what the different units of measure are in the U.S. system and how to convert measures from one unit to another.

The **Objectives** in Section 6.1 will help you

- A Understand the U.S. system of measure for length and time.
- B Understand the U.S. system of measure for weight and volume.
- C Convert units of measure in the U.S. system.

VOCABULARY PREVIEW *Check the box that applies.*

	Got It	Must Study
U.S. System of Measure: The **U.S. system of measure** is the primary system of measurement used in the United States. Very few countries use the U.S. system of measure.		
Units Fraction: A **units fraction** contains units of equal measure in the numerator and denominator. Units fractions are equal to 1.		

Study these words when they appear in the text. After the section, test your understanding by completing the Vocabulary Review in the section exercises.

Objective A Understand the U.S. System of Measure for Length and Time

The Concept Many units of measure in the U.S. system arose from measurement tools that were easily accessible. For example, units of feet came from the length of a person's foot. The U.S. measures of length shown below are related to feet.

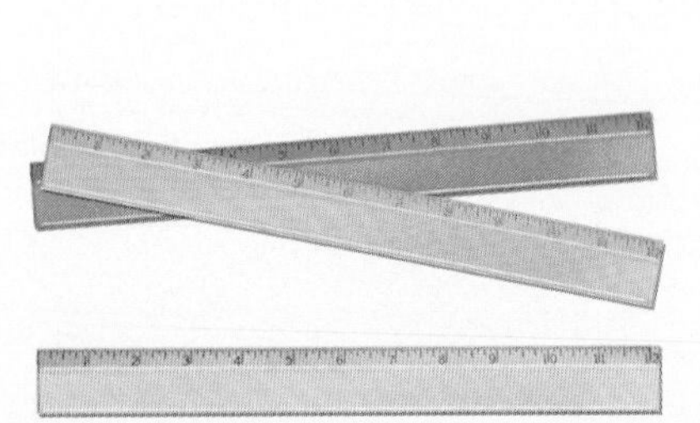

1 foot = 12 inches

1 yard = 3 feet

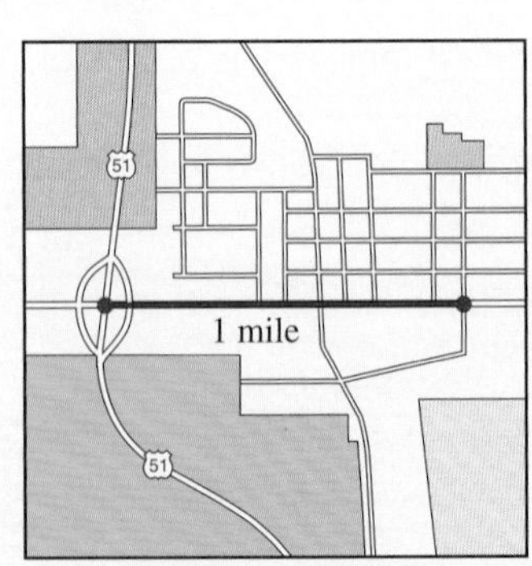

1 mile = 5,280 feet

Table 6.1 lists conversion factors for units of length and units of time in the U.S. system. It is useful to memorize the equal measures in this table.

TABLE 6.1 U.S. UNITS AND THEIR EQUIVALENTS

Length	Time
12 inches (in.) = 1 foot (ft)	60 seconds (sec) = 1 minute (min)
3 feet = 1 yard (yd)	60 minutes = 1 hour (hr)
5,280 feet = 1 mile (mi)	24 hours = 1 day (d)
	7 days = 1 week (wk)
	365 days = 1 year (yr)
	12 months (mos) = 1 year

▶ **NOTE** The same units of time are used in both the U.S. and metric systems.

Concept Check

Identify the U.S. unit of measure that most conveniently measures each item.

A1. distance across the state of Georgia

A2. length of a notebook

A3. time you spend in a typical class meeting

A4. width of a classroom

A5. time span of a college semester

Objective A Practice

Fill in each blank.

A6. Write the correct number in each blank.
- **a.** 1 yard = ______ feet
- **b.** 1 minute = ______ seconds
- **c.** 7 days = ______ week
- **d.** 1 mile = ______ feet
- **e.** 1 year = ______ days
- **f.** 60 minutes = ______ hour
- **g.** 1 day = ______ hours
- **h.** 12 inches = ______ foot

A7. Write the correct abbreviation in each blank.
- **a.** 1 ______ = 24 hr
- **b.** 365 d = 1 ______
- **c.** 1 ______ = 5,280 ft
- **d.** 1 ______ = 60 sec
- **e.** 12 ______ = 1 ft
- **f.** 1 ______ = 7 d
- **g.** 60 ______ = 1 hr
- **h.** 1 ______ = 3 ft

Objective B Understand the U.S. System of Measure for Weight and Volume

The Concept Many objects can be described using a U.S. system unit of weight or volume.

Weight

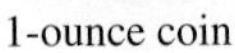
1-ounce coin

1-pound lobster

1-ton bison

Volume

1 fluid ounce of perfume

1 cup of coffee

1 pint of liquid

1 gallon of milk

Table 6.2 lists conversion factors for units of weight and units of volume in the U.S. system. Memorizing the equal measures in this table makes these units easy to use in everyday situations.

TABLE 6.2 U.S. UNITS AND THEIR EQUIVALENTS

Weight	Volume
16 ounces (oz) = 1 pound (lb)	8 fluid ounces (fl oz) = 1 cup (c)
2,000 pounds = 1 ton (T)	2 cups = 1 pint (pt)
	2 pints = 1 quart (qt)
	4 quarts = 1 gallon (gal)

▶ **NOTE**

- Fluid ounces (fl oz) measure volume.
- Ounces (oz) measure weight.

Concept Check

Identify the U.S. unit of measure that most conveniently measures each item.

B1. weight of the human body

B2. weight of a pencil

B3. volume of an aquarium

B4. volume of a single-serving soda bottle

B5. weight of an iceberg

Objective B Practice

Fill in each blank.

B6. Write the correct number in each blank.

a. 1 ton = ______ pounds
b. 1 pint = ______ cups
c. ______ ounces = 1 pound
d. 1 gallon = ______ quarts
e. 1 quart = ______ pints
f. ______ fluid ounces = 1 cup

B7. Write the correct abbreviation in each blank.

a. 4 qt = 1 ______
b. 1 qt = 2 ______
c. 1 T = 2,000 ______
d. 1 lb = 16 ______
e. 1 c = 8 ______
f. 1 pt = 2 ______

Objective C Convert Units of Measure in the U.S. System

The Concept Converting from one unit of measure to another can help you solve everyday problems. For example, converting cups into pints can help you determine the amount of milk needed to double a recipe. To convert units of measure, multiply by a units fraction. A **units fraction** contains equal measures in the numerator and denominator.

TABLE 6.3 U.S. UNITS OF MEASURE AND THEIR EQUIVALENTS

Time	Volume
60 seconds (sec) = 1 minute (min)	8 fluid ounces (fl oz) = 1 cup (c)
60 minutes = 1 hour (hr)	2 cups = 1 pint (pt)
24 hours = 1 day (d)	2 pints = 1 quart (qt)
7 days = 1 week (wk)	4 quarts = 1 gallon(gal)
52 weeks = 1 year (yr)	
365 days = 1 year (yr)	
Length	**Weight**
12 inches (in.) = 1 foot (ft)	16 ounces (oz) = 1 pound (lb)
3 feet = 1 yard (yd)	2,000 pounds = 1 ton (T)
5,280 feet = 1 mile (mi)	

Because the units of measure are equal in a units fraction, units fractions equal 1. For example, seven days and one week are equal measures of time. If we make a fraction with these measures, the fraction is equal to 1 and is a units fraction.

$$\frac{7 \text{ days}}{1 \text{ week}} = 1$$

INTERACTIVE DEFINITION Units Fraction

A **units fraction** contains units of equal measure in the numerator and denominator. Units fractions are equal to 1.

EXAMPLE

1. Identify the two units fractions that convert between feet and yards.

Since 3 feet = 1 yard, the two unit fractions are

$$\frac{3\text{ ft}}{1\text{ yd}} = 1 \quad \text{and} \quad \frac{1\text{ yd}}{3\text{ ft}} = 1.$$

GUIDED PRACTICE

1. Identify the two units fractions that convert between inches and feet.

Since _____ inches = _____ foot, the two unit fractions are

$$\frac{\quad}{\quad} = 1 \quad \text{and} \quad \frac{\quad}{\quad} = 1.$$

DO YOU UNDERSTAND how to build units fractions? Got It Get Help

In Example 1, we identified the two units fractions that convert between feet and yards. When converting between feet and yards, we will use one of the following units fractions:

$$\frac{3\text{ ft}}{1\text{ yd}} \quad \text{or} \quad \frac{1\text{ yd}}{3\text{ ft}}$$

The units fraction we use will depend on whether we are converting from yards to feet or feet to yards.

DETAILED EXAMPLES Using Units Fractions to Convert U.S. System Units of Measure

Convert 18 feet to yards.

Step 1 Write 18 feet as a fraction by writing it over 1.

Step 2 3 feet = 1 yard. Multiply by the units fraction that will divide out the feet and introduce yards.

Step 3 Simplify to convert the units.

$$18\text{ ft} = \frac{18\text{ ft}}{1}$$

$$= \frac{18\text{ ft}}{1} \cdot \left(\frac{1\text{ yd}}{3\text{ ft}}\right)$$

To introduce yards, write yards in the numerator.

To divide out feet, write feet in the denominator.

$$= \frac{18}{1} \cdot \left(\frac{1\text{ yd}}{3}\right)$$

$$= \frac{3 \cdot 6 \cdot 1\text{ yd}}{1 \cdot 3}$$

$$= 6\text{ yards}$$

Convert 12 yards to feet.

Step 1 Write 12 yards as a fraction by writing it over 1.

Step 2 3 feet = 1 yard. Multiply by the units fraction that will divide out the yards and introduce feet.

Step 3 Simplify to convert the units.

$$12\text{ yd} = \frac{12\text{ yd}}{1}$$

$$= \frac{12\text{ yd}}{1} \cdot \left(\frac{3\text{ ft}}{1\text{ yd}}\right)$$

To introduce feet, write feet in the numerator.

To divide out yards, write yards in the denominator.

$$= \frac{12}{1} \cdot \left(\frac{3\text{ ft}}{1}\right)$$

$$= 36\text{ feet}$$

▶ **NOTE** Only units that measure the same type of thing can be converted. For example, miles (distance) can be converted to inches (distance). Inches (distance) cannot be converted to minutes (time).

Procedure Convert Units of Measure in the U.S. System

Step 1 Identify equal measures for the units involved in the conversion.

Step 2 Write the original quantity as a fraction over 1.

Step 3 Multiply by a units fraction to divide out the original units and introduce the new units.

Step 4 Simplify.

EXAMPLES

2. Convert 10 quarts to pints.

Step 1 Identify equal measures. 1 quart = 2 pints

Step 2 Write the original quantity as a fraction over 1. $10 \text{ qt} = \frac{10 \text{ qt}}{1}$

Step 3 Multiply by a units fraction. $= \frac{10 \text{ qt}}{1} \cdot \left(\frac{2 \text{ pt}}{1 \text{ qt}}\right)$

$= \frac{10}{1} \cdot \left(\frac{2 \text{ pt}}{1}\right)$

Step 4 Simplify. = 20 pints

3. Convert 2 gallons to quarts.

Step 1 Identify equal measures. 1 gallon = 4 quarts

Step 2 Write the original quantity as a fraction over 1. $2 \text{ gal} = \frac{2 \text{ gal}}{1}$

Step 3 Multiply by a units fraction. $= \frac{2 \text{ gal}}{1} \cdot \left(\frac{4 \text{ qt}}{1 \text{ gal}}\right)$

$= \frac{2}{1} \cdot \left(\frac{4 \text{ qt}}{1}\right)$

Step 4 Simplify. = 8 quarts

GUIDED PRACTICE

2. Convert 6 pints to quarts.

_____ pint(s) = _____ quart(s)

$6 \text{ pt} = \frac{\quad}{1}$

$= \frac{\quad}{1} \cdot \left(\frac{\quad}{\quad}\right)$

$= \frac{\quad}{\quad}$

=

3. Convert 6 pounds to ounces.

_____ pound(s) = _____ ounce(s)

$6 \text{ lb} = \frac{\quad}{1}$

$= \frac{\quad}{1} \cdot \left(\frac{\quad}{\quad}\right)$

$= \frac{\quad}{1} \cdot \left(\frac{\quad}{\quad}\right)$

=

4. Convert 60 inches to feet.

Step 1 Identify equal measures. 1 foot = 12 inches

Step 2 Write the original quantity as a fraction over 1.

$$60 \text{ in.} = \frac{60 \text{ in.}}{1}$$

Step 3 Multiply by a units fraction.

$$= \frac{60 \text{ in.}}{1} \cdot \left(\frac{1 \text{ ft}}{12 \text{ in.}}\right)$$

$$= \frac{60}{1} \cdot \left(\frac{1 \text{ ft}}{12}\right)$$

Step 4 Simplify.

$$= \frac{5 \cdot 12 \text{ ft}}{12}$$

$$= 5 \text{ feet}$$

4. Convert 28 cups to pints.

_____ pint = _____ cups

$$28 \text{ c} = \frac{\quad}{1}$$

$$= \frac{\quad}{1} \cdot \left(\frac{\quad}{\quad}\right)$$

$$= \frac{\quad}{1} \cdot \left(\frac{\quad}{\quad}\right)$$

$$= \frac{\quad}{\quad}$$

$$=$$

Sometimes we will need to perform two or more unit conversions in one exercise. When that happens, perform the conversion in two steps.

EXAMPLE

5. Convert 120 miles per hour to feet per minute.

Step 1 Write the original quantity as a fraction. The word *per* indicates a fraction bar.

$$120 \text{ miles per hour} = \frac{120 \text{ mi}}{1 \text{ hr}}$$

Step 2 Multiply by two units fractions.

- One units fraction divides out the miles and introduces feet.
- The other units fraction divides out the hours and introduces minutes.

$$= \frac{120 \text{ mi}}{1 \text{ hr}} \cdot \left(\frac{5{,}280 \text{ ft}}{1 \text{ mi}}\right)$$

$$= \frac{120 \text{ mi}}{1 \text{ hr}} \cdot \left(\frac{5{,}280 \text{ ft}}{1 \text{ mi}}\right) \cdot \left(\frac{1 \text{ hr}}{60 \text{ min}}\right)$$

Step 3 Simplify.

$$= \frac{2 \cdot 60 \cdot 5{,}280 \text{ ft}}{60 \text{ min}}$$

$$= \frac{10{,}560 \text{ ft}}{1 \text{ min}}$$

$$= 10{,}560 \text{ feet per minute}$$

GUIDED PRACTICE

5. Convert 2,640 feet per minute to miles per hour.

Step 1 Write the original quantity as a fraction. The word *per* indicates a fraction bar.

$2{,}640 \text{ ft per min} = \underline{\qquad}$

Step 2 Multiply by two units fractions.

- One units fraction divides out the feet and introduces miles.
- The other units fraction divides out the minutes and introduces hours.

$= \underline{\qquad} \cdot \left(\underline{\qquad}\right)$

$= \underline{\qquad} \cdot \left(\underline{\qquad}\right) \cdot \left(\underline{\qquad}\right)$

$= \underline{\qquad}$

Step 3 Simplify.

$= \underline{\qquad}$

$=$ miles per hour

Concept Check

C1. What are the two units fractions that convert between fluid ounces and cups?

C2. What are the two units fractions that convert between days and weeks?

C3. Simplify.

$$6 \text{ quarts} = \frac{6 \text{ quarts}}{1} \cdot \frac{2 \text{ pints}}{1 \text{ quart}}$$
$$= \underline{\qquad} \text{ pints}$$

C4. Simplify.

$$8{,}000 \text{ pounds} = \frac{8{,}000 \text{ pounds}}{1} \cdot \frac{1 \text{ ton}}{2{,}000 \text{ pounds}}$$
$$= \underline{\qquad} \text{ tons}$$

Objective C Practice

Write the appropriate units fraction and perform the conversion.

C5. $\frac{10 \text{ ft}}{1} \cdot \left(\frac{\text{in.}}{\quad}\right) = \underline{\quad} \text{ in.}$

C6. $\frac{10 \text{ yd}}{1} \cdot \left(\frac{\text{ft}}{\quad}\right) = \underline{\quad} \text{ ft}$

C7. $\frac{16 \text{ c}}{1} \cdot \left(\frac{\text{pt}}{\quad}\right) = \underline{\quad} \text{ pt}$

C8. $\frac{14 \text{ pt}}{1} \cdot \left(\frac{\text{qt}}{\quad}\right) = \underline{\quad} \text{ qt}$

Use units fractions to perform each conversion.

C9. How many feet are in 9 yards?

C10. How many days are in 11 weeks?

C11. 240 ounces is equal to how many pounds?

C12. 48 fluid ounces is equal to how many cups?

C13. Convert 280 yards per minute to feet per second.

C14. Convert 1,320 feet per minute to miles per hour.

Combining Concepts and Applications

CONCEPT I Multistep Exercises

EXAMPLES

6. No interest is charged on a $4,800 loan that must be paid back in monthly payments over 2.5 years. What is the monthly payment?

Step 1 Find the number of months needed to pay back the loan.

$$2.5 \text{ years} = \frac{2.5 \text{ yr}}{1}$$
$$= \frac{2.5 \cancel{\text{yr}}}{1} \cdot \left(\frac{12 \text{ mos}}{1 \cancel{\text{yr}}}\right)$$
$$= 30 \text{ months}$$

Step 2 Divide the loan amount by the number of monthly payments.

$$\text{Payment} = \text{Loan amount} \div \text{Months}$$
$$= 4{,}800 \div 30$$
$$= 160$$

The monthly payment is $160.

7. A car loan of $1,000 must be paid back, plus interest, over two years. If the monthly payment is $50, how much total interest will be paid back?

Step 1 Determine how much money will be paid back in total.

$$2 \text{ years} = \frac{2 \cancel{\text{yr}}}{1} \cdot \left(\frac{12 \text{ mos}}{1 \cancel{\text{yr}}}\right)$$
$$= 24 \text{ months}$$
$$\text{Total} = \text{Monthly payment} \cdot \text{Months}$$
$$= \$50 \cdot 24$$
$$= \$1{,}200$$

Step 2 Find the total interest by subtracting the loan amount from the total amount paid back.

$$\text{Total interest} = \text{Total amount paid} - \text{Loan amount}$$
$$= \$1{,}200 - \$1{,}000$$
$$= \$200$$

$200 will be paid back in total interest.

GUIDED PRACTICE

6. No interest is charged on a $1,500 loan that must be paid back in monthly payments over 1.5 years. What is the monthly payment?

Step 1 Find the number of months needed to pay back the loan.

$$1.5 \text{ years} = \frac{\quad}{1}$$
$$= \frac{\quad}{1} \cdot \left(\frac{\quad}{\quad}\right)$$
$$= \quad \text{months}$$

Step 2 Divide the loan amount by the number of monthly payments.

$$\text{Payment} = \underline{\qquad\qquad} \div \underline{\qquad}$$
$$= \quad \div$$
$$=$$

The monthly payment is ________.

7. A car loan of $2,000 must be paid back, plus interest, over three years. If the monthly payment is $70, how much total interest will be paid back?

Step 1 Determine how much money will be paid back in total.

$$3 \text{ years} = \frac{\quad}{1} \cdot \left(\frac{\quad}{\quad}\right)$$
$$= \quad \text{months}$$
$$\text{Total} = \text{Monthly payment} \cdot \text{Months}$$
$$=$$
$$=$$

Step 2 Find the total interest by subtracting the loan amount from the total amount paid back.

$$\text{Total interest} = \text{Total amount paid} - \text{Loan amount}$$
$$=$$
$$=$$

$_____ will be paid back in total interest.

Section 6.1 Exercises

FOR EXTRA HELP     

To understand the U.S. system of measure for length and time:

1. Answer the Objective A Concept Checks.
2. Answer the odd Objective A Practice Exercise.
3. Answer the even Objective A Practice Exercise.

To understand the U.S. system of measure for weight and volume:

4. Answer the Objective B Concept Checks.
5. Answer the odd Objective B Practice Exercise.
6. Answer the even Objective B Practice Exercise.

To convert units of measure in the U.S. system:

7. Answer the Objective C Concept Checks.
8. Answer the odd Objective C Practice Exercises.
9. Answer the even Objective C Practice Exercises.

VOCABULARY REVIEW *Review the Vocabulary Preview for Section 6.1 and the tables of equivalent measures on pages 6-4 and 6-5. Study the definitions until you can check* Got It *for every word.*

Fill in each blank.

	Got It	Get Help		Got It	Get Help
10. a. 7 days = ____ week			**11. a.** 4 quarts = ____ gallon		
b. 1 pint = ____ cups			**b.** 365 days = ____ year		
c. 16 ounces = ____ pound			**c.** 1 ton = ____ pounds		
d. 1 year = ____ days			**d.** 60 minutes = ____ hour		
e. 1 mile = ____ feet			**e.** 1 yard = ____ feet		
f. 8 fluid ounces = ____ cup			**f.** ____ pint = 2 cups		

Use the abbreviations of measurements to fill in each blank.

	Got It	Get Help		Got It	Get Help
12. a. 1 ____ = 3 ft			**13. a.** 5,280 ft = 1 ____		
b. 1 ____ = 60 sec			**b.** 1 ____ = 12 in.		
c. 1 ____ = 2,000 lb			**c.** 1 qt = 2 ____		
d. 1 ____ = 5,280 ft			**d.** 1 min = 60 ____		
e. 1 ____ = 4 qt			**e.** 1 wk = 7 ____		
f. 60 ____ = 1 hr			**f.** 1 ____ = 16 oz		
g. 1 d = 24 ____			**g.** 1 ____ = 8 fl oz		

How will you get help for any vocabulary that you are unsure about?

Your instructor ____ MyMathLab ____ A classmate ____ A tutor ____ Other ____

Write the appropriate units fraction and perform the conversion.

14. $\frac{10\text{ c}}{1} \cdot \left(\frac{\text{fl oz}}{\quad}\right) = \quad \text{fl oz}$

15. $\frac{6\text{ d}}{1} \cdot \left(\frac{\text{hr}}{\quad}\right) = \quad \text{hr}$

16. $\frac{48\text{ oz}}{1} \cdot \left(\frac{\text{lb}}{\quad}\right) = \quad \text{lb}$

17. $\frac{24\text{ qt}}{1} \cdot \left(\frac{\text{gal}}{\quad}\right) = \quad \text{gal}$

Use units fractions to perform each conversion.

18. 21,120 feet = _____ miles

19. 10,560 feet = _____ miles

20. 960 seconds = _____ minutes

21. 1,500 seconds = _____ minutes

22. 10 gallons = _____ quarts

23. 14 gallons = _____ quarts

24. Mount Aconcagua in South America is 22,835 ft tall. How many miles high is this mountain? Round your answer to the nearest tenth of a mile. *Source:* www.britannica.com

25. Mount McKinley in North America is 20,320 ft tall. How many miles high is this mountain? Round your answer to the nearest tenth of a mile. *Source:* www.britannica.com

26. Annabel bought 3 pounds of tomatillos to make salsa. If a recipe calls for 6 ounces of tomatillos, how many batches of the recipe can she make?

27. Dehron bought 4 pounds of bok choy to make soup. If a recipe calls for 7 ounces of bok choy, how many batches of the recipe can he make?

28. 3 miles = ______ feet

29. 7 miles = ______ feet

30. How many tons are in 1,600 pounds?

31. How many tons are in 1,400 pounds?

32. 14 cups = _____ pints

33. 24 cups = _____ pints

34. 6 feet is equal to how many inches?

35. 8 feet is equal to how many inches?

36. 3.25 pounds = _____ ounces

37. 5.5 pounds = _____ ounces

38. How many feet are in 75 inches?

39. How many feet are in 87 inches?

Answer each question.

40. The New York City Marathon is 26.2 miles. *Source:* www.nycmarathon.org
- **a.** How long is the marathon in feet?
- **b.** If an average step is 1 yard, how many steps will it take to walk the New York City Marathon?
- **c.** If an average stride is 2 yards, how many strides will it take to jog the New York City Marathon?

41. The Binder Park Road Race is 6.2 miles. *Source:* www.binderparkzoo.org
- **a.** How long is the race in feet?
- **b.** If an average step is 1 yard, how many steps will it take to walk the Binder Park Road Race?
- **c.** If an average stride is 2 yards, how many strides will it take to jog the Binder Park Road Race?

42. Does it take more cups or more gallons to measure the same amount of stew in a pot?

43. Does it take more ounces or more pounds to measure the same amount of bananas?

Smith To divide out the original units, write them again in the denominator of the units fraction.
For more tips and tweets, go to twitter.com/gstbasicmath

44. Does it take more yards or more miles to measure the distance from Albuquerque to Dallas?

45. Does it take more quarts or more pints to measure the amount of water in Lake Okeechobee?

Perform each conversion. You will need to use more than one units fraction. Express your answer as a decimal if necessary.

46. 15 miles per hour = ____ feet per minute

47. 45 miles per hour = ____ feet per minute

48. 24 quarts per day = ____ gallons per week

49. 60 quarts per day = ____ gallons per week

50. 25 feet per minute = ____ inches per second

51. 45 feet per minute = ____ inches per second

52. 100 pounds per minute = ____ tons per hour

53. 250 pounds per minute = ____ tons per hour

Solve each application exercise.

54. Mandy and Hector are putting new trim in their living room. The perimeter of the living room is 100 feet and 4 inches. If the trim costs $0.30 per inch, how much will the new trim cost?

55. Aretha is replacing the sidewalk at her home. The sidewalk is 5 yards and 2 feet long. If new sidewalk cost $19 per foot, how much will the new sidewalk cost?

56. To stay hydrated, a runner wants to drink 4 fluid ounces of water for every mile of the New York City Marathon. If the race is 26.2 miles, how many cups of water will he need to drink during the marathon?

57. To stay hydrated, a runner wants to drink 4 fluid ounces of water for every mile of the Binder Park Road Race. If the race is 6.2 miles, how many cups of water will she need to drink during the road race?

Use the conversion 660 acres = 1 square mile to answer 58 and 59.

58. A fire burning near Yosemite National Park burned 7,920 acres. How many square miles is this?

59. A fire in Newberry, Michigan, burned 25.5 square miles. How many acres is this?

Solve each multistep exercise.

60. Shanelle's parents loaned her $15,600 to purchase a car. They are not charging her interest. To pay back the loan, she will make monthly payments for 5 years. How much is each payment?

61. Shania loaned her son $95,400 to purchase a house. She is not charging him interest. To pay back the loan, he will make monthly payments for 30 years. How much is each payment?

62. Andrew purchased a car for $15,500. His monthly car payment is $320, which will take him 5 years to pay off. How much will he pay in total interest?

63. Antwan purchased a house for $95,000. His monthly mortgage payment is $630, which will take him 30 years to pay off. How much will he pay in total interest?

SELF-ASSESSMENT

	YES	NO
1. On your last test, were you satisfied with your score? *If you answered yes, good job. Do not continue here.* *If you answered no, answer the following.*		
2. Did you preview each section of the textbook before your instructor taught it in class?		
3. Did you complete all of the homework before it was due?		
4. Did you get help with the homework exercises that you didn't understand?		

5. Set two goals that will help you become better prepared for your next test.

Goal 1: ______________________________

Goal 2: ______________________________

Section 6.1 Question Log

Use this space to write questions. Be sure to get these answered and revisit them when you prepare for your exam.

Page ____ **Answered** ☐	**Page** ____ **Answered** ☐
Page ____ **Answered** ☐	**Page** ____ **Answered** ☐

Section 6.2 Metric System Units of Measure

Most countries use the metric system of measurement. Because we work in a global economy, it is important to understand the metric system. In this section, you will study the metric system of measure and learn how to convert measures from one unit to another.

The only countries that do not use the metric system are the United States, Liberia, and Myanmar.

The **Objectives** in Section 6.2 will help you

A Understand the metric system units of measure for length, volume, and weight.
B Convert units of measure in the metric system.

VOCABULARY PREVIEW *Check the box that applies.*

	Got It	Must Study
Metric System of Measure: The **metric system of measure** is the most widely used measurement system in the world. Conversions in the metric system are based on place values.		
Fundamental Units: Fundamental units are the metric system units of measure that all other metric units are based on. Metric prefixes are added to fundamental units to result in larger or smaller measurements. Three fundamental units are meters, liters, and grams.		
Meter: The **meter (m)** is the fundamental unit of length in the metric system.		
Liter: The **liter (L)** is the fundamental unit of volume in the metric system.		
Gram: The **gram (g)** is the fundamental unit of mass (weight) in the metric system.		
Metric Prefixes: Metric prefixes are added to a fundamental unit (meters, liters, or grams) to create units that are larger or smaller than the fundamental unit. Six common prefixes and their values are listed below. kilo- (k) $= 1{,}000$; hecto- (h) $= 100$; deka- (da) $= 10$; deci- (d) $= \frac{1}{10}$; centi- (c) $= \frac{1}{100}$; milli- (m) $= \frac{1}{1{,}000}$		

Study these words when they appear in the text. After the section, test your understanding by completing the Vocabulary Review in the section exercises.

Objective A Understand the Metric System Units of Measure for Length, Volume, and Weight

The Concept In the metric system, three fundamental units are used to measure length, volume, and weight. They are meters, liters, and grams, respectively.

Meters measure length

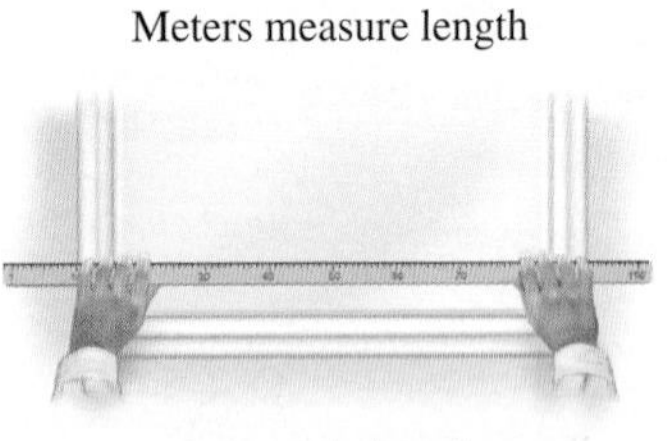
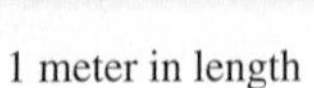

1 meter in length

Liters measure volume.

1 liter of water

Grams measure weight[1].

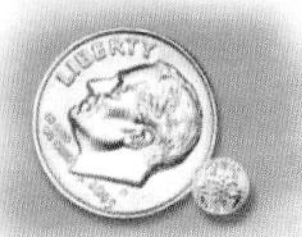

1 gram of gold

Metric prefixes are added to a fundamental unit (meters, liters, or grams) to create metric units that are larger or smaller than the fundamental unit. Each prefix indicates a power of ten, matching our own number system. Table 6.4 matches the place values in our number system: large values greater than 1 are on the left, and small values less than 1 are on the right.

TABLE 6.4 VALUES OF METRIC PREFIXES

(k) kilo- thousand	(h) hecto- hundred	(da) deka- ten	Fundamental Unit	(d) deci- tenth	(c) centi- hundredth	(m) milli- thousandth
1,000	100	10	1	$\frac{1}{10}$	$\frac{1}{100}$	$\frac{1}{1,000}$

The sentence ***K**ing **H**enry **D**ied **U**nceremoniously **D**rinking **C**hocolate **M**ilk* can help you remember the order of the prefixes.

To identify a measurement in the metric system, write a metric prefix followed by a fundamental unit. The following chart shows the relative size of some commonly used metric measurements.

	Length: Meters	Volume: Liters	Weight: Grams
Kilos or 1,000's	1 kilometer = 1000 meters 1 km	1 kiloliter = 1,000 liters 1 kL ≈ 5 barrels	1 kilogram = 1,000 grams 1 kg
Fundamental Unit	1 meter 1 m	1 liter 1 L	1 gram 1 g

(Continued)

[1] The gram actually measures an object's mass. The weight of an object is the force acting on the object due to gravity. An astronaut's mass is the same standing on Earth's surface or floating in space, where the astronaut is weightless. Since all the problems in this text take place on Earth, we will refer to a gram as a unit of weight, not mass.

	Length: Meters	Volume: Liters	Weight: Grams
Centis or $\frac{1}{100}$'s	1 centimeter = $\frac{1}{100}$ meter 1 cm	1 centiliter = $\frac{1}{100}$ liter 1 cL	1 centigram = $\frac{1}{100}$ gram 1 cg ≈ weight of 1 mustard seed
Millis or $\frac{1}{1,000}$'s	1 millimeter = $\frac{1}{1,000}$ meter 1 mm The thickness of a dime is about 1 mm.	1 milliliter = $\frac{1}{1,000}$ liter 1 mL	1 milligram = $\frac{1}{1,000}$ gram Twenty grains of sand is about 1 mg.

Concept Check

A1. What fundamental unit is used to measure length in the metric system?

A2. What fundamental unit is used to measure volume in the metric system?

A3. What fundamental unit is used to measure weight in the metric system?

A4. Complete the saying used to remember the metric prefixes.
King Henry __________ __________ __________ __________ milk.

A5. Write the number that each metric prefix represents.

a. centi- **b.** deci- **c.** hecto-
d. deka- **e.** milli- **f.** kilo-

Objective A Practice

Identify the metric measurement that most conveniently measures each quantity.

A6. Weight of an apple

A7. Weight of a person

A8. Length of a notebook

A9. Width of a classroom

A10. Perimeter of California

A11. Thickness of lead in a pencil

A12. Volume of a large soda bottle

A13. Volume of a syringe

(Continued)

For each exercise, identify three objects that could be conveniently measured using the given metric measure.

A14. Milliliters

A15. Liters

A16. Kilograms

A17. Milligrams

A18. Centimeters

A19. Kilometers

Objective B Convert Units of Measure in the Metric System

The Concept To convert 6,000 meters to kilometers, we can use the units fraction $\frac{1 \text{ kilometer (km)}}{1{,}000 \text{ meters (m)}}$:

$$\begin{aligned} 6{,}000 \text{ m} &= \frac{6{,}000 \text{ m}}{1} \cdot \frac{1 \text{ km}}{1{,}000 \text{ m}} \\ &= \frac{6{,}000 \not{\text{m}}}{1} \cdot \frac{1 \text{ km}}{1{,}000 \not{\text{m}}} \\ &= \frac{6{,}000 \text{ km}}{1{,}000} \\ &= 6 \text{ km} \end{aligned}$$

Since the metric system was designed to work with our decimal number system, we can also perform conversions by moving the decimal point. Listing the metric prefixes from largest to smallest will tell us how many places and in which direction to move the decimal point.

Kilometers are 3 places to the left of meters.

km hm dam m dm

Move the decimal 3 places to the left.

$$\begin{aligned} 6{,}000 \text{ m} &= 6.000 \text{ km} \\ &= 6 \text{ km} \end{aligned}$$

Whether you are converting meters, liters, or grams, all metric units are converted using the same method. Move the decimal point so that it matches the movement shown in the metric conversion table.

Metric Conversion Table

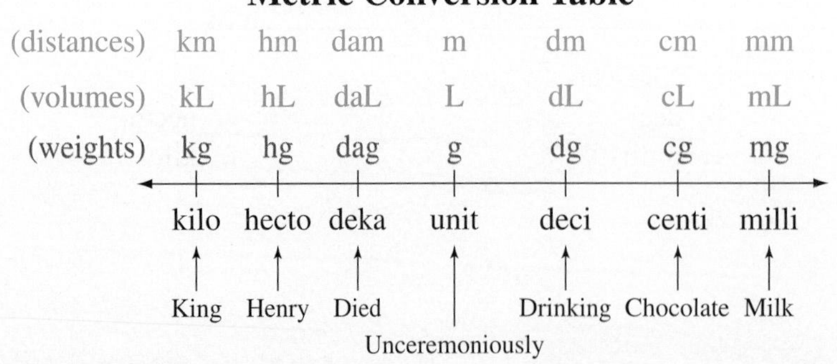

Procedure Convert Units of Measure in the Metric System

Step 1 Identify the direction and the number of places to move on the metric conversion table.

Step 2 Move the decimal point in the same direction and number of places as in step 1.

EXAMPLES

1. Convert 600 grams (g) to kilograms (kg).

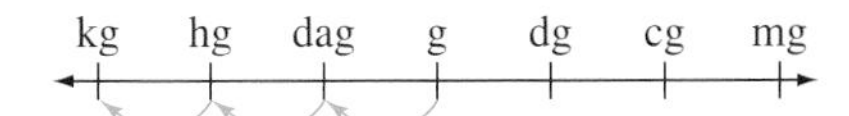

Step 1 To move from grams to kilograms on the metric conversion table, we must move three places to the left.

Step 2 Move the decimal point three places to the left.

600. g = 0.6 kg

The decimal point is on the right side of a whole number.

2. Convert 60.1 kilometers (km) to centimeters (cm).

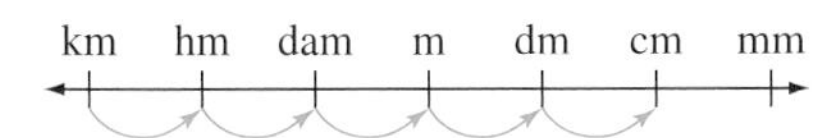

Step 1 To move from kilometers to centimeters on the metric conversion table, we must move five places to the right.

Step 2 Move the decimal point five places to the right.

60.1 km = 60.10000 cm

= 6,010,00 cm

Write 4 zeros to the right of 60.1

GUIDED PRACTICE

1. Convert 23 meters (m) to kilometers (km).

km hm dam m dm cm mm

Step 1 To move from _____ to _________ on the metric conversion table, we must move _____ places to the _____.

Step 2 Move the decimal point _____ places to the _____.

23. m =

The decimal point is on the right side of a whole number.

2. Convert 28.2 kiloliters (kL) to milliliters (mL).

kL hL daL L dL cL mL

Step 1 To move from _________ to _________ on the metric conversion table, we must move ______ places to the _____.

Step 2 Move the decimal point ___ places to the _____.

28.2 kL = mL

Write _____ zeros to the right of 28.2

Concept Check

Exercises B1 and B2 will help you understand why "moving the decimal point" is just a shortcut for multiplying by a units fraction to convert units.

B1. Convert 5.25 kilometers to meters.

a. What number does "kilo" represent?

1 km = _____ m

b. Use the answer from part a to create a units fraction and complete the conversion.

$$\frac{5.25 \text{ km}}{1} \cdot \frac{\quad}{\quad} = \quad \text{m}$$

c. To move from kilometers to meters on the metric conversion table, we must move _____ place(s) to the _____.

d. Move the decimal point _____ place(s) to the _____ to convert 5.25 kilometers to meters.

e. Do your answers from parts b and d match?

B2. Convert 3,253 milligrams to grams.

a. What number does "milli" represent?

_____ mg = 1 g

b. Use the answer from part a to create a units fraction and complete the conversion.

$$\frac{3{,}253 \text{ mg}}{1} \cdot \frac{\quad}{\quad} = \quad \text{g}$$

c. To move from milligrams to grams on the metric conversion table, we must move _____ place(s) to the _____.

d. Move the decimal point _____ place(s) to the _____ to convert 3,253 milligrams to grams.

e. Do your answers from parts b and d match?

Objective B Practice

Perform each conversion.

B3. 244 kilometers to meters

B4. 110 liters to kiloliters

B5. 329 milliliters to deciliters

B6. 564 hectograms to decigrams

B7. 1,264 dekagrams to decigrams

B8. 1,327 kilometers to millimeters

Combining Concepts and Applications

CONCEPT I Applications

EXAMPLE

3. A jeweler is making a custom necklace for a client. To make each link in the necklace, she needs 8 mm of platinum wire. If the necklace needs 500 links, will the 5 meters of wire in stock be enough to make the necklace?

Step 1 Determine how much wire is needed.

$$8 \text{ mm per link} = \frac{8 \text{ mm}}{\text{link}}$$

$$= \frac{8 \text{ mm}}{\cancel{1 \text{ link}}} \cdot \frac{500 \cancel{\text{ links}}}{1}$$

$$= 4{,}000 \text{ mm}$$

Step 2 Convert millimeters to meters to compare this amount to the 5 meters of wire in stock.

$$4{,}000. \text{ mm} = 4 \text{ m}$$

Since 4 m < 5 m, there is enough wire to make the necklace.

GUIDED PRACTICE

3. A pharmaceutical company is filling an order for 8,000 doses of placebos. To make each dose, the company requires 1.3 mL of glucose solution. Will the 10 liters of glucose solution in stock be enough to make the necessary doses of placebos?

Step 1 Determine how much glucose solution is needed.

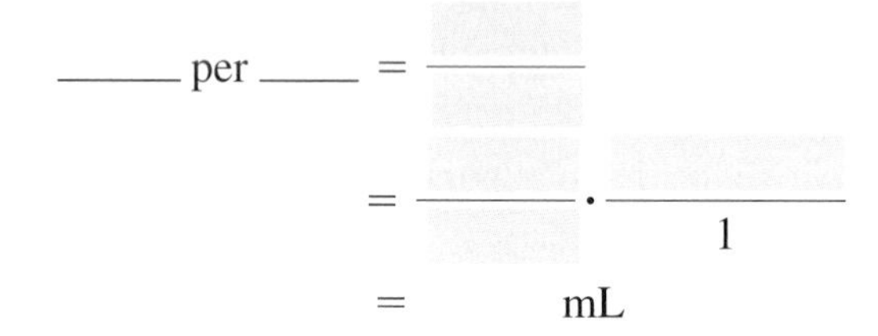

_____ per _____ = $\frac{\_\_\_\_\_}{\_\_\_\_\_}$

$= \frac{\_\_\_\_\_}{\_\_\_\_\_} \cdot \frac{\_\_\_\_\_}{1}$

= _____ mL

Step 2 Convert milliliters to liters to compare this amount to the 10 liters of glucose solution in stock.

_____ mL = _____

Since _____ L ☐ 10 L, there is/is not enough glucose solution to make the necessary doses of placebos.

Section 6.2 Exercises

FOR EXTRA HELP

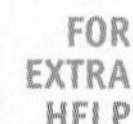

MyMathLab MathXL PRACTICE WATCH DOWNLOAD READ REVIEW

To understand the metric system units of measure for length, volume, and weight:

1. Answer the Objective A Concept Checks.
2. Answer the odd Objective A Practice Exercises.
3. Answer the even Objective A Practice Exercises.

To convert units of measure in the metric system:

4. Answer the Objective B Concept Checks.
5. Answer the odd Objective B Practice Exercises.
6. Answer the even Objective B Practice Exercises.

VOCABULARY REVIEW *Review the Vocabulary Preview for Section 6.2. Study the definitions until you can check* Got It *for every word.*

Use these words to complete each sentence.

metric system of measure • liter • meter • gram • metric prefixes • fundamental units

	Got it	Get help
7. ____________ are added to fundamental units to result in larger or smaller measurements.		
8. The fundamental unit of length in the metric system is ____________.		
9. The fundamental unit of volume in the metric system is ____________.		
10. The fundamental unit of weight in the metric system is ____________.		
11. Write the number that each metric prefix represents. **a.** hecto- ____________ **b.** kilo- ____________ **c.** centi- ____________ **d.** deka- ____________ **e.** deci- ____________ **f.** milli- ____________		
12. Write the metric prefix that corresponds to each place value. **a.** thousands ________ **b.** thousandths ________ **c.** hundreds ________ **d.** hundredths ________ **e.** tenths ________ **f.** tens ________		
13. Write six metric prefixes, in order, from largest to smallest. ____________, ____________, ____________, Fundamental Unit, ____________, ____________, ____________		

How will you get help for any vocabulary that you are unsure about?

Your instructor ____ MyMathLab ____ A classmate ____ A tutor ____ Other ____

For each exercise, identify the most appropriate measurement for the situation. Explain your reasoning.

14. Width of a kitchen
- **a.** 32 millimeters
- **b.** 3.2 meters
- **c.** 3 kilometers
- **d.** 102 centimeters

15. Volume of a soda can
- **a.** 2.1 hectoliters
- **b.** 320 milliliters
- **c.** 5.9 liters
- **d.** 2.9 kiloliters

16. Weight of an adult person.
- **a.** 10 grams
- **b.** 300 centigrams
- **c.** 5 kilograms
- **d.** 65 kilograms

17. Height of a classroom door
- **a.** 2 meters
- **b.** 50 centimeters
- **c.** 5.5 millimeters
- **d.** 7 kilometers

18. Weight of a bowling ball
a. 8 kg
b. 3000 mL
c. 2 m

19. Weight of a tomato
a. 0.5 kg
b. 12 hm
c. 42 L

20. Length of a cruise ship
a. 0.3 km
b. 25,000 g
c. 500 liters

21. Weight of an elephant
a. 5,000 kg
b. 4,000 mL
c. 1,000 mm

Perform each conversion.

22. 546 meters to kilometers

23. 116 kiloliters to liters

24. 1,264,000 decigrams to dekagrams

25. 32.4 decigrams to dekagrams

26. 4.5 centiliters to milliliters

27. 1.327 kilometers to millimeters

28. 5.64 hectograms to decigrams

29. 6.8 milligrams to kilograms

30. 712 milliliters to deciliters

31. 244 kilometers to meters

32. 14 km = ________ dm

33. 22 hL = ________ cL

34. 7 g = ________ dag

35. 5 m = ________ km

36. 16.4 hg = ________ cg

37. 12.3 kg = ________ dag

38. 550 cm = ________ dam

39. 4,092 mg = ________ kg

40. 123,546 g = ________ hg

41. 456,678 cm = ________ hm

42. 5,600 mL = ________ daL

43. 6,700 cL = hL________

44. 0.0876 kg = ________ g

45. 0.965 hm = ________ mm

46. 6 mL = ________ daL

47. 9 g = ________ hg

48. 0.0005 mm = ________ cm

49. 0.00008 cL = ________ dL

Solve each application.

50. To stop a disease outbreak, veterinarians need 3 mL of vaccine for every dog. How many liters of solution are needed to treat 27,000 dogs?

51. A chemistry teacher has 3 liters of sodium hydroxide. To perform an experiment, each student needs 60 mL of sodium hydroxide. Will the teacher have enough solution for 200 students?

52. Main Street Coffee sells 500-gram bags of coffee for $7.99. If Pat bought 2 kg of coffee, how much did he spend?

53. Tujax Pizza is placing an order for 52 kg of cheese. If shredded mozzarella costs $6.79 for 2,000 grams, how much will this order cost?

54. A librarian has estimated that each book in the library is 2.5 cm thick. If the library has 85,000 books, about how many kilometers of shelves does the library need to store all the books?

55. A brick road is made from rows of paving stones that are each about 16 centimeters long. If a road is 2 kilometers long, about how many rows of bricks are in the road?

tobey Converting metric units of length is easier with a chart with the notations km, hm, dam, m, dm, cm, mm to check your work.
For more tips and tweets, go to twitter.com/gstbasicmath

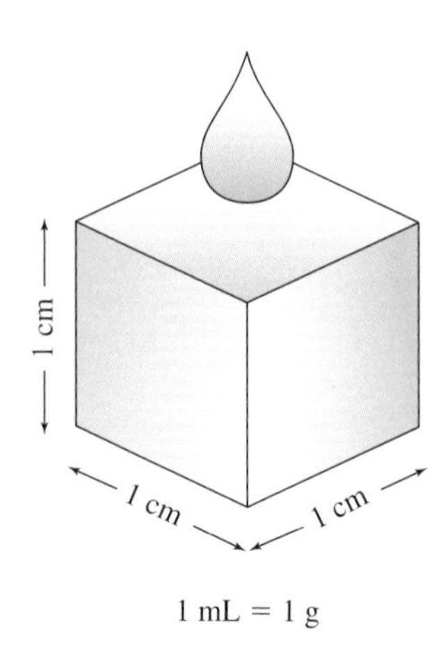

The metric units for length, volume, and weight are all related by water. One cubic centimeter (1 cc) of water is a cube with 1 cm sides. 1 cc of water has a volume of 1 mL and a weight of 1 g. You can use $1\ mL_{water} = 1\ g_{water}$ to perform each of the following conversions.

56. a. What is the volume of 2.1 kg of water in milliliters?
b. Convert your answer from part a to liters.

57. a. What is the volume of 4,300 grams of water in milliliters?
b. Convert your answer from part a to liters.

58. a. What is the weight of 356 mL of water in grams?
b. Convert your answer from part a to kilograms.

59. a. What is the weight of 8.56 liters of water in grams?
b. Convert your answer from part a to kilograms.

Solve each application.

60. Commercial kitchens often measure ingredients using weight instead of volume. If one loaf of bread requires 400 grams of flour, how many kilograms of flour are needed to make 2,500 loaves?

61. A commercial bakery is going to make 2,500 loaves of French bread. If each loaf requires 180 mL of water, how many liters of water are needed?

62. In a manufacturing plant, a shipping crate can hold 800 parts that weigh 250 grams each. Can a container designed to hold 1,000 kg be used to ship 4 full crates?

63. A microbrewery has three 500-liter tanks that are used to make beer. If one bottle holds 375 mL of beer, do the three tanks hold enough beer to fill a 3,600-bottle order?

Section 6.2 Question Log

Use this space to write questions. Be sure to get these answered and revisit them when you prepare for your exam.

Page ____ **Answered** ☐	**Page** ____ **Answered** ☐
Page ____ **Answered** ☐	**Page** ____ **Answered** ☐

Section 6.3 Converting Between the U.S. System and the Metric System

As we move toward a more global economy, converting between the metric system and the U.S. system is becoming more important. Many occupations use both systems.

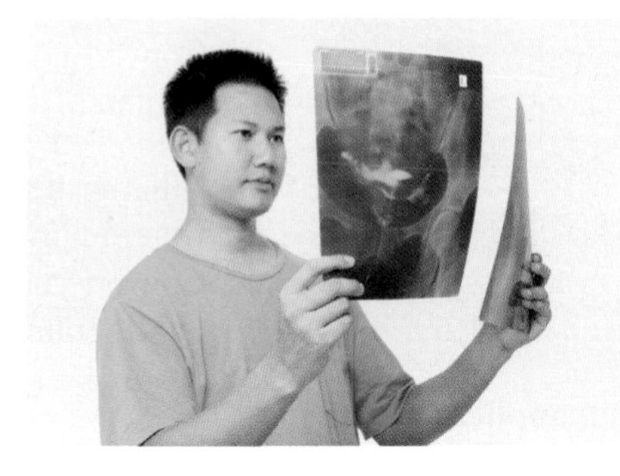

Chemists | Nurses | X-ray technicians

If you are studying chemistry, nursing, X-ray technology, machine technology, mechanics, or any science, you will probably need to convert between the U.S. system and the metric system.

The **Objectives** in Section 6.3 will help you

- **A** Convert between units of measure in the metric and U.S. systems.
- **B** Convert between Celsius and Fahrenheit degrees of temperature.

VOCABULARY PREVIEW *Check the box that applies.*	Got It	Must Study
Units Fraction: A **units fraction** contains units of equal measures in the numerator and denominator. Units fractions are equal to one.		
Celsius: Celsius is a measure of temperature based on the Celsius scale of degrees. Celsius is generally used in countries that use the metric system.		
Fahrenheit: Fahrenheit is a measure of temperature based on the Fahrenheit scale of degrees. Fahrenheit is generally used in countries that use the U.S. system.		

Study these words when they appear in the text. After the section, test your understanding by completing the Vocabulary Review in the section exercises.

Objective A Convert Between Units of Measure in the Metric and U.S. Systems

The Concept To convert between U.S. and metric units, it is helpful to have equivalent values. Several of the most commonly used conversion equivalents are listed in the following table.

TABLE 6.5 U.S. AND METRIC SYSTEM UNITS OF EQUIVALENT MEASURES

	U.S. to Metric	Metric to U.S.
Length	1 mile (mi) ≈ 1.61 kilometers (km) 1 yard (yd) ≈ 0.914 meter (m) 1 foot (ft) ≈ 0.305 meter (m) 1 inch (in.) = 2.54 centimeters (cm)	1 kilometer (km) ≈ 0.62 mile (mi) 1 meter (m) ≈ 1.09 yards (yd) 1 meter (m) ≈ 3.28 feet (ft) 1 centimeter (cm) ≈ 0.394 inch (in.)
Volume	1 gallon (gal) ≈ 3.79 liters (L) 1 quart (qt) ≈ 0.946 liter (L)	1 liter (L) ≈ 0.264 gallon (gal) 1 liter (L) ≈ 1.06 quarts (qt)
Weight	1 pound (lb) ≈ 0.454 kilogram (kg) 1 ounce (oz) ≈ 28.35 grams (g)	1 kilogram (kg) ≈ 2.2 pounds (lb) 1 gram (g) ≈ 0.0353 ounce (oz)

Procedure **Convert Between Units of Measure in the Metric and U.S. Systems**

Step 1 Write the original quantity as a fraction over 1.

Step 2 Multiply by a units fraction to divide out the original units and introduce the new units.

Step 3 Simplify.

When choosing a units fraction, your calculations will be easier if you pick the units fraction that contains "1 unit" in the denominator. For example, it is easier to use $\frac{3.79 \text{ L}}{1 \text{ gal}}$ than $\frac{1 \text{ L}}{0.264 \text{ gal}}$ when converting from gallons to liters.

EXAMPLES

1. Convert 6 meters to feet.

Step 1 Write the original quantity as a fraction over 1.

$$6 \text{ m} = \frac{6 \text{ m}}{1}$$

Step 2 Multiply by a units fraction to divide out the original units.

$$\approx \frac{6 \cancel{\text{m}}}{1} \cdot \left(\frac{3.28 \text{ ft}}{1 \cancel{\text{m}}}\right)$$

Step 3 Simplify.

$$= \frac{6}{1} \cdot \left(\frac{3.28 \text{ ft}}{1}\right)$$

$$= 19.68 \text{ feet}$$

GUIDED PRACTICE

1. Convert 4 pounds to kilograms.

$$4 \text{ lb} = \frac{\phantom{4 \text{ lb}}}{1}$$

$$\approx \frac{\phantom{4 \text{ lb}}}{1} \cdot \left(\frac{\phantom{0.454 \text{ kg}}}{1 \text{ lb}}\right)$$

$$= \frac{}{1} \cdot \left(\frac{\phantom{0.454 \text{ kg}}}{}\right)$$

$$=$$

2. Convert 60 quarts to liters.

Step 1 Write the original quantity as a fraction over 1.

$$60 \text{ qt} = \frac{60 \text{ qt}}{1}$$

Step 2 Multiply by a units fraction to divide out the original units.

$$\approx \frac{60 \text{ qt}}{1} \cdot \left(\frac{0.946 \text{ L}}{1 \text{ qt}}\right)$$

Step 3 Simplify.

$$= \frac{60}{1} \cdot \left(\frac{0.946 \text{ L}}{1}\right)$$

$$= 56.76 \text{ liters}$$

2. Convert 200 grams to ounces.

$$200 \text{ g} = \frac{\quad}{1}$$

$$\approx \frac{\quad}{1} \cdot \left(\frac{\quad}{1 \text{ g}}\right)$$

$$= \frac{\quad}{1} \cdot \left(\frac{\quad}{\quad}\right)$$

$$=$$

3. To calculate a medical dose, a nurse needs to convert 60 pounds to kilograms. How many kilograms are equal to 60 pounds?

Step 1 Write the original quantity as a fraction over 1.

$$60 \text{ lb} = \frac{60 \text{ lb}}{1}$$

Step 2 Multiply by a units fraction to divide out the original units.

$$\approx \frac{60 \text{ lb}}{1} \cdot \left(\frac{0.454 \text{ kg}}{1 \text{ lb}}\right)$$

Step 3 Simplify.

$$= \frac{60}{1} \cdot \left(\frac{0.454 \text{ kg}}{1}\right)$$

$$= 27.24 \text{ kilograms}$$

3. To calculate the diameter of a part, a machinist needs to convert 10 inches to centimeters. How many centimeters are equal to 10 inches?

$$10 \text{ in.} = \frac{\quad}{1}$$

$$\approx \frac{\quad}{1} \cdot \left(\frac{\quad}{\quad}\right)$$

$$= \frac{\quad}{1} \cdot \left(\frac{\quad}{\quad}\right)$$

$$=$$

EXAMPLE

4. A tourist from North Dakota is vacationing in Canada. In kilometers per hour, how fast is the tourist traveling if he is driving 60 miles per hour?

Step 1 Write the original quantity as a fraction over 1. The word *per* indicates a fraction bar.

$$60 \text{ miles per hour} = \frac{60 \text{ mi}}{1 \text{ hr}}$$

Step 2 Multiply by a units fraction to divide out the original unit of miles.

$$= \frac{60 \text{ mi}}{1 \text{ hr}} \cdot \left(\frac{1.61 \text{ km}}{1 \text{ mi}}\right)$$

Step 3 Simplify.

$$= \frac{60}{1 \text{ hr}} \cdot \left(\frac{1.61 \text{ km}}{1}\right)$$

$$= \frac{96.6 \text{ km}}{1 \text{ hr}}$$

$$= 96.6 \text{ kilometers per hour}$$

GUIDED PRACTICE

4. A Spanish couple is vacationing in the United States. In miles per hour, how fast is the couple traveling if they are driving 120 kilometers per hour?

Step 1 Write the original quantity as a fraction over 1. The word *per* indicates a fraction bar.

Step 2 Multiply by a units fraction to divide out the original unit of miles.

Step 3 Simplify.

120 km per hour = ______

= ______

= ______ (______)

= ______

= ______ miles per hour

Concept Check

A1. Which is longer, 1 mile or 1 kilometer?

A2. Which has more volume, 1 quart or 1 liter?

A3. Which weighs more, 1 pound or 1 kilogram?

A4. Choose the units fraction (a) or (b) that will make the conversion easier.
Hint: People usually find multiplication easier to perform than division.

a. $\frac{5 \text{ mi}}{1} \cdot \left(\frac{1 \text{ km}}{0.62 \text{ mi}}\right)$ **b.** $\frac{5 \text{ mi}}{1} \cdot \left(\frac{1.61 \text{ km}}{1 \text{ mi}}\right)$

A5. Choose the units fraction (a) or (b) that will make the conversion easier.
Hint: People usually find multiplication easier to perform than division.

a. $\frac{8 \text{ m}}{1} \cdot \left(\frac{1 \text{ ft}}{0.305 \text{ m}}\right)$ **b.** $\frac{8 \text{ m}}{1} \cdot \left(\frac{3.28 \text{ ft}}{1 \text{ m}}\right)$

Objective A Practice

Perform each conversion.

A6. 100 yards to meters

A7. 10 liters to gallons

A8. 5 ounces to grams

A9. 7 inches to centimeters

A10. 3 meters per second to feet per second

A11. 4 feet per second to meters per second

Objective B Convert Between Celsius and Fahrenheit Degrees of Temperature

The Concept The temperature scales of Celsius and Fahrenheit are used in many places around the world. On the Celsius scale, water boils at 100 degrees Celsius (100°C) and freezes at 0°C. On the Fahrenheit scale, water boils at 212 degrees Fahrenheit (212°F) and freezes at 32°F. To convert between Celsius and Fahrenheit degrees of temperature, use the following formulas:

Formulas Used to Convert Between Degrees Celsius and Degrees Fahrenheit

Celsius to Fahrenheit	Fahrenheit to Celsius
$F = \frac{9}{5} \cdot C + 32$	$C = \frac{5 \cdot F - 160}{9}$

Procedure Convert Between Celsius and Fahrenheit Degrees of Temperature

Step 1 Write the appropriate formula.

Step 2 Substitute the known temperature into the formula.

Step 3 Simplify.

EXAMPLES

5. Convert 45°C to degrees Fahrenheit.

Step 1 Write the appropriate formula. $F = \frac{9}{5} \cdot C + 32$

Step 2 Substitute the known temperature. $= \frac{9}{5} \cdot (45) + 32$

Step 3 Write the value as a fraction. $= \frac{9}{5} \cdot \frac{45}{1} + 32$

Multiply. $= \frac{9 \cdot \cancel{5} \cdot 9}{\cancel{5} \cdot 1} + 32$

Add. $= 81 + 32$

$= 113$

45°C = 113°F

6. Convert 50°F to degrees Celsius.

Step 1 Write the appropriate formula. $C = \frac{5 \cdot F - 160}{9}$

Step 2 Substitute the known temperature. $= \frac{5 \cdot (50) - 160}{9}$

Step 3 Multiply. $= \frac{250 - 160}{9}$

Subtract. $= \frac{90}{9}$

Divide. $= 10$

50°F = 10°C

GUIDED PRACTICE

5. Convert 30°C to degrees Fahrenheit.

$F = \frac{9}{5} \cdot C + 32$

$= \frac{9}{5} \cdot (\quad) + 32$

$= \frac{9}{5} \cdot \frac{\quad}{\quad} + 32$

$= \frac{\quad}{\quad} + \quad$

$=$

$=$

30°C = ___°F

6. Convert 86°F to degrees Celsius.

$C = \frac{5 \cdot F - 160}{9}$

$= \frac{5 \cdot (\quad) - 160}{9}$

$= \frac{\quad - 160}{9}$

$= \frac{\quad}{9}$

$=$

86°F = ___

Concept Check

B1. At what degrees Celsius does water freeze?

B2. Write the formula used to convert from degrees Celsius to degrees Fahrenheit.

Objective B Practice

Perform each conversion.

B3. 15°C to degrees Fahrenheit

B4. 40°C to degrees Fahrenheit

B5. 95°F to degrees Celsius

B6. 140°F to degrees Celsius

Combining Concepts and Applications

CONCEPT I **Comparing Two Measurements** Many situations require a comparison of units between the metric system and the U.S. system. Sometimes the comparison doesn't require a conversion between the systems.

EXAMPLE

7. Which is greater, 300 gallons or 300 liters?

Since the quantities differ only in the units, see which unit is larger using a conversion equivalent.

1 gallon $\approx$ 3.79 liters

Since 1 gallon is greater than 1 liter, 300 gallons is greater than 300 liters.

GUIDED PRACTICE

7. Which is greater, 200 pounds or 200 kilograms?

Since the quantities differ only in the units, see which unit is larger using a conversion equivalent.

_____ pound $\approx$ _____ kilogram

Since 1 _______ is greater than 1 _____, 200 ________, is greater than 200 ______.

In other cases, a comparison of units between the metric system and the U.S. system requires a conversion.

EXAMPLE

8. Use > or < to compare 123 quarts to 115 liters.

Step 1 Find a conversion equivalent. 1 quart (qt) $\approx$ 0.946 liter (L)

Step 2 Convert one unit to match the other.

$$123 \text{ qt} = \frac{123 \text{ qt}}{1} \cdot \frac{0.946 \text{ L}}{1 \text{ qt}}$$

$$= 116.34 \text{ L}$$

Step 3 Use the converted value to compare the quantities.

123 qt [>] 115 L

GUIDED PRACTICE

8. Use > or < to compare 12 inches to 31 centimeters.

____ $\approx$ ______

$$= \frac{\quad}{1} \cdot \frac{\quad}{\quad}$$

$=$

12 in. [ ] 31 cm

Section 6.3 Exercises

FOR EXTRA HELP      

To convert between units of measure in the metric and U.S. systems:

1. Answer the Objective A Concept Checks.
2. Answer the odd Objective A Practice Exercises.
3. Answer the even Objective A Practice Exercises.

To convert between Celsius and Fahrenheit degrees of temperature:

4. Answer the Objective B Concept Checks.
5. Answer the odd Objective B Practice Exercises.
6. Answer the even Objective B Practice Exercises.

VOCABULARY REVIEW *Review the Vocabulary Preview for Section 6.3. Study the definitions until you can check* Got It *for every word.*

Use these words to complete each sentence.

Celsius • Fahrenheit • units fraction

	Got It	Get Help
7. Most countries that use the U.S. system measure temperature with the ________ scale.		
8. Most countries that use the metric system measure temperature with the ________ scale.		
9. Use the numbers 1 (smallest) through 8 (largest) to order the following lengths. ________ meter ________ inch ________ centimeter ________ mile ________ height of a person ________ length of a pencil ________ kilometer ________ yard		
10. Use the numbers 1 (smallest) through 7 (largest) to order the following volumes. ________ fluid ounce ________ liter ________ quart ________ gallon ________ glass of water ________ volume of a bathtub ________ cup		
11. Use the numbers 1 (smallest) through 7 (largest) to order the following weights. ________ ounce ________ gram ________ pound ________ kilogram ________ ton ________ weight of a cruise ship ________ weight of a grain of sand		

How will you get help for any vocabulary that you are unsure about?

Your instructor ____ MyMathLab ____ A classmate ____ A tutor ____ Other ____

Compare U.S. and metric units of measure.

12. Which metric measure is a little longer than one yard?

13. Which U.S. measure is a little smaller than a liter?

14. Which U.S. measure is about twice the length of a centimeter?

15. Which metric measure is a little more than twice the weight of a pound?

Perform each conversion. Round answers to the nearest tenth when necessary.

16. 6 cm to in.

17. 10 cm to in.

18. 9 in. to cm

19. 8 in. to cm

20. 12 gal to L

21. 11 gal to L

22. 17.2 lb to kg

23. 14.6 lb to kg

24. 95 mi to km

25. 85 mi to km

26. 18 m to yd

27. 20 m to yd

28. 1.2 oz to g

29. 0.3 oz to g

30. 11 kg to lb

31. 32 kg to lb

32. 100 km to mi

33. 5 km to mi

34. 50 g to oz

35. 100 g to oz

36. Convert 24 ft to cm in two steps.
a. Convert 24 feet to meters.
b. Convert the answer from part a to centimeters.

37. Convert 6 gal to L in two steps.
a. Convert 6 gallons to quarts.
b. Convert the answer from part a to liters.

38. Convert 250 g to pounds in two steps.
a. Convert 250 g to kilograms.
b. Convert the answer from part a to pounds.

39. Convert 321 mm to inches in two steps.
a. Convert 321 mm to centimeters.
b. Convert the answer from part a to inches.

Answer each question. Round answers to the nearest tenth when necessary.

40. Toivo has a 120 cm board. Is the board long enough to cut a piece that is 3 ft long?

41. Eino needs a 13-inch board to complete a project. Will a 28.2 cm piece of scrap from another board be long enough?

42. Will a 75-liter container hold 25 gallons of water?

43. Will a 35-gallon container hold 100 liters of water?

44. 34 quarts equals how many liters?

45. 46 quarts equals how many liters?

46. Abbot has run a distance of 100 m. Umberto has run a distance of 326 ft. Who has run the longest distance?

47. Costello has run a distance of 300 m. Antuan has run a distance of 312 yd. Who has run the longest distance?

48. How many gallons equal 15 liters?

49. How many gallons equal 21 liters?

50. 5.7 liters equal how many quarts?

51. 4.5 liters equal how many quarts?

52. What is 55 mi per hr when it is converted to km per hr?

53. What is 70 mi per hr when it is converted to km per hr?

goetz When converting units, remember that a units fraction will have the form of "new units over original units."
For more tips and tweets, go to twitter.com/gstbasicmath

54. Convert 105 km per hr to mi per hr.

55. Convert 85 km per hr to mi per hr.

56. Can a 2-meter couch fit a 6-foot space?

57. Can 4 liters be poured into a 1-gallon jug?

58. A contestant on a game show can choose one of two prizes, 400 grams of gold or 1 pound of gold. Which should she choose?

59. To win a reality show, a contestant must answer this question: Can a robin flying 19 miles per hour catch a pig flying 25 kilometers per hour? What should he answer?

Answer each question about the Fahrenheit and Celsius scales.

60. At what temperature does water boil on the Fahrenheit scale?

61. At what temperature does water boil on the Celsius scale?

62. Write the formula used to convert from Fahrenheit to Celsius degrees of temperature.

63. Write the formula used to convert from Celsius to Fahrenheit degrees of temperature.

Convert between degrees Celsius and degrees Fahrenheit. Round to the nearest tenth if necessary.

64. 10°C to °F

65. 105°C to °F

66. 68°F to °C

67. 41°F to °C

68. If it is 140°F outside, what is the temperature in degrees Celsius?

69. If it is 95°F outside, what is the temperature in degrees Celsius?

70. If an object is 20°C, what is its temperature in degrees Fahrenheit?

71. If an object is 85°C, what is its temperature in degrees Fahrenheit?

Solve each application.

72. Sharon is driving 65 miles per hour on a Canadian highway. If the speed limit is 80 kilometers per hour, is Sharon speeding?

73. Kevin is driving 100 kilometers per hour on a U.S. highway. If the speed limit is 70 miles per hour, is Kevin speeding?

74. Juan is a guide for a boy's hiking trip. He purchased 3 kg of peanuts to share with 20 boys. His scale measures pounds and ounces. Are there enough peanuts for Juan and each boy to get 5 ounces each? Recall that 1 lb = 16 oz.

75. A grocery store is selling 20-oz steaks. Toshi has been asked to pick up 5 kilograms of steak for his father's "Texas Barbeque." Will 8 of the 20-oz steaks provide enough steak for his father's party? Recall that 1 lb = 16 oz.

76. Stanley is not from the United States and does not fully understand the measurement gallons. After driving to a hockey game, he needed to refuel his car at a station selling gas for $4.25 per gallon. He guessed that it would take 40 liters to fill the tank.
a. How many gallons is 40 liters?
b. How much will it cost to fill the tank?

77. Sandra is taking 9 Girl Scouts to Death Valley. To have the recommended 2 gallons of drinking water per person, she needs to purchase 20 gallons of water for $0.69 a liter.
a. How many liters of water are needed?
b. How much will the water cost?

Section 6.3 Question Log

Use this space to write questions. Be sure to get these answered and revisit them when you prepare for your exam.

Page ____ **Answered** ☐	**Page** ____ **Answered** ☐
Page ____ **Answered** ☐	**Page** ____ **Answered** ☐

Chapter 6 Organizer

VOCABULARY

Use the following steps to review the vocabulary for Chapter 6.

1. *Write the definition for each word from memory.*
2. *Compare the definitions you have written to the definitions in the Vocabulary Preview.*
3. *Study any definitions that you could not remember or that you defined incorrectly.*

6.1
U.S. System of Measure • Units Fraction

6.2
Metric System of Measure • Fundamental Units • Meter • Liter • Gram • Metric Prefixes

6.3
Units Fraction • Celsius • Fahrenheit

PROCEDURES

Procedure/Topic	Steps	Example
Convert Units of Measure in the U.S. System (Section 6.1)	**Step 1** Identify equal measures for the units involved in the conversion. **Step 2** Write the original quantity as a fraction over 1. **Step 3** Multiply by a units fraction to divide out the original units and introduce the new units. **Step 4** Simplify.	Convert 13 quarts to pints. 1 quart = 2 pints $13 \text{ qt} = \frac{13 \text{ qt}}{1}$ $= \frac{13 \cancel{\text{qt}}}{1} \cdot \left(\frac{2 \text{ pt}}{1 \cancel{\text{qt}}}\right)$ $= \frac{13}{1} \cdot \left(\frac{2 \text{ pt}}{1}\right)$ $= 26$ pints
Convert Units of Measure in the Metric System (Section 6.2)	**Step 1** Identify the direction and the number of places to move on the metric conversion table. **Step 2** Move the decimal point in the same direction and number of places as in step 1.	Convert 2.36 km to centimeters. kilo hecto deka unit deci centi milli Centi is 5 places to the right of kilo. Move the decimal 5 places to the right. 2.36 km = 236,000 cm **Write 3 zeros to the right of 2.36**
Convert Between Units of Measure in the Metric and U.S. Systems (Section 6.3)	**Step 1** Write the original quantity as a fraction over 1. **Step 2** Multiply by a units fraction to divide out the original units and introduce the new units. **Step 3** Simplify.	Convert 5 meters to feet. $5 \text{ m} = \frac{5 \text{ m}}{1}$ $\approx \frac{5 \cancel{\text{m}}}{1} \cdot \left(\frac{3.28 \text{ ft}}{1 \cancel{\text{m}}}\right)$ $= \frac{5}{1} \cdot \left(\frac{3.28 \text{ ft}}{1}\right)$ $= 16.4$ feet

Procedure/Topic	Steps	Example
Convert Between Celsius and Fahrenheit Degrees of Temperature (Section 6.3)	Step 1 Write the appropriate formula. Step 2 Substitute the known temperature into the formula. Step 3 Simplify.	Convert 45°C to degrees Fahrenheit. $F = \frac{9}{5} \cdot C + 32$ $= \frac{9}{5} \cdot 45 + 32$ $= \frac{9 \cdot \cancel{5} \cdot 9}{\cancel{5} \cdot 1} + 32$ $= 81 + 32$ $= 113$ 45°C = 113°F

Chapter 6 Review Exercises

6.1

Identify the U.S. unit of measure that most conveniently measures each item.

1. Distance from New York to Texas
2. Distance from your eyes to your lips
3. Amount of time needed to do your math homework each week
4. Amount of time it takes to get a college degree
5. Weight of a car
6. Weight of a turkey
7. Volume of a soda can
8. Weight of a toad
9. Length of a whale
10. Weight of a whale
11. Diameter of a quarter
12. Volume of a quarter

Perform each conversion. Round to the nearest tenth when necessary.

13. 4 feet to inches
14. 1.5 miles to feet
15. 3 weeks to hours
16. 2.5 hours to seconds
17. 52 yards to feet
18. 320 feet to yards
19. 2.5 qt to pt
20. 3 lb to oz
21. 8 gal to pt
22. 4.532 tons to lb
23. 36 inches per day to inches per hour
24. $231 per week to dollars per day

Perform each conversion. You will need to use more than one units fraction. Round to the nearest tenth when necessary.

25. 60 miles per hour to feet per minute
26. 3 tons per day to pounds per hour
27. Avery is shoveling stone at a rate of 100 pounds per minute. How many tons of stone will he shovel in 4 hours?
28. A faucet is leaking at a rate of 1 ounce per minute, How many gallons does the faucet leak in one day?

6.2

Identify the metric unit of measure that most conveniently measures each item.

29. Weight of a person

30. Weight of a mouse

31. Distance from Miami to Seattle

32. Thickness of a quarter

33. Volume of an aquarium

34. Volume of cough syrup in one dose

35. Volume of a large sports bottle

36. Distance someone can run in 30 seconds

Perform each conversion. Round to the nearest tenth when necessary.

37. 1,234 meters to kilometers

38. 0.523 meter to centimeters

39. 3.89 liters to milliliters

40. 350 milligrams to grams

41. 349 mL to cL

42. 750 g to kg

43. 1.2 m to mm

44. 2 mg to kg

Perform each conversion. You will need to use more than one units fraction. Round to the nearest tenth when necessary.

45. 50 milliliters per minute to liters per hour

46. 320 meters per minute to kilometers per hour

47. A nurse gives 0.5 liter of solution to a patient over 15 minutes. How many milliliters are given per second?

48. A race car is traveling 300 kilometers per hour. How many meters per second is this?

6.3

Order each item or measurement from largest to smallest.

49. 1 kilogram, the weight of an apple, 1 ounce, 1 gram, 1 pound

50. 1 gallon, 1 kiloliter, 1 fluid ounce, the volume of a mug of coffee, 1 centiliter

51. 1 fluid ounce, 1 pint, 1 liter, 1 cup, 1 milliliter, 1 gallon, 1 quart, the volume of water in a bathtub, the volume of a 2-liter soda bottle

52. the width of this page, 1 meter, 1 inch, 1 foot, 1 kilometer, 1 mile, the width of a classroom

Perform each conversion. Round to the nearest tenth when necessary.

53. 12 inches to centimeters

54. 5 meters to feet

55. 500 kilograms to pounds

56. 3 pounds to grams

57. 35° Fahrenheit to degrees Celsius

58. 35° Celsius to degrees Fahrenheit

59. 2 quarts to liters

60. 7 liters to gallons

Perform each conversion in two steps.

61. 2 feet to centimeters
   a. Convert 2 ft to m.
   b. Convert the answer from part a to cm.

62. 2,640 feet to kilometers
   a. Convert 2,640 ft to mi.
   b. Convert the answer from part a to km.

63. 400 milliliters to gallons
   a. Convert 400 mL to L.
   b. Convert the answer from part a to gal.

64. 5 liters to cups
   a. Convert 5 L to qt.
   b. Convert the answer from part a to cups.

65. While traveling in Canada, Dale sees that the price of gas is \$0.98 per liter. If gas costs \$3.70 per gallon in the United States, where is gas more expensive and by how much? Assume that \$1 is the same value in both the United States and Canada.

66. Dale is driving 65 miles per hour in Canada, where the speed limit is 100 kilometers per hour. By how much is he speeding, in kilometers per hour?

67. Which is a better buy—5 pounds of sugar for \$1.99 or 2 kilograms of sugar for \$1.49?

68. Who runs faster—an athlete who runs 100 meters in 15 seconds or an athlete who runs 100 yards in 13 seconds?

69. Geraldo sees that milk costs \$2.90 per gallon in the United States, but it is \$0.89 per liter in Canada. Where is milk cheaper? Assume that \$1 is the same value in both the United States and Canada.

70. Which is a better buy—2 pints of orange juice for \$1.25 or 1 liter of orange juice for \$1.17?

71. In a race, which boat is traveling faster—*Tranquility* at 440 feet per minute or *Carpe Carp* at 2.68 meters per second?

72. While visiting Cozumel, Mexico, during spring break, Shauntel received a ticket for traveling 25 kilometers per hour over the speed limit. By how many miles per hour was she speeding?

Chapter 6 Practice Test

In each situation, what unit would be most appropriate for each system of measure?

1. Weight of a large dog
  a. U.S. system
  b. metric system

2. Amount of liquid in a small bottle
  a. U.S. system
  b. metric system

3. Height of an oak tree
  a. U.S. system
  b. metric system

4. Weight of a ping-pong ball
  a. U.S. system
  b. metric system

Perform each conversion. Round to the nearest tenth when necessary.

5. 8 yards to feet

6. 32 millimeters to meters

7. 1.4 kiloliters to liters

8. 13 ounces to cups

9. 85° Fahrenheit to degrees Celsius

10. 115° Celsius to degrees Fahrenheit

11. 43 centigrams to grams

12. 2.5 pounds to ounces

13. 15 feet per second to miles per hour

14. 31 kilometers per hour to meters per second

15. 25 milliliters per second to liters per hour

16. 8 ounces per hour to gallons per day

17. 400 meters to yards

18. 100 miles to kilometers

19. 10 feet to meters

20. 30 kilometers to miles

21. Convert 3 inches to millimeters in two steps.
a. Convert 3 in. to cm.
b. Convert the answer from part a to mm.

22. Convert 250 grams to pounds in two steps.
a. Convert 250 g to kg.
b. Convert the answer from part a to lb.

23. Convert 1 quart to liters in two steps.
a. Convert 1 qt to gal.
b. Convert the answer from part a to L.

24. Convert 1,000 yards to kilometers in two steps.
a. Convert 1,000 yd to m.
b. Convert the answer from part a to km.

Answer each question. Round to the nearest tenth when necessary.

25. A tortoise walks 480 feet in one hour. How many feet does the tortoise walk per minute?

26. A hummingbird flaps its wings 50 times per second. How many times does the hummingbird flap its wings in one minute?

27. Which is a better buy—flour at $0.58 per pound or at $1.25 per kilogram?

28. Which is a better buy—gas at $4.23 per gallon or at $1.22 per liter?

29. An avalanche is traveling at 120 kilometers per hour. Can a snowmobile traveling 25 meters per second outrun the avalanche?

30. A wildfire is advancing at 12 feet per second. Can a firefighter running 10 miles per hour outrun the fire?

31. A crane at a quarry loaded 22 tons of rock onto a truck in one scoop. How many kilograms is 22 tons?

32. If 17 gallons of water flow over a waterfall every second, how many liters flow over the waterfall per minute?

CHAPTER 7

Geometry

Geometry is a branch of mathematics that addresses the size, shape, and properties of objects. These properties include the length of sides, angles between sides, perimeter, area, and volume.

To build a house, workers use many geometry concepts.

- They measure the *length* of every board and the *angles* at which to cut them.
- To order materials, they must know the *perimeter* of each window, the *area* of the roof, and the *volume* of concrete in the foundation.
- They use *similarity* to translate blueprint dimensions to actual building dimensions.

These geometry concepts are presented in exercises throughout Chapter 7.

Section 7.1 Angles

Angles describe the relative position and orientation of the sides of a figure.

The **Objectives** in Section 7.1 will help you

- **A** Understand angles and their measures.
- **B** Classify angles.
- **C** Solve for unknown angles using angle relationships.

VOCABULARY PREVIEW *Check the box that applies.*	Got It	Must Study
Line: A **line** extends forever in two directions.		
Line Segment: A **line segment** is a part of a straight line and has a starting point and an endpoint.		
Ray: A **ray** is a part of a straight line. It has one endpoint and goes on forever in one direction.		
Angle: An **angle** is made of two rays that begin at a common endpoint, called the vertex.		
Vertex: The **vertex** of an angle is the common endpoint of the two rays that form the angle.		
Measure (of an Angle): The **measure** of an angle indicates the size of the opening between the two rays of the angle. The measure of angle A is represented by $m\angle A$.		
Acute Angle: Acute angles have measures greater than 0° and less than 90°.		
Right Angle: Any angle that measures 90° is a **right angle.**		
Obtuse Angle: Obtuse angles have measures greater than 90° and less than 180°.		
Straight Angle: Any angle that measures 180° is a **straight angle.**		
Complementary Angles: Two angles are **complementary** if their measures add to 90°.		
Supplementary Angles: Two angles are **supplementary** if their measures add to 180°.		

Study these words when they appear in the text. After the section, test your understanding by completing the Vocabulary Review in the section exercises.

Objective A Understand Angles and Their Measures

The Concept An **angle** is made of two **rays** that begin at a common endpoint, called the **vertex.** The red and blue rays below form the sides of an angle. The arrows indicate that each ray extends forever in the direction in which it is pointing.

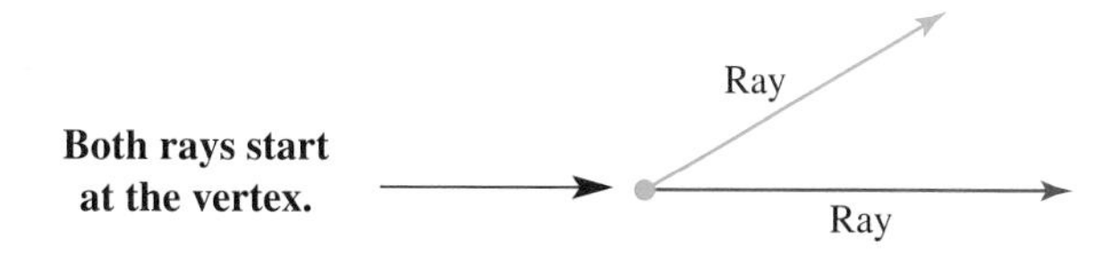

Angles can also be made of **lines** and **line segments**. The next angle shown is made of a ray and a line. It has a measure of 30°, which is read as "thirty degrees." The arrows indicate that the ray extends forever in one direction and the line extends forever in two directions.

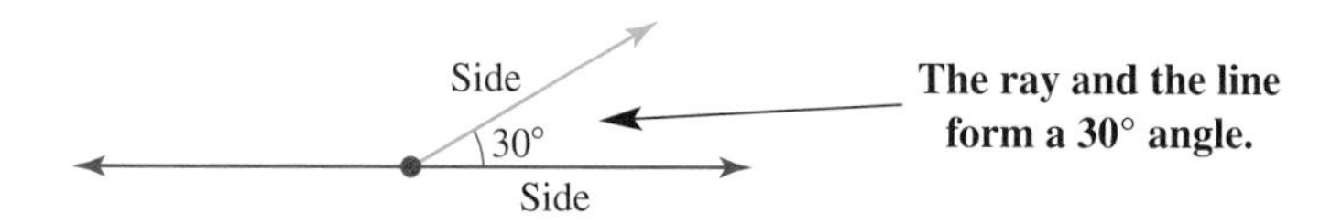

A protractor can be used to check that the angle has a measure of 30°. To use a protractor, place the center of the protractor at the angle's vertex and align the bottom of the protractor with one of the angle's rays. Then read the measure between the rays.

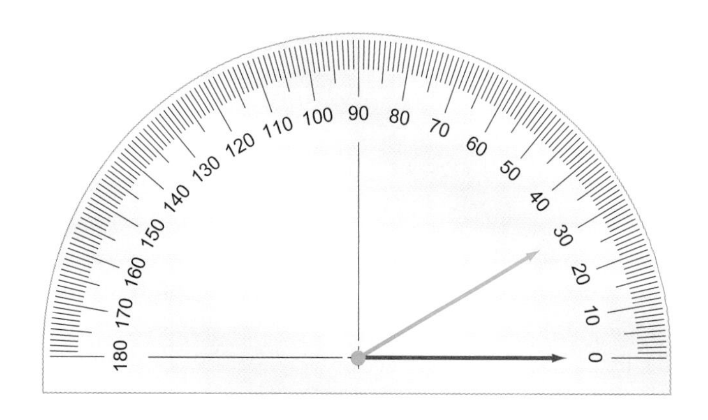

The angle is 30°, or the size of the opening between the rays is 30°.

Procedure Measure an Angle

Use a protractor to determine the measure of the angle between two rays.

Procedure Draw an Angle

Use a protractor to draw two rays with the given angle between them.

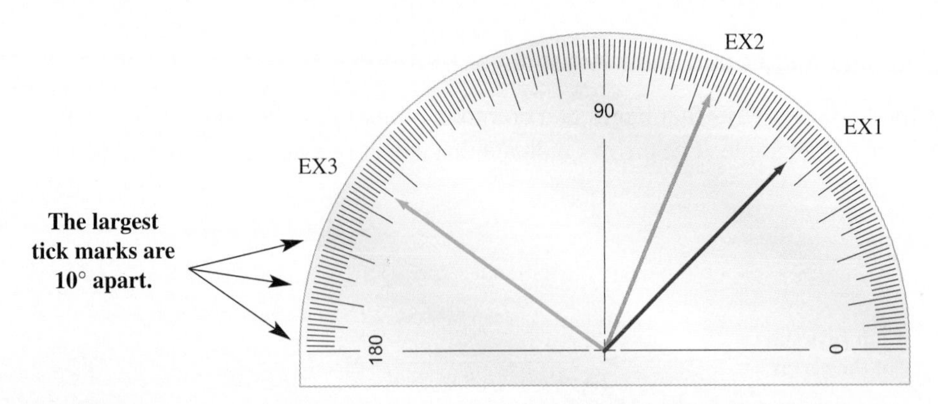

EXAMPLES	GUIDED PRACTICE
1. Draw a ray labeled EX1 at 45°.	**1.** Draw a ray labeled GP1 at 110°.
To draw ray EX1 at 45°, count four and a half big tick marks from zero.	To draw ray GP1 at 110°, count ____ big tick marks from ____.
2. What is the measure of the angle formed by EX1 and EX2?	**2.** What is the measure of the angle between GP1 and EX3?
EX1 is drawn at 45°. EX2 is drawn at 67°. The angle between them is the difference in those values. The angle has a measure of 67° − 45° = 22°.	GP1 is drawn at ____. EX3 is drawn at ____. The angle between them is the difference in those values. The angle has a measure of ____ − ____ = ____.
3. Draw a ray named EX3 that forms an angle of 100° with EX1.	**3.** Draw a ray labeled GP3 that forms an angle of 90° with GP1.
An angle could be drawn to the right or left of EX1. Since 100° isn't shown to the right of EX1, it must be drawn to the left. Draw EX3 at 45° + 100° = 145°.	An angle could be drawn to the right or left of GP1. Since 90° isn't shown to the ____ of GP1, it must be drawn to the ____. Draw GP3 at 110° +/− ____ = ____ **Choose either + or −.**

Concept Check

A1. What is the difference between a ray and a line? Compare the definitions given in the Vocabulary Preview.

A2. The point at which two rays meet to form an angle is called the ______ of the angle.

A3. What measuring tool can be used to measure angles?

Objective A Practice

In the following exercises, each angle starts at 0° on the protractor.

A4. Draw a ray labeled A4 at 170°.

A5. Draw a ray labeled A5 at 80°.

A6. What is the measure of the angle between the blue and orange rays?

A7. What is the measure of the angle between A4 and the green ray?

A8. Draw a ray labeled A8 that is 70° from the green ray.

A9. What is the measure of the angle between A8 and A4?

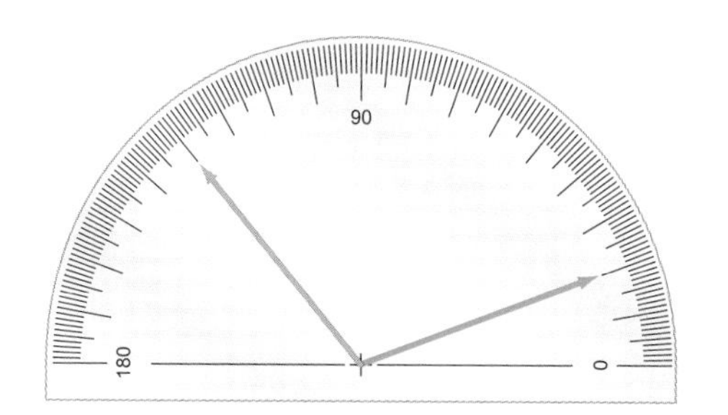

Objective B Classify Angles

The Concept To work with angles, we must be able to classify them according to their size or measure.

Angle Classifications

The words *acute, right, obtuse,* and *straight* describe the size of an angle.

Acute angles measure between 0° and 90° but not 0° or 90°.	**Right angles** measure exactly 90°.	**Obtuse angles** measure between 90° and 180° but not 90° or 180°.	**Straight angles** measure exactly 180°.
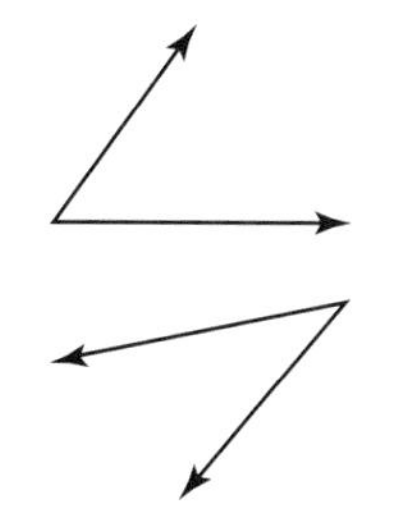	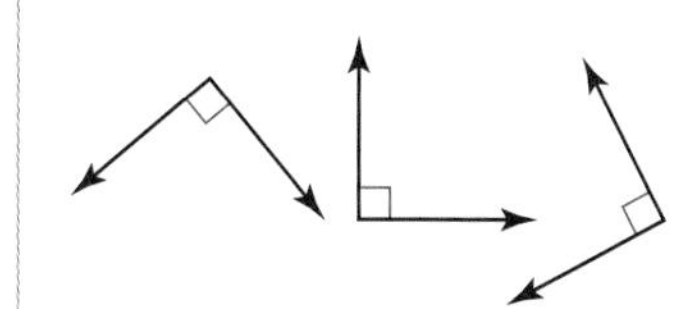 Right angles are often shown with a square drawn at the vertex.	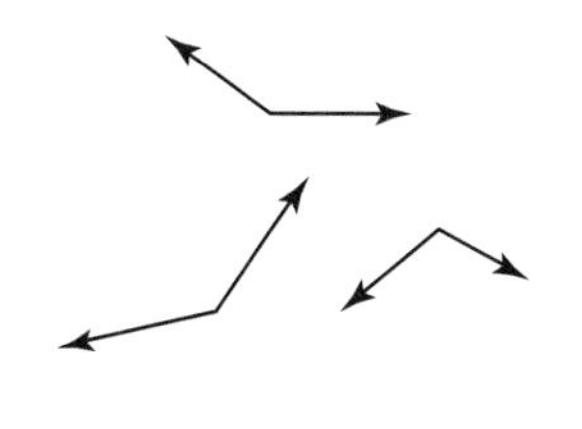	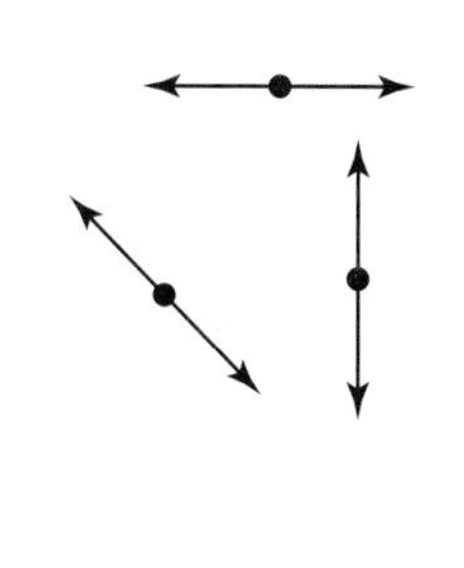

How to Name an Angle

Any of the following three methods can be used to name an angle.

1. The **vertex** can be used to name an angle if the vertex is not used for another angle.	2. **Three points** on the rays can be used to name an angle.	3. **A variable** for the measure of an angle can be used to name the angle.
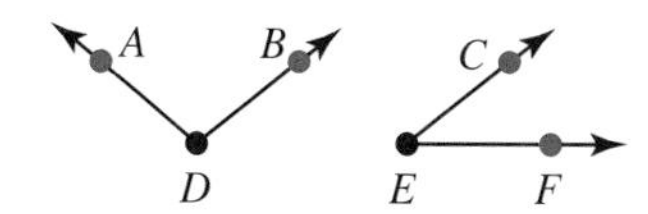 Two angles are shown: $\angle D$ and $\angle E$	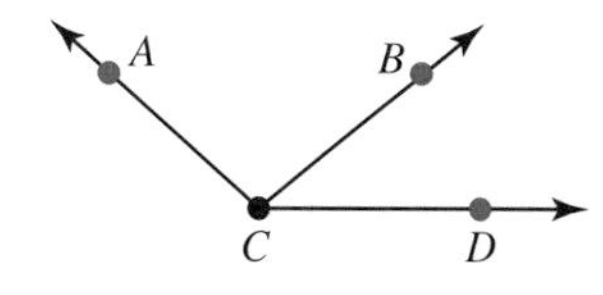 Three angles are shown: $\angle ACB$, $\angle BCD$, and $\angle ACD$ 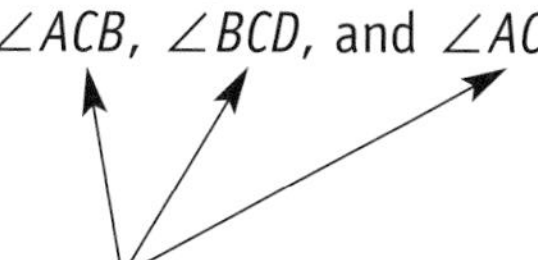 **The vertex, C, *must* be written in the middle.**	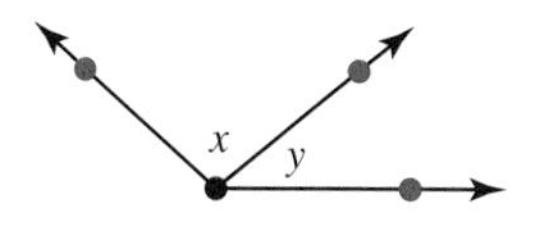 Three angles are shown: $\angle x$, $\angle y$, and $\angle(x + y)$

Use the following two diagrams to answer Examples/Guided Practice 4–8.

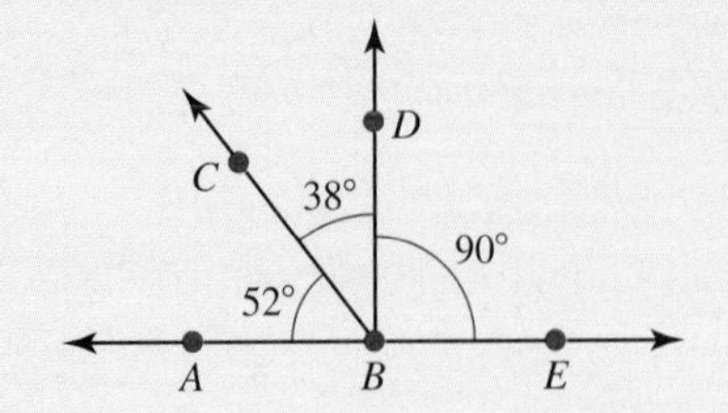

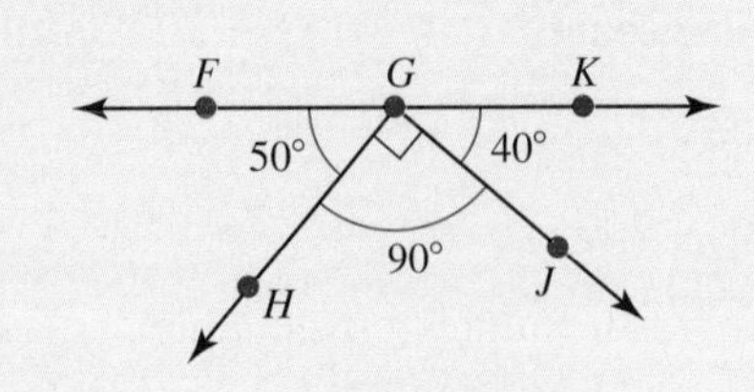

▶ **NOTE** The measure of an angle such as $\angle CBD$ is indicated using the notation $m\angle CBD$.

EXAMPLES

4. What is the name of the 90° angle? Is the angle acute, right, obtuse, or straight?

The 90° angle can be named as $\angle DBE$ or $\angle EBD$.
The angle is a right angle.

5. Find the measure of $\angle CBD$, written as $m\angle CBD$. Is the angle acute, right, obtuse, or straight?

$m\angle CBD = 38°$, an acute angle.

6. Find $m\angle ABD$. Is the angle acute, right, obtuse, or straight?

$$\begin{aligned} m\angle ABD &= m\angle ABC + m\angle CBD \\ &= 52° + 38° \\ &= 90° \end{aligned}$$

$\angle ABD$ is a right angle.

7. Find $m\angle ABE$. Is the angle acute, right, obtuse, or straight?

$$\begin{aligned} m\angle ABE &= m\angle ABC + m\angle CBD + m\angle DBE \\ &= 52° + 38° + 90° \\ &= 180° \end{aligned}$$

$\angle ABE$ is a straight angle.

8. Find $m\angle CBE$. Is the angle acute, right, obtuse, or straight?

$$\begin{aligned} m\angle CBE &= m\angle CBD + m\angle DBE \\ &= 38° + 90° \\ &= 128° \end{aligned}$$

$\angle CBE$ is an obtuse angle.

GUIDED PRACTICE

4. What is the name of the 50° angle? Is the angle acute, right, obtuse, or straight?

The 50° angle can be named as $\angle$ ____ or $\angle$ ____.
The angle is a(n) ____ angle.

5. Find the measure of $\angle HGJ$, written as $m\angle HGJ$. Is the angle acute, right, obtuse, or straight?

$m\angle HGJ =$ ____°, a(n) ____ angle.

6. Find $m\angle HGK$. Is the angle acute, right, obtuse, or straight?

$$\begin{aligned} m\angle HGK &= m\angle \qquad + m\angle \\ &= \\ &= \end{aligned}$$

$\angle HGK$ is a(n) ______ angle.

7. Find $m\angle FGJ$. Is the angle acute, right, obtuse, or straight?

$m\angle FGJ =$

$\angle FGJ$ is a(n) ______ angle.

8. Find $m\angle FGK$. Is the angle acute, right, obtuse, or straight?

$m\angle FGK =$

$\angle FGK$ is a(n) ______ angle.

Concept Check

B1. Why would it be confusing to describe $\angle BCD$ as $\angle C$?

B2. Is $\angle BCD$ an acute, obtuse, or right angle?

B3. Is $\angle BCA$ an acute, obtuse, or right angle?

B4. What is often drawn on a right angle to show that it measures 90°?

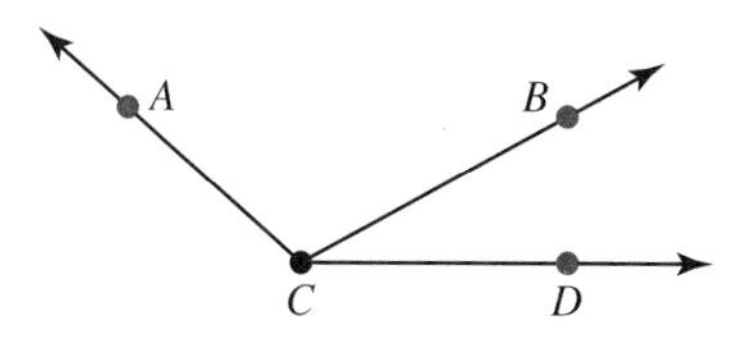

Objective B Practice

Determine the measure of each angle. $\angle JOK$ is a straight angle. $\angle JOL$ is a right angle.

B5. $m\angle NOL$ **B6.** $m\angle LOM$ **B7.** $m\angle JON$

B8. $m\angle MOK$ **B9.** $m\angle JOM$ **B10.** $m\angle LOK$

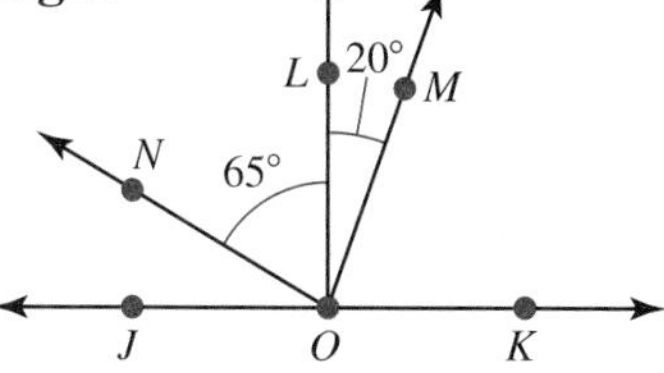

A total of ten different angles are shown in the figure.

B11. Name the five acute angles in the figure.

B12. Name the two obtuse angles in the figure.

B13. Name the straight angle in the figure.

B14. Name the two right angles in the figure.

Objective C Solve for Unknown Angles Using Angle Relationships

The Concept When the measures of two angles add to 180°, the angles are called **supplementary angles.** When the measures of two angles add to 90°, the angles are called **complementary angles.** Using those definitions, we can solve for unknown angles.

INTERACTIVE DEFINITION Supplementary Angles

Two angles are **supplementary** if their measures add to 180°. One angle is the *supplement* of the other angle.

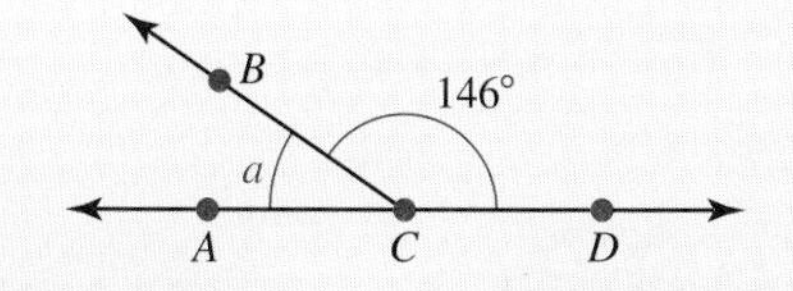

Since $\angle ACD$ is a straight angle, $\angle a$ and the angle of 146° are **supplementary angles.**

$$m\angle a + 146° = 180°$$
$$m\angle a = 180° - 146°$$
$$m\angle a = 34°$$

EXAMPLE

9. Find $m\angle a$.

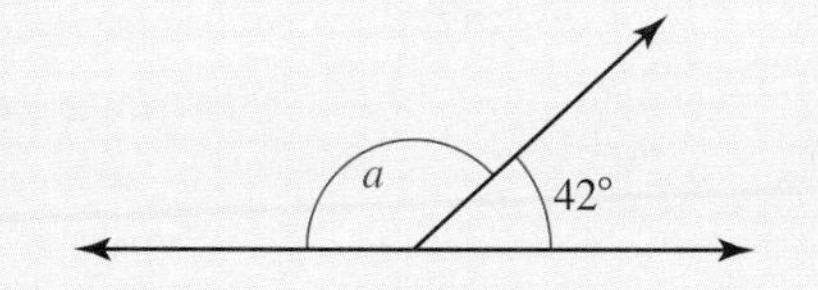

$\angle a$ is supplementary to 42°.

$$m\angle a + 42° = 180°$$
$$m\angle a = 180° - 42°$$
$$m\angle a = 138°$$

GUIDED PRACTICE

9. Find $m\angle a$.

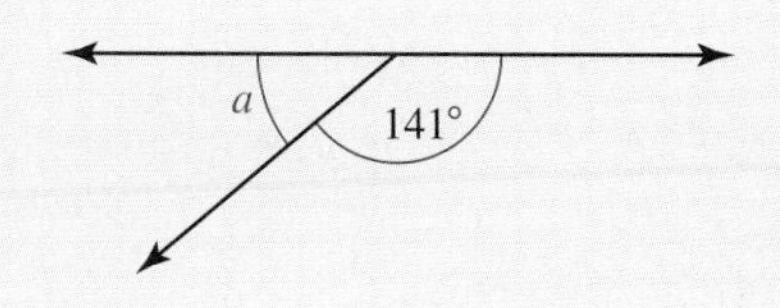

$\angle a$ is __________ to 141°.

$$m\angle a + 141° = \_\_\_°$$
$$m\angle a = \_\_\_° - \_\_\_°$$
$$m\angle a = \_\_\_°$$

DO YOU UNDERSTAND how to find the measure of supplementary angles? Got It Get Help

INTERACTIVE DEFINITION Complementary Angles

Two angles are **complementary** if their measures add to 90°. One angle is the *complement* of the other angle.

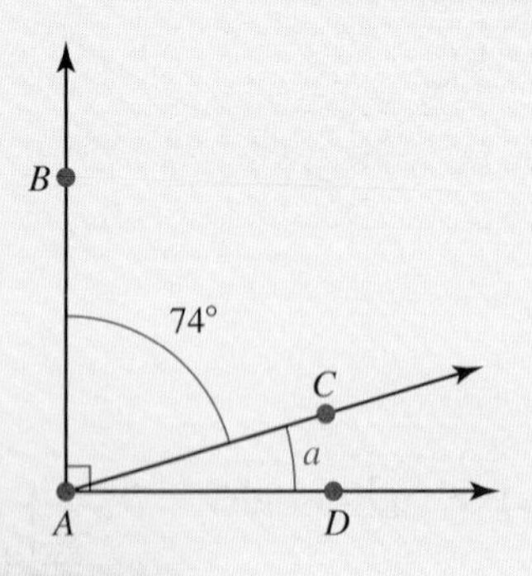

Since $\angle BAD$ is a right angle, $\angle a$ and the angle of 74° are **complementary angles.**

$$m\angle a + 74° = 90°$$
$$m\angle a = 90° - 74°$$
$$m\angle a = 16°$$

EXAMPLE

10. Find $m\angle a$.

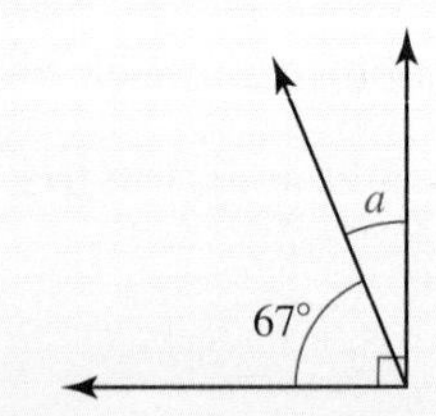

$\angle a$ is complementary to 67°.

$$m\angle a + 67° = 90°$$
$$m\angle a = 90° - 67°$$
$$m\angle a = 23°$$

GUIDED PRACTICE

10. Find $m\angle a$.

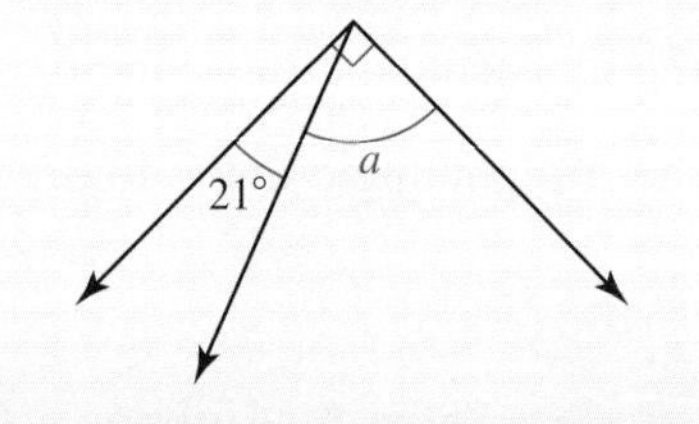

$\angle a$ is ____________ to _____°.

$$m\angle a + 21° = \_\_\_\_°$$
$$m\angle a = \_\_\_\_° - \_\_\_\_°$$
$$m\angle a = \_\_\_\_°$$

DO YOU UNDERSTAND how to find the measure of complementary angles? Got It Get Help

We do not need to have supplements or complements to solve for unknown angles. Often, we can find unknown angles if we know the total measure of the angles.

EXAMPLE

11. Find $m\angle a$.

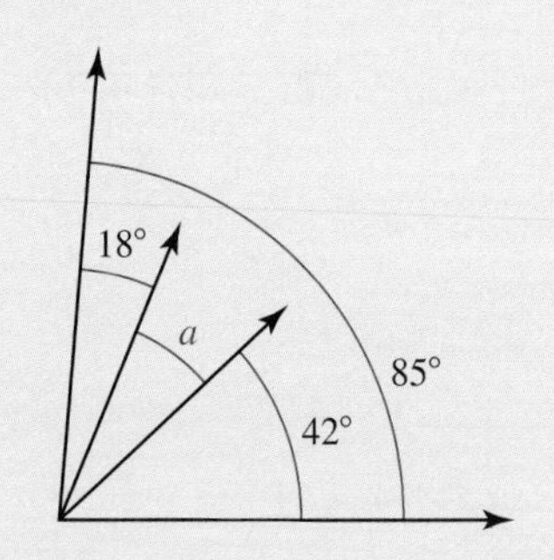

GUIDED PRACTICE

11. Find $m\angle a$.

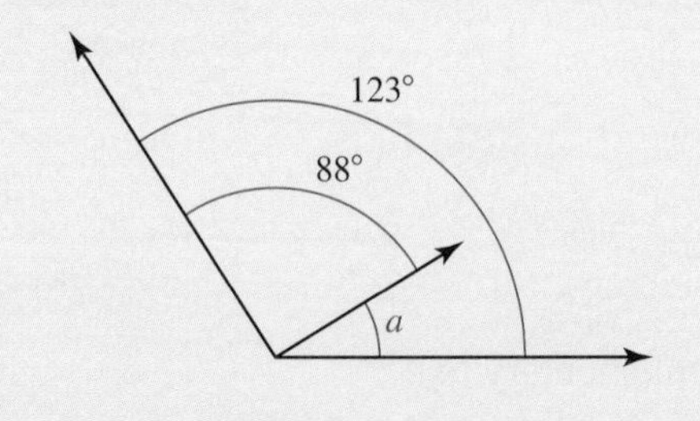

There are no 90° or 180° angles, so we won't be able to use complements or supplements.	There are no 90° or 180° angles, so we won't be able to use ________ or ________.
However, the angle 85° has the same measure as the angles 18°, $\angle a$, and 42° added together.	However, the angle _____ has the same measure as the angles _____ and _____ added together.
The smaller angles added = the larger angle.	The smaller angles added = the larger angle.
$18° + m\angle a + 42° = 85°$ $m\angle a = 85° - 18° - 42°$ $m\angle a = 67° - 42°$ $m\angle a = 25°$	= = =

Concept Check

C1. The measure of an angle and its supplement add to ______.

C2. The measure of an angle and its complement add to ______.

Objective C Practice

Find the measure of each angle.

C3. the supplement of a 35° angle

C4. the complement of a 35° angle

C5. the complement of a 45° angle

C6. the supplement of a 90° angle

C7. the complement of $\angle NOL$

C8. $m\angle MOK$

C9. the supplement of $\angle NOK$

C10. the supplement of $\angle MON$

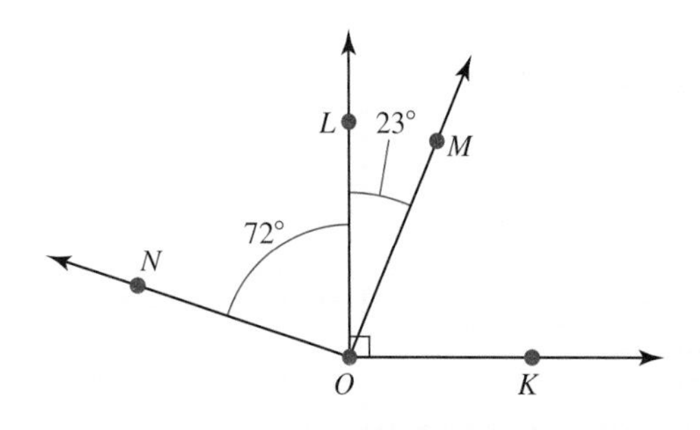

C11. Find $m\angle a$.

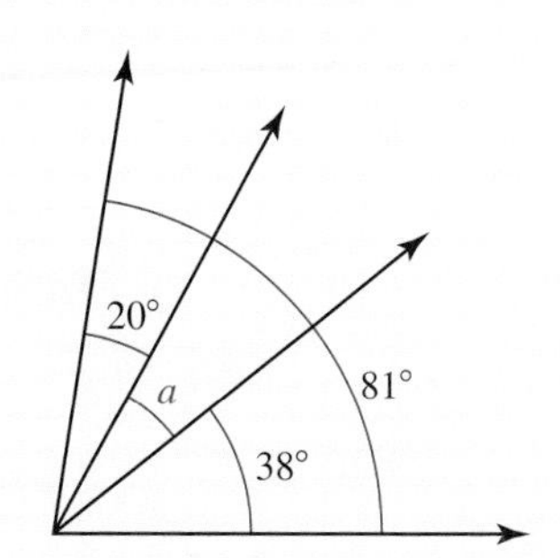

C12. Find $m\angle a$.

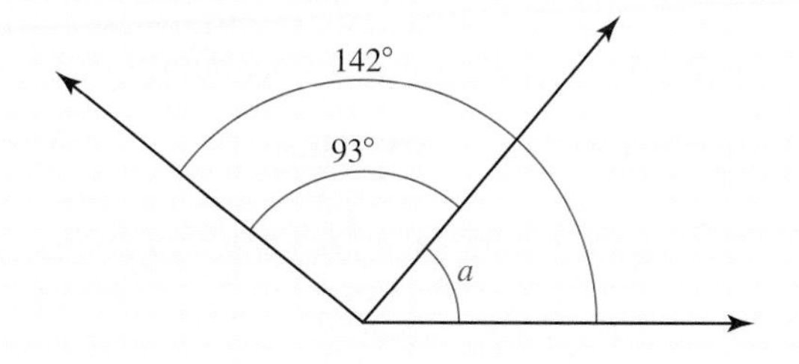

Section 7.1 Exercises

FOR EXTRA HELP 

 PRACTICE  WATCH  DOWNLOAD  READ  REVIEW

To understand angles and their measures:

1. Answer the Objective A Concept Checks.
2. Answer the odd Objective A Practice Exercises.
3. Answer the even Objective A Practice Exercises.

To classify angles:

4. Answer the Objective B Concept Checks.
5. Answer the odd Objective B Practice Exercises.
6. Answer the even Objective B Practice Exercises.

To solve for unknown angles using angle relationships:

7. Answer the Objective C Concept Checks.
8. Answer the odd Objective C Practice Exercises.
9. Answer the even Objective C Practice Exercises.

VOCABULARY REVIEW *Review the Vocabulary Preview for Section 7.1. Study the definitions until you can check* Got It *for every word.*

Use these words to complete each sentence.

line • vertex • obtuse angle • supplementary angles • line segment • measure • acute angle • rays • straight angle • angle • right angle • complementary angles

	Got It	Get Help
10. $m\angle a$ represents the ____________ of angle a.		
11. An angle is made of two ____________ that begin at a common endpoint called the ____________.		
12. If the measures of two angles add to 180°, the angles are ____________.		
13. An angle that measures 90° is a(n) ____________.		
14. Any angle that measures 180° is a(n) ____________.		
15. ____________ have measures less than 90°, and ____________ have measures greater than 90°.		
16. A(n) ____________ extends forever in two directions.		
17. If the measures of two angles add to 90°, then the angles are ____________.		

How will you get help for any vocabulary that you are unsure about?

Your instructor ____ MyMathLab ____ A classmate ____ A tutor ____ Other ____

Draw each ray on the protractor.

18. ray A at 45°

19. ray B at 135°

20. ray C at 87°

21. ray D at 120°

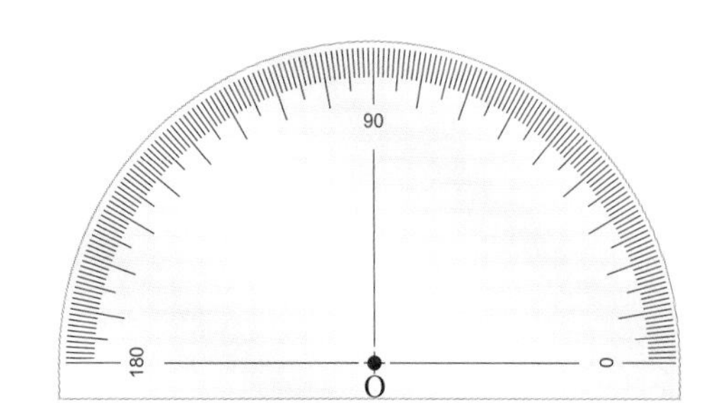

Use the rays from Exercises 18–21 to find the measures of the indicated angles.

22. Find $m\angle DOC$.

23. Find $m\angle AOD$.

24. Find $m\angle AOC$.

25. Find $m\angle BOC$.

26. a. What is the measure of the largest angle formed between any two of the rays A, B, C, and D?

b. What is the name of the largest angle formed between any two of the rays A, B, C, and D?

27. a. What is the measure of the smallest angle formed between any two of the rays A, B, C, and D?

b. What is the name of the smallest angle formed between any two of the rays A, B, C, and D?

Use the figure to answer each question.

28. Name the two obtuse angles that include ray B.

29. Name the two obtuse angles that do not include ray B.

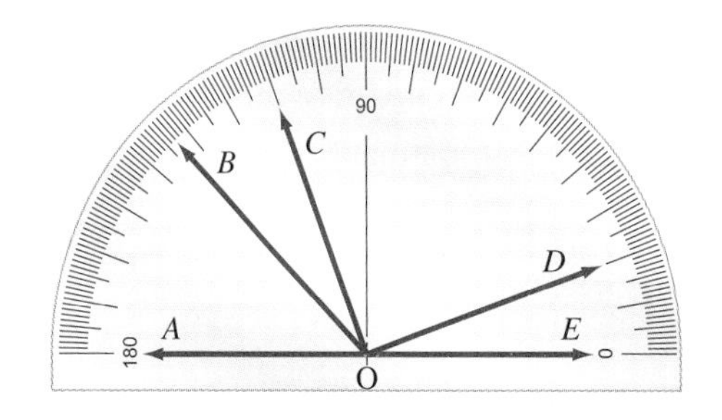

30. Name the two acute angles that do not include ray B.

31. Name the two acute angles that include ray B.

32. Name all the right angles.

33. Name all the straight angles.

34. What is the measure of the complement of $\angle BOC$?

35. What is the measure of the complement of $\angle AOB$?

36. What is the measure of the supplement of $\angle EOC$?

37. What is the measure of the supplement of $\angle BOD$?

For each figure, find the measures of the indicated angles.

38. $\angle AOD$ is a straight angle.

a. Find $m\angle AOC$.

b. Find $m\angle BOC$.

c. Find $m\angle AOB$'s supplement.

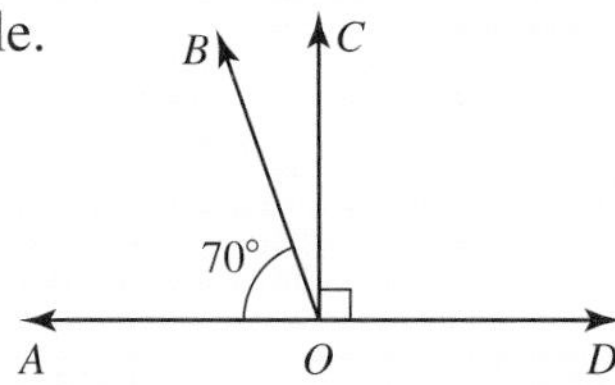

39. $\angle AOD$ is a straight angle.

a. Find $m\angle COD$.

b. Find $m\angle AOB$.

c. Find $m\angle COD$'s supplement.

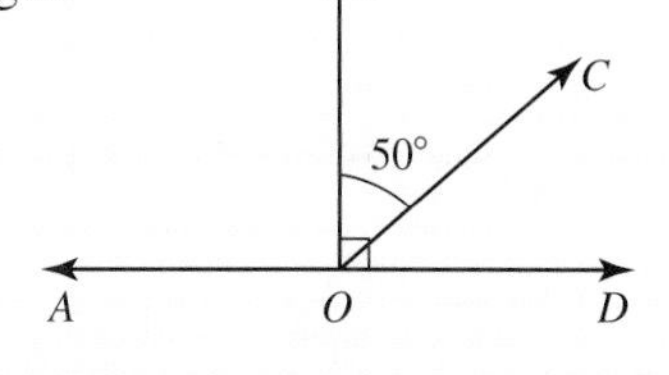

Smith To get help with #38 and #39, revisit the Interactive Definitions in this section for supplementary and complementary angles.

For more tips and tweets, go to twitter.com/gstbasicmath

40. $\angle AOE$ is a straight angle.
a. Find $m\angle COD$.
b. Find $m\angle AOC$.
c. Find $m\angle AOB$.
d. What are the names of the 3 right angles?

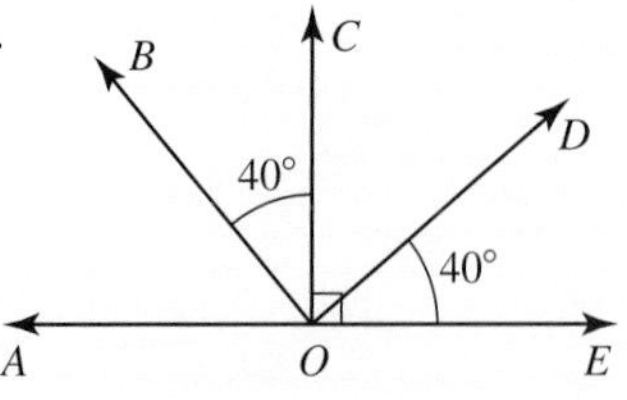

41. $\angle AOE$ is a straight angle.
a. Find $m\angle BOC$.
b. Find $m\angle DOE$.
c. Find $m\angle BOE$.
d. What are the names of the 3 right angles?

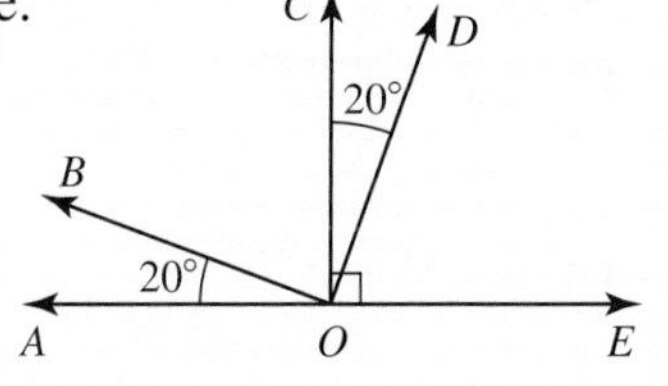

42. a. $m\angle BOC = ?$
b. $m\angle BOD = ?$
c. $m\angle AOC = ?$
d. Which is greater—$\angle BOD$'s supplement or $\angle AOC$'s supplement?

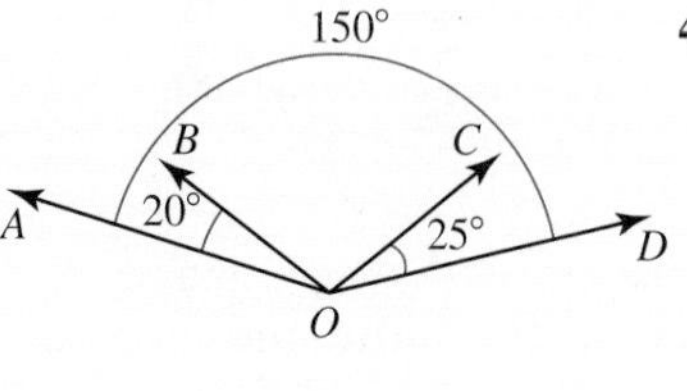

43. a. $m\angle AOB = ?$
b. $m\angle AOC = ?$
c. $m\angle BOC = ?$
d. Which is less—$\angle AOB$'s complement or $\angle BOC$'s complement?

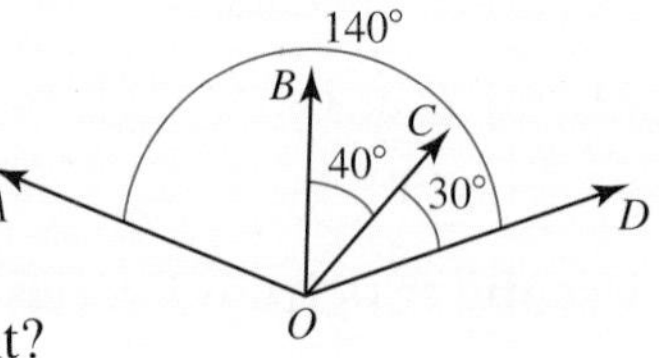

44. A carpenter cuts two pieces of wood to form a joint in an octagonal window. At what angle a should she cut the wood to make the two angles equal to each other?

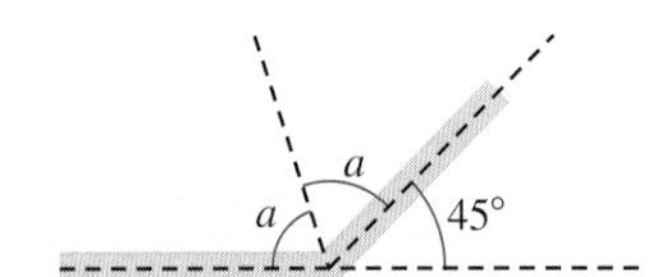

45. A quilter cuts four pieces of cloth to make the pattern shown. If the angle a on each piece of cloth is the same, what is that angle?

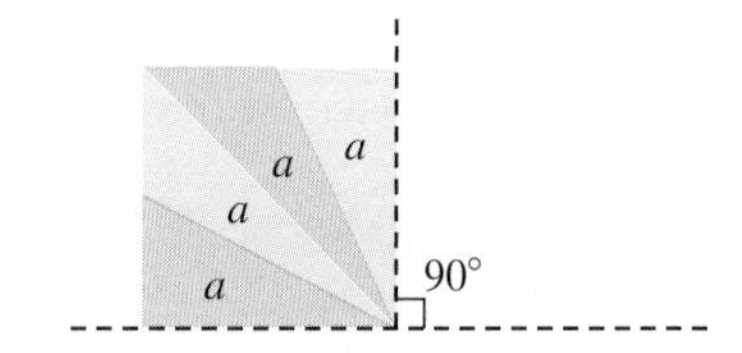

For each exercise, decide if the statement is always true, sometimes true, or never true.
- **If the answer is always true or never true, state why.**
- **If the answer is sometimes true, give an example of when it is true and when it is false.**

46. An angle is equal to its supplement.

47. An angle is equal to its complement.

48. If two angles are complementary, then at most one is acute.

49. If two angles are supplementary, then one of the angles is obtuse.

50. If the measure of an angle's complement is subtracted from 90°, then the result is the same as the measure of the angle.

51. If the measure of an angle's supplement is subtracted from 180°, then the result is the same as the measure of the angle.

SELF ASSESSMENT

	YES	NO
1. On your last test, were you satisfied with your score? *If you answered yes, good job. Do not continue here.* *If you answered no, answer the following.*		
2. Did you preview each section of the textbook before your instructor taught them in class?		
3. Did you complete all of the homework before it was due?		
4. Did you get help with the homework exercises that you didn't understand?		

5. Set two goals that will help you become better prepared for your next test.

Goal 1: ______

Goal 2: ______

Section 7.1 Question Log

Use this space to write questions. Be sure to get these questions answered and revisit them when you prepare for your exam.

Page ____ Answered ☐	Page ____ Answered ☐
Page ____ Answered ☐	Page ____ Answered ☐

Section 7.2 Polygons

Polygons are figures that can be drawn on a flat piece of paper using line segments. The name of any particular polygon depends on how many sides it has and the angles between its sides.

Some polygons you have seen before are as follows:

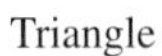
Triangle

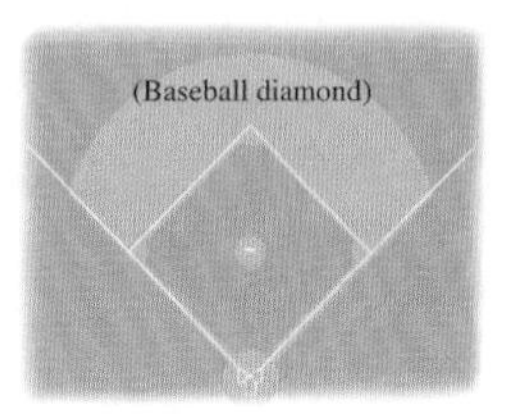

Square

Rectangle

Octagon

The **Objectives** in Section 7.2 will help you

- **A** Classify polygons and find their perimeters.
- **B** Classify quadrilaterals.
- **C** Classify triangles.

VOCABULARY PREVIEW *Check the box that applies.*

Got It | Must Study

Line Segment: A **line segment** is a part of a straight line that has a start point and an endpoint.

Polygon: A **polygon** can be drawn using only line segments that do not cross and form a closed, flat figure.

Polygons are named according to the number of sides and the measures of their angles.

Triangle 3 Sides	Quadrilateral 4 Sides	Pentagon 5 Sides	Hexagon 6 Sides
Heptagon 7 Sides	Octagon 8 Sides	Nonagon 9 Sides	Decagon 10 Sides

Regular Polygon: A polygon with equal sides and equal angles is called a **regular polygon.**

Perimeter: The distance around a figure is its **perimeter.**

Special Quadrilaterals:

Trapezoid: One pair of opposite sides is parallel.
Parallelogram: Two pairs of opposite sides are parallel.
Rectangle: All angles measure 90°.
Rhombus: All sides are equal, and opposite sides are parallel.
Square: All angles measure 90°, and all sides have equal length.

	Got It	Must Study
Special Triangles: **Scalene:** None of the sides or angles are equal. **Isosceles:** Two sides and two angles are equal. **Equilateral:** All sides and all angles are equal. **Right:** One angle is 90°. **Obtuse:** One of the angles is obtuse (more than 90°). **Acute:** All of the angles are acute (less than 90°).		

Study these words when they appear in the text. After the section, test your understanding by completing the Vocabulary Review in the section exercises.

Objective A Classify Polygons and Find Their Perimeters

The Concept A **polygon** is drawn with line segments that do not cross and form a closed, flat figure. A general way to classify a polygon is to count its sides. Notice that the sides of polygons can have different lengths and the measures of the angles may not be equal.

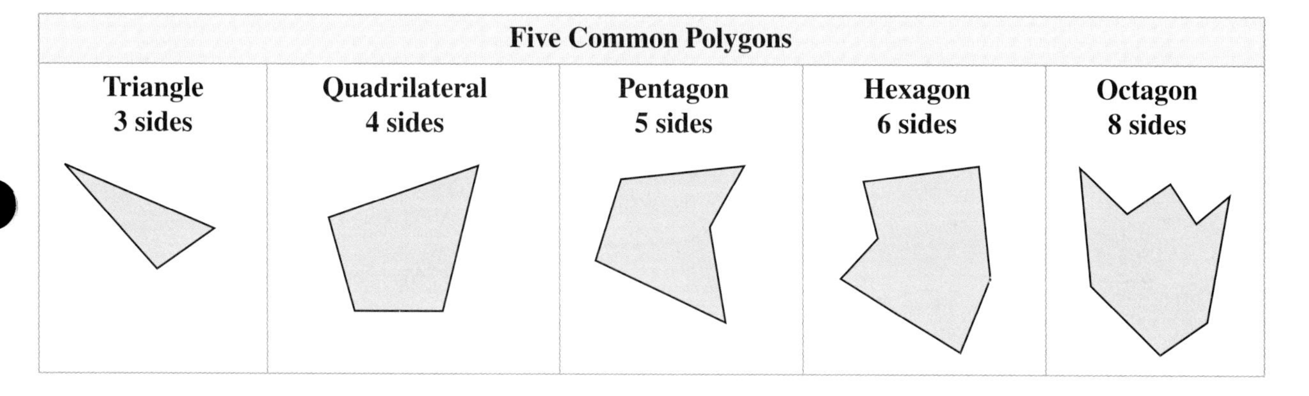

Five Common Polygons				
Triangle **3 sides**	**Quadrilateral** **4 sides**	**Pentagon** **5 sides**	**Hexagon** **6 sides**	**Octagon** **8 sides**

Regular polygons have all sides of the same length and all angles of the same measure.

Five Regular Polygons				
Regular **Triangle**	**Regular** **Quadrilateral**	**Regular** **Pentagon**	**Regular** **Hexagon**	**Regular** **Octagon**

Three Reasons Why a Figure Is Not a Polygon		
One of the sides is not straight.	The sides cross each other.	The sides do not form a closed figure.

Procedure Classify Polygons

Step 1 Count the number of sides and use the appropriate name for the figure.

Step 2 If every side has the same length and every angle has the same measure, state that the polygon is regular.

EXAMPLES

1. Classify the figure.

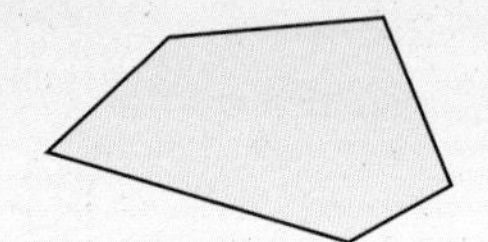

There are five sides.
The angles between sides are different.

The figure is a pentagon.

2. Classify the figure.

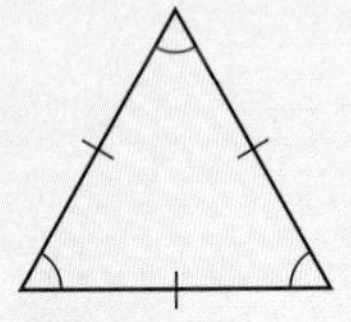

There are three sides.
The angles between sides are the same, and the lengths of the sides are the same.

The figure is a regular triangle.

3. Classify the figure.

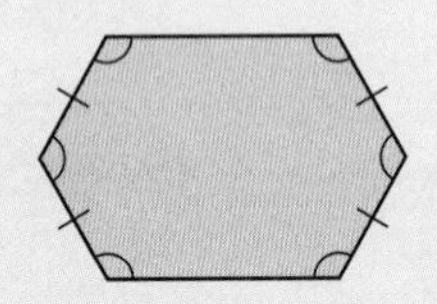

There are six sides.
Opposite sides are parallel.
The angles between sides are the same, but the lengths of the sides are different.

The figure is a hexagon.

GUIDED PRACTICE

1. Classify the figure.

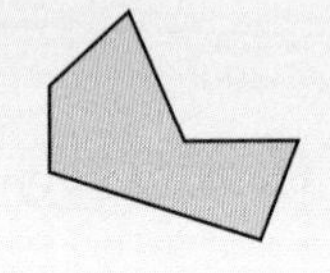

There are ______ sides.
The angles between sides are the same/different.

The figure is a(n) ________.

2. Classify the figure.

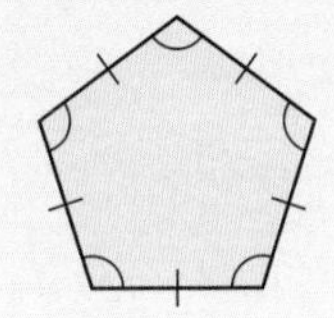

There are _____ sides.
The angles between sides are the same/different, and the lengths of the sides are the same/different.

The figure is a(n) _______________.

3. Classify the figure.

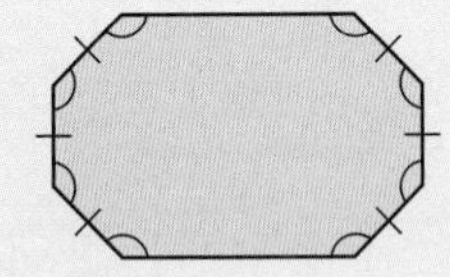

There are _____ sides.
The angles between sides are the same/different, and the lengths of the sides are the same/different .

The figure is a(n) ________.

The **perimeter** of a polygon is the distance around the figure. To find the perimeter of a polygon, add the lengths of the sides.

EXAMPLE

4. Find the perimeter of the polygon.

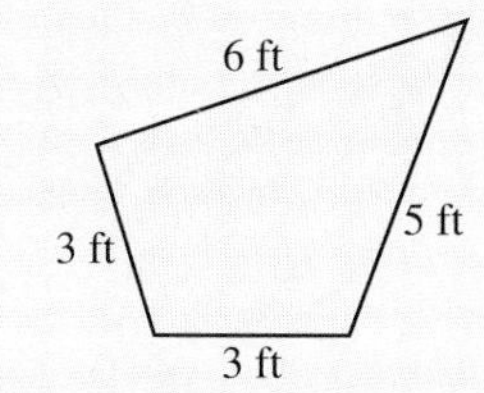

The perimeter is the distance around the figure. Add the lengths of the sides.

$$
\begin{aligned}
P &= 3 + 3 + 6 + 5 \\
&= 6 + 6 + 5 \\
&= 12 + 5 \\
&= 17 \text{ ft}
\end{aligned}
$$

GUIDED PRACTICE

4. Find the perimeter of the polygon.

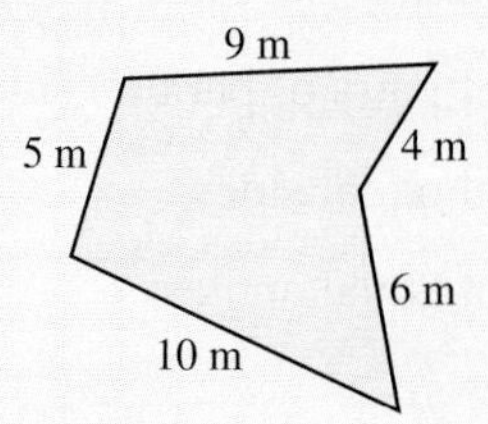

The perimeter is the distance around the figure. Add the lengths of the sides.

$P =$

Concept Check

A1. In your own words, when is a polygon a regular polygon?

A2. **a.** How many sides does a quadrilateral have?
b. What figure has five sides?
c. How many sides does an octagon have?
d. What figure has six sides?

Objective A Practice

Classify each figure as accurately as possible using only the names from this objective.

A3.

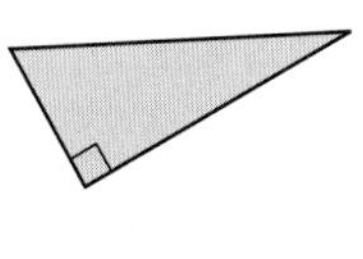

A4.

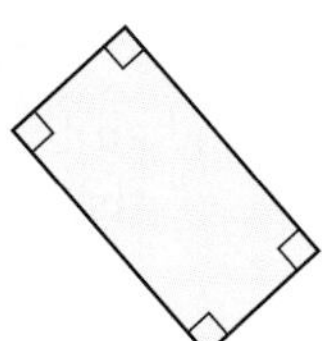

A5.

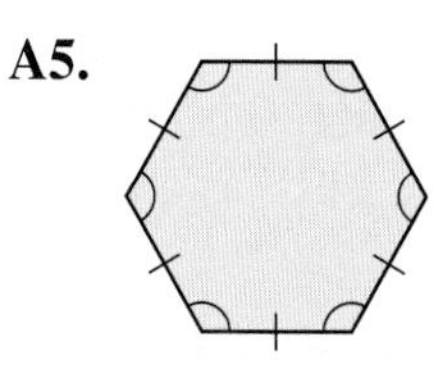

A6.

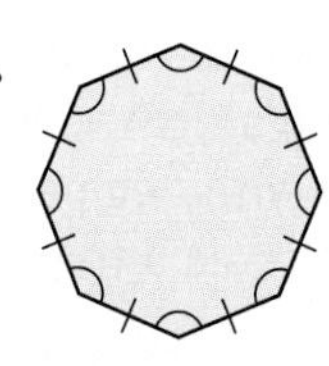

A7.

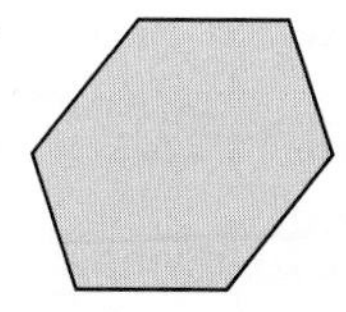

A8.

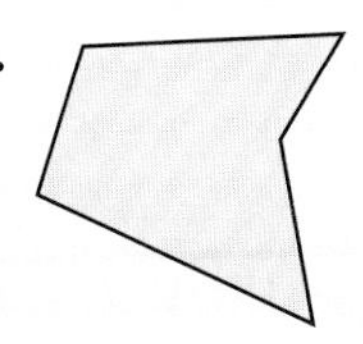

A9.

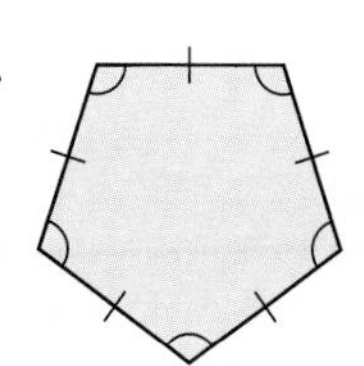

A10.

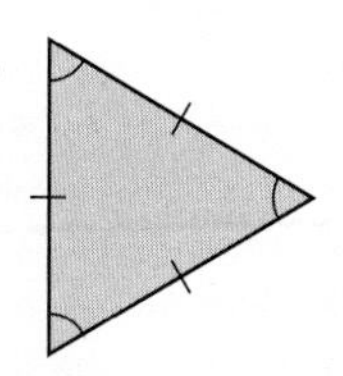

A11. What is the perimeter of the figure?

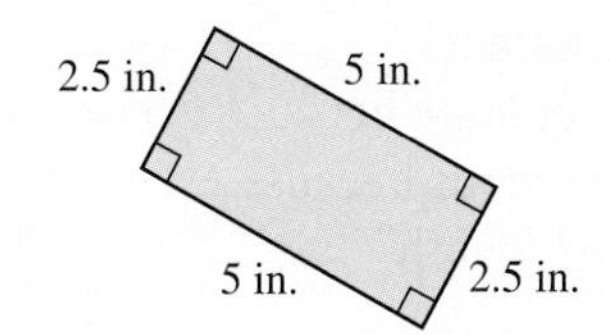

A12. What is the perimeter of the figure?

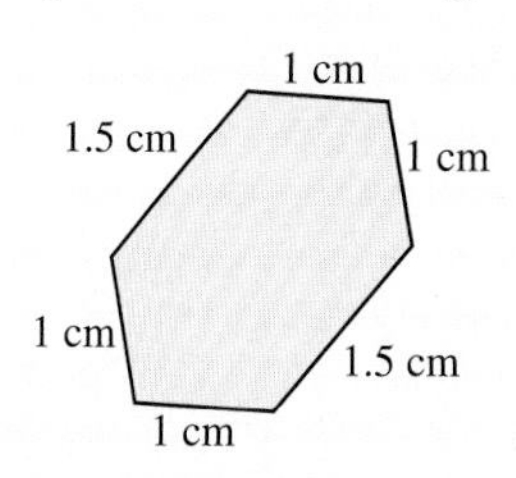

Objective B Classify Quadrilaterals

The Concept Every polygon with four sides is a *quadrilateral*. Many quadrilaterals are classified with special names. To identify the most descriptive name of a quadrilateral, determine the following:

- The number of pairs of parallel sides
- The number of equal sides
- The number of equal angles

Quadrilateral Types

Plain Quadrilaterals

If no sides of the quadrilateral are parallel, then the figure is a quadrilateral.

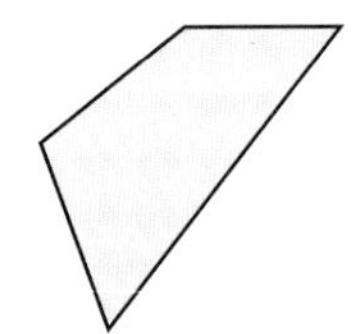

Trapezoids

If only one pair of sides is parallel, then the quadrilateral is a trapezoid.

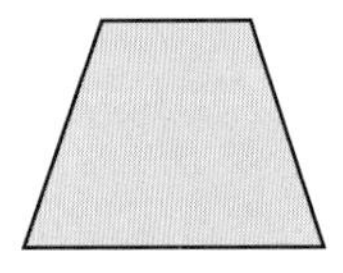

Parallelograms

If two pairs of opposite sides are parallel, then the quadrilateral is a parallelogram.

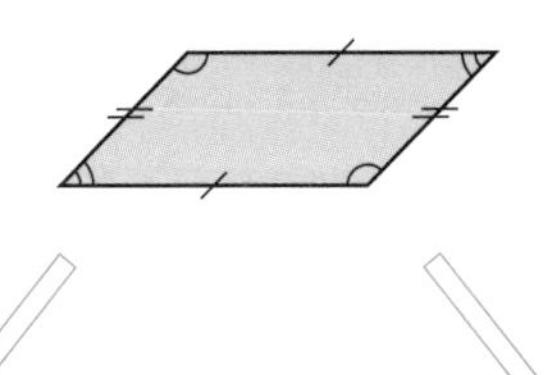

Rectangles

If the angles are all 90°, then the parallelogram is also a rectangle.

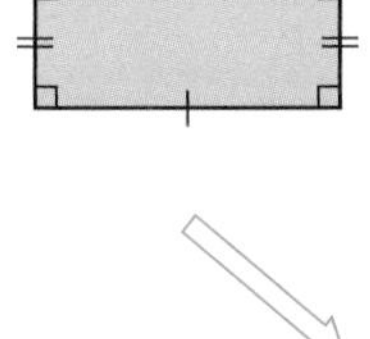

Rhombuses

If all the sides are the same length, then the parallelogram is also a rhombus.

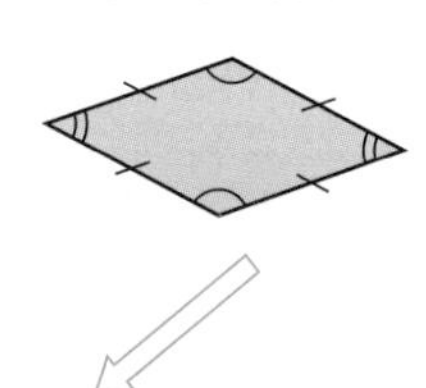

Squares

If all the angles are 90° and all the sides are the same length, then the parallelogram is also a square.

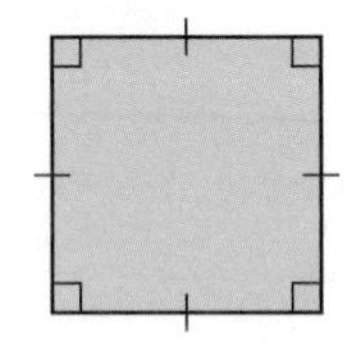

▶ **NOTE** A square is all of the following:

- A quadrilateral because it has four sides
- A parallelogram because opposite sides are parallel
- A rectangle because every angle is 90°
- A rhombus because every side has the same length

▶ **NOTE** In mathematics, sometimes different definitions exist for the same word. This happens with trapezoids.

In addition to the definition above, a trapezoid can be defined as having at least one pair of parallel sides. This means that a figure with two pairs of parallel sides, such as a rectangle, would also be a trapezoid.

In this text, a trapezoid is defined as having exactly one set of parallel sides.

EXAMPLES

GUIDED PRACTICE

The following figures are needed for Examples/Guided Practice 5–7.

a. 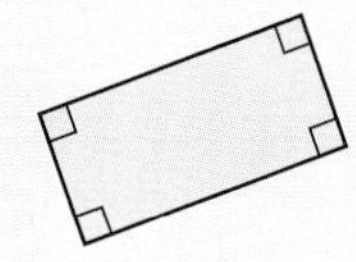

b. 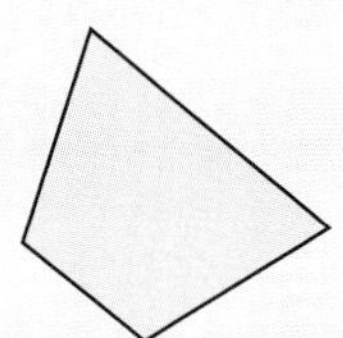

c. 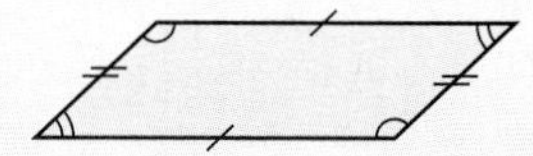

d. 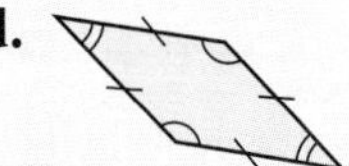

e. 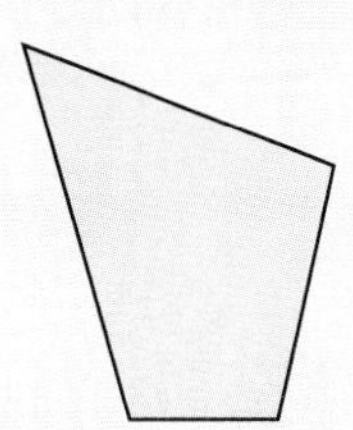

f. 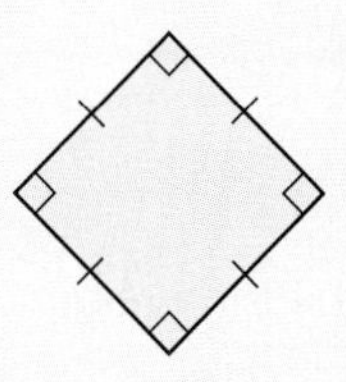

5. Which figures are squares?

Figure f is a square because:

- Each side has the same length.
- All angles measure 90°

6. Which figures are quadrilaterals?

All of the figures are quadrilaterals because all of the figures have four sides.

7. Which figures are trapezoids?

Figure b is a trapezoid because exactly one pair of opposite sides is parallel.

5. Which figures are rhombuses?

Figures ____ and ____ are rhombuses because all ____ are equal.

6. Which figures are parallelograms?

Figures ____, ____, ____, and ____ are parallelograms because the ______ sides are ______.

7. Which figures are rectangles?

Figures ____ and ____ are rectangles because all of the _____ are _____.

Concept Check

B1. Is a rectangle also a parallelogram? Explain.

B2. Must a parallelogram be a rectangle? Explain.

B3. Can a rectangle be a rhombus? Explain.

B4. Why is a rectangle a quadrilateral? Explain.

Objective B Practice

Classify each figure as accurately as possible using only the names from this section.

B5. 

B6. 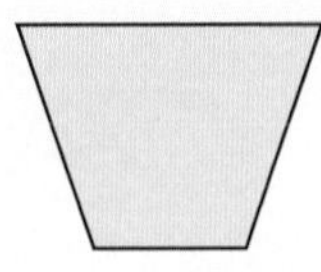

B7. 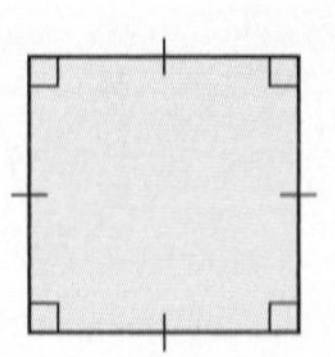

B8. 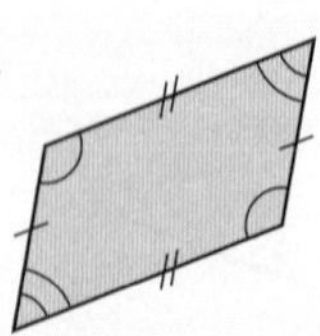

B9. 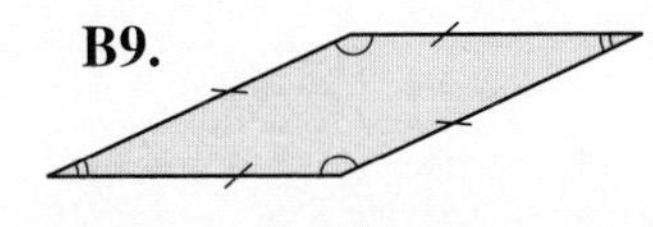

B10. 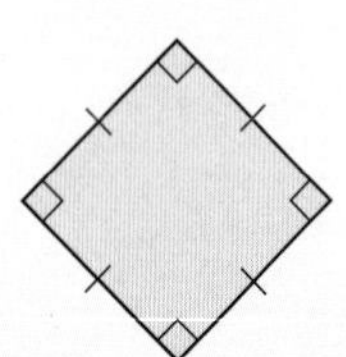

B11. 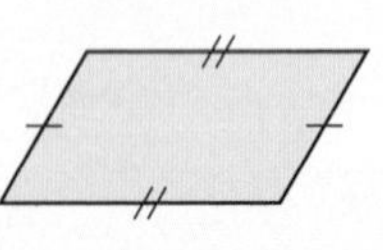

B12. 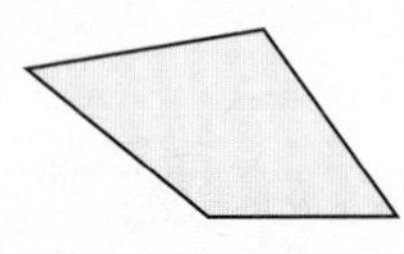

For each exercise, what type of quadrilateral has the given properties?

B13.
- No sides are parallel.
- No sides have equal length.

B14.
- Opposite sides are parallel.
- All sides are equal.
- Angles are not equal.

B15.
- Angles between sides are 90°.
- Sides have different length.

B16.
- One pair of sides is parallel.
- Other two sides are not parallel.

Objective C Classify Triangles

The Concept Every polygon with three sides is a **triangle.** Depending on how many sides have equal length, a triangle can be **scalene, isosceles,** or **equilateral.**

Scalene, Isosceles, and Equilateral Triangles

Scalene Triangle	**Isosceles Triangle**	**Equilateral Triangle**
No sides are equal. No angles are equal.	Two sides are equal. Two angles are equal.	Three sides are equal. Three angles are equal.
	▶ **NOTE** Equal sides and equal angles will be opposite each other.	60° 60° 60° **These marks indicate equal sides or angles**

A triangle with a **right angle** (90°) is also called a **right triangle.** Scalene and isosceles triangles can be right triangles. Equilateral triangles cannot be right triangles.

Right Triangles		
Right Scalene Triangle	**Right Isosceles Triangle**	
• One angle is 90°. • The other angles are not equal. • No two sides are the same length.	• One angle is 90°. • The other two angles are equal (45°). • Two sides are equal in length.	It is impossible to draw a right equilateral triangle because every angle of an equilateral triangle must be 60°.

If a triangle is not a right triangle, then it is either an acute triangle or an obtuse triangle.

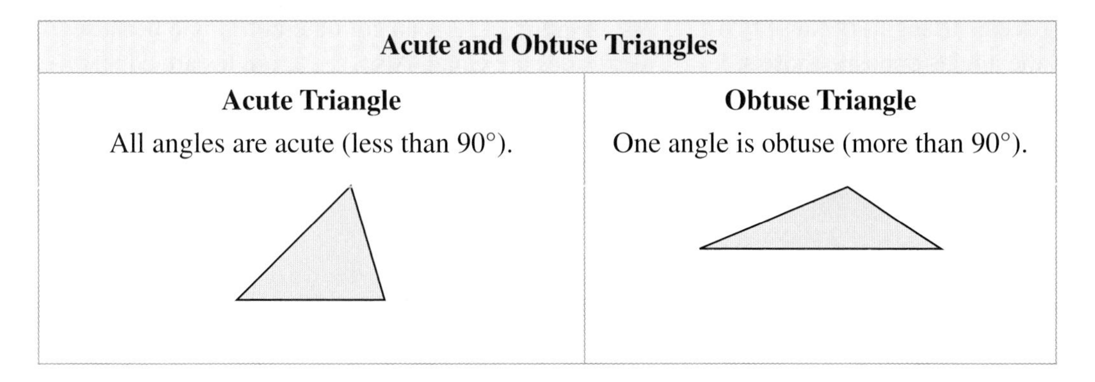

Acute and Obtuse Triangles	
Acute Triangle	**Obtuse Triangle**
All angles are acute (less than 90°).	One angle is obtuse (more than 90°).

To classify triangles, do the following:

- Determine the number of equal sides or the number of equal angles.
- Determine if one of the angles is a right angle (90°).
- If the triangle is not a right triangle, decide if it is acute or obtuse.

The following figures are needed for Examples/Guided Practice 8–11.

a. **b.** **c.** **d.** **e.**

EXAMPLES

8. Which figures are right triangles?

Figures a and e are right triangles because one of their angles measures 90°.

9. Which figures are isosceles triangles?

Figures b and e are isosceles triangles because two sides are equal. Also, the angles opposite those sides are equal.

10. Which figures are right isosceles triangles?

Only figures a and e are right triangles. Of those, only figure e has two equal sides. Therefore, only figure e is a right isosceles triangle.

11. Which figures are acute triangles?

In an acute triangle, every angle is less than 90°. This is true only of figures b and d.

Figures b and d are acute triangles.

12. What name best describes the following triangle?

- Two of the sides have the same length. The triangle is isosceles.
- There is no right angle. The triangle is not a right triangle.
- Since the triangle is not a right triangle, it is either acute or obtuse. Since an angle of the triangle is more than 90°, it is obtuse.

The triangle is an obtuse isosceles triangle.

GUIDED PRACTICE

8. Which figures are scalene triangles?

Figures ____ and ____ are scalene triangles because ____ sides are equal.

9. Which figures are equilateral triangles?

Figure ____ is an equilateral triangle because ____ sides are equal. Also, ____ angles are equal.

10. Which figures are right scalene triangles?

Only figures ____ and ____ are scalene triangles. Of those two, only figure ____ has an angle of ____°. Therefore, only figure ____ is a right scalene triangle.

11. Which figures are obtuse triangles?

In an obtuse triangle, one of the angles will be ________ 90°. This is true only of figure ____.

Figure ____ is an obtuse triangle.

12. What name best describes the following triangle?

- ____ of the sides have the same length. The triangle is/is not isosceles.
- There is/is not a right angle. The triangle is/is not a right triangle.
- If the triangle is not a right triangle, it is either acute or obtuse. The triangle is acute/obtuse because ____________________.

The triangle is a(n) ____________.

Concept Check

C1. What is the difference between an isosceles triangle and an equilateral triangle?

C2. What properties will a right scalene triangle have?

C3. Why can't an equilateral triangle have a right angle?

Objective C Practice

Classify each figure as accurately as possible using only the names from this section.

C4. 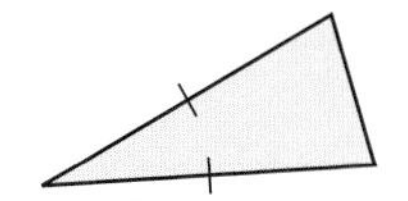

C5. 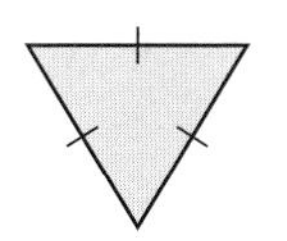

C6. 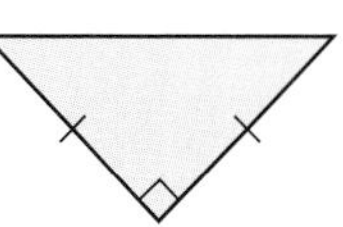

C7. 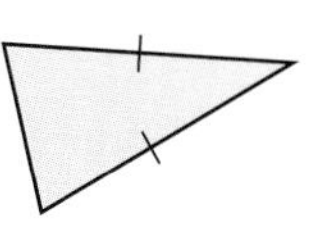

C8. 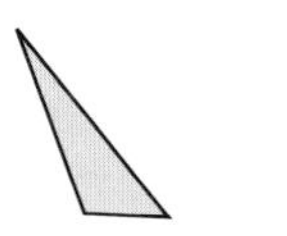

C9. 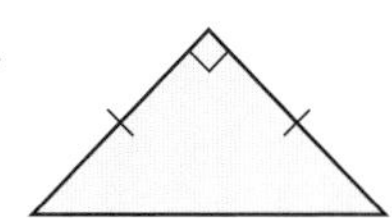

C10. 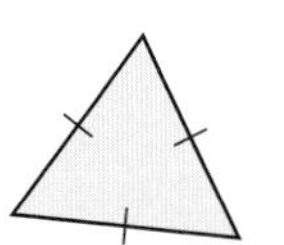

C11. 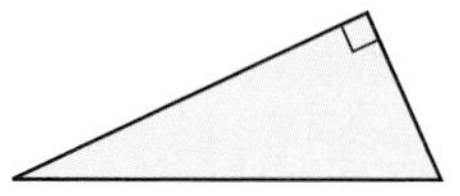

Section 7.2 Exercises

To classify polygons and find their perimeters:

1. Answer the Objective A Concept Checks.
2. Answer the odd Objective A Practice Exercises.
3. Answer the even Objective A Practice Exercises.

To classify quadrilaterals:

4. Answer the Objective B Concept Checks.
5. Answer the odd Objective B Practice Exercises.
6. Answer the even Objective B Practice Exercises.

To classify triangles:

7. Answer the Objective C Concept Checks.
8. Answer the odd Objective C Practice Exercises.
9. Answer the even Objective C Practice Exercises.

VOCABULARY REVIEW *Review the Vocabulary Preview for Section 7.2. Study the definitions until you can check* Got It *for every word.*

Use these words to complete each sentence.

line segment • polygon • perimeter • regular polygon

Polygons: triangle • quadrilateral • pentagon • hexagon • heptagon • octagon • nonagon • decagon

Special Triangles: scalene • isosceles • equilateral • right • obtuse • acute

Special Quadrilaterals: trapezoid • parallelogram • rectangle • rhombus • square

(Continued)

	Got It	Get Help
10. A three-sided polygon with only two sides of the same length is called a(n) ____________.		
11. When every side of a polygon is the same length, the polygon is called a(n) ____________.		
12. A four-sided polygon with all sides of the same length is called a(n) ____________.		
13. A six-sided polygon is called a(n) ____________.		
14. A triangle with three equal sides is a(n) ____________.		

How will you get help for any vocabulary that you are unsure about?

Your instructor ____ MyMathLab ____ A classmate ____ A tutor ____ Other ____

15. True or false. A polygon can have a curved side.

16. True or false. The sides of a polygon can cross.

Classify each polygon as accurately as possible. Indicate if it is regular. If the figure is not a polygon, then write "not a polygon."

17. 

18. 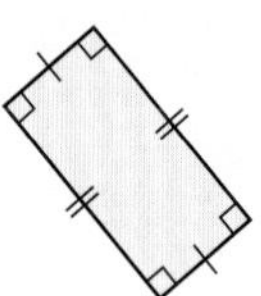

19. 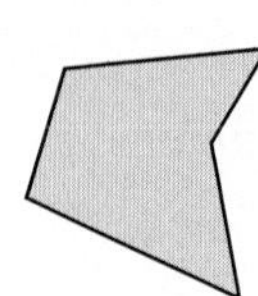

20. 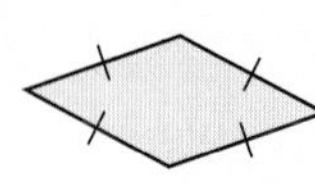

21. 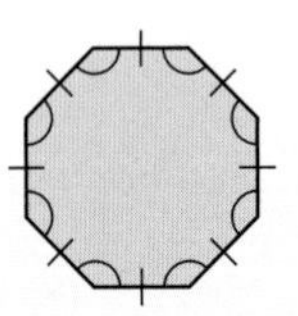

22. 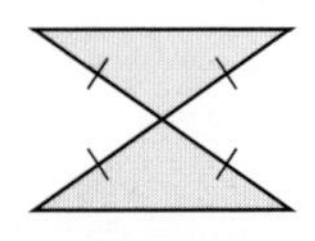

23. 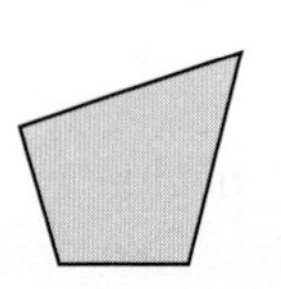

24. 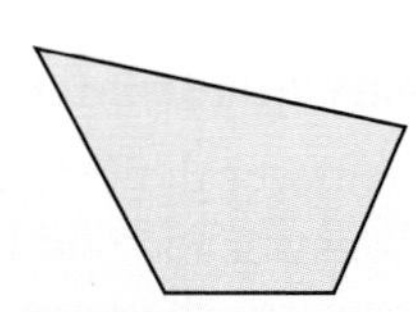

25. 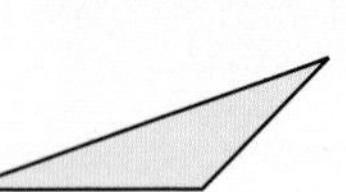

26. 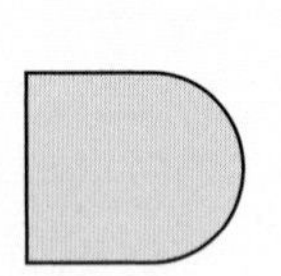

27. 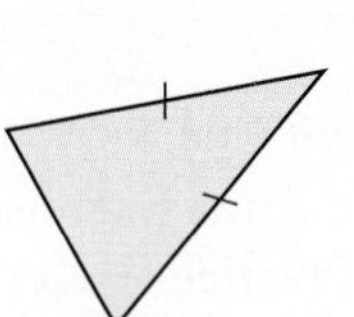

28. 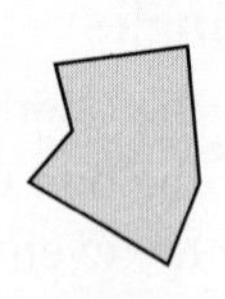

29. 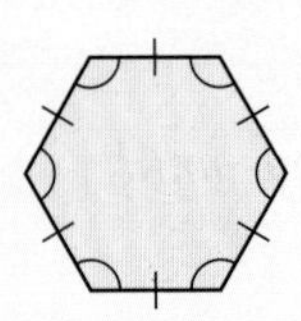

30. 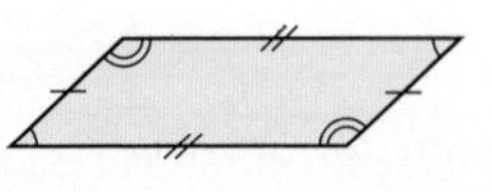

31. 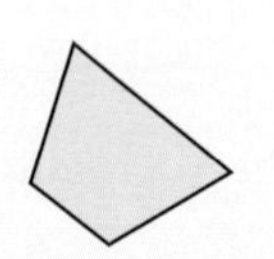

32. 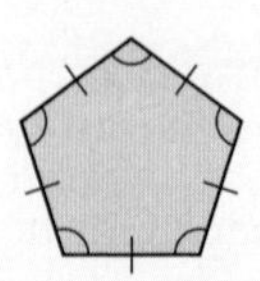

33. 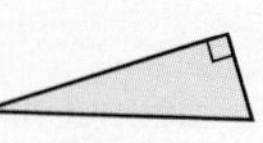

34. 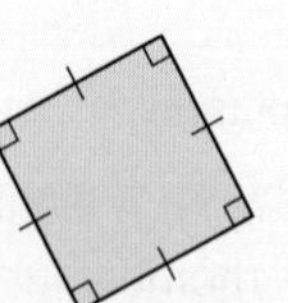

35. 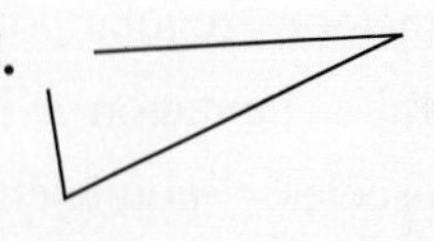

36. 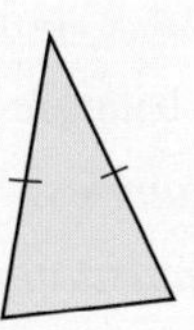

tobey Be sure you can identify which polygons are regular polygons. This is an essential skill.

For more tips and tweets, go to twitter.com/gstbasicmath

37. Which of the triangles in Exercises 17–36 are acute triangles?

38. Which of the triangles in Exercises 17–36 are obtuse triangles?

For each exercise, several properties of a polygon are listed. Classify each polygon as accurately as possible using only the names from this section.

39. 4 sides, all with the same length

40.
- 4 sides, all with the same length
- All angles equal 90°.

41.
- 3 sides
- Two sides have equal length.

42.
- 5 sides, all with the same length
- All angles measure 36°.

43.
- 4 sides
- Each pair of opposite sides is parallel.
- Pairs of opposite sides have different lengths.

44.
- 3 sides
- Every angle has a different measure.

45. 6 sides

46.
- 8 sides, all with the same length
- All angles are equal in measure.

The following figures are drawn using two regular polygons, with one placed on top of the other. What are those two polygons? Outline them on the figure.

47.

48.

The following figures are drawn using identical regular polygons that do not overlap. Draw the outlines of the polygons. The number of polygons used is written in parentheses.

49. (3 polygons)

50. (3 polygons)

51. (6 polygons)

52. (4 polygons)

What polygon best approximates each image? Give a reason for your answer.

53. The head of a giraffe

54. One cell of a honeycomb

55. The striking face of a croquet mallet

56. The windshield of a bus

57. The wheels on the bicycle shown

This bike actually rides smoothly . . . provided it's on the right ground.

58. The entire kite's perimeter

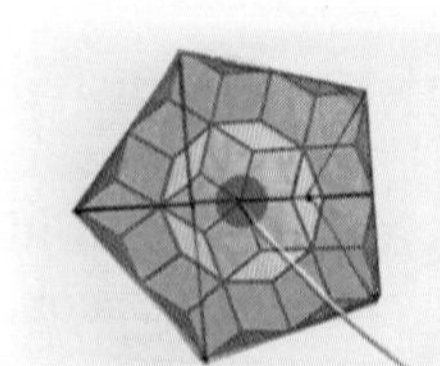

59. The Beatles plate shown

60. The star shown

61. Each side of the rear wheel from Exercise 57 measures 2.2 feet. What is the wheel's perimeter?

62. Each side of the kite in Exercise 58 measures 4.3 feet. What is the kite's perimeter?

63. Each side of the plate in Exercise 59 measures $2\frac{1}{2}$ inches. What is the plate's perimeter?

64. Each side of the star in Exercise 60 measures $5\frac{1}{5}$ inches. What is the star's perimeter?

Section 7.2 Question Log

Use this space to write questions. Be sure to get these questions answered and revisit them when you prepare for your exam.

Page ____ **Answered** ☐	**Page** ____ **Answered** ☐
Page ____ **Answered** ☐	**Page** ____ **Answered** ☐

Section 7.3 Perimeter and Area

To install new carpet and trim in a room, a contractor must find the area and perimeter of the room. The area, shown using blue squares, indicates the amount of carpet to install. The perimeter, shown with the dashed red line, indicates the length of trim to install around the edge of the room.

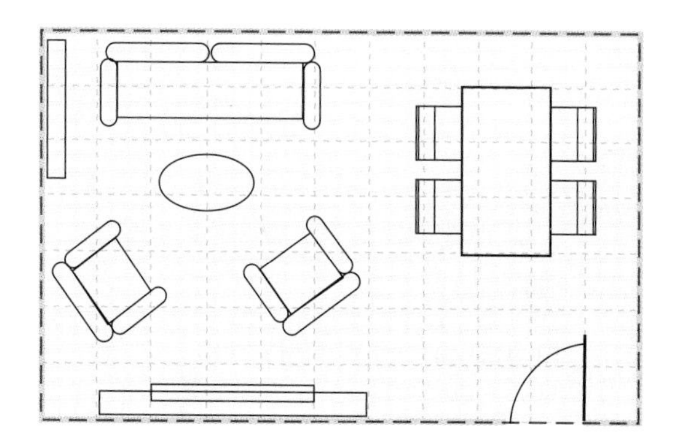

The **Objectives** in Section 7.3 will help you

- **A** Understand perimeter.
- **B** Understand area.
- **C** Use formulas to find area.

VOCABULARY PREVIEW *Check the box that applies.*

	Got It	Must Study
Perimeter: The distance around a figure is the **perimeter.**		
Area: The amount of surface inside a figure is the **area.**		

Study these words when they appear in the text. After the section, test your understanding by completing the Vocabulary Review in the section exercises.

Objective A Understand Perimeter

The Concept The **perimeter** of any figure is the distance around the figure. To find the perimeter of a figure, we could wrap a string along the outside of a figure. The perimeter of the figure would be the length of the string.

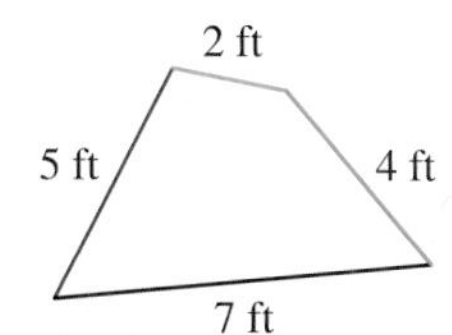

To find the perimeter, add the length of each side.

5 ft + 2 ft + 4 ft + 7 ft = 18 ft

The perimeter of the quadrilateral is $P = 5 + 2 + 4 + 7$

$= 18$ ft

Procedure Find the Perimeter of a Figure

Add the length of the sides.

You may need to find the unknown lengths of some sides.

EXAMPLES

1. Find the perimeter of the polygon.

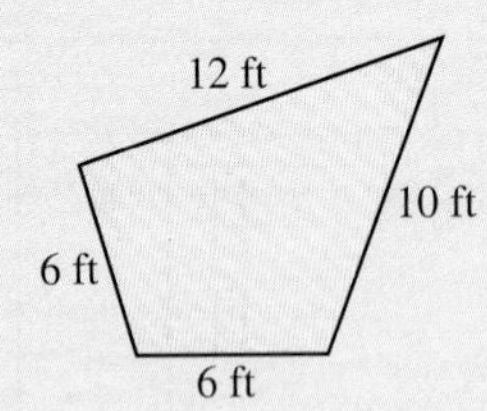

The perimeter is the distance around the polygon. Add the length of the sides.

$$P = 6 + 12 + 10 + 6$$
$$= 18 + 10 + 6$$
$$= 28 + 6$$
$$= 34 \text{ ft}$$

2. Find the perimeter of the figure.

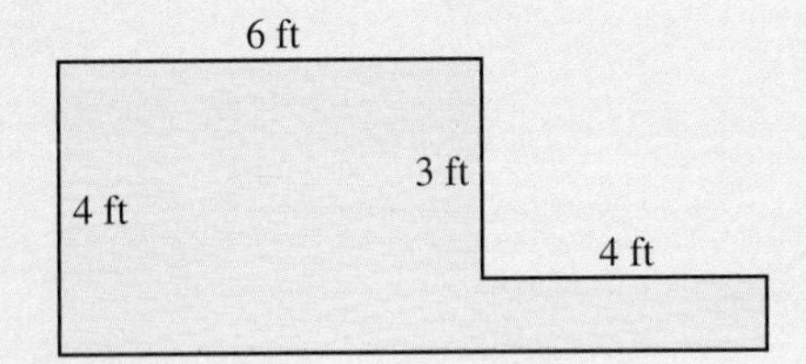

Before beginning, make sure that the length of every side is known. The length of two sides is unknown.

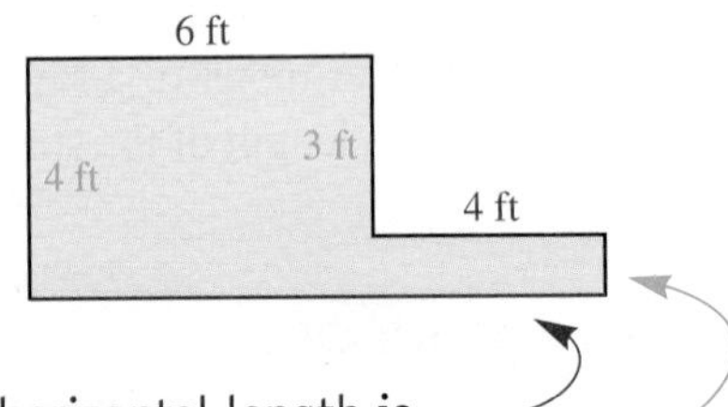

The missing horizontal length is
$6 + 4 = 10$ ft.
The missing vertical length is
$4 - 3 = 1$ ft.

The perimeter is:

$$P = 4 + 6 + 3 + 4 + 1 + 10$$
$$= 28 \text{ ft}$$

GUIDED PRACTICE

1. Find the perimeter of the polygon.

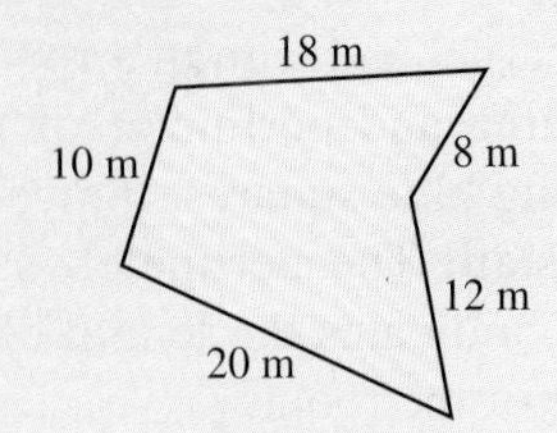

The perimeter is the distance around the polygon. Add the length of the sides.

$P =$

2. Find the perimeter of the figure.

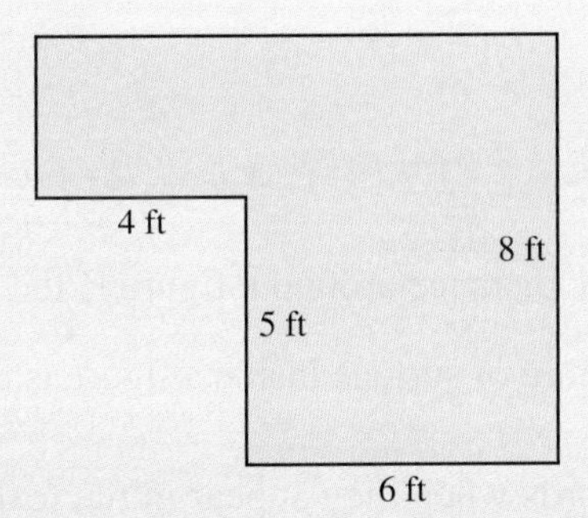

Before beginning, make sure that the length of every side is known. The length of two sides is unknown.

The missing horizontal length is
( ) + ( ) = (   ).

The missing vertical length is
( ) − ( ) = (   ).

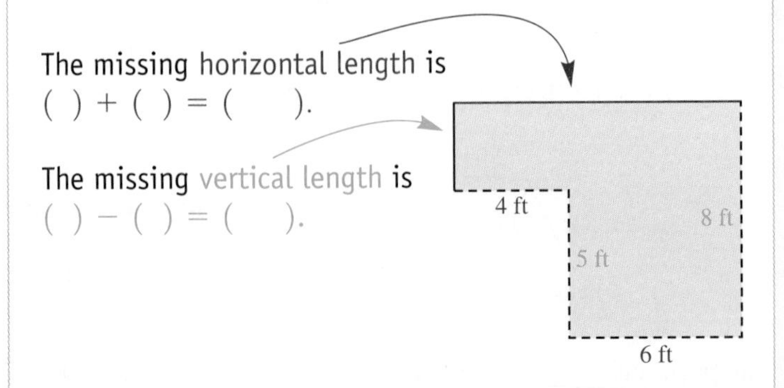

The perimeter is:

$P =$

$=$

Concept Check

A1. In your own words, what is the perimeter of a figure?

A2. Give three examples of units that can be used to measure perimeter.

Objective A Practice

Find the perimeter of each figure. Be sure to find any unknown lengths of sides.

A3. 3 ft, 2 ft, 3 ft, 4 ft, 3 ft

A4. 8 ft, 3 ft

A5. 4 ft, 6 ft

A6. 9 ft, 3 ft, 3 ft, 4 ft, 4 ft, 9 ft

A7. 7 ft, 17 ft, 20 ft, 12 ft, 24 ft

A8. 21 ft, 13 ft, 13 ft, 6 ft, 10 ft, 4 ft

A9. 6 ft, 6 ft, 6 ft, 20 ft, 20 ft, 18 ft

A10.

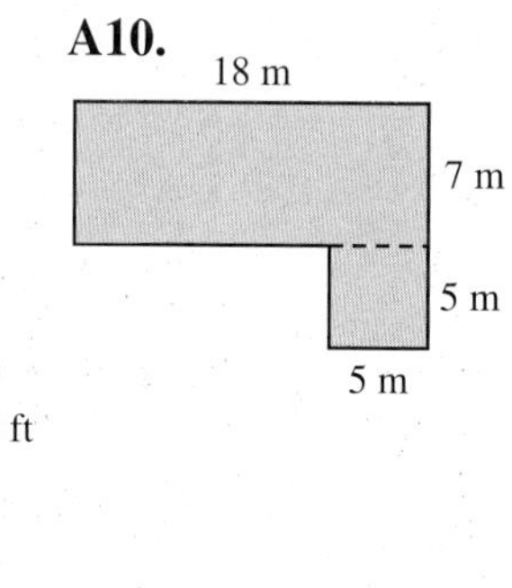

A11. What is the perimeter of the figure if each short side is 2.5 feet long and each long side is 5 feet long?

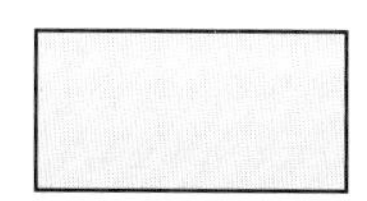

A12. What is the perimeter of the figure if each short side is 1 foot long and each long side is 1.5 feet long?

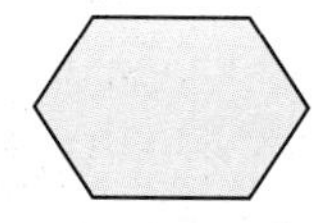

Objective B Understand Area

The Concept **Area** measures how much surface a figure covers. Area is measured in square units, or units^2, and can be found by counting the number of square units that the figure covers.

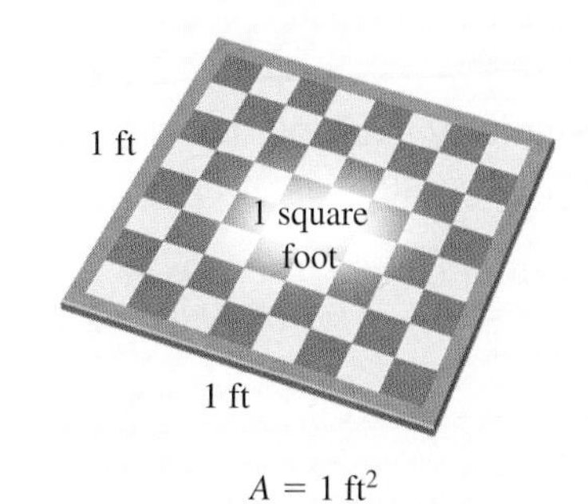

$A = 1 \text{ ft}^2$

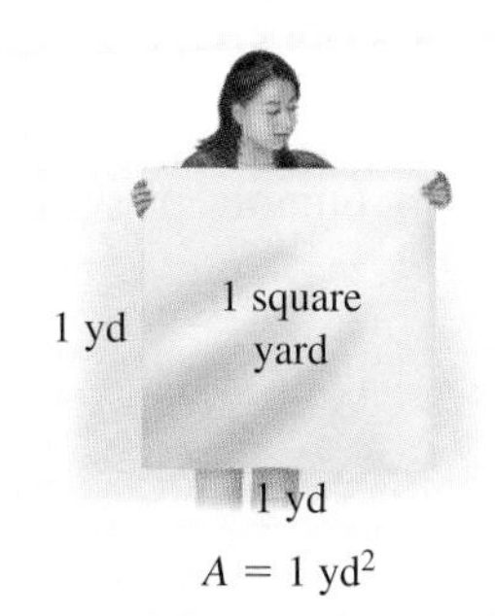

$A = 1 \text{ yd}^2$

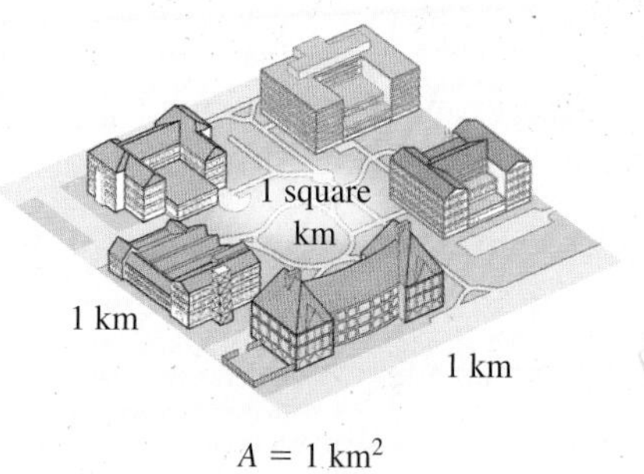

$A = 1 \text{ km}^2$

In the following figure, a grid of squares that are 1 cm by 1 cm is drawn over the triangle. Since it takes 8 squares to cover the triangle, the triangle's area is 8 cm^2.

Four $\frac{1}{2}$ squares are covered, making 2 full squares, or 2 cm^2.

6 full squares are covered, 6 cm^2

Area =

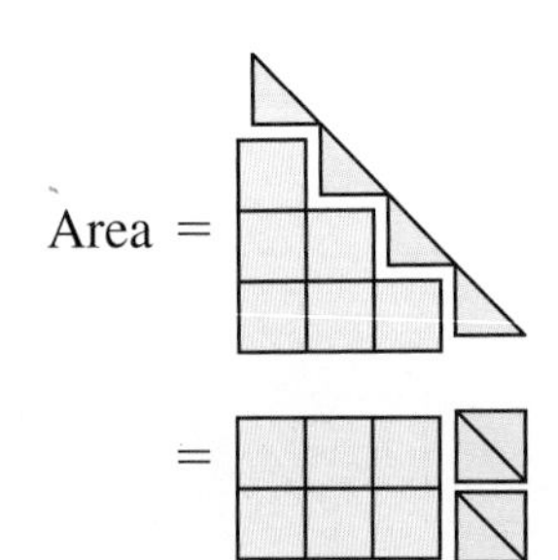

Area = 6 squares + 2 squares
= 6 cm^2 + 2 cm^2
= 8 cm^2, or 8 square centimeters

EXAMPLE

3. Find the area of the rectangle by counting squares. Each square has a side length of 1 foot.

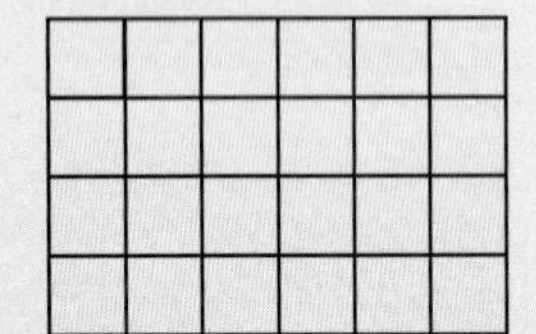

Counting, there are a total of 24 squares.

The area is 24 square feet.

To count the squares more quickly, multiply the number of squares per row by the number of rows.

Area = (squares per row) · (number of rows)
= 6 · 4
= 24 square feet

GUIDED PRACTICE

3. Find the area of the rectangle by counting squares. Each square has a side length of 1 meter.

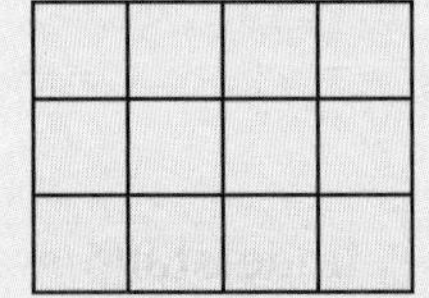

Counting, there are a total of ______ squares.

The area is ______ square ______.

To count the squares more quickly, multiply the number of squares per row by the number of rows.

Area = (squares per row) · (number of rows)
= ______ · ______
= ______ square ______

Fortunately, formulas exist for many figures, which makes finding area quicker than counting squares.

Area of Rectangles and Triangles

The **area of a rectangle** is the product of its length and width.	The **area of a triangle** is one-half the product of its base and height.
$\text{Area} = \text{Length} \cdot \text{Width}$	$\text{Area} = \frac{1}{2} \cdot \text{Base} \cdot \text{Height}$
$A = l \cdot w$	$A = \frac{1}{2} \cdot b \cdot h$
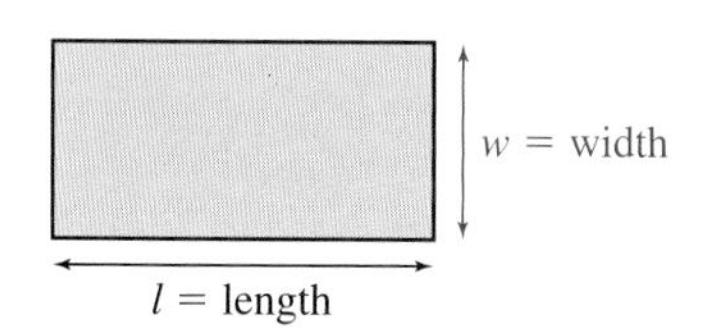	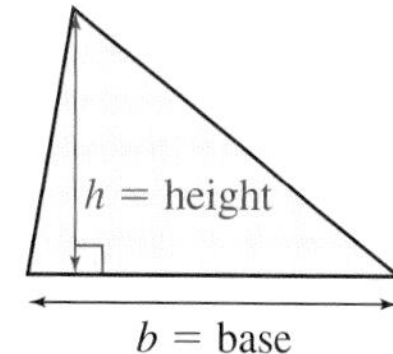 **Note: The base and height *must* make a 90° angle.**

▶ **NOTE** To use the formula for a triangle, the base and height must make a 90° angle. To see the height in some triangles, it may help to rotate the triangle.

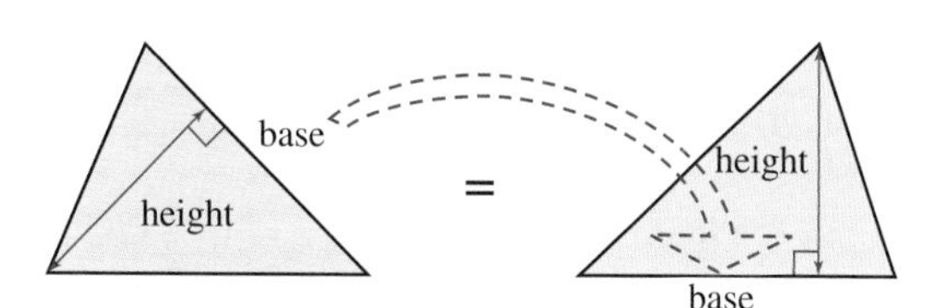

EXAMPLES

4. Find the area of the triangle.

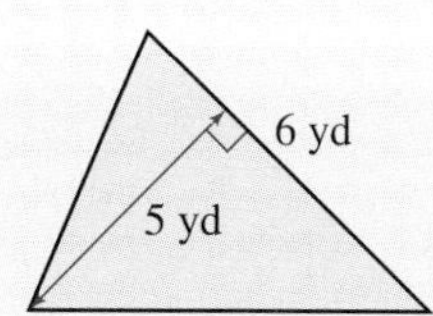

From the diagram, the height is 5 yd and the base is 6 yd.

$$
\begin{aligned}
A &= \frac{1}{2} \cdot b \cdot h \\
&= \frac{1}{2} \cdot (6) \cdot (5) \\
&= 3 \cdot 5 \\
&= 15 \text{ square yards}
\end{aligned}
$$

GUIDED PRACTICE

4. Find the area of the rectangle.

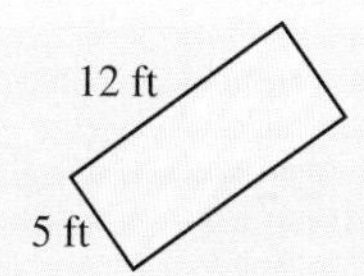

From the diagram, the width is ______ and the length is ______.

$$
\begin{aligned}
A &= l \cdot w \\
&= (\quad)(\quad) \\
&= \_\_\_\_\_\_ \text{ square } \_\_\_\_\_\_
\end{aligned}
$$

5. The following figure can be formed by combining two other basic figures. Find the area of the entire figure by adding the areas of the two figures that make it.

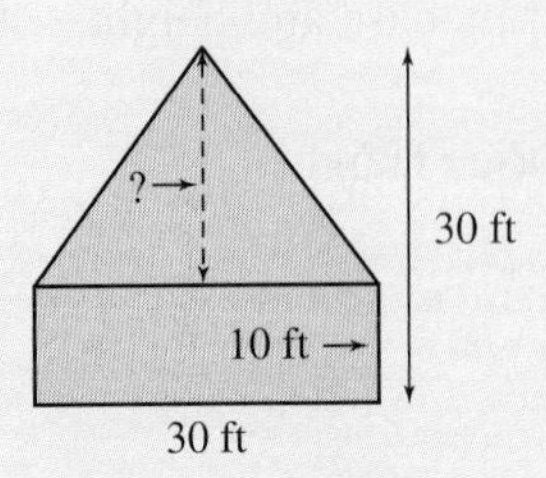

The figure is made of a rectangle and a triangle.

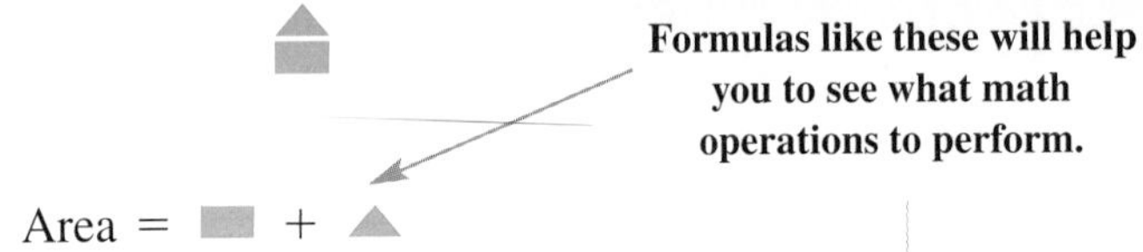

Area = ▬ + ▲

Rectangle	**Triangle**
length = 30 ft	base = 30 ft
width = 10 ft	height = 30 − 10
	= 20 ft

$A_{rectangle} = l \cdot w$
$= 30 \cdot 10$
$= 300$ square feet

$A_{triangle} = \frac{1}{2} \cdot b \cdot h$
$= \frac{1}{2} \cdot 30 \cdot 20$
$= 15 \cdot 20$
$= 300$ square feet

Area = ▬ + ▲
= 300 + 300
= 600 square feet

5. The following figure can be formed by combining two other basic figures. Find the area of the entire figure by adding the areas of the two figures that make it.

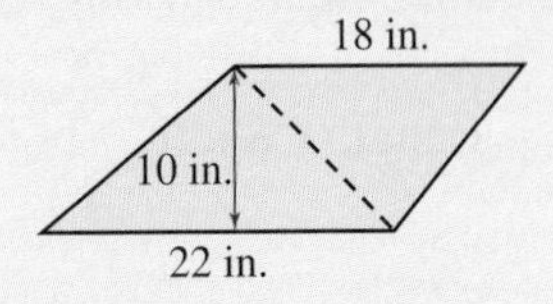

The figure is made of two triangles.

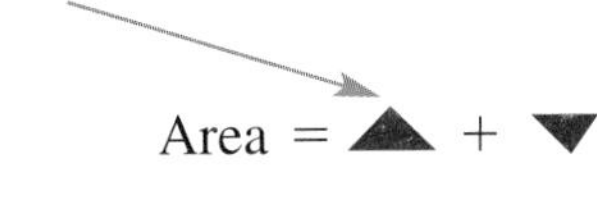

Area = ▲ + ▼

1st Triangle	**2nd Triangle**
base = ____	base = ____
height = ____	height = ____

$A_{triangle} = \frac{1}{2} \cdot b \cdot h$
$= \frac{1}{2} \cdot$ ____ $\cdot$ ____
$=$ ____ $\cdot$ ____
$=$ ____ square ______

$A_{triangle} = \frac{1}{2} \cdot b \cdot h$
$= \frac{1}{2} \cdot$ ____ $\cdot$ ____
$=$ ____ $\cdot$ ____
$=$ ____ square ______

Area = ▲ + ▼
= ______ + ______
= ____ square ____

Concept Check

B1. In your own words, why are the units of area square units, or units^2?

B2. Find the area of the figure shown by counting the squares inside the figure.

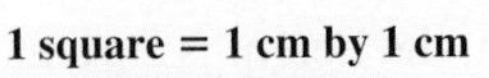

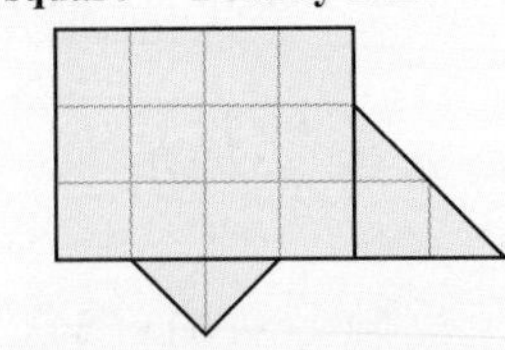

B3. A figure is measured in inches. Explain why the perimeter of the figure has units of inches and why the area of the figure has units of square inches (inches^2).

B4. Use the rectangle shown to explain why the formula for the area of a triangle is $A = \frac{1}{2} b \cdot h$.

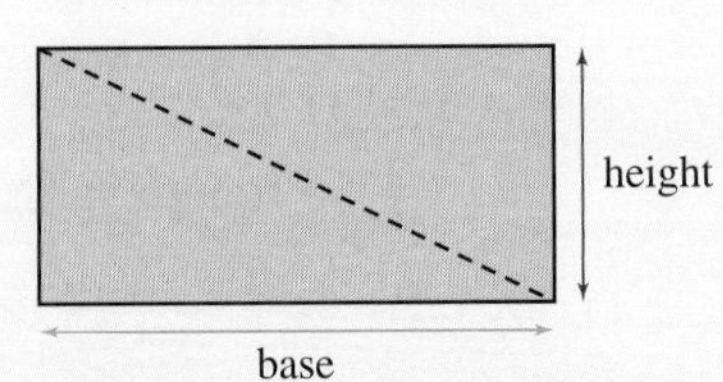

Objective B Practice

Find the area of each basic figure.

B5.

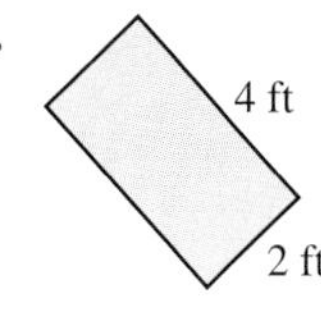

B6.

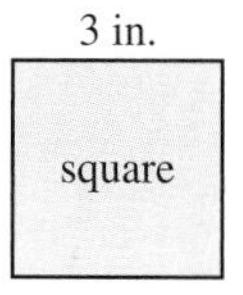

B7.

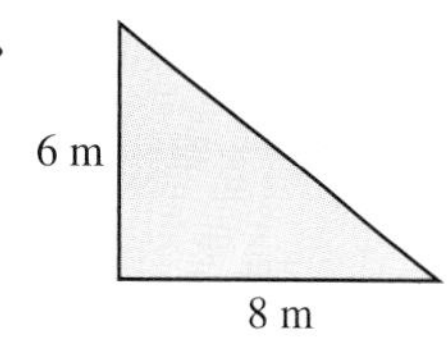

B8.

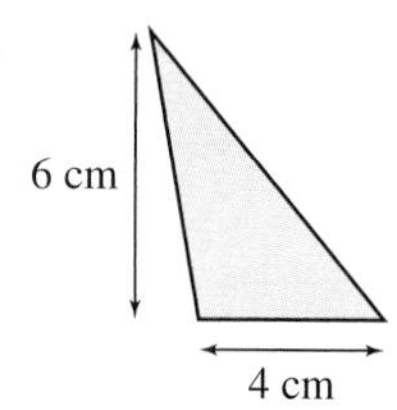

Draw one line on each figure to split it into two basic figures (rectangles or triangles).

B9.

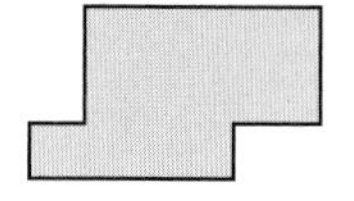

B10.

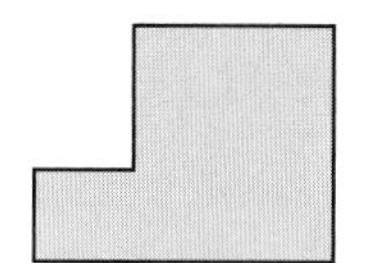

B11.

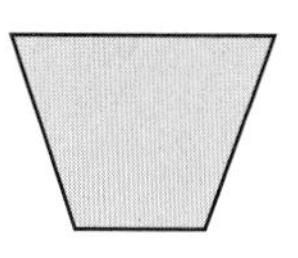

B12.

Draw one line on each figure to split it into two basic figures (rectangles or triangles). Then find the area of the figure by adding the areas of the basic figures that make up the larger figure.

B13.

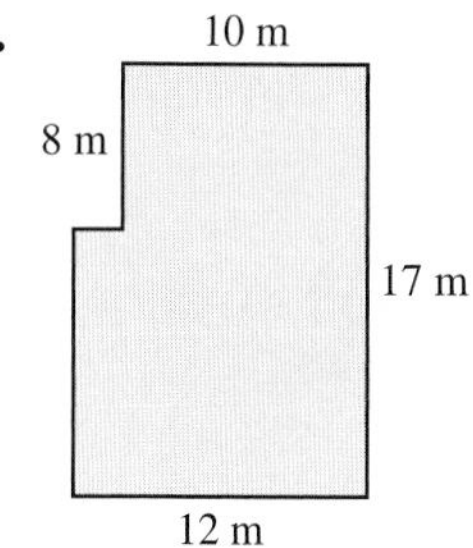

B14.

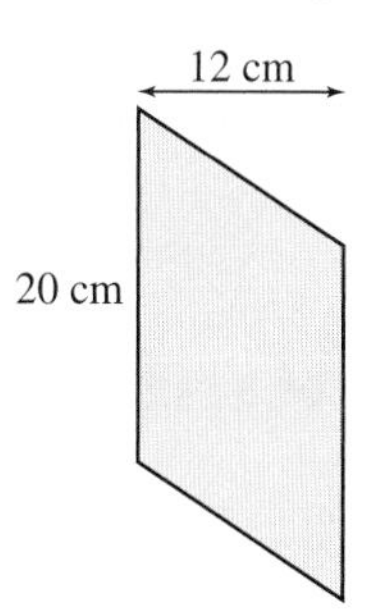

B15.

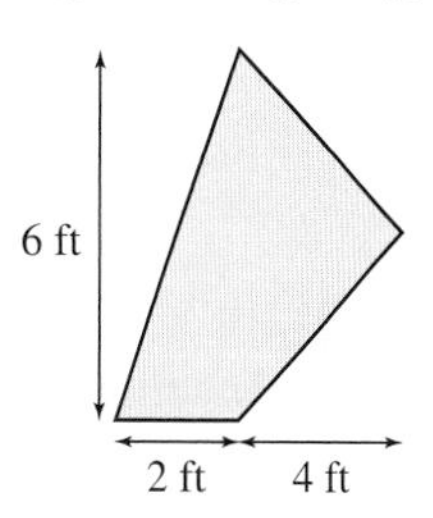

B16.

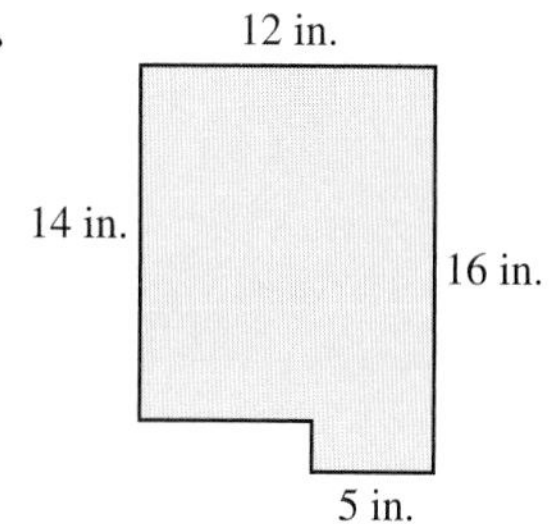

Objective C Use Formulas to Find Area

The Concept The area of many figures can be found using a formula. Two such figures are a parallelogram and a trapezoid.

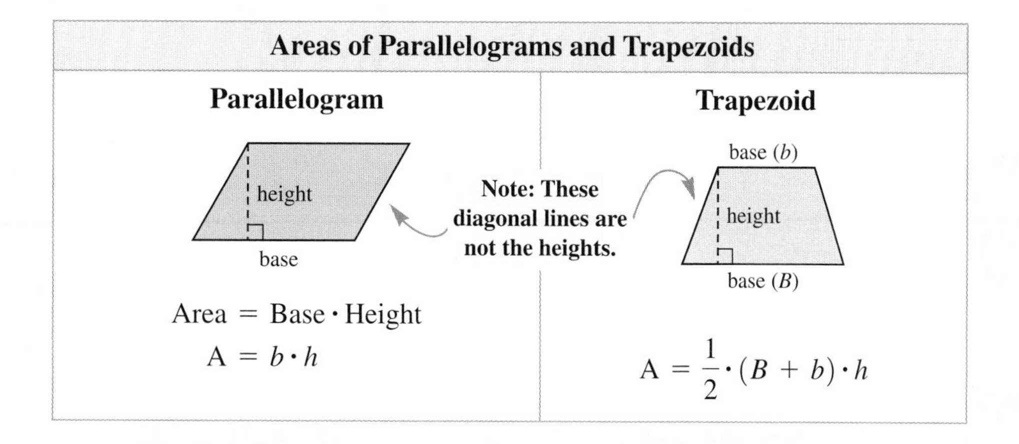

WHY IT WORKS Parallelogram and Trapezoid Formulas

If we cut the triangle off the left side of the parallelogram and reattach it to the right side, we form a rectangle. The formula will be consistent with the formula for a rectangle.

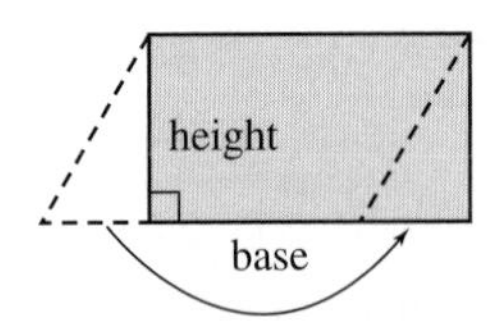

Since the area of a parallelogram is the same as the area of a rectangle, $A = b \cdot h$.

If we attach a second trapezoid, upside down, to the right of a trapezoid, we form a parallelogram. Notice that the length of the parallelogram is the sum of the two bases.

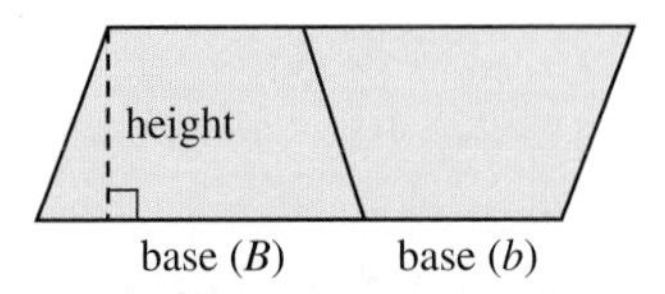

The area of this parallelogram is

$$\begin{aligned} A &= \text{base} \cdot \text{height} \\ &= (B + b) \cdot h \end{aligned}$$

Since the parallelogram is made of two identical trapezoids, multiply the formula above by one-half.

$$A = \frac{1}{2} \cdot (B + b) \cdot h$$

Procedure Use a Formula to Find the Area of a Figure

Step 1 Write the appropriate formula.

Step 2 Substitute the values for each variable.

Step 3 Simplify.

EXAMPLES

6. What is the area of the figure?

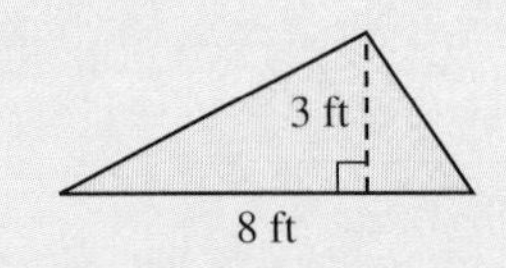

Step 1 Write the appropriate formula. $A = \frac{1}{2} \cdot b \cdot h$

Step 2 Substitute the values for each variable. $= \frac{1}{2} \cdot 8 \cdot 3$

Step 3 Simplify. $= 4 \cdot 3$

$= 12$

State the area, with units. The area is 12 ft ^{2}.

GUIDED PRACTICE

6. What is the area of the figure ?

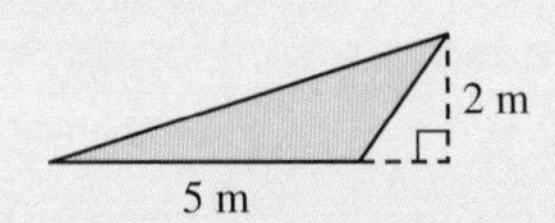

$A =$

$=$

$=$

$=$

The area is ______.

7. What is the area of the figure?

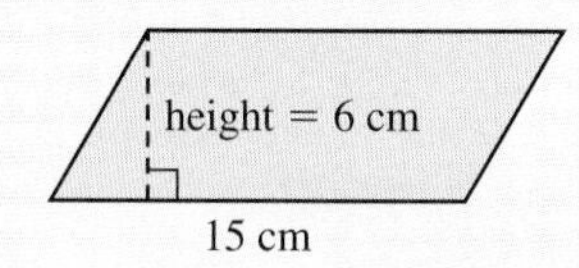

Step 1 Write the appropriate formula. $A = b \cdot h$

Step 2 Substitute the values for each variable. $= 15 \cdot 6$

Step 3 Simplify. $= 90$

The area is 90 cm^2.

7. What is the area of the figure?

height = 3 in.

2 in.

$A =$

$=$

$=$

The area is ______.

8. What is the area of the figure?

base = 6 yd

height = 2 yd

base = 12 yd

Step 1 Write the appropriate formula. $A = \frac{1}{2} \cdot (B + b) \cdot h$

Step 2 Substitute the values for each variable. $= \frac{1}{2} \cdot (12 + 6) \cdot 2$

Step 3 Simplify. $= \frac{1}{2} \cdot (18) \cdot 2$

$= 9 \cdot 2$

$= 18$

The area is 18 yd^2.

8. What is the area of the figure?

base = 20 ft

height = 7 ft

base = 14 ft

$A =$

$=$

$=$

$=$

$=$

The area is ______.

Concept Check

C1. From memory, write the formula for the area of each figure.

a. Triangle
b. Rectangle
c. Parallelogram
d. Trapezoid

C2. Fill in the blank.
A rectangle that measures 4 feet by 5 feet has an area of 20 ______ feet.

Objective C Practice

Find the area of each figure.

C3.

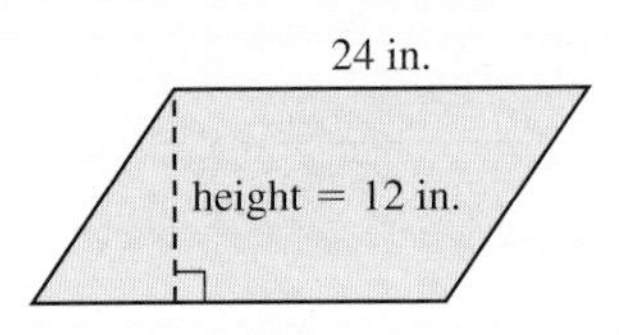

C4.

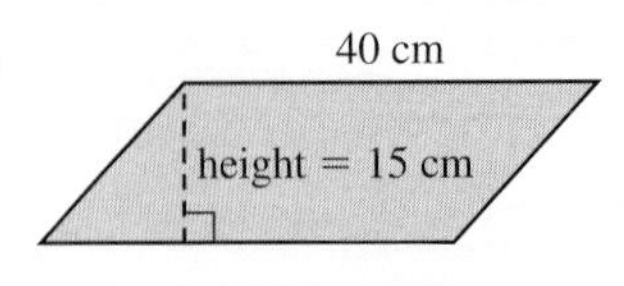

C5.

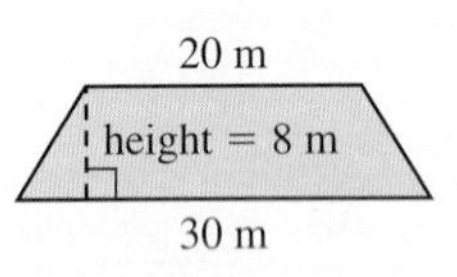

C6.

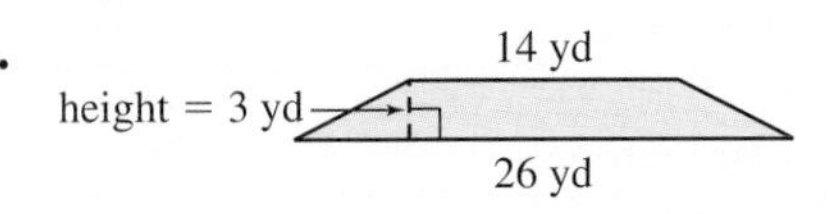

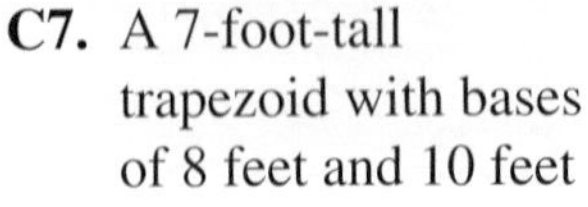

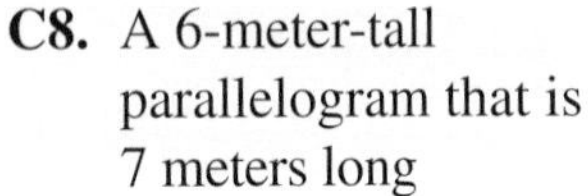

C7. A 7-foot-tall trapezoid with bases of 8 feet and 10 feet

C8. A 6-meter-tall parallelogram that is 7 meters long

C9. A 3-inch-tall trapezoid with bases of 6 inches and 8 inches

C10. An 8-yard-tall trapezoid with bases of 3 yards and 6 yards

Section 7.3 Exercises

To understand perimeter:

1. Answer the Objective A Concept Checks.
2. Answer the odd Objective A Practice Exercises.
3. Answer the even Objective A Practice Exercises.

To understand area:

4. Answer the Objective B Concept Checks.
5. Answer the odd Objective B Practice Exercises.
6. Answer the even Objective B Practice Exercises.

To use formulas to find area:

7. Answer the Objective C Concept Checks.
8. Answer the odd Objective C Practice Exercises.
9. Answer the even Objective C Practice Exercises.

VOCABULARY REVIEW *Review the Vocabulary Preview for Section 7.3. Study the definitions until you can check* Got It *for every word.*

Use these words to complete each sentence.

perimeter • area • lengths • lengths^2

	Got It	Get Help
10. The units used to determine the perimeter of a figure is ______________.		
11. The amount of surface inside a figure is the ______________.		
12. The distance around a figure is the ______________.		
13. The units used to determine the area of a figure is ______________.		

How will you get help for any vocabulary that you are unsure about?

Your instructor ____ MyMathLab ____ A classmate ____ A tutor ____ Other ____

Find the perimeter of each figure.

14.
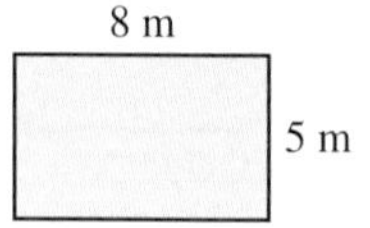

15.
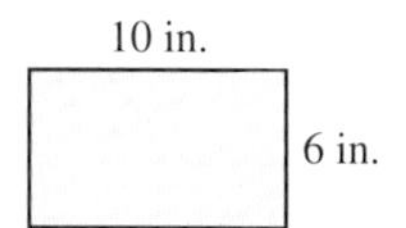

16.
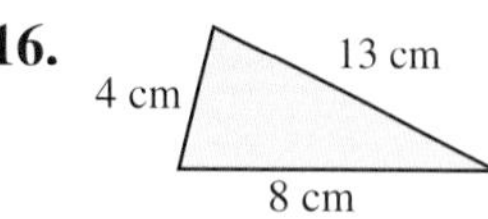

17.
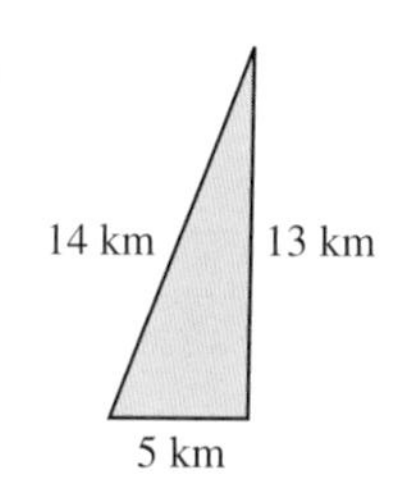

18.
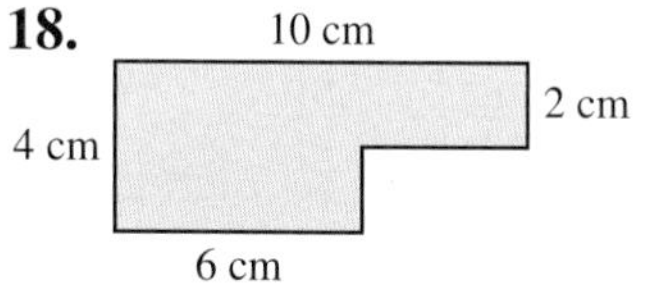

19.
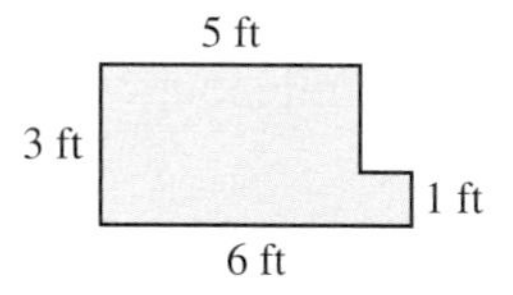

20.
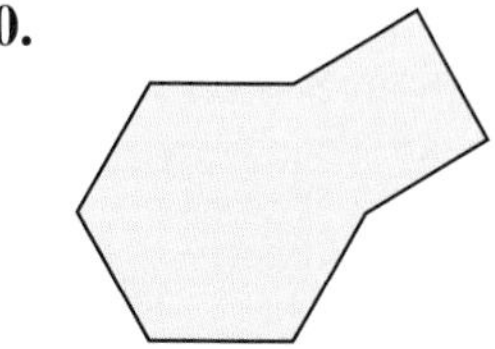

This figure is formed by attaching a regular hexagon to a square. Each side is 3 yd long.

21.
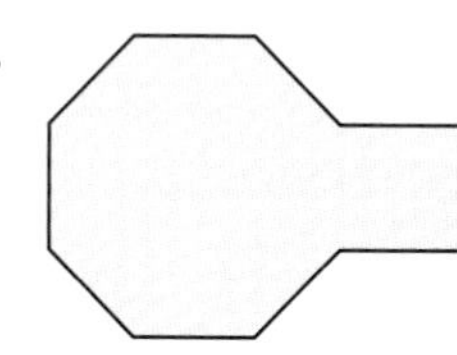

This figure is formed by attaching a regular octagon to a square. Each side is 4 m long.

Find the area of each figure

22.
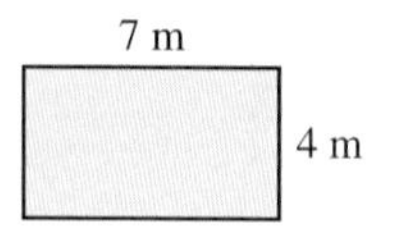

23.
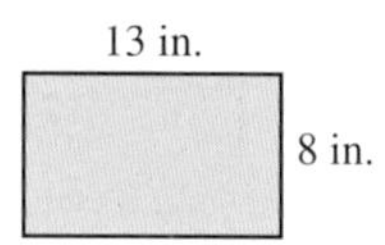

24.
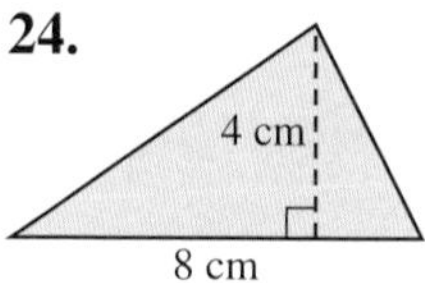

25.
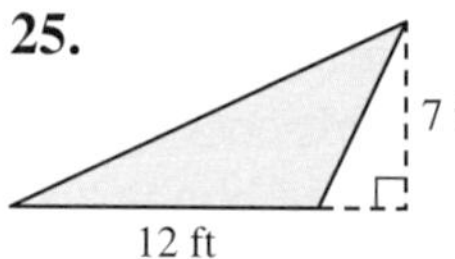

26.
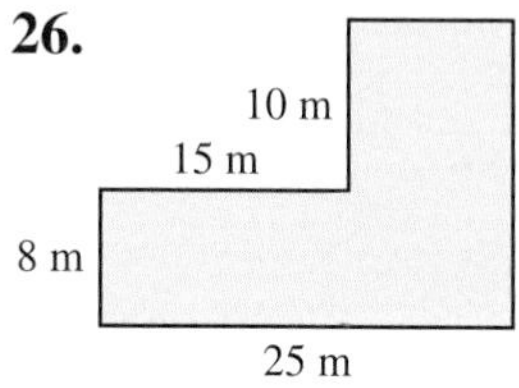

27.
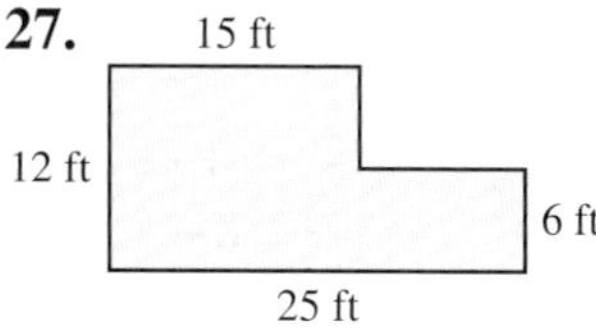

28.
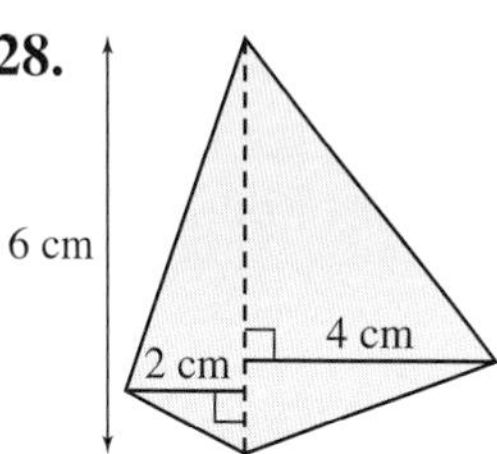

29.
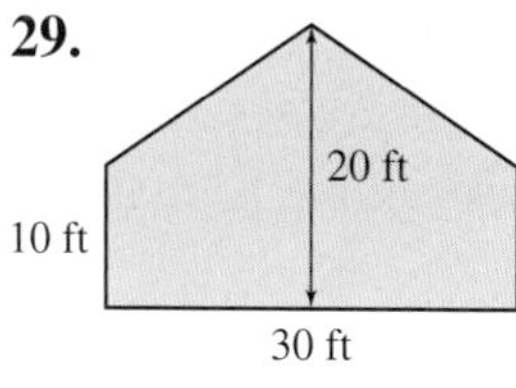

30.
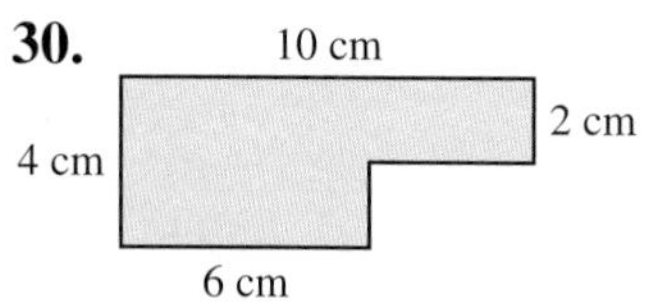

31.
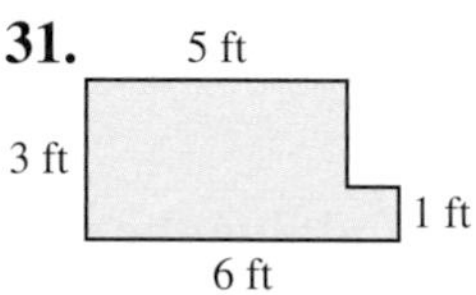

32.
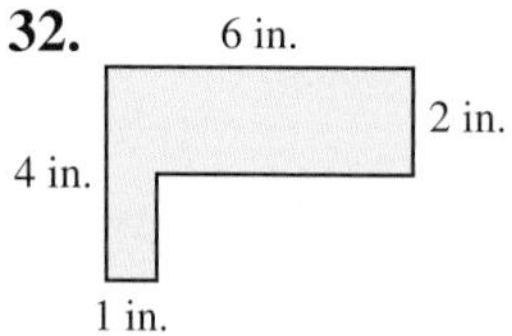

33.
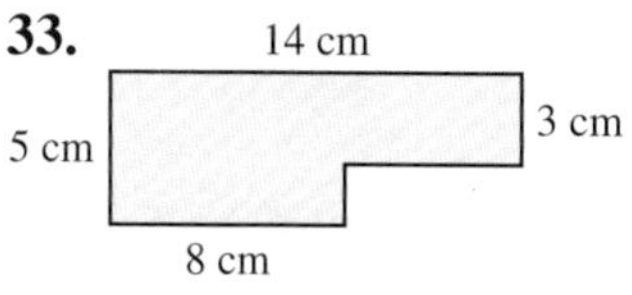

Use the appropriate formula to find the area of the figure shown or described.

34.
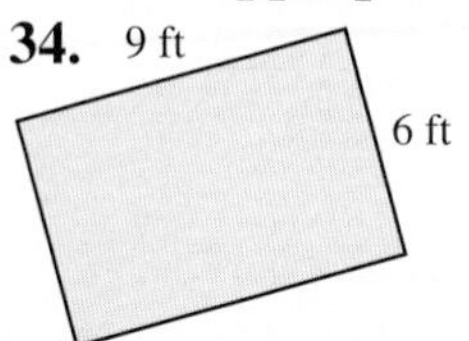

35.
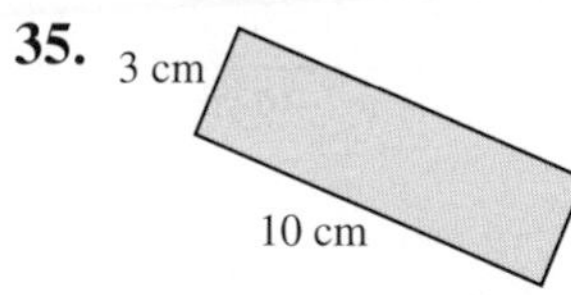

36.
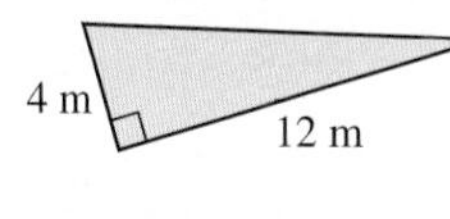

37.
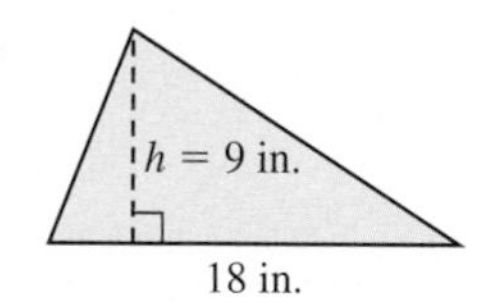

goetz For increased accuracy, draw a figure formula for each figure's area. It's easy to do and will help.

For more tips and tweets, go to twitter.com/gstbasicmath

38.

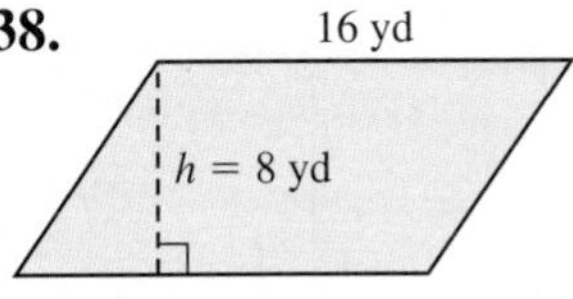

39.

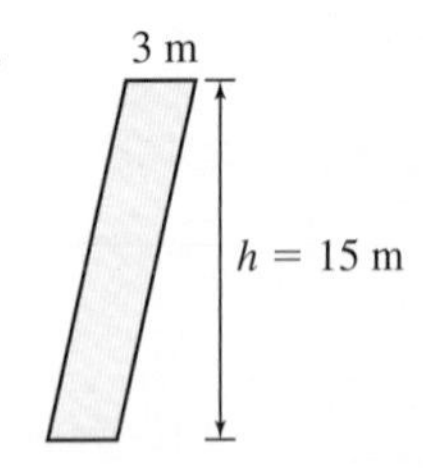

40.

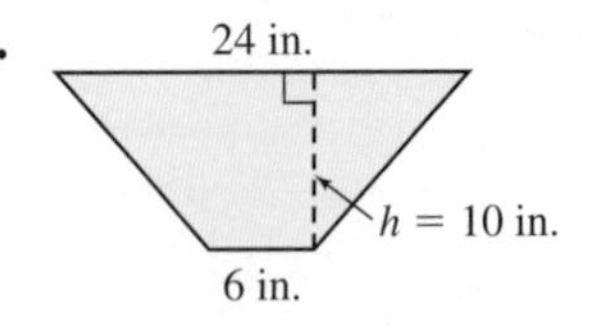

41.

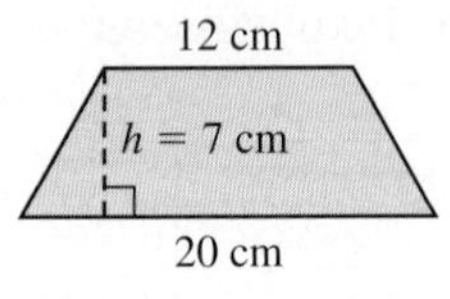

42. What is the area of a 3-foot-tall trapezoid that has bases of 4 feet and 6 feet?

43. What is the area of a 3-inch-tall triangle with an 8-inch base?

44. What is the area of a square with sides that are $\frac{1}{4}$ mile long?

45. What is the area of a rectangle that is $\frac{3}{8}$ mile long by $\frac{1}{2}$ mile wide?

Use the following floor plan to answer each question.

46. a. What are the areas of Office 1 and Office 2?

b. If carpet costs $3 per square foot, what is the total cost for carpeting both offices?

47. a. If each window is 5 feet tall, what is the combined area of all the windows?

b. If blinds cost $2 per square foot, what is the total cost of installing blinds?

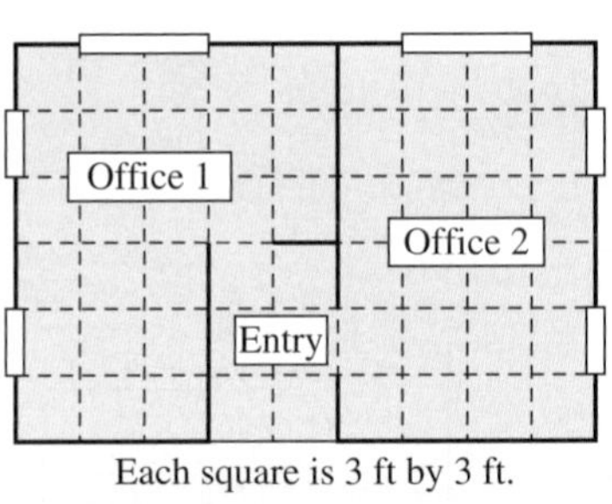

Each square is 3 ft by 3 ft.

= Windows

48. a. What is the combined perimeter of Office 1 and Office 2? Do not include door openings in the perimeter.

b. If trim costs $1.50 per foot, what is the total cost for installing trim along the perimeter of both offices?

49. a. If each window is 5 feet tall, what is the combined perimeter of all the windows?

b. If window trim costs $2.25 per foot, how much will it cost to install trim around the perimeter of all the windows?

Find the area of each figure by "cutting" it into two triangles and adding the areas of the two triangles.

50.

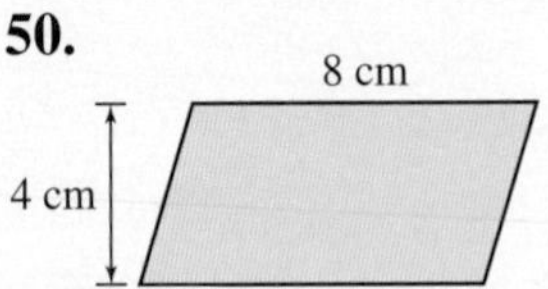

51.

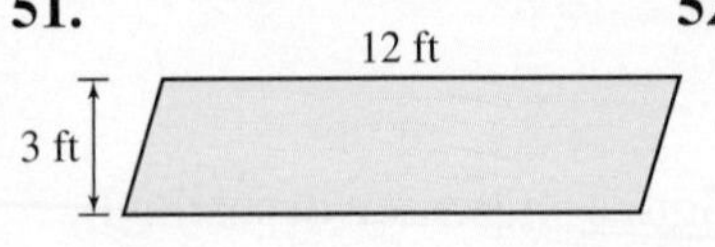

52.

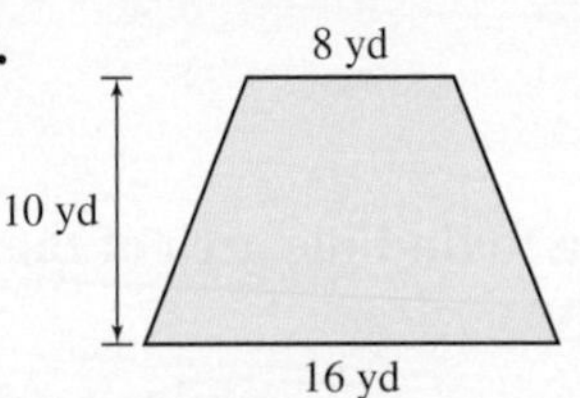

53.

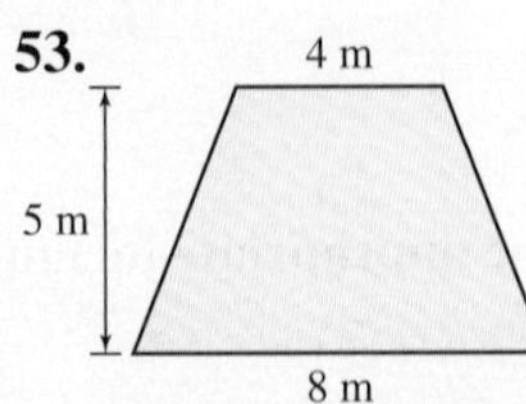

Answer each question.

54. An open box is made by cutting 2-inch square corners from an 8.5-inch by 11-inch sheet of cardboard and folding the paper on the dotted lines, as shown.

a. What is the area of the original sheet of cardboard?

b. What is the area of each square corner cut from the box?

c. What is the area of the cardboard used to make the box?

d. Find the percentage of material that is "waste" by dividing the combined area of the four squares by the area of the original cardboard. Then convert this number to a percent.

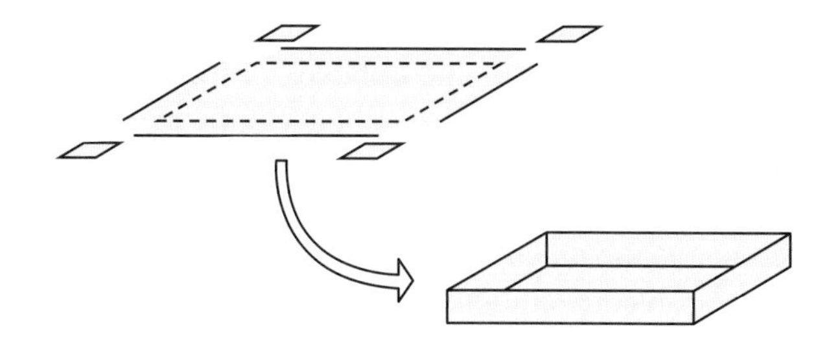

55. a. A student said that he could use triangles to estimate the area of a circle and drew Figure a. What was his estimate for the area of the circle if the base of each blue triangle is 1 inch and the height is 1.3 inches?

b. The student repeated the experiment using the red triangles in Figure b. Which would be more accurate, the estimate with the blue triangles or the estimate with the red triangles? Why?

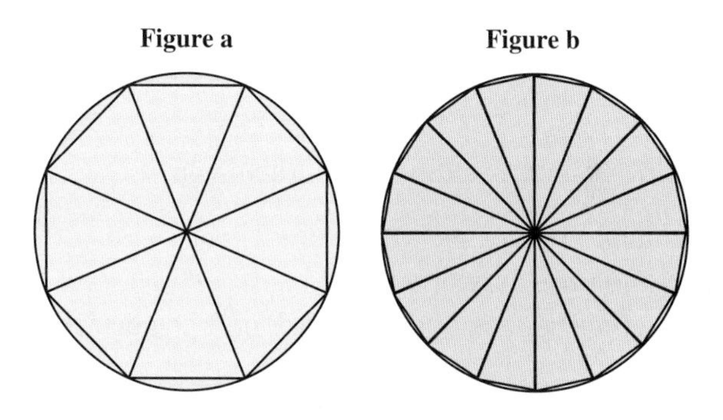

56. a. To estimate the area of the red region in Figure c, a student drew the three rectangles shown. What is the estimated area of the red region?

b. Show how two triangles can be added to the rectangles to improve the estimate. What is the new estimate?

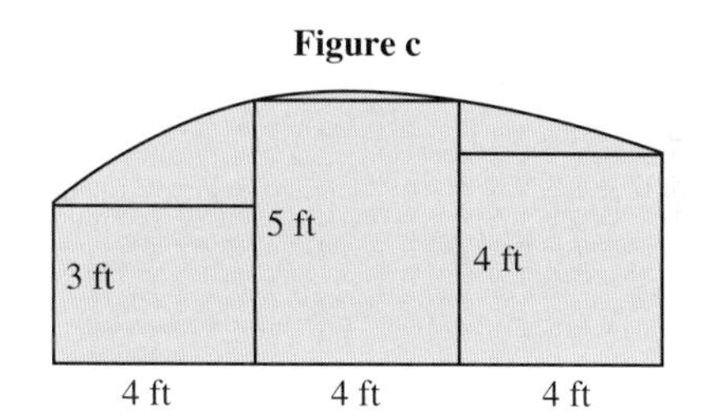

Section 7.3 Question Log

Use this space to write questions. Be sure to get these questions answered and revisit them when you prepare for your exam.

Page ____ Answered ☐	Page ____ Answered ☐
Page ____ Answered ☐	Page ____ Answered ☐

Section 7.4 Circles

Since circles appear in many everyday situations, it is important to understand the related terminology. For example, when designing a car, an engineer uses her knowledge of a circle's radius, diameter, and the number pi to determine the circumference of the wheels. The circumference is the distance the car travels every time the wheel makes one revolution.

- Two dimensions can be used to define a circle, the **radius** and the **diameter.**
- The perimeter of a circle is called the **circumference.**
- **Pi** (π) is a special number that relates the diameter and the circumference of a circle. $\pi \approx 3.14$ ($\approx$ means "is approximately equal to.")

The **Objectives** in Section 7.4 will help you

A Find the circumference of a circle.
B Find the area of a circle.
C Find the area and perimeter of a composite figure.

VOCABULARY PREVIEW *Check the box that applies.*

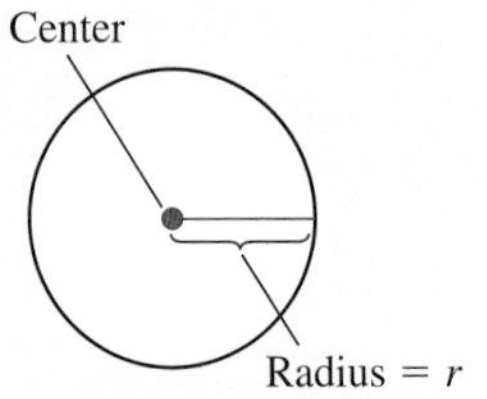

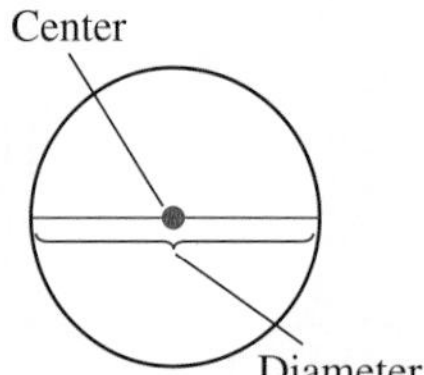

	Got It	Must Study
Radius: The **radius** is the distance from the center of a circle to its outer edge.		
Diameter: The **diameter** is the distance across a circle, through the center.		
Circumference: The perimeter of a circle is the **circumference.**		
π (Pi): π is a number that is approximately equal to 3.14. π is defined as the ratio of a circle's circumference to its diameter. Because the digits of pi never end and never repeat, it is approximated with 3.14. Many calculators approximate pi as $\pi \approx 3.141592654$.		

Study these words when they appear in the text. After the section, test your understanding by completing the Vocabulary Review in the section exercises.

Objective A Find the Circumference of a Circle

The Concept To determine the circumference (perimeter) of a circle, we could wrap a string around the outside of the circle. The length of the string would be the circumference of the circle.

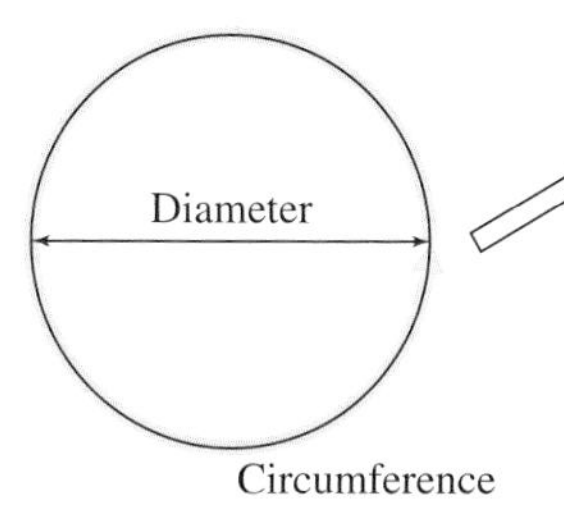

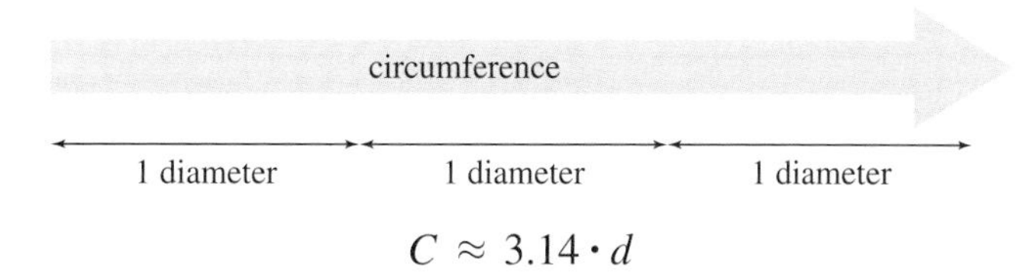

Comparing the circumference to the diameter, we see that the circumference is a little more than 3 times the diameter. In fact, the string is π times as long as 1 diameter.

circumference

1 diameter 1 diameter 1 diameter

$$C \approx 3.14 \cdot d$$

The exact circumference is given by the formula $C = \pi \cdot d$, where $\pi \approx 3.14$. Note that $\approx$ means "is approximately equal to."

Because a circle's diameter is twice the radius, we also can use the formula $C = 2\pi r$ to find the circumference.

Circumference of a Circle

$C = \pi d$ or $C = 2\pi r$,

where $\pi \approx 3.14$.

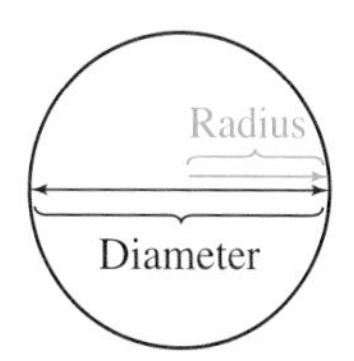

Procedure Find the Circumference of a Circle

Determine the diameter or radius. Then use the appropriate formula.

EXAMPLES

1. Find the circumference. Use $\pi \approx 3.14$.

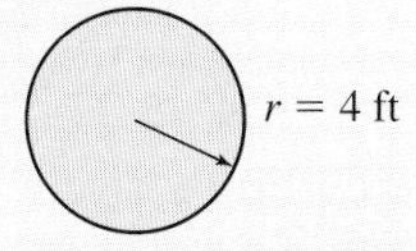

Because the distance from the center to the outside edge is given, we know the radius. $r = 4$ ft

$$C = 2\pi r$$
$$= 2 \cdot \pi \cdot (4)$$
$$\approx 8 \cdot 3.14$$
$$\approx 25.12 \text{ ft}$$

You can use the commutative property to multiply the whole numbers first.

GUIDED PRACTICE

1. Find the circumference. Use $\pi \approx 3.14$.

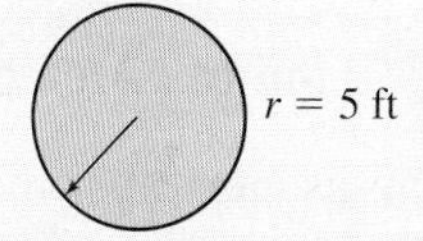

Because the distance from the center to the outside edge is given, we know the radius. $r =$ ______

$$C = 2\pi r$$
$$= 2 \cdot \pi \cdot (\ )$$
$$\approx \quad \cdot 3.14$$
$$\approx$$

(Continued)

2. Find the circumference. Use $\pi \approx 3.14$.

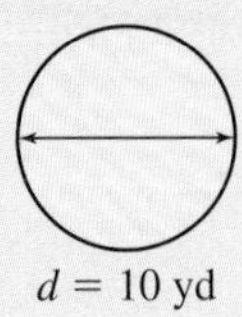

$d = 10$ yd

Because the distance across the circle is given, we know the diameter. $d = 10$ yd

$$\begin{aligned} C &= \pi d \\ &= \pi \cdot (10) \\ &\approx 3.14 \cdot 10 \\ &\approx 31.4 \text{ yd} \end{aligned}$$

2. Find the circumference. Use $\pi \approx 3.14$.

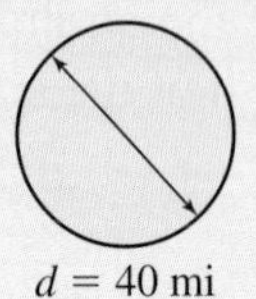

$d = 40$ mi

Because the distance across the circle is given, we know the diameter. $d =$ ______

$$\begin{aligned} C &= \pi d \\ &= \pi \cdot (\quad) \\ &\approx \quad \cdot \\ &\approx \end{aligned}$$

Concept Check

A1. If the radius of a circle is 4 miles, what is the circle's diameter?

A2. What is the name of the measure that describes the distance from the center of a circle to its outer edge?

A3. What is the name used for the perimeter of a circle?

Objective A Practice

Find the circumference of each circle. Use $\pi \approx 3.14$.

A4.

$r = 12$ cm

A5.

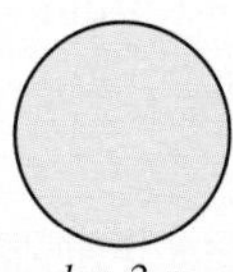

$d = 3$ m

A6.

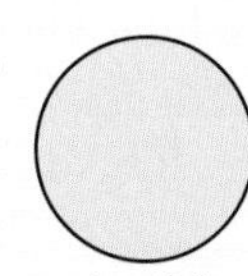

$d = 8$ ft

A7.

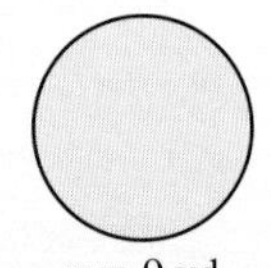

$r = 9$ yd

Objective B Find the Area of a Circle

The Concept We can use an experiment to estimate the area of a circle. Imagine cutting a circle to form 8 pie-shaped sectors. Rearranging the pie-shaped sectors, we can make a figure that approximates a parallelogram with the given base and height.

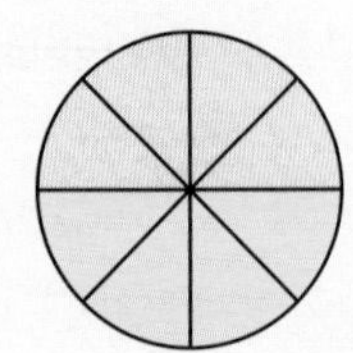

1 radius

$h \approx r$

$\frac{\text{Circumference}}{2}$

Since the green sectors form a half circle, the parallelogram's base is about half of the circumference.

We can approximate the area of the circle using the formula for the area of a parallelogram, $A_{\text{parallelogram}}$.

$$\begin{aligned} A_{\text{parallelogram}} &= \text{Base} \cdot \text{Height} \\ &\approx \frac{\text{Circumference}}{2} \cdot \text{Height} \\ &\approx \frac{\cancel{2}\pi r}{\cancel{2}} \cdot r \\ &\approx \pi \cdot r \cdot r \\ &\approx \pi r^2 \end{aligned}$$

This approximation becomes more accurate when more sectors are used. In later math courses, it can be shown that this approximation gives the exact area of a circle.

Area of a Circle

$$A = \pi \cdot r^2$$

Procedure Find the Area of a Circle

Determine the radius. Then use the formula $A = \pi \cdot r^2$.

EXAMPLES

3. Find the area of the circle. Use $\pi \approx 3.14$.

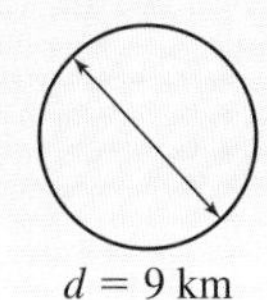

To find a radius, divide the diameter by 2. Because the diameter is 9 km, the radius is $9 \div 2 = 4.5$ km.

$$\begin{aligned} A &= \pi \cdot r^2 \\ &= \pi \cdot (4.5)^2 \\ &\approx 3.14 \cdot (20.25) \\ &\approx 63.585 \text{ square km} \end{aligned}$$

Important:

- Follow the order of operations. Perform exponents before multiplication.
- Describe area using square units.

GUIDED PRACTICE

3. Find the area of the circle. Use $\pi \approx 3.14$.

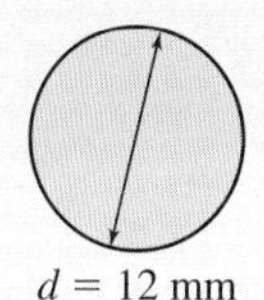

To find the radius, divide the diameter by 2. Because the diameter is ______, the radius is ____ ÷ ____ = ______.

$$\begin{aligned} A &= \pi \cdot r^2 \\ &= \pi \cdot (\quad)^2 \\ &\approx 3.14 \cdot (\quad) \\ &\approx \underline{\qquad} \text{ square } \underline{\qquad} \end{aligned}$$

Important:

- Follow the order of operations. Perform exponents before multiplication.
- Describe area using square units.

4. Find the area inside the square but outside the circle. Use $\pi \approx 3.14$.

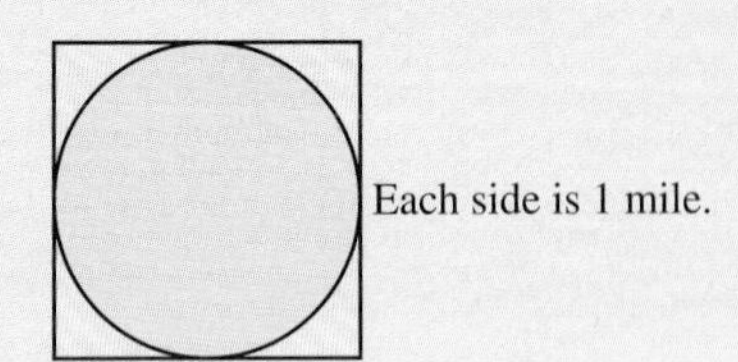

The diameter of the circle is the same as the length of the side of the square, so the diameter is 1 mi. The radius is 0.5 mi.

Area = 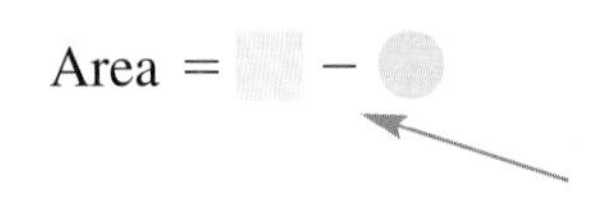

"Figure formulas" like these will help you see the math operations that you must perform.

Square

$A = l \cdot w$

$= 1 \cdot 1$

$= 1 \text{ mi}^2$

Circle

$A = \pi \cdot r^2$

$= \pi \cdot (0.5)^2$

$\approx 3.14 \cdot (0.25)$

$\approx 0.785 \text{ mi}^2$

Area = ▢ − ◯

$\approx 1 - 0.785$

$\approx 0.215 \text{ mi}^2$

4. Find the area inside the rectangle but outside the circle. Use $\pi \approx 3.14$.

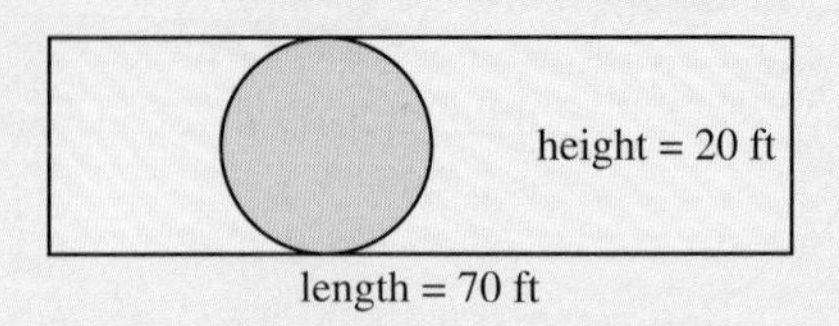

The diameter is the same as the length of the ______ of the rectangle, so the diameter is ______. The radius is ______.

Area = 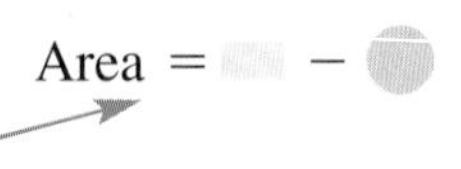

Rectangle

$A = \quad \cdot$

$= \quad \cdot$

$=$

Circle

$A = \pi \cdot r^2$

$= \pi \cdot (\quad)^2$

$\approx \quad \cdot$

$\approx$

Area = ▭ − ◯

$\approx \quad -$

$\approx$

Concept Check

B1. Write a figure formula to find the area of the orange region: the area inside the triangle but outside the circle. You do not need to find the actual area, just the figure formula.

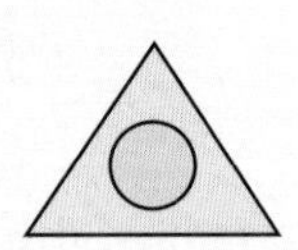

B2. If the radius of a circle is given in inches, then the circle's area will have units of ______ inches.

Objective B Practice

Find the area of each circle. Use $\pi \approx 3.14$.

B3.

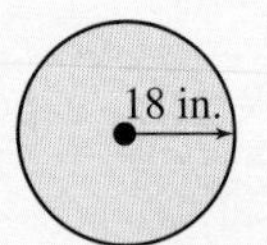

B4.

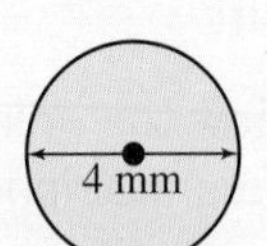

B5.

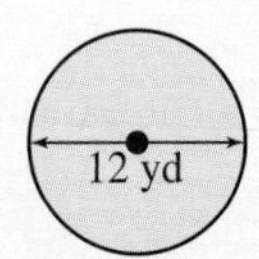

B6.

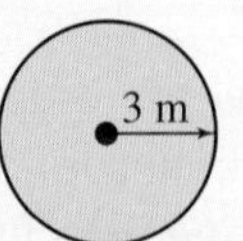

Draw each figure described. Then find the area outside the circle but inside the other figure. The circle must be drawn as large as possible. Use $\pi \approx 3.14$.

B7. A circle is drawn inside a square with sides of 10 m.

B8. A circle is drawn inside a parallelogram with a height of 4 inches and a base of 9 inches.

B9. A circle is drawn inside a trapezoid with a height of 3 mm and base lengths of 5 mm and 7 mm.

B10. A circle is drawn inside a rectangle with a length of 3 inches and a width of 5 inches.

Objective C Find the Area and Perimeter of a Composite Figure

The Concept Composite figures are made up of two or more simpler figures. The word *composite* means "a collection of parts or components." In this case, the components are two or more simpler figures.

- To find the area of a composite figure, find the area of each simpler figure, then add or subtract the areas.
- To find the perimeter of a composite figure, add the lengths of the outside edges.

EXAMPLES

5. Find the area of the composite figure. Use $\pi \approx 3.14$.

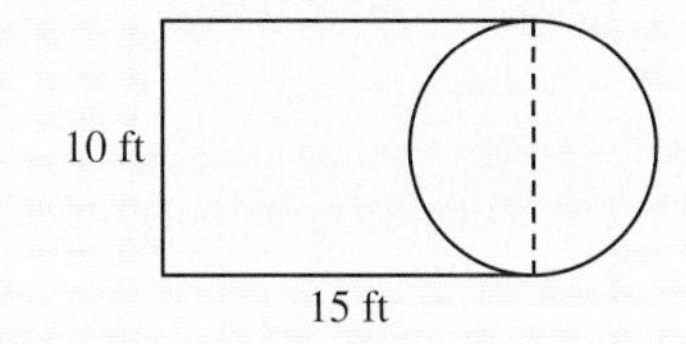

Start by writing a figure formula.

Area = +

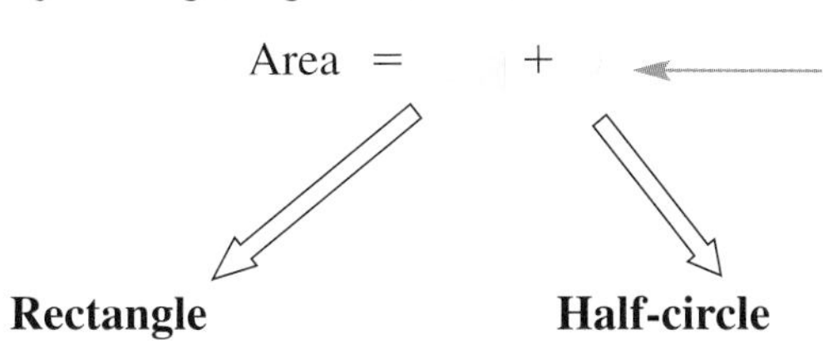

Rectangle

$$\text{Area} = l \cdot w$$
$$= 10 \cdot 15$$
$$= 150 \text{ ft}^2$$

Half-circle

$$\text{Area} = \frac{1}{2}\pi r^2$$
$$\approx (0.5) \cdot (3.14) \cdot 5^2$$
$$\approx (0.5) \cdot (3.14) \cdot 25$$
$$\approx 39.25 \text{ ft}^2$$

$$\text{Area} \approx 150 + 39.25$$
$$\approx 189.25 \text{ ft}^2$$

Figure formulas like these will help you see the math operations you must perform.

Multiply by $\frac{1}{2}$ to find half of the circle.

GUIDED PRACTICE

5. Find the area of the composite figure. Use $\pi \approx 3.14$.

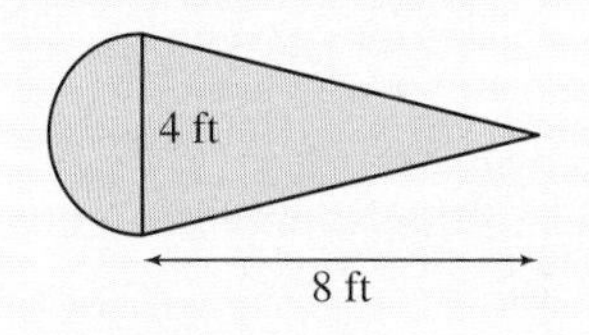

Start by writing a figure formula.

Area = +

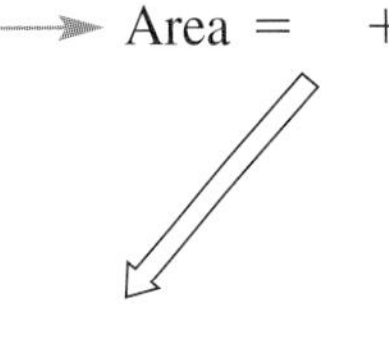

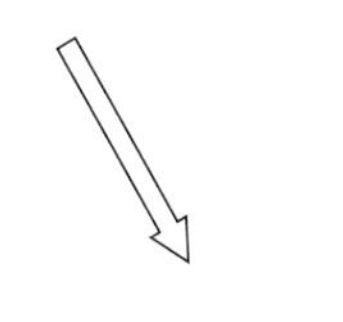

________ ________

Area =

Area =

Area ≈ +

≈

6. Find the perimeter of the composite figure.
Use $\pi \approx 3.14$.

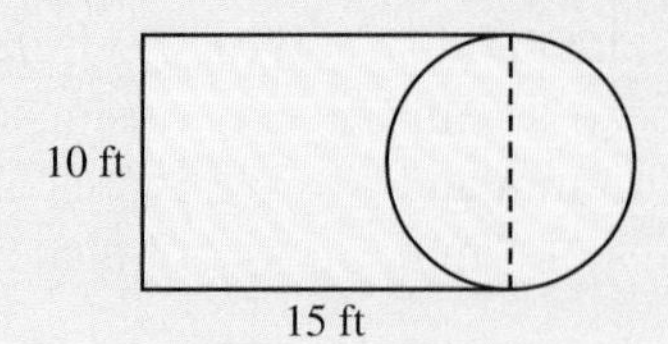

Start by writing a figure formula for the outside edges of the composite figure.

Perimeter =

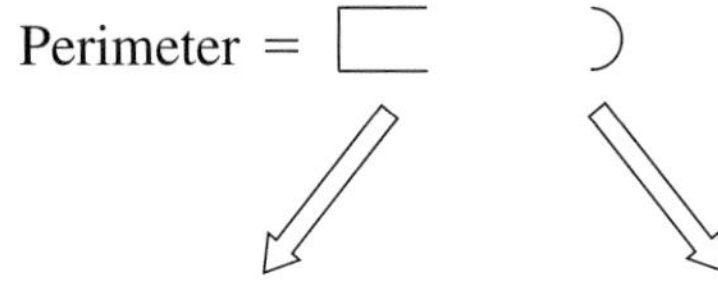

3 sides of the rectangle

$$P = 15 + 10 + 15$$
$$= 40 \text{ ft}$$

Half-circle

Multiply by $\frac{1}{2}$ to find half of the circle.

$$P = \frac{1}{2}(2\pi r)$$
$$= \pi r$$
$$= \pi \cdot 5$$
$$\approx 15.7 \text{ ft}$$

$$\text{Perimeter} \approx 40 + 15.7$$
$$\approx 55.7 \text{ ft}$$

6. Find the perimeter of the composite figure.
Use $\pi \approx 3.14$.

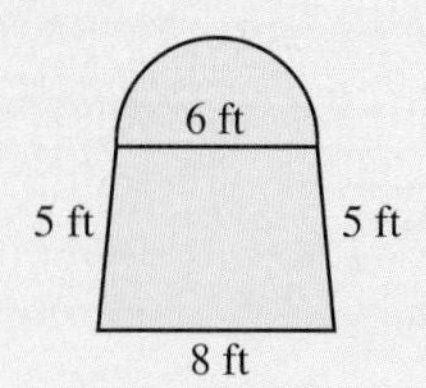

Start by writing a figure formula for the outside edges of the composite figure.

Perimeter = +

______ ______

$P =$ $P =$

Perimeter ≈ +

≈

7. Find the area of the composite figure.
Use $\pi \approx 3.14$.

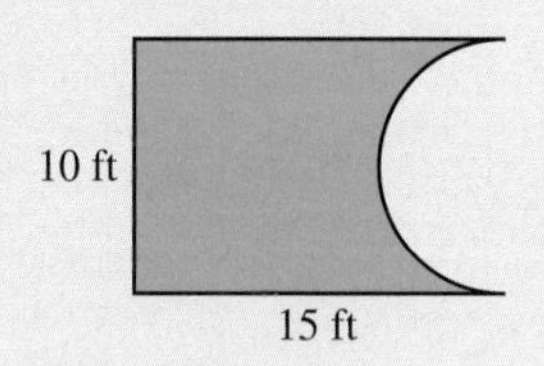

Start by writing a figure formula.

Area = ▬ − ◖

$$\text{Area} = A_{\text{rectangle}} - \frac{1}{2}A_{\text{circle}}$$
$$= 10 \cdot 15 - \frac{1}{2}\pi \cdot (5)^2$$
$$\approx 150 - 39.25$$
$$\approx 110.75 \text{ ft}^2$$

7. Find the area of the composite figure.
Use $\pi \approx 3.14$.

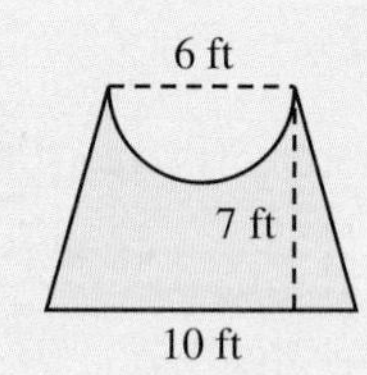

Start by writing a figure formula.

Area = −

Area = −

8. Why is the perimeter of each composite figure the same even though the areas are different?

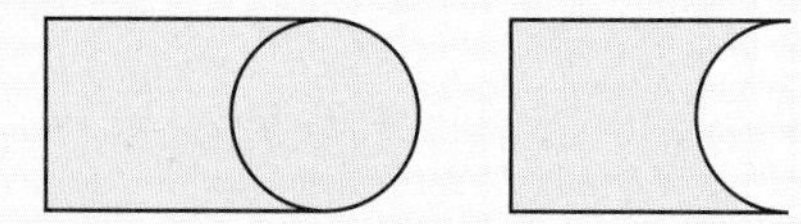

The perimeter of both composite figures includes three sides from the rectangle and half the circumference of the circle. It doesn't matter that the circular edge is drawn inside or outside the rectangular base.

8. Why is the perimeter of each composite figure the same even though the areas are different?

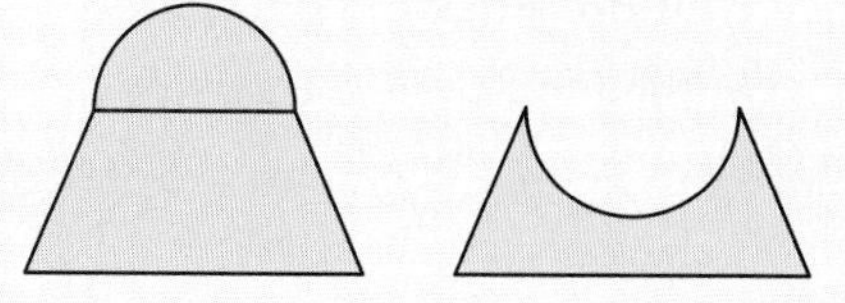

Answer this question with one or two complete sentences.

Concept Check

C1. Write a figure formula that could be used to find the area of the red composite figure. Do not find the area.

C2. Write a figure formula that could be used to find the perimeter of the orange composite figure. Do not find the perimeter.

C3. Write a figure formula that could be used to find the area of the blue composite figure. Do not find the area.

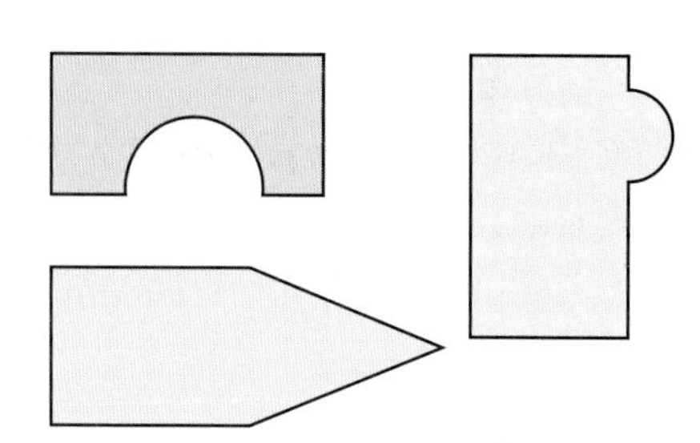

Objective C Practice

Find the area of each shaded composite figure. Use $\pi \approx 3.14$.

C4.

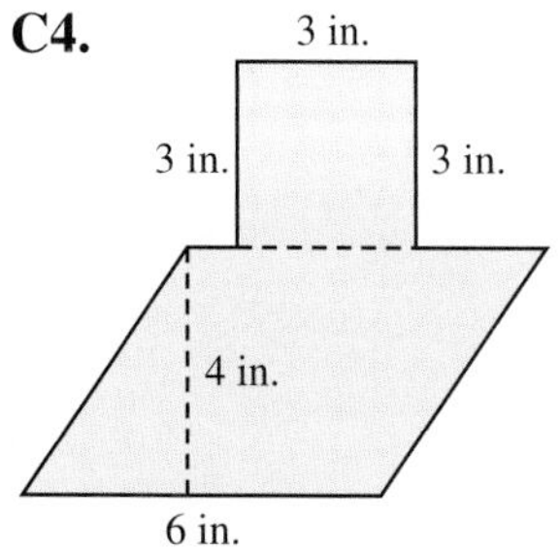

C5.

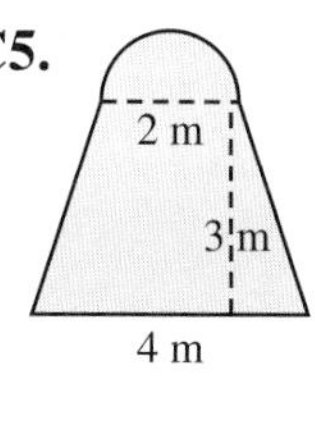

C6.

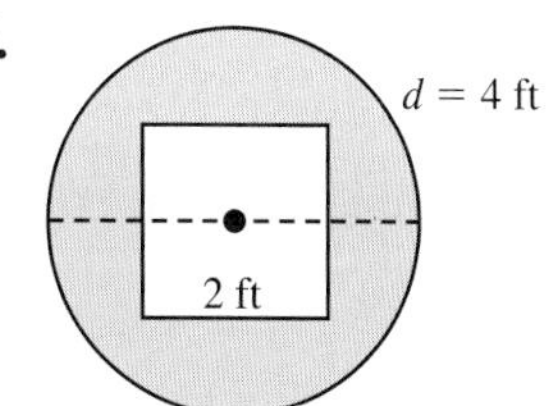

C7.

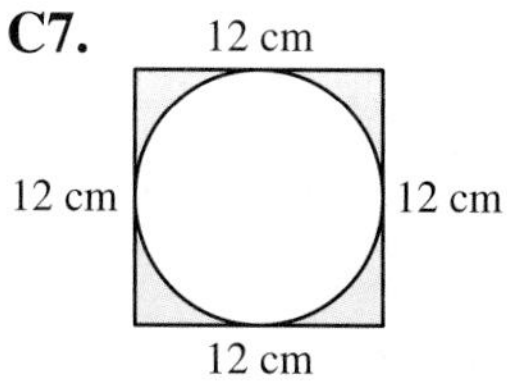

Find the perimeter of each composite figure. Use $\pi \approx 3.14$. Be careful to add only the lengths on the outside of each figure.

C8.

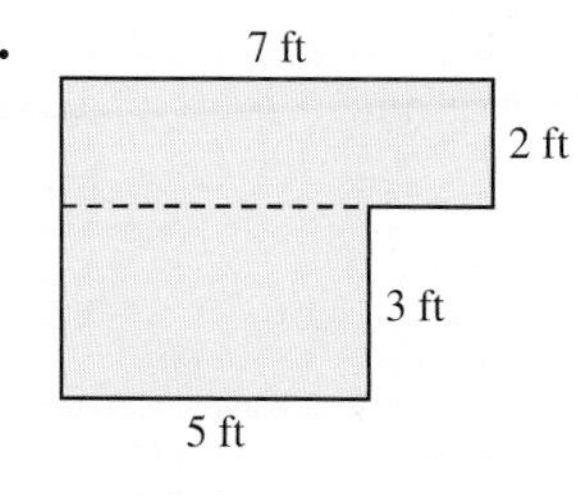

C9.

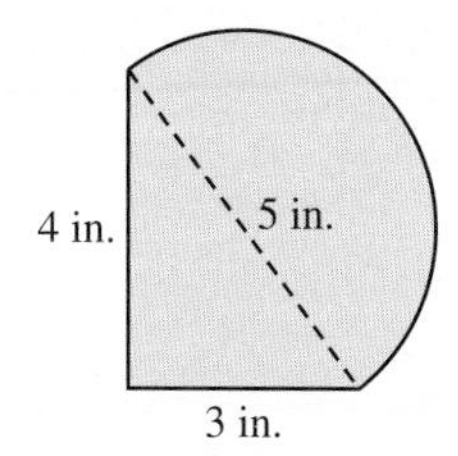

C10.

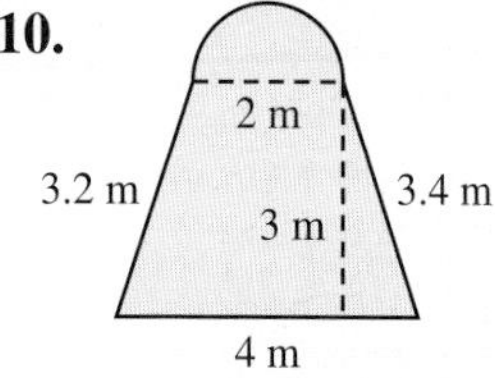

C11.

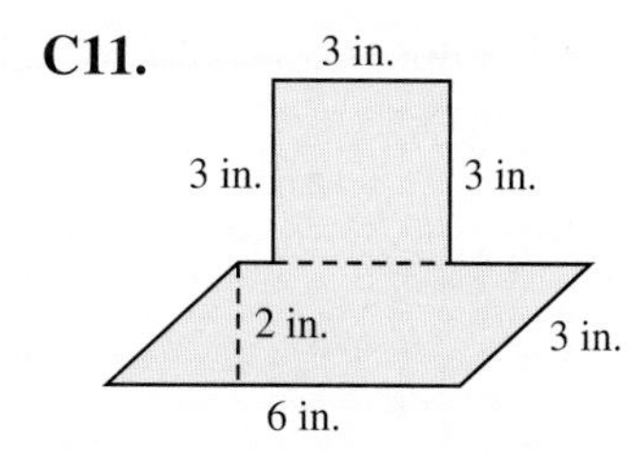

Section 7.4 Exercises

FOR EXTRA HELP      

To find the circumference of a circle:

1. Answer the Objective A Concept Checks.
2. Answer the odd Objective A Practice Exercises.
3. Answer the even Objective A Practice Exercises.

To find the area of a circle:

4. Answer the Objective B Concept Checks.
5. Answer the odd Objective B Practice Exercises.
6. Answer the even Objective B Practice Exercises.

To find the area and perimeter of a composite figure:

7. Answer the Objective C Concept Checks.
8. Answer the odd Objective C Practice Exercises.
9. Answer the even Objective C Practice Exercises.

VOCABULARY REVIEW *Review the Vocabulary Preview for Section 7.4. Study the definitions until you can check* Got It *for every word.*

Use the formulas from this section and these words to complete each sentence.

radius • diameter • circumference • pi

	Got It	Get Help
10. ____________ is the special name given to the perimeter of a circle.		
11. ____________ is approximately equal to 3.14.		
12. The ____________ of a circle is the distance across a circle, through the center.		
13. The ____________ of a circle is the distance from the center of a circle to its outer edge.		
14. The area of a circle is given by the formula $A =$ ____________.		
15. The two formulas for the circumference of a circle are as follows: $C =$ ____________ and $C =$ ____________		

How will you get help for any vocabulary that you are unsure about?

Your instructor _____ MyMathLab _____ A classmate _____ A tutor _____ Other _____

A circle's diameter or radius is given. Find the value of the other quantity.

16. radius = 5 ft
diameter = ?

17. radius = 7 meters
diameter = ?

18. diameter = 5 inches
radius = ?

19. diameter = 7 cm
radius = ?

Find the circumference of each circle. Use $\pi \approx 3.14$.

20.

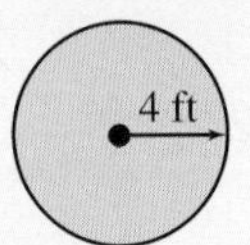

21.

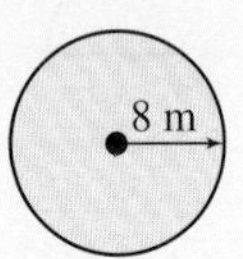

22.

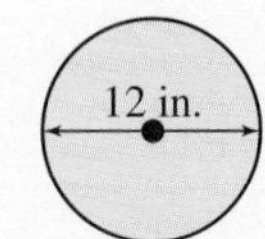

23.

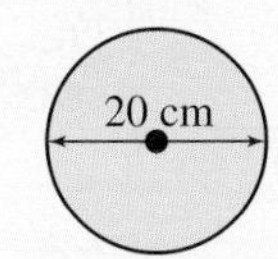

Find the area of each circle. Use $\pi \approx 3.14$.

24. 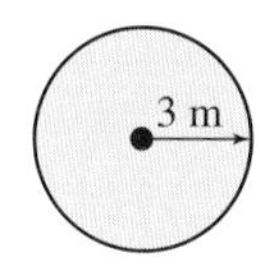

25. 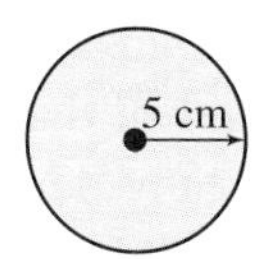

26. 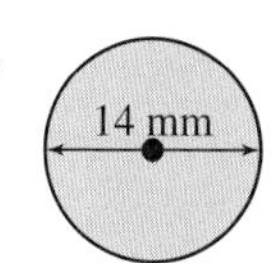

27. 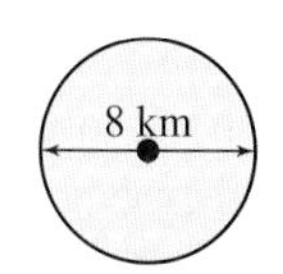

Draw each figure described. Then find the area outside the circle but inside the other figure. The circle must be drawn as large as possible. Use $\pi \approx 3.14$.

28. A circle is drawn inside a rectangle with a length of 3 meters and a width of 5 meters.

29. A circle is drawn inside a rectangle with a length of 7 feet and a width of 9 feet.

30. A circle is drawn inside a trapezoid with bases of 8 inches and 6 inches in length and a height of 1 inch.

31. A circle is drawn inside a trapezoid with bases of 12 cm and 10 cm in length and a height of 3 cm.

For each exercise:

a. Write a figure formula to describe the composite figure's perimeter.

b. Find the perimeter of each composite figure. Use $\pi \approx 3.14$.

32. 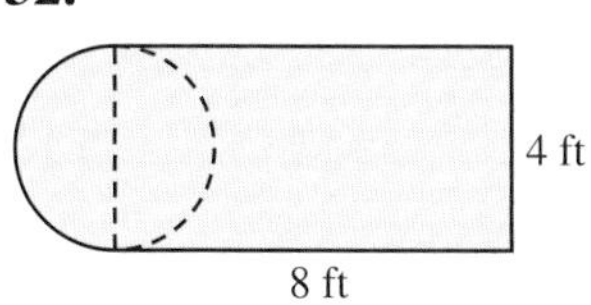

33. 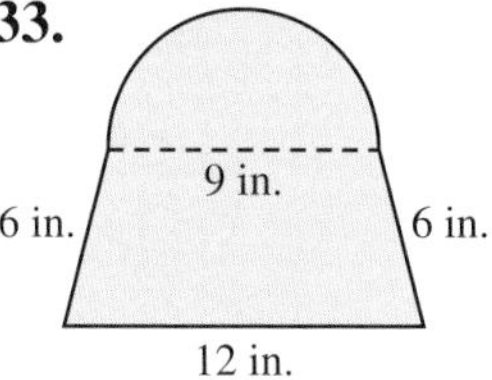

34. 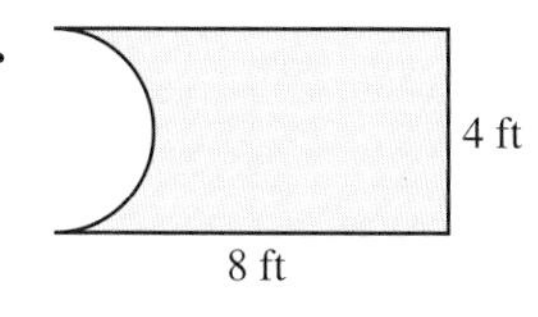

35. 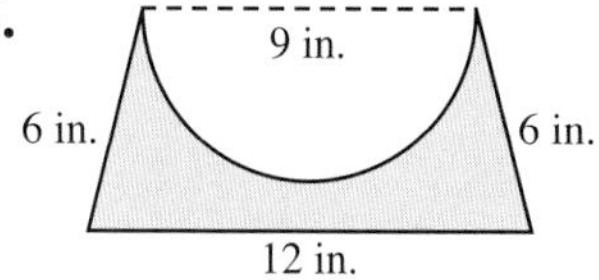

36. **37.** 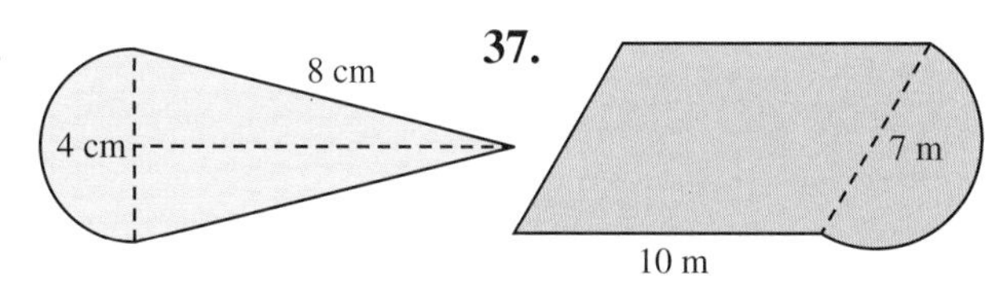

38. 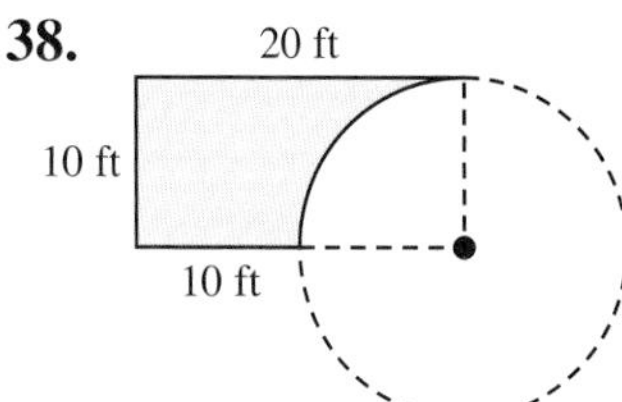

39. 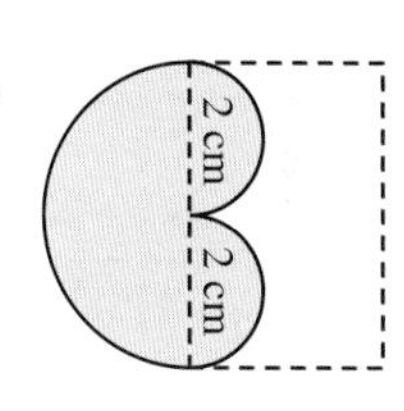

For each exercise:

a. Write a figure formula to describe the composite figure's area.

b. Find the area of each composite figure. Use $\pi \approx 3.14$.

40. 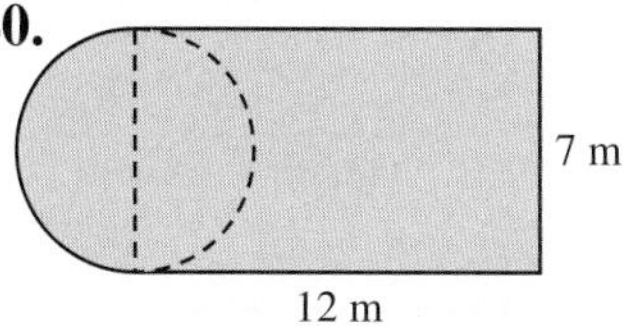

41. 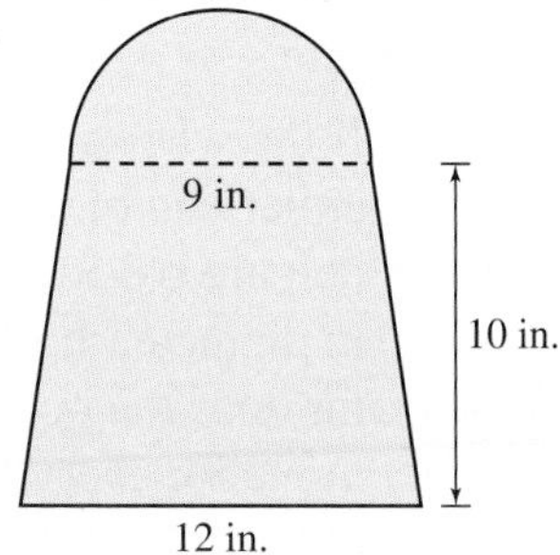

42. 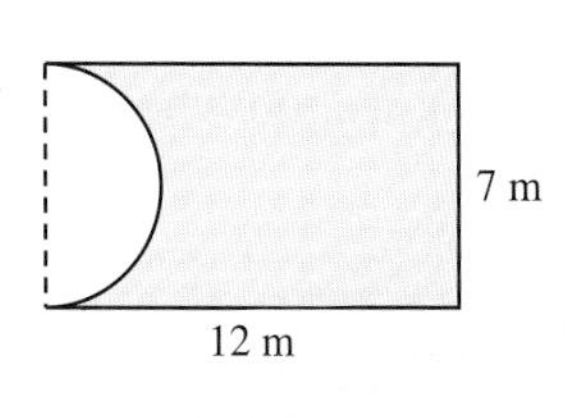

43. 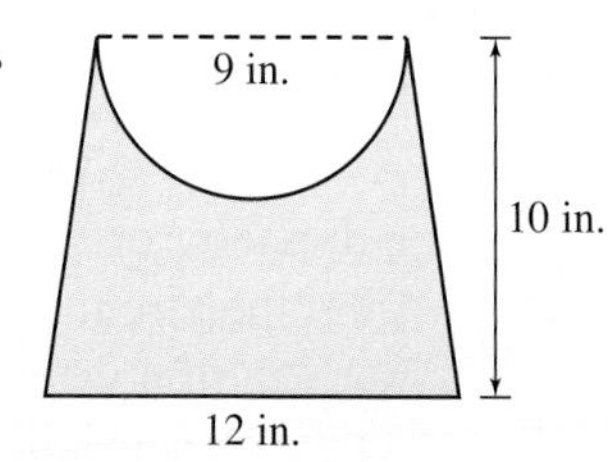

Smith To find the perimeter of half a circle, find the perimeter of the whole circle and then multiply it by 1/2 or divide it by 2.
For more tips and tweets, go to twitter.com/gstbasicmath

Answer each question.

44. a. New railings will be installed along the perimeter of an oil tanker's deck. The deck is a rectangle with two half-circles at either end. The rectangle measures 300 meters by 50 meters. How much railing must be installed?

b. The deck on the oil tanker will be painted. If one gallon of paint covers 40 square meters, how many gallons of paint are needed?

45. a. A rubber gasket will be installed around the perimeter of the door and window unit shown. The rectangular portion of the unit is 6 feet wide and 8 feet tall. How long must the rubber gasket be?

b. A set of curtains will be installed on the door and window unit shown. The curtains will be made using an area of fabric that is three times the area of the entire unit. How much fabric is needed?

46. What is the area inside the square but outside the circle?

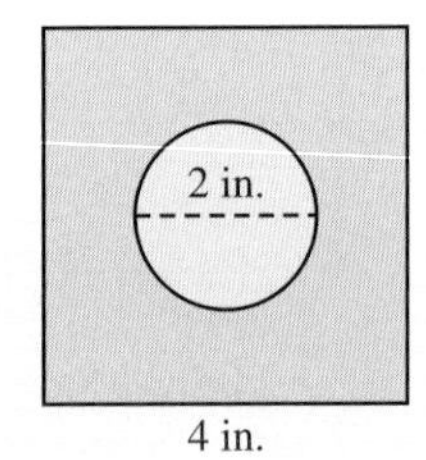

47. What is the area inside the circle but outside the triangle?

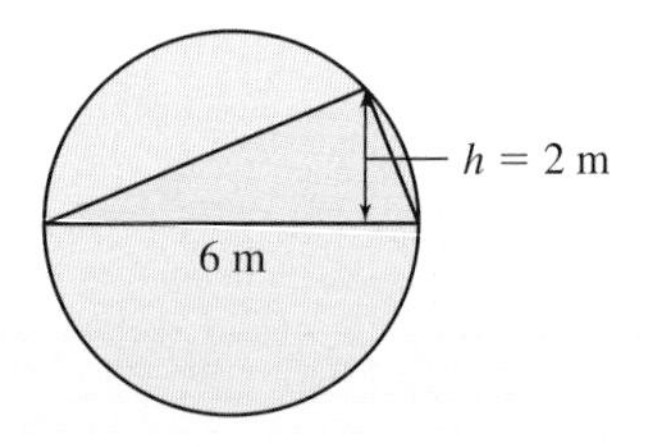

48. What is the area inside the circle but outside the parallelogram?

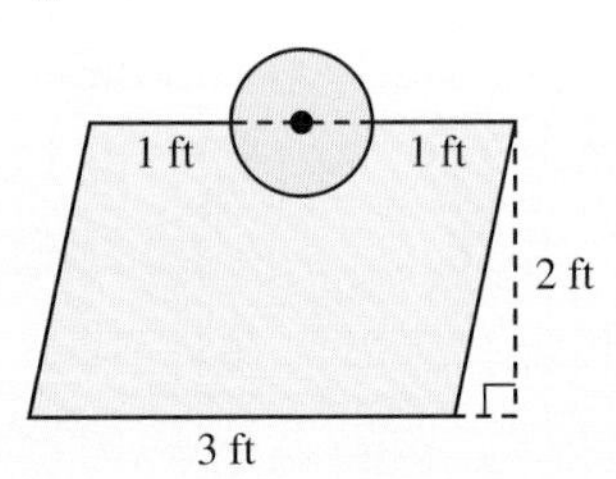

49. What is the area inside the parallelogram and inside the circle?

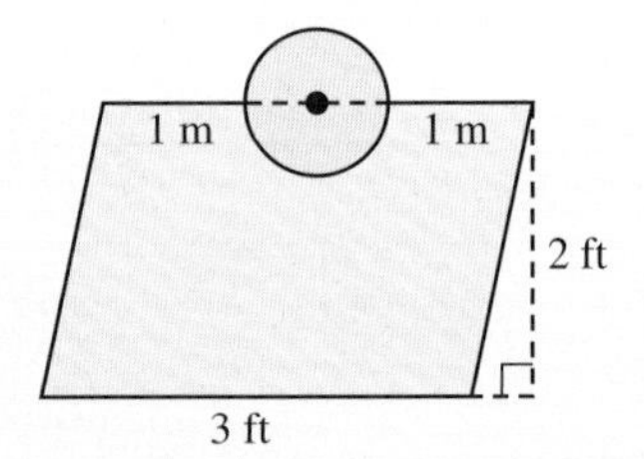

50. Early settlers could measure how far they traveled by counting the revolutions of a wagon wheel.

a. A wagon's wheel has a diameter of 4 feet. Find the circumference of a wagon wheel to determine how far the wagon goes in one revolution.

b. A mile is 5,280 feet. About how many revolutions does the wheel make per mile?

51. A conveyor belt must move a box 15.7 feet in one minute.

a. If the wheels at either end of the belt are 0.5 foot in diameter, how far does the box move for every revolution of the wheel?

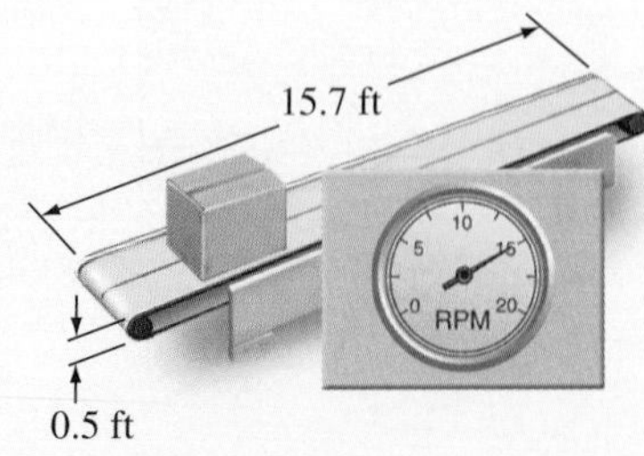

b. How many revolutions must occur in one minute (rpm) to move the one part 15.7 feet?

c. Based on the rpm reading, is the conveyor belt set too fast or too slow?

A circle can be stretched to form an ellipse, as shown by the beige figure. An ellipse is defined by measuring the longest and shortest distances from the center. The longer distance has a length of a units. The shorter distance has a length of b units.

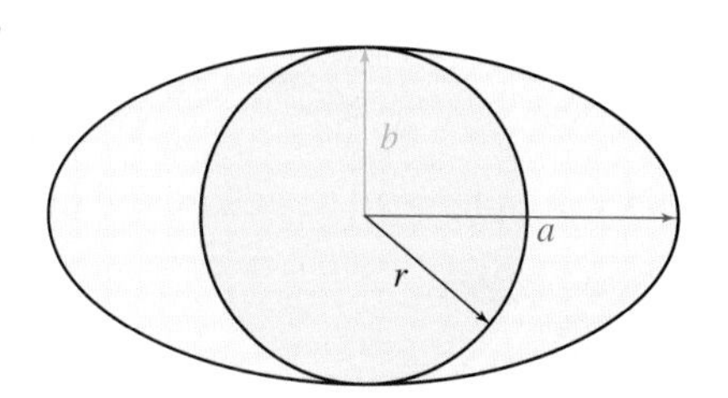

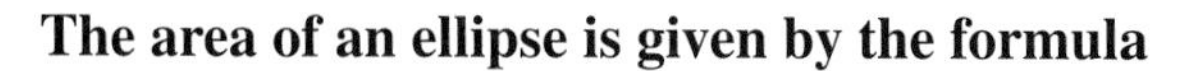

The area of an ellipse is given by the formula

$$A = \pi a \cdot b$$

52. Use the information about an ellipse to fill in the table.
Suppose that we want to find how changing the longer distance a and the shorter distance b will change the area of the ellipse. The answer in each box within the table will be "increases," "decreases," "remains the same," or "not enough information."

Example: The center box represents what will happen if nothing is changed. If neither a nor b are changed, then the ellipse remains the same. Therefore, the area remains the same.

What happens to the area of an ellipse when you change its dimensions?		Change in longer distance, a		
		Increase a	**Same** a	**Decrease** a
Change in shorter distance, b	**Increase** b			
	Same b		remains the same	
	Decrease b			

53. Using the given formula for the area of an ellipse, imagine decreasing the longer distance a until it is the same as the shorter distance b.

a. What is the resulting figure?

b. What is the formula for the area of that figure?

c. Show that the area of the ellipse is the same as your answer for part b when a is the same as b.

Section 7.4 Question Log

Use this space to write questions. Be sure to get these questions answered and revisit them when you prepare for your exam.

Page ____ Answered ☐	Page ____ Answered ☐
Page ____ Answered ☐	Page ____ Answered ☐

Section 7.5 Volume

Many people need to calculate volume as part of their job.

Chefs measure volumes of ingredients using cups.

NASCAR pit crews measure volumes of gasoline using gallons.

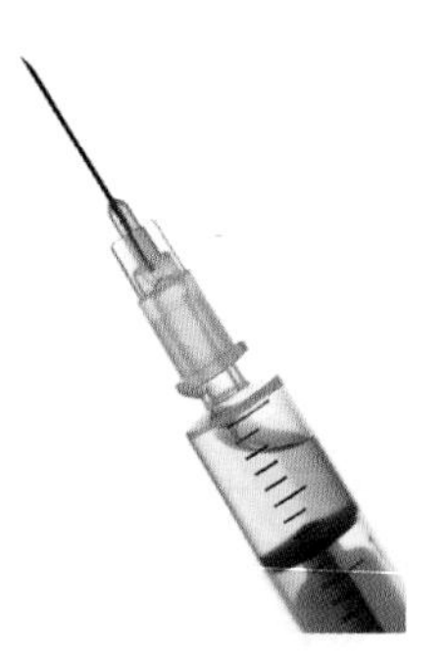

Nurses and doctors measure volumes of medication using milliliters.

Each of the above examples shows a volume that is measured using a cylinder. To design each cylinder, a formula was used to determine the container's dimensions so that it would have the right volume.

The **Objectives** in Section 7.5 will help you

A Understand volume.
B Find the volume of common figures.

VOCABULARY PREVIEW *Check the box that applies.*	Got It	Must Study
Perimeter: The distance around a figure is its **perimeter.** The units of perimeter are expressed in terms of length.		
Area: The **area** of a figure is the amount of surface that the figure covers. The units of area are expressed in terms of square lengths, or length^2.		
Volume: The **volume** of a figure is the amount of space that the figure fills. The units of volume are expressed in terms of cubic lengths, or length^3.		
Cylinder: A **cylinder** is a three-dimensional figure with an identical circular base and top.		
Cone: A **cone** is a three-dimensional figure with a circular base that tapers to a point.		
Pyramid: A **pyramid** is a three-dimensional figure with a rectangular base and sides that taper to a point.		

Study these words when they appear in the text. After the section, test your understanding by completing the Vocabulary Review in the section exercises.

Objective A Understand Volume

The Concept A figure's **volume** indicates how much space it takes to fill the figure. The units of volume are expressed in terms of cubic lengths, or length^3. To understand volume and its units, it is helpful to see how volume relates to perimeter and area.

INTERACTIVE DEFINITION Units for Perimeter, Area, and Volume

- **Perimeter:** The distance around a figure is its **perimeter.** The units of perimeter are expressed in terms of length.
- **Area:** The **area** of a figure is the amount of surface that the figure covers. The units of area are expressed in terms of square lengths, or length^2.
- **Volume:** The **volume** of a figure is the amount of space that the figure fills. The units of volume are expressed in terms of cubic lengths, or length^3.

EXAMPLES

1. Use the following quadrilateral to understand perimeter.

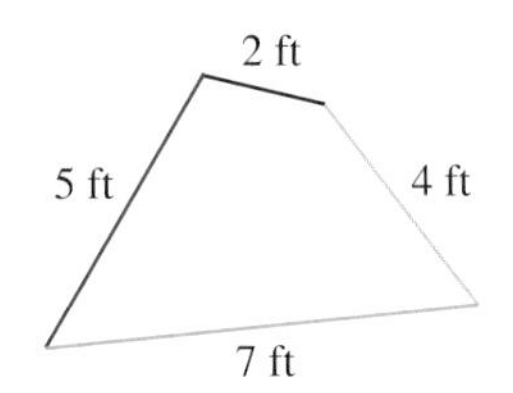

To find the perimeter, add the lengths of the sides that go around the figure.

5 ft + 2 ft + 4 ft + 7 ft = 18 ft

Perimeter is a 1-dimensional measure. We measure perimeter with line segments, so perimeter has units of length such as feet.

2. Use the following square to understand area.

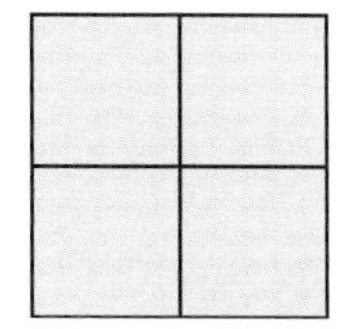

To find area, add the number of squares that cover the figure. Each square is 1 ft^2.

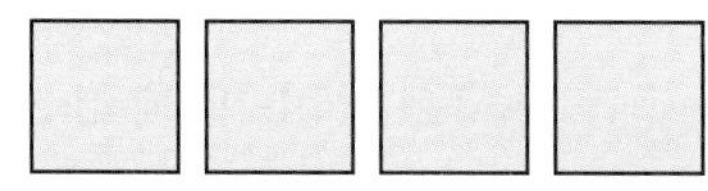

$1\text{ ft}^2 + 1\text{ ft}^2 + 1\text{ ft}^2 + 1\text{ ft}^2 = 4\text{ ft}^2$

Area is a 2-dimensional measure. We measure area with squares, so area has units of square length such as square feet or ft^2.

GUIDED PRACTICE

1. Use the following triangle to understand perimeter.

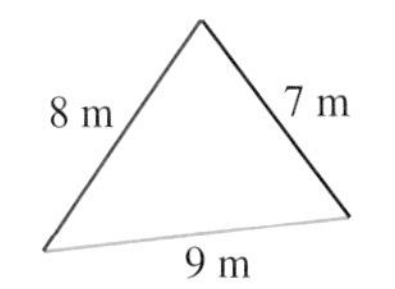

To find the perimeter, add the lengths of the sides that go around the figure.

_____ + _____ + _____ = _____

Perimeter is a __________ measure. We measure perimeter with __________, so perimeter has units of ______ such as ______.

2. Use the following rectangle to understand area.

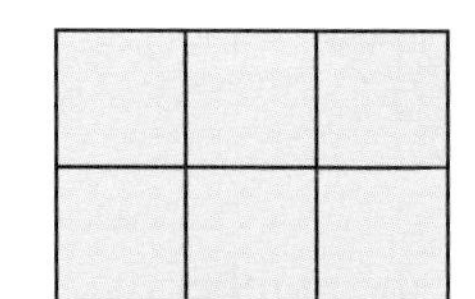

To find area, add the number of squares that cover the figure. Each square is 1 ft^2.

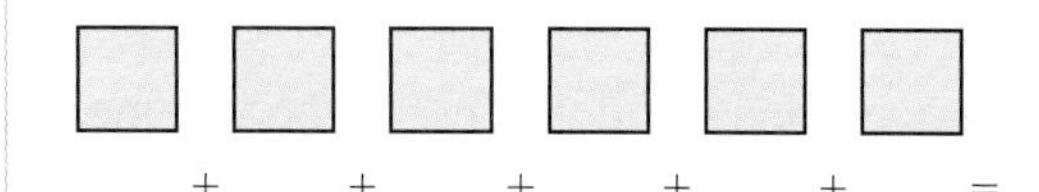

+ + + + + =

Area is a __________ measure. We measure area with ______, so area has units of __________ such as __________ or _____.

(Continued)

3. Use the following box to understand volume.

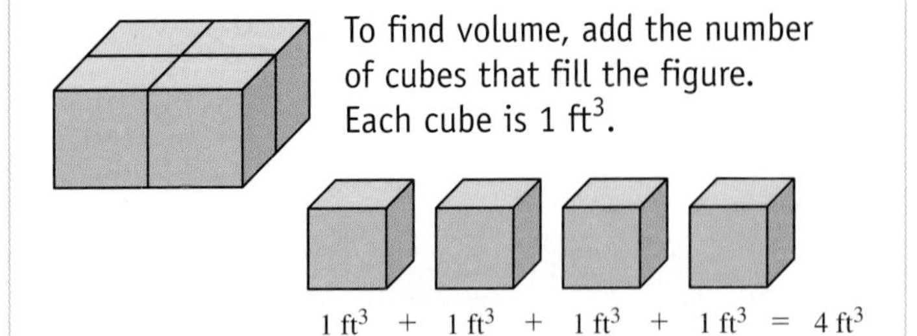

To find volume, add the number of cubes that fill the figure. Each cube is 1 ft^3.

$1\ ft^3 + 1\ ft^3 + 1\ ft^3 + 1\ ft^3 = 4\ ft^3$

Volume is a 3-dimensional measure. We measure volume with cubes, so volume has units of cubic length such as cubic feet, or ft^3.

3. Use the following box to understand volume.

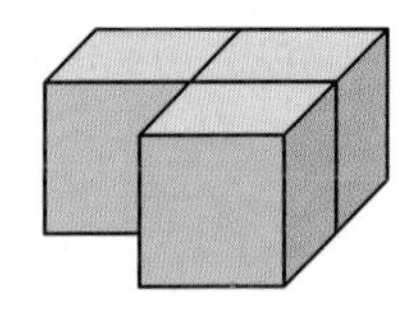

To find volume, add the number of cubes that fill the figure. Each cube is 1 m^3.

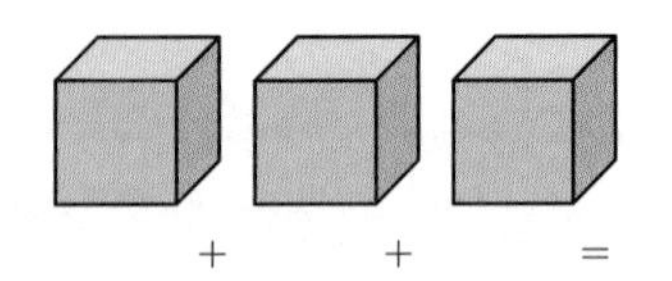

_____ + _____ + _____ = _____

Volume is a __________ measure. We measure volume with _____, so volume has units of __________ such as __________, or _____.

DO YOU UNDERSTAND the units for perimeter, area, and volume? Got It Get Help

Procedure Find the Volume of a Figure

Count the number of cubes needed to fill the figure.

Remember that units of volume are $length^3$ (cubic lengths).

EXAMPLES

4. Find the volume of the figure. Each cube is 1 cm^3.

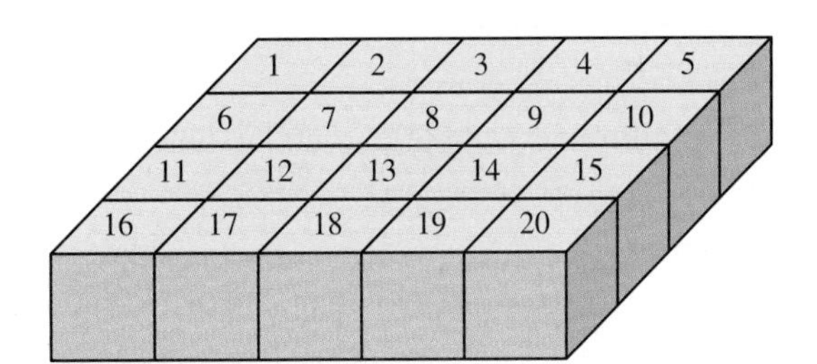

To find the volume of the figure, count the number of cubes needed to fill the figure.

Since 20 cubes are needed to fill the figure, the volume is 20 cm^3, or 20 cubic centimeters.

GUIDED PRACTICE

4. Find the volume of the figure. Each cube is 1 in^3.

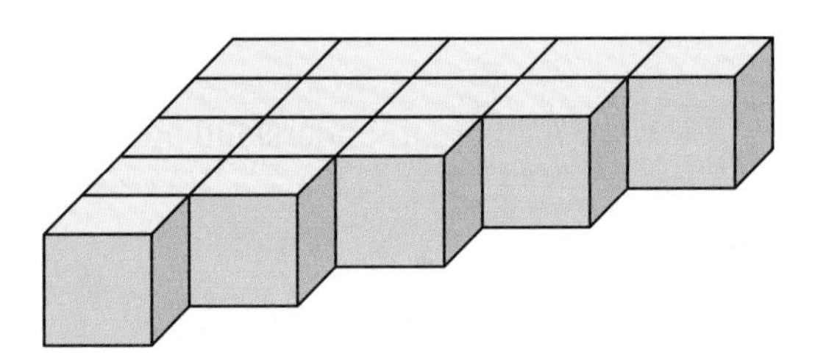

To find the volume of the figure, count the number of cubes needed to fill the figure.

Since _____ cubes are needed to fill the figure, the volume is _____, or _______ inches.

5. Find the volume of the box.
Each cube is 1 ft $\times$ 1 ft $\times$ 1 ft.

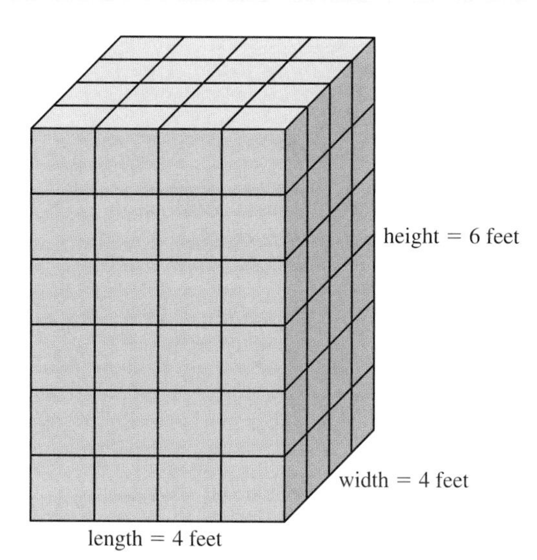

To find the number of cubes needed to fill the box, multiply the length, width, and height.

$$
\begin{aligned}
\text{Volume} &= \text{Length} \cdot \text{Width} \cdot \text{Height} \\
&= 4 \cdot 4 \cdot 6 \\
&= 16 \cdot 6 \\
&= 96
\end{aligned}
$$

The volume is 96 ft^3, or 96 cubic feet.

5. Find the volume of the box.
Each cube is 1 m $\times$ 1 m $\times$ 1 m.

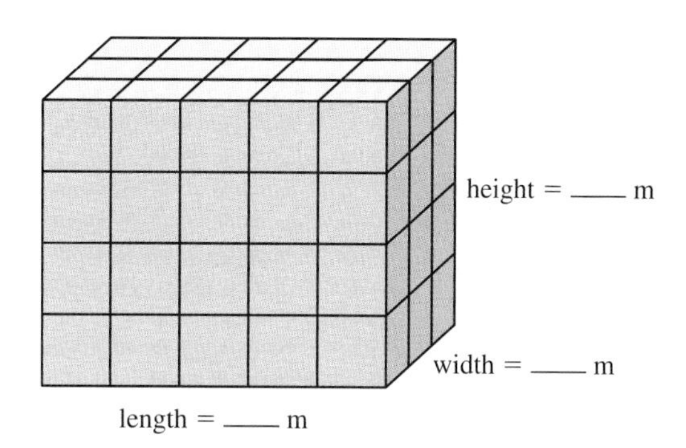

To find the number of cubes needed to fill the box, multiply the ______, ______, and ______.

Volume = ______ · ______ · ______
= ______ · ______ · ______
= ______ · ______
= ______

The volume is ______, or ______ cubic ______.

Concept Check

A1. In order, what units are used to measure perimeter, area, and volume?

A2. Explain why the units used for volume are length^3 (cubic lengths).

Objective A Practice

Find the volume of each figure by counting the number of cubes needed to fill the figure.

A3.

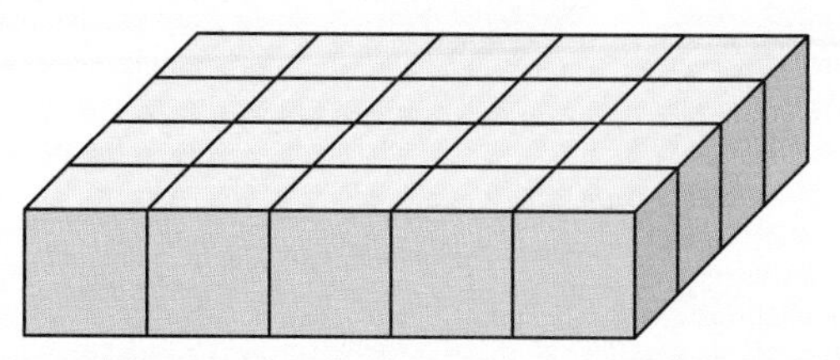

Each cube has sides that measure 1 foot.

A4.

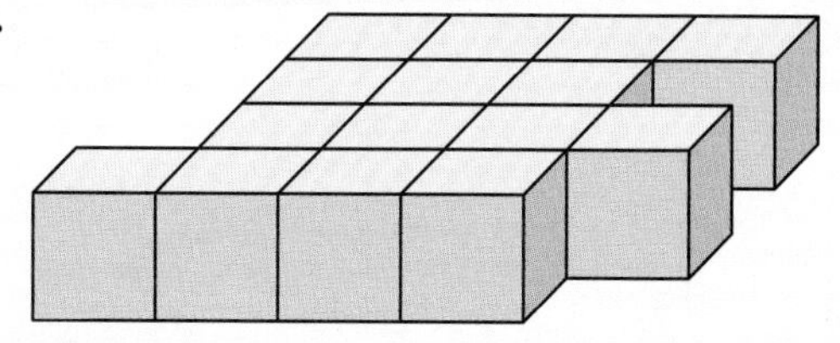

Each cube has sides that measure 1 centimeter.

(Continued)

Answer each question.

A5.

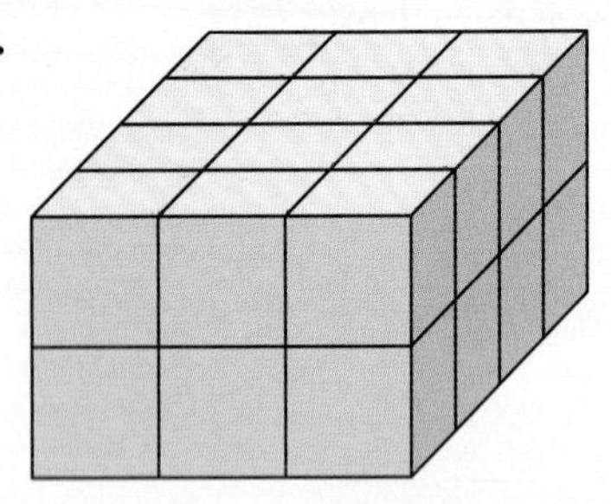

Each cube has sides that measure 1 meter.

a. What is the box's length?
b. What is the box's width?
c. What is the box's height?
d. Find the volume of the box by multiplying its dimensions.

A6.

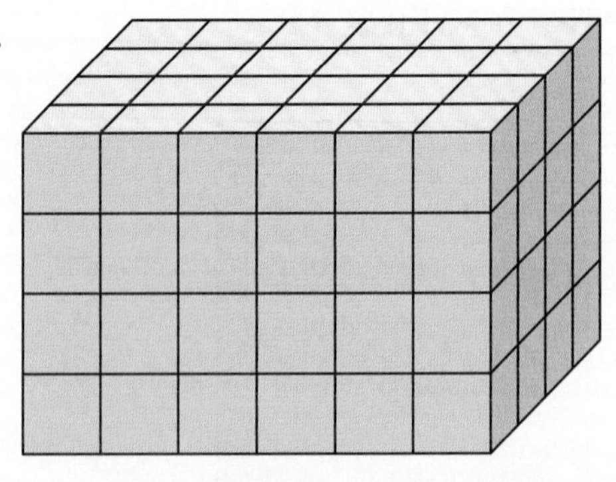

Each cube has sides that measure 1 yard.

a. What is the box's length?
b. What is the box's width?
c. What is the box's height?
d. Find the volume of the box by multiplying its dimensions.

A7. This exercise will help explain why we can find the volume of a box by multiplying the length, width, and height. Use the box in Exercise A6 to help you answer the following questions.

a. True or false. Each layer of the box has the same number of cubes. Explain.
b. True or false. The number of cubes in a layer of the box can be found by multiplying length and width. Explain.
c. True or false. Multiplying the number of cubes in a layer of the box by the number of layers will give the number of cubes in the box. Explain.
d. Explain why volume = length · width · height.

Objective B Find the Volume of Common Figures

The Concept It is very difficult to count the number of cubes needed to fill some figures. Fortunately, the volume of many figures can be found using formulas.

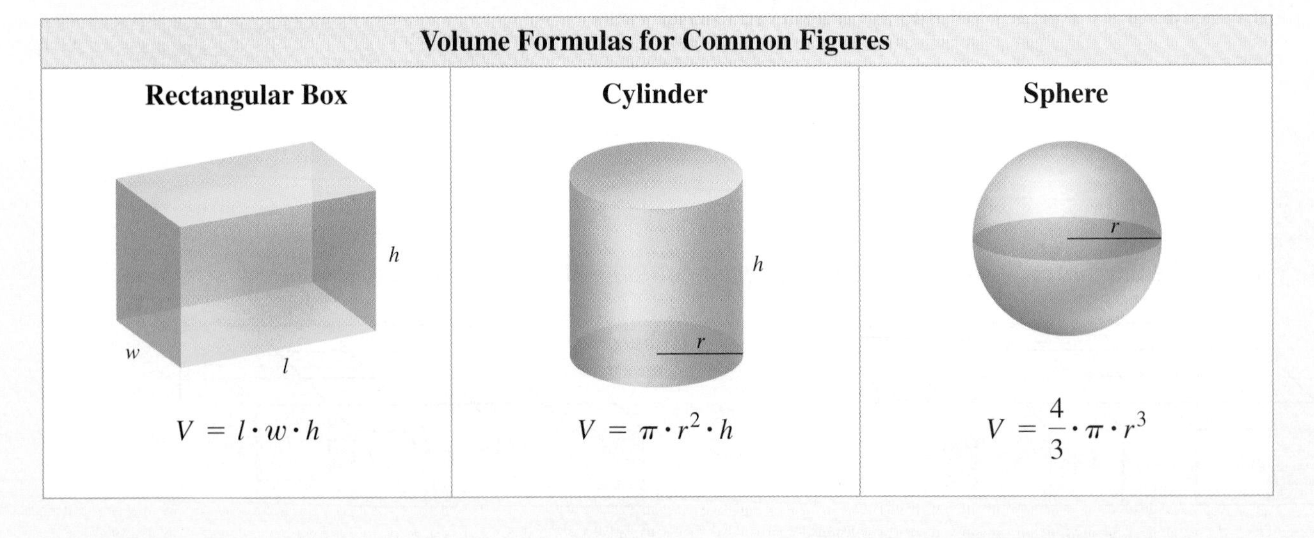

Volume Formulas for Common Figures		
Rectangular Box	**Cylinder**	**Sphere**
$V = l \cdot w \cdot h$	$V = \pi \cdot r^2 \cdot h$	$V = \frac{4}{3} \cdot \pi \cdot r^3$

Pyramid	Cone
 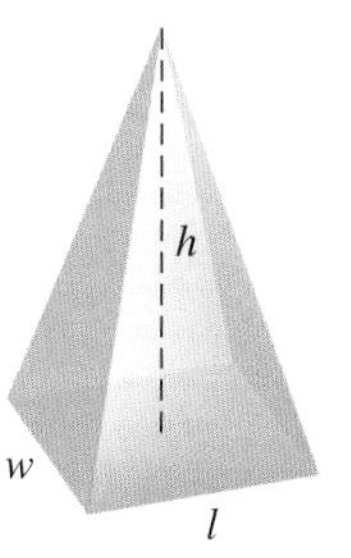	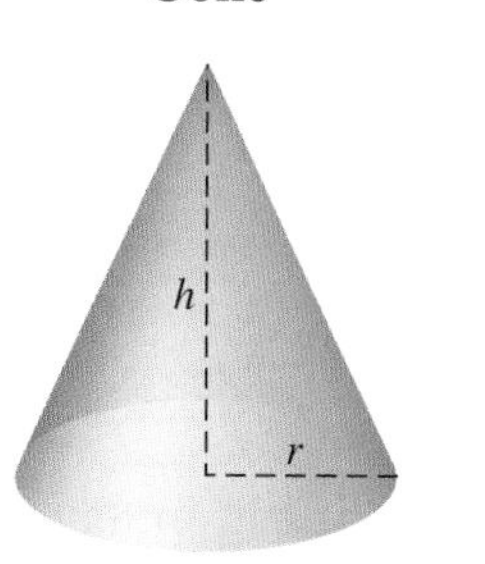
$V = \frac{1}{3} \cdot l \cdot w \cdot h$	$V = \frac{1}{3} \cdot \pi \cdot r^2 \cdot h$

The following observations can help you remember these formulas.

- The volume of a rectangular box and a cylinder can be found by multiplying the area of the base ($l \cdot w$ or πr^2) by the height h of the figure.
- The volume of a pyramid is one-third the volume of a rectangular box with the same dimensions.
- The volume of a cone is one-third the volume of a cylinder with the same dimensions.

Procedure Use a Formula to Find the Volume of a Figure

Step 1 Write the appropriate formula.
Step 2 Substitute the values into the formula.
Step 3 Simplify.

EXAMPLES

6. A cone has a radius of 1 inch and a height of 4 inches. What is its volume?

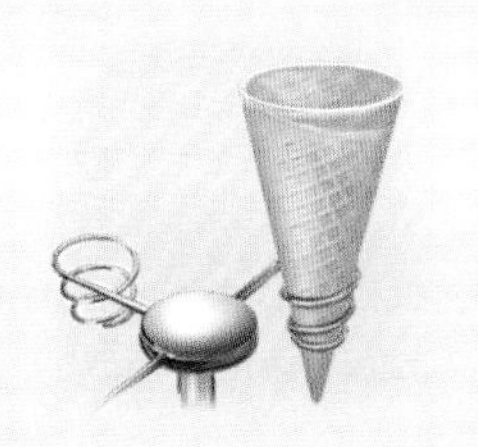

Step 1 Write the appropriate formula. $V = \frac{1}{3} \cdot \pi \cdot r^2 \cdot h$

Step 2 Substitute the values. $\approx \frac{1}{3} \cdot 3.14 \cdot 1^2 \cdot 4$

Step 3 Simplify. $\approx \frac{1}{3} \cdot 3.14 \cdot 1 \cdot 4$

$\approx 1.05 \cdot 1 \cdot 4$

$\approx 1.05 \cdot 4$

≈ 4.2

Since an approximation for π was used, the calculation is not exact. The volume of the cone is about 4.2 in.^{3}.

GUIDED PRACTICE

6. A pyramid-shaped steeple has a square base that is 4 feet by 4 feet. It is 12 feet tall. What is its volume?

$V = \frac{\_\_}{\_\_} \cdot \_\_ \cdot \_\_ \cdot \_\_$

$= \_\_ \cdot \_\_ \cdot \_\_ \cdot \_\_$

$= \_\_ \cdot \_\_ \cdot \_\_$

$= \_\_ \cdot \_\_$

$= \_\_$

The volume of the steeple is ______^{3}.

(Continued)

7. A cylindrical can has a diameter of 8 cm and a height of 12 cm. Its volume, in cm^3, is the same as the can's volume in milliliters. How many milliliters are in the can?

If the diameter is 8 cm, the radius is half of 8 cm.
Radius = 4 cm

Step 1 Write the appropriate formula.

Step 2 Substitute the values.

Step 3 Simplify.

$$V = \pi \cdot r^2 \cdot h$$
$$\approx 3.14 \cdot 4^2 \cdot 12$$
$$\approx 3.14 \cdot 16 \cdot 12$$
$$\approx 3.14 \cdot 192$$
$$\approx 602.88$$

Because multiplication is commutative, we can wait and multiply by π at the last step.

The can's volume is about 602.88 milliliters.

7. A rubber ball has a diameter of 6 cm. Its volume, in cm^3, is the same as the ball's volume in milliliters. How many milliliters of rubber are required to make the ball?

If the diameter is 6 cm, the radius is half of 6 cm.
Radius = ____ cm

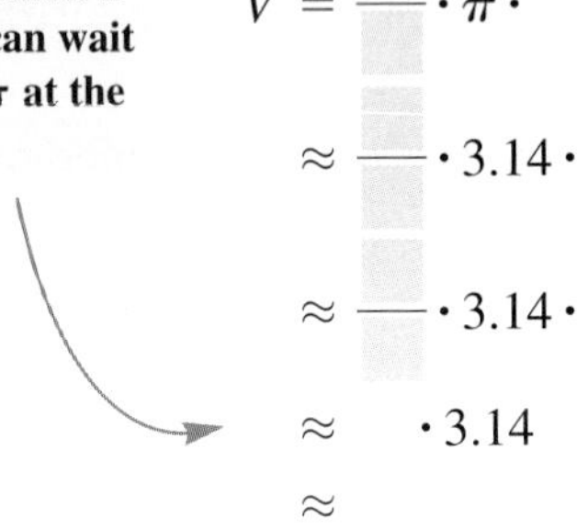

$V = \frac{\quad}{\quad} \cdot \pi \cdot$

$\approx \frac{\quad}{\quad} \cdot 3.14 \cdot$

$\approx \frac{\quad}{\quad} \cdot 3.14 \cdot$

$\approx \quad \cdot 3.14$

$\approx$

The ball's volume is about ____________ milliliters.

Concept Check

B1. What decimal approximation is used for π?

B2. If you know a diameter, how do you find the corresponding radius?

B3. A student made the comment that "writing out the formula helps me avoid silly mistakes." Why do you suppose she feels this way?

Objective B Practice

Find the volume of each figure.

B4. What is the volume of a cylinder with a diameter of 2 feet and a height of 6 feet?

B5. What is the volume of a 5-foot-tall pyramid with a rectangular base that is 6 feet by 4 feet?

B6. The roaster pan shown is approximately a rectangular box. If it is 3 inches tall, 9 inches wide, and 14 inches long, what is its volume?

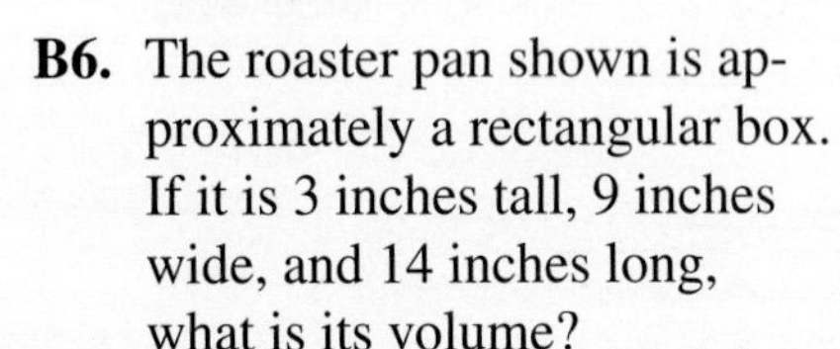

B7. The tank shown is a cylinder that is 50 feet tall and 20 feet across. What is its volume?

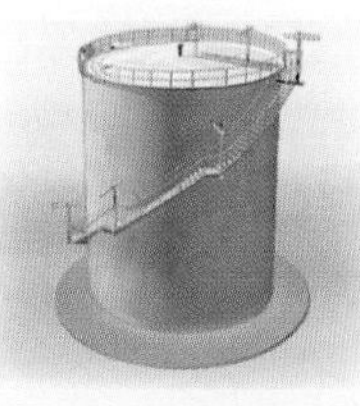

The grain hopper shown is made of an upside-down cone positioned under a cylinder. The diameter of the hopper is 2 meters.

B8. If the cone is 1.5 meters high, how much grain fits in the cone?

B9. If the cylinder is about 5 meters high, how much grain fits in the cylinder?

B10. Add your answers from Exercises B8 and B9 to find the total volume of the hopper.

Section 7.5 Exercises

FOR EXTRA HELP      

To understand volume:

1. Answer the Objective A Concept Checks.
2. Answer the odd Objective A Practice Exercises.
3. Answer the even Objective A Practice Exercises.

To find the volume of common figures:

4. Answer the Objective B Concept Checks.
5. Answer the odd Objective B Practice Exercises.
6. Answer the even Objective B Practice Exercises.

VOCABULARY REVIEW *Review the Vocabulary preview for Section 7.5. Study the definitions until you can check* Got It *for every word.*

Use these words to complete each sentence.

perimeter • area • volume • cylinder • cone • pyramid • lengths • lengths^2 • lengths^3

	Got It	Get Help
7. A(n) ____________ is a three-dimensional figure with a circular base, and it comes to a point.		
8. The ____________ of a figure is the amount of surface that the figure covers. The units are expressed in terms of ____________.		
9. A(n) ____________ is a three-dimensional figure with a rectangular base, and it comes to a point.		
10. The ____________ of a figure is the amount of space that the figure fills. The units are expressed in terms of ____________.		
11. A(n) ____________ is a three-dimensional figure with an identical circular base and top.		
12. The distance around a figure is its ____________. The units are expressed in terms of ____________.		

How will you get help for any vocabulary that you are unsure about?

Your instructor ____ MyMathLab ____ A classmate ____ A tutor ____ Other ____

Find the volume of each figure by counting the number of cubes needed to fill the figure.

13. The edge of each cube is 1 ft long.

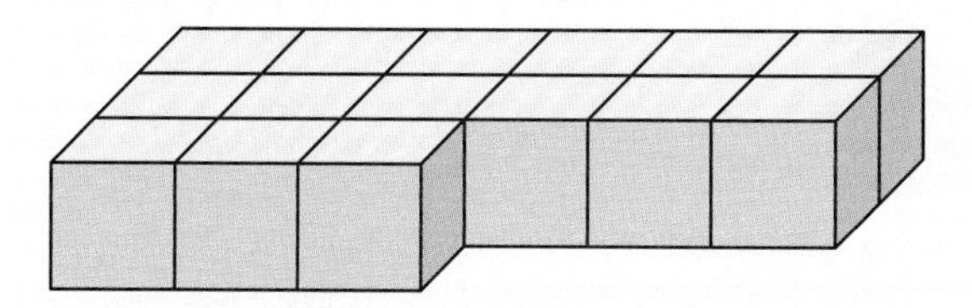

14. The edge of each cube is 1 meter long.

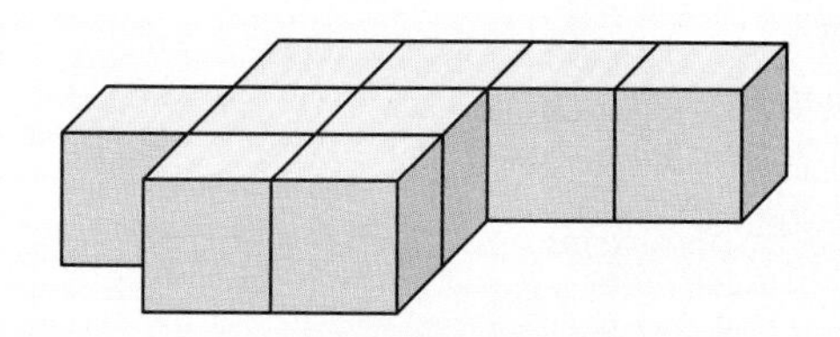

15. The edge of each cube is 1 cm long.

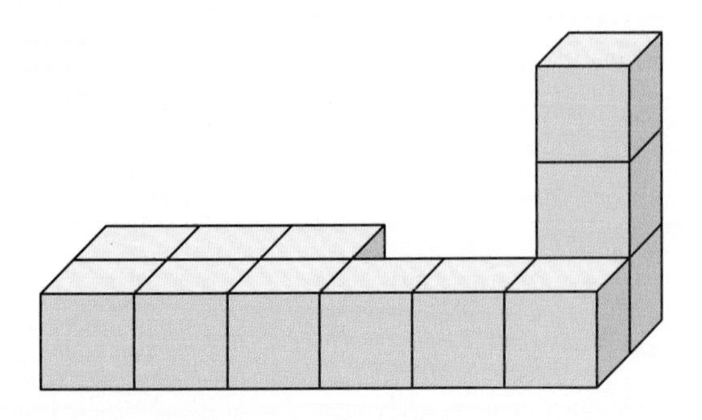

16. The edge of each cube is 1 yard long.

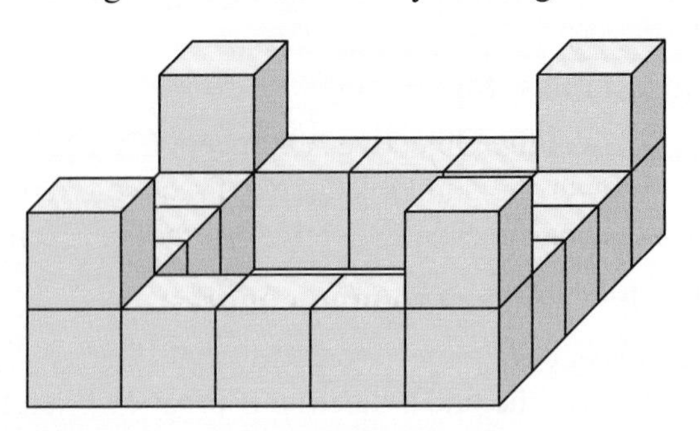

17. The edge of each cube is 1 mm long.

When finding the volume of this shape, assume that the cubes in the upper layer are supported by cubes in the lower layer with no space between them.

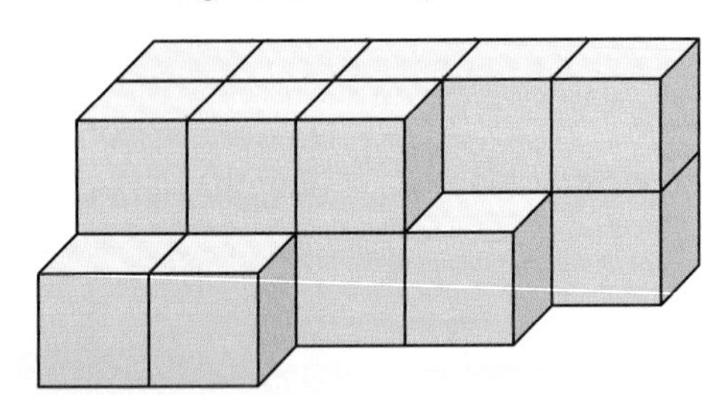

18. The edge of each cube is 1 in. long.

When finding the volume of this shape, assume that the cubes in the upper layer are supported by cubes in the lower layer with no space between them.

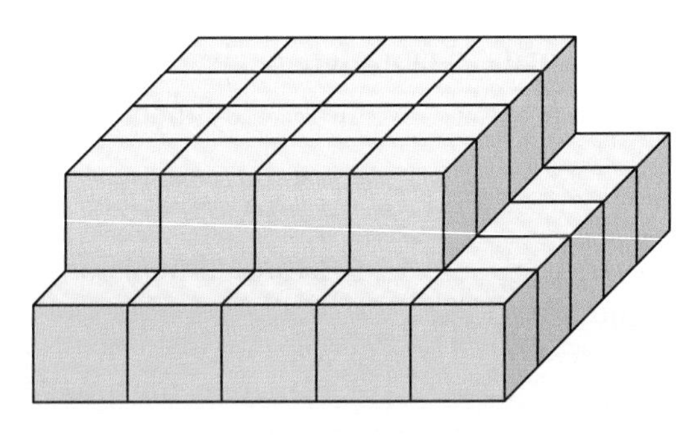

Use a formula to find the volume of each figure.

19. The edge of each cube is 1 ft long.

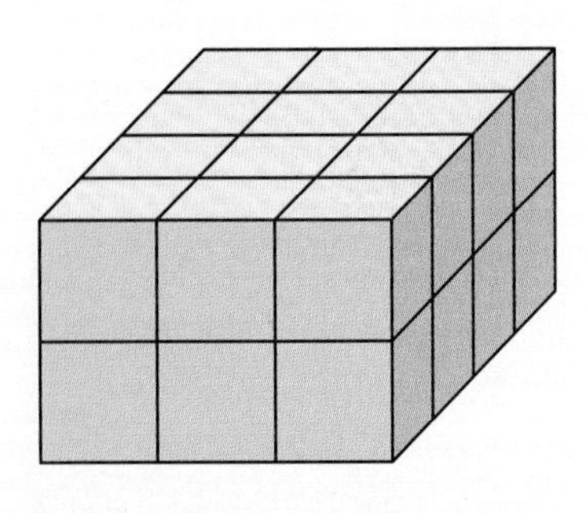

20. The edge of each cube is 1 m long.

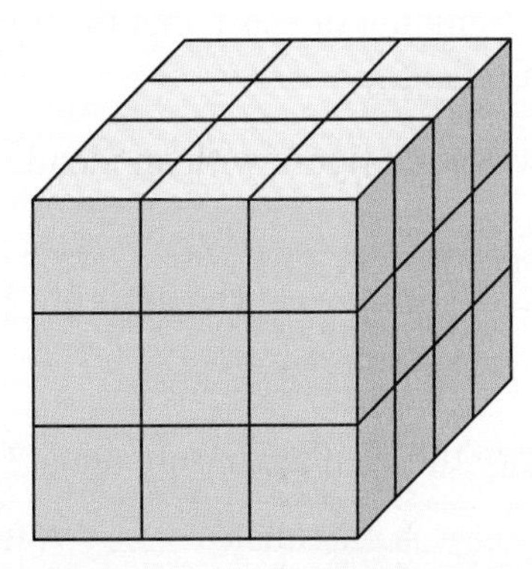

21. A cone with a radius of 6 inches and a height of 15 inches

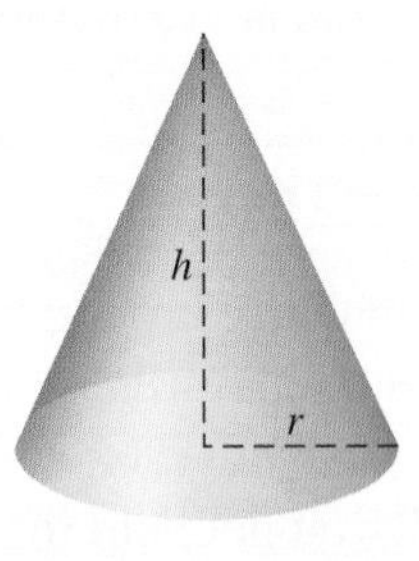

22. A cylinder with a 3-inch radius and height of 5 inches

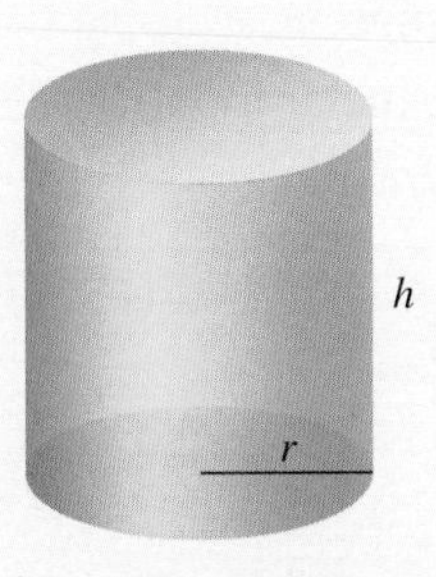

23. A sphere with a diameter of 12 inches

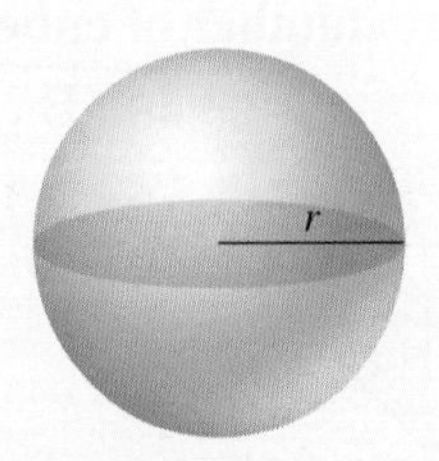

24. An 8-inch-tall pyramid with a 4-inch by 6-inch rectangular base

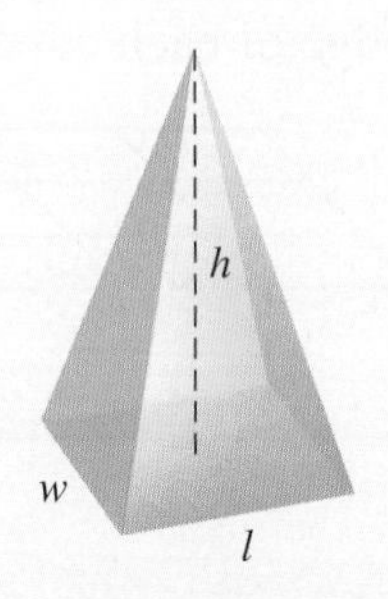

25. The moon's radius is approximately 1,000 miles long. What is its approximate volume?

26. Earth's radius is approximately 4,000 miles long. What is its approximate volume?

27. The oil tank is 8 feet long and 6 feet in diameter. What is its approximate volume?

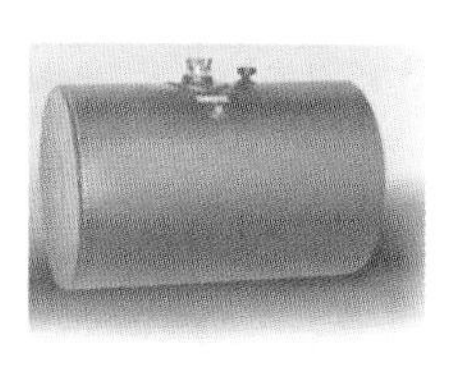

28. The pyramid at La Louvre has a square base that is about 116 feet wide and a height that is about 71 feet. What is the approximate volume?

29. The great pyramid at Giza has a square base that is about 750 feet wide and a height that is about 450 feet. What is the approximate volume?

30. A volcano is approximately a cone. It is about 2,000 feet high and has a radius of about 2,000 feet. What is the approximate volume?

31. A Kansas City skyscraper is approximately a rectangular solid. The skyscraper measures about 630 feet high with a square base that is about 150 feet long. What is its approximate volume?

32. a. What shape best approximates the nitrogen truck shown?

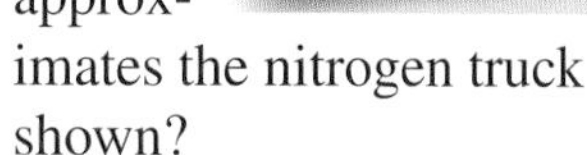

b. Use an appropriate formula to estimate the volume. The truck is 18 meters long and 3 meters wide.

33. a. What shape best approximates the juice container shown?

b. Use an appropriate formula to estimate the volume. The container is about 9 inches tall and 4 inches wide and long.

Answer each question.

34. The box and pyramid shown have the same dimensions.

a. The box has a length of 2 feet, a width of 1 foot, and a height of 3 feet. What is the volume?

b. The pyramid has a length of 2 feet, a width of 1 foot, and a height of 3 feet. What is the volume?

c. How many pyramids full of water would you have to pour into the box to fill the box?

d. Your answer from part c demonstrates how the volume of a pyramid and a box are related. Explain the relationship.

e. Compare the formulas for the volume of a box and a pyramid. How does your answer from part d explain the relationship between the formulas?

35. The cylinder and cone shown have the same dimensions.

a. The cylinder has a radius of 1 inch and a height of 4 inches. What is the volume?

b. The cone has a radius of 1 inch and a height of 4 inches. What is the volume?

c. How many cones full of water would you have to pour into the cylinder to fill the cylinder?

d. Your answer from part c demonstrates how the volume of a cone and a cylinder are related. Explain the relationship.

e. Compare the formulas for the volume of a cylinder and a cone. How does your answer from part d explain the relationship between the formulas?

tobey Take your time to reason and follow the steps of these problems. You will find the results to be interesting.

For more tips and tweets, go to twitter.com/gstbasicmath

36. a. What is the volume of the deck of cards shown in figure a?

b. Figure (b) shows the same deck of cards. Has the volume of any single card changed by shifting the deck? Does the volume of the entire deck change by shifting the deck? Why or why not?

c. Figure b is called a *parallelepiped*. Use your answers from parts a and b to write the formula for the volume of a parallelepiped.

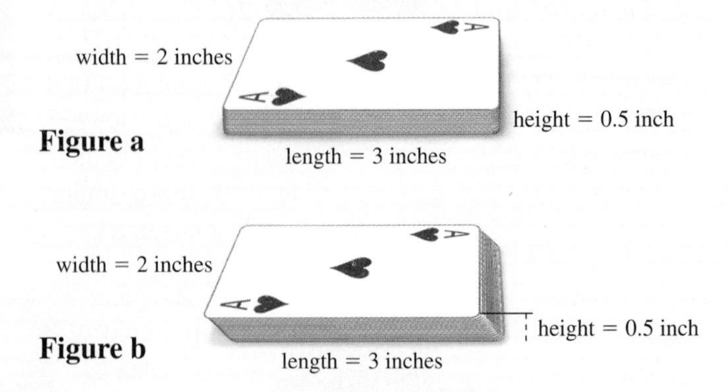

Section 7.5 Question Log

Use this space to write questions. Be sure to get these questions answered and revisit them when you prepare for your exam.

Page ______ **Answered**	**Page** ______ **Answered**
Page ______ **Answered**	**Page** ______ **Answered**

Section 7.6 Square Roots and the Pythagorean Theorem

The Pythagorean theorem allows us to find the length of one side of a right triangle when the lengths of the other two sides are known. Using the Pythagorean theorem, we can find the unknown length (x) shown in each right triangle below.

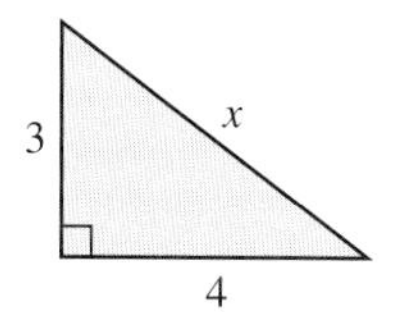

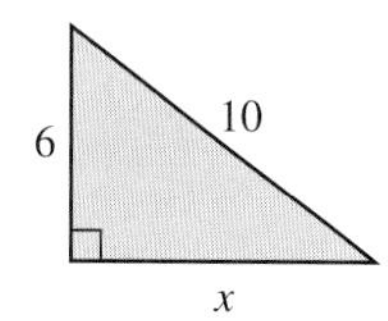

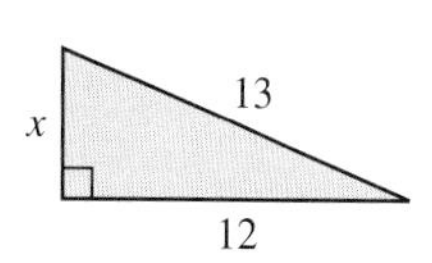

To use the Pythagorean theorem, we must find the square root of a number. For example, we will learn that $\sqrt{25} = 5$, which is read as "the square root of twenty-five is equal to 5."

The **Objectives** in Section 7.6 will help you

- **A** Evaluate the square root of a perfect square.
- **B** Approximate the square root of a number.
- **C** Use the Pythagorean theorem.

VOCABULARY PREVIEW *Check the box that applies.*	Got It	Must Study
Square: To **square** a number, multiply the number by itself.		
Perfect Square: A **perfect square** is the result of squaring a whole number.		
Square Root: The square root of a number n is written as $\sqrt{n}$. The **square root** of n is the number that, when multiplied by itself, gives n.		
Right Triangle: A **right triangle** has a 90° angle.		
Hypotenuse: The longest side of a right triangle is called the **hypotenuse.**		
Legs: The **legs** of a right triangle are the two shorter sides.		
Pythagorean Theorem: Given a right triangle, the sum of the squares of the legs is equal to the square of the hypotenuse. 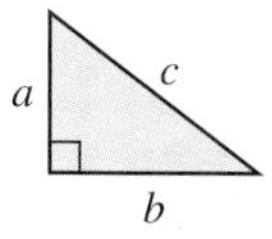 (triangle with legs a, b and hypotenuse c) $a^2 + b^2 = c^2$		

Study these words when they appear in the text. After the section, test your understanding by completing the Vocabulary Review in the section exercises.

Objective A Evaluate the Square Root of a Perfect Square

The Concept Given a number n, the square root of n is written as $\sqrt{n}$. The **square root** of n is the number that, when multiplied by itself, gives n.

$$\sqrt{4} = 2 \quad \text{because } (2)\cdot(2) = 4.$$
$$\sqrt{36} = 6 \quad \text{because } (6)\cdot(6) = 36.$$

If a number can be written as a product of two identical factors, then the factor is a square root.

Procedure **Evaluate the Square Root of a Number**

Determine what number, when multiplied by itself, gives the number inside the square root symbol.

EXAMPLES

1. Find $\sqrt{81}$.

$$\sqrt{81} = 9$$

because $9 \cdot 9 = 81$.

2. Find $\sqrt{121}$.

$$\sqrt{121} = 11$$

because $11 \cdot 11 = 121$.

3. Find $\sqrt{\frac{16}{25}}$.

We must find the fraction, times itself, that gives $\frac{16}{25}$.

Find the square root of the numerator, 16: $4 \cdot 4 = 16$

Find the square root of the denominator, 25: $5 \cdot 5 = 25$

$$\sqrt{\frac{16}{25}} = \frac{4}{5}$$

because $\frac{4}{5} \cdot \frac{4}{5} = \frac{16}{25}$.

GUIDED PRACTICE

1. Find $\sqrt{49}$.

$\sqrt{49} =$ ____

because ____ $\cdot$ ____ $= 49$.

2. Find $\sqrt{64}$.

$\sqrt{64} =$ ____

because ____ $\cdot$ ____ $= 64$.

3. Find $\sqrt{\frac{9}{4}}$.

We must find the fraction, times itself, that gives $\frac{9}{4}$.

Find the square root of the numerator, 9: $(\ )\cdot(\ ) = 9$

Find the square root of the denominator, 4: $(\ )\cdot(\ ) = 4$

$$\sqrt{\frac{9}{4}} = \frac{\quad}{\quad}$$

because $\frac{\quad}{\quad} \cdot \frac{\quad}{\quad} = \frac{9}{4}$.

Square roots are evaluated during the exponent step of the order of operations.

EXAMPLE

4. Find $\sqrt{25} \cdot \sqrt{81}$.

This is an order of operations exercise. First, evaluate the square roots. Then multiply.

$$\sqrt{25} \cdot \sqrt{81} = 5 \cdot 9$$
$$= 45$$

GUIDED PRACTICE

4. Find $\sqrt{100} - \sqrt{64}$.

This is an order of operations exercise. First, ________ ________. Then ________.

$\sqrt{100} - \sqrt{64} =$

$=$

Concept Check

A1. In your own words, what is the square root of a number?

A2. Why does finding the square root of a fraction involve finding the square root of two numbers?

A3. When finding square roots, it is useful to know some perfect squares.

Complete the table below. Take time to study this table to help you learn how to recognize perfect squares.

Number, n	1	2	3	4	5	6	7	8	9	10	11	12	13	14	15
Perfect square, n^2	1	4	9											196	225

A4. Square roots must be evaluated during the ________ step of the order of operations.

Objective A Practice

Evaluate each square root.

A5. $\sqrt{9}$

A6. $\sqrt{4}$

A7. $\sqrt{0}$

A8. $\sqrt{1}$

A9. $\sqrt{\frac{49}{36}}$

A10. $\sqrt{\frac{1}{64}}$

A11. $\sqrt{\frac{225}{49}}$

A12. $\sqrt{\frac{196}{25}}$

Follow the order of operations to simplify each expression.

A13. $\sqrt{4} + \sqrt{9}$

A14. $\sqrt{144} - \sqrt{36}$

A15. $\sqrt{25} \cdot \sqrt{16}$

A16. $\sqrt{49} \cdot \sqrt{1}$

Objective B Approximate the Square Root of a Number

The Concept Often, the square root of a whole number will not be another whole number.

$$\sqrt{4} = 2$$
$$\sqrt{5} = ?$$
$$\sqrt{9} = 3$$

Since 5 is between 4 and 9, $\sqrt{5}$ must be between the square roots of 4 and 9, or 2 and 3. $\sqrt{5}$ cannot be a whole number because there aren't two whole numbers between 2 and 3 that multiply to 5. If we used a calculator, we would see that $\sqrt{5} \approx 2.23606798$, a decimal number. In the following table, we see that $\sqrt{5}$ is about 2.236. This value is consistent with our estimate since 2.236 is between 2 and 3. That is, $2 < 2.236 < 3$.

Before using a square root table, always estimate the square root by first determining the two whole numbers between which the square root falls. Estimation will help you avoid mistakes.

Number n	Square root of the number, $\sqrt{n}$	Number n	Square root of the number, $\sqrt{n}$	Number n	Square root of the number, $\sqrt{n}$
1	**1**	8	2.828	15	3.873
2	1.414	**9**	**3**	**16**	**4**
3	1.732	10	3.162	17	4.123
4	**2**	11	3.317	18	4.243
5	2.236	12	3.464	19	4.359
6	2.449	13	3.606	20	4.472
7	2.646	14	3.742	21	4.583

A version of this table that includes up to $\sqrt{200}$ is available in Appendix B on page APP-8 of this text.

EXAMPLES

5. $\sqrt{51}$ falls between which two whole numbers?

Since $7^2 = 49$ and $8^2 = 64$, $\sqrt{51}$ is between 7 and 8.

6. Approximate $\sqrt{13}$ to the thousandths place using the square root table.

$$\sqrt{13} \approx 3.606$$

7. Approximate $\sqrt{7} + \sqrt{25}$ to the thousandths place.

Square roots are evaluated during the exponent step of the order of operations. They are evaluated before multiplication, division, addition, or subtraction.

$$\sqrt{7} + \sqrt{25} \approx 2.646 + 5$$
$$\approx 7.646$$

We use the square root table for $\sqrt{7}$.

GUIDED PRACTICE

5. $\sqrt{33}$ falls between which two whole numbers?

Since $5^2 =$ __ and __$^2 = 36$, $\sqrt{33}$ is between __ and __.

6. Approximate $\sqrt{11}$ to the thousandths place using the square root table.

$$\sqrt{11} \approx \_\_\_\_$$

7. Approximate $\sqrt{6} \cdot \sqrt{4}$ to the thousandths place.

Square roots are evaluated during the ________ step of the order of operations. They are evaluated before ________, ________, ________, or ________.

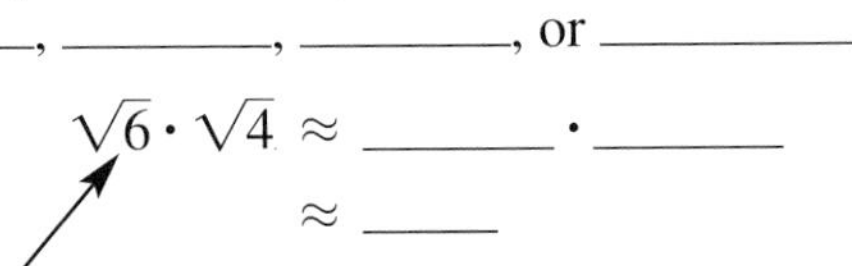

$$\sqrt{6} \cdot \sqrt{4} \approx \_\_\_\_ \cdot \_\_\_\_$$
$$\approx \_\_\_\_$$

Use the square root table for $\sqrt{6}$.

Concept Check

B1. The square root of a number that is a(n)__________ will have no decimal part.

B2. If a number is not a(n)__________, its square root will have a decimal part.

B3. Write the first nine perfect squares.
$(1)^2 = 1$ $(2)^2 = 4$ $(\ \ )^2 =$ __ $(\ \ )^2 =$ __ $(\ \ )^2 =$ __ $(\ \ )^2 =$ __
$(\ \ )^2 =$ __ $(\ \ )^2 =$ __ $(\ \ )^2 =$ __

B4. Write a set of steps to help a friend understand how to determine the 2 whole numbers between which $\sqrt{31}$ falls.

Objective B Practice

Each square root falls between which two whole numbers? Do not find the actual square root.

B5. $\sqrt{3}$ **B6.** $\sqrt{19}$ **B7.** $\sqrt{17}$ **B8.** $\sqrt{5}$

Find each square root rounded to the thousandths place.

B9. $\sqrt{3}$ **B10.** $\sqrt{19}$ **B11.** $\sqrt{17}$ **B12.** $\sqrt{5}$

Use the order of operations to evaluate each expression.

B13. $\sqrt{36} + \sqrt{3}$ **B14.** $\sqrt{49} - \sqrt{20}$ **B15.** $\sqrt{12} - \sqrt{3}$

B16. $\sqrt{8} + \sqrt{8}$ **B17.** $\sqrt{18} \cdot \sqrt{2}$ **B18.** $\sqrt{5} \cdot \sqrt{3}$

Objective C Use the Pythagorean Theorem

The Concept When two sides of a right triangle are known, we can use the Pythagorean theorem to find the length of the third side.

INTERACTIVE DEFINITION The Pythagorean Theorem

The **Pythagorean theorem** states that for any right triangle,

$$(\text{1st leg})^2 + (\text{2nd leg})^2 = (\text{hypotenuse})^2.$$

The Pythagorean theorem is used to find an unknown side of a right triangle using one of the following formulas:

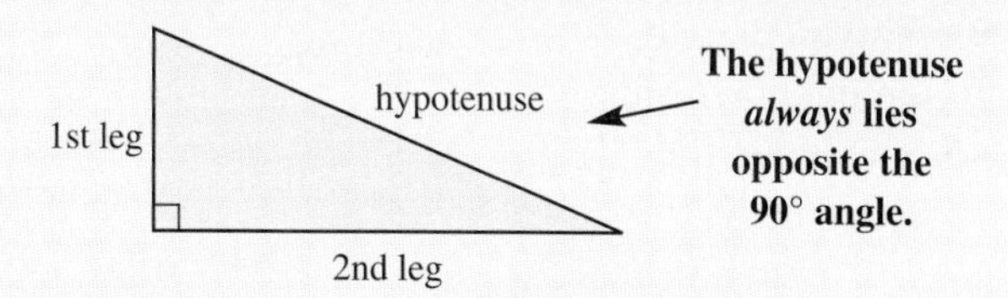

Formula for a Leg

$$\text{leg} = \sqrt{(\text{hypotenuse})^2 - (\text{known leg})^2}$$

Formula for the Hypotenuse

$$\text{hypotenuse} = \sqrt{(\text{1st leg})^2 + (\text{2nd leg})^2}$$

EXAMPLE

8. a. Identify the lengths of the legs and the hypotenuse in the right triangle.

b. Substitute the values into the appropriate formula.

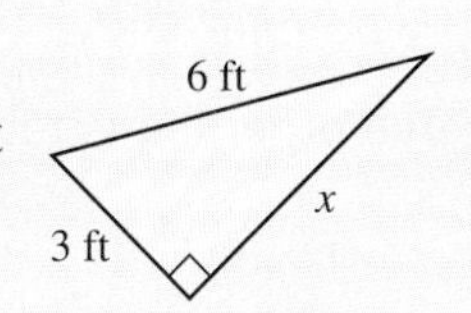

a. The hypotenuse is opposite the right angle.

Hypotenuse = 6 ft
1st leg = 3 ft
2nd leg = x ft

b. Since a leg is unknown, use the formula

$$\text{leg} = \sqrt{(\text{hypotenuse})^2 - (\text{known leg})^2}$$

Substitute the values:

$$x = \sqrt{6^2 - 3^2}$$

GUIDED PRACTICE

8. a. Identify the lengths of the legs and hypotenuse in the right triangle.

b. Substitute the values into the appropriate formula.

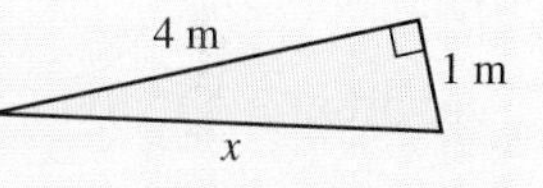

a. The hypotenuse is opposite the ________.

Hypotenuse = _____.
1st leg = _____.
2nd leg = _____

b. Since the ________ is unknown, use the formula

$$\_\_\_\_\_ = \sqrt{(\_\_)^2 \_\_ (\_\_)^2}$$

Substitute the values: +/−?

$$x = \sqrt{(\ )^2 \_\_ (\ )^2}$$

DO YOU UNDERSTAND how to identify the legs and hypotenuse of a right triangle?	Got It	Get Help
DO YOU UNDERSTAND how to substitute values into the Pythagorean theorem correctly?	Got It	Get Help

Prodecure Use the Pythagorean Theorem

Step 1 Use the appropriate formula.
Step 2 Substitute the values.
Step 3 Simplify.

After substituting the known lengths into the Pythagorean theorem, you must follow the order of operations. The inside of a square root symbol should be evaluated in the groupings step of the order of operations. The square root itself is evaluated during the exponent step of the order of operations. Always simplify the inside of a square root before evaluating the square root itself.

EXAMPLES

9. Find the length of the unknown side, rounded to the thousandths place.

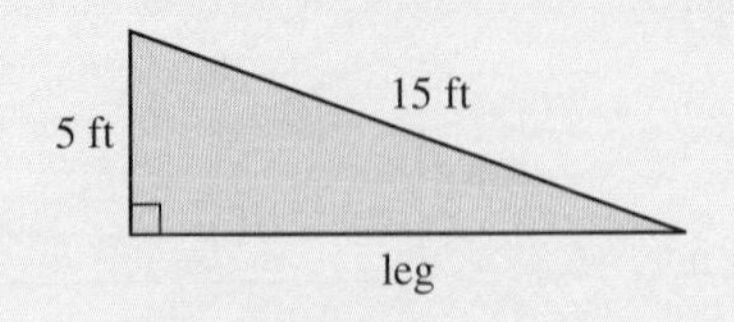

Since the triangle is a right triangle, we can use the Pythagorean theorem.

One leg $= 5$ ft
Hypotenuse $= 15$ ft

The other leg is found using the formula

$$\begin{aligned} \text{leg} &= \sqrt{(\text{hypotenuse})^2 - (\text{known leg})^2} \\ &= \sqrt{15^2 - 5^2} \\ &= \sqrt{225 - 25} \\ &= \sqrt{200} \\ &\approx 14.142 \text{ ft} \end{aligned}$$

Always simplify inside the square root before evaluating the square root.

▶ **NOTE** This value came from the square root table in Appendix B.

10. Find the length of the unknown side, rounded to the thousandths place.

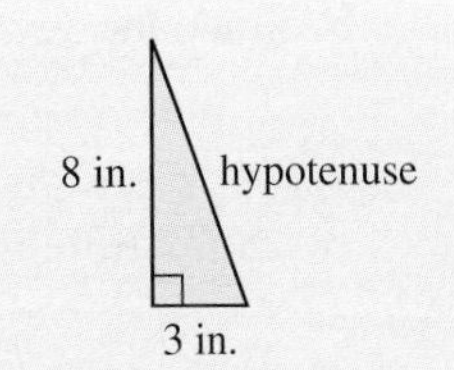

Since the triangle is a right triangle, we can use the Pythagorean theorem.

1st leg $= 8$ in.
2nd leg $= 3$ in.

The hypotenuse is found using the formula

$$\begin{aligned} \text{hypotenuse} &= \sqrt{(\text{1st leg})^2 + (\text{2nd leg})^2} \\ &= \sqrt{8^2 + 3^2} \\ &= \sqrt{64 + 9} \\ &= \sqrt{73} \\ &\approx 8.544 \text{ inches} \end{aligned}$$

Always simplify inside the square root before evaluating the square root.

GUIDED PRACTICE

9. Find the length of the unknown side, rounded to the thousandths place.

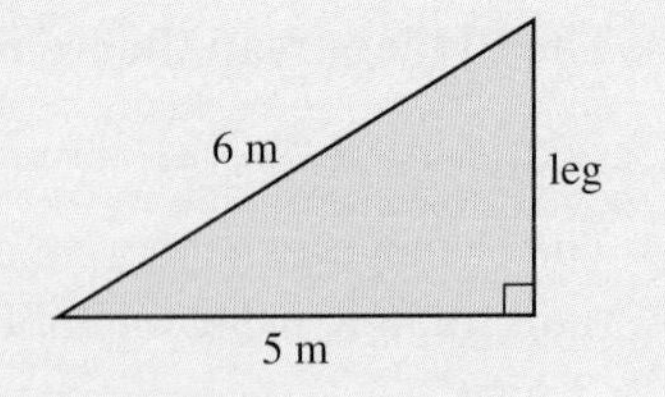

Since the triangle is a(n) ____ triangle, we can use the Pythagorean theorem.

One leg = ____
Hypotenuse = ____

The other leg is found using the formula

$$\begin{aligned} \text{leg} &= \sqrt{(\text{hypotenuse})^2 - (\text{known leg})^2} \\ &= \sqrt{(\quad)^2 - (\quad)^2} \\ &= \sqrt{\quad - \quad} \\ &= \sqrt{\quad} \\ &\approx \end{aligned}$$

10. Find the length of the unknown side, rounded to the thousandths place.

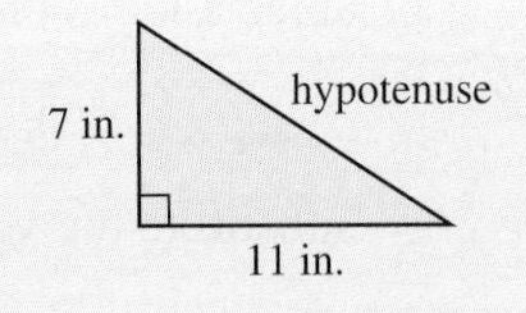

Since the triangle is a(n) ____________, we can use the ________________.

1st leg = ____
2nd leg = ____

The hypotenuse is found using the formula

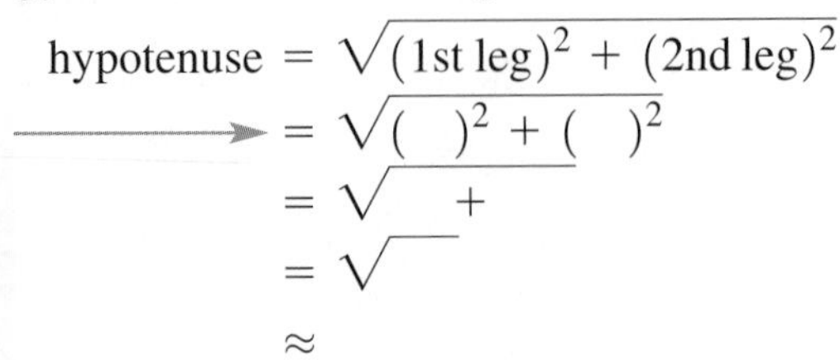

$$\begin{aligned} \text{hypotenuse} &= \sqrt{(\text{1st leg})^2 + (\text{2nd leg})^2} \\ &= \sqrt{(\quad)^2 + (\quad)^2} \\ &= \sqrt{\quad + \quad} \\ &= \sqrt{\quad} \\ &\approx \end{aligned}$$

11. A 25 ft ladder is placed against a building and touches the building at a point 22 ft above ground. How far is the base of the ladder from the building? Round your answer to the nearest tenth.

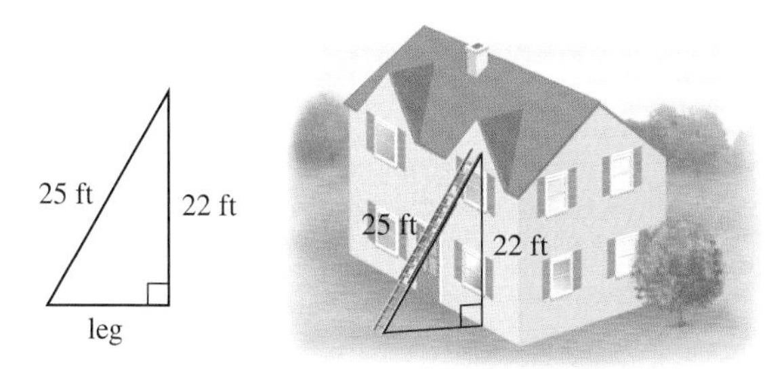

1st leg = 22 ft
2nd leg = unknown
Hypotenuse = 25 ft

Be sure to choose the appropriate formula to solve for a leg.

$$\begin{aligned}\text{leg} &= \sqrt{(\text{hypotenuse})^2 - (\text{known leg})^2}\\ &= \sqrt{25^2 - 22^2}\\ &= \sqrt{625 - 484}\\ &= 141\\ &\approx 11.9 \text{ ft}\end{aligned}$$

11. A kite is flying on a 30 yd string. The kite is directly above a rock. The rock is 27 yd from the child flying the kite. How far is the kite above the ground? Round your answer to the nearest tenth.

Read the problem and fill in the missing values on the diagram.

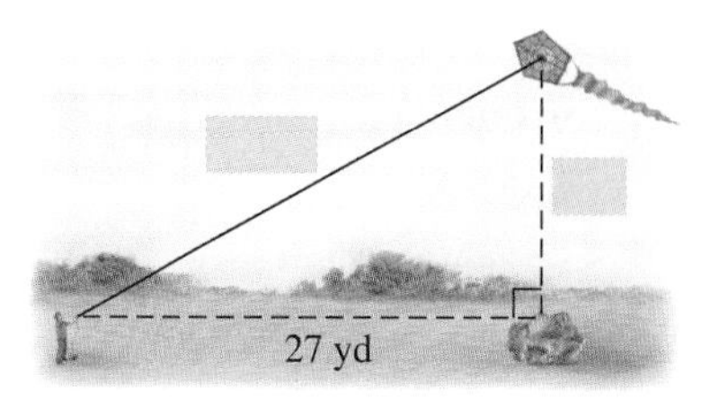

1st leg = ________
2nd leg = ____________
Hypotenuse = ________

Be sure to choose the appropriate formula to solve for a(n) _____.

$$= \sqrt{(\qquad)^2 - (\qquad)^2}$$
$$=$$

As you work on the exercises below, you will need to use a calculator or the square roots table in Appendix B on page APP-8 of this text. Be sure to ask your instructor what is expected in your course.

Concept Check

C1. The Pythagorean theorem can be used only with _____ triangles.

C2. What is the name of the longest side of a right triangle?

C3. **a.** Write the formula used to find the length of the hypotenuse in a right triangle.
b. Write the formula used to find the length of a leg in a right triangle.

Objective C Practice

Use the Pythagorean theorem to find the unknown side of each right triangle. Round your answer to the nearest tenth.

C4.

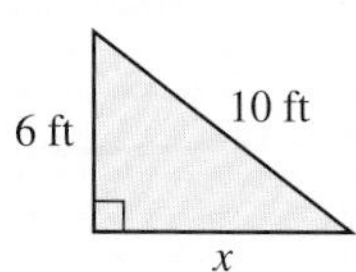

C5.

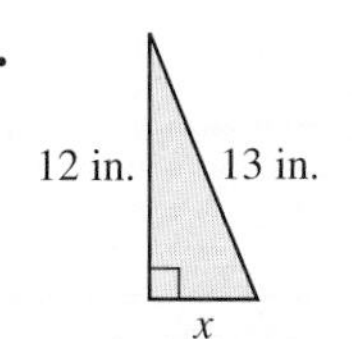

C6.

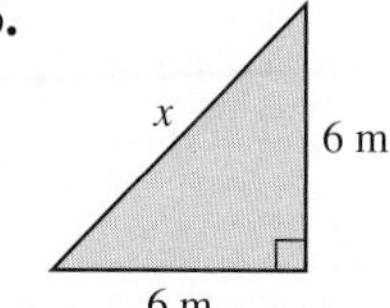

C7.

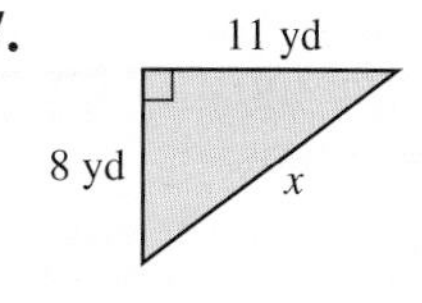

(Continued)

C8.
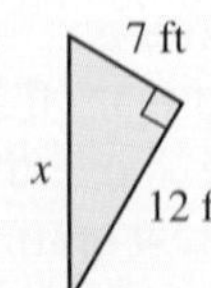

C9.
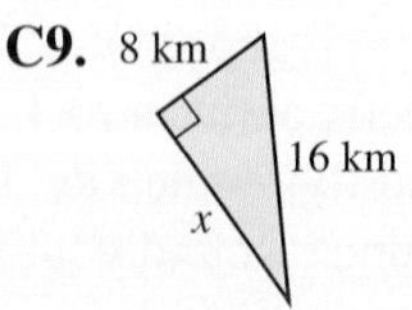

C10.
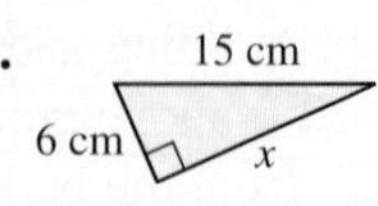

C11.
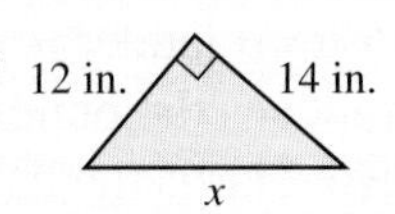

FOCUS ON ALGEBRA The Pythagorean Theorem

The Pythagorean Theorem

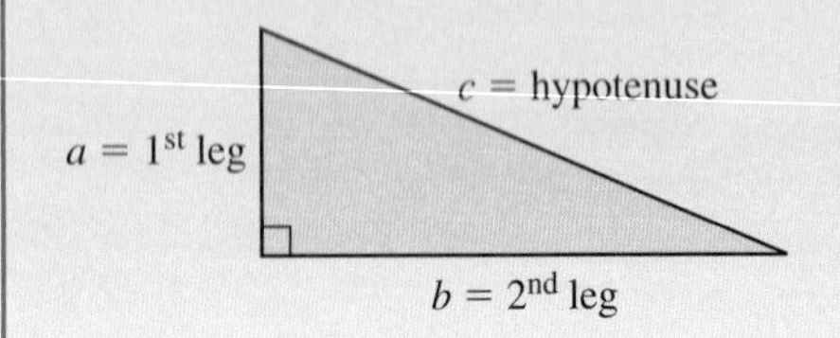

$$a^2 + b^2 = c^2$$

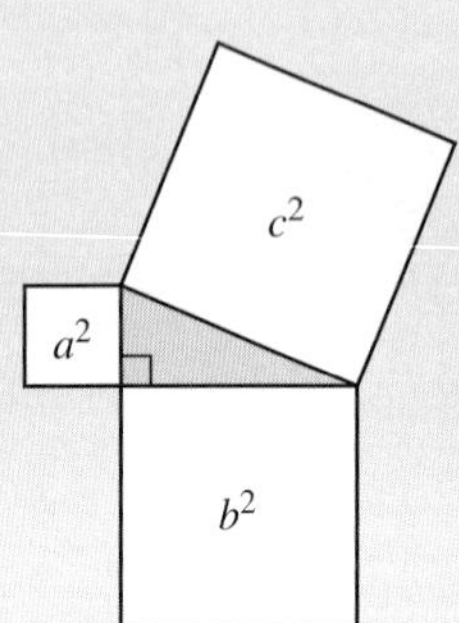

In a right triangle, the sum of the squares of the legs equals the square of the hypotenuse.

Graphical interpretation:
The area of the large square, c^2, is equal to the total area of the small squares, $a^2 + b^2$.

Procedure Use the Pythagorean Theorem

Step 1 Substitute the given information into the formula $a^2 + b^2 = c^2$.

Step 2 Simplify.

Step 3 Isolate the variable.

- Isolate a^2, b^2, or c^2.
- Undo the square by taking the square root of both sides.

EXAMPLES

12. Find the length of the unknown side.

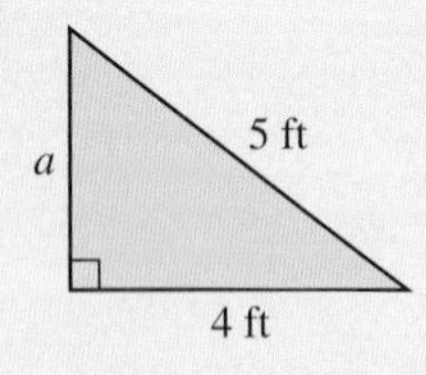

$a = 4,\ b = \text{unknown},\ c = 5$

Step 1 Substitute the values.

$a^2 + b^2 = c^2$

$(4)^2 + b^2 = (5)^2$

Step 2 Simplify.

$16 + b^2 = 25$

Step 3 Isolate the variable.

$16 - 16 + b^2 = 25 - 16$

$b^2 = 9$

Undo the square with a square root.

$\sqrt{b^2} = \sqrt{9}$

$b = 3\text{ ft}$

GUIDED PRACTICE

12. Find the length of the unknown side.

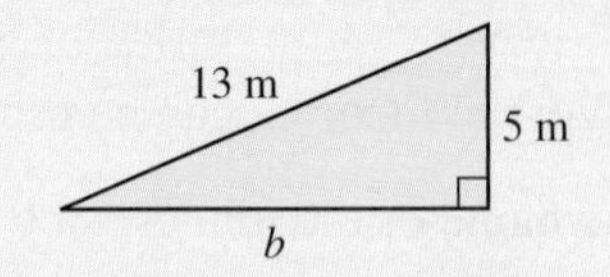

$a =$ ____, $b =$ ______, $c =$ ____

$a^2 + b^2 = c^2$

$(\quad)^2 + b^2 = (\quad)^2$

$\quad + b^2 = $

$- \quad + b^2 = \quad -$

$b^2 =$

$\sqrt{b^2} = \sqrt{\quad}$

$b =$

13. Find the length of the unknown side.

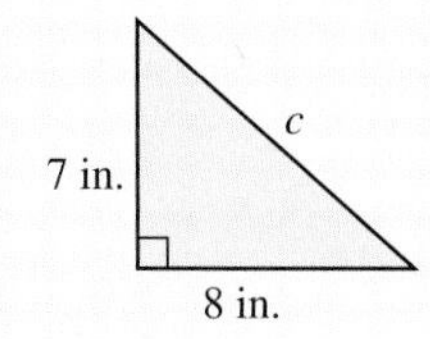

$a = 7,\ b = 8,\ c = \text{unknown}$

Step 1 Substitute the values.

$$a^2 + b^2 = c^2$$
$$(7)^2 + (8)^2 = c^2$$

Step 2 Simplify.

$$49 + 64 = c^2$$

Step 3 The variable is isolated.

$$113 = c^2$$

Undo the square with a square root.

$$\sqrt{113} = \sqrt{c^2}$$
$$10.630 \text{ in.} = c$$

13. Find the length of the unknown side.

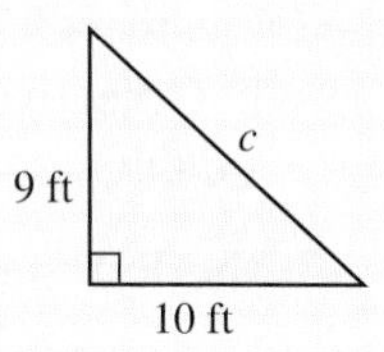

$a =$ ____, $b =$ ____, $c =$ ______

$$a^2 + b^2 = c^2$$
$$(\quad)^2 + (\quad)^2 = c^2$$
$$\quad + \quad = c^2$$
$$\quad = c^2$$
$$\sqrt{\quad} = \sqrt{\quad}$$
$$\quad = c$$

PRACTICE

Use the Pythagorean theorem to find the length of the unknown side. Round each answer to the thousandths place.

1.

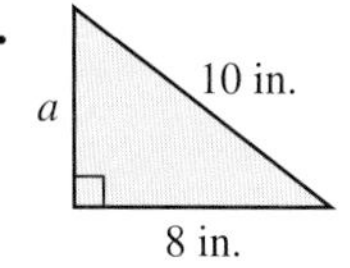

2.

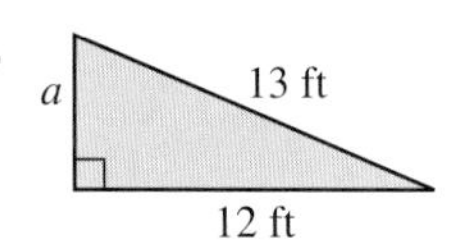

3.

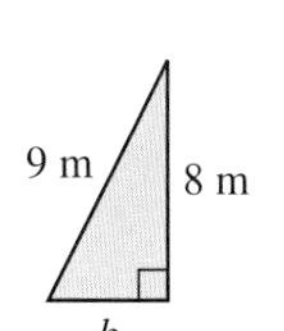

4.

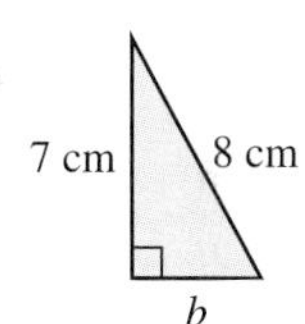

5.

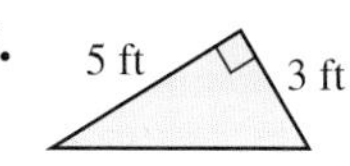

6.

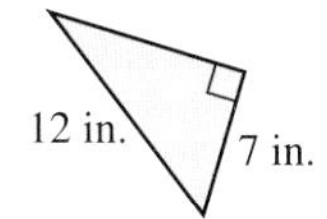

7.

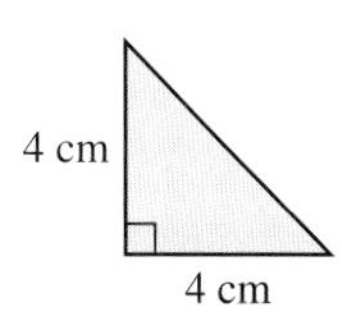

8.

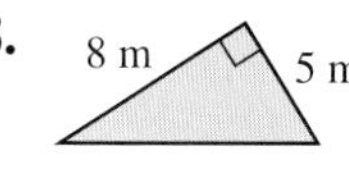

Section 7.6 Exercises

FOR EXTRA HELP

To evaluate the square root of a perfect square:

1. Answer the Objective A Concept Checks.
2. Answer the odd Objective A Practice Exercises.
3. Answer the even Objective A Practice Exercises.

To approximate the square root of a number:

4. Answer the Objective B Concept Checks.
5. Answer the odd Objective B Practice Exercises.
6. Answer the even Objective B Practice Exercises.

To use the Pythagorean theorem:

7. Answer the Objective C Concept Checks.
8. Answer the odd Objective C Practice Exercises.
9. Answer the even Objective C Practice Exercises.

To use algebra with the Pythagorean theorem:

10. Answer the odd Focus On Algebra Exercises.
11. Answer the even Focus On Algebra Exercises.

VOCABULARY REVIEW *Review the Vocabulary Preview for Section 7.6. Study the definitions until you can check* Got It *for every word.*

Use these words to complete each sentence.

perfect square • square root • hypotenuse • legs • square • Pythagorean theorem • right triangle

	Got It	Get Help
12. A ____________ is the longest side of a right triangle.		
13. The result of squaring a whole number is a ____________.		
14. In the ____________, the sum of the squares of the ____________ is equal to the ____________ of the hypotenuse.		
15. In the equation $\sqrt{9} = 3$, 3 is the ____________ of 9.		

How will you get help for any vocabulary that you are unsure about?

Your instructor ____ MyMathLab ____ A classmate ____ A tutor ____ Other ____

16. $\sqrt{33}$ is read as the ____________ of ____.

17. How do you know that $\sqrt{33}$ is between 5 and 6?

Evaluate each square root.

18. $\sqrt{36}$ **19.** $\sqrt{49}$ **20.** $\sqrt{100}$ **21.** $\sqrt{144}$

22. $\sqrt{\frac{4}{81}}$ **23.** $\sqrt{\frac{64}{25}}$ **24.** $\sqrt{\frac{121}{9}}$ **25.** $\sqrt{\frac{4}{169}}$

Each square root falls between which two whole numbers?

26. $\sqrt{3}$ **27.** $\sqrt{7}$ **28.** $\sqrt{30}$ **29.** $\sqrt{55}$

30. $\sqrt{95}$ **31.** $\sqrt{65}$ **32.** $\sqrt{102}$ **33.** $\sqrt{77}$

Find the square root of each square's area to determine the length of each side.

34. 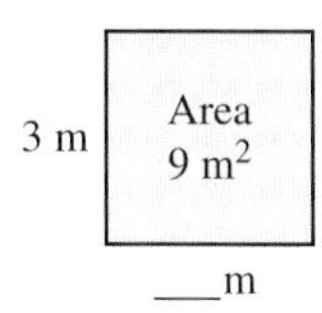

35. 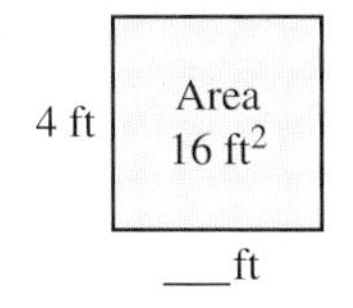

36. 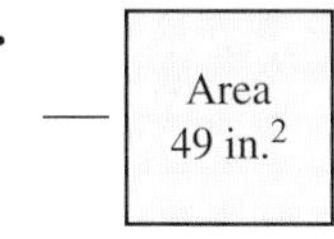

37. 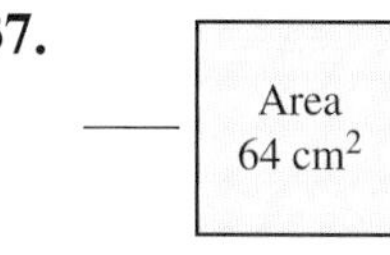

38. 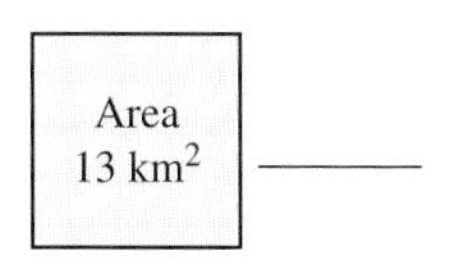

39. 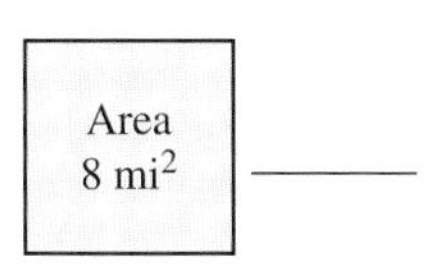

40. 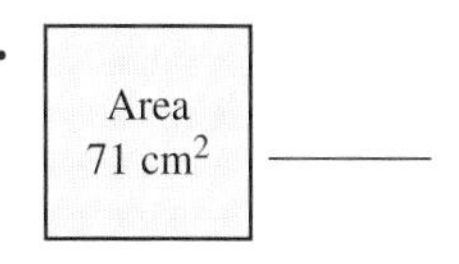

41. 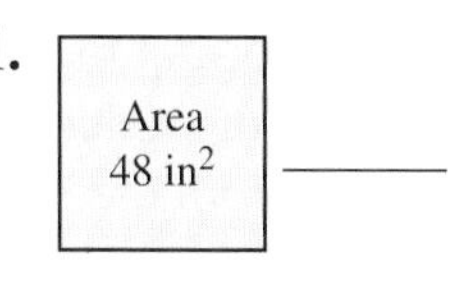

42. Based on Exercises 34–41, explain how finding the length of the side of each square is related to finding the square root of the square's area.

43. A square is attached to each side of a right triangle. Use the given side length of each square to show that the sum of the squares of the legs is equal to the square of the hypotenuse.

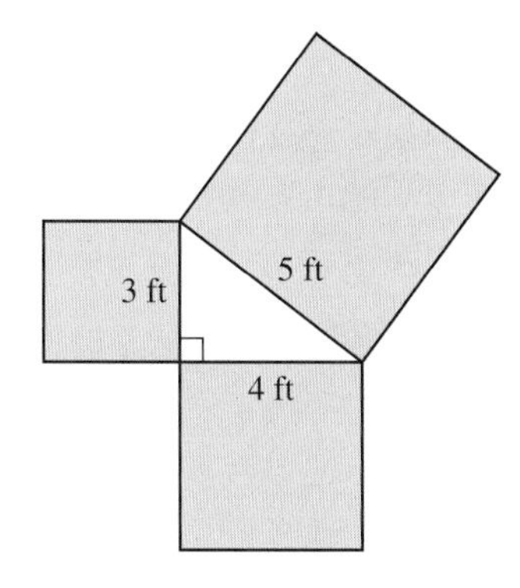

44. A square is attached to each side of a right triangle. Use the given side length of each square to show that the sum of the squares of the legs is equal to the square of the hypotenuse.

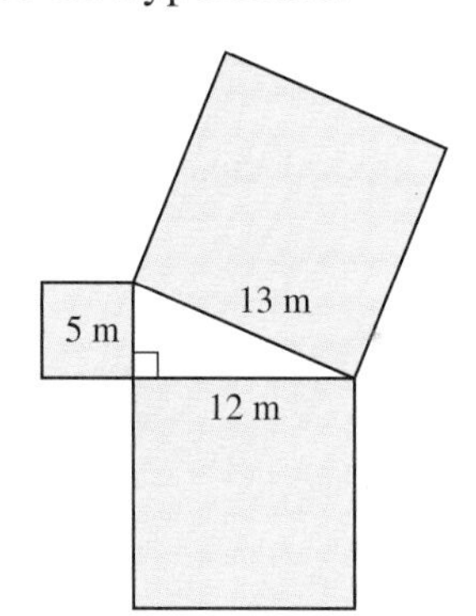

Use the square root table (p. APP-8) to approximate each value. Round to the thousandths place.

45. $\sqrt{3}$

46. $\sqrt{7}$

47. $\sqrt{30}$

48. $\sqrt{55}$

49. $\sqrt{83}$

50. $\sqrt{33}$

51. $\sqrt{8}$

52. $\sqrt{18}$

53. $\sqrt{4} + \sqrt{36}$

54. $\sqrt{25} + \sqrt{9}$

55. $\sqrt{22} - \sqrt{2}$

56. $\sqrt{39} - \sqrt{9}$

Find the length of the unknown side of each right triangle. Round your answer to the nearest hundredths place.

57. 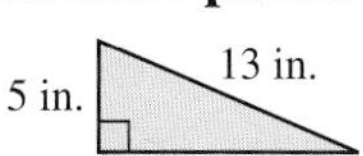

58. 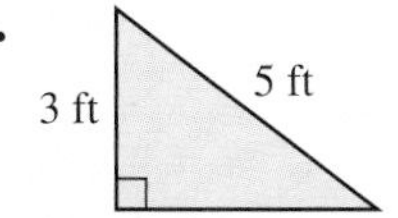

59. 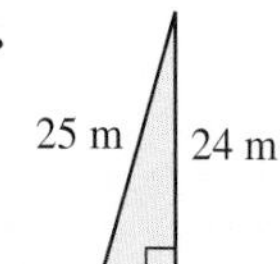

60. 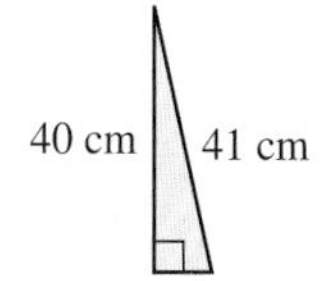

goetz Some right triangles have whole number measures, including 3-4-5, 6-8-10, and 5-12-13. Remembering these will help you work quicker. *For more tips and tweets, go to twitter.com/gstbasicmath*

61. 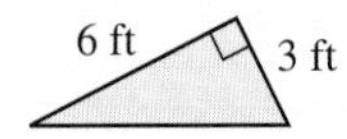

62. 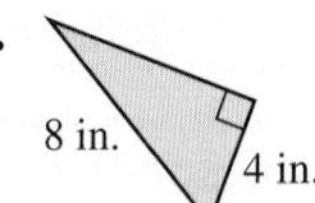

63. 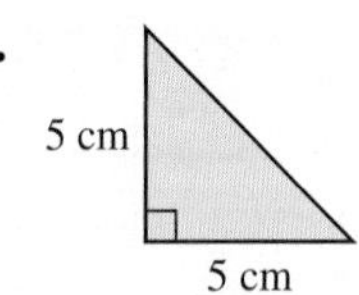

64. 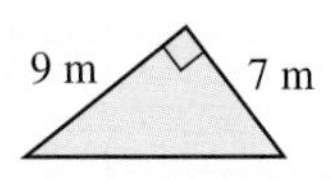

Answer each question. Round your answer to the thousandths place.

65. A 10-foot-tall basketball hoop is lowered onto a 6-foot-tall ladder. How far must the ladder be placed from the base of the hoop?

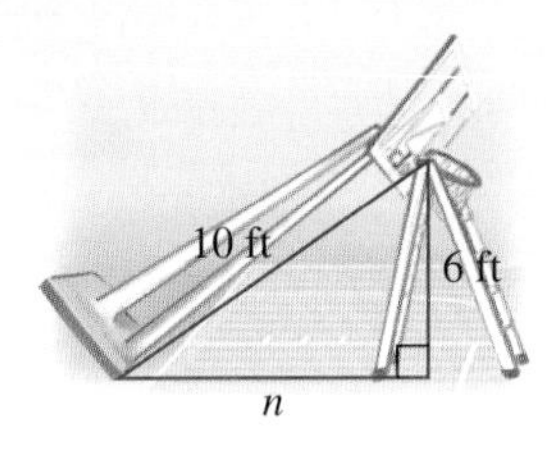

66. A 15-foot ladder is leaning against a wall. The base of the ladder is 6 feet from the wall. How high does the ladder touch the wall?

67. Antonio must find the area of a roof (front and back) to purchase the correct number of shingles.

a. What are the length and width of each side of the roof?

b. What is the area of the entire roof, front and back?

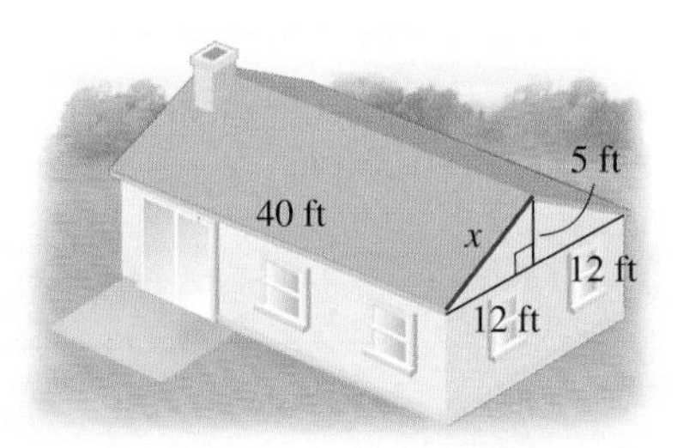

68. The ridge of a tent is 7 feet long. The triangular front of the tent is 4 feet wide and 4 feet tall.

a. What is the width of one side of the tent's roof?

b. What is the total area of the fabric used to make the roof of the tent, front and back?

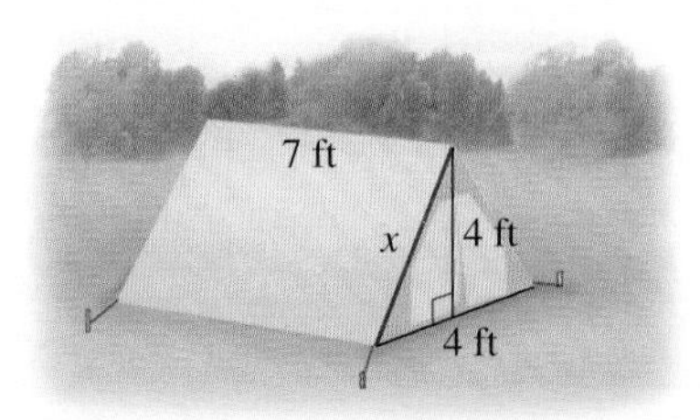

69. A carpenter installs a window with two panes of glass. One pane is a square with a side of 2 ft, and the other pane is an isosceles right triangle.

a. What is the area of the glass used in the window?

b. How many feet of trim must be used on the perimeter of the window?

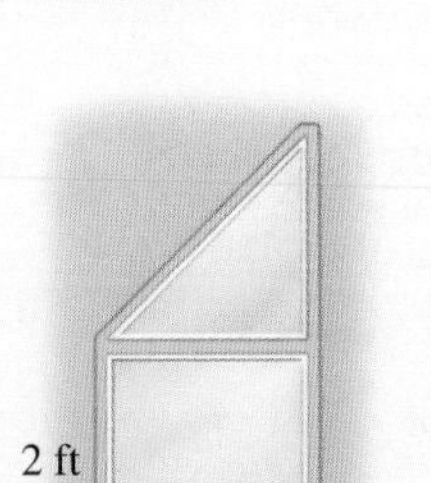

70. Rudolph makes a radio tower for his model train, using five 2-inch sections. The sections alternate between red and black, as shown.

a. The wire is attached to the top of the fourth section. How high is this?

b. How long is the wire if it is anchored 5 inches from the model's base?

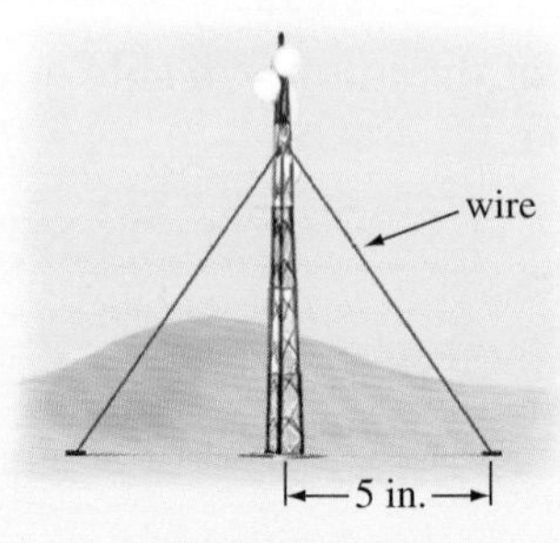

71. Each side of the square shown is 1 meter long.

a. Use the Pythagorean theorem to find the diameter of the circle.

b. What is the area of the circle?

c. What is the area inside the circle but outside the square?

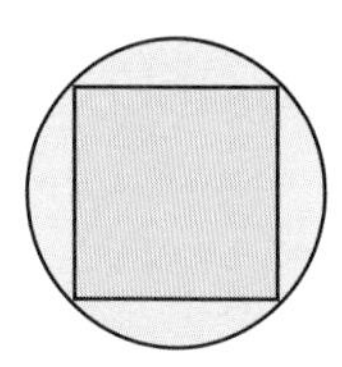

Section 7.6 Question Log

Use this space to write questions. Be sure to get these questions answered and revisit them when you prepare for your exam.

Page ____ **Answered**	**Page** ____ **Answered**
Page ____ **Answered**	**Page** ____ **Answered**

Section 7.7 Similarity

The concept of similarity is used to represent objects with smaller or larger models. Similar objects are proportional to each other, so we can use proportions to solve for unknown lengths in similar figures.

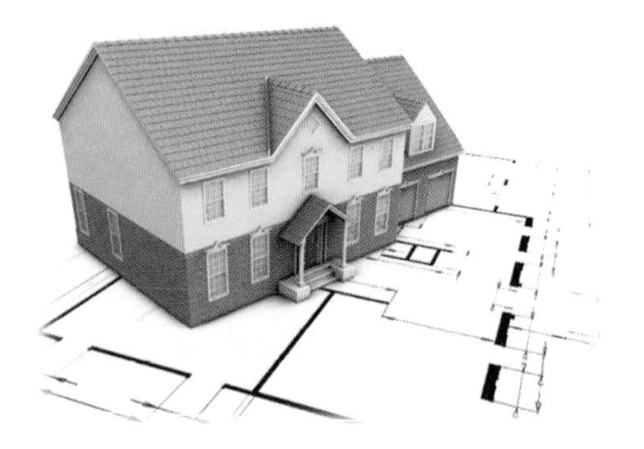

A blueprint is similar to the object that it represents.

The **Objectives** in this section will help you

- A Identify similar figures.
- B Find unknown dimensions of similar figures.
- C Find the area of a similar figure.

VOCABULARY PREVIEW *Check the box that applies.*	Got It	Must Study
Similar Objects: Similar objects have the same shape but may have different sizes.		
Corresponding Angles: The **corresponding angles** of similar figures have equal measure.		
Corresponding Sides: The **corresponding sides** of similar figures form equivalent ratios.		
Proportion: A **proportion** states that two ratios or rates are equal.		

Study these words then they appear in the text. After the section, test your understanding by completing the Vocabulary Review in the section exercises.

Objective A Identify Similar Figures

The Concept **Similar objects** have the same shape but may have different sizes. One similar figure can be thought of as a photocopy, perhaps an enlargement, of the other figure. Because the following two triangles have identical shapes, they are similar. The larger triangle is an enlargement of the smaller triangle.

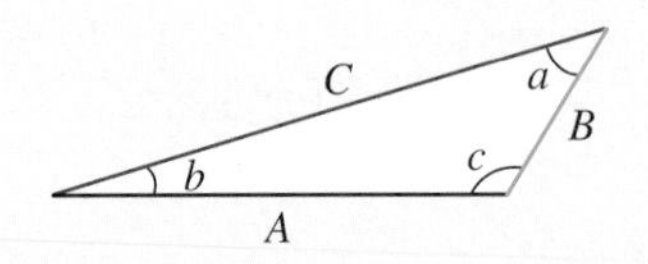

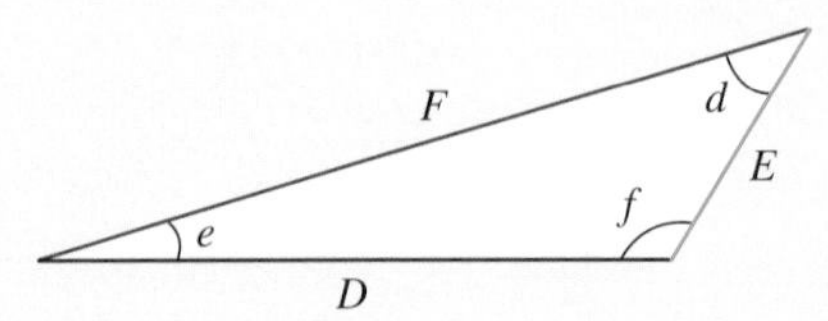

The **corresponding angles** in each triangle have equal measure and are located directly opposite the **corresponding sides.**

Corresponding Angles			Corresponding Sides	
These pairs of angles are equal in measure.	$a = d$ $b = e$ $c = f$	$\Rightarrow$	A and D B and E C and F	**The corresponding sides of similar figures form equivalent ratios.** $\frac{A}{D} = \frac{B}{E} = \frac{C}{F}$

The following triangles are used in Examples/Guided Practice 1–3.

a. A B C

b.

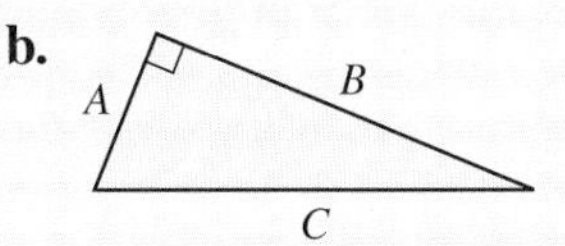

c. A B C

d.

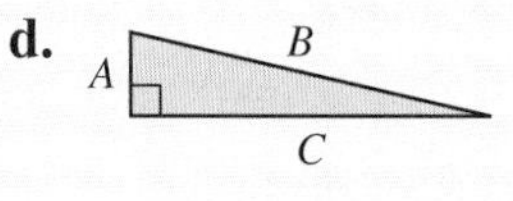

e. A B C

f.

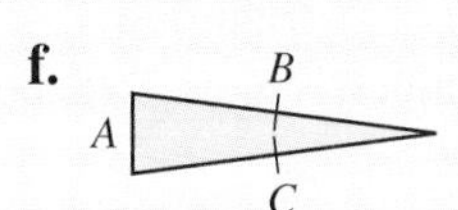

g.

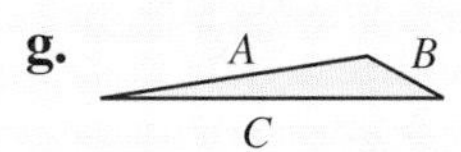

EXAMPLES

1. The following triangle is similar to exactly one of the triangles (a)–(g).

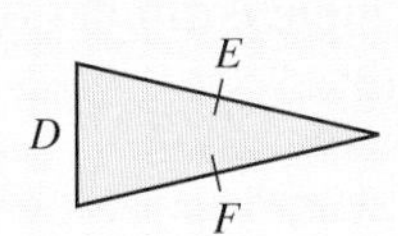

a. Which of the triangles, (a)–(g), is similar to the triangle above?
b. What are the corresponding sides?

a. Since the triangle is isosceles, it can be similar only to the other isosceles triangles: (a), (c), or (f). The longest side is about 2 times the shortest side.

This matches (c). The above triangle is similar to (c).

b. D corresponds to A.
E corresponds to B.
F corresponds to C.

GUIDED PRACTICE

1. The following triangle is similar to exactly one of the triangles (a)–(g).

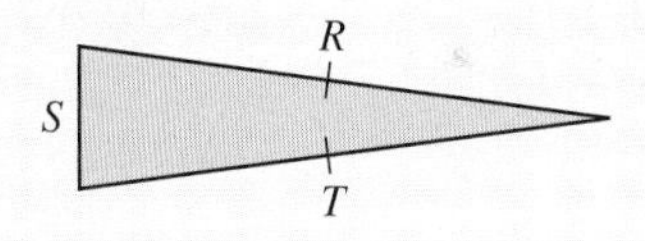

a. Which of the triangles, (a)–(g), is similar to the triangle above?
b. What are the corresponding sides?

a. Since the triangle is ______ it can be similar only to the other ______ triangles: ____, ____, or ____. The longest side is about ____ times the shortest side. That matches ____.

The above triangle is similar to ____

b. ____ corresponds to C.
____ corresponds to B.
____ corresponds to A.

(Continued)

2. The following triangle is similar to exactly one of the triangles (a)–(g).

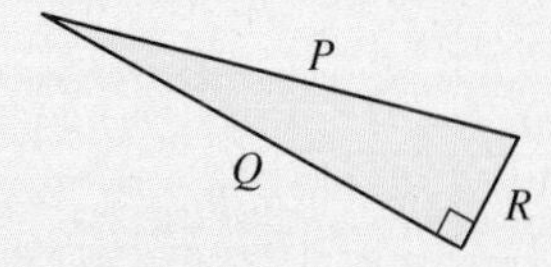

a. Which of the triangles, (a)–(g), is similar to the above triangle?
b. What are the corresponding sides?

a. Since the triangle is a right triangle, it can be similar only to the other right triangles: (a), (b), or (d). The longest side is much longer than the shortest side. This matches (d).

The triangle is similar to (d).

b. Identify corresponding sides by matching the longest, shortest, and medium length sides.

Shortest sides: R corresponds to A.
Longest sides: P corresponds to B.
Medium sides: Q corresponds to C.

3. The following triangle is similar to exactly one of the triangles (a)–(g).

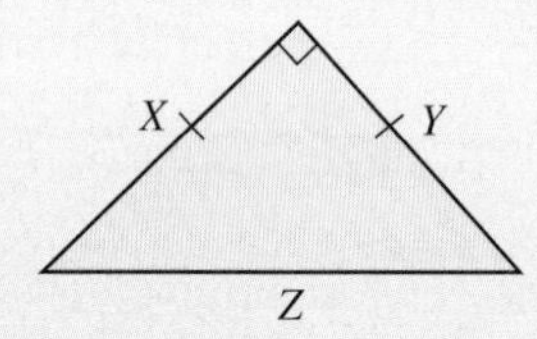

a. Which of the triangles, (a)–(g), is similar to the above triangle?
b. What are the corresponding sides?

a. Since the triangle is right isosceles, it can be similar only to the other right isosceles triangle. This matches (a).

The triangle is similar to (a).

b. Since the triangle is right isosceles, the hypotenuses correspond, as do the other legs.

Z corresponds to B.
X corresponds to C.
Y corresponds to A.

2. The following triangle is similar to exactly one of the triangles (a)–(g).

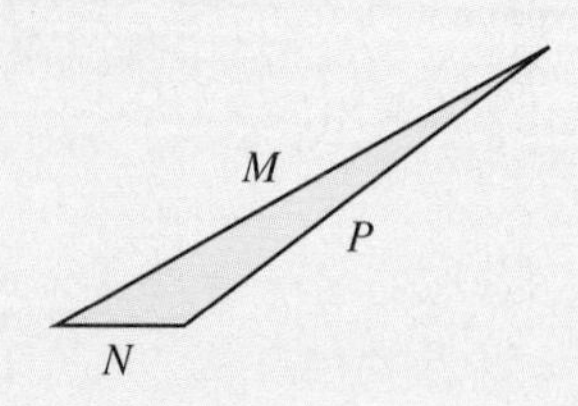

a. Which of the triangles, (a)–(g), is similar to the above triangle?
b. What are the corresponding sides?

a. Since the triangle is ______ it can be similar only to the other ______ triangles: ______ or ______. The longest side is much longer than the shortest side. This matches ______.

The triangle is similar to ______.

b. Identify corresponding sides by matching the longest, shortest, and medium length sides.

Shortest sides: ______ corresponds to B.
Longest sides: ______ corresponds to C.
Medium sides: ______ corresponds to A.

3. The following triangle is similar to exactly one of the triangles (a)–(g).

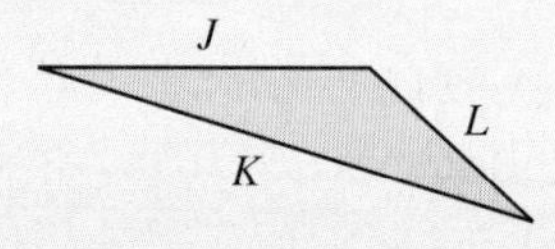

a. Which of the triangles, (a)–(g), is similar to the above triangle?
b. What are the corresponding sides?

a. Since the triangle is ______, it can be similar only to the other ______ triangles: ______ or ______. The two shortest sides are close in length. This matches ______.

The triangle is similar to ______.

b. Identify corresponding sides by matching the longest, shortest, and medium length sides.

______ corresponds to ______.
______ corresponds to ______.
______ corresponds to ______.

Concept Check

A1. When two figures are similar, the corresponding sides are located directly opposite the __________ angles.

A2. Similar figures may have different __________, but they must have the same __________.

Objective A Practice

The figures in Exercises A3–A8 are similar to exactly one figure, (A)–(F), drawn at right. Determine which figures are similar and explain why the figures are similar.

A3. 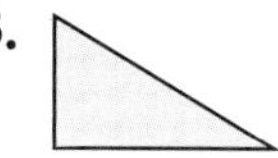

A4. 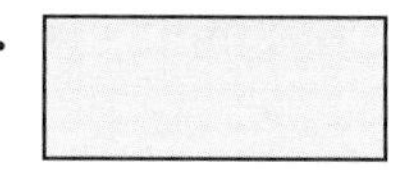

A5. 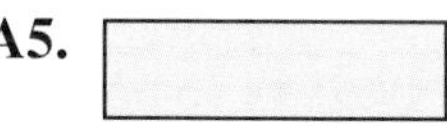

A6. 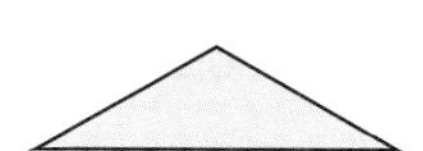

A7. 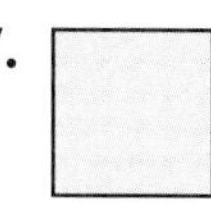

A8. 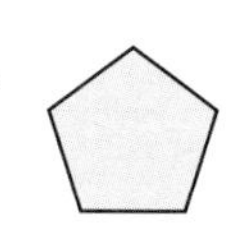

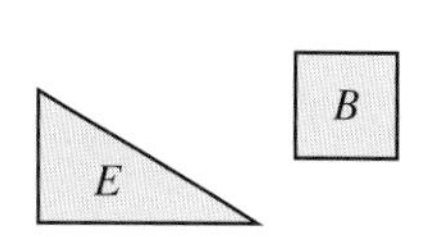

For each pair of similar figures, list all pairs of corresponding sides.

A9.

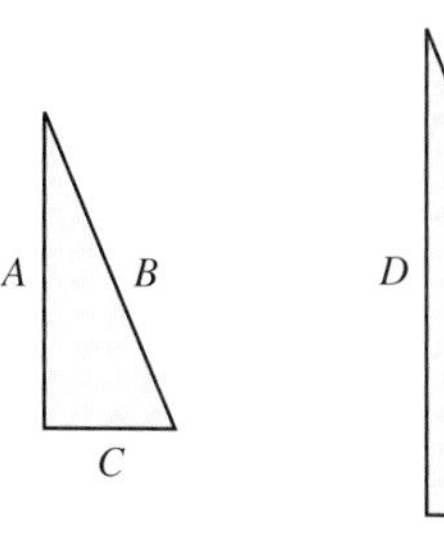

A10.

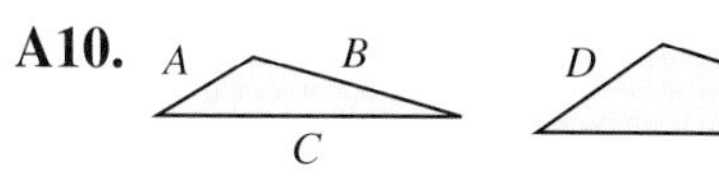

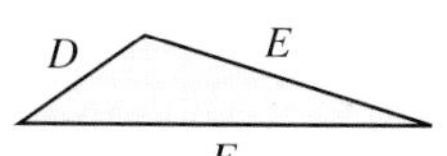

A11.

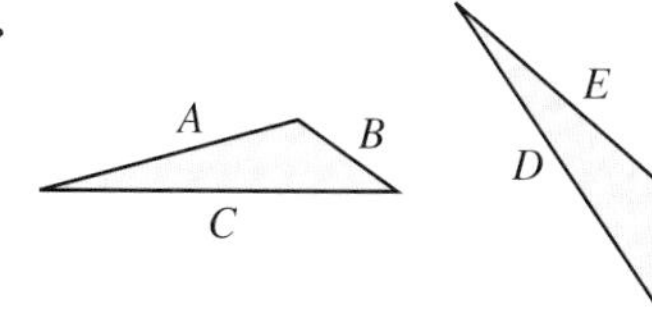

A12.

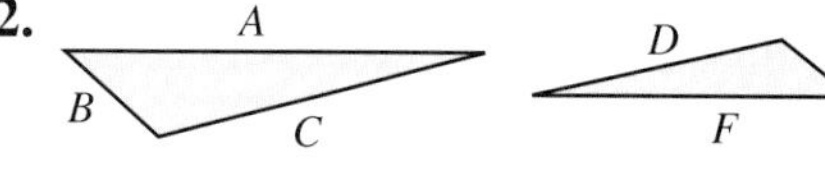

Objective B Find Unknown Dimensions of Similar Figures

The Concept When two figures are similar, the corresponding sides form equivalent ratios. The following triangles are similar, so each pair of corresponding sides forms an equivalent ratio. The equivalent ratios may be used to write a proportion, which is then used to solve the length of the unknown side, x.

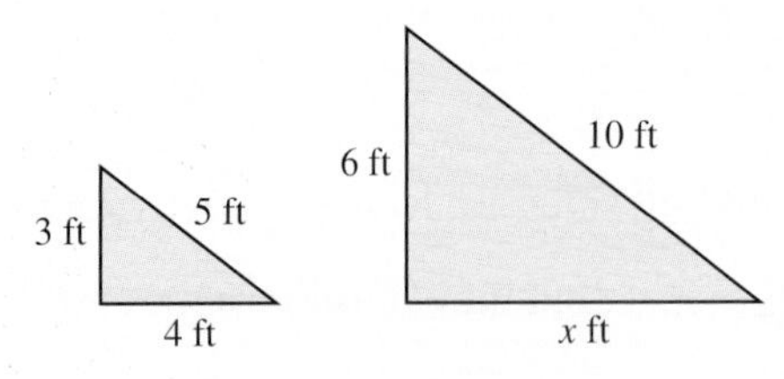

Three Proportions

$$\frac{3}{6} = \frac{4}{x} \quad or \quad \frac{10}{5} = \frac{x}{4} \quad or \quad \frac{x}{4} = \frac{6}{3}$$

The length of the unknown side, x, can be found using any of these proportions.

> **Procedure Find the Unknown Dimensions of Similar Figures**
>
> **Step 1** Set up a proportion.
>
> **Step 2** Solve the proportion.
> Get the variable into the numerator and then isolate it.

EXAMPLES

4. The following two triangles are similar. Write a proportion that could be used to solve for the length of the unknown side.

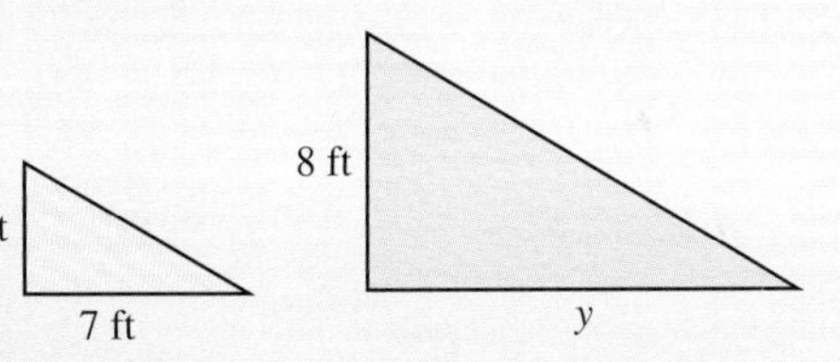

Form ratios using corresponding sides.

Find the ratio of the horizontal sides. $\frac{y}{7}$

Find the ratio of the vertical sides. $\frac{8}{4}$

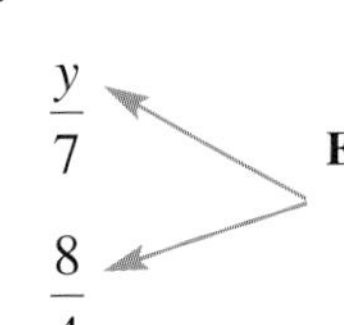

Each ratio is of the form $\frac{\text{large shape's side}}{\text{small shape's side}}$.

Set up a proportion. $\frac{y}{7} = \frac{8}{4}$

5. The following two triangles are similar. Solve for the length of the unknown side.

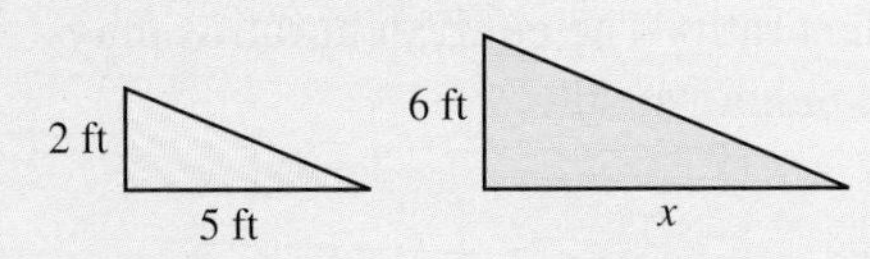

GUIDED PRACTICE

4. The following two rectangles are similar. Write a proportion that could be used to solve for the length of the unknown side.

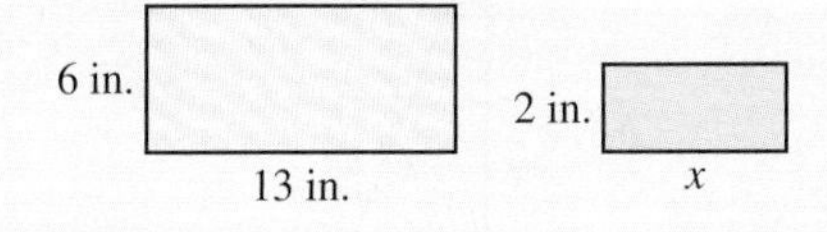

Form ratios using ____________ sides.

$\frac{\quad}{\quad}$

$\frac{\quad}{\quad}$

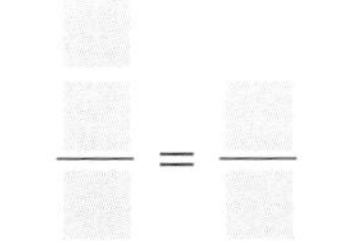

$\frac{\quad}{\quad} = \frac{\quad}{\quad}$

5. The following two rectangles are similar. Solve for the length of the unknown side.

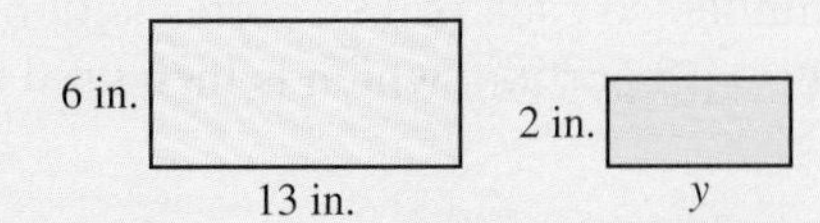

Set up a proportion.

$$\frac{6}{2} = \frac{x}{5}$$

Solve the proportion.
Division is undone with multiplication.

$$\frac{6}{2} \cdot \frac{5}{1} = \frac{x}{5} \cdot \frac{5}{1}$$

$$\frac{30}{2} = x$$

$$15 \text{ ft} = x$$

The unknown side is 15 ft long.

$$\frac{\quad}{\quad} = \frac{\quad}{\quad}$$

=

=

=

The unknown side is ______ long.

6. A flagpole casts a shadow of 36 ft. A nearby tree that is 3 ft tall has a 5-foot shadow. How tall is the flagpole?

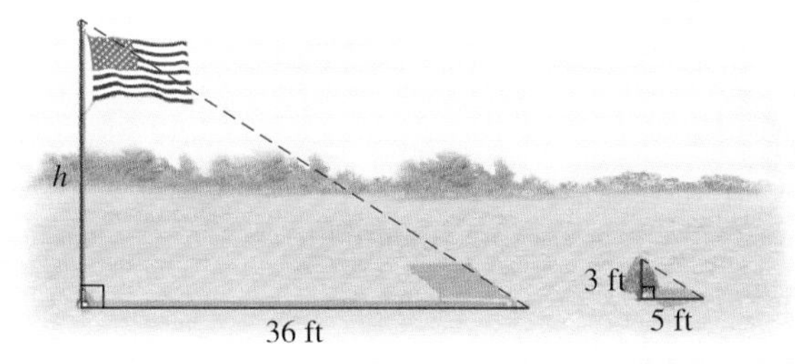

Set up a proportion.

$$\frac{36}{5} = \frac{h}{3}$$

Solve the proportion.
Division is undone with multiplication.

$$\frac{36}{5} \cdot \frac{3}{1} = \frac{h}{3} \cdot \frac{3}{1}$$

$$\frac{108}{5} = h$$

$$21.6 \text{ ft} = h$$

The flagpole is 21.6 feet tall.

6. A 5-foot-tall person casts a shadow that is 2 feet long. A nearby building casts a 20-foot shadow. How tall is the building?

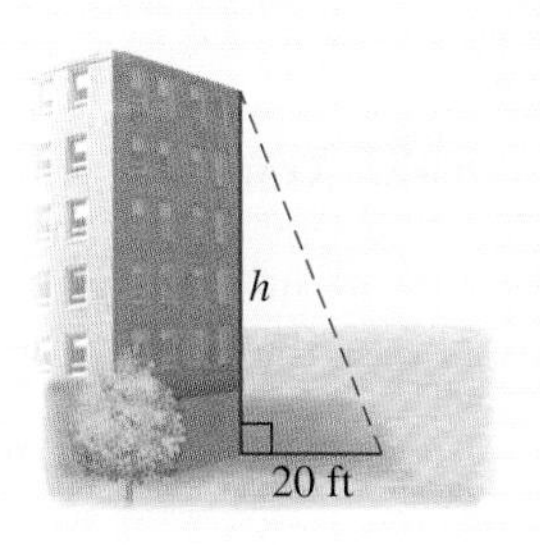

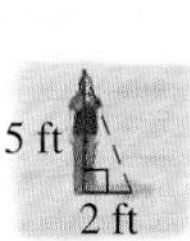

$$\frac{\quad}{\quad} = \frac{\quad}{\quad}$$

=

=

=

The building is _____ tall.

Concept Check

B1. The corresponding sides of similar figures form an equivalent _____.

B2. A proportion is an equation stating that two _____ or _____ are equal.

Objective B Practice

Find the unknown length in each figure. Round to the tenths place. For each exercise, the two figures are similar.

B3.

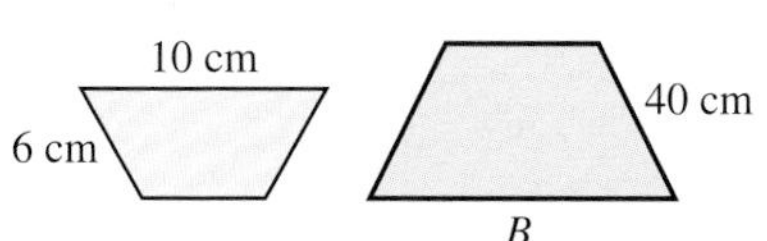

B4.

2 in.
7 in.
30 in.
D

(Continued)

Complete the sketch to show two similar triangles. Include the dimensions on your drawing.

B5. A flagpole casts a shadow that is 15 feet long. A 6-foot-tall person casts a 4-foot shadow.

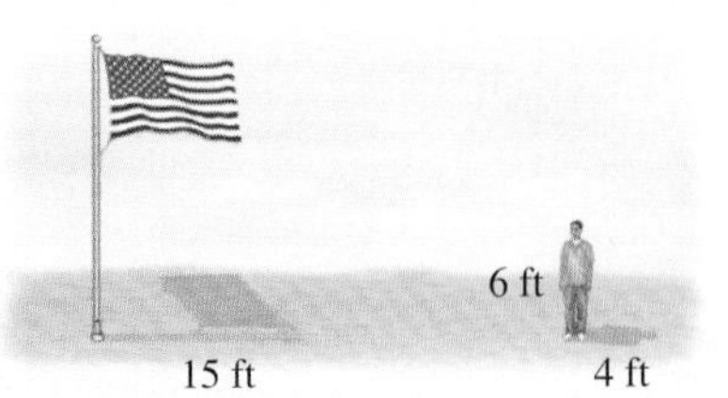

B6. A 20-foot-tall house casts a 25-foot shadow. A radio tower casts a 120-foot shadow.

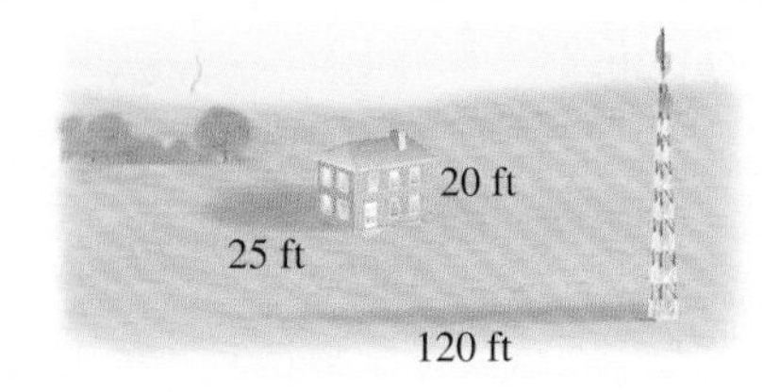

Answer each question.

B7. How tall is the flagpole in Exercise B5?

B8. How tall is the radio tower in Exercise B6?

Objective C Find the Area of a Similar Figure

The Concept To find the area of similar figures, we use similarity to determine the unknown lengths of any sides needed to find the area. Once the dimensions are known, we can use the appropriate formula to find area.

Procedure Find the Area of a Similar Figure

Step 1 Find any unknown lengths.

Step 2 Use an appropriate formula.

EXAMPLES

7. The two triangles are similar. Find the area of the green triangle.

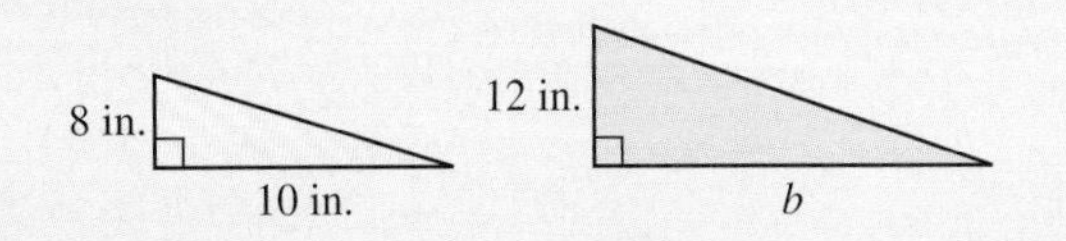

Step 1 Find the unknown length of b.

$$\frac{b}{10} = \frac{12}{8}$$

$$\frac{\cancel{10}}{1} \cdot \frac{b}{\cancel{10}} = \frac{10}{1} \cdot \frac{12}{8}$$

$$b = \frac{120}{8}$$

$$b = 15 \text{ in.}$$

GUIDED PRACTICE

7. The two rectangles are similar. Find the area of the green rectangle.

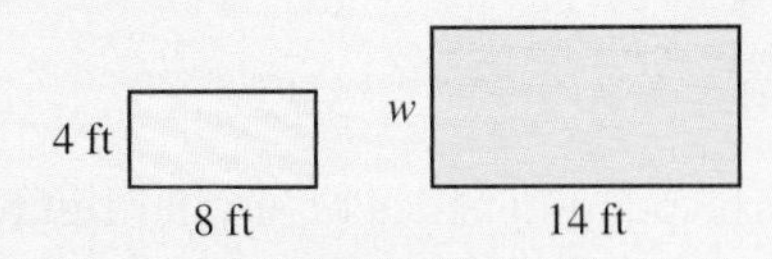

Step 1 Find the unknown length of w.

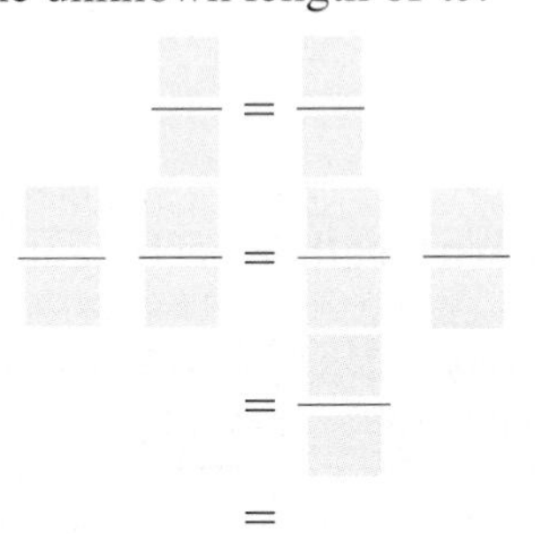

Step 2 Use an appropriate formula.

$$A = \frac{1}{2} \cdot b \cdot h$$
$$= \frac{1}{2} \cdot 15 \cdot 12$$
$$= 6 \cdot 15$$
$$= 90 \text{ in.}^2$$

8. A blueprint of a rectangular room is 3 inches by 4 inches. If the longest wall of the actual room is 12 feet long, what is the area of the actual room?

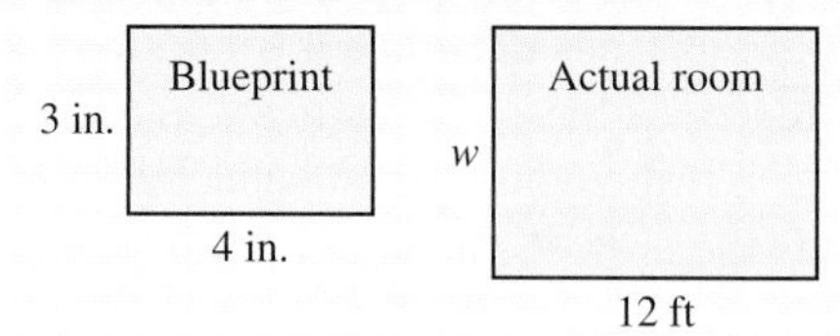

Step 1 Find the unknown length of w

$$\frac{w}{3} = \frac{12}{4}$$
$$\frac{\cancel{3}}{1} \cdot \frac{w}{\cancel{3}} = \frac{3}{1} \cdot \frac{12}{4}$$
$$w = \frac{36}{4}$$
$$w = 9 \text{ ft}$$

Step 2 Use an appropriate formula.

$$A = l \cdot w$$
$$= 12 \cdot 9$$
$$= 108 \text{ ft}^2$$

Step 2 Use an appropriate formula.

$A =$

$=$

$=$

8. The triangular wing on the model of a jet is 8 cm long and extends 4 cm out from the body of the plane. If the length of the actual wing is 18 ft, what is the area of the top of the actual wing?

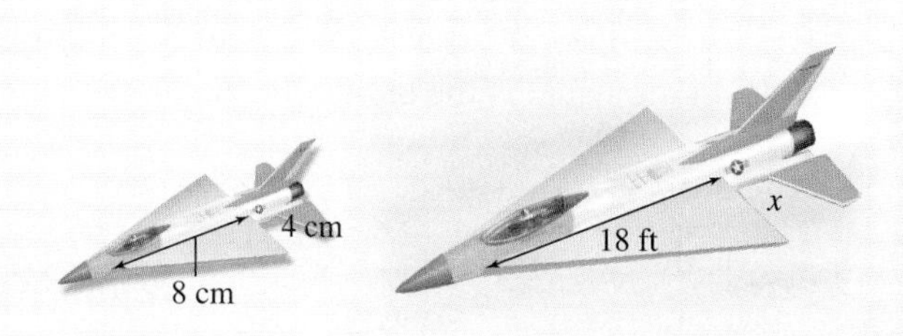

Step 1 Find the unknown length of x.

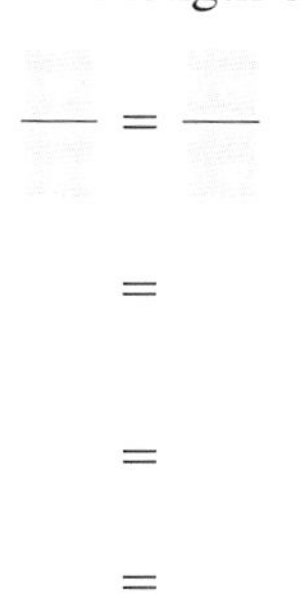

$=$

$=$

$=$

Step 2 Use an appropriate formula.

$A =$

$=$

$=$

Concept Check

C1. If an object is measured in meters, what units are used to write the object's area?

C2. Write three proportions that could be used to solve for the unknown length x. Write each proportion with the variable in the numerator.

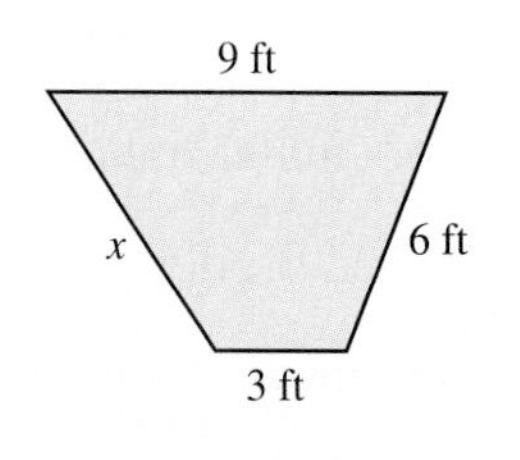

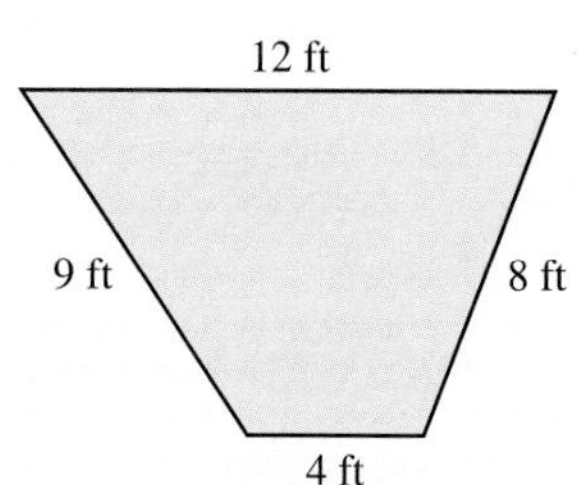

Objective C Practice

For each exercise, the two figures are similar. Round to the tenths place. Find the area of the blue figure.

C3.

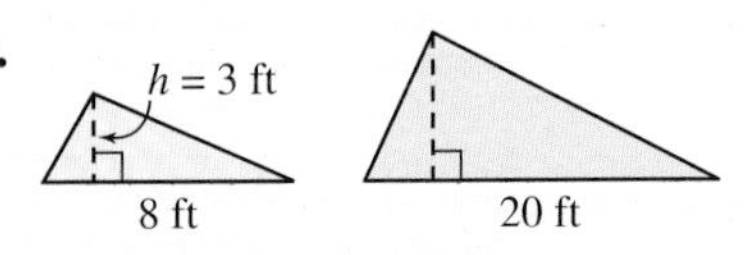

C4.

12 ft
24 ft
h
3 ft

C5.

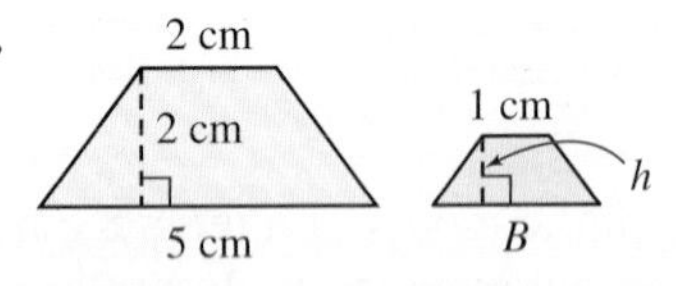

C6.

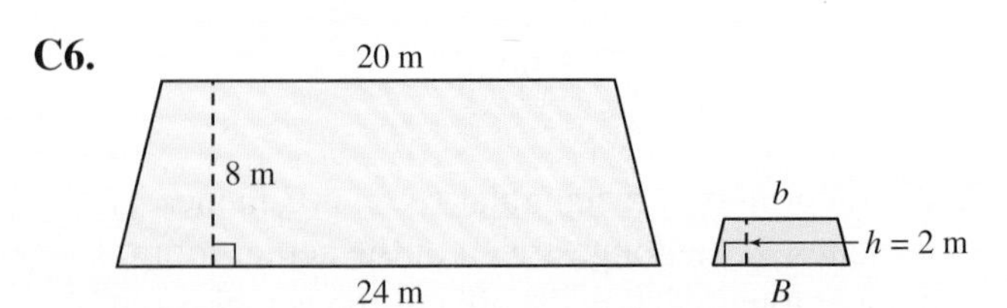

C7. A blueprint of a rectangular room is 7 inches by 12 inches. If the longest wall in the actual room is 50 feet long, what is the area of the actual room?

C8. A landscaper has designed a triangular patio. Her drawing has a base of 8 inches and a height of 6 inches. If the actual patio has a base of 20 feet, what is the area of the actual patio?

Section 7.7 Exercises

To identify similar figures:

1. Answer the Objective A Concept Checks.
2. Answer the odd Objective A Practice Exercises.
3. Answer the even Objective A Practice Exercises.

To find unknown dimensions of similar figures:

4. Answer the Objective B Concept Checks.
5. Answer the odd Objective B Practice Exercises.
6. Answer the even Objective B Practice Exercises.

To find the area of a similar figure:

7. Answer the Objective C Concept Checks.
8. Answer the odd Objective C Practice Exercises.
9. Answer the even Objective C Practice Exercises.

VOCABULARY REVIEW *Review the Vocabulary Preview for Section 7.7. Study the definitions until you can check* Got It *for every word.*

Use these words to complete each sentence.

similar objects • corresponding angles • corresponding sides • proportion

	Got It	Get Help
10. The __________ of similar figures have equal measure.		
11. Two objects that have the same shape but different sizes are __________.		
12. The __________ of similar figures always form equivalent ratios.		
13. A __________ is used to solve for an unknown length in similar figures.		

How will you get help for any vocabulary that you are unsure about?

Your instructor ____ MyMathLab ____ A classmate ____ A tutor ____ Other ____

The figures in Exercises 14–19 are similar to exactly one of the figures (a)–(f).

a. Match each figure with its similar figure.

b. Identify the corresponding sides.

14.

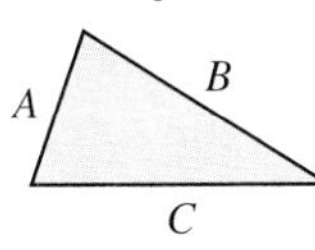

15.

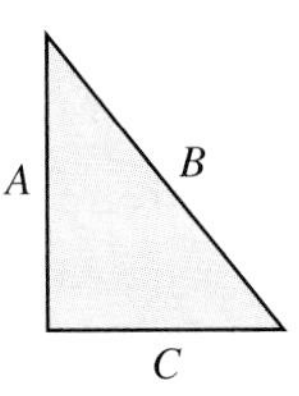

(a) J K L

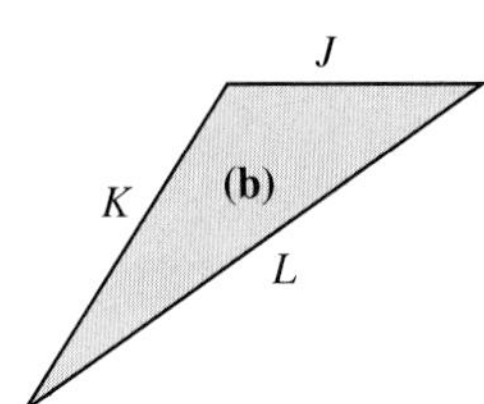

16.

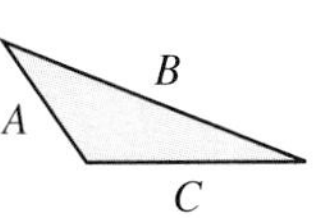

17.

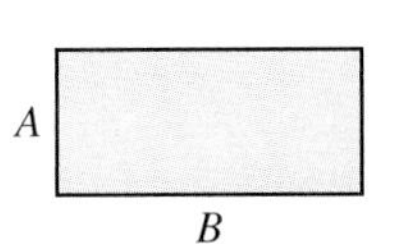

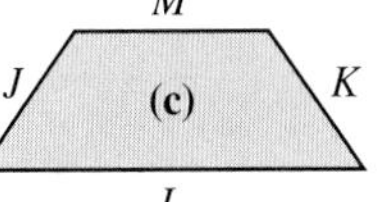

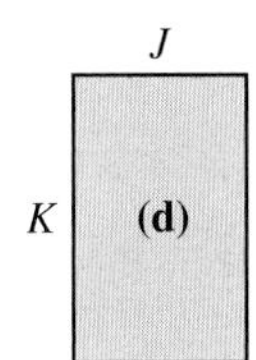

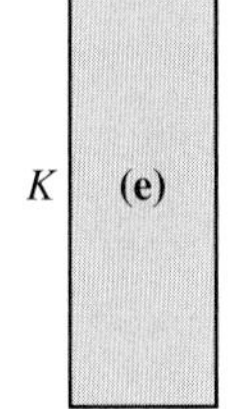

18. A B

19. A B C D

(f) J K L

For each pair of similar figures, find the unknown length x. Round each answer to the nearest tenth.

20.

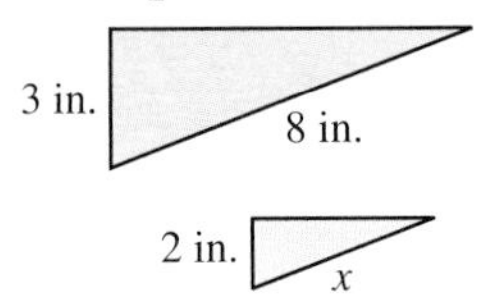

21.

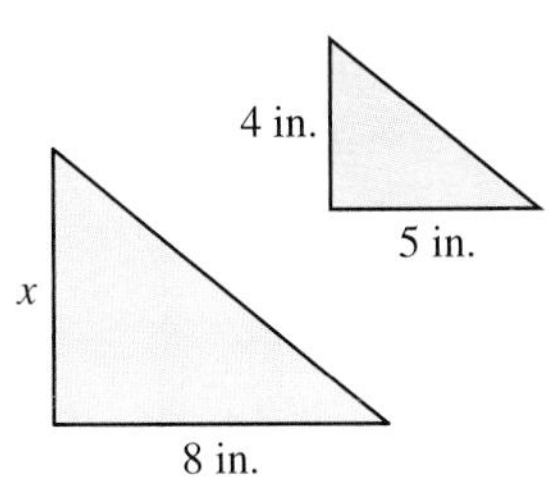

22.

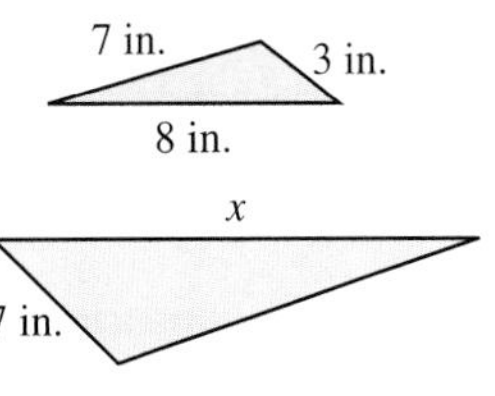

23.

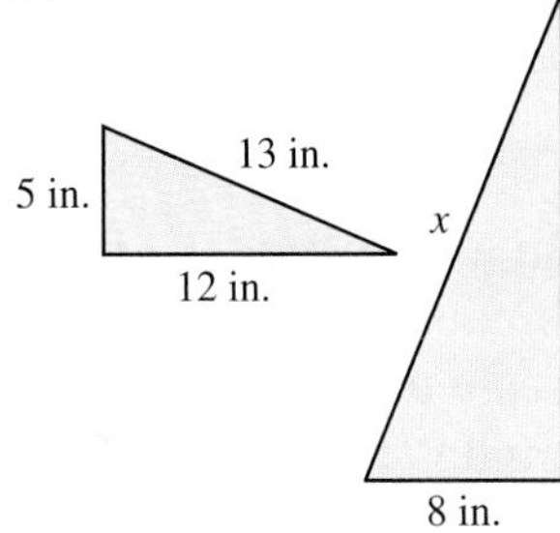

Shadows cast from the flagpoles and trees form similar triangles. Find the height of each flagpole.

24.

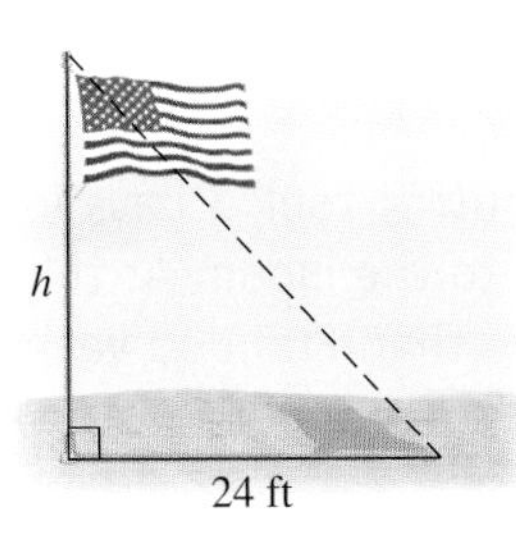

25.

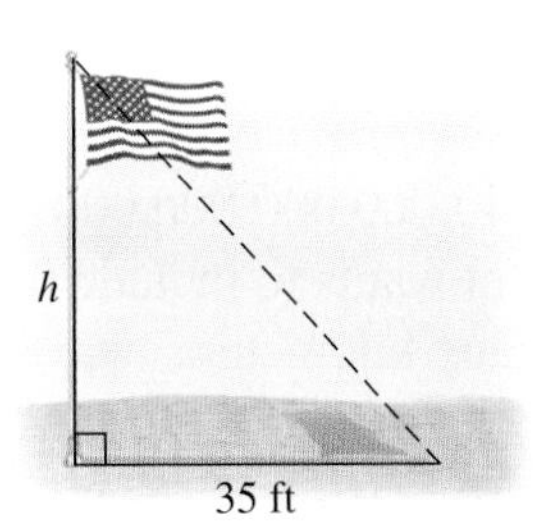

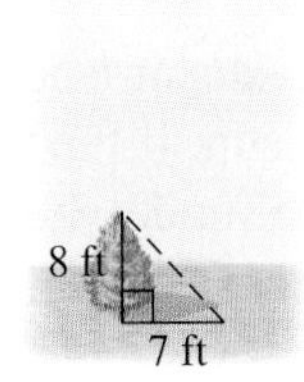

Each pair of triangles is similar. Round answers to the nearest tenth. Find the area of the blue triangle.

26.

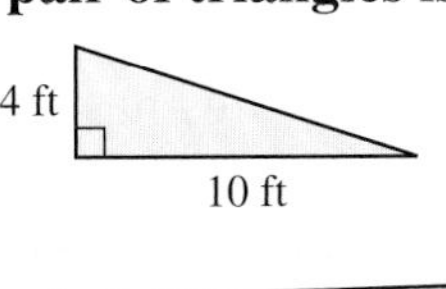

27.

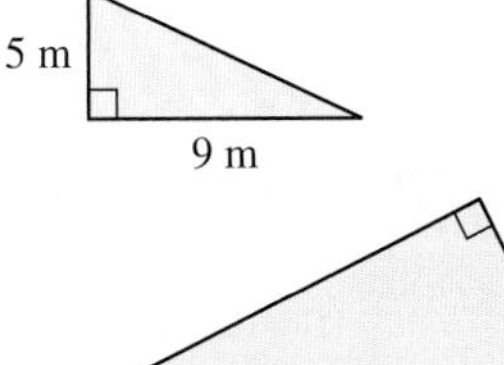

28.

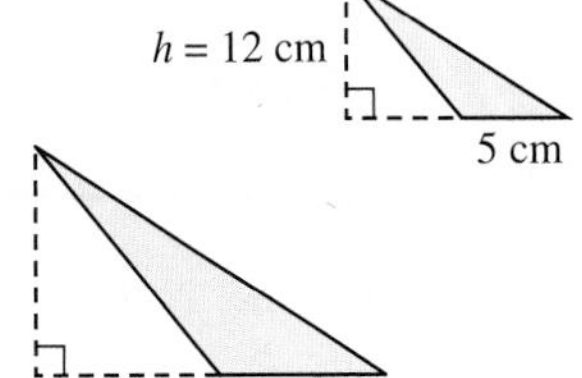

29.

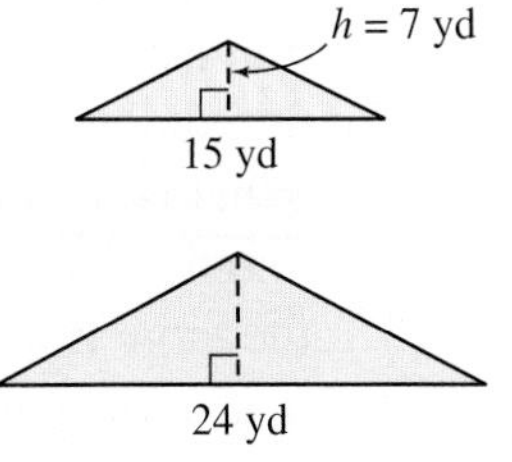

Smith Using your height, the length your shadow, and this technique, you can find the height of any object from the length of its shadow. *For more tips and tweets, go to twitter.com/gstbasicmath*

Answer each question. Round answers to the nearest tenth.

30. A drawing of a sculpture has a base of 10 inches and a height of 18 inches. If the actual sculpture is 8 feet tall, how wide is the actual base?

31. A plastic model of a van measures 4 inches wide by 7 inches long. If the actual van is 7 feet wide, how long is the actual van?

32. A blueprint shows a bathtub that is 2 inches wide and 5 inches long. If the actual tub is 6 feet long, what area does the actual tub occupy in the bathroom?

33. A blueprint shows a room that is 5 inches wide and 6 inches long. If the actual room is 15 feet long, how many square feet of carpet are needed to carpet the room?

34. A surveyor used similar triangles to find the distance across a river, as shown. How wide is the river?

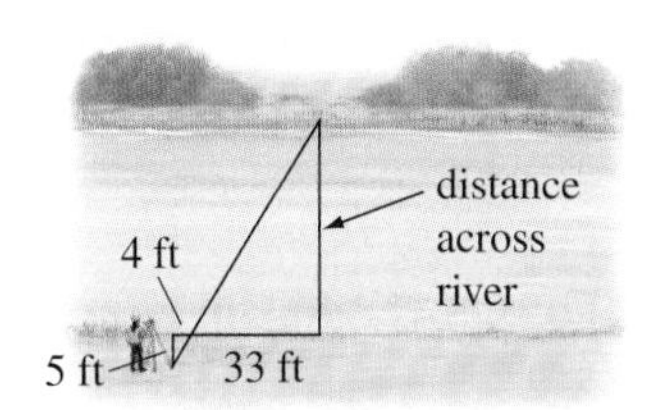

35. A boat captain wants to know the distance to an island. A 3-foot ruler is placed at the front of the boat, 20 feet from the captain. The volcano on the island's shore is about 1.3 miles high. How far away is the island?

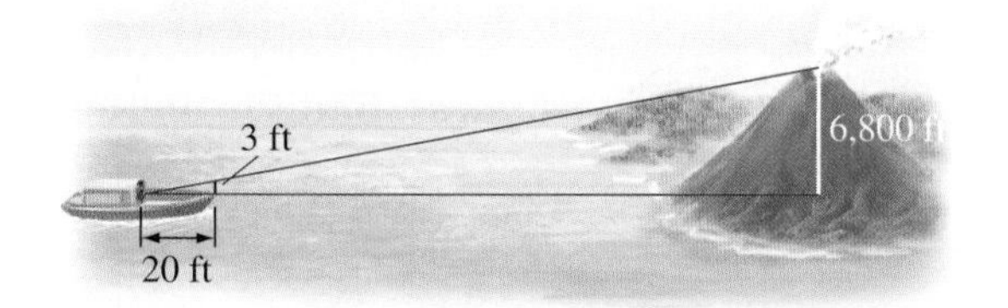

In a scale model, all actual figures are similar to the model objects. A student at a science fair builds a scale model to represent the solar system. In the model, the earth is 1 centimeter from the moon. The actual distance to the moon is about 400,000 km.

36. If Earth is about 150,000,000 km from the sun, how far is Earth from the sun in the model? Give your answer in meters.

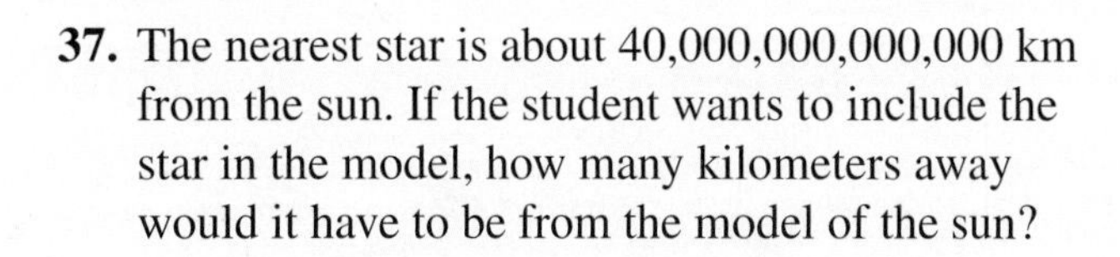

37. The nearest star is about 40,000,000,000,000 km from the sun. If the student wants to include the star in the model, how many kilometers away would it have to be from the model of the sun?

Something to think about: It took 3 days for the Apollo Lander to get to the moon. At that speed, it would take more than 800,000 years to get to the nearest star! Use a proportion to verify this fact.

38. Small equilateral triangles are joined to form a second larger equilateral triangle.

Each of the small triangles has an area of 1 ft^2.

a. What is the ratio of the length of the sides of the large triangle to that of the small triangle?

b. What is the ratio of the large triangle's perimeter to that of the small triangle?

c. What is the ratio of the area of the large triangle to that of the small triangle?

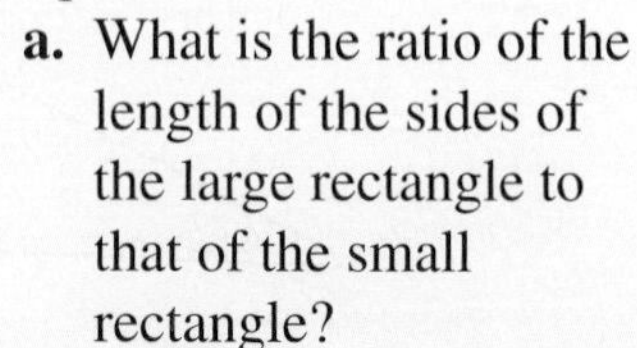

39. Small squares are joined to form a second larger square.

Each of the small squares has an area of 1 ft^2.

a. What is the ratio of the length of the sides of the large rectangle to that of the small rectangle?

b. What is the ratio of the large rectangle's perimeter to that of the small rectangle?

c. What is the ratio of the area of the large rectangle to that of the small rectangle?

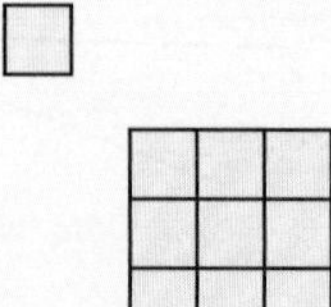

40. Small equilateral triangles are joined to form a second larger equilateral triangle.

Each of the small triangles has an area of 1 ft^2.

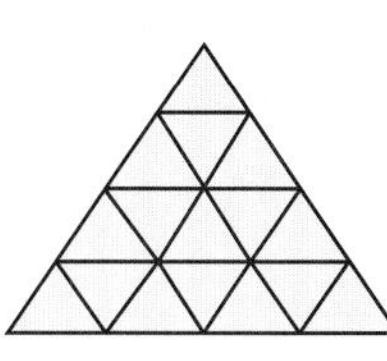

a. What is the ratio of the length of the sides of the large triangle to that of the small triangle?

b. What is the ratio of the large triangle's perimeter to that of the small triangle?

c. What is the ratio of the area of the large triangle to that of the small triangle?

41. Small squares are joined to form a second larger square.

Each of the small squares has an area of 1 ft^2.

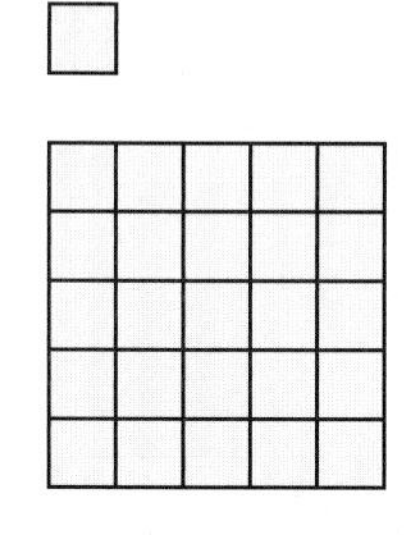

a. What is the ratio of the length of the sides of the large rectangle to that of the small rectangle?

b. What is the ratio of the large rectangle's perimeter to that of the small rectangle?

c. What is the ratio of the area of the large rectangle to that of the small rectangle?

42. Based on Exercises 38–41, if you enlarge a figure by multiplying each side by the number n, what will happen to the perimeter? (By what number will the perimeter be multiplied?)

43. Based on Exercises 38–41, if you enlarge a figure by multiplying each side by the number n, what will happen to the area? (By what number will the area be multiplied?)

Use your answers from Exercises 42 and 43 to answer the following questions.

44. The following are similar figures.

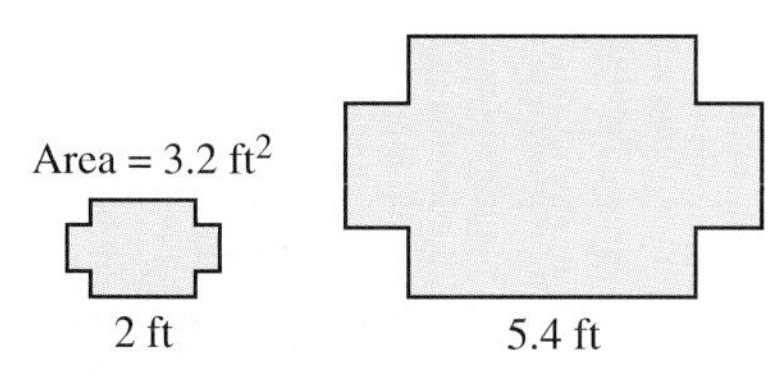

a. What is the ratio of the sides of the two figures?

b. What is the area of the enlarged figure?

45. The following are similar figures.

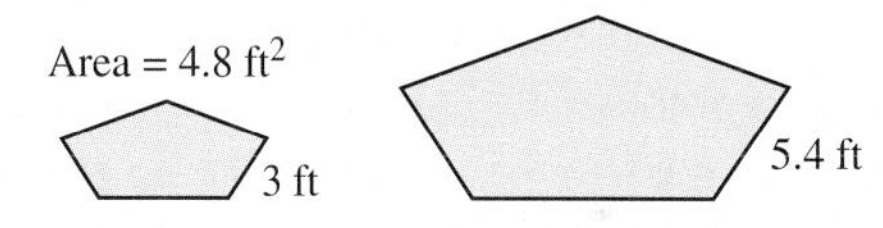

a. What is the ratio of the sides of the two figures?

b. What is the area of the enlarged figure?

46. When ordering school portraits, a "full-size" photo fills an entire sheet of paper. A "half-size" photo is half as wide and half as tall as a full-size photo. How many half-size photos can be printed on one sheet of paper?

47. When ordering school photos, a "full-size" photo fills an entire sheet of paper. A "one-fourth-size" photo is one-fourth as wide and one-fourth as tall as a full-sized photo. How many one-fourth-sized photos can be printed on one sheet of paper?

Section 7.7 Question Log

Use this space to write questions. Be sure to get these questions answered and revisit them when you prepare for your exam.

Page ______ Answered ☐	Page ______ Answered ☐
Page ______ Answered ☐	Page ______ Answered ☐

Chapter 7 Organizer

VOCABULARY

Use the following steps to review the vocabulary for Chapter 7.

1. *Write the definition for each word from memory.*
2. *Compare the definitions you have written with the definitions in the Vocabulary Preview.*
3. *Study any definitions that you could not remember or that you defined incorrectly.*

7.1

Line • Line Segment • Ray • Angle • Vertex • Measure (of an Angle) • Acute Angle • Right Angle • Obtuse Angle • Straight Angle • Complementary Angles • Supplementary Angles

7.2

Line Segment • Polygon • Triangle • Quadrilateral • Pentagon • Hexagon • Heptagon • Octagon • Nonagon • Decagon • Regular Polygon • Perimeter • Trapezoid • Parallelogram • Rectangle • Rhombus • Square • Scalene Triangle • Isosceles Triangle • Equilateral Triangle • Right Triangle • Obtuse Triangle • Acute Triangle

7.3

Perimeter • Area

7.4

Radius • Diameter • Circumference • π (Pi)

7.5

Perimeter • Area • Volume • Cylinder • Cone • Pyramid

7.6

Square • Perfect Square • Square Root • Right Triangle • Hypotenuse • Legs • Pythagorean Theorem

7.7

Similar Objects • Corresponding Angles • Corresponding Sides • Proportion

PROCEDURES

Procedure/Topic	Steps	Examples
Measure an Angle (Section 7.1)	Use a protractor to determine the measure of the angle between two rays.	Find the measure of the angle between the red and blue rays. The angle between the red and blue rays is 120°.
Draw an Angle (Section 7.1)	Use a protractor to draw two rays with the given angle between them.	Draw an angle that has a measure of 50°. 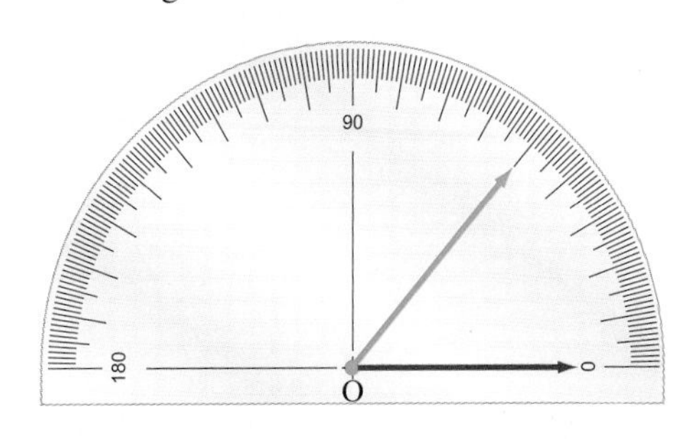 To draw a 50° angle, we draw a red ray at 0° and a green ray at 50°. The measure difference between the rays is 50°.

Procedure/Topic	Steps	Examples
Classify Polygons (Section 7.2)	**Step 1** Count the number of sides and use the appropriate name for the figure. **Step 2** If every side has the same length and every angle has the same measure, state that the polygon is regular.	Classify each figure. Parallelogram: 4 sides, opposite sides parallel Regular hexagon: 6 sides, same length, angles of same measure
Find the Perimeter of a Figure (Section 7.3)	Add the length of the sides.	Find the perimeter of the polygon. 7 ft, 4 ft, 7 ft, 3 ft The perimeter is the distance around the figure. Add the lengths of the sides. $P = 4 + 7 + 7 + 3$ $= 11 + 7 + 3$ $= 18 + 3$ $= 21$ ft
Use a Formula to Find the Area of a Figure (Section 7.3)	**Step 1** Write the appropriate formula. **Step 2** Substitute the values for each variable. **Step 3** Simplify.	What is the area of the figure? height = 6 m, 13 m In the parallelogram, the base is 13 m and the height is 6 m. $A = \text{base} \cdot \text{height}$ $= 13 \cdot 6$ $= 78 \text{ m}^2$
Find the Circumference of a Circle (Section 7.4)	Determine the diameter or radius. Then use the appropriate formula: $C = \pi d$ *or* $C = 2\pi r$	Find the circumference of a circle with a 4-foot diameter. $C = \pi d$ $\approx (3.14) \cdot 4$ ≈ 12.56 ft

Procedure/Topic	Steps	Examples
Find the Area of a Circle (Section 7.4)	Determine the radius. Then use the formula $A = \pi \cdot r^2$.	Find the area of the circle. Use $\pi \approx 3.14$. $d = 12$ km To find the radius, divide the diameter by 2. Because the diameter is 12 km, the radius is $12 \div 2 = 6$ km. $A = \pi \cdot r^2$ $= \pi \cdot (6)^2$ $\approx 3.14 \cdot (36)$ ≈ 113.04 square km
Use a Formula to Find the Volume of a Figure (Section 7.5)	**Step 1** Write the appropriate formula. **Step 2** Substitute the values into the formula. **Step 3** Simplify.	Find the volume of a cylinder with a radius of 3 feet and a height of 6 feet. $V = \pi r^2 h$ $= \pi \cdot 3^2 \cdot 6$ $= \pi \cdot 9 \cdot 6$ $= \pi \cdot 54$ $\approx 169.56\text{ ft}^3$
Evaluate the Square Root of a Number (Section 7.6)	Determine what number, when multiplied by itself, gives the number inside the square root symbol.	Find $\sqrt{36}$. $\sqrt{36} = 6$ because $6 \cdot 6 = 36$.
Use the Pythagorean Theorem (Section 7.6)	**Step 1** Use the appropriate formula. **Step 2** Substitute the values. **Step 3** Simplify.	Find the length of the unknown side, rounded to the hundredths place. 4 m, 8 m, hypotenuse $\text{hypotenuse} = \sqrt{(\text{1st leg})^2 + (\text{2nd leg})^2}$ $= \sqrt{4^2 + 8^2}$ $= \sqrt{16 + 64}$ $= \sqrt{80}$ $= 8.94$ m
Find the Unknown Dimension of Similar Figures (Section 7.7)	**Step 1** Set up a proportion. **Step 2** Solve the proportion. Get the variable into the numerator and then isolate it.	These triangles are similar. Find the unknown height. 4 m, 8 m; h, 12 m $\frac{h}{4} = \frac{12}{8}$ $\frac{4}{1} \cdot \frac{h}{4} = \frac{4}{1} \cdot \frac{12}{8}$ $h = \frac{\cancel{4}}{1} \cdot \frac{\cancel{2} \cdot 6}{\cancel{2} \cdot \cancel{4}}$ $h = 6$ m

Procedure/Topic	Steps	Examples
Find the Area of a Similar Figure (Section 7.7)	**Step 1** Find any unknown lengths. **Step 2** Use an appropriate formula.	The two triangles are similar. Find the area of the orange triangle. 6 in. 12 in. 8 in. b $\frac{b}{12} = \frac{8}{6}$ $\frac{\cancel{12}}{1} \cdot \frac{b}{\cancel{12}} = \frac{12}{1} \cdot \frac{8}{6}$ $b = \frac{96}{6}$ $b = \frac{\cancel{6} \cdot 16}{\cancel{6}}$ $b = 16$ in. $A = \frac{1}{2} \cdot b \cdot h$ $= \frac{1}{2} \cdot 16 \cdot 8$ $= 8 \cdot 8$ $= 64$ in.^{2}

Chapter 7 Review Exercises

7.1

Use the protractor with the given measurements to answer each exercise.

1. Find $m\angle AOD$.
2. Find $m\angle BOD$.
3. What are the names of the three acute angles?
4. What are the names of the two obtuse angles?
5. What is the name of the one right angle?
6. Draw ray E on the protractor so that is $m\angle BOE$ is 100°. At what value on the protractor should you draw ray E?

Use the following figure to find the measure of the indicated angles. $\angle AOE$ is a straight angle.

7. Find $m\angle DOE$.
8. Find $m\angle BOE$.

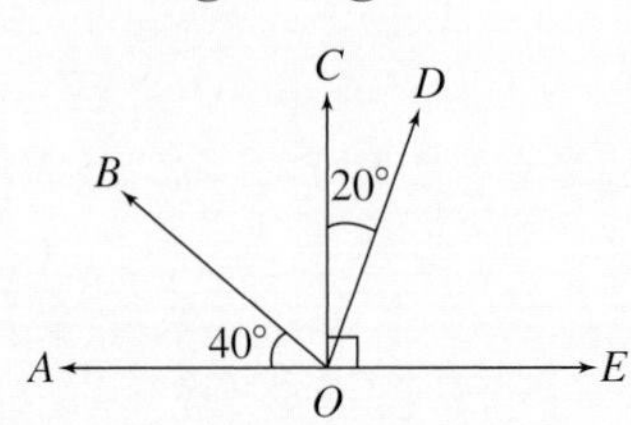

9. Find $m\angle AOD$.
10. Find $m\angle BOD$.

7.2

Classify each polygon as accurately as possible. If the figure is not a polygon, explain why.

11.

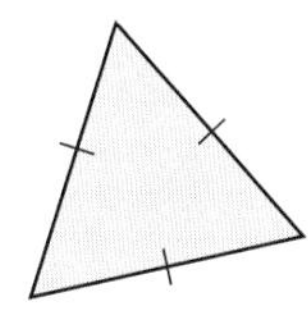

12.

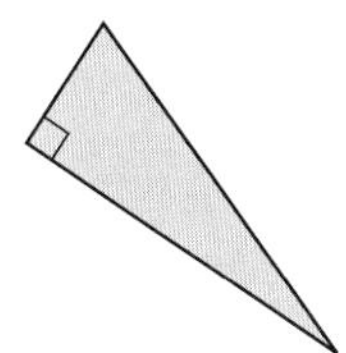

13.

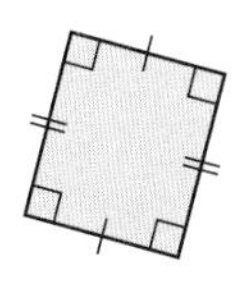

14.

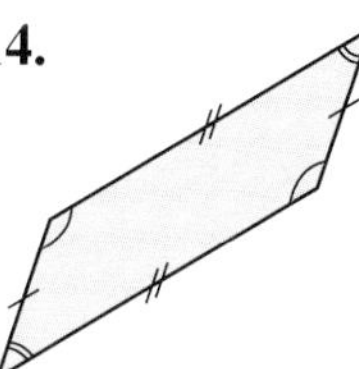

15.

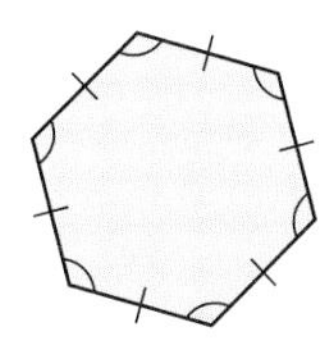

16.

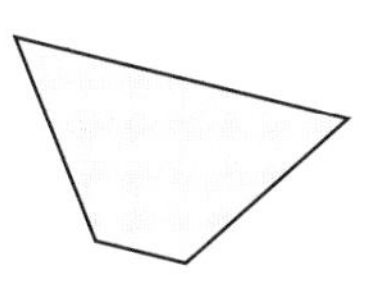

17.

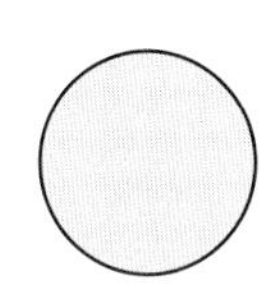

18.

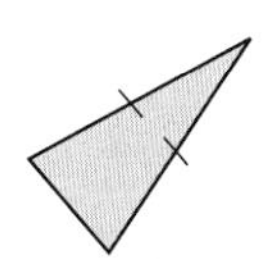

19. What is the most accurate name for a polygon with 8 sides of equal length and 8 equal angles?

20. What is the most accurate name for a triangle with all angles less than 90° and all sides of different lengths?

7.3

Find the perimeter of each figure.

21. 8 in. 5 in.

22. 3 cm 12 cm 9 cm

23. 15 cm 2 cm 6 cm 8 cm

24. 7 ft 5 ft 2 ft 9 ft

Find the area of each figure.

25.

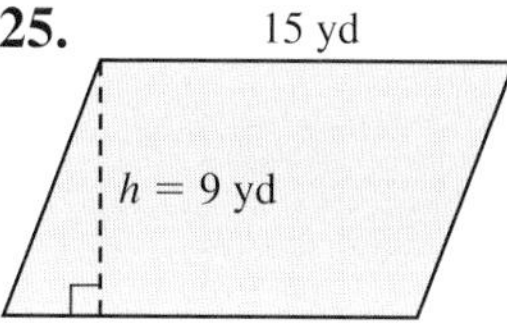

26.

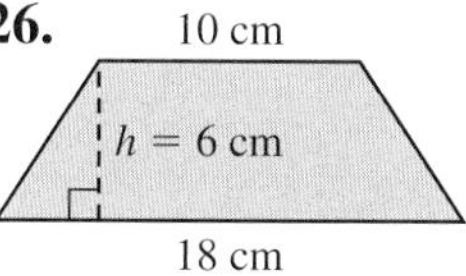

27.

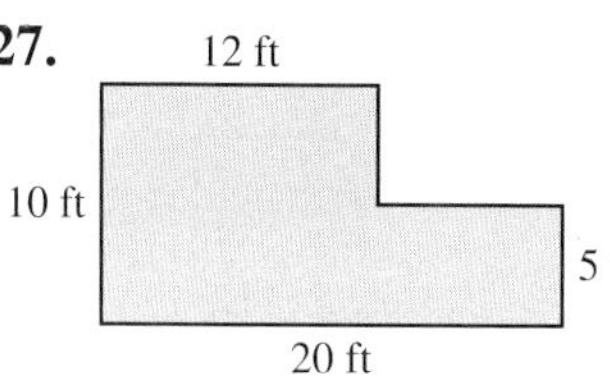

28.

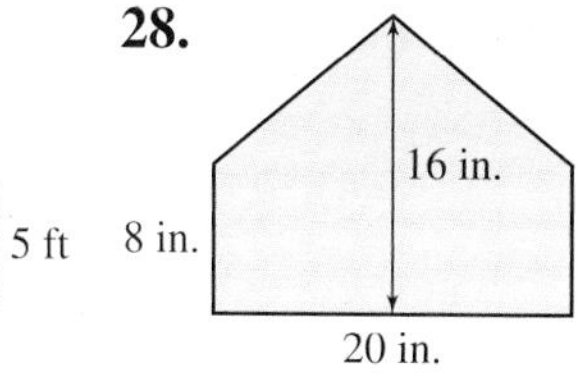

7.4

Answer each question.

29. What is the area of a circle with a radius of 5 inches?

30. What is the area of a circle with a diameter of 6 inches?

31. What is the circumference of a circle with a diameter of 20 feet?

Find the circumference of each figure.

32.

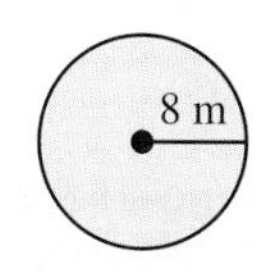

33. A semi-circle.

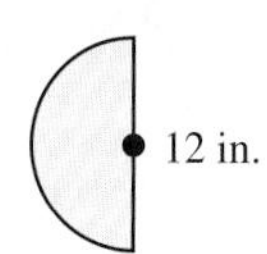

Find the area of each figure.

34.

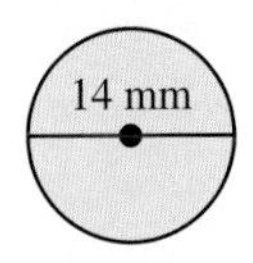

35. The area outside the square but inside the circle.

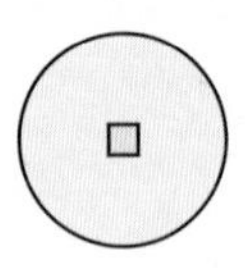

$r = 7$ cm; the sides of the square measure 2 cm.

Find the perimeter of each figure.

36.

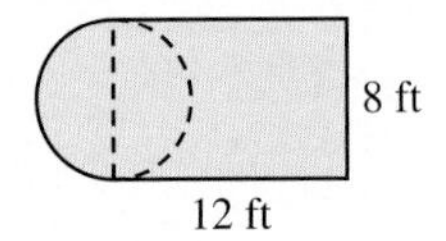

37.

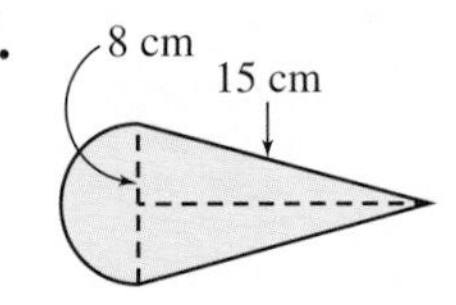

▶ **NOTE** The triangle is isosceles.

Find the area of each figure.

38.

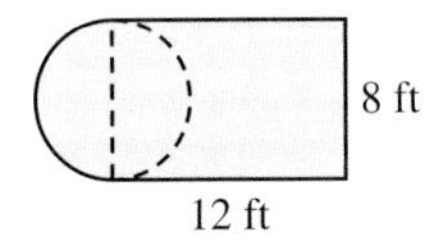

39.

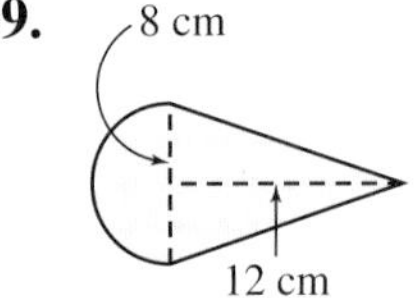

▶ **NOTE** The triangle is isosceles.

7.5

Find the volume of each figure. When boxes are used to make the figure, assume that each upper layer is supported by boxes underneath.

40. Each box has sides that are 1 ft long.

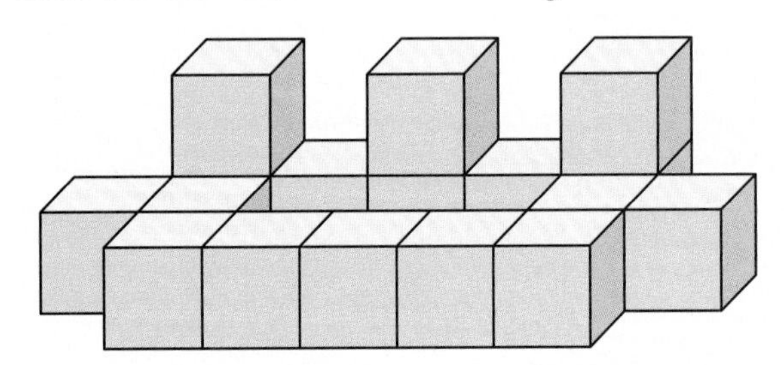

41. Each box has sides that are 1 cm long.

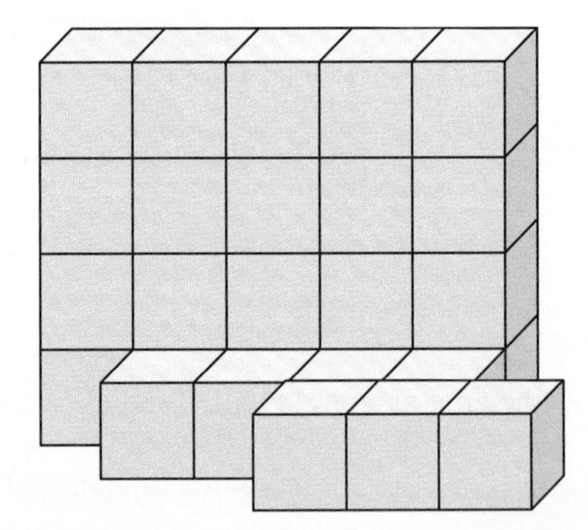

42. Each cylindrical column is approximately 40 feet high and 3 feet across. What is the volume of one column?

43. The sphere measures 20 cm across. What is its volume?

44. A pyramid is 120 feet tall and has a square base with 140-foot sides. What is the volume of the pyramid?

45. A pile of sand at a quarry has formed a cone. If the cone is 15 yards high and 40 yards across at its base, how many cubic yards of sand are in the pile of sand?

7.6

Between which two whole numbers is each square root? Do not find the actual value of the square root.

46. $\sqrt{13}$ **47.** $\sqrt{29}$ **48.** $\sqrt{85}$ **49.** $\sqrt{56}$

Evaluate each square root.

50. $\sqrt{36}$ **51.** $\sqrt{49}$ **52.** $\sqrt{\frac{4}{81}}$ **53.** $\sqrt{\frac{64}{25}}$

Find the perimeter or area of each right triangle shown. Round your answers to the nearest tenth. You may need to use the Pythagorean theorem to find an unknown length.

54. What is the area?

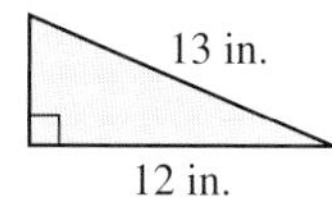

55. What is the area?

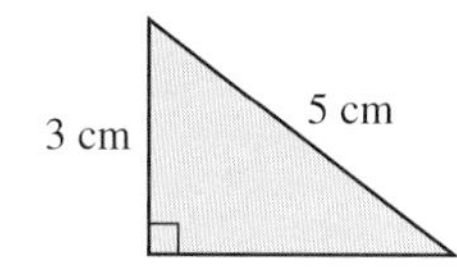

56. What is the perimeter?

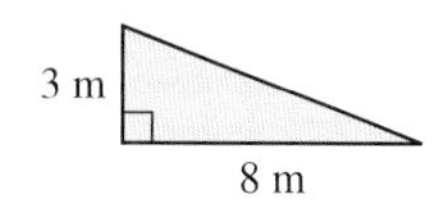

57. What is the perimeter?

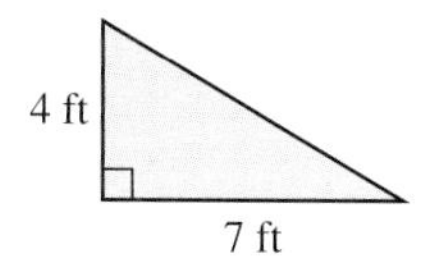

Find the unknown length for each pair of images. Assume that the sun is casting a shadow off each object at the same angle.

58.

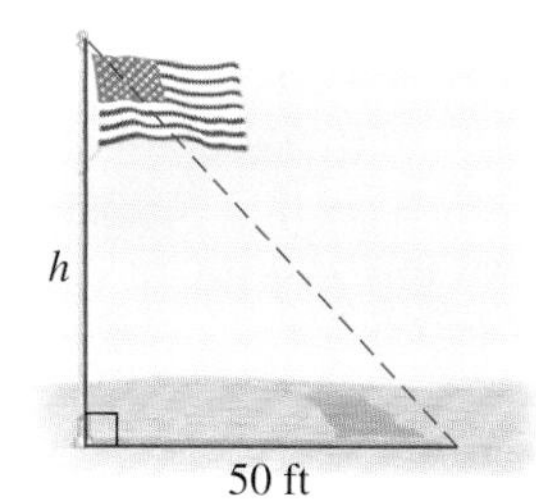

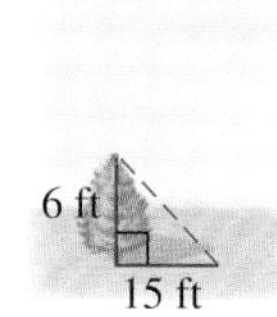

59.

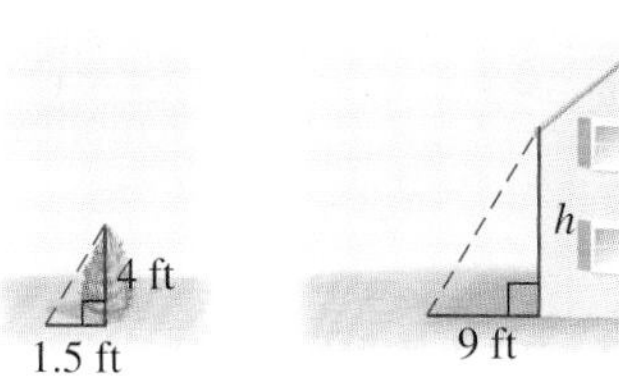

Chapter 7 Practice Test

Use the protractor with the given measurements to answer each question.

1. For $\angle BOC$
  a. What is its measure?
  b. Is the angle acute, obtuse, or right?

2. For $\angle AOC$
  a. What is its measure?
  b. Is the angle acute, obtuse, or right?

3. For $\angle AOD$
  a. What is its measure?
  b. Is the angle acute, obtuse, or right?

4. What is the measure of $\angle BOC$'s complement?

5. What is the measure of $\angle BOC$'s supplement?

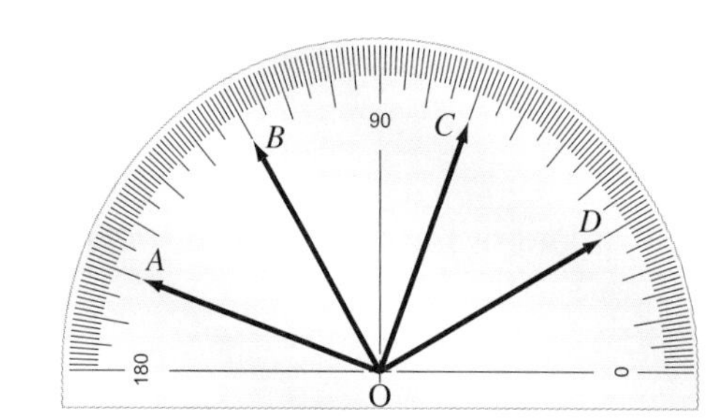

Classify each polygon as accurately as possible. If the figure is not a polygon, explain why.

6.

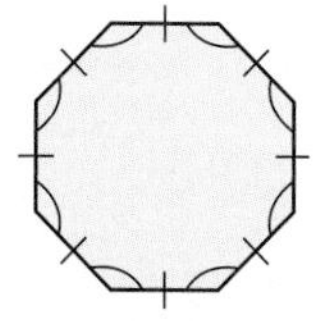

7.

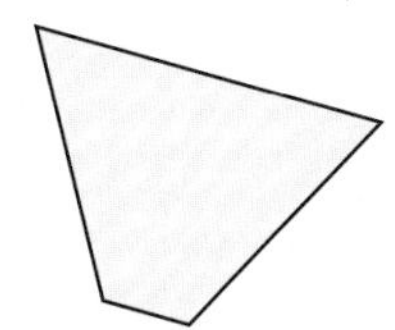

8.

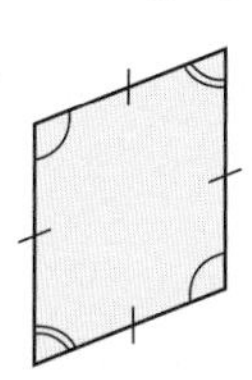

9.

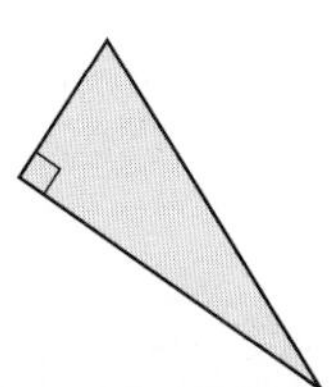

Find the perimeter of each figure.

10.

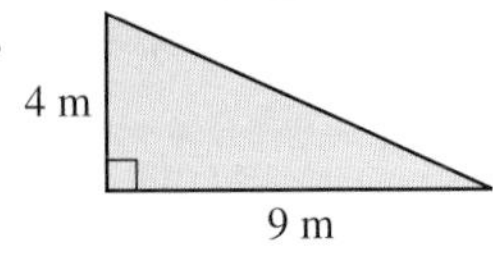

11.

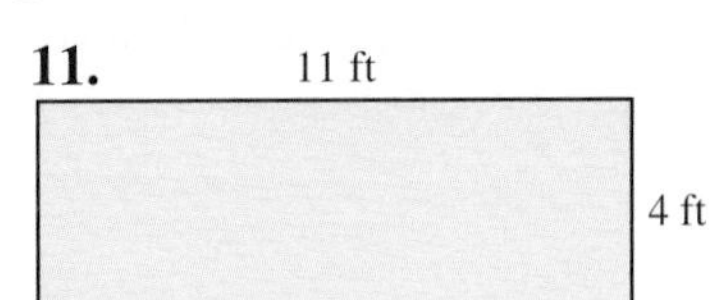

12. A regular octagon with sides that are 7 centimeters long

13. A parallelogram with lengths of 7 feet and widths of 4 feet

Find the area of each figure.

14.

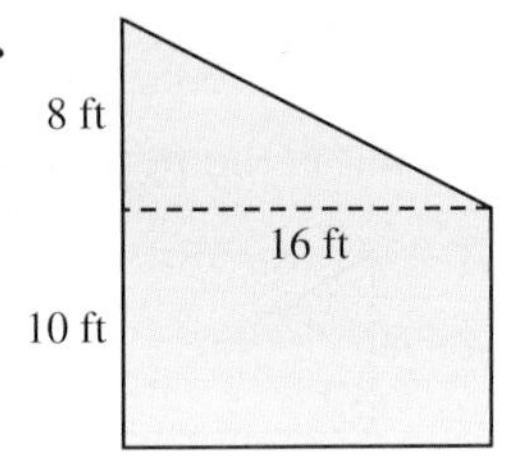

15.

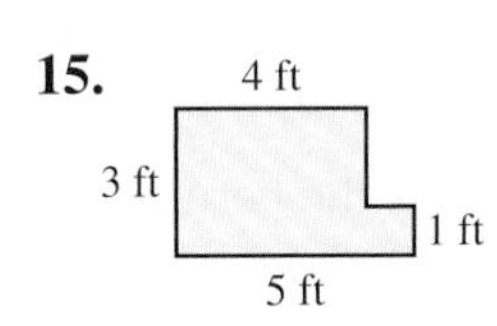

16. A trapezoid with a height of 4 inches and bases with lengths of 7 inches and 11 inches

17.

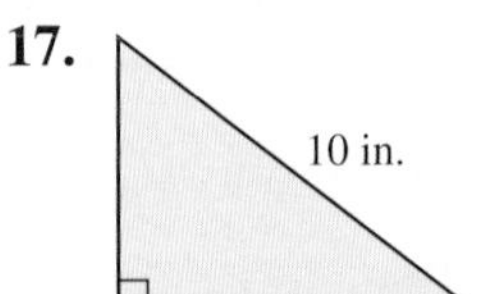

Find the volume of each figure (described or shown). Round to the tenths place when necessary.

18. The sides of the boxes are 1 meter long.

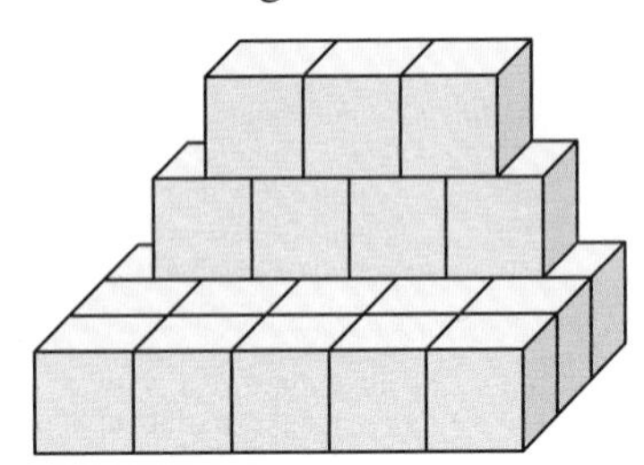

19. A softball with a diameter of 4 inches

20. A cylinder with a diameter of 10 meters and a height of 5 meters

21. A 12-foot-tall pyramid with a rectangular base that measures 20 feet by 14 feet

Between which two whole numbers is each square root? Do not find the actual value of the square root.

22. $\sqrt{73}$

23. $\sqrt{3}$

Evaluate each square root. Round to the hundredths place when necessary.

24. $\sqrt{\frac{9}{25}}$

25. $\sqrt{\frac{81}{16}}$

26. $\sqrt{31}$

27. $\sqrt{21}$

Find the unknown length for each pair of images. Assume that the sun is casting a shadow off each object at the same angle.

28.

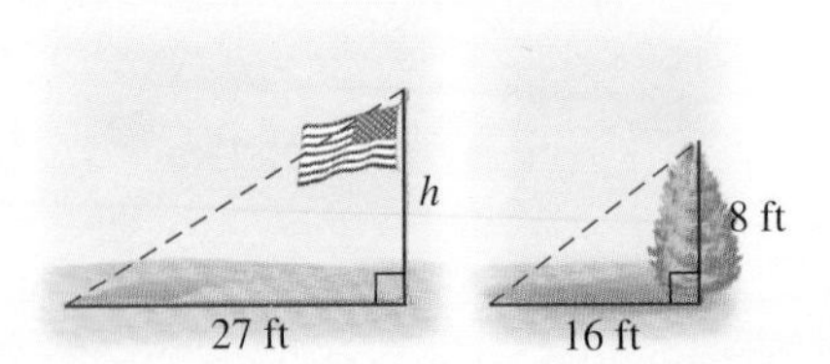

29.

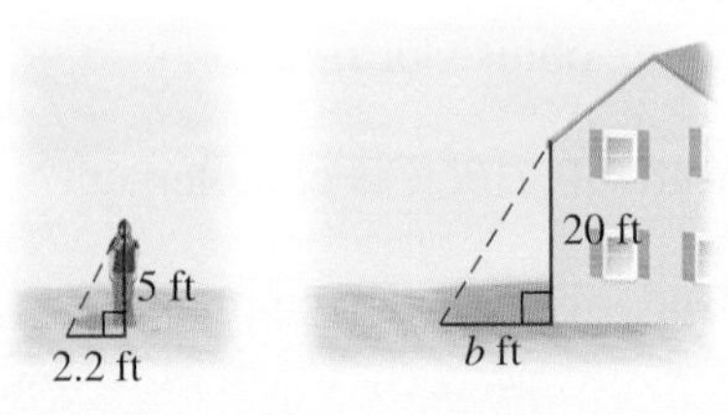

CHAPTER 8

Statistics

Median Salary, by City, for Statisticians in the U.S.

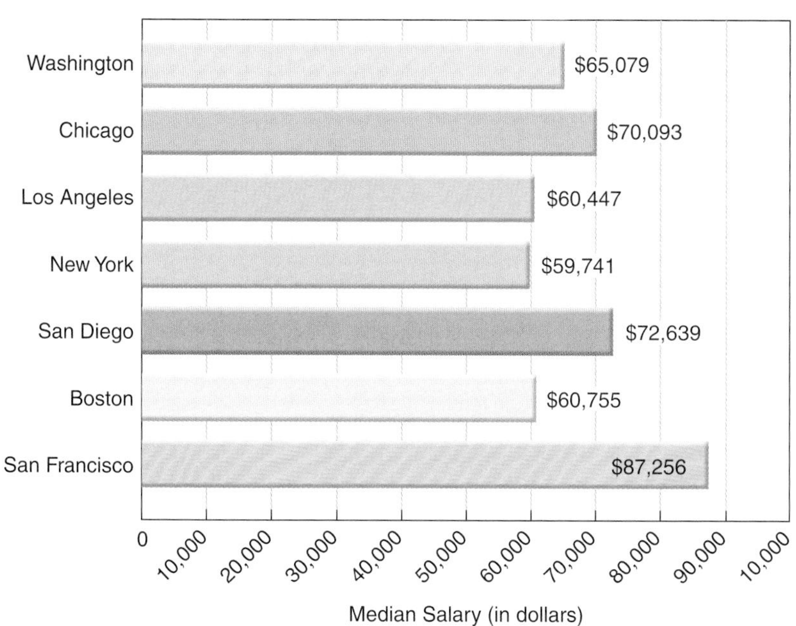

Source: PayScale, Inc.

Statistics is a branch of mathematics that is used to collect and study data. You have probably seen the results of statistical studies in newspapers, on Internet sites, and on television. Statisticians make their living studying and reporting statistical information. At left is a graph of statistician salaries in cities throughout the United States.

Based on the graph, we might guess that a typical statistician earns around $65,000 per year.

In Exercise 47, Section 8.2, you will use statistical measures to identify a typical statistician's salary with more certainty.

Section 8.1 Reading Graphs

Graphs are useful in displaying large amounts of information in a form that is easy to interpret. This is why newspapers, magazines, and television programs frequently use graphs to present data. There are many types of graphs, including circle graphs, bar graphs, and histograms. Each will be discussed in this section.

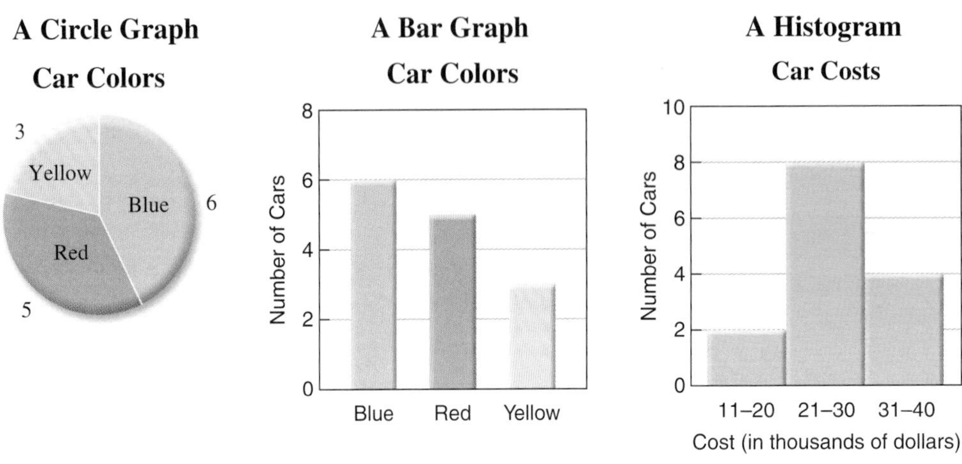

The **Objectives** in Section 8.1 will help you

- **A** Read a circle graph.
- **B** Read a bar graph and a line graph.
- **C** Construct a histogram.

VOCABULARY PREVIEW *Check the box that applies.*

	Got It	Must Study
Data: Data is a collection of facts, usually numbers, from which conclusions may be drawn.		
Circle Graph: A **circle graph** is a graph in the form of a circle that is divided into pie-shaped pieces, where each piece represents one part of the data set.		
Sectors: The **sectors** of a circle graph are pie-shaped pieces that represent each part of the data set.		
Bar Graph: A **bar graph** uses rectangular bars to make a visual comparison of data.		
Line Graph: Line graphs display data points that are connected by lines to show a trend.		
Histogram: A **histogram** is a type of bar graph that shows the frequency of data in class intervals.		
Class Interval: The interval size in a histogram is called the **class interval.**		
Class Frequency: The number of times that data points appear in a particular class interval is called the **class frequency.**		

Study these words when they appear in the text. After the section, test your understanding by completing the Vocabulary Review in the section exercises.

Objective A Read a Circle Graph

The Concept A **circle graph** is a graph in the form of a circle that is divided into pie-shaped pieces, where each piece represents one part of the data set. The entire circle represents the whole, or 100% of the data. The **sectors** of a circle graph are pie-shaped pieces that represent each part of the data set.

Circle graphs are often used to show the results of statistical polls. For example, a poll of 10,000 students at East Community College was performed to collect information about student majors. Each major is represented by a sector in the circle. Notice that the numbers in the sectors add to 10,000, the total amount of data in the survey.

East Community College Student Majors

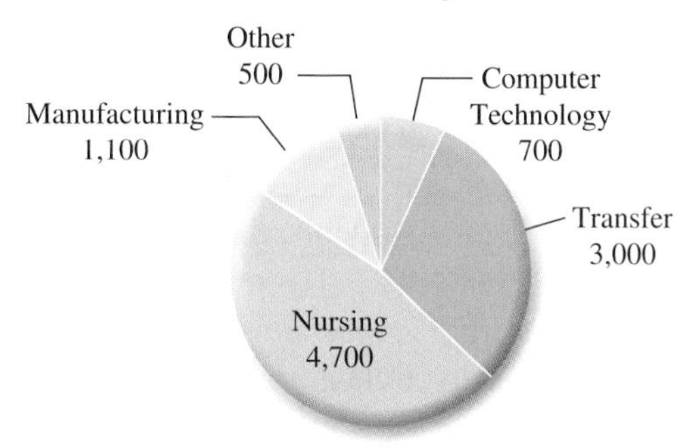

EXAMPLES

1. What field of study has the most students?

Since the largest pie-shaped sector of the circle is labeled "Nursing," that field of study has the most students.

2. How many students are studying manufacturing?

The number next to the manufacturing sector is 1,100. That is the number of students studying manufacturing.

3. What percent of the students are studying manufacturing?

1,100 out of 10,000 students are studying manufacturing.

$$\frac{\text{Manufacturing students}}{\text{Total students}} \Rightarrow \frac{1{,}100}{10{,}000} = \frac{11 \cdot \cancel{100}}{100 \cdot \cancel{100}} = \frac{11}{100} = 11\%$$

11% of the students are studying manufacturing.

GUIDED PRACTICE

1. What field of study has the fewest students?

Since the ______ pie-shaped sector of the circle is labeled "______," that field of study has the fewest students.

2. How many students plan to transfer?

The number next to the transfer sector is ____. That is the number of students planning to transfer.

3. What percent of the students are studying nursing?

______ out of 10,000 students are studying nursing.

$$\frac{\text{Nursing students}}{\text{Total students}} \Rightarrow \underline{\qquad} = \underline{\qquad} = \underline{\qquad} =$$

______ of the students are studying nursing.

(Continued)

4. What is the ratio of computer technology students to transfer students? Write the ratio in lowest terms.

Number of computer technology students: 700
Number of transfer students: 3,000

$$\frac{\text{Number of computer technology students}}{\text{Number of transfer students}} = \frac{700}{3{,}000} = \frac{7 \cdot \cancel{100}}{30 \cdot \cancel{100}} = \frac{7}{30}$$

The ratio of computer technology students to transfer students is 7 to 30. For every 7 computer technology students, there are 30 transfer students.

4. What is the ratio of manufacturing students to nursing students? Write the ratio in lowest terms.

Number of manufacturing students: ______
Number of nursing students: ______

$$\frac{\text{Number of manufacturing students}}{\text{Number of nursing students}} = \frac{\quad}{\quad} = \frac{\quad}{\quad} = \frac{\quad}{\quad}$$

The ratio of manufacturing students to nursing students is ______ to ______. For every ______ manufacturing students, there are ______ nursing students.

Because it is easy to compare the relative size of sectors, circle graphs are sometimes used to display parts of data that are written as percents. The percents in the following circle graph show the locations of TV viewers during the 2006 *American Idol* season.

Where *American Idol* Viewers Live

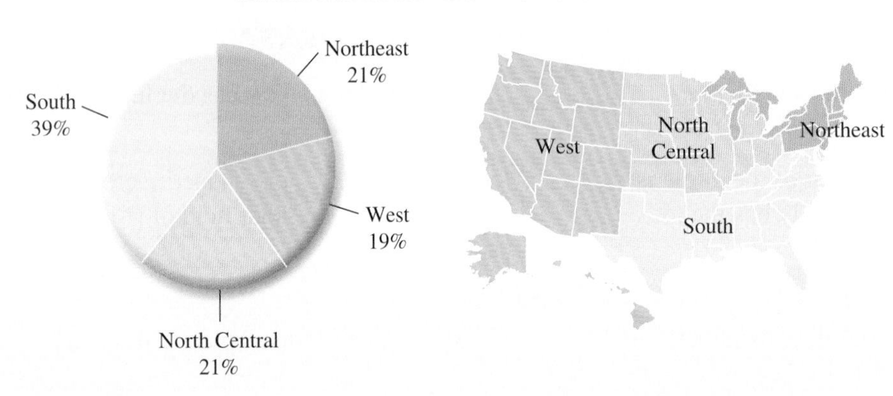

Source: Reality TV Magazine

EXAMPLES

5. Which of the four regions had the fewest viewers?

Since the smallest sector of the circle is labeled "West," that region contributed the fewest viewers. The West contributed 19% of the overall viewers.

6. What percent of the total viewers are from the Northeast or West regions?

Northeast: 21%
West: 19%

$$21\% + 19\% = 40\%$$

40% of the total viewers are from the Northeast or West regions.

GUIDED PRACTICE

5. Which of the four regions had the most viewers?

Since the ______ sector of the circle is labeled "______," that region contributed the most viewers. The ______ contributed ______% of the overall viewers.

6. What percent of the total viewers are from the North Central or South regions?

North Central: ______
South: ______

______% + ______% = ______%

______% of the total viewers are from the North Central or South regions.

7. If 1,045 people participated in the poll, how many viewers polled were from the West? Round your answer to the nearest whole number.

19% of 1,045 people were from the West.

Percent	·	Base	=	Amount
↓		↓		↓
(0 .19)	·	(1,045)	=	198.55
			≈	199

199 of the people polled were from the West.

7. If 1,045 people participated in the poll, how many viewers polled were from the South? Round your answer to the nearest whole number.

_____% of 1,045 people were from the South.

Percent	·	Base	=	Amount
↓		↓		↓
( )	·	(1,045)	=	
			≈	

_____ of the people polled were from the South.

Concept Check

A1. Why is a circle graph useful for showing how the parts in a data set relate to the whole data set?

A2. What is the name for the pie-shaped pieces that make up a circle graph?

Objective A Practice

Use the circle graph to answer the following questions.

U.S. Medals won at the 2006 Winter Olympics

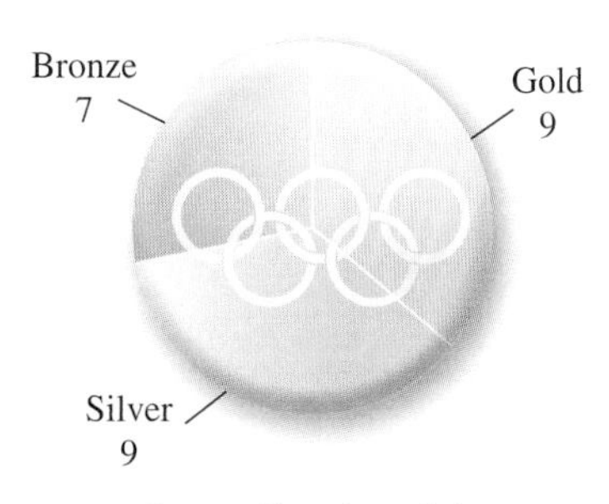

Source: Olympics website

A3. How many medals did the United States win?

A4. How many more silver medals were won than bronze medals?

A5. What percent of the medals won were silver? Round your answer to the nearest percent.

A6. What percent of the medals won were bronze? Round your answer to the nearest percent.

A7. What was the ratio of bronze medals won to total medals won?

A8. What was the ratio of gold medals won to bronze medals won?

1,045 people were asked which judge, if any, influenced their voting on *American Idol*. Use the circle graph to answer each question. Round answers to the nearest whole person.

Percent of Votes Influenced by Each Judge on *American Idol*, 2007

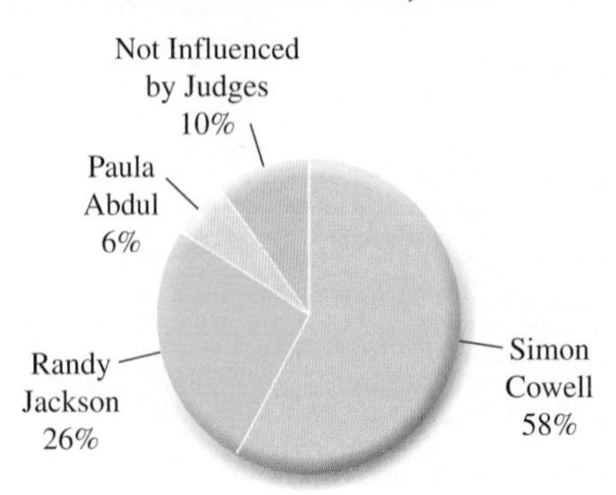

Source: *Reality TV Magazine*

A9. What percent of the total votes were influenced by either Randy Jackson or Paula Abdul?

A10. As a percent, how many more votes were influenced by Simon Cowell than by Paula Abdul?

A11. Of the 1,045 people surveyed, how many said that they were influenced by Paula Abdul?

A12. Of the 1,045 people surveyed, how many said that the judges did not influence their opinion?

Objective B Read a Bar Graph and a Line Graph

The Concept **Bar graphs** are useful for making visual comparisons of data. The following bar graph shows recording artists and the number of gold singles they earned in the United States.

Gold Singles by Artist

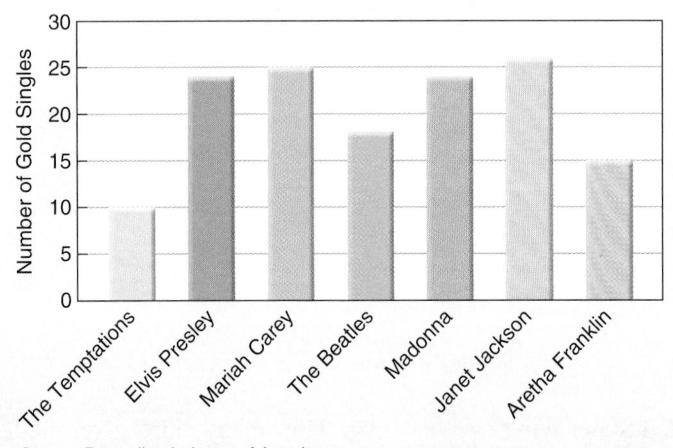

Source: Recording Industry of America

EXAMPLES

8. Which of the artists has earned 15 gold singles in the United States?

The green bar has a height of 15. Looking at the graph, we see that the green bar represents Aretha Franklin.

Aretha Franklin has earned 15 gold singles in the United States.

GUIDED PRACTICE

8. Which of the artists has earned the most gold singles in the United States?

The _____ bar has the highest height. Looking at the graph, we see that the _____ bar represents ___________.

___________ has earned _____ gold singles in the United States.

9. How many more gold singles does Madonna have than The Beatles?

Madonna has 24 gold singles.
The Beatles have 18 gold singles.

$$24 - 18 = 6$$

Madonna has 6 more gold singles than The Beatles.

9. How many more gold singles does Mariah Carey have than The Temptations?

Mariah Carey has _____ gold singles.
The Temptations have _____ gold singles.

Mariah Carey has _____ more gold singles than The Temptations.

Line graphs display data points that are connected by lines to show a trend. In a line graph, points are plotted to indicate each specific value. The points are then connected with a line to show a trend. The following line graph shows the number of customers per month at Rudy's Texas Barbecue restaurant.

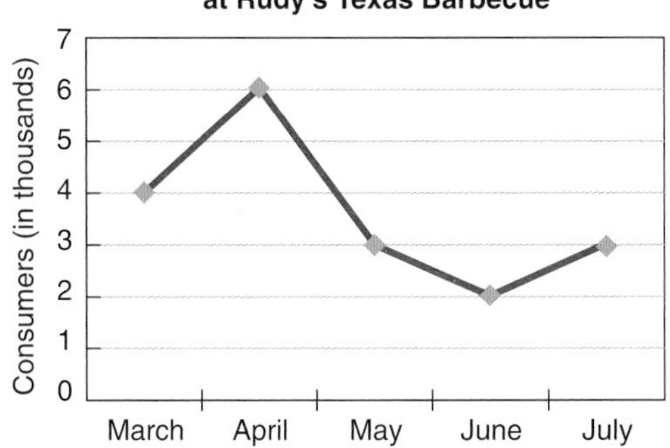

▶ **NOTE** Although the line on this graph is useful for visualizing trends, only the blue squares can be used to read data values.

EXAMPLES

10. In which month did the fewest customers visit the restaurant?

Since the lowest point on the graph occurred in June, that was the month with the fewest customers.

11. What was the approximate decrease in customers from April to May?

In April, there were approximately 6,000 customers.
In May, there were approximately 3,000 customers.

$$6{,}000 - 3{,}000 = 3{,}000$$

There was a decrease of approximately 3,000 customers from April to May.

12. Between what two months did the largest increase in customers occur?

The line goes up at the steepest angle between the months of March and April, so this represents the largest increase in customers.

GUIDED PRACTICE

10. In which month did the most customers visit the restaurant?

Since the highest point on the graph occurred in _____, that was the month with the most customers.

11. What was the approximate increase in customers from March to April?

In March, there were approximately _____ customers.
In April, there were approximately _____ customers.

There was an increase of approximately _____ customers from March to April.

12. Between what two months did the largest decrease in customers occur?

The line goes down at the _____ angle between the months of _____ and _____, so this represents the largest decrease in customers.

Concept Check

B1. What type of graph includes data points connected by lines to show a trend?

B2. Give an example of data that could be graphed using a bar graph.

Objective B Practice

Below are the results of a 2006 salary survey of practicing nurses.

Average Salaries of Practicing Nurses

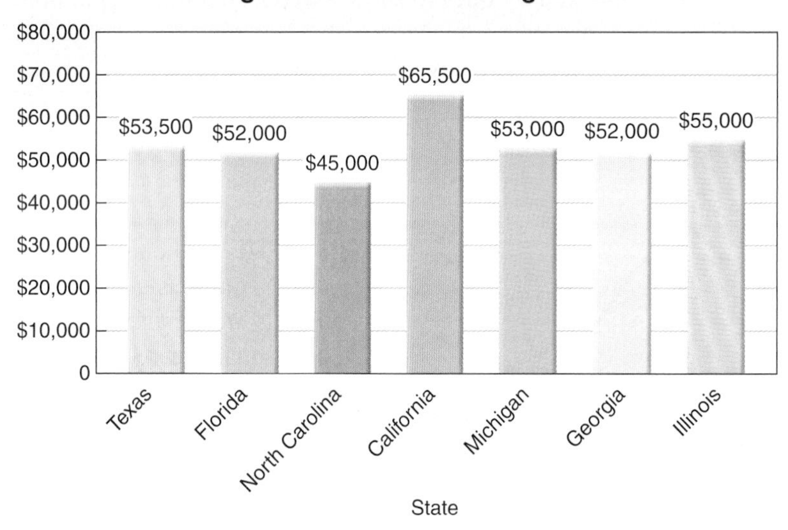

Source: PayScale, Inc.

B3. What is the average salary of a nurse working in Illinois?

B4. What is the average salary of a nurse working in Florida?

B5. How much more is the average nursing salary in Texas than in Georgia?

B6. How much more is the average nursing salary in Michigan than in North Carolina?

B7. Identify the two states that have the largest difference in nursing salaries.

B8. Identify the two states that have the smallest difference in nursing salaries.

Passive solar heating uses south-facing windows to capture the sun's heat. To demonstrate how well her garage is heated using passive solar heating, Anaïs made a graph comparing the inside and outside temperature over the course of one day.

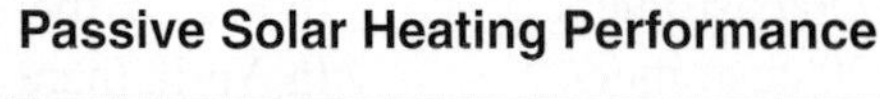

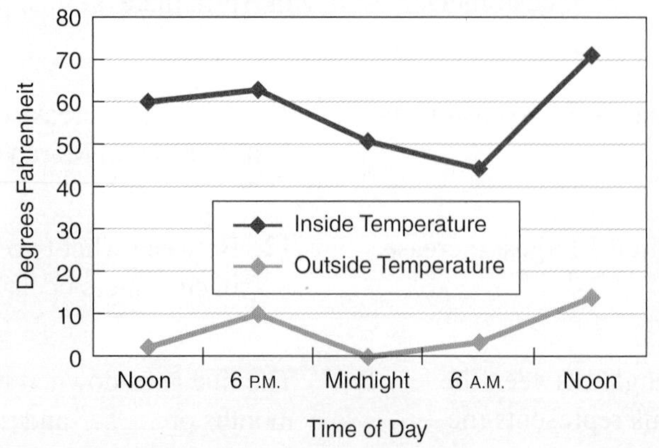

Source: Home Power magazine

B9. What was the approximate outside temperature at 6 P.M.?

B10. What was the approximate inside temperature at 6 P.M.?

B11. About how much warmer was the inside temperature than the outside temperature at midnight?

B12. About how much warmer was the inside temperature than the outside temperature at 6 A.M.?

Objective C Construct a Histogram

The Concept A **histogram** is a type of bar graph that shows the frequency of data in **class intervals.** The number of data entries in each class interval is the **class frequency** of data in the interval. To find each class frequency below, a teacher made a *tally* of the grade data by making a mark for each score in the corresponding class interval. A score of 90–99 is an A, 80–89 is a B, and so on. The class intervals are of size 10 (90–99, for instance), and the tally for each class interval gives us the class frequency.

GRADES IN A HISTORY CLASS

Grade	Tally	Class Frequency
90–99	𝍸 IIII	9
80–89	𝍸 𝍸	10
70–79	𝍸 I	6
60–69	II	2
50–59	IIII	4

The grades are divided into 5 class intervals.

The number of students in each class interval is the class frequency. For example, 9 students received a grade between 90 and 99.

Using this table, we can make a histogram.

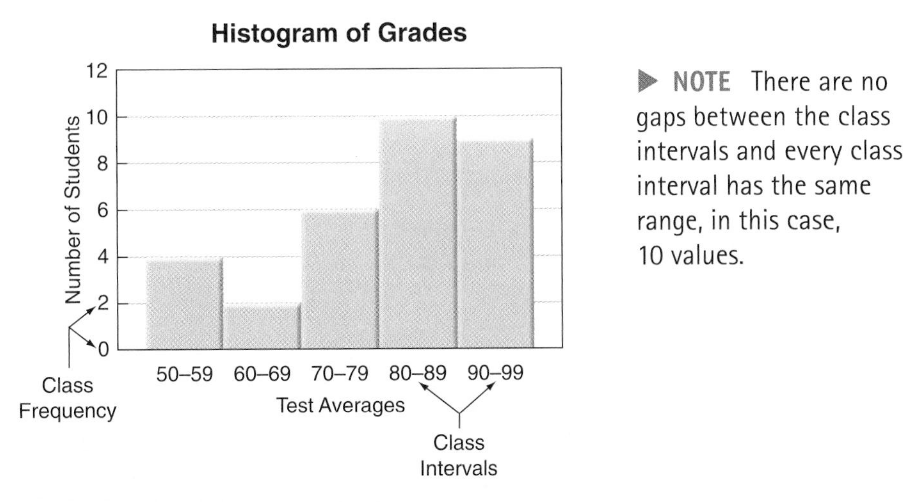

▶ **NOTE** There are no gaps between the class intervals and every class interval has the same range, in this case, 10 values.

- In a histogram, the class intervals always have the same width. Here, each class interval has ten numbers (50–59, 60–69, and so on).
- The height of each bar is the class frequency of the class interval. In this case, the height of each bar corresponds to the number of students that received a grade that was within the specific class interval.

Procedure Create a Histogram

Step 1 Tally the data using equal-sized class intervals.

Step 2 Create a histogram.

- The width of each bar must be the same.
- There should be no gaps between the class intervals.
- The height of each bar is the class frequency of data in that class interval.

EXAMPLES

GUIDED PRACTICE

An instructor has given an exam and wants to make a histogram of the grades. A score of 90–99 is an A, 80–89 is a B, and so on. Organize the data into a table and represent the data as a histogram.

13. The exam scores were 75, 60, 82, 95, 72, 70, 86, 88, 89, 91, 74, 92, 88, 69, and 80.

Step 1 Use class intervals of size 10 to tally the data.

Grade/Score	Tally	Class Frequency
90–99	III	3
80–89	~~IIII~~ I	6
70–79	IIII	4
60–69	II	2
50–59		0

Step 2 Create a histogram.

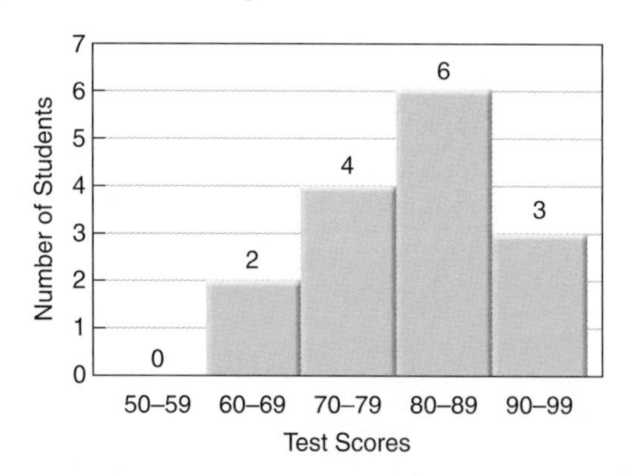

14. Based on the histogram, what was the most frequently assigned grade?

Because the class interval 80–89 has the highest frequency, the grade of B was assigned the most. There were more Bs than any other grade.

13. The exam scores were 76, 86, 80, 94, 72, 70, 86, 88, 80, 91, 74, and 65.

Step 1 Use class intervals of size 10 to tally the data.

Grade/Score	Tally	Class Frequency

Step 2 Create a histogram.

14. Based on the histogram, what was the most frequently assigned grade?

Because the class interval ______ has the ______ frequency, the grade of _____ was assigned the most. There were more _____s than any other grade.

15. A football team won all 12 of its games in a season. The margins of victory were 14, 7, 10, 12, 13, 15, 21, 19, 3, 18, 18, and 8 points. Make a histogram of the margins of victory.

Step 1 Use class intervals of size 5 to tally the data.

Margin of Win (points)	Tally	Frequency
1–5	|	1
6–10	|||	3
11–15	||||	4
16–20	|||	3
21–25	|	1

Step 2 Create a histogram.

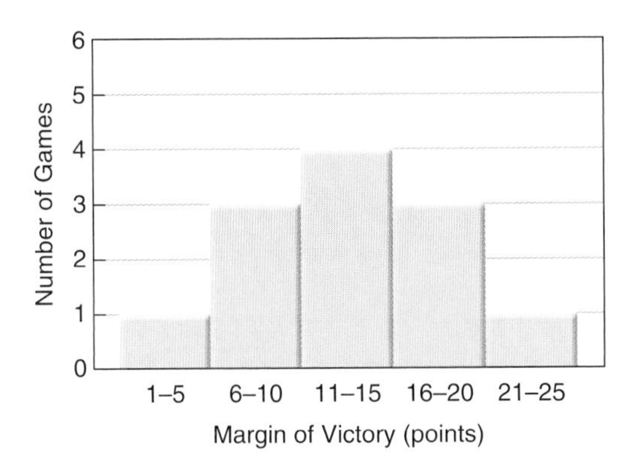

16. Based on the histogram, in what class interval were the most victories?

The class frequency is highest for the class interval of 11 to 15 points. The margin of victory was more likely to be in this interval than in any other.

15. Raphael received 13 bonuses at work. They were $208, $158, $119, $209, $259, $176, $225, $131, $230, $120, $322, $299, and $233. Make a histogram of his bonuses.

Step 1 Use class intervals of size 50 to tally the data.

Work Bonus (dollars)	Tally	Frequency
101–150		
151–200		
201–		
–		
–		

Step 2 Create a histogram.

16. Based on the histogram, in what class interval were the most bonuses?

The class frequency is highest for the class interval of $______ to $______. Raphael was more likely to get a bonus in this class interval than in any other.

Concept Check

C1. Explain why the following statement is true or false: A class interval of 6–10 includes five whole numbers.

C2. Explain why the following statement is true or false: A class interval of 7–12 includes five whole numbers.

C3. A data set includes whole numbers from 31–60. To make a histogram with class intervals of size 10, how many class intervals will be needed?

Objective C Practice

Fifteen daily high temperatures, in degrees Fahrenheit, were recorded in Anchorage, Alaska. The temperatures were 34, 42, 47, 60, 56, 31, 58, 40, 44, 52, 35, 41, 53, 47, and 45.

C4. What are the lowest and highest points in the data set?

C5. Complete the following table. Use class intervals of size 6.

Class Intervals	Tally	Class Frequency
31–36		
37–		
–		
–		
–		

C6. Construct a histogram for the daily high temperatures using the information in the table.

Section 8.1 Exercises

FOR EXTRA HELP

To read a circle graph:

1. Answer the Objective A Concept Checks.
2. Answer the odd Objective A Practice Exercises.
3. Answer the even Objective A Practice Exercises.

To read a bar graph and a line graph:

4. Answer the Objective B Concept Checks.
5. Answer the odd Objective B Practice Exercises.
6. Answer the even Objective B Practice Exercises.

To construct a histogram:

7. Answer the Objective C Concept Checks.
8. Answer the odd Objective C Practice Exercise.
9. Answer the even Objective C Practice Exercises.

VOCABULARY REVIEW *Review the Vocabulary Preview for Section 8.1. Study the definitions until you can check* Got It *for every word.*

Use these words to complete the sentences on the next page.

data • circle graph • sectors • bar graph • line graph • histogram • class interval • class frequency

	Got It	Get Help
10. In a histogram, the top of each bar indicates the ___________.		
11. A ___________ is a graph that is divided into pie-shaped pieces, where each piece represents one part of a data set.		
12. A ___________ uses rectangular bars to make a visual comparison of data.		
13. A ___________ is a type of bar graph that shows the frequency of data in class intervals.		
14. A ___________ displays data points that are connected by lines to show a trend.		
15. A circle graph uses ___________, in the shape of pie pieces, to represent the parts of a data set.		
16. ___________ is a collection of facts, usually numbers, from which conclusions may be drawn.		

How will you get help for any vocabulary that you are unsure about?

Your instructor ____ MyMathLab ____ A classmate ____ A tutor ____ Other ____

Use the circle graph to answer each question.

Due to millions of poinsettia sales, the Christmas/Hanukkah period is the top holiday period for floral sales. The circle graph shows the number of floral sales, by holiday.

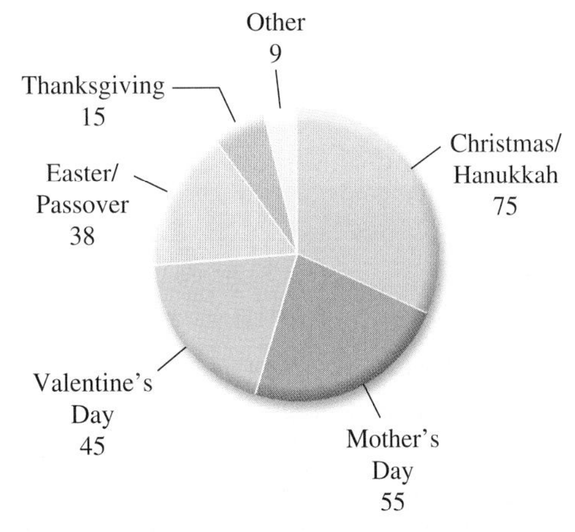

17. Which holiday has the third most floral sales?

18. Which holiday has the second most floral sales?

19. How many more floral sales take place around Christmas than Valentine's Day?

20. What is the combined number of floral sales that take place around Easter and Mother's Day?

21. What percent of floral sales take place around Valentine's Day?

22. What percent of floral sales take place around Mother's Day?

Use the following circle graph to answer each question.

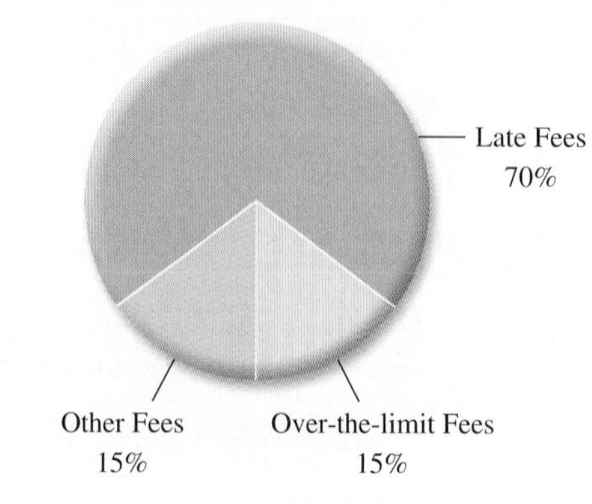

23. Most of the penalty fees are in which category?

24. What percent of the penalty fees are for over-the-limit fees?

25. What percent of the penalty fees were for late fees and over-the-limit fees combined?

26. What percent of the penalty fees were not late fees?

27. If the credit card companies collected \$17.1 billion in 2006, what amount was collected in late fees?

28. If the credit card companies collected \$17.1 billion in 2006, what amount was collected in over-the-limit fees?

Use the following bar graph to answer each question.

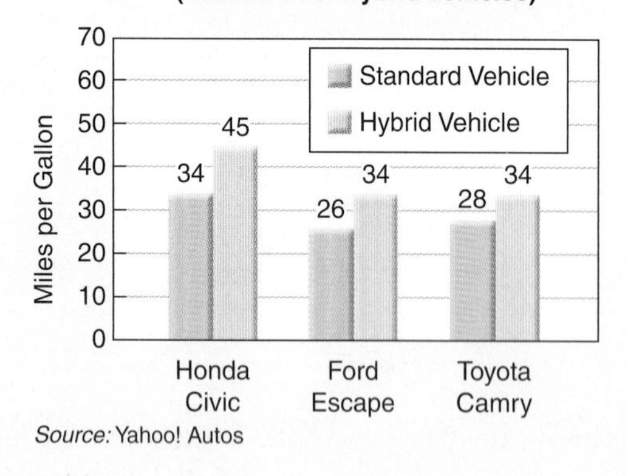

29. How many more miles per gallon does the hybrid Honda Civic get than the standard Ford Escape?

30. How many more miles per gallon does the hybrid Ford Escape get than the standard Toyota Camry?

31. Comparing the standard and hybrid vehicles, which model has the greatest difference in fuel economy?

32. Comparing the standard and hybrid vehicles, which model has the smallest difference in fuel economy?

33. Determine how many gallons of gas it would take to drive each vehicle 495 miles. Round each answer to the nearest gallon.

a. Ford Escape (hybrid)
b. Ford Escape (standard)

34. Determine how many gallons of gas it would take to drive each vehicle 650 miles. Round each answer to the nearest gallon.

a. Honda Civic (hybrid)
b. Honda Civic (standard)

Use the following bar graph to answer each question.

Researchers from the Pew Research Center surveyed 2,000 adults by phone in 2006. They read a list of items and asked, "Do you think of this item as a necessity or as a luxury you could do without?"

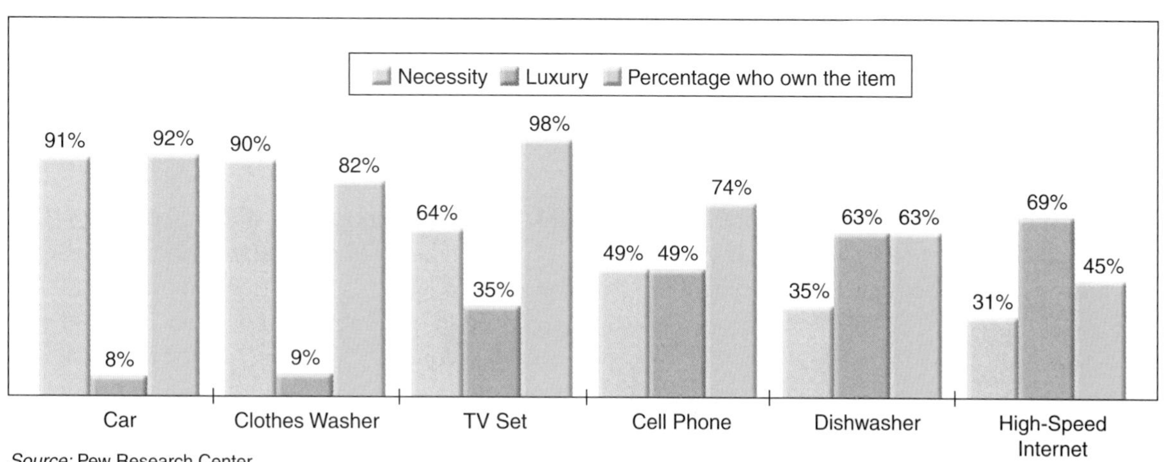

35. What percent of respondents said that a cell phone is a necessity?

36. What percent of respondents said that a dishwasher is a luxury?

37. As a percent, what is the difference between respondents who said that they owned a TV and those who said that TV is a luxury?

38. As a percent, what is the difference between respondents who said that high-speed Internet is a luxury and those who said that it is a necessity?

39. List the top three items that respondents thought were necessities.

40. List the top three items that respondents thought were luxuries.

41. Compare the differences between the data showing that an item is a necessity and the percentage of people that own the item. Which item do most people find to be necessary but do not own? There is no right answer to this question. State your opinion and support it using evidence from the bar graph.

42. Compare the differences between the data showing that an item is a luxury and the percentage of people that own the item. Which item do most people find to be a luxury but still own? There is no right answer to this question. State your opinion and support it using evidence from the bar graph.

Use the bar graph to answer each question.

American Deena Kastor ran the world's fastest women's marathon in 2006, becoming the first U.S. woman to set the record since 1983. The bar graph shows American women who have the fastest marathon times for the given years. Time is shown as hours: minutes: seconds.

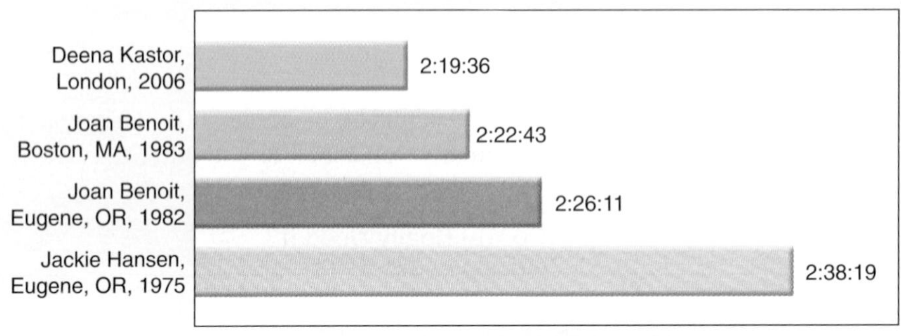

Source: World Marathon Majors

43. By how much time did Joan Benoit reduce her best marathon time from 1982 to 1983? If you need to borrow, remember that 1 minute equals 60 seconds.

44. How much more time did it take Jackie Hansen to finish the 1975 Eugene marathon compared to Deena Kastor's 2006 London marathon? If you need to borrow, remember that 1 minute equals 60 seconds.

Use the line graph to answer each question.

U.S. spending on energy efficiency fell with power industry deregulation in the mid-1990s but has nearly doubled since 1999.

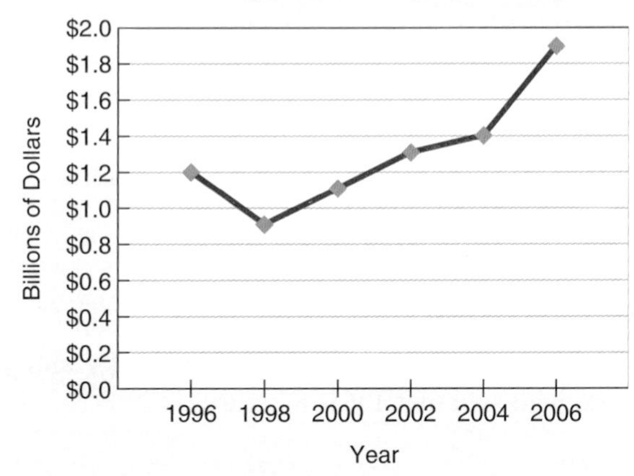

Source: USA Today

45. What trend in energy-efficiency spending do you see from 1996 to 1998?

46. What trend in energy-efficiency spending do you see from 2000 to 2006?

47. What is the approximate change in energy-efficient spending from 1996 to 2004? Round data to the nearest $0.1 billion.

48. What is the approximate change in energy-efficient spending from 1998 to 2006? Round data to the nearest $0.1 billion.

tobey When borrowing, it is helpful to use 1 minute = 60 seconds to rewrite the time. Thus 3:18:22 could be written as 3:17:82.

For more tips and tweets, go to twitter.com/gstbasicmath

Use the bar graph to answer each question.

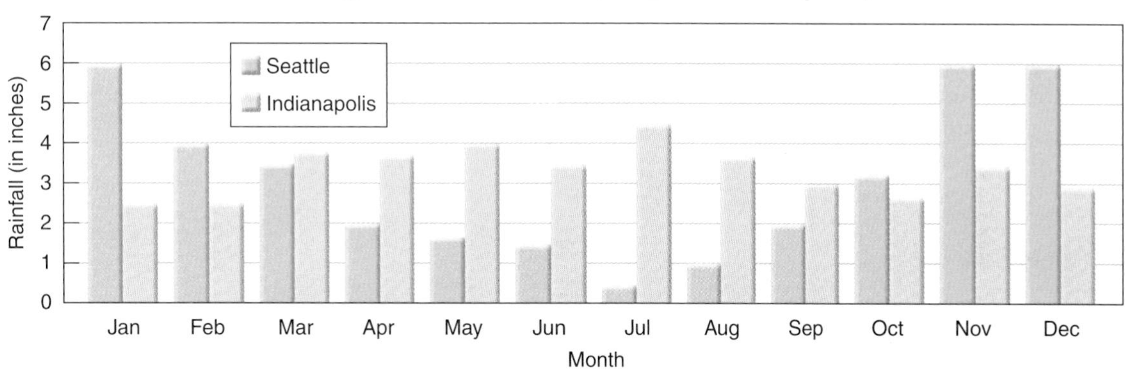

49. What was the approximate rainfall in Seattle during the month of September?

50. What was the approximate rainfall in Indianapolis during the month of December?

51. During July, how much more rain fell in Indianapolis than in Seattle?

52. During November, how much more rain fell in Seattle than in Indianapolis?

53. For how many months was the rainfall in Indianapolis greater than the rainfall in Seattle?

54. During which months was the rainfall in Seattle greater than the rainfall in Indianapolis?

Use the histogram to answer each question.

Student Ages in Algebra Class

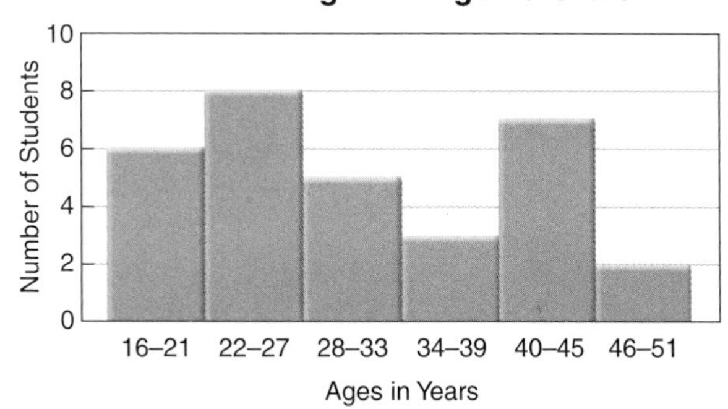

55. How many students in this class are between the ages of 28 to 33?

56. How many students in this class are between the ages of 40 to 45?

57. How many students in this class are less than 34 years old?

58. How many students in this class are more than 27 years old?

Twenty-four students in a Health Dynamics course recorded their heights to the nearest inch as follows:

65, 60, 64, 71, 77, 61, 65, 71, 71, 66, 67, 72, 76, 63, 70, 62, 67, 68, 68, 69, 74, 68, 74, 61

59. **a.** What is the range of heights, in inches?

b. Tally the data into six 3-inch class intervals, starting with the lowest height.

c. Make a histogram showing the heights of the students. Be sure to label the histogram completely.

d. Based on the histogram, most of the students fall within what range of heights?

The Young family likes to dine out at restaurants and have the following restaurant bills for one year:

$76, $54, $52, $115, $46, $40, $73, $64, $58, $109, $82, $85, $64, $74, $68, $92, $64, $71, $60, $50

60. **a.** What is the range of the family's restaurant bills, in dollars?
b. Tally the data into four $20 class intervals, starting with the lowest bill.
c. Make a histogram showing the Young's restaurant bills. Be sure to label the histogram completely.
d. Based on the histogram, most of the bills fall within what range of dollars?

SELF-ASSESSMENT

	YES	NO
1. On your last test, were you satisfied with your score? *If you answered yes, good job. Do not continue here.* *If you answered no, answer the following.*		
2. Did you preview each section of the textbook before your instructor taught it in class?		
3. Did you complete all of the homework before it was due?		
4. Did you get help with the homework exercises that you didn't understand?		

5. Set two goals that will help you become better prepared for your next test.

Goal 1: ____________________

Goal 2: ____________________

Section 8.1 Question Log

Use this space to write questions. Be sure to get these answered and revisit them when you prepare for your exam.

Page ____ **Answered** ☐	**Page** ____ **Answered** ☐
Page ____ **Answered** ☐	**Page** ____ **Answered** ☐

Section 8.2 Mean, Median, and Mode

The following graph shows the results of a survey for statistician salaries in seven U.S. cities. When data is collected in a survey, people often use the data to determine a *typical response*. For the data in the bar graph shown, the typical response is the salary that someone could expect to earn as a statistician.

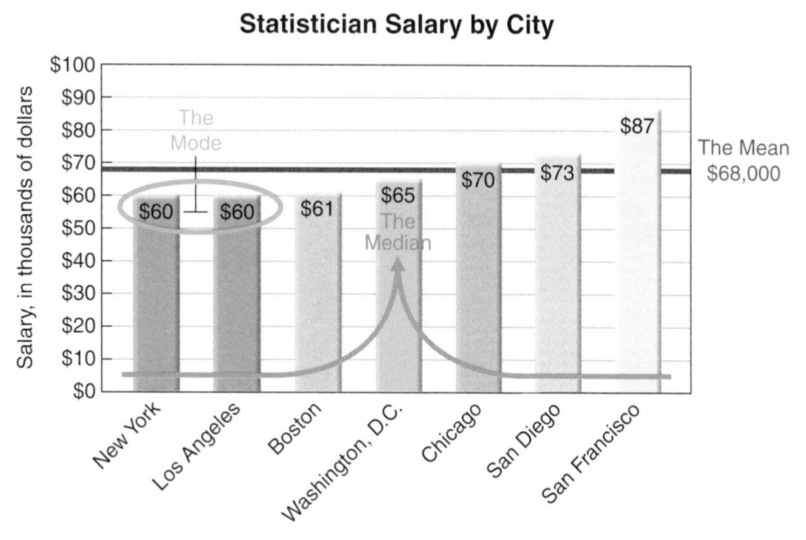

This section explores three values that are used to identify the typical response, or central numbers, in a set of data. These values are called **measures of central tendency** because they identify the center of the data set. Each of the three measures is shown on the bar graph.

- The mean (or average) of a data set is found by adding the values and then dividing by the number of values. Above, the mean is $68,000.
- The median of a data set is in the middle of the data set when the data is ordered, from least to greatest. Above, the median is $65,000.
- The mode of a data set is the response that occurs most frequently. Above, the mode is $60,000.

The **Objectives** in Section 8.2 will help you

A Find the mean of a data set.
B Find the median of a data set.
C Find the mode of a data set.

VOCABULARY PREVIEW *Check the box that applies.*

	Got It	Must Study
Measures of Central Tendency: Measures of central tendency are statistical measures that give us information about the center of a data set. They include the mean, median, and mode.		
Mean: The **mean** is the average of a data set. It is calculated by dividing the sum of the values by the number of values.		
Median: The **median** is the middle number in a data set that has been ordered from least to greatest.		
Mode: The **mode** is the number (or numbers) that occurs most often in a data set.		

Study these words when they appear in the text. After the section, test your understanding by completing the Vocabulary Review in the section exercises.

Objective A Find the Mean of a Data Set

The Concept One commonly used measure of central tendency is the mean. The **mean** is the average of a data set. You have already calculated the mean in other sections of this text. The mean is often used to calculate grades in college courses. The following bar graph shows Saeed's scores for five exams. The red line represents the mean, which is calculated to the right of the bar graph.

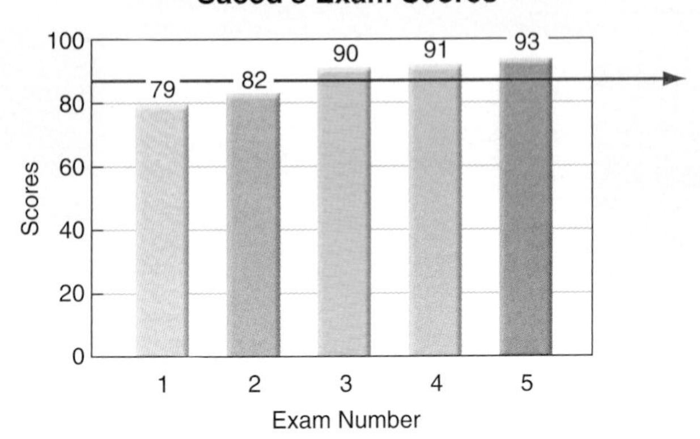

To find the mean, add the scores and divide by the number of exams.

$$\text{Mean} = \frac{79 + 82 + 90 + 91 + 93}{5}$$
$$= \frac{435}{5}$$
$$= 87$$

$5\overline{)435}$ = 87

Procedure Find the Mean of a Data Set

Step 1 Find the sum of the values.

Step 2 Divide the sum by the number of values.

INTERACTIVE DEFINITION Mean

The **mean** is the average of a set of data. It is calculated by dividing the sum of the values by the number of values.

EXAMPLE

1. Find the mean of the data set.

$\{5, 8, 9, 4, 4\}$

$$\frac{\text{Sum of values}}{\text{Number of values}} = \frac{5 + 8 + 9 + 4 + 4}{5}$$
$$= \frac{30}{5}$$
$$= 6$$

The mean of the set of data is 6.

GUIDED PRACTICE

1. Find the mean of the data set.

$\{2, 8, 3, 6, 3, 2\}$

$$\frac{\text{Sum of values}}{\text{Number of values}} = \frac{\qquad}{\quad}$$
$$= \frac{\quad}{\quad}$$
$$=$$

The mean of the set of data is ____.

DO YOU UNDERSTAND how to find the mean of a data set? Got It Get Help

EXAMPLES

2. Find the mean of the data set.

$$\{6, 3, 2, 9\}$$

$$\frac{\text{Sum of values}}{\text{Number of values}} = \frac{6 + 3 + 2 + 9}{4} = \frac{20}{4} = 5$$

The mean of the data set is 5.

3. Find the mean of the data set. Round the answer to the nearest tenth.

$$\{10, 6, 15, 7, 3, 3\}$$

$$\frac{\text{Sum of values}}{\text{Number of values}} = \frac{10 + 6 + 15 + 7 + 3 + 3}{6} = \frac{44}{6} \approx 7.33$$

The mean rounded to the nearest tenth is 7.3.

GUIDED PRACTICE

2. Find the mean of the data set.

$$\{5, 7, 4, 2, 2\}$$

$$\frac{\text{Sum of values}}{\text{Number of values}} = \underline{\qquad\qquad} = \frac{\quad}{\quad} = $$

The mean of the data set is ____.

3. Find the mean of the data set. Round the answer to the nearest tenth.

$$\{8, 6, 11, 7, 1, 2, 2\}$$

$$\frac{\text{Sum of values}}{\text{Number of values}} = \underline{\qquad\qquad} = \frac{\quad}{\quad} \approx $$

The mean rounded to the nearest tenth is ____.

Concept Check

A1. What word could you use to help yourself remember the definition of the mean?

A2. If the largest and smallest values in a data set are 80 and 42, the mean must fall between the values ____ and ____.

A3. How do you calculate the mean?

Objective A Practice

Find the mean for each data set. Round to the nearest tenth if necessary.

A4. $\{5, 7, 12, 8\}$

A5. $\{22, 26, 29\}$

A6. $\{8, 5, 10, 0, 3\}$

A7. $\{12, 4, 7, 2, 1, 6\}$

A8. $\{16, 9, 5, 13, 7\}$

A9. $\{24, 2, 5, 35, 0, 16, 3\}$

Objective B Find the Median of a Data Set

The Concept The median is another commonly used measure of central tendency. The **median** is the middle number in a set of data when it is ordered from least to greatest. The following bar graph shows Saeed's scores for five exams. The arrow on the blue lines represents the median.

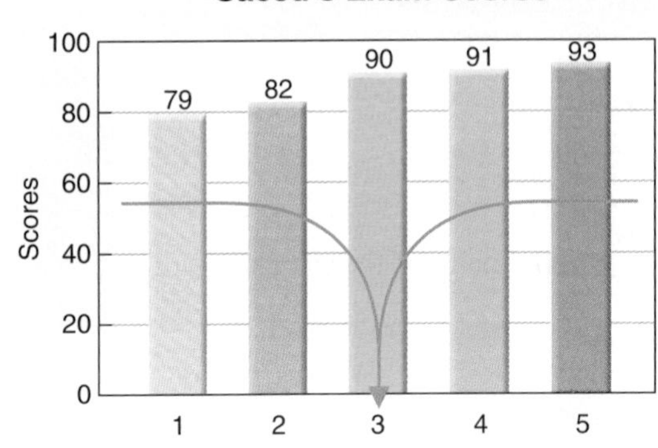

The median is the middle data value.

79, 82, 90, 91, 93

The value 90 is the median because it is in the middle of the data set. There are two values on either side of 90.

Generally speaking, the median will be a different value than the mean. Students often ask, "Is it better to use the mean or median to find the center of a data set?" Depending on the particular data set, one measure may be better than the other. You will explore data sets in the section exercises to help you understand when the mean or the median is a better indicator of the center of a data set.

Procedure Find the Median of a Data Set

Step 1 Arrange the data in numerical order.

Step 2 Find the data element in the middle of the list.

If there are two middle numbers, the median is the average of the two numbers.

INTERACTIVE DEFINITION Median

The **median** is the middle number in a data set that is organized from least to greatest.

EXAMPLE

4. Find the median of the data set.

$\{5, 8, 9, 4, 4\}$

Arrange in numerical order. $\Rightarrow$ 4, 4, 5, 8, 9

Find the middle. $\Rightarrow$ ~~4~~, ~~4~~, 5, ~~8~~, ~~9~~

The median of the data set is 5.

GUIDED PRACTICE

4. Find the median of the data set.

$\{2, 8, 3, 6, 3\}$

Arrange in numerical order. $\Rightarrow$

Find the middle. $\Rightarrow$

The median of the data set is ______.

DO YOU UNDERSTAND how to find the median of a data set? Got It Get Help

If a data set has an even number of elements, the middle of the data set will be between two numbers.

{1, [3, 5,] 7}

the middle

The median is 4 because it is in the middle of 3 and 5. To find the middle of two numbers, calculate the average of the two numbers.

EXAMPLE

5. Find the median of the data set.

$\{7, 3, 2, 9, 10, 4\}$

If there are two middle numbers, the median is the average of these two numbers.

Arrange in numerical order. ⇒ 2, 3, 4, 7, 9, 10

Find the middle. ⇒ 2, 3, [4, 7,] 9, 10

the middle

Calculate the mean of the two middle numbers.

$$\frac{4+7}{2} = \frac{11}{2}$$
$$= 5.5$$

The median is 5.5.

GUIDED PRACTICE

5. Find the median of the data set.

$\{11, 0, 4, 2, 15, 17\}$

If there are two middle numbers, the median is the average of these two numbers.

Arrange in numerical order. ⇒

Find the middle. ⇒

Calculate the mean of the two middle numbers.

$$\frac{\quad}{\quad} = \frac{\quad}{\quad}$$
$$=$$

The median is ____.

Concept Check

B1. What word could you use to help yourself remember the definition of the median?

B2. If there are two middle numbers in a data set, how do you find the median?

Objective B Practice

Find the median for each data set.

B3. $\{5, 7, 12, 8, 2\}$

B4. $\{16, 9, 5, 13, 7\}$

B5. $\{8, 5, 10, 0, 3, 3\}$

B6. $\{2, 4, 7, 2, 1, 6\}$

Objective C Find the Mode of a Data Set

The Concept The **mode** is the number that occurs most often in a data set. The mode gives us another way to identify a central number in a set of data.

The set $\{9, 7, 12, 7, 8\}$ lists the sizes of shoes sold one morning. The mode of this data set would be useful to a salesperson who wants to know what size shoe is sold most often. The mode of the data set is 7. Knowing the mode, a salesperson can better meet the needs of her clients.

$$\{9, \boxed{7,}\ 12, \boxed{7,}\ 8\}$$

Size 7 is the mode because it is listed most frequently.

INTERACTIVE DEFINITION Mode

The **mode** is the number (or numbers) that occurs most often in a data set.

EXAMPLE

6. Find the mode of the data set.

$$\{5, 8, 9, 4, 4\}$$

Since 4 occurs most often, the mode is 4.

GUIDED PRACTICE

6. Find the mode of the data set.

$$\{2, 8, 3, 6, 3\}$$

Since ____ occurs most often, the mode is ____.

DO YOU UNDERSTAND how to find the mode of a data set? Got It | Get Help

A DATA SET CAN HAVE ONE MODE, SEVERAL MODES, OR NO MODE

One Mode	Several Modes	No Mode
One number occurs most frequently.	Two or more numbers are tied for "most frequent," and they appear more frequently than another number.	Every number occurs equally as often.
$\{2, 2, 3, 3, 3\}$ Mode $= 3$	$\{1, 1, 2, 2, 3\}$ Modes $= 1$ and 2	$\{1, 1, 2, 2\}$ No mode
3 occurs most frequently.	1 and 2 occur the most number of times, and 3 appears less frequently.	No number occurs more frequently than any other number.

Procedure Find the Mode of a Data Set

Identify the number (or numbers) that occurs most often in a data set.

EXAMPLES

7. Find the mode(s) of the data set.

$$\{1, 3, 2, 6, 2\}$$

$$\{1, 3, 2, 6, 2\} = \{1, 2, 2, 3, 6\}$$

Since 2 occurs most often, the mode is 2.

GUIDED PRACTICE

7. Find the mode(s) of the data set.

$$\{6, 3, 6, 2, 3, 3\}$$

$$\{6, 3, 6, 2, 3, 3\} = \{\ \_,\ \_,\ \_,\ \_,\ \_,\ \_\ \}$$

Since ____ occurs ________, the mode is ____.

8. Find the mode(s) of the data set.

$$\{6, 3, 6, 9, 10, 3\}$$

$$\{6, 3, 6, 9, 10, 3\} = \{3, 3, 6, 6, 9, 10\}$$

The data has two modes, 3 and 6.

9. Find the mode(s) of the data set.

$$\{1, 3, 5, 6, 2, 7\}$$

$$\{1, 3, 5, 6, 2, 7\} = \{1, 2, 3, 5, 6, 7\}$$

The data has no mode since every value appears the same number of times.

8. Find the mode(s) of the data set.

$$\{17, 0, 4, 2, 0, 17\}$$

$$\{17, 0, 4, 2, 0, 17\} = \{\ \ ,\ \ ,\ \ ,\ \ ,\ \ ,\ \ \}$$

The data has ____ modes, ____ and ____.

9. Find the mode(s) of the data set.

$$\{1, 0, 7, 4\}$$

$$\{1, 0, 7, 4\} = \{\ \ ,\ \ ,\ \ ,\ \ \}$$

The data has ____ mode since ________ appears the ____ number of times.

Concept Check

C1. What word could you use to help yourself remember the definition of the mode?

C2. If each number appears equally often in a data set, is there a mode?

Objective C Practice

Find the mode for each data set.

C3. $\{5, 7, 12, 5, 2\}$

C4. $\{7, 9, 5, 13, 7\}$

C5. $\{8, 5, 8, 0, 3, 3\}$

C6. $\{8, 2, 6, 4, 2, 8\}$

C7. $\{1, 6, 4, 5, 9, 0\}$

C8. $\{2, 4, 7, 3, 1, 6\}$

Section 8.2 Exercises

FOR EXTRA HELP

To find the mean of a data set:

1. Answer the Objective A Concept Checks.
2. Answer the odd Objective A Practice Exercises.
3. Answer the even Objective A Practice Exercises.

To find the median of a data set:

4. Answer the Objective B Concept Checks.
5. Answer the odd Objective B Practice Exercises.
6. Answer the even Objective B Practice Exercises.

To find the mode of a data set:

7. Answer the Objective C Concept Checks.
8. Answer the odd Objective C Practice Exercises.
9. Answer the even Objective C Practice Exercises.

VOCABULARY REVIEW *Review the Vocabulary Preview for Section 8.2. Study the definitions until you can check* Got It *for every word.*

Use these words to complete the definitions below.

mean • median • mode • measures of central tendency

	Got It	Get Help
10. The ____________ of a data set is calculated by dividing the sum of the data values by the number of values.		
11. The ____________ of the data set is found by locating the middle number in a data set.		
12. The ____________ of a data set indicates which data value occurs most often.		

How will you get help for any vocabulary that you are unsure about?

Your instructor ____ MyMathLab ____ A classmate ____ A tutor ____ Other ____

Find the mean for each data set. Round your answer to the tenths place if necessary.

13. $\{2, 7, 8, 8, 5\}$

14. $\{9, 4, 6, 4, 2\}$

15. $\{54, 68, 32, 91\}$

16. $\{34, 61, 67, 82\}$

17. $\left\{\frac{1}{4}, \frac{3}{8}, \frac{1}{8}\right\}$

18. $\left\{\frac{7}{12}, \frac{5}{6}, \frac{1}{12}\right\}$

Find the median for each data set. Round your answer to the tenths place if necessary.

19. $\{5, 8, 2, 4, 8\}$

20. $\{3, 7, 9, 3, 4\}$

21. $\{5, 7, 123, 634, 43, 64\}$

22. $\{99, 43, 67, 43, 23, 65\}$

23. $\{34, 25, 67, 84, 21, 84, 23\}$

24. $\{45, 92, 1, 0, 93, 121, 50\}$

Find the mode for each data set. If a set has no mode, write "no mode."

25. $\{1, 6, 3, 5, 1\}$

26. $\{4, 3, 7, 9, 3\}$

27. $\{7, 3, 0, 3, 0, 7\}$

28. $\{2, 7, 4, 4, 2, 7\}$

29. $\{1, 6, 1, 6, 7\}$

30. $\{5, 8, 4, 3, 4, 8\}$

Find the mean, median, and mode for each set of data.

31. $\{10, 10, 16, 13, 11\}$

32. $\{18, 19, 16, 10, 19\}$

Answer each question. Round to the tenths place if necessary.

33. The number of pizzas delivered by Tom's Pizza over the last seven days was 32, 20, 17, 24, 26, 38, and 37.

a. What is the mean number of pizza deliveries per day?

b. What is the median?

34. The number of customers at Hugh's Pets has increased over the last 5 days. On those days, there were 37, 44, 53, 65, and 86 customers.

a. What is the mean number of customers per day?

b. What is the median?

goetz Study the graph on page 8-19. It can help you remember the differences between the mean, median, and mode.

For more tips and tweets, go to twitter.com/gstbasicmath

35. The annual salaries of financial officers at a company are \$135,000, \$112,000, \$72,500, \$83,000, \$72,700, and \$65,000. What is the mean of these salaries, to the nearest dollar?

36. The annual salaries of college presidents at six universities are \$167,000, \$145,000, \$174,000, \$112,000, \$145,500, and \$110,500. What is the mean of these salaries, to the nearest dollar?

37. a. Find the mean, median, and mode of the test scores recorded by a teacher.
{88, 77, 98, 87, 82, 75, 45, 99, 92, 77, 59, 62, 81, 85, 83, 90}

b. Make a histogram of the data, using class intervals of 40–49, 50–59, and so on.

c. In your opinion, does the mean, median, or mode best indicate how the class did? There is no correct answer to this question, so be sure to defend your answer with evidence.

38. a. Find the mean, median, and mode of the vehicle speeds recorded by a police officer.
{92, 62, 64, 67, 73, 66, 85, 64, 79, 85, 77, 75, 65, 85, 72, 74, 73}

b. Make a histogram of the data, using class intervals of 60–64, 65–69, and so on.

c. In your opinion, does the mean, median, or mode best indicate vehicle speeds on the road? There is no correct answer to this question, so be sure to defend your answer with evidence.

Each exercise demonstrates a common error. An average of averages will generally give bad data. Answer each question to see why an average of averages (miles per gallon) can be inaccurate.

39. Kendra has been recording the number of miles she drives and the gas she uses. Her data from a recent trip was recorded in the table.

	Day 1	Day 2	Day 3	Day 4
Miles Driven	252	286	42	290
Gallons Used	9	11	2	10

a. Calculate the mileage (miles per gallon) for each day of her trip.

b. Some students mistakenly calculate the total gas mileage by finding the average of the daily gas mileages. What value do they get?

c. Calculate the actual gas mileage for the trip by dividing total miles driven by the total amount of gas used. Round your answer to the nearest tenth.

40. Gilbert has been recording the number of miles he drives and the gas he uses. His data from a recent trip was recorded in the table.

	Day 1	Day 2	Day 3	Day 4
Miles Driven	484	336	34	456
Gallons Used	22	16	2	19

a. Calculate the mileage (miles per gallon) for each day of his trip.

b. Some students mistakenly calculate the total gas mileage by finding the average of the daily gas mileages. What value do they get?

c. Calculate the actual gas mileage for the trip by dividing total miles driven by the total amount of gas used. Round your answer to the nearest tenth.

41. Repeat Exercise 39 but change the day 3 data to show Kendra driving 210 miles using 10 gallons of gas.

42. Repeat Exercise 40 but change the day 3 data to show Gilbert driving 255 miles using 15 gallons of gas.

43. Notice that the mileages for each day are identical in Exercises 39 and 41. Explain why increasing the number of miles driven on day three decreases the overall mileage in Exercise 41.

44. Notice that the mileages for each day are identical in Exercises 40 and 42. Explain why increasing the number of miles driven on day three decreases the overall mileage in Exercise 42.

Answer each question. The purpose of these exercises is to investigate the effectiveness of the mean, median, and mode as measures of central tendency.

45. Six people work at a restaurant and earn the amounts listed in the table.

Job	Hourly Wage
Line cook	$9.50
Host	$8.75
Server	$6.85
Server	$6.85
Server	$6.85
Manager	$23.54

a. What are the mean, median, and mode for their wages?

b. In your opinion, which measure of central tendency best describes the center of the data set (the wage someone can expect to earn at the restaurant)?

46. Julia and Javier kept the following scores for their practice round in a couples' bowling tournament.

Player	Score
Julia	232
Julia	220
Julia	220
Javier	183
Javier	196
Javier	161

a. What are the mean, median, and mode of the scores?

b. In your opinion, which measure of central tendency best describes the center of the data set (the score they can expect to get as a team)?

Use the bar graph to answer each question.

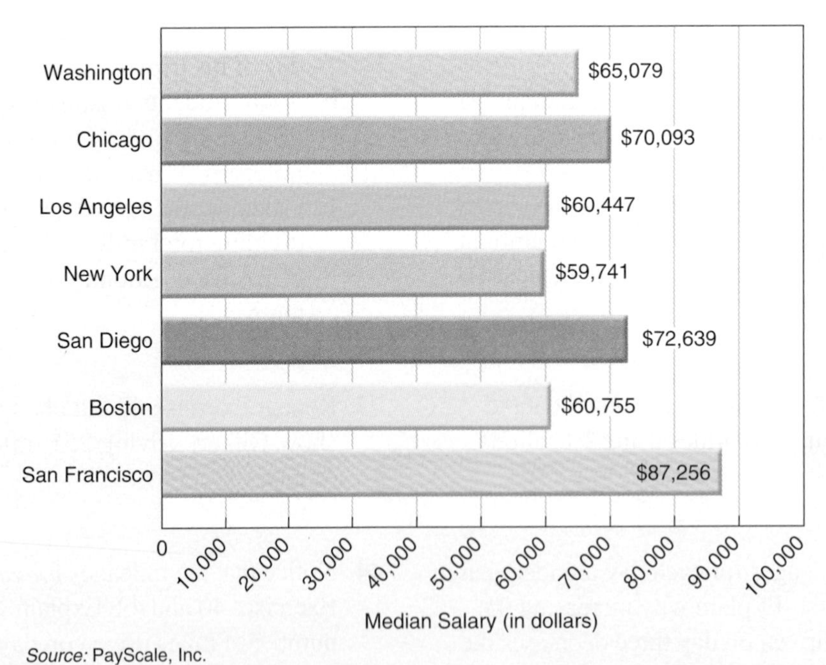

Source: PayScale, Inc.

47. a. Calculate the mean, median, and mode for the statistician salaries in these U.S. cities. Round to the nearest cent.

b. In your opinion, which measure of central tendency best describes the center of the data set (the wage someone can expect to earn as a statistician)?

For the following exercises, complete the table.

48. One data element in a data set is changed from 16 to 21, as shown below. Without performing any calculations, explain why each measure of central tendency will increase, decrease, or stay the same.

$\{10, 10, 16, 13, 11\}$

↓

$\{10, 10, 21, 13, 11\}$

	Increases/Decreases/Stays the Same?
Mean	
Median	
Mode	

49. One data element in a data set is changed from 10 to 5, as shown below. Without performing any calculations, explain why each measure of central tendency will increase, decrease, or stay the same.

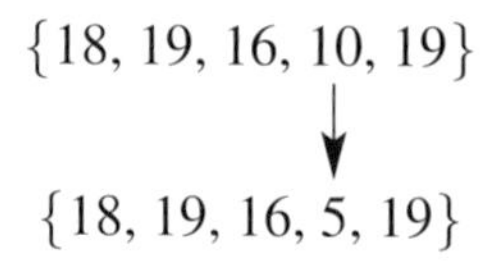

$\{18, 19, 16, 10, 19\}$

↓

$\{18, 19, 16, 5, 19\}$

	Increases/Decreases/Stays the Same?
Mean	
Median	
Mode	

50. One data element in a data set is changed from 10 to 15, as shown below. Without performing any calculations, explain why each measure of central tendency will increase, decrease, or stay the same.

$\{10, 10, 16, 13, 11\}$

↓

$\{10, 15, 16, 13, 11\}$

	Increases/Decreases/Stays the Same?
Mean	
Median	
Mode	

51. One data element in a data set is changed from 19 to 14, as shown below. Without performing any calculations, explain why each measure of central tendency will increase, decrease, or stay the same.

$\{18, 19, 16, 10, 19\}$

↓

$\{18, 19, 16, 10, 14\}$

	Increases/Decreases/Stays the Same?
Mean	
Median	
Mode	

52. Using the data set $\{3, 4, 5\}$, change the largest element from 5 to 8.

a. What change do you notice when you calculate the mean of both data sets?

b. What change do you notice when you calculate the median of both data sets?

c. Do you think the results of parts a and b will always occur when you increase the largest element in a data set that has three or more elements?

53. If the value of one element in a data set is changed, the mode may or may not change. Why?

Answer each question.

54. In a company, ten employees earn \$25,000 each.

Mean = \$25,000

Median = \$25,000

If the owner, who earns \$1,000,000, is counted, the data changes as follows:

Mean = \$113,636.36

Median = \$25,000

a. Use the given information to construct two data sets.

b. Why doesn't the median change when the owner is counted in the data?

c. The owner's salary is called an *outlier* because it is so far from all the other salaries. Why does including the outlier salary make such a large change on the mean but not the median?

55. In a company, ten people earn \$35,000 and one owner earns \$90,000.

Mean = \$40,000

Median = \$35,000

To help her employees weather hard times, the owner decides to pay each employee a one-time bonus of \$5,000 from her own salary. The data changes as follows:

Mean = \$40,000

Median = \$40,000

a. Use the given information to construct two data sets.

b. Why does the median change when the owner pays her employees out of her own salary?

c. Why doesn't the mean change when the owner pays her employees out of her own salary?

Section 8.2 Question Log

Use this space to write questions. Be sure to get these answered and revisit them when you prepare for your exam.

Page ____ Answered	Page ____ Answered
Page ____ Answered	Page ____ Answered

Chapter 8 Organizer

VOCABULARY

Use the following steps to review the vocabulary for Chapter 8.

1. *Write the definition for each word from memory.*
2. *Compare the definitions you have written with the definitions in the Vocabulary Preview.*
3. *Study any definitions that you could not remember or that you defined incorrectly.*

8.1

Data • Circle Graph • Sectors • Bar Graph • Line Graph • Histogram • Class Interval • Class Frequency

8.2

Measures of Central Tendency • Mean • Median • Mode

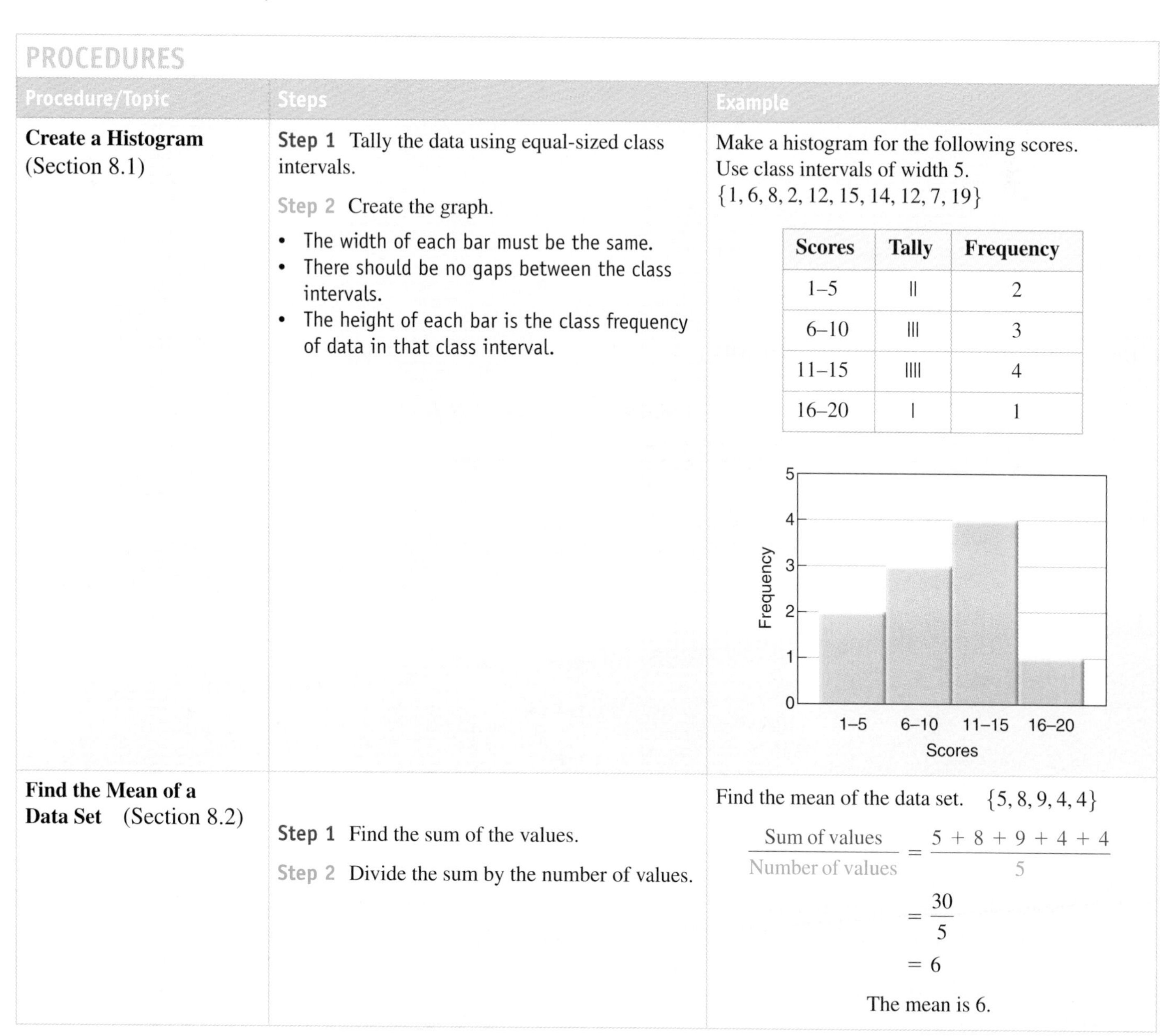

PROCEDURES

Procedure/Topic	Steps	Example
Create a Histogram (Section 8.1)	**Step 1** Tally the data using equal-sized class intervals. **Step 2** Create the graph. • The width of each bar must be the same. • There should be no gaps between the class intervals. • The height of each bar is the class frequency of data in that class interval.	Make a histogram for the following scores. Use class intervals of width 5. $\{1, 6, 8, 2, 12, 15, 14, 12, 7, 19\}$
Find the Mean of a Data Set (Section 8.2)	**Step 1** Find the sum of the values. **Step 2** Divide the sum by the number of values.	Find the mean of the data set. $\{5, 8, 9, 4, 4\}$ $\frac{\text{Sum of values}}{\text{Number of values}} = \frac{5+8+9+4+4}{5} = \frac{30}{5} = 6$ The mean is 6.

Scores	Tally	Frequency
1–5	II	2
6–10	III	3
11–15	IIII	4
16–20	I	1

Procedure/Topic	Steps	Example
Find the Median of a Data Set (Section 8.2)	**Step 1** Arrange the data in numerical order. **Step 2** Find the data element in the middle of the list. If there are two middle numbers, the median is the average of the two numbers.	Find the median of the data set. $\{6, 3, 2, 9, 10, 4\}$ 2, 3, 4, 6, 9, 10 ~~2~~, ~~3~~, 4, 6, ~~9~~, ~~10~~ The middle $\frac{4+6}{2} = \frac{10}{2}$ $= 5$ The median is 5.
Find the Mode of a Data Set (Section 8.2)	Identify the number (or numbers) that occurs most often in a data set.	Find the mode of the data set. $\{5, 8, 9, 4, 4\}$ Since 4 occurs most often, the mode is 4.

Chapter 8 Review Exercises

8.1

Use the circle graph to answer each question.

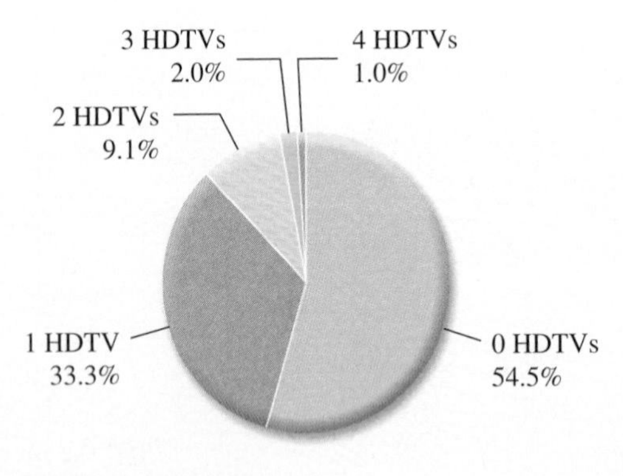

1. What was the most common response to the survey question, "How many HDTVs do you own?"

2. As a percent, how many more households own one HDTV than own two HDTVs?

3. What percent of U.S. homes have two or more HDTVs?

4. What percent of U.S. homes have fewer than two HDTVs?

5. If 1,059 people were surveyed to collect this data, how many had one HDTV?

6. If 1,059 people were surveyed to collect this data, how many had two HDTVs?

Use the bar graph to answer each question.

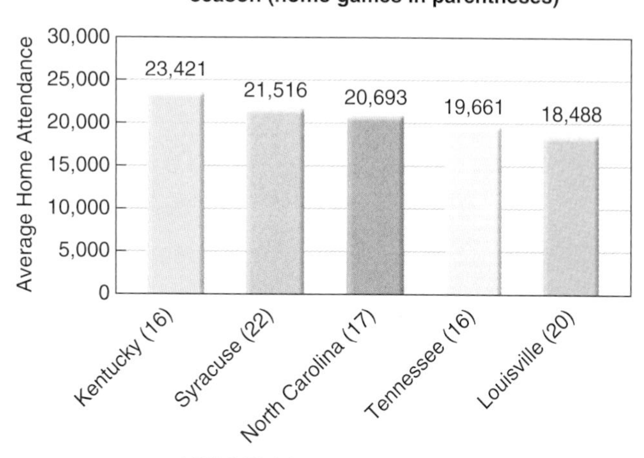

7. What was the average home attendance for a Kentucky basketball game in 2006–2007?

8. What was the average home attendance for a North Carolina basketball game in 2006–2007?

9. How much more was the average home attendance at a Syracuse basketball game than at a Tennessee basketball game?

10. How much more was the average home attendance at a North Carolina basketball game than at a Louisville basketball game?

11. Which two teams had the smallest difference in average home attendance? What was the difference?

12. Which two teams had the largest difference in average home attendance? What was the difference?

Use the line graph to answer each question.

Crunching Numbers
Average Number of Nextel Cup Accidents or Spinouts per Race

Average Number of Accidents/Spinouts (0–10); Year: 2001, 2002, 2003, 2004, 2005, 2006, 2007

Source: *USA Today*

13. During which year did the least number of accidents/spinouts occur?

14. Between which two years did the greatest increase in accidents/spinouts occur?

15. About how many more accidents/spinouts occurred per race in 2005 than in 2004?

16. Based on this graph, what overall trend in the Nextel Cup do you see happening for these years?

Use the histogram to answer each question.

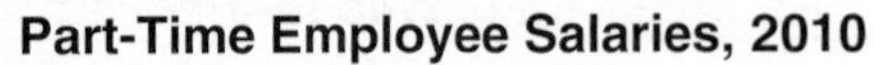

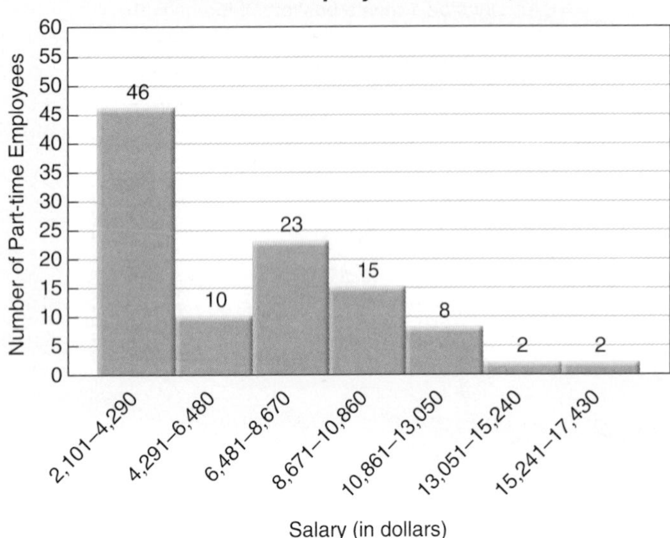

17. What range of salaries describes the most part-time employees?

18. How many part-time employees make between \$6,481 and \$13,050?

19. How many part-time employees make less than \$15,241?

20. How many part-time employees make more than \$10,860?

8.2

21. What word could help you remember the definition of mean?

22. What word could help you remember the definition of median?

23. What word could help you remember the definition of mode?

Find the mean of each data set.

24. $\{10, 5, 12, 5\}$

25. $\{8, 7, 7, 2, 1\}$

26. $\{13, 15, 12, 12, 10\}$

27. $\{18, 10, 11, 14\}$

28. The Arugala family's gas bills last winter were \$145, \$198, \$230, \$210, \$176, and \$138. Find the mean of these bills. Round your answer to the nearest hundredth of a dollar.

29. Knight Watch Security is buying new vehicles for the company. The vehicle costs are \$28,500, \$29,300, \$21,690, \$35,000, \$37,000, \$43,600, \$45,300, and \$38,600. Find the mean of the vehicle costs. Round your answer to the nearest dollar.

Find the median of each data set.

30. $\{3, 15, 11, 12, 6\}$

31. $\{8, 19, 12, 14, 24\}$

32. For marketing purposes, the ages of customers were taken as they entered Disneyland. Find the median of the following customer ages:

$$\{6, 7, 43, 40, 3, 5, 35, 35\}$$

33. A track coach timed his athletes running the 100-meter dash. Find the median of the following times:

$$\{12.1, 11.8, 13.0, 11.6, 11.2, 11.1\}$$

Find the mode of each data set.

34. $\{3, 5, 11, 5, 3\}$

35. $\{7, 16, 12, 10, 25\}$

36. The high temperatures, in degrees Fahrenheit, for Miami over the last seven days were 82°, 84°, 78°, 74°, 75°, 79°, and 87°. Find the mode of these temperatures.

37. Valencia Community College surveyed incoming freshmen and recorded their ages. The first five freshmen surveyed had the following ages: 19, 18, 17, 17, 22, and 18. Find the mode of these ages.

Find the mean, median, and mode of each data set. Round to the nearest tenth when necessary.

38. $\{25, 31, 15, 11, 12, 11\}$

39. $\{18, 19, 18, 14, 24\}$

40. The number of pizzas delivered by North Side Pizza over 7 days:

$$\{21, 16, 15, 19, 24, 13, 18\}$$

41. The number of lunch customers at Big City Subs over 8 days:

$$\{36, 46, 25, 28, 34, 20, 17, 28\}$$

Make a histogram for each data set.

42. Rudy wrote 12 checks last month. Make a histogram of the following check values using class intervals of width \$75.

{\$390, \$480, \$320, \$250, \$125, \$42, \$35, \$38, \$46, \$89, \$150, \$290}

43. During a fund-raiser, 14 students collected the following amounts. Make a histogram of these amounts using class intervals of width \$20.

{\$25, \$43, \$32, \$85, \$15, \$25, \$125, \$49, \$58, \$20, \$10, \$75, \$90, \$10}

Chapter 8 Practice Test

Use the circle graph to answer each question.

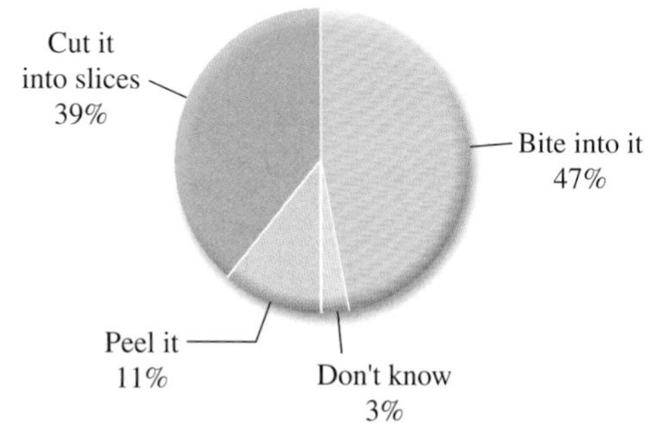

1. Based on this survey, what percent of people answered "Don't know" when asked how they eat an apple?

2. As a percent, how many more people answered "Bite into it" than "Peel it"?

3. What percent of people answered either "Cut it into slices" or "Peel it"?

4. Out of 2,000 people surveyed, how many people answered "Peel it"?

Use the bar graph to answer each question.

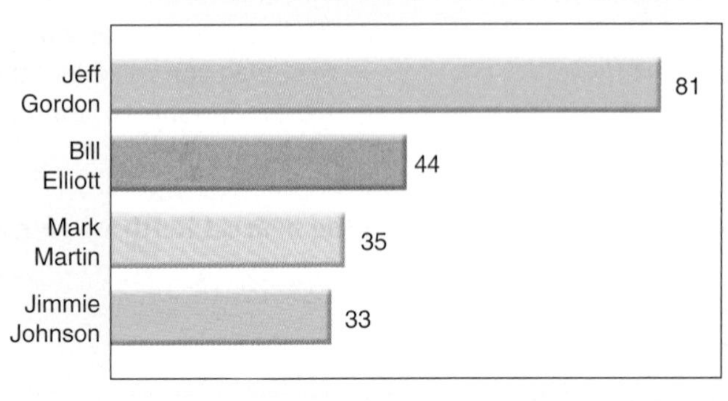

5. How many wins did Bill Elliott have in 2007?

6. How many more wins did Mark Martin have than Jimmie Johnson in 2007?

7. Write a ratio that compares Jimmie Johnson's number of wins to Jeff Gordon's number of wins.

8. Jeff Gordon's number of wins is what percent of Bill Elliott's number of wins?

Use the line graph to answer each question. Estimate the data from the graph to the nearest 50 resorts.

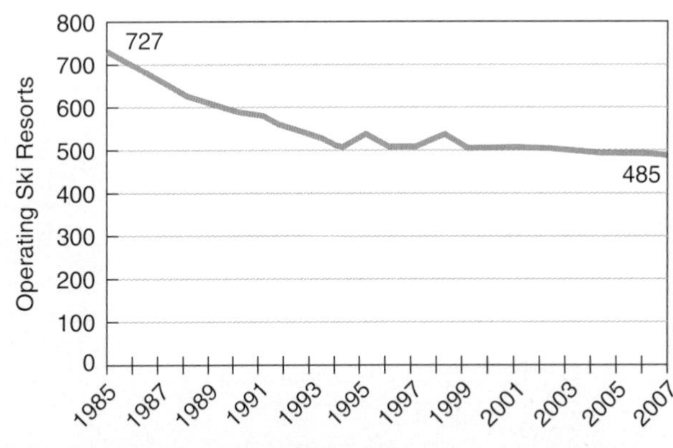

9. During the 22 years shown, in what year did the number of U.S. ski resorts in operation first drop below 700?

10. During the years shown, the number of U.S. ski resorts in operation decreased. However, there were two years when a slight increase occurred. In which two years did those increases occur?

11. About how many fewer U.S. ski resorts were in operation in 1990 than in 1985?

12. The number of U.S. ski resorts in operation in 2007 is what percent of the U.S. ski resorts in operation in 1985? Round to the nearest tenth.

Use the histogram to answer each question.

13. What range of salaries describes the most employees?

14. How many employees make between \$26,001 and \$38,000?

15. What is the size of the class intervals?

16. Which class interval has the lowest class frequency?

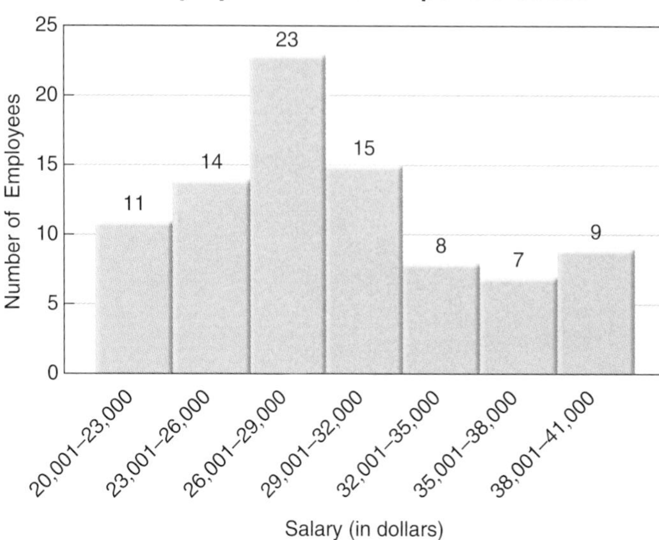

Answer each question about mean, median, and mode.

17. Terry received the following grades in her classes last semester: History, A; English, B; Mathematics, A; Psychology, A.

Using the scale A = 4, B = 3, C = 2, D = 1, and F = 0, calculate Terry's grade point average by finding the mean. Assume that all of her classes have the same number of credit hours.

18. Your study group contains 5 students. On the last test, your group members scored 74, 87, 72, 93, and 94 points.

a. Is the score 87 above or below the mean?

b. Is the score 87 above or below the median?

c. In your opinion, did the person who earned the 87 do better, worse, or about the same as the overall group? There is no right answer to this question. Support your opinion with evidence from the problem.

19. For marketing purposes, the ages of customers were taken as they entered Meadowlands Stadium in New Jersey. The ages were 16, 72, 43, 40, 13, 50, and 35 years. What was the median age?

20. A track coach timed his athletes running the 400-meter dash. The times were 52.1, 61.3, 56.0, 57.6, 53.2, and 60.1 seconds. What was the median time?

21. The high temperatures in Eugene, Oregon, were $62°, 64°, 73°, 64°, 75°, 70°$, and $73°$ Fahrenheit over the last 7 days. What is the mode of these temperatures?

22. Houston Community College surveyed incoming freshmen and recorded their ages. The ages of the first six freshmen surveyed were 19, 18, 27, 22, 38, and 18 years. What is the mode of these ages?

Find the mean, median, and mode of each data set. Round to the nearest tenth when necessary.

23. $\{15, 11, 15, 24, 4\}$

24. $\{8, 35, 18, 23, 24\}$

25. The number of cars sold at Timeless Imports was recorded for 6 days. Over these days, it sold 8, 11, 10, 0, 7, and 13 cars.

26. The number of lunch customers at Nina's Taqueria was recorded for 6 days. Over these days, there were 48, 47, 35, 38, 39, and 47 customers.

27. The following data set represents Andrea's daily green bean harvest (in pounds) over a two-week period. Make a histogram using class intervals of 5 pounds.

$\{65, 71, 78, 84, 83, 72, 61, 75, 74, 81, 90, 60, 83, 84\}$

CHAPTER

9 Signed Numbers

The signed numbers include the positive numbers, negative numbers, and zero. Numbers that are greater than zero are *positive*. Numbers that are less than zero are *negative*. Zero is neither positive nor negative.

The New York Stock Exchange (NYSE) uses signed numbers to indicate the increase or decrease in stock values. On the stock ticker, a display of −1.16% means that a stock has lost 1.16% of its value. A display of +0.16% means that a stock has gained 0.16% of its value.

In Exercises 56 and 57 in Section 9.4, signed numbers are used to calculate the change in value of a stock portfolio.

Section 9.1 Understanding Signed Numbers

For every positive number, there is a corresponding negative number. Together, positive numbers, negative numbers, and zero are called the **signed numbers.**

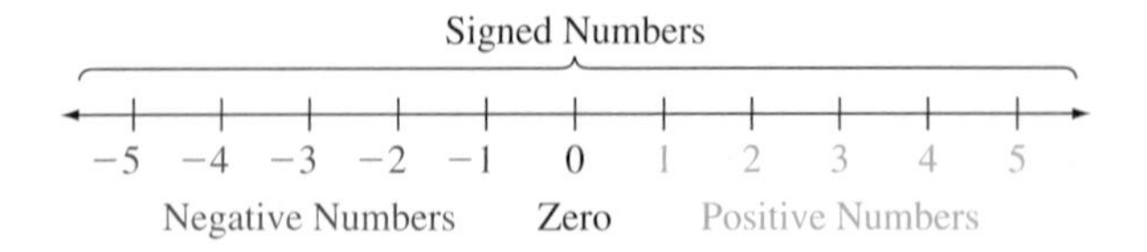

In this section, you will learn the relationship between positive and negative numbers and use them to describe real-life events.

The **Objectives** in Section 9.1 will help you

- **A** Represent real-world quantities with signed numbers.
- **B** Graph signed numbers on a number line.
- **C** Compare signed numbers.
- **D** Simplify opposites of numbers.
- **E** Find the absolute value of a number.

VOCABULARY PREVIEW *Check the box that applies.*

	Got It	Must Study
Positive Number: A number that is greater than zero is a **positive number.**		
Negative Number: A number that is less than zero is a **negative number.**		
Signed Numbers: Together, positive numbers, negative numbers, and zero are called the **signed numbers.** Signed numbers are also called *real numbers.*		
Less Than: Given two numbers, the number that is graphed to the left on a number line is **less than** the other number.		
Greater Than: Given two numbers, the number that is graphed to the right on a number line is **greater than** the other number.		
Opposites: Two numbers are **opposites** if they are the same distance from zero on a number line and are on opposite sides of zero.		
Absolute Value: The **absolute value** of a number is the number's distance from zero on a number line.		

Study these words when they appear in the text. After the section, test your understanding by completing the Vocabulary Review in the section exercises.

Objective A Represent Real-World Quantities with Signed Numbers

The Concept To describe the amount of money in a checking account, we need to give two pieces of information.

- Use a positive sign to show that an account has money or a negative sign to show that the account owes money to the bank.
- Write how many dollars are in the account or are owed to the bank.

First, we find the sign of the number; then we describe the size of the number. If an account owes \$20.32 to the bank, the balance is −\$20.32. If \$49.50 is in the account, the balance is \$49.50, or +\$49.50.

When a + sign or no sign is written in front of a number, the number is positive. When a − sign is written in front of a number, the number is negative. The following phrases are typically used to show positive or negative quantities.

KEY WORDS USED TO IDENTIFY SIGN			
Negative		**Positive**	
Move Down	Lose	Move Up	Gain
Move Left	Delete	Move Right	Add
Decrease	Descend	Increase	Ascend

Procedure Represent Real-World Quantities with Signed Numbers

Step 1 Determine if the quantity is positive or negative.

Step 2 Assign the appropriate value.

EXAMPLES

1. What signed number best describes this statement?

A plane descended 10,000 feet.

Step 1 Determine if the quantity is positive or negative.

Since a descending plane's altitude is decreasing, use a negative number.

Step 2 Assign the appropriate value.

Descending 10,000 feet can be represented with the signed number −10,000.

GUIDED PRACTICE

1. What signed number best describes this statement?

Dan deleted 317 songs from his MP3 player.

Step 1 Determine if the quantity is positive or negative.

Since deleting songs increases/decreases the amount of music on an MP3 player, use a positive/negative number.

Step 2 Assign the appropriate value.

Deleting 317 songs can be represented with the signed number ______.

2. What signed number best describes this statement? A company had a loss of $2.3 million.	**2.** What signed number best describes this statement? Carmella added 2.5 GB of RAM to her laptop.
Step 1 Determine if the quantity is positive or negative. Since a loss is a decrease, use a negative number.	**Step 1** Determine if the quantity is positive or negative. Since adding ram is a(n) ________, use a ________ number.
Step 2 Assign the appropriate value. A loss of $2.3 million can be represented with the signed number −2.3.	**Step 2** Assign the appropriate value. Adding 2.5 GB of RAM can be represented with the signed number ________.

Concept Check

A1. In your own words, how do you determine if a real-world quantity should be represented with a positive or a negative number?

A2. Represent each statement with a signed number.

a. A child lost 48 marbles to a friend.

b. A child's friend won 48 marbles.

c. Is it possible that the same event can be described with either a positive or a negative number?

d. Give an example of another event that different people could represent with either a positive or a negative number.

Objective A Practice

Use a signed number to represent each real-world quantity.

A3. The temperature was fifteen degrees below zero.

A4. The flash drive holds five hundred songs.

A5. Startled, the man backed up $3\frac{1}{2}$ feet.

A6. A carpenter cut $5\frac{1}{2}$ inches off a board.

A7. A checking account owes $25.34 to the bank.

A8. A checking account has a balance of $267.15.

A9. A teller's cash drawer had $4.56 more than the recorded receipts.

A10. A teller's cash drawer had $3.56 less than the recorded receipts.

Objective B Graph Signed Numbers on a Number Line

The Concept To graph negative numbers on a number line, we extend the number line to the left of zero.

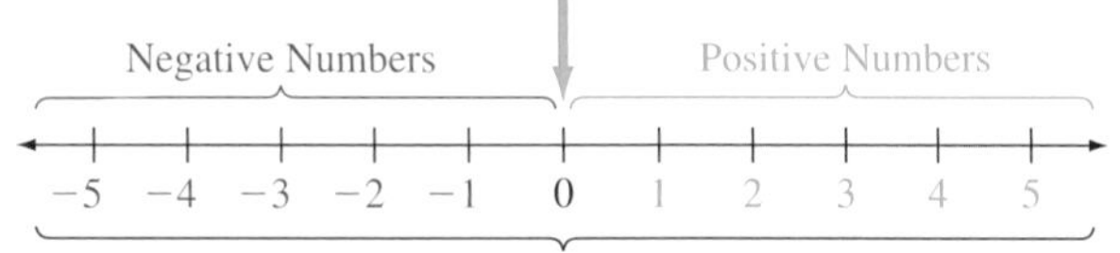

The following number line shows the graph of signed numbers −200, 0, 157, and 350.

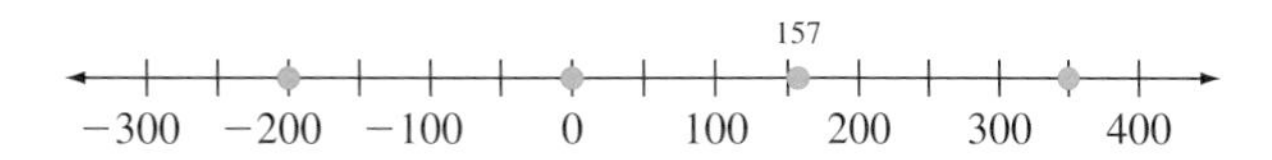

Notice the following details about this graph:

- The range of the number line extends to the left of the smallest value, −200.
- The range of the number line extends to the right of the largest value, 350.
- The tick marks are evenly spaced.
- Since 157 does not fall on a tick mark, a label is included.
- Since 350 falls on a tick mark, a label is not included.

Procedure Graph Signed Numbers on a Number Line

Step 1 Draw a number line with an appropriate range.

Step 2 Graph each signed number on the number line.

EXAMPLE

3. Graph 3, −2, 1.5, and $3\frac{1}{2}$ on a number line.

Step 1 The range of the number line goes beyond the smallest value, −2, and the largest value, $3\frac{1}{2}$.

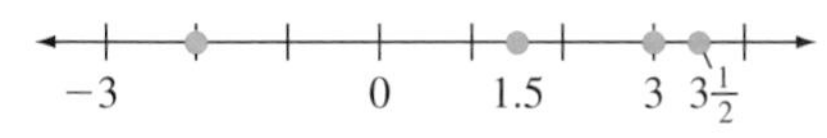

Step 2 We graph each number on the number line.

Since 1.5 and $3\frac{1}{2}$ do not fall on tick marks, we include labels for these signed numbers.

GUIDED PRACTICE

3. Graph $2\frac{1}{3}$, −1, 3, −2.5, and 0 on a number line.

Step 1 The range of the number line goes beyond the smallest value, _____, and the largest value, _____.

Step 2 We graph each number on the number line.

Make sure that every dot is on a tick mark or includes a label.

Concept Check

B1. Which signed number is neither positive nor negative?

B2. Why is it incorrect to draw a number line without any labels?

Objective B Practice

Graph each set of signed numbers on a number line.

B3. $\{3, -8, 7.2, -4\frac{1}{5}, 0, -3\}$

B4. $\{20, -18, 17.5, -16\frac{2}{3}, 35, -12\}$

B5. $\left\{2\frac{1}{2}, -2\frac{3}{4}, -1\frac{1}{2}, \frac{3}{4}, -1, 1\right\}$

B6. $\left\{-3\frac{1}{2}, \frac{3}{4}, 1\frac{1}{2}, -1\frac{1}{4}, 2\frac{3}{4}, 2\right\}$

B7. $\{-0.079, 0.7, -0.07, 0.79, -0.008, 0.8\}$

B8. $\{0.056, -0.65, 0.05, -0.6, -0.068, 0.1\}$

Objective C Compare Signed Numbers

The Concept Visualizing a number line is a good way to understand the relative size of a signed number. In Section 1.1, you learned that when two numbers are graphed on a number line, the number to the right of the other is the greater number.

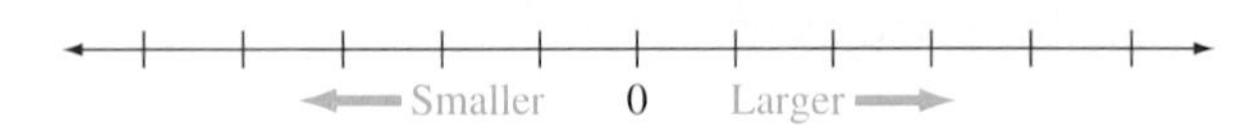

Which number is greater, −8 or 2? To order these signed numbers, graph them on a number line. The larger number will be on the right.

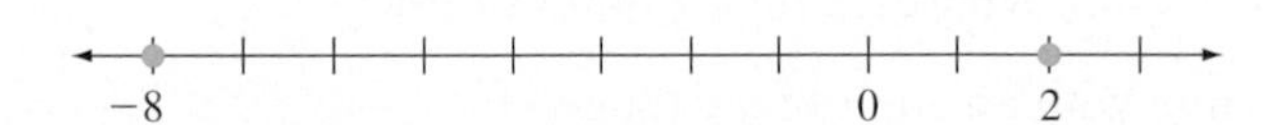

Since 2 is to the right of −8 on the number line, 2 is greater than −8. This can be written in two ways.

$2 > -8$	$-8 < 2$
2 is greater than −8.	−8 is less than 2.

We can also use money to help us understand why 2 is greater than −8. Adding \$10 to a debt of \$8 increases the balance from −\$8 to \$2. \$2 is greater than −\$8.

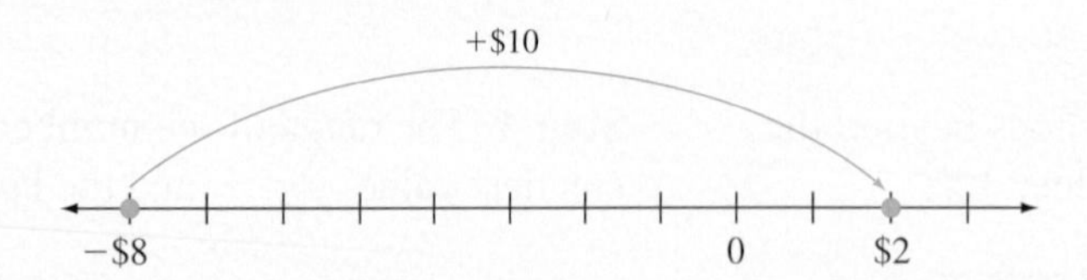

Be sure to draw number lines as you practice this material. After a short while, you will be able to "see" number lines in your head.

Procedure Compare Signed Numbers

Step 1 Visualize each number on a number line.

Step 2 Insert an inequality symbol that opens toward the larger number.

EXAMPLES

4. Insert $<$ or $>$ to make a true statement.

$$-5 \square -7$$

Step 1 Visualize each number on a number line.

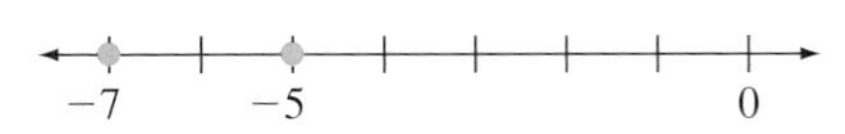

Step 2 Insert an inequality symbol that opens toward the larger number.

$$-5 \boxed{>} -7$$

-5 is greater than -7.

5. Insert $<$ or $>$ to make a true statement.

$$-\frac{3}{10} \square -\frac{1}{5}$$

To make the graph, it helps to have a common denominator.

$$\frac{3}{10} = \frac{3}{10} \qquad \frac{1}{5} \cdot \frac{2}{2} = \frac{2}{10}$$

Step 1 Visualize each number on a number line.

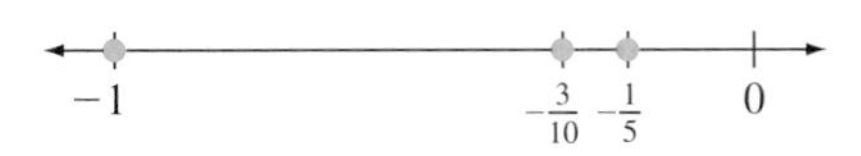

Step 2 Insert an inequality symbol that opens toward the larger number.

$$-\frac{3}{10} \boxed{<} -\frac{1}{5}$$

$-\frac{3}{10}$ is less than $-\frac{1}{5}$.

GUIDED PRACTICE

4. Insert $<$ or $>$ to make a true statement.

$$-9 \square -12$$

Step 1 Visualize each number on a number line.

Step 2 Insert an inequality symbol that opens toward the larger number.

$$-9 \square -12$$

____ is greater/less than ____.

5. Insert $<$ or $>$ to make a true statement.

$$-\frac{5}{12} \square -\frac{1}{4}$$

To make the graph, it helps to have a common denominator.

$$\frac{5}{12} = \frac{5}{12} \qquad \frac{1}{4} \cdot \frac{\quad}{\quad} = \frac{\quad}{12}$$

Step 1 Visualize each number on a number line.

Step 2 Insert an inequality symbol that opens toward the larger number.

$$-\frac{5}{12} \square -\frac{1}{4}$$

$-\frac{5}{12}$ is less/greater than $-\frac{1}{4}$.

Concept Check

C1. In your own words, explain why a negative number is always less than a positive number.

C2. In your own words, explain why zero is always greater than a negative number.

Objective C Practice

Insert < or > between each pair of signed numbers to make a true statement.

C3. $-5 \square 2$ **C4.** $-13 \square 8$ **C5.** $10 \square -12$ **C6.** $3 \square -15$

C7. $-0.054 \square -0.05$ **C8.** $0.093 \square 0.09$ **C9.** $-\frac{7}{16} \square -\frac{3}{8}$ **C10.** $\frac{5}{3} \square \frac{14}{9}$

Objective D Simplify Opposites of Numbers

The Concept It is often necessary to find the opposite of a number. Two numbers that are the same distance from zero on a number line but are on opposite sides of zero are called **opposites.** For example, 3 and −3 are opposites.

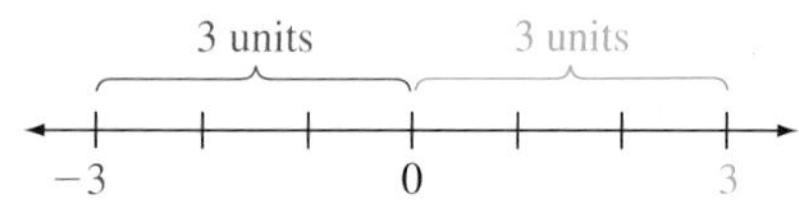

To find the opposite of a number, change the sign of the number.

- The opposite of a positive number is negative.
- The opposite of a negative number is positive.

When simplifying an opposite, it helps to read the phrase in words.

−(7) reads as "the opposite of seven."

−(−3) reads as "the opposite of negative 3."

Procedure Simplify Opposites

Step 1 Write the math phrase in words.

Step 2 Simplify.

EXAMPLE	GUIDED PRACTICE
6. Write −(6) in words and simplify.	**6.** Write −(−4) in words and simplify.
Step 1 Write −(6) in words. −(6) reads as "the opposite of six." **Step 2** Simplify. $-(6) = -6$	**Step 1** Write −(−4) in words. −(−4) reads as "the ________ of ________." **Step 2** Simplify. $-(-4) =$

Concept Check

D1. Write each math phrase in words and simplify.

a. $-(-5)$ reads as "   "; so $-(-5) =$ ______.

b. $-(2)$ reads as "   "; so $-(2) =$ ______.

c. $-\left(\frac{1}{2}\right)$ reads as "   "; so $-\left(\frac{1}{2}\right) =$ ______.

Objective D Practice

Simplify.

D2. $-(-2)$ **D3.** $-(+4)$ **D4.** $-(+5)$ **D5.** $-(-6)$

D6. $-(11)$ **D7.** $-(-14)$ **D8.** $-(-13.5)$ **D9.** $-(2.59)$

Objective E Find the Absolute Value of a Number

The Concept Vertical bars appearing on both sides of a number indicate the *absolute value* of that number. In words,

$|-5|$ reads as "the absolute value of negative five."

$|16|$ reads as "the absolute value of sixteen."

INTERACTIVE DEFINITION Absolute Value

The **absolute value** of a number is the number's distance from zero on a number line.
To *evaluate* the absolute value of a number, find the number's distance from zero on a number line.

EXAMPLE

7. Evaluate $|-3|$ (the absolute value of negative three).

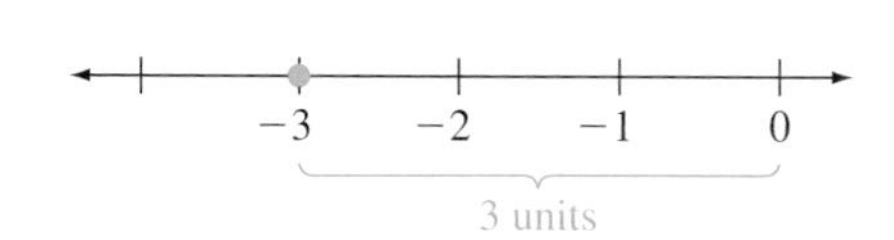

Since -3 is 3 units from zero, $|-3| = 3$.

Since the absolute value is a measure of distance, the answer is never negative.

GUIDED PRACTICE

7. Evaluate $|4|$ (the absolute value of four).

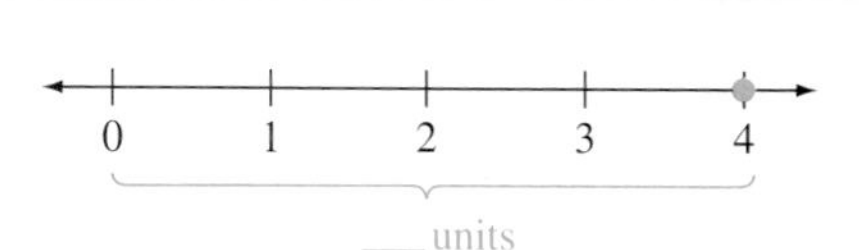

Since 4 is ______ units from zero, $|4| =$ ______.

Since the absolute value is a measure of ________, the answer is never ________.

DO YOU UNDERSTAND how to identify and evaluate the absolute value of a number? Got It Get Help

▶ **NOTE** If someone mentions a distance such as -3 meters, it means two things. First, the distance is 3 meters. Second, the distance is measured in the negative direction.

Procedure Find the Absolute Value of a Number

Determine the number's distance from zero.

EXAMPLES

8. Write $|-5|$ in words. Then evaluate.

$|-5|$ reads as "the absolute value of negative five."

$|-5| = 5$ because -5 is 5 units from zero.

9. Write $\left|\frac{5}{3}\right|$ in words. Then evaluate.

$\left|\frac{5}{3}\right|$ reads as "the absolute value of five-thirds."

$\left|\frac{5}{3}\right| = \frac{5}{3}$ because $\frac{5}{3}$ is $\frac{5}{3}$ units from zero.

10. Evaluate $|0|$.

$|0| = 0$ because 0 is 0 units from zero.

GUIDED PRACTICE

8. Write $|-12|$ in words. Then evaluate.

$|-12|$ reads as "________________."

$|-12| =$ ______ because -12 is ______ units from zero.

9. Write $|45|$ in words. Then evaluate.

$|45|$ reads as "________________."

$|45| =$ ______ because 45 is ______ units from zero.

10. Evaluate $\left|-\frac{2}{3}\right|$.

$\left|-\frac{2}{3}\right| = \frac{\quad}{\quad}$ because $-\frac{2}{3}$ is $\frac{\quad}{\quad}$ unit from zero.

Important! $|10|$ and (10) mean very different things even though they are equal in value. An absolute value symbol indicates an operation. Parentheses do not indicate an operation. A set of parentheses is used to organize information. To understand the difference between $-|-2|$ and $-(-2)$, write each expression in words.

INTERACTIVE DEFINITION Absolute Value and Parentheses

Absolute value is an operation.
Parentheses are not an operation. They are used to organize information.

EXAMPLES

11. Write $-|-2|$ in words. Then evaluate.

$-|-2|$ reads as "the opposite of the absolute value of negative two."

Evaluate the absolute value. $-|-2| = -(2)$

Find the opposite. $= -2$

Since we had to evaluate the absolute value before finding the opposite, this exercise required 2 steps.

GUIDED PRACTICE

11. Write $-|-6|$ in words. Then evaluate.

$-|-6|$ reads as "________________
________."

Evaluate the absolute value. $-|-6| = -(\quad)$

Find the opposite. $=$

Since we had to evaluate the absolute value before finding the opposite, this exercise required ______ steps.

12. Write $-(-2)$ in words. Then evaluate.

$-(-2)$ reads as "the opposite of negative two."

Find the opposite. $-(-2) = 2$

Since a grouping is not an operation, we find the opposite of negative two in one step.

12. Write $-(-6)$ in words. Then evaluate.

$-(-6)$ reads as "______________________."

Find the opposite. $-(-6) =$ ______

Since a grouping is not an ____________, we find the opposite of negative six in ______ step.

DO YOU UNDERSTAND the difference between absolute value and parentheses? Got It | Get Help

Concept Check

E1. In your own words, how does the absolute value of a number relate to the number's position on a number line?

E2. Why can't the absolute value of a number be negative?

For questions E3–E6, do the following.
a. Determine how many steps are required. See Examples 11 and 12.
b. Evaluate.

E3. $-|7| =$ **E4.** $-|-3| =$ **E5.** $-(-7) =$ **E6.** $-(3) =$

Objective E Practice

Evaluate.

E7. $|-5|$ **E8.** $|16|$ **E9.** $\left|\frac{7}{9}\right|$ **E10.** $\left|\frac{-1}{10}\right|$

E11. $-(-5)$ **E12.** $-(-31)$ **E13.** $-|-42|$ **E14.** $-|-15|$

Combining Concepts and Applications

CONCEPT I Ordering Signed Numbers To order signed numbers, it helps to visualize them on a number line.

Procedure Order Signed Numbers

Step 1 Simplify each number.
Step 2 Visualize the numbers on a number line.
Step 3 Order the original numbers.

EXAMPLE

13. Order the numbers from least to greatest.

$|-5|, -3, -(-8), 0, -|-2|$

Step 1 Simplify. $|-5| = 5$ $-3 = -3$ $-(-8) = 8$ $0 = 0$ $-|-2| = -2$

Step 2 Graph.

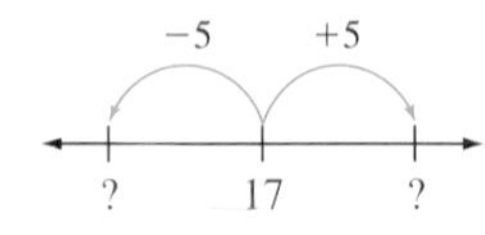

Step 3 Order. $-3, -|-2|, 0, |-5|, -(-8)$

GUIDED PRACTICE

13. Order the numbers from least to greatest.

$|-6|, -1, -(-4), 0, -|-4|$

$|-6| =$ $-1 =$ $-(-4) =$ $0 =$ $-|-4| =$

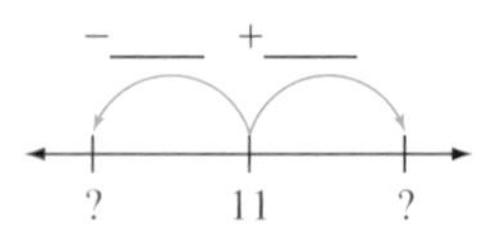

CONCEPT II Getting Ready for Algebra The following Example and Guided Practice use concepts that you may encounter in an algebra course.

EXAMPLE

14. What two numbers are 5 units from 17?

Step 1 Visualize the two numbers on the number line.

-5 $+5$

? 17 ?

The two numbers are 5 units to the left and right of 17.

Step 2 Use addition and subtraction to find the numbers.

$17 - 5 = 12$

$17 + 5 = 22$

Step 3 Check your answers.

17 and 12 are $17 - 12 = 5$ units apart.
17 and 22 are $22 - 17 = 5$ units apart.

Both 12 and 22 are 5 units from 17.

GUIDED PRACTICE

14. What two numbers are 7 units from 11?

$-$___ $+$___

? 11 ?

The two numbers are ____ units to the left and right of ____.

$11 - (\quad) =$

$11 + (\quad) =$

____ and ____ are $(\quad) - (\quad) = (\quad)$ units apart.
____ and ____ are $(\quad) - (\quad) = (\quad)$ units apart.

Both ____ and ____ are ____ units from ____.

Section 9.1 Exercises

FOR EXTRA HELP

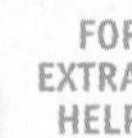

To represent real-world quantities with signed numbers:

1. Answer the Objective A Concept Checks.
2. Answer the odd Objective A Practice Exercises.
3. Answer the even Objective A Practice Exercises.

To graph signed numbers on a number line:

4. Answer the Objective B Concept Checks.
5. Answer the odd Objective B Practice Exercises.
6. Answer the even Objective B Practice Exercises.

To compare signed numbers:

7. Answer the Objective C Concept Checks.
8. Answer the odd Objective C Practice Exercises.
9. Answer the even Objective C Practice Exercises.

To simplify opposites of numbers:

10. Answer the Objective D Concept Check.
11. Answer the odd Objective D Practice Exercises.
12. Answer the even Objective D Practice Exercises.

To find the absolute value of a number:

13. Answer the Objective E Concept Checks.
14. Answer the odd Objective E Practice Exercises.
15. Answer the even Objective E Practice Exercises.

VOCABULARY REVIEW *Review the Vocabulary Preview for Section 9.1. Study the definitions until you can check* Got It *for every word.*

Use these words to complete each sentence.

positive number • negative number • signed numbers • less than • greater than • opposites • absolute value

	Got it	Get help
16. When two numbers are graphed on the same number line, the number that is graphed to the right is ____________ the number to the left.		
17. Two numbers that are the same distance from zero are called ____________.		
18. A number's distance from zero is its ____________.		
19. A number that is less than zero is called a(n) ____________.		
20. ____________ include all numbers, whether positive, negative, or zero.		

How will you get help for any vocabulary that you are unsure about?

Your instructor ____ MyMathLab ____ A classmate ____ A tutor ____ Other ____

Represent each real-world quantity with a signed number.

21. Randy earned 89 points on an exam.

22. A parachutist opened her parachute 1,235 feet above the ground.

23. The car's speed decreased by 15 mph.

24. The hiker descended 1,389 feet into a valley.

25. The number of DVDs available at Sunni's Video Mart has increased by 1,453 in the last six months.

26. The number of videocassettes available at Sunni's Video Mart has decreased by 892 in the last six months.

27. During the Ice Bowl, a 1967 football game between the Packers and the Cowboys, the temperature was thirteen degrees below zero.

28. The windchill during the Ice Bowl was forty degrees below zero.

29. In the last year, we have all become one year older.

30. The average speed of a new computer has increased by 1.8 Ghz in the last two years.

Graph each set of signed numbers on a number line.

31. $\{-5, 8, -12, 10, 3, -1\}$

32. $\{3, -4, 2, -8, -1, 5\}$

33. $\{-120, 80, -60, 60, -53, 5\}$

34. $\{-50, 250, -100, 125, -45, 12\}$

35. $\left\{5, -3, 8, -1\frac{1}{2}, -4\right\}$

36. $\{-8, 2, 0, -4.3, 5\}$

37. $\left\{\frac{6}{5}, 1.8, -3, \frac{-8}{5}, 4, -\frac{1}{5}\right\}$

38. $\left\{-\frac{3}{4}, 2.75, -5, -1.5, 3, \frac{4}{4}\right\}$

Insert < or > between each pair of signed numbers to make a true statement.

39. $-8 \square 3$

40. $15 \square -8$

41. $-6 \square -3$

42. $-12 \square -100$

43. $-6 \square -18$

44. $6 \square 18$

45. $42 \square -50$

46. $50 \square -42$

47. $-\frac{1}{3} \square -\frac{1}{4}$

48. $-\frac{1}{3} \square -\frac{1}{2}$

49. $-4.18 \square -4.2$

50. $-6.4 \square -6.399$

51. In your own words, why is every negative number less than every positive number? Be sure to discuss the number line in your answer.

52. In your own words, why is every positive number greater than every negative number? Be sure to discuss the number line in your answer.

53. What number is greater than every negative number and less than every positive number? Be sure to explain your answer.

54. In your own words, what is the procedure to find the absolute value of a number?

Write each absolute value expression in words and then evaluate.

55. a. $|-12|$ reads as "________________________."
b. $|-12| =$

56. a. $|50|$ reads as "________________________."
b. $|50| =$

57. a. $-(-3)$ reads as "________________________."
b. $-(-3) =$

58. a. $-(4)$ reads as "________________________."
b. $-(4) =$

59. **a.** $-(7)$ reads as "____________________."
b. $-(7) =$

60. **a.** $-(-9)$ reads as "____________________."
b. $-(-9) =$

61. **a.** $-|-10|$ reads as "____________________."
b. $-|-10| =$

62. **a.** $-|-11|$ reads as "____________________."
b. $-|-11| =$

Simplify.

63. $|-8|$

64. $|12|$

65. (-5)

66. (13)

67. $-(-7)$

68. $-(7)$

69. $-|-13|$

70. $-|62|$

71. $-(91)$

72. $-(-91)$

73. $-|75|$

74. $-|-100|$

75. What numbers are equal to their own absolute value?

76. In each expression below, n can be any signed number except zero.
a. Is $|n|$ equal to a positive or negative number?
b. Is $-|n|$ equal to a positive or negative number?

77. What is the only number that is neither positive nor negative?

Insert <, =, or > between each pair of signed numbers to make a true statement.

78. $|-33| \square (25)$

79. $|25| \square (-33)$

80. $|-16| \square |16|$

81. $|-45| \square |-100|$

82. $-|1| \square -|0|$

83. $-|2| \square -|-8|$

84. $-(-15) \square (16)$

85. $-(-2) \square -(3)$

Order the numbers from least to greatest.

86. $|-6|, -(6), -|-7|, |-7|$

87. $|-10|, -(10), -|-4|, |-4|$

88. $-(-9), -|-9|, -(8), |-8|$

89. $-(-12), -|-12|, -(15), |-15|$

90. What two numbers are 4 units from 7?

91. What two numbers are 5 units from -3?

92. What two numbers are 7 units from 4?

93. What two numbers are 3 units from -5?

94. Solve the equation $|n| = 4$ by finding the two numbers whose absolute value is 4.

95. Solve the equation $|n| = 7$ by finding the two numbers whose absolute value is 7.

smith Do exercises 69 and 70 in two steps. Find the absolute value of the number, and then write the opposite of this absolute value.
For more tips and tweets, go to twitter.com/gstbasicmath

SELF-ASSESSMENT

	YES	NO
1. On your last test, were you satisfied with your score? *If you answered yes, good job. Do not continue here.* *If you answered no, answer the following.*		
2. Did you preview each section of the textbook before your instructor taught it in class?		
3. Did you complete all of the homework before it was due?		
4. Did you get help with the homework exercises that you didn't understand?		

5. Set two goals that will help you become better prepared for your next test.

Goal 1: ____________________

Goal 2: ____________________

Section 9.1 Question Log

Use this space to write questions. Be sure to get these answered and revisit them when you prepare for your exam.

Page ____ **Answered** ☐	**Page** ____ **Answered** ☐
Page ____ **Answered** ☐	**Page** ____ **Answered** ☐

Section 9.2 Adding and Subtracting Signed Numbers

Positive and negative numbers occur frequently in the real world. Knowing how to add and subtract signed numbers will help you balance your checkbook, understand the news, and work in a world where signed numbers occur everywhere.

The **Objectives** in Section 9.2 will help you

- A Add numbers with the same sign.
- B Add numbers with opposite signs.
- C Subtract signed numbers.
- D Add and subtract signed numbers using a number line.

VOCABULARY PREVIEW *Check the box that applies.*	Got It	Must Study
Zero Pair: Two numbers that add to zero make a **zero pair.** A zero pair consists of any number and its opposite.		
Additive Inverse: A number's **additive inverse** is its opposite.		
Commutative Property of Addition: The **commutative property of addition** lets you reorder addition exercises. $2 + 3 = 3 + 2$		
Associative Property of Addition: The **associative property of addition** lets you regroup addition exercises. $(5 + 12) + 3 = 5 + (12 + 3)$		

Study these words when they appear in the text. After the section, test your understanding by completing the Vocabulary Review in the section exercises.

Objective A Add Numbers with the Same Sign

The Concept When finding sums such as $(-3) + (-2)$, we add numbers with the same sign. To find this sum, it is useful to visualize *signed discs.*

INTERACTIVE DEFINITION Adding Numbers with the Same Sign

A sum of two numbers can be visualized using signed discs.

EXAMPLE

1. Find the sum. $(-3) + (-2)$

Draw 3 negative discs for -3 and 2 negative discs for -2.

$(-3) + (-2)$

(−1) (−1)
(−1) (−1)
(−1)

Together, there are a total of 5 negative discs.

$(-3) + (-2) = -5$

GUIDED PRACTICE

1. Find the sum. $(-4) + (-2)$

Draw ________ discs for -4 and ________ discs for -2.

$(-4) + (-2)$

Together, there are a total of ______ negative discs.

$(-4) + (-2) =$

DO YOU UNDERSTAND how to use signed discs to add numbers with the same sign? Got It Get Help

This visualization demonstrates a formal procedure for adding two numbers with the same sign. When both numbers have the same sign, the sum is the total number of discs with that sign.

Procedure Add Numbers with the Same Sign

Step 1 Add the absolute values of the numbers.

Step 2 Notice that the sign of the answer matches the sign of the original numbers.

▶ **NOTE** First, visualize the addition using signed discs. It will help you understand the steps in the procedure.

When adding large numbers, it is not convenient to draw many discs. Instead, visualize the discs to count the total number.

EXAMPLES

2. Add. $(-12) + (-17)$

Visualization:

$(-12) + (-17) = -29$

12 negative discs + 17 negative discs = 29 negative discs

GUIDED PRACTICE

2. Add. $(-34) + (-21)$

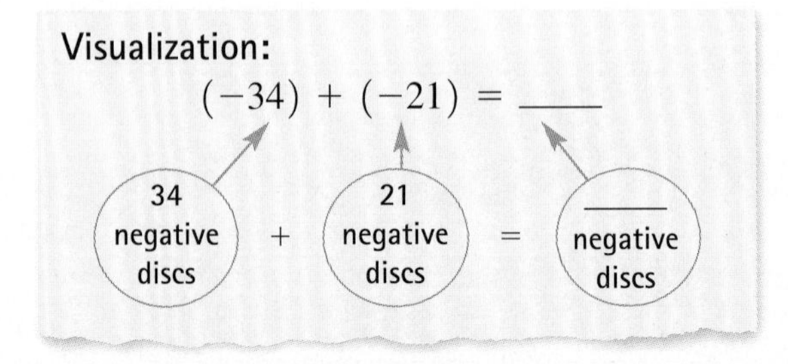

Step 1 Add the absolute values of the numbers.

$$|(-12)| = 12 \quad \text{and} \quad |(-17)| = 17$$
$$12 + 17 = 29$$

Step 2 The sign of the answer matches the sign of the original numbers.

$$(-12) + (-17) = -29$$

When adding two negative numbers, the answer is negative.

3. Add. $\frac{-9}{30} + \frac{-8}{30}$

Because we have like fractions, we can add the numerators.

Visualization:

Add the numerators.

9 negative discs + 8 negative discs = 17 negative discs

$$\frac{-9}{30} + \frac{-8}{30} = \frac{-17}{30}$$

Since both numbers are negative, the answer is negative.

Step 1 Add the absolute values of the numbers.

$$\left|\frac{-9}{30}\right| = \frac{9}{30} \quad \text{and} \quad \left|\frac{-8}{30}\right| = \frac{8}{30}$$
$$\frac{9}{30} + \frac{8}{30} = \frac{17}{30}$$

Step 2 The sign of the answer matches the sign of the original numbers.

$$\frac{-9}{30} + \frac{-8}{30} = \frac{-17}{30}$$

Step 1 Add the absolute values of the numbers.

$$|-34| = \_\_\_\_ \qquad |-21| = \_\_\_\_$$
$$\_\_\_\_ + \_\_\_\_ = \_\_\_\_$$

Step 2 The sign of the answer matches the sign of the original numbers.

$$(-34) + (-21) = \_\_\_\_$$

When adding two ________ numbers, the answer is ________.

3. Add. $\frac{-3}{18} + \frac{-4}{18}$

Because we have like fractions, we can add the numerators.

Visualization:

Add the numerators.

3 negative discs + 4 negative discs = ____ negative discs

$$\frac{-3}{18} + \frac{-4}{18} = \frac{\_\_\_\_}{18}$$

Since both numbers are negative, the answer is ________.

Step 1 Add the absolute values of the numbers.

$$\left|\frac{-3}{18}\right| = \frac{\_\_}{\_\_} \quad \text{and} \quad \left|\frac{-4}{18}\right| = \frac{\_\_}{\_\_}$$
$$\frac{\_\_}{\_\_} + \frac{\_\_}{\_\_} = \frac{\_\_}{\_\_}$$

Step 2 The sign of the answer matches the sign of the original numbers.

$$\frac{-3}{18} + \frac{-4}{18} = \frac{\_\_}{\_\_}$$

Concept Check

A1. Explain why each of the following statements is true or false.

a. When adding two positive numbers, the answer will be positive.

b. When adding two negative numbers, the answer will be negative.

Objective A Practice

Add.

A2. $(-8) + (-9)$ **A3.** $(-5) + (-6)$ **A4.** $-12 + -16$ **A5.** $-19 + -8$

A6. $(-17.3) + (-3.2)$ **A7.** $(-8.7) + (-9.5)$ **A8.** $\left(\frac{-7}{28}\right) + \left(\frac{-12}{28}\right)$ **A9.** $\left(\frac{-3}{10}\right) + \left(\frac{-4}{10}\right)$

Objective B Add Numbers with Opposite Signs

The Concept To add numbers with opposite signs, such as $(3) + (-2)$, we can visualize the numbers as signed discs. When a positive disc and a negative disc are added together, the result is zero.

A Positive Disc Added to a Negative Disc Is Zero.

(+1)(−1) = 0

INTERACTIVE DEFINITION Adding Numbers with Opposite Signs

When **adding numbers with opposite signs,** you can visualize the sum using signed discs. A positive disc and a negative disc add to zero.

EXAMPLE

4. Add. $3 + (-2)$

Draw 3 positive discs and 2 negative discs. $3 + (-2)$

(+1)(−1) = 0
(+1)(−1) = 0
(+1)

(1) + (−1) = 0, so two pairs of discs are removed.

One positive disc remains. $3 + (-2) = 1$

GUIDED PRACTICE

4. Add. $(-4) + 2$

Draw ____ negative discs and ____ positive discs. $(-4) + 2$

(1) + (−1) = 0, so ____ pairs of discs are removed.

____ positive/negative disc(s) remain(s). $(-4) + 2 =$

DO YOU UNDERSTAND how to use signed discs to add numbers with opposite signs? Got It Get Help

The visualization above uses the concept of *zero pairs*. A **zero pair** consists of two numbers that are opposites. The "zero" comes from the fact that the two numbers add to zero.

INTERACTIVE DEFINITION Identifying Zero Pairs

Two numbers that add to zero make a **zero pair.** A zero pair consists of any number and its opposite.

EXAMPLE

5. Identify the zero pairs.

Zero Pair	Reason
1 and −1	$1 + (-1) = 0$
3 and −3	$3 + (-3) = 0$
352 and −352	$352 + (-352) = 0$

GUIDED PRACTICE

5. Identify the zero pairs.

Zero Pair	Reason
4 and ____	4 + (____) = 0
____ and −12	____ + (____) = 0
231 and ____	____ + (____) = 0

DO YOU UNDERSTAND how to identify zero pairs? Got It Get Help

DETAILED EXAMPLE Visualizing to Add Numbers with Opposite Signs

Add. $(15) + (-20)$

Form the largest zero pair possible.

$(15) + (-20) = ?$

(15) + (−15) =0
(−?)

- All 15 positives are used to form the zero pair.
- Only 15 out of the 20 negatives are used to form the zero pair.
- Since there are more negatives than positives, the answer is negative.

The answer is the remaining signed discs.

$(15) + (-20) = -5$

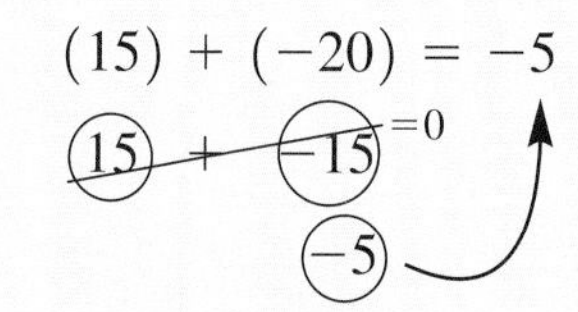

Since 5 negatives remain, the answer is −5.

EXAMPLE

6. Add using signed discs. $(17) + (-9)$

The largest zero pair is 9 and −9.

The answer will be positive since there are more positives than negatives.

Visualization:

$(17) + (-9) = 8$

(+9) + (−9) =0
(+8)

Subtract 17 − 9 = 8 to find how many positives remain.

GUIDED PRACTICE

6. Add using signed discs. $(-14) + (8)$

The largest zero pair is ____ and ____.

The answer will be positive/negative since there are more ______ than ______.

Visualization:

$(-14) + (8) =$

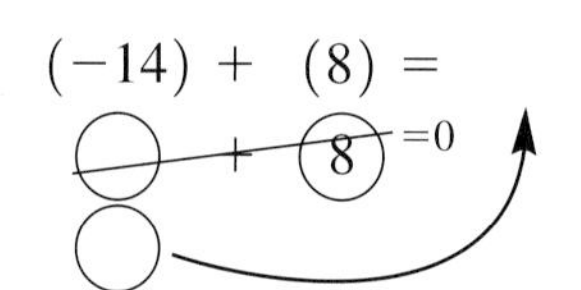

Subtract 14 − 8 = ____ to find how many positives/negatives remain.

Procedure Add Numbers with Opposite Signs

Step 1 Find the difference between the absolute values of the numbers.

Step 2 Use the sign of the number with the larger absolute value.

▶ **NOTE** Visualizing the addition of signed numbers will help you understand the steps in the formal procedures.

EXAMPLES

7. Add. $(23) + (-32)$

Visualization:

$(23) + (-32) = -9$

$(+23) + (-23) = 0$

(-9)

Subtract 32 − 23 = 9 to find how many negatives remain.

Step 1 Find the difference between the absolute values of the numbers.

$$|(23)| = 23 \qquad |(-32)| = 32$$

$$32 - 23 = 9$$

Step 2 Use the sign of the number with the larger absolute value.

$$(23) + (-32) = -9$$

Since $|-32|$ is larger than $|23|$, the answer is negative.

8. Add. $\frac{1}{3} + \left(\frac{-11}{12}\right)$

To add, build like fractions.

$$\frac{1}{3} + \left(\frac{-11}{12}\right) = \frac{1}{3}\left(\frac{4}{4}\right) + \left(\frac{-11}{12}\right)$$

$$= \frac{4}{12} + \left(\frac{-11}{12}\right)$$

GUIDED PRACTICE

7. Add. $(-14) + (30)$

Visualization:

$(-14) + (30) =$

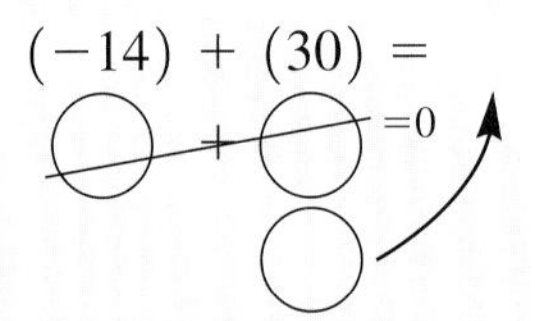

Subtract 30 − 14 = ____ to find how many positives/negatives remain.

Step 1 Find the difference between the absolute values of the numbers.

$$|(-14)| = \_\_\_\_ \qquad |(30)| = \_\_\_\_$$

$$\_\_\_\_ - \_\_\_\_ = \_\_\_\_$$

Step 2 Use the sign of the number with the larger absolute value.

$$(-14) + (30) = \_\_\_\_$$

Since |_____| is larger than |_____|, the answer is positive/negative.

8. Add. $\frac{-4}{15} + \frac{1}{3}$

To add, build like fractions.

$$\frac{-4}{15} + \frac{1}{3} = \frac{-4}{15} + \frac{1}{3}\left(\frac{\quad}{\quad}\right)$$

$$= \frac{-4}{15} + \frac{\quad}{\quad}$$

Visualization:

$$\left(\frac{4}{12}\right) + \left(\frac{-11}{12}\right) = -\frac{7}{12}$$

Subtract 11 − 4 = 7 to find how many negative discs remain.

Step 1 Find the difference between the absolute values of the numbers.

$$\left|\frac{4}{12}\right| = \frac{4}{12} \qquad \left|\frac{-11}{12}\right| = \frac{11}{12}$$

$$\frac{11}{12} - \frac{4}{12} = \frac{7}{12}$$

Step 2 Use the sign of the number with the larger absolute value.

$$\frac{4}{12} + \left(\frac{-11}{12}\right) = -\frac{7}{12}$$

Since $\left|\frac{-11}{12}\right|$ is larger than $\left|\frac{4}{12}\right|$, the answer is negative.

Visualization:

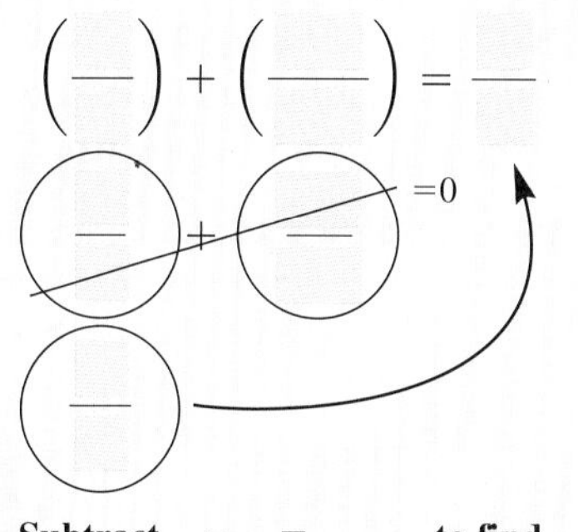

Subtract – = ____ to find how many positive/negative discs remain.

Step 1 Find the difference between the absolute values of the numbers.

$$\left|\frac{-4}{15}\right| = \frac{\quad}{\quad} \qquad \left|\frac{5}{15}\right| = \frac{\quad}{\quad}$$

$$\frac{\quad}{\quad} - \frac{\quad}{\quad} = \frac{\quad}{\quad}$$

Step 2 Use the sign of the number with the larger absolute value.

$$\frac{-4}{15} + \frac{5}{15} = \frac{\quad}{\quad}$$

Since $\left|\frac{\quad}{\quad}\right|$ is larger than $\left|\frac{\quad}{\quad}\right|$, the answer is positive/negative.

Concept Check

B1. When numbers with opposite signs are added, the sum will have the same sign as the number with the larger absolute value. In your own words, why is this true?

B2. When adding numbers with opposite signs, you must subtract the absolute values of the numbers. In your own words, why is this done?

Objective B Practice

Draw positive and negative discs to show a visualization for each sum.

B3. $(-6) + (3)$

B4. $(5) + (-2)$

Add.

B5. $(-11) + (21)$

B6. $(50) + (-43)$

B7. $(-19) + (15)$

B8. $(-29) + (16)$

B9. $\left(\frac{4}{5}\right) + \left(\frac{-7}{10}\right)$

B10. $\left(\frac{-3}{8}\right) + \left(\frac{1}{2}\right)$

B11. $(4.9) + (-1.5)$

B12. $(-15.3) + (21.8)$

FOCUS ON Personal Finance

Alfonso has been trying to keep up with his finances. On January 10, he was in a hurry and did not calculate his balance. His account became overdrawn, resulting in a debt. Balance his checkbook to make sure that Alfonso's balance is the same as the bank's calculation. Don't forget to check for other bank fees or fines.

Alfonso's Check Ledger:

Trans Type	Date	Description of Transaction	Payment/ Debit (–)	Fee (if any)	Deposit/ Credit (+)	Balance
	1-Jan	Balance forward				$512.73
Check 343	1-Jan	Rent	$375.00			$137.73
ATM	5-Jan	Needed cash	$20.00	$3.00		$114.73
Check 344		VOID				$114.73
Check 345	8-Jan	Groceries	$78.95			$35.78
ATM	8-Jan	Needed cash	$20.00	$3.00		$12.78
	8-Jan	Paycheck			$438.24	$451.02
Check 346	10-Jan	Credit card bill	$100.00			
Check 348	10-Jan	Cable	$59.99			
Check 349	10-Jan	Phone	$48.73			
ATM	11-Jan	Needed cash	$20.00	$3.00		
ATM	13-Jan	Needed cash	$20.00	$3.00		
Check 350	15-Jan	Car payment	$185.00			
Check 351	16-Jan	Owed brother money	$150.00			
		Tried to make ATM withdrawal				
	18-Jan	Bank says insufficient funds!				

The Bank's Ledger:

Starting Balance	$512.73
Debits	$1,077.67
Fees	$62.00
Credits	$438.24
Ending Balance	−$188.70

Debits

1/1	Ck 343	$375.00	1/10	Ck 346	$100.00	1/13	ATM Fee	$3.00
1/5	ATM	$20.00	1/10	Ck 348	$59.99	1/15	Ck 350	$185.00
1/5	ATM Fee	$3.00	1/10	Ck 349	$48.73	1/16	Ck 351	$150.00
1/8	Ck 345	$78.95	1/11	ATM	$20.00	1/16	Overdraft Fee	$20.00
1/8	ATM	$20.00	1/11	ATM Fee	$3.00	1/18	Returned Check Fee	$30.00
1/8	ATM Fee	$3.00	1/13	ATM	$20.00			

Credits

1/8 Draft Deposit $438.24		

PRACTICE

1. How much money must Alfonso pay in ATM fees in January? What are some ways that he can minimize his ATM fees in the future?
2. In your opinion, is Alfonso being careful enough in how he uses his checks and documents his transactions? Explain.
3. Alfonso's carelessness cost him $50 in bank fees. He figures that it would have taken him only 3 minutes to keep his balance up to date. How much did his lapse in bookkeeping cost him as a unit rate, given in dollars per hour?

Objective C Subtract Signed Numbers

The Concept The steps to subtract signed numbers may seem surprising. To subtract a signed number, add its opposite. The following example explains how this works.

Subtracting Signed Numbers	
Subtract a Positive	**Subtract a Negative**
To subtract a positive, add its opposite.	**To subtract a negative, add its opposite.**
$5 - (+2) = 5 + (-2)$	$5 - (-2) = 5 + (+2)$
Both expressions equal the same number.	Both expressions equal the same number.
$5 - (+2) = 3$ and $5 + (-2) = 3$	$5 - (-2) = 7$ and $5 + (+2) = 7$

The following application demonstrates why subtracting a negative number is the same as adding a positive number.

WHY IT WORKS Subtracting a Negative Is the Same as Adding a Positive

Suppose you used a credit card to purchase \$50 worth of merchandise from a store. If you return a \$10 item, the store will credit your account. There are two ways of looking at this situation:

1. The store says that it removed a \$10 debt from your account.
2. You say that the store added a \$10 credit to your account.

The store's perspective	must be the same as	your perspective.
Removing a debt	is the same as	adding a credit.
$-\$50 - (-\$10)$	is the same as	$-\$50 + (+\$10)$
$-50 - (-10)$	$=$	$-50 + (+10)$
Subtracting a negative	**is the same as**	**adding a positive.**

It is important to distinguish between the two meanings of the symbols $+$ and $-$. These symbols can be used to represent an operation, addition or subtraction. They can also be used to indicate a number's sign, positive or negative.

INTERACTIVE DEFINITION The two meanings of + and −

The symbols $+$ and $-$ can represent an operation, addition or subtraction, or the sign of a number, positive or negative.

When two signs are written next to each other in an addition or subtraction exercise:

- Treat the first sign as an operation (add or subtract).
- Treat the second sign as the sign of the number (positive or negative).

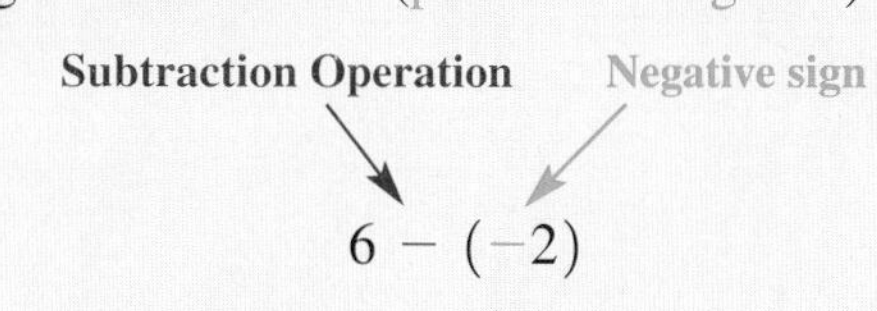

(Continued)

EXAMPLE

9. Translate each exercise into words.

$6 - (-4)$	Six subtract negative four
$4 + (-2)$	Four plus negative two
$-5 - (+2)$	Negative five subtract positive two

GUIDED PRACTICE

9. Translate each exercise into words.

$3 - (-2)$ ____________________.

$9 + (-5)$ ____________________.

$-1 - (+3)$ ____________________.

DO YOU UNDERSTAND the two meanings of "+" and "−?"

Got It | Get Help

Procedure Subtract Signed Numbers

Step 1 Change the subtraction symbol to an addition symbol.

Step 2 Change the sign of the second number.

Step 3 Add.

This procedure can be stated in one sentence: To subtract a number, add its opposite.

DETAILED EXAMPLE Subtracting Signed Numbers

Subtract. $(-3) - (5)$

Step 1 Change the subtraction symbol to an addition symbol.

$(-3) - (5) = (-3) +$

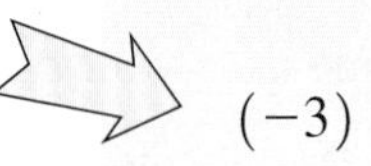

Step 2 Change the sign of the second number.

$(-3) - (5) = (-3) + (-5)$

Step 3 Add.

$(-3) - (5) = (-3) + (-5)$
$= -8$

EXAMPLES

10. Subtract. $-7 - 12$

Step 1 Change the subtraction symbol to an addition symbol.

Step 2 Change the sign of the second number.

Step 3 Add.

$-7 - 12 = -7 + (-12)$
$= -19$

11. Subtract. $-21 - (-15)$

Step 1 Change the subtraction symbol to an addition symbol.

Step 2 Change the sign of the second number.

Step 3 Add.

$-21 - (-15) = -21 + (+15)$
$= -6$

GUIDED PRACTICE

10. Subtract. $-13 - 17$

$-13 - 17 = -13$
$=$

11. Subtract. $-32 - (-40)$

$-32 - (-40) =$
$=$

Concept Check

C1. Write each subtraction exercise as an addition exercise. Do not solve.

a. $10 - 16$

b. $-12 - 5$

c. $15 - (-2)$

d. $-4 - (-8)$

C2. In your own words, why can $5 - (3)$ be solved using $5 + (-3)$?

Objective C Practice

Subtract.

C3. $16 - 20$ **C4.** $15 - 19$ **C5.** $-12 - 13$ **C6.** $-15 - 10$

C7. $8 - (-3)$ **C8.** $6 - (-2)$ **C9.** $(-3) - (-6)$ **C10.** $(-2) - (-8)$

Objective D Add and Subtract Signed Numbers Using a Number Line

The Concept The procedures in Objectives A, B, and C work well for adding or subtracting signed numbers. However, in an algebra course, you may want to simplify exercises by eliminating extra signs. Changing double signs to a single operation will help you write an exercise more simply.

CHANGING DOUBLE SIGNS TO A SINGLE OPERATION

Subtraction			Addition		
Adding a negative	⟶	$3 + (-2) = 1$	Subtracting a negative	⟶	$3 - (-2) = 5$
or subtracting a positive	⟶	$3 - (+2) = 1$	or adding a positive	⟶	$3 + (+2) = 5$
can be replaced with a single subtraction.	⟶	$3 - 2 = 1$	can be replaced with a single addition.	⟶	$3 + 2 = 5$

INTERACTIVE DEFINITION Changing Double Signs to a Single Operation

If the signs are different, replace them with a single subtraction.
If the signs are the same, replace them with a single addition.

EXAMPLE

12. Rewrite each exercise using a single operation.

Same signs: add. $3 - (-2) = 3 + 2$

Different signs: subtract. $4 + (-2) = 4 - 2$

Different signs: subtract. $-5 - (+3) = -5 - 3$

GUIDED PRACTICE

12. Rewrite each exercise using a single operation.

______ signs: ______. $3 + (-5) = 3$ ____ 5

______ signs: ______. $-7 - (+5) = -7$ ____ 5

______ signs: ______. $2 - (-3) = 2$ ____ 3

DO YOU UNDERSTAND how to change double signs to a single operation? Got It Get Help

Once we have rewritten double signs using a single operation, we can use a number line to visualize sums and differences.

Procedure **Add and Subtract Signed Numbers Using a Number Line**

Step 1 Graph the first number.

Step 2 Move left if subtracting. Move right if adding.

Step 3 Move a number of units equal to the second number.

DETAILED EXAMPLE Using a Number Line to Add or Subtract Signed Numbers

Subtract. $10 - 21$

Step 1 Graph the first number, 10.

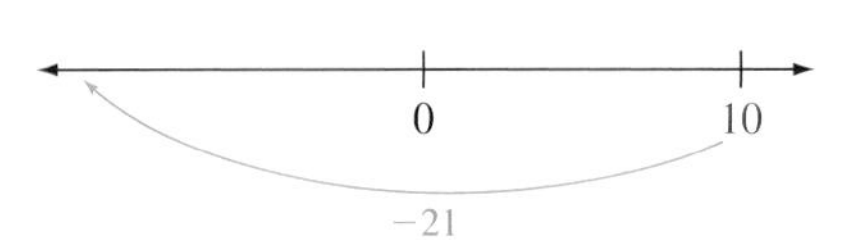

Step 2 To subtract, move left.

Step 3 Move 21 units to the left.

- We pass 0, so the answer is negative.
- $21 - 10$ is 11, so $10 - 21 = -11$.

EXAMPLES

13. Subtract. $12 - 15$

Graph 12.
To subtract 15, move 15 units to the left.

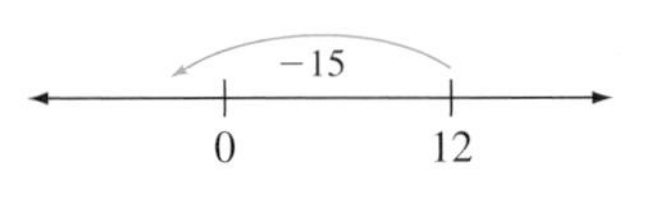

- We pass 0, so the answer is negative.
- $15 - 12$ is 3, so $12 - 15 = -3$.

14. Add. $-5 + 3$

Graph −5.
To add 3, move 3 units to the right.

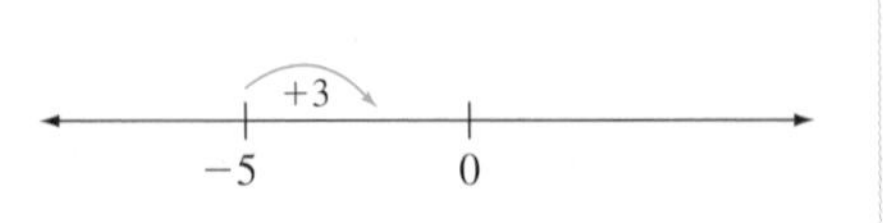

- We do not pass 0, so the answer is negative.
- $5 - 3$ is 2, so $-5 + 3 = -2$.

GUIDED PRACTICE

13. Subtract. $13 - 19$

Graph ____.
To subtract 19, move ____ units to the ____.

0

- We do/do not pass ____, so the answer is ____.
- ____ − ____ is ____, so $13 - 19 =$ ____.

14. Add. $-7 + 5$

Graph ____.
To add 5, move ____ units to the ____.

0

- We do/do not pass ____, so the answer is ____.
- ____ − ____ is ____, so $-7 + 5 =$ ____.

Procedure **Add and Subtract Signed Numbers Using a Number Line**

Step 1 Change double signs to a single operation.

Step 2 Graph the first number.

Step 3 Move left if subtracting. Move right if adding.

Step 4 Move a number of units equal to the second number.

EXAMPLES

15. Add. $24 + (-42)$

$24 + (-42) = 24 - 42$

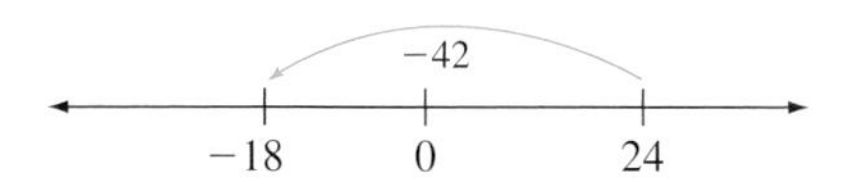

We pass 0, so the answer is negative.

$24 + (-42) = -18$

16. Add. $(-24) + (+16)$

$(-24) + (+16) = -24 + 16$

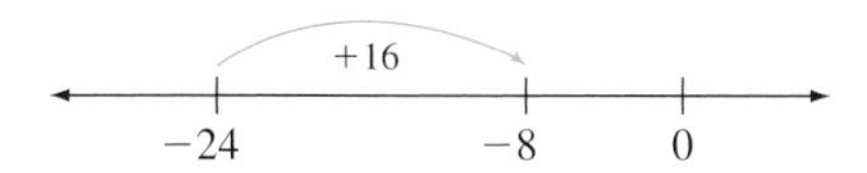

We do not pass zero, so the answer is negative.

$(-24) + (+16) = -8$

17. Subtract. $13 - (-16)$

$13 - (-16) = 13 + 16$

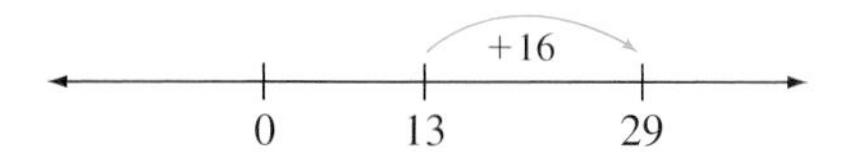

We do not pass zero, so the answer is positive.

$13 - (-16) = 29$

18. Subtract. $-31 - (+18)$

$-31 - (+18) = -31 - 18$

−18

−49 −31 0

We do not pass zero, so the answer is negative.

$-31 - (+18) = -49$

GUIDED PRACTICE

15. Add. $30 + (-51)$

$30 + (-51) = 30$ ___ 51

0 30

We do/do not pass ____, so the answer is ______.

$30 + (-51) =$

16. Add. $(-13) + (+20)$

$(-13) + (+20) = -13$ ___ 20

−13 0

We do/do not pass ____, so the answer is ______.

$(-13) + (+20) =$

17. Subtract. $21 - (-30)$

$(21) - (-30) = 21$ ___ 30

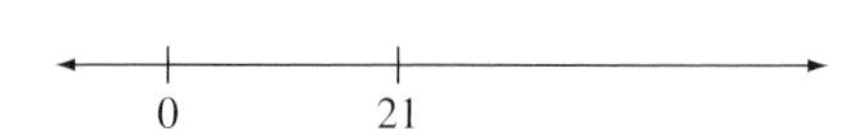

We do/do not pass ____, so the answer is ______.

$(21) - (-30) =$

18. Subtract. $(-61) - (+75)$

$(-61) - (+75) = -61$ ___ 75

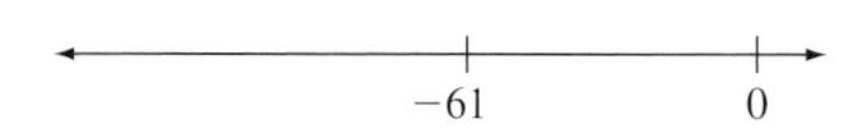

We do/do not pass ____, so the answer is ______.

$(-61) - (+75) =$

Concept Check

D1. In your own words, why is subtracting a negative the same as adding a positive?

D2. Examine both visualizations for $3 + (-4) = -1$ to answer the following questions.

Number Line Visualization:

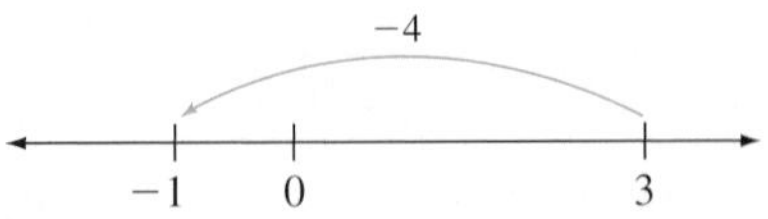

Zero Pair Visualization:

$(3) + (-4) = -1$

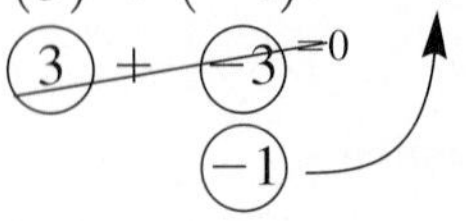

a. In the number line visualization, how many units to the left of the starting point, 3, do you need to move to reach zero?
b. In the zero pair visualization, what is the largest zero pair that can be made from the numbers?
c. What are the similarities between these two visualizations?
d. What are the differences between these two visualizations?

Objective D Practice

Exercises D3–D6 have three parts.
a. Change any double signs to a single operation.
b. Show a visualization using a number line.
c. Write the result.

D3. a. $10 - (+16) = 10 \quad 16$

b.

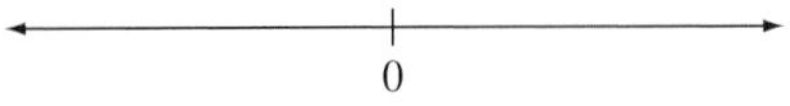

c. $10 - (+16) =$

D4. a. $15 - (+18) = 15 \quad 18$

b.

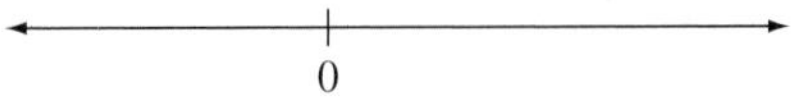

c. $15 - (+18) =$

D5. a. $-12 - (-13) = -12 \quad 13$

b.

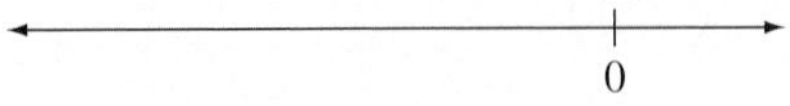

c. $-12 - (-13) =$

D6. a. $-7 - (-5) = -7 \quad 5$

b.

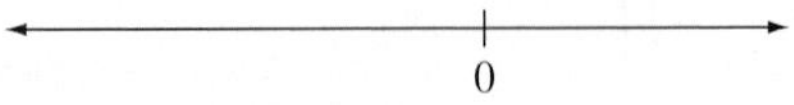

c. $-7 - (-5) =$

Change any double signs to a single operation and write the result.

D7. $5 - (-6)$

D8. $4 - (-8)$

D9. $-6 - (-9)$

D10. $-40 - (-50)$

D11. $51 - (+16)$

D12. $40 + (-65)$

D13. $-19 - (-30)$

D14. $-50 + (+75)$

Combining Concepts and Applications

CONCEPT I Using the Order of Operations Recall the order of operations.

First: Perform operations in groupings and evaluate absolute values.
Second: Perform operations with exponents.
Third: Perform multiplication and division as they occur from left to right.
Fourth: Perform addition and subtraction as they occur from left to right.

EXAMPLE	GUIDED PRACTICE
19. Evaluate. $(-5) - \lvert -3 \rvert + 12$	**19.** Evaluate. $(-3) - \lvert -6 \rvert + 8$
Groupings: Evaluate the absolute value. $(-5) - \lvert -3 \rvert + 12 = \underbrace{(-5) - 3}_{-8} + 12$	$(-3) - \lvert -6 \rvert + 8 = (-3) - \_\_\_\_ + 8$
Add and subtract. $= \underbrace{-8 + 12}_{+4}$	$=$
$= 4$	$=$

CONCEPT II Adding Several Signed Numbers Two algebraic properties will make adding and subtracting several signed numbers easier.

The commutative property lets you reorder addition exercises.

Commutative Property of Addition
Given any two numbers a and b, $a + b = b + a$.

For example, $2 + 3 = 3 + 2$ and $3 + (-2) = (-2) + 3$.

The associative property lets you regroup addition exercises.

Associative Property of Addition
Given any three numbers a, b, and c, $(a + b) + c = a + (b + c)$.

For example, $(5 + 12) + 3 = 5 + (12 + 3)$ and $(3 + (-2)) + (-8) = 3 + ((-2) + (-8))$.

▶ **NOTE** These properties do not work for subtraction. You can neither reorder nor regroup subtraction exercises.

EXAMPLE

20. Evaluate. $3 - (-2) + (-5) - 8 + 16 - 7$

Step 1 Change subtraction to the addition of an opposite.

$$3 - (-2) + (-5) - 8 + 16 - 7$$
$$= 3 + 2 + (-5) + (-8) + 16 + (-7)$$

positives listed first — negatives listed second

Step 2 Use the commutative property of addition.

$$= \underbrace{3 + 2 + 16}_{\text{positives listed first}} + \underbrace{(-5) + (-8) + (-7)}_{\text{negatives listed second}}$$

Step 3 Use the associative property of addition.

$$= (3 + 2 + 16) + ((-5) + (-8) + (-7))$$

Step 4 Add the quantities in the groupings.

$$= 21 + (-20)$$

Add the remaining numbers.

$$= 1$$

GUIDED PRACTICE

20. Evaluate. $5 + (-17) - 3 + 4 - 2 - (-5)$

Step 1 Change subtraction to the addition of an opposite.

$$5 + (-17) - 3 + 4 - 2 - (-5)$$
$$= \quad + (\quad) + (\quad) + \quad + (\quad) + (\quad)$$

list positives first — list negatives second

Step 2 Use the commutative property of addition.

$$= \quad\quad + \quad\quad$$

Step 3 Use the associative property of addition.

$$= (\quad + \quad + \quad) + ((-\quad) + (-\quad) + (-\quad))$$

Step 4 Add the quantities in the groupings.

$$= \quad + (-\quad)$$

Add the remaining numbers.

$$=$$

Regrouping and reordering addition problems reduces the chances of making a sign error when adding many signed numbers.

CONCEPT III Application Exercises

When solving application exercises with signed numbers, be sure to read carefully to decide if a value is positive or negative.

EXAMPLE

21. A hiker on the Appalachian Trail started hiking at an altitude of 3,482 feet. Throughout the day, the hiker used her GPS unit to keep track of her change in altitude. She noted the change only at the high and low points on the trail. What were the highest and lowest points of her path on the trail?

Because the highest and lowest points are needed, we find her altitude at each time.

Altitude Log		Change	Altitude
8:00 A.M.	started hiking		3,428 ft
9:35 A.M.	climbed 748 feet	+748	4,176 ft
11:20 A.M.	descended 830 feet	−830	3,346 ft
12:08 P.M.	climbed 200 feet	+200	3,546 ft
2:00 P.M.	descended 90 feet	−90	3,456 ft
4:30 P.M.	climbed 693 feet	+693	4,149 ft
6:00 P.M.	descended 420 feet	−420	3,729 ft

The hiker's highest altitude was 4,176 ft at 9:35 A.M. and her lowest altitude was 3,346 feet at 11:20 A.M.

GUIDED PRACTICE

21. The same hiker from Example 21 took a different trail the next day to get back to her starting point. When she made her last entry at 4:08 P.M., she saw her car parked in a valley below her as seen above. How much farther down the valley is her car?

Altitude Log		Change	Altitude
8:00 A.M.	started hiking	Start	3,729 ft
9:20 A.M.	climbed 250 feet		
10:35 A.M.	descended 490 feet		
12:35 P.M.	climbed 172 feet		
2:15 P.M.	descended 100 feet		
4:08 P.M.	climbed 82 feet		
Car is parked at 3,428 feet.		?	3,428 ft

To determine how far down the valley she must still hike, subtract the altitude of her car from her altitude at 4:08 P.M.

The hiker must descend ______ feet to get to her car.

Section 9.2 Exercises

FOR EXTRA HELP       

To add numbers with the same sign:

1. Answer the Objective A Concept Check.
2. Answer the odd Objective A Practice Exercises.
3. Answer the even Objective A Practice Exercises.

To add numbers with opposite signs:

4. Answer the Objective B Concept Checks.
5. Answer the odd Objective B Practice Exercises.
6. Answer the even Objective B Practice Exercises.

To use signed numbers to balance a checkbook:

7. Balance the checkbook in Focus On Personal Finance.
8. Answer the odd and even Focus On Personal Finance Exercises.

To subtract signed numbers:

9. Answer the Objective C Concept Checks.
10. Answer the odd Objective C Practice Exercises.
11. Answer the even Objective C Practice Exercises.

To add and subtract signed numbers using a number line:

12. Answer the Objective D Concept Checks.
13. Answer the odd Objective D Practice Exercises.
14. Answer the even Objective D Practice Exercises.

VOCABULARY REVIEW *Review the Vocabulary Preview for Section 9.2. Study the definitions until you can check* Got It *for every word.*

Use these words to complete each sentence.

zero pair • additive inverse • commutative property of addition • associative property of addition

	Got It	Get Help
15. Two opposites form a(n) ____________.		
16. A(n) ____________ is the opposite of a number.		

How will you get help for any vocabulary that you are unsure about?

Your instructor ____ MyMathLab ____ A classmate ____ A tutor ____ Other ____

Add.

17. $(-4) + (-6)$

18. $(-12) + (-9)$

19. $73 + 28$

20. $25 + 27$

21. $14 + (-9)$

22. $23 + (-12)$

23. $(-6) + 5$

24. $(-15) + 13$

25. $18 + (-23)$

26. $54 + (-72)$

27. $(-8) + 20$

28. $(-17) + 30$

29. $(-2.5) + 3.8$

30. $4.7 + (-3.9)$

31. $(-2.2) + (-8.8)$

32. $(-6.5) + (7.5)$

33. $\frac{3}{5} + \left(-\frac{7}{10}\right)$

34. $\left(-\frac{1}{12}\right) + \frac{1}{4}$

35. $\left(-\frac{8}{9}\right) + \left(-\frac{2}{9}\right)$

36. $\left(-\frac{8}{15}\right) + \left(-\frac{3}{15}\right)$

Subtract.

37. $4 - 12$

38. $16 - 30$

39. $21 - 8$

40. $38 - 29$

41. $(-17) - 14$

42. $(-50) - 25$

43. $38 - (-12)$

44. $24 - (-51)$

45. $(-13) - (-18)$

46. $(-50) - (-108)$

47. $(-71) - (-86)$

48. $(-14) - (-8)$

49. $\frac{3}{5} - \left(-\frac{1}{10}\right)$

50. $\left(-\frac{1}{8}\right) - \frac{1}{2}$

51. $(-5.5) - (-6.3)$

52. $(-3.5) - (9.5)$

Answer each question about the commutative and associative properties.

53. a. Evaluate. $17 - 32$
b. Evaluate. $32 - 17$
c. Based on this exercise, is subtraction commutative?

54. a. Evaluate. $10 - 16$
b. Evaluate. $16 - 10$
c. Based on this exercise, is subtraction commutative?

55. a. Evaluate. $(15 + 2) + -8$
b. Evaluate. $15 + (2 + (-8))$
c. Based on this exercise, is addition associative?

56. a. Evaluate. $(10 - 21) - 8$
b. Evaluate. $10 - (21 - 8)$
c. Based on this exercise, is subtraction associative?

Evaluate using the order of operations.

57. $7 - 3 + 5$

58. $8 + 3 - 14$

59. $8 + 3 - 6$

60. $10 - 5 + 5$

61. $-4 + |-3| - (-2)$

62. $-5 + (-8) + |12|$

63. $3 - (-3) - 4$

64. $-10 - (-8) + (-4)$

65. $7 + (-5) - 9 + (-15)$

66. $13 - (-5) + 7 - 12$

67. $2.1 - (-3.8) + 4.5 + (-1.1)$

68. $-0.43 + 0.98 - 1.1 - (-2)$

69. $1 + 2 - 3 + 4 - 5 + 6 - 7$

70. $7 - 6 + 5 - 4 + 3 - 2 + 1$

71. $|5 - (-3) - 15|$

72. $|1 - (-3) + (-4)|$

tobey Don't forget to change subtraction to the addition of an opposite. Thus, $7 - (-3) = 7 + 3$ and $-8 - (-4) = -8 + 4$.

For more tips and tweets, go to twitter.com/gstbasicmath

73. A freight plane lands in New York City, unloading 3,149 pounds and then loading 2,987 pounds of freight. What is the change in the weight of freight?

74. A freight plane lands in Newark, New Jersey, unloading 1,582 pounds and then loading 1,793 pounds of freight. What is the change in the weight of freight?

75. Yesterday, the nighttime temperature in Seoul, South Korea, was 43°F and the daytime temperature was 85°F. What was the change in temperature from day to night?

76. The average daytime temperature in Battle Creek, Michigan, is 34°F in December and 78°F in June. What is the change in average daytime temperature from June to December?

77. On New Year's Eve, the temperature was −3°C. On New Year's Day, the temperature was 9°C. What was the change in temperature from New Year's Eve to New Year's Day?

78. A hiker was climbing Mount Everest. If the temperature at the mountain's base was 45°F and the temperature at the peak was −4°F, what was the temperature change from base to peak?

79. The lowest point in the United States is in Death Valley, California, with an altitude of −282 feet (below sea level). The surface of the Caspian Sea is 170 feet higher than the floor of Death Valley. What is the altitude of the Caspian Sea?

80. To freeze microwave dinners, each dinner is dipped in liquid nitrogen at a temperature of −320°F and then placed in a freezer at 23°F. What is the change in temperature from the liquid nitrogen bath to the freezer?

Use the following table to answer Exercises 81–84.

81. If the company had $113,200 on the first of January, how much money did it have at the end of March?

82. What is the total profit/loss for January and February?

83. What is the total profit/loss for February and March?

84. If the company had $113,200 on the first of January, how much money did it have at the end of May?

Month	Profit	Loss
January	$18,700	
February		$34,700
March		$6,300
April	$43,600	
May		$12,400

▶ **NOTE** Dollar amounts in the table are recorded at the end of the month.

85. When they are keeping score in a golf match, golfers sometimes use symbols to indicate how many shots they are above or below par. Negative values are "below par." Positive values are "above par."

◎ $= -2$ ○ $= -1$ empty $= 0$ □ $= +1$ ⧈ $= +2$

a. How many shots above or below par did each golfer score by the end of the match?

b. Order the golfers from 1st to 4th place, using the fact that a low score wins in golf.

	1	2	3	4	5	6	7	8	9	10	11	12	13	14	15	16	17	18
Chante			□			⧈		□	□		□	○			⧈	○	○	
Angelica		○	□	□		○			○	○	□	◎	□			○	○	○
Sergio	□	○				□	⧈	□		□		□	□	⧈	○			
Mike		○	○	□			◎	⧈		⧈			○	○			⧈	⧈

86. A high score on a hole suggests that the hole is more difficult.

a. Which hole was most difficult for the golfers as a group?

b. Which hole was the easiest?

Section 9.2 Question Log

Use this space to write questions. Be sure to get these answered and revisit them when you prepare for your exam.

Page ______ Answered

Page ______ Answered

Page ______ Answered

Page ______ Answered

Section 9.3 Multiplying and Dividing Signed Numbers

Multiplying and dividing signed numbers involve two steps:

1. Use the same multiplication and division procedures from earlier chapters.
2. Determine the sign of the product or quotient.

Since you can already multiply and divide, the only new concept in this section is to learn how to determine the sign of a product or quotient.

The **Objectives** in Section 9.3 will help you

- **A** Multiply two signed numbers.
- **B** Divide two signed numbers.
- **C** Multiply or divide several signed numbers.
- **D** Evaluate an exponential expression involving signed numbers.

VOCABULARY PREVIEW *Check the box that applies.*	Got It	Must Study
Product: A **product** is the result of multiplying.		
Quotient: A **quotient** is the result of dividing.		
Repeated Addition: Multiplication can be performed using **repeated addition.** $4 \cdot 3 = 3 + 3 + 3 + 3$		
Repeated Multiplication: Exponents indicate **repeated multiplication.** $4^3 = 4 \cdot 4 \cdot 4$		
Factors: Factors are numbers that are multiplied to give a product. 2 and 5 are factors of 10 because $2 \cdot 5 = 10$.		
Base and Exponent: The **base** is the symbol that appears immediately before the **exponent.** $(-4)^3$ has a base of (-4) and an exponent of 3. base → $(-4)^3$ ← exponent		

Study these words when they appear in the text. After the section, test your understanding by completing the Vocabulary Review in the section exercises.

Objective A Multiply Two Signed Numbers

The Concept To multiply two signed numbers, use the multiplication procedures from earlier chapters and use the following rules to determine the sign of the product.

Opposite signs result in a negative product.

$(+) \cdot (-) =$ negative product
$(-) \cdot (+) =$ negative product

Same signs result in a positive product.

$(+) \cdot (+) =$ positive product
$(-) \cdot (-) =$ positive product

WHY IT WORKS Using Sign Rules for Multiplication

To understand why factors with opposite signs result in a negative product, study the following pattern.

Moving down the left side, the first factor keeps decreasing by one.

$$\begin{aligned} 2 \cdot 4 &= 8 \\ 1 \cdot 4 &= 4 \\ 0 \cdot 4 &= 0 \\ -1 \cdot 4 &= -4 \\ -2 \cdot 4 &= -8 \end{aligned}$$

Moving down the right side, the product keeps decreasing by four.

Continuing this pattern, we see that factors with different signs result in a negative product.

To understand why two negative factors result in a positive product, study the following pattern.

Moving down the left side, the first factor keeps decreasing by one.

$$\begin{aligned} 2 \cdot (-4) &= -8 \\ 1 \cdot (-4) &= -4 \\ 0 \cdot (-4) &= 0 \\ (-1) \cdot (-4) &= 4 \\ (-2) \cdot (-4) &= 8 \end{aligned}$$

Moving down the right side, the product keeps increasing by four.

Continuing this pattern, we see that two negative factors result in a positive product.

Procedure Multiply Two Signed Numbers

Step 1 Determine the sign of the product.

- **a.** If the factors have opposite signs, the product is negative.
- **b.** If the factors have the same sign, the product is positive.

Step 2 Multiply the absolute values of the factors.

EXAMPLES

1. Multiply. $7 \cdot (-5)$

Step 1 Since the factors have opposite signs, the product is negative.

Step 2 Multiply the absolute values of the factors.

$$\begin{aligned} 7 \cdot (-5) &= -(7 \cdot 5) \\ &= -35 \end{aligned}$$

GUIDED PRACTICE

1. Multiply. $(-9) \cdot (8)$

$$\begin{aligned} (-9) \cdot (8) &= \quad (\quad \cdot \quad) \\ &= \end{aligned}$$

(Continued)

2. Multiply. $\left(-\frac{5}{8}\right)\cdot\left(-\frac{10}{5}\right)$

Step 1 Since the factors have the same sign, the product is positive.

Step 2 Multiply the absolute values of the factors.

$$\left(-\frac{5}{8}\right)\cdot\left(-\frac{10}{15}\right) = +\left(\frac{5}{8}\cdot\frac{10}{15}\right)$$
$$= \frac{5\cdot 10}{8\cdot 15}$$
$$= \frac{5\cdot \cancel{2}\cdot \cancel{5}}{\cancel{2}\cdot 2\cdot 2\cdot 3\cdot \cancel{5}}$$
$$= \frac{5}{12}$$

2. Multiply. $\left(-\frac{3}{4}\right)\cdot\left(-\frac{8}{1}\right)$

$$\left(-\frac{3}{4}\right)\cdot\left(-\frac{8}{1}\right) = \square\left(\frac{\square}{\square}\cdot\frac{\square}{\square}\right)$$
$$=$$
$$=$$
$$=$$

Several symbols are used to indicate multiplication, including $\cdot$ and $\times$. Two sets of parentheses next to each other also indicate multiplication. For example, the product $(-2)\cdot(-3)$ can be written as $(-2)(-3)$.

EXAMPLES

3. Multiply. $(-3)(-6)$

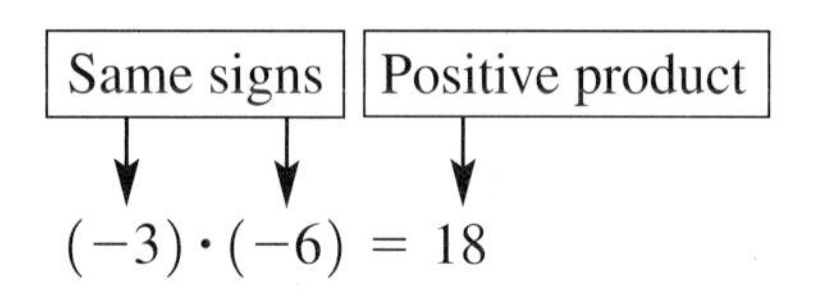

4. Multiply. $(1.1)(-4)$

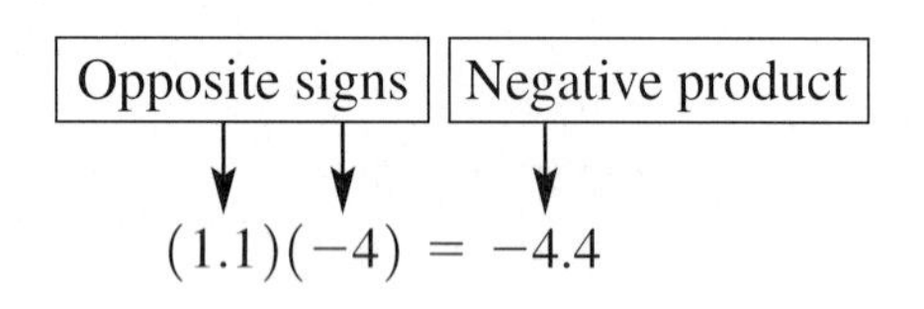

GUIDED PRACTICE

3. Multiply. $(-3)(2)$

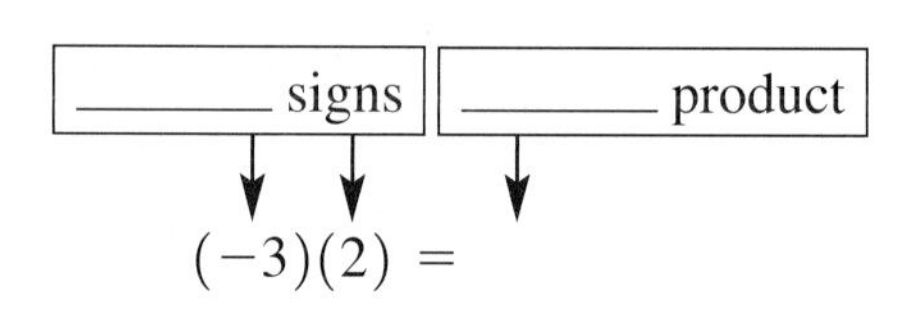

4. Multiply. $(-1.6)(-1)$

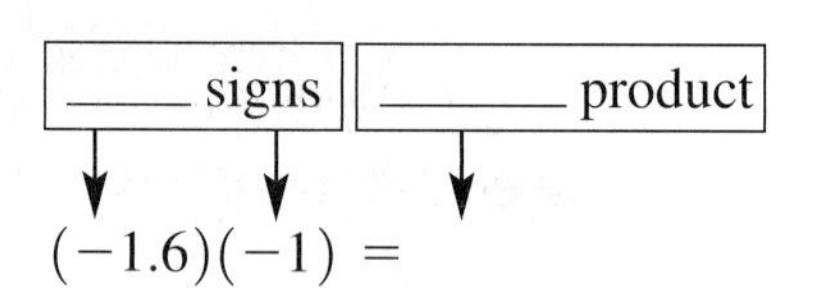

Concept Check

A1. Use the fact that "multiplication can be performed by repeated addition" to explain why $3\cdot(-4) = -12$. *Hint:* The equation $3\cdot(-4) = -12$ can be read as "Three negative fours equal negative twelve."

A2. a. Complete the following pattern.

$$1\cdot(-2) = -2$$
$$0\cdot(-2) = 0$$
$$(-1)\cdot(-2) = 2$$
$$(\quad)\cdot(-2) = \square$$
$$(\quad)\cdot(\quad) = \square$$

b. Based on the pattern, two negatives result in a ______ product.

Objective A Practice

Multiply.

A3. $13 \cdot (-8)$ **A4.** $21 \cdot (-20)$ **A5.** $(-7)(-15)$ **A6.** $(-8)(-22)$

A7. $(-3.3)(2)$ **A8.** $(-5.5)(-10)$ **A9.** $(-1.1)(-100)$ **A10.** $(-7.1)(3)$

Objective B Divide Two Signed Numbers

The Concept The sign rules for division are the same as the sign rules for multiplication.

Opposite signs result in a negative quotient.

$$\frac{(+)}{(-)} = \text{negative quotient}$$

$$\frac{(-)}{(+)} = \text{negative quotient}$$

Same signs result in a positive quotient.

$$\frac{(+)}{(+)} = \text{positive quotient}$$

$$\frac{(-)}{(-)} = \text{positive quotient}$$

WHY IT WORKS Using Sign Rules for Division

Sign rules for division are the same as the sign rules for multiplication.

Division exercises can be solved using corresponding multiplication exercises.
For example, $6 \div 2 = 3$ corresponds to $3 \cdot 2 = 6$.

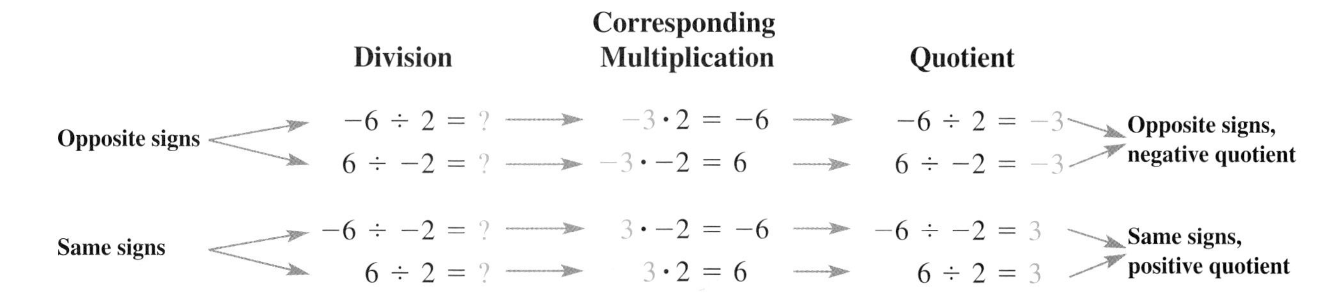

The sign rules for multiplication and division are identical.

- Opposite signs result in a negative product or quotient.
- Same signs result in a positive product or quotient.

Procedure Divide Two Signed Numbers

Step 1 Determine the sign of the quotient.

a. If the numbers have opposite signs, the quotient is negative.
b. If the numbers have the same sign, the quotient is positive.

Step 2 Divide the absolute values of the numbers.

EXAMPLES

5. Divide. $(-8) \div (-2)$

Same signs | Positive quotient

$(-8) \div (-2) = 4$

6. Divide. $(-4.2) \div (7)$

Opposite signs | Negative quotient

$(-4.2) \div (7) = -0.6$

7. Divide. $\left(-\frac{3}{8}\right) \div \left(-\frac{9}{4}\right)$

Same signs | Positive quotient

$$\left(-\frac{3}{8}\right) \div \left(-\frac{9}{4}\right) = +\left(\frac{3}{8}\right) \cdot \left(\frac{4}{9}\right)$$

$$= \frac{\cancel{3} \cdot \cancel{4}}{2 \cdot \cancel{4} \cdot 3 \cdot \cancel{3}}$$

$$= \frac{1}{6}$$

GUIDED PRACTICE

5. Divide. $(-20) \div (5)$

______ signs | ______ quotient

$(-20) \div (5) =$

6. Divide. $(-2.7) \div (-9)$

_____ signs | _______ quotient

$(-2.7) \div (-9) =$

7. Divide. $\left(-\frac{5}{4}\right) \div \left(-\frac{35}{8}\right)$

_____ signs | _______ quotient

$$\left(-\frac{5}{4}\right) \div \left(-\frac{35}{8}\right) = \left(\frac{\quad}{\quad}\right) \cdot \left(\frac{\quad}{\quad}\right)$$

$$=$$

$$=$$

Concept Check

B1. Will the product or quotient of two numbers with the same sign be positive or negative?

B2. Will the product or quotient of two numbers with opposite signs be positive or negative?

Objective B Practice

Divide.

B3. $(27) \div (-3)$

B4. $(14) \div (-7)$

B5. $(-30) \div (-12)$

B6. $(-40) \div (-15)$

B7. $\left(\frac{5}{8}\right) \div \left(\frac{-3}{4}\right)$

B8. $\left(-\frac{21}{1}\right) \div \left(-\frac{1}{2}\right)$

B9. $(-4.4) \div (.02)$

B10. $(-6.4) \div (-0.8)$

Objective C Multiply or Divide Several Signed Numbers

The Concept When multiplying or dividing several signed numbers, we can use two rules to determine the sign of the answer.

Rule 1: An even number of negatives gives a positive product or quotient.

Two negatives

$(-2)\cdot(3) \div (-5) =$ Positive answer

Four negatives

$(-20) \div (-5)\cdot(-2) \div (-1) =$ Positive answer

Rule 2: An odd number of negatives gives a negative product or quotient.

One negative

$(-16) \div (2)\cdot(1) =$ Negative answer

Three negatives

$(-1)\cdot(-3) \div (-4) =$ Negative answer

WHY IT WORKS Using Sign Rules for Multiplying or Dividing Several Signed Numbers

Rule 1: An even number of negatives gives a positive product or quotient.
Rule 2: An odd number of negatives gives a negative product or quotient.

In the following examples, negatives are paired to form positive products or quotients.

$\underbrace{(-3) \div (-1)}_{+}$ ← **The positive sign on this bracket means that the pair gives a positive quotient.**

- **Why does an even number of negatives give a positive product or quotient?**

When two negative numbers are multiplied or divided, the result is positive. If there is an even number of negatives, all the negatives are paired to form positive products or quotients. The answer is positive.

$\underbrace{(-10) \div (-2)}_{+}\cdot\underbrace{(-3) \div (-1)}_{+} =$ Positive answer

$\underbrace{(-6) \div (-3)}_{+}\cdot\underbrace{(-1) \div (-2)}_{+}\cdot\underbrace{(-8) \div (-2)}_{+} =$ Positive answer

Each pair of negatives gives a positive result. Since every negative is paired, the answer is positive.

- **Why does an odd number of negatives give a negative product or quotient?**

If there is an odd number of negatives, all but one negative are paired to form positive products or quotients. The answer is negative.

$\underbrace{(-10) \div (-2)}_{+}\cdot(-3) =$ Negative answer

$\underbrace{(-6) \div (-3)}_{+}\cdot\underbrace{(-1) \div (-2)}_{+}\cdot(-8) =$ Negative answer

There is always one negative that is not paired. The one leftover negative gives a negative answer.

Based on these sign rules, the product of $(2) \cdot (-3) \cdot (-1) \cdot (2) \cdot (-2)$ is negative because three of the factors (an odd number) are negative.

$$\begin{aligned}(2) \cdot (-3) \cdot (-1) \cdot (2) \cdot (-2) &= \underbrace{(2) \cdot (-3)}_{-6} \cdot (-1) \cdot (2) \cdot (-2) \\ &= \underbrace{(-6) \cdot (-1)}_{6} \cdot (2) \cdot (-2) \\ &= \underbrace{(6) \cdot (2)}_{12} \cdot (-2) \\ &= \underbrace{(12) \cdot (-2)}_{-24} \\ &= -24\end{aligned}$$

Procedure Multiply or Divide Several Signed Numbers

Step 1 Determine the sign of the product or quotient.

- An odd number of negatives results in a negative answer.
- An even number of negatives results in a positive answer.

Step 2 Multiply or divide using the absolute value of the numbers.

Follow the order of operations. Multiply or divide from left to right.

EXAMPLES

8. Evaluate. $(-2) \cdot (10) \div (-5)$

Step 1 Determine the sign of the product or quotient.

There is an even number of negatives, so the answer is positive.

Step 2 Multiply or divide the absolute value of the numbers.

Sign of the answer ↓

$$\begin{aligned}(-2) \cdot (10) \div (-5) &= +\underbrace{(2) \cdot (10)}_{20} \div (5) \\ &= +20 \div (5) \\ &= +4\end{aligned}$$

GUIDED PRACTICE

8. Evaluate. $(-14) \div (-2) \cdot (-7)$

Step 1 Determine the sign of the product or quotient.

There is an ________ number of negatives, so the answer is ________.

Step 2 Multiply or divide the absolute value of the numbers.

Sign of the answer ↓

$$\begin{aligned}(-14) \div (-2) \cdot (-7) &= \square(14) \div (2) \cdot (7) \\ &= \\ &= \end{aligned}$$

9. Evaluate. $(-3)\cdot(-5)\cdot(2) \div 6 \div (-2)$

Step 1 Determine the sign of the product or quotient.

There is an odd number of negatives, so the answer is negative.

Step 2 Multiply or divide the absolute value of the numbers.

$$\begin{aligned}(-3)\cdot(-5)\cdot(2) \div 6 \div (-2) &= -\underbrace{(3)\cdot(5)}_{15}\cdot(2) \div 6 \div (2)\\ &= -\underbrace{(15)\cdot(2)}_{30} \div 6 \div (2)\\ &= -\underbrace{30 \div 6}_{5} \div (2)\\ &= -5 \div (2)\\ &= -2\frac{1}{2}\end{aligned}$$

Sign of the answer (arrow pointing to the minus sign)

9. Evaluate. $(4)\cdot(-3)\cdot(2) \div (3) \div (-5)$

Step 1 Determine the sign of the product or quotient.

There is an ________ number of negatives, so the answer is ________.

Step 2 Multiply or divide the absolute value of the numbers.

$$\begin{aligned}(4)\cdot(-3)\cdot(2) \div (3) \div (-5) &= \quad (4)\cdot(3)\cdot(2) \div (3) \div (5)\\ &= \quad \cdot \quad \div \quad \div\\ &=\\ &=\\ &=\end{aligned}$$

Sign of the answer (arrow pointing to the blank)

Concept Check

C1. In your own words, when is the product of several signed numbers negative?

C2. When multiplying or dividing several signed numbers, how do you determine which operation to perform first?

Objective C Practice

Follow the order of operations to evaluate each expression.

C3. $(-3)\cdot(2)\cdot 1\cdot(-1)$

C4. $(-3)\cdot(-2)\cdot 2\cdot(-1)$

C5. $(-2)\cdot(3)\cdot(3)$

C6. $6\cdot(-4) \div (-2)\cdot 3$

C7. $(-12)\cdot 2 \div 4\cdot(-2)$

C8. $18 \div (-3)\cdot(-3)$

C9. $\dfrac{2\cdot(-3)\cdot 5}{(-20)}$

C10. $3\cdot\left(-\dfrac{8}{9}\right)\cdot(-6)$

Objective D Evaluate an Exponential Expression Involving Signed Numbers

The Concept Exponents indicate repeated multiplication. To evaluate 4^3 (4 to the third power), multiply three 4's together.

$$\begin{aligned}4^3 &= 4\cdot 4\cdot 4\\ &= 64\end{aligned}$$

When evaluating exponents, it is important to identify whether the base is positive or negative.

INTERACTIVE DEFINITION Identify an Exponent and Its Base

The **base** is the symbol that appears immediately before the **exponent.** In $(-4)^3$, the exponent touches the grouping; so the exponent is 3 and the base is (-4).

base → $(-4)^3$ ← exponent

EXAMPLES

10. Identify the exponent and base in $(-2)^3$.

The exponent "touches" the grouping. The base includes the negative sign.

The exponent is 3.

The base is (-2).

$(-2)^3 = (-2)\cdot(-2)\cdot(-2)$

11. Identify the exponent and base in -3^4.

The exponent "touches" only the 3, not the negative sign. The base does not include the negative sign.

The exponent is 4.

The base is 3.

$-3^4 = -(3\cdot3\cdot3\cdot3)$

GUIDED PRACTICE

10. Identify the exponent and base in $(-4)^2$.

The exponent "touches" the __________. The base includes the __________.

The exponent is _____.

The base is _____.

$(-4)^2 =$ __________

11. Identify the exponent and base in -5^2.

The exponent "touches" only the _____, not the negative sign. The base does not include the __________.

The exponent is _____.

The base is _____.

$-5^2 =$ __________

DO YOU UNDERSTAND how to identify the base and exponent when working with signed numbers? Got It | Get Help

Before you evaluate an exponential expression involving signed numbers, it is worth restating the rules for finding the sign of a product.

Rule 1: An odd number of negative factors gives a negative product.
Rule 2: An even number of negative factors gives a positive product.

Procedure Evaluate an Exponential Expression Involving Signed Numbers

Step 1 Write the repeated multiplication.
Identify the base and exponent carefully.

Step 2 Determine the sign of the product and multiply the absolute value of the factors.

EXAMPLES

12. Evaluate. $(-2)^3$

Step 1 Write the repeated multiplication.
Step 2 Three negative factors result in a negative product.

$$(-2)^3 = (-2)\cdot(-2)\cdot(-2)$$
$$= -(2)\cdot(2)\cdot(2)$$
$$= -8$$

13. Evaluate. -4^2

Step 1 Write the repeated multiplication. The negative sign is not part of the base.
Step 2 One negative factor results in a negative product.

$$-4^2 = -4\cdot 4$$
$$= -(4\cdot 4)$$
$$= -16$$

14. Evaluate. $(-5)^2$

Step 1 Write the repeated multiplication.
Step 2 Two negative factors result in a positive product.

$$(-5)^2 = (-5)\cdot(-5)$$
$$= +(5)\cdot(5)$$
$$= 25$$

15. Evaluate. -5^2

Step 1 Write the repeated multiplication. The negative sign is not part of the base.
Step 2 One negative factor results in a negative product.

$$-5^2 = -(5\cdot 5)$$
$$= -25$$

GUIDED PRACTICE

12. Evaluate. $(-1)^3$

$(-1)^3 =$
$=$
$=$

13. Evaluate. -3^2

$-3^2 =$
$=$
$=$

14. Evaluate. $(-7)^2$

$(-7)^2 =$
$=$
$=$

15. Evaluate. -7^2

$-7^2 =$
$=$

Concept Check

D1. Complete the table.

	Result
a. 3^2 has a base of 3 and an exponent of 2.	$3^2 = 9$
b. $(-5)^3$ has a base of ____ and an exponent of ____.	$(-5)^3 =$
c. -8^2 has a base of ____ and an exponent of ____.	$-8^2 =$
d. $(-1)^9$ has a base of ____ and an exponent of ____.	$(-1)^9 =$

Objective D Practice

Evaluate.

D2. $(5)^3$ **D3.** 10^3 **D4.** $(-3)^2$ **D5.** $(-7)^2$

D6. -2^5 **D7.** -2^4 **D8.** $(-2)^5$ **D9.** $(-2)^4$

Combining Concepts and Applications

CONCEPT I Finding the Average of Signed Numbers In earlier chapters, you learned how to find an average. Use the same procedure to find the average of signed numbers.

EXAMPLE

16. In Nome, Alaska, the average daily temperatures are recorded in degrees Fahrenheit.

Mon	Tues	Wed	Thurs	Fri
−21°F	−11°F	−16°F	4°F	0°F

What is the average temperature for these days?

Step 1 Find the sum of the temperatures.
Step 2 Divide by the total number of days.

$$\text{Average} = \frac{\text{Sum of the data values}}{\text{Number of data values}}$$

$$= \frac{(-21) + (-11) + (-16) + (4) + (0)}{5}$$

$$= \frac{-44}{5}$$

$$= -8.8$$

The average daily temperature is −8.8°F.

GUIDED PRACTICE

16. In Moskva, Russia, the average monthly temperatures are recorded in degrees Celsius.

Jan	Feb	Mar	Apr
−10.2°C	−8.9°C	−4.0°C	4.5°C

What is the average temperature for these months?

Step 1 Find the sum of the temperatures.
Step 2 Divide by the total number of months.

$$\text{Average} = \frac{\text{Sum of the data values}}{\text{Number of data values}}$$

$$= \frac{(\quad) + (\quad) + (\quad) + (\quad)}{\quad}$$

$$= \frac{\quad}{\quad}$$

$$=$$

The average monthly temperature is ______.

CONCEPT II Getting Ready for Algebra In an algebra class, you may be asked to identify signed numbers that multiply to give a certain value and that add to give another value. The following exercises will help you practice this skill.

EXAMPLES

17. Find two numbers whose product is 10 and whose sum is -7.

Step 1 List the factor pairs whose product is 10.

Step 2 Add the pairs and look for a sum of -7.

Products that = 10	Sums of factors
$1 \cdot 10$	$1 + 10 = 11$
$2 \cdot 5$	$2 + 5 = 7$
Now check the opposites.	
$(-1) \cdot (-10)$	$(-1) + (-10) = -11$
$(-2) \cdot (-5)$	$(-2) + (-5) = -7$

The two numbers are (-2) and (-5).
$(-2)(-5) = 10$ and $(-2) + (-5) = -7$

18. Find two numbers whose product is -20 and whose sum is 1.

Step 1 List the factor pairs whose product is -20.

Step 2 Add the pairs and look for a sum of 1.

Products that = (-20)	Sums of factors
$1 \cdot (-20)$	$1 + (-20) = -19$
$2 \cdot (-10)$	$2 + (-10) = -8$
$4 \cdot (-5)$	$4 + (-5) = -1$
Now check the opposites.	
$(-1) \cdot 20$	$(-1) + 20 = 19$
$(-2) \cdot 10$	$(-2) + 10 = 8$
$(-4) \cdot 5$	$(-4) + 5 = 1$

The two numbers are (-4) and (5).
$(-4)(5) = -20$ and $(-4) + 5 = 1$

GUIDED PRACTICE

17. Find two numbers whose product is 6 and whose sum is -5.

Step 1 List the factor pairs whose product is ____.

Step 2 Add the pairs and look for a sum of ____.

Products that = ( )	Sums of factors
Now check the opposites.	

The two numbers are (   ) and (   ).
(   )(   ) = ____ and (   ) + (   ) = ____

18. Find two numbers whose product is -12 and whose sum is -4.

Step 1 List the factor pairs whose product is ____.

Step 2 Add the pairs and look for a sum of ____.

Products that = ( )	Sums of factors
Now check the opposites.	

The two numbers are (   ) and (   ).
(   )(   ) = ____ and (   ) + (   ) = ____

Section 9.3 Exercises

FOR EXTRA HELP       

To multiply two signed numbers:

1. Answer the Objective A Concept Checks.
2. Answer the odd Objective A Practice Exercises.
3. Answer the even Objective A Practice Exercises.

To divide two signed numbers:

4. Answer the Objective B Concept Checks.
5. Answer the odd Objective B Practice Exercises.
6. Answer the even Objective B Practice Exercises.

To multiply or divide several signed numbers:

7. Answer the Objective C Concept Checks.
8. Answer the odd Objective C Practice Exercises.
9. Answer the even Objective C Practice Exercises.

To evaluate an exponential expression involving signed numbers:

10. Answer the Objective D Concept Check.
11. Answer the odd Objective D Practice Exercises.
12. Answer the even Objective D Practice Exercises.

VOCABULARY REVIEW *Review the Vocabulary Preview for Section 9.3. Study the definitions until you can check* Got It *for every word.*

Use these words to complete each sentence.

product • quotient • repeated addition • repeated multiplication • factors • base • exponent

	Got It	Get Help
13. A(n) ____________ is the result of dividing.		
14. ____________ are numbers that are multiplied to give a product.		
15. A(n) ____________ indicates repeated multiplication.		
16. The ____________ is the symbol appearing immediately before an exponent.		

How will you get help for any vocabulary you are unsure about?

Your instructor ____ MyMathLab ____ A classmate ____ A tutor ____ Other ____

Multiply.

17. $-2 \cdot 12$ **18.** $5 \cdot -8$ **19.** $(-13) \cdot (-3)$ **20.** $(-7) \cdot (-20)$

21. $(2.5) \cdot (8)$ **22.** $(3.4) \cdot (5)$ **23.** $(-10.1) \cdot (9)$ **24.** $7 \cdot (-5.3)$

25. $\left(-\frac{3}{5}\right) \cdot \left(-\frac{10}{9}\right)$ **26.** $\left(-\frac{7}{8}\right) \cdot \left(-\frac{16}{21}\right)$ **27.** $\left(-\frac{3}{12}\right) \cdot \left(\frac{4}{17}\right)$ **28.** $\left(-\frac{2}{3}\right) \cdot \left(-\frac{9}{6}\right)$

Divide.

29. $50 \div (-25)$

30. $(-16) \div (4)$

31. $(-65) \div (-13)$

32. $(-84) \div (-7)$

33. $(-1.6) \div (4)$

34. $(-4.8) \div (-12)$

35. $(-0.35) \div (-0.07)$

36. $(0.23) \div (-0.46)$

37. $\frac{-100}{-20}$

38. $\frac{-64}{-16}$

39. $\left(-\frac{3}{5}\right) \div \left(-\frac{9}{10}\right)$

40. $\left(\frac{6}{5}\right) \div \left(-\frac{12}{20}\right)$

Follow the order of operations to evaluate each expression.

41. $(-14)\cdot(2) \div (-7)\cdot(2)$

42. $(6)\cdot(-8) \div (-12)\cdot(-2)$

43. $(60) \div (-2) \div (3)\cdot(10)$

44. $(-84) \div (-7) \div (4)\cdot(3)$

45. $\left(-\frac{3}{4}\right)\cdot\left(-\frac{1}{3}\right) \div (-7)$

46. $\left(-\frac{8}{3}\right)\cdot\left(\frac{18}{2}\right) \div (-3)$

47. $(16) \div (8)\cdot(2) \div (-0.1)$

48. $(25) \div (5)\cdot(5) \div (-0.1)$

49. $(13)\cdot(-2) \div (-26)\cdot(0)$

50. $(-17)\cdot(-3) \div (-51)\cdot(0)$

51. $(-100) \div (-25) \div (2)$

52. $(120) \div (-40) \div (3)$

Evaluate.

53. 9^2

54. 4^3

55. $(-3)^3$

56. $(-2)^5$

57. -2^4

58. -7^2

59. $(6)^2$

60. $(12)^2$

61. $\left(-\frac{3}{5}\right)^2$

62. $\left(-\frac{6}{7}\right)^2$

63. $-\left(\frac{2}{3}\right)^3$

64. $-\left(\frac{3}{2}\right)^3$

65. 0^4

66. 0^5

67. $(-1)^{10}$

68. $(-1)^9$

Identify the signed numbers whose product and sum result in the given values.

69. Find two numbers whose product is 3 and whose sum is 4.

70. Find two numbers whose product is 7 and whose sum is 8.

71. Find two numbers whose product is -8 and whose sum is 2.

72. Find two numbers whose product is -27 and whose sum is 6.

73. Find two numbers whose product is -18 and whose sum is -7.

74. Find two numbers whose product is -25 and whose sum is -24.

75. Find two numbers whose product is 40 and whose sum is -13.

76. Find two numbers whose product is 42 and whose sum is -13.

Answer the following exercises.

77. In your own words, why are these three fractions equal?

$$\frac{-6}{2} = \frac{6}{-2} = -\frac{6}{2}$$

78. Explain how the rules for determining the sign of a product are related to the rules for determining the sign of a quotient.

goetz Be careful. An exponent applies only to a negative sign when the sign is part of the base.

For more tips and tweets, go to twitter.com/gstbasicmath

79. A friend asked the following question: "What number, when multiplied by itself, is equal to itself?" There are two correct answers to the question. What are they?

80. Dani uses her debit card to withdraw $25 from her savings account every week for 7 weeks. What is the change in the account balance at the end of 7 weeks?

81. During landing, a jet descends 30 feet per second. What is the jet's change in altitude in 1 minute?

82. Paul owns 40 shares of stock whose price increased by $0.75 per share. He also owns 100 shares of a stock whose price decreased by $0.20 per share. What is the total change in the value of Paul's stocks?

83. Greta owns 80 shares of a stock whose price increased by $1.50 per share. She also owns 30 shares of a stock whose price decreased by $2.20 per share. What is the total change in the value of Greta's stocks?

Ions are atoms or molecules with positive or negative electrical charges. The charges of some ions are given in the following box.

Aluminium	+3	Choloride	−1	Magnesium	+2
Oxide	−2	Phosphate	−3	Silver	+1

84. What is the total charge of 6 phosphate ions?

85. What is the total charge of 13 oxide ions?

86. What is the charge of the sum of 8 aluminum ions and 12 oxide ions?

87. What is the charge of the sum of 7 silver ions and 6 magnesium ions?

88. Eight chloride ions are removed from a substance. What is the change in the charge of the remaining substance?

89. Six oxide ions are removed from a substance. What is the change in the charge of the remaining substance?

90. Four ions have a total charge of −11. What are the ions?

91. Four ions have a total charge of +11. What are the ions?

Find the average of each set of signed numbers.

92. 7, −8, −10, 3

93. 20, −30, −40, 10

94. 2, −4, 6, −8, 10, −12

95. 1, −3, 5, −7, 9, −11

96. 10, 20, 30, −5, −15, −40

97. 3, 7, 30, −12, −8, −20

Section 9.3 Question Log

Use this space to write questions. Be sure to get these answered and revisit them when you prepare for your exam.

Page ____ Answered

Page ____ Answered

Page ____ Answered

Page ____ Answered

Mid-Chapter 9 Review Exercises

Before studying Section 9.4, you may find it useful to work the Mid-Chapter Review Exercises in Appendix A.

Section 9.4 The Order of Operations and Signed Numbers

In every chapter that we have learned about a new class of numbers (whole numbers in Chapter 1, fractions in Chapter 2, and decimals in Chapter 3), we have concluded the chapter with the order of operations. We do this again with signed numbers.

The **Objective** for Section 9.4 will help you

A. Apply the order of operations to signed numbers.

VOCABULARY PREVIEW *Check the box that applies.*

	Got It	Must Study
Order of Operations: The **order of operations** is the order in which mathematical operations must be performed. The order of operations is as follows. **First:** Perform operations in groupings and calculate absolute values. **Second:** Perform operations with exponents. **Third:** Perform multiplication and division as they occur from left to right. **Fourth:** Perform addition and subtraction as they occur from left to right.		

Objective A Apply the Order of Operations to Signed Numbers

The Concept When performing operations with signed numbers, we must follow the order of operations and be careful of the signs in our calculations.

Procedure Apply the Order of Operations

First: Perform operations in groupings and calculate absolute values.
Second: Perform operations with exponents.
Third: Perform multiplication and division as they occur from left to right.
Fourth: Perform addition and subtraction as they occur from left to right.

EXAMPLES

1. Evaluate. $8 - (-4) + (-3)$

Write subtraction as addition. $8 - (-4) + (-3) = 8 + (+4) + (-3)$

Add from left to right. $= 12 + (-3)$

$= 9$

2. Evaluate. $16 + (-4)\cdot(-8)$

Multiply. $16 + (-4)\cdot(-8) = 16 + 32$

Add. $= 48$

3. Evaluate. $\dfrac{17 + (-3)}{-7\cdot(-4)}$

Numerators and denominators are groupings. $\dfrac{17 + (-3)}{-7\cdot -4} = \dfrac{14}{28}$

Divide out common factors to simplify the fraction. $= \dfrac{\cancel{14}}{2\cdot\cancel{14}}$

$= \dfrac{1}{2}$

GUIDED PRACTICE

1. Evaluate. $(-3) - (4) + (-7)$

$(-3) - (4) + (-7) = (\quad) + (\quad) + (\quad)$

$= \quad + \quad$

$=$

2. Evaluate. $35 - 15\cdot(-2)$

$35 - 15\cdot(-2) =$

$=$

3. Evaluate. $\dfrac{-42 \div (-21)}{6 - (-8)}$

$\dfrac{-42 \div (-21)}{6 - (-8)} = \dfrac{\quad}{\quad}$

$= \dfrac{\quad}{\quad}$

$= \dfrac{\quad}{\quad}$

When performing the order of operations, use the three *C*'s to keep your work organized.

Choose Copy Calculate

DETAILED EXAMPLE Organizing an Order of Operations Exercise

Evaluate. $(-5 + 3)^2 \div 16$

Choose which operation will be performed.

$(-5 + 3)^2 \div 16$

The addition inside the grouping is calculated first.

Copy everything in the expression that will not change.

$(-5 + 3)^2 \div 16 = (\quad)^2 \div 16$

Copy everything except the calculation in the grouping.

Calculate the operation.

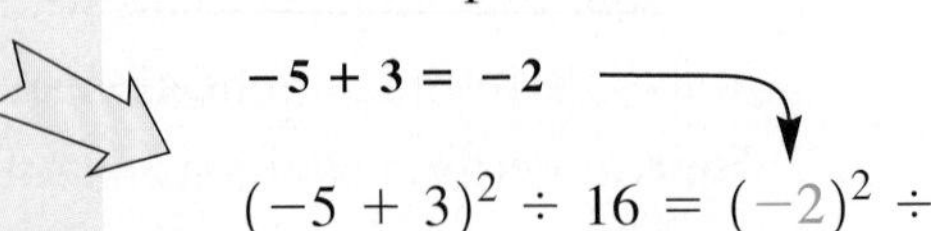

EXAMPLE

4. Evaluate. $(-3)\cdot(-4)^2 + (-4 + -5) \div 3$

Perform operations in groupings.	$(-3)\cdot(-4)^2 + (-4 + -5) \div 3 = (-3)\cdot(-4)^2 + (-9) \div 3$
Evaluate exponents.	$= (-3)\cdot 16 + (-9) \div 3$
Multiply/divide, left to right.	$= -48 + (-9) \div 3$
	$= -48 + (-3)$
Add/subtract, left to right.	$= -51$

GUIDED PRACTICE

4. Evaluate. $(1 - 3)^2 + (5 - 7)^3 \div 2\cdot(-4)$

Perform operations in groupings.	$(1 - 3)^2 + (5 - 7)^3 \div 2\cdot(-4) = (\quad)^2 + (\quad)^3 \div 2\cdot(-4)$
Evaluate exponents.	$= (\quad) + (\quad) \div 2\cdot(-4)$
Multiply/divide, left to right.	$= (\quad) + (\quad)\cdot(-4)$
	$= (\quad) + (\quad)$
Add/subtract, left to right.	$=$

You may have found Guided Practice 4 manageable because the work was organized to perform one operation at a time. If you organize your work this way, you may find the exercises easier and will probably make fewer mistakes.

Concept Check

A1. State the order of operations from memory.

Objective A Practice

Evaluate.

A2. $-7 - 3 + 2$

A3. $-5 - 8 + 4$

A4. $-10 \div 2\cdot 5$

A5. $-6 \div 2\cdot 3$

A6. $(-3) + (-2)\cdot 6 - 5$

A7. $(4) - (6) \div (-2) + (-3)$

A8. $(-1)^3 - (-1)^2 + (-5)\cdot 2$

A9. $(-1)^4 + (-1)^3 - (-3)\cdot(-4)$

A10. $\frac{1}{2} \div \frac{1 + 2}{4}$

A11. $\frac{3}{4} \div \frac{5 + (-2)}{8}$

A12. $\left(\frac{5 - (7 - 12)}{(-5) + (7 - 12)}\right)^2$

A13. $\left(\frac{(-4) + (13 - 8)}{4 - (13 - 8)}\right)^2$

Combining Concepts and Applications

CONCEPT I Getting Ready for Algebra In an algebra course, formulas and equations are often used to solve exercises. Writing several operations into a single formula or equation can help you organize the work necessary for solving the exercise.

EXAMPLE

5. Use one multiplication and one addition operation to write a single equation that will solve the following exercise.

On Monday, the high temperature was 14°C. The high temperature fell 3 degrees a day for the next 4 days. What was the high temperature on Friday?

First: Write a word equation to describe the situation.

New temperature = Old temperature + Change

Second: Find the total change.

Change = Number of days · Change per day

Change $= 4 \cdot (-3)$

Third: Substitute the values into the word equation.

New temperature = Old temperature + Change

$= 14 + 4 \cdot (-3)$

The answer is an expression that requires one multiplication and one addition operation.

GUIDED PRACTICE

5. Use one multiplication and one addition operation to write a single equation that will solve the following exercise.

A child was playing video games at an arcade. He had $3.45 in his pocket at noon. An hour later, he had played 7 games at 25 cents each. How much money did he have left?

First: Write a word equation to describe the situation.

Money left =

Second: Find the total change.

Change = ______________ · ______________ ← Word Equation

Change = · ← Number Equation

Third: Substitute the values into the word equation.

Money left =

=

The answer is an expression that requires one multiplication and one addition operation.

CONCEPT II Adding Parentheses to Make an Equation True Adding a set of parentheses to an expression can change the order in which operations are performed. This can change the value of the expression.

EXAMPLE

6. If needed, add a set of parentheses to make the equation $5 \cdot 2 - 13 = -55$ true.

Step 1 Check to see if the equation is true.

$$5 \cdot 2 - 13 = -55$$
$$10 - 13 = -55$$
$$-3 \neq -55$$

The equation is not true.

Step 2 Add a set of parentheses to change the calculation order.

$$(5 \cdot 2) - 13 = -55$$

The calculation order is not changed with parentheses here. Do not check this equation.

$$5 \cdot (2 - 13) = -55$$

The calculation order is changed with parentheses here. Check this equation.

Step 3 Check to see if the resulting equation is true.

$$5 \cdot (2 - 13) = -55$$
$$5 \cdot (-11) = -55$$
$$-55 = -55$$

The equation $5 \cdot (2 - 13) = -55$ is true.

GUIDED PRACTICE

6. If needed, add a set of parentheses to make the equation $16 \div 2 - 6 = -4$ true.

Step 1 Check to see if the equation is true.

$$16 \div 2 - 6 = -4$$
$$= -4$$
$$\stackrel{?}{=} -4$$

Is the equation true?

Step 2 Add a set of parentheses to change the calculation order.

$$16 \div 2 \quad - 6 = -4$$

Does the calculation order change with parentheses here? Do/do not check this equation.

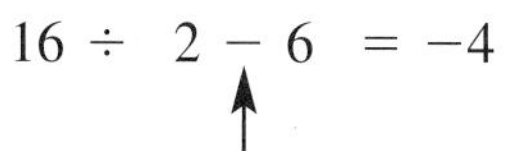

$$16 \div \quad 2 - 6 \quad = -4$$

Does the calculation order change with parentheses here? Do/do not check this equation.

Step 3 Check to see if the resulting equation is true.

$$16 \div \quad 2 - 6 \quad = -4$$
$$= -4$$
$$= -4$$

The equation ________________ is true.

This flowchart is presented to help you organize your thoughts about operations with signed numbers.

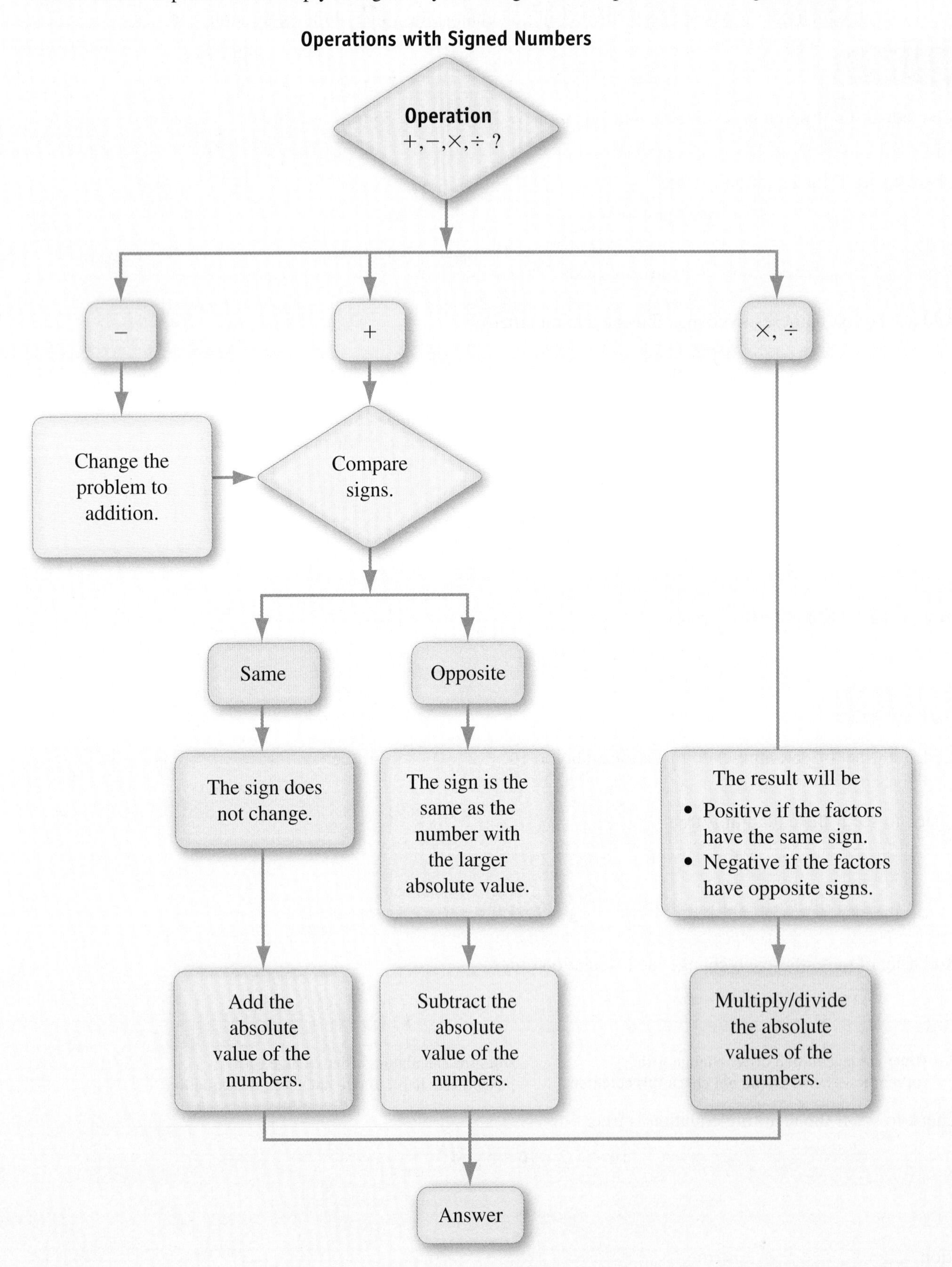

Section 9.4 Exercises

FOR EXTRA HELP  MyMathLab     

To apply the order of operations to signed numbers:

1. Answer the Objective A Concept Check.
2. Answer the odd Objective A Practice Exercises.
3. Answer the even Objective A Practice Exercises.

Follow the order of operations to evaluate each expression.

4. $(4) - 3 + (-7)$

5. $(-5) + 12 - (8)$

6. $14 \div 7 \cdot (-2)$

7. $3 \cdot (-8) \div (4)$

8. $16 - (-8) \cdot 2$

9. $15 \cdot (-2) \div (-1)$

10. $6 \cdot (-5) - 2 \cdot (4)$

11. $-7 \cdot (-7) + 3 \cdot (-8)$

12. $4 + (-3) \cdot (2) + 12$

13. $5 - (7) \cdot (3) + (-35)$

14. $\frac{3}{4} \div \frac{2 + (-5)}{8}$

15. $\frac{3}{4} \cdot \frac{1 + (-7)}{5}$

16. $21 \div (-13 + 6) + 2 \cdot (-3)$

17. $28 \div (-8 - 6) + 2 \cdot (-3)$

18. $(3^3 - 5^2) \cdot 2 \div 4$

19. $(6^2 - 2^3) \div 4 \cdot 2$

20. $3(4 - 7) + 1$

21. $3(6 - 8) + 5$

22. $|7 - 9| - 15 + 21 \div (-7)$

23. $|15 - 17| - 2 + (-12) \div (6)$

24. $5 - 4 + (7 - 9)^2$

25. $6 - 7 + (9 - 5)^2$

26. $13 - (-7 + 3)$

27. $10 - (-30 - 17)$

Evaluate. Treat the numerator and denominator as groupings.

28. $\frac{2 \cdot 3 - (-4)}{7 - 5}$

29. $\frac{(-3) \cdot 4 - (-2)}{-2 + 4}$

30. $\frac{8 \div 2 \cdot (-3)}{15 - 3}$

31. $\frac{(-9) \div 3 \cdot 4}{4 - (-8)}$

32. $\frac{6 + (-3) \cdot 4 - 18}{2^4 + 8 - 4 \cdot 3}$

33. $\frac{(-7) - 5 \cdot 2 + 13}{(-1)^5 + 9 - 8 \cdot 2}$

34. $\frac{(-1)^2 + 3^2}{2^2 + (-4)^2}$

35. $\frac{3^2 - 5^2}{6^2 - 8^2}$

36. $\frac{4(-3) \div 6 + 2}{16 - 5 + 3}$

37. $\frac{5 \cdot 4 \div (-2) + 15}{-13 + 8 \cdot 1}$

38. $\frac{49 - 25 + 1}{10 + (-15)}$

39. $\frac{64 + (-36) + 8}{-21 + 3}$

Find the average in each situation.

40. A bouncer was keeping track of how many people were inside a club. In five-minute intervals, he saw that 12 people entered, 7 left, 9 entered, and 11 entered. What was the average number of people to enter or leave over the four intervals?

41. A basketball team played 6 games. In those games, the team won by 8 points, lost by 12, won by 7, won by 9, lost by 1, and won by 7. What was the average difference in points over the six games?

42. In the last few years, Craig bought and sold several houses. He earned \$12,000, earned \$26,000, lost \$2,500, and earned \$18,000. What was the average amount that Craig earned per house?

43. Delores has four investment accounts, each with a \$1,000 balance. The rates of return were 12%, 5%, −3%, and 2%. What was the average rate of return for the four accounts?

Evaluate.

44. $(-2)^2 \div \frac{4}{3} - \frac{1}{8}$

45. $(-3)^2 \div \frac{9}{2} + \frac{-1}{2}$

46. $(-3) \cdot \frac{5}{6} + \frac{-7}{8}$

47. $(-4) \cdot \frac{3}{20} + \frac{-3}{10}$

48. $\left(\frac{1}{3} - \frac{1}{4}\right) + \left(\frac{1}{6} - \frac{1}{2}\right)$

49. $\left(\frac{1}{4} + \frac{-1}{3}\right) + \left(\frac{1}{8} - \frac{1}{6}\right)$

50. $(0.2)^2 - \frac{4}{100}$

51. $(0.3)^2 - \frac{9}{100}$

52. $-(-7)^2 + 35 \div (-7)$

53. $-(6)^2 + 24 \div (-4)$

54. $-(-1)^4 + (13 - 11)^3$

55. $-(-1)^6 + (12 - 14)^3$

Using one or two multiplication and one addition operation, write an arithmetic expression for each situation. Do not evaluate the expression.

56. Phillipe owns 300 shares of a stock that fell \$1.20 per share and 500 shares that rose \$0.35 per share. What is the total change in Phillipe's stock value?

57. Sandra owns 150 shares of a stock that rose \$3.21 per share and 200 shares that fell \$1.75 per share. What is the total change in Sandra's stock value?

58. Lenny budgeted \$50 to buy lunch for the next 7 days. If he buys a soda each day for \$1.75, what amount of his lunch budget will remain?

59. Malaya makes a monthly mortgage payment of \$850. If she budgeted \$11,000 of her salary for mortgage payments this year, what amount of this budget will remain after 12 months?

If needed, add a set of parentheses to make each equation true.

60. $3 \cdot 4 - 5 = -3$

61. $2 - 3 \cdot 4 = -4$

62. $4 - 4 + 2 = 2$

63. $1 - 2 + 3 = 2$

64. $12 \div 6 \cdot 2 = 1$

65. $18 \div 3 \cdot 2 = 3$

66. $3 \cdot 10 \div 15 = 2$

67. $28 \div 4 \cdot 3 = 21$

68. $14 + 6 \div 2 = 10$

69. $8 + 8 \div 4 = 4$

smith Be careful! Some of the values are negative, even though they don't have a negative sign written in front.

For more tips and tweets, go to twitter.com/gstbasicmath

Before you begin the remaining exercises, you may want to review the table of Key Words in Mathematics in Section 1.9.

Match each word expression with its corresponding math expression. Each expression is used only once.

70. The average of negative four and ten

71. The sum of twice ten and four

72. Half the difference between ten and negative four

73. The product of two and negative four, multiplied by ten

74. Twice the sum of ten and four

75. The quotient of ten and two, decreased by four

76. The quotient of ten and two less than negative four

77. The quotient of ten and negative four, increased by two

78. The product of two and negative four, less ten

79. Negative ten, decreased by half of negative four

a. $\frac{10 - (-4)}{2}$

b. $2(-4) \cdot 10$

c. $2(10) + 4$

d. $\frac{10}{-4} + 2$

e. $\frac{-4 + 10}{2}$

f. $2(-4) - 10$

g. $2(10 + 4)$

h. $-10 - \frac{-4}{2}$

i. $\frac{10}{2} - 4$

j. $\frac{10}{-4 - 2}$

Section 9.4 Question Log

Use this space to write questions. Be sure to get these answered and revisit them when you prepare for your exam.

Page ____ **Answered** ☐	**Page** ____ **Answered** ☐
Page ____ **Answered** ☐	**Page** ____ **Answered** ☐

Chapter 9 Organizer

VOCABULARY

Use the following steps to review the vocabulary for Chapter 9.

1. *Write the definition for each word from memory.*
2. *Compare the definitions you have written with the definitions in the Vocabulary Preview.*
3. *Study any definitions that you could not remember or that you defined incorrectly.*

9.1

Positive Number • Negative Number • Signed Numbers • Less Than • Greater Than • Opposites • Absolute Value

9.2

Zero Pair • Additive Inverse • Commutative Property of Addition • Associative Property of Addition

9.3

Product • Quotient • Repeated Addition • Repeated Multiplication • Factors • Base • Exponent

9.4

Order of Operations

PROCEDURES

Procedure/Topic	Steps	Example
Represent Real-World Quantities with Signed Numbers (Section 9.1)	**Step 1** Determine if the quantity is positive or negative. **Step 2** Assign the appropriate value.	What signed number best describes this statement? A plane descended 8,000 feet. Since a descending plane's altitude is decreasing, use a negative number. Descending 8,000 feet can be represented with the signed number −8,000.
Graph Signed Numbers on a Number Line (Section 9.1)	**Step 1** Draw a number line with an appropriate range. **Step 2** Graph each signed number on the number line.	Graph 3, −2, 1.5, and $3\frac{1}{2}$ on a number line. The range goes beyond the smallest value −2 and largest value $3\frac{1}{2}$. [number line: −3, 0, 1.5, 3, $3\frac{1}{2}$] Since 1.5 and $3\frac{1}{2}$ did not fall on tick marks, we were sure to include labels for them.
Compare Signed Numbers (Section 9.1)	**Step 1** Visualize each number on a number line. **Step 2** Insert an inequality symbol that opens towards the larger number.	Insert < or > to make a true statement. −50 □ −70 [number line: −70, −50, 0] −50 [>] −70 −50 is greater than −70.

Procedure/Topic	Steps	Example
Simplify Opposites (Section 9.1)	**Step 1** Write the math phrase in words. **Step 2** Simplify.	Simplify. $-(-13)$ $-(-13)$ is read as "the opposite of negative thirteen." $-(-13) = 13$
Find the Absolute Value of a Number (Section 9.1)	Determine the number's distance from zero.	What is the value of $\lvert -4 \rvert$? −4 0 -4 is 4 units from zero, so $\lvert -4 \rvert = 4$.
Order Signed Numbers (Section 9.1)	**Step 1** Simplify each number. **Step 2** Visualize the numbers on a number line. **Step 3** Order the original numbers.	Order the numbers from least to greatest. $\lvert -8 \rvert, -3, -(-5), 0$ $\lvert -8 \rvert = 8 \quad -3 = -3 \quad -(-5) = 5 \quad 0 = 0$ −3 0 5 8 $-3, 0, -(-5), \lvert -8 \rvert$
Add Numbers with the Same Sign (Section 9.2)	▶ **NOTE** First, visualize the addition using signed discs. It will help you understand the steps in the procedure. **Step 1** Add the absolute values of the numbers. **Step 2** Notice that the sign of the answer matches the sign of the original numbers.	Add. $(-11) + (-15)$ Visualization: $(-11) + (-15) = -26$ 11 negative discs + 15 negative discs = 26 negative discs $\lvert(-11)\rvert = 11$ and $\lvert(-15)\rvert = 15$ $11 + 15 = 26$ $(-11) + (-15) = -26$ When adding two negative numbers, the answer is negative.

Procedure/Topic	Steps	Example
Add Numbers with Opposite Signs (Section 9.2)	▶ **NOTE** First, visualize the addition using signed discs. It will help you understand the steps in the procedure. **Step 1** Find the difference between the absolute values of the numbers. **Step 2** Use the sign of the number with the larger absolute value.	Add. $(17) + (-28)$ Visualization: $(17) + (-28) = -11$ (+17) + (−17) =0 (−11) **Subtract 28 − 17 = 11 to find how many negatives remain.** $\|(17)\| = 17 \quad \|(-28)\| = 28$ $28 - 17 = 11$ $(17) + (-28) = -11$ Because $\|-28\|$ is larger than $\|17\|$, the answer is negative.
Subtract Signed Numbers (Section 9.2)	**Step 1** Change the subtraction symbol to an addition symbol. **Step 2** Change the sign of the second number. **Step 3** Add.	Subtract. $-5 - 13$ $-5 - 13 = -5 + (-13)$ $= -18$
Add and Subtract Signed Numbers Using a Number Line (Section 9.2)	**Step 1** Change double signs to a single operation. **Step 2** Graph the first number. **Step 3** Move left if subtracting. Move right if adding. **Step 4** Move a number of units equal to the second number.	Add. $18 + (-23)$ $18 + (-23) = 18 - 23$ [number line: 0, 18; arrow −23] We pass 0, so the answer is negative. $18 + (-23) = -5$
Multiply Two Signed Numbers (Section 9.3)	**Step 1** Determine the sign of the product. **a.** If the factors have opposite signs, the product is negative. **b.** If the factors have the same sign, the product is positive. **Step 2** Multiply the absolute values of the factors.	Multiply. $7 \cdot (-5)$ The factors have opposite signs. The product is negative. $7 \cdot (-5) = -(7 \cdot 5)$ Multiply the absolute values of the factors. $= -35$
Divide Two Signed Numbers (Section 9.3)	**Step 1** Determine the sign of the quotient. **a.** If the numbers have opposite signs, the quotient is negative. **b.** If the numbers have the same sign, the quotient is positive. **Step 2** Divide the absolute values of the numbers.	Divide. $(-12) \div (-3)$ Same signs → Positive quotient $(-12) \div (-3) = 4$

Procedure/Topic	Steps	Example
Multiply or Divide Several Signed Numbers (Section 9.3)	**Step 1** Determine the sign of the product or quotient. • An odd number of negatives results in a negative answer. • An even number of negatives results in a positive answer. **Step 2** Multiply or divide using the absolute value of the numbers. • Follow the order of operations. Multiply or divide from left to right.	Evaluate. $(-3)\cdot(8) \div (-6)$ There is an even number of negatives. The answer is positive. $(-3)\cdot(8) \div (-6) = +\underbrace{(3)\cdot(8)}_{24} \div (6)$ $= +24 \div (6)$ $= +4$
Evaluate an Exponential Expression Involving Signed Numbers (Section 9.3)	**Step 1** Write the repeated multiplication. Identify the base and exponent carefully. **Step 2** Determine the sign of the product and multiply the absolute value of the factors.	Evaluate. $(-2)^3$. The exponent is 3, and the base is (-2). $(-2)^3 = (-2)\cdot(-2)\cdot(-2)$ $= -(2)\cdot(2)\cdot(2)$ $= -8$
Apply the Order of Operations (Section 9.4)	**First:** Perform operations in groupings and calculate absolute values. **Second:** Perform operations with exponents. **Third:** Perform multiplication and division as they occur from left to right. **Fourth:** Perform addition and subtraction as they occur from left to right.	Evaluate. $12 + (-3)\cdot(-5)$ $12 + (-3)\cdot(-5) = 12 + 15$ $= 27$

Chapter 9 Review Exercises

9.1

Represent each real-world quantity with a signed number.

1. A diver's altitude when she is 75 feet under water

2. A falcon's speed when it is flying at 175 miles per hour

3. Pikes Peak has an elevation of 7,400 feet above sea level.

4. Shelia's checking account is overdrawn \$48.50.

Insert < or > between each pair of signed numbers to make a true statement.

5. $-11 \square -21$ **6.** $-24 \square -19$ **7.** $-7.89 \square -7.8$ **8.** $-3.4 \square -3.46$

Simplify.

9. $-(24)$ **10.** $-(-52)$ **11.** $-|-6|$ **12.** $-|56|$

Graph each set of numbers on the number line. Then order the numbers from least to greatest.

13. $\{-9, -4\}$ **14.** $\{3, -7\}$ **15.** $\{-(-9), |-8|, -5\}$ **16.** $\{|-3|, -(-2), 5\}$

17. In your own words, why is every negative number less than every positive number? Be sure to discuss the number line in your answer.

18. In your own words, why is every positive number greater than every negative number? Be sure to discuss the number line in your answer.

9.2

Use signed discs to draw a visualization of each sum or difference.

19. $-5 + (-6)$ **20.** $-3 + (-2)$ **21.** $-5 + 7$ **22.** $2 + (-6)$

Use a number line to draw a visualization of each sum or difference.

23. $-7 + (-5)$ **24.** $18 - (+34)$ **25.** $-5 - 10$ **26.** $-4 - 7$

Find each sum or difference.

27. $-12 + (-4)$ **28.** $-6 + -18$ **29.** $10 - 16$

30. $8 - 24$ **31.** $-11 - (-9)$ **32.** $-13 - (-2)$

33. $37 + (-24)$ **34.** $-28 + 24$ **35.** $-\frac{3}{8} + \frac{3}{2}$

36. $\frac{2}{3} - \left(-\frac{3}{2}\right)$ **37.** $|21| + (-3) - 30 - (-2)$ **38.** $|-15| - 17 + (-4) - (-8)$

Find each of the missing balances in the following check ledger.

39.

Trans Type	Date	Description of Transaction	Payment/ Debit (–)	Fee (if any)	Deposit/ Credit (+)	Balance
	4-May	Balance Forward				$132.85
Check 212	5-May	Credit card pmt	$48.24			
ATM	8-May	Needed cash	$20.00	$3.00		
Deposit	17-May	Pay day!			$324.62	
Check 213	18-May	Bill repayment for roommate	$117.43			
ATM	24-May	Needed cash	$40.00	$2.00		

9.3

Follow the order of operations to evaluate each expression.

40. $4 \cdot (-3)$ **41.** $(-5) \cdot (-7)$ **42.** $-15 \div 3$ **43.** $-20 \div (-4)$

44. $3 \cdot (-2) \cdot (-8)$ **45.** $-2 \cdot (-14) \div (-4)$ **46.** $(-3)^2$ **47.** -4^2

48. $\frac{4}{15} \cdot (-3)^2$ **49.** $\frac{-3}{14} \cdot (-7)^2$ **50.** $\left(3 \cdot \frac{-2}{3}\right)^2$ **51.** $\left(-2 \div \frac{1}{3}\right)^2$

Answer the following exercises.

52. Find two numbers whose product is -20 and whose sum is -8.

53. Find two numbers whose product is 14 and whose sum is -9.

9.4

Follow the order of operations to evaluate each expression.

54. $-8 + 12 \div 4$

55. $18 - 9 \div 3$

56. $2 \cdot (-3)^2 - 6$

57. $-10 - 3^2 \cdot 2$

58. $3 - 2(4 - 6)$

59. $4 + 3(5 - 8)$

60. $\dfrac{2 - 7}{8 + (-6)}$

61. $\dfrac{3 + 2}{15 - 8}$

62. $\dfrac{4 - 7 + 3}{7^2}$

63. $\dfrac{5 - 8 + 3}{5^2}$

Find the average of each set of signed numbers.

64. $\{-10, 4, -8, 3, -6, 5\}$

65. $\{20, -33, 15, -17, -2, 7\}$

Use one multiplication operation and one addition operation to write a single equation that will solve each exercise.

66. Byron is washing his car. He has \$8.45 in his pocket when he starts washing his car. If he spends 18 quarters to wash his car, how much money does he have left?

67. Tania has budgeted \$450 for entertainment this year. If she subscribes to Netflix, which costs her \$20 a month, how much will she have left for entertainment this year?

Chapter 9 Practice Test

Use a signed number to represent each real-world quantity.

1. Xailu deleted 54 songs from his MP3 player.

2. Carrie added 2.5 GB of ram to her laptop.

Insert $<$ or $>$ between each pair of signed numbers to make a true statement.

3. $-18 \; \square \; -125$

4. $-5.61 \; \square \; -5.6$

Graph each set of signed numbers on a number line. Then order the numbers from least to greatest.

5. $\{|-4|, -(-8), -5\}$

6. $\{-4, -(-3), |-2|\}$

Evaluate.

7. $5 + (-12)$

8. $-4.8 + (-12)$

9. $(-5)(-6)$

10. $21 \div (-7)$

11. $-\dfrac{3}{4} + \dfrac{7}{12}$

12. $130 - 175$

13. $-3 - (-9)$

14. $-(-3)^3$

15. Before the holiday season, Alexis's checking account had a balance of \$284. After the holidays, the balance was $-\$193$. What is the change in the balance?

16. A truck has spilled concrete on a road. The concrete is flowing down a hill at a rate of 3.4 feet per second. How far will the concrete flow in 8 seconds?

Follow the order of operations to evaluate each expression.

17. $(-12) \div 2 + 4$

18. $7 - 9 + 11$

19. $10 - 3^2$

20. $12 - 4^2$

21. $5(-7 + 3)$

22. $\dfrac{-4 + 5}{-15 - 10}$

23. $(-2)(-1) + (4 - 7)^2$

24. $-\dfrac{1}{8} - \dfrac{1}{2} \cdot \dfrac{1}{4}$

Find the average of each set of signed numbers.

25. $\{-10, 4, -9\}$

26. $\{20, -3, 15, -4\}$

Use the graph to answer each question.

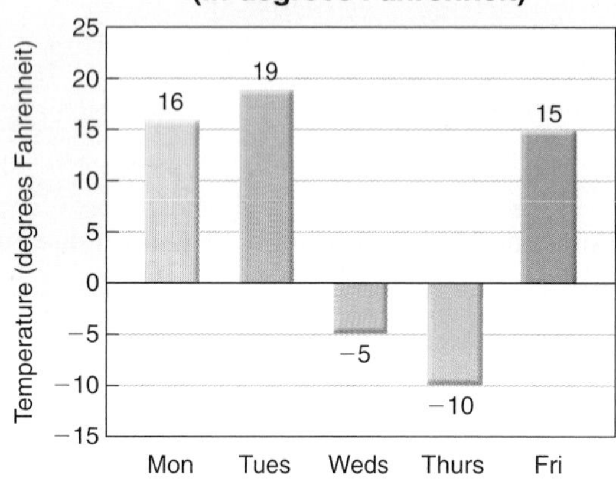

27. What was the average high temperature from Monday to Friday?

28. a. What was the greatest increase in high temperature from one day to the next?

b. What was the greatest decrease in high temperature from one day to the next?

c. What was the greatest change, either positive or negative, in high temperature from one day to the next?

29. Water freezes at 32°F. How many degrees below freezing was the high temperature on Thursday?

30. What three-day period had an average high temperature of 0° F?

CHAPTER

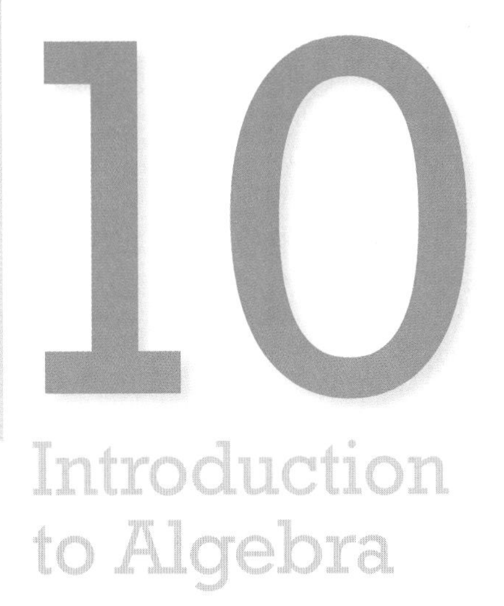

Introduction to Algebra

By representing unknown quantities with variables, algebra makes it easy to organize and solve many real-world problems.

Monthly Cost of a Phone Plan Without Using Algebra:
The cost is \$60 plus \$0.12 for each minute over 500.

Monthly Cost of a Phone Plan Using Algebra:

$$c = 60 + 0.12m,$$

where c = monthly cost and
m = number of minutes over 500

Algebra allows us to describe "monthly cost" and "number of minutes over 500" with the variables c and m.

In Exercises 58 and 59 in Section 10.1, we use algebra to compare the costs of two products and determine the best buy.

Section 10.1 Introduction to Variables

To build a foundation for algebra, it is important to understand variables. This section will help you to reinforce what you have already learned, as well as introduce new concepts related to variables.

The **Objectives** in Section 10.1 will help you

- A Write variable expressions.
- B Evaluate variable expressions.
- C Understand the vocabulary of variable expressions.

VOCABULARY PREVIEW *Check the box that applies.*

	Got It	Must Study
Variable: A **variable** is a symbol, usually a letter of the alphabet, that represents an unknown number.		
Variable Expression: A **variable expression** is a mathematical phrase that contains at least one variable.		
Evaluating an Expression: Finding the value of a variable expression by replacing each variable with a given value is called **evaluating an expression.**		
Term: A **term** can be a number, variable, or product of a number and one or more variables. Terms are separated by addition or subtraction.		
Variable Term: Any term with a variable is a **variable term.**		
Constant Term: Any term without a variable is a **constant term.** Without a variable, the term cannot change; so it is a constant value.		
Coefficient: The **coefficient** is the numerical factor of a term. • A *coefficient of a variable term* is the number that multiplies the variable part of the term. • The *coefficient of a constant term* is the constant term itself. A coefficient includes the positive or negative sign written in front of the term.		
Like Terms: Like terms have identical variable parts. Identical variable parts have the same variables with the same exponents.		

Study these words when they appear in the text. After the section, test your understanding by completing the Vocabulary Review in the section exercises.

Objective A Write Variable Expressions

The Concept Anya is having a party and needs to know if she will have enough soda for her guests. She has two bottles of soda. A friend will bring 3 cases of soda. Since Anya does not know how many bottles are in a case, she does not know how many total bottles of soda she will have for the party. However, she can describe the amount of soda using a variable and a variable expression.

INTERACTIVE DEFINITION Variable and Variable Expression

A **variable** is a symbol, usually a letter of the alphabet, that represents an unknown number.
A **variable expression** is a mathematical phrase that contains at least one variable.

EXAMPLE

1. Write a variable expression that describes the number of soda bottles at Anya's party.

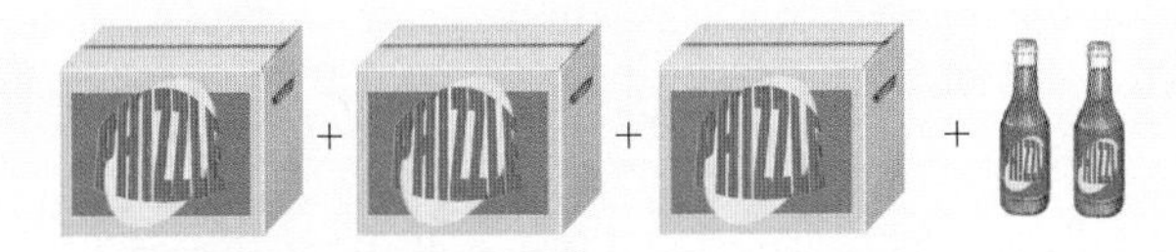

Step 1 Choose a variable.

Since the number of bottles in a case is unknown, we let b = the number of bottles per case.

Step 2 Write a variable expression.

There are 3 cases and 2 bottles of soda. We can describe the number of bottles using the variable expression $3b + 2$.

GUIDED PRACTICE

1. Write a variable expression that describes the number of lollipops shown.

Step 1 Choose a variable.

Since the number of ______ in a bag is unknown, we let ____ = the number of lollipops per bag.

Step 2 Write a variable expression.

There are ____ bags and ____ lollipops. We can describe the number of lollipops using the variable expression ____ + ____.

DO YOU UNDERSTAND how to choose a variable and write a variable expression? Got It Get Help

▶ **NOTE** It may seem strange to have both variables and numbers in your answer. Answers to algebra exercises often include both variables and numbers.

Procedure Write Variable Expressions

Step 1 Choose a variable to represent the unknown.

Step 2 Write a variable expression that describes the situation.

EXAMPLES

2. Write a variable expression that describes the number of toothpicks in the picture.

Step 1 Choose a variable. Let t = number of toothpicks in one box.

Step 2 Write a variable expression. Number of toothpicks = $3t + 6$

GUIDED PRACTICE

2. Write a variable expression that describes the number of batteries in the picture.

Let ____ = ______________

Number of batteries =

3. John used 5 binder clips from 3 boxes that he bought last week. Write a variable expression that describes the number of binder clips remaining.	**3.** A chef purchased 3 cases of corn and used 14 cans in one night. Write a variable expression that describes the number of cans of corn remaining.
Step 1 Choose a variable. Let c = number of binder clips in each box. **Step 2** Write a variable expression. Number of binder clips $= 3c - 5$	Let ____ = ______________ Number of cans of corn = ______

Concept Check

A1. In your own words, what is a variable?

A2. The expression $4b$ represents the total number of batteries in four boxes. Why is b multiplied by 4?

Objective A Practice

Write a variable expression to describe the number of objects shown or described.

A3.

A4.

A5. A college bookstore has six boxes of notebooks in stock and 31 individual notebooks on display. How many total notebooks are in the bookstore?

A6. A gas station has eight boxes of jerky sticks in stock and 14 jerky sticks on display. How many total jerky sticks are in the gas station?

A7. Candice withdraws $1,500 from an account. How much money is still in the account?

A8. Caleb deposits $3,500 in an escrow account that already has money in it. What is the new balance of the escrow account?

Objective B Evaluate Variable Expressions

The Concept Finding the value of a variable expression by replacing each variable with its given value is called **evaluating an expression.** The variable expression $3b + 2$ describes the quantity of soda pictured. If each case contains 12 bottles, you can evaluate $3b + 2$ when $b = 12$.

DETAILED EXAMPLE Evaluating Variable Expressions

Evaluate $3b + 2$ when $b = 12$.

$$\text{Number of soda bottles} = 3b + 2$$

Replace b with 12

$$= 3(\quad) + 2$$
$$= 3(12) + 2$$
$$= 36 + 2$$
$$= 38$$

Use empty parentheses when evaluating variable expressions. This will help you stay organized and perform the arithmetic correctly.

Procedure Evaluate Variable Expressions

Step 1 Replace the variable with the given number.
Write ( ) in place of the variable and insert the given value within the parentheses.

Step 2 Follow the order of operations to simplify the expression.

EXAMPLES

4. Evaluate $2x - 7$ when $x = 5$.

Step 1 Replace the variable with the given number.
Step 2 Simplify.

$$2x - 7 = 2(5) - 7$$
$$= 10 - 7$$
$$= 3$$

5. Evaluate $10 - x$ when $x = -6$.

Step 1 Replace the variable with the given number.
Step 2 Simplify.

$$10 - x = 10 - (-6)$$
$$= 10 + 6$$
$$= 16$$

6. Evaluate $\frac{2x - y}{2y}$ when $x = 3, y = -6$.

Step 1 Replace the variables with the given number.
Step 2 Simplify.

$$\frac{2x - y}{2y} = \frac{2(3) - (-6)}{2(-6)}$$
$$= \frac{6 - (-6)}{-12}$$
$$= \frac{6 + 6}{-12}$$
$$= \frac{12}{-12}$$
$$= -1$$

GUIDED PRACTICE

4. Evaluate $3x - 8$ when $x = 4$.

$$3x - 8 = 3(\quad) - 8$$
$$=$$
$$=$$

5. Evaluate $8 - y$ when $y = -7$.

$$8 - y = 8 - (\quad)$$
$$=$$
$$=$$

6. Evaluate $\frac{5x - y}{y}$ when $x = 8, y = -5$.

$$\frac{5x - y}{y} = \frac{5(\quad) - (\quad)}{(\quad)}$$
$$=$$
$$=$$
$$=$$
$$=$$

Concept Check

B1. In your own words, what does it mean to evaluate a variable expression?

B2. When evaluating $2x$ for $x = -3$, which of the following is correct, a or b? Explain why.
a. $2x = 2 - 3$ or **b.** $2x = 2(-3)$

B3. In your own words, why is replacing a variable with a set of parentheses useful? In your answer, explain parts a and b in Exercise B2.

Objective B Practice

Evaluate each expression using $x = 4, y = 8$, and $z = -2$.

B4. $2x + 5$ **B5.** $3x + 4$ **B6.** $10 - z$ **B7.** $25 - z$

B8. $\frac{2x + y}{y}$ **B9.** $\frac{3x - y}{y}$ **B10.** $\frac{4xz}{y}$ **B11.** $\frac{4y}{xz}$

B12. $y^2 - 2z$ **B13.** $x^2 - 3z$ **B14.** $\frac{4x}{y} - (2x + 1)$ **B15.** $\frac{3y}{x} - (4y + 6)$

Objective C Understand the Vocabulary of Variable Expressions

The Concept To communicate ideas in algebra, it is important to know the vocabulary. Each of the following Example and Guided Practice pairs will help you learn and use the vocabulary for this chapter.

INTERACTIVE DEFINITION Terms in a Variable Expression

A **term** can be a number, variable, or product of a number and one or more variables. Terms are separated by addition or subtraction.

EXAMPLE

7. Identify the terms. $3x - 4y - 7$

Separate the terms with a line.

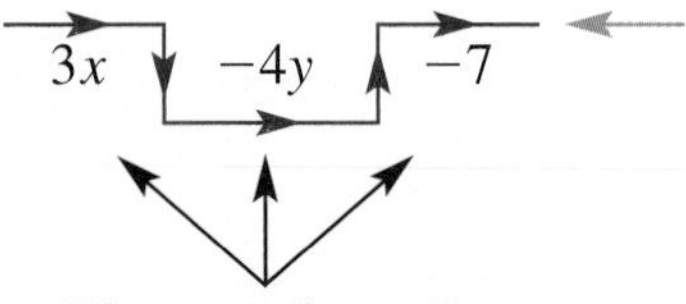

There are 3 terms.

The terms are $3x$, $-4y$, and -7.

To separate the terms, draw the vertical parts of the line in front of each + and −.

GUIDED PRACTICE

7. Identify the terms. $5z - 2y - 3 + z$

Separate the terms with a line.

$5z \quad -2y \quad -3 \quad +z$

There are ____ terms.

The terms are ________________.

DO YOU UNDERSTAND how to identify terms in a variable expression? Got It Get Help

INTERACTIVE DEFINITION Constant Terms and Variable Terms

There are two types of terms: *constant terms* and *variable terms*.

Any term without a variable is a **constant term.**
Any term with a variable is a **variable term.**

EXAMPLES

8. Identify the constant terms.

$$2x + 7 - 2y - 8$$

Separate the terms with a line.

$$2x \mid +7 \mid -2y \mid -8$$

Constant terms (arrows point to $+7$ and -8)

Of the four terms, only 7 and -8 have no variable. The constant terms are 7 and -8.

9. Identify the variable terms.

$$3x^2 + 4xy - 5y + 2$$

Separate the terms with a line.

$$3x^2 \mid +4xy \mid -5y \mid +2$$

Variable terms (arrows point to $3x^2$, $+4xy$, and $-5y$)

Of the four terms, $3x^2$, $4xy$, and $-5y$ have at least one variable. The variable terms are $3x^2$, $4xy$, and $-5y$.

GUIDED PRACTICE

8. Identify the constant terms.

$$2y - 3 + 15 - x$$

Separate the terms with a line.

$$2y \quad -3 \quad +15 \quad -x$$

Of the ____ terms, only ____ and ____ have no variable. The constant terms are ____ and ____.

9. Identify the variable terms.

$$4y^2 - 6 + 5yz - 3x$$

Separate the terms with a line.

$$4y^2 \quad -6 \quad +5yz \quad -3x$$

Of the four terms, ____, ____, and ____ have at least one variable. The variable terms are ____, ____, and ____.

DO YOU UNDERSTAND how to identify a constant term?	Got It	Get Help
DO YOU UNDERSTAND how to identify a variable term?	Got It	Get Help
DO YOU UNDERSTAND the difference between a constant term and a variable term?	Got It	Get Help

INTERACTIVE DEFINITION The Coefficient and Variable Part of a Term

Variable terms have two parts: the *coefficient* and the *variable part.*

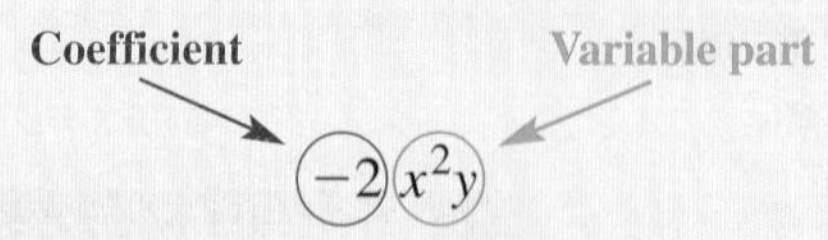

Coefficient: The **coefficient** is the numerical factor of a term.

- A coefficient of a variable term is the number that multiplies the variable part of the term.
- The coefficient of a constant term is the constant term itself.

A coefficient includes the positive or negative sign written in front of the term.

The **variable part** of a term consists of the variables and their exponents.

EXAMPLE

10. Identify the coefficients and variable parts of each term.

$$3x^2 - xy - 7$$

Separate the terms with a line.

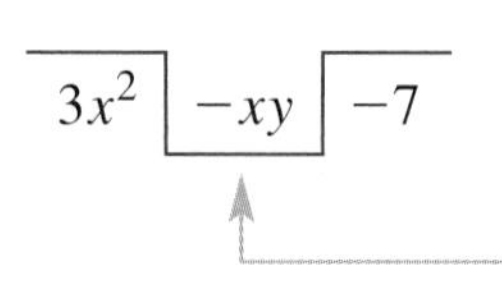

When a negative sign is written in front of a variable, it means that the coefficient is −1.

Terms	$3x^2$	$-xy$	-7
Coefficients	3	-1	-7
Variable Parts	x^2	xy	

GUIDED PRACTICE

10. Identify the coefficients and variable parts of each term.

$$-y^2 + 5xy - 3$$

Separate the terms with a line.

$$-y^2 \quad +5xy \quad -3$$

Terms			
Coefficients			
Variable Parts			

DO YOU UNDERSTAND how to identify the coefficient of a term? Got It | Get Help

DO YOU UNDERSTAND how to identify the variable part of a term? Got It | Get Help

INTERACTIVE DEFINITION Like Terms

Like terms have identical variable parts.

Identical variable parts have the same variables with the same exponents.

EXAMPLE

11. Identify all like terms.

$$2x^2y - 4x + 3 - x^2 + 2x - 4$$

Draw arrows that point to the pairs of like terms.

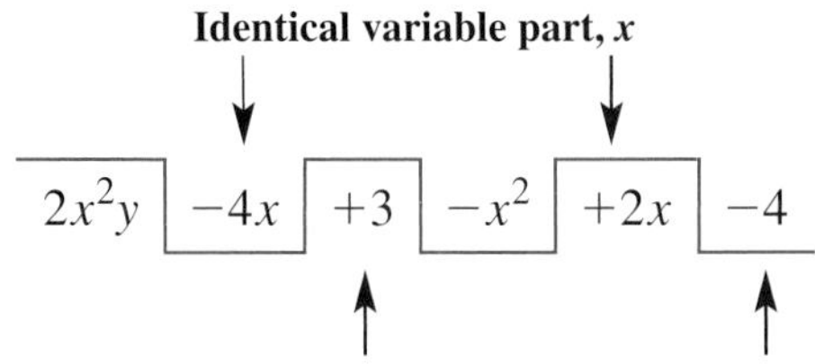

$-4x$ and $2x$ are like terms.

3 and -4 are like terms.

$2x^2y$ and $-x^2$ are unlike terms because the variable parts are different.

GUIDED PRACTICE

11. Identify all like terms.

$$-xy^2 + 5xy + 2 - 2x - 3xy - 7$$

Draw arrows that point to the pairs of like terms.

Identical variable part, ____

$-xy^2$ $+5xy$ $+2$ $-2x$ $-3xy$ -7

Constant terms have ____ variable parts.

____ and ____ are like terms.

____ and ____ are like terms.

$-xy^2$ and $-2x$ are ____ terms because the variable parts are different.

DO YOU UNDERSTAND how to identify like terms? Got It Get Help

Concept Check

C1. What is the difference between a variable term and a constant term?

C2. Sometimes it is helpful to draw a vertical line to separate terms. Why might this be useful?

C3. In your own words, what are like terms?

Objective C Practice

C4. $2x - 5y - 6x + 3$
- **a. Identify the number of terms.**
- **b. List any constant terms.**
- **c. List any variable terms.**
- **d. List any coefficients.**
- **e. List any sets of like terms.**

C5. $3r - 8y - 6r + 5$
- **a. Identify the number of terms.**
- **b. List any constant terms.**
- **c. List any variable terms.**
- **d. List any coefficients.**
- **e. List any sets of like terms.**

C6. $-7 + 3y - y^2 - 4y$
- **a. Identify the number of terms.**
- **b. List any constant terms.**
- **c. List any variable terms.**
- **d. List any coefficients.**
- **e. List any sets of like terms.**

C7. $-z + 4z^2 - 2z^2 + 14$
- **a. Identify the number of terms.**
- **b. List any constant terms.**
- **c. List any variable terms.**
- **d. List any coefficients.**
- **e. List any sets of like terms.**

Combining Concepts and Applications

CONCEPT I Using Formulas Juanita is deciding between two cell phone plans. To compare the costs, she will write formulas to help her find the cost of each plan based on the number of minutes used. Juanita expects to use about 500 minutes each month.

The cost of a basic Spirit phone plan is \$30.00 per month plus \$0.45 per minute for each minute over 200 minutes.

The cost of a basic Horizon phone plan is \$60.00 per month plus \$0.45 per minute for each minute over 450 minutes.

EXAMPLE

12. a. Write a formula to find Juanita's monthly cost for a Spirit plan.

b. Use the formula from part a to find the cost if she uses 500 minutes in one month.

a. Write a formula.

Let c = cost per month.
Let m = minutes over 200.

Cost equals \$30 plus \$0.45 per minute over 200.

$$c = 30 + 0.45m$$

b. Evaluate the formula.

Find the number of minutes used that go over 200 minutes.

$$\begin{aligned} m &= 500 - 200 \\ &= 300 \end{aligned}$$

Replace m with this value and evaluate the formula.

$$\begin{aligned} c &= 30 + 0.45m \\ &= 30 + 0.45(300) \\ &= 30 + 135 \\ &= 165 \end{aligned}$$

It will cost Juanita \$165 to talk 500 minutes on the Spirit plan.

GUIDED PRACTICE

12. a. Write a formula to find Juanita's monthly cost for a Horizon plan.

b. Use the formula from part a to find the cost if she uses 500 minutes in one month.

a. Write a formula.

Let c = cost per month.
Let m = minutes over 450.

Cost equals \$60 plus \$0.45 per minute over 450.

$$c =$$

b. Evaluate the formula.

Find the number of minutes used that go over 450 minutes.

$$\begin{aligned} m &= \\ &= \end{aligned}$$

Replace m with this value and evaluate the formula.

$$\begin{aligned} c &= \\ &= \\ &= \\ &= \end{aligned}$$

It will cost Juanita ______ to talk 500 minutes on the Horizon plan.

Section 10.1 Exercises

FOR EXTRA HELP      

To write variable expressions:

1. Answer the Objective A Concept Checks.
2. Answer the odd Objective A Practice Exercises.
3. Answer the even Objective A Practice Exercises.

To evaluate variable expressions:

4. Answer the Objective B Concept Checks.
5. Answer the odd Objective B Practice Exercises.
6. Answer the even Objective B Practice Exercises.

To understand the vocabulary of variable expressions:

7. Answer the Objective C Concept Checks.
8. Answer the odd Objective C Practice Exercises.
9. Answer the even Objective C Practice Exercises.

VOCABULARY REVIEW *Review the Vocabulary Preview for Section 10.1. Study the definitions until you can check* Got It *for every word.*

Use these words to complete each sentence.

variable • variable expression • evaluating an expression • term • variable term
constant term • coefficient • like terms

	Got It	Get Help
10. Any term with a variable is a ____________.		
11. The numerical factor of a term is called the ____________.		
12. Finding the value of a variable expression by replacing each variable with a given value is called ____________.		
13. ____________ have identical variable parts.		
14. A ____________ is a symbol, usually a letter of the alphabet, that represents an unknown number.		
15. Any term without a variable is a ____________.		
16. A ____________ is a mathematical phrase that contains at least one variable.		
17. A ____________ can be a number, variable, or product of a number and one or more variables.		

How will you get help for any vocabulary that you are unsure about?

Your instructor ____ MyMathLab ____ A classmate ____ A tutor ____ Other ____

Write a variable expression to describe each situation.

18. A college bookstore has 5 boxes of T-shirts and 23 individual T-shirts on display. How many T-shirts does the bookstore have?

19. A gas station has 18 boxes of foam cups and 40 individual foam cups on display. How many foam cups does the gas station have?

20. Angel withdraws \$500 from an account with a balance of b. How much money does Angel have left in her account?

21. Buffy deposits \$900 in a college fund with a balance of b. How much money does Buffy have left in her account?

22. a. Write a variable expression that describes the number of chocolate candies.

b. If there are 125 chocolate candies in each bag, how many candies are there in total?

23. a. Write a variable expression that describes the number of candy bars.

b. If there are 36 candy bars in each box, how many candy bars are there in total?

Evaluate each variable expression when $x = 3, y = 9$, and $z = -4$.

24. $5x - 7$

25. $4x - 5$

26. $x - z$

27. $y - z$

28. $4xz$

29. $5yz$

30. $\dfrac{y}{2x - y}$

31. $\dfrac{2y}{3x + y}$

32. $6(z + 3) - 10$

33. $7(x - 4) - 8$

34. $2z^2$

35. $3z^2$

36. $8 - 5(x - 2)$

37. $7 - 2(3y - 4)$

38. $\dfrac{yz}{x}$

39. $\dfrac{24x}{yz}$

40. $\dfrac{1}{2}x + \dfrac{1}{2}$

41. $\dfrac{1}{4}y + \dfrac{1}{4}$

42. $3x - (4x + 1)$

43. $2y - (y + 6)$

Identify the following parts of each variable expression.

44. $z + 5z + 2x + 3$

a. Identify the number of terms.
b. List any constant terms.
c. List any variable terms.
d. List any coefficients.
e. List any sets of like terms.

45. $4r + y + r + 5$

a. Identify the number of terms.
b. List any constant terms.
c. List any variable terms.
d. List any coefficients.
e. List any sets of like terms.

46. $-9 + 3y^2 - y - 4$

a. Identify the number of terms.
b. List any constant terms.
c. List any variable terms.
d. List any coefficients.
e. List any sets of like terms.

47. $3x^3 + 4 - x + 14$

a. Identify the number of terms.
b. List any constant terms.
c. List any variable terms.
d. List any coefficients.
e. List any sets of like terms.

48. $4x^4 + 5x - x^4 - 2x$
a. Identify the number of terms.
b. List any constant terms.
c. List any variable terms.
d. List any coefficients.
e. List any sets of like terms.

49. $10m - 4m^3 + 6m + m^3$
a. Identify the number of terms.
b. List any constant terms.
c. List any variable terms.
d. List any coefficients.
e. List any sets of like terms.

50. $4r + 5t - 9 - r^5 - 2r$
a. Identify the number of terms.
b. List any constant terms.
c. List any variable terms.
d. List any coefficients.
e. List any sets of like terms.

51. $10l - 4m^3 + 6l + m - 8$
a. Identify the number of terms.
b. List any constant terms.
c. List any variable terms.
d. List any coefficients.
e. List any sets of like terms.

52. Using the formula $A = lw$, find the area of a rectangular TV screen that is 17 inches by 23 inches.

53. Using $A = lw$, find the area of a rectangular laptop screen that is 22 cm by 29 cm.

54. Sloane and Melissa are installing a wallpaper border around their rectangular living room. If the room measures 30 feet by 25 feet, how much wallpaper do they need for the border? Use $P = 2l + 2w$ to find the room's perimeter.

55. The Winicks want to put a fence around their garden. If the garden is 13 feet by 21 feet, how much fencing will they need to enclose the garden? Use $P = 2l + 2w$ to find the garden's perimeter.

56. Use the formula $d = rt$, where d = distance, r = rate (speed), and t = time. Find the distance that Kumar traveled if he drove 55 miles per hour for 4 hours.

57. Use the formula $d = rt$, where d = distance, r = rate (speed), and t = time. Find the distance that Ralucca traveled if she drove 35 miles per hour for 9 hours.

58. Diana is deciding between two cell phone plans. Spirit charges $58 per month plus $0.05 per minute for each minute over 300 minutes. Horizon charges $39 per month plus $0.25 per minute for each minute over 300 minutes. Diana usually talks about 400 minutes per month.

a. Write a variable expression that describes the charges for each plan.
b. Use the variable expressions in part a to find the monthly bill for each plan.
c. Which plan is a better buy for Diana?

59. A customer is deciding between two car rental companies. Clifton's car rental charges $32 per day plus $0.35 per mile for each mile over 100 miles. Kyle's car rental charges $37 per day plus $0.15 per mile for each mile over 100 miles. The customer plans to drive 140 miles on his trip.

a. Write a variable expression that describes the charges for each car rental company.
b. Use the variable expressions in part a to find the cost of a 140-mile trip that lasts one day.
c. Which plan is a better buy for the customer?

tobey These types of situations are often encountered. The use of variables here will help you understand the facts more clearly.
For more tips and tweets, go to twitter.com/gstbasicmath

SELF-ASSESSMENT

	YES	NO
1. On your last test, were you satisfied with your score? *If you answered yes, good job. Do not continue here.* *If you answered no, answer the following.*		
2. Did you preview each section of the textbook before your instructor taught it in class?		
3. Did you complete all of the homework before it was due?		
4. Did you get help with the homework exercises that you didn't understand?		

5. Set two goals that will help you become better prepared for your next test.

Goal 1: ______________________

Goal 2: ______________________

Section 10.1 Question Log

Use this space to write questions. Be sure to get these answered and revisit them when you prepare for your exam.

Page ____ **Answered** ☐	**Page** ____ **Answered** ☐
Page ____ **Answered** ☐	**Page** ____ **Answered** ☐

Section 10.2 Operations with Variable Expressions

To use variable expressions efficiently, you must learn how to simplify them. Simplifying variable expressions involves the same rules as simplifying numerical expressions. For example, simplifying requires following the order of operations and adding like terms.

The **Objectives** in Section 10.2 will help you

- A Simplify variable expressions by combining like terms.
- B Simplify variable expressions using multiplication.
- C Simplify variable expressions using the order of operations.

VOCABULARY PREVIEW *Check the box that applies.*

	Got It	Must Study
Term: A **term** is a number, variable, or product of a number and one or more variables. Terms are separated by addition or subtraction operations.		
Coefficient: The **coefficient** is the numerical factor of a term. • A coefficient of a variable term is the number that multiplies the variable part of the term. • The coefficient of a constant term is the constant term itself. A coefficient includes the positive or negative sign written in front of the term.		
Like Terms: Like terms have identical variable parts. Identical variable parts have the same variables with the same exponents.		
Collect Like Terms: To **collect like terms,** use the commutative property of addition to reorder and group like terms together.		
Combine Like Terms: To **combine like terms,** add or subtract their coefficients.		
Distributive Property: When a factor multiplies a grouping, the **distributive property** states that the factor must multiply every term in the grouping.		

Study these words when they appear in the text. After the section, test your understanding by completing the Vocabulary Review in the section exercises.

Objective A Simplify Variable Expressions by Combining Like Terms

The Concept Combining like terms is a process used to organize and simplify an expression. To simplify variable expressions with like terms, we group the like terms together and then add or subtract them. By adding or subtracting the coefficients of like terms, those terms with identical variable parts are combined into a single term. The result is an expression that is written more simply.

Combining like terms is similar to cleaning a messy stockroom. The following pictures represent cases of myPods, CDs, and RAM.

To clean the stockroom, the cases are organized and grouped together.

A messy stockroom . . .	. . . is organized by collecting like items together	. . . and simplified by adding like items into single piles.
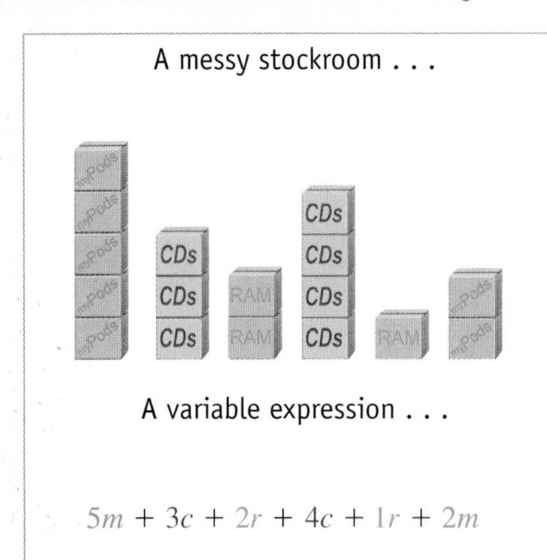	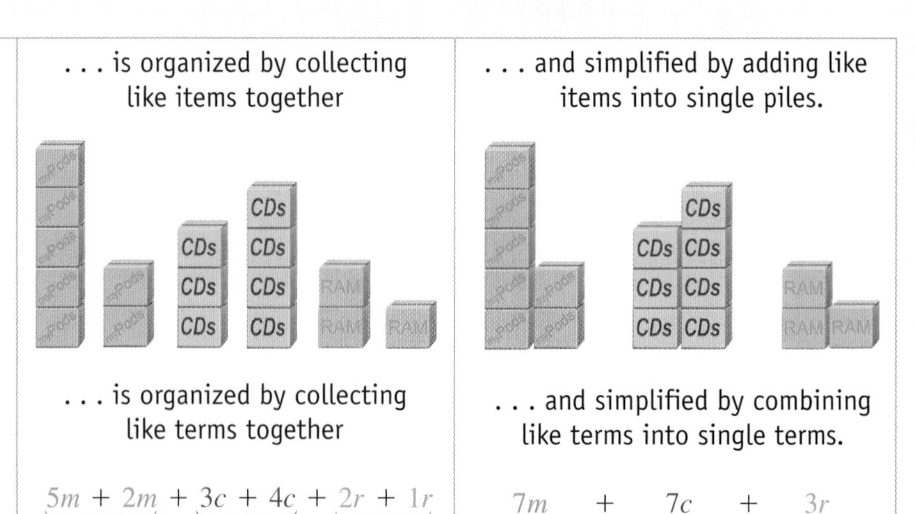	
A variable expression . . .	. . . is organized by collecting like terms together	. . . and simplified by combining like terms into single terms.
$5m + 3c + 2r + 4c + 1r + 2m$	$\underbrace{5m + 2m} + \underbrace{3c + 4c} + \underbrace{2r + 1r}$	$7m \quad + \quad 7c \quad + \quad 3r$

To simplify a variable expression, we collect and combine like terms.

Collect like terms: To collect like terms, we use the commutative property of addition to reorder the terms and group like terms together.

$$5m + 3c + 2r + 4c + 1r + 2m = \underbrace{5m + 2m} + \underbrace{3c + 4c} + \underbrace{2r + 1r}$$

Combine like terms: To combine like terms, we add or subtract their coefficients. In the variable expression $5m + 2m$, we add $5m$ to $2m$.

$$\begin{aligned} 5m + 2m &= (5 + 2)m \\ &= 7m \end{aligned}$$

Procedure Simplify by Combining Like Terms

Step 1 Collect like terms.
Use the commutative property to reorder and group like terms.

Step 2 Combine like terms.
Add the coefficients of the like terms.

EXAMPLES

1. Simplify. $5x + 10 + 2x + 3$

Step 1 Collect like terms. $5x + 10 + 2x + 3 = \underbrace{5x + 2x}_{x\text{-terms}} + \underbrace{10 + 3}_{\text{constants}}$

Step 2 Combine like terms: $= 7x \quad + \quad 13$

Add the variable terms.
Add the constant terms.

GUIDED PRACTICE

1. Simplify. $6x + 14 + 4x + 9$

$6x + 14 + 4x + 9 = \underbrace{\qquad\qquad}_{x\text{-terms}} \underbrace{\qquad\qquad}_{\text{constants}}$

$=$

2. Simplify. $\frac{1}{4}x + 1 + \frac{1}{2}x + 8$

Step 1 Collect like terms.

$$\frac{1}{4}x + 1 + \frac{1}{2}x + 8 = \underbrace{\frac{1}{4}x + \frac{1}{2}x}_{x\text{-terms}} + \underbrace{1 + 8}_{\text{constants}}$$

Step 2 Combine like terms.

$$= \frac{1}{4}x + \frac{2}{4}x + 1 + 8$$

$$= \frac{3}{4}x + 9$$

2. Simplify. $\frac{1}{3}x + 11 + \frac{5}{6}x + 18$

$$\frac{1}{3}x + 11 + \frac{5}{6}x + 18 = \underbrace{\qquad}_{x\text{-terms}} \; \underbrace{\qquad}_{\text{constants}}$$

Remember to build like fractions. →

$=$

$=$

When collecting or combining negative terms in a variable expression, remember that the sign of the term is part of the coefficient.

DETAILED EXAMPLES Collecting and Combining Negative Terms

Simplify. $5x + 1 - 2x$

Step 1 Collect like terms. $5x + 1 - 2x = 5x - 2x + 1$ ← **Notice that we include the negative sign when we move $-2x$.**

Step 2 Combine like terms. $= 5x + (-2x) + 1$

$= 3x + 1$

To combine like terms, add. Since a negative sign is in front of $2x$, we add $-2x$.

EXAMPLES

3. Simplify. $3n - 10 - 5n + 3$

Step 1 Collect like terms.

$$3n - 10 - 5n + 3 = \underbrace{3n + (-5n)}_{n\text{-terms}} + \underbrace{(-10) + 3}_{\text{constants}}$$

Step 2 Combine like terms.

$$= -2n + (-7)$$

$$= -2n - 7$$

4. Simplify. $-4 - 7y + 2 + 9y$

Step 1 Collect the like terms.

$$-4 - 7y + 2 + 9y = \underbrace{(-7y) + 9y}_{y\text{-terms}} + \underbrace{(-4) + 2}_{\text{constants}}$$

Step 2 Combine like terms.

$$= 2y + (-2)$$

$$= 2y - 2$$

GUIDED PRACTICE

3. Simplify. $7z - 2 - 3z + 5$

$$7z - 2 - 3z + 5 = \underbrace{\qquad}_{x\text{-terms}} \; \underbrace{\qquad}_{\text{constants}}$$

$=$

4. Simplify. $-8 - 8t + 4 + 6t$

$$-8 - 8t + 4 + 6t = \underbrace{\qquad}_{t\text{-terms}} \; \underbrace{\qquad}_{\text{constants}}$$

$=$

$=$

Concept Check

A1. Simplify.

a. like objects: 3 pencils + 2 pencils =

b. like fractions: $\frac{3}{5} + \frac{1}{5} =$

c. like terms: $5x - 2x =$

A2. In mathematics, what does the word *like* mean?

A3. Write two unlike variable terms. Describe the reason(s) your terms are unlike.

Objective A Practice

Simplify each variable expression.

A4. $2x + 3x$

A5. $7z - 3z$

A6. $6 - 5y - 10$

A7. $4 - 7t - 26$

A8. $-6z + 9z - 14$

A9. $-2x + 17 + 4x$

A10. $11 + 7y - y$

A11. $15 + 8n - n$

A12. $3 + 2x - 3x + 13$

A13. $4 + 5z - 6z + 4$

A14. $13y - 9 + y - 14$

A15. $7h - 8 + h - 17$

A16. $-\frac{1}{4}x + \frac{3}{4}x - 3$

A17. $\frac{4}{5}y - \frac{2}{5}y - 23$

A18. $2.4r - 10s + 3.2s + 4r$

A19. $10t - 4.5v - 3.6t + 5v$

Objective B Simplify Variable Expressions Using Multiplication

The Concept Variable expressions can be simplified using the associative and commutative properties of multiplication.

ASSOCIATIVE AND COMMUTATIVE PROPERTIES OF MULTIPLICATION	
The associative property of multiplication lets us *regroup* factors.	The commutative property of multiplication lets us *reorder* factors.
$5(12x) = (5 \cdot 12)x$ $= 60x$	$3x \cdot 5 = 3 \cdot 5 \cdot x$ $= 15x$

When a factor multiplies a grouping, we use the distributive property to simplify the expression. In $2(x + 3)$, 2 multiplies the grouping $(x + 3)$. We can use the distributive property to simplify this variable expression to $2x + 6$.

INTERACTIVE DEFINITION The Distributive Property

When a factor multiplies a grouping, the **distributive property** states that it must multiply every term in the grouping. $a(b + c) = a \cdot b + a \cdot c$

EXAMPLE

5. Distribute the 2 in $2(x + 3)$.

$$2(x + 3) = 2 \cdot (x) + 2 \cdot (3)$$
$$= 2x + 6$$

GUIDED PRACTICE

5. Distribute the 4 in $4(2x + 7)$.

$$4(2x + 7) = 4 \cdot (\quad) + 4 \cdot (\quad)$$
$$=$$

DO YOU UNDERSTAND how to distribute a factor? Got It | Get Help

Procedure Simplify Using Multiplication

Multiply.

Use the distributive, associative, and commutative properties as appropriate.

EXAMPLES

6. Simplify. $2(5y)$

Regroup the factors using the associative property. Then multiply.

$$2(5y) = (2 \cdot 5)y$$
$$= 10y$$

7. Simplify. $10x \cdot 4$

Reorder the factors using the commutative property. Then multiply.

$$10x \cdot 4 = 10 \cdot 4 \cdot x$$
$$= 40x$$

8. Simplify. $4(6y - 2)$

Distribute the 4 to remove the parentheses.

$$4(6y - 2) = 4 \cdot (6y) + 4 \cdot (-2)$$
$$= 24y - 8$$

GUIDED PRACTICE

6. Simplify. $8(7x)$

$$8(7x) = (\quad)$$
$$=$$

7. Simplify. $11z \cdot 6$

$$11z \cdot 6 =$$
$$=$$

8. Simplify. $4(5x - 1)$

$$4(5x - 1) =$$
$$=$$

9. Simplify. $-3(y - 2)$	**9.** Simplify. $-3(x - 8)$
Distribute the -3 to remove the parentheses. $-3(y - 2) = -3 \cdot (y) - 3 \cdot (-2)$ $= -3y + 6$ **Since $-3(-2)$ is positive 6, we write + in front of the 6.**	$-3(x - 8) =$ $=$ **Since $-3(-8)$ is positive 24, we write _____ in front of the 24.**

Concept Check

B1. In the variable expression $5(5x + 4)$, why can't we perform the addition inside the parentheses?

B2. When a factor multiplies a grouping, it must multiply _____ term in the grouping.

B3. Find the value of $2(3 + 4)$ using the two methods in parts a and b.
- **a.** Use the order of operations. Simplify the grouping and then multiply by two.
- **b.** Distribute the 2 and then add.
- **c.** What do you notice about the answers from parts a and b?

Objective B Practice

Simplify each variable expression.

B4. $2(4x)$ **B5.** $5(7y)$ **B6.** $3x \cdot 3$ **B7.** $12z \cdot 4$

B8. $2(x + 6)$ **B9.** $4(y + 3)$ **B10.** $7(4x - 7)$ **B11.** $8(3x - 6)$

B12. $-6(6x + 1)$ **B13.** $-5(2x + 2)$ **B14.** $-8(3x - 1)$ **B15.** $-1(4x - 1)$

Objective C Simplify Variable Expressions Using the Order of Operations

The Concept When using the distributive property or combining like terms, we must follow the order of operations.

- **Distributive property:** Multiplication is used in the distributive property. Apply the distributive property in the multiplication/division step of the order of operations.
- **Combining like terms:** Addition is used to combine like terms. Combine like terms in the addition/subtraction step of the order of operations.

Procedure Use the Order of Operations

First: Perform operations in groupings.

Second: Evaluate exponents.

Third: Multiply and divide from left to right (distribute).

Fourth: Add and subtract from left to right (combine like terms).

EXAMPLES

10. Simplify. $5x + 3(2x) - 10$

Multiply. $5x + 3(2x) - 10 = 5x + 6x - 10$

Combine like terms. $= 11x - 10$

11. Simplify. $\frac{24z}{2} + 11z$

Perform the division by dividing out the common factor. $\frac{24z}{2} + 11z = \frac{12 \cdot \cancel{2}z}{\cancel{2}} + 11z$

$= 12z + 11z$

Combine like terms. $= 23z$

12. Simplify. $-7(z^2 + 3) + 2(3)$

Since z^2 and 3 are *unlike* terms, we cannot simplify inside the grouping.

Distribute. $-7(z^2 + 3) + 2(3) = -7z^2 - 21 + 2(3)$

Multiply. $= -7z^2 - 21 + 6$

Combine like terms. $= -7z^2 - 15$

GUIDED PRACTICE

10. Simplify. $4(2x) + 6(3x) - 114$

$4(2x) + 6(3x) - 114 =$

$=$

11. Simplify. $\frac{10z}{5} + 28z$

$\frac{10z}{5} + 28z =$

$=$

$=$

12. Simplify. $-2(n - 8) + 5(6)$

Since n and -8 are __________, we cannot simplify inside the grouping.

$-2(n - 8) + 5(6) =$

$=$

$=$

Concept Check

C1. At what step of the order of operations can the distributive property be applied?

C2. At what step of the order of operations can combining like terms be performed?

Objective C Practice

Simplify each variable expression.

C3. $6(3t) + 5$

C4. $6(5y) + 10$

C5. $\frac{4z}{2} - 3z$

C6. $\frac{8x}{4} + 3x$

C7. $2(n + 8) + 4n$

C8. $4(n + 2) + 7n$

C9. $-6(y - 4) + 5(6) + 4y$

C10. $-3(m - 9) + 9(3) + 7m$

C11. $2(n - 8) - 5(n + 1)$

C12. $7(t^2 - 2) - 4(t^2 + 1)$

Combining Concepts and Applications

CONCEPT I Finding Perimeter and Area When a figure has side lengths that are variable expressions, we can still find the perimeter and area of the figure.

EXAMPLES

13. Find the perimeter.

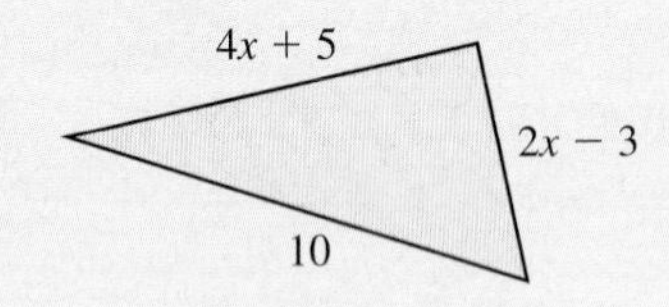

Add the lengths of all sides.

$$P = \text{Side}_1 + \text{Side}_2 + \text{Side}_3$$
$$P = (4x + 5) + (2x - 3) + (10)$$
$$P = 4x + 5 + 2x - 3 + 10$$
$$P = 4x + 2x + 5 - 3 + 10$$
$$P = 6x + 12 \text{ units}$$

14. Find the area.

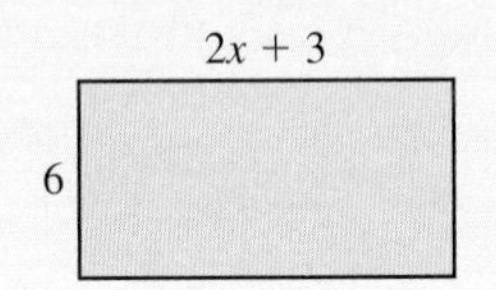

Multiply the length and width of the rectangle.

$$A = \text{Length} \cdot \text{Width}$$
$$A = (6) \cdot (2x + 3)$$
$$A = 12x + 18 \text{ square units}$$

GUIDED PRACTICE

13. Find the perimeter.

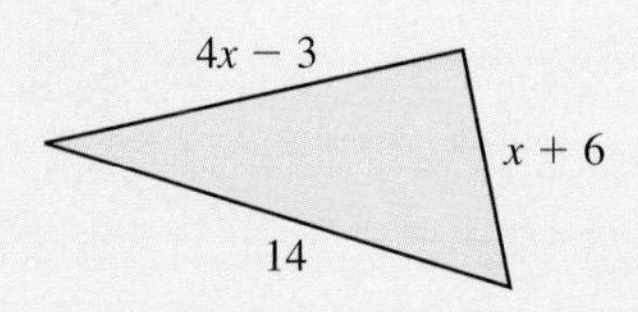

Add the lengths of all sides.

$$P = \text{Side}_1 + \text{Side}_2 + \text{Side}_3$$
$$P = (\qquad) + (\qquad) + (\qquad)$$
$$P =$$
$$P =$$
$$P =$$

14. Find the area.

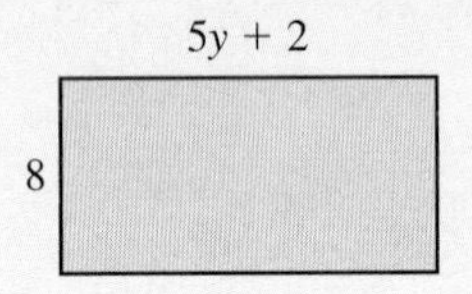

Multiply the length and width of the rectangle.

$$A = \text{Length} \cdot \text{Width}$$
$$A =$$
$$A =$$

Section 10.2 Exercises

To simplify variable expressions by combining like terms:

1. Answer the Objective A Concept Checks.
2. Answer the odd Objective A Practice Exercises.
3. Answer the even Objective A Practice Exercises.

To simplify variable expressions using multiplication:

4. Answer the Objective B Concept Checks.
5. Answer the odd Objective B Practice Exercises.
6. Answer the even Objective B Practice Exercises.

To simplify variable expressions using the order of operations:

7. Answer the Objective C Concept Checks.
8. Answer the odd Objective C Practice Exercises.
9. Answer the even Objective C Practice Exercises.

VOCABULARY REVIEW *Review the Vocabulary Preview for Section 10.2. Study the definitions until you can check* Got It *for every word.*

Use these words to complete each sentence.

term • coefficient • like terms • collect like terms • combine like terms • distributive property

	Got It	Get Help
10. To __________, add their coefficients.		
11. The numerical factor of a term is called its __________.		
12. A __________ is a number, variable, or product of a number and one or more variables.		
13. When a factor multiplies a grouping, the __________ states that it must multiply every term in the grouping.		
14. To __________, use the commutative property of addition to reorder and group like terms together.		
15. __________ have identical variable parts.		

How will you get help for any vocabulary that you are unsure about?

Your instructor ____ MyMathLab ____ A classmate ____ A tutor ____ Other ____

If the items are like terms, combine them. If the items are unlike terms, write *unlike*.

16. a. 6 students + 2 students =
b. 4 pens + 2 students =
c. $7z + 2z =$
d. $8x + 3y =$

17. a. 8 pencils − 2 pencils =
b. 5 teachers − 4 pens =
c. $9x - 4x =$
d. $6z - 1t =$

Simplify each variable expression by combining like terms.

18. $7x + 3x$

19. $8z - 4z$

20. $6x - 10x$

21. $4t - 7t$

22. $3x + 2x - 3x + 13x$

23. $4z + 2z - z + 4z$

24. $3y - 9 + y - 4$

25. $5h - 6 + 2h - 12$

26. $4x + 8 - 10x - 2$

27. $5x + 3 - 12x - 8$

28. $\frac{1}{5}y + 4 - \frac{3}{5}y$

29. $\frac{3}{7}y + 5 - \frac{6}{7}y$

30. $-\frac{1}{2}x + \frac{3}{4}x - 6$

31. $\frac{2}{5}y - \frac{2}{3}y - 23$

32. $6.2x - 5s + 3.2x + 4s$

33. $10x - 4v - 3.6x + 5v$

Answer each question.

34. a. Write a variable expression that represents the following pictures. Let h = one hippo and z = one zebra.

b. What are the like animals? Why can they be combined?

c. Simplify the variable expression from part a so that it represents the pictures as simply as possible.

35. a. Write a variable expression that represents the following pictures. Let t = one turtle and f = one frog.

b. What are the like animals? Why can they be combined?

c. Simplify the variable expression from part a so that it represents the pictures as simply as possible.

Simplify each variable expression using multiplication.

36. $5(7x)$ **37.** $4(3x)$ **38.** $2y \cdot 6$ **39.** $4z \cdot 8$

40. $2(x + 5)$ **41.** $4(y + 6)$ **42.** $3(4x - 3)$ **43.** $6(3x - 6)$

44. $-9(2x + 1)$ **45.** $-8(3x + 2)$ **46.** $-6(3x - 1)$ **47.** $-8(4x - 1)$

Simplify each variable expression using the order of operations.

48. $3(2x) - 3x + 13$ **49.** $4(2z) - z + 4$ **50.** $\frac{2y}{2} + y - 14$ **51.** $\frac{3h}{3} + 2h - 17$

52. $4y - y - x + 24$ **53.** $-11x + 3y - y - 23$

54. $4s - 10 + 3(2r) + 4r$ **55.** $10t - 45 - 3v + 5(4v)$

56. $5n + 2(n + 7)$ **57.** $7n + 4(n + 1)$ **58.** $2x - 3(4x + 5)$

59. $9x - 5(x + 6)$ **60.** $8z - 2(6z - 3)$ **61.** $6z - 3(4z - 7)$

62. $4(3) + 7m - 2(m - 9)$ **63.** $5(6) + 3m - 5(m - 2)$ **64.** $-3(t - 2) + 4(t + 1)$

65. $-7(t - 3) + 6(2t - 5)$ **66.** $-2(x + 4) - 4(3 - 2x)$ **67.** $6(x + 1) - 6(2 + x)$

goetz The distributive property deals with multiplication, not addition. In $2(x + 3) + 4$, the 2 distributes, the 4 doesn't because 4 is added.

For more tips and tweets, go to twitter.com/gstbasicmath

Solve each application.

68. Find the perimeter of the triangle.

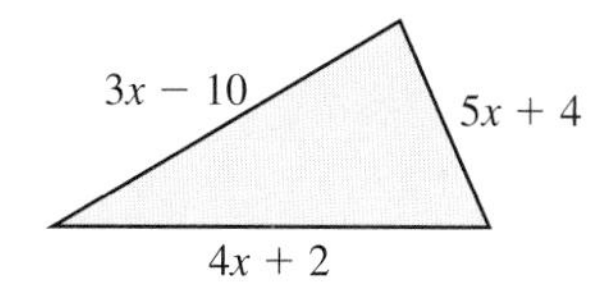

69. Find the perimeter of the pentagon.

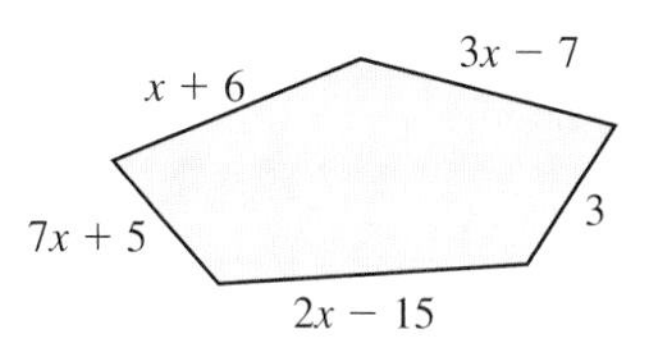

70. a. Find the perimeter of the rectangle.
b. Find the area of the rectangle.

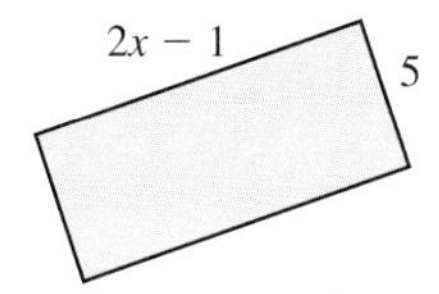

71. a. Find the perimeter of the rectangle.
b. Find the area of the rectangle.

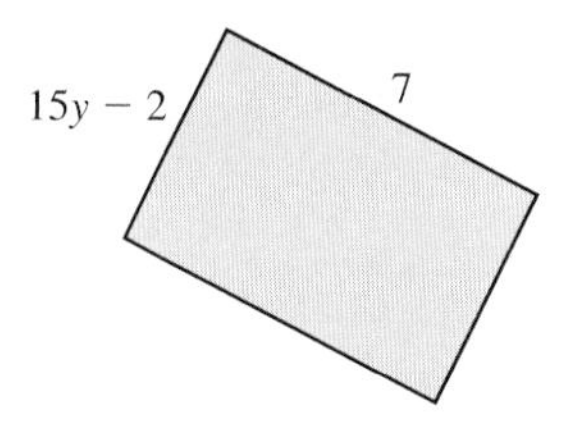

72. Find the area of the triangle. The formula for the area of a triangle is $A = \frac{1}{2}bh$.

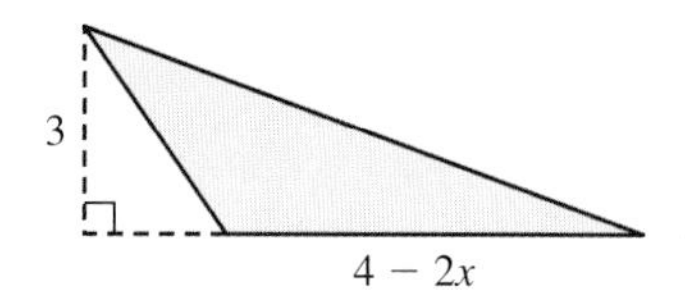

73. Find the area of the triangle. The formula for the area of a triangle is $A = \frac{1}{2}bh$.

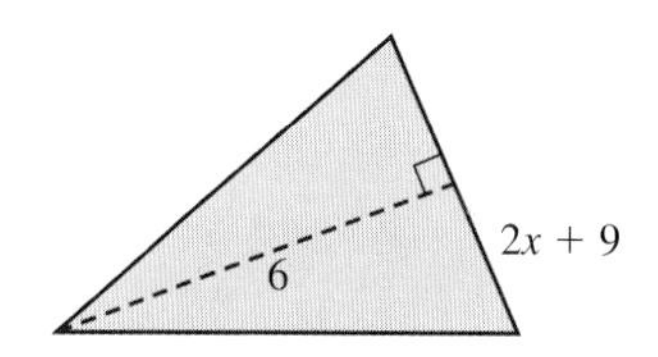

74. A rectangle, measured in feet, has a width of $2x + 5$ and a length of 9. A square that measures 2 feet by 2 feet is cut out of the rectangle.

a. Draw a picture of the resulting figure.
b. What is the remaining area of the rectangle?

75. A triangle, measured in feet, has a base of $2x + 12$ and a height of 10. A rectangle that is 5 feet by 6 feet is cut out of the triangle.

a. Draw a picture of the resulting figure.
b. What is the remaining area of the triangle?

Section 10.2 Question Log

Use this space to write questions. Be sure to get these answered and revisit them when you prepare for your exam.

Page ______ **Answered** ☐	**Page** ______ **Answered** ☐
Page ______ **Answered** ☐	**Page** ______ **Answered** ☐

Section 10.3 Solving One-Step Equations

Solving an equation is one of the most important skills in algebra. When we solve an equation, we are finding the value of a variable that will make the equation true.

What must x equal to make the equation $x + 3 = 4$ true?

$$x + 3 = 4$$
$$x = 1$$

We solved this equation using the basic fact $1 + 3 = 4$.

Most equations are too complicated to solve using basic facts. To solve most equations, we use a series of steps that get the variable alone on one side of the equal sign. When the variable is alone, or isolated, the solution is easily identified.

Isolate the variable to solve an equation.

$$y + 35 = 115$$
$$y + 35 - 35 = 115 - 35$$
$$y = 80$$

Subtracting 35 from each side **of the equation isolates the variable to identify the solution.**

In this section, you will learn the steps for isolating a variable and solving an equation.

The **Objectives** in Section 10.3 will help you

- **A** Determine if a number is a solution to an equation.
- **B** Use inverse operations to isolate a variable.
- **C** Solve equations of the form $x + a = b$.
- **D** Solve equations of the form $ax = b$.

VOCABULARY PREVIEW *Check the box that applies.*

	Got It	Must Study
Equation: An **equation** is a mathematical statement stating that two expressions are equal.		
Solution: A **solution** to an equation is a value of the variable that makes the equation true.		
Inverse Operations: Inverse operations are operations that undo each other.		
Addition Property of Equations: The **addition property of equations** lets us add or subtract the same number to each side of an equation to get an equivalent equation. For example, if $a = b$, then $a + c = b + c$.		
Multiplication Property of Equations: The **multiplication property of equations** lets us multiply or divide each side of an equation by the same nonzero number to get an equivalent equation. For example, if $a = b$, then $a \cdot c = b \cdot c$.		
Isolate a Variable: To **isolate a variable,** use a series of steps that will get the variable by itself on one side of the equal sign.		
Equivalent Equations: Equivalent equations are equations with the same solution.		
Solve an Equation: To **solve an equation,** isolate the variable on one side of the equal sign.		

Study these words when they appear in the text. After the section, test your understanding by completing the Vocabulary Review in the section exercises.

Objective A Determine If a Number Is a Solution to an Equation

The Concept An **equation** is a mathematical statement stating that two expressions are equal. The equation $x + 10 = 15$ states that the variable expression $x + 10$ is equal to the numerical expression 15.

$$x + 10 = 15$$

Depending on the value that is substituted for the variable, $x + 10 = 15$ can be true or false. If a value is substituted for the variable to make $x + 10 = 15$ true, then that value is a *solution* to the equation.

INTERACTIVE DEFINITION Solution to an Equation

A **solution** to an equation is a value of the variable that makes the equation true.

EXAMPLE

1. Is $x = 2$ a solution to $x + 10 = 15$?

Replace the variable with the given number to see if it makes the equation true.

$$x + 10 = 15$$
$$(2) + 10 \stackrel{?}{=} 15$$
$$12 \neq 15$$

$x = 2$ is not a solution to $x + 10 = 15$.

GUIDED PRACTICE

1. Is $x = 5$ a solution to $x + 10 = 15$?

Replace the variable with the given number to see if it makes the equation true.

$$x + 10 = 15$$
$$(\quad) + 10 \stackrel{?}{=} 15$$
$$15$$

$x = 5$ is/is not a solution to $x + 10 = 15$.

DO YOU UNDERSTAND when the value of a variable is a solution to an equation? Got It Get Help

Procedure Determine If a Number Is a Solution to an Equation

Step 1 Using parentheses, replace the variable with the given number.

Step 2 Check to see if the equation is true.

Follow the order of operations to simplify each side of the equation.

EXAMPLE

2. Is 13 a solution to $y + 12 = 25$?

Step 1 Using parentheses, replace the variable with the given number.

Step 2 Follow the order of operations to simplify each side.

$$y + 12 = 25$$
$$(13) + 12 \stackrel{?}{=} 25$$
$$25 = 25$$

Since $25 = 25$ is true, 13 is a solution.

GUIDED PRACTICE

2. Is 6 a solution to $4x = 24$?

$$4x = 24$$
$$4(\quad) \stackrel{?}{=} 24$$
$$= 24$$

Since ____ = ____ is true/false, 6 is/isn't a solution.

3. Is -4 a solution to $3x + 2 = x - 4$?	**3.** Is -3 a solution to $2x - 4 = x - 8$?
Step 1 Using parentheses, replace the variable with the given number. $3x + 2 = x - 4$ $3(-4) + 2 \stackrel{?}{=} (-4) - 4$ **Step 2** Follow the order of operations to simplify each side. $-12 + 2 \stackrel{?}{=} -8$ $-10 = -8$	$2x - 4 = x - 8$ $2(\quad) - 4 \stackrel{?}{=} (\quad) - 8$ $\stackrel{?}{=}$ $=$
Since $-10 = -8$ is a false statement, -4 is not a solution.	Since _____ = _____ is true/false, -3 is/isn't a solution.

Concept Check

A1. In your own words, what is a solution to an equation?

A2. An equation is a statement where one ____________ is equal to another ____________.

Objective A Practice

Determine if the given number is a solution to the equation.

A3. Is $x = 26$ a solution to $x - 16 = 10$?

A4. Is $z = 10$ a solution to $24 = z + 14$?

A5. Is $y = 5$ a solution to $3y = 18$?

A6. Is $y = 4$ a solution to $27 = 9y$?

A7. Is $x = -7$ a solution to $-53 = 8x + 3$?

A8. Is $z = -9$ a solution to $-112 = 12z - 4$?

A9. Is $y = 40$ a solution to $\frac{y}{5} + 1 = 9$?

A10. Is $y = 80$ a solution to $11 = \frac{y}{10} + 3$?

Objective B Use Inverse Operations to Isolate a Variable

The Concept **Inverse operations** are operations that undo each other.

INVERSE OPERATIONS	
Addition and subtraction are inverse operations.	**Multiplication and division are inverse operations.**
Since addition and subtraction are inverse operations, they undo each other. $5 + 4 = 9$ 5 9 $9 - 4 = 5$ **If we add 4 and then subtract 4, we'll end up where we started.**	Since multiplication and division are inverse operations, they undo each other. 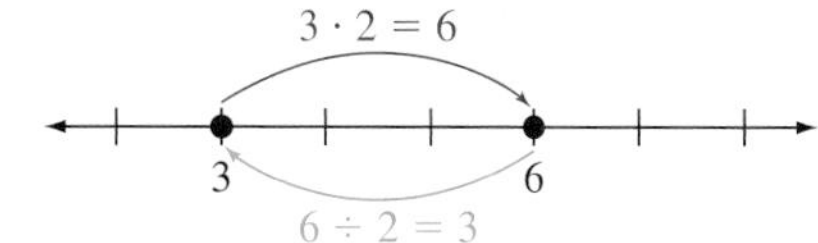 **If we multiply by 2 and then divide by 2, we'll end up where we started.**

The primary goal when solving an equation is to isolate the variable.

Procedure Isolate the Variable

Step 1 Determine what action is being done to the variable.

Step 2 Identify the inverse operation that will undo that action.

EXAMPLES	GUIDED PRACTICE
4. Consider the variable expression $x - 7$.	**4.** Consider the variable expression $5 + y$.
a. What action is being done to the variable? 7 is being subtracted from x. **b.** What inverse operation will undo that action? Subtraction of 7 is undone by the addition of 7.	**a.** What action is being done to the variable? ___ is being ________ y. **b.** What inverse operation will undo that action? ________ of ___ is undone by ________ of ___.
5. Consider the variable expression $-4x$.	**5.** Consider the variable expression $-10y$.
a. What action is being done to the variable? x is being multiplied by -4. **b.** What inverse operation will undo that action? Multiplication by -4 is undone with division by -4.	**a.** What action is being done to the variable? y is being ________ by ___. **b.** What inverse operation will undo that action? ________ by ___ is undone with ________ by ___.
6. Consider the variable expression $\frac{x}{3}$.	**6.** Consider the variable expression $\frac{z}{13}$.
a. What action is being done to the variable? x is being divided by 3. **b.** What inverse operation will undo that action? Division by 3 is undone with multiplication by 3.	**a.** What action is being done to the variable? z is being ________ by ___. **b.** What inverse operation will undo that action? ________ by ___ is undone with ________ by ___.

Concept Check

B1. List two pairs of inverse operations.

B2. For each of the following, write the inverse operation and fill in the correct number to make each equation true. Part a has been completed for you.

a. $3 + 4 \quad\quad = 3$

b. $2 \cdot 6 \quad\quad = 2$

c. $4 \div 2 \quad\quad = 4$

d. $8 - 5 \quad\quad = 8$

Objective B Practice

For each exercise:

a. Determine what action is being done to the variable.

b. Identify the inverse operation that will undo that action.

B3. $17 + x$ **B4.** $x - 12$ **B5.** $\frac{y}{5}$ **B6.** $\frac{z}{-2}$

B7. $-12 + x$ **B8.** $-15 + z$ **B9.** $-9y$ **B10.** $-x$

Objective C Solve Equations of the Form $x + a = b$

The Concept Modeling an equation with a balanced scale can help you understand how to solve equations. To keep a scale balanced, always add or subtract the same weight from both sides. To keep an equation true, always add or subtract the same number from both sides.

The Balanced Scale Solution	The Equation Solution
How many objects are in the box? To answer that question, isolate the box on one side of the balance.	**What is the value of x?** To answer that question, isolate x on one side of the equation.
Equal weights 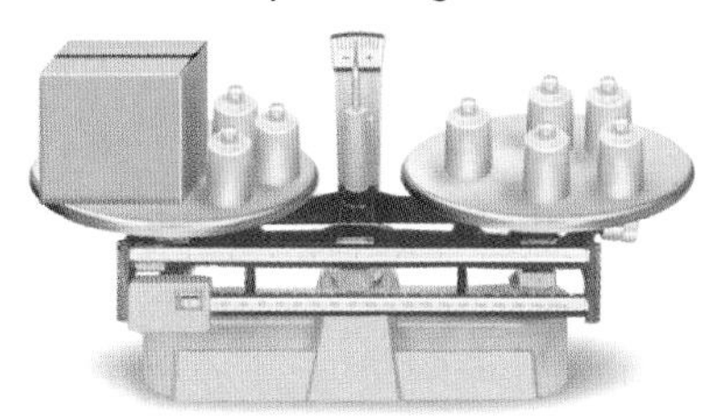	Equal Expressions $x + 3 = 5$
The box is not isolated because 3 objects are being added to it.	The variable x is not isolated because 3 is being added to it.
Isolate the box by removing 3 objects from each side.	Isolate x by subtracting 3 from each side.
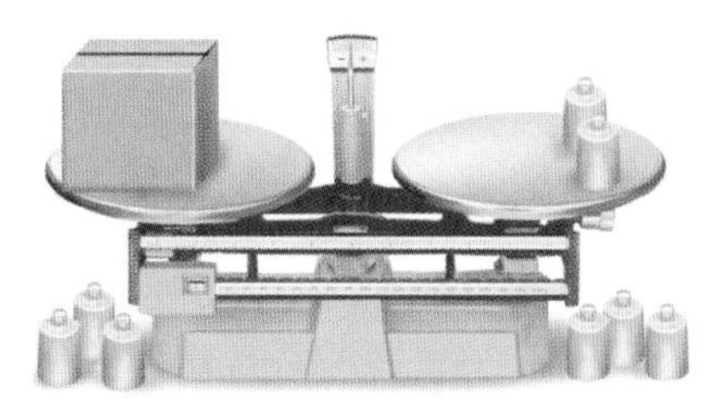	$x + 3 - 3 = 5 - 3$
The weights are still equal.	The expressions are still equal.
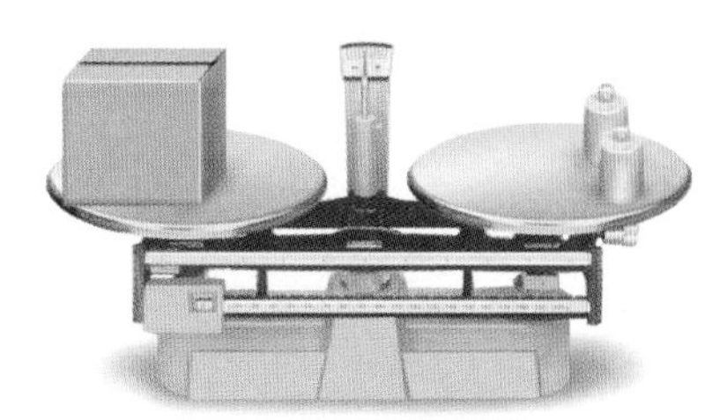	$x = 2$
The box must have 2 objects in it for the original scale to balance.	$x = 2$ is the solution to the original equation $x + 3 = 5$.

To solve the equation $x + 3 = 5$, we **isolated** the variable by undoing addition and getting x alone on one side of the equal sign. The following steps show the techniques used to solve $x + 3 = 5$.

To isolate x, we must undo the addition of 3.	$x + 3 = 5$
We use the inverse operation, subtraction of 3.	$x + 3 - 3 = 5 - 3$
This isolates x and solves the equation.	$x = 2$

Notice that we subtracted 3 from *both sides* of the equation. An equation must be kept balanced as if it were a scale. Any operation you do to one side of the equation you must to the other side as well.

INTERACTIVE DEFINITION Addition Property of Equations

The **addition property of equations** lets us add or subtract the same number to each side of an equation to get an equivalent equation. For example, if $a = b$, then $a + c = b + c$.

EXAMPLES

7. Subtract 7 from both sides of the equation $10 = 10$ to demonstrate the addition property of equations.

Subtract 7 from both sides of the equation.

$$10 = 10$$
$$10 - 7 = 10 - 7$$
$$3 = 3$$

Subtract 7 from both sides, and the equation remains true.

8. Add 3 to both sides of $x - 3 = 10$ to solve the equation.

To undo the subtraction of 3, add 3 to both sides of the equation.

$$x - 3 = 10$$
$$x - 3 + 3 = 10 + 3$$
$$x = 13$$

Adding 3 to both sides of the equation isolates the variable and solves the equation.

GUIDED PRACTICE

7. Add 2 to both sides of the equation $5 = 5$ to demonstrate the addition property of equations.

Add 2 to both sides of the equation.

$$5 = 5$$
$$5 + \_\_ = 5 + \_\_$$
$$\_\_ = \_\_$$

Add 2 to both sides, and the equation remains _____.

8. Subtract 5 from both sides of $y + 5 = 15$ to solve the equation.

To undo the addition of 5, subtract _____ from both sides of the equation.

$$y + 5 = 15$$
$$y + 5 - \_\_ = 15 - \_\_$$
$$y = \_\_$$

Subtracting 5 from both sides of the equation _____ the variable and _____ the equation.

DO YOU UNDERSTAND how to use the addition property of equations? Got It Get Help

Procedure Solve Equations of the Form $x + a = b$

Isolate the variable.

Determine what action is being done to the variable.

Undo that action with its inverse operation.

Check your result.

EXAMPLES

9. Solve. $12 = y - 18$

Solution:

A number is subtracted from the variable.

$$12 = y - 18$$

To undo subtraction, add the number to both sides.

$$12 + 18 = y - 18 + 18$$
$$30 = y$$

Check:

Write the original equation. $12 = y - 18$

Replace the variable with the solution. $12 \stackrel{?}{=} (30) - 18$

Simplify. $12 = 12$ ✓

The solution checks.

10. Solve. $y + \frac{2}{3} = \frac{3}{4}$

Solution:

A number is added to the variable.

$$y + \frac{2}{3} = \frac{3}{4}$$

To undo addition, subtract the number from both sides.

$$y + \frac{2}{3} - \frac{2}{3} = \frac{3}{4} - \frac{2}{3}$$

Build like fractions to simplify the right side.

$$y = \frac{9}{12} - \frac{8}{12}$$
$$y = \frac{1}{12}$$

Check:

Write the original equation. $y + \frac{2}{3} = \frac{3}{4}$

Replace the variable with the solution. $\left(\frac{1}{12}\right) + \frac{2}{3} \stackrel{?}{=} \frac{3}{4}$

Simplify. $\frac{1}{12} + \frac{8}{12} \stackrel{?}{=} \frac{3}{4}$

$$\frac{9}{12} \stackrel{?}{=} \frac{3}{4}$$
$$\frac{3}{4} = \frac{3}{4}$$ ✓

GUIDED PRACTICE

9. Solve. $-23 = x - 8$

$$-23 = x - 8$$
$$-23 + \quad = x - 8 +$$
$$=$$

$-23 = x - 8$

$-23 \stackrel{?}{=} (\quad) - 8$

$\stackrel{?}{=}$

Does the solution check?

10. Solve. $x + \frac{1}{3} = \frac{1}{2}$

$$x + \frac{1}{3} = \frac{1}{2}$$
$$x + \frac{1}{3} - \quad = \frac{1}{2} -$$
$$x = \frac{\quad}{\quad} - \frac{\quad}{\quad}$$
$$x = \frac{\quad}{\quad}$$

$x + \frac{1}{3} = \frac{1}{2}$

$\left(\frac{\quad}{\quad}\right) + \frac{1}{3} \stackrel{?}{=} \frac{1}{2}$

$\stackrel{?}{=}$

$\stackrel{?}{=}$

Does the solution check?

Concept Check

C1. If you add or subtract a number to one side of an equation, why must you also do the same to the other side of the equation?

C2. For the equation $x - 4 = 17$, answer each question.
a. What action is being done to the variable?
b. What inverse operation will undo that action?

C3. In your own words, what does it mean to isolate the variable?

C4. In the equation $x + \frac{1}{8} = 2$, what action will isolate the variable?

Objective C Practice

Solve each equation and check your solution.

C5. $y + 10 = 14$

C6. $z + 24 = 29$

C7. $-14 = t - 25$

C8. $-16 = z + 24$

C9. $x - 10 = -10$

C10. $x - 30 = -30$

C11. $-4 = t - 16$

C12. $7 = t - 11$

C13. $x - \frac{1}{2} = 6$

C14. $y + \frac{1}{5} = 4$

C15. $z + \frac{1}{3} = \frac{3}{4}$

C16. $x - \frac{1}{6} = \frac{1}{2}$

Objective D Solve Equations of the Form $ax = b$

The Concept In the equation $5n = 20$, we must undo multiplication by 5 to isolate n. Because multiplication and division are inverse operations, dividing both sides by 5 isolates n and solves the equation.

$$5n = 20$$

Divide both sides by 5 to isolate n.

$$\frac{\cancel{5}n}{\cancel{5}} = \frac{20}{5}$$

$$n = 4$$

Whatever is done to one side of the equation must be done to the other side as well.

To undo multiplication or division, we use the multiplication property of equations.

INTERACTIVE DEFINITION The Multiplication Property of Equations

The **multiplication property of equations** lets us multiply or divide each side of an equation by the same nonzero number to get an equivalent equation. For example, if $a = b$, then $a \cdot c = b \cdot c$ $(c \neq 0)$.

EXAMPLES

11. Divide each side of the equation $14 = 14$ by 7 to demonstrate the multiplication property of equations.

Divide both sides of the equation by 7.

$$14 = 14$$

$$\frac{14}{7} = \frac{14}{7}$$

$$2 = 2$$

Divide both sides by 7, and the equation remains true.

GUIDED PRACTICE

11. Multiply each side of the equation $6 = 6$ by 4 to demonstrate the multiplication property of equations.

Multiply both sides of the equation by 4.

$$6 = 6$$

$$6 \cdot \square = 6 \cdot \square$$

$$\square = \square$$

Multiply both sides by 4, and the equation remains ______.

12. Multiply both sides by 3 to solve the equation $\frac{y}{3} = 10$.

$$\frac{y}{3} = 10$$

$$\frac{y}{3} \cdot 3 = 10 \cdot 3$$

$$\frac{y}{\cancel{3}} \cdot \frac{\cancel{3}}{1} = 10 \cdot 3$$

$$y = 30$$

Multiplying both sides of the equation by 3 isolates the variable and solves the equation.

12. Divide both sides by 5 to solve the equation $5x = 20$.

$$5x = 20$$

$$\frac{5x}{} = \frac{20}{}$$

$$x = \underline{\qquad}$$

$$x =$$

Dividing both sides of the equation by 5 ______ the variable and ______ the equation.

DO YOU UNDERSTAND how to use the multiplication property of equations? Got It Get Help

Procedure Solve Equations of the Form $ax = b$

Isolate the variable.

Determine what action is being done to the variable.

Undo that action with its inverse operation.

Check your result.

EXAMPLES

13. Solve. $-3y = 18$

Solution:

A number is multiplying the variable.

To undo multiplication, divide both sides by the number.

$$-3y = 18$$

$$\frac{\cancel{-3}y}{\cancel{-3}} = \frac{18}{-3}$$

$$y = -6$$

Check:

Write the original equation.	$-3y = 18$
Replace the variable with the solution.	$-3(-6) \stackrel{?}{=} 18$
Simplify the left side.	$18 = 18$ ✓

GUIDED PRACTICE

13. Solve. $-9z = 90$

$$-9z = 90$$

$$\frac{-9z}{} = \frac{90}{}$$

$$z =$$

$$-9z = 90$$

$$-9(\quad) \stackrel{?}{=} 90$$

$$\stackrel{?}{=}$$

Does the solution check?

14. Solve. $\frac{y}{4} = 8$

Solution:

A number is dividing the variable. $\frac{y}{4} = 8$

To undo division, multiply both sides by the number. $4 \cdot \frac{y}{4} = 8 \cdot 4$

Write the number as a fraction to simplify the left side. $\frac{4}{1} \cdot \frac{y}{4} = 8 \cdot 4$

$\frac{\cancel{4}}{1} \cdot \frac{y}{\cancel{4}} = 32$

$y = 32$

Check:

Write the original equation. $\frac{y}{4} = 8$

Replace the variable with the solution. $\frac{(32)}{4} \stackrel{?}{=} 8$

Simplify the left side. $8 = 8$ ✓

14. Solve. $\frac{x}{7} = 6$

$\frac{x}{7} = 6$

$\quad \cdot \frac{x}{7} = 6 \cdot \quad$

$\frac{\quad}{\quad} \cdot \frac{\quad}{\quad} = \quad$

$\frac{\quad}{\quad} \cdot \frac{\quad}{\quad} = \quad$

$x =$

$\frac{x}{7} = 6$

$\frac{(\quad)}{7} \stackrel{?}{=} 6$

$\stackrel{?}{=} 6$

Does the solution check?

15. Solve. $3y = 0$

Solution:

A number is multiplying the variable. $3y = 0$

To undo multiplication, divide both sides by the number. $\frac{3y}{3} = \frac{0}{3}$

$y = 0$

Check:

Write the original equation. $3y = 0$

Replace the variable with the solution. $3(0) \stackrel{?}{=} 0$

Simplify the left side. $0 = 0$ ✓

15. Solve. $6x = 0$

$6x = 0$

$\frac{6x}{\quad} = \frac{0}{\quad}$

$x =$

$6x = 0$

$6(\quad) \stackrel{?}{=} 0$

$0 \stackrel{?}{=} 0$

Does the solution check?

16. Solve. $\frac{2}{5}x = \frac{4}{3}$.

Solution:

The variable is multiplied by a fraction. $\frac{2}{5}x = \frac{4}{3}$

Multiply both sides by the fraction's reciprocal. $\frac{5}{2} \cdot \frac{2}{5}x = \frac{5}{2} \cdot \frac{4}{3}$

Simplify each side to solve the equation. $\frac{\cancel{5}}{\cancel{2}} \cdot \frac{\cancel{2}}{\cancel{5}}x = \frac{5}{\cancel{2}} \cdot \frac{\cancel{2} \cdot 2}{3}$

$x = \frac{10}{3}$

It might feel wrong to use multiplication to undo multiplication. However, to divide by a fraction, we multiply by its reciprocal.

16. Solve. $\frac{2}{3}x = \frac{10}{3}$.

$\frac{2}{3}x = \frac{10}{3}$

$\frac{\quad}{\quad} \cdot \frac{2}{3}x = \frac{\quad}{\quad} \cdot \frac{10}{3}$

$\frac{\quad}{\quad} \cdot \frac{\quad}{\quad}x = \frac{\quad}{\quad} \cdot \frac{\quad}{\quad}$

$x =$

Check:

Write the original equation. $\frac{2}{5}x = \frac{4}{3}$

Replace the variable with the solution. $\frac{2}{5}\left(\frac{10}{3}\right) \stackrel{?}{=} \frac{4}{3}$

Simplify each side. $\frac{2 \cdot 2 \cdot \cancel{5}}{\cancel{5} \cdot 3} \stackrel{?}{=} \frac{4}{3}$

$\frac{4}{3} = \frac{4}{3}$

$\frac{2}{3}x = \frac{10}{3}$

$\frac{2}{3}\left(\underline{\quad}\right) \stackrel{?}{=} \frac{10}{3}$

$\underline{\quad} \stackrel{?}{=} \underline{\quad}$

$\underline{\quad} \stackrel{?}{=} \underline{\quad}$

Does the solution check?

Concept Check

D1. Use the multiplication property of equations to demonstrate that dividing both sides of $100 = 100$ by the same number will result in a true equation.

D2. In the equation $\frac{3}{8}x = 6$, what action must be done to isolate the variable?

Objective D Practice

Solve and check each equation.

D3. $4x = 32$

D4. $5y = 45$

D5. $-4z = 20$

D6. $-7y = 28$

D7. $-55 = -5t$

D8. $48 = -6t$

D9. $5x = 0$

D10. $7z = 0$

D11. $\frac{x}{2} = -3$

D12. $\frac{y}{3} = -7$

D13. $\frac{4}{5}y = \frac{12}{5}$

D14. $\frac{3}{4}x = \frac{9}{4}$

Section 10.3 Exercises

FOR EXTRA HELP **MyMathLab** MathXL PRACTICE

WATCH

DOWNLOAD

READ

REVIEW

To determine if a number is a solution to an equation:

1. Answer the Objective A Concept Checks.
2. Answer the odd Objective A Practice Exercises.
3. Answer the even Objective A Practice Exercises.

To use inverse operations to isolate a variable:

4. Answer the Objective B Concept Checks.
5. Answer the odd Objective B Practice Exercises.
6. Answer the even Objective B Practice Exercises.

To solve equations of the form $x + a = b$:

7. Answer the Objective C Concept Checks.
8. Answer the odd Objective C Practice Exercises.
9. Answer the even Objective C Practice Exercises.

To solve equations of the form $ax = b$:

10. Answer the Objective D Concept Checks.
11. Answer the odd Objective D Practice Exercises.
12. Answer the even Objective D Practice Exercises.

VOCABULARY REVIEW

Review the Vocabulary Preview for Section 10.3. Study the definitions until you can check Got It *for every word.*

Use these words to complete each sentence.

equation • solution • inverse operations • addition property of equations
multiplication property of equations • isolate a variable • equivalent equations •
solve an equation

	Got It	Get Help
13. To ____________, isolate the variable on one side of the equal sign.		
14. The ____________ lets us add or subtract the same number to each side of an equation to get an equivalent equation.		
15. To ____________, use a series of steps that will get the variable by itself on one side of the equal sign.		
16. Two equations with the same solution are called ____________.		
17. The ____________ allows us to multiply or divide each side of an equation by the same nonzero number to get an equivalent equation.		
18. Operations that undo each other are called ____________.		
19. A(n) ____________ is a mathematical statement stating that two expressions are equal.		
20. A(n) ____________ of an equation is a value of the variable that makes the equation true.		

How will you get help for any vocabulary that you are unsure about?

Your instructor ____ MyMathLab ____ A classmate ____ A tutor ____ Other ____

Determine if the given number is a solution to the equation.

21. Is $y = 19$ a solution to $y - 14 = 5$?

22. Is $x = 11$ a solution to $x + 13 = 24$?

23. Is $t = 5$ a solution to $35 = 6t$?

24. Is $y = 7$ a solution to $56 = 9y$?

25. Is $x = 3$ a solution to $16 = 8x - 8$?

26. Is $z = 4$ a solution to $28 = 9z - 8$?

27. Is $y = 30$ a solution to $\frac{y}{2} - 10 = 8$?

28. Is $y = 54$ a solution to $8 = \frac{y}{6} + 1$?

For each expression:
a. Determine what action is being done to the variable.
b. Undo that action with its inverse operation.

29. $7 + t$ **30.** $y - 19$ **31.** $\frac{z}{15}$ **32.** $\frac{n}{-9}$

33. $\frac{2}{3} + x$ **34.** $\frac{3}{4} + z$ **35.** $-2y$ **36.** $-5t$

Solve and check each equation.

37. $y + 12 = 23$ **38.** $z + 13 = 25$ **39.** $-18 = t - 14$ **40.** $-16 = z - 22$

41. $x - 15 = -15$ **42.** $x - 38 = -38$ **43.** $-9 = t - 12$ **44.** $5 = t - 14$

45. $18 + v = -4$ **46.** $10 + v = -8$ **47.** $x - \frac{1}{3} = 2$ **48.** $y - \frac{1}{8} = 6$

49. $\frac{3}{4} = z + \frac{2}{3}$ **50.** $\frac{5}{6} = x + \frac{4}{5}$ **51.** $x - 4 = 5.8$ **52.** $z - 5 = 3.4$

53. $7z = 56$ **54.** $8y = 64$ **55.** $-3z = -21$ **56.** $-6y = -42$

57. $3 = -2x$ **58.** $7 = -3y$ **59.** $\frac{v}{4} = 10$ **60.** $\frac{t}{5} = 11$

61. $-25 = -5t$ **62.** $18 = -6t$ **63.** $0 = 3x$ **64.** $0 = 4z$

65. $\frac{2}{5}y = \frac{1}{2}$ **66.** $\frac{2}{3}x = \frac{6}{7}$ **67.** $15 = -\frac{x}{3}$ **68.** $14 = -\frac{y}{2}$

Identify the unknown dimension in each figure.

69. The area of this rectangle is 60 square meters.

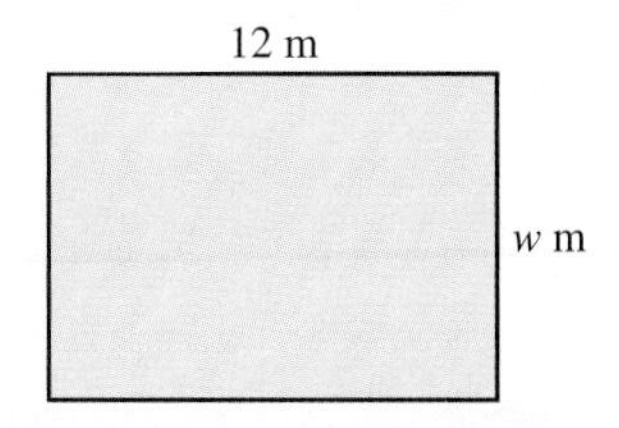

a. Substitute the given values into the equation $A = l \cdot w$.
b. Solve the equation for w to find the width of this rectangle.

70. The area of this rectangle is 77 square inches.

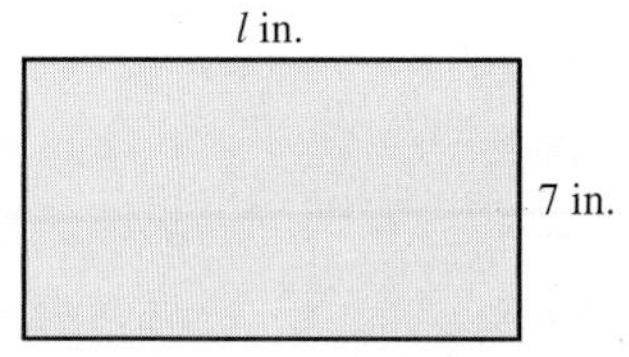

a. Substitute the given values into the equation $A = l \cdot w$.
b. Solve the equation for l to find the length of this rectangle.

smith To solve Exercises 49 and 50, subtract a fraction from both sides of the equal sign. When subtracting, make sure you have a common denominator. *For more tips and tweets, go to twitter.com/gstbasicmath*

71. The area of this triangle is 20 square inches.

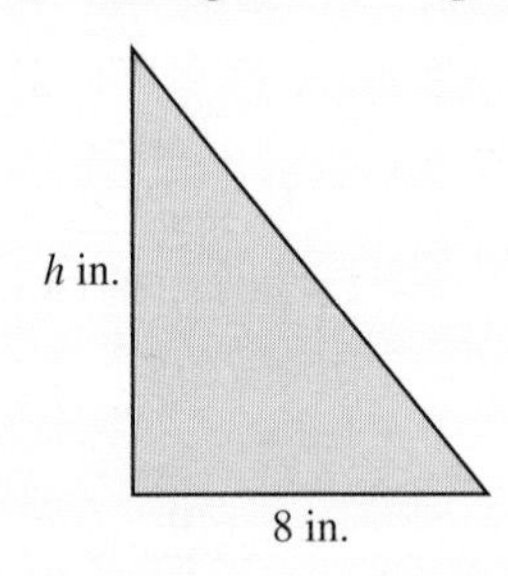

a. Substitute the given values into the equation $A = \frac{1}{2}b \cdot h$.

b. Solve the equation for h to find the height of this triangle.

72. The area of this triangle is 27 square inches.

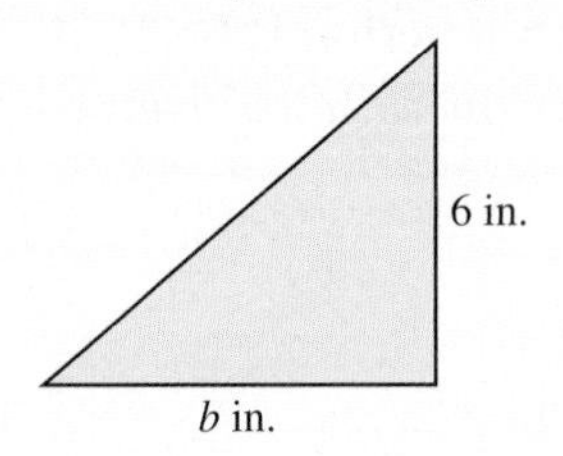

a. Substitute the given values into the equation $A = \frac{1}{2}b \cdot h$.

b. Solve the equation for b to find the base of this triangle.

A common mistake was made in each exercise.
a. Find the mistake and explain why it is wrong.
b. Solve the equation correctly.

73. Solve. $20 = 4 + y$

Incorrect Solution:

$$20 = y + 4$$
$$20 \div 4 = y + 4 \div 4$$
$$5 = y$$

74. Solve. $-3x = 9$

Incorrect Solution:

$$-3x + 3 = 9 + 3$$
$$x = 12$$

75. Solve. $-5z = 10$

Incorrect Solution:

$$\frac{-5z}{-5z} = \frac{10}{-5}$$
$$z = 5$$

76. Solve. $3 = 6 + y$

Incorrect Solution:

$$3 - 3 = 6 - 3 + y$$
$$3 = y$$

Section 10.3 Question Log

Use this space to write questions. Be sure to get these answered and revisit them when you prepare for your exam.

Page ____ Answered ☐	Page ____ Answered ☐
Page ____ Answered ☐	Page ____ Answered ☐

Section 10.4 Solving Multistep Equations

In Section 10.3, the equations required only one step to solve. In this section, you will learn how to solve equations that require several steps. To do this, you will follow a specific procedure that is organized into three main goals.

Three Goals Used for Solving Equations

Goal 1 Simplify each side of the equation.
Goal 2 Gather the variable terms on one side of the =. Gather constant terms on the other side.
Goal 3 Undo multiplication or division to isolate the variable.

Depending on the equation, you may need to use one, two, or three of these goals. You should treat the three goals as a list that you check whenever solving an equation.

The **Objectives** in Section 10.4 will help you

A Solve equations using the addition property of equations.
B Solve equations using the addition and multiplication properties of equations.
C Solve equations that can be simplified.

VOCABULARY PREVIEW *Check the box that applies.*	Got It	Must Study
Variable Term: Any term *with* a variable is a **variable term.**		
Constant Term: Any term *without* a variable is a **constant term.**		
Coefficient: The **coefficient** is the numerical factor of a term. • A coefficient of a variable term is the number that multiplies the variable part of the term. • The coefficient of a constant term is the constant term itself. A coefficient includes the positive or negative sign written in front of the term.		
Solve: To **solve** an equation, the variable must be isolated on one side of the equal sign, giving the solution on the other side of the equal sign.		

Study these words when they appear in the text. After the section, test your understanding by completing the Vocabulary Review in the section exercises.

Objective A Solve Equations Using the Addition Property of Equations

The Concept When variable terms and constant terms are on both sides of the equal sign, we use the second goal.

Goal 2 Gather the variable terms on one side of the =. Gather constant terms on the other side.

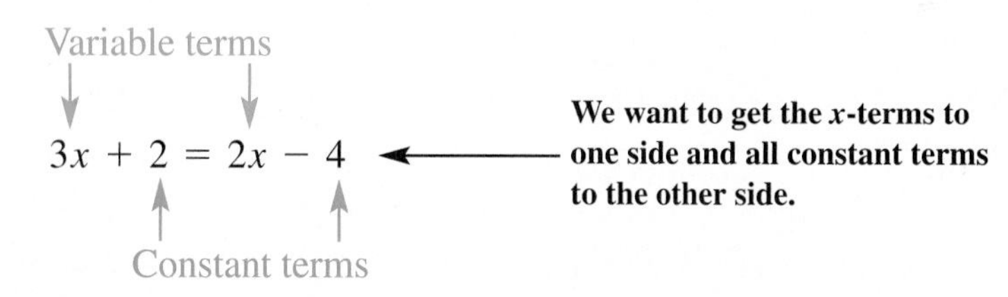

Procedure Solve Equations Using the Addition Property of Equations

Goal 1 Not used in this objective.

Goal 2 Gather the variable terms on one side of the =.
Gather constant terms on the other side.

Goal 3 Not used in this objective.

DETAILED EXAMPLE Gathering Variable Terms and Constant Terms on Opposite Sides

Solve. $-3x + 5 = -4x + 8$

First, move $-4x$ to the left side. $-3x + 5 = -4x + 8$

To move $-4x$ to the left side, add $4x$ to both sides. $-3x + 5 + 4x = -4x + 4x + 8$

Now move $+5$ to the right side. $x + 5 = 8$

To move $+5$ to the right side, subtract 5 from both sides. $x + 5 - 5 = 8 - 5$

$x = 3$

Check:

Write the original equation. $-3x + 5 = -4x + 8$

Replace the variable with the solution. $-3(3) + 5 \stackrel{?}{=} -4(3) + 8$

Simplify. $-9 + 5 \stackrel{?}{=} -12 + 8$

$-4 = -4$ ✓

Students often ask, "Which variable term should I move?" It doesn't matter. If you are careful, you will get the correct solution either way. However, moving the term with the smaller coefficient often makes the exercise easier because the result is a positive coefficient on the variable that is being isolated.

EXAMPLES

1. Solve. $5x + 3 = 4x - 7$

Since 4 is less than 5, we move $4x$ to the left by subtracting $4x$ from both sides.

$5x + 3 = 4x - 7$

$5x - 4x + 3 = 4x - 4x - 7$

$x + 3 = -7$

Move $+3$ to the other side by subtracting 3 from both sides.

$x + 3 - 3 = -7 - 3$

$x = -10$

GUIDED PRACTICE

1. Solve. $8x + 1 = 9x + 4$

Since ____ is less than ____, we move ____ x to the ____ by subtracting ____ x from both sides.

$8x + 1 = 9x + 4$

$8x - \_\_\_ + 1 = 9x - \_\_\_ + 4$

____ = ____

Move ____ to the ________ by subtracting ____ from both sides.

____ = ____

____ = ____

Check your solution on scratch paper.

2. Solve. $-2x + 4 = -x + 6$	**2.** Solve. $-3x - 3 = -4x + 1$
Since -2 is less than -1, we move $-2x$ to the right by adding $2x$ to both sides. $-2x + 4 = -x + 6$; $-2x + 2x + 4 = -x + 2x + 6$; $4 = x + 6$	Since ____ is less than ____, we move ____ x to the ____ by adding ____ x to both sides. $-3x - 3 = -4x + 1$; $-3x + \quad - 3 = -4x + \quad + 1$; $=$
Move $+6$ to the other side by subtracting 6 from both sides. $4 - 6 = x + 6 - 6$; $-2 = x$	Move ____ to the ________ by adding ____ to both sides. $=$; $=$
	Check your solution on scratch paper.

Concept Check

A1. What mathematical operations are used to gather the variable and constant terms on opposite sides of an equal sign?

A2. When there are variable terms on both sides of an equation, moving the term with the ______________ often makes the exercise easier because the result is a positive coefficient on the variable that is being isolated.

Objective A Practice

Solve each equation.

A3. $2x + 4 = 3x - 5$

A4. $5x - 3 = 4x + 2$

A5. $9x + 5 = 8x - 1$

A6. $6x - 13 = 5x - 2$

A7. $-3x - 3 = -2x + 4$

A8. $-6x + 2 = -5x - 4$

A9. $-17x + 3 = -16x + 1$

A10. $-9x + 7 = -8x + 2$

Objective B Solve Equations Using the Addition and Multiplication Properties of Equations

The Concept To isolate the variable in many equations, you will need to use both the second and third goal for solving equations.

Goal 2 Gather the variable terms on one side of the =. Gather constant terms on the other side.

Goal 3 Undo multiplication or division to isolate the variable.

Goal 2 uses the addition property of equations to gather the variable terms on one side and the constant terms on the other. →

$$5x = 2x + 6$$
$$5x - 2x = 2x - 2x + 6$$
$$3x = 6$$
$$\frac{3x}{3} = \frac{6}{3}$$
$$x = 2$$

← **Goal 3 uses the multiplication property of equations to isolate the variable.**

DETAILED EXAMPLE Solving Equations Using the Addition and Multiplication Properties of Equations

Solve. $4x - 1 = x + 8$

$$4x - 1 = x + 8$$

Move x to the left side by subtracting x from both sides.

$$4x - x - 1 = x - x + 8$$

$$3x - 1 = 8$$

Move -1 to the other side by adding 1 to both sides.

$$3x - 1 + 1 = 8 + 1$$

$$3x = 9$$

To undo multiplication by 3, divide both sides by 3.

$$\frac{\cancel{3}x}{\cancel{3}} = \frac{9}{3}$$

$$x = 3$$

Check:

Write the original equation.	$4x - 1 = x + 8$
Replace the variable with the solution.	$4(3) - 1 \stackrel{?}{=} (3) + 8$
Simplify.	$12 - 1 \stackrel{?}{=} 3 + 8$
	$11 = 11$ ✓

Procedure Solve Equations Using the Addition and Multiplication Properties

Goal 1 Not used in this objective.

Goal 2 Gather the variable terms on one side of the =.
Gather constant terms on the other side.

Goal 3 Undo multiplication or division to isolate the variable.

EXAMPLES

3. Solve. $5x + 2 = 3x - 8$

$$5x + 2 = 3x - 8$$

Move $3x$ to the left side.

$$5x - 3x + 2 = 3x - 3x - 8$$

$$2x + 2 = -8$$

Move $+2$ to the other side.

$$2x + 2 - 2 = -8 - 2$$

$$2x = -10$$

Divide both sides by 2 to undo the multiplication.

$$\frac{\cancel{2}x}{\cancel{2}} = \frac{-10}{2}$$

$$x = -5$$

GUIDED PRACTICE

3. Solve. $9x + 2 = 6x + 5$

$$9x + 2 = 6x + 5$$

Move ____ x to the ____ side.

=

=

Move ____ to the ____ side.

=

=

Divide both sides by ____ to undo the multiplication.

=

=

Check your solution on scratch paper.

4. Solve. $-7x + 3 = 4x - 8$

	$-7x + 3 = 4x - 8$
Move $-7x$ to the right side.	$-7x + 7x + 3 = 4x + 7x - 8$
	$3 = 11x - 8$
Move -8 to the other side.	$3 + 8 = 11x - 8 + 8$
	$11 = 11x$
Divide both sides by 11 to undo the multiplication.	$\frac{11}{11} = \frac{11x}{11}$
	$1 = x$

4. Solve. $-6x + 18 = 4x - 2$

	$-6x + 18 = 4x - 2$
Move ____ x to the ____ side.	=
	=
Move ____ to the ____ side.	=
	=
Divide both sides by ____ to undo the multiplication.	=
	=

Check your solution on scratch paper.

Concept Check

B1. What is the first step you would do to solve each equation?
a. $2x - 6 = 3x$ **b.** $-6 = 4x + 4$ **c.** $6x = -7x + 2$

B2. When solving the equation $-2x = 4$, why is it incorrect to add 2 to both sides to isolate the variable?

Objective B Practice

Solve each equation.

B3. $6x + 5 = 2x - 3$

B4. $4x + 7 = 8x - 5$

B5. $y + 6 = 4y - 3$

B6. $7y - 16 = y + 8$

B7. $4x - 12 = 3x - 12$

B8. $8x - 22 = 5x - 22$

B9. $-2z + 8 = z + 5$

B10. $4x - 10 = -3x + 4$

Objective C Solve Equations That Can Be Simplified

The Concept An equation states that two expressions are equal. Equations are often easier to solve when both expressions are simplified. Therefore, our first goal is to simplify each side of the equation.

Goal 1 Simplify each side of the equation.

$$3(x - 2) = -2x + 6x$$

The expression on the left side is simplified by distributing the 3.

The expression on the right side is simplified by combining like terms.

$$3(x - 2) = -2x + 6x$$
$$3x - 6 = 4x$$

$$3x - 6 = 4x$$
$$3x - 3x - 6 = 4x - 3x$$
$$-6 = x$$

Procedure **Use the Three Goals to Solve an Equation**

Goal 1 Simplify each side of the equation.

- Use the distributive property to remove any grouping symbols.
- Combine like terms.

Goal 2 Gather the variable terms on one side of the =.
Gather constant terms on the other side.

- Use the addition property of equations.

Goal 3 Undo multiplication or division to isolate the variable.

- Use the multiplication property of equations.

EXAMPLES

5. Solve. $6(x - 2) = 10 + 2x + 2$

Distribute.	$6(x - 2) = 10 + 2x + 2$
Combine like terms.	$6x - 12 = 10 + 2x + 2$
	$6x - 12 = 2x + 12$
Gather the variable terms.	$6x - 2x - 12 = 2x - 2x + 12$
	$4x - 12 = 12$
Gather the constant terms.	$4x - 12 + 12 = 12 + 12$
	$4x = 24$
Isolate the variable.	$\frac{4x}{4} = \frac{24}{4}$
	$x = 6$

GUIDED PRACTICE

5. Solve. $3(2x + 5) = -3x + 5x + 1$

$$3(2x + 5) = -3x + 5x + 1$$
$$= -3x + 5x + 1$$
$$=$$
$$=$$
$$=$$
$$=$$
$$=$$
$$=$$
$$=$$

Check your solution on scratch paper.

Combining Concepts and Applications

CONCEPT I Converting Temperatures In the United States, temperature is usually measured in degrees Fahrenheit (°F). In other countries, temperature is often measured in degrees Celsius (°C).

The relationship between Celsius and Fahrenheit is given in the following equation.

$$F = \frac{9}{5}C + 32$$

To convert between measurements, follow these steps:

1. Substitute the known measurement into the equation.
2. Solve the formula to find the unknown measurement.

Some History About °F and °C

Fahrenheit: Daniel Fahrenheit invented the Fahrenheit scale of temperature, but there is some uncertainty about how he created it. One theory is that 0°F was "the coldest winter day" and 100°F was his own body temperature. Other theories refer to the thought that people could freeze to death below 0°F or die of heat stroke above 100°F.

Celsius: Anders Celsius created the Celsius scale. Water freezes at 0°C and boils at 100°C.

EXAMPLE

7. On a particularly hot day, the temperature was 104°F. What was this temperature in degrees Celsius?

$$F = \frac{9}{5}C + 32$$

Substitute the known measurement.

$$(104) = \frac{9}{5}C + 32$$

Solve the equation.

$$104 - 32 = \frac{9}{5}C + 32 - 32$$

$$72 = \frac{9}{5}C$$

$$\frac{5}{9} \cdot \frac{72}{1} = \frac{\cancel{5}}{\cancel{9}} \cdot \frac{\cancel{9}}{\cancel{5}}C$$

$$\frac{5 \cdot 8 \cdot \cancel{9}}{\cancel{9}} = C$$

$$40 = C$$

104°F is the same as 40°C.

GUIDED PRACTICE

7. On a particularly cold day, the temperature was −4°F. What was this temperature in degrees Celsius?

$$F = \frac{9}{5}C + 32$$

=

=

=

=

=

=

_____°F is the same as _____°C.

Section 10.4 Exercises

FOR EXTRA HELP       

To solve equations using the addition property of equations:

1. Answer the Objective A Concept Checks.
2. Answer the odd Objective A Practice Exercises.
3. Answer the even Objective A Practice Exercises.

To solve equations using the addition and multiplication properties of equations:

4. Answer the Objective B Concept Checks.
5. Answer the odd Objective B Practice Exercises.
6. Answer the even Objective B Practice Exercises.

To solve equations that can be simplified:

7. Answer the Objective C Concept Checks.
8. Answer the odd Objective C Practice Exercises.
9. Answer the even Objective C Practice Exercises.

VOCABULARY REVIEW *Review the Vocabulary Preview for Section 10.4. Study the definitions until you can check* Got It *for every word.*

Use these words to complete each sentence.

variable term • constant term • coefficient • solve

	Got It	Get Help
10. The numerical factor of a term is the term's ____________.		
11. Any term without a variable is a ____________.		
12. To ____________ an equation, the variable must be isolated on one side of the equal sign, giving the solution on the other side of the equal sign.		
13. Any term with a variable is a ____________.		

How will you get help for any vocabulary that you are unsure about?

Your instructor ____ MyMathLab ____ A classmate ____ A tutor ____ Other ____

Three Goals Used For Solving Equations

Goal 1 Simplify each side of the equation.

- Use the distributive property to remove any grouping symbols.
- Combine like terms.

Goal 2 Gather the variable terms on one side of the =. Gather the constant terms on the other side.

- Use the addition property of equations.

Goal 3 Undo multiplication or division to isolate the variable.

- Use the multiplication property of equations.

Use only goal 2 to solve each equation. Check your solution.

14. $4x + 4 = 3x - 6$ **15.** $5x - 7 = 6x + 2$ **16.** $6x - 5 = 7x + 9$ **17.** $4x - 11 = 5x + 2$

18. $-3x - 2 = -4x + 1$ **19.** $-4x - 3 = -5x - 4$ **20.** $-7x + 6 = -6x + 4$ **21.** $-7x + 5 = -8x + 3$

Use goals 2 and 3 to solve each equation. Check your solution.

22. $5x - 16 = 4$ **23.** $6x - 18 = 24$ **24.** $3 = 7 - 2x$ **25.** $5 = 8 - 3x$

26. $\frac{t}{5} - 4 = 8$ **27.** $\frac{c}{4} - 6 = 4$ **28.** $18 - 2z = 18 - 8z$ **29.** $20 - 5v = 20 - 3v$

30. $5x - 2 = -3x + 3$ **31.** $6x - 25 = -7x + 1$ **32.** $\frac{3}{7}y - 10 = -\frac{4}{7}y + 3$ **33.** $\frac{3}{4}v - 6 = -\frac{1}{4}v + 5$

Use all three goals to solve each equation. Check your solution.

34. $6(x + 5) = 6$ **35.** $5(x + 7) = 10$ **36.** $x - 2 - 3x = 3x - 12$

37. $-3x + 19 = 3x - 1 - x$ **38.** $7 = 2(y - 1) + y$ **39.** $10 = 3(z - 2) + z$

40. $-4z + 2 + 2z = 4(z + 5)$ **41.** $-2c + 12 + 4c = 4(c + 6)$ **42.** $3 - 2(x - 5) = 17$

43. $4 - 3(t - 8) = 37$ **44.** $5(z + 6) = -6(z + 5)$ **45.** $-5(c - 9) = -9(c - 5)$

Solve each equation. Check your solution.

46. $6x = -3$ **47.** $4x = -2$ **48.** $2(x - 5) = 4x + 2$

49. $3(t - 8) = 7t + 4$ **50.** $2(x + 9) = 5x - 2(x + 6)$ **51.** $4(y + 2) = 5y - 3(y + 4)$

52. $y + 21 = 3y - 3$ **53.** $y - 32 = 5y + 8$ **54.** $4 + 3(2x - 1) = 1 + 2x + 2$

55. $9x - 2 + x = 3 + 2(x + 1)$ **56.** $5 - (x + 3) = 5x - 10$ **57.** $3x - (2 + 2x) = 7 + 2x$

Use the formula $F = \frac{9}{5}C + 32$ to answer each question. Round to the nearest tenth.

58. What is the temperature in degrees Fahrenheit if it is 12°C?

59. What is the temperature in degrees Fahrenheit if it is 21°C?

60. What is the temperature in degrees Celsius if it is 48°F?

61. What is the temperature in degrees Celsius if it is 84°F?

tobey Be sure to remove grouping symbols and combine like terms before doing any other steps. This is the key to solving these equations. *For more tips and tweets, go to twitter.com/gstbasicmath*

A rectangle has a perimeter of 60 inches. The perimeter can be found using the formula $60 = 2l + 2w$. The area can be found using the formula $A = l \cdot w$. Use these formulas to answer each question.

62. **a.** What is the width of the rectangle if its length is 10 inches?
b. What is the area of the rectangle?

63. **a.** What is the length of the rectangle if its width is 12 inches?
b. What is the area of the rectangle?

The total number of points that an NFL kicker scores in a season can be described by the equation $T = 3f + p$. Use this equation to answer each question.

T = total points

f = number of field goals (3 points for each field goal)

p = number of extra points (1 point for each extra point)

64. Jason Hanson of the Detroit Lions scored 84 total points. If 27 of those were extra points, how many field goals did Hanson kick?

65. Matt Bryant of the Tampa Bay Buccaneers scored 94 total points. If 31 of those were extra points, how many field goals did Bryant kick?

66. Jason Elam of the Denver Broncos scored 90 total points. If he kicked 24 field goals, how many extra points did Elam kick?

67. Sebastian Janikowski of the Oakland Raiders scored 90 total points. If he kicked 20 field goals, how many extra points did Janikowski kick?

Car rental companies often use equations to calculate the cost of renting a car. Use the equations for each car rental company to answer each question.

Axis Rental calculates the total cost of a car rental with the equation $C = 20d + 0.32m$.

C = total cost

d = number of rental days

m = number of miles driven

Harts Rental calculates the total cost of a car rental with the equation $C = 40d + 0.18m$.

C = total cost

d = number of rental days

m = number of miles driven

68. Doug is planning a four-day trip in which he travels a total of 500 miles. Which rental company has a better deal?

69. Deanna is planning a one-day trip in which she travels a total of 190 miles. Which rental company has a better deal?

70. Alexis has a $300 travel budget. She must travel 1,000 miles in two days. Can she afford either company's plan? If the answer is yes, which plan?

71. Andréa has a $90 travel budget. She must travel 120 miles in two days. Can she afford either company's plan? If the answer is yes, which plan?

A common mistake was made in each exercise.
a. Find the mistake and explain why it is wrong.
b. Solve the equation correctly.

72. $3x + 6 = 24$

Incorrect steps:

$$3x + 6 = 24$$
$$\frac{3x}{3} + 6 = \frac{24}{3}$$
$$x + 6 = 8$$
$$x + 6 - 6 = 8 - 6$$
$$x = 2$$

73. $3x - 2(x + 4) = 2x$

Incorrect steps:

$$3x - 2(x + 4) = 2x$$
$$3x - 2x + 8 = 2x$$
$$x + 8 = 2x$$
$$x - x + 8 = 2x - x$$
$$8 = x$$

74. $5 - 3(x + 2) = 12$

Incorrect steps:

$$5 - 3(x + 2) = 12$$
$$2\,(x + 2) = 12$$
$$2x + 4 = 12$$
$$2x + 4 - 4 = 12 - 4$$
$$2x = 8$$
$$\frac{2x}{2} = \frac{8}{2}$$
$$x = 4$$

75. $-5x + 2(x + 1) = 8$

Incorrect steps:

$$-5x + 2(x + 1) = 8$$
$$-5x + 2x + 2 = 8$$
$$-3x + 2 = 8$$
$$-3x + 2 - 2 = 8 - 2$$
$$-3x = 6$$
$$\frac{-3x}{3} = \frac{6}{3}$$
$$x = 2$$

Consecutive numbers can be described with variable expressions. For example, the consecutive numbers 3, 4, and 5 can be described using the variable expressions x, $x + 1$, and $x + 2$ if $x = 3$. Solve for x in each exercise to determine the set of consecutive numbers.

76. x, $x - 1$, and $x - 2$ are three consecutive whole numbers. If their sum is 216, what are the numbers?

77. x, $x + 1$, and $x + 2$ are three consecutive whole numbers. If their sum is 168, what are the numbers?

78. x, $x + 2$, and $x + 4$ are three consecutive odd numbers. If their sum is 219, what are the numbers?

79. x, $x + 2$, and $x + 4$ are three consecutive even numbers. If their sum is 252, what are the numbers?

Section 10.4 Question Log

Use this space to write questions. Be sure to get these answered and revisit them when you prepare for your exam.

Page ____ **Answered** ☐	**Page** ____ **Answered** ☐
Page ____ **Answered** ☐	**Page** ____ **Answered** ☐

Chapter 10 Organizer

VOCABULARY

Use the following steps to review the vocabulary for Chapter 10.

1. *Write the definition for each word from memory.*
2. *Compare the definitions you have written with the definitions in the Vocabulary Preview.*
3. *Study any definitions that you could not remember or that you defined incorrectly.*

10.1

Variable • Variable Expression • Evaluating an Expression • Term • Variable Term • Constant Term • Coefficient • Like Terms

10.2

Term • Coefficient • Like Terms • Collect Like Terms • Combine Like Terms • Distributive Property

10.3

Equation • Solution • Inverse Operations • Addition Property of Equations • Multiplication Property of Equations • Isolate a Variable • Equivalent Equations • Solve an Equation

10.4

Variable Term • Constant Term • Coefficient • Solve

PROCEDURES

Procedure/Topic	Steps	Example
Write Variable Expressions (Section 10.1)	**Step 1** Choose a variable to represent the unknown. **Step 2** Write a variable expression that describes the situation.	Write a variable expression that describes the number of jewel cases in the picture. Let c = the number of cases in one box. Number of jewel cases = $2c + 5$
Evaluate Variable Expressions (Section 10.1)	**Step 1** Replace the variable with the given number. Write ( ) in place of the variable and insert the given value within the parentheses. **Step 2** Follow the order of operations to simplify the expression.	Evaluate $3x - 2$ when $x = 4$. $3x - 2 = 3(4) - 2$ $= 12 - 2$ $= 10$
Use the Vocabulary of Variable Expressions (Section 10.1)	**Term:** A term can be a number, variable, or product of a number and one or more variables. Terms are separated by addition or subtraction. **Constant term:** Any term without a variable is a constant term. **Variable term:** Any term with a variable is a variable term. **Coefficient:** The coefficient is the numerical factor of a term. **Like terms:** Like terms have identical variable parts.	Identify the terms, constant terms, variable terms, coefficients, and like terms in $2x - 5y - x + 3$. The four terms are $2x$, $-5y$, $-x$, and $+3$. The one constant term is $+3$. The three variable terms are $2x$, $-5y$, and $-x$. The four coefficients are 2, -5, -1, and 3. The like terms are $2x$ and $-x$.

Procedure/Topic	Steps	Example
Simplify by Combining Like Terms (Section 10.2)	**Step 1** Collect like terms. Use the commutative property to reorder and group like terms. **Step 2** Combine like terms. Add the coefficients of the like terms.	Simplify. $6x + 13 + 3x + 1$ $6x + 13 + 3x + 1 = \underbrace{6x + 3x}_{x\text{-terms}} + \underbrace{13 + 1}_{\text{constants}}$ $= 9x + 14$
Simplify Using Multiplication (Section 10.2)	Multiply. Use the distributive, associative, and commutative properties as appropriate.	Simplify. $3(5y - 4)$ $3(5y - 4) = 3 \cdot 5y + 3 \cdot (-4)$ $= 15y - 12$
Use the Order of Operations (Section 10.2)	First: Perform operations in groupings. Second: Evaluate exponents. Third: Multiply and divide from left to right (distribute). Fourth: Add and subtract from left to right (combine like terms).	Simplify. $-5(z + 2) + z + 3(5)$ $-5(z + 2) + z + 3(5) = -5z - 10 + z + 3(5)$ $= -5z - 10 + z + 15$ $= -5z + z - 10 + 15$ $= -4z + 5$
Determine If a Number Is a Solution to an Equation (Section 10.3)	**Step 1** Using parentheses, replace the variable with the given number. **Step 2** Check to see if the equation is true. Follow the order of operations to simplify each side of the equation.	Is 11 a solution to $y + 16 = 27$? $y + 16 = 27$ $(11) + 16 \stackrel{?}{=} 27$ $27 = 27$ Yes, 11 is a solution.
Solve Equations of the Form $x + a = b$ (Section 10.3)	Isolate the variable. Determine what action is being done to the variable. Undo that action with its inverse operation.	Solve. $17 = y - 9$ $17 = y - 9$ $17 + 9 = y - 9 + 9$ $26 = y$
Solve Equations of the Form $ax = b$	Isolate the variable. Determine what action is being done to the variable. Undo that action with its inverse operation.	Solve. $-3y = 18$ $-3y = 18$ $\frac{-3y}{-3} = \frac{18}{-3}$ $y = -6$
Use the Three Goals to Solve an Equation (Section 10.4)	**Goal 1** Simplify each side of the equation. • Use the distributive property to remove any grouping symbols. • Combine like terms. **Goal 2** Gather the variable terms on one side of the =. Gather constant terms on the other side. • Use the addition property of equations. **Goal 3** Undo multiplication or division to isolate the variable. • Use the multiplication property of equations.	Solve. $3(x + 3) = 4x + 2x - 3$ $3(x + 3) = 4x + 2x - 3$ $3x + 9 = 4x + 2x - 3$ $3x + 9 = 6x - 3$ $3x - 3x + 9 = 6x - 3x - 3$ $9 = 3x - 3$ $9 + 3 = 3x - 3 + 3$ $12 = 3x$ $\frac{12}{3} = \frac{3x}{3}$ $4 = x$

Chapter 10 Review Exercises

10.1

Write a variable expression to describe each situation.

1. Describe the number of candy bars.

2. Describe the number of lollipops.

3. A college bookstore has five boxes of baseball caps and 15 baseball caps on display.

4. Chip's Computer Sales has 10 boxes of DVDs and 14 DVDs on display.

Fill in each blank.

5. The two types of terms are ________ terms and ________ terms.

6. The numerical factor of a term is called its ________.

7. Like terms have identical ___________.

Identify the following parts of each variable expression.

8. $12x - 4y - 1x + 3$
   a. Identify the number of terms.
   b. List any constant terms.
   c. List any variable terms.
   d. List any coefficients.
   e. List any sets of like terms.

9. $3z^2 - 5y - 6r^2 + 5$
   a. Identify the number of terms.
   b. List any constant terms.
   c. List any variable terms.
   d. List any coefficients.
   e. List any sets of like terms.

10. $-17 + 3y^3 - y^2 - 14y^3$
   a. Identify the number of terms.
   b. List any constant terms.
   c. List any variable terms.
   d. List any coefficients.
   e. List any sets of like terms.

11. $3z^2 + 9z - 2z^2 + 35$
   a. Identify the number of terms.
   b. List any constant terms.
   c. List any variable terms.
   d. List any coefficients.
   e. List any sets of like terms.

Use the formula $d = rt$ to find the distance in each situation.

12. Harvey drove his car for 3 hours at a rate of 50 miles per hour. How far did he drive?

13. Karel jogged for $1\frac{1}{2}$ hours at a rate of 6 miles per hour. How far did she jog?

10.2

Simplify each variable expression.

14. $5 - 2z + 7$

15. $3x + 5 - 2x + 6$

16. $10z^2 + 8 - 4z - 5z^2$

17. $-9x - 12 - 7x^2 - 11x$

18. $3(4x)$

19. $5(2y)$

Answer each of the following questions.

20. In your own words, why can we add "like things" together? Use the examples below in your explanation.
a. 2 carrots + 3 carrots = 5 carrots
b. $2x + 3x = 5x$

21. In your own words, why can't we add "unlike things" together? Use the examples below in your explanation.
a. 2 carrots + 3 potatoes = 2 carrots + 3 potatoes
b. $2x + 3y = 2x + 3y$

Use the distributive property to simplify each variable expression.

22. $5(x + 7)$ **23.** $7(y - 8)$ **24.** $-4(z + 8)$ **25.** $-2(y - 3)$

Follow the order of operations to simplify each variable expression.

26. $3(4x) + 5x$ **27.** $\frac{5h}{5} + 7h$ **28.** $7n + 2(n + 1)$

29. $5t - 3(t + 3)$ **30.** $5(x - 1) + 3(2x - 2)$ **31.** $6(x + 4) - 3(x + 4)$

10.3

Determine if the given number is a solution to the equation.

32. Is $y = 10$ a solution to $3y = y + 20$? **33.** Is $x = 12$ a solution to $5x - 16 = 4x$?

34. In your own words, what is a solution to an equation?

Solve each equation. Check your solution.

35. $x - 6 = 10$ **36.** $15 = y + 7$ **37.** $-12 = z + 5$ **38.** $x + \frac{1}{3} = \frac{3}{5}$

39. $56 = 8y$ **40.** $-5x = 55$ **41.** $\frac{x}{5} = 9$ **42.** $\frac{2}{3}x = \frac{1}{2}$

A common mistake was made in each exercise.
a. Identify the mistake and explain what was done wrong.
b. Solve the equation correctly.

43.
$$x - 3 = 10$$
$$x - 3 + 3 = 10$$
$$x = 10$$

44.
$$12 = -2x$$
$$12 + 2 = -2x + 2$$
$$14 = x$$

45.
$$4 = 4x$$
$$\frac{4}{4} = \frac{4x}{4}$$
$$0 = x$$

46. When adding or subtracting a number from one side of an equation, why must you do the same thing to the other side of the equation?

10.4

Use goal 2 to solve each equation.

Goal 2 Gather the variable terms on one side of the =. Gather constant terms on the other side.

- Use the addition property of equations.

47. $5x + 3 = 4x - 7$

48. $-2x + 4 = -x + 6$

Use goals 2 and 3 to solve each equation.

Goal 2 Gather the variable terms on one side of the =. Gather constant terms on the other side.

- Use the addition property of equations.

Goal 3 Undo multiplication or division to isolate the variable.

- Use the multiplication property of equations.

49. $5x + 2 = 3x - 8$

50. $-7x - 8 = 4x - 8$

Use the three goals to solve each equation.

Goal 1 Simplify each side of the equation.

- Use the distributive property to remove any grouping symbols.
- Combine like terms.

Goal 2 Gather the variable terms on one side of the =. Gather constant terms on the other side.

- Use the addition property of equations.

Goal 3 Undo multiplication or division to isolate the variable.

- Use the multiplication property of equations.

51. $3(x + 4) = 5x - 3x$

52. $6(x - 2) = 10 + 2x + 2$

Solve each equation. Check your result.

53. $-15x = 45$

54. $\frac{x}{7} = 9$

55. $-\frac{2}{3}y = \frac{1}{4}$

56. $10y + 3 = 9y - 17$

57. $x + 45 = 45 + 3x$

58. $-3x + 13 = 4x - 8$

59. $-5x = 2$

60. $7(x - 2) = 8 + 2x + 3$

61. $4 - 2(x + 8) = 2x$

62. $6 - 3(x + 6) = 3x$

63. $9y = 5$

64. $5x + 10 = 4x + 13$

Chapter 10 Practice Test

Write a variable expression to describe each situation.

1. Describe the number of cans of soda.

2. A sports store has 4 boxes of jerseys and 25 jerseys on display. Describe the number of jerseys.

Fill in each blank.

3. When simplifying an expression, ____________ terms can be combined.

4. Terms that do not contain a variable part are called ____________ terms.

Identify the following parts of each variable expression.

5. $-5 + 4y - 2y^2 - 14$
- **a.** Identify the number of terms.
- **b.** List any constant terms.
- **c.** List any variable terms.
- **d.** List any coefficients.
- **e.** List any sets of like terms.

6. $3 + 5z + 4z^2 - z - 8$
- **a.** Identify the number of terms.
- **b.** List any constant terms.
- **c.** List any variable terms.
- **d.** List any coefficients.
- **e.** List any sets of like terms.

Use the formula $d = rt$ to find the distance.

7. Mark drove his car for 6 hours at a rate of 70 miles per hour. How far did he drive?

Simplify each variable expression.

8. $5x - 12 + 17x$

9. $7(8x)$

10. $x^2 + 8x - 4 - x^2$

11. $4x - 7 - 2x - 12$

Use the distributive property to simplify each variable expression.

12. $6(x + 9)$

13. $-5(y - 7)$

Follow the order of operations to simplify each variable expression.

14. $n + 9(n + 4)$

15. $\frac{3x}{3} + 5x$

16. $9y - 2(y + 3)$

17. $2(x + 1) - 7(x + 2)$

Determine if the given number is a solution to the equation.

18. Is $y = 22$ a solution to $2y = y + 20$?

Solve each equation. Check your solution.

19. $x - 12 = 40$

20. $121 = 11y$

21. $\frac{1}{2}x = \frac{4}{5}$

22. $y + \frac{1}{4} = \frac{4}{5}$

In your own words, describe the steps that you would use to isolate each variable.

23. $y - 10 = 12$

24. $\frac{x}{-6} = 5$

25. $-7z = -28$

26. $x + \frac{2}{3} = \frac{5}{3}$

27. Without looking in the book, list the three goals used to solve equations.

Goal 1: ________________________________

Goal 2: ________________________________

Goal 3: ________________________________

Solve each equation. Check your solution.

28. $5(x + 2) = 4x - 2x + 25$

29. $20 = -10y$

30. $7y + 13 = 6y - 12$

31. $2 + 2x + 3 = 7(x - 5)$

32. $\frac{x}{4} = 8$

33. $-10x + 11 = 4x - 17$

34. $3y + 14 = 2y + 14$

35. $5x - 4(2x - 6) = x$

36. $8y - 1 = 4$

Photograph Credits

1 Photos.com; **1-42** Adisa / Shutterstock; **1-55** Chris Hondros / Getty Images; **1-126** Shutterstock / Sarunas Krivickas; **2-1** Getty Images / Staff / Getty Images; **2-16** iStockphoto / Jani Bryson; **2-19** Courtesy Brian Goetz; **2-21 (l)** aceshot1 / Shutterstock; **(m)** bsankow / Shutterstock; **(r)** 4774344sean / Dreamstime.com; **2-91** Shutterstock / Timmary; **2-94** Shutterstock / Feng Yu; **3-1** Pierre Ducharme / Reuters / Landov; **3-17 (l)** Courtesy NASA; **(r)** Walid Nohra / Shutterstock; **3-19** Superstock Royalty Free; **3-21** Shutterstock / Scott Pehrson; **3-39 (t)** Shutterstock / Trinacria Photo; **(b)** iStockphoto / bluebird13; **3-42 (l)** Corbis Super RF / Alamy; **(r)** Stephen Mcsweeny / Shutterstock; **4-1** WARNER BROS TV / AMBLIN TV / THE KOBAL COLLECTION; **4-32** Chris Steele-Perkins / Magnum Photos; **4-35 (l)** Rmarmion / Dreamstime.com; **(m)** Zhudifeng / Dreamstime.com; **(r)** Ron Chapple Photography Inc. / MediaMagnet / SuperStock; **4-39 (tr)** Sarah Jojnson / Shutterstock; **(bl)** Manfred Steinback / Shutterstock; **(br)** Zina Seletskaya / Shutterstock; **4-57, 4-59** Reproduced with Permission of Abbott Laboratories; **4-50** Used with permission of Roxane Laboratories, Inc.; **5-25** Design Pics Inc. / Alamy; **6-4 (t)** Shutterstock / Morgan Lane Photography; **(b)** iStockphoto / Aleaimage; **6-16 (tl)** iStockphoto / Abel Leao; **(tr)** iStockphoto / funkypoodle; **(b)** iStockphoto / technotr; **6-17 (t)** iStockphoto.com; **(b)** Maxwellartandphoto.com / Pearson Math; **624 (l)** Yurchyks / Shutterstock; **(m)** Dnadigital / Dreamstime.com; **(r)** iStockphoto.com; **7-1** Elemental Imaging / Shutterstock; **7-25 (l)** iStockphoto / Warwick Lister-Kaye; **(m)** iStockphoto / Viktor Fischer; **(r)** Shutterstock / omers; **7-26** Stan Wagon on the square-wheel bike at Macalester College, St. Paul, Minnesota /; **7-40** Serp / Shutterstock; **7-50** Shutterstock / Marek Slusarczyk; **7-52 (l)** PhotoBliss / Alamy Images Royalty Free; **(m)** David Alan Harvey / Magnum Photos; **(r)** Csaba Peterdi / Shutterstock; **7-61 (Ex. 29)** Shutterstock / Luciano Mortula; **(Ex. 32)** Shutterstock / Rob Wilson; **7-76** iStockphoto / Kirsty Pargeter; **7-94 (l)** iStockphoto / Alexei Zarubin; **(r)** iStockphoto / Asher Welstead; **9-1** iStockphoto; **9-36** Mikael Damkier / Shutterstock; **10-1** Davidarts / Dreamstime.com; **10-24 (Ex. 34a)** iStockphoto / Graeme Purdy; **(Ex. 34b)** iStockphoto / Peter Fuchs; **(Ex. 34c)** iStockphoto / Editorial12; **(Ex. 35a)** iStockphoto / Tammy Peluso; **(Ex. 35b)** iStockphoto / Pavel Mozzhukhin; **10-51** Duane Burleson / AP Wide World; **Cover** Carsten Reisinger / Shutterstock; Lee Waters / Jupiter Images; Albert Campbell / Shutterstock; Robert Adrian Hillman / Shutterstock; Elena Snow / Shutterstock

Appendix A

Additional Practice and Review

Section 1.2 Extra Practice Exercises; Addition Facts

Add.

1. $\begin{array}{r} 23 \\ +11 \\ \hline \end{array}$ **2.** $\begin{array}{r} 21 \\ +48 \\ \hline \end{array}$ **3.** $\begin{array}{r} 52 \\ +33 \\ \hline \end{array}$ **4.** $\begin{array}{r} 55 \\ +44 \\ \hline \end{array}$

5. $\begin{array}{r} 83 \\ +18 \\ \hline \end{array}$ **6.** $\begin{array}{r} 48 \\ +35 \\ \hline \end{array}$ **7.** $67 + 73$ **8.** $93 + 66$

9. $\begin{array}{r} 6 \\ 7 \\ +5 \\ \hline \end{array}$ **10.** $\begin{array}{r} 9 \\ 5 \\ +3 \\ \hline \end{array}$ **11.** $17 + 14 + 13$ **12.** $15 + 11 + 17$

13. $\begin{array}{r} 453 \\ +\ 54 \\ \hline \end{array}$ **14.** $\begin{array}{r} 667 \\ +\ 38 \\ \hline \end{array}$ **15.** $652 + 49$ **16.** $645 + 58$

17. $763 + 528$ **18.** $878 + 381$ **19.** $675 + 378$ **20.** $983 + 658$

21. $\begin{array}{r} 25 \\ 42 \\ +81 \\ \hline \end{array}$ **22.** $\begin{array}{r} 27 \\ 74 \\ +67 \\ \hline \end{array}$ **23.** $75 + 52 + 63$ **24.** $87 + 41 + 36$

Section 1.3 Extra Practice Exercises; Subtraction Facts

Subtract.

1. $\begin{array}{r} 23 \\ -11 \\ \hline \end{array}$ **2.** $\begin{array}{r} 26 \\ -15 \\ \hline \end{array}$ **3.** $\begin{array}{r} 52 \\ -33 \\ \hline \end{array}$ **4.** $\begin{array}{r} 53 \\ -44 \\ \hline \end{array}$

5. $\begin{array}{r} 83 \\ -18 \\ \hline \end{array}$ **6.** $\begin{array}{r} 41 \\ -35 \\ \hline \end{array}$ **7.** $73 - 28$ **8.** $93 - 66$

9. $26 - 7 - 5$ **10.** $19 - 5 - 3$ **11.** $37 - 14 - 13$ **12.** $45 - 11 - 17$

13. $\begin{array}{r} 453 \\ -\ 54 \\ \hline \end{array}$ **14.** $\begin{array}{r} 677 \\ -\ 78 \\ \hline \end{array}$ **15.** $652 - 69$ **16.** $645 - 58$

17. $763 - 578$ **18.** $878 - 389$ **19.** $675 - 378$ **20.** $983 - 688$

21. $\begin{array}{r} 25{,}002 \\ -\ 8{,}145 \\ \hline \end{array}$ **22.** $\begin{array}{r} 27{,}006 \\ -\ 6{,}734 \\ \hline \end{array}$ **23.** $\begin{array}{r} 45{,}600 \\ -\ 8{,}765 \\ \hline \end{array}$ **24.** $\begin{array}{r} 870{,}003 \\ -\ 7{,}453 \\ \hline \end{array}$

Section 1.4 Extra Practice Exercises; Multiplication Facts

Multiply.

1. 45×6

2. 52×5

3. 67×7

4. 83×9

5. 526×3

6. 987×4

7. 2463×8

8. 3351×7

Multiply by the given power of 10.

9. 67×10

10. 92×10

11. 234×1000

12. 542×1000

Find each product.

13. 45×30

14. 93×40

15. 6300×5000

16. 1800×6000

17. 65×28

18. 73×45

19. $71 \cdot 64$

20. $51 \cdot 76$

21. $5 \cdot 3 \cdot 4$

22. $6 \cdot 6 \cdot 5$

23. $13 \cdot 14 \cdot 15$

24. $12 \cdot 20 \cdot 14$

25. 332×92

26. 561×85

27. $643 \cdot 39$

28. $265 \cdot 74$

29. 537×657

30. 210×734

31. $705 \cdot 408$

32. $803(650)$

Mid-Chapter 1 Review Exercises

Round each whole number to the given place value.

1. 32,894; thousands

2. 121; tens

3. 349; hundreds

4. 1,332,984; ten thousands

Write each whole number in words as you would on a check.

5. \$4,908

6. \$25,045

Perform each operation.

7. $398 + 215$

8. $467 + 241$

9. $212 - 78$

10. $452 - 383$

11. $48 \cdot 27$

12. $56 \cdot 94$

13. $154 \div 8$

14. $213 \div 9$

15. $14\overline{)367}$

16. $13\overline{)298}$

17. $600 - 126$

18. $800 - 135$

19. $81 \cdot 100$

20. $45 \cdot 1{,}000$

Answer each question.

21. If Stephania's starting balance is \$792 and she writes a check for \$325, what is her new balance?

22. Alexis bought an \$840 computer and will pay for it in 24 monthly payments. How much is each monthly payment?

23. Andre bought his brother's car and agreed to pay \$55 monthly for 36 months. How much will Andre pay in total?

24. Raj went on a road trip. He drove 63 miles to pick up a friend, 128 miles to Busch Gardens amusement park, and then 16 miles to the beach. How far did he drive in total?

Round each whole number to its largest place value and then estimate each product or quotient.

25. $32 \cdot 675$ **26.** $84 \cdot 512$ **27.** $831 \div 39$ **28.** $976 \div 45$

Answer each question.

29. Raymond and Monique invited two hundred forty-six guests to their wedding. If one hundred eighty-six guests attended, how many invited guests did not attend?

30. If Micah and Jamie drove 448 miles in 7 hours, how fast were they traveling in miles per hour?

31. A painter must find the area of the gable on a house in order to determine how much orange paint to use. Use the figure to find the area of the gable.

32. A painter must find the perimeter of the gable on a house in order to determine how much white paint to use. Use the figure to find the perimeter of the gable.

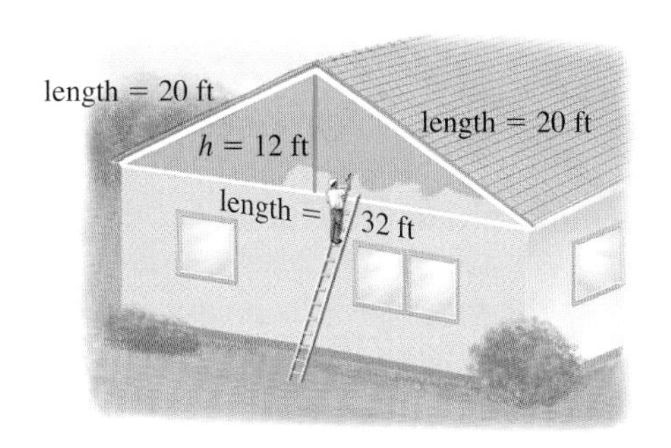

Solve for the unknown number.

33. $7 \cdot x = 42$ **34.** $x + 18 = 53$ **35.** $x \div 5 = 8$ **36.** $x - 14 = 7$

Mid-Chapter 2 Review Exercises

To help you study, use the flowchart on the page 2-86 to guide you through these problems. Recite the steps used to solve an exercise before you begin working it.

Add, subtract, multiply, or divide the fractions.

1. $\frac{1}{12} + \frac{1}{14}$ **2.** $\frac{1}{18} + \frac{3}{81}$ **3.** $\frac{3}{15} - \frac{1}{10}$ **4.** $\frac{7}{8} - \frac{1}{14}$

5. $\frac{5}{4} \cdot \frac{3}{10}$ **6.** $\frac{5}{12} \cdot \frac{4}{5}$ **7.** $\frac{13}{15} \div \frac{7}{10}$ **8.** $\frac{1}{24} \div \frac{1}{10}$

9. $\frac{3}{5} + \frac{3}{10} - \frac{1}{3}$ **10.** $\frac{7}{8} + \frac{4}{5} - \frac{1}{10}$ **11.** $\frac{5}{6} - \frac{1}{18} + \frac{4}{36}$ **12.** $\frac{3}{2} - \frac{2}{3} + \frac{1}{9}$

13. $\frac{9}{10} \div \frac{5}{12} \cdot \frac{2}{4}$ **14.** $\frac{4}{25} \div \frac{6}{15} \cdot \frac{15}{2}$ **15.** $\frac{6}{25} \cdot \frac{15}{4} \div \frac{9}{10}$ **16.** $\frac{9}{18} \cdot \frac{1}{24} \div \frac{9}{12}$

17. Which operations (+, −, ·, ÷) require you to have like fractions?

18. Which operations (+, −, ·, ÷) can be performed with unlike fractions?

19. Give an example of a pair of fractions that are like but not equivalent.

20. Give an example of a pair of fractions that are equivalent but not like.

21. A fraction will be equal to one if its denominator is ______ as the numerator.

Match each numbered exercise with the lettered step that could be used to evaluate it. Be careful when working these exercises. We have set this up to include common mistakes that students make.

22. $\frac{1}{2}\cdot\frac{1}{3}$ **a.** $\frac{1}{2}+\frac{3}{1}$ **b.** $\frac{1}{6}$ **23.** $\left(\frac{1}{3}\right)^2$

24. $\frac{6}{7}-\frac{1}{3}$ **c.** $\frac{1}{81}$ **d.** $\frac{3}{8}\cdot\frac{6}{5}$ **25.** $\frac{2+1}{8}\cdot\frac{6}{5}$

26. $\frac{1}{9}\cdot\frac{1}{9}$ **e.** $\frac{1}{2}\cdot\frac{3}{1}$ **f.** $\frac{3}{8}\cdot\frac{5}{6}$ **27.** $\frac{3}{8}\div\frac{6}{5}$

28. $\frac{1}{2}\div\frac{1}{3}$ **g.** $\frac{1}{9}$ **h.** $\frac{18}{21}-\frac{7}{21}$ **29.** $\frac{1}{2}+3$

Insert <, >, or = to make a true statement.

30. $\frac{5}{9}\square\frac{12}{19}$ **31.** $\frac{7}{8}\square\frac{7}{13}$ **32.** $\frac{3}{10}\square\frac{2}{5}$ **33.** $\frac{5}{12}\square\frac{7}{12}$

34. $\frac{3}{7}\square\frac{10}{21}$ **35.** $\frac{1}{2}\square\frac{20}{39}$ **36.** $\frac{11}{34}\square\frac{5}{13}$ **37.** $\frac{13}{20}\square\frac{51}{80}$

38. $\frac{1}{2}+\frac{1}{2}\square\frac{1}{2}\cdot\frac{5}{6}$ **39.** $\frac{7}{8}+\frac{3}{4}\square\frac{8}{7}\cdot\frac{5}{2}$

40. $\frac{15}{4}-\frac{5}{2}\square\frac{8}{3}\cdot\frac{5}{3}$ **41.** $\frac{2}{3}+\frac{3}{4}\square\frac{5}{6}\cdot\frac{1}{13}$

42. $\frac{15}{16}+\frac{7}{4}\square\frac{8}{3}\cdot\frac{6}{5}$ **43.** $\frac{21}{5}-\frac{5}{2}\square\frac{2}{1}\cdot\frac{3}{5}$

44. There are 5,280 feet in one mile. Convert 6 miles into feet.

45. There are 16 ounces in a pound. Convert 80 ounces to pounds.

46. If there are about 500 jelly beans in a 1-pint jar, approximately how many jelly beans will be in a 5-gallon container?

Mid-Chapter 9 Review Exercises

Order each set of numbers from least to greatest.

1. $\{-5, 7, -3, 4\}$

2. $\{-13, 12, 7, -8\}$

3. $\{|-8|, -|-4|, -(-3), -(7)\}$

4. $\{-(-2), 0, |-9|, -|-2|\}$

Evaluate.

5. $-8 + (-16)$

6. $-5 + 12$

7. $(-3)(-12)$

8. $-32 \div 8$

9. $7 - (-8)$

10. $-13 - 7$

11. $(-2)(-5)(7)$

12. $(-12) \div 3 \cdot (-2)$

13. $-6 - 5 + 8$

14. $6 + (-15) - 4$

15. -2^4

16. $(-3)^2$

17. $\frac{-1}{5} + \frac{-7}{10}$

18. $\frac{1}{7} + \frac{-3}{14}$

19. $\frac{-3}{5} \div \frac{9}{10}$

20. $\frac{-12}{25} \cdot \frac{-35}{16}$

Follow the order of operations to evaluate each expression.

21. $4 \cdot (-6) - 3 \cdot (5)$

22. $8 - (5) \cdot (7) + (-21)$

23. $8 - 10 + (4 - 6)^2$

24. $(2^3 - 3^2) \cdot 8 \div (-2)$

25. $\frac{4 \cdot 5 - (-7)}{9 - 6}$

26. $\frac{(-4) \cdot 3 - (-2)}{-4 + 2}$

Answer each question.

27. A youth baseball team just won the Junior World Series. In the 3 games played, they won by 1, lost by 8, and won by 4. What was their average margin of victory?

28. Elaine has five investment accounts. The rates of return are 8%, −2.3%, −4.2%, 7%, and 1%. What is the average rate of return for the five accounts?

29. Find two numbers whose product is −15 and whose sum is 2.

30. Find two numbers whose product is 24 and whose sum is −11.

31. Doris owns 100 shares of a stock whose value fell by $0.32 per share and 150 shares of a stock whose value increased by $0.80 per share. What is the total change in the value of these two stocks?

32. Eight coal miners enter an elevator into a mine that is 4,500 feet deep. If the elevator descends for 1 minute at a rate of 36 feet per second, how far is the elevator from the entrance of the mine after 1 minute?

Appendix B

Addition, Multiplication, and Square Root Tables

TABLE OF BASIC ADDITION FACTS										
+	**0**	**1**	**2**	**3**	**4**	**5**	**6**	**7**	**8**	**9**
0	0	1	2	3	4	5	6	7	8	9
1	1	2	3	4	5	6	7	8	9	10
2	2	3	4	5	6	7	8	9	10	11
3	3	4	5	6	7	8	9	10	11	12
4	4	5	6	7	8	9	10	11	12	13
5	5	6	7	8	9	10	11	12	13	14
6	6	7	8	9	10	11	12	13	14	15
7	7	8	9	10	11	12	13	14	15	16
8	8	9	10	11	12	13	14	15	16	17
9	9	10	11	12	13	14	15	16	17	18

TABLE OF BASIC MULTIPLICATION FACTS													
×	**0**	**1**	**2**	**3**	**4**	**5**	**6**	**7**	**8**	**9**	**10**	**11**	**12**
0	0	0	0	0	0	0	0	0	0	0	0	0	0
1	0	1	2	3	4	5	6	7	8	9	10	11	12
2	0	2	4	6	8	10	12	14	16	18	20	22	24
3	0	3	6	9	12	15	18	21	24	27	30	33	36
4	0	4	8	12	16	20	24	28	32	36	40	44	48
5	0	5	10	15	20	25	30	35	40	45	50	55	60
6	0	6	12	18	24	30	36	42	48	54	60	66	72
7	0	7	14	21	28	35	42	49	56	63	70	77	84
8	0	8	16	24	32	40	48	56	64	72	80	88	96
9	0	9	18	27	36	45	54	63	72	81	90	99	108
10	0	10	20	30	40	50	60	70	80	90	100	110	120
11	0	11	22	33	44	55	66	77	88	99	110	121	132
12	0	12	24	36	48	60	72	84	96	108	120	132	144

TABLE OF SQUARE ROOTS									
n	$\sqrt{n}$	n	$\sqrt{n}$	n	$\sqrt{n}$	n	$\sqrt{n}$	n	$\sqrt{n}$
1	1.000	**41**	6.403	**81**	9.000	**121**	11.000	**161**	12.689
2	1.414	**42**	6.481	**82**	9.055	**122**	11.045	**162**	12.728
3	1.732	**43**	6.557	**83**	9.110	**123**	11.091	**163**	12.767
4	2.000	**44**	6.633	**84**	9.165	**124**	11.136	**164**	12.806
5	2.236	**45**	6.708	**85**	9.220	**125**	11.180	**165**	12.845
6	2.449	**46**	6.782	**86**	9.274	**126**	11.225	**166**	12.884
7	2.646	**47**	6.856	**87**	9.327	**127**	11.269	**167**	12.923
8	2.828	**48**	6.928	**88**	9.381	**128**	11.314	**168**	12.961
9	3.000	**49**	7.000	**89**	9.434	**129**	11.358	**169**	13.000
10	3.162	**50**	7.071	**90**	9.487	**130**	11.402	**170**	13.038
11	3.317	**51**	7.141	**91**	9.539	**131**	11.446	**171**	13.077
12	3.464	**52**	7.211	**92**	9.592	**132**	11.489	**172**	13.115
13	3.606	**53**	7.280	**93**	9.644	**133**	11.533	**173**	13.153
14	3.742	**54**	7.348	**94**	9.695	**134**	11.576	**174**	13.191
15	3.873	**55**	7.416	**95**	9.747	**135**	11.619	**175**	13.229
16	4.000	**56**	7.483	**96**	9.798	**136**	11.662	**176**	13.266
17	4.123	**57**	7.550	**97**	9.849	**137**	11.705	**177**	13.304
18	4.243	**58**	7.616	**98**	9.899	**138**	11.747	**178**	13.342
19	4.359	**59**	7.681	**99**	9.950	**139**	11.790	**179**	13.379
20	4.472	**60**	7.746	**100**	10.000	**140**	11.832	**180**	13.416
21	4.583	**61**	7.810	**101**	10.050	**141**	11.874	**181**	13.454
22	4.690	**62**	7.874	**102**	10.100	**142**	11.916	**182**	13.491
23	4.796	**63**	7.937	**103**	10.149	**143**	11.958	**183**	13.528
24	4.899	**64**	8.000	**104**	10.198	**144**	12.000	**184**	13.565
25	5.000	**65**	8.062	**105**	10.247	**145**	12.042	**185**	13.601
26	5.099	**66**	8.124	**106**	10.296	**146**	12.083	**186**	13.638
27	5.196	**67**	8.185	**107**	10.344	**147**	12.124	**187**	13.675
28	5.292	**68**	8.246	**108**	10.392	**148**	12.166	**188**	13.711
29	5.385	**69**	8.307	**109**	10.440	**149**	12.207	**189**	13.748
30	5.477	**70**	8.367	**110**	10.488	**150**	12.247	**190**	13.784
31	5.568	**71**	8.426	**111**	10.536	**151**	12.288	**191**	13.820
32	5.657	**72**	8.485	**112**	10.583	**152**	12.329	**192**	13.856
33	5.745	**73**	8.544	**113**	10.630	**153**	12.369	**193**	13.892
34	5.831	**74**	8.602	**114**	10.677	**154**	12.410	**194**	13.928
35	5.916	**75**	8.660	**115**	10.724	**155**	12.450	**195**	13.964
36	6.000	**76**	8.718	**116**	10.770	**156**	12.490	**196**	14.000
37	6.083	**77**	8.775	**117**	10.817	**157**	12.530	**197**	14.036
38	6.164	**78**	8.832	**118**	10.863	**158**	12.570	**198**	14.071
39	6.245	**79**	8.888	**119**	10.909	**159**	12.610	**199**	14.107
40	6.325	**80**	8.944	**120**	10.954	**160**	12.649	**200**	14.142

Note: Square root values are rounded to the nearest thousandth unless the answer ends in .000

Answers to Selected Exercises

CHAPTER 1

Section 1.1 Guided Practice

1. two; thousands; four hundred twelve thousand, seven hundred six
2. three; millions; twelve million, three
3. two; thousands; seventeen thousand, four hundred
4. millions; hundreds; hundred millions
5. ones; 890; hundreds; 890; hundreds
6. 6; 8; 3; 1; 5; 4; 451,386 = 400,000 + 50,000 + 1,000 + 300 + 80 + 6
7. 5,300,010 = 5,000,000 + 300,000 + 10

0 2 4 6 8 10 12

8. 43 67
0 10 20 30 40 50 60 70

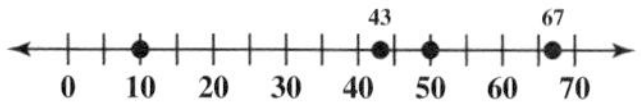

9. 0 100 200 300 400 500 600 700

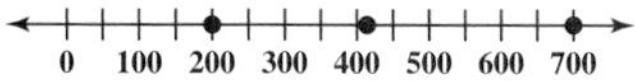

10.

11. 2; $130; $120; 130; 130; $130
12. 900; 800; 800; down; 800
13. 6; 5; 6; up; 6,000
14. 140; 130; halfway; 130; 140; up; 140
15. 10; 9; 9,000; down; 9,000
16. 3[4], 821; 35,000; 34,000; 8; up; up; 35,000
17. 17, [6]83; 17,700; 17,600; 8; up; up; 17,700
18. [5]2, 908; 60,000; 50,000; 2; down; 50,000
19. 3 years; 15,000 **20.** new; 33,000
21. new; 1 year **22.** 4; $13,000; 4

Concept Check and Objective Practice

A1. 0, 1, 2, 3, 4, 5, 6, 7, 8, 9 **A2.** ten thousands
A3. ones, thousands, millions, billions, trillions
A4. one hundred fifty-three **A5.** four hundred ninety-two
A6. seven thousand, five **A7.** three thousand, eighty
A8. one hundred thousand, three hundred nine
A9. fifty thousand, three **A10.** four billion, five
A11. eight billion, five thousand
A12. 4 is in the thousands place.
A13. 4 is in the thousands place.
A14. 4 is in the hundreds place.
A15. 4 is in the hundred thousands place.
A16. 4 is in the hundreds place.
A17. 4 is in the hundred millions place.
A18. 4 is in the ten thousands place.
A19. 4 is in the billions place.
B1. **a.** yes
b. 505 = 500 + 5
550 = 500 + 50
c. Answers may vary. In 550, the second 5 represents 5 tens, while in 505, it represents 5 ones.
B2. **a.** yes **b.** 560 = 500 + 60 and 506 = 500 + 6
c. Answers may vary. In 560, the 6 represents 6 tens, while in 506, it represents 6 ones.
B3. 1,000 + 500 + 30 + 4 **B4.** 5,000 + 400 + 90 + 2
B5. 6,000 + 2 **B6.** 3,000 + 40
B7. 200,000,000 + 5,000,000 + 100,000 + 300 + 9
B8. 400,000,000 + 20,000,000 + 50,000 + 3
B9. 4,000,000,000 + 5 **B10.** 8,000,000,000 + 5,000
C1. Answers may vary. Example: If a tick mark was placed at every unit, the graph would be too large to fit on the paper.
C2. Answers may vary. Example: The graphed number will already be identified.
C3. 0 2 4 6 8 10 12
C4. 0 2 4 6 8 10 12 14
C5. 70 88 105
0 20 40 60 80 100 120
C6. 5 30 107
0 20 40 60 80 100
C7. 120
0 300 600 900 1,200
C8. 570
0 300 600 900 1,200
D1. **a.** 11,500 **b.** If the number is less than 11,500, it is closer to 11,000.
D2. If the number is less than 500,000, it is closer to 0.
D3. Answers may vary. Example: Rounding is a way to approximate a number to a certain place value.
D4. 2,000 **D5.** 1,000 **D6.** 320 **D7.** 170
D8. 433,000 **D9.** 237,000 **D10.** 60 **D11.** 90
D12. 700 **D13.** 800 **D14.** 16,000 **D15.** 64,000
D16. 900,000 **D17.** 100,000 **D18.** $190,000
D19. $3,800,000 **D20.** $1,500 **D21.** $12,000
E1. bar graph
E2. Answers may vary. Example: A collection of facts that may be used to draw conclusions.
E3. about 120 mph **E4.** about 170 mph
E5. about 140 mph **E6.** about 155 mph **E7.** 6 **E8.** 2

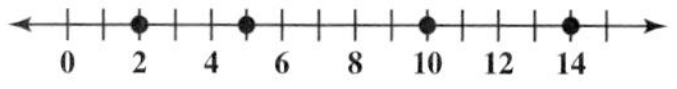

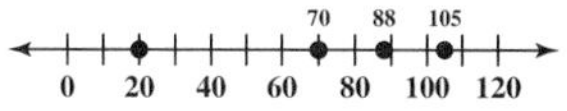

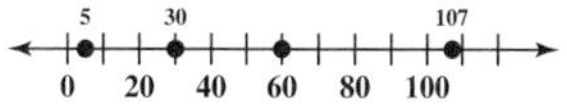

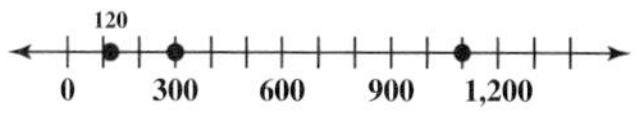

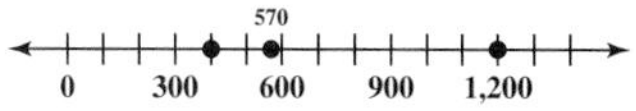

Exercises

For 1.–15.: See preceding Concept Check and Objective Practice answers.

17. digits **19.** rounding **21.** 0, 1, 2, 3, 4, 5, 6, 7, 8, 9
23. sixty-seven thousand, four hundred fifty-six
25. three thousand, one hundred twenty-seven
27. three hundred fifteen
29. one million, one hundred twenty-five thousand, five hundred sixty-eight
31. five hundred sixty million, two hundred thousand, one hundred seven
33. twelve million, five thousand **35.** 808,088
37. two hundred one thousand, forty-one
39. 30 + 7 **41.** 3,000 + 800
43. 80,000 + 5,000 + 200 + 90
45. 8,000,000,000 + 100,000

47. 200 + 20 + 2 **49.** 340 **51.** 654
53. 2,720 **55.** 3,009
57. 0 20 22 24 26 28 30 32
59. 0 10 20 30 40 50 60 70
61. 0 200 400 600 800 1,000 1,200 1,400 1,600
63. 90 **65.** 900 **67.** 85,000 **69.** 700,000
71. 5,300 lb **73.** 80 hrs
75. a. $123,600 **b.** $100,000 **c.** $120,000
77. a. 3,564,300 **b.** 3,600,000 **c.** 4,000,000
79. a. California **b.** California, 53; Texas, 32; Florida, 25; Ohio, 18; Michigan, 15; North Dakota, 1 **c.** yes, North Dakota

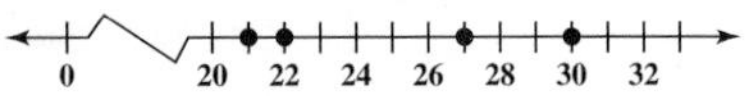

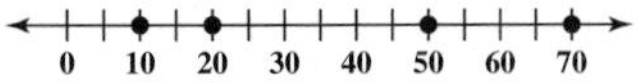

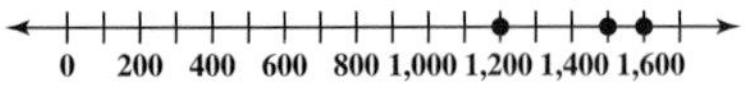

Section 1.2 Guided Practice

1. +3; 9 12
2. 6 + 5 = 11; 5 + 6 = 11; order **3.** 59 **4.** 63
5. 111 **6.** 126 **7.** 4719 **8.** 2287
9. 1,520 + 730 + 1,250
= 2,250 + 1,250
= 3,500
10. 6; 6; 10 **11.** 45, 55, 65; 3; 3; 3; 45, 55, 65, 66, 67; 3; 3; 3; 3; 33; 33

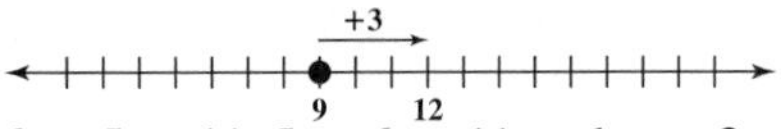

Concept Check and Objective Practice

A1. Answers may vary. Example: When you count on the number line, you add 1 for each tick mark.
A2. +2; 0 4 6
A3. +2; 0 6 8
A4. +8; 7 15
A5. +6; 9 15
A6.

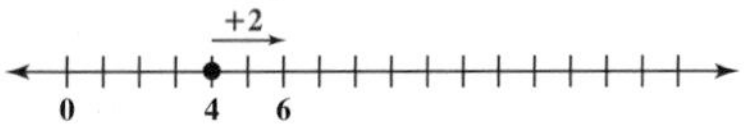

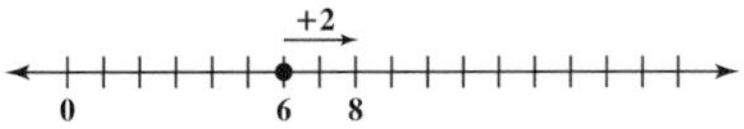

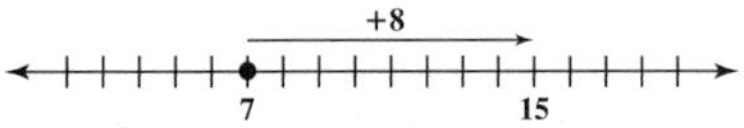

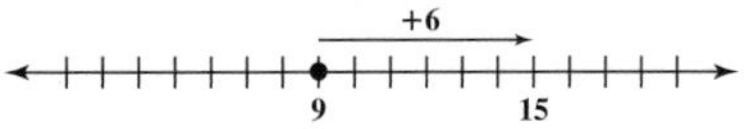

	7	3	9	8
5	**12**	**8**	**14**	**13**
7	**14**	**10**	**16**	**15**
8	**15**	**11**	**17**	**16**
6	**13**	**9**	**15**	**14**

A7.

	4	9	5	6
4	**8**	**13**	**9**	**10**
9	**13**	**18**	**14**	**15**
3	**7**	**12**	**8**	**9**
6	**10**	**15**	**11**	**12**

A8. a. 6 **b.** 3 **c.** 11 **d.** 9 **e.** 13 **f.** 13 **g.** 7 **h.** 13 **i.** 16 **j.** 16
A9. a. 10 **b.** 15 **c.** 7 **d.** 10 **e.** 9 **f.** 13 **g.** 12 **h.** 12 **i.** 12 **j.** 15
A10. a. 11 **b.** 13 **c.** 10 **d.** 13 **e.** 8 **f.** 14 **g.** 17 **h.** 5 **i.** 18 **j.** 14
A11. a. 17 **b.** 6 **c.** 14 **d.** 8 **e.** 15 **f.** 9 **g.** 14 **h.** 11 **i.** 9 **j.** 13
B1. If you have 13 ones, you can exchange them for 1 ten and 3 ones.
B2. 86 **B3.** 99 **B4.** 81 **B5.** 93 **B6.** 121
B7. 129 **B8.** 143 **B9.** 141 **B10.** 142 **B11.** 169
C1. 2 + 3 is 5; 5 + 5 is 10; 10 + 7 is 17 **C2.** 966
C3. 997 **C4.** 871 **C5.** 742 **C6.** 8,569 **C7.** 9,143
C8. 91,893 **C9.** 78,825 **C10.** $1,126 **C11.** $4,179

Focus on Adding Several Numbers Efficiently

1. 28 **2.** 24 **3.** 25 **4.** 21 **5.** 29 **6.** 24

Exercises

For 1.–11.: See preceding Concept Check and Objective Practice answers.

13. Commutative Property of Addition
15. variable **17.** sum
19. +2; 5 7
21. +3; 9 12
23. +19; 3 22
25. 8
27.

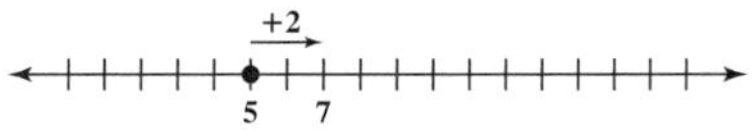

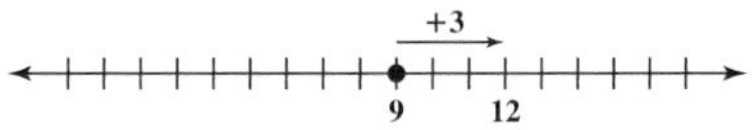

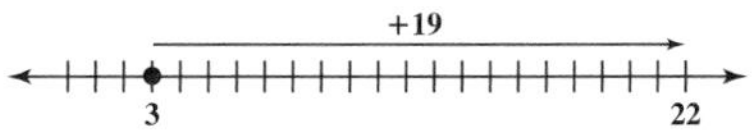

	9	7	5	4
3	**12**	**10**	**8**	**7**
6	**15**	**13**	**11**	**10**
5	**14**	**12**	**10**	**9**
9	**18**	**16**	**14**	**13**

29.

	8	4	9	0
2	**10**	**6**	**11**	**2**
7	**15**	**11**	**16**	**7**
6	**14**	**10**	**15**	**6**
9	**17**	**13**	**18**	**9**

31. a. 10 **b.** 15 **c.** 7 **d.** 12 **e.** 17 **f.** 13 **g.** 13
33. a. 17 **b.** 6 **c.** 14 **d.** 9 **e.** 15 **f.** 15 **g.** 16
35. 69 **37.** 79 **39.** 93 **41.** 159 **43.** 14 **45.** 53
47. 599 **49.** 293 **51.** 929 **53.** 1,241 **55.** $153
57. 182 miles **59.** 1,031 **61.** 730 **63.** 11,139
65. 67,825 **67.** 1,057,117 **69.** 1,423,277 **71.** 20
73. 23 **75.** 20 **77.** 44 inches **79.** 28 feet
81. 390 feet **83. a.** $492 **b.** no **85.** 450 miles
87. Answers may vary. Example: Too low because the curve of the shoreline appears longer than the straight lines estimating its length.
89. Answers may vary. **91.** 991 **93.** 189 **95.** 1,000
97. $46,500,000 **99.** 864 feet **101.** 7 **103.** 9
105. 7 **107.** 9 **109.** 21 **111.** 36

Section 1.3 Guided Practice

1. 10 − 6 = 4 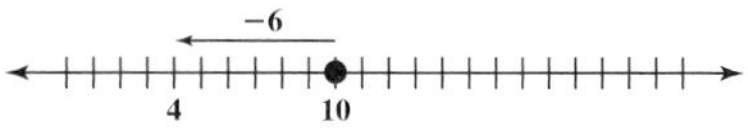

2. 35 **3.** 45 **4.** 126 **5.** 297 **6.** 3,173 **7.** 5,188

8. $38 - 2 - 5 = 36 - 5$
$= 31$

9. $47 - 5 - 6 - 9 = 42 - 6 - 9$
$= 36 - 9$
$= 27$

10. $976 - 428 - 35 = 548 - 35$
$= 513$

11. Asia's highest point: 29,028 feet
Australia's highest point: 7,310 feet
Difference = 29,028 − 7,310 = 21,718 feet
The difference between the highest mountain in Asia and the highest mountain is Australia is 21,718 feet.

12. 12; 12; 3; 12; 3; 3; 15

13. 15; 15; 34; 15; *y*; 34; 34; 19

14. Total order = Items ordered + items needed
Total order = Items ordered + x
$300 = 265 + x$
$300 - 265 = 265 - 265 + x$
$35 = x$
The store must order 35 more items to earn the discount.

Concept Check and Objective Practice

A1. Answers may vary. Example: Start at the first number given and move to the left the number of tick marks to be subtracted.

A2. 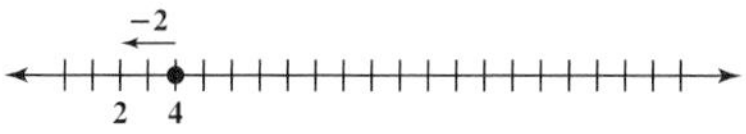

A3. 

A4. 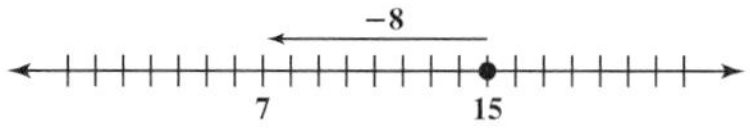

A5. 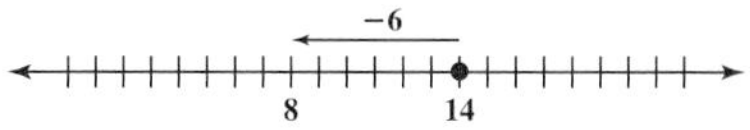

A6.

	7	**4**	8	10
5	12	**9**	**13**	**15**
2	**9**	**6**	**10**	12
4	**11**	**8**	12	**14**
9	**16**	13	**17**	19

A7.

	2	**6**	**7**	**3**
5	7	11	**12**	**8**
6	**8**	**12**	13	**9**
7	**9**	13	**14**	10
8	10	**14**	15	**11**

A8. **a.** 5 **b.** 8 **c.** 3 **d.** 7 **e.** 6 **f.** 2 **g.** 5 **h.** 4 **i.** 7 **j.** 6

A9. **a.** 9 **b.** 6 **c.** 7 **d.** 4 **e.** 9 **f.** 7 **g.** 8 **h.** 8 **i.** 3 **j.** 2

A10. **a.** 6 **b.** 5 **c.** 3 **d.** 8 **e.** 0 **f.** 9 **g.** 4 **h.** 7 **i.** 9 **j.** 2

A11. **a.** 9 **b.** 5 **c.** 9 **d.** 3 **e.** 8 **f.** 9 **g.** 7 **h.** 7 **i.** 2 **j.** 9

B1. Answers may vary. Example: Change one $10 bill for ten $1 bills. Changing the bills is the same as borrowing from the tens place.

B2. Answers may vary. Example: Borrowing is unnecessary when subtracting a smaller digit from a larger digit.

B3. 12 **B4.** 13 **B5.** 25 **B6.** 13 **B7.** 119

B8. 34 **B9.** 91 **B10.** 394 **B11.** $22 **B12.** $110

C1. Answers may vary. Example: If you remember this, you can write 100 directly as 9 tens and 10 ones.

C2. Answers may vary. Example: We borrow the same number of times for each problem. We just subtract a few more columns in the second problem.

C3. 270 **C4.** 228 **C5.** 1,095 **C6.** 2,090 **C7.** 144

C8. 322 **C9.** 1,479 **C10.** 27,669 **C11.** $3,950

C12. $895

D1. Answers may vary. Example: Subtract the first two numbers. From that answer, subtract the next value. Continue in this manner until all numbers have been subtracted

D2. 13 **D3.** 20 **D4.** 62 **D5.** 42 **D6.** 165

D7. 332 **D8.** $468 **D9.** $118 **D10.** $368

D11. no

Exercises

For 1.–12.: *See preceding Concept Check and Objective Practice answers.*

13. borrow **15.** difference

17. 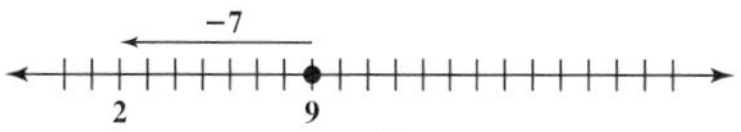

19. 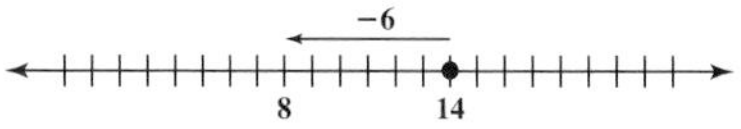

21.

	3	**9**	**7**	**5**
8	11	**17**	15	**13**
9	12	18	**16**	**14**
7	**10**	**16**	14	**12**
4	**7**	13	**11**	9

23.

	5	**8**	9	**6**
4	**9**	**12**	13	**10**
6	11	14	**15**	12
7	**12**	**15**	16	**13**
1	**6**	**9**	10	**7**

25. **a.** 6 **b.** 7 **c.** 8 **d.** 3 **e.** 6 **f.** 2 **g.** 9

27. **a.** 4 **b.** 8 **c.** 7 **d.** 5 **e.** 7 **f.** 8 **g.** 8

29. 2 **31.** 27 **33.** 29 **35.** 7 **37.** 16 **39.** 17

41. 483 **43.** 177 **45.** $367 **47.** 135 pounds

49. 27 **51.** 112 **53.** 345 **55.** 431 **57.** 685

59. 121 **61.** 7,668 **63.** 7,118 **65.** 893,847

67. 254,199 **69.** 25 **71.** 43 **73.** 28 **75.** 24

77. 40 **79.** 353 miles **81.** 656 pages **83.** $1,804

85. 8 hours 15 minutes **87.** $502 **89.** $452

91. 10,927 feet **93.** 28 **95.** 41 **97.** 30 **99.** 19
101. 42 **103.** 52

Section 1.4 Guided Practice

1. product; factors; 63; 7 and 9
2. $4 \cdot 5 = 20$; $5 \cdot 4 = 20$; order **3.** 96 **4.** 371
5. 540 **6.** 4; 630,000 **7.** 7; 65,130,000,000
8. 1,365; 1,638; 17,745; 17,745 **9.** 6,552; 7,488; 1,872; 268,632; 268,632
10. $212 \approx 200$ $285 \approx 300$
$212 \cdot 285 \approx 200 \cdot 300 \approx 60{,}000$
11. $1{,}183 \approx 1{,}000$ $299 \approx 300$
$1{,}183 \cdot 299 \approx 1{,}000 \cdot 300 \approx 300{,}000$
12. 15; 105; 210 **13.** 24; 72; 360; 720
14. $A = 15 \cdot 16$
$= 240$
240 square inches
15. 6 **16.** 41; 205; 41, 82, 123, 164, 205; Five; 5

Concept Check and Objective Practice

A1. Answers may vary. Example: Multiplication is a quick way to perform repeated addition.
A2. $2 \cdot 5 = 10$ and $5 \cdot 2 = 10$
A3. 3 and 4 are the factors. 12 is the product.
A4. 15 **A5.** 28 **A6.** 18 **A7.** 32
A8.

	1	2	3	4	5	6	7	8	9	10
1	**1**	**2**	**3**	**4**	**5**	**6**	**7**	**8**	**9**	**10**
2	**2**	**4**	**6**	**8**	**10**	**12**	**14**	**16**	**18**	**20**
3	**3**	**6**	**9**	**12**	**15**	**18**	**21**	**24**	**27**	**30**
4	**4**	**8**	**12**	**16**	**20**	**24**	**28**	**32**	**36**	**40**
5	**5**	**10**	**15**	**20**	**25**	**30**	**35**	**40**	**45**	**50**
6	**6**	**12**	**18**	**24**	**30**	**36**	**42**	**48**	**54**	**60**
7	**7**	**14**	**21**	**28**	**35**	**42**	**49**	**56**	**63**	**70**
8	**8**	**16**	**24**	**32**	**40**	**48**	**56**	**64**	**72**	**80**
9	**9**	**18**	**27**	**36**	**45**	**54**	**63**	**72**	**81**	**90**
10	**10**	**20**	**30**	**40**	**50**	**60**	**70**	**80**	**90**	**100**

A9. **a.** 0 **b.** 48 **c.** 54 **d.** 3 **e.** 30 **f.** 42 **g.** 28 **h.** 40 **i.** 64 **j.** 16
A10. **a.** 4 **b.** 54 **c.** 0 **d.** 49 **e.** 50 **f.** 56 **g.** 45 **h.** 20 **i.** 72 **j.** 6
A11. **a.** 12 **b.** 30 **c.** 72 **d.** 32 **e.** 35 **f.** 63 **g.** 0 **h.** 56 **i.** 18 **j.** 42
A12. **a.** 9 **b.** 48 **c.** 40 **d.** 0 **e.** 63 **f.** 36 **g.** 40 **h.** 36 **i.** 81 **j.** 8
B1. Answers may vary. Example: When multiplying $4 by 9, the answer is $36, or 3 tens and 6 ones. The 6 goes in the ones column, and the three must be "carried" to the tens column.
B2. Answers may vary. Example: We may need to carry digits to the next higher place value.
B3. 82 **B4.** 39 **B5.** 80 **B6.** 90 **B7.** 686
B8. 776 **B9.** 420 **B10.** 420 **B11.** $224 **B12.** $112
C1. Answers may vary. Example: $3 \cdot 100$ gives 3 hundreds, so appending two zeros is a quick way to multiply by 100.
C2. Commutative Property of Multiplication
C3. 140 **C4.** 230 **C5.** 523,000 **C6.** 1,200
C7. 7,300 **C8.** 879,000 **C9.** 9,870,000
C10. 8,700,000
D1. Answers may vary. Example: In multiplying 34 by 12, we need to multiply $1 \cdot 34$ and $2 \cdot 34$.
D2. Answers may vary. Example: In multiplying 34 by 12, we must add the products of $1 \cdot 34$ and $2 \cdot 34$, making sure to align the place values correctly.
D3. 5,076 **D4.** 2,982 **D5.** 6,570 **D6.** 17,616
D7. 47,424 **D8.** 119,884 **D9.** 696,877 **D10.** 50,964
D11. 7,215 calories **D12.** 10,335 calories

Focus on Estimation

1. correct **2.** correct **3.** incorrect; 85,544
4. incorrect; 79,506 **5.** incorrect; 12,206
6. incorrect; 32,361 **7.** incorrect; 933,696
8. incorrect; 7,167,404

E1. Answers may vary. Multiply the first pair of factors. Multiply this product by the next factor. Continue until you have multiplied by all of the factors.
E2. 90 **E3.** 70 **E4.** 108 **E5.** 336 **E6.** 420
E7. 560 **E8.** 288 **E9.** 648

Exercises

***For 1.–17.:** See preceding Concept Check and Objective Practice answers.*

19. factors **21.** factor · factor; product **23.** 8
25. 8
27.

	7	8	4	3
5	**35**	**40**	**20**	**15**
8	**56**	**64**	**32**	**24**
6	**42**	**48**	**24**	**18**
9	**63**	**72**	**36**	**27**

29. **a.** 36 **b.** 18 **c.** 36 **d.** 63 **e.** 28 **f.** 72 **g.** 30
31. **a.** 20 **b.** 24 **c.** 56 **d.** 18 **e.** 32 **f.** 81 **g.** 40
33. 92 **35.** 294 **37.** 2,715 **39.** 2,235 **41.** $7,389
43. $257,360 **45.** 840 **47.** 85,400 **49.** 561,000
51. 578,900 **53.** 900 **55.** 277,800 **57.** $120,000
59. $260,000 **61.** 2,352 **63.** 4,446 **65.** 224
67. 2,184 **69.** 17,928 **71.** 28,938 **73.** $884,282
75. 29,602,490 **77.** 305,505 **79.** 365,670
81. 78,645 **83.** 118,665 **85.** 80 square inches
87. 338,843 square feet **89.** 1,610 **91.** 2,862
93. 738 **95.** $465 **97.** $5,100 **99.** 8,100 square feet
101. 546 miles **103.** **a.** $5,500 **b.** less
105. **a.** 4,700 square feet **b.** 4,000 square feet **c.** no
107. $850 **109.** 3 **111.** 4 **113.** 8 **115.** 3

Section 1.5 Guided Practice

1. $10 \div 5 = \boxed{?}$

$\boxed{?} \cdot 5 = 10$

2. 6; 24; 4; 6; 24; 4 **3.** $3\overline{)12}$ with quotient 4; 12; 3; 4

4. 4; 5; 4; 4; 24; 24; 5; 5; 5; 29; four; 5; 4R5

5.
```
      2 4 R1
  3) 7 3
   - 6
     1 3
   - 1 2
       1
```
6.
```
      7 6 R3
  6) 4 5 9
   - 4 2
       3 9
     - 3 6
         3
```
73 ÷ 3 = 24R1 459 ÷ 6 = 76R3

7. quotient = 5; remainder = 2; divisor = 3 $5\frac{2}{3}$

8. 4R1; $4\frac{1}{6}$ **9.** 10; tens; 200; hundreds; $10\overline{)200}$ with quotient 20; 20

10. 30; tens; 600; hundreds; $30\overline{)600}$ with quotient 20; 20

11.
```
       0 1 7
  21) 3 7 2
    - 2 1
      1 6 2
      1 4 7
        1 5
```
17R15

12.
```
       0 3 8
  12) 4 5 6
    - 3 6
        9 6
      - 9 6
          0
```
38

73 ÷ 3 = 24R1 459 ÷ 6 = 76R3

13.
```
       0 3 8 5 R2
  22) 8 4 7 2
    - 6 6
      1 8 7
    - 1 7 6
        1 1 2
      - 1 1 0
            2
```
385R2

14. 7; *y*; 7; 6; 6

15. 13; 65; 13; 26; 39; 52; 65; five; 5

Focus on Division Involving Zero

1. a. ? × 0 = 10 **b.** undefined

2. a. ? × 0 = 10 **b.** undefined

3. a. ? × 10 = 0 **b.** 0

4. a. ? × 0 = 10 **b.** undefined

5. a. ? × 10 = 0 **b.** 0

6. a. ? × 10 = 0 **b.** 0

Concept Check and Objective Practice

A1. ? · 9 = 45

A2. There is no multiplication problem for which ? · 0 = 6.

A3. 10 **A4.** 10 **A5.** 6 **A6.** 6 **A7.** 9 **A8.** 9

A9. 4 **A10.** 2

A11.

	7	**6**	8	**2**
5	**35**	**30**	40	10
9	**63**	**54**	**72**	18
6	**42**	36	48	**12**
4	28	**24**	**32**	**8**

A12.

	3	5	9	**6**
9	**27**	**45**	81	**54**
7	**21**	35	**63**	42
4	**12**	**20**	**36**	**24**
8	24	**40**	72	**48**

A13. a. 2 **b.** 6 **c.** 9 **d.** 1 **e.** 5 **f.** 7 **g.** 4 **h.** undefined **i.** 8 **j.** 4

A14. a. 4 **b.** 6 **c.** 0 **d.** 7 **e.** undefined **f.** 8 **g.** 6 **h.** 4 **i.** 9 **j.** 3

A15. a. 4 **b.** 10 **c.** 8 **d.** 4 **e.** 5 **f.** 9 **g.** 0 **h.** 7 **i.** undefined **j.** 6

A16. a. 3 **b.** 8 **c.** 5 **d.** 5 **e.** 7 **f.** 6 **g.** 4 **h.** 4 **i.** 9 **j.** undefined

B1. Answers may vary. Example: It will help keep the digits aligned.

B2. Answers may vary. Example: The remainder is what is left over after the divisor is subtracted as many times as possible.

B3. 18R3 **B4.** 41R1 **B5.** 8R3 **B6.** 6R6

B7. 163R2 **B8.** 281R2 **B9.** 87R1 **B10.** 89R2

B11. \$300 **B12.** \$250

Focus on the Meaning of Remainders

1. $4\frac{2}{3}$ **2.** $2\frac{5}{8}$ **3.** $8\frac{2}{3}$ **4.** $2\frac{1}{8}$ **5.** $6\frac{2}{5}$ **6.** $4\frac{1}{7}$

7. $8\frac{1}{5}$ **8.** $7\frac{4}{7}$

C1. Answers may vary. Example: Estimating can help determine whether the answer is correct.

C2. a. 5,000 **b.** 90,000 **c.** 20 **d.** 900 **C3.** 20

C4. 20 **C5.** 20 **C6.** 30 **C7.** 250 **C8.** 400

C9. 100 **C10.** 30 **C11.** 450 pounds

C12. 300 pounds **D1.** 300 ÷ 30 = 10

D2. 31 · 9 = 279, which is greater than 271. Therefore, 31 goes into 271 less than 9 times.

D3. Answers may vary. Example: The 24 really represents 24 tens. Because 24 ÷ 12 = 2, the 2 belongs in the quotient above the 4.

D4. Answers may vary. Example: It would have made it very clear where the 2 belonged.

D5. 6R3 **D6.** 7R49 **D7.** 43R10 **D8.** 27

D9. 159R6 **D10.** 189R4 **D11. a.** \$632 **b.** \$793

Exercises

For 1.–16.: See preceding Concept Check and Objective Practice answers.

17. quotient **19.** divisor **21.** dividend; divisor; quotient

23. $\text{divisor}\overline{)\text{dividend}}$ with quotient above

25. a. $y \cdot 4 = 16$ **b.** 4

27. a. $x \cdot 7 = 28$ **b.** 4

29. a. $z \cdot 9 = 54$ **b.** 6

31. a. $y \cdot 12 = 60$ **b.** 5

33.

	6	8	9	3
5	**30**	40	**45**	**15**
4	24	**32**	**36**	12
7	42	**56**	63	**21**
8	**48**	64	**72**	**24**

35. a. 4 **b.** 3 **c.** 8 **d.** 8 **e.** 7 **f.** 9 **g.** 0
37. a. 7 **b.** 6 **c.** 9 **d.** undefined **e.** 6 **f.** 7 **g.** 7
39. 13 **41.** 12 **43.** 16 **45.** 54 **47.** 5R1
49. 3R3 **51.** 21R6 **53.** 32 **55.** 37
57. \$216 **59.** 61R4 **61.** 261 **63.** 28,436
65. 1,933R3 **67.** 411 **69.** undefined **71.** 47
73. 21 **75.** 5 **77.** 93 pages **79. a.** 2,000 **b.** 2,086
81. a. 20 **b.** 16 **83. a.** 3,500 **b.** 3,769R1
85. a. 1,000 **b.** 745 **87.** 32 **89.** 82R1 **91.** 48
93. \$60 **95.** \$5,600 **97.** 3 hours
99. a. \$328 **b.** 5 months **101. a.** \$1,400 **b.** \$1,500
103. a. \$185 **b.** no **105.** 6 **107.** 6 **109.** 10
111. 200

Section 1.6 Guided Practice

1. 9; 8
2. 2; 4; 4; 2; $2^4 = 2 \cdot 2 \cdot 2 \cdot 2 = 4 \cdot 2 \cdot 2 = 8 \cdot 2 = 16$
3. 7; 3; 3; 7; $3 \cdot 3 \cdot 3 \cdot 3 \cdot 3 \cdot 3 \cdot 3 = 3^7$
4. 3; four; $3 \cdot 3 \cdot 3 \cdot 3 = 9 \cdot 3 \cdot 3 = 27 \cdot 3 = 81$
5. 12; two; $12 \cdot 12 = 144$ **6.** zero; 1; $3{,}589^0 = 1$
7. $15 - 3 = 12$ **8.** $3[6] = 18$
9. 10; $\frac{10}{5}$; $10 \div 5 = 2$
10. $3 \cdot (\text{1st highest} + \text{2nd highest})$
$3 \cdot (12 + 9)$; $3(12 + 9) = 3(21) = 63$
Candace scored 63 points.
11. 18; 30 **12.** 5; 20; 10 **13.** 18; 48; 40
14. $16 + 10 \cdot 2 = 16 + 20 = 36$
15. $5 + 5^2 - 20 = 5 + 25 - 20$
$= 30 - 20$
$= 10$
16. $27 \div 3^3 + 8 \cdot 2 - 7 = 27 \div 27 + 8 \cdot 2 - 7$
$= 1 + 8 \cdot 2 - 7$
$= 1 + 16 - 7$
$= 10$
17. a. 55; 83; 55; 83; 70
b. 5; $\frac{55 + 58 + 72 + 83 + 62}{5} = \frac{330}{5} = 66°\text{F}$

Concept Check and Objective Practice

A1. The exponent is written above and to the right of the base.
A2. The exponent indicates how many of the base must be multiplied.
A3. base: 3 exponent: 7
A4. base: 7 exponent: 2
A5. base: 13 exponent: 2
A6. base: 4 exponent: 10
A7. $6 \cdot 6 \cdot 6$; 216 **A8.** $5 \cdot 5 \cdot 5$; 125 **A9.** $3 \cdot 3 \cdot 3 \cdot 3$; 81
A10. $3 \cdot 3 \cdot 3 \cdot 3 \cdot 3$; 243 **A11.** $16 \cdot 16$; 256
A12. $21 \cdot 21$; 441 **A13.** 7^0; 1 **A14.** 9^0; 1
B1. Answers may vary. Example: A grouping indicates that the operations inside must be done first.
B2. Operations with numbers in parentheses are calculated before a number with exponents.
B3. 9 **B4.** 1 **B5.** 7 **B6.** 27 **B7.** 15 **B8.** 2
B9. 1 **B10.** 1 **B11.** made \$1,404 **B12.** \$44
C1. Answers may vary. Example: When everyone follows the same order, the results will always be the same.
C2. Considering only the operations of multiplication and division, they are performed as encountered from left to right.
C3. Considering only the operations of addition and subtraction, they are performed as encountered from left to right.
C4. 7 **C5.** 9 **C6.** 8 **C7.** 18 **C8.** 4 **C9.** 14
C10. 5 **C11.** 12 **C12.** 18 **C13.** 32 **C14.** 30
C15. 8

Exercises

For 1.–9.: See preceding Concept Check and Objective Practice answers.

11. Exponents **13.** grouping
15. When considering only multiplication and division, the operation appearing first when moving from left to right should be performed first.
17. 4^3 **19.** 6^5 **21.** $6 \cdot 6 \cdot 6 \cdot 6 \cdot 6$ **23.** $9 \cdot 9 \cdot 9$
25. a. 1 **b.** 9 **c.** 8 **d.** 16 **e.** 49 **f.** 36 **g.** 81
27. a. 25 **b.** 4 **c.** 125 **d.** 9 **e.** 36 **f.** 27 **g.** 81
29. 81 **31.** 5 **33.** 1 **35.** 1,000 **37.** 16 **39.** 25
41. 64 **43.** 64 **45.** 100 **47.** 72 **49.** 28
51. 16 **53.** 2 **55.** 11 **57.** 40 **59.** 3 **61.** 2
63. 65 **65.** 54 **67.** d **69.** c **71.** g **73.** a
75. 29 **77.** 2 **79.** 5 **81.** \$358,000 **83.** 16
85. 180 **87.** 75 **89.** 160
91. a. 40°F Answers may vary. **b.** 42°F
93. 12

Section 1.7 Guided Practice

1. 2, 3, 4, 5, 6; 1; prime
2. 3; 5; $15 \div 3 = 5$; $15 \div 5 = 3$; itself; 1; composite
3. a. even; odd; is not; even; is; odd; is not
b. 3; 15; 15; is; is; $7 + 2 + 0 = 9$; 9; is; is; $8 + 3 + 7 = 18$; 18; is; is
c. 0; 5; 5; is; 0; is; 7; is not
4. odd; does not; is not; 1; is not; is not; does; prime; it is divisible by only itself and 1.
5. odd; is not; 8; is not; does not; 5; does; composite; it is divisible by a number other than itself and 1.
6. 87; 4; 9; 25; 49; 121; 169; 2, 3, 5, and 7
7. 109; 4; 9; 25; 49; 121; 169; 2, 3, 5, and 7
8. 4; 9; 25; 49; 121; 121; 7; odd; is not; $5 + 3 = 8$; is not; is not; 3; is not; is not; prime
9. 1; 12; 2; 6; 3; 4; 4; 3; 6; 2; 12 ;1; 1, 2, 3, 4, 6, 12
10. 1 and 12, 2 and 6, 3 and 4 **11.** $1 \cdot 12$; $2 \cdot 6$; $3 \cdot 4$
12. 6; 6; $6 = 2 \cdot 3$; $12 = 2 \cdot 6 = 2 \cdot 2 \cdot 3$; $12 = 2 \cdot 2 \cdot 3$; prime; 12
13.

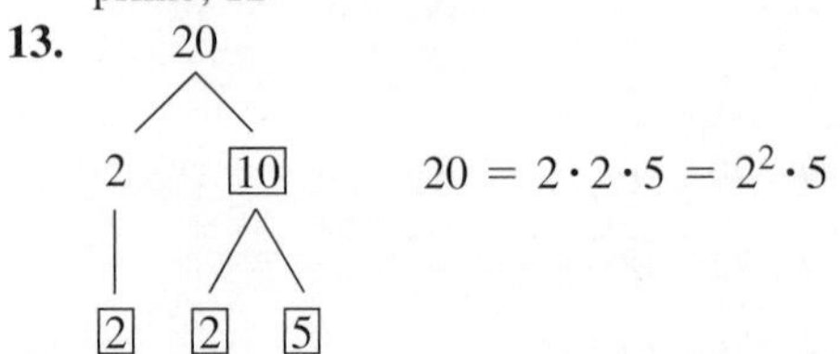

14. 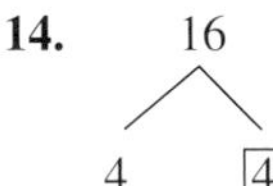

16

4 [4]

[2] [2] [2] [2]

$16 = 2 \cdot 2 \cdot 2 \cdot 2 = 2^4$

15. $1 \cdot 4 = 4$
$2 \cdot 4 = 8$
$3 \cdot 4 = 12$
$4 \cdot 4 = 16$
$5 \cdot 4 = 20$

16. $1 \cdot 12 = 12$
$2 \cdot 12 = 24$
$3 \cdot 12 = 36$
$4 \cdot 12 = 48$
$5 \cdot 12 = 60$

Concept Check and Objective Practice

A1. A prime number is divisible only by itself and 1.
A2. A composite number is divisible by a whole number other than itself and 1. **A3.** 12; 34; 220
A4. 12; 75; 87; 111; 213 **A5.** 75; 220
A6. You can stop at 3 because $5^2 = 25$, which is greater than 23.
A7. **a.** 2, 3, 5, 7 **b.** composite
A8. **a.** 2, 3, 5, 7 **b.** composite
A9. **a.** 2, 3, 5, 7 **b.** composite
A10. **a.** 2, 3, 5, 7 **b.** prime
A11. **a.** 2, 3, 5, 7, 11 **b.** composite
A12. **a.** 2, 3, 5, 7, 11 **b.** composite
A13. **a.** 2, 3, 5, 7 **b.** prime
A14. **a.** 2, 3, 5, 7 **b.** composite
B1. Answers may vary. Example: A prime factorization is a way to write a number as a product of prime numbers.
B2. Answers may vary. Example: The use of exponents enables us to write prime factorization more concisely. We can easily see how many times each factor is repeated in the factorization.
B3. **a.** 1, 36, 2, 18, 3, 12, 4, 9, 6
b. $1 \cdot 36$; $2 \cdot 18$; $3 \cdot 12$; $4 \cdot 9$; $6 \cdot 6$ **c.** $2^2 \cdot 3^2$
B4. **a.** 1, 40, 2, 20, 4, 10, 5, 8 **b.** $1 \cdot 40$; $2 \cdot 20$; $4 \cdot 10$; $5 \cdot 8$
c. $2^3 \cdot 5$
B5. $5 \cdot 7$ **B6.** $3 \cdot 11$ **B7.** $2 \cdot 5^2$ **B8.** $3^2 \cdot 5$
B9. $2^4 \cdot 3$ **B10.** $2^2 \cdot 5^2$ **B11.** $2^3 \cdot 3^2$ **B12.** $2^3 \cdot 7$
C1. Answers may vary. Example: 10, 20, 30, 40
C2. 1, 2, 5, 10 **C3.** factors **C4.** multiples
C5. 8, 16, 24, 32, 40 **C6.** 7, 14, 21, 28, 35
C7. 30, 60, 90, 120, 150 **C8.** 20, 40, 60, 80, 100
C9. 16, 32, 48, 64, 80 **C10.** 18, 36, 54, 72, 90
C11. 24, 48, 72, 96, 120 **C12.** 32, 64, 96, 128, 160

Exercises

For 1.–9.: See preceding Concept Check and Objective Practice answers.

11. prime number
13. multiple **15.** divisible
17. prime factorization
19. the digits of the number add to a value that is divisible by 3.
21. prime
23. composite
25. composite **27.** composite
29. 1, 2, 3, 6 **31.** 1, 11
33. 1, 2, 4, 7, 14, 28 **35.** 1, 2, 4, 8, 16, 32
37. 1, 3, 5, 15, 25, 75 **39.** 1, 2, 17, 34
41. **a.** 2, 3 **b.** composite **43.** **a.** 2, 3 **b.** prime
45. **a.** 2, 3, 5 **b.** composite **47.** **a.** 2, 3, 5 **b.** prime
49. **a.** 2, 3, 5, 7 **b.** prime **51.** **a.** 2, 3, 5, 7 **b.** composite
53. $2 \cdot 3^2$ **55.** $2 \cdot 3 \cdot 7$ **57.** $2^2 \cdot 3$ **59.** prime
61. $2 \cdot 3^2$ **63.** $2 \cdot 3 \cdot 7$ **65.** $2^3 \cdot 3^2$ **67.** prime
69. $2^2 \cdot 5^2$ **71.** $2^2 \cdot 3 \cdot 5$ **73.** 6, 12, 18, 24, and 30
75. 25, 50, 75, 100, and 125 **77.** 14, 28, 42, 56, and 70
79. 17, 34, 51, 68, and 85 **81.** 150, 300, 450, 600, and 750
83. 35, 70, 105, 140, and 175 **85.** 1, 2, 3, and 6
87. 8, 12, 16, 20, 24, and 80 **89.** 1, 2, 3, 6, and 12
91. 5, 10, 20, 75, and 80 **93.** 1, 2, and 8 **95.** 20 and 80
97. factor **99.** multiple

Section 1.8 Guided Practice

1. Factors of 12: [1], [2], [3], 4, [6], 12
Factors of 18: [1], [2], [3], [6], 9, 18
1, 2, 3, and 6; 1; 2; 3; 6
2. 1, 2, 3, and 6; 6; 6
3. Factors of 20: [1], [2], [4], 5, 10, 20
Factors of 8: [1], [2], [4], 8
1, 2, and 4; 4; 4
4. Factors of 14: [1], [2], [7], [14]
Factors of 42: [1], [2], 3, 6, [7], [14], 21, 42
1, 2, 7, and 14; 14; 14
5. Factors of 8: [1], 2, 4, 8
Factors of 27: [1], 3, 9, 27
1; 1; 1
6. Multiples of 4: 4, 8, 12, 16, 20, 24
Multiples of 6: 6, 12, 18, 24
First two common multiples: 12, 24
7. 12; 24; 12; 12
8. Multiples of 5: 5, 10, 15, 20, 25, 30
Multiples of 6: 6, 12, 18, 24, 30
30
9. Multiples of 12: 12, 24, 36, 48
Multiples of 16: 16, 32, 48
48
10. **a.** turkey, tomato, swiss
b. ham, lettuce, peppers, swiss **c.** swiss
11. $4 = 2 \cdot 2$
$6 = 2 \cdot 3$
2; 2; 3

Factors of 4: 2 | both: 2 | Factors of 6: 3

12. $20 = 2 \cdot 2 \cdot 5$
$12 = 2 \cdot 2 \cdot 3$
2; 2

Factors of 20: 5 | both: 2, 2 | Factors of 12: 3

13. 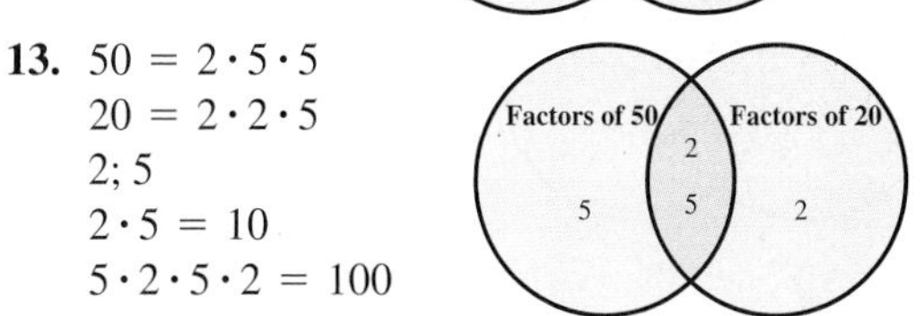

$50 = 2 \cdot 5 \cdot 5$
$20 = 2 \cdot 2 \cdot 5$
2; 5
$2 \cdot 5 = 10$
$5 \cdot 2 \cdot 5 \cdot 2 = 100$

6. Solve. $2(x + 4) = 5x - x + 8$		**6.** Solve. $4 + 8t - 5 = 3(t + 8)$
Distribute.	$2(x + 4) = 5x - x + 8$	$4 + 8t - 5 = 3(t + 8)$
Combine like terms.	$2x + 8 = 5x - x + 8$	$4 + 8t - 5 =$
	$2x + 8 = 4x + 8$	$=$
Gather the variable terms.	$2x - 2x + 8 = 4x - 2x + 8$	$=$
	$8 = 2x + 8$	$=$
Gather the constant terms.	$8 - 8 = 2x + 8 - 8$	$=$
	$0 = 2x$	$=$
Isolate the variable.	$\frac{0}{2} = \frac{\cancel{2}x}{\cancel{2}}$	$=$
	$0 = x$	$t =$
		Check your solution on scratch paper.

An answer of $x = 0$ might seem wrong to many students. Remember that 0 is just another number, and it can be a solution to an equation.

Concept Check

C1. List the three goals used to solve an equation.

C2. When simplifying an expression like $3 - 2(x - 5)$, students often forget to distribute the -2 to both the x and -5. What advice would you offer a friend to help her remember that she must distribute -2 to both terms?

Objective C Practice

Solve each equation.

C3. $6(x + 5) = 9 + 2x - 3$

C4. $5(x + 7) = 10 + 3x - 3$

C5. $6 = 4(y - 1) + y$

C6. $2 = 8(z - 2) + z$

C7. $4x - 1 - 2x = 3x - 12$

C8. $3x + 14 = 3x - 1 - x$

C9. $3 - 2(x - 5) = 4x + 1$

C10. $4 - 3(t - 8) = 2t - 7$

14. $18 = 2 \cdot 3 \cdot 3$
$54 = 2 \cdot 3 \cdot 3 \cdot 3$
$2 \cdot 3 \cdot 3 = 18$
$2 \cdot 3 \cdot 3 \cdot 3 = 54$

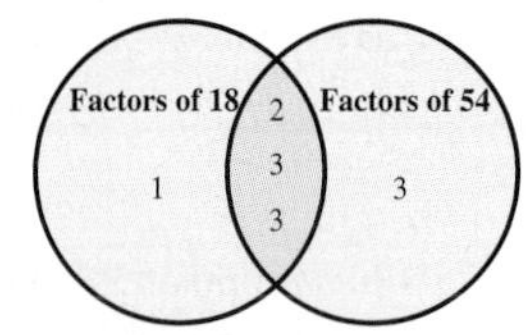

15. Multiples of 5: 5, 10, 15, 20, 25, 30
Multiples of 6: 6, 12, 18, 24, 30; 30 days

Concept Check and Objective Practice

A1. 8 is a factor of 16 but not of 18.
A2. 5 is a common factor, but so is 10. Because 10 is greater than 5, 10 is the greatest common factor.
A3. Example: 12 and 18
A4. **a.** Factors of 6: 1, 2, 3, 6
Factors of 12: 1, 2, 3, 4, 6, 12
b. 1, 2, 3, and 6 **c.** 6
A5. **a.** Factors of 12: 1, 2, 3, 4, 6, and 12
Factors of 36: 1, 2, 3, 4, 6, 9, 12, 18, and 36
b. 1, 2, 3, 4, 6, 12 **c.** 12
A6. 2 **A7.** 5 **A8.** 3 **A9.** 2 **A10.** 1 **A11.** 1
A12. 28 **A13.** 16
B1. The least common multiple is the smallest number that is divisible by the two numbers.
B2. The greatest common factor is the largest number that divides the two numbers exactly.
B3. LCM **B4.** GCF **B5.** 16 **B6.** 25 **B7.** 60
B8. 36 **B9.** 150 **B10.** 60 **B11.** 1,200 **B12.** 2,400
B13. LCM **B14.** GCF **B15.** GCF **B16.** LCM
C1. GCF because it is less than either number.
C2. LCM because it is greater than either number.
C3. **a.** LCM **b.** GCF **C4.** 1
C5. **a.** 18 **b.** 6 **c.** 72
C6. **a.** 1 **b.** 11 **c.** 55

In Exercises C7–C18, the first value is the GCF and the second value is the LCM.

C7. 6; 24 **C8.** 9; 36 **C9.** 5; 75 **C10.** 7; 105
C11. 11; 110 **C12.** 12; 144 **C13.** 2; 448 **C14.** 2; 204
C15. 1; 144 **C16.** 1; 300 **C17.** 10; 140 **C18.** 10; 450

Exercises

***For 1.–9.:** See preceding Concept Check and Objective Practice answers.*

11. factor **13.** least common multiple
15. greatest common factor **17.** 3 **19.** 6 **21.** 7
23. 10 **25.** 10 **27.** 33 **29.** 66 **31.** 90
33. **a.** 1 **b.** 14 **c.** 210
35. **a.** **b.** 5 **c.** 75

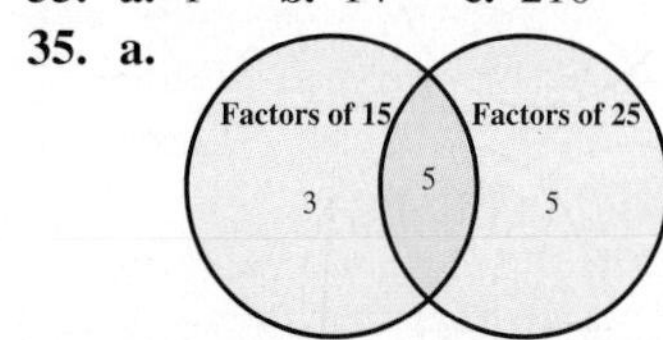

37. **a.** **b.** 6 **c.** 36

Factors of 12 | Factors of 18
2 | 2, 3 | 3

39. **a.** **b.** 5 **c.** 175

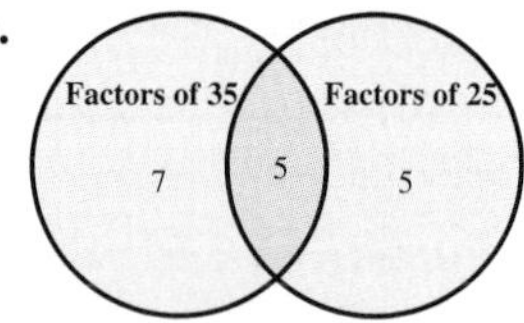

41. **a.** **b.** 2 **c.** 312

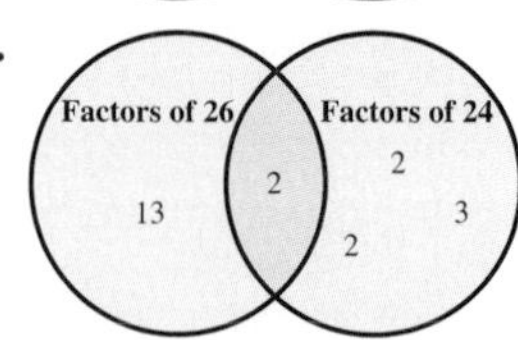

43. **a.** **b.** 7 **c.** 105

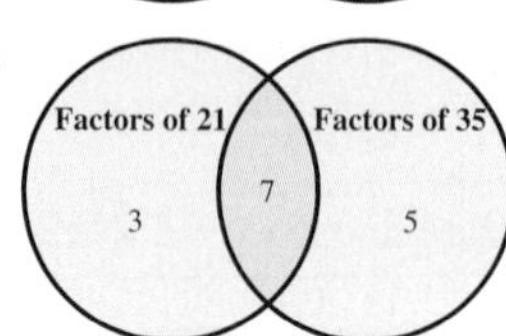

45. **a.** **b.** 9 **c.** 90

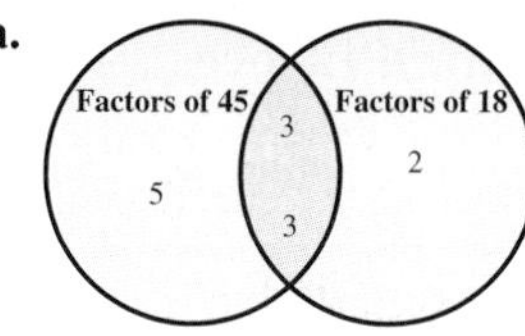

47. **a.** **b.** 18 **c.** 216

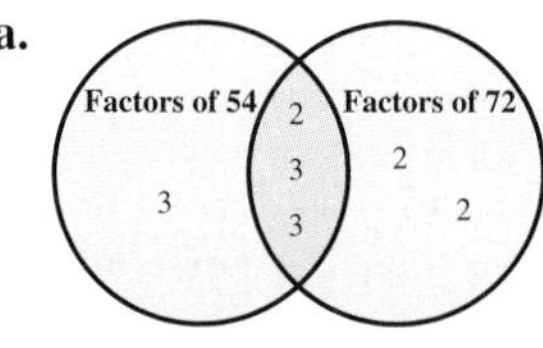

49. on the 11th
51.

	9	7	4	8
7	63	**49**	**28**	56
5	**45**	35	**20**	40
6	54	**42**	24	**48**
3	**27**	21	**12**	**24**

53. **a.** 700 **b.** 14 revolutions **c.** 5 revolutions

Section 1.9 Guided Practice

1. "Tripling" indicates multiplying by 3.
The result $= 3 \cdot 40$
2. "Go into" indicates division.
The result $= 32 \div 4$
3. "Deposit" indicates addition.
Balance $= \$50 + \$45$
4. "Withdrew" indicates subtraction; subtract 25 from 76.
New balance = Old balance − Withdrawal
$= \$76 - \$25 = \$51$
Jim's new balance is $51.
5. # cakes = (cakes per hour) · (hours)
$= 8 \cdot 4$
$= 32$
Buggs can make 32 carrot cakes.
6. # bags = (lb carrots) ÷ (lb per bag)
$= 45 \div 3$
$= 15$
Daniel fills 15 bags of carrots.

7. $\text{Avg temp} = \dfrac{\text{Sum of the daily temps}}{\text{Number of days}}$

$= \dfrac{64 + 61 + 72 + 59}{4}$

$= \dfrac{256}{4}$

$= 64$

The four-day average of 64° was lower than 65°.

8. Average amount of water wasted

$= \dfrac{\text{Total water wasted}}{\text{Number of hrs}}$

$= \dfrac{\text{Change in meter}}{\text{Number of hrs}}$

$= \dfrac{13{,}037 - 12{,}983}{2}$

$= \dfrac{54}{2}$

$= 27$

The average amount of water wasted per hour was 27 gallons.

Concept Check and Objective Practice

A1. **a.** Examples include sum, add, increased by, more than, plus, total.
b. Examples include difference, less than, subtracted from, fewer than, withdrawal, decreased by.
c. Examples include times, product, of, doubled.
d. Examples include ratio, divided by, quotient, goes into.
e. Examples include is, was, equals, is the same as.

A2. Example: A child has 10 pennies. She is given 20 more. How many pennies does she have?

A3. Example: A soccer coach has $20 to spend on his 10 team members. How much can he spend on each member of the team?

A4. $15 - 7$ **A5.** $8 + 9$ **A6.** $2 \cdot 15$ **A7.** $42 \div 3$

A8. $\$80 - \$40$ **A9.** $5 \cdot 6$ **A10.** $3 \cdot 4 \cdot 5$

A11. $3 + 4 + 5$

B1. Answers may vary. Example: Sarah pays $50 a month for her wireless service. How much does she pay for this service in a year?

B2. Answers may vary. Example: When Simon purchased a rug for his bedroom for $600, he was given the opportunity to purchase it over 12 months same as cash. If he chose to take advantage of this plan, what were his monthly payments?

B3. It will take Sergio 5 hours to paint 75 birdhouses.

B4. The bookstore ordered 720 pens.

B5. The repairs cost $9,312.

B6. There is a difference of 29 students in the class sizes.

B7. Gumby's business collected $141.

B8. Terrence had $470 left.

B9. There are 1,001,880 sq ft in 23 acres of land.

B10. Their average income is $44,000.

C1. Answers may vary. Example: It will help focus on the words that lead directly to the equation.

C2. Answers may vary. Example: There is a great deal of difference between 1 inch and 1 mile. Indicating the units gives greater clarity to the answer.

C3. The average life expectancy of men is 72 years.

C4. The average life expectancy of women is 77 years.

C5. 63 freshmen or sophomores entered the competition.

C6. Yanis' bill was $49.

C7. The total bill after he used the gift card was $242.

C8. Janise spent $218 on the project.

Exercises

For 1.–9.: *See preceding Concept Check and Objective Practice answers.*

11. division **13.** division **15.** equality
17. addition **19.** subtraction **21.** subtraction
23. equality **25.** 855 **27.** 11 **29.** 46 **31.** 58
33. $991 **35.** 81,973 **37.** $484 **39.** 144 eggs
41. 52° **43.** 2,125 gallons **45.** $235 **47.** 56°
49. 2,265,120 square feet **51.** 26 miles per gallon
53. $94 **55.** 32 mpg **57.** $1,764

Chapter 1 Review Exercises

1. ten thousands **2.** millions
3. one million, ninety-eight thousand, three hundred forty-two
4. thirty-four thousand, nine hundred eight
5. $30{,}000 + 400 + 2$ **6.** $100{,}000 + 5{,}000 + 30$
7. 938,000 **8.** 100

9.
16 32 43
0 10 20 30 40 50 60 70

10.
80
0 100 200 300

11. 70 **12.** 50 miles **13.** 58 **14.** 97
15. 172 **16.** 472 **17.** 2,079 **18.** 3,065
19. 15,534 **20.** 16,677 **21.** $705, yes
22. $850, yes **23.** 32 **24.** 33 **25.** 134
26. 437 **27.** 1,092 **28.** 4,909 **29.** 4,792
30. 5,018 **31.** year 4; Elaine earned $7,000 less.
32. Elaine earned $122,000. Blair earned $125,000.
33. Blair earned $3,000 more than Elaine.
34. Addition: right; carry; subtract
Subtraction: left; borrow; add
35. 800; 861 **36.** 900; 864 **37.** 1,400; 1,608
38. 3,200; 3,002 **39.** 10,000; 9,752
40. 24,000; 21,660 **41.** 420 **42.** 336
43. short: 420 square feet
tall: 399 square feet
The area of the short rectangle is 21 square feet larger than the tall rectangle.
44. $5,580; $7,276 **45.** 30 points **46.** $396
47. 8 **48.** 7 **49.** 10; 10R3 **50.** 20; 19R1
51. 20; 12R3 **52.** 10;13R3 **53.** 350; 312R20
54. 200; 209R16 **55.** 9 sheds **56.** 3 hours
57. grouping symbols; exponents; multiplication/division in order from left to right; addition/subtraction in order from left to right
58. 8 **59.** 16 **60.** 29 **61.** 51 **62.** 14 **63.** 9
64. 9 **65.** 24 **66.** 3 **67.** 28 **68.** 2 **69.** 2
70. Answers may vary. Example: 85 **71.** 84; B
72. Answers may vary. Example: The lowest grade Alex received on any one exam was 71, a C–. The average will not be lower than the lowest grade.
73. composite **74.** prime **75.** 1 and 20; 2 and 10; 4 and 5
76. 1 and 36; 2 and 18; 3 and 12; 4 and 9; 6 and 6 **77.** $2^3 \cdot 5$
78. $2 \cdot 3 \cdot 11$ **79.** 6, 12, 18, 24, 30, 36, 42, 48
80. 8, 16, 24, 32, 40, 48
81. 33 is the greatest common factor because the GCF must be less than or equal to the other numbers.

82. 198 is the least common multiple because the LCM must be greater than or equal to the other numbers.
83. 8 **84.** 7 **85.** 60 **86.** 36

In Exercises 87–90, the LCM is the first value and the GCF is the second value.

87. 120; 4 **88.** 224; 4 **89.** 40; 20 **90.** 210; 1
91. 384 **92.** 4 **93.** 26 **94.** 26 **95.** $7 **96.** $218
97. 115 kwh; making money **98.** 876 kwh; 276 kwh

Chapter 1 Practice Test

1. hundred thousands **2.** sixty thousand, eight hundred nine
3. 830,000 **4.** 100,000 + 7,000 + 60
5. 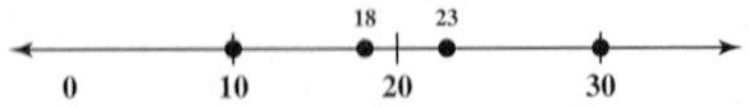

6. He can purchase a 0.3-carat ruby. **7.** 179 **8.** 241
9. 289 **10.** 673 **11.** 5,018 **12.** 6,361 **13.** 4,693
14. 7,328 **15.** $113,000,000 **16.** $20,000,000
17. $35,000,000 **18.** $49,000,000 **19.** 3,500; 3,456
20. 40; 38R5 **21.** 20; 21R14 **22.** 14,000; 10,854
23. 3,000,000; 2,815,716 **24.** 140; 131R22
25. 2,850 pounds **26.** 172 square inches **27.** 32 books
28. $532 **29.** 50 **30.** 35 **31.** 13 **32.** 100
33. 8 **34.** 2 **35.** 1 and 40; 2 and 20; 4 and 10; 5 and 8
36. $2 \cdot 3 \cdot 7$ **37.** 75 **38.** 10

In Exercises 39 and 40, the LCM is the first value and the GCF is the second value.

39. 168; 6 **40.** 112; 4 **41.** $223 **42.** 15 gallons

CHAPTER 2

Section 2.1 Guided Practice

1. 3; 3; 5; 5; Fraction $= \frac{3}{5}$ **2.** 6; 6; 4; 4; Fraction $= \frac{4}{6}$
3. 3 + 1 + 2 + 1; 7; reading; 3; $\frac{3}{7}$
4. denominator = 8; 8 equal-sized parts; numerator = 3; 3 of those parts 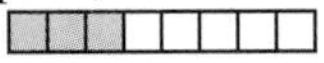
5. Proper fractions: a; smaller; e; less than Improper fractions: d; f; equal to or greater than; b; c; at least; whole
6. 3 equal-sized parts; denominator = 3
7 shaded parts; numerator = 7

fraction $= \frac{7}{3}$

7. denominator = 5; 5 equal-sized parts numerator = 11; shade 11 of those parts; 3 whole objects; 11

8. 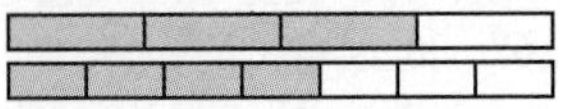 $\frac{3}{4}, \frac{3}{4}, \frac{4}{7}$
9. 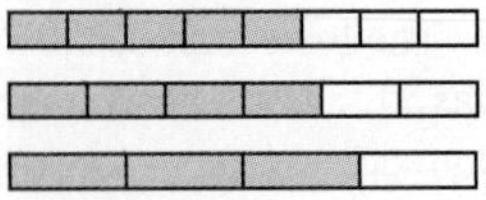 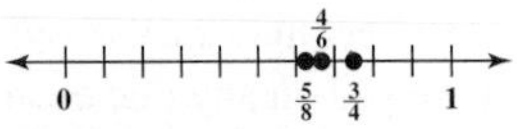

$\frac{5}{8}, \frac{4}{6}$, and $\frac{3}{4}$
10. $\frac{1}{3} < \frac{1}{2}$; $\frac{2}{6} > \frac{1}{6}$ **11.** larger pieces; 5; 6; $\frac{4}{5} > \frac{4}{6}$
12. fewer pieces; $\frac{3}{5} < \frac{4}{5}$ **13.** 8; 13; $\frac{8}{13}$
14. 25; 25 + 38 = 63; $\frac{25}{63}$
15. 4; 4; 4; 4 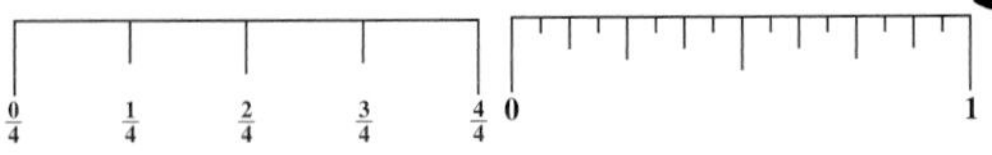

16. 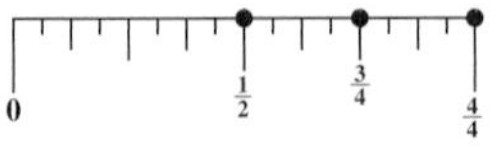

17. 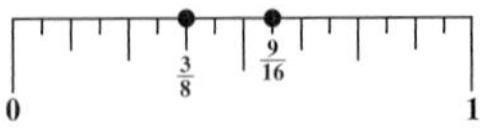

Concept Check and Objective Practice

A1. 2; 7 **A2.** denominator; numerator **A3.** $\frac{3}{5}$ **A4.** $\frac{3}{4}$
A5. $\frac{1}{2}$ **A6.** $\frac{2}{2}$ **A7.** $\frac{4}{6}$ **A8.** $\frac{2}{4}$ **A9.** $\frac{4}{13}$ **A10.** $\frac{5}{8}$
B1. the number of parts to shade
B2. the number of parts in the whole **B3.** c **B4.** a
B5. d **B6.** b **B7.** 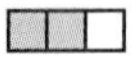 **B8.** 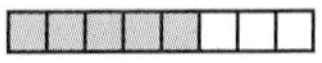
B9. 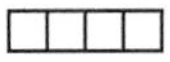 **B10.** 
B11. 
B12. 
C1. more than **C2.** Its numerator and denominator are equal. **C3.** c **C4.** b **C5.** d **C6.** a
C7.  **C8.** 
C9.  **C10.** 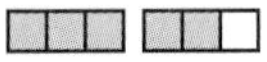
C11. $\frac{6}{3}$; improper **C12.** $\frac{6}{4}$; improper **C13.** $\frac{1}{3}$; proper
C14. $\frac{3}{4}$; proper
C15. Answers may vary. Example: It's not possible to draw a whole with 0 parts.
D1. Answers may vary. Example: Given that the "whole" pictures are the same size, the larger fraction picture will have more area shaded than the smaller fraction picture.
D2. $\frac{1}{5}$ **D3.** $\frac{4}{6}$ **D4.** $\frac{3}{4}$ **D5.** $\frac{6}{8}$
D6. $\frac{1}{5}, \frac{1}{4}, \frac{1}{3}$ 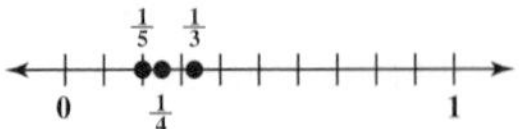

D7. $\frac{2}{3}, \frac{5}{7}, \frac{3}{4}$ 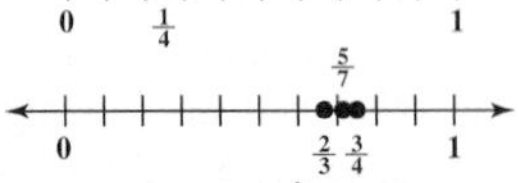

D8. $\frac{3}{8}, \frac{3}{7}, \frac{3}{6}, \frac{3}{5}$ 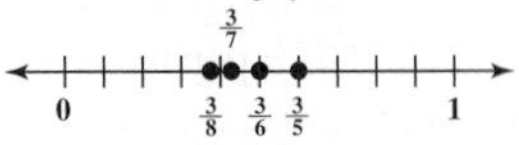

D9. $\frac{2}{7}, \frac{2}{6}, \frac{2}{5}, \frac{2}{4}$ 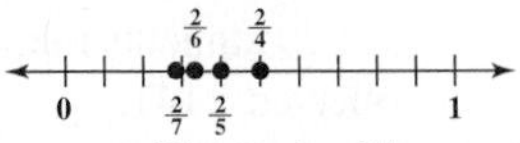

E1. Answers may vary. Example: There will be more shaded parts. 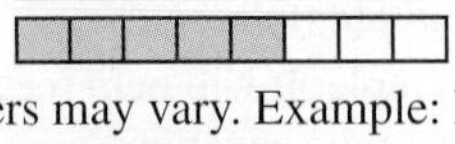
E2. Answers may vary. Example: Each individual part will be larger. 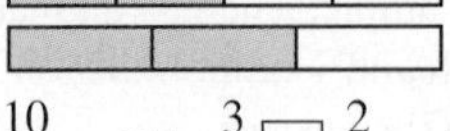
E3. $\frac{9}{7}$ [<] $\frac{10}{7}$ **E4.** $\frac{3}{3}$ [>] $\frac{2}{3}$ **E5.** $\frac{15}{31}$ [<] $\frac{15}{19}$

E6. $\frac{21}{52} \boxed{<} \frac{21}{25}$ **E7.** $\frac{6}{2} \boxed{>} \frac{6}{4}$ **E8.** $\frac{7}{10} \boxed{<} \frac{7}{4}$

E9. $\frac{8}{90} \boxed{<} \frac{10}{90}$ **E10.** $\frac{70}{120} \boxed{>} \frac{20}{120}$

Exercises

For 1.–15.: *See preceding Concept Check and Objective Practice answers.*

17. fraction **19.** proper fraction **21.** denominator

23. Answers may vary. Example: It tells how many parts are shaded.

25. $\frac{7}{4}$ 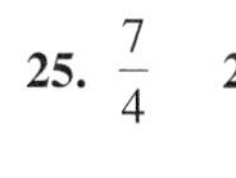 **27.** 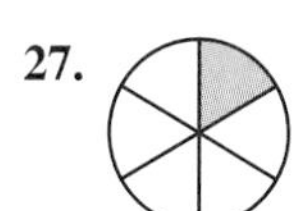 **29.** 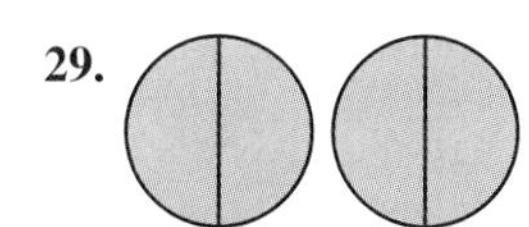

31. $\frac{2}{6}$ 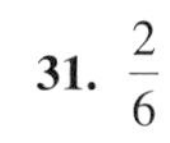 **33.** $\frac{0}{6}$ **35.** $\frac{3}{4}$ **37.**  $\frac{4}{8}$

39. 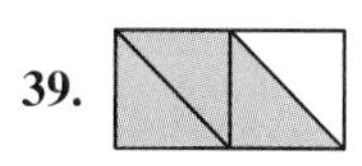 $\frac{3}{4}$ 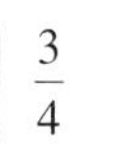 **41.** 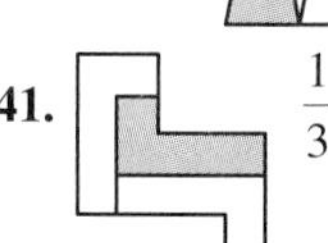 $\frac{1}{3}$

43. Subtraction; there are fewer shaded parts at the conclusion of the "problem" than at the start.

45. a. 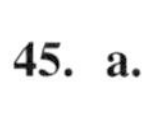 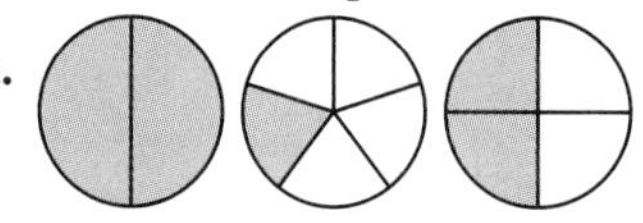 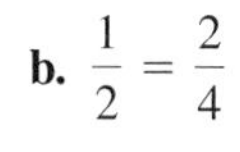 **b.** $\frac{1}{2} = \frac{2}{4}$

47. a. 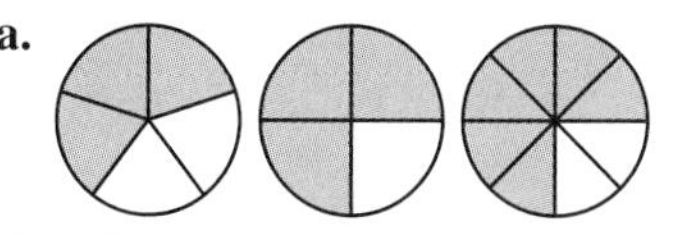 **b.** $\frac{3}{4} = \frac{6}{8}$

49. $\frac{4}{7} > \frac{3}{7}$ With like denominators, the larger fraction is the one with the larger numerator.

51. $\frac{1}{20} > \frac{1}{100}$ With like numerators, the larger fraction is the one with the smaller denominator.

53. $\frac{5}{12} > \frac{5}{20}$ With like numerators, the larger fraction is the one with the smaller denominator.

55. $\frac{12}{11} < \frac{13}{11}$ With like denominators, the larger fraction is the one with the larger numerator.

57. 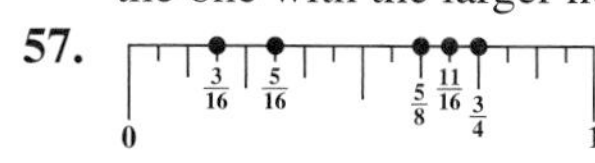

59. 0, $\frac{1}{16}$, $\frac{3}{16}$, $\frac{1}{4}$, $\frac{5}{16}$, $\frac{3}{8}$, 1

61. a. Answers may vary. Example: Since each object has different sized pieces, they require different denominators.

b. 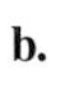 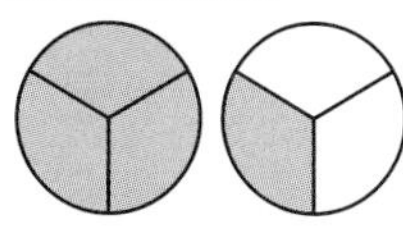

63. $\frac{8}{12}$ **65. a.** 25 **b.** $\frac{14}{25}$ **c.** $\frac{11}{25}$

67. a. $\frac{7}{10}$ **b.** $\frac{3}{10}$

69. a. $\frac{12}{31}$ **b.** $\frac{15}{31}$ **c.** $\frac{27}{31}$

71. a. $\frac{8}{10}$ **b.** $\frac{7}{10}$ **c.** $\frac{15}{10}$

73. a. $\frac{7}{13}$ **b.** $\frac{6}{13}$ **c.** $\frac{4}{13}$ **d.** $\frac{9}{13}$

Section 2.2 Guided Practice

1. a. $\frac{4}{6}; \frac{2}{3}$; equivalent **b.** equivalent; $\frac{4}{6}; \frac{2}{3}; \frac{2}{3}$

2. 3; three; three; 3; $\frac{6}{9} = \frac{2 \cdot \cancel{3}^1}{3 \cdot \cancel{3}_1} = \frac{2}{3}; \frac{2}{3}$

3. $\frac{4}{10} = \frac{2 \cdot 2}{2 \cdot 5} = \frac{\cancel{2}^1 \cdot 2}{\cancel{2}_1 \cdot 5} = \frac{2}{5}; \frac{2}{5}$

4. $\frac{18}{36} = \frac{2 \cdot 3 \cdot 3}{2 \cdot 2 \cdot 3 \cdot 3} = \frac{\cancel{2}^1 \cdot \cancel{3}^1 \cdot \cancel{3}^1}{\cancel{2}_1 \cdot 2 \cdot \cancel{3}_1 \cdot \cancel{3}_1} = \frac{1}{2}; \frac{1}{2}$

5. $\frac{240}{16} = \frac{\cancel{2}^1 \cdot \cancel{2}^1 \cdot \cancel{2}^1 \cdot \cancel{2}^1 \cdot 3 \cdot 5}{\cancel{2}_1 \cdot \cancel{2}_1 \cdot \cancel{2}_1 \cdot \cancel{2}_1} = \frac{15}{1} = 15; 15$

9. $\frac{12}{5} \cdot \frac{10}{9} = \frac{12 \cdot 10}{5 \cdot 9} = \frac{\cancel{3}^1 \cdot 4 \cdot 2 \cdot \cancel{5}^1}{\cancel{5}_1 \cdot \cancel{3}_1 \cdot 3} = \frac{8}{3}$

10. $\frac{14}{9} \cdot \frac{3}{35} = \frac{14 \cdot 3}{9 \cdot 35} = \frac{2 \cdot \cancel{7}^1 \cdot \cancel{3}^1}{\cancel{3}_1 \cdot 3 \cdot 5 \cdot \cancel{7}_1} = \frac{2}{15}$

11. $6 \cdot \frac{5}{12} = \frac{6}{1} \cdot \frac{5}{12} = \frac{6 \cdot 5}{1 \cdot 12} = \frac{\cancel{6}^1 \cdot 5}{1 \cdot \cancel{6}_1 \cdot 2} = \frac{5}{2}$

12. $\frac{3}{5} \cdot 5{,}000 = \frac{3}{5} \cdot \frac{5{,}000}{1} = \frac{3 \cdot 5{,}000}{5 \cdot 1} = \frac{3 \cdot \cancel{5}^1 \cdot 1{,}000}{\cancel{5}_1 \cdot 1}$

$= \frac{3{,}000}{1} = 3{,}000$; \$3,000

They will receive \$3,000.

13. The missing multiplier is 7 because $5 \cdot 7 = 35$.

$\frac{3}{5} \cdot \frac{7}{7} = \frac{21}{35}; \frac{21}{35}$; equivalent

14. The missing multiplier is 3 because $3 \cdot 7 = 21$.

$\frac{3}{7} \cdot \frac{3}{3} = \frac{9}{21}; \frac{9}{21}$; equivalent **15.** 12; 1; $\frac{12\text{ in.}}{1\text{ ft}}; \frac{1\text{ ft}}{12\text{ in.}}$

16. 1; 2; $\frac{1\text{ pt}}{2\text{ c.}}; \frac{2\text{ c.}}{1\text{ pt}}; \frac{2\text{ cups}}{1\text{ pint}}; \frac{5\text{ }\cancel{\text{pints}}}{1} \cdot \frac{2\text{ cups}}{1\text{ }\cancel{\text{pint}}} = \frac{10\text{ cups}}{1}$

$= 10$ cups

17. 1; 5,280 $\frac{4\text{ }\cancel{\text{miles}}}{1} \cdot \frac{5280\text{ ft}}{1\text{ }\cancel{\text{mile}}} = \frac{21{,}120\text{ ft}}{1} = 21{,}120$ ft

18. 28 c. $= \frac{28\text{ c.}}{1}$

$\frac{^{14}\cancel{28}\text{ }\cancel{c}}{1} \cdot \frac{1\text{ pt}}{_1\cancel{2}\text{ }\cancel{c}} = 14$ pt

19. pounds; ounces; 6 lb $\frac{6\text{ }\cancel{\text{lb}}}{1} \cdot \frac{16\text{ oz}}{1\text{ }\cancel{\text{lb}}} = 96$ oz No, will not

20. $\frac{2}{3} \cdot \frac{2}{3} \cdot \frac{2}{3} \cdot \frac{2}{3} = \frac{2 \cdot 2 \cdot 2 \cdot 2}{3 \cdot 3 \cdot 3 \cdot 3} = \frac{16}{81}$

21. $\frac{2}{3}\cdot\left(\frac{9}{8}\cdot\frac{4}{3}\right)^3 = \frac{2}{3}\cdot\left(\frac{\cancel{3}^1\cdot 3\cdot {}^1\cancel{2}\cdot\cancel{2}^1}{{}^1\cancel{2}\cdot\cancel{2}^1 2\cdot\cancel{3}^1}\right)^3$

$= \frac{2}{3}\cdot\left(\frac{3}{2}\right)^3$

$= \frac{2}{3}\cdot\left(\frac{3}{2}\cdot\frac{3}{2}\cdot\frac{3}{2}\right)$

$= \frac{\cancel{2}^1}{{}^1\cancel{3}}\cdot\frac{{}^1\cancel{3}\cdot 3\cdot 3}{\cancel{2}^1\, 2\cdot 2}$

$= \frac{9}{4}$

22. Area of yard $= 30\text{ ft}\cdot 60\text{ ft} = 1{,}800\text{ ft}^2$
One-ninth; the back yard area
Area of garden $= \frac{1}{9}\cdot\frac{1{,}800\text{ ft}^2}{1} = \frac{1{,}800\text{ ft}^2}{9}$

$= \frac{1}{{}^1\cancel{9}}\cdot\frac{{}^1\cancel{9}\cdot 200\text{ ft}^2}{1} = 200\text{ ft}^2$

Concept Check and Objective Practice

A1. Answers may vary. Example: Fractions in their simplest form are easy to understand. It's easier to visualize 1 part out of 2 than 9 parts out of 18.

A2. a. $\frac{2}{6}$ and $\frac{1}{3}$ are equivalent **b.** $\frac{1}{3}$ is in simplest form

A3. a. $\frac{3}{9}$ and $\frac{1}{3}$ are equivalent **b.** $\frac{1}{3}$ is in simplest form

A4.  $= \frac{1}{2}$ **A5.** 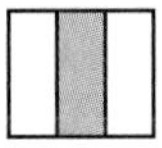 $= \frac{1}{3}$

A6. $\frac{2}{3}$ **A7.** $\frac{3}{4}$ **A8.** $\frac{2}{5}$ **A9.** $\frac{2}{5}$ **A10.** $\frac{4}{3}$

A11. $\frac{10}{7}$ **A12.** 6 **A13.** 7 **A14.** $\frac{1}{8}$ **A15.** $\frac{1}{13}$

Focus on Estimating

1. $\frac{11}{30} \approx \frac{10}{30} \approx \frac{1}{3}$

$\frac{1}{3}$ 

$\frac{11}{30}$ 

2. $\frac{9}{40} \approx \frac{10}{40} \approx \frac{1}{4}$

$\frac{1}{4}$ 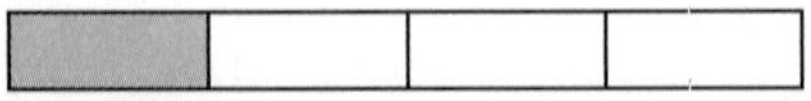

$\frac{9}{40}$ 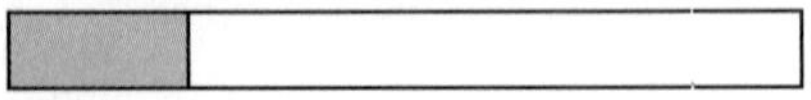

3. $\frac{20}{35} \approx \frac{20}{40} \approx \frac{1}{2}$

$\frac{1}{2}$ 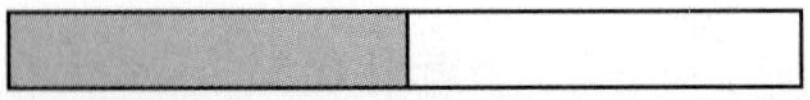

$\frac{20}{35}$ 

4. $\frac{60}{94} \approx \frac{60}{90} \approx \frac{2}{3}$

$\frac{2}{3}$ 

$\frac{60}{94}$ 

5. $\frac{60}{83} \approx \frac{60}{80} \approx \frac{3}{4}$

$\frac{3}{4}$ 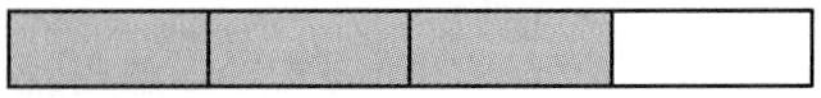

$\frac{60}{83}$ 

6. $\frac{18}{20} \approx \frac{20}{20} \approx 1$

1 

$\frac{18}{20}$ 

B1. 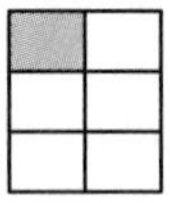 **B2.** 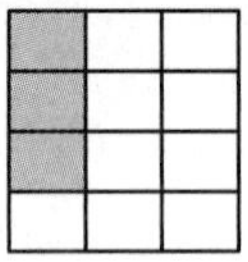

B3. $\frac{15}{28}$ **B4.** $\frac{18}{55}$ **B5.** $\frac{1}{12}$ **B6.** 20 **B7.** 20

B8. $\frac{5}{3}$ **B9.** $\frac{9}{25}$ **B10.** $\frac{6}{17}$

B11. Step 1: Multiply the numerators.
Step 2: Multiply the denominators
Step 3: Simplify if possible.

B12. \$8,000 **B13.** \$4,000

C1. Answers may vary. Example: $\frac{3}{3}; \frac{7}{7}; \frac{10}{10}$

C2. Answers may vary. Example: The number of shaded parts is the same as the number of parts in the whole.

C3. Answers may vary. Example: Any number divided by itself is one.

C4. True. $\frac{5}{5} = 1$ Any number multiplied by one is the same number.

C5. False. For the value to be the same after multiplying, the multiplier must equal 1. $\frac{9}{5}$ does not equal 1.

C6. $\frac{1}{4}\cdot\frac{3}{3} = \frac{3}{12}$ **C7.** $\frac{1}{3}\cdot\frac{5}{5} = \frac{5}{15}$ **C8.** $\frac{12}{16}$ **C9.** $\frac{35}{21}$

C10. $\frac{21}{27}$ **C11.** $\frac{22}{24}$ **C12.** $\frac{64}{84}$ **C13.** $\frac{10}{46}$ **C14.** $\frac{30}{39}$

C15. $\frac{60}{64}$ **C16. a.** $\frac{1}{12}$ **b.** $\frac{6}{12}$ **C17. a.** $\frac{1}{6}$ **b.** $\frac{4}{6}$

D1. $\frac{60\text{ min}}{1\text{ hr}}; \frac{1\text{ hr}}{60\text{ min}}$ **D2.** $\frac{4\text{ qt}}{1\text{ gal}}; \frac{1\text{ gal}}{4\text{ qt}}$

D3. $\frac{8\,\cancel{\text{qt}}}{1}\cdot\frac{2\text{ pt}}{1\,\cancel{\text{qt}}} = 16\text{ pt}$ **D4.** $\frac{4{,}000\,\cancel{\text{lb}}}{1}\cdot\frac{1\text{ ton}}{2{,}000\,\cancel{\text{lb}}} = 2\text{ tons}$

D5. $\frac{5\,\cancel{\text{ft}}}{1}\cdot\frac{12\text{ in.}}{1\,\cancel{\text{ft}}} = 60\text{ in.}$ **D6.** $\frac{4\,\cancel{\text{yd}}}{1}\cdot\frac{3\text{ ft}}{1\,\cancel{\text{yd}}} = 12\text{ ft}$

D7. $\frac{{}^4\cancel{8}\,\cancel{\text{cups}}}{1}\cdot\frac{1\text{ pt}}{{}^1\cancel{2}\,\cancel{\text{cups}}} = 4\text{ pt}$ **D8.** $\frac{{}^6\cancel{12}\,\cancel{\text{pt}}}{1}\cdot\frac{1\text{ qt}}{{}^1\cancel{2}\,\cancel{\text{pt}}} = 6\text{ qt}$

D9. 21 ft **D10.** 63 d **D11.** 12 lb **D12.** 14 d

D13. yes **D14.** no

Exercises

For 1.–14.: See preceding Concept Check and Objective Practice answers.

15. simplest form **17.** equivalent

19. factor **21.** $\frac{2}{3}$ **23.** $\frac{5}{7}$ **25.** $\frac{2}{3}$ **27.** $\frac{1}{21}$

29. $\frac{11}{13}$ **31.** $\frac{42}{85}$

33. Answers may vary. Example:
Step 1: Factor the numerator and denominator.
Step 2: Divide out all common factors.

35. $\frac{3}{20}$ **37.** $\frac{5}{4}$ **39.** $\frac{9}{10}$ **41.** $\frac{1}{4}$ **43.** $\frac{3}{2}$ **45.** 2

47. $\frac{1}{12}$ **49.** $\frac{3}{10}$

51. Answers may vary. Example:
Step 1: Multiply numerators.
Step 2: Multiply denominators.
Step 3: Simplify the product fraction if possible.

53. $\frac{2\cdot2\cdot2}{7\cdot7\cdot7}$ **55.** $\frac{36}{49}$ **57.** $\frac{1}{60}$ **59.** $\frac{1}{64}$

61. $\frac{45}{4}$ **63.** $\frac{18}{81}$ **65.** $\frac{56}{64}$ **67.** $\frac{44}{48}$ **69.** $\frac{21}{48}$

71. $\frac{10\text{ ft}}{1}\cdot\frac{12\text{ in.}}{1\text{ ft}} = 120\text{ in.}$

73. $\frac{^{14}98\text{ d}}{1}\cdot\frac{1\text{ wk}}{^{1}7\text{ d}} = 14\text{ wk}$

75. 48 oz **77.** 120 oz **79.** 20 lb **81.** $\frac{1}{10}$

83. **a.** 72 **b.** 36 **c.** 108

85. **a.** \$77 **b.** \$154 **87.** 1 sq yd

89. **a.** \$1,400 **b.** \$600

Section 2.3 Guided Practice

1. $\frac{7}{5};\frac{7}{5};\frac{7}{5}$ **2.** $\frac{23}{65};\frac{23}{65}$ **3.** $\frac{96}{1};\frac{1}{96}$

4. $\frac{4}{7}\cdot\frac{35}{16} = \frac{^{1}4\cdot5\cdot7^{1}}{7^{1}\cdot4\cdot{}^{1}4} = \frac{5}{4}$

5. $\frac{3}{8}\cdot\frac{2}{5} = \frac{3\cdot2^{1}}{2^{1}\cdot4\cdot5} = \frac{3}{20}$

6. $\frac{4}{1}\div\frac{3}{5} = \frac{4}{1}\cdot\frac{5}{3} = \frac{4\cdot5}{1\cdot3} = \frac{20}{3}$

7. $\frac{3}{4}$; 3; division $3\div\frac{3}{4} = \frac{3}{1}\div\frac{3}{4} = \frac{3^{1}}{1}\cdot\frac{4}{3^{1}} = 4$
It will take the child 4 years to grow three inches.

8. $\left(\frac{1}{3}\cdot\frac{2}{3}\right)^2\cdot7 = \left(\frac{2}{9}\right)^2\cdot7$

$= \left(\frac{2}{9}\right)\cdot\left(\frac{2}{9}\right)\cdot7$

$= \left(\frac{4}{81}\right)\cdot\frac{7}{1}$

$= \frac{28}{81}$

9. $\frac{1}{16}$; 7; 7; $\frac{1}{16}$; $7\div\frac{1}{16} = \frac{7}{1}\cdot\frac{16}{1} = \frac{112}{1} = 112$
The 7 cakes have a total of 112 pieces.

Concept Check and Objective Practice

A1. its reciprocal **A2.** The answer to all of the problems is 1.

A3. $\frac{7}{9}$ **A4.** $\frac{2}{5}$ **A5.** $\frac{13}{3}$ **A6.** $\frac{52}{21}$ **A7.** 3

A8. 9 **A9.** $\frac{1}{22}$ **A10.** $\frac{1}{16}$

B1. **a.** $2\overline{)8}$ with quotient 4; $8\cdot\frac{1}{2} = \frac{8}{1}\cdot\frac{1}{2} = \frac{2^{1}\cdot4\cdot1}{1\cdot2^{1}} = \frac{4}{1} = 4$

b. $4\overline{)12}$ with quotient 3; $12\cdot\frac{1}{4} = \frac{12}{1}\cdot\frac{1}{4} = \frac{3\cdot4^{1}\cdot1}{1\cdot4^{1}} = \frac{3}{1} = 3$

c. $9\overline{)36}$ with quotient 4; $36\cdot\frac{1}{9} = \frac{36}{1}\cdot\frac{1}{9} = \frac{4\cdot9^{1}\cdot1}{1\cdot9^{1}} = \frac{4}{1} = 4$

B2. reciprocal **B3.** reciprocal

B4. Answers may vary. Example:
a. Rewrite the fraction $\frac{4}{3}$.
b. Change the division sign to a multiplication sign.
c. Write the reciprocal of $\frac{5}{6}$, which is $\frac{6}{5}$.
d. Multiply the numerators to get the new numerator of 24.
e. Multiply the denominator to get the new denominator of 15.
f. Factor both the numerator and denominator
g. Divide out any common factors.
h. Multiply the remaining factors.

B5. $\frac{2}{3}$ **B6.** $\frac{5}{7}$ **B7.** $\frac{1}{3}$ **B8.** $\frac{9}{28}$ **B9.** $\frac{2}{11}$

B10. $\frac{1}{54}$ **B11.** 10 **B12.** 20 **B13.** 40 yr **B14.** yes

Exercises

For 1.–6.: *See preceding Concept Check and Objective Practice answers.*

7. dividend **9.** reciprocal **11.** $\frac{2}{3}$ **13.** $\frac{5}{6}$ **15.** 2

17. undefined **19.** 0 **21.** $\frac{49}{144}$ **23.** 32 **25.** $\frac{5}{2}$

27. $\frac{5}{2}$ **29.** $\frac{10}{7}$ **31.** $\frac{15}{2}$ **33.** $\frac{1}{36}$ **35.** $\frac{9}{64}$ **37.** $\frac{1}{81}$

39. $\frac{1}{9}$ **41.** $\frac{5}{3}$ **43.** 0 **45.** 8

47. Division by 0 is undefined.

49. **a.** multiplied instead of divided **b.** $\frac{27}{8}$

51. **a.** multiplied the denominator by 3 **b.** $\frac{9}{4}$

53. Answers may vary. Identical pairings are B and E and D and F. You may choose any one of those pairings for answers to part a and a different one for part b.

55. 56 patients **57.** yes **59.** 4

61. 9 in. **63.** 20 mpg

65. 21 in. **67.** 8 laps

Section 2.4 Guided Practice

1. $\frac{1+7}{6} = \frac{8}{6} = \frac{^{1}2\cdot4}{^{1}2\cdot3} = \frac{4}{3}$

2. $\frac{9-5}{12} = \frac{4}{12} = \frac{^{1}4\cdot1}{^{1}4\cdot3} = \frac{1}{3}$

3. 10; 8; 10; Keep going; 20; Keep going; 30; Keep going; 40; stop; 40

4. 6, 12; 12; 12

$$\frac{3}{4}\cdot\frac{3}{3}=\frac{9}{12}$$
$$\frac{1}{6}\cdot\frac{2}{2}=\frac{2}{12}$$

5. 14, 21, 28, 35; 35; 35

$$\frac{4}{7}\cdot\frac{5}{5}=\frac{20}{35}$$
$$\frac{2}{5}\cdot\frac{7}{7}=\frac{14}{35}$$

6. 12, 24, 36; 36; 36

$$\frac{1}{12}\left(\frac{3}{3}\right)=\frac{3}{36}$$
$$\frac{11}{9}\left(\frac{4}{4}\right)=\frac{44}{36}$$

7. The LCD is 20.

$$\frac{1}{4}\left(\frac{5}{5}\right)=\frac{5}{20}$$
$$\frac{3}{5}\left(\frac{4}{4}\right)=\frac{12}{20}$$
$$\frac{5}{10}+\frac{12}{20}=\frac{17}{20}$$

$\frac{17}{20}$ is in simplest form.

8. The LCD is 30.

$$\frac{7}{15}-\frac{1}{6}=\frac{7}{15}\left(\frac{2}{2}\right)-\frac{1}{6}\left(\frac{5}{5}\right)$$
$$=\frac{14}{30}-\frac{5}{30}$$
$$=\frac{9}{30}$$
$$=\frac{{}^{1}\cancel{3}\cdot 3}{{}^{1}\cancel{3}\cdot 10}$$
$$=\frac{3}{10}$$

9. The LCD is 36.

$$\frac{4}{9}-\frac{1}{12}=\frac{4}{9}\left(\frac{4}{4}\right)-\frac{1}{12}\left(\frac{3}{3}\right)$$
$$=\frac{16}{36}-\frac{3}{36}$$
$$=\frac{13}{36}$$

$\frac{13}{36}$ is in simplest form.

10. 4; $\frac{6}{1}+\frac{1}{4}=\frac{6}{1}\cdot\frac{4}{4}+\frac{1}{4}$

$$=\frac{24}{4}+\frac{1}{4}$$
$$=\frac{25}{4}$$

11. $18 = 2\cdot3\cdot3$
$27 = 3\cdot3\cdot3$

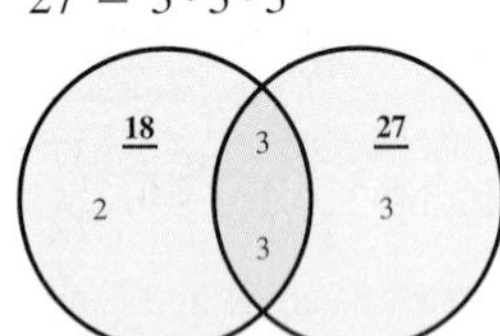

LCD $= 2\cdot3\cdot3\cdot3 = 54$

12. $14 = 2\cdot7$
$10 = 2\cdot5$

14
10
7
2
5

LCD $= 2\cdot5\cdot7 = 70$

$$\frac{3}{14}\left(\frac{5}{5}\right)=\frac{15}{70}$$
$$\frac{1}{10}\left(\frac{7}{7}\right)=\frac{7}{70}$$

13. $12 = 2\cdot2\cdot3$
$54 = 2\cdot3\cdot3\cdot3$

12
54
2
2
3
3
3

LCD $= 2\cdot2\cdot3\cdot3\cdot3 = 108$

$$\frac{13}{12}\left(\frac{9}{9}\right)=\frac{117}{108}$$
$$\frac{17}{54}\left(\frac{2}{2}\right)=\frac{34}{108}$$

14. $3\cdot2\cdot5\cdot5 = 150$

6
50
3
2
5
5

$$\frac{1}{6}+\frac{3}{50}=\frac{1}{6}\left(\frac{25}{25}\right)+\frac{3}{50}\left(\frac{3}{3}\right)$$
$$=\frac{25}{150}+\frac{9}{150}=\frac{34}{150}$$
$$=\frac{\cancel{2}\cdot 17}{\cancel{2}\cdot 75}=\frac{17}{75}$$

15.

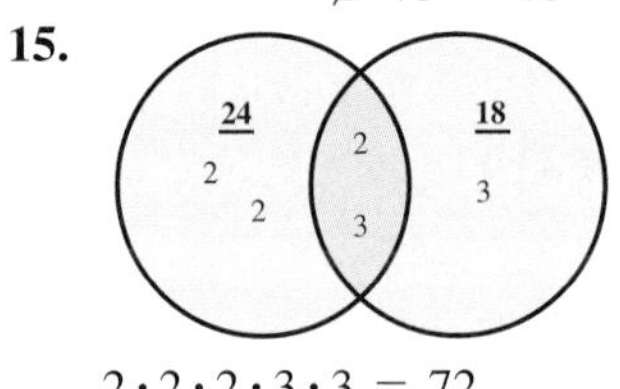

$2\cdot2\cdot2\cdot3\cdot3 = 72$

$$\frac{17}{24}-\frac{3}{18}=\frac{17}{24}\left(\frac{3}{3}\right)-\frac{3}{18}\left(\frac{4}{4}\right)$$
$$=\frac{51}{72}-\frac{12}{72}$$
$$=\frac{39}{72}$$
$$=\frac{{}^{1}\cancel{3}\cdot 13}{{}^{1}\cancel{3}\cdot 24}$$
$$=\frac{13}{24}$$

16. 12; 12^{ths} and 8^{ths}; 24; 24

$$\frac{3}{4}-\frac{1}{3}+\frac{1}{8}=\frac{3}{4}\left(\frac{6}{6}\right)-\frac{1}{3}\left(\frac{8}{8}\right)+\frac{1}{8}\left(\frac{3}{3}\right)$$
$$=\frac{18}{24}-\frac{8}{24}+\frac{3}{24}$$
$$=\frac{10}{24}+\frac{3}{24}$$
$$=\frac{13}{24}$$

17. Subtract the board thickness from the length of the nail.

$$\frac{3}{4}-\frac{5}{8}=\frac{3}{4}\left(\frac{2}{2}\right)-\frac{5}{8}$$
$$=\frac{6}{8}-\frac{5}{8}$$
$$=\frac{1}{8}\text{ in.}$$

The nail will stick out $\frac{1}{8}$ in.

Concept Check and Objective Practice

A1. The denominators are the same.
A2. Answers may vary. Example: $\frac{1}{6}, \frac{5}{6}; \frac{2}{7}, \frac{4}{7}; \frac{3}{10}, \frac{7}{10}$
A3. 3 is the numerator. 10 is the denominator.
A4. **a.** 14 oranges **b.** 4 fifths **c.** 4 gumballs **d.** 5 tenths
A5. $\frac{2}{3}$ **A6.** $\frac{7}{11}$ **A7.** $\frac{1}{5}$ **A8.** $\frac{2}{7}$ **A9.** $\frac{6}{8} = \frac{3}{4}$
A10. $\frac{18}{24} = \frac{3}{4}$ **A11.** $\frac{13}{26} = \frac{1}{2}$ **A12.** $\frac{11}{22} = \frac{1}{2}$
A13. 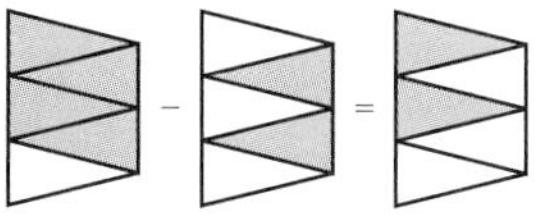 $\frac{4}{5} - \frac{2}{5} = \frac{2}{5}$
A14. 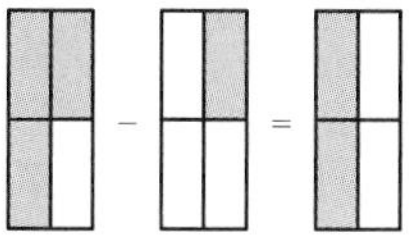 $\frac{3}{4} - \frac{1}{4} = \frac{2}{4}$
A15. 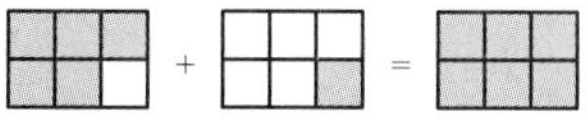 $\frac{5}{6} + \frac{1}{6} = \frac{6}{6}$
A16. 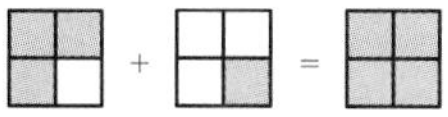 $\frac{3}{4} + \frac{1}{4} = \frac{4}{4}$
A17.  $\frac{1}{2} + \frac{1}{2} = \frac{2}{2}$
A18. 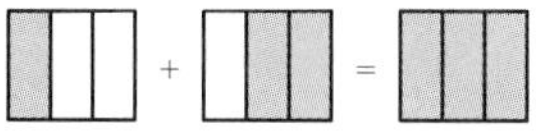 $\frac{1}{3} + \frac{2}{3} = \frac{3}{3}$
B1. Answers may vary. Example: The pieces were not the same size.
B2. Answers may vary. Example: The pieces were the same size.
B3. **a.** 1 **b.** $\frac{4}{4}$ and 1 are equal **c.** True; true
d. Answers may vary. Example: Both fractions were multiplied by one, which gives the beginning value.
e. Answers may vary. Example: In order not to change the value of the fraction, it can be multiplied only by 1.
B4. The LCD is 6.
$\frac{1}{2}\left(\frac{3}{3}\right) = \frac{3}{6}$
$\frac{1}{6}\left(\frac{1}{1}\right) = \frac{1}{6}$
B5. The LCD is 15.
$\frac{2}{5}\left(\frac{3}{3}\right) = \frac{6}{15}$
$\frac{4}{15}\left(\frac{1}{1}\right) = \frac{4}{15}$
B6. The LCD is 40.
$\frac{3}{8}\left(\frac{5}{5}\right) = \frac{15}{40}$
$\frac{7}{10}\left(\frac{4}{4}\right) = \frac{28}{40}$
B7. The LCD is 24.
$\frac{3}{8}\left(\frac{3}{3}\right) = \frac{9}{24}$
$\frac{2}{3}\left(\frac{8}{8}\right) = \frac{16}{24}$
B8. $\frac{3}{20}, \frac{12}{20}$ **B9.** $\frac{1}{6}, \frac{4}{6}$ **B10.** $\frac{16}{36}, \frac{15}{36}$ **B11.** $\frac{20}{36}, \frac{27}{36}$
B12. $\frac{51}{75}, \frac{25}{75}$ **B13.** $\frac{15}{24}, \frac{6}{24}$ **B14.** $\frac{22}{52}, \frac{13}{52}$ **B15.** $\frac{7}{42}, \frac{9}{42}$
C1. **a.** $\frac{1}{2} =$ 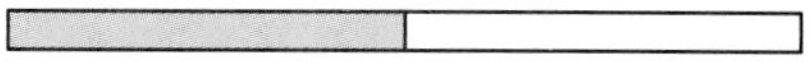
$\frac{1}{3} =$ 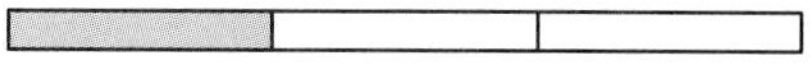
$\frac{2}{5} =$ 
b. no **c.** no
d. Answers may vary. Example: She added the numerators and added the denominators.
C2. To make sure we are adding the same size pieces
C3. $\frac{3}{4}$ **C4.** $\frac{4}{9}$ **C5.** $\frac{13}{30}$ **C6.** $\frac{19}{20}$ **C7.** $\frac{13}{30}$
C8. $\frac{7}{18}$ **C9.** $\frac{5}{24}$ **C10.** $\frac{7}{12}$
C11. Answers may vary. Example:
a. Find the LCD.
b. Multiply the numerator and denominator of each fraction by the missing multiplier.
C12. Answers may vary. Example:
a. Add or subtract the numerators.
b. Keep the denominators the same.
D1. Answers may vary. **D2.** Answers may vary.
D3. Answers may vary.
D4. $\frac{7}{13}\left(\frac{11}{11}\right) = \frac{77}{143}; \frac{5}{11}\left(\frac{13}{13}\right) = \frac{65}{143}$
D5. $\frac{3}{10}\left(\frac{8}{8}\right) = \frac{24}{80}; \frac{5}{16}\left(\frac{5}{5}\right) = \frac{25}{80}$
D6. $\frac{3}{14}\left(\frac{3}{3}\right) = \frac{9}{42}; \frac{1}{6}\left(\frac{7}{7}\right) = \frac{7}{42}$
D7. $\frac{7}{8}\left(\frac{11}{11}\right) = \frac{77}{88}; \frac{5}{22}\left(\frac{4}{4}\right) = \frac{20}{88}$
D8. $\frac{21}{140}; \frac{24}{140}$ **D9.** $\frac{7}{147}; \frac{6}{147}$ **D10.** $\frac{35}{126}; \frac{15}{126}$
D11. $\frac{19}{589}; \frac{31}{589}$
E1. Answers may vary. Example: As fractions become more complicated, LCDs will be difficult to find using the listing multiples technique.
E2. $\frac{37}{120}$ **E3.** $\frac{2}{15}$ **E4.** $\frac{47}{84}$ **E5.** $\frac{43}{110}$ **E6.** $\frac{37}{72}$
E7. $\frac{121}{156}$ **E8.** $\frac{7}{60}$ **E9.** $\frac{7}{48}$

Exercises

For 1.–15.: See preceding Concept Check and Objective Practice answers.

17. building like fractions **19.** missing multiplier
21. $\frac{8}{7}$ **23.** $\frac{3}{5}$ **25.** $\frac{4}{5}$ **27.** $\frac{4}{7}$
29. 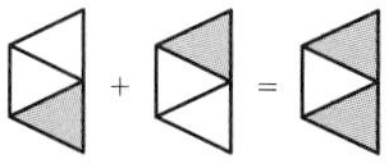 $\frac{1}{3} + \frac{1}{3} = \frac{2}{3}$
31. 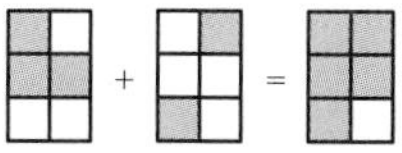 $\frac{7}{8} - \frac{6}{8} = \frac{1}{8}$
33. 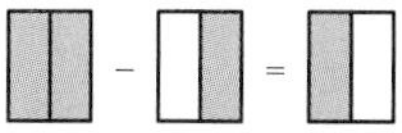 $\frac{2}{2} - \frac{1}{2} = \frac{1}{2}$

35. 10 **37.** 18 **39.** 105 **41.** 84 **43.** $\frac{7}{6}$

45. $\frac{19}{30}$ **47.** $\frac{14}{45}$ **49.** $\frac{31}{42}$ **51.** $\frac{41}{60}$ **53.** $\frac{19}{42}$

55. $\frac{35}{54}$ **57.** $\frac{7}{60}$ **59.** $\frac{83}{144}$ **61.** $\frac{7}{120}$ **63.** $\frac{7}{12}$

65. $\frac{1}{8}$ **67.** $\frac{133}{15}$ **69.** $\frac{3}{8}$ **71. a.** $\frac{5}{8}; \frac{6}{8}$ **b.** $\frac{5}{8}; \frac{3}{4}$

73. a. $\frac{13}{16}; \frac{12}{16}$ **b.** $\frac{3}{4}; \frac{13}{16}$ **75. a.** $\frac{25}{30}; \frac{24}{30}$ **b.** $\frac{4}{5}; \frac{5}{6}$

77. a. $\frac{20}{30}; \frac{21}{30}$ **b.** $\frac{2}{3}; \frac{7}{10}$ **79.** $\frac{11}{32}; \frac{1}{2}; \frac{5}{8}$ **81.** $\frac{3}{8}; \frac{11}{32}; \frac{5}{16}$

83. Answers may vary. Example: The denominator tells the type of fraction being added. The type of fraction does not change.

85. $\frac{11}{12}$ lb **87.** $\frac{7}{6}$ in. **89.** 42 questions

91. a. Rosie saved more. **b.** $\frac{22}{35}$ **c.** $\frac{13}{35}$

93. $\frac{1}{2}$ turn **95.** $\frac{59}{100}$ **97.** 1 km **99.** $\frac{1}{12}$ d

Section 2.5 Guided Practice

1. $\frac{5}{8} \div \frac{3}{9-5} = \frac{5}{8} \div \frac{3}{4} = \frac{5}{8}\cdot\frac{4}{3} = \frac{5\cdot \cancel{4}^1}{2\cdot\cancel{4}^1\cdot 3} = \frac{5}{6}$

2. $\left(\frac{3}{5}+\frac{1}{10}\right) - \frac{7}{15} = \left(\frac{6}{10}+\frac{1}{10}\right) - \frac{7}{15} = \frac{7}{10} - \frac{7}{15} = \frac{21}{30} - \frac{14}{30} = \frac{7}{30}$

3. $\frac{13}{20} - \frac{3}{4}\cdot\frac{9-5}{5} = \frac{13}{20} - \frac{3}{4}\cdot\frac{4}{5} = \frac{13}{20} - \frac{3\cdot\cancel{4}^1}{\cancel{4}^1\cdot 5} = \frac{13}{20} - \frac{3}{5} = \frac{13}{20} - \frac{12}{20} = \frac{1}{20}$

4. $\left(\frac{11}{6}\cdot\frac{1}{5}+\frac{17}{30}\right)\cdot\frac{6}{7} = \left(\frac{11}{30}+\frac{17}{30}\right)\cdot\frac{6}{7} = \frac{28}{30}\cdot\frac{6}{7} = \frac{4\cdot{}^1\cancel{7}\cdot\cancel{6}^1}{5\cdot\cancel{6}^1\cdot{}^1\cancel{7}} = \frac{4}{5}$

5. before lunch: $\frac{3}{5}+\frac{3}{10} = \frac{6}{10}+\frac{3}{10} = \frac{9}{10}$

after lunch: $1 - \frac{9}{10} = \frac{10}{10} - \frac{9}{10} = \frac{1}{10}$

$300\cdot\frac{1}{10} = \frac{\cancel{300}^{30}}{1}\cdot\frac{1}{\cancel{10}^1} = 30$

Alex will need to set up 30 chairs after lunch.

6. $n - \frac{2}{3} = \frac{1}{2}$

$n - \frac{2}{3} + \frac{2}{3} = \frac{1}{2} + \frac{2}{3}$

$n = \frac{3}{6} + \frac{4}{6}$

$n = \frac{7}{6}$

7. $\frac{3}{2}\cdot n = \frac{4}{5}$

$\frac{2}{3}\cdot\frac{3}{2}\cdot n = \frac{2}{3}\cdot\frac{4}{5}$

$n = \frac{8}{15}$

Concept Check and Objective Practice

A1. Violation of the order of operations: The subtraction needs to be performed first since it is to the left of the addition.

A2. Computation error: The square of $\frac{3}{5}$ is $\frac{3}{5}\cdot\frac{3}{5}$, or $\frac{9}{25}$.

A3. Numerators and denominators are groupings.

A4. $\frac{1}{20}$ **A5.** $\frac{5}{7}$ **A6.** $\frac{3}{8}$ **A7.** $\frac{1}{15}$ **A8.** 1 **A9.** $\frac{13}{10}$

A10. $\frac{1}{3}$ **A11.** 10 **A12.** $\frac{1}{7}$ **A13.** $\frac{1}{8}$ **A14.** $\frac{13}{8}$

A15. $\frac{11}{10}$

Focus on Algebra

1. $\frac{1}{2}$ **2.** $\frac{1}{2}$ **3.** 2 **4.** $\frac{1}{2}$ **5.** $\frac{5}{8}$ **6.** $\frac{4}{9}$ **7.** $\frac{12}{5}$

8. $\frac{18}{7}$ **9.** $\frac{5}{18}$ **10.** $\frac{1}{12}$ **11.** $\frac{7}{6}$ **12.** $\frac{1}{2}$ **13.** $\frac{11}{30}$

14. $\frac{11}{18}$ **15.** $\frac{3}{16}$ **16.** $\frac{10}{7}$

Exercises

For 1.–5.: See preceding Concept Check and Objective Practice answers.

7. $\frac{19}{120}$ **9.** $\frac{5}{8}$ **11.** $\frac{3}{4}$ **13.** $\frac{3}{8}$ **15.** $\frac{1}{8}$ **17.** $\frac{2}{7}$

19. $\frac{1}{3}$ **21.** $\frac{5}{8}$ **23.** $\frac{1}{25}$ **25.** $\frac{1}{4}$ **27.** $\frac{2}{7}$ **29.** $\frac{1}{8}$

31. 3 **33.** $\frac{43}{70}$ **35.** $\frac{8}{3}$ spools **37.** $\frac{3}{4}$ **39.** $\frac{1}{6}$

41. $\frac{1}{2}$ **43.** $\frac{37}{45}$

45. a. The equation expresses the fact that Victor began with some oranges, used $\frac{7}{2}$ oranges, and finished with $\frac{5}{2}$.

b. 6 oranges

47. a. 280 ft **b.** 245 ft **c.** $\frac{1}{8}$

Section 2.6 Guided Practice

1. $3; \frac{1}{4}; 3\frac{1}{4}$

2.

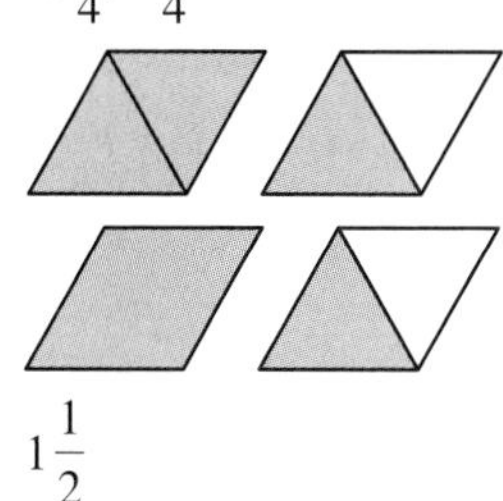

$1\frac{1}{2}$

3. $1; 1; 2; 1\frac{1}{2}$ **4.** $2R5; 2\frac{5}{6}$ **5.** $1R7; 1\frac{7}{8}$

6. 12R0; 12

7.

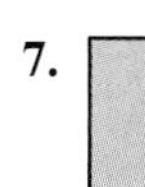 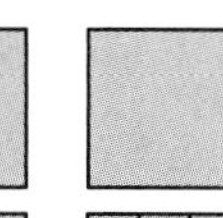 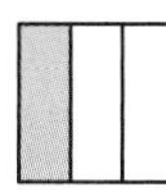

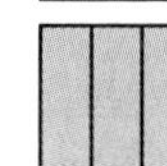  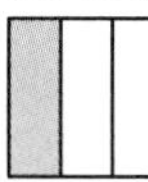

$$= \frac{3}{3} + \frac{3}{3} + \frac{1}{3}$$
$$= \frac{6}{3} + \frac{1}{3}$$
$$= \frac{7}{3}$$

8. **a.** 2; 3; 6 **b.** 1 **c.** 6; 1; 7; 7

9. **a.** 5; 8; 40 **b.** 7 **c.** 40; 7; 47; 47

10. $4; 3; 1; 13; \frac{13}{4}$ **11.** $9; 5; 7; 52; \frac{52}{9}$ **12.** $>$; up; 6

13. $<$; down; 12

14. $7 \cdot 7 \approx 49$

$$\frac{50}{7} \cdot \frac{33}{5} = \frac{\cancel{5}^1 \cdot 10 \cdot 3 \cdot 11}{7 \cdot \cancel{5}^1} = \frac{330}{7} = 47\frac{1}{7}; \text{ yes}$$

15. $16 \div 8 \approx 2$

$$\frac{65}{4} \div \frac{25}{3} = \frac{65}{4} \cdot \frac{3}{25} = \frac{\cancel{5}^1 \cdot 13 \cdot 3}{4 \cdot \cancel{5}^1 \cdot 5} = \frac{39}{20} = 1\frac{19}{20}; \text{ yes}$$

16. $3\frac{1}{8} + 4\frac{3}{4} \approx 3 + 5 = 8$

$$3\frac{1}{8} + 4\frac{3}{4} = 3\frac{1}{8} + 4\frac{6}{8} = 3 + 4 + \frac{7}{8} = 7\frac{7}{8}; \text{ yes}$$

17. $7\frac{5}{6} + 5\frac{2}{3} \approx 8 + 6 = 14$

$$7\frac{5}{6} + 5\frac{2}{3} = 7\frac{5}{6} + 5\frac{4}{6} = 7 + 5 + \frac{9}{6}$$
$$= 7 + 5 + 1\frac{1}{2} = 13\frac{1}{2}; \text{ yes}$$

18. $5\frac{5}{8} - 2\frac{1}{16} \approx 6 - 2 = 4$

$$5\frac{5}{8} - 2\frac{1}{16} = 5\frac{10}{16} - 2\frac{1}{16} = 3\frac{9}{16}; \text{ yes}$$

19. **a.** $\frac{5}{6}; \frac{1}{6}; 7$ **b.** $1; \frac{6}{6}; 7$ **c.** $6\frac{7}{6}; 3\frac{2}{6}; 3\frac{1}{3}$

20. $\frac{1}{5}; \frac{2}{5}; 3; \frac{5}{5}; 2 + \frac{5}{5}; 2 + \frac{5}{5}; 2\frac{6}{5}; 1\frac{4}{5}$

21. $6 - 3 = 3$

$$5\frac{5}{8} - 2\frac{7}{8} = 4 + \frac{8}{8} + \frac{5}{8} - 2\frac{7}{8}$$
$$= 4\frac{13}{8} - 2\frac{7}{8}$$
$$= 2\frac{6}{8}$$
$$= 2\frac{3}{4}; \text{ yes}$$

22. $9\frac{1}{4} - 4\frac{1}{3} \approx 9 - 4 = 5$

$$9\frac{1}{4} - 4\frac{1}{3} = 9\frac{3}{12} - 4\frac{4}{12}$$
$$= 8 + \frac{12}{12} + \frac{3}{12} - 4\frac{4}{12}$$
$$= 8\frac{15}{12} - 4\frac{4}{12}$$
$$= 4\frac{11}{12}; \text{ yes}$$

23. $4 - 2\frac{1}{5} \approx 4 - 2 \approx 2$

$$3 + \frac{5}{5} - 2\frac{1}{5} = 3 + \frac{5}{5} - 2\frac{1}{5}$$
$$= 1\frac{4}{5}; \text{ yes}$$

24.

$$\left(3\frac{1}{6} - 1\frac{1}{3}\right)^2 \div \frac{1}{2} = \left(\frac{19}{6} - \frac{8}{6}\right)^2 \cdot \frac{2}{1}$$
$$= \left(\frac{11}{6}\right)^2 \cdot \frac{2}{1}$$
$$= \frac{121}{36} \cdot \frac{2}{1}$$
$$= \frac{11 \cdot 11 \cdot \cancel{2}^1}{2 \cdot \cancel{2}^1 \cdot 9 \cdot 1}$$
$$= \frac{121}{18}$$
$$= 6\frac{13}{18}$$

25. Total packaged $= 5 \cdot \frac{3}{4} + 1 \cdot 1\frac{1}{2} + 2 \cdot 1$

$$= \frac{5}{1} \cdot \frac{3}{4} + \frac{1}{1} \cdot \frac{3}{2} + 2 \cdot 1$$
$$= \frac{15}{4} + \frac{3}{2} + 2$$
$$= \frac{15}{4} + \frac{6}{4} + \frac{8}{4}$$
$$= \frac{29}{4} = 7\frac{1}{4} \text{ lb}$$

Leftover coffee $= 10 - 7\frac{1}{4}$

$$= 9\frac{4}{4} - 7\frac{1}{4}$$
$$= 2\frac{3}{4} \text{ lb}$$

$2\frac{3}{4}$ lb of coffee will be left over.

Concept Check and Objective Practice

A1. Answers may vary. Example: The quotient is the whole number, the remainder is the numerator, and the divisor is the denominator.

A2. $\frac{15}{8}; 1\frac{7}{8}$ **A3.** $\frac{8}{5}; 1\frac{3}{5}$ **A4.** $\frac{19}{8}; 2\frac{3}{8}$ **A5.** $\frac{32}{9}; 3\frac{5}{9}$

A6. $1\frac{1}{2}; \frac{3}{2}$ **A7.** $2\frac{3}{4}; \frac{11}{4}$ **A8.** $2\frac{2}{5}$ **A9.** $5\frac{2}{3}$ **A10.** $4\frac{1}{5}$

A11. $4\frac{1}{3}$ **A12.** 6 **A13.** 13

B1. Answers may vary. Example:
a. Multiplying the whole number by the denominator gives the numerator of a fraction equal to the whole number with the same denominator as the fraction.
b. The improper fraction can be formed by adding the "like fractions."

B2. a. 48 **b.** 7 **c.** 55 **B3. a.** 27 **b.** 2 **c.** 29

B4. $\frac{13}{7}$ **B5.** $\frac{17}{5}$ **B6.** $\frac{5}{2}$ **B7.** $\frac{33}{5}$ **B8.** $\frac{31}{3}$

B9. $\frac{31}{10}$ **B10.** $\frac{82}{9}$ **B11.** $\frac{61}{8}$

Focus on Rounding Mixed Numbers

1. 6 **2.** 4 **3.** 7
4. 17 **5.** 13 **6.** 2 **7.** 6 **8.** 44 **9.** 8 **10.** 1
11. 12 **12.** 14

C1. Answers may vary. Example: Determine if the fraction is less than or more than $\frac{1}{2}$. If it's less, the mixed number approximately equals the whole number. If it's more than or equal to $\frac{1}{2}$, the mixed number approximately equals the next larger whole number.

C2. Answers may vary. Example: It indicates if the final answer is reasonable.

C3. 12; $14\frac{1}{7}$ **C4.** 6; 4 **C5.** 72; $72\frac{1}{3}$ **C6.** 30; $32\frac{4}{7}$

C7. 2; $1\frac{1}{8}$ c. **C8.** 2; $1\frac{1}{8}$ lb **C9.** 6; $4\frac{3}{8}$ **C10.** $1\frac{1}{3}; 1\frac{13}{72}$

C11. $\frac{1}{3}; \frac{21}{64}$ **C12.** $\frac{1}{3}; \frac{9}{32}$

D1. Answers may vary. Example: Add the fractions first in case it is necessary to carry a whole number.

D2. $7\frac{31}{56}$ **D3.** $3\frac{5}{6}$ **D4.** $16\frac{14}{15}$ **D5.** $16\frac{9}{10}$

D6. $3\frac{3}{8}$ lb **D7.** $8\frac{5}{12}$ lb **D8.** $8\frac{1}{18}$ **D9.** $20\frac{1}{3}$

D10. $20\frac{11}{14}$ **D11.** $100\frac{2}{5}$ **E1.** $4\frac{7}{6}$ **E2.** $7\frac{9}{7}$

E3. Answers may vary. Example: You are borrowing "one" from the whole number. For 1 to be added to the fraction, it must be in fraction form with the same denominator as the fraction to which 1 is being added. One in fraction form has the same numerator as denominator. Thus, the value of the denominator has been added to the existing numerator.

E4. $4\frac{1}{10}$ **E5.** $9\frac{7}{16}$ **E6.** $1\frac{1}{15}$ **E7.** $4\frac{1}{3}$ **E8.** $5\frac{4}{9}$

E9. $12\frac{17}{21}$ **E10.** $12\frac{14}{15}$ **E11.** $19\frac{8}{15}$ **E12.** $4\frac{2}{3}$

E13. $6\frac{5}{6}$ **E14.** $2\frac{1}{6}$ **E15.** $3\frac{4}{5}$

Exercises

For 1.–17.: See preceding Concept Check and Objective Practice answers.

19. proper fraction **21.** mixed number

23. a. $\frac{15}{4}$ **b.** $3\frac{3}{4}$ **25. a.** $\frac{22}{7}$ **b.** $3\frac{1}{7}$

27. a. $\frac{35}{8}$ **b.** $4\frac{3}{8}$ **29. a.** $\frac{8}{6}$ **b.** $1\frac{1}{3}$

31. $\frac{25}{8}$ **33.** $5\frac{1}{3}$ **35.** $\frac{40}{3}$ **37.** $14\frac{4}{7}$ **39.** $2\frac{13}{14}$

41. a. 12 **b.** $10\frac{1}{12}$ **43. a.** $\frac{3}{4}$ **b.** $\frac{8}{13}$

45. a. $3\frac{2}{3}$ **b.** $3\frac{11}{14}$ **47. a.** 12 **b.** $12\frac{1}{7}$

49. a. 26 **b.** $25\frac{2}{3}$ **51. a.** 11 **b.** $10\frac{6}{11}$

53. a. 25 **b.** $24\frac{7}{10}$ **55. a.** 5 **b.** $4\frac{9}{10}$

57. a. 2 lb **b.** $2\frac{3}{20}$ lb **59. a.** 14 **b.** $13\frac{3}{10}$

61. a. 26 **b.** $26\frac{2}{5}$ **63. a.** 2 **b.** $1\frac{11}{12}$

65. a. 127 **b.** $126\frac{11}{16}$ **67. a.** 8 **b.** $6\frac{1}{3}$

69. a. 1 **b.** $\frac{29}{30}$ **71. a.** 12 **b.** $10\frac{1}{12}$

73. a. $1\frac{4}{5}$ **b.** $1\frac{17}{20}$ **75. a.** $51\frac{3}{4}$ km **b.** $17\frac{1}{4}$ km

77. $2\frac{1}{2}$ **79.** $10\frac{1}{2}$ **81.** $13\frac{1}{3}$ **83.** $22\frac{1}{3}$ **85.** $2\frac{25}{32}$

87. $\frac{9}{23}$ **89.** $1\frac{1}{8}$ lb **91.** $\frac{1}{2}$ **93.** $550

Chapter 2 Review Exercises

1. $\frac{5}{15}$ **2.** $\frac{14}{5}$ **3.** 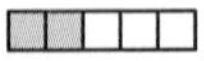 **4.**

5. 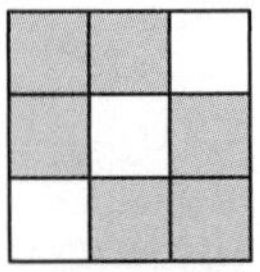 $\frac{6}{9}$ **6.** 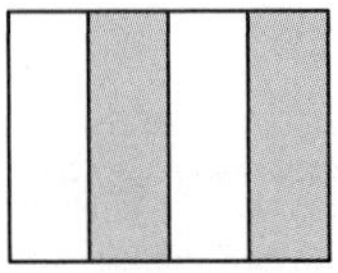 $\frac{2}{4}$

7. $\frac{5}{8} > \frac{5}{9}$ When the numerators are the same, the larger fraction is the one with the smaller denominator. Five pieces of a pie that has been cut into 8 pieces is more pie than five pieces of a pie that has been cut into 9 pieces.

8. $\frac{5}{12} < \frac{7}{12}$ When the denominators are the same, the larger fraction is the one with the larger numerator. Five pieces of a pie that has been cut into 12 pieces is less pie than seven pieces.

9. $\frac{6}{13}$ **10.** $\frac{8}{15}$ **11. a.** B; C **b.** C

12. 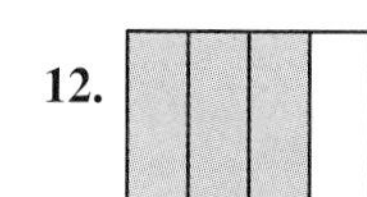 **13.** $\frac{3}{4}$ **14.** $\frac{3}{7}$ **15.** $\frac{21}{4}$ **16.** $\frac{3}{5}$

17. $\frac{12}{5}$ **18.** $\frac{3}{4}$ **19.** 6 **20.** $\frac{25}{9}$ **21.** $\frac{4}{5}$ **22.** \$180

23. 180 in. **24.** 88 oz **25.** $\frac{9}{5}$ **26.** 45 **27.** $\frac{1}{3}$

28. $\frac{5}{4}$ **29.** 4 **30.** $\frac{25}{36}$ **31.** 4 **32.** 1 **33.** 0

34. $\frac{1}{4}$ **35.** $\frac{6}{5}$ **36.** 5 pieces **37.** $17\frac{1}{2}$ bags **38.** 24 mpg

39. 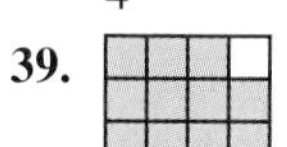 – 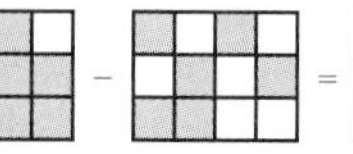 = $\frac{11}{12} - \frac{6}{12} = \frac{5}{12}$

40. 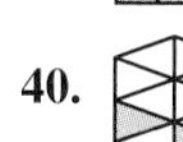 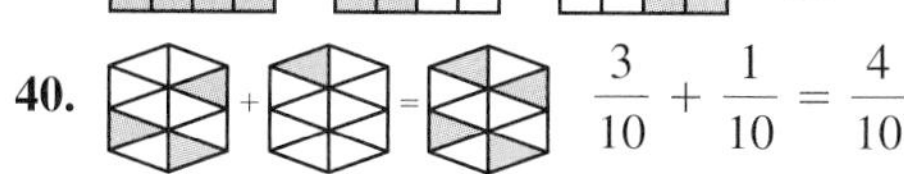 $\frac{3}{10} + \frac{1}{10} = \frac{4}{10}$

41. a. 24 **b.** $\frac{10}{24}; \frac{3}{24}$ **42. a.** 24 **b.** $\frac{21}{24}; \frac{20}{24}$

43. $\frac{8}{9}$ **44.** $\frac{4}{7}$ **45.** $\frac{11}{18}$ **46.** $\frac{4}{15}$ **47.** $\frac{5}{42}$ **48.** $\frac{41}{50}$

49. $\frac{1}{2}$ **50.** $\frac{5}{4}$ **51.** $\frac{7}{16}; \frac{3}{8}; \frac{1}{4}$ **52.** $\frac{9}{15}; \frac{4}{5}; \frac{5}{6}$

53. $1\frac{1}{8}$ gal **54.** $\frac{1}{2}$ **55.** $\frac{9}{20}$ **56.** $\frac{8}{21}$ **57.** $\frac{1}{9}$

58. $\frac{64}{105}$ **59.** $\frac{25}{36}$ **60.** $\frac{11}{6}$ **61.** $\frac{1}{6}$ **62.** $\frac{5}{6}$

63. $\frac{1}{15}$ **64.** $\frac{5}{3}$ **65.** $\frac{7}{6}$ **66.** 200 shelves **67.** $4\frac{3}{8}$ c.

68. a. $\frac{19}{5}$ **b.** $3\frac{4}{5}$ **69. a.** $\frac{7}{4}$ **b.** $1\frac{3}{4}$ **70.** $\frac{43}{6}$

71. $6\frac{5}{8}$ **72.** 8 **73.** 45 **74.** $5\frac{11}{14}$ **75.** $\frac{49}{76}$ **76.** $6\frac{1}{2}$

77. 10 **78.** $8\frac{4}{9}$ **79.** $5\frac{1}{6}$ **80.** $9\frac{3}{5}$ **81.** $1\frac{1}{4}$

82. $89\frac{13}{15}$ **83.** $10\frac{9}{16}$

Chapter 2 Practice Test

1. 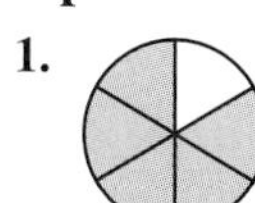 **2. a.** A; C **b.** C

3. $\frac{5}{6} - \frac{2}{6} = \frac{3}{6}$

4. $\frac{3}{8} + \frac{2}{8} = \frac{5}{8}$ **5.** $\frac{1}{15}$ **6.** $\frac{1}{2}$

7. $\frac{5}{6}$ **8.** $\frac{11}{12}$ **9.** $\frac{55}{84}$ **10.** $\frac{3}{4}$ **11.** $\frac{23}{80}$ **12.** $\frac{5}{2}$

13. $\frac{9}{14}$ **14.** $\frac{26}{15}$ **15.** $\frac{17}{56}$ **16.** 2 **17.** $\frac{9}{8}$

18. 2 **19.** $\frac{3}{4}; \frac{15}{24}; \frac{7}{12}$ **20.** $\frac{2}{5}$ **21.** 33 documents

22. 6,986 people **23.** $2\frac{3}{4}$ **24.** $\frac{22}{7}$ **25.** $\frac{20}{6}$ **26.** $6\frac{3}{5}$

27. $\frac{11}{16}$ **28.** $4\frac{5}{8}$ **29.** $\frac{5}{6}$ **30.** $14\frac{2}{5}$ **31.** $2\frac{1}{8}$ **32.** $\frac{1}{9}$

33. $3\frac{3}{8}$ **34.** $\frac{1}{2}$ **35.** 8 **36.** $\frac{9}{16}$ **37.** 600 points

CHAPTER 3

Section 3.1 Guided Practice

1. third; tenths, hundredths, thousandths; thousandths
2. fifth; tenths, hundredths, thousandths, ten thousandths, hundred thousandths; hundred thousandths
3. $8.19 = 8 + \frac{1}{10} + \frac{9}{100}$
4. $300.40012 = 300 + \frac{4}{10} + \frac{1}{10{,}000} + \frac{2}{100{,}000}$
5. 43; 3; forty-three hundredths
6. 987; 7; nine hundred eighty-seven hundred thousandths
7. thirteen and ninety-four thousandths
8. five hundred three and one thousand six ten thousandths
9. six and eight hundred thirty-nine thousandths; $6.839 = 6\frac{839}{1{,}000}$
10. three hundred twenty and four ten thousandths; $320.0004 = 320\frac{4}{10{,}000}$
11. twenty-three hundredths; 0.23
12. seventy-four and one hundred three hundred thousandths; 74.00103
13. 1.35; up **14.** 3.4; down **15.** tie; round up; up
16. 2; 6.873; 6.872; 6; up; 6.873
17. 7; 89.08; 89.07; 2; down; 89.07
18. 5; 18.6; 18.5; 2; down; 18.5
19. 0.073
0.200
thousandths
$200 > 73$
$0.2 > 0.073$
20. 0.400
0.039
0.380
thousandths
$39 < 380 < 400$
$0.039 < 0.38 < 0.4$

Concept Check and Objective Practice

A1. Answers may vary. Example: $1 \cdot \frac{1}{10} = \frac{1}{10}$

A2. Three; ones **A3.** $4 + \frac{5}{10} + \frac{6}{100}$

A4.

ten thousands	thousands	hundreds	tens	ones	decimal point	tenths	hundredths	thousandths	ten thousandths
10,000	1,000	100	10	1	.	$\frac{1}{10}$	$\frac{1}{100}$	$\frac{1}{1{,}000}$	$\frac{1}{10{,}000}$

A5. thousandths **A6.** hundredths **A7.** tenths
A8. ten thousandths **A9.** ten thousandths
A10. hundred thousandths **A11.** tenths **A12.** thousandths
A13. $\frac{8}{10}$ **A14.** $\frac{9}{10}$ **A15.** $\frac{7}{10} + \frac{6}{100} + \frac{8}{1,000}$
A16. $\frac{4}{10} + \frac{7}{100} + \frac{5}{1,000}$ **A17.** $\frac{8}{1,000} + \frac{7}{10,000}$
A18. $\frac{5}{100} + \frac{6}{1,000} + \frac{7}{10,000}$ **A19.** $4 + \frac{9}{10} + \frac{9}{1,000}$
A20. $7 + \frac{7}{100} + \frac{6}{1,000}$ **B1.** 1 d; 2 a; 3 b; 4 c
B2. and **B3.** They are the same. **B4.** eight tenths
B5. nine tenths **B6.** eighty-seven ten thousandths
B7. five hundred sixty-seven ten thousandths
B8. one and fifty-three hundredths
B9. five and forty-nine hundredths
B10. sixty and two hundredths
B11. fifty-five and five hundredths
B12. five and eight hundred thirty-four thousandths
B13. four and eight hundred ninety-two thousandths
B14. five and nine thousand six ten thousandths
B15. six and ninety thousand three hundred thousandths
C1. three hundred ten and sixty-seven hundredths
C2. Answers may vary. Example: The word *and* places the decimal point incorrectly.
C3. ten thousandths **C4.** $4\frac{9}{100}$ **C5.** $3\frac{12}{100}$
C6. $54\frac{4}{10}$ **C7.** $23\frac{2}{10}$ **C8.** $3,000\frac{3}{1,000}$
C9. $4,000\frac{4}{1,000}$ **C10.** $230\frac{43}{100,000}$ **C11.** $837\frac{439}{10,000}$
C12. 0.3 **C13.** 0.7 **C14.** 0.37 **C15.** 0.73
C16. 5.01 **C17.** 3.013 **C18.** 36.0027 **C19.** 25.00029
D1. Answers may vary. Example: 300 is 310.143 rounded to hundreds not hundredths.
D2. Answers may vary. Example: The digit in the place value directly to the right of the given place value indicates whether the number is less than half way to the next larger number. Knowing this, it's possible to state which number it is approximately equal to.
D3. Answers may vary. Example: If the 0 was not included in 240, the number would be 24, which is not a good approximation of 241.
D4. 5.68 **D5.** 2.6 **D6.** 2 **D7.** 0.4 **D8.** ≈ 8.99
D9. ≈ 37.4 **D10.** ≈ 100.009 **D11.** ≈ 782.1286
D12. $\approx 13,000$ **D13.** $\approx 54,800$ **D14.** ≈ 11 **D15.** ≈ 20
D16. 120 mph **D17.** 130 mph
E1. **a.** $\frac{3}{1,000}, \frac{4}{100}, \frac{2}{10}$ **b.** $\frac{3}{1,000}, \frac{40}{1,000}, \frac{200}{1,000}$
c. 0.2, 0.04, 0.003
E2. **a.** $>$; $\frac{2}{10} = \frac{20}{100}$; $\frac{5}{100}$; $\frac{2}{10} > \frac{5}{100}$
b. $>$; $0.2 = 0.20$; $0.05 = 0.05$; $0.2 > 0.05$
c. place value
E3. $0.4 > 0.06$ **E4.** $0.01 > 0.004$ **E5.** $0.07 < 0.6$
E6. $0.024 < 0.15$ **E7.** 0.12, 0.042, 0.04
E8. 0.53, 0.087, 0.056 **E9.** Mobile Metro
E10. Motorola L6

Exercises

For 1.–15.: See preceding Concept Check and Objective Practice answers.

17. like decimals **19.** digits **21.** thousandths
23. hundredths **25.** tenths **27.** ones
29. $1,000 + 4 + \frac{1}{100} + \frac{2}{1,000} + \frac{3}{10,000}$
31. $\frac{3}{100} + \frac{2}{1,000} + \frac{4}{100,000}$ **33.** sixty-five hundredths
35. two hundred three and one hundred twenty-nine thousandths
37. nine and two ten thousandths
39. twelve thousand and twelve ten thousandths
41. fifteen and 67/100 **43.** one hundred two and 45/100
45. five hundred four and 98/100
47. $\frac{12}{100}$ **49.** $1\frac{2}{10}$ **51.** $12\frac{9}{1,000}$ **53.** $102\frac{908}{1,000}$
55. $92\frac{12,902}{100,000}$ **57.** $\frac{1,004}{10,000}$ **59.** 0.3 **61.** 2.5
63. 14.15 **65.** 3.07 **67.** 54.005 **69.** 10,000.011
71. 0.00001 **73.** 165.1674 **75.** 163.5 **77.** 160
79. 236.98 **81.** 200 **83.** 872.826 **85.** 7,327.828
87. 864.5° **89.** $345.69 **91.** 0.7, 0.45, 0.103, 0.08
93. 1.83, 1.8, 1.38, 1.308, 1.083
95. 149.601 mph; 149.335 mph; 148.295 mph
97. Gypsum; Fingernail; Glass **99.** no **101.** 155 lb

Section 3.2 Guided Practice

1.
$$\begin{array}{r} {}^{1}3.879 \\ +\ 5.710 \\ \hline 9.589 \end{array}$$
2.
$$\begin{array}{r} 0.002378 \\ +\ 3.392000 \\ \hline 3.394378 \end{array}$$
3. 3;
$$\begin{array}{r} \overset{5}{\cancel{6}}.\overset{9}{\cancel{0}}\,{}^{1}02378 \\ -\ 3.\ 3\ 92000 \\ \hline 2.\ 6\ 10378 \end{array}$$
4. 2;
$$\begin{array}{r} \overset{3}{\cancel{4}}\,\overset{9}{{}^{1}\cancel{0}}.\ \overset{9}{{}^{1}\cancel{0}}\ \overset{1}{0} \\ -\ 7.\ 3\ 2 \\ \hline 3\ 2.\ 6\ 8 \end{array}$$
5. addition; subtraction; subtraction
New value = Starting value + Rose − Decreased − Fell
= 13,562.33 + 129.43 − 39.87 − 67.13
= 13,691.76 − 39.87 − 67.13
= 13,651.89 − 67.13
= 13,584.76
6. 2007; difference; (47.5 − 41.2 = $6.3); $6.3; $6,300; 2008; 2009; difference; (36.3 − 19.5 = 16.8); $16.8; $16,800; 2006

Concept Check and Objective Practice

A1. decimal points
A2. Answers may vary. Example: Both represent tenths, and adding the numbers is similar to adding numerators.
A3. $0.3 = \frac{3}{10}$; $0.30 = \frac{30}{100} = \frac{\cancel{10} \cdot 3}{\cancel{10} \cdot 10} = \frac{3}{10}$
A4. Answers may vary. Example: It is the same as building equivalent fractions.
A5. 3.8 **A6.** 5.8 **A7.** 16.98 **A8.** 5.99 **A9.** 7.4
A10. 17.21 **A11.** 9.32 **A12.** 7.22 **A13.** 4.08
A14. 11.19 **A15.** 4.7208 **A16.** 69.7101 **A17.** 3.86
A18. 6.13 **A19.** 13.2 million **A20.** 0.5 million

Exercises

For 1.–3.: See preceding Concept Check and Objective Practice answers.

5. sum **7.** difference **9.** 11.7 **11.** 70.05
13. 8.32 **15.** 10.712 **17.** $78.91 **19.** $85.35
21. 4.1 **23.** 8.83 **25.** 13.45 **27.** 81.817
29. $71.49 **31.** $13.76 **33.** 427.8184 **35.** 923.877
37. 136.83 **39.** 307.55 **41.** 51.0227 **43.** 122.86
45. 4.97 lb **47. a.** 1.05 mi **b.** 4.15 mi **49.** 18.7876
51. 100 **53. a.** $680 **b.** yes **c.** $666.85 **d.** $31.58
55. $364.13, $89.13, $61.13, $11.18, $268.16
57. 0.9 million square miles less
59. more than 5.6 million square miles

Section 3.3 Guided Practice

1.
3; 0.003
2; × 0.02
$3 + 2 = 5$ 0.00006

2.
1; 0.7
3; × 0.009
$1 + 3 = 4$ 0.0063

3.
0; 5
7; × 0.0000003
$0 + 7 = 7$ 0.0000015

4.
2; 50.67
2; × 0.03
$2 + 2 = 4$ 1.5201
$50.67 \times 0.03 = 1.5201$

5.
5; 0.00345
0; × 27
2415
+ 690
$5 + 0 = 5$ 0.09315
$0.00345 \times 27 = 0.09315$

6.
0; 1500
5; × 0.00073
4500
+ 10500
$0 + 5 = 5$ 1.09500
$1{,}500 \times 0.00073 = 1.095$

7. 0.01; smaller; two; left; $501 \times 0.01 = 5.01$
8. not; 0.001; smaller; three; left;
$3.01 \times 0.001 = 0.00301$
9. 100,000; larger; five; right;
$23.2 \times 100{,}000 = 2{,}320{,}000$
10. 1,000,000; larger; six; right;
$3.76 \times 1{,}000{,}000 = 3{,}760{,}000$
11. $0.004 \times 0.07 \approx 0.00028$, $4 \cdot 7 = 28$; thousandths; three; left; 28; 0.028; hundredths; two; left; 0.028; 0.00028
12. $9 \times 0.08 \approx 0.72$, $9 \cdot 8 = 72$; ones; zero; left; 72; 72; hundredths; two; left; 72; 0.72
13. $5{,}000 \times 0.05 \approx 250$, $5 \cdot 5 = 25$; thousands; three; right; 25; 25,000; hundredths; two; left; 25,000; 250
14. half; 2; $4.8 \div 2 = 2.4$
The radius is 2.4 cm.
15. twice; 2; $3.75 \times 2 = 7.5$
The diameter is 7.5 in.
16.
$C = 2\pi r$
$C = 2(3.14)(20 \text{ meters})$
$C = (6.28)(20 \text{ meters})$
$C = 125.6 \text{ meters}$
17.
$A = \pi r^2$
$A = (3.14)(4 \text{ cm})^2$
$A = (3.14)(16 \text{ cm}^2)$
$A = 50.24 \text{ cm}^2$

Concept Check and Objective Practice

A1. a. $0.3 = \frac{3}{10}$; $0.5 = \frac{5}{10}$
b. It indicates the number of zeros. **c.** $\frac{3}{10} \cdot \frac{5}{10} = \frac{15}{100}$
d. 0.15
e. The total number of decimal places for 0.3 and 0.5 is equal to the number of decimal places in the answer.
A2. 0.00012 **A3.** 0.000016 **A4.** 0.00000048
A5. 0.0003 **A6.** 0.0036 **A7.** 0.003 **A8.** 72
A9. 16 **A10.** 0.32 in. **A11.** 12 in.
B1. Answers may vary. Example: Count the total number of decimal places in the factors. The answer must have the same number of decimal places.
B2. Answers may vary. Example: The zero does not add value to the number. It is simply a placeholder.
B3. 0.04011 **B4.** 0.001378 **B5.** 29.93 **B6.** 11.832
B7. 0.147088 **B8.** 1.964269 **B9.** 8,056 **B10.** 8,056
B11. $262.50 **B12.** $454.50
C1. Answers will vary. Example:
a. 42.2 **b.** 4.22 **c.** smaller
C2. 1,000 **C3.** 1,360 **C4.** 3,710 **C5.** 0.010001
C6. 10.1 **C7.** 980 **C8.** 8.54982 **C9.** 0.0000987
C10. 8,700,000 **D1.** left; 4 **D2.** right; 3
D3. neither; 0 **D4.** 0.24 **D5.** 0.16 **D6.** 28
D7. 36 **D8.** 90 **D9.** 35 **D10.** 0.00018
D11. 0.0028 **D12.** $8,000 **D13.** $240
E1. Circumference is the perimeter of a circle.
E2. 18.86 square centimeters and 78.5 ft^2
E3. Divide the diameter of the circle by 2.
E4. 28.26 sq km **E5.** 12.56 sq mi **E6.** 131.88 in.
E7. 182.12 cm **E8.** 153.86 sq in. **E9.** 615.44 sq m
E10. 23.864 mm **E11.** 15.7 in.

Exercises

For 1.–15.: See preceding Concept Check and Objective Practice answers.

17. Pi (π) **19.** product **21.** circumference
23. 0.000025 **25.** 0.00063 **27.** 0.924 **29.** 7.14
31. 1,126.78 **33.** 0.048246 **35.** $9.50 **37.** $112
39. 4.32 **41.** 1,265 **43.** 0.00000078 **45.** 2.1
47. a. 100; 3 **b.** 300 **c.** 400
49. a. 40; 0.006 **b.** 0.24 **c.** 0.26
51. a. 2; 0.007 **b.** 0.014 **c.** 0.013
53. a. 900; 0.05 **b.** 45 **c.** 40
55. 25.1 km **57.** 7.1 sq yd **59.** 28.3 in.
61. 2,826 sq cm **63.** $30.31 **65.** $50.69
67. a. $52.47 **b.** $82.46 **c.** no

Section 3.4 Guided Practice

1. 0.91; two; two
9 1.)3 4 8 6.

2. 1.2; one; one
1 2)3 2.1

3. three; zeros
1 2.)1 2 3 0 0.

4.
0 5 5.
4.)2 2 0.
2 0
2 0
2 0
0

5. hundredths; hundredths; two

6. $5\overline{)0.060}$ quotient 0.012; -5, 10, -10

$0.006 \div 0.05 \approx 0.1$

7. $6\overline{)70.000}$ quotient 11.666; -6, 10, -6, 40, -36, 40

$7 \div 0.06 \approx 11.67$

8. $\frac{3}{4} \Rightarrow 4\overline{)3.00}$ quotient 0.75; -28, 20, -20, 0 $\quad \frac{3}{4} = 0.75$

9. $\frac{1}{6} \Rightarrow 6\overline{)1.000}$ quotient 0.166; -6, 40, -36, 40, -36, 4 $\quad \frac{1}{6} = 0.1\overline{6}$

10. $7\overline{)3.000}$ quotient 0.428; -28, 20, -14, 60 $\quad \frac{3}{7} \approx 0.43$

11. $\frac{1}{5} \Rightarrow 5\overline{)1.0}$ quotient 0.2; -10, 0 $\quad 4\frac{1}{5} = 4.2$

12. $9.32 \approx 9.0$
$0.0425 \approx 0.04$
$04\overline{)900}$ quotient 225
$9.32 \div 0.0425 \approx 225$

13. $0.0784 \approx 0.08$
$8.726 \approx 9$
$9\overline{)0.0800}$ quotient 0.0088
$0.0784 \div 8.726 \approx 0.009$

14. $0.004643 \approx 0.005$
$0.0835 \approx 0.08$
$8\overline{)0.5000}$ quotient 0.0625
$0.004643 \div 0.0835 \approx 0.063$

15. $16\overline{)280} \Rightarrow 20\overline{)300}$ quotient 15
$2.8 \div 0.16 \approx 15$
The quotient is close to the estimate.

$16\overline{)280.0}$ quotient 017.5; -160, 120, -112, 80, -80, 0

$2.8 \div 0.16 = 17.5$

16. $52\overline{)863.4} \Rightarrow 50\overline{)900}$ quotient 18
$8.634 \div 0.52 \approx 18$
The quotient is close to the estimate.

$52\overline{)863.40}$ quotient 16.60; -52, 343, -312, 314, -312 $\approx$ **16.6**

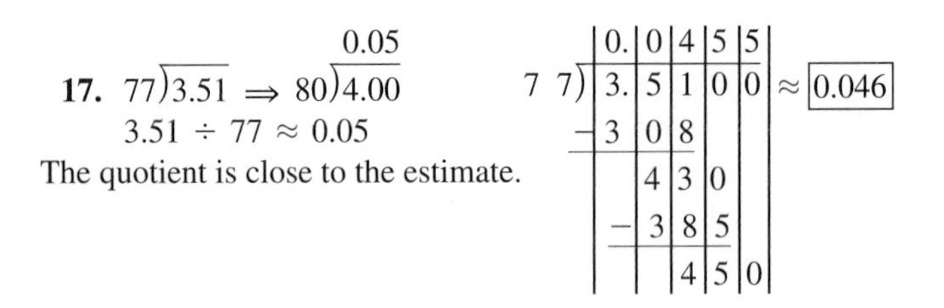

17. $77\overline{)3.51} \Rightarrow 80\overline{)4.00}$ quotient 0.05
$3.51 \div 77 \approx 0.05$
The quotient is close to the estimate.

$77\overline{)3.5100}$ quotient 0.0455; -308, 430, -385, 450 $\approx$ **0.046**

Concept Check and Objective Practice

A1. a. 1.2 **b.** 0.06 **A2.** three places

A3. $4\overline{)125.}$ **A4.** $5\overline{)132.}$ **A5.** $2\overline{)71.45}$

A6. $3\overline{)10.97}$ **A7.** $2\overline{)3100.}$ **A8.** $2\overline{)4300.}$

A9. $1332\overline{)3560.}$ **A10.** $2672\overline{)4260.}$

B1. hundredths

B2. Answers may vary. Example:
a. Place a zero at the end of the number.
b. They do not change the value, but rather are the same as writing equivalent fractions.

B3. 57.5 **B4.** 662.5 **B5.** 0.15 **B6.** 0.012 **B7.** 600
B8. 2,800 **B9.** 19.648 **B10.** 535.95 **B11.** 84 cases
B12. $7.20 **C1. a.** 0.5 **b.** 0.25 **c.** 0.2 **d.** 0.1
C2. a. 3.25 **b.** 4
C3. Answers may vary. Example: The whole number of the mixed number becomes the whole number portion of the decimal.
C4. 0.4 **C5.** 0.7 **C6.** 0.625 **C7.** 0.125 **C8.** $0.\overline{3}$
C9. $0.8\overline{3}$ **C10.** 7.6 **C11.** 8.3 **C12.** 0.38
C13. 0.67 **C14.** 1.42 **C15.** 3.88
D1. Answers may vary. **D2.** Answers may vary.
D3. Answers may vary. Example: The rest of the answers have the decimal in an incorrect position.
D4. 15 **D5.** 25 **D6.** 0.00035 **D7.** 0.003
D8. 2.5 **D9.** 0.15 **D10.** 17.5 **D11.** 45
D12. 3 hr **D13.** 20 hr

E1. thousandths **E2.** $40\overline{)400}$ quotient 10 **E3.** $80\overline{)400}$ quotient 5
E4. $60\overline{)6.0}$ quotient 0.1 **E5.** $30\overline{)3.0}$ quotient 0.1 **E6.** 0.03 **E7.** 0.05
E8. 16.09 **E9.** 13.68 **E10.** 130 **E11.** 11
E12. 11.6 **E13.** 15.6 **E14.** 6.7 hr **E15.** 47 months

Exercises

For 1.–15.: See preceding Concept Check and Objective Practice answers.

17. divisor

19. 6 → dividend; 3 → divisor; 2 → quotient
5 → divisor; 35 → dividend; 7 → quotient
16 → dividend; 8 → divisor; 2 → quotient

21. $21\overline{)\,3\,|\,4.\,|\,5\,|}$ **23.** $12\overline{)\,3.\,|\,4\,|}$ **25.** 0.39 **27.** 10.95
29. 0.254 **31.** 83.333 **33.** 5,001.25 **35.** 220
37. \$22.45 **39.** 1.2 **41.** 0.56 **43.** 4.25 **45.** 0.44
47. 1.16 **49.** 3.29 **51.** 2.31 **53.** 49.41 **55.** 41.22
57. 0.13 **59.** 90.33 **61.** 3.12 **63.** \$321.25 **65.** 12.8
67. 20.3 **69.** 1,312.5 **71.** 849.4 **73.** 1,772.63
75. 0.18 **77.** The package is missing an iPod. **79.** 68 parts

Chapter 3 Review Exercises

1. one and two-tenths; both represent the same quantity.
2. tenths; hundredths; thousandths
3. rounding **4.** and **5.** thousands **6.** thousandths
7. hundred thousandths **8.** ten thousandths
9. seventy-six and thirty-four hundredths
10. two and fifty-seven thousandths
11. twelve and six thousand three ten thousandths
12. five and seven thousand two hundred thousandths
13. $\frac{35}{1,000}$ **14.** $\frac{45}{100}$ **15.** 0.014 **16.** 0.0008 **17.** 67.8
18. 578.88 **19.** 3.567 **20.** 1,400 **21.** 0.0435; 0.0436; 0.14
22. 0.0987; 0.567; 7.003 **23.** the decimal point
24. Answers may vary. Example: It's the same as writing equivalent fractions.
25. 3.11 **26.** 734.4 **27.** 39.461 **28.** 129.29
29. 28.72 **30.** 360.9 **31.** 12.69 **32.** 2.11
33. a. 0.8 mi **b.** 1.1 mi **34. a.** \$403.45 **b.** \$1,052.80
35. product **36.** 1,000 **37.** Multiply the radius by 2.
38. 0.00018 **39.** 0.000018 **40.** 0.32 **41.** 0.035
42. 6.23 **43.** 0.007 **44.** 41 **45.** 87 **46.** 0.48 in.
47. \$5.25 **48. a.** 0.018 **b.** 0.0184032
49. a. 0.00021 **b.** 0.00017784 **50. a.** 8 **b.** 9.071
51. a. 35 **b.** 32.21242 **52.** 50.24 m **53.** 18.84 mm
54. 28.26 sq ft **55.** 78.5 sq in. **56.** two; right
57. ten thousandths **58.** 200 **59.** 40,000 **60.** 0.4
61. 14 **62.** 44.5 **63.** 48 **64.** 26.63 **65.** 44.60
66. 3.3 hr **67.** \$18.96, \$18.96, and \$18.95 **68.** 0.6
69. 0.625 **70.** $1.1\overline{6}$ **71.** $1.8\overline{3}$ **72.** $5.\overline{3}$ **73.** $7.\overline{6}$

Chapter 3 Practice Test

1. hundredths; thousandths; ten thousandths
2. Answers may vary. Example: Count the total number of decimal places in the factors.
3. three; right
4. hundredths
5. two hundred sixteen and 67/100
6. $\frac{16}{100}$ **7.** 0.009 **8.** 145.3 **9.** 5.87 **10.** 10.548
11. 1,000 **12.** 0.0499; 0.0655; 0.101 **13.** 44.01
14. 21.4 **15.** 2,708.371 **16.** 341.911
17. a. 1.2 mi **b.** 4.9 mi
18. a. \$397.67 **b.** \$87,162.33
19. a. 1.2 **b.** 1.0065 **20. a.** 0.63 **b.** 0.65415
21. a. 1.8 **b.** 1.477368 **22.** \$286.84 **23.** 43.96 ft
24. 200.96 sq in. **25.** 1.829 **26.** 0.428 **27.** 19 weeks
28. 28.3 points **29.** 0.375 **30.** 4.4 **31.** $0.8\overline{3}$
32. 2.25

CHAPTER 4

Section 4.1 Guided Practice

1. 5 to 7; $\frac{5}{7}$; 5 : 7 **2.** $\frac{10¢}{1¢} = \frac{10}{1}$; cents; ratio; cents; $\frac{10}{1}$

3. $\frac{3}{7}$ **4.** to; $\frac{6}{13}$

5. Julian's hot dogs to Mikito's hot dogs
$$= \frac{\text{Julian's hot dogs}}{\text{Mikito's hot dogs}} = \frac{40\ \cancel{\text{hot dogs}}}{13\ \cancel{\text{hot dogs}}} = \frac{40}{13}$$

6. State taxes to Federal taxes
$$= \frac{\text{State taxes}}{\text{Federal taxes}} = \frac{\cancel{\$}\,23}{\cancel{\$}\,88} = \frac{23}{88}$$

7. $\frac{42\ \cancel{\text{kg}}}{30\ \cancel{\text{kg}}} = \frac{{}^{1}\cancel{6}\cdot 7}{{}^{1}\cancel{6}\cdot 5} = \frac{7}{5}$ **8.** $\frac{0.13\ \cancel{L}}{0.5\ \cancel{L}} = \frac{0.13}{0.5}\cdot\frac{100}{100} = \frac{13}{50}$

9. $\frac{1\frac{1}{2}}{2\frac{1}{4}} = \frac{\frac{3}{2}}{\frac{9}{4}} = \frac{3}{2} \div \frac{9}{4} = \frac{3}{2}\cdot\frac{4}{9} = \frac{{}^{1}\cancel{3}}{\cancel{2}^{1}}\cdot\frac{\cancel{2}^{1}\cdot 2}{{}^{1}\cancel{3}\cdot 3} = \frac{2}{3}$

10. different; 2 days $= \frac{2\ \cancel{\text{days}}}{1}\cdot\frac{24\text{ hours}}{1\ \cancel{\text{day}}} = 48$ hours

$\frac{14.4\ \cancel{\text{hours}}}{48\ \cancel{\text{hours}}} = \frac{14.4}{48}\cdot\frac{10}{10} = \frac{144}{480} = \frac{3\cdot\cancel{48}^{1}}{10\cdot\cancel{48}^{1}} = \frac{3}{10}$

11. $\frac{5\text{ miles}}{42\text{ minutes}}$ **12.** $\frac{\$200}{15\text{ hours}} = \frac{\cancel{5}^{1}\cdot\$40}{\cancel{5}^{1}\cdot 3\text{ hours}} = \frac{\$40}{3\text{ hours}}$

13. $\frac{40\text{ mg}}{120\text{ lb}} = \frac{{}^{1}\cancel{40}\cdot 1\text{ mg}}{{}^{1}\cancel{40}\cdot 3\text{ lb}} = \frac{1\text{ mg}}{3\text{ lb}}$

14. $\frac{\$42}{5\text{ hours}} = \frac{\$8.40}{1\text{ hour}} = \$8.40$ per hour

15. $\frac{160\text{ miles}}{4\text{ gallons}}$; $\frac{40\text{ miles}}{1\text{ gallon}}$; 40 miles; gallon

16. $\frac{\$138}{12\text{ hours}}$; $\frac{\$11.50}{1\text{ hour}}$; \$11.50; hour

17. $\frac{20\text{ milligrams}}{100\text{ pounds}}$; $\frac{0.2\text{ milligrams}}{1\text{ pound}}$; 0.2 milligram per pound

18. a. $\frac{20\text{ oz}}{\$5.00} = 4$ oz per dollar
b. $\frac{27\text{ oz}}{\$5.40} = 5$ oz per dollar; the larger; cereal; dollar; 27 oz

19. a. $\frac{\$1.50}{6\text{ rutabagas}} = \$0.25$ per rutabaga
b. $\frac{\$2.60}{10\text{ rutabagas}} = \$0.26$ per rutabaga; rutabaga; Hector's

Concept Check and Objective Practice

A1. same **A2.** Answers may vary. Example: The units are not the same nor can one unit be converted into the other.
A3. $\frac{2}{3}$ **A4.** $\frac{1}{2}$ **A5.** $\frac{11}{5}$ **A6.** $\frac{20}{27}$ **A7.** $\frac{14}{17}$

A8. $\frac{1}{2}$ **A9.** $\frac{6}{355}$ **A10.** $\frac{18}{265}$ **A11.** $\frac{10}{3}$ **A12.** $\frac{1,753}{375}$
A13. $\frac{1}{2}$ **A14.** $\frac{21}{2}$ **A15.** $\frac{300}{1}$ **A16.** $\frac{100}{1}$
B1. Answers may vary. Example: A ratio compares quantities with like units, while a rate compares quantities with unlike units.
B2. Answers may vary. Example: A unit rate is a rate whose denominator is always one unit.
B3. **a.** Trevor: 21 mpg; Valerie: 23 mpg **b.** Valerie's **c.** Answers may vary. Example: It makes comparison easier.
B4. 1 mi per 6 min **B5.** 16 mi per 5 min
B6. $32 per 3 hr **B7.** $23 per 2 hr
B8. $6.25 per hr **B9.** $5.50 per hr
B10. $0.26 per oz **B11.** $0.24 per oz
B12. 5 mg per lb **B13.** 5 mg per lb
B14. 53 miles per hr **B15.** 45 miles per hr

Exercises

For 1.–6.: See preceding Concept Check and Objective Practice answers.

7. unit rate **9.** ratio
11. **a.** ratio **b.** With no units indicated, the assumption is that they are the same. **c.** 1:5
13. **a.** ratio **b.** With no units indicated, the assumption is that they are the same. **c.** 4:7
15. **a.** rate **b.** The units are not the same. **c.** 14 girls to 27 students
17. **a.** rate **b.** The units are not the same. **c.** 5 yards per dollar
19. **a.** rate **b.** The units are not the same. **c.** $45 per credit hour
21. **a.** rate **b.** The units are not the same. **c.** $28 per 5 hours
23. **a.** ratio **b.** The units are the same. **c.** 20:37
25. 3 to 50 **27.** 3 to 7
29. Answers may vary. Example: 120 to 78 (measure of blood pressure); 1 cup oatmeal to 2 cups water (recipe for oatmeal cereal); 16:9 is the aspect ratio of an HDTV
31. 58 miles per hour **33.** 24 students per class
35. 20 cents per dollar **37.** 200 square feet per hour
39. 8.6 mg per minute **41.** 40 people per bus
43. **a.** factor the numerator and denominator **b.** divide out all common factors
45. **a.** divide the numerator by the denominator and use the units "miles per gallon" in the answer
47. 12 issues **49.** **a.** 15 oz for $2.32 **b.** 20 oz for $4.89
51. 50 miles per hour; 24 miles per gallon
20 miles per hour; 15 miles per gallon
53. $31 per PDA **55.** 2,400 teeth per year
57. 10,000 to 1,000, or 10 to 1
59. **a.** Mark McGwire **b.** Mark McGwire **c.** Answers may vary.
61. **a.** 30 parts per hour **b.** $1.50 per part
63. Kareem Abdul-Jabbar **65.** Michael Jordan
67. Answers may vary. **69.** Answers may vary.

Section 4.2 Guided Practice

1. good parts; total parts;
$\frac{\text{Good parts}}{\text{Total parts}}$; $\frac{480 \text{ Good parts}}{500 \text{ Total parts}} = \frac{384 \text{ Good parts}}{400 \text{ Total parts}}$

2. price; number of boxes
$\frac{\text{Price}}{\text{Number of boxes}}$ $\frac{\$1.50}{1 \text{ box}} = \frac{\$4.50}{3 \text{ boxes}}$

3. amount of medication; weight of the bull
$\frac{\text{Milligrams}}{\text{Pounds}}$ $\frac{4,000 \text{ milligrams}}{1,500 \text{ pounds}} = \frac{x \text{ milligrams}}{1,200 \text{ pounds}}$

4. $15 \cdot 98 = 14 \cdot 105$
$1,470 = 1,470$; is; are

5. $175 \cdot 40 = 50 \cdot 140$
$7,000 = 7,000$; is; are

6. $360 \cdot 8 = 280 \cdot 10.1$
$2,880 \neq 2,828$; isn't; aren't

7.
$$\left(\frac{2}{2}\right)\frac{3}{6} = \frac{n}{12}$$
$$\frac{6}{12} = \frac{n}{12}$$
$$6 = n$$

8.
$$24 = 8 \cdot 3$$
$$\text{Denominator} = 8 \cdot \text{numerator}$$
$$\text{Denominator} = 8 \cdot \text{numerator}$$
$$n = 8 \cdot 13$$
$$n = 104$$

9.
$$\frac{\cancel{28}^{1}}{1} \cdot \frac{n}{\cancel{28}_{1}} = \frac{28}{1} \cdot \frac{3}{14}$$
$$n = \frac{2 \cdot \cancel{14}^{1}}{1} \cdot \frac{3}{\cancel{14}^{1}}$$
$$n = 6$$
Check:
$$\frac{6}{28} \stackrel{?}{=} \frac{3}{14}$$
$$6 \cdot 14 \stackrel{?}{=} 3 \cdot 28$$
$$84 = 84$$

10.
$$\frac{n}{4} = \frac{7}{16}$$
$$\frac{\cancel{4}^{1}}{1} \cdot \frac{n}{\cancel{4}^{1}} = \frac{\cancel{4}^{1}}{1} \cdot \frac{7}{16}$$
$$n = \frac{\cancel{4}^{1}}{1} \cdot \frac{7}{\cancel{4} \cdot 4}$$
$$n = \frac{7}{4}$$
Check:
$$\frac{4}{\frac{7}{4}} \stackrel{?}{=} \frac{16}{7}$$
$$4 \cdot 7 \stackrel{?}{=} 16 \cdot \frac{7}{4}$$
$$4 \cdot 7 \stackrel{?}{=} \frac{\cancel{4}^{1} \cdot 4}{1} \cdot \frac{7}{\cancel{4}^{1}}$$
$$28 = 28$$

11.
$$\frac{1.5}{4} = \frac{n}{12}$$
$$\frac{12}{1} \cdot \frac{1.5}{4} = \frac{n}{12} \cdot \frac{12}{1}$$
$$\frac{3 \cdot {}^{1}\cancel{4}}{1} \cdot \frac{1.5}{{}^{1}\cancel{4}} = n$$
$$4.5 = n$$
Check:
$$\frac{1.5}{4} \stackrel{?}{=} \frac{4.5}{12}$$
$$1.5 \cdot 12 \stackrel{?}{=} 4 \cdot 4.5$$
$$18 = 18$$

A1. Proportion
A2. Answers may vary. Example: A rate is a comparison of two quantities with unlike units. A proportion compares two rates (or ratios).
A3. Answers may vary. Example: The units in the numerators of the two rates are not the same. It could become a proportion by interchanging the numerator and denominator of one of the rates. For example: 14 pounds is to 4 bricks as 28 pounds is to 8 bricks.
A4. $\frac{5 \text{ cups}}{3 \text{ pounds}} = \frac{15 \text{ cups}}{9 \text{ pounds}}$
A5. $\frac{8 \text{ liters}}{7 \text{ kilograms}} = \frac{32 \text{ liters}}{28 \text{ kilograms}}$
A6. $\frac{9 \text{ boards}}{10 \text{ feet}} = \frac{18 \text{ boards}}{20 \text{ feet}}$ **A7.** $\frac{3 \text{ pounds}}{8 \text{ vases}} = \frac{9 \text{ pounds}}{24 \text{ vases}}$

A8. $\dfrac{5 \text{ copies}}{45 \text{ pieces of paper}} = \dfrac{50 \text{ copies}}{450 \text{ pieces of paper}}$

A9. $\dfrac{1 \text{ packet}}{1 \text{ row}} = \dfrac{30 \text{ packets}}{30 \text{ rows}}$ **A10.** $\dfrac{\$8}{1 \text{ hour}} = \dfrac{n \text{ dollars}}{7.5 \text{ hours}}$

A11. $\dfrac{\$18}{1 \text{ hour}} = \dfrac{n \text{ dollars}}{16.4 \text{ hours}}$

B1. A proportion is true if the cross products are equal.
B2. Answers may vary.
B3. no **B4.** yes **B5.** yes **B6.** no
B7. yes **B8.** yes **B9.** no **B10.** yes

B11. a. $\dfrac{24 \text{ points}}{10 \text{ minutes}}; \dfrac{50 \text{ points}}{25 \text{ minutes}}$ **b.** no

B12. a. $\dfrac{5 \text{ driveways}}{2 \text{ hours}}; \dfrac{7 \text{ driveways}}{3 \text{ hours}}$ **b.** no

Focus on Proportional Reasoning

1. $n = 15$ **2.** $n = 30$ **3.** $n = 6$ **4.** $n = 1$
5. $n = 20$ **6.** $n = 49$ **7.** $n = 3$ **8.** $n = 9$
9. $n = 32$ **10.** $n = 6$ **11.** $n = 36$ **12.** $n = 55$
13. $n = 5$ **14.** $n = 5$ **15.** $n = 28$ **16.** $n = 36$
17. $n = 3$ **18.** $n = 2$ **19.** $n = 1\frac{1}{2}$ **20.** $n = 1\frac{1}{2}$
21. 43 is greater than 32, but 15 is less than 23.
22. 15 is three times 5, but 38 is not three times 13.
23. 88 is very close to 90, but 41 is not very close to 63.
24. 100 is five times 20, but 36 is not five times 6.
25. 1 is less than half of 3, but 2 is exactly half of 4.
26. 2 is half of 4, but 7 is not half of 10.
27. 53 is not very close to 30, but 62 is close to 60.
28. 2 is larger than 1, but 500 is smaller than 1,000.

C1. Answers may vary. Example: A woman on a diet ate 3 cookies. According to the cookie's package, one cookie has 250 calories. How many calories did the woman consume?

C2. $\dfrac{250 \text{ calories}}{1 \text{ cookie}} = \dfrac{x \text{ calories}}{3 \text{ cookies}}$

C3. Answers may vary. Example: The first ratio, $\frac{4}{5}$, is cloudy days to all days. The second ratio, $\frac{30}{C}$, has "all days" in the numerator.

C4. $n = 9$ **C5.** $n = 14$ **C6.** $n = 17\frac{1}{2}$ **C7.** $n = 1\frac{9}{13}$

C8. $n = 3\frac{1}{3}$ **C9.** $n = 1\frac{4}{5}$ **C10.** $n = 1\frac{5}{7}$

C11. $n = 6\frac{1}{3}$ **C12.** 16 qt **C13.** 50 min

Exercises

For 1.–11.: See preceding Concept Check and Objective Practice answers.

13. proportion **15.** $\frac{5}{3} = \frac{n}{7}$ **17.** $\frac{8}{11} = \frac{0.06}{n}$

19. $\dfrac{53}{100} = \dfrac{n}{4{,}300}$, where n represents the number of people who oppose the proposition

21. $\dfrac{\$0.261}{1 \text{ square inch}} = \dfrac{n}{766 \text{ square inches}}$, where n represents the approximate purchase price

23. $\dfrac{\$8.55}{1 \text{ hour}} = \dfrac{n}{30 \text{ hours}}$, where n represents Sansui's earnings

25. $\dfrac{250 \text{ words}}{1 \text{ page}} = \dfrac{3{,}000 \text{ words}}{n}$, where n represents the number of pages in Jessica's report

27. $\dfrac{\frac{1}{4} \text{ cup}}{1 \text{ batch}} = \dfrac{1\frac{1}{2} \text{ cups}}{n}$, where n represents the number of batches of chili

29. yes **31.** no **33.** no **35.** no **37.** yes **39.** no

41. a. $\dfrac{30 \text{ mph}}{5°\text{F}} \stackrel{?}{=} \dfrac{5 \text{ mph}}{33°\text{F}}$ **b.** no

43. a. $\dfrac{3 \text{ inches}}{\frac{1}{3} \text{ inch}}; \dfrac{9 \text{ square inches}}{\frac{1}{9} \text{ square inch}}$

b. $\dfrac{3 \text{ inches}}{\frac{1}{3} \text{ inch}} \stackrel{?}{=} \dfrac{9 \text{ square inches}}{\frac{1}{9} \text{ square inch}}$ **c.** no

45. no **47. a.** $\dfrac{9.6 \text{ inches}}{320 \text{ feet}} \stackrel{?}{=} \dfrac{1.8 \text{ inches}}{60 \text{ feet}}$ **b.** yes

49. $n = 2$ **51.** $n = 12$ **53.** $n = 6.4$

55. $n = 270$ **57.** $n = 1\frac{2}{9}$ **59.** $n = 21$ **61.** $500

63. 100 lb **65.** $62.50 **67.** 20 m **69.** 3 hr

Section 4.3 Guided Practice

1. 5

$$\frac{5 \text{ ft}}{1 \text{ in.}} = \frac{n \text{ ft}}{5 \text{ in.}}$$

$$\frac{5 \text{ ft}}{1 \cancel{\text{in.}}} \cdot \frac{5 \cancel{\text{in.}}}{1} = \frac{n \text{ ft}}{\cancel{5 \text{ in.}}} \cdot \frac{\cancel{5 \text{ in.}}}{1}$$

$$25 \text{ ft} = n$$

The width of the cabin is 25 ft.

2. Length of the Bathroom:

$$\frac{5 \text{ ft}}{1 \text{ in.}} = \frac{n \text{ ft}}{3.5 \text{ in.}}$$

$$\frac{5 \text{ ft}}{1 \cancel{\text{in.}}} \cdot \frac{3.5 \cancel{\text{in.}}}{1} = \frac{n \text{ ft}}{\cancel{3.5 \text{ in.}}} \cdot \frac{\cancel{3.5 \text{ in.}}}{1}$$

$$17.5 \text{ ft} = n$$

Width of the Bathroom:

$$\frac{5 \text{ ft}}{1 \text{ in.}} = \frac{n \text{ ft}}{2 \text{ in.}}$$

$$\frac{5 \text{ ft}}{1 \cancel{\text{in.}}} \cdot \frac{2 \cancel{\text{in.}}}{1} = \frac{n \text{ ft}}{\cancel{2 \text{ in.}}} \cdot \frac{\cancel{2 \text{ in.}}}{1}$$

$$10 \text{ ft} = n$$

Area of the Bathroom: $= 17.5 \text{ ft} \cdot 10 \text{ ft}$
$= 175$ square feet

The area of the bathroom is 175 ft^2

3. rate $= \dfrac{86 \text{ white jelly beans}}{1.3 \text{ lb}}$

$$\frac{86 \text{ white jelly beans}}{1.3 \text{ lb}} = \frac{n}{100 \text{ lb}}$$

$$\frac{86 \text{ white jelly beans}}{1.3 \cancel{\text{lb}}} \cdot \frac{100 \cancel{\text{lb}}}{1} = \frac{n}{\cancel{100 \text{ lb}}} \cdot \frac{\cancel{100 \text{ lb}}}{1}$$

6,615 white jelly beans $\approx n$

There are approximately 6,615 white jelly beans in the 100 lb box.

4. $\dfrac{50 \text{ mg}}{1 \text{ kg}}; \dfrac{25 \text{ mg}}{1 \text{ kg}}; \dfrac{300 \text{ mg}}{5 \text{ kg}} = \dfrac{60 \text{ mg}}{1 \text{ kg}}; \dfrac{60 \text{ mg}}{1 \text{ kg}} > \dfrac{25 \text{ mg}}{1 \text{ kg}};$ is; don't

5. 300; is

$$\frac{1\text{ mL}}{50\text{ mg}} = \frac{n\text{ mL}}{125\text{ mg}}$$

$$\frac{125\text{ mg}}{1} \cdot \frac{1\text{ mL}}{50\text{ mg}} = \frac{n\text{ mL}}{{}^{1}125\text{ mg}} \cdot \frac{{}^{1}12.5\text{ mg}}{1}$$

$$\frac{125\text{ mL}}{50} = n$$

$$2.5\text{ mL} = n$$

2.5 mL must be given.

Concept Check and Objective Practice

A1. 4 ft **A2.** 30 ft **A3.** 25 ft **A4.** 17.5 ft
A5. $656.25 **A6.** $656.25 **A7.** 112.5 mi **A8.** 696 in.
A9. 6.5 ft
A10. a. Answers may vary.
b. Answers may vary
B1. a. students at the college **b.** students in the dining hall
B2. a. 500 trees **b.** 20 trees
B3. Answers may vary. Example: Sampling techniques do not give an exact amount.
B4. Approximately 1,800 students take night classes.
B5. He receives approximately 1,120 pieces of junk mail.
B6. Approximately 19,200 fish swam upstream in 24 hours.
B7. There are approximately 104,720 words in the dictionary.
C1. 20 mg/kg **C2.** 30–50 mg/kg/d in divided doses
C3. Adults: 250 mg administered once every 24 hours or 300 mg given in divided doses at 8- or 12-hour intervals
Children: 15–25 mg/kg up to a maximum of 250 mg per single daily injection
C4. a. 25 mg per kg **b.** yes
C5. a. 15 mg per kg **b.** yes
D1. a. 100 mg **b.** 150 mg
c. There is 125 mg of the drug in a 2 mL injection.
D2. a. 30–50 mg/kg/day in two divided doses **b.** yes
c. 200 mg per 5 mL **d.** 8.75 mL
e.

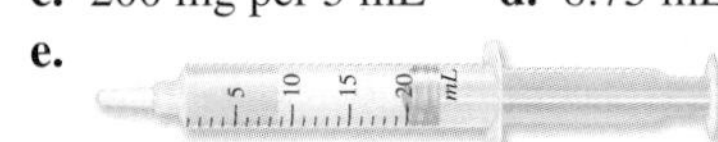

D3. a. Adults: 250 mg administered once every 24 hours or 300 mg in divided doses Children: 15–25 mg/kg
b. yes **c.** 50 mg per mL **d.** 8.8 mL
e.

Exercises

For 1.–12.: See preceding Concept Check and Objective Practice answers.

13. prescribed dose **15.** usual dose **17.** drug concentration
19. about 19.5 ft **21.** about 3 ft **23.** about 7.5 ft
25. about 2.25 ft **27.** 15 ft by 20 ft **29.** about 300 sq ft
31. about 160 sq ft **33.** about 90 ft
35. No, the ratio is 2 to 1. **37.** about 40,000,000
39. about 750 **41.** about 4,950,000
43. Answers may vary.
45. a. no
c.

47. a. yes **b.** 4 mL
c.

49. a. yes **b.** 2 mL
c.

51. a. yes **b.** 7.5 mL
c.

Chapter 4 Review Exercises

1. like **2.** $\frac{7}{5}$ **3.** $\frac{15}{32}$ **4.** $\frac{7}{250}$ **5.** $\frac{7}{80}$
6. A ratio is a comparison of items with the same units, while a rate is a comparison of items with different units.
7. A unit rate has a denominator of 1 unit.
8. $9.33 per hr **9.** $7.20 per hr **10.** $0.15 per oz
11. $0.13 per oz **12.** 68 oz for $1.56 **13.** 128 oz for $2.55
14. $0.01 per oz **15.** $0.08 per oz
16. proportion
17. Answers may vary. Example: Since the numerator of the first rate is money, the numerator of the second rate also must be money for the comparison to be a proportion.
18. $\frac{15}{45} = \frac{1}{3}$ **19.** $\frac{11}{55} = \frac{22}{110}$ **20.** $\frac{4\text{ d}}{60\text{ m}} = \frac{x\text{ d}}{100\text{ m}}$
21. $\frac{3\text{ hr}}{5\text{ pages}} = \frac{h\text{ hr}}{8\text{ pages}}$ **22.** no **23.** no **24.** yes
25. no **26.** 12 **27.** 28 **28.** $18\frac{2}{3}$ **29.** $2\frac{4}{9}$
30. $119 **31.** $33.33 **32.** about 62.5 mi
33. about 56 mi **34.** about 20 in. **35.** about 28 in.
36. about 14 in. **37.** about 8 in. **38.** about 1,125 students
39. about 692 pieces **40.** about 1,050 fish
41. about 246,500 words
42. a. no **b.**
c.

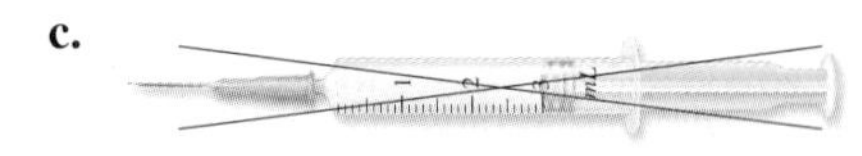

43. a. yes **b.** 6.25 mL
c.

44. a. yes **b.** 1.4 mL
c.

Chapter 4 Practice Test

1. $\frac{2\text{ cups}}{3\text{ loaves}}$ is a rate because the units are different
2. $\frac{3}{2}$ is a ratio because the units are the same
3. $\frac{1.24}{41.27}$ is a ratio because the units are the same
4. $\frac{2\text{ cartridges}}{1\text{ box of paper}}$ is a rate because the units are different
5. $0.16 per oz **6.** 24 mpg **7.** 58 mph

8. \$8.20 per hr **9.** \$2.69 for 15 oz **10.** \$1.99 for 5 lb
11. \$67.50 **12. a.** 3.75 mg per kg **b.** 187.5 mg
13. yes **14.** yes **15.** no **16.** no **17.** 12.5
18. 105 m **19.** 4 gal **20.** 4.5 hr **21.** 280 sq ft
22. $3\frac{1}{2} \times 1\frac{1}{2}$ gridlines **23.** about 211 clients
24. about 3,125 fish **25.** yes
26. 12.5 ml

CHAPTER 5

Section 5.1 Guided Practice

1. 37

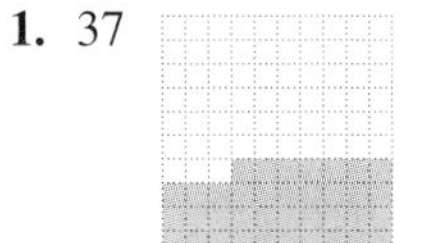

2. 63

3. 171; 71

4. $60\% = \frac{60\%}{1} \cdot \frac{1}{100\%} = \frac{60}{100} = \frac{3 \cdot 20}{5 \cdot 20} = \frac{3}{5}$

5. $16\frac{2}{3}\% = \frac{50\%}{3} = \frac{50\%}{3} \cdot \frac{1}{100\%} = \frac{50}{300} = \frac{1 \cdot 50}{6 \cdot 50} = \frac{1}{6}$

6. $220\% = \frac{220\%}{1} \cdot \frac{1}{100\%} = \frac{220}{100} = \frac{11 \cdot 20}{5 \cdot 20} = \frac{11}{5} \text{ or } 2\frac{1}{5}$

7. $\frac{18}{25} = \frac{18}{25} \cdot \frac{100\%}{1} = \frac{18}{{}^{1}25} \cdot \frac{{}^{4}100\%}{1} = \frac{72\%}{1} = 72\%$

8. $\frac{3}{8} = \frac{3}{8} \cdot \frac{100\%}{1} = \frac{300\%}{8} = 37\frac{1}{2}\%$

9. $1\frac{1}{12} = \frac{13}{12} = \frac{13}{12} \cdot \frac{100\%}{1} = \frac{13}{4 \cdot 3} \cdot \frac{4 \cdot 25\%}{1} = \frac{325}{3}\% = 108\frac{1}{3}\%$

10. The missing multiplier is 20 because $5 \times 20 = 100$.
$\frac{4}{5} \cdot \left(\frac{20}{20}\right) = \frac{80}{100} = 80\%$

11. $2\frac{6}{25} = 2 + \frac{6}{25}$
$2 = 2 \cdot 100\% = 200\%$
$\frac{6}{25} = \frac{6}{25} \cdot \frac{4}{4} = \frac{24}{100} = 24\%$
$2\frac{6}{25} = 200\% + 24\% = 224\%$

12. $62\% = \frac{62\%}{1} \cdot \frac{1}{100\%} = \frac{62}{100} = 0.62$

13. $0.0031 = 0.0031 \times 100\% = 0.31\%$

14. $340\% = \frac{340\%}{1} \cdot \frac{1}{100\%} = \frac{340}{100} = 3.4$

15. $10.3 = 10.3 \times 100\% = 1{,}030\%$

16. $\frac{3}{5} = 0.60 = 0.60 \cdot 100\% = 60\%$

17. $\frac{3}{5} = \frac{3}{5} \cdot \frac{20}{20} = \frac{3 \cdot 5 \cdot 20\%}{5} = 60\%$

18. $0.045 = \frac{45}{1{,}000} = \frac{45}{1{,}000} \cdot \frac{100\%}{1} = \frac{45}{10 \cdot 100} \cdot \frac{100\%}{1} = \frac{45}{10}\% = 4.5\%$

19. $0.045 = 0.045 \times 100\% = 4.5\%$

20. a. red; $\frac{1}{2}; \frac{1}{2} \cdot \frac{100\%}{1} = \frac{100\%}{2} = 50\%; 50\%$

b. $\frac{1}{4}; \frac{1}{3}; \frac{1}{4}; \frac{1}{3}; \frac{1}{4} \cdot \frac{100\%}{1} = \frac{100\%}{4} = 25\%;$
$\frac{1}{3} \cdot \frac{100\%}{1} = \frac{100\%}{3} = 33\frac{1}{3}\%; 30; 30$

c. 100; 100; 50; 30; 50; 30; 20; 20

Concept Check and Objective Practice

A1. f **A2.** h **A3.** e **A4.** b **A5.** i **A6.** g
A7.
A8.
A9.
A10.
A11.
A12.
A13.
A14.

A15. 95% **A16.** 20% **A17.** 120% **A18.** 200%

B1. Answers may vary. Example: A percent indicates the number of parts out of a whole that is divided into one hundred parts.

B2. Answers may vary. Example: Both represent parts of a whole.

B3. $\frac{3}{4}$ **B4.** $\frac{1}{5}$ **B5.** $\frac{89}{100}$ **B6.** $\frac{2}{25}$ **B7.** $\frac{1}{3}$ **B8.** $\frac{3}{8}$

B9. $1\frac{3}{5}$ **B10.** $1\frac{3}{20}$ **C1.** one

C2. a quantity multiplied by a mixed number

C3. a. Each value is increased by 100%. $3\frac{1}{4} = 325\%$; $4\frac{1}{4} = 425\%$
b. The whole-number part of the mixed number becomes a percent that is a multiple of 100%.

C4. $41\frac{2}{3}\%$ **C5.** $42\frac{1}{2}\%$ **C6.** $12\frac{1}{2}\%$ **C7.** $83\frac{1}{3}\%$

C8. 130% **C9.** 260% **C10.** 150% **C11.** 335%

Focus on Converting Fractions

1. 90% **2.** 30% **3.** 20% **4.** 80% **5.** 275%
6. 325% **7.** 205% **8.** 105% **9.** 200% **10.** 500%
11. 650% **12.** 350%
13. May vary. The whole number portion corresponds to hundreds of percent.

D1. a. $\frac{1}{100\%}$ **b.** Answers may vary. Example: This multiplication will divide out the percent symbol.
D2. a. ii **b.** This has the same effect as multiplying by $\frac{1}{100\%}$.
D3. 0.85 **D4.** 0.36 **D5.** 92% **D6.** 12% **D7.** 8.23
D8. 5.55 **D9.** 650% **D10.** 420% **D11.** 0.003
D12. 0.87% **D13.** 0.12% **D14.** 0.005 **D15.** 10
D16. 400,000% **D17.** 100,000% **D18.** 40

Exercises

For 1.–14.: See preceding Concept Check and Objective Practice answers.

15. Percent **17.** Per

19.
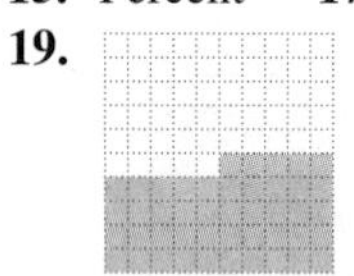

21.
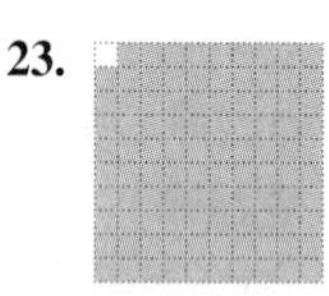

23.

25. 57% **27.** $\frac{51}{100}$ **29.** $\frac{4}{5}$ **31.** 1 **33.** $\frac{1}{3}$ **35.** $\frac{7}{8}$

37. $1\frac{3}{4}$ **39.** 30 **41.** $\frac{1}{5{,}000}$ **43.** $\frac{11}{20}$ **45.** $1\frac{1}{10}$

47. Answers may vary. **49.** 30% **51.** $44\frac{4}{9}\%$

53. $166\frac{2}{3}\%$ **55.** 320% **57.** 85% **59.** 100%

61. 330% **63.** 600% **65.** 25% **67.** $44\frac{4}{9}\%$

69. 0% **71.** 300% **73.** 0.6 **75.** 81% **77.** 2.5
79. 0.76% **81.** 0.0008 **83.** 450% **85.** 4.5
87. 400% **89.** 0.06 **91.** 8.2% **93.** 0.00001
95. a. 0.05; 5% **b.** 5%
97. a. $\frac{34}{1{,}000}$; 3.4% **b.** 3.4%

99. a. 2.2; $\frac{11}{5} = 2\frac{1}{5}$ **b.** $2\frac{1}{5}$

101. decimal: 0.0025; fraction: $\frac{25}{10{,}000} = \frac{1}{400}$

103. percent: 85% decimal: 0.85

105. Multiply by $\frac{1}{100\%}$.

107. Multiply the fraction by 100%.

109. a. 40% **b.** 60% **c.** Answers may vary. Example: All of the parts of the whole must add to 100%. Since there are only “two parts” in this problem’s whole, their sum must equal 100%.

111. a. 25% **b.** 40% **c.** Answers may vary. Example: These represent only two of the possible outcomes, not all of them.

		Decimal	Fraction	Percent
113. Halves	**a.**	0.5	$\frac{1}{2}$	50%
	b.	1	$\frac{2}{2}$	100%
115. Fourths	**a.**	0.25	$\frac{1}{4}$	25%
	b.	0.5	$\frac{2}{4}$	50%
	c.	0.75	$\frac{3}{4}$	75%
	d.	1	$\frac{4}{4}$	100%
117. Fifths	**a.**	0.2	$\frac{1}{5}$	20%
	b.	0.4	$\frac{2}{5}$	40%
	c.	0.6	$\frac{3}{5}$	60%
	d.	0.8	$\frac{4}{5}$	80%
	e.	1	$\frac{5}{5}$	100%
119.	**a.**	$0.\overline{5}$	$\frac{5}{9}$	$55\frac{5}{9}\%$
	b.	1.28	$1\frac{7}{25}$	128%
	c.	2.9	$2\frac{9}{10}$	290%
	d.	0.03125	$\frac{1}{32}$	3.125%

Section 5.2 Guided Practice

1. 10; 20; 2 **2.** percent; 20; 50; $\frac{P}{100}=\frac{50}{20}$

3. 120; 35; amount; $\frac{120}{100}=\frac{A}{35}$

4. 45; B; total number of customers; 10; $\frac{45}{100}=\frac{10}{B}$

5. Henry and Anais are each writing a 40-page term paper. After printing a draft, Anais ran out of paper. Henry gave her 232 pages, which was 58% of his paper. How much paper did Henry have before he gave paper to Anais?

58; B; total number of sheets of paper; 232; $\frac{58}{100}=\frac{232}{B}$

6. percent; $\frac{P}{100}=\frac{16}{40}$

$$\frac{P}{\cancel{100}}\cdot\frac{\cancel{100}}{1}=\frac{16}{40}\cdot\frac{100}{1}$$
$$P=\frac{16\cdot 100}{40}$$
$$P=\frac{4\cdot\cancel{4}\cdot\cancel{10}\cdot 10}{\cancel{4}\cdot\cancel{10}}$$
$$P=40$$

40% of 40 is 16.

7. 16; the number; 64

$$\frac{16}{100}=\frac{64}{B}$$
$$\frac{100}{16}=\frac{B}{64}$$
$$\frac{100}{16}\cdot\frac{64}{1}=\frac{B}{{}^1\cancel{64}}\cdot\frac{{}^1\cancel{64}}{1}$$
$$\frac{100\cdot 4\cdot\cancel{16}}{\cancel{16}}=B$$

64 is 16% of 400.

8. 40; 20; pages to be completed

$$\frac{40}{100}=\frac{A}{20}$$
$$\frac{40}{100}\cdot\frac{20}{1}=\frac{A}{{}^1\cancel{20}}\cdot\frac{{}^1\cancel{20}}{1}$$
$$\frac{8\cdot\cancel{5}\cdot\cancel{20}}{\cancel{5}\cdot\cancel{20}}=A$$
$$8=A$$

Kato must write 8 pages by tomorrow.

9. What percent of the bikes are nonracing bikes?

Nonracing bikes = Total bikes − Racing bikes
= 125 − 5
= 120

P; the percent; 125; A = nonracing bikes = 120

$$\frac{P}{100}=\frac{120}{125}$$
$$\frac{{}^1\cancel{100}}{1}\cdot\frac{P}{{}^1\cancel{100}}=\frac{120}{125}\cdot\frac{100}{1}$$
$$P=\frac{{}^1\cancel{5}\cdot 24\cdot 4\cdot\cancel{25}^1}{{}^1\cancel{5}\cdot\cancel{25}^1}$$
$$P=96$$

96% are nonracing bikes.

10. How much more does she need to save?
= 100% − 60%
= 40%

40; 1,500; amount needed

$$\frac{40}{100}=\frac{A}{1{,}500}$$
$$\frac{1{,}500}{1}\cdot\frac{40}{100}=\frac{A}{\cancel{1{,}500}^1}\cdot\frac{\cancel{1{,}500}^1}{1}$$
$$\frac{15\cdot 100\cdot 40}{100}=A$$
$$15\cdot 40=A$$
$$600=A$$

Samantha still needs to save $600.

Concept Check and Objective Practice

A1. **a.** the horses **b.** black
A2. **a.** percent; % **b.** percent of **c.** is or equal
A3. $\frac{P}{100}=\frac{18}{30}$ **A4.** $\frac{P}{100}=\frac{12}{40}$ **A5.** $\frac{80}{100}=\frac{A}{20}$
A6. $\frac{20}{100}=\frac{A}{50}$ **A7.** $\frac{250}{100}=\frac{90}{B}$ **A8.** $\frac{125}{100}=\frac{50}{B}$
A9. $\frac{4}{100}=\frac{8.80}{B}$ **A10.** $\frac{5}{100}=\frac{3{,}750}{B}$
A11. $\frac{P}{100}=\frac{110}{150}$ **A12.** $\frac{P}{100}=\frac{79}{80}$

Focus on Writing Percent Proportions

1. b **2.** e **3.** a **4.** d **5.** c **6.** f **7.** d
8. c **9.** f **10.** a **11.** e **12.** b
B1. 24; 22; some number
B2. **a.** Answers may vary. Example: A must be more than 30 because 120 is more than 100. **b.** Answers may vary. Example: P must be less than 100% because 20 is less than 25.
B3. 18 **B4.** 44 **B5.** 125% **B6.** 200% **B7.** 300
B8. 2,000 **B9.** 189 **B10.** 80% **B11.** 64
B12. 60%

Exercises

For 1.–8.: See preceding Concept Check and Objective Practice answers.

9. percent proportion **11.** is **13.** $\frac{35}{100}=\frac{A}{70}$
15. $\frac{40}{100}=\frac{60}{B}$ **17.** $\frac{P}{100}=\frac{33}{50}$
19. $\frac{1.2}{100}=\frac{A}{300}$ **21.** $\frac{P}{100}=\frac{2}{450}$
23. $\frac{40}{100}=\frac{A}{300}$ **25.** 12 **27.** 70,000 **29.** 77.78
31. 200% **33.** 97% **35.** 83.33% **37.** 6%
39. 72% **41.** 633% **43.** 278% **45.** 160%
47. about 129% **49.** about 238% **51.** about 67 grams
53. 73% **55.** 70 mg **57.** 0 **59.** $1,950
61. $27 **63.** $11,750 **65.** 76
67. 200% = 2; 50% = $\frac{1}{2}$; Answers may vary. Example: They are reciprocals, and their product is one.

Section 5.3 Guided Practice

1. 25; 200; multiplied; 50; multiplying the percent and the base
2. 16; multiplication; points scored; multiplied; percent; the number of points scored; equal sign; 8; $16\% \cdot B = 8$
3. 40; 200; the number; $40\% \cdot 200 = A$
4. percent; 700; 35; $P \cdot 700 = 35$
5. 15; B; the tiger's weight; 61; $15\% \cdot B = 61$
6. Oprah Winfrey earns more money than almost any other celebrity. In 2005, she earned about 668% of Shaquille O'Neal's earnings. If Shaquille O'Neal earned \$33.4 million in 2005, how much did Oprah earn that year? 668%; Shaquille O'Neal's earnings; 33.4 million; Oprah's earnings; A; Oprah; $668\% \cdot 33.4 = A$
7. 20%; 0.2; 50; what number
   $P \cdot B = A$
   $0.2 \cdot 50 = A$
   $10 = A$
   The number is 10.
8. 6%; 0.06; what number; 4.2
   $P \cdot B = A$
   $0.06 \cdot B = 4.2$
   $\frac{0.06 \cdot B}{0.06} = \frac{4.2}{0.06}$
   $B = 70$
   The number is 70.
9. What percent; 55; 11
   $P \cdot B = A$
   $P \cdot 55 = 11$
   $\frac{P \cdot 55}{55} = \frac{1}{5}$
   $P = \frac{1}{5}$
   $P = 0.2$
   The percent is 20%.
10. 40%; 0.4; 10; what number
   $P \cdot B = A$
   $0.4 \cdot 10 = A$
   $4 = A$
   The number is 4.
11. 26%; 0.26; what number, 104
   $P \cdot B = A$
   $0.26 \cdot B = 104$
   $\frac{0.26 \cdot B}{0.26} = \frac{104}{0.26}$
   $B = 400$
   The number is 400.
12. What number; 52; 13
   $P \cdot B = A$
   $P \cdot 52 = 13$
   $\frac{P \cdot 52}{52} = \frac{13}{52}$
   $P = \frac{13}{4 \cdot 13}$
   $P = 0.25$
   Angel repaid 25% of her debt.
13. 30; 10%; 30; 30 (10%, 25) 25; is not; 10%; 30; is not
14. 120; 100%; 230%; 2; 120

100% | 100% | 30%

Since each 100% is the same as 120, 2 bases is 240. The difference between 240 and 276 is 36, which is about one-third of 120 as 30% is about one-third of 100%. 276 is reasonable.

15. 60; 45%; 60; 60

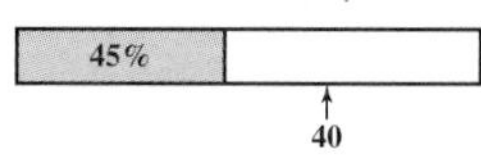

40 is more than half of 60, but 45% is less than half of 100%. The answer is not reasonable.

16. 90; 30%; 90; 90

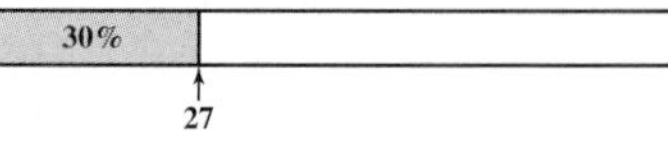

30% is about one-third of 100%, and 27 is about one-third of 90. The answer is reasonable.

17. 30%; 0.3; $9{,}200 - 1{,}400 = 7{,}800$; the down payment
   $0.3 \cdot 7{,}800 = A$
   $2{,}340 = A$
   He needs \$2,340 for the down payment.

Concept Check and Objective Practice

A1. of
A2. Answers may vary. Examples include is, are, was, were, equals, and results.
A3. $P \cdot B = A$ **A4.** $P \cdot 15 = 12$ **A5.** $P \cdot 20 = 7$
A6. $30\% \cdot 70 = A$ **A7.** $70\% \cdot 30 = A$
A8. $35\% \cdot B = 60$ **A9.** $25\% \cdot B = 60$
A10. $6\% \cdot \text{Cost} = \$900$ **A11.** $20\% \cdot 45.50 = \text{Tip}$
A12. $P \cdot 120 = 90$ **A13.** $P \cdot 40 = 37$

Focus on Writing Percent Equations

1. f **2.** a **3.** d **4.** e **5.** b **6.** c **7.** c
8. a **9.** e **10.** d **11.** f **12.** b
B1. **a.** percent, decimal **b.** percent, decimal
c. decimal, percent
B2. division
B3. **a.** more; Answers may vary. Example: Since 110% is more than the whole (100%), the amount must be more than the starting value. **b.** less; Answers may vary. Example: Since 90% is less than the whole (100%), the amount must be less than the starting value.
B4. 3.6 **B5.** 225 **B6.** 50 **B7.** 200 **B8.** 60%
B9. $33\frac{1}{3}\%$ **B10.** \$121.60 **B11.** 3.6 hr

Exercises

For 1.–8.: See preceding Concept Check and Objective Practice answers.

9. percent **11.** is **13.** base **15.** $13\% \cdot 312 = A$
17. $P \cdot 20 = 54$ **19.** $30\% \cdot B = 30$
21. $0.05\% \cdot 10{,}000 = A$ **23.** $P \cdot 10{,}000 = 2$
25. $135\% \cdot B = 297$ **27.** $150\% \cdot 70 = A$
29. $P \cdot 600 = 120$ **31.** $350\% \cdot B = 21$ **33.** 125%
35. 159 **37.** 12.25 **39.** 72 **41.** 175 **43.** 120%
45. 105 **47.** 20% **49.** 6 **51.** reasonable
53. not reasonable **55.** reasonable **57.** not reasonable
59. not reasonable **61.** reasonable **63.** 32
65. 37.5% **67.** 80 **69.** 25% **71.** 17 **73.** 30
75. \$0.83 **77.** about 29% **79.** 80% **81.** yes

Section 5.4 Guided Practice

1. 5%; 0.05; $19,000
   $c = 0.05 \cdot 19{,}000$
   $c = 950$
   The salesperson made $950 in commission.
2. $45,000; 5%; 0.05; amount of total sales
   $45{,}000 = 0.05 \cdot s$
   $\frac{45{,}000}{0.05} = \frac{\cancel{0.05}^1 \cdot s}{\cancel{0.05}^1}$
   $900{,}000 = s$
   Hank's total sales were $900,000.
3. r; $2,500,000
   $12{,}500 = r \cdot 2{,}500{,}000$
   $\frac{12{,}500}{2{,}500{,}000} = \frac{r \cdot \cancel{2{,}500{,}000}^1}{\cancel{2{,}500{,}000}^1}$
   $\frac{12{,}500}{2{,}500{,}000} = r$
   $0.005 = r$
   Gloria earned a 0.5% commission rate.
4. $3,500; 1.5%; 0.015; month; 2 months
   $I = 3{,}500 \cdot 0.015 \cdot 2$
   $I = 52.5 \cdot 2$
   $I = 105$
   Antonio was charged $105.00 interest.
5. $12,000; $420; interest rate; year; 1 year
   $420 = 12{,}000 \cdot r \cdot 1$
   $420 = 12{,}000 \cdot r$
   $\frac{420}{12{,}000} = \frac{{}^1\cancel{12{,}000} \cdot r}{{}^1\cancel{12{,}000}}$
   $0.035 = r$
   $3.5\% = r$
   Samantha earned an interest rate of 3.5%.
6. $9,500; I; interest; 9%; 9%; 0.75%; 0.0075; 1 month
   $I = 9{,}500 \cdot 0.0075 \cdot 1$
   $I = 9{,}500 \cdot 0.0075$
   $I = 71.25$
   Andrea owes $71.25 interest after 1 month.
7. **a.** increase **b.** decrease
8. 30%; 0.30; c; 36 in.
   $\text{Percent change} = \frac{\text{Change in amount of snow}}{\text{Original amount of snow}}$
   $0.30 = \frac{c}{36}$
   $0.30 \times 36 = \frac{c}{\cancel{36}^1} \times \frac{\cancel{36}^1}{1}$
   $10.80 = c$
   The amount of snow increased by 10.8 inches.
9. p = the percent change; decrease; 70,000 balls; 500,000 balls
   $\text{Percent change} = \frac{\text{Change in production}}{\text{Original production}}$
   $p = \frac{70{,}000}{500{,}000}$
   $p = 0.14$
   $p = 14\%$
   Tennis ball production decreased by 14%.
10. p = the percent decrease; $32,000 − $28,000; $4,000; $32,000
    $\text{Percent change} = \frac{\text{Change in salary}}{\text{Original salary}}$
    $p = \frac{4{,}000}{32{,}000}$
    $p = 0.125$
    $p = 12.5\%$
    Daniel's salary decreased by 12.5%.
11. $100; 2% = 0.02; 0.02; 1.02
    $(1.02) \cdot \$100 = \$102$
    $(1.02) \cdot \$102 = \$104.04$
    $(1.02) \cdot \$104.04 = \$106.12.$
    The account balance after three years is $106.12.
12. **a.** $7,500 − $7,000 ≈ $500
    b. $23,000 − $13,000 ≈ $10,000
    c. the interest earned in a compound interest account also earns interest.

Concept Check and Objective Practice

A1. Commission = Sales · Commission rate
A2. Answers may vary. Example: This would mean that she earned more than she sold.
A3. $72 **A4.** $21,000 **A5.** 5% **A6.** 15%
A7. $3,150 **A8.** $25,000
B1. Multiply the monthly rate by 12.
B2. Divide the annual rate by 12. **B3.** 7.5% **B4.** 6%
B5. $24.57 **B6.** $158 **B7.** $4,000 **B8.** $1,620
C1. no **C2.** $735 **C3.** $1,140 **C4.** 23%
C5. 60% **C6.** 12.5% **C7.** 3.75%

Exercises

For 1.–9.: *See preceding Concept Check and Objective Practice answers.*

11. percent increase **13.** commission rate
15. interest rate **17.** Commission **19.** $63,250
21. 2.5% **23.** $275,000 **25.** $300 **27.** $1,290
29. $100,000 **31.** $180 **33.** $32.25 **35.** $1,853
37. 66% **39.** 200% **41.** 11.25% **43.** $1,800
45. 80% **47.** 80% **49.** 20% **51.** 144%
53. $164,000 **55.** 17.27 sec **57.** $6.65 **59.** $2,160
61. $9,720,000 **63.** 300% **65.** 20% **67.** 25%
69. 0%

Chapter 5 Review Exercises

1. % **2.** cent
3. a. 350%; 450% **b.** 100%; 200%; 300%; 900%; 1,000%

4. **5.** **6.**

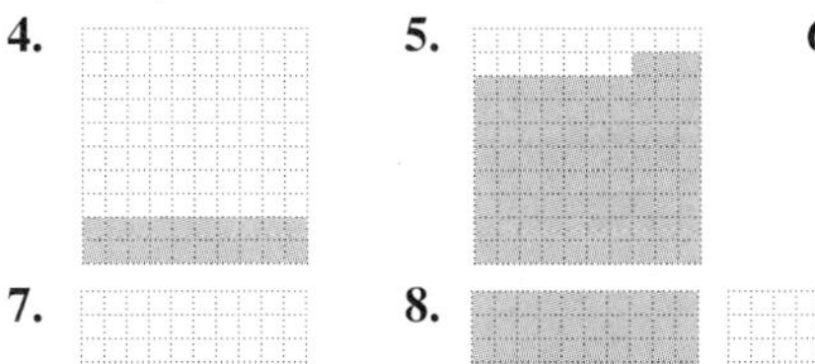

7. **8.**

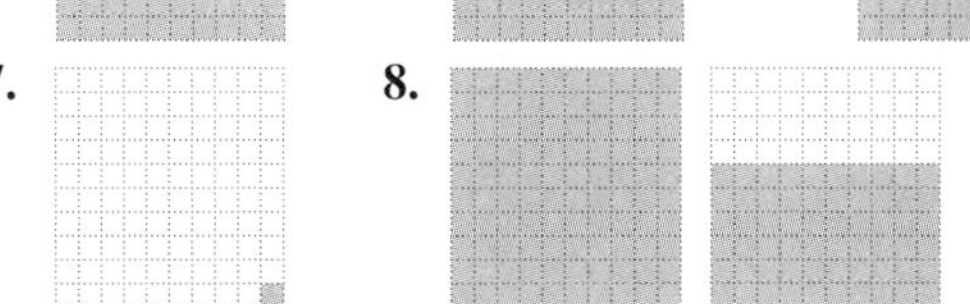

9.

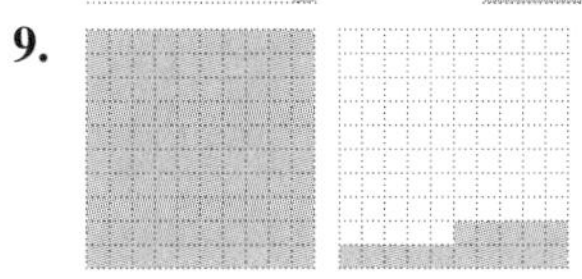

10. $\frac{1}{4}$ **11.** $\frac{2}{5}$ **12.** $\frac{39}{50}$ **13.** $\frac{41}{50}$ **14.** $\frac{201}{400}$ **15.** $\frac{81}{200}$
16. $1\frac{4}{5}$ **17.** $1\frac{1}{5}$ **18.** 70% **19.** 55% **20.** 37.5%
21. 62.5% **22.** 410% **23.** 520% **24.** $211.\overline{1}\%$
25. $383.\overline{3}\%$ **26.** 0.57 **27.** 0.48 **28.** 67%
29. 52% **30.** 2.35 **31.** 360% **32.** 870% **33.** 4.34
34. 0.54% **35.** 0.002 **36.** 0.009 **37.** 0.25%
38. 60 **39.** 700,000% **40.** 300,000% **41.** 30

	Decimal	Fraction	Percent
42.	0.75	$\frac{3}{4}$	75%
43.	$0.\overline{3}$	$\frac{1}{3}$	$33\frac{1}{3}\%$
44.	0.2	$\frac{1}{5}$	20%
45.	1	$\frac{3}{3}$	100%

	Decimal	Fraction	Percent
46.	0.125	$\frac{1}{8}$	12.5%
47.	1.22	$1\frac{11}{50}$	122%
48.	2.6	$2\frac{3}{5}$	260%
49.	0.7	$\frac{7}{10}$	70%

50. $\frac{P}{100} = \frac{A}{B}$ **51.** is **52.** of **53.** 24 **54.** 10
55. 250% **56.** 200% **57.** 72 **58.** 100 **59.** 140%
60. 51 **61.** $26,500 **62.** 60% **63.** 2,500 mg
64. about 23.5% **65.** 6 g **66.** of **67.** is, are
68. $P \cdot B = A$ **69.** percent **70.** base **71.** amount
72. 5.25 **73.** 17.6 **74.** 62.5 **75.** 230 **76.** $433.\overline{3}\%$
77. 700% **78.** 84% **79.** $12 **80.** $21,800
81. Commission **82.** interest rate **83.** principal
84. Interest **85.** $96 **86.** $22,000 **87.** 9%
88. 12% **89.** $260,000 **90.** $14,500 **91.** $25.80
92. $81 **93.** 2.5% **94.** 1.6% **95.** $34 **96.** $22.50
97. $1,350 **98.** $720 **99.** 19% **100.** 11%

Chapter 5 Practice Test

1. Percent **2.** percent **3.** base **4.** amount **5.** is
6. of **7.** commission **8.** principal **9.** decrease
10. **11.**

12. $\frac{7}{25}$ **13.** $1\frac{3}{5}$ **14.** $\frac{161}{400}$ **15.** 90% **16.** 12.5%
17. 340% **18.** 0.72 **19.** 7.8% **20.** 45% **21.** 1.23
22. 30 **23.** 125% **24.** 2,000 **25.** 33% **26.** 38
27. 6.16 **28.** 125 **29.** 325% **30.** 63% **31.** $68
32. $44 **33.** 7% **34.** $13,600 **35.** $695.25
36. 2% **37.** $26 **38.** $520 **39.** 25%

CHAPTER 6

Section 6.1 Guided Practice

1. 12; 1; $\frac{12 \text{ in.}}{1 \text{ ft}} = 1$; $\frac{1 \text{ ft}}{12 \text{ in.}} = 1$

2. 2; 1; $6 \text{ pt} = \frac{6 \text{ pt}}{1}$
$= \frac{6 \cancel{\text{pt}}}{1} \cdot \left(\frac{1 \text{ qt}}{2 \cancel{\text{pt}}}\right)$
$= \frac{6 \text{ qt}}{2}$
$= 3 \text{ qt}$

3. 1; 16; $6 \text{ lb} = \frac{6 \text{ lb}}{1}$
$= \frac{6 \cancel{\text{lb}}}{1} \cdot \left(\frac{16 \text{ oz}}{1 \cancel{\text{lb}}}\right)$
$= \frac{6}{1} \cdot \left(\frac{16 \text{ oz}}{1}\right)$
$= 96 \text{ oz}$

4. 1; 2 $28 \text{ c} = \frac{28 \text{ c}}{1}$
$= \frac{28 \cancel{\text{c}}}{1} \cdot \left(\frac{1 \text{ pt}}{2 \cancel{\text{c}}}\right).$
$= \frac{28}{1} \cdot \left(\frac{1 \text{ pt}}{2}\right)$
$= \frac{\cancel{2} \cdot 14}{\cancel{2}}$
$= 14 \text{ pt}$

5. $\frac{2{,}640 \text{ ft}}{1 \text{ min}} = \frac{2{,}640 \cancel{\text{ft}}}{1 \text{ min}} \cdot \left(\frac{1 \text{ mi}}{5{,}280 \cancel{\text{ft}}}\right)$
$= \frac{2{,}640}{1 \cancel{\text{min}}} \cdot \left(\frac{1 \text{ mi}}{5{,}280}\right) \cdot \left(\frac{60 \cancel{\text{min}}}{1 \text{ hr}}\right)$
$= \frac{\cancel{2{,}640} \cdot \cancel{2} \cdot 30 \text{ mi}}{\cancel{2} \cdot \cancel{2{,}640} \text{ hr}}$
$= \frac{30 \text{ mi}}{1 \text{ hr}}$
$= 30$ miles per hour

6. $1.5 \text{ years} = \frac{1.5 \text{ yr}}{1}$
$= \frac{1.5 \cancel{\text{yr}}}{1} \times \left(\frac{12 \text{ mo}}{1 \cancel{\text{yr}}}\right)$
$= 18 \text{ mos}$
Payment = Loan amount ÷ Months
= 1,500 ÷ 18
= 83.33
The monthly payment is $83.33.

7. $3 \text{ years} = \frac{3 \cancel{\text{yr}}}{1} \cdot \frac{12 \text{ mos}}{1 \cancel{\text{yr}}} = 36 \text{ mos}$
= $70 · 36
= $2,520
= $2,520 − $2,000
= $520
$520 will be paid back in total interest.

Concept Check and Objective Practice

A1. miles **A2.** inches **A3.** minutes
A4. feet **A5.** weeks
A6. a. 3 feet **b.** 60 seconds **c.** 1 week **d.** 5,280 feet **e.** 365 days **f.** 1 hour **g.** 24 hours **h.** 1 foot
A7. a. d **b.** yr **c.** mi **d.** min **e.** in. **f.** wk **g.** min **h.** yd
B1. pounds **B2.** ounces **B3.** gallons
B4. fluid ounces **B5.** tons
B6. a. 2,000 pounds **b.** 2 cups **c.** 16 ounces **d.** 4 quarts **e.** 2 pints **f.** 8 fluid ounces
B7. a. gal **b.** pt **c.** lb **d.** oz **e.** fl oz **f.** c
C1. $\frac{8\text{ fl oz}}{1\text{ c}}; \frac{1\text{ c}}{8\text{ fl oz}}$ **C2.** $\frac{7\text{ d}}{1\text{ wk}}; \frac{1\text{ wk}}{7\text{ d}}$ **C3.** 12 **C4.** 4
C5. $\frac{10\cancel{\text{ ft}}}{1} \cdot \left(\frac{12\text{ in.}}{1\cancel{\text{ ft}}}\right) = 120\text{ in.}$
C6. $\frac{10\cancel{\text{ yd}}}{1} \cdot \left(\frac{3\text{ ft}}{1\cancel{\text{ yd}}}\right) = 30\text{ ft}$ **C7.** $\frac{16\cancel{\text{ c}}}{1} \cdot \left(\frac{1\text{ pt}}{2\cancel{\text{ c}}}\right) = 8\text{ pt}$
C8. $\frac{14\cancel{\text{ pt}}}{1} \cdot \left(\frac{1\text{ qt}}{2\cancel{\text{ pt}}}\right) = 7\text{ qt}$ **C9.** 27 ft **C10.** 77 d
C11. 15 lb **C12.** 6 c **C13.** 14 ft per sec **C14.** 15 mi per hr

Exercises

For 1.–9.: See preceding Concept Check and Objective Practice answers.

11. a. 1 gallon **b.** 1 year **c.** 2,000 pounds **d.** 1 hour **e.** 3 feet **f.** 1 pint
13. a. mi **b.** ft **c.** pt **d.** sec **e.** d **f.** lb **g.** c
15. $\frac{6\cancel{\text{ d}}}{1} \cdot \frac{24\text{ hr}}{1\cancel{\text{ d}}} = 144\text{ hr}$ **17.** $\frac{24\cancel{\text{ qt}}}{1} \cdot \left(\frac{1\text{ gal}}{4\cancel{\text{ qt}}}\right) = 6\text{ gal}$
19. 2 miles **21.** 25 minutes **23.** 56 quarts
25. 3.8 miles **27.** 9 batches **29.** 36,960 feet
31. 0.7 ton **33.** 12 pints **35.** 96 inches
37. 88 ounces **39.** 7.25 feet
41. a. 32,736 feet **b.** 10,912 steps **c.** 5,456 strides
43. ounces **45.** pints **47.** 3,960 ft per min
49. 105 gal per wk **51.** 9 in. per sec
53. 7.5 T per hr **55.** \$323 **57.** 3.1 c **59.** 16,830 acres
61. \$265 **63.** \$131,800

Section 6.2 Guided Practice

1. km hm dam m dm cm mm

meters; kilometers; three; left; three; left; 0.023 km

2. kL hL daL L dL cL mL

kiloliters; milliliters; six; right; six; right; five; 28,200,000 mL

3. $1.3\text{ mL per dose} = \frac{1.3\text{ mL}}{\text{dose}}$
$= \frac{1.3\text{ mL}}{\cancel{\text{dose}}} \times \frac{8{,}000\cancel{\text{ doses}}}{1}$
$= 10{,}400\text{ mL}$
$10{,}400\text{ mL} = 10.4\text{ L}$
$10.4\text{ L} > 10\text{ L}$; is not

Concept Check and Objective Practice

A1. meter **A2.** liter **A3.** gram
A4. King Henry died unceremoniously drinking chocolate milk.
A5. a. hundredths **b.** tenths **c.** hundreds **d.** tens **e.** thousandths **f.** thousands
A6. grams **A7.** kilograms **A8.** centimeters
A9. meters **A10.** kilometers **A11.** millimeters
A12. liters **A13.** milliliters **A14.** Answers will vary.
A15. Answers will vary. **A16.** Answers will vary.
A17. Answers will vary. **A18.** Answers will vary.
A19. Answers will vary.
B1. a. 1,000 **b.** $\frac{5.25\cancel{\text{ km}}}{1} \cdot \frac{1{,}000\text{ m}}{1\cancel{\text{ km}}} = 5{,}250\text{ km}$
c. three; right **d.** three; right; 5,250 m **e.** yes
B2. a. 1,000 **b.** $\frac{3{,}253\cancel{\text{ mg}}}{1} \cdot \frac{1\text{ g}}{1{,}000\cancel{\text{ mg}}} = 3.253\text{ g}$
c. three; left **d.** three; left; 3.253 g **e.** yes
B3. 244,000 m **B4.** 0.11 kL **B5.** 3.29 dL
B6. 564,000 dg **B7.** 126,400 dg **B8.** 1,327,000,000 mm

Exercises

For 1.–6.: See preceding Concept Check and Objective Practice answers.

7. Metric prefixes **9.** liter
11. a. hundreds **b.** thousands **c.** hundredths **d.** tens **e.** tenths **f.** thousandths
13. kilo, hecto, deka, deci, centi, milli
15. b; the other values are too large for the volume of a soda can.
17. a; the other values are too small or too large for the height of a door.
19. a; kilograms is the only unit listed that measures weight.
21. a; kilograms is the only unit listed that measures weight.
23. 116,000 L **25.** 0.324 dag **27.** 1,327,000 mm
29. 0.0000068 kg **31.** 244,000 m **33.** 220,000 cL
35. 0.005 km **37.** 1,230 dag **39.** 0.004092 kg
41. 45.6678 hm **43.** 0.67 hL **45.** 96,500 mm
47. 0.09 hg **49.** 0.000008 dL **51.** no
53. \$176.54 **55.** 12,500 rows
57. a. 4,300 mL **b.** 4.3 L
59. a. 8,560 g **b.** 8.56 kg
61. 450 L **63.** yes

Section 6.3 Guided Practice

1. $\frac{4\text{ lb}}{1} \approx \frac{4\cancel{\text{ lb}}}{1} \cdot \left(\frac{0.454\text{ kg}}{1\cancel{\text{ lb}}}\right) = \frac{4}{1} \cdot \left(\frac{0.454\text{ kg}}{1}\right) = 1.816\text{ kg}$

2. $\frac{200\text{ g}}{1} = \frac{200\cancel{\text{ g}}}{1} \cdot \left(\frac{0.0353\text{ oz}}{1\cancel{\text{ g}}}\right)$
$= \frac{200}{1} \cdot \left(\frac{0.0353\text{ oz}}{1}\right) = 7.06\text{ oz}$

3. $\frac{10\text{ in.}}{1} = \frac{10\cancel{\text{ in.}}}{1} \cdot \left(\frac{2.54\text{ cm}}{1\cancel{\text{ in.}}}\right)$
$= \frac{10}{1} \cdot \left(\frac{2.54\text{ cm}}{1}\right) = 25.4\text{ cm}$

4. $\frac{120\text{ km}}{\text{hr}} = \frac{120\cancel{\text{ km}}}{1} \cdot \left(\frac{0.62\text{ mi}}{1\cancel{\text{ km}}}\right) = \frac{120}{1} \cdot \left(\frac{0.62\text{ mi}}{1}\right)$
$= \frac{74.4\text{ mi}}{\text{hr}}$
$= 74.4$ miles per hour

5. $\frac{9}{5}\cdot 30 + 32 = \frac{9}{5}\cdot\frac{30}{1} + 32$
$= \frac{9\cdot 6\cdot \cancel{5}}{\cancel{5}\cdot 1} + 32 = 54 + 32 = 86$
$30°C = 86°F$

6. $\frac{5\cdot(86) - 160}{9} = \frac{430 - 160}{9} = \frac{270}{9} = 30$
$86°F = 30°C$

7. 1; 0.454; kilogram; pound; kilograms; pounds

8. 1 in. ≈ 2.54 cm
$12 \text{ in.} = \frac{12 \cancel{\text{in.}}}{1}\cdot\frac{2.54 \text{ cm}}{1 \cancel{\text{in.}}} = 30.48 \text{ cm}$
12 in. $\boxed{<}$ 31 cm

Concept Check and Objective Practice

A1. 1 mile **A2.** 1 liter **A3.** 1 kilogram
A4. $\frac{1.61 \text{ km}}{1 \text{ mi}}$ **A5.** $\frac{3.28 \text{ ft}}{1 \text{ m}}$ **A6.** 91.4 m **A7.** 2.64 gal
A8. 141.75 g **A9.** 17.78 cm **A10.** 9.84 ft per sec
A11. 1.22 m per sec **B1.** 0°C **B2.** $F = \frac{9}{5}C + 32$
B3. 59°F **B4.** 104°F **B5.** 35°C **B6.** 60°C

Exercises

For 1.–6.: See preceding Concept Check and Objective Practice answers.

7. Fahrenheit **9.** 5, 2, 1, 8, 6, 3, 7, 4
11. 3, 2, 4, 5, 6, 7, 1 **13.** quart **15.** kilogram
17. 3.9 in. **19.** 20.3 cm **21.** 41.7 L **23.** 6.6 kg
25. 136.9 km **27.** 21.8 yd **29.** 8.5 g **31.** 70.4 lb
33. 3.1 mi **35.** 3.5 oz **37. a.** 24 qt **b.** 22.7 L
39. a. 32.1 cm **b.** 12.6 in. **41.** no **43.** yes
45. 43.5 L **47.** Costello **49.** 5.5 gal **51.** 4.8 qt
53. 112.7 km/hr **55.** 52.7 mi/hr **57.** no **59.** yes
61. 100°C **63.** $F = \frac{9}{5}C + 32$ **65.** 221°F **67.** 5°C
69. 35°C **71.** 185°F **73.** no **75.** no
77. a. 75.8 L **b.** $52.30

Chapter 6 Review Exercises

1. miles **2.** inches **3.** hours **4.** years
5. tons **6.** pounds **7.** fluid ounces **8.** ounces
9. yards **10.** tons **11.** inches **12.** fluid ounces
13. 48 in. **14.** 7,920 ft **15.** 504 hr **16.** 9,000 sec
17. 156 ft **18.** 106.7 yd **19.** 5 pt **20.** 48 oz
21. 64 pt **22.** 9,064 lb **23.** 1.5 in. per hr
24. $33 per d **25.** 5,280 ft per min **26.** 250 lb per hr
27. 12 T **28.** 11.3 gal per d **29.** kilograms
30. grams **31.** kilometers **32.** millimeters
33. liters **34.** milliliters **35.** liters **36.** meters
37. 1.234 km **38.** 52.3 cm **39.** 3,890 mL
40. 0.35 g **41.** 34.9 cL **42.** 0.75 kg **43.** 1,200 mm
44. 0.000002 kg **45.** 3 L per hr **46.** 19.2 km per hr
47. 0.6 mL per sec **48.** 83.3 m per sec
49. 1 kilogram, 1 pound, the weight of an apple, 1 ounce, 1 gram
50. 1 kiloliter, 1 gallon, the volume of a mug of coffee, 1 fluid ounce, 1 centiliter
51. the volume of water in a bathtub, 1 gallon, the volume of a 2-liter soda bottle, 1 liter, 1 quart, 1 pint, 1 cup, 1 fluid ounce, 1 milliliter
52. 1 mile, 1 kilometer, the width of a classroom, 1 meter, 1 foot, the width of this page, 1 inch
53. 30.5 cm **54.** 16.4 ft **55.** 1,100 lb **56.** 1,360.8 g
57. 1.7°C **58.** 95°F **59.** 1.9 L **60.** 1.8 gal
61. a. 0.61 m **b.** 61 cm **62. a.** 0.5 mi **b.** 0.8 km
63. a. 0.4 L **b.** 0.1 gal **64. a.** 5.3 qt **b.** 21.2 c
65. Canadian gas by $0.01 per gal **66.** 4.7 km per hr
67. 2 kg for $1.49 **68.** 100 yd in 13 sec
69. United States **70.** 1 L for $1.17 **71.** Carpe Carp
72. 15.5 mi per hr

Chapter 6 Practice Test

1. a. pound **b.** kilogram
2. a. fluid ounce **b.** milliliter
3. a. foot **b.** meter **4. a.** ounce **b.** gram
5. 24 ft **6.** 0.032 m **7.** 1,400 L **8.** 1.6 c
9. 29.4°C **10.** 239°F **11.** 0.43 g **12.** 40 oz
13. 10.2 mi per hr **14.** 8.6 m per sec
15. 90 L per hr **16.** 1.5 gal per d **17.** 436 yd
18. 161 km **19.** 3.1 m **20.** 18.6 mi
21. a. 7.62 cm **b.** 76.2 mm **22. a.** 0.25 kg **b.** 0.55 lb
23. a. 0.25 gal **b.** 0.9475 L
24. a. 914 m **b.** 0.914 km **25.** 8 ft per min
26. 3,000 times **27.** $1.25 per kg **28.** $4.23 per gal
29. no **30.** yes **31.** 19,976 kg **32.** 3,865.8 L per min

CHAPTER 7

Section 7.1 Guided Practice

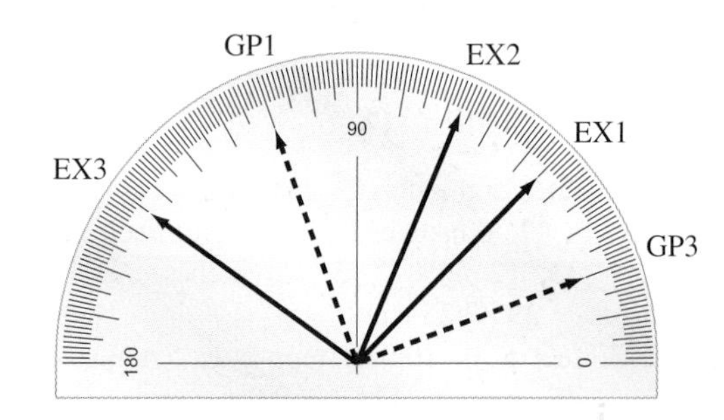

1. 11; zero **2.** 110°; 145°; 145° − 110° = 35°
3. left; right; 110° − 90° = 20° **4.** FGH; HGF; acute
5. 90°; right
6. $m\angle HGK = m\angle HGJ + m\angle JGK$
$= 90° + 40°$
$= 130°$
obtuse
7. $m\angle FGJ = m\angle FGH + m\angle HGJ$
$= 50° + 90°$
$= 140°$
obtuse
8. $m\angle FGK = m\angle FGH + m\angle HGJ + m\angle JGK$
$= 50° + 90° + 40°$
$= 180°$
straight
9. supplementary; 180°; 180° − 141°; 39°
10. complementary; 21°; 90°; 90° − 21°; 69°
11. complements; supplements; 123°; 88°; $m\angle a$
$m\angle a + 88° = 123°$
$m\angle a = 123° - 88°$
$m\angle a = 35°$

Concept Check and Objective Practice

A1. A ray has one endpoint and extends forever in one direction. A line has no endpoints and extends forever in both directions.
A2. vertex
A3. protractor

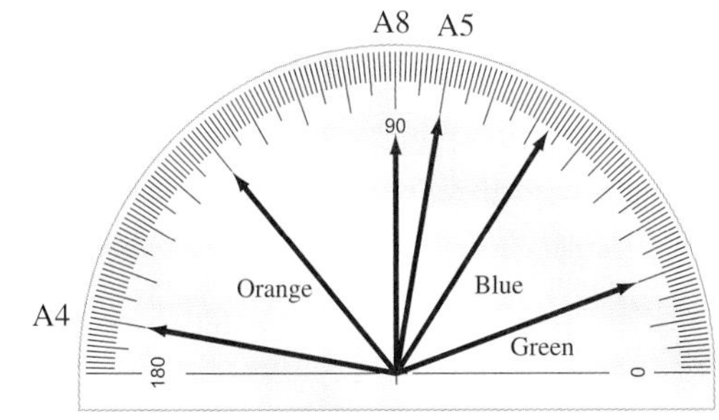

A6. 73° **A7.** 150° **A9.** 80°
B1. Answers may vary. There are three angles at C. It would be hard to discern which angle was being referred to.
B2. acute **B3.** obtuse **B4.** a square **B5.** 65°
B6. 20° **B7.** 25° **B8.** 70° **B9.** 110° **B10.** 90°
B11. $\angle JON$; $\angle NOL$; $\angle NOM$; $\angle LOM$; $\angle MOK$
B12. $\angle JOM$; $\angle NOK$ **B13.** $\angle JOK$
B14. $\angle JOL$; $\angle LOK$ **C1.** 180° **C2.** 90° **C3.** 145°
C4. 55° **C5.** 45° **C6.** 90° **C7.** 18° **C8.** 67°
C9. 18° **C10.** 85° **C11.** 23° **C12.** 49°

Exercises

For 1.–9.: *See preceding Concept Check and Objective Practice answers.*

11. rays; vertex **13.** right angle
15. Acute angles; obtuse angles
17. complementary angles **19.** See B in figure.
21. See D in figure.

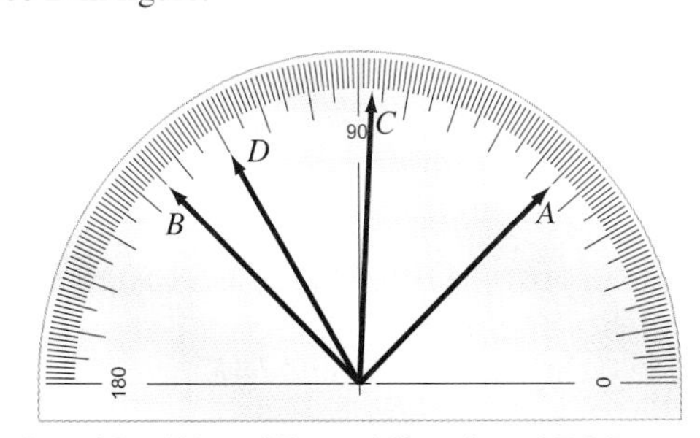

23. 75° **25.** 48° **27. a.** 15° **b.** $\angle BOD$
29. $\angle AOD$; $\angle COE$ **31.** $\angle AOB$; $\angle BOC$
33. $\angle AOE$ **35.** 43° **37.** 67°
39. a. 40° **b.** 90° **c.** 140°
41. a. 70° **b.** 70° **c.** 160° **d.** $\angle AOC$; $\angle BOD$; $\angle COE$
43. a. 70° **b.** 110° **c.** 40° **d.** $\angle AOB$'s complement
45. 22.5°
47. sometimes true; true example: $m\angle a = 45°$; false example: $m\angle a = 30°$
49. sometimes true; true example: 130° and 50°; false example: 90° and 90°.
51. always true; Answers may vary. Example: Pick any angle, say 100°. The supplement of 100° is 80°. To find the supplement of 80°, subtract it from 180°. $180° - 80° = 100°$ or the original angle. This is true for any angle chosen.

Section 7.2 Guided Practice

1. six; different; hexagon
2. five; the same; the same; regular pentagon
3. eight; the same; different; octagon
4. $P = 9 + 4 + 6 + 10 + 5$
$= 13 + 6 + 10 + 5$
$= 19 + 10 + 5$
$= 29 + 5$
$= 34$ m
5. d; f; sides **6.** a; c; d; f; opposite; parallel
7. a; f; angles; 90° **8.** a; c; no **9.** d; all; all
10. a; c; a; 90°; a **11.** greater than; c; c
12. None; is not; is not; is not; acute; all angles measure less than 90°; an acute scalene triangle

Concept Check and Objective Practice

A1. Answers may vary. Example: A polygon is regular when all the angles have the same measure and all the sides are the same length.
A2. a. 4 **b.** pentagon **c.** 8 **d.** hexagon
A3. triangle **A4.** quadrilateral **A5.** regular hexagon
A6. regular octagon **A7.** hexagon **A8.** pentagon
A9. regular pentagon **A10.** regular triangle
A11. 15 in. **A12.** 7 cm
B1. yes; Answers may vary. Example: In a parallelogram, opposite sides are equal and parallel. This is also true of a rectangle.
B2. no; Answers may vary. Example: In a rectangle, the angles must measure 90°. This is not true of a parallelogram.
B3. yes; A square is both a rectangle and a rhombus.
B4. Answers may vary. A rectangle is a polygon with four sides, which is the definition of a quadrilateral.
B5. rectangle **B6.** trapezoid **B7.** square
B8. parallelogram **B9.** rhombus **B10.** square
B11. parallelogram **B12.** quadrilateral
B13. quadrilateral **B14.** rhombus
B15. rectangle **B16.** trapezoid
C1. Answers may vary. Example: In an equilateral triangle, all three sides are equal in length and all three angles have the same measure. In an isosceles triangle, this is true of only two sides and two angles.
C2. Answers may vary. Example: A right scalene triangle will have one angle measuring 90° and two unequal acute angles. None of the sides will be equal.
C3. Every angle of an equilateral triangle must be 60°.
C4. acute isosceles triangle **C5.** equilateral triangle
C6. right isosceles triangle **C7.** acute isosceles triangle
C8. obtuse scalene triangle **C9.** right isosceles triangle
C10. equilateral triangle **C11.** right scalene triangle

Exercises

For 1.–9.: *See preceding Concept Check and Objective Practice answers.*

11. regular polygon **13.** hexagon **15.** false
17. octagon **19.** pentagon **21.** regular hexagon
23. quadrilateral **25.** obtuse scalene triangle
27. acute isosceles triangle **29.** regular hexagon
31. trapezoid **33.** right scalene triangle
35. not a polygon **37.** 27 and 36 **39.** rhombus

41. isosceles triangle **43.** parallelogram **45.** hexagon

47. ; equilateral triangles

49.

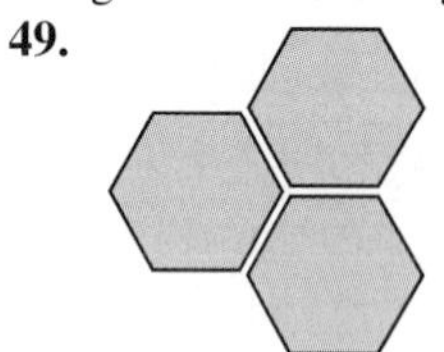

51.

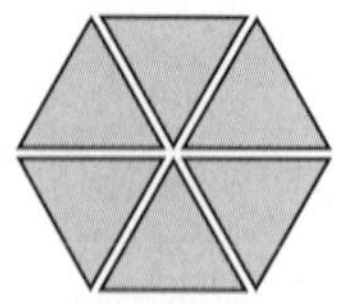

53. triangle; Answers may vary. The giraffe's head is a three sided figure.

55. square; Answers may vary. The mallet's end is a four-sided polygon that appears to have sides that are the same length with right corners.

57. square; Answers may vary. The wheels are four-sided polygons whose angles appear to be right angles and whose sides appear to be equal in length.

59. regular octagon; Answers may vary. The plate has eight sides and each side and angle appears to be the same.

61. 8.8 ft **63.** 20 in.

Section 7.3 Guided Practice

1. $P = 18 + 8 + 12 + 20 + 10$
$= 26 + 12 + 20 + 10$
$= 38 + 20 + 10$
$= 58 + 10$
$= 68$ m

2. $4 + 6 = 10$ ft; $8 - 5 = 3$ ft
$P = 6 + 5 + 4 + 3 + 10 + 8$
$= 36$ ft

3. 12; 12; meters; $4 \cdot 3$; 12; meters

4. 5 ft; 12 ft; 12; 5; 60 square feet

5. 1st Triangle:
22 in.; 10 in.; 22 in.; 10 in.; 11; 10; 110 square inches
2nd Triangle:
18 in.; 10 in.; 18 in.; 10 in.; 9; 10; 90 square inches
Area $= 110 + 90$
$= 200$ square inches

6. $A = \frac{1}{2} \cdot b \cdot h$
$= \frac{1}{2} \cdot 5 \cdot 2$
$= \frac{5}{2} \cdot 2$
$= 5$
5 m^2

7. $A = b \cdot h$
$= 2 \cdot 3$
$= 6$
6 $in.^2$

8. $A = \frac{1}{2} \cdot (B + b) \cdot h$
$= \frac{1}{2} \cdot (20 + 14) \cdot 7$
$= \frac{1}{2}(34) \cdot 7$
$= 17 \cdot 7$
$= 119$
119 ft^2

Concept Check and Objective Practice

A1. Answers may vary. Example: the distance around the outside of the object

A2. Answers may vary. Example: inches, feet, centimeters

A3. 15 ft **A4.** 22 ft **A5.** 20 ft **A6.** 32 ft

A7. 88 ft **A8.** 80 ft **A9.** 88 ft **A10.** 60 m

A11. 15 ft **A12.** 7 ft

B1. Answers may vary. Example: To find area, the shape needs to be "covered." Linear measures are only one-dimensional. To cover a region, a two-dimensional measure is needed. Square units are two-dimensional.

B2. 15 sq cm

B3. Answers may vary. Example: Perimeter is the distance around and is measured with a one-dimensional measure (i.e., inches). Area is the amount needed to "cover" the region and is measured with a two-dimensional measure (i.e., square inches).

B4. Answers may vary. Example: By drawing a diagonal, the rectangle has been "cut in half," forming two triangles. Thus, the area of a triangle is half that of the corresponding rectangle.

B5. 8 ft^2 **B6.** 9 $in.^2$ **B7.** 24 m^2 **B8.** 12 cm^2

B9. **B10.**

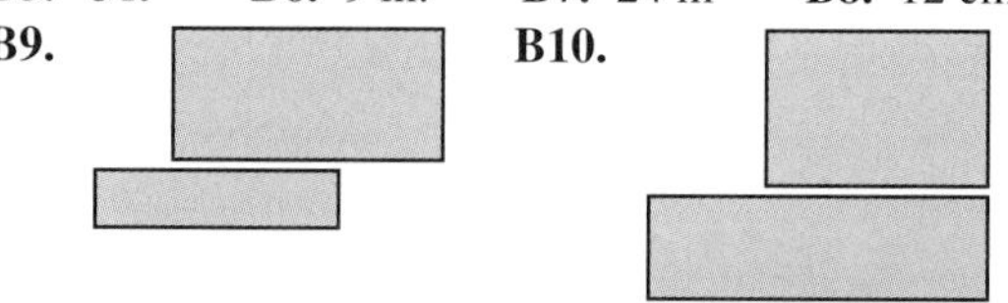

Answers may vary.

B11.

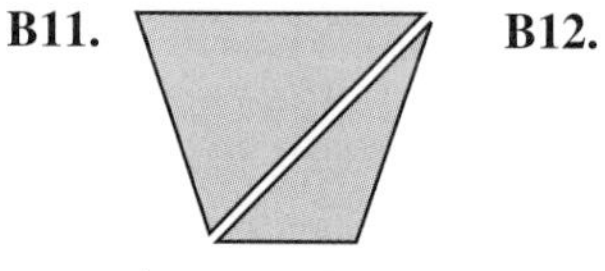

Answers may vary.

B12.

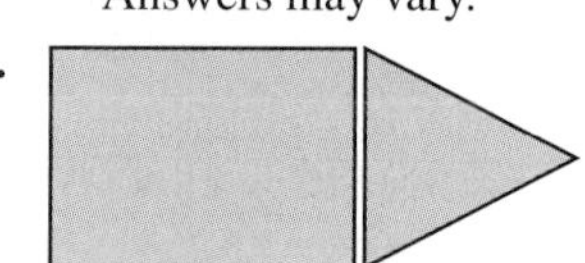

B13. **B14.**

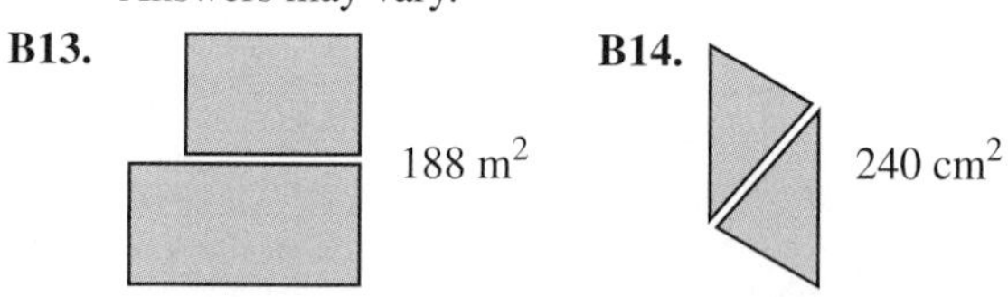

188 m^2 240 cm^2

Answers may vary. Answers may vary.

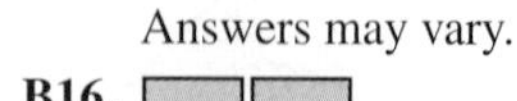

B15.

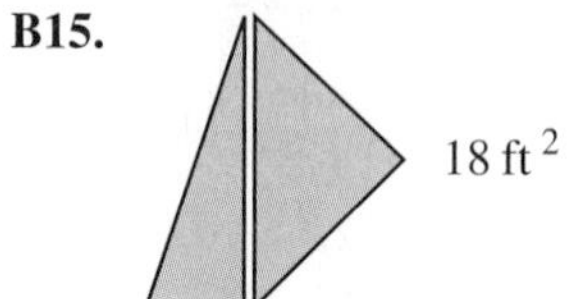

18 ft^2

B16.

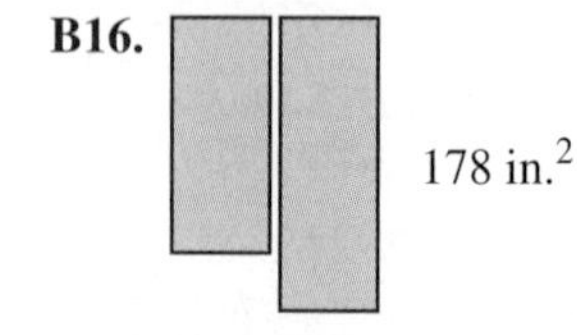

178 $in.^2$

Answers may vary.

C1. **a.** $A = \frac{1}{2}bh$ **b.** $A = lw$ **c.** $A = bh$
d. $A = \frac{1}{2}(B + b)h$ **C2.** square **C3.** 288 $in.^2$

C4. 600 cm^2 **C5.** 200 m^2 **C6.** 60 yd^2 **C7.** 63 ft^2

C8. 42 m^2 **C9.** 21 $in.^2$ **C10.** 36 yd^2

Exercises

***For 1.–9.:** See preceding Concept Check and Objective Practice answers.*

11. area **13.** $lengths^2$ **15.** 32 in. **17.** 32 km

19. 18 ft **21.** 40 m **23.** 104 $in.^2$ **25.** 42 ft^2

27. 240 ft^2 **29.** 450 ft^2 **31.** 16 ft^2 **33.** 58 cm^2

35. 30 cm^2 **37.** 81 in.^{2} **39.** 45 m^2 **41.** 112 cm^2
43. 12 in.^{2} **45.** $\frac{3}{16}$ mi^2 **47. a.** 120 ft^2 **b.** $240
49. a. 108 ft **b.** $243 **51.** 51.36 ft^2 **53.** 30 m^2
55. a. 5.2 in.^{2}
b. Answers may vary. Example: the estimate with the red triangle because the length of the base is shorter and will create less "waste" in the circle.

Section 7.4 Guided Practice

1. 5 ft; 5; 10; 31.4 ft
2. 40 mi; 40; 3.14 · 40 ; 125.6 mi
3. 12 mm; 12 ÷ 2 = 6 mm; 6; 36; 113.04 square mm
4. height; 20 ft; 10 ft

Rectangle:
$A = l \cdot w$
$= 70 \cdot 20$
$= 1{,}400 \text{ ft}^2$

Circle:
$A = \pi \cdot r^2$
$= \pi \cdot (10)^2$
$= 3.14 \cdot 100$
$= 314 \text{ ft}^2$

Area = [rectangle] − [circle]
$= 1{,}400 - 314$
$= 1{,}086 \text{ ft}^2$

5. Area = [triangle] + [half-circle]

Triangle
$\text{Area} = \frac{1}{2}bh$
$= 0.5(4)(8)$
$= 16 \text{ ft}^2$

Half-circle
$\text{Area} = \frac{1}{2}\pi r^2$
$\approx 0.5(3.14)(2)^2$
$\approx 0.5(3.14)(4)$
$\approx 6.28 \text{ ft}^2$

$\text{Area} \approx 6.28 + 16$
$\approx 22.28 \text{ ft}^2$

6. Perimeter =

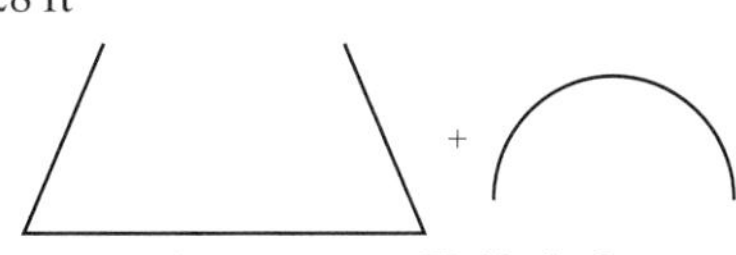

3 sides of the trapezoid
$P = 5 + 8 + 5$
$= 18 \text{ ft}$

Half-circle
$P = \frac{1}{2}\pi d$
$\approx \frac{1}{2}(3.14) \cdot 6$
$\approx 9.42 \text{ ft}$

$\text{Perimeter} \approx 18 + 9.42$
$\approx 27.42 \text{ ft}$

7. Area = [trapezoid] − [half-circle]

$\text{Area} = A_{\text{trapezoid}} - A_{\text{half-circle}}$
$= \frac{1}{2} \cdot (10 + 6) \cdot 7 - \frac{1}{2}\pi \cdot (3)^2$
$= \frac{1}{2} \cdot 16 \cdot 7 - \frac{1}{2} \cdot \pi \cdot 9$
$\approx 56 - 14.13$
$\approx 41.87 \text{ ft}^2$

8. Answers may vary. Example: The perimeter of both figures includes the lengths of the same three sides of the trapezoid and half the circumference of the circle.

Concept Check and Objective Practice

A1. 8 mi **A2.** radius **A3.** circumference
A4. 75.36 cm **A5.** 9.42 m **A6.** 25.12 ft
A7. 56.52 yd **B1.** [triangle] − [circle] **B2.** square
B3. 1,017.36 in.^{2} **B4.** 12.56 mm^2 **B5.** 113.04 yd^2
B6. 28.26 m^2 **B7.** [square with circle] 21.5 m^2
B8. [parallelogram with circle] 23.44 in.^{2} **B9.** [trapezoid with circle] 10.935 mm^2
B10. [rectangle with circle] 7.935 in.^{2} **C1.** [rectangle] − [half-circle]
C2. [open rectangle] + [open triangle] **C3.** [rectangle] + [half-circle]
C4. 33 in.^{2} **C5.** 10.57 m^2 **C6.** 8.56 ft^2
C7. 30.96 cm^2 **C8.** 24 ft **C9.** 14.85 in.
C10. 13.74 m **C11.** 24 in.

Exercises

For 1.–9.: See preceding Concept Check and Objective Practice answers.

11. Pi **13.** radius **15.** $C = 2\pi r$; $C = \pi d$
17. 14 m **19.** 3.5 cm **21.** 50.24 m **23.** 62.8 cm
25. 78.5 cm^2 **27.** 50.24 km^2 **29.** 24.535 ft^2
31. 25.935 cm^2 **33.** 38.13 in. [half-circle] + [open rectangle]
35. 38.13 in. [half-circle] + [open rectangle] **37.** 37.99 m [trapezoid] + [half-circle]
39. 12.56 cm [half-circle] + [half-circle] + [half-circle] **41.** 136.7925 in.^{2} [half-circle] + [square]
43. 73.2075 in.^{2} [square] − [half-circle]
45. a. 31.42 ft **b.** 186.39 ft^2
47. 22.26 m^2 **49.** 0.3925 m^2
51. a. 1.57 ft **b.** 10 revolutions **c.** too fast
53. a. circle **b.** $A = \pi r^2$
c. Answers may vary. Example: Use the formula for the area of an ellipse with a and b equal to the same value; for instance, 5. Then the area is $\pi \cdot 5 \cdot 5$, or $\pi \cdot 25$. Using the formula for the area of a circle with the radius equal to 5, we get $\pi \cdot 5^2$, or $\pi \cdot 25$, which is the same as the area of the ellipse with $a = b$.

Section 7.5 Guided Practice

1. 8 m + 7 m + 9 m = 24 m; 1-dimensional; line segments; length; meters
2. 1 in.^{2} + 1 in.^{2} + 1 in.^{2} + 1 in.^{2} + 1 in.^{2} + 1 in.^{2} = 6 in.^{2}
   2-dimensional; squares; square length; square inches; in.^{2}
3. 1 m^3 + 1 m^3 + 1 m^3 = 3 m^3; 3-dimensional; cubes; cubic length; cubic meter; m^3
4. 15; 15 in.^{3}; 15 cubic
5. 4; 3; 5; length; width; height

Volume = Length · Width · Height
$= 5 \cdot 3 \cdot 4$
$= 15 \cdot 4$
$= 60$

60 m^3, or 60 cubic meters

6. $V = \frac{1}{3} \cdot l \cdot w \cdot h$
$= \frac{1}{3} \cdot 4 \cdot 4 \cdot 12$
$= \frac{4}{3} \cdot 4 \cdot 12$
$= \frac{16}{3} \cdot 12$
$= 64$
64 ft^3

7. 3; $V = \frac{4}{3} \cdot \pi \cdot r^3$
$\approx \frac{4}{3} \cdot 3.14 \cdot 3^3$
$\approx \frac{4}{3} \cdot 3.14 \cdot 27$
$\approx 36 \cdot 3.14$
≈ 113.04
113.04 milliliters

Concept Check and Objective Practice

A1. lengths; square lengths; cubic lengths
A2. Answers may vary. Example: Volume is measured using a three-dimensional unit or cube.
A3. 20 ft^3 **A4.** 15 cm^3
A5. 3 m; 4 m; 2 m; 24 m^3
A6. 6 yd; 4 yd; 4 yd; 96 yd^3
A7. **a.** true; Answers may vary. There are no "indentations" in the top layer to indicate that any layer has fewer cubes than the top layer.
b. true; Answers may vary. When the number of cubes in the layer is counted, the answer is the same as multiplying the number of cubes in the length by the number of cubes in the width.
c. true; Answers may vary. When the number of cubes in the entire object is counted, the answer is the same as multiplying the number of layers by the number of cubes in each layer.
d. Answers may vary. The number of cubes in a layer is the same as multiplying the number of cubes in the width of the layer by the number of cubes in the length. Multiplying the number of cubes in the layer by the number of layers is the same as multiplying the layer by the height of the object.
B1. 3.14 **B2.** Divide the diameter by 2.
B3. Answers may vary. Example: She can easily see what values must be substituted into the formula.
B4. 18.84 ft^3 **B5.** 40 ft^3 **B6.** 378 in.^3
B7. 15,700 ft^3 **B8.** 1.57 m^3 **B9.** 15.7 m^3
B10. 17.27 m^3

Exercises

For 1.–6.: See preceding Concept Check and Objective Practice answers.

7. cone **9.** pyramid **11.** cylinder **13.** 15 ft^3
15. 12 cm^3 **17.** 19 mm^3 **19.** 30 ft^3 **21.** 565.2 in.^3
23. 904.32 in.^3 **25.** about 4,186,666,667 mi^3
27. about 226.08 ft^3 **29.** 84,375,000 ft^3
31. 14,175,000 ft^3
33. **a.** rectangular solid **b.** $V \approx 72 \text{ in.}^3$
35. **a.** 12.56 in.^3 **b.** about 4.19 in.^3 **c.** 3
d. Answers may vary. Example: The volume of a cone is one-third that of a cylinder.
e. Answers may vary. Example: The volume for a cone is the formula for a cylinder multiplied by one-third.

Section 7.6 Guided Practice

1. 7; 7; 7 **2.** 8; 8; 8 **3.** 3; 3; 2; 2; $\frac{3}{2}$; $\frac{3}{2} \cdot \frac{3}{2} = \frac{9}{4}$
4. evaluate the square roots; subtract
$\sqrt{100} - \sqrt{64} = 10 - 8$
$= 2$
5. 25; 6; 5; 6 **6.** 3.317
7. exponent; multiplication; division; addition; subtraction; $\sqrt{6} \cdot \sqrt{4} \approx 2.449 \cdot 2 \approx 4.898$
8. **a.** right angle; x; 1 m; 4 m
b. hypotenuse;
hypotenuse $= \sqrt{(\text{leg})^2 + (\text{leg})^2}$
$x = \sqrt{(1)^2 + (4)^2}$
9. right; 5 m; 6 m
$= \sqrt{(6)^2 - (5)^2}$
$= \sqrt{36 - 25}$
$= \sqrt{11}$
≈ 3.317 m
10. right triangle; Pythagorean theorem; 7 in.; 11 in.
$= \sqrt{(7)^2 + (11)^2}$
$= \sqrt{49 + 121}$
$= \sqrt{170}$
≈ 13.038 in.
11. 27 yd; unknown; 30 yd; leg
leg $= \sqrt{(\text{hypotenuse})^2 - (\text{known leg})^2}$
$= \sqrt{(30)^2 - (27)^2}$
$= \sqrt{900 - 729}$
$= \sqrt{171}$
≈ 13.1 yd
12. 5; unknown; 13
$(5)^2 + b^2 = (13)^2$
$25 + b^2 = 169$
$25 - 25 + b^2 = 169 - 25$
$b^2 = 144$
$\sqrt{b^2} = \sqrt{144}$
$b = 12$ m
13. 9; 10; unknown
$(9)^2 + (10)^2 = c^2$
$81 + 100 = c^2$
$181 = c^2$
$\sqrt{181} = \sqrt{c^2}$
$c = 13.454$ ft

Concept Check and Objective Practice

A1. Answers may vary. Example: It's the number that, when multiplied by itself, gives the original value.
A2. Answers may vary. Example: To find the square root of a fraction, you must find the square root of the numerator and the square root of the denominator.
A3. 16; 25; 36; 49; 64; 81; 100; 121; 144; 169
A4. exponent **A5.** 3 **A6.** 2 **A7.** 0 **A8.** 1
A9. $\frac{7}{6}$ **A10.** $\frac{1}{8}$ **A11.** $\frac{15}{7}$ **A12.** $\frac{14}{5}$ **A13.** 5
A14. 6 **A15.** 20 **A16.** 7 **B1.** perfect square
B2. perfect square
B3. $(3)^2 = 9$; $(4)^2 = 16$; $(5)^2 = 25$; $(6)^2 = 36$; $(7)^2 = 49$; $(8)^2 = 64$; $(9)^2 = 81$
B4. Answers may vary. Example: Determine the square roots of the perfect squares just less than 31 and just greater than 31. The square root of 31 will lie between those two values.

B5. 1 and 2 **B6.** 4 and 5 **B7.** 4 and 5 **B8.** 2 and 3
B9. 1.732 **B10.** 4.359 **B11.** 4.123 **B12.** 2.236
B13. 7.732 **B14.** 2.528 **B15.** 1.732 **B16.** 5.656
B17. 6.000 **B18.** 3.873 **C1.** right **C2.** hypotenuse
C3. a. hypotenuse $= \sqrt{(\text{leg})^2 + (\text{leg})^2}$
b. leg $= \sqrt{(\text{hypotenuse})^2 + (\text{leg})^2}$
C4. 8 ft **C5.** 5 in. **C6.** 8.5 m **C7.** 13.6 yd
C8. 13.9 ft **C9.** 13.9 km **C10.** 13.7 cm **C11.** 18.4 in.

Focus on Algebra

1. 6 in. **2.** 5 ft **3.** 4.123 m **4.** 3.873 cm
5. 5.831 ft **6.** 9.747 in. **7.** 5.657 cm **8.** 9.434 m

Exercises

For 1.–11.: See proceding Concept Check and Objective Practice answers.

13. perfect square **15.** square root
17. Answers may vary. Example: $5^2 = 25$; $6^2 = 36$. Since 33 lies between 25 and 36, the square root of 33 will lie between 5 and 6.
19. 7 **21.** 12 **23.** $\frac{8}{5}$ **25.** $\frac{2}{13}$ **27.** 2 and 3
29. 7 and 8 **31.** 8 and 9 **33.** 8 and 9 **35.** 4 ft
37. 8 cm **39.** 2.828 mi **41.** 6.928 in.
43. $3^2 = 9$; $4^2 = 16$; $5^2 = 25$; $9 + 16 = 25$
45. 1.732 **47.** 5.477 **49.** 9.110 **51.** 2.828 **53.** 8
55. 3.276 **57.** 12 in. **59.** 7 m **61.** 6.71 ft
63. 7.07 cm **65.** 8 ft **67. a.** 40 ft × 13 ft **b.** 1,040 ft^2
69. a. 6 ft^2 **b.** 10.828 ft
71. a. 1.414 m **b.** 1.570 m^2 **c.** 0.570 m^2

Section 7.7 Guided Practice

1. a. isosceles; isosceles; (a); (c); (f); 4; (f); (f) **b.** *T*; *R*; *S*
2. a. obtuse; obtuse; (e); (g); (g); (g) **b.** *N*; *M*; *P*
3. a. obtuse; obtuse; (e); (g); (e); (e)
b. *C* to *K*; *B* to *L*; *A* to *J*
4. corresponding; $\frac{13}{x}$; $\frac{6}{2}$; $\frac{13}{x} = \frac{6}{2}$

5.
$$\frac{2}{6} = \frac{y}{13}$$
$$\frac{13}{1} \cdot \frac{2}{6} = \frac{y}{\cancel{13}^1} \cdot \frac{\cancel{13}^1}{1}$$
$$\frac{26}{6} = y$$
$$4.33 \text{ in.} = y$$

6.
$$\frac{h}{5} = \frac{20}{2}$$
$$\frac{\cancel{5}^1}{1} \cdot \frac{h}{\cancel{5}^1} = \frac{20}{2} \cdot \frac{5}{1}$$
$$h = \frac{100}{2}$$
$$h = 50 \text{ ft}$$
50 ft

7.
$$\frac{w}{4} = \frac{14}{8}$$
$$\frac{\cancel{4}^1}{1} \cdot \frac{w}{\cancel{4}^1} = \frac{14}{\cancel{8}^2} \cdot \frac{\cancel{4}^1}{1}$$
$$w = \frac{14}{2}$$
$$w = 7 \text{ ft}$$
$$A = bh$$
$$= 14 \cdot 7$$
$$= 98 \text{ ft}^2$$

8.
$$\frac{x}{4} = \frac{18}{8}$$
$$\frac{\cancel{4}^1}{1} \cdot \frac{x}{\cancel{4}^1} = \frac{18}{\cancel{8}^2} \cdot \frac{\cancel{4}^1}{1}$$
$$x = \frac{18}{2}$$
$$x = 9 \text{ ft}$$
$$A = \frac{1}{2} \cdot b \cdot h$$
$$= \frac{1}{2} \cdot 9 \cdot 18$$
$$= 81 \text{ ft}^2$$

Concept Check and Objective Practice

A1. corresponding **A2.** sizes; shape
A3. E; right triangle **A4.** A; short rectangle
A5. C; long rectangle **A6.** F; obtuse triangle
A7. B; square **A8.** D; pentagon
A9. *A to D; B to E; C to F* **A10.** *A to D; B to E; C to F*
A11. *A to E; B to F; C to D* **A12.** *A to F; B to E; C to D*
B1. ratio **B2.** ratios; rates
B3. 66.7 cm **B4.** 8.6 in.
B5. **B6.**

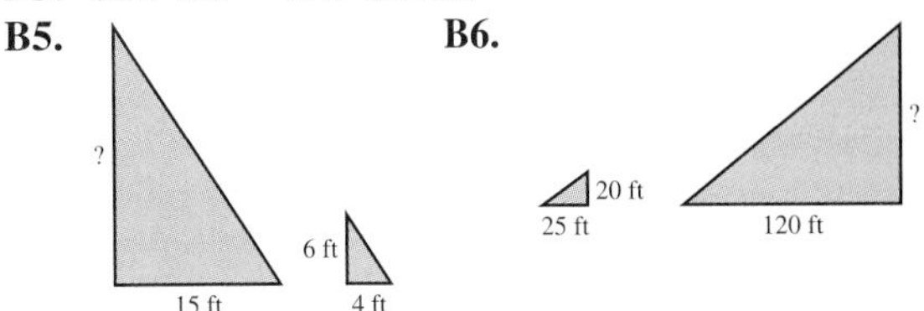

B7. 22.5 ft **B8.** 96 ft **C1.** square meters
C2. $\frac{x}{9} = \frac{9}{12}$; $\frac{x}{9} = \frac{6}{8}$; $\frac{x}{9} = \frac{3}{4}$ **C3.** 75 ft^2 **C4.** 2.3 ft^2
C5. 1.8 cm^2 **C6.** 11 m^2 **C7.** 1,458.3 ft^2 **C8.** 150 ft^2

Exercises

For 1.–9.: See preceding Concept Check and Objective Practice answers.

11. similar objects **13.** proportion
15. (a); *A* to *K*; *B* to *L*; *C* to *J* **17.** (d); *A* to *J*; *B* to *K*
19. (c); *A* to *M*; *B* to *J*; *C* to *K*; *D* to *L* **21.** 6.4 in.
23. 20.8 in. **25.** 40 ft **27.** 57.6 m^2 **29.** 134.4 yd^2
31. 12.3 ft **33.** 187.5 ft^2 **35.** about 8.7 mi
37. 1,000 km **39. a.** 3 : 1 **b.** 3 : 1 **c.** 9 : 1
41. a. 5 : 1 **b.** 5 : 1 **c.** 25 : 1
43. The area will be multiplied by a factor equal to n^2.
45. a. 1 : 1.8 **b.** 15.552 ft^2 **47.** 16

Chapter 7 Review Exercises

1. 160° **2.** 95° **3.** $\angle AOB$; $\angle BOC$; $\angle COD$
4. $\angle AOD$; $\angle BOD$ **5.** $\angle AOC$
6. 5°

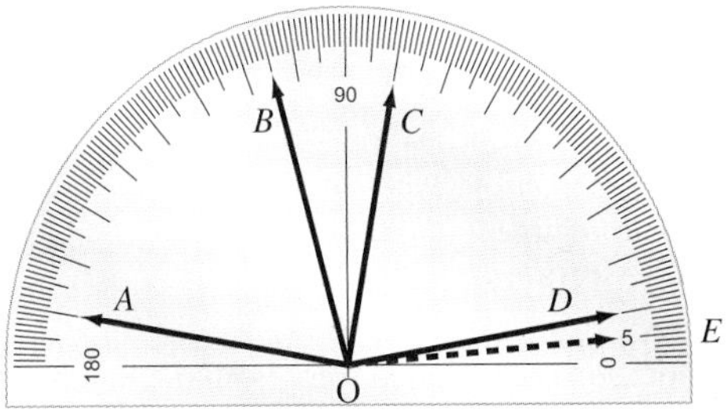

7. 70° **8.** 140° **9.** 110° **10.** 70°
11. equilateral triangle **12.** right scalene triangle
13. rectangle **14.** parallelogram **15.** regular hexagon
16. trapezoid **17.** circle **18.** acute isosceles triangle
19. regular octagon **20.** acute scalene triangle **21.** 26 in.
22. 24 cm **23.** 42 cm **24.** 28 ft **25.** 135 yd^2
26. 84 cm^2 **27.** 160 ft^2 **28.** 240 in.^2 **29.** 78.5 in.^2
30. 28.26 in.^2 **31.** 62.8 ft **32.** 50.24 m **33.** 30.84 in.
34. 153.86 mm^2 **35.** 149.86 cm^2 **36.** 44.56 ft
37. 42.56 cm **38.** 121.12 ft^2 **39.** 73.12 cm^2
40. 17 ft^3 **41.** 27 cm^3 **42.** 282.6 ft^3 **43.** 4,187 cm^3
44. 784,000 ft^3 **45.** 6,280 yd^3 **46.** 3 and 4 **47.** 5 and 6
48. 9 and 10 **49.** 7 and 8 **50.** 6 **51.** 7 **52.** $\frac{2}{9}$

53. $\frac{8}{5}$ **54.** 30 in.^{2} **55.** 6 cm^2 **56.** 19.5 m **57.** 19.1 ft
58. 20 ft **59.** 24 ft

Chapter 7 Practice Test

1. a. 50° **b.** acute **2. a.** 90° **b.** right
3. a. 130° **b.** obtuse **4.** 40° **5.** 130°
6. regular octagon **7.** trapezoid **8.** rhombus
9. right scalene triangle **10.** 22.8 m **11.** 30 ft
12. 56 cm **13.** 22 ft **14.** 224 ft^2 **15.** 13 ft^2
16. 36 in.^{2} **17.** 24 in.^{2} **18.** 22 m^3 **19.** 33.5 in.^{3}
20. 392.5 m^3 **21.** 1,120 ft^3 **22.** 8 and 9 **23.** 1 and 2
24. $\frac{3}{5}$ **25.** $\frac{9}{4}$ **26.** 5.57 **27.** 4.58 **28.** 13.5 ft
29. 8.8 ft

CHAPTER 8

Section 8.1 Guided Practice

1. smallest; Other **2.** 3,000
3. 4,700

$$\frac{\text{Nursing students}}{\text{Total students}} \Rightarrow \frac{4{,}700}{10{,}000} = \frac{47 \cdot \cancel{100}^1}{100 \cdot \cancel{100}^1} = \frac{47}{100} = 47\%$$

47% of the students are studying nursing.

4. 1,100; 4,700

$$\frac{\text{Number of manufacturing students}}{\text{Number of nursing students}} \Rightarrow \frac{1{,}100}{4{,}700} = \frac{11 \cdot \cancel{100}^1}{47 \cdot \cancel{100}^1} = \frac{11}{47}$$

11; 47; 11; 47

5. largest; South; South; 39%
6. 21%; 39%; 21% + 39% = 60%
60% of the total viewers are from the North Central or South regions.
7. 39%; $(0.39) \cdot (1{,}045) = 407.55 \approx 408$
408 of the people polled were from the South.
8. yellow; yellow; Janet Jackson; Janet Jackson; 26
9. 25; 10; 25 − 10 = 15; 15 **10.** April
11. 4,000; 6,000; 6,000 − 4,000 = 2,000; 2,000
12. steepest; April; May
13.

Grade/Score	Tally	Frequency
A 90–99	||	2
B 80–89	𝍸	5
C 70–79	||||	4
D 60–69	|	1
F 50–59		0

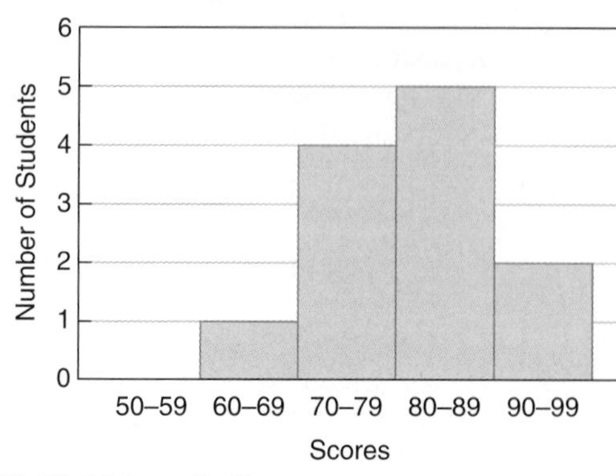

14. 80–89; highest; B; B
15.

Work Bonus	Tally	Frequency
101–150	|||	3
151–200	||	2
201–250	𝍸	5
251–300	||	2
301–350	|	1

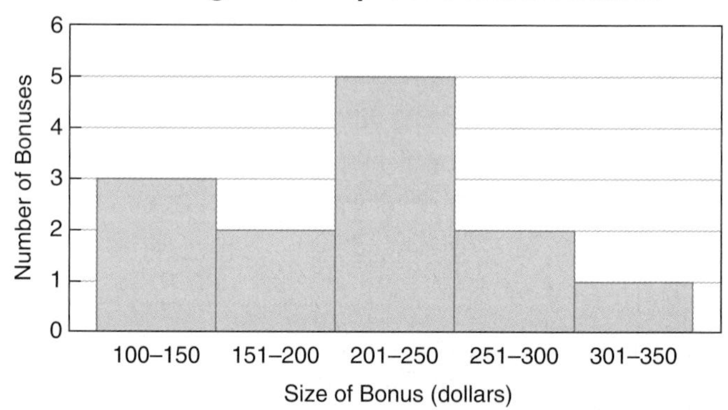

16. 201; 250

Concept Check and Objective Practice

A1. Answers may vary. Example: The reader can clearly see that the parts, when "glued" together, will be a whole.
A2. sectors **A3.** 25 **A4.** 2 **A5.** 36% **A6.** 28%
A7. 7 to 25 **A8.** 9 to 7 **A9.** 32% **A10.** 52%
A11. about 63 **A12.** about 105 **B1.** line graph
B2. Answers may vary. Example: enrollment at several area colleges.
B3. \$55,000 **B4.** \$52,000 **B5.** \$1,500 **B6.** \$8,000
B7. North Carolina and California **B8.** Georgia and Florida
B9. 10° **B10.** 62° **B11.** about 50° **B12.** about 40°
C1. True: There are 5 whole numbers in the interval. They are 6, 7, 8, 9, and 10.
C2. False: There are 6 values in the interval: 7, 8, 9, 10, 11, and 12.
C3. 3 **C4.** 31; 60

C5.

Intervals	Tally	Frequency
31–36	III	3
37–42	III	3
43–48	IIII	4
49–54	II	2
55–60	III	3

C6. Histogram of Daily High Temperatures in Anchorage, Alaska

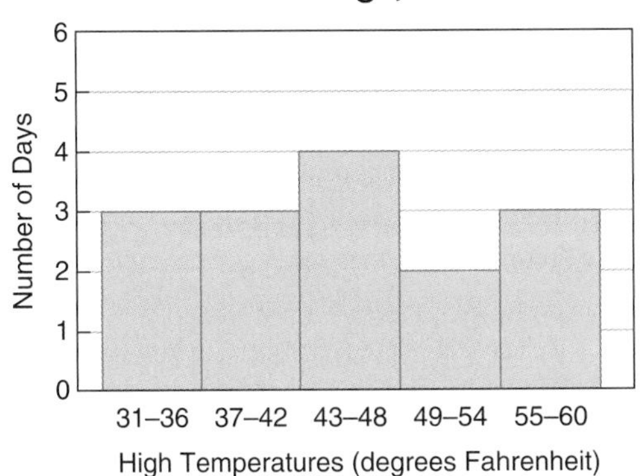

Exercises

For 1.–9.: See preceding Concept Check and Objective Practice answers.

11. circle graph **13.** histogram **15.** sectors
17. Valentine's Day **19.** 30 million **21.** about 19%
23. late fees **25.** 85% **27.** about 12 billion
29. 19 miles per gal **31.** Honda Civic
33. a. 15 gal **b.** 19 gal
35. 49% **37.** 63% **39.** car, clothes washer, TV set
41. Answers may vary. **43.** 3 min 28 sec **45.** decrease
47. \$0.2 billion **49.** about 2 in. **51.** about 4 in.
53. 7 months **55.** 5 students **57.** 19 students
59. a. 60–77 **b.**

Height	Tally	Frequency
60–62	IIII	4
63–65	IIII	4
66–68	~~IIII~~ I	6
69–71	~~IIII~~	5
72–74	III	3
75–77	II	2

c. Histogram of Students' Heights

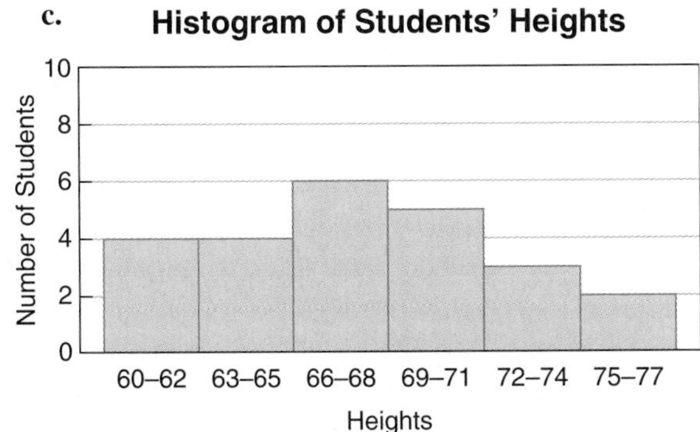

d. 66–68 in.

Section 8.2 Guided Practice

1. $\dfrac{2+8+3+6+3+2}{6} = \dfrac{24}{6} = 4; 4$

2. $\dfrac{5+7+4+2+2}{5} = \dfrac{20}{5} = 4; 4$

3. $\dfrac{8+6+11+7+1+2+2}{7} = \dfrac{37}{7} \approx 5.29; 5.3$

4. 2, 3, 3, 6, 8; ~~2~~, ~~3~~, 3, ~~6~~, ~~8~~; 3

5. 0, 2, 4, 11, 15, 17; ~~0~~, ~~2~~, $\boxed{4, 11}$, ~~15~~, ~~17~~
$\dfrac{4+11}{2} = \dfrac{15}{2} = 7.5; 7.5$

6. 3; 3 **7.** {2, 3, 3, 3, 6, 6}; 3; most often; 3
8. {0, 0, 2, 4, 17, 17}; two; 0; 17
9. {0, 1, 4, 7}; no; every value; the same

Concept Check and Objective Practice

A1. average **A2.** 80 and 42
A3. Add the values and divide by the number of values.
A4. 8 **A5.** 25.7 **A6.** 5.2 **A7.** 5.3 **A8.** 10
A9. 12.1 **B1.** middle
B2. Find the mean of the two numbers.
B3. 7 **B4.** 9 **B5.** 4 **B6.** 3 **C1.** most **C2.** no
C3. 5 **C4.** 7 **C5.** 3 and 8 **C6.** 2 and 8
C7. no mode **C8.** no mode

Exercises

For 1.–9.: See Preceding Concept Check and Objective Practice answers

11. median **13.** 6 **15.** 61.3 **17.** $\frac{1}{4}$ **19.** 5
21. 53.5 **23.** 34 **25.** 1 **27.** no mode **29.** 1 and 6
31. mean: 12; median: 11; mode: 10 **33. a.** 27.7 **b.** 26
35. \$90,033
37. a. mean: 80; median: 82.5; mode: 77
b. Histogram of Test Scores

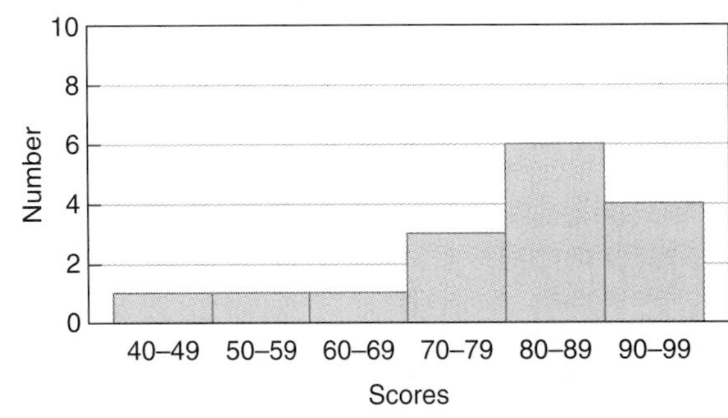

c. Answers may vary. Example: Choosing the median would tell about the distribution of the grades on the test.

39. a. Day 1: 28 mi per gal Day 2: 26 mi per gal
Day 3: 21 mi per gal Day 4: 29 mi per gal
b. 26 mi per gal **c.** 27.2 mi per gal

41. a. Day 3: 21 mi per gal **b.** 26 mi per gal
c. 26.0 mi per gal

43. In Exercise 41, the miles driven on day three increases, raising its influence on the mean. Since the mileage of day three is low (21 mpg), it lowers the overall average.

45. a. mean: \$10.39; median: \$7.80; mode: \$6.85
b. Answers may vary. Example: The most meaningful measure is the mode. An individual is more likely to be hired as a server than any other position.

47. **a.** mean: $68,001.43; median: $65,079; mode: none
b. Answers may vary.

49. mean decreases: the total of the data points decreases; median stays the same: the order has not changed; mode stays the same: the value of the most data points has not changed

51. mean decreases: the total of the data points decreases; median decreases: the new value is smaller than the old value, thereby affecting the order of the values; mode no longer exists: the value replaced was the mode value

53. Answers may vary. Example: The mode will change if the change is with a value that is the current mode or if the change duplicates a value in the set of numbers.

55. **a.**

Data Set 1	Data Set 2
$35,000	$40,000
$35,000	$40,000
$35,000	$40,000
$35,000	$40,000
$35,000	$40,000
$35,000	$40,000
$35,000	$40,000
$35,000	$40,000
$35,000	$40,000
$35,000	$40,000
$90,000	$40,000

b. Answers may vary. Example: The median changed because all the salary amounts changed.
c. Answers may vary. Example: The mean didn't change because the total amount of the salaries paid and the number of salaries did not change.

Chapter 8 Review Exercises

1. 0 **2.** 24.2% **3.** 12.1% **4.** 87.8%
5. about 353 **6.** about 96 **7.** 23,421 **8.** 20,693
9. 1,855 **10.** 2,205
11. Syracuse and North Carolina; 823
12. Kentucky and Louisville; 4,933
13. 2002 **14.** 2002 and 2003 **15.** 1
16. The average number of accidents and spinouts is increasing.
17. $2,101–$4,290 **18.** 46 **19.** 104 **20.** 12
21. average **22.** middle **23.** most **24.** 8
25. 5 **26.** 12.4 **27.** 13.25 **28.** $182.83 **29.** $34,874
30. 11 **31.** 14 **32.** no mode **33.** 11.7 **34.** 3 and 5
35. none **36.** no mode **37.** 17 and 18
38. mean: 17.5; median: 13.5; mode: 11
39. mean: 18.6; median: 18; mode: 18
40. mean: 18; median: 18; mode: none
41. mean: 29.3; median: 28; mode: 28

42. Histogram of Rudy's Checks

Number of Checks; 1–75, 76–150, 151–225, 226–300, 301–375, 376–450, 451–525; Amount (in dollars)

43.

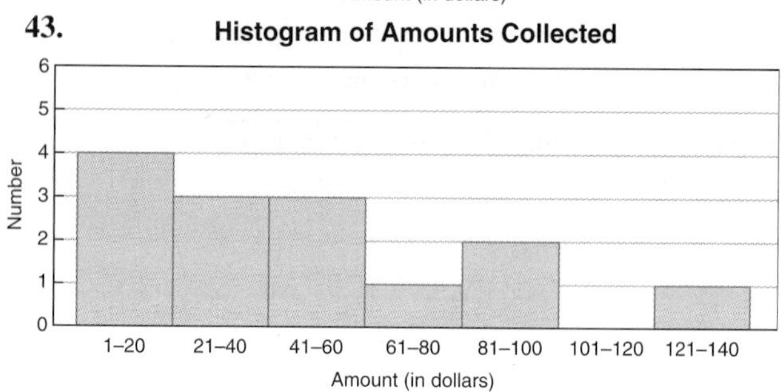

Chapter 8 Practice Test

1. 3% **2.** 36% **3.** 50% **4.** 220 **5.** 44 **6.** 2
7. 11 to 27 **8.** about 184% **9.** 1987
10. 1995 and 1998 **11.** about 127 **12.** 66.7%
13. $26,001–$29,000 **14.** 53 **15.** $3,000
16. $35,001–$38,000 **17.** 3.75
18. **a.** above **b.** neither; 87 is the median.
c. Answers may vary.
19. 40 **20.** 56.8 seconds **21.** 64° and 73° **22.** 18
23. mean: 13.8; median: 15; mode: 15
24. mean: 21.6; median: 23; mode: none
25. mean: 8.2; median: 9; mode: none
26. mean: 42.3; median: 43; mode: 47
27. mean: 75.8; median: 76.5; mode: 83 and 84

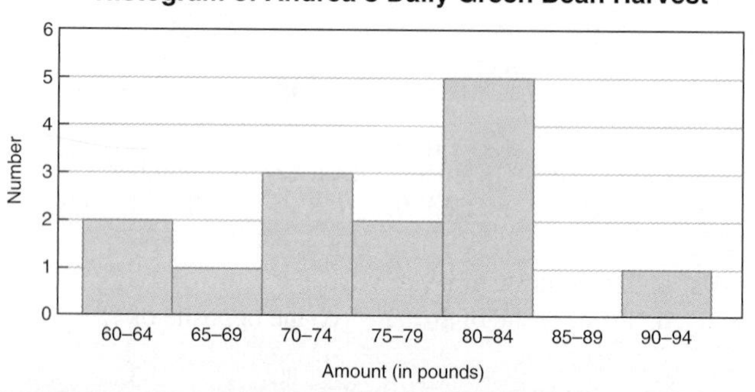

CHAPTER 9

Section 9.1 Guided Practice

1. decreases; negative; −317 **2.** increase; positive; 2.5

3.

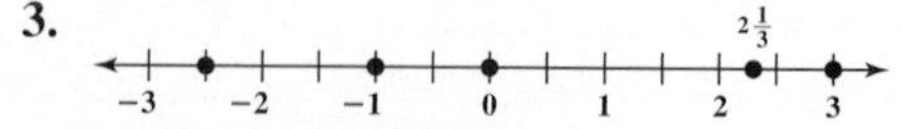

−2.5; 3

4.

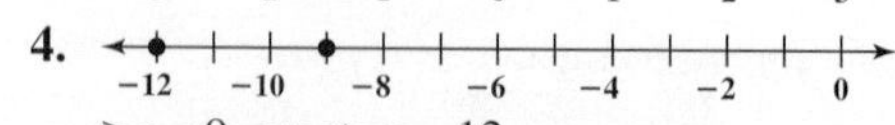

>; −9; greater; −12

5.

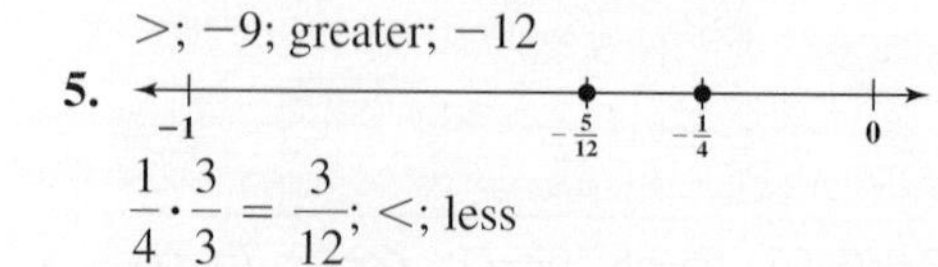

$\frac{1}{4} \cdot \frac{3}{3} = \frac{3}{12}$; <; less

6. opposite; negative four; 4 7. 4; 4; 4; distance; negative
8. the absolute value of negative twelve; 12; 12
9. the absolute value of forty-five; 45; 45 10. $\frac{2}{3}; \frac{2}{3}$
11. the opposite of the absolute value of negative six; 6; −6; 2
12. the opposite of negative six; 6; operation; one
13. 6; −1; 4; 0; −4

$-|-4|$ −1 0 $-(-4)$ $|-6|$

$-|-4|, -1, 0, -(-4), |-6|$

14.

? 11 ?

−7; +7; 7; 11; 11 − 7 = 4; 11 + 7 = 18; 11; 4; 11 − 4 = 7; 11; 18; 18 − 11 = 7; 4; 18; 7; 11

Concept Check and Objective Practice

A1. Answers will vary.
A2. **a.** −48 **b.** 48 **c.** yes **d.** Answers will vary.
A3. −15 **A4.** 500 **A5.** $-3\frac{1}{2}$ **A6.** $-5\frac{1}{2}$
A7. −25.34 **A8.** 267.15 **A9.** 4.56 **A10.** −3.56
B1. Zero
B2. Answers may vary. Example: There would be no indication of the values of the points on the line.
B3. $-4\frac{1}{5}$ 7.2

−10 −8 −6 −4 −2 0 2 4 6

B4. $-16\frac{2}{3}$ −12 17.5 35

−18 −10 0 10 20 30 40

B5. $-2\frac{3}{4}$ $-1\frac{1}{2}$ $\frac{3}{4}$ $2\frac{1}{2}$

−3 −2 −1 0 1 2 3

B6. $-1\frac{1}{4}$ $\frac{3}{4}$ $2\frac{3}{4}$

−3 −2 −1 0 1 2 3

B7. −0.079 −0.008 0.79

−0.3 0 0.1 0.3 0.5 0.7 1

−0.07

B8. −0.65 −0.6 −0.068 0.05 0.056

−0.7 −0.5 −0.3 −0.1 0 0.1 0.3 0.5

C1. Answers will vary. Example: A negative number is always further to the left on the number line than a positive number.
C2. Answers will vary. Example: Zero is always further to the right on the number line than a negative number.
C3. < **C4.** < **C5.** > **C6.** > **C7.** < **C8.** >
C9. < **C10.** >
D1. **a.** the opposite of negative five; 5
b. the opposite of 2; −2 **c.** the opposite of $\frac{1}{2}$; $-\frac{1}{2}$
D2. 2 **D3.** −4 **D4.** −5 **D5.** 6 **D6.** −11
D7. 14 **D8.** 13.5 **D9.** −2.59
E1. Answers may vary. Example: The absolute value of a number indicates its distance from 0 on the number line.
E2. Answers may vary. Example: Distance is always measured with a positive value.
E3. **a.** two **b.** −7 **E4.** **a.** two **b.** −3
E5. **a.** one **b.** 7 **E6.** **a.** one **b.** −3
E7. 5 **E8.** 16 **E9.** $\frac{7}{9}$ **E10.** $\frac{1}{10}$ **E11.** 5
E12. 31 **E13.** −42 **E14.** −15

Exercises

For 1.–15.: See preceding Concept Check and Objective Practice answers.

17. opposites **19.** negative number **21.** 89 **23.** −15
25. 1,453 **27.** −13 **29.** 1
31. −5 −1 3

−12 −8 −4 0 4 8 12

33. −53 5

−120 −80 −40 0 40 80

35. $-1\frac{1}{2}$

−10 −8 −6 −4 −2 0 2 4 6 8 10

37. $-\frac{8}{5}$ $-\frac{1}{5}$ $\frac{6}{5}$ 1.8

−4 −2 0 2 4 6 8

39. < **41.** < **43.** > **45.** > **47.** < **49.** >
51. Answers will vary. All negative numbers are to the left of all positive numbers on the number line.
53. Answers will vary. Zero is greater than every negative number and less than every positive number because it lies to the right of all negative numbers and to the left of all positive numbers on the number line.
55. **a.** the absolute value of negative twelve **b.** 12
57. **a.** the opposite of negative three **b.** 3
59. **a.** the opposite of seven **b.** −7
61. **a.** the opposite of the absolute value of negative ten **b.** −10
63. 8 **65.** −5 **67.** 7
69. −13 **71.** −91 **73.** −75
75. Zero and all positive numbers are equal to their own absolute value.
77. zero **79.** > **81.** < **83.** > **85.** >
87. $-(10), -|-4|, |-4|, |-10|$
89. $-(15), -|-12|, -(-12), |-15|$
91. −8 and 2 **93.** −8 and −2 **95.** −7 and 7

Section 9.2 Guided Practice

1. 4 negative; 2 negative; 6
$(-4) + (-2)$

(−1) (−1)
(−1) (−1)
(−1)
(−1)

$(-4) + (-2) = -6$

2. −55; 55; 34; 21; 34 + 21 = 55; −55; negative; negative
3. 7; −7; negative; $\frac{3}{18}; \frac{4}{18}; \frac{3}{18} + \frac{4}{18} = \frac{7}{18}; \frac{-7}{18}$
4. 4; 2; two; Two; negative
$(-4) + 2$

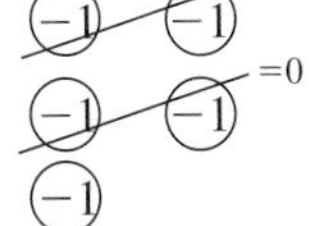

(−1)
(−1)

$(-4) + 2 = -2$

5. -4; -4; 12; $12 + (-12)$; -231; $231 + (-231)$

6. 8; -8; negative; negatives; positives
$(-14) + (8) = -6$

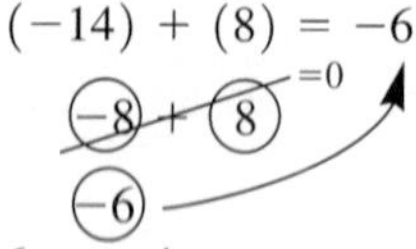

6; negatives

7. $(-14) + (30) = 16$

14; 30; $30 - 14 = 16$; 16; 30; -14; positive

8. $\frac{5}{5}$; $\frac{5}{15}$;
$\left(-\frac{4}{15}\right) + \left(\frac{5}{15}\right) = \frac{1}{15}$

1; positive; $\frac{4}{15}$; $\frac{5}{15}$; $\frac{5}{15} - \frac{4}{15} = \frac{1}{15}$; $\frac{1}{15}$; $\frac{5}{15}$; $\frac{-4}{15}$; positive

9. Three subtract negative two
Nine plus negative five
Negative one subtract positive three

10. $-13 - 17 = -13 + (-17)$
$= -30$

11. $-32 - (-40) = -32 + (+40)$
$= 8$

12. Different; subtract; $3 - 5$
Different; subtract; $-7 - 5$
Same; add; $2 + 3$

13. 13; 19; left; do; 0; negative; $19 - 13$; 6; $13 - 19 = -6$

14. -7; 5; right; do not; 0; negative; $7 - 5$; 2;
$-7 + 5 = -2$;

15. $30 + (-51) = 30 - 51$ do; 0; negative
$= -21$

16. $(-13) + (+20) = -13 + 20$; do; 0; positive
$= 7$

17. $(21) - (-30) = 21 + 30$; do not; 0; positive
$= 51$

18. $(-61) - (+75) = -61 - 75$; do not; 0; negative
$= -136$

19. $(-3) - |-6| + 8 = (-3) - 6 + 8$
$= -9 + 8$
$= -1$

20. $5 + (-17) - 3 + 4 - 2 - (-5)$
$= 5 + (-17) + (-3) + 4 + (-2) + 5$
$= 5 + 4 + 5 + (-17) + (-3) + (-2)$
$= (5 + 4 + 5) + ((-17) + (-3) + (-2))$
$= 14 + (-22)$
$= -8$

21.

Change	Altitude
Start	3,729 ft
+250	3,979 ft
−490	3,489 ft
+172	3,661 ft
−100	3,561 ft
+82	3,643 ft
−215	3,428 ft

−215 ft

Concept Check and Objective Practice

A1. a. & b. When adding numbers with the same sign, the total number of positives or negatives will increase, so the sign remains the same.

A2. -17 **A3.** -11 **A4.** -28 **A5.** -27

A6. -20.5 **A7.** -18.2 **A8.** $\frac{-19}{28}$ **A9.** $\frac{-7}{10}$

B1. Answers will vary. Example: There will be "more" of that type of number.

B2. Answers will vary. Example: Subtracting tells us how many positives or negatives remain.

B3. $(-6) + (3) = -3$ **B4.** $(5) + (-2) = 3$

B5. 10 **B6.** 7 **B7.** -4 **B8.** -13 **B9.** $\frac{1}{10}$ **B10.** $\frac{1}{8}$

B11. 3.4 **B12.** 6.5

Focus on Personal Finance

Date	Payment Debit	Fee (if any)	Deposit Credit	Balance
1-Jan				\$512.73
1-Jan	\$375.00			\$137.73
5-Jan	\$20.00	\$3.00		\$114.73
8-Jan	\$78.95			\$35.78
8-Jan	\$20.00	\$3.00		\$12.78
8-Jan			\$438.24	\$451.02
10-Jan	\$100.00			\$351.02
10-Jan	\$59.99			\$291.03
10-Jan	\$48.73			\$242.30
11-Jan	\$20.00	\$3.00		\$219.30
13-Jan	\$20.00	\$3.00		\$196.30
15-Jan	\$185.00			\$11.30
16-Jan	\$150.00			−\$138.70
16-Jan		\$20.00	Fee	−\$158.70
18-Jan		\$30.00	Fee	−\$188.70

1. In the first half of the month, Alfonso paid \$12 in ATM fees. He should try to determine the amount of money he will need for a couple of weeks and withdraw it all at the same time.
2. Answers may vary. **3.** It cost him \$1,000 per hour.

C1. a. $10 + (-16)$ **b.** $-12 + (-5)$
c. $15 + (+2)$ **d.** $-4 + (+8)$
C2. Answers may vary. Example: Both problems have the same answer.
C3. -4 **C4.** -4 **C5.** -25 **C6.** -25 **C7.** 11
C8. 8 **C9.** 3 **C10.** 6
D1. Answers may vary. Example: Subtraction is the opposite of addition. A negative value is the opposite of a positive value. By changing both the operation and the sign of the number to its opposite, the same problem has been maintained. It is important to remember that *BOTH* the operation and the sign must be changed to the opposite.
D2. a. 3 units **b.** 3 and -3
c. Answers may vary. **d.** Answers may vary.
D3. a. $10 - 16$
b.
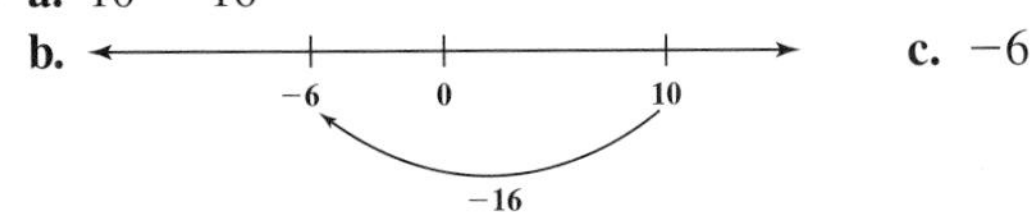

c. -6
D4. a. $15 - 18$
b. (−3, 0, 15; −18) **c.** -3
D5. a. $-12 + 13$
b. (−12, 0 1; +13) **c.** 1
D6. a. $-7 + 5$
b. (−7, −2 0; +5) **c.** -2
D7. $5 + 6 = 11$ **D8.** $4 + 8 = 12$ **D9.** $-6 + 9 = 3$
D10. $-40 + 50 = 10$ **D11.** $51 - 16 = 35$
D12. $40 - 65 = -25$ **D13.** $-19 + 30 = 11$
D14. $-50 + 75 = 25$

Exercises

For 1.–14.: See preceding Concept Check and Objective Practice answers.

15. zero pair **17.** -10 **19.** 101 **21.** 5 **23.** -1
25. -5 **27.** 12 **29.** 1.3 **31.** -11.0 **33.** $\frac{-1}{10}$
35. $-1\frac{1}{9}$ **37.** -8 **39.** 13 **41.** -31 **43.** 50
45. 5 **47.** 15 **49.** $\frac{7}{10}$ **51.** 0.8
53. a. -15 **b.** 15 **c.** no
55. a. 9 **b.** 9 **c.** yes **57.** 9 **59.** 5 **61.** 1
63. 2 **65.** -22 **67.** 9.3 **69.** -2 **71.** 7
73. -162 lb **75.** -42°F **77.** 12°C **79.** -112 ft
81. \$90,900 **83.** $-\$41,000$
85. 1st place: Angelica; 5 under par
2nd place: Mike; 3 above par
3rd place: Chante; 5 above par
4th place: Sergio; 8 above par

Section 9.3 Guided Practice

1. $(-9)(8) = -(9 \cdot 8) = -72$
2. $\left(-\frac{3}{4}\right)\left(-\frac{8}{1}\right) = +\left(\frac{3}{4}\right)\cdot\left(\frac{8}{1}\right) = \frac{3 \cdot 8}{4 \cdot 1} = \frac{3 \cdot 2 \cdot \cancel{4}^1}{\cancel{4}^1 \cdot 1} = 6$
3. Opposite; Negative; -6 **4.** Same; Positive; 1.6
5. Opposite; Negative; -4 **6.** Same; Positive; 0.3
7. Same; Positive
$$\left(-\frac{5}{4}\right) \div \left(-\frac{35}{8}\right) = +\left(\frac{5}{4}\right)\cdot\left(\frac{8}{35}\right) = \frac{{}^1\cancel{5} \cdot 2 \cdot \cancel{4}^1}{\cancel{4}^1 \cdot {}^1\cancel{5} \cdot 7} = \frac{2}{7}$$
8. odd; negative
$$(-14) \div (-2)\cdot(-7) = -(14) \div (2)\cdot(7) = -7 \cdot 7 = -49$$
9. even; positive
$$(4)\cdot(-3)\cdot(2) \div (3) \div (-5) = +(4)\cdot(3)\cdot(2) \div (3) \div (5) = +(12)\cdot(2) \div (3) \div (5) = +(24) \div (3) \div (5) = +(8) \div (5) = +1\frac{3}{5}$$

10. grouping; negative sign; 2; −4; $(-4)\cdot(-4)$
11. 5; negative sign; 2; 5; $-(5\cdot 5)$
12. $(-1)^3 = (-1)(-1)(-1) = -(1\cdot 1\cdot 1) = -1$
13. $-3^2 = -3\cdot 3 = -(3\cdot 3) = -9$
14. $(-7)^2 = (-7)(-7) = +(7\cdot 7) = 49$
15. $-7^2 = -(7\cdot 7) = -7\cdot 7 = -49$
16. $\dfrac{(-10.2)+(-8.9)+(-4.0)+(4.5)}{4} = \dfrac{-18.6}{4} = -4.65$

−4.65°C
17. 6; −5;

Products that = 6	Sums of factors
$1\cdot 6$	$1+6=7$
$2\cdot 3$	$2+3=5$
Now check the opposites	
$(-1)(-6)$	$(-1)+(-6)=-7$
$(-2)(-3)$	$(-2)+(-3)=-5$

−2; −3; $(-2)(-3)=6$; $(-2)+(-3)=-5$
18. −12; −4

Products that = −12	Sums of factors
$(1)(-12)$	$1+(-12)=-11$
$(2)(-6)$	$2+(-6)=-4$
$(3)(-4)$	$3+(-4)=-1$
Now check the opposites	
$(-1)(12)$	$(-1)+12=11$
$(-2)(6)$	$(-2)+6=4$
$(-3)(4)$	$-3+4=1$

2; −6; $(2)(-6)=-12$; $(2)+(-6)=-4$

Concept Check and Objective Practice

A1. Answers will vary. Example: $(-4)+(-4)+(-4)=-12$. Therefore, adding −4 three times gives −12.
A2. **a.** −2; 4; −3; −2; 6 **b.** positive
A3. −104 **A4.** −420 **A5.** 105 **A6.** 176
A7. −6.6 **A8.** 55 **A9.** 110 **A10.** −21.3
B1. positive **B2.** negative **B3.** −9 **B4.** −2
B5. $\frac{5}{2} = 2\frac{1}{2}$ **B6.** $\frac{8}{3} = 2\frac{2}{3}$ **B7.** $-\frac{5}{6}$ **B8.** 42
B9. −220 **B10.** 8
C1. Answers will vary. Example: The answer is negative when there is an odd number of negative signed numbers.
C2. If there are no grouping symbols, perform the operations in the order encountered from left to right.
C3. 6 **C4.** −12 **C5.** −18 **C6.** 36 **C7.** 12
C8. 18 **C9.** $\frac{3}{2} = 1\frac{1}{2}$ **C10.** 16
D1. **b.** −5; 3; −125 **c.** 8; 2; −64 **d.** −1; 9; −1
D2. 125 **D3.** 1,000 **D4.** 9 **D5.** 49 **D6.** −32
D7. −16 **D8.** −32 **D9.** 16

Exercises

For 1.–12.: See preceding Concept Check and Objective Practice answers.

13. quotient **15.** exponent **17.** −24 **19.** 39
21. 20 **23.** −90.9 **25.** $\frac{2}{3}$ **27.** $-\frac{1}{17}$ **29.** −2
31. 5 **33.** −0.4 **35.** 5 **37.** 5 **39.** $\frac{2}{3}$ **41.** 8
43. −100 **45.** $-\frac{1}{28}$ **47.** −40 **49.** 0 **51.** 2
53. 81 **55.** −27 **57.** −16 **59.** 36 **61.** $\frac{9}{25}$
63. $-\frac{8}{27}$ **65.** 0 **67.** 1 **69.** 1 and 3 **71.** −2 and 4
73. −9 and 2 **75.** −8 and −5
77. Answers will vary. Example: Each of these fractions is equal to negative three.
79. 0 and 1 **81.** −1,800 ft **83.** +$54 **85.** −26
87. 19 **89.** 12 **91.** 3 aluminum and 1 magnesium
93. −10 **95.** −1 **97.** 0

Section 9.4 Guided Practice

1. $(-3)-(4)+(-7) = (-3)+(-4)+(-7)$
$= -7+(-7)$
$= -14$
2. $35 - 15\cdot(-2) = 35 + 30 = 65$
3. $\dfrac{-42 \div -21}{6-(-8)} = \dfrac{2}{14} = \dfrac{{}^{1}\not{2}}{{}^{1}\not{2}\cdot 7} = \dfrac{1}{7}$
4. $(1-3)^2 + (5-7)^3 \div 2\cdot(-4)$
$= (-2)^2 + (-2)^3 \div 2\cdot(-4)$
$= 4 + (-8) \div 2\cdot(-4)$
$= 4 + (-4)\cdot(-4)$
$= 4 + 16$
$= 20$
5. First: Money left = Original amount + Change
Second:
Change = Number of games · Cost per game
Change = 7(−$0.25)
Third: Money left = Original amount + Change
= $3.45 + 7(−$0.25)
6. $8-6$; 2; no; $(16 \div 2) - 6 = -4$; do not;
$16 \div (2-6) = -4$; Do;
$16 \div (2-6) = -4$
$16 \div (-4) = -4$
$-4 = -4$
$16 \div (2-6) = -4$

Concept Check and Objective Practice

A1. **a.** Perform all operations inside grouping symbols and calculate absolute values.
b. Perform operations with exponents.
c. Multiply/Divide in order from left to right.
d. Add/Subtract in order from left to right.
A2. −8 **A3.** −9 **A4.** −25 **A5.** −9 **A6.** −20
A7. 4 **A8.** −12 **A9.** −12 **A10.** $\frac{2}{3}$ **A11.** 2
A12. 1 **A13.** 1

Exercises

For 1.–3.: See preceding Concept Check and Objective Practice answers.

5. -1 **7.** -6 **9.** 30 **11.** 25 **13.** -51
15. $-\frac{9}{10}$ **17.** -8 **19.** 14 **21.** -1 **23.** -2
25. 15 **27.** 57 **29.** -5 **31.** -1 **33.** $\frac{1}{2}$
35. $\frac{4}{7}$ **37.** -1 **39.** -2 **41.** 3 points
43. 4% **45.** $1\frac{1}{2}$ **47.** $-\frac{9}{10}$ **49.** $-\frac{1}{8}$
51. 0 **53.** -42 **55.** -9
57. $150 \cdot (\$3.21) + 200 \cdot (-\$1.75)$
59. $\$11{,}000 + 12(-\$850)$ **61.** $(2 - 3) \cdot 4 = -4$
63. Parentheses not needed. **65.** $18 \div (3 \cdot 2) = 3$
67. Parentheses not needed. **69.** $(8 + 8) \div 4 = 4$
71. c **73.** b **75.** i **77.** d **79.** h

Chapter 9 Review Exercises

1. -75 **2.** 175 **3.** \$48.50 **4.** 7,400 **5.** $>$ **6.** $<$
7. $<$ **8.** $>$ **9.** -24 **10.** 52 **11.** -6 **12.** -56

13. −10 −9 −8 −7 −6 −5 −4 −3 −2 −1 0 1 2

14. −8 −7 −6 −5 −4 −3 −2 −1 0 1 2 3 4

15. −5 −4 −3 −2 −1 0 1 2 3 4 5 6 7 |−8| −(−9)

16. −2 −1 0 1 −(−2) |−3| 4 5

17. All negative numbers are to the left of all positive numbers on the number line.
18. All positive numbers are to the right of all negative numbers on the number line.
19. $(-5) + (-6) = (-11)$
20. $-3 + -2 = -5$
21. $-5 + 7 = 2$
22. $2 + (-6) = -4$

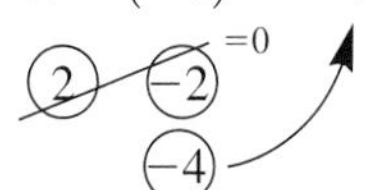

23. $-7 + (-5) = -12$ (−12, −7, 0; −5)
24. $18 - (+34) = -16$ (−16, 0, 18; −34)
25. $-5 - 10 = -15$ (−15, −5, 0; −10)
26. $-4 - 7 = -11$ (−11, −4, 0; −7)

27. -16 **28.** -24 **29.** -6 **30.** -16 **31.** -2
32. -11 **33.** 13 **34.** -4 **35.** $\frac{9}{8} = 1\frac{1}{8}$
36. $\frac{13}{6} = 2\frac{1}{6}$ **37.** -10 **38.** 2
39.

Type	Date	Desc	Payment/ Debit (−)	Fee (if any)	Deposit/ Credit (+)	Balance
	4-May	Bal Fwd				\$132.85
212	5-May	Credit card	\$48.24			\$84.61
ATM	8-May	Needed cash	\$20.00	\$3.00		\$61.61
Dep	17-May	Pay day!			\$324.62	\$386.23
213	18-May	Bill	\$117.43			\$268.80
ATM	24-May	Needed cash	\$40.00	\$2.00		\$226.80

40. -12 **41.** 35 **42.** -5 **43.** 5 **44.** 48
45. -7 **46.** 9 **47.** -16 **48.** $\frac{12}{5} = 2\frac{2}{5}$
49. $\frac{-21}{2} = -10\frac{1}{2}$ **50.** 4 **51.** 36 **52.** -10 and 2
53. -7 and -2 **54.** -5 **55.** 15 **56.** 12 **57.** -28
58. 7 **59.** -5 **60.** $\frac{-5}{2} = -2\frac{1}{2}$ **61.** $\frac{5}{7}$ **62.** 0
63. 0 **64.** -2 **65.** $\frac{-10}{6} = -1\frac{2}{3}$
66. $8.45 + 18(-0.25)$ **67.** $450 + 12(-20)$

Chapter 9 Practice Test

1. -54 **2.** 2.5 **3.** $>$ **4.** $<$

5. |−4| −5 −4 −3 −2 −1 0 1 2 3 5 6 7 −(−8)

6.

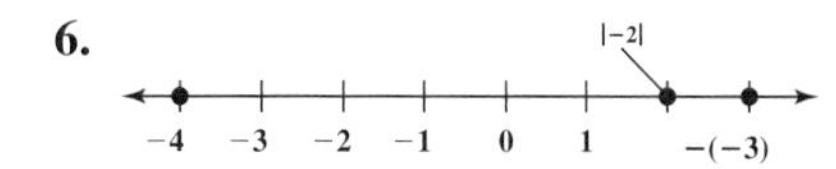

7. -7 **8.** -16.8 **9.** 30 **10.** -3 **11.** $-\frac{1}{6}$
12. -45 **13.** 6 **14.** 27 **15.** $-\$477$ **16.** 27.2 ft
17. -2 **18.** 9 **19.** 1 **20.** -4 **21.** -20
22. $-\frac{1}{25}$ **23.** 11 **24.** $-\frac{1}{4}$ **25.** -5 **26.** 7
27. 7°F **28. a.** 25°F **b.** -24°F **c.** 25°F
29. -42°F **30.** Wednesday to Friday

CHAPTER 10

Section 10.1 Guided Practice

1. lollipops l; 2; 4; $2l + 4$
2. b; number of batteries per case; $3b + 4$
3. c; number of cans of corn in each case; $3c - 14$
4. $3x - 8 = 3(4) - 8$
$= 12 - 8$
$= 4$
5. $8 - y = 8 - (-7)$
$= 8 + 7$
$= 15$

6. $\frac{5x - y}{y} = \frac{5(8) - (-5)}{(-5)}$

$= \frac{40 + 5}{-5}$

$= \frac{45}{-5}$

$= -9$

7. $5z \mid -2y \mid -3 \mid +z$ 4; $5z$; $-2y$; -3; z

8. $2y \mid -3 \mid +15 \mid -x$ 4; -3; 15; -3; 15

9. $4y^2 \mid -6 \mid +5yz \mid -3x$ $4y^2$; $5yz$; $-3x$; $4y^2$; $5yz$; $-3x$

10. $-y^2 \mid +5xy \mid -3$

Terms	$-y^2$	$5xy$	-3
Coefficients	-1	5	-3
Variable parts	y^2	xy	

11. $-xy^2 \mid +5xy \mid +2 \mid -2x \mid -3xy \mid -7$ xy; no; $5xy$; $-3xy$; 2; -7; unlike

12. a. $c = 60 + 0.45m$

b. $m = 500 - 450 = 50$

$c = 60 + 0.45m$

$c = 60 + 0.45(50)$

$= 60 + 22.5$

$= 82.5$

$82.50

Concept Check and Objective Practice

A1. Answers may vary. Example: A variable is a symbol or letter representing a numerical quantity.

A2. Answers may vary. Example: b represents the number of batteries per box, and there are four boxes. To find the total number of batteries, multiply the number of boxes (4) by the number of batteries in each box (b).

A3. Let c represent the number of pieces of chalk per box. $3c + 6$

A4. Let b represent the number of pieces of candy per case. $4b + 5$

A5. Let n represent the number of notebooks per case. $6n + 31$

A6. Let j represent the number of jerky sticks per box. $8j + 14$

A7. Let a represent the amount in the account. $a - 1{,}500$

A8. Let a represent the amount in the account. $a + 3{,}500$

B1. Answers may vary. Example: Substitute the values for the variables and calculate the answer.

B2. b is correct. When a coefficient sits directly to the left of a variable with no spaces or other signs of punctuation or operation, it indicates multiplication.

B3. Answers may vary. Example: By placing the open ( ) in the expression, the operation can easily be identified.

B4. 13 **B5.** 16 **B6.** 12 **B7.** 27 **B8.** 2 **B9.** $\frac{1}{2}$

B10. -4 **B11.** -4 **B12.** 68 **B13.** 22 **B14.** -7

B15. -32

C1. Answers may vary. Example: A variable term has a variable part; a constant term does not.

C2. Answers may vary. Example: The lines clearly identify the terms.

C3. Answers may vary. Example: Like terms have identical variable parts.

C4. a. 4 **b.** 3 **c.** $2x, -5y, -6x$ **d.** $2, -5, -6, 3$ **e.** $2x, -6x$

C5. a. 4 **b.** 5 **c.** $3r, -8y, -6r$ **d.** $-7, 3, -8, -6, 5$ **e.** $3r, -6r$

C6. a. 4 **b.** -7 **c.** $3y, -y^2, -4y$ **d.** $3, -1, -4$ **e.** $3y, -4y$

C7. a. 4 **b.** 14 **c.** $-z, 4z^2, -2z^2$ **d.** $-1, 4, -2, 14$ **e.** $4z^2, -2z^2$

Exercises

***For 1.–9.:** See preceding Concept Check and Objective Practice answers.*

11. coefficient **13.** Like terms

15. constant term **17.** term

19. Let c represent the number of foam cups per box. $18c + 40$

21. $b + 900$

23. a. Let k represent the number of candy bars in a box. $3k + 2$

b. $3k + 2 = 3(36) + 2 = 108 + 2 = 110$ There are a total of 110 candy bars.

25. 7 **27.** 13 **29.** -180 **31.** 1 **33.** -15

35. 48 **37.** -39 **39.** -2 **41.** $\frac{5}{2}$ **43.** 3

45. a. 4 **b.** 5 **c.** $4r, y, r$ **d.** $4, 1, 1, 5$ **e.** $4r$ and r

47. a. 4 **b.** 4, 14 **c.** $3x^3, -x$ **d.** $3, 4, -1, 14$ **e.** 4 and 14

49. a. 4 **b.** none **c.** $10m, -4m^3, 6m, m^3$ **d.** $10, -4, 6, 1$ **e.** $10m$ and $6m$, $-4m^3$ and m^3

51. a. 5 **b.** -8 **c.** $10l, -4m^3, 6l, m$ **d.** $10, -4, 6, 1, -8$ **e.** $10l$ and $6l$

53. 638 sq cm **55.** 68 ft **57.** 315 mi

59. a. Let c represent the total charges. Let d represent the number of days the car is rented. Let m represent the number of miles driven over the limit.
Clifton: $c = 32d + 0.35m$ Kyle: $c = 37d + 0.15m$

b. Clifton: $46 Kyle: $43 **c.** Kyle

Section 10.2 Guided Practice

1. $6x + 14 + 4x + 9 = 6x + 4x + 14 + 9$
$= 10x + 23$

2. $\frac{1}{3}x + 11 + \frac{5}{6}x + 18 = \frac{1}{3}x + \frac{5}{6}x + 11 + 18$
$= \frac{2}{6}x + \frac{5}{6}x + 11 + 18$
$= \frac{7}{6}x + 29$

3. $7z - 2 - 3z + 5 = 7z + (-3z) + (-2) + 5$
$= 4z + 3$

4. $-8 - 8t + 4 + 6t = (-8t) + 6t + (-8) + 4$
$= -2t + (-4)$
$= -2t - 4$

5. $4(2x + 7) = 4 \cdot 2x + 4 \cdot 7$
$= 8x + 28$

6. $(8 \cdot 7)x = 56x$ **7.** $(11 \cdot 6)z = 66z$

8. $4(5x) + 4(-1) = 4 \cdot 5x + 4 \cdot (-1)$
$= 20x - 4$

9. $-3(x) + (-3) \cdot (-8) = -3x + 24$

10. $4(2x) + 6(3x) - 114 = 8x + 18x - 114 = 26x - 114$

11. $\frac{10z}{5} + 28z = \frac{2 \cdot \cancel{5}^1 z}{_1\cancel{5}} + 28z = 2z + 28z = 30z$

12. unlike terms $-2(n - 8) + 5(6) = -2n + 16 + 5(6)$
$= -2n + 16 + 30$
$= -2n + 46$

13. $P = \text{Side}_1 + \text{Side}_2 + \text{Side}_3$
$= (4x - 3) + (x + 6) + (14)$
$= 4x - 3 + x + 6 + 14$
$= 4x + x - 3 + 6 + 14$
$= 5x + 17$ units

14. $A = \text{Length} \cdot \text{Width}$
$= (8) \cdot (5y + 2)$
$= 40y + 16$ square units

Concept Check and Objective Practice

A1. **a.** 5 pencils **b.** $\frac{4}{5}$ **c.** $3x$
A2. *Like* indicates the same type of item; that is, pencils, fifths, and x's.
A3. Answers may vary. $3y$ and $3x$; The variable parts are not identical.
A4. $5x$ **A5.** $4z$ **A6.** $-5y - 4$ **A7.** $-7t - 22$
A8. $3z - 14$ **A9.** $2x + 17$ **A10.** $6y + 11$
A11. $7n + 15$ **A12.** $-x + 16$ **A13.** $8 - z$
A14. $14y - 23$ **A15.** $8h - 25$
A16. $\frac{1}{2}x - 3$ **A17.** $\frac{2}{5}y - 23$
A18. $6.4r - 6.8s$ **A19.** $6.4t + 0.5v$
B1. The terms inside the parentheses are not like terms.
B2. every
B3. **a.** $2(3 + 4) = 2(7) = 14$
b. $2(3 + 4) = 2 \cdot 3 + 2 \cdot 4 = 6 + 8 = 14$
c. They are the same.
B4. $8x$ **B5.** $35y$ **B6.** $9x$ **B7.** $48z$ **B8.** $2x + 12$
B9. $4y + 12$ **B10.** $28x - 49$ **B11.** $24x - 48$
B12. $-36x - 6$ **B13.** $-10x - 10$ **B14.** $-24x + 8$
B15. $-4x + 1$ **C1.** at the multiplication step
C2. at the addition/subtraction step
C3. $18t + 5$ **C4.** $30y + 10$ **C5.** $-z$ **C6.** $5x$
C7. $6n + 16$ **C8.** $11n + 8$ **C9.** $-2y + 54$
C10. $4m + 54$ **C11.** $-3n - 21$ **C12.** $3t^2 - 18$

Exercises

For 1.–9.: See preceeding Concept Check and Objective Practice answers.

11. coefficient **13.** distributive property **15.** Like terms
17. **a.** 6 pencils **b.** unlike **c.** $5x$ **d.** unlike
19. $4z$ **21.** $-3t$ **23.** $9z$ **25.** $7h - 18$ **27.** $-7x - 5$
29. $-\frac{3}{7}y + 5$ **31.** $-\frac{4}{15}y - 23$ **33.** $6.4x + v$
35. **a.** $t + f + t$
b. The turtles are like animals.They can be combined because they can be described using the same name.
c. $2t + f$
37. $12x$ **39.** $32z$ **41.** $4y + 24$ **43.** $18x - 36$
45. $-24x - 16$ **47.** $-32x + 8$ **49.** $7z + 4$
51. $3h - 17$ **53.** $-11x + 2y - 23$
55. $10t - 45 + 17v$ **57.** $11n + 4$ **59.** $4x - 30$
61. $-6z + 21$ **63.** $-2m + 40$ **65.** $5t - 9$ **67.** -6
69. $13x - 8$ units
71. **a.** $30y + 10$ units **b.** $105y - 14$ sq units
73. $6x + 27$
75. **a.**

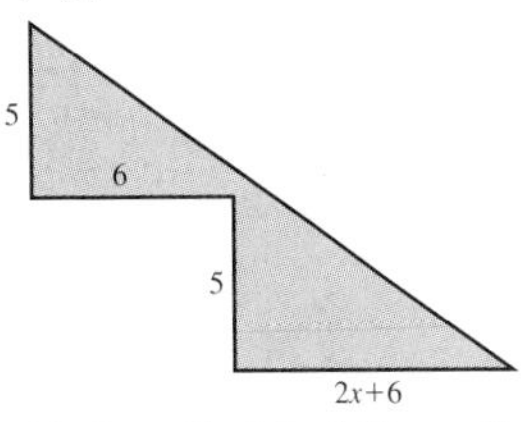

b. $10x + 30$ sq ft

Section 10.3 Guided Practice

1. 5; 15 =; is **2.** 6; 24; 24; 24; true; is

3.
$2x - 4 \stackrel{?}{=} x - 8$
$2(-3) - 4 \stackrel{?}{=} (-3) - 8$
$-6 - 4 \stackrel{?}{=} -3 - 8$
$-10 \neq -11$
-10; -11; false; isn't

4. **a.** 5; added to **b.** Addition; 5; subtraction; 5
5. **a.** multiplied; -10.
b. Multiplication; -10; division; -10
6. **a.** divided; 13 **b.** Division; 13; multiplication; 13
7. 2; 2; 7; 7; true
8. 5; 5; 5; 10; isolates; solves

9. $-23 + (8) = x - 8 + (8)$
$-15 = x$
Check: $-23 \stackrel{?}{=} (-15) - 8$
$-23 \stackrel{?}{=} -23$

10. $x + \frac{1}{3} - \frac{1}{3} = \frac{1}{2} - \frac{1}{3}$
$x = \frac{3}{6} - \frac{2}{6}$
$x = \frac{1}{6}$
yes

Check: $\frac{1}{6} + \frac{1}{3} = \frac{1}{2}$
$\frac{1}{6} + \frac{2}{6} = \frac{1}{2}$
$\frac{3}{6} = \frac{1}{2}$

11. 4; 4; 24; 24; true

12. $\frac{5x}{5} = \frac{20}{5}$
$\frac{{}^1\cancel{5}x}{{}^1\cancel{5}} = \frac{4 \cdot \cancel{5}^1}{\cancel{5}^1}$
$x = 4$
isolates; solves

13. -9; -9; -10; -10; 90; 90; yes

14. $7 \cdot \frac{x}{7} = 6 \cdot 7$
$\frac{{}^1\cancel{7}}{1} \cdot \frac{x}{\cancel{7}_1} = 6 \cdot 7$
$x = 42$
yes

Check: $\frac{x}{7} = 6$
$\frac{42}{7} \stackrel{?}{=} 6$
$6 \stackrel{?}{=} 6$

15. $\frac{6x}{6} = \frac{0}{6}$
$x = 0$
yes

Check: $6x = 0$
$6(0) \stackrel{?}{=} 0$
$0 \stackrel{?}{=} 0$

16. $\frac{3}{2} \cdot \frac{2}{3}x = \frac{3}{2} \cdot \frac{10}{3}$
$\frac{{}^1\cancel{3}}{{}^1\cancel{2}} \cdot \frac{{}^1\cancel{2}}{{}^1\cancel{3}}x = \frac{{}^1\cancel{3}}{{}^1\cancel{2}} \cdot \frac{{}^1\cancel{2} \cdot 5}{{}^1\cancel{3}}$
$x = 5$
yes

Check: $\frac{2}{3}x = \frac{10}{3}$
$\frac{2}{3} \cdot \frac{5}{1} \stackrel{?}{=} \frac{10}{3}$
$\frac{2 \cdot 5}{3 \cdot 1} \stackrel{?}{=} \frac{10}{3}$
$\frac{10}{3} \stackrel{?}{=} \frac{10}{3}$

Concept Check and Objective Practice

A1. Answers may vary. Example: A solution is the numerical value that makes the equation a true statement.
A2. expression; expression **A3.** yes **A4.** yes
A5. no **A6.** no **A7.** yes **A8.** yes **A9.** yes
A10. yes
B1. Addition and subtraction are inverse operations. Multiplication and division are inverse operations.
B2. a. -4 **b.** $\div 6$ **c.** $\cdot 2$ **d.** $+5$
B3. a. addition of 17 **b.** subtraction of 17
B4. a. subtraction of 12 **b.** addition of 12
B5. a. division by 5 **b.** multiplication by 5
B6. a. division by -2 **b.** multiplication by -2
B7. a. subtraction of 12 **b.** addition of 12
B8. a. subtraction of 15 **b.** addition of 15
B9. a. multiplication by -9 **b.** division by -9
B10. a. multiplication by -1 **b.** division by -1
C1. Answers may vary. Example: to keep the equation in balance **C2. a.** subtraction of 4 **b.** addition of 4
C3. Answers may vary. Example: to get the variable alone on one side of the equal sign
C4. subtracting $\frac{1}{8}$ from both sides of the equation
C5. $y = 4$ **C6.** $z = 5$ **C7.** $11 = t$ **C8.** $-40 = z$
C9. $x = 0$ **C10.** $x = 0$ **C11.** $12 = t$
C12. $18 = t$ **C13.** $x = \frac{13}{2}$ **C14.** $y = \frac{19}{5}$
C15. $z = \frac{5}{12}$ **C16.** $x = \frac{2}{3}$
D1.
$$\begin{aligned} 100 &= 100 \\ 100 \div 2 &= 100 \div 2 \\ 50 &= 50 \end{aligned}$$
D2. Multiply both sides of the equation by $\frac{8}{3}$.
D3. $x = 8$ **D4.** $y = 9$ **D5.** $z = -5$ **D6.** $y = -4$
D7. $t = 11$ **D8.** $t = -8$ **D9.** $x = 0$ **D10.** $z = 0$
D11. $x = -6$ **D12.** $y = -21$ **D13.** $y = 3$ **D14.** $x = 3$

Exercises

For 1.–12.: See preceeding Concept Check and Objective Practice answers.

13. solve an equation **15.** isolate a variable
17. multiplication property of equations **19.** equation
21. yes **23.** no **25.** yes **27.** no
29. a. addition of 7 **b.** subtraction of 7
31. a. division by 15 **b.** multiplication by 15
33. a. addition of $\frac{2}{3}$ **b.** subtraction of $\frac{2}{3}$
35. a. multiplication by -2 **b.** division by -2
37. $y = 11$ **39.** $t = -4$ **41.** $x = 0$ **43.** $t = 3$
45. $v = -22$ **47.** $x = \frac{7}{3}$ **49.** $z = \frac{1}{12}$ **51.** $x = 9.8$
53. $z = 8$ **55.** $z = 7$ **57.** $x = -\frac{3}{2}$ **59.** $v = 40$
61. $t = 5$ **63.** $x = 0$ **65.** $y = \frac{5}{4}$ **67.** $x = -45$
69. a. $60 = 12 \cdot w$ **b.** 5 m
71. a. $20 = \frac{1}{2} \cdot 8 \cdot h$ **b.** 5 in.
73. a. divided by 4 instead of subtracting 4 **b.** $y = 16$
75. a. division fact error $10 \div (-5) = -2$ **b.** $z = -2$

Section 10.4 Guided Practice

1. 8; 9; 8; right; 8; 4; other side; 4
$$\begin{aligned} 8x - 8x + 1 &= 9x - 8x + 4 \\ 1 &= x + 4 \\ 1 - 4 &= x + 4 - 4 \\ -3 &= x \end{aligned}$$

2. -4; -3; -4; left; 4; -3; other side; 3
$$\begin{aligned} -3x + 4x - 3 &= -4x + 4x + 1 \\ x - 3 &= 1 \\ x - 3 + 3 &= 1 + 3 \\ x &= 4 \end{aligned}$$

3. 6; left; 2; other side; 3
$$\begin{aligned} 9x - 6x + 2 &= 6x - 6x + 5 \\ 3x + 2 &= 5 \\ 3x + 2 - 2 &= 5 - 2 \\ 3x &= 3 \\ \frac{{}^{1}\cancel{3}x}{{}_{1}\cancel{3}} &= \frac{{}^{1}\cancel{3}}{{}_{1}\cancel{3}} \\ x &= 1 \end{aligned}$$

4. -6; right; -2; other side; 10
$$\begin{aligned} -6x + 6x + 18 &= 4x + 6x - 2 \\ 18 &= 10x - 2 \\ 18 + 2 &= 10x - 2 + 2 \\ 20 &= 10x \\ \frac{20}{10} &= \frac{\cancel{10}x}{\cancel{10}} \\ 2 &= x \end{aligned}$$

5.
$$\begin{aligned} 6x + 15 &= -3x + 5x + 1 \\ 6x + 15 &= 2x + 1 \\ 6x - 2x + 15 &= 2x - 2x + 1 \\ 4x + 15 &= 1 \\ 4x + 15 - 15 &= 1 - 15 \\ 4x &= -14 \\ \frac{4x}{4} &= \frac{-14}{4} \\ x &= -\frac{7}{2} \end{aligned}$$

6.
$$\begin{aligned} 4 + 8t - 5 &= 3t + 24 \\ 8t - 1 &= 3t + 24 \\ 8t - 3t - 1 &= 3t - 3t + 24 \\ 5t - 1 &= 24 \\ 5t - 1 + 1 &= 24 + 1 \\ 5t &= 25 \\ \frac{5t}{5} &= \frac{25}{5} \\ t &= 5 \end{aligned}$$

7.
$$\begin{aligned} (-4) &= \frac{9}{5}C + 32 \\ -4 - 32 &= \frac{9}{5}C + 32 - 32 \\ -36 &= \frac{9}{5}C \\ \frac{5}{9} \cdot \frac{-36}{1} &= \frac{\cancel{5}}{\cancel{9}} \cdot \frac{\cancel{9}}{\cancel{5}}C \\ \frac{5 \cdot -4 \cdot \cancel{9}}{\cancel{9}} &= C \\ -20 &= C \end{aligned}$$
-4°F is the same as -20°C.

Concept Check and Objective Practice

A1. addition and subtraction **A2.** smaller coefficient
A3. $9 = x$ **A4.** $x = 5$ **A5.** $x = -6$ **A6.** $x = 11$
A7. $-7 = x$ **A8.** $6 = x$ **A9.** $2 = x$ **A10.** $5 = x$
B1. a. Subtract $2x$ from both sides.
b. Subtract 4 from both sides. **c.** Add $7x$ to both sides.
B2. The "doing" operation is multiplication. Multiplication is "undone" using division, not addition.
B3. $x = -2$ **B4.** $3 = x$ **B5.** $3 = y$ **B6.** $y = 4$
B7. $x = 0$ **B8.** $x = 0$ **B9.** $1 = z$ **B10.** $x = 2$
C1. Simplify each side, gather variable terms and constant terms to opposite sides, isolate the variable.
C2. Answers may vary. Example: Tell her to write $-2(x - 5)$ as the sum of two multiplication problems, $-2(x) + (-2)(-5)$.
C3. $x = -6$ **C4.** $x = -14$ **C5.** $2 = y$ **C6.** $2 = z$
C7. $11 = x$ **C8.** $x = -15$ **C9.** $2 = x$ **C10.** $7 = t$

Exercises

For 1.–9.: *See preceeding Concept Check and Objective Practice answers.*

11. constant term **13.** variable term **15.** $-9 = x$
17. $-13 = x$ **19.** $x = -1$ **21.** $x = -2$ **23.** $x = 7$
25. $x = 1$ **27.** $c = 40$ **29.** $v = 0$ **31.** $x = 2$
33. $v = 11$ **35.** $x = -5$ **37.** $x = 4$ **39.** $z = 4$
41. $c = -6$ **43.** $t = -3$ **45.** $c = 0$ **47.** $x = -\frac{1}{2}$
49. $-7 = t$ **51.** $y = -10$ **53.** $-10 = y$ **55.** $x = \frac{7}{8}$
57. $-9 = x$ **59.** 69.8°F **61.** $\frac{260}{9} = 28.9$°C
63. a. 18 in. **b.** 216 sq in. **65.** 21 field goals
67. 30 extra points **69.** Harts **71.** yes, Axis
73. a. did not distribute the -2 correctly **b.** $-8 = x$
75. a. divided by 3 instead of -3 **b.** $x = -2$
77. 55, 56, and 57 **79.** 82, 84, and 86

Chapter 10 Review Exercises

1. Let b represent the number of candy bars per box. $2b + 2$
2. Let b represent the number of lollipops per box. $b + 5$
3. Let b represent the number of baseball caps per box. $5b + 15$
4. Let b represent the number of DVDs per box. $10b + 14$
5. variable; constant **6.** coefficient
7. variable parts
8. a. 4 **b.** 3 **c.** $12x, -4y, -1x$
d. $12, -4, -1, 3$ **e.** $12x$ and $-1x$
9. a. 4 **b.** 5 **c.** $3z^2, -5y, -6r^2$ **d.** $3, -5, -6, 5$
e. none
10. a. 4 **b.** -17 **c.** $3y^3, -y^2, -14y^3$
d. $-17, 3, -1, -14$ **e.** $3y^3$ and $-14y^3$
11. a. 4 **b.** 35 **c.** $3z^2, 9z, -2z^2$
d. $3, 9, -2, 35$ **e.** $3z^2$ and $-2z^2$
12. 150 mi **13.** 9 mi **14.** $12 - 2z$ **15.** $x + 11$
16. $5z^2 - 4z + 8$ **17.** $-7x^2 - 20x - 12$
18. $12x$ **19.** $10y$
20. Answers may vary. Example: When adding things together, there must be a way to describe the combined thing. Adding carrots to carrots results in carrots. Adding x to x results in x.
21. Answers may vary. Example: In order to add things together, there must be a way to describe the combined thing. There is not a name for the combination of a carrot and a potato. Also, there is not a name for addition of x and y.
22. $5x + 35$ **23.** $7y - 56$ **24.** $-4z - 32$
25. $-2y + 6$ **26.** $17x$ **27.** $8h$ **28.** $9n + 2$
29. $2t - 9$ **30.** $11x - 11$ **31.** $3x + 12$
32. yes **33.** no
34. Answers may vary. Example: A solution is a value that, when substituted for the variable, creates a true statement.
35. $x = 16$ **36.** $8 = y$ **37.** $-17 = z$ **38.** $x = \frac{4}{15}$
39. $7 = y$ **40.** $x = -11$ **41.** $x = 45$ **42.** $x = \frac{3}{4}$
43. a. 3 should have been added to both sides of the equation.
b. $x = 13$
44. a. Both sides should have been divided by -2.
b. $-6 = x$
45. a. division fact error **b.** $1 = x$
46. Answers may vary. Example: An equal sign indicates that both sides are in balance. To be an equation, both sides must be in balance. Adding to or subtracting from only one side will cause the equation to become unbalanced.
47. $x = -10$ **48.** $-2 = x$ **49.** $x = -5$
50. $0 = x$ **51.** $x = -12$ **52.** $x = 6$ **53.** $x = -3$
54. $x = 63$ **55.** $y = -\frac{3}{8}$ **56.** $y = -20$
57. $0 = x$ **58.** $3 = x$ **59.** $x = -\frac{2}{5}$
60. $x = 5$ **61.** $-3 = x$ **62.** $-2 = x$ **63.** $y = \frac{5}{9}$
64. $x = 3$

Chapter 10 Practice Test

1. Let c represent the number of cans of soda in each box. $2c + 3$
2. Let j represent the number of jerseys in each box. $4j + 25$ **3.** like **4.** constant
5. a. 4 **b.** $-5, -14$ **c.** $4y, -2y^2$
d. $-5, 4, -2, -14$ **e.** -5 and -14
6. a. 5 **b.** $3, -8$ **c.** $5z, 4z^2, -z$ **d.** $3, 5, 4, -1, -8$
e. 3 and -8; $5z$ and $-z$
7. 420 miles **8.** $22x - 12$ **9.** $56x$ **10.** $8x - 4$
11. $2x - 19$ **12.** $6x + 54$ **13.** $-5y + 35$
14. $10n + 36$ **15.** $6x$ **16.** $7y - 6$ **17.** $-5x - 12$
18. no **19.** $x = 52$ **20.** $11 = y$ **21.** $x = \frac{8}{5}$
22. $y = \frac{11}{20}$ **23.** Add 10 to both sides of the equation.
24. Multiply both sides of the equation by -6.
25. Divide both sides of the equation by -7.
26. Subtract $\frac{2}{3}$ from both sides of the equation.
27. Goal 1: Simplify both sides of the equation.
Goal 2: Gather variable terms on one side of the equation. Gather constant terms on the other side of the equation.
Goal 3: Isolate the variable.
28. $x = 5$ **29.** $-2 = y$ **30.** $y = -25$ **31.** $8 = x$
32. $x = 32$ **33.** $2 = x$ **34.** $y = 0$ **35.** $6 = x$
36. $y = \frac{5}{8}$

APPENDIX A

Section 1.2 Extra Practice Exercises; Addition Facts

1. 34 **2.** 69 **3.** 85 **4.** 99 **5.** 101 **6.** 83 **7.** 140 **8.** 159 **9.** 18 **10.** 17 **11.** 44 **12.** 43 **13.** 507 **14.** 705 **15.** 701 **16.** 703 **17.** 1,291 **18.** 1,259 **19.** 1,053 **20.** 1,641 **21.** 148 **22.** 168 **23.** 190 **24.** 164

Section 1.3 Extra Practice Exercises; Subtraction Facts

1. 12 **2.** 11 **3.** 19 **4.** 9 **5.** 65 **6.** 6 **7.** 45 **8.** 27 **9.** 14 **10.** 11 **11.** 10 **12.** 17 **13.** 399 **14.** 599 **15.** 583 **16.** 587 **17.** 185 **18.** 489 **19.** 297 **20.** 295 **21.** 16,857 **22.** 20,272 **23.** 36,835 **24.** 862,550

Section 1.4 Extra Practice Exercises; Multiplication Facts

1. 270 **2.** 260 **3.** 469 **4.** 747 **5.** 1,578 **6.** 3,948 **7.** 19,704 **8.** 23,457 **9.** 670 **10.** 920 **11.** 234,000 **12.** 542,000 **13.** 1,350 **14.** 3,720 **15.** 31,500,000 **16.** 10,800,000 **17.** 1,820 **18.** 3,285 **19.** 4,544 **20.** 3,876 **21.** 60 **22.** 180 **23.** 2,730 **24.** 3,360 **25.** 30,544 **26.** 47,685 **27.** 25,077 **28.** 19,610 **29.** 352,809 **30.** 154,140 **31.** 287,640 **32.** 521,950

Mid-Chapter 1 Review Exercises

1. 33,000 **2.** 120 **3.** 300 **4.** 1,330,000 **5.** Four thousand, nine hundred eight **6.** Twenty-five thousand, forty-five **7.** 613 **8.** 708 **9.** 134 **10.** 69 **11.** 1,296 **12.** 5,264 **13.** 19R2 **14.** 23R6 **15.** 26R3 **16.** 22R12 **17.** 474 **18.** 665 **19.** 8,100 **20.** 45,000 **21.** $467 **22.** $35 **23.** $1,980 **24.** 207 miles **25.** 21,000 **26.** 40,000 **27.** 20 **28.** 20 **29.** 60 guests **30.** 64 miles per hour **31.** 192 square feet **32.** 72 feet **33.** $x = 6$ **34.** $x = 35$ **35.** $x = 40$ **36.** $x = 21$

Mid-Chapter 2 Review Exercises

1. $\frac{13}{84}$ **2.** $\frac{5}{54}$ **3.** $\frac{1}{10}$ **4.** $\frac{45}{56}$ **5.** $\frac{3}{8}$ **6.** $\frac{1}{3}$ **7.** $\frac{26}{21}$ **8.** $\frac{5}{12}$ **9.** $\frac{17}{30}$ **10.** $\frac{63}{40}$ **11.** $\frac{8}{9}$ **12.** $\frac{17}{18}$ **13.** $\frac{27}{25}$ **14.** 3 **15.** 1 **16.** $\frac{1}{36}$ **17.** Addition and Subtraction **18.** Multiplication and Division **19.** Answers may vary. Example: $\frac{1}{5}$ and $\frac{3}{5}$ **20.** Answers may vary. Example: $\frac{1}{2}$ and $\frac{2}{4}$ **21.** the same **22.** b **23.** g **24.** h **25.** d **26.** c **27.** f **28.** e **29.** a **30.** $<$ **31.** $>$ **32.** $<$ **33.** $<$ **34.** $<$ **35.** $<$ **36.** $<$ **37.** $>$ **38.** $>$ **39.** $<$ **40.** $<$ **41.** $>$ **42.** $<$ **43.** $>$ **44.** 31,680 feet **45.** 5 pounds **46.** about 20,000 jelly beans

Mid-Chapter 9 Review Exercises

1. $-5, -3, 4, 7$ **2.** $-13, -8, 7, 12$ **3.** $-(7), -|-4|, -(-3), |-8|$ **4.** $-|-2|, 0, -(-2), |-9|$ **5.** -24 **6.** 7 **7.** 36 **8.** -4 **9.** 15 **10.** -20 **11.** 70 **12.** 8 **13.** -3 **14.** -13 **15.** -16 **16.** 9 **17.** $-\frac{9}{10}$ **18.** $-\frac{1}{14}$ **19.** $-\frac{2}{3}$ **20.** $\frac{21}{20}$ **21.** -39 **22.** -48 **23.** 2 **24.** 4 **25.** 9 **26.** 5 **27.** -1 **28.** 1.9% **29.** -3 and 5 **30.** -8 and -3 **31.** $88 **32.** 2,160 feet

Index

U.S./Metric Conversion

METRIC PREFIXES FOR DISTANCE, VOLUME, AND WEIGHT

(distances, meters)	km	hm	dam	m	dm	cm	mm
(volumes, liters)	kL	hL	daL	L	dL	cL	mL
(weights, grams)	kg	hg	dag	g	dg	cg	mg
	kilo	hecto	deka	unit	deci	centi	milli
	King	**H**enry	**D**ied	**U**nceremoniously	**D**rinking	**C**hocolate	**M**ilk

U.S. UNITS OF MEASURE AND THEIR EQUIVALENTS

Time	Volume
60 seconds (s) = 1 minute (min) 60 minutes = 1 hour (h) 24 hours = 1 day (d) 7 days = 1 week (wk) 52 weeks = 1 year (yr) 365 days = 1 year (yr)	8 fluid ounces (fl oz) = 1 cup (c) 2 cups = 1 pint (pt) 2 pints = 1 quart (qt) 4 quarts = 1 gallon (gal)
Length	**Weight**
12 inches (in.) = 1 foot (ft) 3 feet = 1 yard (yd) 5,280 feet = 1 mile (mi)	16 ounces (oz) = 1 pound (lb) 2,000 pounds = 1 ton (T)

U.S. AND METRIC SYSTEM: UNITS OF EQUIVALENT MEASURES

	U.S. to Metric	Metric to U.S.
Length	1 mile (mi) ≈ 1.61 kilometers (km) 1 yard (yd) ≈ 0.914 meter (m) 1 foot (ft) ≈ 0.305 meter (m) 1 inch (in.) = 2.54 centimeters (cm)	1 kilometer (km) ≈ 0.62 mile (mi) 1 meter (m) ≈ 1.09 yard (yd) 1 meter (m) ≈ 3.28 feet (ft) 1 centimeter (cm) ≈ 0.394 inch (in.)
Volume	1 gallon (gal) ≈ 3.79 liters (L) 1 quart (qt) ≈ 0.946 liter (L)	1 liter (L) ≈ 0.264 gallon (gal) 1 liter (L) ≈ 1.06 quarts (qt)
Weight	1 pound (lb) ≈ 0.454 kilogram (kg) 1 ounce (oz) ≈ 28.35 grams (g)	1 kilogram (kg) ≈ 2.2 pounds (lb) 1 gram (g) ≈ 0.0353 ounce (oz)